Spread Spectrum Systems for GNSS and Wireless Communications

For a listing of recent titles
in the *Artech House GNSS Technology and Applications Series*,
turn to the back of this book.

Spread Spectrum Systems for GNSS and Wireless Communications

Jack K. Holmes

ARTECH
HOUSE
BOSTON | LONDON
artechhouse.com

Library of Congress Cataloguing-in-Publication Data
A catalog record for this book is available from the U.S. Library of Congress.

British Library Cataloguing in Publication Data
A catalogue record for this book is available from the British Library.

ISBN-13: 978-1-59693-083-4

Cover design by Igor Valdman

The opinions expressed in this book are solely those of the author and do not reflect those of the publisher or any other party.

685 Canton Street
Norwood, MA 02062

10 9 8 7 6 5 4 3 2 1

To my wife Fontayne, my daughters Julie and Rachel, my grandchildren Sarah, Danielle, Emily, and Andrew, and my sons-in-law Scott and Vince

Contents

CHAPTER 1

An Introduction to Spread Spectrum Systems

1.0 INTRODUCTION

A spread spectrum communications system is a communication system that purposely spreads the spectrum of the signal (considerably) beyond the required bandwidth needed for the information content of the message. Spread spectrum signals are transmitted by electromagnetic waves in free space with applications to both nonmilitary and military systems. There are at least six reasons that one would consider using spread spectrum signals. The first reason is that these signals reject hostile jamming as well as unintentional interference, so information can be communicated effectively. A second reason is that spread spectrum signals have a low probability of intercept (LPI), because the signal is spread over a large bandwidth. This makes detectability more difficult than for a nonspread signal, since oftentimes the signal lies below the noise. The third reason for using spread spectrum signals is that an unauthorized listener who lacks knowledge of the spreading code and precise timing cannot readily demodulate these signals and thus spread spectrum signaling has inherent message privacy. The fourth reason that spread spectrum signaling is used is that it provides good rejection of multipath signals (although not complete rejection). A high degree of ranging precision can be obtained with direct sequence spread spectrum signaling and is the fifth reason for using spread spectrum signals. The sixth reason for using spread spectrum signaling is its compatibility with code division multiple access capability, which allows simultaneous signaling on the same frequency.

The Milstar military communication system is an example of a system in use now that was designed for jamming protection (first reason). Another example of an existing system that uses spread spectrum signals for ranging (fifth reason) is the Global Positioning System (GPS). In addition, GPS spread spectrum signaling provides jamming protection (first reason), multipath protection (fourth reason), and a high degree of range precision (fifth reason).

In today's world, there are probably more applications on the commercial side than on the military side, which is a considerable change from the 1970s and 1980s when the primary applications were to military systems with many of the attributes indicated earlier. Milstar was mentioned as an example of a military spread spectrum system, which was (originally) based on six geosynchronous satellites that provide voice and data communications from virtually any location on the Earth to any other location on Earth. Milstar utilizes frequency-hopping spread spectrum communications over a very large bandwidth, providing a considerable jamming immunity from unfriendly (jamming) signals.

Consider now some examples of commercial (actually some have military applications as well) systems that have launched or will be launched shortly. GPS utilizes direct sequence spread spectrum modulation and has a constellation of 24 satellites (with three spares) located at about one half the altitude of a geostationary satellite. Two direct sequence signals at L1 frequency (L-band) include a clear access (C/A) code and a precision (P(Y)) code. In addition, a new military code (M-code) and some additional civilian signals are in the process of being added. Another (P(Y)) code is located at L2 (a somewhat lower frequency in the L-band). Also a new C/A like code has been added to L2 and a new civil signal L5 has been added. Any four GPS satellites in view enable a commercial user (e.g., a soldier, an airplane, a hiker, a boat, an auto) to determine its location with an accuracy less than 50 feet in each dimension, although it is possible to obtain a much higher accuracy if desired. GPS has demonstrated that it can automate airplane landings with a very high degree of accuracy using a form of differential GPS.

OmniTRACS is a mobile satellite communications system providing two-way messaging, paging and position location/reporting operating in the United States. It is owned and operated by QUALCOMM, Inc. Typical users are long-haul transportation, utilities, construction, and agriculture. In order to minimize interference with other satellite users, the signals use a hybrid direct sequence and frequency-hopping scheme with frequency shift keying (FSK) modulation. User location is obtained by ranging through both satellites, and the location is sent to the operator.

GLOBALSTAR is an example of a mobile satellite service (MSS). This system is a proposed low Earth orbit MSS operating in L-band and S-band utilizing 48 satellites in eight different orbital planes 700 miles above Earth. This system offers voice and data service, facsimile, worldwide paging, and messaging. GLOBALSTAR utilizes code division multiple access (CDMA) spread spectrum techniques and can provide its own position location capability with an accuracy of about 30 meters. A unique feature of the GLOBALSTAR system is its use of the CDMA digital cellular standard IS-95 developed by QUALCOMM, which, along with Loral, are principals of GLOBALSTAR.

Cellular radio is another example in which at least one firm (QUALCOMM) is using CDMA spread spectrum techniques to provide superior performance in an interference environment and to provide more capacity for cellular radio. Competitors include the European system known as the Global System for Mobile Communications (GSM), which is a time division multiple access (TDMA) system, and IS-54, which is also a TDMA system and is a North American standard.

The common codeless phone utilizes spread spectrum technology to provide a telephonic interference-rejecting communication link. Both direct sequence and frequency hopping (FH) SS schemes are used in cordless telephones in the 900-MHz and 2.4-GHz bands. Within the FH phones, both narrowband and wideband hopping schemes are used. Direct sequence phones have 20 channels to select from (usually automatically) so that hopefully one channel will have good reception.

In addition to these applications, spread spectrum communication methods have been considered an alternative to hard wiring for indoor private branch offices, laboratories, or factories where the transmission medium involves severe multipath fading. Spread spectrum methods have also been proposed for improvement of the throughput performance in packet radio networks. A home security system using spread spectrum techniques imposed on the AC power line has been used in Japan. It appears that the applications of spread spectrum techniques to nonmilitary applications are virtually limitless!

1.1 A VERY BRIEF HISTORY OF SPREAD SPECTRUM COMMUNICATIONS

It is not easy to state what the first spread spectrum system was, but it is possible to outline some of the more important ones. Perhaps Hedy Lamarr and George Antheil made one of the most interesting patents filed in mid-1941. This was after Hedy had immigrated to the United States of America, from her native Austria. Hedy later became a well-known movie star. George was a volatile symphony conductor whom Hedy met in this country.

Basically they conceived of a frequency-hopping guidance system for controlling a torpedo to its target [1], even though neither one had electrical engineering backgrounds! The transmitted carrier would change frequency according to a randomized nonrepeating code. Hedy and George thought that this technique had applications to guided missiles also; however, the American government wasn't very interested.

One of the earlier uses of direct sequence signals occurred for the purpose of ranging for the tracking-range radar system at the Jet Propulsion Laboratory (JPL) for use in the Corporal guidance system link [2]. Frank Lehan of JPL noted that radar signal correlation function was of prime importance in determining the accuracy of the range estimate. The term "pseudonoise" with its abbreviation PN was consistently used at JPL from 1953 onward in relation to the spread spectrum signals used in a direct sequence system [2].

The pseudonoise codes (actually *m*-sequences) were investigated starting around the early 1950s. One of the earliest works started with Gilbert [3] at Bell Labs. However his report had limited circulation. The work of Golomb [4] and Zierler [5, 6] appeared slightly later. Around this time period Birdsall and Ristenbatt [7] released their work while at the University of Michigan.

One of the earliest coherent spread spectrum systems was constructed and designed by Magnavox, and the codes used for spreading in the spread spectrum system were Gold codes [2], named after the inventor Dr. Robert Gold. This was probably the first used of Gold codes in a spread spectrum system. This project was instrumental in making Magnavox one of the early leaders in spread spectrum technology. One of the earliest reports of a tutorial nature was written by Dr. Charlie Cahn and published by Magnavox [8] on the subject of spread spectrum systems.

1.2 A DIGITAL SPREAD SPECTRUM COMMUNICATION SYSTEMS MODEL

In a spread spectrum system, the system designer has to be concerned with a number of factors in the design of the system. One important factor is interference, either intentional as in military systems, or unintentional, which is encountered in commercial systems. The system designer must build into the design a resistance to interference so as to mitigate the effects of interference. This also occurs in code division multiple access systems where all signals, except the desired one, act as a noise-like signal that interferes with the desired signal. One important factor in the design of spread spectrum systems is the ratio of spread bandwidth to the data bandwidth. The larger this factor is, the better the rejection of narrowband interference. Typically the bandwidth is spread by a pseudorandom code sequence that modulates the frequency or phase so as to increase the bandwidth over what is needed to send just the data modulation.

The basic elements of a spread spectrum system are shown in Figure 1.2-1. A communication channel is any medium over which information can be transmitted. Communication channel examples include free space, ionospheric propagation, fiber optic cables, microwave waveguides, and coaxial cables. In general these media accept signals and deliver them to their destination. A channel may attenuate, distort, corrupt with additive noise or multiplicative effects, and in general make the reception of the signal a challenge. Channel encoding and decoding are typical in communication systems, and in spread spectrum systems the modulation process provides both data modulation as well as a form of spectrum widening modulation. Both the modulator and the demodulator contain a pseudorandom sequence generator, which is used to spread the transmitted signal spectrum and to despread it at the demodulator. Typically the spread spectrum process is removed first at the demodulator and then the data is demodulated and then decoded.

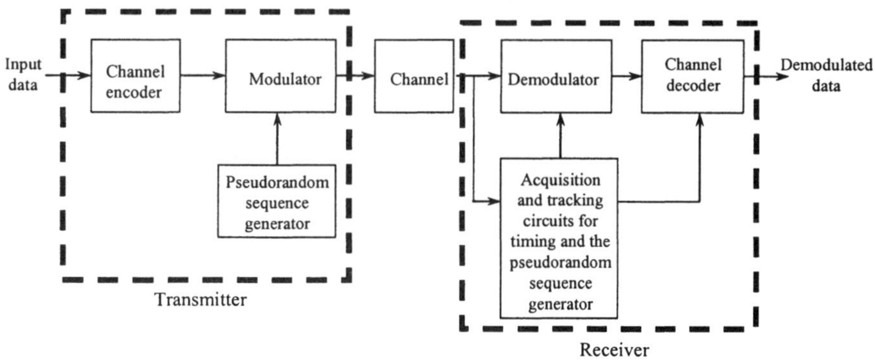

Figure 1.2-1 Model of a spread spectrum system.

A major requirement for the process to work is that the receiver be synchronized with the received signal. Acquisition circuits, which will be addressed later in the book, search the time and frequency uncertainty regions to obtain time and frequency (and sometimes phase) synchronization. After acquisition is accomplished, tracking is provided to maintain synchronization of the pseudorandom generator. The despread signal then can be data demodulated and decoded. Both the demodulation process and the decoding process require timing information. The demodulation process may need phase or frequency information as well.

Interference is introduced in the transmission of the spread spectrum signal through the effects of the channel. Various types of interference can be observed, such as other information bearing signals, pulse interferers, narrowband interferers, and broadband interferers. In this context, broadband means on the order of the bandwidth of the spread signal spectrum, and narrowband means narrow compared to the spread signal spectrum. Other forms of interference can occur also, such as in CDMA systems, in which all the other signals in the same frequency band produce a controlled noise-like interference to the desired signal. The despreading process collapses the signal, but it also changes narrowband interference to broadband interference to the data and thus has a reduced effect on the data demodulation. This phenomenon of spreading the interference results in a processing gain of the receiver against the narrowband interferer, and it will be discussed in more detail later in this chapter.

1.3 NARROWBAND SIGNALS

In dealing with spread spectrum signals, it is often advantageous to use complex envelopes to describe the signal when it is narrowband. Then filtering operations can be described by operations on the complex envelope only, rather than the complete signal. At this point a short discussion of complex signals will be presented and then applied to the various modulations including frequency hopping and direct sequence signals.

1.3.1 Narrowband Processes Via the Complex Envelope

Let the received signal plus noise be denoted by $x(t)$. Then the received signal can be written as

$$x(t) = \mathrm{Re}\left\{v(t)e^{i\omega_0 t}\right\} \tag{1.3.1-1}$$

where Re{ }denotes the "real part" of quantity in braces, $v(t)$ is the *complex envelope* of $x(t)$ [9, 10], and ω_0 is the angular carrier frequency. For a complex variable z, it is true that

$$\mathrm{Re}\{z\} = \frac{1}{2}z + \frac{1}{2}z^* \tag{1.3.1-2}$$

where the asterisk denotes the complex conjugate. From (1.3.1-1) it follows that

$$x(t) = \frac{1}{2}v(t)e^{i\omega_0 t} + \frac{1}{2}v^*(t)e^{-i\omega_0 t} \tag{1.3.1-3}$$

Let $V(f)$ denote the Fourier transform of $v(t)$; then the Fourier transform of the second term produces

$$F\left[v^*(t)\right] = \left(\int_{-\infty}^{\infty} v^*(t)e^{-j\omega t}dt\right) = \left(\int_{-\infty}^{\infty} v(t)e^{j\omega t}dt\right)^* = V^*(-f) \tag{1.3.1-4}$$

where $V^*(-f)$ is the Fourier transform of $v^*(t)$. Hence it follows that the Fourier transform of $x(t)$, from (1.3.1-4), is written as

$$X(f) = F\{x(t)\} = \frac{1}{2}V(f - f_0) + \frac{1}{2}V^*(-f - f_0) \tag{1.3.1-5}$$

If $v(t)$ is real, then $V^*(f) = V(-f)$ and then (1.3.1-5) can be simplified to

$$X(f) = F\{x(t)\} = \frac{1}{2}V(f - f_0) + \frac{1}{2}V(f + f_0) \tag{1.3.1-6}$$

For a *narrowband signal* (one in which the bandwidth is small compared to the center frequency) it is true that

$$V(f) \cong 0 \quad \text{for} \quad f >> B/2 \ \text{ and } \ B/2 << f_0 \tag{1.3.1-7}$$

where B is the nominal bandwidth of the signal $x(t)$, and $B/2$ is the nominal bandwidth of the complex envelope of the signal $v(t)$; both of which are the positive frequency bandwidths.

1.3.2 Narrowband Signals Through Narrowband Systems

Consider now the result when a narrowband signal is passed through a linear narrowband system. Let the linear narrowband system under consideration have a transfer function $H(f)$ and impulse response $h(t)$. A system is a *narrowband system* when the transfer function can be written as

$$H(f) = H_L(f - f_0) + H_L^*(-f - f_0) \tag{1.3.2-1}$$

with

$$|H(f)| \cong 0 \quad \text{for} \quad |f| >> W/2 \ \text{ and } \ W/2 << f_0 \tag{1.3.2-2}$$

with W the system radio frequency bandwidth and $H_L(f)$ the *baseband* (*or low pass*) *equivalent transfer function* (having bandwidth $W/2$). The impulse response of the narrowband system is given by

$$h(t) = \int_{-\infty}^{\infty} \left[H_L(f - f_0) + H_L^*(-f - f_0) \right] e^{i2\pi ft} df \tag{1.3.2-3}$$

Changing the variables in the two integrals produces

$$h(t) = e^{i2\pi f_0 t} \int_{-\infty}^{\infty} H_L(f) e^{i2\pi ft} df + e^{-i2\pi f_0 t} \int_{-\infty}^{\infty} H_L^*(f) e^{-i2\pi ft} df \tag{1.3.2-4}$$

or

$$h(t) = h_L(t) e^{i2\pi f_0 t} + h_L^*(t) e^{-i2\pi f_0 t} \tag{1.3.2-5}$$

where $h_L(t)$ is the *low pass equivalent impulse response*. Using (1.3.1-2) it follows that

$$h(t) = 2\,\text{Re}\left[h_L(t) e^{i2\pi f_0 t} \right] \tag{1.3.2-6}$$

where $h_L(t)$ is the Fourier transform of $H_L(f)$, thus

$$h_L(t) = F^{-1}\left[H_L(f) \right] = \int_{-\infty}^{\infty} H_L(f) e^{i2\pi ft} df \tag{1.3.2-7}$$

where $F^{-1}(g(t))$ is the inverse transform of $g(t)$. In problem 1 it is shown that the input complex envelope is related to the output complex envelope, by what one would perhaps expect:

$$v_0(t) = \int_{-\infty}^{\infty} v_i(u) h_L(t-u) du \tag{1.3.2-8}$$

And in addition, the Fourier transform of the output spectrum is related to the input spectrum by

$$V_0(f) = V_i(f) H_L(f) \tag{1.3.2-9}$$

where

$$\begin{aligned} V_0(f) &= F[v_0(t)] \\ V_i(f) &= F[v_i(t)] \end{aligned} \tag{1.3.2-10}$$

Figure 1.3-1 illustrates the signals and their transforms, along with the complex envelope and its transform.

Example 1 Consider a simple resonant circuit as shown in Figure 1.3-2. It can be shown that the transfer function is given by

$$H(f) = \frac{1}{1 + iQ\left(\dfrac{f}{f_0} - \dfrac{f_0}{f}\right)}$$

where the system is narrowband if $Q >> 1$. The transfer can be written approximately as

$$H(f) \cong \frac{1}{1 + i2Q\left(\dfrac{f - f_0}{f_0}\right)} + \frac{1}{1 - i2Q\left(\dfrac{-f - f_0}{f_0}\right)}$$

so that one can conclude from (1.3.2-1) that

$$H_L(f) = \frac{1}{1 + i2Q\left(\dfrac{f}{f_0}\right)}$$

and hence the baseband equivalent transfer function has been determined. Note that a second-order bandpass transfer function produced a first-order baseband equivalent transfer function.

To complete our characterization of bandpass signals, one more point is of interest. That is the power spectral density of the radio frequency signal. It may be obtained from the complex envelope power spectral density from the fact that when the complex envelope is a wide sense stationary random process then the autocorrelation function of the radio frequency signal $x(t)$ is given by

$$R_x(\tau) = \mathrm{Re}\left\{R_v(\tau) e^{i\omega_0 \tau}\right\} \tag{1.3.2-11}$$

It then follows that the power spectral density of the radio frequency signal is given by

$$S_x(f) = \frac{1}{2} S_v(f - f_0) + \frac{1}{2} S_v(-f - f_0) \tag{1.3.2-12}$$

where $S_v(f)$ is the power spectral density of the complex envelope and $S_x(f)$ is the power spectral density of the radio frequency signal. Note that the result of converting from the complex envelope of the radio frequency signal is to shift ½ of the spectrum up to f_0 and ½ of the spectrum down to $-f_0$.

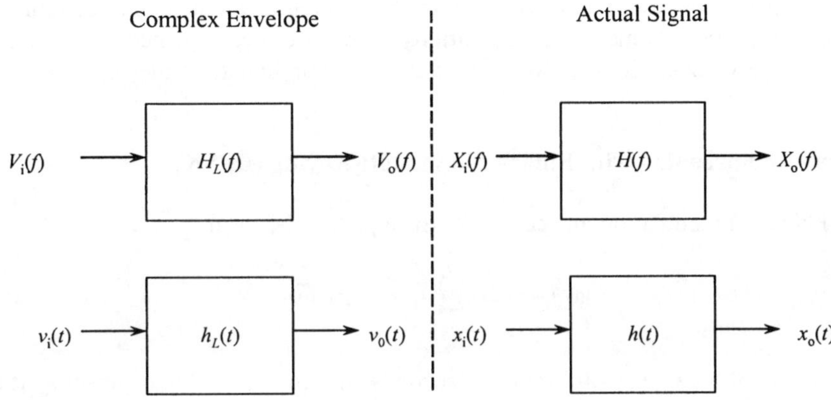

Figure 1.3-1 Relationship of the actual and complex envelope of a signal.

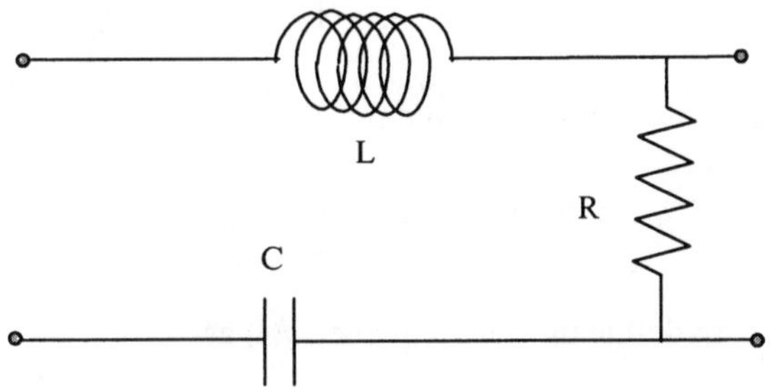

Figure 1.3-2 A simple resonant LCR (inductance, capacitance, and resistance) circuit.

1.3.3 Complex Envelope Characterization for Direct Sequence and Frequency-Hopping Signals

In the following sections the concept of the complex envelope will be applied to various forms of direct sequence and frequency-hopping signals [10, 11]. As mentioned in Section 1.3.1, the actual transmitted signal will be assumed to be of the form

$$s(t) = \mathrm{Re}\left\{v(t)e^{i\omega_0 t}\right\} \tag{1.3.3-1}$$

where $v(t)$ is the complex envelope of the code and data modulation. Thus any narrowband digital modulation process $s(t)$ can represented in this manner for any digital data and code modulation used.

In the following sections $p_T(t)$, will denote a unit rectangular pulse of duration T seconds. In addition, it will be assumed that the chip rate is much higher than the data rate. It will also be assumed for direct sequence signals that when data is present, an integer number of direct sequence chips will fit exactly into one data bit.

1.4 DIRECT SEQUENCE SPREAD SPECTRUM SYSTEMS

There are three primary ways to spread the spectrum of a signal. The first is direct sequence spread spectrum techniques, the second is frequency hopping, and the third is time hopping. The first two are much more common in practice than time hopping. In addition to these three techniques, any combination of the three types can be used in a hybrid scheme. There are various forms of direct sequence spreading, and they will be elaborated on in the following sections, followed by frequency hopping and time hopping.

1.4.1 Direct Sequence Spreading with Binary Phase Shift Keying (BPSK)

For BPSK chip (PN code) modulation, the complex envelope has the form

$$v(t) = \sqrt{2P}\sum_k a_k p_T(t-kT) \tag{1.4.1-1}$$

where $P_T(t)$ is a unit amplitude pulse of duration T seconds, starting at $t = 0$ and finishing at $t = T$ seconds. It is seen that $v(t)$ is real. In (1.4.1-1), P denotes the spread spectrum signal power, and the summation represents the direct sequence chip sequence in which the coefficients take on the values

$$a_k \in \{1, \ -1\} \tag{1.4.1-2}$$

For convenience it will sometimes be useful to denote the chip sequence by the notation $PN(t)$, so that

$$PN(t) = \sum_k a_k p_T(t-kT) \tag{1.4.1-3}$$

In addition it will be convenient to define the data sequence d(t) as

$$d(t) = \sum_j d_j p_{MT}(t-jMT) \tag{1.4.1-4}$$

where $p_{MT}(t)$ is a pulse of amplitude 1 and duration of MT seconds, when time is within the time period 0 to MT seconds, and is zero otherwise. Thus it is assumed that there are exactly M chips per bit or coded bit. The sequence d_j are the individual data symbols, which take on the values of either 1 or –1. In our modeling it will be assumed that data bits take on the value of 1 or –1 with the probability of 1/2, and each symbol is assumed to be statistically independent of each other. When BPSK data modulation is employed with BPSK chip modulation, the complex envelope can be written in terms of the two series as

$$v(t) = \sqrt{2P}\sum_j d_j p_{MT}(t-jMT)\sum_k a_k p_T(t-kT) \tag{1.4.1-5}$$

where it is assumed that there are exactly M PN chips per data bit (or coded symbol), where M is an integer. Alternatively the complex envelope can be written compactly as

$$v(t) = \sqrt{2P}d(t)PN(t) \tag{1.4.1-6}$$

Thus, from (1.3.3-1) and (1.4.1-5), the BPSK direct sequence modulated, BPSK data modulated, spread spectrum signal is of the form:

$$s(t) = \sqrt{2P}\sum_j d_j p_{MT}(t-jMT)\sum_k a_k p_T(t-kT)\cos(\omega_0 t) \tag{1.4.1-7}$$

This can alternatively be written more compactly as

$$s(t) = \sqrt{2P}d(t)PN(t)\cos(\omega_0 t) \tag{1.4.1-8}$$

Let us define a *random sequence* to be a sequence in which each symbol value outcome is statistically independent from all other symbol outcomes, and the probability of a "1" or a "–1" is one half. The power spectral density of the complex envelope of a BPSK chip and data modulated direct sequence signal, when the sequence and the data can be modeled as a random sequence, is given by

$$S(f) = PT\text{sinc}^2(fT) \tag{1.4.1-9}$$

where P is the power of the baseband waveform, T is the chip duration, and

$$\text{sinc}(x) = \frac{\sin(\pi x)}{(\pi x)} \tag{1.4.1-10}$$

Notice in this special case the spectral density is not dependent on the bit duration as long as M is greater than 1 and an integer. For narrowband processes that can be modeled as a wide-sense stationary random process, the power spectral density of the complex envelope of a signal and that of the radio frequency version of the signal is essentially the same, except for the shift in frequency. That is, the radio frequency power spectral density of a *PN* chip and data modulated BPSK signal, modulated by random chip and data sequences, is given by

$$S(f) = \frac{PT}{2}\text{sinc}^2((f - f_0)T) + \frac{PT}{2}\text{sinc}^2((f + f_0)T) \tag{1.4.1-11}$$

This translation from baseband to bandpass at frequency f_0 with half the power located at f_0 and half the power at $-f_0$ will be assumed to apply to all the modulations to follow, with the appropriate complex envelope power spectral density inserted. If the chip sequence cannot be modeled as a random sequence, then these expressions do not apply exactly but may approximate the spectral envelope. All finite length codes will have a discrete line structure; however, long maximal length code spectral densities can be well approximated by (1.4.1-9). Short codes, especially nonmaximal length codes, may still roughly follow the envelope of the spectral density expression of (1.4.1-9), but will have a discrete line spectrum structure. If the continuous spectral density of (1.4.1-11) is not a good approximation, then the actual spectral density would have to be used to obtain a more accurate spectral density representation.

Figure 1.4-1 illustrates an example of a transmitter for a BPSK DSSS system. Mathematically the multiplication operations can be interchanged; the goal is to produce the signal indicated at the output of the last multiplier. Example 2 discusses the waveforms at various places in the signal path.

Example 2 Figure 1.4-2(a) illustrates a baseband nonreturn to zero (NRZ) bit sequence in which there are three data bits "1, –1,1." Figure 1.4-2(b) illustrates a *PN* sequence composed of six chips taking on the values "–1,1,1, –1,1, –1," over the duration of three data bits; hence the processing gain (see Section 1.4.1.1) is 2 in this example. When the direct sequence code or pseudonoise code waveform *PN*(t) is phase shift keyed onto the carrier, the effect of the modulation can again be expressed as a multiplicative factor so that the data and pseudonoise coded signal, when BPSK modulation is used for both, can be simply expressed by

$$s(t) = d(t)PN(t)\cos(\omega_0 t) \tag{1.4.1-12}$$

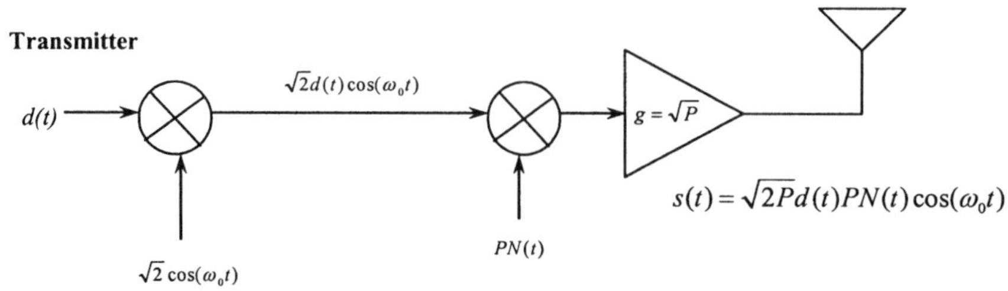

Figure 1.4-1 Direct sequence spread spectrum BPSK transmitter.

where the power is 1/2, since the amplitude is unity. The code symbols (chips) are of duration $T = 0.5$. The cosine wave ($\cos(2\pi 5t)$) is illustrated in Figure 1.4-2(c), and the resulting signal $s(t)$ (P =1/2) is shown in Figure 1.4-2(d). Notice that whenever the product $d(t)PN(t)$ changes algebraic sign, the phase of the carrier is flipped (the phase changes by π radians), as shown in the Figure 1.4-2(d) by the four inner vertical lines in the figure. There are in fact four phase changes of magnitude π radians in Figure 1.4-2(d). The noise has been neglected in these plots to clarify the signal operations.

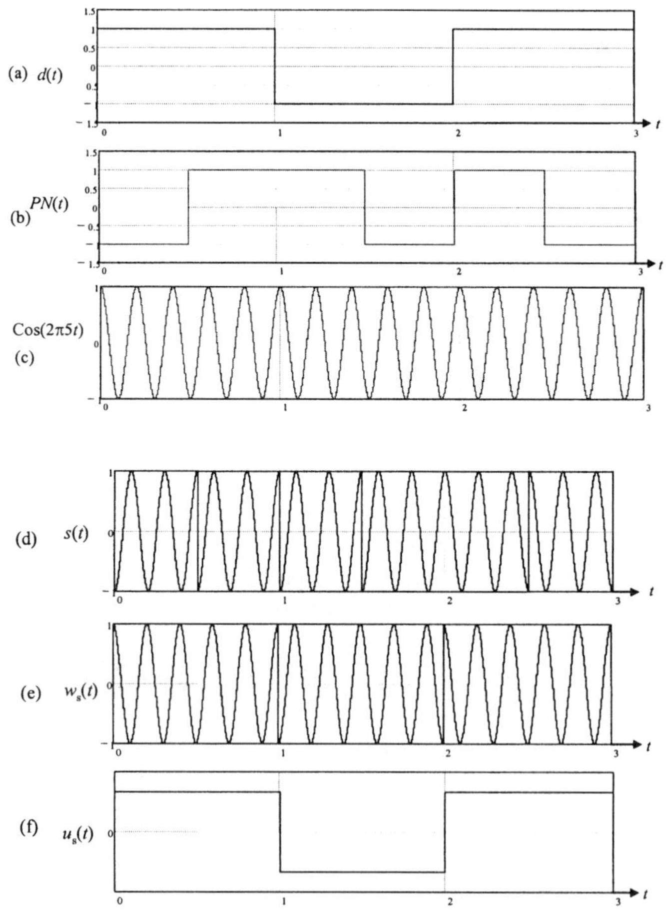

Figure 1.4-2 (a–f) An example of a BPSK data and PN modulated signal.

Figure 1.4-3 illustrates a model of a DSSS receiver and Figure 1.4-2(d–f) applies to the receiver processing. After *despreading* (multiplying by the receiver code $PN(t)$ at the proper timing), the code is removed, as shown in Figure 1.4-2(e) (again the noise has been neglected). Finally after coherent demodulation, the carrier is removed from the bit stream and the data is recovered in the data demodulator shown in Figure 1.3-2(f) (actually an estimate is obtained, $\hat{d}(t)$).

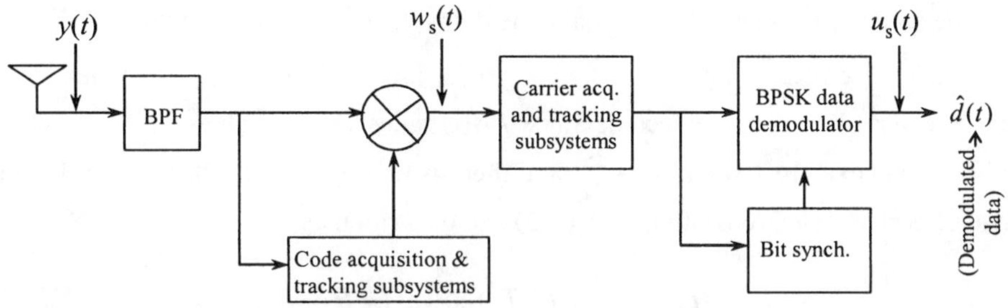

Figure 1.4-3 Direct sequence spread spectrum BPSK receiver.

For the transmitter depicted in Figure 1.4-1, the primary spectral lobe of the transmitted signal is about $2/T$ Hz wide, where T is the chip time in the pseudonoise code sequence. If a pseudonoise code for spreading were not used with the data, so that only the data were transmitted, then the bandwidth would only be about $2/T_d$ Hz, where T_d is the data (or symbol if coded) duration. This is true if one takes the bandwidth as the mainlobe frequency width (of the signal spectral density). Thus the spread spectrum modulation increases the bandwidth considerably if $T_d/T >> 1$. The ratio $PG = T_d/T$ for BPSK modulation is called the *processing gain*. In this discussion it was assumed that both the data and the spread spectrum modulation were BPSK. However it is not necessary that the data also be BPSK; it could be any type of phase modulation in general, such as DPSK, but most commonly BPSK is found in applications for both the data and the spreading process. If the data modulation process is not BPSK, the transmitted signal can be written as

$$s(t) = \sqrt{2P_t}PN(t)\cos(\omega_0 t + \theta_d(t)) \tag{1.4.1-13}$$

where $\theta_d(t)$ represents the data phase modulation and P_t is the transmitted power. Clearly if the phase modulation is DPSK or BPSK, then the data can be factored out as expressed in (1.4.1-12).

Returning to the case that the data and code are BPSK modulated, it will be assumed that the transmission process attenuates the signal strength and causes a carrier phase delay as well as a time delay to the data and the spreading sequence. Then the received signal can be written as

$$y(t) = \sqrt{2P}d(t - T_p)PN(t - T_p)\cos(\omega_0 t + \theta) + n(t) + J(t) \tag{1.4.1-14}$$

where $P << P_t$ is the received (attenuated) signal amplitude and the delay is $T_p > 0$ seconds, the carrier phase θ is modeled as a uniform random variable, uniformly distributed between 0 and 2π radians, and the last two terms are the naturally occurring receiver noise $n(t)$ and any interference $J(t)$, including jamming that may be present at the pass band of the receiver front end. For now, only the signal plus thermal noise $n(t)$ will be considered in what follows.

The receiver for BPSK modulated signals must be able to obtain both time synchronization and phase synchronization with the received signal. As noted earlier, Figure 1.4-3 illustrates a model of a BPSK receiver. The first function of the receiver is to align the receiver timing with the received code. This process is called *code acquisition*. When the code of the receiver is aligned in time with the received signal code, then the receiver timing must be continually maintained with the code tracking circuit. This process of maintaining the receiver time aligned with the received code is called *code tracking*. In the tracking mode, the code is removed from the received signal by the mathematical process of multiplying the local code with the received waveform. Thus the despread signal waveform (neglecting the noise and interference terms for now) is of the form

$$w(t) = \sqrt{2P}d(t - T_p)PN(t - T_p)\cos(\omega_0 t + \theta)PN(t - \hat{T}_p) + PN(t - \hat{T}_p)n(t) \qquad (1.4.1\text{-}15)$$

Let $w_s(t)$ denote the signal part of the despread waveform $w(t)$, which is the first term in (1.4.1-15). The parameter $\hat{T}_p$ is the receiver estimate of the signal code delay T_p from the transmitter.

The locally generated estimate $PN(t - \hat{T}_p)$ is identical to $PN(t-T_p)$ when $T_p = \hat{T}_p$ if the transmitter and receiver filtering is neglected. Since $PN(t)$ takes on the values of +1 or –1 the product of $PN(t-\hat{T}_p)$ $PN(t-T_p)$ is exactly 1 when $T_p = \hat{T}_p$ and there is no appreciable filtering on the received signal. Thus in this case of perfect synchronization (1.4.1-15) can be written as

$$w(t) = \sqrt{2P}d(t - T_p)\cos(\omega_0 t + \theta) + n'(t) \qquad (1.4.1\text{-}16)$$

where $n'(t)$ is the product of $n(t)$ and $PN(t-\hat{T}_p)$. Figure 1.4-2(e) illustrates the signal part of $w(t)$, which we shall call $w_s(t)$, with the delay suppressed for drawing convenience. Note that the transitions only occur at the data transitions. At this point the signal portion of the received waveform is a despread (the spreading is removed) signal containing only the BPSK data modulated signal.

To demodulate BPSK data, it is necessary to provide a local carrier in phase synchronism with the received carrier. This entails acquiring the carrier phase, tracking the carrier phase, and using the regenerated carrier to demodulate the data. The product of the locally generated unit power carrier and the despread signal produces

$$u(t) = \sqrt{2}\cos(\omega_0 t + \hat{\theta})\left[\sqrt{2P}d(t - T_p)\cos(\omega_0 t + \theta) + n'(t)\right] \qquad (1.4.1\text{-}17)$$

When $\theta = \hat{\theta}$ (that is, when the receiver phase is the same as the received carrier phase), the signal is reduced to a delayed and attenuated version of $d(t)$, as shown in Figure 1.4-2(f), plus a noise term; however, the noise is not shown for clarification purposes. It has been assumed that the term at $2\omega_0$ has been neglected, since it would normally be negligible due to the filter action of the detector filter or a filter would be used in the design to attenuate the $2\omega_0$ term. Equation (1.4.1-16) yields the result of the coherent demodulation

$$u(t) = \sqrt{P}d(t - T_p) + \sqrt{2}\cos(\omega_0 t + \hat{\theta})n'(t) \qquad (1.4.1\text{-}18)$$

where the first term of $u(t)$ is denoted as $u_s(t)$.

After carrier acquisition and tracking are completed, a bit synchronizer is used to obtain data bit timing; the data detector, with this timing, then detects the data sequence, outputting a (delayed) data estimate of the transmitted data.

It is instructive to view the same process in the frequency domain, since it provides additional insight into the whole process. From (1.4.1-12) it is possible to approximate the spectral density of the waveform under some reasonable conditions. First it will be assumed that the PN code has a long period and is a random sequence in its generation of ones and minus ones. If the data operates at a submultiple of the chip rate and has synchronous time epochs, then the power spectral density of the BPSK modulated spread spectrum signal of (1.4.1-12) is given by

$$S_{PN}(f) = \frac{PT}{2}\text{sinc}^2\left[(f - f_0)T\right] + \frac{PT}{2}\text{sinc}^2\left[(f + f_0)T\right] \qquad (1.4.1\text{-}19)$$

where T is the chip duration, P is the received power, and the sinc(x) function is defined by (1.4.1-10). The spectral density of the despread signal, having only data modulated onto the carrier, is given by

$$S_d(f) = \frac{PT_d}{2}\text{sinc}^2\left[(f - f_0)T_d\right] + \frac{PT_d}{2}\text{sinc}^2\left[(f + f_0)T_d\right] \quad (1.4.1\text{-}20)$$

in which T_d is the bit duration in seconds.

Both power spectral densities are sketched in Figure 1.4-4. Notice that the spectral density is located about both f_0 and $-f_0$ frequencies in both cases. The spectral width of the spread signal mainlobe is $2/T$ Hz, or twice the chip rate. The spectral density of the demodulated data is illustrated in the bottom of Figure 1.4-4. The data spectral density is centered at baseband and has a spectral width of the mainlobe, of $2/T_d$. The power spectral density of the spread signal is $PT/2$ at the peaks' ($\pm f_0$) spectral level for the spread signal and PT_d for the despread signal. Furthermore, when $T_d >> T$ then the bandwidth of the spread signal is much greater than the bandwidth of the despread signal.

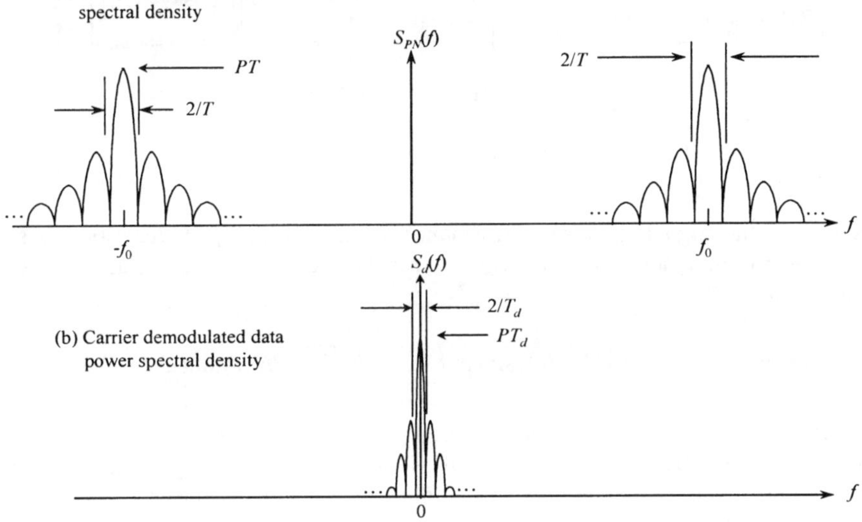

Figure 1.4-4 (a, b) A sketch of a BPSK spread spectrum signal before and after despreading.

1.4.1.1 Processing Gain for a BPSK Spread Spectrum Signal

It is convenient at this point to discuss the concept of processing gain of a BPSK direct sequence spread spectrum system for an uncoded communication system. Processing gain is a commonly used term for spread spectrum systems because it is a measure of how effective the spread spectrum system design is against a jammer. For a BPSK direct sequence system, the *processing gain* is defined as the ratio of the chip rate to the bit rate, which is the processing gain, and is given by

$$PG = \frac{R_c}{R_b} = \frac{T_b}{T_c} \quad (1.4.1.1\text{-}1)$$

as mentioned earlier.

Consider the coherent receiver for BPSK direct sequence signaling in Figure 1.4-5(a). Figure 1.4-5(b) will be discussed in the problem set. The input, $y(t)$, is a direct sequence spread spectrum BPSK signal with a tone jammer at the same frequency having power P_J. Hence the received waveform can be written as

$$y(t) = \sqrt{2P}d(t)PN(t)\sin\left(\omega_0 t + \theta_s\right) + \sqrt{2P_J}\sin\left(\omega_0 t + \theta_J\right) \quad (1.4.1.1\text{-}2)$$

where P is the signal power, P_J is the jammer power, θ_s and θ_J are the signal and jammer carrier phases, respectively, $PN(t)$ is the spread code, and $d(t)$ (± 1) is the bit sequence. The tilde over the $PN(t)$ term denotes filtering by the front end of the receiver. However we will neglect this filtering until Problem 4.

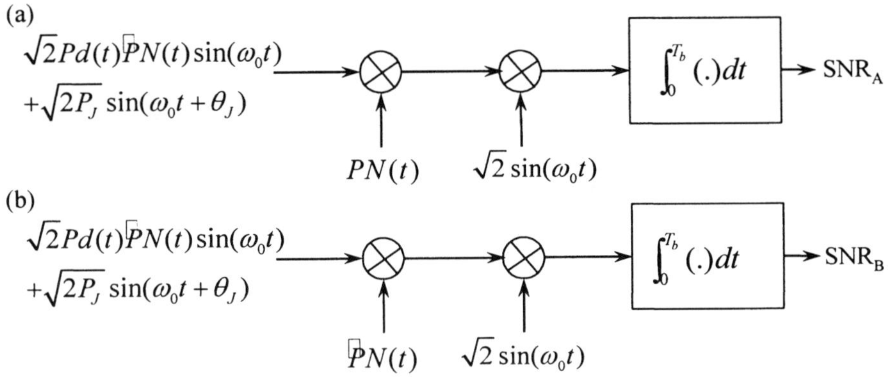

Figure 1.4-5 (a, b) Two possible ways to despread the received waveform.

Now consider the signal-to-noise performance of the model just described. Assuming that the receiver code generator is in synchronization with the received code, the signal, after the code despreading operation, is of the form

$$W_s(t) + W_n(t) = y(t)PN(t) = \sqrt{2P}d(t)\sin(\omega_0 t + \theta_S) + \sqrt{2P_J}PN(t)\sin(\omega_0 t + \theta_J) \quad (1.4.1.1\text{-}3)$$

where the despread signal term is $W_s(t)$ and the despread jammer term is $W_n(t)$.

It is to be noticed that the desired signal is despread (spectrally speaking) down to the data bandwidth, and in addition the tone jammer is spread out to the original bandwidth (since it is multiplied by the PN code). Thus the despreading operation has spread the undesired jammer and collapsed the desired signal. It should be clear that the "spread" jammer power would have less effect on the data detector than a narrowband jammer, since much of the jammer power is spread out beyond the detector data bandwidth. The carrier is demodulated down to baseband by multiplication by a coherent reference of the form

$$c(t) = \sqrt{2}\sin(\omega_0 t + \theta_r) \quad (1.4.1.1\text{-}4)$$

The result is given by

$$e_i(t) = \sqrt{P}d(t)\cos(\theta_S - \theta_r) + \sqrt{P_J}PN(t)\cos(\theta_J - \theta_r) + O(2\omega_0 t) \quad (1.4.1.1\text{-}5)$$

where the $O(2\omega_0 t)$ term denotes a term that is of the order of $2\omega_0 t$ and will be filtered out by the bit detector or a combination of a low-pass filter plus the bit detector. Now in order to continue, it is necessary to make some assumptions on the phases. It will be assumed that the receiver is coherent with the incoming signal so that $\theta_S = \theta_r$. In addition it will be assumed that the jammer is coherent with the signal since that is the most advantageous phase relationship for the jammer. Thus with these assumptions one obtains

$$e_i(t) = \sqrt{P}d(t) + \sqrt{P_J}PN(t) + O(2\omega_0 t) \quad (1.4.1.1\text{-}6)$$

Now let us compute the SNR out of the integrate-and-dump bit detector. First consider the signal part. Thus, as the signal output, at the end of a bit time, neglecting the $O(2\omega_0 t)$ term since it will be filtered out at the output of the data detector, one has

$$E\{e_o(T_b)\} = \sqrt{P}d(T_b)T_b \tag{1.4.1.1-7}$$

with $d(T_b)$ denoting the bit value during the time from zero to T_b, and $E\{x\}$ denoting the ensemble average of x. The $O(2\omega_0 t)$ will integrate to a value dependent on the inverse of $(2\omega_0)$ and consequently is negligible. Now the variance of the output is due to the despread jammer, which will be modeled as a random *PN* code, given by the product of the power spectral density at $f = 0$ and the noise bandwidth; thus

$$Var\{e_0(T_b)\} = E\left\{\left(e_0(T_b) - E\{e_0(T_b)\}\right)^2\right\} = P_J T_c NBW \tag{1.4.1.1-8}$$

The term $P_J T_c$ is the two-sided power spectral density of a random waveform *PN* code having symbol duration of T_c seconds and having amplitude 1 or –1 (a good approximate model for a long *PN* code) at $f = 0$, and *NBW* is the noise bandwidth of the integrate-and-dump filter. The two-sided noise bandwidth of the integrate-and-dump filter can be calculated as

$$NBW = \int_{-\infty}^{\infty} |H(f)|^2 df = \int_{-\infty}^{\infty} T_b^2 \frac{\sin(\pi f T_b)^2}{(\pi f T_b)^2} df = T_b \tag{1.4.1.1-9}$$

since the transfer function of an integrating filter is given by [12]

$$H(f) = T_b \frac{\sin(\pi f T_b)}{(\pi f T_b)} e^{-j\omega T_b/2} \tag{1.4.1.1-10}$$

Thus the variance can be written as

$$Var\{e_0(T_b)\} = P_J T_c T_b \tag{1.4.1.1-11}$$

The signal to noise ratio is defined to be the mean signal squared divided by the variance of the noise, so that

$$SNR = \frac{P}{P_J}\frac{T_b}{T_c} = \frac{P}{P_J}\frac{R_c}{R_b} = \frac{P}{P_J} PG \tag{1.4.1.1-12}$$

where $R_b = 1/T_b$ is the bit rate and $R_c = 1/T_c$ is the chip rate, and $PG = R_c/R_b$. In this case it is seen that the output signal to noise ratio is the signal to jammer power times the ratio R_c/R_b, which is the processing gain for a direct sequence BPSK SS and BPSK data modulated spread spectrum system when the system is uncoded. The processing gain is the reduction in the jammer power due to the spreading action on the narrowband jammer, which spreads the jammer out to a wideband process, that is much wider than the data bandwidth. The processing gain is an approximate measure of how effective the spread spectrum system is in the environment of a narrowband jammer.

It should be cautioned that when coding is employed in this same SS system, it is tempting to argue that the processing gain is diminished by the act of coding since the code rate is greater than the bit rate (for example, by 3 dB for rate 1/2 convolutional coding). Following this logic, one would conclude that therefore if the coding gain did not yield more gain than the code rate loss, coding would not be beneficial! This argument is fallacious [13] and in fact coding does improve the performance of spread spectrum systems, and it should be emphasized that *the processing gain is defined to be the ratio of the chip rate to bit rate* (and not the coded bit rate). Since R_c/R_b is usually much bigger than unity, the output SNR is enhanced by the processing gain.

The actual bit error rate (BER) performance of the communication system depends upon the actual code, its length, and the bandwidths involved. Short codes that have spectral lines that are relatively wide apart can have more or less effect on the detector than estimated from the processing gain, depending on where the lines fall in relationship to the detector passband. Codes that are very long, so that code length in time is large compared to the bit time, can be approximated by continuous spectra. Hence for long codes the continuous interference spectra may be used (the processing gain is a reasonable estimate of the improvement), whereas the discrete line nature of the spectra must be taken into account for short codes and will thus affect the effective processing gain.

It is also important to note that a wideband (relative to the data bandwidth) jammer, having a bandwidth equal to the *PN* code, is not reduced by the processing gain, since it is already of the *PN* code bandwidth. In other words the processing gain in a direct sequence system comes about from the spreading of a narrowband interference signal. And if it is already at least as wide as the direct sequence signal, spreading will, to first order, not appreciably reduce its spectral density in the data bandwidth, although there will be some reduction due to the convolution of the receiver code and the input jammer spectral density yields a lower spectral density at band center than the original spectral density at band center.

1.4.1.2 Correlation Loss of a RF Filtered BPSK Direct Sequence Signal

In this section an expression will be derived for the cross correlation that one obtains when the input signal is filtered and correlated with the local unfiltered code. Figure 1.4-6 illustrates the model that will be used. This discussion follows [8, 14]. This result will be useful to determine the loss in a correlated signal and the effect of a bandpass filter on the discriminator curve in code tracking analysis. In addition the effect of notch filters, as well as other chip symbols, such as Manchester symbols, can be readily investigated with these results. The received BPSK modulated direct sequence signal may be modeled as

$$s(t) = \sqrt{2P}\,\mathrm{Re}\left[PN(t)e^{j\omega_0 t}\right] = \sqrt{2P}PN(t)\cos(\omega_0 t) \tag{1.4.1.2-1}$$

where the signal power is P, and $PN(t)$ is the baseband direct sequence spread spectrum signal, with the carrier centered at ω_0, radians/second. The term Re[z] denotes the real part of the complex number z. Initially the *PN* sequence will be truncated to length $2TI$ seconds. It will be assumed that the *PN* sequence can be approximated by a sequence in which each chip is randomly picked with probability of 0.5, and each chip is statistically independent from each other. From (1.4.1.2-1) (data has been neglected), the finite time duration signal, in terms of the Fourier transform, produces

$$s(t) = \sqrt{2P}\,\mathrm{Re}\left[e^{j\omega_o t}\frac{1}{2\pi}\int_{-\infty}^{\infty}A_{2TI}(j\omega)e^{j\omega t}d\omega\right] \tag{1.4.1.2-2}$$

where $A_{2TI}(\omega)$ denotes the $2TI$ second segment Fourier transform of $PN(t)$ for $(|t| \le TI)$.

After filtering by the bandpass filter centered at ω_0, and denoting the low-pass equivalent filter of the bandpass filter $H(\omega)$ as $H_L(\omega)$, one can write for the low pass equivalent filter (see (1.3.2-1))

$$\begin{aligned} H_L(\omega-\omega_0) &= H(\omega) \quad \omega > 0 \\ &\quad 0 \qquad\quad \omega \le 0 \end{aligned} \tag{1.4.1.2-3}$$

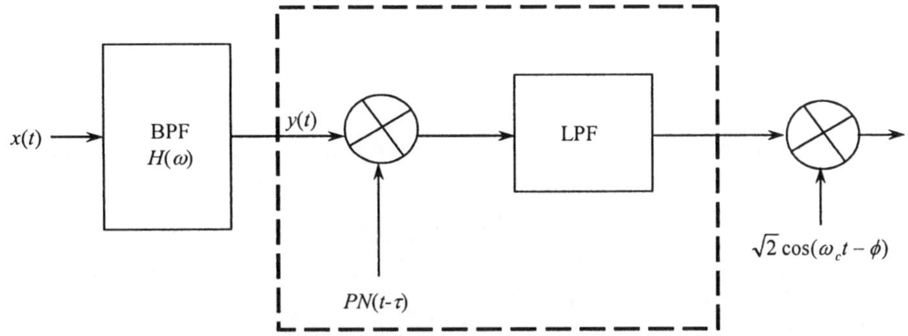

Figure 1.4-6 System model for the filtered correlation function.

It therefore follows from (1.3.2-9) that the response of a bandpass system to a bandpass signal can be obtained from the input complex envelope, which is real in this case (i.e., $PN(t)$) and the low-pass equivalent filter transfer function via

$$y_{2TI}(t) = \sqrt{2P}\,\mathrm{Re}\left[e^{j\omega_0 t}\frac{1}{2\pi}\int_{-\infty}^{\infty} A_{2TI}(j\omega)H_L(\omega)e^{j\omega t}d\omega\right] \tag{1.4.1.2-4}$$

In general the reference signal, composed of the code and the carrier, will be shifted in time and RF phase from $s(t)$ so that it will be modeled as

$$r(t) = \sqrt{2}\,\mathrm{Re}\left[PN(t-\tau)e^{j\omega_0 t - j\phi}\right] \tag{1.4.1.2-5}$$

With τ, the PN code time shift, and ϕ the RF carrier phase difference, the $2TI$ second portion of $r(t)$ may be written as

$$r_{2TI}(t) = \sqrt{2}\,\mathrm{Re}\left[e^{j\omega_0 t - j\phi}\frac{1}{2\pi}\int_{-\infty}^{\infty} A_{2TI}(j\omega)e^{j\omega(t-\tau)}d\omega\right] \tag{1.4.1.2-6}$$

Now define the cross correlation function between the two random functions $y_{2TI}(t)$ and $r_{2TI}(t)$ by

$$R(\tau,\phi) = \lim_{TI\to\infty}\frac{1}{2TI}E\left\{\int_{-\infty}^{\infty} y_{2TI}(t)r_{2TI}(t)dt\right\} \tag{1.4.1.2-7}$$

where $E\{x\}$ denotes the ensemble average of the random variable x. It will be assumed that the PN sequence can be modeled as a random sequence. Using the fact that

$$\mathrm{Re}\{z_1\}\mathrm{Re}\{z_2\} = \frac{1}{2}\mathrm{Re}\{z_1 z_2\} + \frac{1}{2}\mathrm{Re}\{z_1 z_2^*\} \tag{1.4.1.2-8}$$

and letting z_1 denote $r_{2TI}(t)$ and z_2 denote $y_{2TI}(t)$, and keeping only the baseband term (neglecting the term at $2\omega_0$) one obtains the result

$$\frac{R(\tau,\phi)}{\sqrt{P}} = \lim_{T\to\infty}\frac{1}{2TI}\mathrm{Re}\,E\left\{e^{j\phi}\left[\frac{1}{2\pi}\int_{-TI}^{TI} A_{2TI}(j\omega)H_L(\omega)e^{j\omega t}d\omega\right]\left[\frac{1}{2\pi}\int_{-TI}^{TI} A_{2TI}^*(j\omega')e^{-j\omega'(t-\tau)}d\omega'\right]\right\} \tag{1.4.1.2-9}$$

Now using the fact that

$$\frac{1}{2\pi}\int_{-\infty}^{\infty} e^{j(\omega-\omega')t}dt = \delta(\omega-\omega') \qquad (1.4.1.2\text{-}10)$$

It is well known that the power spectral density [14] of an ergodic random process is given by

$$S_{PN}(f) = \lim_{TI\to\infty} E\left\{\left[\frac{A_{2TI}(j\omega)A_{2TI}^{*}(j\omega)}{2TI}\right]\right\} \qquad (1.4.1.2\text{-}11)$$

Using (1.4.1.2-11) and (1.4.1.2-10) in (1.4.1.2-9) produces the basic result

$$R(\tau,\phi) = \max_{\phi}\sqrt{P}\,\mathrm{Re}\left[\frac{1}{2\pi}\int_{-\infty}^{\infty} S_{PN}(\omega)H_L(\omega)e^{j(\omega\tau+\phi)}d\omega\right] \qquad (1.4.1.2\text{-}12)$$

with the maximum taken to account for coherent phase tracking effect. The maximum value of (1.4.1.2-12) occurs by searching over τ. $R(\tau,\ \phi)$ is the cross correlation function between the bandpass filtered *PN* code and the local reference code, which is not filtered. It was assumed in this calculation that the correlation time was infinite. This calculation also assumed that the *PN* code could be modeled by a random binary valued (± 1) sequence. In practice, with an actual *PN* code, this result is an approximation. However, it is a very useful result that can be used to determine the effects of filtering on a spread spectrum direct signal. This includes the cross correlation function due to a band-limiting filter and other types of filters such as notch filters. In addition it is not limited to continuous spectra; discrete power spectral densities can be used for periodic codes, which include all "real" codes.

Let us consider the value of ϕ that makes $R(\tau,\phi)$ be a maximum when $H_L(\omega)$ *is complex conjugate symmetric* (that is, the baseband equivalent impulse response is real) and equivalently that

$$H_L^{*}(\omega) = H_L(-\omega) \qquad (1.4.1.2\text{-}13)$$

Using the fact that for z, a complex number that

$$\mathrm{Re}(z) = \frac{z+z^{*}}{2} \qquad (1.4.1.2\text{-}14)$$

in (1.4.1.2-12) and the fact that $S_{PN}(\omega)$ is real, produces

$$\frac{R(\tau,\phi)}{\sqrt{P}} = \left[\frac{1}{2\pi}\int_{-\infty}^{\infty} S_{PN}(\omega)\frac{H_L(\omega)e^{j(\omega\tau+\phi)} + H_L^{*}(\omega)e^{-j(\omega\tau+\phi)}}{2}d\omega\right] \qquad (1.4.1.2\text{-}15)$$

Since $H_L(\omega)$ is complex conjugate symmetric one obtains, with a change of variable on the second integral of (1.4.1.2-15), the result

$$R(\tau,\phi) = \frac{\sqrt{P}}{2\pi}\int_{-\infty}^{\infty} S_{PN}(\omega)H_L(\omega)e^{j\omega\tau}\cos\phi\, d\omega \qquad (1.4.1.2\text{-}16)$$

clearly $R(\tau,\phi)$ is maximized when $\phi = 0$, mod 2π. A maximum of $R(\tau,\phi)$ occurs at $\phi = 0$, so that the result for the maximum is given by

$$R(\tau,\phi)=\frac{\sqrt{P}}{2\pi}\int_{-\infty}^{\infty}S_{PN}(\omega)H_L(\omega)e^{j\omega\tau}d\omega \tag{1.4.1.2-17}$$

Therefore we conclude that when the baseband equivalent filter is complex conjugate symmetric that $R(\tau,\phi)$ is given by (1.4.1.2-16) and the maximum occurs at $\phi = 0 \bmod 2\pi$.

In order to obtain a feel for the effects of the filter bandwidth on the cross correlation function (actually the cross correlation function between the filtered and unfiltered codes), consider an ideal filter that is complex conjugate symmetric and a model for a multipole filter, having an RF bandwidth of B Hz and therefore a baseband equivalent bandwidth of $B/2$ Hz. Let

$$H_L(\omega)=\begin{cases} e^{-j\omega\alpha} & |\omega|<2\pi(B/2) \\ 0 & otherwise \end{cases} \tag{1.4.1.2-18}$$

where α is the delay of the filter. Assume that the *PN* code can be modeled by a random sequence. From (1.4.1.2-16) it follows that the cross correlation function at $\tau = \alpha$ is given by

$$R(\alpha,0)=\frac{\sqrt{P}}{2\pi}\int_{-\pi B}^{\pi B}T\left[\frac{\sin(\omega T/2)}{(\omega T/2)}\right]^2 d\omega \tag{1.4.1.2-19}$$

This value is plotted in Figure 1.4-7 showing the unfiltered case ($BT = 40$), the BT=1 and the BT= 2 cases. It is also of interest to obtain a plot of the correlation loss as a function of the BT product. This corresponds to the reduction in the maximum correlation value as a function of the BT product.

As another example of the correlation loss obtained from passing a spread signal through a filter, consider an ideal filter bandpass filter (BPF), which approximates a multipole bandpass filter. Figure 1.4-8 illustrates the correlation loss due a receiver filter having a predetection filter bandwidth B and a PN code of chip duration T, with an ideal bandpass front end filter. The loss is plotted in decibels, for various BT values from 20 log($L(BT)$), where $L(\infty)$=1 is the loss at BT=∞ and is of course 0 dB.

Example 3 Assume that a system designer is going to design a direct sequence spread spectrum system. Assume that the requirements are to use a multipole front-end filter and that the specifications only allow 0.5 dB loss due to this filter. What bandwidth should the designer select? Referring to Figure 1.4-8, it is seen that BT needs to be about 3.5 to keep the correlation loss to 0.5 dB. Assuming that the power spectral density is given by (1.4.1.2-19), then the RF bandwidth would be such that it would extend slightly less than between the second nulls.

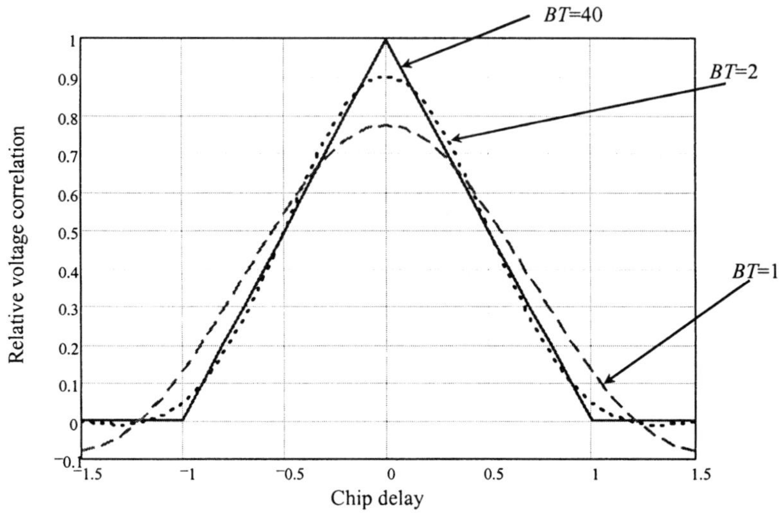

Figure 1.4-7 Voltage cross-correlation curve for various BT products.

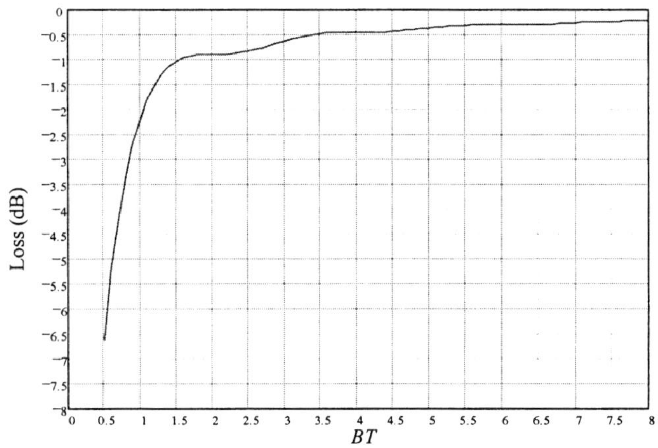

Figure 1.4-8 Correlation loss versus the BT product.

It should be emphasized that other filter types can be appraised with (1.4.1.2-12). For example, notch filters and low-pass filters can be considered for their effect on the correlation function. And the results apply to other ± 1 valued modulation formats besides NRZ, such as Manchester encoded symbols. In addition the power spectral density need not be continuous as has been assumed in these calculations; however, in this case the integral then collapses to a discrete sum.

1.4.1.3 Noise Spectral Density Reduction for the BPSK Despreading Process

In accessing the performance of a BPSK direct sequence spread spectrum system, it is noted that the despread spectral density of the noise is lower at the carrier frequency than without despreading. This calculation relates to code acquisition and data detection. Consider the model illustrated in Figure 1.4-9. Band-limited bandpass white noise (WGN) may be represented by

$$n(t) = \sqrt{2}n_c(t)\cos(\omega_0 t) + \sqrt{2}n_s(t)\sin(\omega_0 t) \tag{1.4.1.3-1}$$

where ω_0 is the center angular frequency and $n_c(t)$ and $n_s(t)$ are two baseband, statistically independent, band-limited white Gaussian random noise processes. The despread noise process, which is denoted by $n'(t)$, and is indicated in the figure, is given by

$$n'(t) = n(t)PN(t) \tag{1.4.1.3-2}$$

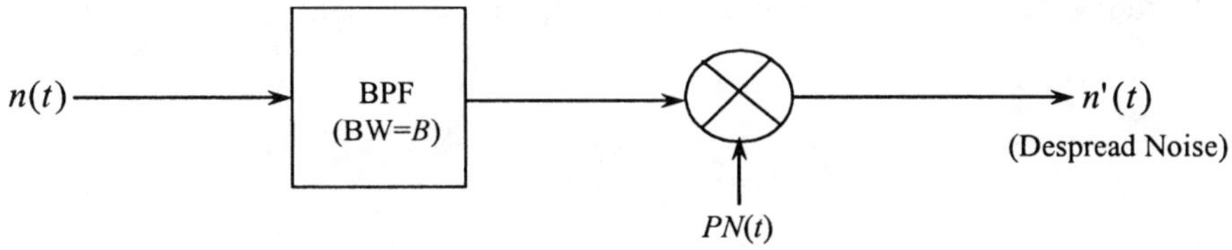

Figure 1.4-9 Despread noise model.

If the epoch time of the *PN* sequence is randomized to make it stationary, then the autocorrelation function of $n'(t)$ is given by

$$R_{n'}(\tau) = R_{PN}(\tau)R_{n_c}(\tau)\cos(\omega_0\tau) + R_{PN}(\tau)R_{n_s}(\tau)\cos(\omega_0\tau) \tag{1.4.1.3-3}$$

The power spectral density at ω is given by

$$S_{n'}(\omega) = 2\int_{-\infty}^{\infty} R_{PN}(\tau)R_{n_c}(\tau)\cos(\omega_0\tau)e^{-j\omega\tau}d\tau \tag{1.4.1.3-4}$$

The two quadrature noise correlation functions are identical. Now the cosine function can be expanded into the sum of two exponentials to yield

$$S_{n'}(\omega) = \int_{-\infty}^{\infty} R_{PN}(\tau)R_{n_c}(\tau)\left[e^{j\omega_0\tau} + e^{-j\omega_0\tau}\right]e^{-j\omega\tau}d\tau \tag{1.4.1.3-5}$$

Since the only nonnegligible component occurs at ω_0, one has

$$S_{n'}(\omega_0) = \int_{-\infty}^{\infty} R_{PN}(\tau)R_{n_c}(\tau)d\tau \tag{1.4.1.3-6}$$

By Parseval's theorem the integral of the product of the autocorrelation functions is equal to the integral of the product of the corresponding power spectral densities, so that

$$S_{n'}(\omega_0) = \frac{1}{2\pi}\int_{-\infty}^{\infty} S_{PN}(\omega)S_{n_c}(\omega)d\omega \tag{1.4.1.3-7}$$

Again, using the assumption of a random binary sequence valued *PN* code model, the spectral density of the random chip code model is given by

$$S_{PN}(f) = T_c\left(\frac{\sin\pi fT_c}{\pi fT_c}\right)^2 \tag{1.4.1.3-8}$$

where T_c is the chip duration in seconds. Changing the variable to f from ω in (1.4.1.3-7) produces our result

$$S_{n'}(f_0)=\frac{N_0}{2}\int_{-\infty}^{\infty}T_c\left(\frac{\sin\pi fT_c}{\pi fT_c}\right)^2\left|H_L(f)\right|^2 df \qquad (1.4.1.3\text{-}9)$$

where $H_L(f)$ is the low-pass (or baseband) equivalent transfer function of the bandpass filter transfer function. To obtain a feel of the values associated with the bandwidth of the filter let us again assume that the filter is an ideal BPF with bandwidth B. It follows from (1.4.1.3-9), with a change of variable to f from x, and where $x = fT_c$, that

$$S_{n'}(f_0)=\frac{N_0}{2}\int_{-BT_c/2}^{BT_c/2}\left(\frac{\sin(\pi x)}{\pi x}\right)^2 dx \qquad (1.4.1.3\text{-}10)$$

Figure 1.4-10 illustrates the noise power spectral density loss as a function of normalized bandwidth (BT_c). Note that this power spectral density loss is one half that for the signal loss with an ideal bandpass filter, in decibels. The reason is that the signal voltage loss is equal to the noise power spectral density loss squared.

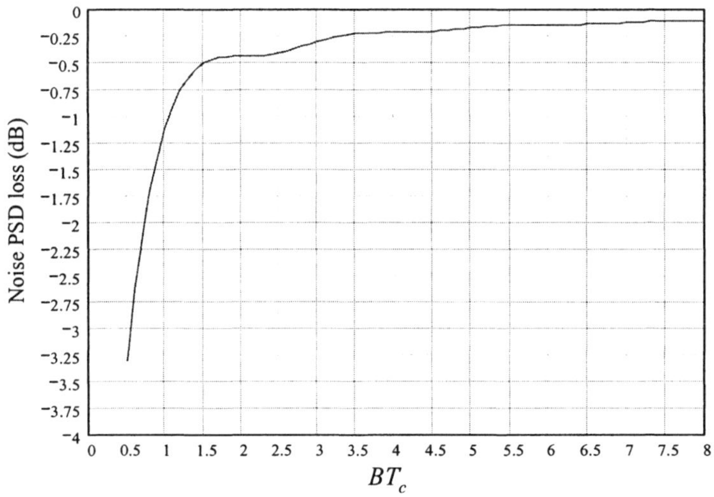

Figure 1.4-10 Noise spectral density decrease by despreading for various BT_c values.

At this point it is possible to account for the loss in C/N_0 as a function of the filtering of the spread signal and the spreading of the noise spectral density from despreading process by combining (1.4.1.3-10) and (1.4.1.2-12) in the form

$$\left(\frac{C}{N_0}\right)_{filtered}=\left(\frac{C}{N_0}\right)_{unfiltered}\frac{\max\limits_{\phi}\left\{\mathrm{Re}\left[\frac{1}{2\pi}\int_{-\infty}^{\infty}S_{PN}(\omega)H_L(\omega)e^{j\omega\tau+\phi}d\omega\right]\right\}^2}{\int_{-\infty}^{\infty}S_{PN}(\omega)\left|H_L(\omega)\right|^2 d\omega/(2\pi)} \qquad (1.4.1.3\text{-}11)$$

where in many cases the assumption of the random binary valued NRZ waveform can be used for the power spectral density, so that

$$S_{PN}(\omega)=T_c\left(\frac{\sin(\omega T_c/2)}{(\omega T_c/2)}\right)^2 \qquad (1.4.1.3\text{-}12)$$

with T_c being the chip duration. Thus, (1.4.1.3-11) accounts for the loss in signal and the loss in the noise spectral density, and therefore is an overall loss. It should be noted, however, that other code chip shapes besides *NRZ* symbols (such as Manchester symbols) could be used in this equation for the power spectral density. Further, there is no requirement to limit the spectral density to continuous ones; discrete spectral densities, derived from known period *PN* sequences, could be used as well.

1.4.2 Quadriphase Direct Sequence Spread Spectrum Systems

Generally quadriphase phase shift keyed (QPSK) data modulation is used for bandwidth efficiency since it takes one half the bandwidth of BPSK modulation. However for spread spectrum applications, bandwidth efficiency of the spread spectrum signal is not normally a concern, but the advantages in low probability of detection and less sensitivity to some types of jamming do offer advantages over BPSK modulated spread spectrum systems.

Two types of quadriphase SS modulators and demodulators will be discussed in this section: one with BPSK data modulation, and one with QPSK data modulation.

1.4.2.1 BPSK Data with QPSK SS Modulation

In this scheme BPSK data is multiplied (BPSK modulated) by an in-phase and quadrature carrier (the 90° hybrid provides the BPSK modulation on two carriers that are 90° apart), each of which is multiplied with a spreading code which is quasi orthogonal. Figure 1.4-11 illustrates the transmitter and the receiver for this scheme. The resulting transmitted complex envelope signal for QPSK spread spectrum modulation with BPSK data modulation (QPSK/BPSK) is given by

$$v_1(t)=\sqrt{P_t}\sum_j d_j p_{MT_c}(t-jMT_c)\sum_k\begin{bmatrix} a_k p_{T_c}(t-kT_c) \\ +ib_k p_{T_c}(t-kT_c)\end{bmatrix} \qquad (1.4.2\text{-}1)$$

where a_k and $b_k \in \{1,-1\}$ and are the in-phase and quadrature code sequences that have chip duration of T_c. seconds. The data sequence is unit valued also so that $d_j \in \{1,-1\}$. Also recall that $p_{T_c}(t)$ is a unit amplitude pulse starting at $t = 0$ and lasting for T_c seconds. Using (1.3.3-1) and denoting the two spreading sequences as $PN_1(t)$ and $PN_2(t)$ and the data sequence by $d(t)$ as

$$d(t)=\sum_j a_j p_{MT_c}(t-jMT_c) \qquad (1.4.2\text{-}2)$$

$$PN_1(t)=\sum_k a_k p_{T_c}(t-kT_c) \qquad (1.4.2\text{-}3)$$

$$PN_2(t)=\sum_k b_k p_{T_c}(t-kT_c) \qquad (1.4.2\text{-}4)$$

The transmitted QPSK SS signal can be written as

$$s(t)=\sqrt{P_t}d(t)\left[PN_1(t)\cos(\omega_0 t)-PN_2(t)\sin(\omega_0 t)\right] \qquad (1.4.2\text{-}5)$$

where P_t is the transmitted power, ω_0 is the radian carrier frequency, and $PN_i(t)$ (i = 1, 2) are the two spreading codes for the SS modulation. The received power in this waveform is P_t since each component contributes $P_t/2$ of the power, and the sine and cosine are orthogonal. Sometimes offset QPSK (OQPSK) is used instead of QPSK since when OQPSK is filtered and amplified in a nonlinear amplifier, the spectral sidelobes don't have as much regrowth, as with QPSK. This is normally a very desirable feature since significant spectral power outside ones allocated frequency band is not allowed. The resulting complex envelope transmitted signal for OQPSK spread spectrum modulation is given by

$$v_2(t) = \sqrt{P_t}\sum_j d_j p_{MT_c}(t - jMT_c)\sum_k \begin{bmatrix} a_k p_{T_c}(t-kT_c) \\ +ib_k p_T(t-kT_c-T_c/2) \end{bmatrix} \quad (1.4.2\text{-}6)$$

where again a_k and $b_k \in \{1, -1\}$ and $d_j \in \{1, -1\}$. Again the total power is P_t. Using (1.3.3-1) and denoting the two spreading sequences by $PN_1(t)$ and $PN_2(t-T_c/2)$ and the data sequence by $d(t)$, one has for the received OQPSK signal

$$s(t) = \sqrt{P_t}d(t)\left[PN_1(t)\cos(\omega_0 t) - PN_2(t-T_c/2)\sin(\omega_0 t)\right] \quad (1.4.2\text{-}7)$$

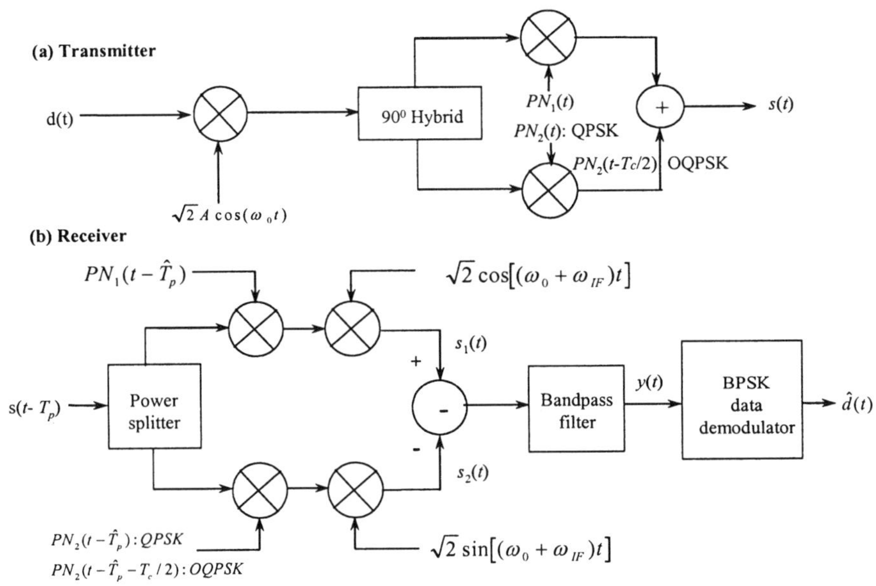

Figure 1.4-11 (a, b) BPSK data modulation with QPSK modulation.

where again $\sqrt{2P_t}$ is the peak voltage, and ω_0 is the radian carrier frequency. The power in this waveform is again P_t, since each component contributes $P_t/2$ of the power, and the sine and cosine are orthogonal, so that the powers add.

Continuing now with the QPSK spread spectrum modulation; the received signal will have been attenuated and delayed, so that the received signal is of the form

$$s(t) = \sqrt{P}d(t-T_p)\begin{bmatrix} PN_1(t-T_p)\cos(\omega_0 t+\theta) \\ -PN_2(t-T_p)\sin(\omega_0 t+\theta) \end{bmatrix} \quad (1.4.2\text{-}8)$$

where $\hat{T}_p$ is the propagation time delay estimate between the transmitter and the receiver, and where T_p is the propagation time delay between the transmitter and the receiver, $\sqrt{2P}$ is the received peak signal voltage, the received power is P, and θ is the carrier time delay expressed as a phase. The receiver is shown in part (b) of Figure 1.4-11. To demodulate the signal it is necessary to despread both phases of the quadriphase signal and then coherently demodulate the signal to baseband. The resulting signals entering into the summer are given by

$$\begin{aligned} s_1(t) &= \frac{\sqrt{P}}{\sqrt{2}} d(t-T_p) PN_1(t-T_p) PN_1(t-\hat{T}_p)\cos(\omega_{IF}t-\theta) \\ &+ \frac{\sqrt{P}}{\sqrt{2}} d(t-T_p) PN_2(t-T_p) PN_1(t-\hat{T}_p)\sin(\omega_{IF}t-\theta) \end{aligned} \tag{1.4.2-9}$$

$$\begin{aligned} s_2(t) &= -\frac{\sqrt{P}}{\sqrt{2}} d(t-T_p) PN_2(t-T_p) PN_2(t-\hat{T}_p)\cos(\omega_{IF}t-\theta) \\ &+ \frac{\sqrt{P}}{\sqrt{2}} d(t-T_p) PN_1(t-T_p) PN_2(t-\hat{T}_p)\sin(\omega_{IF}t-\theta) \end{aligned} \tag{1.4.2-10}$$

For OQPSK spread spectrum modulation in the receiver $PN_2(t-\hat{T}_p)$ is replaced with $PN_2(t-\hat{T}_p-T_c/2)$ and the transmitted signal has $PN_2(t)$ replaced with $PN_2(t-T_c/2)$, see Figure 1.4-11. Therefore $s_1(t)$ and $s_2(t)$, for OQPSK, become

$$\begin{aligned} s_1(t) &= \frac{\sqrt{P}}{\sqrt{2}} d(t-T_p) PN_1(t-T_p) PN_1(t-\hat{T}_p)\cos(\omega_{IF}t-\theta) \\ &+ \frac{\sqrt{P}}{\sqrt{2}} d(t-T_p) PN_2(t-T_p-T_c/2) PN_1(t-\hat{T}_p)\sin(\omega_{IF}t-\theta) \end{aligned} \tag{1.4.2-11}$$

$$\begin{aligned} s_2(t) &= -\frac{\sqrt{P}}{\sqrt{2}} d(t-T_p) PN_2(t-T_p-T_c/2) PN_2(t-\hat{T}_p-T_c/2)\cos(\omega_{IF}t-\theta) \\ &+ \frac{\sqrt{P}}{\sqrt{2}} d(t-T_p) PN_1(t-T_p) PN_2(t-\hat{T}_p-T_c/2)\sin(\omega_{IF}t-\theta) \end{aligned} \tag{1.4.2-12}$$

When the codes are aligned in time, the like-code terms will equal one whereas the cross code products will appear as spread interference noise across the code bandwidth and can often be neglected. After filtering by the bandpass filter and neglecting the cross code terms and summing, one obtains, for either the QPSK or OQPSK spread spectrum modulation, the result

$$y(t) = \sqrt{2P} d(t-T_p)\cos(\omega_{IF}t-\theta) \tag{1.4.2-13}$$

Thus after the SS signal processing, the resulting signal is the original BPSK (data) modulated signal, centered at the IF frequency. A BPSK data demodulator can demodulate this BPSK signal.

1.4.2.2 QPSK Data with QPSK Spread Spectrum Modulation

This form of QPSK allows two independent data streams to be modulated on the SS modulation. Figure 1.4-12 illustrates the transmitter and receiver for QPSK data and QPSK SS modulation. Each data stream is modulated on one of the (two) orthogonal carriers and then modulated by the respective *PN* sequences. Then the two components are summed to form the QPSK data and QPSK SS modulated signal (QPSK/QPSK). The resulting complex envelope signal can be represented by

$$v(t) = \left[A_1 \sum_j d_j p_{MT_c}(t - jMT_c) \sum_l a_l p_{T_c}(t - lT_c) \right] + \left[iA_2 \sum_m c_m p_{MT_c}(t - mMT_c) \sum_k b_k p_{T_c}(t - kT_c) \right] \quad (1.4.2\text{-}14)$$

where d_j is the in-phase data stream and c_m is the quadrature data stream and both have unit amplitude (i.e., d_j and $c_m \in \{1, -1\}$). The coefficients a_l and b_k are the in-phase and quadrature code sequences, respectively, and A_1 and A_2 are the respective in-phase and quadrature peak signal voltages. If the SS signal employs OQPSK modulation instead, then the corresponding complex envelope is changed to

$$v(t) = \left[A_1 \sum_j d_j p_{MT_c}(t - jMT_c) \sum_l a_l p_{T_c}(t - lT_c) \right] + \left[iA_2 \sum_m c_m p_{MT_c}(t - mMT_c) \sum_k ib_k p_T(t - kT_c - T_c/2) \right] \quad (1.4.2\text{-}15)$$

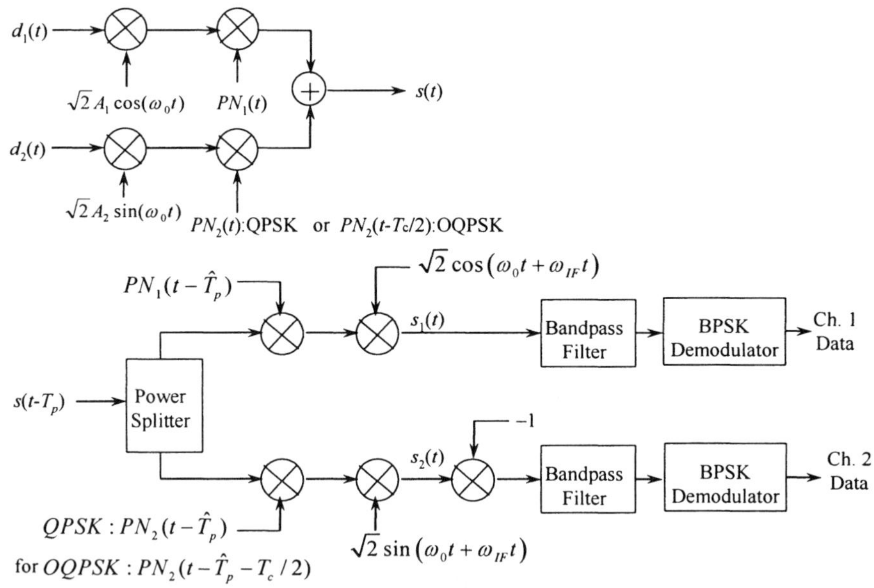

Figure 1.4-12 QPSK data modulation with QPSK SS modulation: (a) the transmitter and (b) the receiver.

The resulting received signal for QPSK or OQPSK is given by

$$s(t) = A_1 d_1(t) PN_1(t)\cos(\omega_0 t) - A_2 d_2(t) PN_2(t)\sin(\omega_0 t) \quad (1.4.2\text{-}16)$$

where the index 1 refers to the cosine (or in-phase) carrier and the index 2 refers to the sine (or quadrature) carrier, and the data summations and the sequence summations have been written in concise notation. The propagation delays and signal reduction have been neglected for convenience in (1.4.2-16). The total received carrier power in this case is given by

$$P = \frac{A_1^2}{2} + \frac{A_2^2}{2} \quad (1.4.2\text{-}17)$$

and they may not be equal, although they often are in practice. If they are equal and A_1=A_2=A, then the power $P = A^2$. At the receiver, the transmitted signal is given by a delayed and attenuated version

$$\begin{aligned} s(t) &= A_1 d_1(t-T_p)PN_1(t-T_p)\cos(\omega_0 t+\theta) \\ &- A_2 d_2(t-T_p)PN_2(t-T_p)\sin(\omega_0 t+\theta) \end{aligned} \tag{1.4.2-18}$$

where the $P_i = A_i^2/2$ $(i = 1 \text{ or } 2)$ are the component received signal powers, T_p is the propagation delay, and θ is the phase equivalent to the carrier time delay. After the receiver PN codes and carrier multiplications are completed in the receiver, the received signals become

$$\begin{aligned} s_1(t) &= \frac{A_1}{2} d_1(t-T_p)PN_1(t-T_p)PN_1(t-\hat{T}_p)\cos(\omega_{IF} t-\theta) \\ &\frac{A_2}{2} d_2(t-T_p)PN_2(t-T_p)PN_1(t-\hat{T}_p)\sin(\omega_{IF} t-\theta) \end{aligned} \tag{1.4.2-19}$$

$$\begin{aligned} s_2(t) &= -\frac{A_2}{2} d_2(t-T_p)PN_2(t-T_p)PN_2(t-\hat{T}_p)\cos(\omega_{IF} t-\theta) \\ &+ \frac{A_1}{2} d_1(t-T_p)PN_1(t-T_p)PN_2(t-\hat{T}_p)\sin(\omega_{IF} t-\theta) \end{aligned} \tag{1.4.2-20}$$

If offset QPSK SS modulation was used instead of regular QPSK SS modulation, then $PN_2(t)$ would be delayed one half of a *PN* chip. Continuing with QPSK, assuming the receiver codes are aligned with the received codes and additionally assuming that the two codes, $PN_1(t)$ and $PN_2(t)$, are essentially orthogonal (that is, the product of the two *PN* codes will integrate to almost zero at the end of a bit time), the result is given by

$$s_1(t) \cong \frac{A_1}{2} d_1(t-T_p)\cos(\omega_{IF} t-\theta) \tag{1.4.2-21}$$

$$s_2(t) \cong -\frac{A_2}{2} d_2(t-T_p)\cos(\omega_{IF} t-\theta) \tag{1.4.2-22}$$

These two signals can be demodulated with BPSK demodulators (after $s_2(t)$ is inverted in sign) to obtain the estimates of a delayed version of $d_1(t)$ and $d_2(t)$.

It is worth noting that the data can be modulated onto the carrier by other modulation means also. One other common form of modulation is differential phase shift keying (DPSK) that can be used with QPSK SS modulation. In fact in Section 1.4.2.1 one could replace BPSK with DPSK as the modulation, and replace the BPSK demodulators with DPSK demodulators in the receivers.

The processing gain for equal chip rate QPSK or OQPSK SS systems with equal data rates in the case of QPSK modulated data is in analogy with BPSK SS systems, given by

$$PG = \frac{R_c}{R_b} \tag{1.4.2-23}$$

where R_c is the chip rate on either phase of the QPSK SS signal, and R_b is the bit rate on either arm.

If the two code sequences are of the same chip rate and can be modeled as two statistically independent random sequences, with each chip taking on the values 1 or –1 with equal probability, and with the number of chips per bit period being an integer number and with the assumption that the data bits form statistically independent sequences, the power spectral density of a QPSK or a OQPSK signal is given by

$$S(f) = \frac{PT_c}{2}\text{sinc}((f - f_0)T_c) + \frac{PT_c}{2}\text{sinc}((f + f_0)T_c) \tag{1.4.2-24}$$

where T_c is the time duration of either chip, and P is defined in (1.4.2-14) and is the total power of the QPSK or OQPSK signal. Note that when the same chip rate on each phase of the QPSK or OQPSK signal is the same as a BPSK chip modulated signal, the power spectral density will be the same also.

Very commonly OQPSK (instead of QPSK) is used to limit the amount of spectral regrowth that occurs when a filtered QPSK (or OQPSK) signal is transmitted through a nonlinear power amplifier such as a TWTA.

1.4.3 Minimum Shift Keying (MSK)

Another method of SS modulation for direct sequence applications involves the use of minimum-shift keying (MSK). One advantage of MSK is based on the fact that the modulated spectral density of pure MSK has sidebands that drop off much faster than BPSK modulation. Also, MSK is more difficult to detect in a low probability of intercept (LPI) environment. An MSK modulator is similar to a QPSK modulator. Figure 1.4-13 illustrates an SS system with BPSK data modulation and MSK SS modulation.

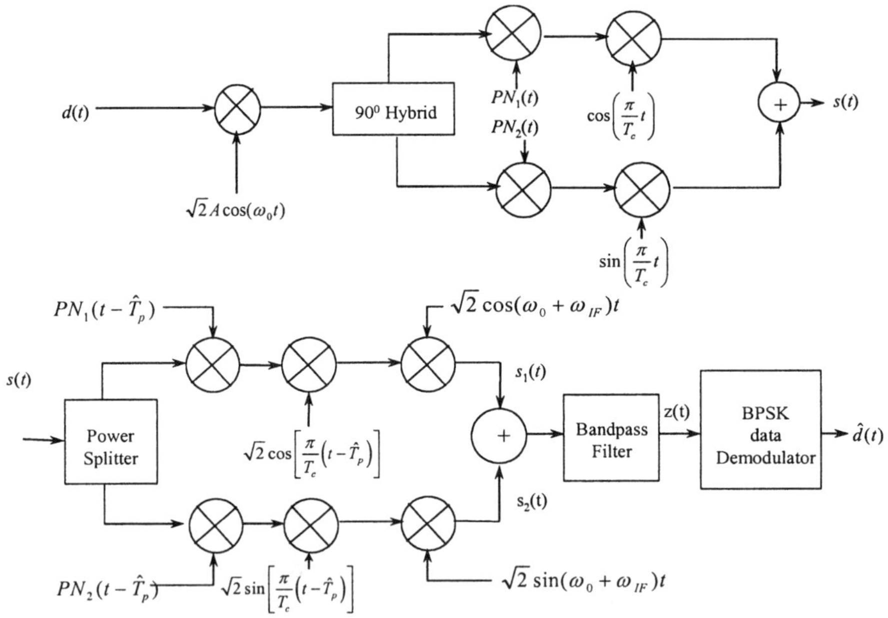

Figure 1.4-13 BPSK data modulation with MSK modulation: (a) the transmitter and (b) the receiver.

The transmitter complex envelope is given by

$$v(t) = \sqrt{2P}\sum_j d_j p_{MT_c}(t - jMT_c)\sum_k \begin{bmatrix} a_k p_T(t - kT_c)\cos(\pi t / T_c) \\ -ib_k p_T(t - kT_c)\sin(\pi t / T_c) \end{bmatrix} \tag{1.4.3-1}$$

Again denoting the two PN sequences a_k and b_k by $PN_1(t)$ and $PN_2(t)$, respectively, and the data by the first series in (1.4.3-1), the transmitted signal can be described by

$$s_t(t)=\sqrt{2P}d(t)\begin{bmatrix} PN_1(t)\cos\left(\dfrac{\pi}{T_c}t\right)\cos(\omega_0 t) \\ +PN_2(t)\sin\left(\dfrac{\pi}{T_c}t\right)\sin(\omega_0 t)\end{bmatrix} \tag{1.4.3-2}$$

where $PN_1(t)$ and $PN_2(t)$ are unit valued PN sequences for which $PN_i(t)\in\{1,-1\}$ for i = 1, 2. Note that the code sequence functions for MSK are not rectangular pulses, but rather are half period sinusoids whose epoch times are aligned with the rectangular chip sequences $PN_1(t)$ and $PN_2(t)$. Since the phase is not changed abruptly, the spectral density for MSK is narrower than either BPSK or QPSK in terms of the 99% power containment bandwidth. If the data can be modeled as a random sequence, and the same is true for the two statistically independent PN codes, and all the sequences are mutually and individually statistically independent, and the data period is an integer multiple of the chip period, then the power spectral density [15] for the transmitted carrier is given by

$$S(f)=\frac{8PT_c}{\pi^2}\left[\frac{\cos^2(2\pi(f-f_0)T_c)}{(1-16(f-f_0)^2T_c^2)^2}+\frac{\cos^2(2\pi(f+f_0)T_c)}{(1-16(f+f_0)^2T_c^2)^2}\right] \tag{1.4.3-3}$$

where P is the total power, and T_c is the chip duration. It should be noted that the power spectral density falls off asymptotically with frequency as f^{-4} whereas QPSK and BPSK fall off asymptotically as f^{-2}, and thus even though it has a wider main-lobe, it is more narrowband overall.

At the receiver, the signal is assumed to be attenuated and delayed by the propagation delay, which is denoted by T_p. Hence the received signal can be written as

$$s(t)=\sqrt{2P}d(t-T_p)\begin{bmatrix} PN_1(t-T_p)\cos\left(\dfrac{\pi}{T_c}(t-T_p)\right)\cos(\omega_0 t+\theta) \\ +PN_2(t-T_p)\sin\left(\dfrac{\pi}{T_c}(t-T_p)\right)\sin(\omega_0 t+\theta)\end{bmatrix} \tag{1.4.3-4}$$

with P denoting the received signal power now, and θ denoting the negative of the frequency delay product, $\theta=-\omega_0 T_p$. To see how the data is obtained from the receiver model of Figure 1.4-13, consider the product of the first code and the received signal.

$$PN_1(t-\hat{T}_p)s(t)=\sqrt{2P}d(t-T_p)\begin{bmatrix} PN_1(t-\hat{T}_p)PN_1(t-T_p) \\ \times\cos\left(\dfrac{\pi}{T_c}(t-T_p)\right)\cos(\omega_0 t+\theta) \\ +PN_1(t-\hat{T}_p)PN_2(t-T_p) \\ \times\sin\left(\dfrac{\pi}{T_c}(t-T_p)\right)\sin(\omega_0 t+\theta)\end{bmatrix} \tag{1.4.3-5}$$

Assuming that $\hat{T}_p=T_p$ and noting that the product of the different codes produce a broad spectral component somewhat similar to each original code spectra, they will be in large part removed by the bandpass filter that occurs after the summation. Assuming that the product of different codes is neglected, one has

$$PN_1(t-\hat{T}_p)s(t) \cong \sqrt{2P}d(t-T_p)\left[\cos\left(\frac{\pi}{T_c}\left(t-T_p\right)\right)\cos(\omega_0 t+\theta)\right] \tag{1.4.3-6}$$

After multiplying by the cosine tone with the assumption that $\hat{T}_p = T_p$, one obtains

$$\sqrt{2}\cos\left[\frac{\pi}{T_c}(t-\hat{T}_p)\right]PN_1(t-T_p)s(t) = \sqrt{P}d(t-T_p)\cos(\omega_0 t+\theta) \tag{1.4.3-7}$$

after neglecting the sum term, which will be filtered out. Now multiplying by the tone at the frequency $\omega_0 + \omega_{IF}$, one obtains the result

$$s_1(t) = \frac{\sqrt{P}}{2}d(t-T_p)\cos(\omega_{IF}t-\theta) \tag{1.4.3-8}$$

Doing the corresponding processing on the lower part of Figure 1.4-13(b) produces

$$s_2(t) = \frac{\sqrt{P}}{2}d(t-T_p)\cos(\omega_{IF}t-\theta) \tag{1.4.3-9}$$

Adding the two signals $s_1(t)$ and $s_2(t)$, and assuming that the bandpass filter has a negligible effect on the data, one obtains

$$z(t) = \sqrt{P}d(t-T_p)\cos(\omega_{IF}t-\theta) \tag{1.4.3-10}$$

And thus the resulting BPSK signal can be demodulated with a BPSK modulator. It should be pointed out that MSK modulation can be implemented in a simpler way involving a frequency offset carrier with an MSK conversion filter. Details can be found in [10]. As a final note it is shown in Problem 10 that the processing gain of a BPSK bit modulated and MSK chip modulated communication system is about 2.1 dB less than when BPSK chip modulation is used. However, it is also true that the bandwidth is less.

1.5 FREQUENCY-HOPPED SPREAD SPECTRUM SYSTEMS

Frequency hopping is a spread spectrum technique in which the center frequency of transmission is changed periodically according to some pseudonoise sequence. The data modulation can be almost any type of modulation including noncoherent and coherent types. Noncoherent types include multiple frequency shift keying (MFSK) and differential phase shift keying (DPSK). Milstar MDR (medium data rate) would be an example of a noncoherent system in existence, and the crosslink of GPS would be an example of a coherent frequency-hopping link. In frequency hopping, when MFSK is employed two or more frequency offsets are possible at each hop time. In slow frequency-hopping DPSK a fixed number of symbols are placed on each hop, with one or more reference symbols preceding the data symbols.

Coherent modulation on a frequency-hopped system is commonly done with the use of a known sync word (or words) placed in the hop to provide a phase and timing estimate. This scheme typically requires two passes (*a two-pass scheme*): the first pass obtains the phase, timing, and possibly frequency, and the second pass demodulates the data.

The hopping frequencies are finite, and the difference of the highest to the lowest carrier frequencies defines the *hopping range.* The bandwidth of the channel needed to pass the whole range of hopping frequencies is the *hopping bandwidth* denoted by W. The bandwidth of the modulation, on a given hop, is denoted by BW_b. After the receiver has synchronized its hop pattern with the received signal and multiplies its reference signal with the received signal, it is referred to as the *dehopped signal.* The ratio of the hopped bandwidth to the data bit bandwidth is defined to be the *processing gain for frequency hopped signals*, which is compatible with the definition for direct sequence spread spectrum systems; that is, the processing gain for a frequency-hopped SS system is given by

$$PG = \frac{W}{BW_b} \tag{1.5-1}$$

where W usually is taken as the first null to first null bandwidth and BW_b is taken as twice the bit rate.

When only one hopping carrier is operating over the bandwidth, it is called *single channel frequency hopping modulation.* If more than one carrier operates over the bandwidth simultaneously, then it is called *multiple channel frequency-hopping modulation.* Figure 1.5-1(a) illustrates the transmitter and Figure 1.5-1(b) shows the receiver for operating a single frequency hopping system. The data modulates the carrier, and that in turn is frequency hopped according to the frequency hopping *PN* code used in the system design. At the receiver the signal is filtered with a bandpass filter to bandwidth W that passes the hopped signal. Then the signal is synchronized with the input signal and dehopped so that it is a fixed frequency feeding into the bandpass filter and demodulator. The bandpass filter eliminates the sum frequency term from the dehopping operation. The demodulator demodulates the dehopped signal to produce the estimate of the input data stream.

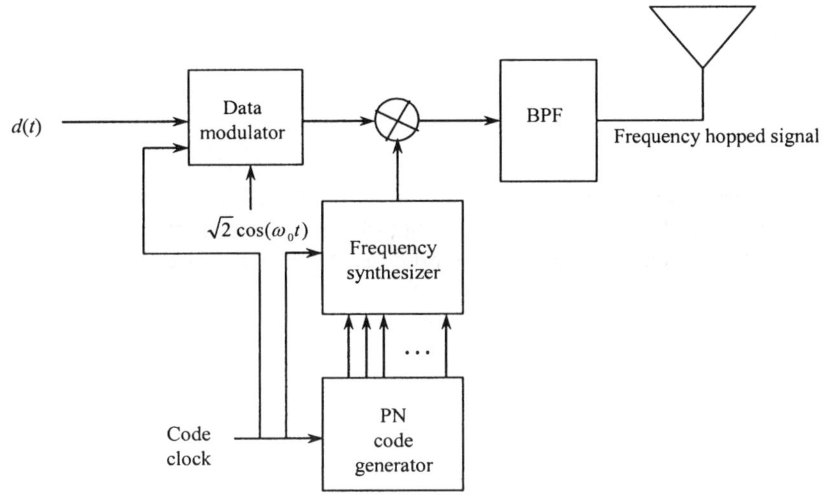

Figure 1.5-1 (a) A single channel frequency hopping modulation transmitter.

There are two types of frequency hopping. *Fast frequency-hopping* occurs when there is one or more (normally an integer number) frequency hops for each transmitted symbol (or, if uncoded, the transmitted bit). *Slow frequency-hopping* occurs when there are two or more symbols per frequency *hop time* T_h. Many frequency-hopped systems use noncoherent or differentially coherent modulation, since it is difficult to maintain phase coherence in the synthesis of the hopping frequencies over the hopping band, especially if it is very wide. However, a number of new systems are considering using coherent modulation on a frequency hopped signal. Only noncoherent frequency hopping will be considered here along with noncoherent, differentially coherent, and coherent data modulation.

Another important parameter in frequency-hopping systems is the frequency-hopping rate. This parameter is very significant when considering the *frequency-follower jamming.* A frequency-follower jammer is a jamming device that intercepts the hopping signal, amplifies it, and retransmits the signal to the receiver with the intention of jamming the receiver. It is important that the jamming signal reach the target receiver within the same hop, or else it will be quite ineffective. Therefore, it is clear that the higher the hop rate, the more resistant a receiver is to a frequency-follower jammer.

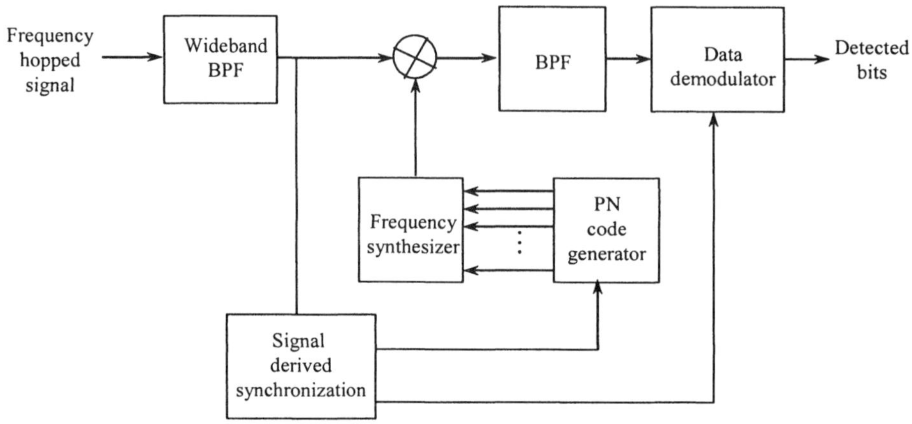

Figure 1.5-1 (b) A single channel modulation receiver.

Figure 1.5-2 illustrates the frequency-follower jammer scenario. The transmitter transmits to the receiver but is also observed by the frequency-follower jammer, which repeats the signal that must traverse both the distance $x2$ plus $x3$, as well as suffer the delay in amplifying the jammer. To be effective in jamming, the jammer must be at the same hop frequency and not delayed more than say, one half of a hop. It follows that to be effective as a follower jammer that

$$\frac{x2 + x3}{c} + T_{proc} \leq \alpha T_h + \frac{x1}{c} \tag{1.5-2}$$

where the xi (i = 1, 2, 3) are the distances involved, c is the speed of light, T_h is the hop duration, T_{proc} is the amplifier processing delay, and α is the fraction of a hop that is *not* hit (so that $(1-\alpha)\ T_{proc}$ is hit (overlaps)), but the jammer can still be effective in the jamming process. Rearranging (1.5-2) produces

$$x2 + x3 \leq x1 + c(\alpha T_h - T_{proc}) \tag{1.5-3}$$

If the inequality is changed to an equality, and it is assumed that the right-hand side is constant, one sees that the region of permissible operation is defined by being inside the ellipse with the location of the receiver and transmitter being the two foci. Rotating the ellipse about the x1 line will produce an ellipsoid. Thus in three dimensions an ellipsoid of revolution is defined as the boundary in which being inside allows the jamming equation to be satisfied, and outside the boundary the jamming equation cannot be satisfied. Clearly as the hop rate increases, the hop time decreases, and the volume of the ellipsoid decreases. Therefore a high hop rate is desirable to guard against a follower jammer.

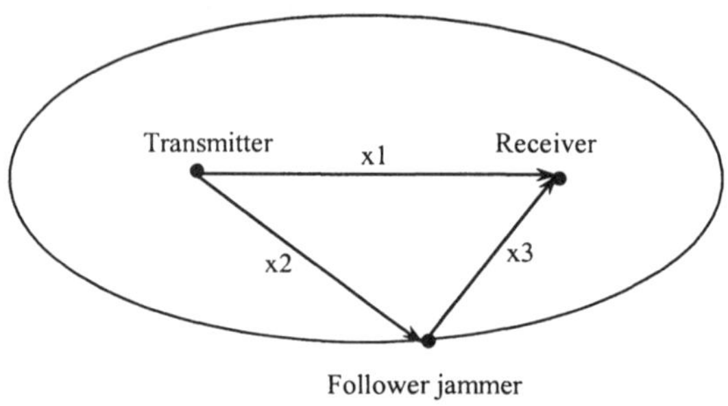

Figure 1.5-2 Geometry of a frequency-follower jammer.

1.5.1 Noncoherent Slow Frequency-Hopped Systems with MFSK Data Modulation

Consider now a slow frequency-hopping signal that has multiple frequency shift keying modulation (SFH/MFSK). Let the F_i be the modulation frequency at the i-th modulation tone time and modeled to be chosen randomly from the M modulation frequencies (tones), and let f_i be the i-th hopping frequency at time i, determined from the spread spectrum code used for the frequency-hopping signal and assumed to have M_h frequencies where M_h=2^k, and k is the number of bits needed to specify the frequency synthesizer.

The hopping frequencies are almost always equally spaced across the hopping band. Assume that the hop frequency separation is Δf, so that the RF bandwidth is approximately $W = M_h\Delta f$. Further let θ_i denote the statistically independent random variable that takes on a value in the range $\{-\pi,\pi\}$ during every modulation tone and changes from modulation tone to modulation tone. This random phase variable θ_i is statistically independent of both f_i and F_i. The random phase is associated with the hopping signal and the frequency modulation process. The complex envelope of a slow frequency-hopped signal, with MFSK data modulation, is given by

$$v(t)=\sqrt{2}A\sum_k e^{i2\pi f_k t}p_{T_h}(t-kT_h)\sum_j e^{i(2\pi F_j t+\theta_j)}p_{T_s}(t-jT_s) \quad (1.5.1\text{-}1)$$

where $p_T(t)$ denotes a rectangular pulse of unit height and duration T seconds that starts at time $t = 0$, and $A^2 = P$, the power. The RF signal is given by

$$x(t)=\text{Re}\left\{\begin{matrix}\sqrt{2}A\sum_k e^{i2\pi f_k t}p_{T_h}(t-kT_h)\\ \times\sum_j e^{i(2\pi F_j t+\theta_j)}p_{T_s}(t-jT_s)e^{i\omega_0 t}\end{matrix}\right\} \quad (1.5.1\text{-}2)$$

This can be simplified somewhat to

$$\begin{aligned}x(t)=&\sqrt{2}A\sum_k p_{T_h}(t-kT_h)\sum_j p_{T_s}(t-jT_s)\\ &\times\cos\left[\left(2\pi f_k+2\pi F_j+\omega_0\right)t+\theta_j\right]\end{aligned} \quad (1.5.1\text{-}3)$$

The power during every tone is $P = A^2$. The hop duration is T_h, the frequency shift keyed modulation (tone modulation) time duration is T_s, and the channel bit time duration is T_{CB}. Note that the complex envelope of the waveform is the product of the hopping complex envelope and the modulation complex envelope, and in general each complex envelope would have a random phase associated with it. However, in this case only one random phase is needed, since the sum of two independent random phases is again a random phase over $\{-\pi,\pi\}$. In the case of *slow frequency-hopping* $T_h > T_s$, and it will be assumed here that $T_h/T_s = N$, an integer greater than one; in other words, there are N MFSK tone symbols per hop ($N>1$).

For the case of an M-ary FSK modulation system, one of M tones is transmitted every T_s seconds, where T_s is the channel symbol (tone) duration. The input data stream is composed of channel bits, which have time duration T_{CB} and $T_s = T_{CB}\log_2 M$. If the channel is not encoded, a bit time would equal a channel-bit time. Normally, the modulation tones are spaced far enough apart so that they are orthogonal. Thus the spacing of the tones would be $1/(T_{CB}\log_2 M) = 1/T_s$ Hz apart. The corresponding modulation bandwidth is therefore approximately $BW_b = M/(T_{CB}\log_2 M) = M/T_s$ Hz. Now after each T_h seconds a new hop frequency is transmitted, and after each T_s seconds a new modulation frequency is transmitted. Normally the hop and modulation tones start at the same clock time and the hop frequency changes every $NT_s = T_h$ seconds. The frequency separation of adjacent hopping frequencies can be on the order of a few Hz or less, or they can be separated as far apart as the modulation bandwidth.

At the receiver the dehopping synthesizer is used to dehop the incoming signal, which leaves the modulation frequencies (possibly offset by an IF) available to demodulate, thereby producing the original data bit sequence (impacted by the noise). It is useful at this point to view the hopping/modulation process pictorially in an example.

Example 4 Now consider Figure 1.5-3, which illustrates a simple example of a 4FSK data modulation scheme with slow hopping. In this example $M = 4$, so that $T_s = \log_2(4)T_{CB} = 2\ T_{CB}$. Further it will be assumed for simplicity that there are eight hopping frequencies, and it will be assumed that the hopping frequencies are spaced apart by the modulation bandwidth. Therefore the total bandwidth is about $W = (8)(4)/(T_s) = 32/T_s$ Hz. In this example it would take three bits to specify each one of eight synthesizer frequencies, and two channel bits at a time are used to specify the 4FSK modulation tones. It will also be assumed that $N = 2$, so that there are two 4FSK tone symbols per hop time. The figure shows a particular modulation frequency each symbol time and a particular hop frequency each hop time. Also the corresponding input channel bit sequence is shown at the bottom of the figure.

At the receiver the dehopping synthesizer is used to dehop the incoming signal that leaves the modulation frequencies (possibly offset by an IF) available to demodulate, thereby producing the original data bit sequence (at least without noise). This process is shown in Figure 1.5-4. The dehopped 4FSK modulation is shown in the figure, and the demodulated channel bits are shown below the figure in the case of no noise.

1.5.2 Noncoherent Fast Frequency-Hopped Systems with MFSK Data Modulation

In contrast to slow frequency hopping, fast frequency-hopping has one or more hops per data symbol (FFH/MFSK). Thus at least one hop, and often more than one hop, occurs during one transmitted data symbol. One major advantage of fast frequency hopping is the fact that it produces frequency diversity on each symbol, which is very advantageous when operating in the environment of a partial band jammer.

The complex envelope of a fast frequency-hopped signal with MFSK data modulation is given by

$$v(t) = \sqrt{2}A\sum_k e^{i(2\pi f_k t+\theta_k)} p_{T_h}(t-kT_h)\sum_j e^{i(2\pi F_j t)} p_{T_S}(t-jT_s) \tag{1.5.2-1}$$

where T_s is the duration of the data frequency modulation (MFSK) with frequencies F_j (assumed to be M equally spaced frequencies), $P = A^2$ is the power, and T_h is the duration of the frequency hops with N frequency hops per one data modulation frequency period (note that the N defined here is the inverse of that defined in Section 1.5.1). It is assumed that the M_h hop frequencies f_k, are pseudorandom variables that are constant from hop to hop, statistically independent from hop to hop, and uniformly distributed over 0 to 2π. The random variables θ_k are statistically independent random variables that are constant from hop to hop, statistically independent from hop to hop, and uniformly distributed over 0 to 2π. In the case of fast frequency hopping, $T_h \leq T_s$.

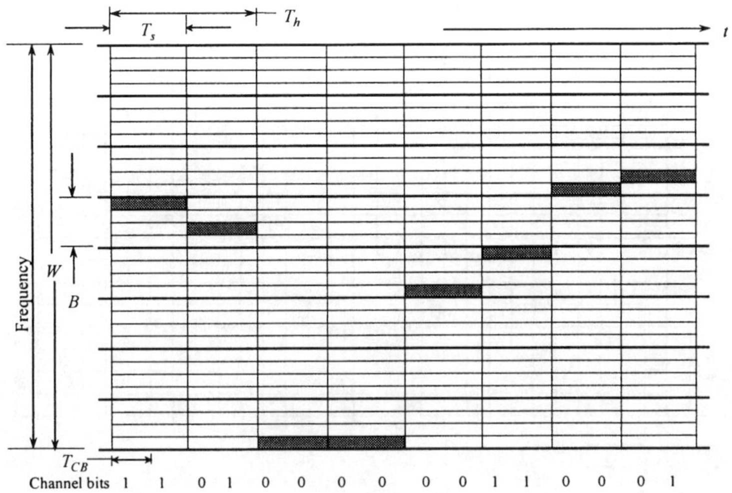

Figure 1.5-3 Slow frequency-hopping transmitter with MFSK modulation.

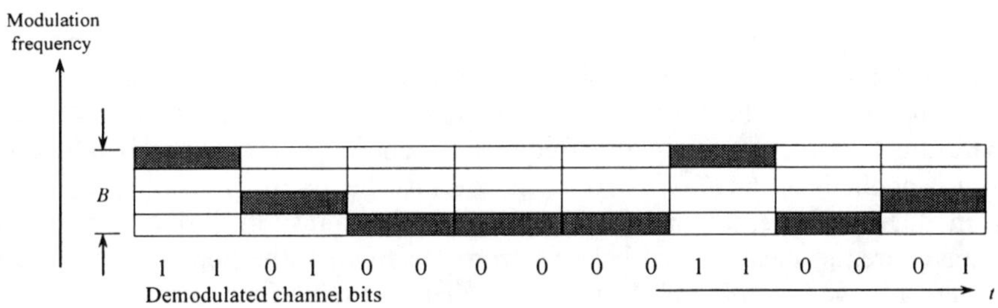

Figure 1.5-4 Slow frequency-hopping receiver with 4FSK modulation showing the dehopped waveform.

The RF signal associated with the indicated complex envelope is given by

$$x(t) = \mathrm{Re}\left\{\sqrt{2}A\sum_k e^{i(2\pi f_k t+\theta_k)} p_{T_h}(t-kT_h)\sum_j e^{i(2\pi F_j t)} p_{Ts}(t-jT_s)e^{i\omega_0 t}\right\} \tag{1.5.2-2}$$

or simplifying somewhat

$$\begin{aligned} x(t) &= \sqrt{2}A\sum_k p_{T_h}(t-kT_h)\sum_j p_{Ts}(t-jT_s) \\ &\times \cos\left[\left(2\pi f_k + 2\pi F_j + \omega_0\right)t + \theta_k\right] \end{aligned} \tag{1.5.2-3}$$

It will be assumed again that MFSK modulation is used, so that the channel bit time T_{CB} is related to the tone time in the same way as in slow hopping; that is, $T_s = \log_2(M)\ T_{CB}$. The approximate required bandwidth for MFSK modulation is $BW_b = M/T_h$, and the total channel bandwidth $W = 2^k BW_b$, where k is number of bits needed to specify the hop frequency, if essentially no overlapping of the modulation bandwidth BW_b is desired. Certainly this is not a requirement and can differ in that the number of hopping frequencies can be a much greater value.

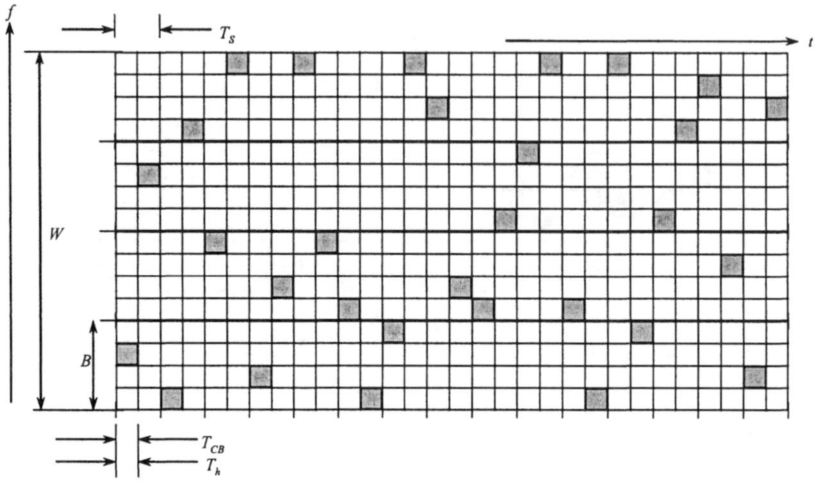

Figure 1.5-5 Fast frequency-hopping transmitter with MFSK modulation.

Example 5 Now consider a very simple example in which a 4FSK modulation system is hopped with four hop frequencies in such a way that the hopping frequencies are spaced apart by the modulation bandwidth. The approximate bandwidth for the four-tone modulation is in this case, $BW_b \cong 4/T_h$. Since it will be assumed that there are four different hopping frequencies, spaced apart by the bandwidth BW_b, the total hopped bandwidth is about $W \cong (4/T_h)4 = 16/T_h$. It will also be assumed that $N = 2$, that is there are two hops per 4-ary modulation tone. Figure 1.5-5 illustrates the transmitter frequencies and Figure 1.5-6 illustrates the receiver frequencies. As can be seen from the figures, the hops occur twice per information tone. Also since there are two bits per each 4-ary tone, a channel bit is of the same duration as a hop time.

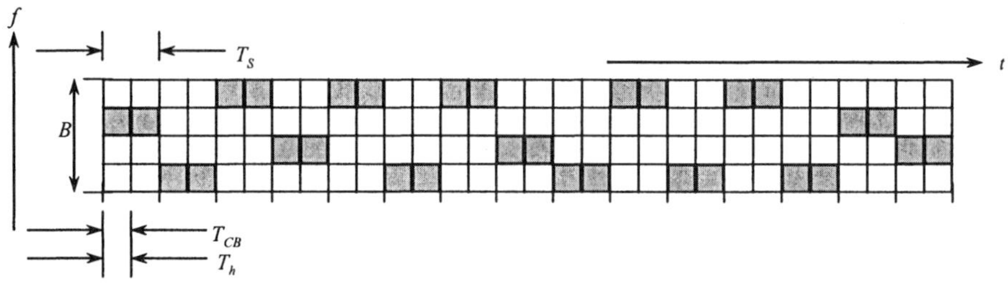

Figure 1.5-6 Fast frequency-hopping receiver with 4FSK modulation showing the dehopped waveform.

1.5.3 Noncoherent Slow Frequency-Hopped Signals with DPSK Data Modulation

The other common data modulation used with slow frequency-hopped signals is differential phase shift keying (SFH/DPSK). On each hop a burst of DPSK channel bits is transmitted, with the first one or more used for the reference bit and the rest used for data. The complex envelope of the frequency hopped signal with DPSK data modulation is given by

$$v(t) = \sqrt{2}A\sum_k e^{i(2\pi f_k t+\theta_k)} p_{T_h}(t-kT_h)\sum_j d_j p_{T_d}(t-jT_d) \tag{1.5.3-1}$$

where f_k are the hop frequencies, T_d is the DPSK data symbol duration, $P = A^2$ is the power, and T_h is the frequency-hop duration. The random phase variables θ_k are assumed to be uniform random phase variables, and independent from hop to hop. In addition the d_j represent the data sequence and are assumed to have equally likely random variables, taking on the value of +1 or −1 with equal probability. It is also assumed that the data, phase, and frequencies are statistically independent. Thus for each hop there are

$$N = T_h / T_d \tag{1.5.3-2}$$

DPSK channel bits per hop time. The signal corresponding to the complex envelope is given by

$$x(t) = \text{Re}\left\{\sqrt{2}A\sum_k e^{i(2\pi f_k t+\theta_k)} p_{T_h}(t-kT_h)\sum_j d_j p_{T_d}(t-jT_d)e^{i\omega_0 t}\right\} \tag{1.5.3-3}$$

which can be somewhat simplified to

$$x(t) = \sqrt{2}A\sum_k p_{T_h}(t-kT_h)\sum_j d_j p_{T_d}(t-jT_d) \times \cos\left[\left(2\pi f_k+\omega_0\right)t+\theta_k\right] \tag{1.5.3-4}$$

Figure 1.5-7 illustrates a picture of the hop with the data chips placed on the hop. Note that the first channel bit is reserved for the reference bit (actually it could be more than one), and the remainder is used for the data. The overhead of one bit causes a loss of efficiency of $(N-1)/N$, when there are N data symbols per hop, which is small when N is large.

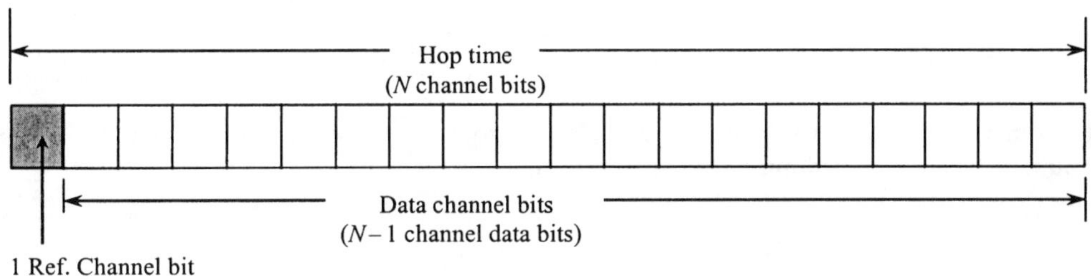

Figure 1.5-7 Hop time showing the reference bit and the data bits.

The spectral density of the resulting waveform can be approximated in the case that N is reasonably large. Note that if the complex envelope, $v(t)$, is viewed as a random process, the RF version of the process, say, $n(t)$, can be written at RF as

$$n(t) = \text{Re}\left[v(t)e^{i\omega_0 t}\right] \tag{1.5.3-5}$$

The auto correlation function of $v(t)$ is given by [9]

$$R_v(\tau) = \frac{1}{2}E\left\{v^*(t)v(t+\tau)\right\} \tag{1.5.3-6}$$

so that the autocorrelation function of the process $n(t)$ is given by

$$R_n(\tau) = \text{Re}\left[R_v(\tau)e^{i\omega_0\tau}\right] \tag{1.5.3-7}$$

and the power spectral density is given by

$$S_n(f)=\frac{1}{2}[S_v(f-f_0)+S_v(f+f_0)] \qquad (1.5.3\text{-}8)$$

In order to obtain the power spectral density of $v(t)$, it is necessary to first take the ensemble average and then the time average, since the hopping process is not a stationary random process. Let $R_v(\tau,t)$ denote the ensemble averaged but not time averaged autocorrelation function. From (1.5.3-1) one can write for $R_v(\tau,t)$

$$R_v(\tau,t)=\frac{1}{2}E\left\{\begin{matrix}\sqrt{2P}\sum_k e^{-i(\omega_k t+\theta_k)}p_{T_h}(t-kT_h)\sum_j d_j p_{T_d}(t-jT_d)\times \\ \sqrt{2P}\sum_l e^{+i(\omega_l(t+\tau)+\theta_l)}p_{T_h}(t+\tau-lT_h)\sum_m d_m p_{T_d}(t+\tau-mT_d)\end{matrix}\right\} \qquad (1.5.3\text{-}9)$$

Since the phase values are independent from hop to hop, this expression can be simplified to

$$R_v(\tau,t)=E\left\{P\sum_k e^{i\omega_k\tau}p_{T_h}(t-kT_h)p_{T_h}(t+\tau-kT_h)\sum_j d_j p_{T_d}(t-jT_d)\sum_m d_m p_{T_d}(t+\tau-mT_d)\right\} \qquad (1.5.3\text{-}10)$$

Approximating the time average by neglecting the hop pulses and only accounting for the data chip pulses, one obtains

$$R_v(\tau)=\frac{P}{M_h}\sum_{k=1}^{M_h}e^{i2\pi f_k\tau}\left[1-\frac{|\tau|}{T_d}\right] \qquad |\tau|\le T_d \qquad (1.5.3\text{-}11)$$

and is zero otherwise. It follows from (1.5.3-8), after grouping the exponentials together along with the triangular function, that the resulting power spectral density is given by

$$S(f)\cong\frac{PT_d}{2M_h}\sum_{k=1}^{M_h}\left\{\left(\text{sinc}((f-f_0-f_k)T_d\right)^2+\left(\text{sinc}((f+f_0+f_k)T_d\right)^2\right\} \qquad (1.5.3\text{-}12)$$

This result assumes that each frequency has the same weighting, $1/M_h$; when the weighting is not the same for each frequency, the spectral density changes to

$$S(f)\cong\frac{PT_d}{2}\sum_{k=1}^{M_h}W_k\left\{\left(\text{sinc}((f-f_0-f_k)T_d\right)^2+\left(\text{sinc}((f+f_0+f_k)T_d\right)^2\right\} \qquad (1.5.3\text{-}13)$$

where W_k is the relative weighting for each frequency.

Example 6 Consider a simplified example of a system that illustrates the spectral density derived in (1.5.3-11). For simplicity let M_h = 8, and the frequencies in GHz to be $\{f_k\}$= {0.2,0.4,0.6,0.8,1,1.2,1.4,1.6}, and let the center frequency be f_0 =1 GHz. The resulting spectral density is illustrated in Figure 1.5-8. Note that the spectral density fills the bandwidth from a little greater than 1.2 to 2.6 GHz.

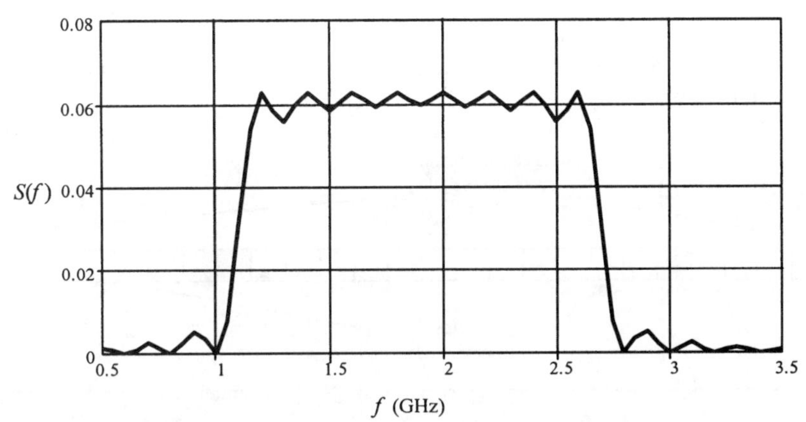

Figure 1.5-8 Power spectral density for a DPSK FH signal.

1.5.4 Noncoherent Slow Frequency-Hopped Signals with BPSK Data Modulation

The other common data modulation used with frequency-hopped signals is coherent phase binary shift keying (SFH/BPSK). On each hop a burst of BPSK channel bits are transmitted along with a group of channel sync bits. The function of the channel sync bits is to provide both phase and channel bit timing necessary to demodulate the data bits. The complex envelope of the frequency hopped signal with BPSK data modulation is given by

$$v(t)=\sqrt{2}A\sum_{k}e^{i(2\pi f_k t+\theta_k)}p_{T_h}(t-kT_h)\sum_{j}d_j p_{T_d}(t-jT_d) \tag{1.5.4-1}$$

where f_k are the hop frequencies, T_d is the BPSK data symbol duration, $A^2 = P$ is the power, and T_h is the frequency hop duration. The θ_k are assumed to be uniform random phase variables, independent from hop to hop. In addition the d_j are assumed to be equally likely random variables representing the data and taking on the value of +1 or –1 with equal likelihood. It is also assumed that the data, phase, and frequencies are mutually statistically independent. The transmitted signal is given by

$$x(t)=\mathrm{Re}\left\{\sqrt{2}A\sum_{k}e^{i(2\pi f_k t+\theta_k)}p_{T_h}(t-kT_h)\sum_{j}d_j p_{T_d}(t-jT_d)e^{i\omega_0 t}\right\} \tag{1.5.4-2}$$

which can be simplified somewhat to

$$\begin{aligned}x(t)=&\sqrt{2}A\sum_{k}p_{T_h}(t-kT_h)\sum_{j}d_j p_{T_d}(t-jT_d)\\&\times\cos\left[\left(2\pi f_k+\omega_0\right)t+\theta_k\right]\end{aligned} \tag{1.5.4-3}$$

It follows that on each hop there are

$$N=T_h/T_d \tag{1.5.4-4}$$

BPSK channel bits per hop time. Figure 1.5-9 illustrates a picture of the hop with the data channel bits and the sync words placed on the hop. If selective pulse jamming is not a concern then the first sync word could be placed at the beginning of the hop and the next one near the end so as to obtain a good estimate of the phase and the phase rate (or frequency). Conversely if selective pulse jamming is a concern then the location of the sync words would have to be located pseudorandomly in the hop. The sync words have to be of sufficient length so as to produce a reasonable estimate of phase and timing. Whether one or more sync words are needed depends on the dynamic environment and the available signal-to-noise ratio. If there are L sync channel bits in the hop, then the relative efficiency is $(N-L)/N$.

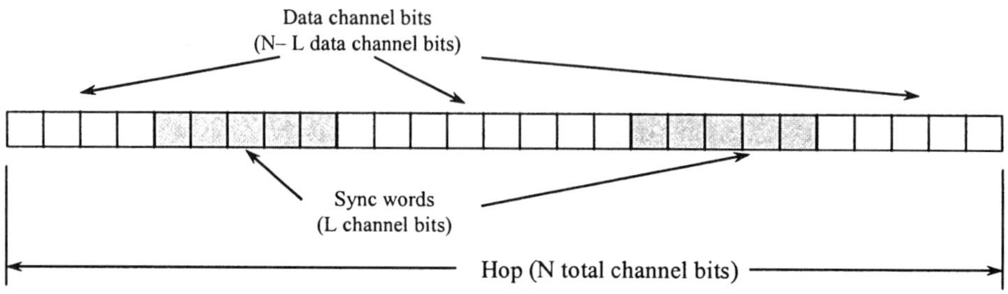

Figure 1.5-9 Hop time showing the sync words and the data bits for noncoherent frequency hopping with BPSK data.

1.6 HYBRID SPREAD SPECTRUM SYSTEMS

In this section the hybrid combination of direct sequence spread spectrum modulation in combination with frequency hopping will be considered. Other possibilities are also possible, of course.

1.6.1 Hybrid DS with Slow Frequency Hopping with BPSK Data

Hybrid spread spectrum direct sequence frequency-hopped systems have some of the advantages of frequency hopping and direct sequence (DS) systems. A hybrid spread spectrum modulation process spreads the signal much more than just frequency hopping does. Each hop, BPSK data, and a BPSK direct sequence signal are imposed on it, so that at any given time the frequency is hopped to an essentially random frequency, and each hop has a spread spectrum sequence spreading that hop. Let T_h denote the hop duration, let T_d denote the data duration, and let T_c denote the chip duration for the direct sequence on the hop. Let the number of channel bits per hop be given by N, where

$$N = \frac{T_h}{T_d} \tag{1.6.1-1}$$

Let M denote the number of direct sequence chips per hop, so that

$$M = \frac{T_h}{T_c} \tag{1.6.1-2}$$

where typically $M >> N$, that is

$$T_d >> T_c \tag{1.6.1-3}$$

with the further assumption that the ratio M/N is an integer. Figure 1.6-1 illustrates the relationship between the chips, hops, and data bits, with N data channel bits and M chips per hop. The complex envelope of the hybrid signal is given by

$$v(t) = \sqrt{2}A\sum_k e^{i(2\pi f_k t+\theta_k)} p_{T_h}(t-kT_h)\sum_j d_j p_{T_d}(t-jT_d)\sum_m a_m p_{T_c}(t-mT_c) \tag{1.6.1-4}$$

where $A^2 = P$ is the signal power, f_k are the hop frequencies which are assumed to be randomly chosen from hop to hop, d_j is the j-th data channel bit that takes on the values 1 and −1, and a_m is the m-th chip in the direct sequence spreading code, which also takes on the values of 1 and −1 with probability of 1/2 and the chips are assumed to be statistically independent. The corresponding RF signal is given by

$$x(t) = \text{Re}\left\{\sqrt{2}A\sum_k e^{i(2\pi f_k t+\theta_k)} p_{T_h}(t-kT_h)\sum_j d_j p_{T_d}(t-jT_d)\sum_m a_m p_{T_c}(t-mT_c)e^{i\omega_0 t}\right\} \quad (1.6.1\text{-}5)$$

which can be somewhat simplified to

$$x(t) = \sqrt{2}A\sum_k p_{T_h}(t-kT_h)\sum_j d_j p_{T_d}(t-jT_d)\sum_m a_m p_{T_c}(t-mT_c)\cos\left[(2\pi f_k+\omega_0)t+\theta_k\right] \quad (1.6.1\text{-}6)$$

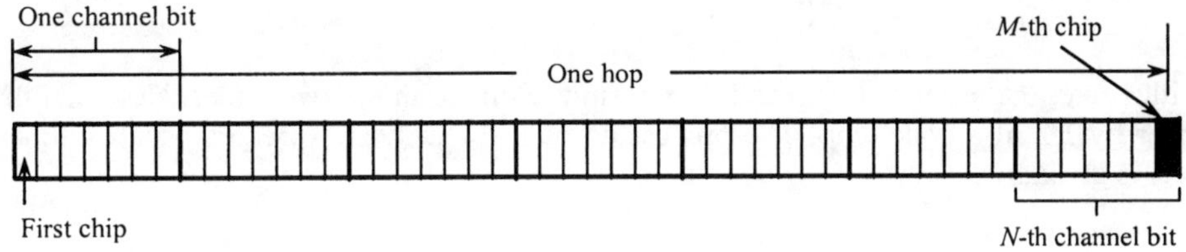

Figure 1.6-1 A hybrid direct sequence frequency-hopped system.

When the hopping bandwidth is not as great as to cause the chip delay to vary significantly from hop to hop, then time tracking can be accomplished by code tracking the direct sequence signal from hop to hop using a noncoherent code tracking loop. If this is not the case, then it is necessary to track the code and data from hop to hop. This could be accomplished by reserving one or more words in the hop to be used for phase (and possibly phase rate) as well as the timing estimates. In this case it may be necessary to store the data, derive the time and phase estimates, and demodulate the data in nonreal time.

The processing gain for hybrid FH/DS SS systems is given by a similar equation as for FH SS systems; that is

$$PG = \frac{W_h + 2R_c}{BW_b} \quad (1.6.1\text{-}7)$$

where usually W_h is the hopped bandwidth and R_c is the chip rate, with BW_b taken as twice the data bit rate. The numerator is the hopping frequency range plus twice the chip rate, which accounts for the direct sequence spectral first nulls on each end of the hopping range.

1.6.2 Hybrid OQPSK DS with SFH with BPSK Data

In addition to BPSK direct sequence spreading modulation, one can also consider (offset or not offset) quadriphase direct sequence spreading with frequency hopping. Although either offset or standard QPSK could be used, typically staggered QPSK is used since it has less spectral regrowth when passed through a nonlinear amplifier such as a solid state or TWT amplifier. In this section we will assume that the data is applied as BPSK modulation (DS-SFH/BPSK). The complex envelope of this waveform is given by

$$v(t) = \sqrt{P}\sum_k e^{i(2\pi f_k t+\theta_k)} p_{T_h}(t-kT_h)\sum_j d_j p_{T_d}(t-jT_d)\left[\sum_m a_m p_{T_c}(t-mT_c) + i\sum_n b_n p_{T_c}(t-nT_c-T_c/2)\right]$$

(1.6.2-1)

where now a_m and b_n are the m-th and n-th elements of the two quadrature sequences used in the spreading process, and P is the signal power. If the data sequence is denoted by $d(t)$ and the two PN sequences are denoted by $PN_1(t)$ and $PN_2(t)$, respectively, then the RF signal is described by

$$x(t) = \text{Re}\left\{\sqrt{P}\sum_k e^{i(2\pi f_k t+\theta_k)} p_{T_h}(t-kT_h)d(t)\left[PN_1(t)+iPN_2(t)\right]e^{i\omega_0 t}\right\} \quad (1.6.2\text{-}2)$$

This can be simplified somewhat to

$$\begin{aligned} x(t) &= \sqrt{P}\sum_k p_{T_h}(t-kT_h)d(t)PN_1(t)\cos\left[(2\pi f_k+\omega_0)t+\theta_k\right] \\ &-\sqrt{P}\sum_k p_{T_h}(t-kT_h)d(t)PN_2(t)\sin\left[(2\pi f_k+\omega_0)t+\theta_k\right] \end{aligned} \quad (1.6.2\text{-}3)$$

Figure 1.6-2 illustrates the baseband chip and data periods along with the two codes. Note that they are offset from each other by 1/2 of a code chip. The processing gain for OQPSK spreading with frequency hopping and BPSK data is given by

$$PG = \frac{W_h + 2R_c}{BW_b} \quad (1.6.2\text{-}4)$$

where W_h is the hopped bandwidth and R_c is the chip rate with BW_b taken as twice the data bit rate. The numerator is the hopping frequency range plus twice the chip rate, which accounts for the direct sequence spectral first nulls on each end of the hopping range, as discussed in the previous section.

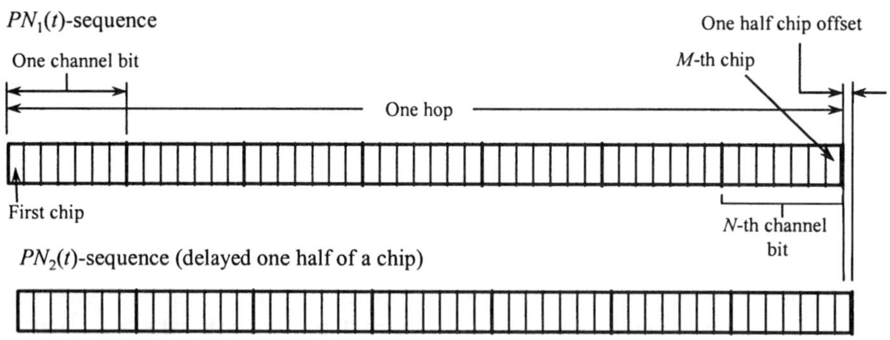

Figure 1.6-2 A hybrid offset quadriphase direct sequence frequency-hopped system.

1.7 TIME HOPPING SPREAD SPECTRUM SIGNALS

Although time hopping is not concurrently used very much, a short discussion of this type of spread spectrum technique will be briefly presented. A time hopping spread spectrum waveform is shown in Figure 1.7-1. Time is broken into M time slots where M satisfies

$$M = 2^m \quad (1.7\text{-}1)$$

During each frame only one time slot will be modulated with a message by any compatible modulation method. A pseudorandom code generator chooses the particular time slot that is chosen for a given frame. All of the channel bits accumulated during the last frame are "burst" into the selected transmitted burst slot. Let

T_f denote the frame duration in seconds, k is the number of channel bits in one frame and is transmitted in one burst of duration, T_f/M, T_b is the duration of one channel bit (nonburst duration), and therefore

$$T_f = kT_b \tag{1.7-2}$$

Now the width of each time slot for the k channel bits is T_f/M. Since there are k channel bits in each burst, the burst channel bit time is given by

$$T_{bBURST} = T_f/(kM) = T_b/M \tag{1.7-3}$$

Thus the transmitted bandwidth is M times the message channel bit bandwidth, so that the processing gain is M, that is

$$PG = M \tag{1.7-4}$$

In the case of time hopping, the processing gain comes from the fact that time is compressed by a factor of M and then expanded, whereas in frequency hopping or direct sequence SS systems it is the frequency that is expanded and then compressed to produce the processing gain.

A time hopping receiver [16] is illustrated in Figure 1.7-2. The receiver accepts the time hopped signal plus noise and is switched via the gate controlled by the PN code generator to pass the correct burst in the particular frame. The demodulator demodulates the signal and with the help of the bit synchronizer the demodulated data is sent to storage and reclocked out at the unburst rate. The bit synchronizer controls the PN code generator to maintain synchronization with the receiver signal.

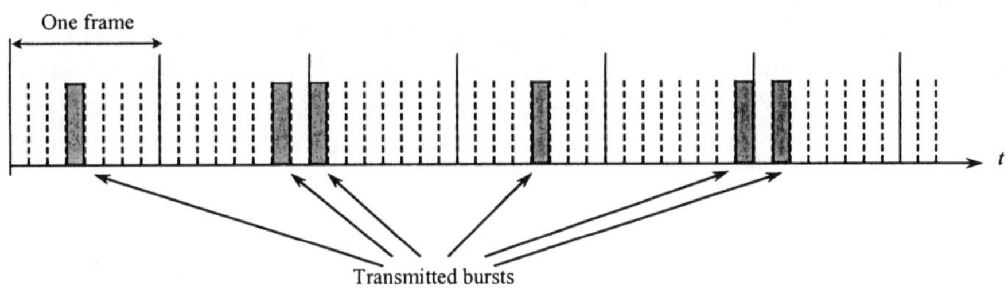

Figure 1.7-1 A time hopping spread spectrum system.

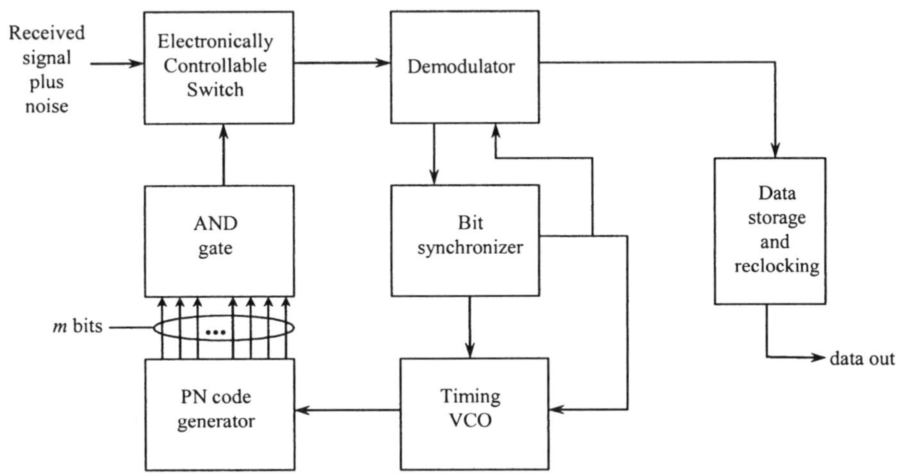

Figure 1.7-2 A time hopping spread spectrum receiver.

1.8 AN INTRODUCTION TO OFDM

Orthogonal frequency division multiplexing (OFDM) is a communication technique in which a large number of modulated carriers are closely spaced in frequency and are selected so that the carriers are orthogonal. To avoid a large number of modulators and filters at the modulator and a similar number of receiver filters and demodulators at the receiver, fast Fourier transform techniques are utilized.

OFDM is utilized in communications for the following reasons [17]. (1) It is an efficient method of combating multipath. For a given delay spread, the implementation complexity is considerably lower compared to a single carrier system with an equalizer. (2) In slow time varying channels, it is possible to considerably increase the capacity by adapting the data rate of each channel according to the signal-to-noise ratio of that carrier (sometimes called a subcarrier) signal. (3) Since narrowband interference affects only a few carriers, OFDM is robust against narrowband interference. OFDM has some disadvantages also. (1) OFDM is more sensitive to frequency offset error and phase noise induced errors. (2) Due to the fact that the sum of the carriers is not constant envelope, the efficiency is reduced compared to a single carrier signal with constant envelope if it is utilized without mitigation.

The basic idea of OFDM is to break the existing bandwidth W Hz into N_s *subchannels* of bandwidth ΔW Hz, so that $W = N_s \Delta W$. Consequently different information symbols can be transmitted on the different subchannels. Thus the data is transmitted via frequency division multiplex (FDM). Each subchannel is associated with a subcarrier signal, at frequency f_n, of the form [18]

$$s_{sc}(t) = \sin(2\pi f_n t) \quad n = 0,1,2...N_s - 1 \tag{1.8-1}$$

and f_n is the center frequency of the n-th ΔW Hz bandwidth. By distributing the data from one channel to N_s channels, the effective symbol rate is reduced to $1/N_s$ of what a single channel would have, if they all have equal rates. For OFDM the symbol rate ($1/T$) of each subchannel is selected to be equal to the separation of adjacent subcarrier channel frequencies. Thus the subcarrier symbols have duration T seconds, and if the subcarrier spacing is set to be $\Delta W = 1/T$, the subcarriers will be orthogonal, and we shall see. Consider two subcarriers at frequencies $f_n = n/T$ and $f_m = m/T$, and let $\theta_n = \theta$ and $\theta_m = 0$ (for computational convenience), then

$$\int_0^T \sin(2\pi f_n t + \theta_n)\sin(2\pi f_m t + \theta_m)dt = \frac{1}{2}\left[\sin\left(2\pi\frac{(n-m)}{T}\right)t + \theta\right]_o^T + \frac{1}{2}\left[\sin\left(2\pi\frac{(n+m)}{T}\right)t + \theta\right]_0^T \tag{1.8-2}$$

Evaluating this integral we find that

$$\int_0^T \sin(2\pi f_n t + \theta_n)\sin(2\pi f_m t + \theta_m)dt = 0, \quad \text{when } n \neq m \tag{1.8-3}$$

And therefore distinct subcarriers are orthogonal, irrespective of their phases. If we let T_s be the symbol period of a single carrier system, then the symbol period of the individual subcarriers is given by

$$T = N_s T_s \tag{1.8-4}$$

This large symbol duration greatly diminishes the intersymbol interference in OFDM.

The modulator and demodulator for OFDM can be constructed from discrete Fourier transform (DFT) processing. When a sufficient number of subcarriers are utilized, the fast Fourier transform (FFT) can be utilized to efficiently modulate and demodulate OFDM. One problem in transmitting a large number of subcarriers is the fact that the envelope is not constant. This can be mitigated somewhat by randomizing the phases of each subcarrier [19] or by clipping the peak power [20]. Without some form of mitigation it is required to back off the input drive level of the transmitter, to minimize the distortion causing power inefficiencies. Lawrey [20] found that OFDM could be heavily clipped with little effect on the receiver BER. As a point of reference, digital TV has somewhere between 2,000–8,000 subcarriers and Hiperlan 2 has about 48 subcarriers [19].

1.8.1 OFDM Communication System Implemented Via the FFT

OFDM often utilizes quadrature amplitude modulation (QAM) or QPSK modulation [17]. We will assume that QAM or QPSK is the modulation used for OFDM. Figure 1.8-1 illustrates the block diagram of the OFDM transmitter and receiver.

The serial to parallel converter subdivides the input data into frames of N_f bits. The N_f bits in each frame are assigned to N_s groups in which the j-th group is assigned n_j bits. The total sum of all the bits in each subcarrier group must be equal to the total number of bits, thus

$$\sum_{j=1}^{N_s} n_j = N_f \tag{1.8.1-1}$$

We can view the multicarrier scheme as generating N_s independent QAM subcarriers for which the subcarrier symbol rate is $1/T$, and therefore the number of bits per symbol may be different if the n_j are not all identical. It follows that the number of signal points, M_j, in the j-th subcarrier channel is given by

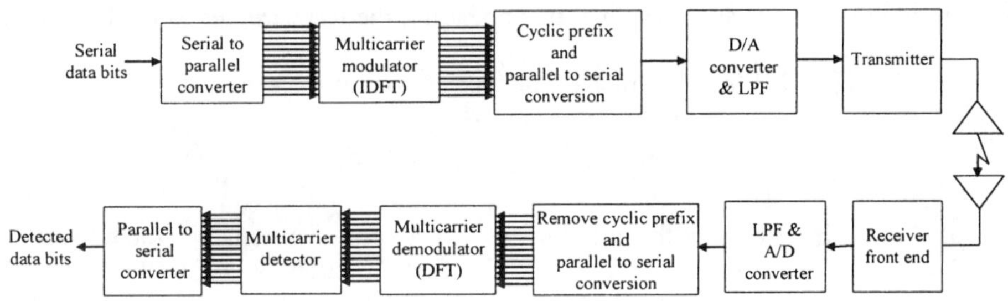

Figure 1.8-1 Block diagram of an OFDM transmitter and receiver.

$$M_j = 2^{n_j} \tag{1.8.1-2}$$

Let d_n ($d_n = a_n + jb_n$) denote the complex valued signal points on the n-th subcarrier channel, where $n = 0,1,2 \ldots N_s - 1$. These data information symbols represent the values of the discrete Fourier transform (DFT) of a multicarrier OFDM signal $s(t)$ for which the data modulation is QAM or QPSK. Hence the N_s point inverse DFT (IDFT) produces the complex valued envelope sequence

$$v_n = \frac{1}{N_s}\sum_{k=0}^{N_s-1} d_k e^{j2\pi nk/(N_s)} \qquad n = 0,1,2,\ldots N_s - 1 \tag{1.8.1-3}$$

The resulting IDFT sequence, v_n ($n = 0,1,2 \ldots N_s-1$), corresponds to samples of the OFDM signal complex envelope, $v(t)$, consisting of N_s subcarriers, and can be written as

$$v(t) = \frac{1}{N_s}\sum_{k=0}^{N_s-1} d_k e^{j2\pi kt/T} \tag{1.8.1-4}$$

in which T is the signal symbol duration and $v_n = v(nT/(N_s))$ for n = 0,1,2,...N_s–1. The subcarrier frequencies are specified by $f_k = k/T$ for k = 0,1,2,...N_s–1. The samples are passed through the D/A with a low pass filter and converted to the continuous OFDM signal complex envelope of the form of (1.8.1-4). The transmitted signal can be obtained from the relationship

$$s(t) = \text{Re}\left\{v(t)e^{j2\pi f_c t}\right\} \tag{1.8.1-5}$$

This can be expanded to the form

$$s(t) = \frac{1}{N_s}\sum_{k=0}^{N_s-1}\left[a_k\cos\left(2\pi(f_c + k/T)t\right) - b_k\sin\left(2\pi(f_c + k/T)t\right)\right] \quad k = 0,1,2,...N_s \tag{1.8.1-6}$$

Thus the transmitter places the modulation on the inphase and quadrature carriers and amplifies the signal and transmits it.

1.8.2 OFDM Intersymbol Interference Reduction Techniques

We have not addressed the cyclic prefix shown in Figure 1.8-1. This is a technique used to prevent intersymbol interference (ISI). Assume that the channel impulse response lasts for $M + 1$ samples and assume that $M << N_s$. The cyclic prefix is accomplished by appending the last samples to the beginning of the block [18].

Therefore, the appended (new) M block sequence consist of the samples

$$d_{N_s-M}, d_{N_s-M+1}, d_{N_s-M+2}, ...d_{N_s-1} \tag{1.8.2-1}$$

These are appended to the beginning of the old block sequence. As a consequence the transmitted sequence is N_s+M samples long, with n = –M to N_s–1, so that the first M samples constitute the cyclic prefix. The meaning of cyclic is now clear, since the first M and the last M samples are the same [17]. The ISI from the previous symbol will expire before the actual (data carrying) samples occur.

An alternative method of preventing ISI is to provide a time guard in between the actual the data samples so that ISI can be avoided on the symbols. This is achieved by appending extra (dummy) samples to each block of N_s signal samples. Then a guard time of $MT/(N_s)$ seconds are appended at the beginning of the block, so that the channel response is negligible when the actual data samples start.

The receiver does the inverse processing on the received signal. First the I and Q components of the transmitted signal are separated and then two components are low pass filtered and analog to digital converted. The cyclic prefix is removed (if it was appended at the transmitter) and the DFT is used to perform the multicarrier detection. Finally the parallel to serial converter puts the data estimates in serial form.

A few comments are in order regarding this discussion. This discussion was intended to be an introduction to how OFDM works in principle, but has not discussed some important issues, such as synchronization. If the synchronization in time or frequency is in error ISI occurs. It is possible to use OFDM training symbols for which the receiver knows the data sequence [17]. Obviously both time and carrier frequency must be obtained by the receiver to demodulate the OFDM signal.

1.8.3 OFDM Power Spectral Density

In order to obtain the power spectral density (PSD) of the OFDM signal we will assume that all orthogonal carriers are modulated by the same rate data with rectangular pulse shapes (not rounded to minimize the sidelobes). Since each subcarrier has a $\sin^2(2\pi fT)/(2\pi fT)^2$ spectral shape, it is assumed that all the subcarriers have statistically independent data (based on the serial input data being statistically independent) so that the PSD is of the form [21]

$$S_{OFDM}(f) = C\sum_{k=0}^{N_s-1}\left|\frac{\sin(\pi(f-f_n)T}{(\pi(f-f_n)T}\right|^2 \quad (1.8.3\text{-}1)$$

and $f_n = n/T$. A plot of the relative PSD (maximum at 0 dB) for the case that N_s =16 is shown in Figure 1.8-2. It can be seen that the bandwidth is about $(N+1)/T$ or

$$BW \cong \frac{N+1}{T} = \frac{N+1}{N_s T_s} \cong \frac{1}{T_s} = R_s \quad (1.8.3\text{-}2)$$

This is the symbol rate of the original single carrier system and is also the approximate bandwidth. Hence from the bandwidth point of view, OFDM offers no advantage or disadvantage.

The spectral shape of OFDM is nearly flat out to last subcarrier (the fifteenth in the figure) and has sidelobes that diminish in amplitude as frequency increases.

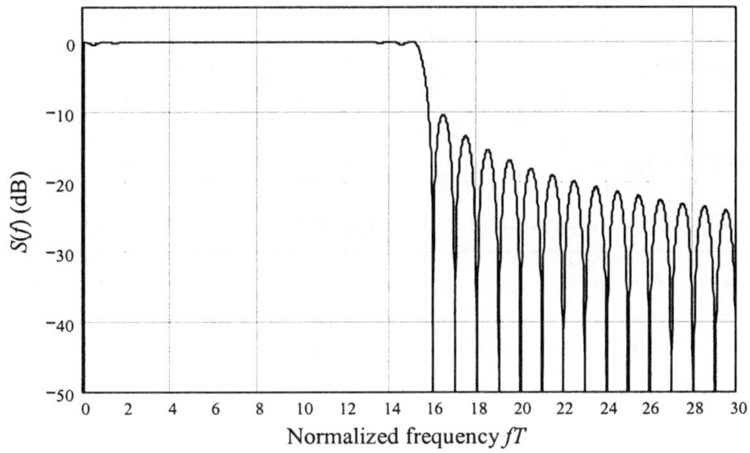

Figure 1.8-2 PSD of the complex envelope of an OFDM signal with $N_s = 16$.

1.9 AN INTRODUCTION TO ULTRAWIDEBAND COMMUNICATIONS

In this section we will present an introduction to ultrawideband (UWB) communications. The term *ultrawideband* has come to denote a group of equivalent terms such as: impulse, carrier-free, baseband, non-sinusoidal, time domain, orthogonal function, and large-relative-bandwidth radio/radar signals [22]. The actual term "ultra wideband" was apparently used by the United States Department of Defense circa 1989. In this section we will introduce the two distinct types of UWB communications.

1.9.1 A Brief Early History of UWB Communications

It may be fair to state that the first application of UWB occurred when Guglielmo Marconi, in 1901, transmitted Morse code across the Atlantic Ocean using a spark gap type radio transmitter. The modern development of what is now known as UWB started around the late 1960s with the pioneering work of H. F.

Harmuth at the Catholic University of America, G. F. Ross and K. W. Robbins at the Sperry Rand Corporation, and P. Van Etten at the United States Air Force's Rome Air Development Center.

The papers and books by Harmuth published in the period of 1969–1984 provide the basic design approach for UWB receivers and transmitters. Independently and at about the same time period (1972–1987), Ross and Robbins pioneered the use of UWB in a number of applications including both communications and radar. An early patent by Ross [23] was a landmark patent in UWB transmitters and receivers. Both Harmuth, and Ross and Robbins applied the principle of matched filtering to UWB systems. The work of Van Etten's testing of UWB radar systems allowed the development of antenna and system design concepts in UWB [24]. R. Morey designed a UWB radar system that could penetrate the Earth, which later became a success in the field of geophysical survey systems in 1974 [25].

In 1978 C. L. Bennett and G. F. Ross [26] published the pulse generation methods known to date. Both industrial and academic research continued unabated. Then in 1990 Los Alamos National Laboratory (LANL) produced a conference that provided a forum for the many developments in the field of UWB systems. Then in 1994 T. E. McEwan [27] invented the micropower impulse radar (MIR), a motion sensor, that provided a practical UWB system that operated at ultra low power and was inexpensive and very compact in size. Barrett [22] has a rather complete history of UWB work, and covers much more than has been covered here.

Many conferences have been successfully done on UWB and at least 14 books have been published on the subject of UWB [28]. In fact a very useful resource on UWB is given in [28]. Currently there are two UWB factions. One led by Intel, Nokia, Texas Instruments, and many others (130 in 2004) are proponents of *The Multiband OFDM Alliance* (MBOA), which is a multiband UWB approach. The other group, led by Freescale, Motorola, Time Domain, and others (40 in 2004), champion the direct sequence UWB approach. Both sides feel their approach is superior; however, in the end only one will probably survive.

1.9.2 Description of UWB Signals

The area of UWB has developed two distinct forms of signaling: (1) one is based on transmitting very short duration pulses (*impulse radio*) to carry information, and (2) the other approach is based on the multiple simultaneous transmission of carriers. The Defense Advanced Research Project Agency (DARPA) in 1990 concluded that in order to be considered a UWB signal the fractional bandwidth must be at least 25%, regardless of the center frequency or the signal time-bandwidth product. The *fractional bandwidth* is defined by (as a fraction)

$$\text{FBW} = \frac{(f_u - f_l)}{((f_u + f_l)/2)} \tag{1.9.2-1}$$

in which f_u is the upper bandwidth limit and f_l is the lower bandwidth limit. Another bandwidth used in UWB work is the *relative bandwidth* [29, 30], defined as

$$\text{RBW} = \frac{(f_u - f_l)}{(f_u + f_l)} \tag{1.9.2-2}$$

The upper and lower limits are presumed to define the bandwidth to say, 95% or 99%.

Unlike classic communications, pure impulse radio does not utilize a modulated carrier to carry the information. Rather the signal is a series of pulses that are extremely short in time duration (nanosecond or less) and the transmit bandwidth is in the gigahertz range, and the FBW is at least 25%.

The propagated impulse radio signal can be represented as

$$s(t) = \sum_{k=-\infty}^{\infty} A_i(t) p(t - kT_f) \tag{1.9.2-3}$$

in which $A_i(t)$ is the amplitude of the pulse and is equal to $\pm\sqrt{E_p}$ according to the sign of the pulse and E_p is the pulse energy. In addition T_f is the repetition time duration, with the knowledge that the frame contains one pulse per frame. The received signal is modified by the channel, and as well by both the transmitter and receiver antenna responses.

The use of multicarrier communications first started in the late 1950s to early 1960s in order to provide greater data rate, high frequency (HF), military communications. The modulation, known as OFDM, has evolved as a densely packed set of subcarriers with overlapping power spectral densities, as was discussed in the previous section. This approach was patented in 1970 [31]. Multicarrier OFDM (MC-OFDM) is quite different from impulse UWB (I-UWB), in that the complex envelope (baseband) model of the modulation is of the form [32]

$$s(t) = \sum_{k=1}^{N} d_k(t) e^{j2\pi k(t/T_s)} \tag{1.9.2-4}$$

where N is the number of carriers (sometimes referred to as subcarriers) in the waveform, T_b is the bit duration, and T_s is the symbol duration and is given by

$$T_s = NT_b \tag{1.9.2-5}$$

The $d_k(t)$ is the symbol stream modulating the kth carrier. Both of these UWB modulations will be discussed in more detail in the following two sections.

1.9.2.1 I-UWB Signal Modeling

Recall that (1.9.2-3) is the propagated signal model (not the envelope) of the impulse UWB (I-UWB) signal. It is to be noted that it is not necessarily true that this model represents the transmitter waveform, since the transmit antenna can distort the pulse shape. The most common pulse shapes for I-UWB work are the Gaussian pulse and its derivatives, since they are easy to describe and work with. The Gaussian pulse is described analytically by

$$p(t) = \frac{A}{\sqrt{2\pi\sigma^2}} e^{-\left(\frac{(t)^2}{2\sigma^2}\right)} \tag{1.9.2-6}$$

where σ^2 is the variance parameter. The pulse width is given by the expression

$$T_p = 2\pi\sigma \tag{1.9.2-7}$$

which defines 99.8% of the pulse area. Another pulse model is the first derivative of the Gaussian pulse. This is used as a model since a UWB antenna may have the effect on the signal to be equivalent to differentiating the pulse with respect to the time variable. Letting the mean value be zero, the first derivative is given by [32, 33]

$$p^{(1)}(t) = -\left(\frac{At}{\sqrt{2\pi\sigma^3}}\right) te^{\left(-\frac{t^2}{2\sigma^2}\right)} \tag{1.9.2-8}$$

where the superscript (n) denotes the n-th derivative. A third model is based on the second derivative of the Gaussian pulse and is given by [32, 34]

$$p^{(2)}(t) = A\left(\frac{t^2}{\sqrt{2\pi\sigma^5}} - \frac{1}{\sqrt{2\pi\sigma^3}}\right)e^{-\frac{(t)^2}{2\sigma^2}} \tag{1.9.2-9}$$

And again the parameter σ determines the pulse width. Figure 1.9-1 illustrates all three pulse types, with amplitude factors changed to make them appear to be of the same peak amplitude with $\sigma = 1$. By changing the value of σ, any time scale can be achieved. The pulses typically exist for a fraction of a nanosecond.

Currently the FCC rules make UWB operation most practical in the 3.1–10.6-GHz frequency band. Consequently the three preceding pulses may not be useful for commercial systems [35]. Thus, another pulse shape has been suggested; it is called the *Gaussian modulated sinusoidal pulse.* The actual pulse shape is given by

$$p(t) = \left(\frac{8k}{\pi}\right)^{1/4} \frac{1}{\sqrt{1+e^{\left(\frac{2\pi^2 f_c^2}{k}\right)}}} e^{-(kt)^2} \cos(2\pi f_c t) \tag{1.9.2-10}$$

where the center frequency is denoted f_c and k is again a spreading parameter. The pulse shape is plotted in Figure 1.9-2.

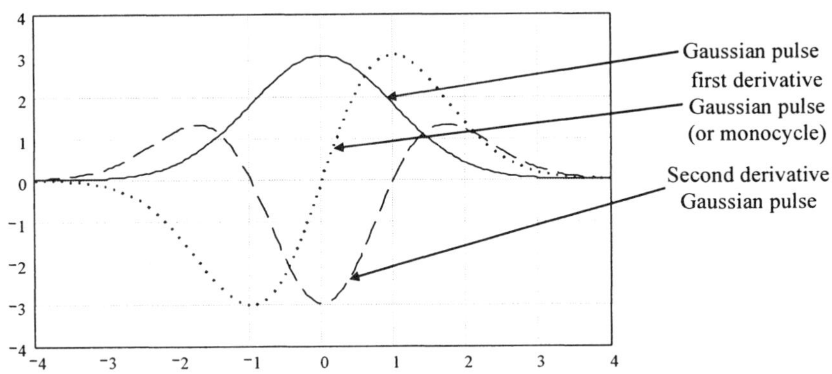

Figure 1.9-1 Gaussian and the first and second derivates of a Gaussian pulse.

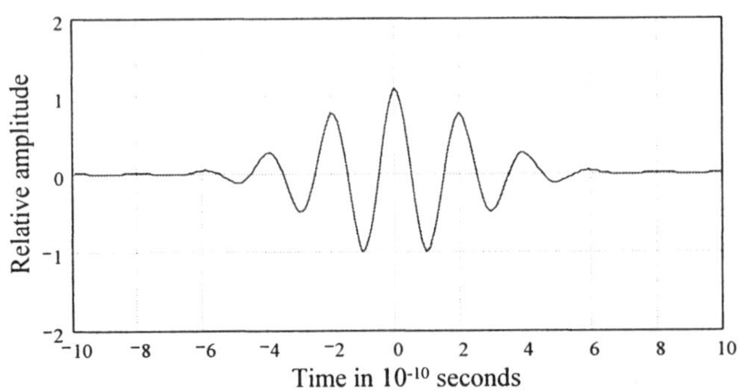

Figure 1.9-2 A plot of the Gaussian modulated sinusoidal pulse with $k = 3*10^9$ and $f_c = 5*10^9$.

As can be seen in the figure, the sine wave is amplitude modulated by a Gaussian shape.

Now consider the power spectral density of some of the derivatives of Gaussian I-UWB signals. Assuming that the probability of a "+1" or a "−1" is 0.5 and the sequence has no memory, then the power spectral density is given by

$$S(f) = \frac{E_p}{T_f}|X_n(f)|^2 \tag{1.9.2-11}$$

where $X_n(f)$ is the Fourier transform of the n-th derivative of the pulse. The Fourier transform of the n-th derivative of a Gaussian pulse is given by [34]

$$X_n(f) = (j2\pi f)^n \exp\left(-\frac{(2\pi f\sigma)^2}{2}\right) \tag{1.9.2-12}$$

So that the amplitude spectrum is given by

$$|X_n(f)| = (2\pi f)^n \exp\left(-\frac{(2\pi f\sigma)^2}{2}\right) \tag{1.9.2-13}$$

The power spectral density of a few pulse cases is shown in Figure 1.9-3.

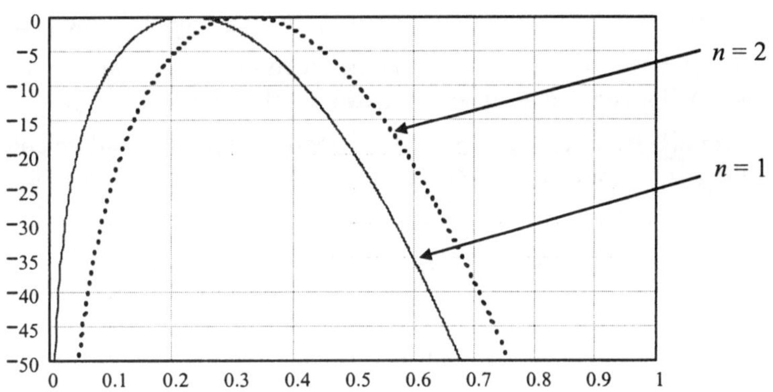

Figure 1.9-3 The power spectral density of the first and second derivative of a Gaussian pulse.

Now let us briefly review a few modulation schemes that are compatible with I-UWB. One is *pulse amplitude modulation* (PAM). In PAM the transmit signal is given by

$$s(t) = \sum_{k=-\infty}^{\infty} Ad_k(t)p(t-kT_f) \tag{1.9.2-14}$$

where $d_k(t)$ represents the k-th amplitude of the pulse and depends on the data and the specific amplitude modulation scheme, and A is the peak amplitude of the pulse when the max ($p(t)$) normalized to unity. For the *pulse position modulation* scheme, the transmit signal is given by

$$s(t) = \sum_{k=-\infty}^{\infty} Ap(t-kT_f-\Delta d_k(t)) \tag{1.9.2-15}$$

where $d_k(t)$ is the time modulation based on the data and Δ is the time base time offset. *On-off keying* (OOK) is also used in I-UWB and is similar to (1.9.2-14) except now the value of $d_k(t)$ is either on or off (1 or 0).

Now let us consider some multiple access methods for I-USB. Multiple access implies that there is more than one signal available at the receiver in the same frequency band. If we let the impulse response between the kth user and the receiver be $h^{(k)}(t)$, then the received complex envelope signal can be represented by

$$r(t) = \sum_{k=1}^{K} s^{(k)}(t)h^{(k)}(t) + n(t) \tag{1.9.2-16}$$

when there are K users. Of course, the actual bandpass received signal is given by

$$r_{BP}(t) = \mathrm{Re}\left[r(t)e^{j2\pi f_c t}\right] \tag{1.9.2-17}$$

First consider CDMA. The k-th signal can be represented as [32]

$$s^{(k)}(t) = \sum_{n=-\infty}^{\infty} PN_n d^{(k)}_{\lfloor n/N_s \rfloor} p(t - nT_f) \tag{1.9.2-18}$$

in which PN_n is the pseudorandom code sequence for the n-th pulse, N_s is the number of pseudonoise chips per data symbol, and $d_i^{(k)}$ is the i-th data symbol for the k-th user. The *floor function* (*greatest integer function*) satisfies, for example, $\lfloor 4.5 \rfloor = 4$ and $\lfloor -4.5 \rfloor = -5$, and $\lfloor x \rfloor$ denotes the greatest integer less than x. The code can have a period of one data symbol or can be much longer than one data symbol. For the case of one period of the pseudorandom code it is equal to the data symbol $PN_n = PN_{n+N_s}$.

PPM can also be utilized as a multiple access method with pseudorandom codes. The transmit signal (before the effects of the channel or antennas) can be modeled as

$$s^{(k)}(t) = \sum_{k=-\infty}^{\infty} Ap\left(t - kT_f - PN_k T_c - \Delta d_{\lfloor k/N_s \rfloor}\right) \tag{1.9.2-19}$$

in which the time hopping is accomplished with the sequence PN_k, and T_c is the hopping offset time or granularity.

1.9.2.2 Multicarrier UWB Signals

Now consider the multicarrier UWB signal (MC-UWB). This UWB signal is also called the frequency domain UWB. Recently this approach has received considerable attention. The transmit signal of MC-UWB has the following complex envelope form

$$v(t) = A\sum_{k=-\infty}^{\infty}\sum_{n=1}^{N} a_n^k p(t - kT_p)e^{j2\pi n f_0 (t - kT_p)} \tag{1.9.2-20}$$

For N carriers, a_n^k is the symbol that is transmitted in the k-th transmission interval over the n-th subcarrier. Also A is the parameter that determines the transmitted power spectral density as well as the energy per bit [32]. The frequency parameter is given by $f_0 = 1/T_p$.

In the multiband OFDM-UWB approach, the available spectrum of 7.5 GHz is divided into multiple 528-MHz bands. In this setup selective band implementation can be utilized while leaving other portions of the spectrum unused. The MBOA proposal has five channels that are, in turn, broken down into 14 bands. Figure 1.9-4 illustrates the MultiBand OFDM Alliance frequency band plan, and the band center frequency is noted under each band. Each band has 128 QPSK modulated OFDM carriers.

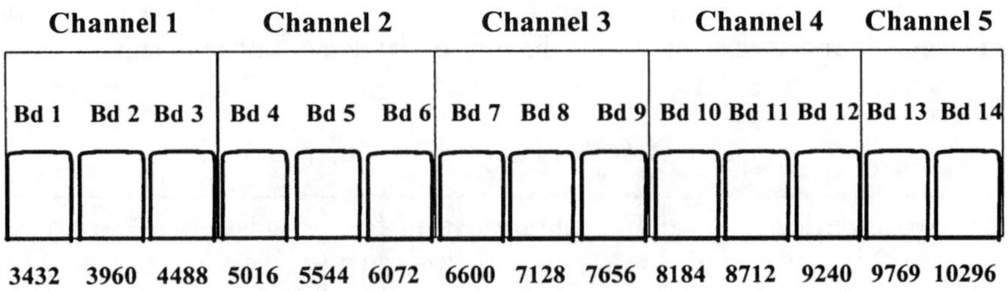

Figure 1.9-4 The frequency plan of the MultiBand OFDM Alliance.

At the time of this writing, bands 1–3 (Bd k denotes band k) are used for Mode 1 (a mandatory mode), but the other four modes are optional. There are up to four time-frequency codes per channel, so that there are a maximum of 20 piconets[1] with the current proposal. Each time frequency code (TFC) corresponds to a logic channel. Logic channels enable simultaneously operating piconets (SOPS). Four SOPS are enabled on band group 1. In the event that interference from 802.11a causes reception problems in channel 2, it may be avoided by not using channel 2.

Orthogonal frequency division multiplex (OFDM) [32] is a multicarrier signaling scheme that allows carriers to overlap in frequency without causing self-interference, thereby improving spectral efficiency. Assigning each user a group of carriers can accommodate multiple users. OFDM-UWB has been proposed as a physical layer for high bit rate, short-range communication networks. The scheme transmits multiple data streams in parallel on separate carriers. As mentioned earlier, it is not necessary to have contiguous carriers operating on the same channel.

1.9.3 Regulatory Constraints and Spectral Masks for Various UWB Applications

Based on a notice of inquiry issued on September 1, 1998, by May 2000 over a thousand documents were received by the Federal Communications Commission (FCC) with the aim of assisting the FCC in developing an appropriate set of specifications for UWB operation. In February 2002 the FCC issued a first report and order [35] that classified UWB operation into three distinct categories [32]: (1) communication and measurement systems; (2) vehicular radar systems; and (3) imaging systems including ground penetrating radar, through-the-wall imaging and surveillance systems, and medical imaging. A summary is provided in Table 1.9-1 [35].

Each category has a unique spectral mask as shown in Figures 1.9-4(a) through 1.9-4(e). The break point frequencies and the emission limits are shown on the five plots. Outside of the United States of America other countries are following a similar approach to licensing UWB technology.

1.9.4 Impact of the Transmit Antenna on the Transmitted Signal

We have alluded to the fact that the antenna can affect the response of the UWB signals, in particular for pulsed UWB systems. I-UWB requires an antenna that covers multioctave bandwidths in order to transmit pulses on the order of a nanosecond in duration, while allowing minimal distortion. It is desired that the

---

[1] A *piconet* is a network of devices connected through Bluetooth technology. It is formed when at least two devices, such as a portable PC and cellular phone, connect. In a piconet one device acts as the master and the other as a slave. Up to eight devices may be connected in a piconet.

antenna pattern will provide a voltage standing wave ratio (VSWR) of no more that 2:1 over the entire frequency band and provide a radiation field that has constant magnitude and linear phase with frequency. An important concept that is applicable to UWB antenna response is the following: The transmitting transient response of an antenna is proportional to the time derivative of the receiving transient response of the same antenna [36, 37]. An antenna that meets the condition on linear phase, constant magnitude, and constant VSWR with frequency will radiate a signal that is the time derivative of the input signal.

Table 1.9-1 FCC Restrictions on UWB Operations

Application	Frequency Band for Operation at Part 15 Limits	User Restrictions
Communications and measurement systems (sensors)	3.1–10.6 GHz (separate emission levels for indoor and outdoor)	None
Vehicular radar for collision avoidance, airbag activation, and suspension system control	24–29 GHz	None
Ground penetrating radar to see or detect buried objects	3.1–10.6 GHz and below 960 MHz	Law enforcement, fire and rescue, research institutes, mining, construction
Wall imaging systems to detect objects contained in walls	3.1–10.6 GHz and below 960 MHz	Law enforcement, fire and rescue, mining, construction
Through-the-wall imaging systems to detect location or movement of objects located on the other side of the wall	1.99–10.6 GHz and below 960 MHz	Law enforcement, fire and rescue
Medical systems for imaging inside people and animals	3.1–10.6 GHz	Medical personnel
Surveillance systems for intrusion detection	1.99–10.6 GHz	Law enforcement, fire and rescue, public utilities, and industry

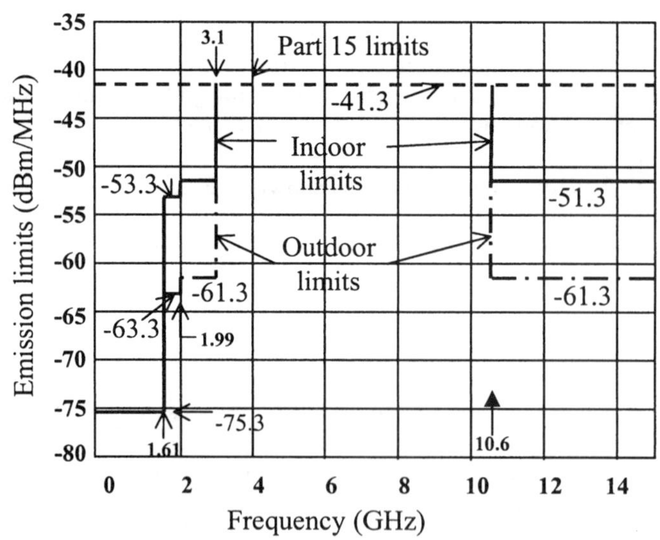

Figure 1.9-4 (a) Indoor UWB communication systems emission limits.

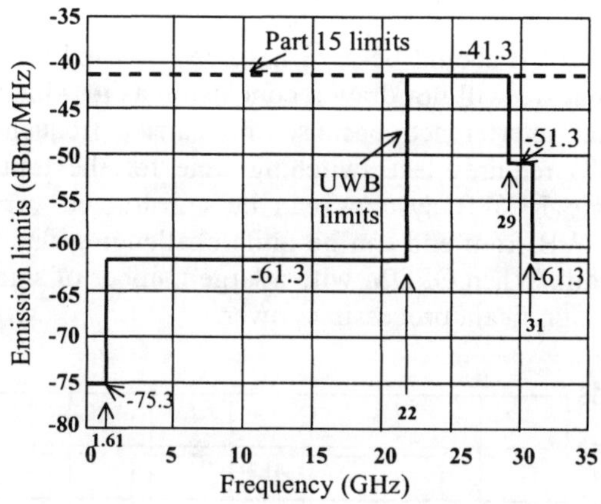

Figure 1.9-4 (b) Vehicular radar emission limits.

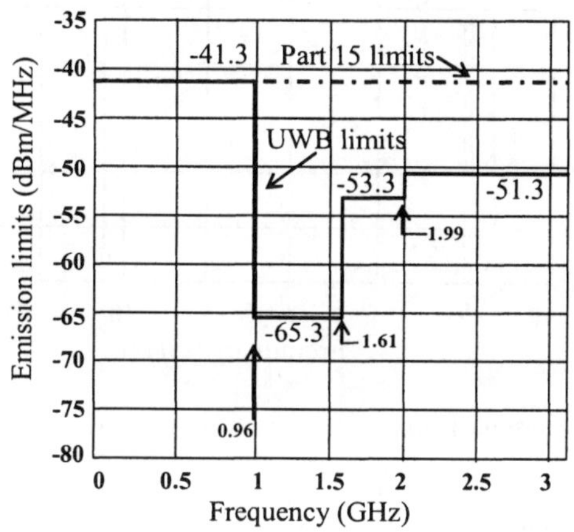

Figure 1.9-4 (c) Low-frequency imaging emission limits.

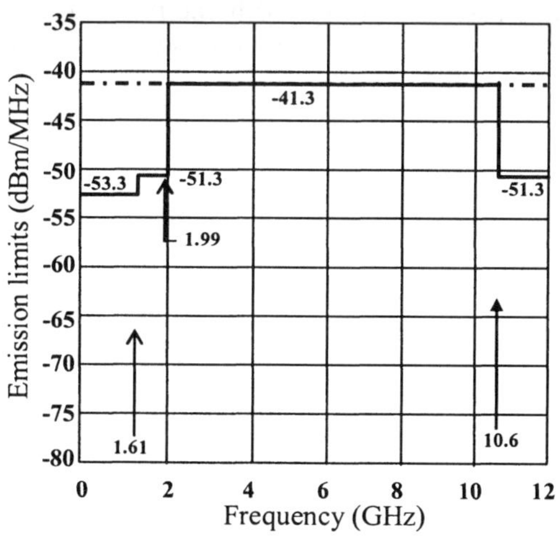

Figure 1.9-4 (d) Mid-frequency imaging emission limits.

1.9.5 The Advantages and the Disadvantages of Impulse Versus Multicarrier UWB

The issue of which approach to UWB should be used has been debated for some time. We will attempt to point out some relative issues, but we will not draw a conclusion as to which is superior. MC-UWB is well suited for avoiding narrowband interference because the carrier frequencies can be chosen to avoid narrowband interference. I-UWB requires fast switching time for the transmitter and receiver and very accurate synchronization. Simple I-UWB systems can be constructed very inexpensively. Alternatively designing and building a MC-UWB front end can be quite challenging due to the continuous variations in power over a very wide bandwidth. When OFDM with a large number of signals is utilized, high-speed FFT processing is needed, leading to significant processing power.

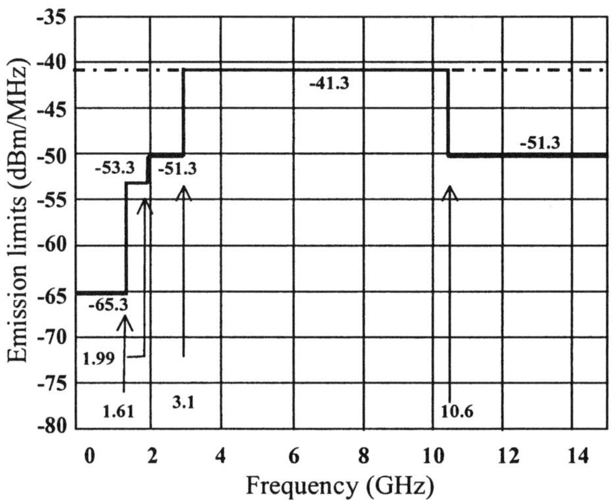

Figure 1.9-4 (e) High-frequency imaging emission limits.

1.9.6 Advantages of UWB Systems

Some of the advantages to UWB are indicated here. (1) High channel capacity achievable by virtue of the large bandwidth, (2) UWB only requires low power and can be manufactured with low cost due to the fact that UWB can directly modulate a baseband pulse, and it can be made nearly all digital with low transmit power levels, and (3) UWB has fading robustness since UWB has a pulse duration of around 1 ns, so only very short multipath delays have any effect.

1.9.7 Applications of UWB

Many new applications of UWB techniques are appearing. Some of the applications include [37]: (1) location of buried objects, (2) location of wooden or steel studs in a wall and steel in concrete, (3) search and rescue under ruble, (4) measurement of explosive velocities up to 3,800 m/s, (5) cargo container intrusion sensing, (6) bridge deck inspection, (7) perimeter security for both military and civilian applications, (8) microwave impulse radar (MIR), (9) wireless home networks, and (10) roadside markers for informing passing cars of weather advisories, road conditions, and road construction.

Obviously we have only presented a very brief summary of the UWB communications. More details can be found in the references.

1.10 THE NEAR-FAR PROBLEM

The *near-far problem* in spread spectrum systems comes about when the received power from each transmitter is not the same at the receiver. One case in which can occur is when equal powered transmitters are at different distances from the receiver. In other words the receiver is closer to some transmitters than others. The closer signals are received at a much higher power level than the signals from the transmitters that are farther away. As a consequence the closer signals mask the farther-away-signals and make them more difficult to detect and process. All the signals except the one that is being tracked add what is sometimes called CDMA noise to the tracked signal. When the tracked signal transmitter is far away from the receiver, it is degraded more by the CDMA noise than are closer transmitted signals. In CDMA systems, capacity is maximized if all the signals are at a similar received power level. In fact, in one commercial CDMA system, received power is maintained to about $\pm$ 0.2 dB, when is it is feasible.

A frequency-hop system is much less susceptible to the near-far problem since normally collisions, and therefore interference, would not be allowed (except occasionally) to occur and therefore interference would not be a significant problem. This is true for single channel or multiple channel frequency-hopped signals. Interference to the desired signal occurs when a collision occurs, and the power of the interference doesn't matter much, as long as the interference signal is stronger.

Another advantage of frequency-hopped systems is the fact that acquisition can be accomplished much quicker that direct sequence systems. This is true since the hop time is normally of much longer duration than the direct sequence chip time, and therefore the number of states needed to search the initial time uncertainty, to obtain acquisition, is much less.

At one time it was believed that a disadvantage to frequency-hopped systems was the fact that it was not considered to be feasible to use coherent modulation for frequency-hopped systems. However there has been considerable recent work showing that coherent demodulation is feasible if one is willing to provide synchronization words or aids in each hop at the expense of some overhead. Noncoherent demodulation is inferior to coherent demodulation in bit error rate performance, so that it is desirable to utilize coherent modulation and demodulation whenever possible. Naturally there is a trade-off to be made for each system design on whether to use coherent or noncoherent data modulation techniques.

1.11 LOW PROBABILITY OF INTERCEPTION

Before concluding this chapter it is worthwhile to discuss one of the advantages of spread spectrum systems that has not been discussed yet, and that is the low probability of interception (LPI) capability. Since SS signals are difficult to detect, special types of detectors in the hands of unfriendly users are needed to detect these signals. There are basically three types of LPI detectors that are used in practice. The first is classified as a transform detector, the second is a radiometer, and the third is the rate line detector. An example of the first class is a spectrum analyzer that displays the power spectral density of the received signal to the operator. The second class is based on the use of a wideband power detector that measures signal and noise power in a given bandwidth. Finally, an example of the third class of detector is based on detecting the chip rate of a direct sequence SS system. These methods will be discussed in more detail in Chapter 11, on LPI detection systems.

Figure 1.11-1 illustrates an example of a radiometer that is designed for detecting SS signals. A bandpass filter limits the spectral range to be searched. The detector detects any signals or noise present in the bandwidth under investigation. The integrator or low-pass filter smoothes the detected output and is compared to a threshold. If the detected signal is above the threshold it is declared detected, otherwise it is not detected.

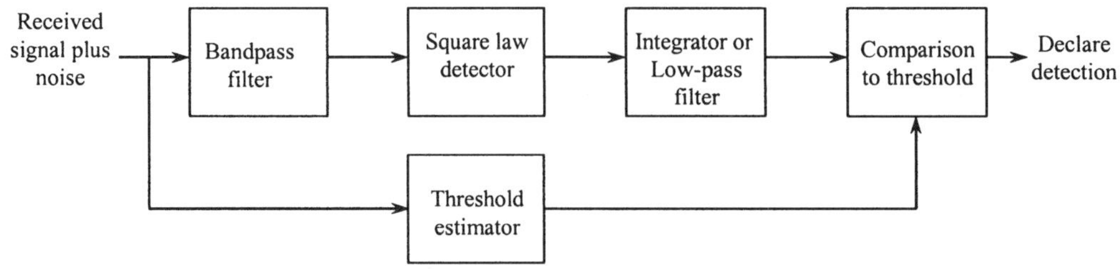

Figure 1.11-1 A classic radiometer for detecting spread spectrum signals.

Many of the topics discussed only briefly in this chapter will be presented in more detail in the following chapters.

1.12 SUMMARY

This chapter introduced the basic concepts of spread spectrum (SS) systems. A very brief introduction to the history of SS systems was presented. A important tool in analyzing these signals is the concept of narrowband signal, which was introduced in this chapter. Direct sequence (DS) systems were presented along with the correlation loss of a signal when filtered. In addition the noise despreading reduction in the despreading process was developed along with the loss in C/N_0 value for a BPSK modulated DS SS system.

Next the various SS narrowband signal models were presented including slow and fast frequency hopping, along with hybrid DS and FH systems, and minimum shift keying systems. Examples of slow and fast frequency hopping were illustrated. This was followed by a discussion of OFDM. Then a brief discussion of UWB was presented. Finally the near-far problem was introduced, followed by a brief discussion of the concept of designing low probability of interception signals.

References

[1] R. Price, "Further Notes and Anecdotes on Spread-Spectrum Origins," *IEEE Trans. on Communications*, Vol. COM-31, No. 1, January 1983.

[2] M. K. Simon, J. K. Omura, R. A. Scholtz, and B. K. Levitt, *Spread Spectrum Communications*, Vol. 1, Rockville, MD: Computer Science Press, 1985.

[3] E. N. Gilbert, "Quasi-Random Binary Sequences," Bell Telephone Laboratories Memorandum MM-53-1400-42, November 27, 1953.

[4] S. W. Golomb, "Sequences with Randomness Properties," Terminal Progress Report under Contract Req. No. 639498, Martin Co. Maryland, June 1955.

[5] N. Zierler," "Several Binary-Sequence Generators," Tech. Report No. 95, Lincoln Labs., MIT, September 12, 1955.

[6] N. Zierler, "Linear Recurring Sequences," *J. Soc. Industrial and Applied Math.*, Vol. 7, March 1959, pp. 31–48.

[7] T. G. Birdsall and M. P. Ristenbatt, "Introduction to Linear Shift-Register Generalized Sequences," University of Michigan Research Institute, Dept. of Elect. Eng. Tech. Report No. 90, October 1958.

[8] C. R. Cahn, "Applications and State-of-the Art Equipments," Paper No. 5, AGARD-NATO Lecture Series, Vol. 58 on "Spread Spectrum Communications," May 28–June 6, 1973.

[9] J. G. Proakis, *Digital Communications*, Section 3.3, New York: McGraw-Hill, 1983.

[10] R. E. Ziemer and R. L. Peterson, *Digital Communications and Spread Spectrum Systems*, New York: Macmillan, 1985.

[11] R. A. Scholtz, "The Spread Spectrum Concept," *IEEE Trans. on Communications*, Vol. COM-25, No. 8, August 1977.

[12] A. B. Carlson, *Communication Systems: An Introduction to Signals and Noise in Electrical Communication*, New York: McGraw-Hill, 1968.

[13] A. J. Viterbi, "Spread Spectrum Communications—Myths and Realities," *IEEE Communication Magazine*, No. 3, May 1979, pp. 11–18.

[14] J. K. Holmes, *Coherent Spread Spectrum Systems*, New York: Wiley-Interscience, 1982 (also published by Krieger Publishing Company, Malibar FL, 1990).

[15] S. A. Gronemeyer and A. L. McBride, "MSK and Offset QPSK Modulation," *IEEE Trans. on Communications*, Vol. COM-24, August 1976, pp. 809–820.

[16] G. R. Cooper and C. D. McGillem, *Modern Communications and Spread Spectrum*, New York: McGraw-Hill, 1986.

[17] R. Van Nee and R. Prasad, *OFDM For Wireless Multimedia Communications*, Norwood, MA: Artech House, 2000.

[18] J. Proakis and M. Salehi, *Fundamentals of Communication Systems*, Englewood Cliffs, NJ: Prentice Hall, 2001.

[19] J. Armstrong, "OFDM–Orthogonal Frequency Division Multiplexing," a La Trobe University Electronic Engineering PowerPoint presentation, circa 2000.

[20] E. Lawrey, B.S. thesis, "The Suitability of OFDM as a Modulation Technique for Wireless Telecommunications, with a CDMA Comparison," Computer Systems Engineering at James Cook University, 1997.

[21] L. Couch III, *Digital and Analog Communications Systems*, 6th ed., Englewood Cliffs, NJ: Prentice Hall, 2001.

[22] T. W. Barrett, "History of Ultra Wideband (UWB) Radar & Communications: Pioneers and Innovators," Progress in *Electromagnetics Symposium 2000 (PIERS2000)*, Cambridge, MA, July 2000.

[23] G. F. Ross, "Transmission and Reception System for Generating and Receiving Base-Band Duration Pulse Signals for Short Base-Band Pulse Communication System," U.S. patent 3,750,025, April 17, 1973.

[24] P. Van Etten, "The Present Technology of Impulse Radars," *International Radar Conference Proc.*, October 1977, pp. 535–539.

[25] R. N. Morey, "Geophysical Survey System Employing Electromagnetic Impulses," U.S. patent 3,806,795, April 1974.

[26] C. L. Bennett and G. F. Ross, "Time-Domain Electromagnetics and Its Application," *Proc. IEEE*, No. 66, 1978, pp. 299–318.

[27] T. E. McEwan, "Ultra-Wideband Radar Motion Sensor," U.S. patent 5,361,070, November 1, 1994.

[28] Spread Spectrum Scene, UWB page 2: Ultra Wideband Resource Information, located at http://www.sss-mag.com/uwbp2.html.

[29] http://www.aetherwire.com/CDROM/General/papers.html.

[30] J. Taylor, editor, *Ultra Wideband Radar Systems*, Boca Raton, FL: CRC Press, 1994.

[31] R. W. Chang, "Orthogonal Frequency Division Multiplexing," U.S. patent 3,488,445, January 6, 1970.

[32] J. Reed, *An Introduction to Ultra Wideband Communications Systems*, Upper Saddle River, NJ: Prentice Hall, 2004.

[33] R. Poisel, *Modern Communication Jamming Principles and Techniques*, Norwood, MA: Artech House, 2004.

[34] H. Sheng, P. Orlik, A. Haimovich, L. Cimini Jr., and J. Zhang, "On the Spectral and Power Requirements for Ultra-Wideband Transmission," Mitsubishi Research Labs., TR2003-66, December 2003.

[35] FCC document entitled "Revision of Part 15 of the Commission's Rules Regarding Ultra-Wideband Transmission Systems," First note and order, Federal Communication Commission, ET-Docket 98-153, adopted February 14, 2002, released April 22, 2002.

[36] K. Motohisa, "Time-Domain Sensors & Radiators," Chapter 5 in E. K. Miller, editor, *Time-Domain Measurements in Electromagnetics*, New York: Van Nostrand Reinhold, 1986.

[37] A. Azevedo, "Exploring the Ultrawideband," September 2004, Lawrence Livermore National Laboratory Magazine S & RT.

Problems

1. Show that complex envelopes satisfy the same relationships as the actual signals themselves; that is show that

$$\begin{aligned}&(1)\quad v_o(t)=v_i(t)*h_L(t)\quad \text{and}\\&(2)\quad V_o(f)=V_i(f)H_L(f)\end{aligned}\qquad \text{(P1-1)}$$

where $V_o(f)$ and $V_i(f)$ are defined in (1.3.2-9) and (1.3.2-10).

2. Using the fact that $R_z(\tau)=(1/2)E\left\{z(t+\tau)z^*(t)\right\}$ is the definition of the autocorrelation function of a complex process $z(t)$, (a) show that the auto correlation function of the radio frequency process can be expressed as $R_x(\tau)=\text{Re}\left\{R_v(\tau)e^{i\omega_0\tau}\right\}$ and (b) show that the radio frequency power spectral density is given by

$$S_x(f)=\frac{1}{2}S_v(f-f_0)+\frac{1}{2}S_v(-f-f_0)\qquad \text{(P1-2)}$$

3. Assume that is it required to send recorded voice which is band limited to 3 kHz. Assuming the voice is transmitted at the Nyquist rate (6 kHz) with a quantization of 256 levels, find the chip rate required to achieve a processing gain of 20 dB.

4. Show that the despreading *PN* code should be filtered in order to maximize the matched filter output signal-to-noise ratio at the data detector output in a BPSK direct sequence spread spectrum system. The SNR isdefined to be the square of the mean output signal divided by the variance of the noise. The system model is shown in Figure 1.4-5, showing both the unfiltered case and the filtered case. Hint: Use the Schwartz inequality in the time domain to show that using a filtered *PN* code to despread the signal is better than using an unfiltered one in the sense that it has a higher SNR. That is show that

$$SNR_A \le SNR_B \qquad \text{(P1-3)}$$

or in other words show that

$$\left[\frac{\left(\int_0^{T_d}\widetilde{PN}(t)PN(t)dt\right)^2}{E\left\{\int_0^{T_d}\int_0^{T_d}n'(t)n'(u)PN(t)PN(u)dt\right\}}\right]\le\left[\frac{\left(\int_0^{T_d}\widetilde{PN}(t)\widetilde{PN}(t)dt\right)^2}{E\left\{\int_0^{T_d}\int_0^{T_d}n'(t)n'(u)\widetilde{PN}(t)\widetilde{PN}(u)dt\right\}}\right]\qquad \text{(P1-4)}$$

where $\widetilde{PN}(t)$ is the filtered version of $PN(t)$. Assume that the product of the carrier and the white noise input is a white Gaussian baseband noise, $n'(t)$, with the same spectral density as the RF process. Also assume, as shown in Figure 1.4-5, that the filtered product of the data and the *PN* code is essentially the data times the filtered *PN* code. Let N_0 denote the one-sided noise spectral density, and T_d denote the data bit time.

5. A designer has chosen to use Manchester symbols, rather than NRZ symbols, for his *PN* code since he is concerned about narrowband interference at the carrier frequency. Assume that the *PN* code is a random sequence. Show that the correlation function of the filtered signal can be written as

$$R(\tau,\phi)=\sqrt{P}\,\mathrm{Re}\left\{\frac{1}{2\pi}\int_{-\infty}^{\infty}T\frac{\sin(\omega T/4)^4}{(\omega T/4)^2}H_L(\omega)e^{j(\omega\tau+\phi)}d\omega\right\} \qquad \text{(P1-5)}$$

6. Show that the total power in the MSK waveform, represented in (1.4.3-2), is given by P.

7. Consider a slow frequency-hopped communication system with 8FSK modulation that uses 10 bits to specify the hopping frequency. Let the hop rate be 12,000 hops per second and let the hopping frequency range be from 20.4 GHz to 21.0 GHz. Assume that there are 40 tones per hop and the synthesizer settling time is 3.333 μs. Determine the channel symbol rate that this design is capable of and the approximate bandwidth needed based on the first null of the spectral sidelobes. Assume that the channel bits are derived from a random sequence.

8. Suppose you are asked to design a frequency-hopped SS communication system and are told to use either DPSK or BPSK. What factors should be considered in making the choice of modulation. Consider BER efficiency first and then any other relevant factors. Hint: Consider the fact that BPSK is more efficient than DPSK and the fact that the overhead channel bits required for BPSK phase ambiguity resolution will be greater than for DPSK, which tends to offset the advantage of BPSK.

9. Consider a hybrid PN/FH communication system that is uncoded and has BPSK *PN* modulation and DPSK data modulation. The data rate is 4,200 bits per second, and there are eight channel bits per hop. At the first of each hop there is one reference bit, used to demodulate the remaining seven data bits. The chip rate is 148,800 chips per second. Also assume that the hops are spaced apart by the chip rate. There are 2,047 hop frequencies, all of which are equally spaced across the bandwidth. (a) Determine the RF bandwidth. (b) Determine the processing gain.

10. By assuming long spreading codes and comparing the spectral densities at $f = f_0$ show that MSK chip modulation with BPSK bit modulation provides 2.1 dB less processing gain than BPSK chip modulation with BPSK bit modulation. Assume the BPSK chip rate is the same as either the I or Q chip rate of the MSK chip modulation. Note that the required transmission bandwidth is reduced also.

CHAPTER 2

Binary Shift Register Codes for Spread Spectrum Systems

2.0 INTRODUCTION

This chapter will cover the essentials of the theory of binary shift register codes. Most of the theory will be directed toward linear codes, which will be defined shortly. Applications of shift register sequences (or codes) include many areas such as code division multiple access applications, ranging, radar systems, spectral spreading, privacy encoding, low probability of detection transmission, interference mitigation, random bit generation, synchronization, and other applications.

Before we delve into the theory of binary shift register codes it is appropriate to mention some of the original work. The first work in the area of maximal length sequences started around the early to mid-1950s. Possibly the earliest work was done by Gilbert [1] at Bell Labs. However his report had limited circulation. The work of Golomb [2] and Zierler [3, 4] appeared slightly later. L. Welch was also active in this area around this time. Slightly later Birdsall and Ristenbatt [5] published their work while at the University of Michigan. Golomb [6] credits the original theory of mod p addition to Lagrange in the eighteenth century, along with a modern mathematical exposition by Hall in 1937. And finally the paper by Sarwate and Pursley [7] provides a good summary up to 1980 and is probably one of the better complete introductions to the subject of linear binary sequences to date.

2.1 FINITE FIELD ARITHMETIC

In order to multiply, divide, add, subtract, and otherwise manipulate the code sequences introduced later in this chapter, it is necessary to have a rudimentary understanding of finite field arithmetic. These manipulations can easily be done if we require the coefficients to be elements of a finite field. Two good references that cover the theory of modern algebra are Birkhoff and MacLane [8] and at a more advanced level Van Der Waerden [9]. In addition, the book by Lin and Costello [10] on coding has a complete introductory treatment.

A finite field, also called a *Galois Field* and denoted by GF(q), has a set of q elements, which follow some special rules of algebra. All finite fields possess the following nine properties. In what follows it is assumed that a, b, and c are elements in the field:

I There are two operations that are defined for combining the elements, and they are addition and multiplication.

II Adding or multiplying any two elements in the field results in the sum or product being in the field.

III The field always contains th additive identity "0" that has the property that if a is in the field then $a+0 = a$.

IV The field always contains the multiplicative identity element "1" and $1 \bullet a = a$.

V For every element a there is an additive inverse element $(-a)$ such that $a + (-a) = 0$.

VI For every element a there is a multiplicative inverse element (a^{-1}) such that $a \bullet a^{-1} = 1$, and the inverse of "0" is not defined.

Note that V and VI allow one to use the customary notions of subtraction and division.

VII The associative law applies, that is, $a + (b + c) = (a + b) + c$, and in addition the following is true $a \bullet (b \bullet c) = (a \bullet b) \bullet c$.

VIII The commutative law applies, that is, $a + b = b + a$ and $a \bullet b = b \bullet a$.

IX The distributive law applies, that is, $a \bullet (b + c) = a \bullet b + a \bullet c$.

Not all values of q, the number of elements in the field, result in a finite field. In general, finite fields exist when q is a prime number, or an integer power of a prime number. When q is prime, the field is called a *prime field*. When there are q^N elements and N is an integer, the field is called an *extension field* over the prime field.

For each value of q there is one and only one unique field. In other words there is only one way to construct the field using the rules for addition and multiplication so that all nine properties are satisfied. If q is a prime number then the field elements are the numbers $\{0, 1, 2, ...q-1\}$. Multiplication in this field is ordinary multiplication, and addition is modulo q. At this point it is worth while considering some examples.

Example 1 Consider GF(5). The field elements are $\{0, 1, 2, 3, 4\}$. The addition and multiplicative tables are shown below.

Addition Table for GF(5)

+	0	1	2	3	4
0	0	1	2	3	4
1	1	2	3	4	0
2	2	3	4	0	1
3	3	4	0	1	2
4	4	0	1	2	3

Multiplication Table for GF(5)

•	0	1	2	3	4
0	0	0	0	0	0
1	0	1	2	3	4
2	0	2	4	1	3
3	0	3	1	4	2
4	0	4	3	2	1

For example, using the addition table, consider $4 + 4 = 8$. When 8 is reduced by 5, it equals 3. As an example of the multiplication table, consider multiplying 2 times 3, which produces 6, and when reduced by 5 we get 1, as seen in the multiplication table. Thus, it is seen that all sums and products are reduced by multiples of 5, so that the result is always in the set $\{0, 1, 2, 3, 4\}$.

Example 2 The smallest field is GF(2). It has the elements $\{0, 1\}$. The addition and multiplication tables are shown next.

Addition and Multiplication Tables for GF(2)

+	0	1
0	0	1
1	1	0

•	0	1
0	0	0
1	0	1

Note that $1+1 = 2$ and when reduced by 2 the sum equals zero as shown in the sum table for addition.

2.1.1 Polynomial Arithmetic

Next we shall consider the arithmetic of polynomials in which the coefficients are from the binary field GF(2). A polynomial $f(x)$ of the variable x, of the form

$$f(x) = f_0 + f_1 x + f_2 x^2 + \cdots f_n x^n \tag{2.1.1-1}$$

with coefficients from GF(2) (i.e., f_i is an element of GF(2)), is of n-th *degree* if the largest nonzero coefficient is f_n. In the literature the term *a polynomial over GF(2)* means "a polynomial with coefficients in GF(2)." Polynomials over GF(2) can be added, subtracted, multiplied, and divided in a similar manner as polynomials with real coefficients except that the coefficients are in GF(2). For example, when adding two polynomials, one of degree n and one of degree m with $m<n$, one has the result

$$f(x) + g(x) = (f_0 + g_0) + (f_1 + g_1)x + \cdots (f_m + g_m)x^m + f_{m+1}x^{m+1} \cdots + f_n x^n \tag{2.1.1-2}$$

where $f_i + g_i$ is combined with modulo-2 addition. When multiplying $f(x)$ with $g(x)$ one obtains the result

$$f(x)g(x) = a_0 + a_1 x + a_2 x^2 + \cdots a_{n+m} x^{n+m} \tag{2.1.1-3}$$

where the coefficients are given by

$$\begin{aligned} a_0 &= f_0 g_0 \\ a_1 &= f_0 g_1 + f_1 g_0 \\ a_2 &= f_0 g_2 + f_1 g_1 + f_2 g_0 \\ &\vdots \\ a_i &= f_0 g_i + f_1 g_{i-1} + f_2 g_{i-2} + \cdots f_i g_0 \\ &\vdots \\ a_{n+m} &= f_n g_m \end{aligned} \tag{2.1.1-4}$$

and the multiplication and addition of the coefficients are in the field GF(2). It is not difficult to show that polynomials over GF(2) are commutative, associative, and distributive.

When $f(x)$ is divided by $g(x)$ (assumed to be not equal to zero), a unique pair of polynomials over GF(2) is obtained, the *quotient* $q(x)$, and the *remainder* $r(x)$, such that

$$f(x) = q(x)g(x) + r(x) \tag{2.1.1-5}$$

with the degree of $r(x)$ is less than $g(x)$. This is known as *Euclid's division algorithm*. Consider an example of Euclid's division algorithm.

Example 3 Let $g(x) = 1 + x + x^4$ and $f(x) = x^6 + 1$. Then using long division one has

$$\begin{array}{r l} & x^2 \quad \text{(quotient)} \\ x^4 + x + 1 & \overline{\big)\, x^6 + 1} \\ & \underline{x^6 + x^3 + x^2} \\ & \quad x^3 + x^2 + 1 \quad \text{(remainder)} \end{array}$$

It therefore follows that $x^6 + 1 = x^2 (x^4 + x + 1) + x^3 + x^2 + 1$, when the addition of the coefficients is modulo-2.

If $f(x)$ is divided by $g(x)$ and has a remainder of zero, we say that $g(x)$ *divides* $f(x)$. If $f(x)$ is divisible by $x–a$, then it is true that $f(a) = 0$. In other words, polynomials over GF(2) have roots and can be factored. It should be pointed out that, in general, the roots of polynomials over GF(2) are not necessarily in GF(2), or even real. A polynomial over GF(2) of degree n is said to be *irreducible* over GF(2) if $f(x)$ is not divisible by any polynomial over GF(2) of degree less than n, but greater than zero. Now consider a polynomial that is factorable over GF(2).

Example 4 As an example of a factorable polynomial over GF(2), consider $f(x) = x^6 + 1$. Since $f(1) = 0$ we know that $(x–1)$ is a root, and therefore so is $(x+1)$. To verify this note that

$$
\begin{array}{r|l}
 & x^3 + x^2 + x + 1 \\
\hline
x+1 & x^4 + 1 \\
 & \underline{x^4 + x^3} \\
 & \quad x^3 + 1 \\
 & \quad \underline{x^3 + x^2} \\
 & \qquad x^2 + 1 \\
 & \qquad \underline{x^2 + x} \\
 & \qquad\quad x + 1 \\
 & \qquad\quad \underline{x + 1} \\
 & \qquad\qquad 0
\end{array}
$$

and thus $x–1$ is a root of x^4+1.

When we discuss m-sequences, we will encounter primitive polynomials which are always irreducible.

2.2 SHIFT REGISTER SEQUENCES

In this section we will consider some different types of shift registers and their corresponding output sequences. When referring to a shift register sequence we shall mean (unless stated otherwise) the resulting sequence of "ones" and "zeros" that emanates from the shift register. Figure 2.2-1 illustrates an output sequence of zeros and ones and its corresponding time waveform, which has values of +1 and –1.

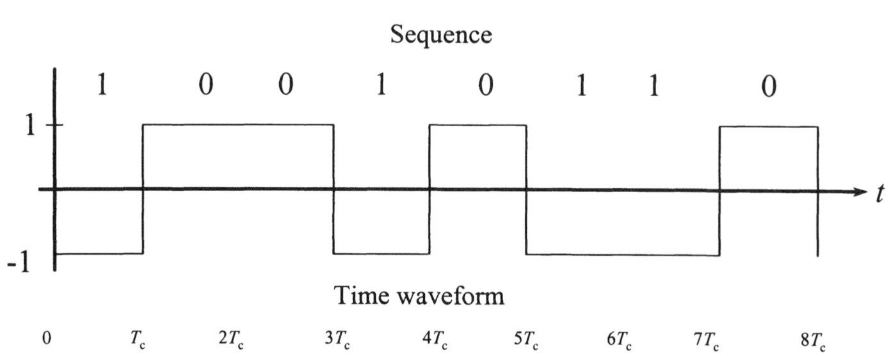

Figure 2.2-1 The PN sequence and the associated time waveform.

The *sequence* takes on the values of "0s" and "1s" from the code generator. The *time waveform* is the continuous time signal associated with the sequence that takes on amplitudes according to the following rule. A binary "1" in the sequence maps to "–1" chip at the waveform level, and a binary "0" in the sequence maps to a "1" chip at the waveform level. The time waveform has a possible change of level every T_c seconds, where T_c is the chip duration. Mathematically the relationship is given by

$$a(t) = \frac{1 - PN(t)}{2} \qquad (2.2\text{-}1)$$

where $a(t)$ takes on the values in $\{0, 1\}$ and $PN(t)$ takes on the values ± 1, In other words "0" $\leftrightarrow +1$ and "1" $\leftrightarrow -1$. The double arrow denotes "maps to each other." This is the only mapping that preserves the multiplication property on the numbers ± 1 with modulo 2 addition of "0" and "1."

Consider a *simple shift register generator* (SSRG). In an SSRG all the feedback signals are returned to a single point. A SSRG contains a shift register for storage (and setting the initial conditions) and shifting its contents, a modulo-2 adder, and a feedback line. An example of an SSRG, in *Fibonacci form*, is illustrated in Figure 2.2-2 showing two common means of depicting the same SSRG.

By a Fibonacci form we mean an SSRG in which the feedback is fed back to the first cell only. Implicit in each version is a clock signal that clocks the register and thereby sets the chip rate. Later we will consider an equivalent form to the SSRG, called the *Galois form,* which has the advantage of operating at greater speed than the Fibonacci form and produces the same output but has multiple feedback points.

Now consider the operation of a simple shift register. Number the shift register cells 1 through 6 in Figures 2.2-2(a) and 2.2-2(b) and note that each cell contains either a "1" or a "0." When a clock pulse arrives the value stored in cell number 1 transfers to cell number 2, the value that was stored in cell number 2 transfers to cell number 3, and so forth. Finally the values of cells 5 and 6 are added modulo 2 to form the new value for cell number 1. Thus a shift register is a device that shifts its stored contents one position to the right for each shift of the clock pulse applied to the shift register. The *shift register sequence* is normally defined to be the output value of the last cell, which changes after each clock pulse. However the output of the first cell may be used as the output also; the difference in the sequence at the first cell and the last cell is just the time delay between the input and output cells.

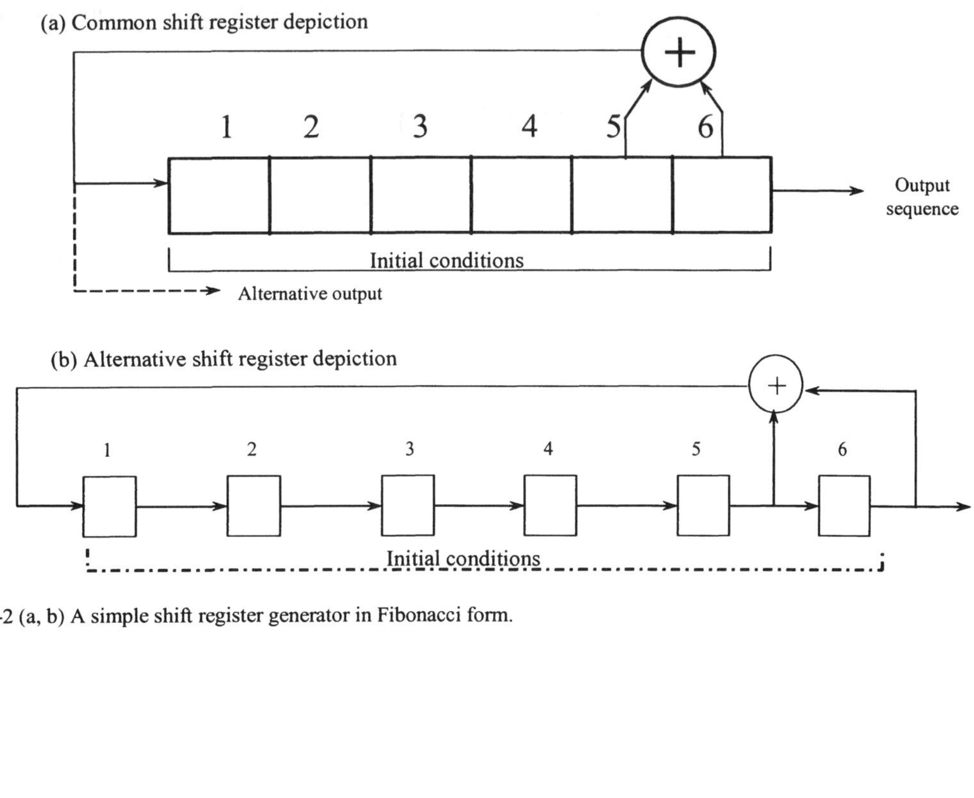

Figure 2.2-2 (a, b) A simple shift register generator in Fibonacci form.

Assume that the initial content of the shift register is 000001, which means all the cells are set to zero except the sixth one. After the first clock pulse it is clear that the register will contain the value 100,000, and the first output value will be "1." The "one" in cell number 1 is the result of the mod 2 sum of the fifth cell a "zero," and the sixth cell a "one." The resulting shift register sequence will be given by

$$100000100001100010100111101000111001001011011101100110101011111 \quad (2.2\text{-}2)$$

This sequence has a period of 63 chips (or bits) in it, and the same sequence will reappear again and again. This sequence is an example of a maximal length sequence and is also called an m-sequence, for which the length (period) $N = 2^n - 1$, where n is the number of cells in the SSRG. This is an example of a Fibonacci shift register, since all the feedback signals are returned to the first cell.

Another type of shift register generator is known as a *multiple return shift register generator* (MRSRG), which has adder outputs to two or more input stages. Figure 2.2-3 illustrates a MRSRG of length 6. Birdsall and Ristenbatt [5] have shown that every multiple return generator that has no transients possesses an equivalent SSRG. As a consequence most of the remaining discussion will deal with SSRGs.

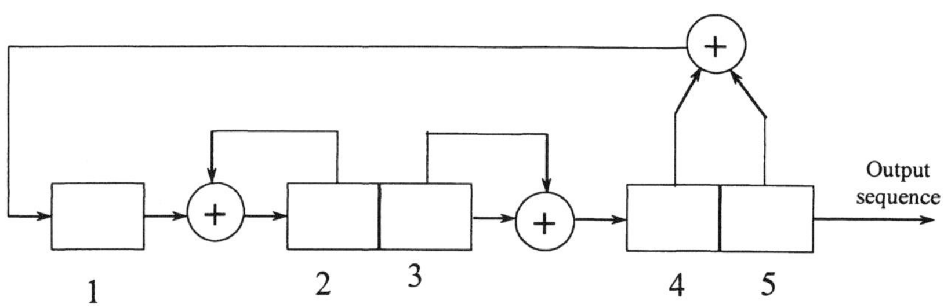

Figure 2.2-3 An example of an MRSRG.

If an SSRG produces a sequence that goes through a transient series of digits before settling into a periodic sequence, then that generator is said to possess *transients*. Figure 2.2-4 illustrates two examples of transient generators and their output sequences. In the (a) portion of the figure the sequence is of length 7 and repeats every seven symbols after an initial "0" is introduced into the sequence. In the (b) portion of the figure the same periodic output occurs with a different configuration but produces the same transient output. The transient symbol is underlined in both output sequences.

(a)

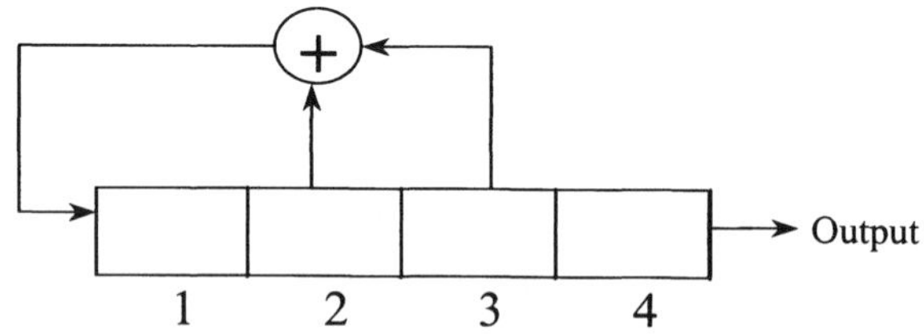

Sequence A with initial condition 1000: 000101110010111

(b)

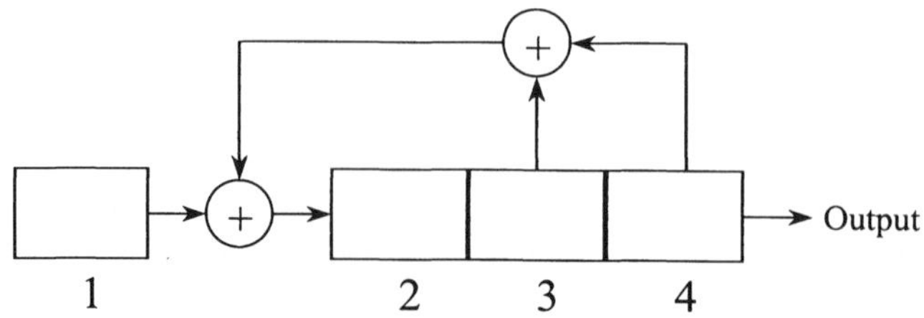

Sequence B with initial condition 1000: 000101110010111

Figure 2.2-4 (a, b) Two examples of transient generators.

Denote the contents of each cell in an SSRG by x_i, with the i-th cell number; then we can say that the shift register is *linear* if the feedback function can be expressed as a modulo 2 sum of the variables x_i; that is, if it can be expressed in the form

$$f(x_1, x_2, x_3, \ldots x_n) = c_1x_1 + c_2x_2 + c_3x_3 + \ldots c_nx_n \tag{2.2-3}$$

where the c_i are either one or zero, and the symbol "+" denotes modulo-2 addition. Figure 2.2-5 illustrates a general shift register generator in Fibonacci form, and if the feedback function satisfies (2.2-3), then it is a linear SSRG.

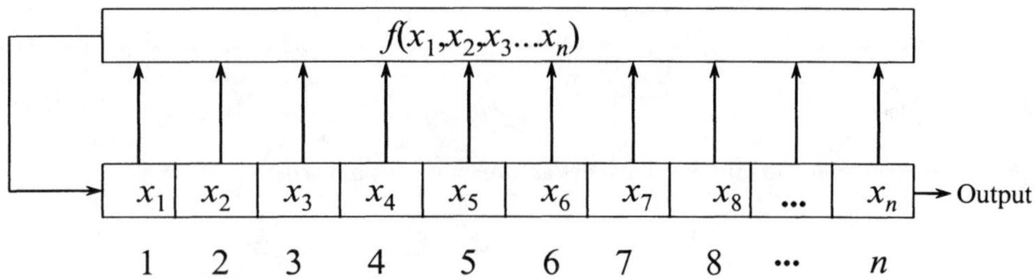

Figure 2.2-5 A general shift register generator.

Consider an example of a nonlinear shift register generator illustrated in Figure 2.2-6. In part (a) of the figure the SSRG is shown and in part (b) the corresponding state diagram is shown. Starting in state 110 (cell 1 contains a "1," cell 2 contains a "1," and cell 3 contains a "0") we see that $x_2x_3 = 0$, so that the new state becomes 011. This process continues until the state 000 is reached. From then on the SSRG remains in state 000. If the SSRG is loaded with the values 111, then it remains in that state from that time on. Another observation can be made about this SSRG. That is, the state 010 can be reached from two distinct predecessors, 101 and 100. Golomb [2] has shown that *for linear shift register generators every state has exactly one predecessor.*

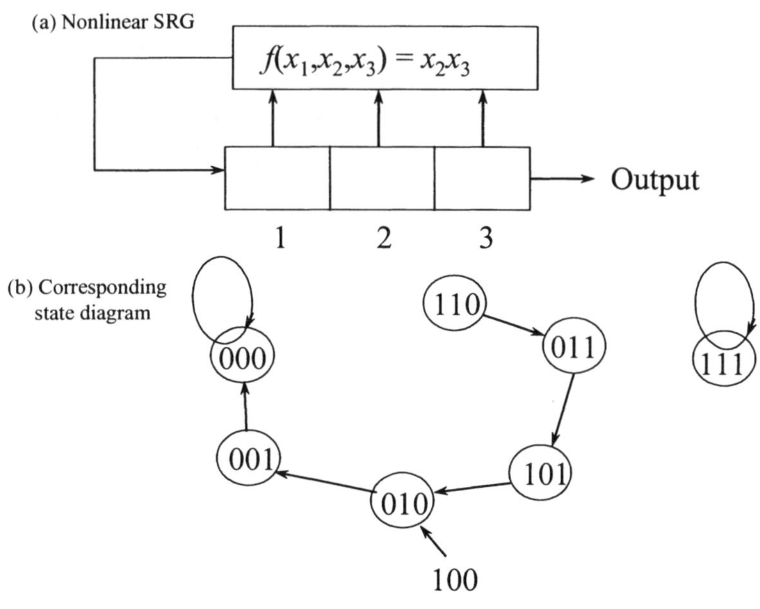

Figure 2.2-6 (a, b) An example of a nonlinear SSRG.

Consider the SSRG in Figure 2.2-7(a) with its tap coefficients $\{c_i\}$ and the n state contents $\{a_{k-1}, a_{k-2}...a_{k-n}\}$, when reading from left to right on the shift register cells.

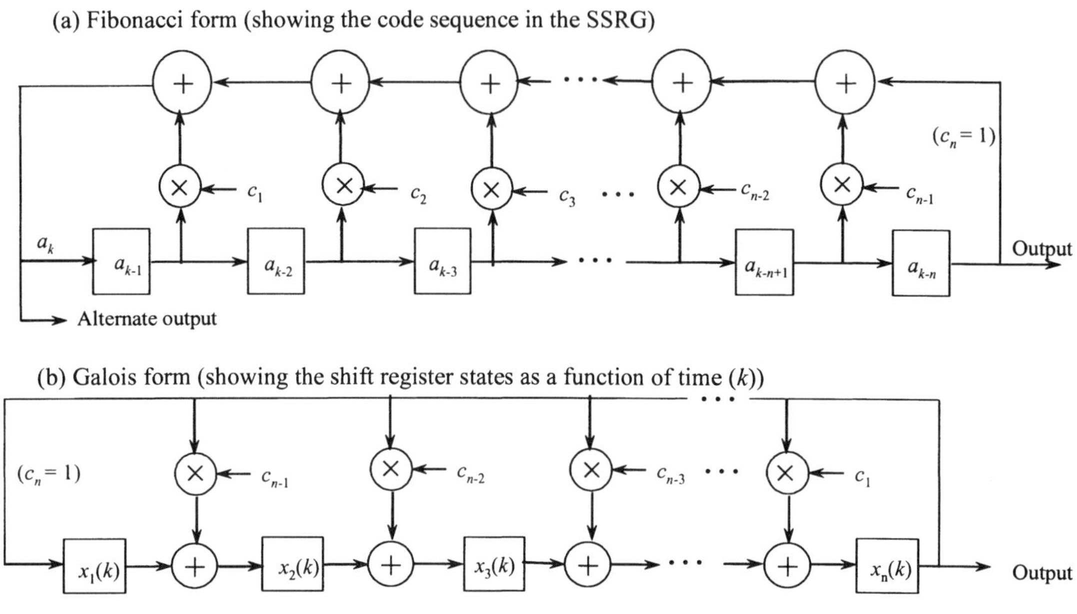

Figure 2.2-7 (a, b) Linear SSRG models.

The *linear recursion* that linear SSRGs satisfy is given directly by

$$a_k = \sum_{i=1}^{n} c_i a_{k-i} \tag{2.2-4}$$

where the c_i take on the value "1" if the feedback connection is connected and takes on the value of "0" if it is open and the sequence $\{a_k\}$ is in GF(2). The summation in (2.2-4) is modulo 2. Note that this is a relationship between the existing state contents, the tap coefficients and the next new state value, which is shifted into the left cell after the clock signal. Note that the all-zeros state corresponds to the initial conditions $a_{-1} = a_{-2} ... = a_{-n} = 0$ and has a period of one. In general the *initial conditions* are given by $a_{-1}, a_{-2},...a_{-n}$ and are not normally all-zero. The output sequence after the initial conditions have evolved, when taken as the output of the rightmost cell, will be $a_0, a_1, a_2,...$. In other words, the output taken from the rightmost cell including the initial conditions will be $a_{-n}, a_{-n+1}, a_{-n+2}, ... a_{-1}, a_0, a_1, a_2, ...$ If instead, the output is taken from the leftmost cell, which is the alternate output, the output would be $a_0, a_1, a_2, ...$ That is the initial conditions will not appear on this output.

2.2.1 Equivalence of the Fibonacci and Galois Forms of a Linear SRG

The Fibonacci and Galois forms of a linear SRG are illustrated in Figures 2.2-7(a) and 2.2-7(b). Now it will be shown that the Fibonacci form and the Galois form of linear SRGs are equivalent, in that they produce the same output sequence. Consider Figure 2.2-7(b), which is a shift register in Galois form, with the states $x_i(j)$, where i is the index of the state when counting from the left, and j is the index of time. We will show that the linear recursion of (2.2-4) is satisfied for both implementations of Figure 2.2-7 [11, 12]. The SRG in Galois form also has n states, with the tap coefficients c_i, now numbered from right to left, which is just the opposite of the Fibonacci form indicated in Figure 2.2-7(a).

By inspection of the output of the Galois SRG form of Figure 2.2-7(b), it is clear that the following is true,

$$x_n(k) = c_1 x_n(k-1) + c_2 x_n(k-2) + \cdots c_{n-1} x_n(k-n+1) + x_1(k-n+1) \quad k \geq n-1 \tag{2.2-5}$$

This follows since the output state $x_n(k)$ is composed of the sum of the previous (one unit earlier) output state $x_n(k–1)$ weighted by c_1 plus $x_n(k–2)$ (the output state two units earlier) weighted by c_2, and so on until the first cell $x_1(k)$ is reached. The first register cell $x_1(k)$ is delayed by $n–1$ units and is added to the sum to yield (2.2-5). This result can be written in summation form as

$$x_n(k) = \sum_{i=1}^{n-1} c_i x_n(k-i) + x_1(k-n+1) \quad ,k \geq n-1 \tag{2.2-6}$$

It is clear from the Figure 2.2-7(b) that the following is true

$$x_1(k) = x_n(k-1) \tag{2.2-7}$$

Define $a_k = x_n(k)$. Inserting the term in (2.2-7) and the definition into (2.2-6) and defining $c_n = 1$, produces the result

$$a_k = \sum_{i=1}^{n-1} c_i a_{k-i} + a_{k-n} = \sum_{i=1}^{n} c_i a_{k-i} \tag{2.2-8}$$

since

$$x_1(k-n+1) = x_n(k-1-n+1) = x_n(k-n) = a_{k-n} \tag{2.2-9}$$

Thus

$$a_k = \sum_{i=1}^{n} c_i a_{k-i} \quad ,k \geq n \tag{2.2-10}$$

We see that the sequence defined by either shift register form produces the same result (2.2-4) and (2.2-10), and thus the two forms are equivalent in that they produce the same sequence. The initial conditions will not be the same for the Fibonacci SRG (FSRG) and the Galois SRG (GSRG) in general, and the states that the two generators move through will not be the same either. Now consider an example that illustrates the two shift register forms.

Example 5 Consider the two shifts registers in the Fibonacci and Galois forms depicted in the figure. Note that the tap connections are just reversed, as indicated in Figure 2.2-7. In part (a) of the figure the output has been determined for the Fibonacci form for both the rightmost cell output and the leftmost cell input. Note that the rightmost cell outputs the initial conditions first and then the sequence, whereas the input to the leftmost cell outputs the sequence directly. In reality there is just a time delay (or phase delay) between the two outputs. In part (b) the Galois form produces the same output as the input to the leftmost cell of the Fibonacci shift register.

One useful parameter in measuring the complexity of the sequence structure is the linear span. The minimum number of memory elements, LS, needed to construct a Fibonacci generator for the sequence a_k is the degree of the minimum-degree recursion that generates the sequence a_k. This number LS is called *the linear span* of the sequence a_k. As an example of this definition the sequence of example 5 has a period of 7 and a linear span of 3, since the smallest linear Fibonacci generator to generate that sequence has three storage elements.

2.3 MATHEMATICAL CHARACTERIZATION OF SRGS

In this section the matrix point of view will be exploited to develop the mathematical basis of SSRGSs [4, 5, 13] and to develop the contents of an SSRG as a function of time.

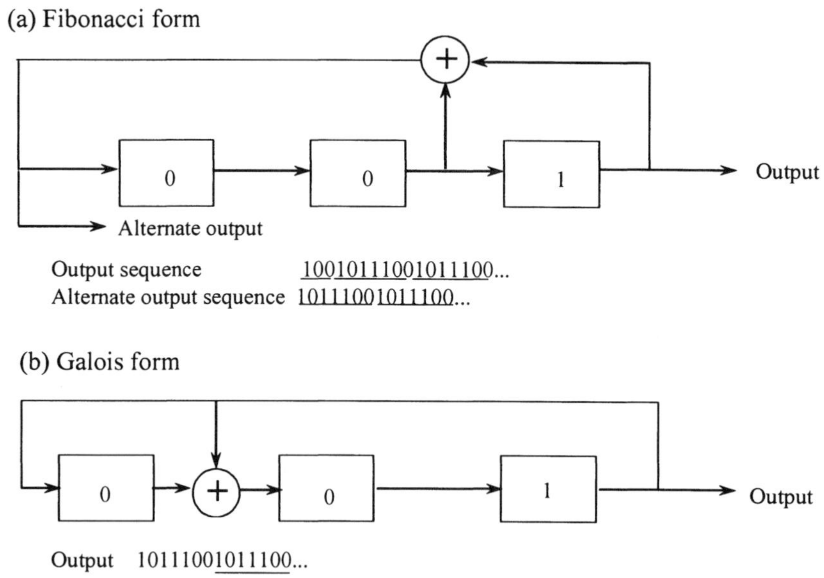

Example 5 figure showing the Fibonacci and Galois forms for the same code generator.

2.3.1 The Shift Register Matrix

The $n \times n$ *shift register matrix* (SRM), when multiplied by the $n \times 1$ column vector $X(j)$ representing the present contents at time j of the SSRG, will produce the contents of the SSRG, in Fibonacci form, in the next state after one clock shift of the SSRG. The SRM, which will be denoted by A, is an $n \times n$ matrix of the form

$$A = \begin{bmatrix} c_1 & c_2 & c_3 & \cdots c_{n\text{-}1} & 1 \\ 1 & 0 & 0 & & 0 & 0 \\ 0 & 1 & 0 & & 0 & 0 \\ 0 & 0 & 1 & & 0 & 0 \\ \vdots & \vdots & \vdots & & \vdots & \vdots \\ 0 & 0 & 0 & \ldots & 1 & 0 \end{bmatrix} \quad n \times n \tag{2.3.1-1}$$

with the n-th register output always connected to the mod 2 adder (hence $c_n = 1$). The notation $n \times n$ denotes the fact that the matrix is of dimensions n by n. Let the column state vector $X(j)$ denote the contents of the shift register at time j, with the first (top) component denoting the first storage element (left side of the

SSRG) and the second element being represented by the second component of the column vector, and so forth. Thus the column state vector $X(j)$ can be represented by

$$X(j) = \begin{Bmatrix} x_1(j) \\ x_2(j) \\ x_3(j) \\ \vdots \\ x_n(j) \end{Bmatrix} n \times 1 \qquad (2.3.1\text{-}2)$$

This notation is the same as that used in the last section on Galois SRGs. The contents of the SSRG, at time $j+1$, is given by

$$X(j+1) = AX(j) \qquad (n \times 1) \qquad (2.3.1\text{-}3)$$

It follows that to obtain the contents of the register after m shifts of the SSRG, one merely has to evaluate

$$X(j+1) = A^m X(j) \qquad (2.3.1\text{-}4)$$

Clearly if the shift register is maximal length then

$$A^N = I \qquad (2.3.1\text{-}5)$$

where N is the length of the maximal sequence ($N = 2^n - 1$, with n the number of shift register cells in the SSRG), and I is the $n \times n$ identity matrix.

Example 6 Consider the following example illustrated in the figure. The A matrix is given by (2.3.1-1), and produces, in this example by inspection, the matrix and vector

$$A = \begin{bmatrix} 1\,0\,1 \\ 1\,0\,0 \\ 0\,1\,0 \end{bmatrix} \text{ and } X(j) = \begin{Bmatrix} x_1(j) \\ x_2(j) \\ x_3(j) \end{Bmatrix}$$

After one shift of the SSRG, the output can be determined by (2.3.1-3) to yield

$$\begin{aligned} x_1(j+1) &= x_1(j) + x_3(j) \\ x_2(j+1) &= x_1(j) \\ x_3(j+1) &= x_2(j) \end{aligned}$$

Thus the state of the SRG can be determined at any number of shifts of the SRG by repeatedly applying (2.3.1-3).

Thus it is seen that shift register sequences can be studied by utilizing matrix theory.

2.3.2 The Characteristic Equation and Characteristic Polynomial

For any $n \times n$ matrix A the *characteristic equation* is formed by evaluating the determinant of the matrix $[A-\lambda I]$ (where I is an $n \times n$ unit matrix and λ is a parameter), and which we denote by

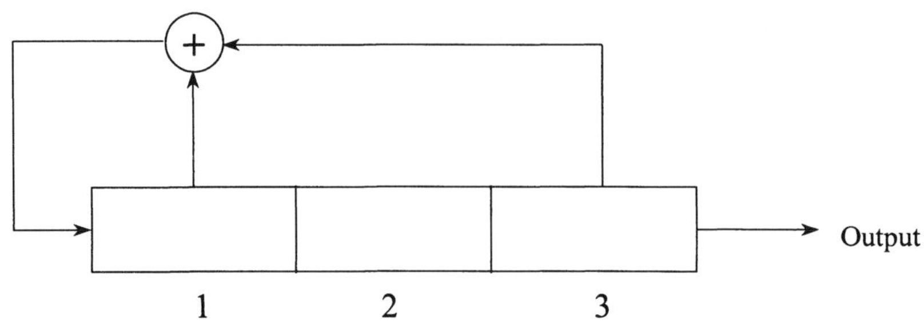

Figure for example 6 illustrating a Fibonacci SSRG.

$$F(\lambda) = |A - \lambda I| \tag{2.3.2-1}$$

Example 7 As an example of the determination of the characteristic equation consider the following A matrix

$$A = \begin{bmatrix} c_1 & c_2 & c_3 & 1 \\ 1 & 0 & 0 & 0 \\ 0 & 1 & 0 & 0 \\ 0 & 0 & 1 & 0 \end{bmatrix}$$

Then the characteristic equation becomes

$$F(\lambda) = |A - \lambda I| = \begin{vmatrix} c_1 - \lambda & c_2 & c_3 & 1 \\ 1 & -\lambda & 0 & 0 \\ 0 & 1 & -\lambda & 0 \\ 0 & 0 & 1 & -\lambda \end{vmatrix}$$

Expanding by minors along the first row one obtains the result

$$F(\lambda) = \lambda^4 + c_1\lambda^3 + c_2\lambda^2 + c_3\lambda + 1$$

Now consider the characteristic equation in the general case of an $n \times n$ matrix A. The characteristic equation $F(\lambda)$ is given by

$$F(\lambda) = \begin{vmatrix} c_1 - \lambda & c_2 & c_3 & c_4 & \cdots & 1 \\ 1 & -\lambda & 0 & 0 & \cdots & 0 \\ 0 & 1 & & & \cdots & 0 \\ 0 & & 1 & & \cdots & 0 \\ \vdots & \vdots & \vdots & \vdots & \vdots & \vdots \\ 0 & 0 & 0 & 0 & \cdots & 1 - \lambda \end{vmatrix} \tag{2.3.2-2}$$

which can be evaluated by expansion of minors to yield the *characteristic equation*:

$$F(\lambda)=\sum_{k=0}^{n} c_k \lambda^{n-k} \quad \text{with } c_0 = c_n = 1 \tag{2.3.2-3}$$

Thus it is possible to write down the characteristic equation directly for a given Fibonacci SSRG via (2.3.2-3).

The characteristic polynomial is defined by[1]

$$f(\lambda)=\sum_{k=0}^{n} c_k \lambda^{k} \quad c_0 = c_n = 1 \tag{2.3.2-4}$$

with each connection corresponding to $c_k = 1$, and each nonconnection corresponding to $c_k = 0$.

Example 8 As an example, suppose that the SSRG under consideration is a six-stage SSRG with connections at the first, third, and sixth cells as shown in the following figure. Then, using (2.3.2.-3), one obtains the result for the characteristic equation

$$F(\lambda)=\lambda^6+\lambda^5+\lambda^3+1$$

The characteristic polynomial is given by (2.3.2-4)

$$f(\lambda)=1+\lambda+\lambda^3+\lambda^6$$

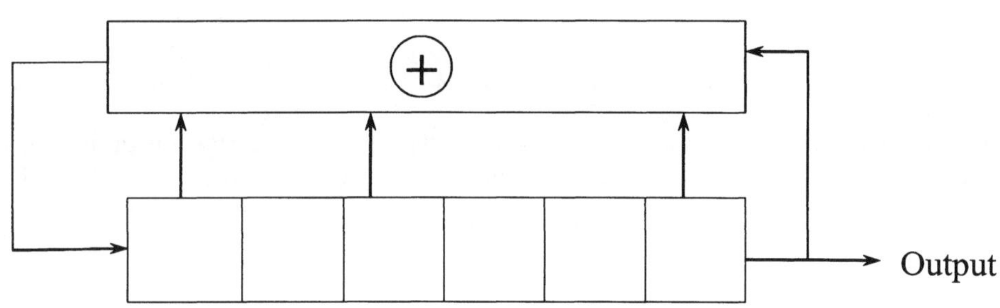

Figure for example 8, illustrating how to determine the characteristic equation and polynomial.

Figure 2.3-1 summarizes both the characteristic equation and the characteristic polynomial in relationship to the SSRG.

2.4 THE GENERATING FUNCTION

A powerful tool in the analysis and characterization of the output sequence is the *generating function*. Denote the output of an SSRG sequence by

$$\{a_m\}=\{a_0,a_1,a_2 ...\} \tag{2.4-1}$$

---

[1] Although some authors use this definition, others use $f(\lambda)=\sum_{k=0}^{n} c_k \lambda^{n-k}$.

where the index m indicates time; that is, a_0 occurs initially, then a_1, then a_2, and so forth. Note that the initial conditions are not present in this output, since it starts after the initial conditions have been shifted out of the shift register.

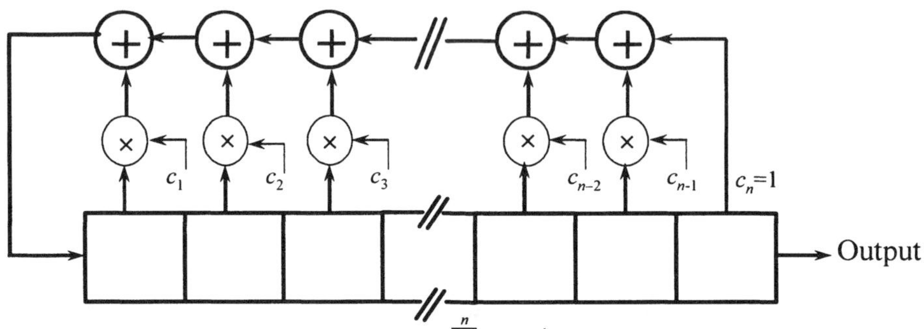

Characteristic equation $F(\lambda)=\sum_{k=0}^{n} c_k \lambda^{n-k}$ with $c_0 = c_n = 1$

Characteristic polynomial: $f(\lambda)=\sum_{k=0}^{n} c_k \lambda^{k}$ with $c_0 = c_n = 1$

Figure 2.3-1 The characteristic polynomial and equation for a Fibonacci SSRG.

Then the generating function of the output sequence is given by

$$G(x)=\sum_{k=0}^{\infty} a_k x^k \tag{2.4-2}$$

where x is a real variable. The *initial state* of the register (reading from left to right) is defined as a_{-1}, a_{-2}, a_{-3}, ..., a_{-n+1}, a_{-n} for an n-stage shift register. Then the output sequence a_0, a_1, a_2, ... is defined by the *linear recursion* relation; see (2.2-3)

$$a_m=\sum_{i=1}^{n} c_i a_{m-i} \qquad m = 0, 1, 2,... \tag{2.4-3}$$

So that the sequence is defined after all of the initial conditions are clocked out of the register.
Using (2.4-3) with (2.4-2) produces

$$G(x)=\sum_{k=0}^{\infty}\sum_{i=1}^{n} c_i a_{k-i} x^k \tag{2.4-4}$$

Changing the order of summation yields

$$G(x)=\sum_{i=1}^{n} c_i x^i \sum_{k=0}^{\infty} a_{k-i} x^{k-i} \tag{2.4-5}$$

or

$$G(x)=\sum_{i=1}^{n} c_i x^i \left[a_{-i}x^{i} + a_{-i+1}x^{-i+1} + ...a_{-1}x^{-1} + \sum_{i=0}^{\infty} a_k x^k \right] \tag{2.4-6}$$

The last term on the right of (2.4-6) is just $G(x)$, so solving for $G(x)$ produces

$$G(x)=\frac{\sum_{i=1}^{n}c_ix^i\left[a_{-i}x^{-i}+a_{-i+1}x^{-i+1}+\cdots+a_{-1}x^{-1}\right]}{1-\sum_{i=1}^{n}c_ix^i} \tag{2.4-7}$$

Now since the denominator is equal to $\sum_{i=0}^{n}c_ix^i$ when the coefficients are in GF(2), one finally obtains one of our main results

$$G(x)=\frac{\sum_{i=1}^{n}c_ix^i\left[a_{-i}x^{-i}+a_{-i+1}x^{-i+1}+\cdots+a_{-1}x^{-1}\right]}{\sum_{i=0}^{n}c_ix^i}=\frac{g(x)}{f(x)} \tag{2.4-8}$$

in which c_0 is assumed to be one (there is no "zeroth" tap of course) and "normally" c_n is one. Thus *G(x)* can be expressed in terms of the initial state a_{-1}, a_{-2}, ..., a_{-n} in the numerator, and the characteristic polynomial in the denominator, via the feedback taps c_1, c_2, ..., c_n.

Note that when we perform long division of (2.4-8) we obtain the series

$$G(x)=a_0+a_1x+a_2x^2+\cdots \tag{2.4-9}$$

so that the coefficient of x^m is the $(m+1)$-th term in the output sequence. Note that when $a_{-1}=a_{-2}=\ldots a_{1-n}=0$ and $a_{-n}=1$, $G(x)$ reduces to ($c_n=1$)

$$G(x)=\frac{1}{f(x)} \tag{2.4-10}$$

Therefore this condition occurs when the initial conditions of a "1" in the last cell (right most) and "0s" in the $n-1$ remaining cells occur located on the left of the "1."

Example 9 Consider the example indicated in the figure, which is a SSRG with three stages with feedback taps on the first and third cells. Using (2.4-8), the function $G(x)$ is given by

$$G(x)=\frac{c_1x(a_{-1}x^{-1})+c_2x^{-2}(a_{-2}x^{-2}+a_{-1}x^{-1})+c_3x^3(a_{-3}x^{-3}+a_{-2}x^{-2}+a_{-1}x^{-1})}{1+x+x^3} \tag{2.4-11}$$

Let the initial conditions be given by $a_{-1}=a_{-2}=0$, and $a_{-3}=1$ so that the numerator is "1" ($c_3=1$), then the division (using the lowest order first in all the polynomials) leads to

$$
\begin{array}{l}
\quad\quad\quad\quad 1\ 1\ 1\ 0\ 1\ 0\ 0\ 1\ 1 \\
\quad\quad\quad\quad 1+x+x^2+x^4+x^7+x^8 \\
1+x+x^3 \overline{)\,1} \\
\quad\quad\quad\quad \underline{1+x+x^3} \\
\quad\quad\quad\quad\quad x+x^3 \\
\quad\quad\quad\quad\quad \underline{x+x^2+x^4} \\
\quad\quad\quad\quad\quad\quad x^2+x^3+x^4 \\
\quad\quad\quad\quad\quad\quad \underline{x^2+x^3+x^5} \\
\quad\quad\quad\quad\quad\quad\quad x^4+x^5 \\
\quad\quad\quad\quad\quad\quad\quad \underline{x^4+x^5+x^7} \\
\quad\quad\quad\quad\quad\quad\quad\quad x^7 \\
\quad\quad\quad\quad\quad\quad\quad\quad \underline{x^7+x^8+x^{10}} \\
\quad\quad\quad\quad\quad\quad\quad\quad\quad x^8+x^{10} \\
\quad\quad\quad\quad\quad\quad\quad\quad\quad \underline{x^8+x^9+x^{11}} \\
\quad\quad\quad\quad\quad\quad\quad\quad\quad\quad x^9+x^{10}+x^{11}
\end{array}
\quad (2.4\text{-}12)
$$

The figure of the SSRG is shown next.

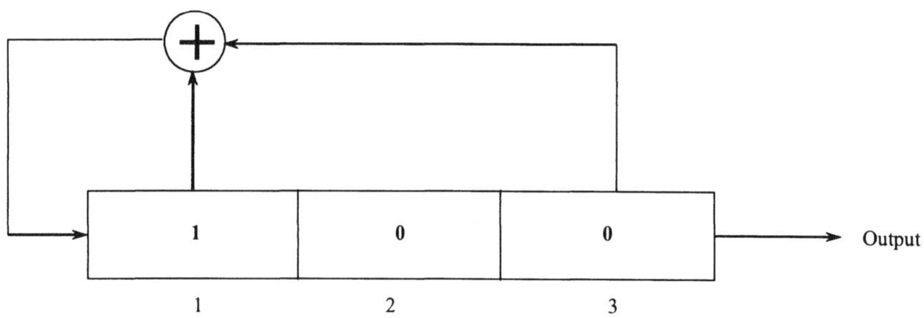

Figure of example 9, showing the SSRG.

By checking the nonzero coefficients of the powers of x one can conclude that the output sequence is given by

$$
\begin{array}{ccccccccc}
a_0 & a_1 & a_2 & a_3 & a_4 & a_5 & a_6 & a_7 & a_8 \\
1 & 1 & 1 & 0 & 1 & 0 & 0 & 1 & 1
\end{array}
\quad (2.4\text{-}13)
$$

Notice that the first digit of the output sequence is a_0, as it should be from the definition of $G(x)$, given by (2.4-2).

2.5 THE CORRELATION FUNCTION OF SEQUENCES

Before discussing the specific types of codes that are available, some definitions related to the various types of correlation functions that are useful when dealing with both sequences and the associated time waveforms will be presented. Figure 2.5-1 illustrates various types of correlations and the associated binary codes, which can be classified by the type of correlation function for which these codes were designed [14].

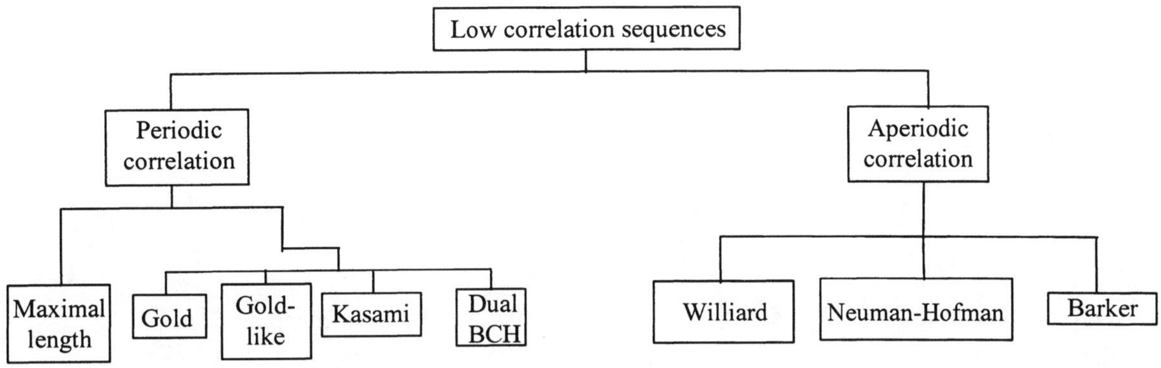

Figure 2.5-1 Correlation function classification of binary codes.

Low correlation sequences (i.e., low autocorrelation and low cross-correlation sequences) can be broken down into different classes of correlations. The first class is that of a *periodic correlation.* This involves the correlation of two codes over the complete code period, with the assumption that each code has the same period. This correlation is very important in code division multiple access (CDMA) applications.

Aperiodic correlation functions typically relate to the correlation over an incomplete period of one code against another code, in which data is often located before and after the code word. Aperiodic correlation functions thus find applications in synchronization schemes.

Within these types of correlations one can further break down the class into autocorrelation and cross-correlation processes. And within these two classes of correlations one can break them down even further into binary and q-ary sequences. These two classes can further be broken down into the individual codes. Only binary codes will be treated here.

It is appropriate and useful at this point to indicate the relationship between the periodic, auto-, and cross-correlation functions of the waveform and the discrete periodic auto- and cross-correlation functions of the sequence. Let the code sequence from a code generator be the sequence $\{a_m\}\ \varepsilon\{0,1\}$ and let the sequence $\{a_m^{'}\}\ \varepsilon\{-1,1\}$ be related by

$$a_m^{'} = 1 - 2a_m \tag{2.5-1}$$

so that a "zero" value of a_m maps to "one" for $a_m^{'}$, and a "one" value for a_m maps to a "minus one" value for $a_m^{'}$. The baseband waveform $u(t)$ is written in terms of the $\{a_m^{'}\}$ as

$$u(t) = \sum_{m=-\infty}^{\infty} a_m^{'} p(t - mT_c) \tag{2.5-2}$$

where T_c is the chip time or duration of each pulse, and $p(t)$ is a unit amplitude rectangular pulse starting at time zero and ending at time T_c seconds, so that $u(t)$ is the binary valued waveform corresponding to the code $\{0,1\}$ sequence $\{a_m\}$. The time autocorrelation function of $u(t)$ is given by

$$R_u(\tau) = \frac{1}{T}\int_0^T u(t)u(t+\tau)dt \tag{2.5-3}$$

where $T = NT_c$ is the code length, and N is the number of chips in the period. Now consider two different binary valued waveforms $u(t)$ and $v(t)$, each with the same chip rate $1/T_c$ and period T. The time cross-correlation function is given by

$$R_{uv}(\tau)=\frac{1}{T}\int_0^T v(t)u(t+\tau)dt \tag{2.5-4}$$

where $v(t)$ is given by

$$v(t)=\sum_{n=-\infty}^{\infty} b_n' p(t-nT_c) \tag{2.5-5}$$

and where $b_n^{'}$ ε$\{-1,1\}$ and b_n ε$\{0,1\}$ and b_n is the code sequence and $b_n^{'}$ is the mapping from $\{0,1\}$ to $\{-1,1\}$ through

$$b_n^{'}=1-2b_n \tag{2.5-6}$$

In general the notation x_n will denote elements in $\{0,1\}$ and the primed version will denote elements in $\{-1,1\}$. Putting the definitions of $u(t)$ and $v(t)$ into (2.5-4) produces

$$R_{uv}(\tau)=\frac{1}{T}\sum_{n=-\infty}^{\infty}\sum_{m=-\infty}^{\infty} a_m^{'}b_n^{'}\int_0^T p(t-mT_c)p(t+\tau-nT_c)dt \tag{2.5-7}$$

Note that the integral is zero when the two pulse functions don't overlap. Now let τ be given by

$$\tau=lT+\Delta\tau \;,\quad 0\le\Delta\tau<T_c \;,\; l \text{ an integer} \tag{2.5-8}$$

the integral in (2.5-7) produces a pulse overlap when $n = l+m$ and when $n = l+m+1$, and it can be broken up into two integrals of the form

$$\begin{aligned} R_{uv}(l,\Delta\tau) &= \frac{1}{N}\sum_{m=0}^{N-1} a_m^{'}b_{m+l}\frac{1}{T_c}\int_0^{T_c-\Delta\tau} p(x)p(x+\Delta\tau)dx + \\ &+\frac{1}{N}\sum_{m=0}^{N-1} a_m^{'}b_{m+l+1}\frac{1}{T_c}\int_{T_c-\Delta\tau}^{T_c} p(x)p(x-T_c+\Delta\tau)dx + \end{aligned} \tag{2.5-9}$$

where the substitution $x = t-mT_c$ has been used in this equation. The *discrete normalized periodic autocorrelation function* and the *discrete normalized periodic cross-correlation function* are defined, respectively, by

$$\begin{aligned} \hat{\theta}_a(l) &= \frac{1}{N}\sum_{n=0}^{N-1} a_n^{'}a_{n+l}^{'} \\ \hat{\theta}_{ab}(l) &= \frac{1}{N}\sum_{n=0}^{N-1} a_n^{'}b_{n+l}^{'} \end{aligned} \tag{2.5-10}$$

The normalization comes from the division by N. The corresponding unnormalized versions follow from (2.5-10) without the division by N, so that

$$\theta_a(l) = \sum_{n=0}^{N-1} a_n' a_{n+l}'$$
$$\theta_{ab}(l) = \sum_{n=0}^{N-1} a_n' b_{n+l}' \quad (2.5\text{-}11)$$

From (2.5-9) it is seen that, in order to analyze the time waveforms it is necessary to analyze the discrete auto and cross-correlation functions. Thus, from (2.5-9) and (2.5-10), one can write (since $\Delta\tau$ is positive)

$$R_{uv}(\tau, \Delta\tau) = \left(1 - \frac{\Delta\tau}{T_c}\right)\hat{\theta}_{ab}(l) + \frac{\Delta\tau}{T_c}\hat{\theta}_{ab}(l+1) \quad (2.5\text{-}12)$$

and when $\Delta\tau = 0$, one has

$$R_{uv}(l) = \theta_{ab}(l) \quad (2.5\text{-}13)$$

It is useful at this point to define the two parameters that measure the peak correlation values for the cross- and autocorrelation functions. Define the *peak periodic cross-correlation function magnitude* by

$$\theta_c = \max\left\{|\theta_{ab}(l)| \text{such that } 0 \le l \le N-1,\ a \ne b\right\} \quad (2.5\text{-}14)$$

In a similar manner one can define the *peak periodic out-of-phase autocorrelation function magnitude* by

$$\theta_a = \max\left\{|\theta_a(l)| \text{such that } 1 \le l \le N-1\right\} \quad (2.5\text{-}15)$$

and finally the maximum of the two correlation parameters by

$$\theta_{\max} = \max\left\{\theta_a, \theta_c\right\} \quad (2.5\text{-}16)$$

Notice that the range of l in θ_c differs from the range of l in θ_a, that is θ_c includes $l = 0$.

Henceforth in this chapter we will drop the primes in (2.5-10) and (2.5-11) with the understanding that the cross-correlation and the autocorrelation functions are based on the sequences with coefficients having the values ± 1 which is related to the 0, 1 sequence through (2.5-1).

2.5.1 Periodic Correlation Functions for Sequences

In this section a few relationships for periodic correlation functions will be presented. Reference [7] was utilized in this development. The discussion will be limited to sequence symbols that are ε GF(2). We define the *shift register generated sequence a*, as

$$a = ..., a_{-3}, a_{-2}, a_{-1}, a_0, a_1, a_2, ...a_{N-1}, a_N, a_{N+1}, ... \quad (2.5.1\text{-}1)$$

with the period being N. If b is also an SSRG generated sequence of period N, it is easy to show that[2]

$$\theta_{ab}(l) = \theta_{ab}(l+N) \quad (2.5.1\text{-}2)$$

It is also easy to verify that

[2] The bold representation of a and b are based on the definition of $\boldsymbol{a}$ and $\boldsymbol{b}$ that are N element vectors, which will be defined shortly.

$$\theta_{ab}(-l) = \theta_{ab}(l) \tag{2.5.1-3}$$

Denote the N component row vector $\boldsymbol{a}$, whose components are εGF(2) as the first N sequence values of an SSRG, that is,

$$\boldsymbol{a} = a_0, a_1, \cdots a_{N-1} \tag{2.5.1-4}$$

and define $\boldsymbol{b}$ in a similar manner. Define the *inner product* of two vectors $\boldsymbol{a}$ and $\boldsymbol{b}$ by

$$\langle \boldsymbol{a}, \boldsymbol{b} \rangle = \sum_{i=0}^{N-1} a_i b_i \tag{2.5.1-5}$$

where, as was mentioned earlier, all components in every summation used for an inner product or discrete correlation function have value ± 1. The *norm of* $\boldsymbol{a}$ is given by

$$\|\boldsymbol{a}\| = \sqrt{\sum_{i=0}^{N-1} a_i^2} \tag{2.5.1-6}$$

The *Cauchy Inequality* is given by

$$|\langle \boldsymbol{a}, \boldsymbol{b} \rangle| \le \|\boldsymbol{a}\| \; \|\boldsymbol{b}\| \tag{2.5.1-7}$$

Using the Cauchy Inequality for binary valued periodic sequences of period N, one has

$$\left| \sum_{i=0}^{N-1} a_i b_{i+l} \right| \le \sqrt{\sum_{i=0}^{N-1} a_i^2} \sqrt{\sum_{i=0}^{N-1} b_i^2} \tag{2.5.1-8}$$

since the norm of $\boldsymbol{b}$ does not depend on where the summation begins, as long as the sequence is periodic.

When $\boldsymbol{b} = \boldsymbol{a}$, the autocorrelation function has the same properties as the cross-correlation function in the sense that

$$\theta_a(l) = \theta_a(l+N) \text{ and } \theta_a(-l) = \theta_a(l) \tag{2.5.1-9}$$

From the results of problem 5 it follows that

$$\left| \sum_{i=0}^{N-1} a_i a_{i+l} \right| \le \sqrt{\sum_{i=0}^{N-1} a_i^2} \sqrt{\sum_{i=0}^{N-1} a_i^2} = \sum_{i=0}^{N-1} a_i^2 = \theta_a(0) \tag{2.5.1-10}$$

If we let $T\boldsymbol{a}$ denote the left cyclic shift operator that shifts the vector cyclically to the left by one place, so that if $\boldsymbol{a}$ is as in (2.5.1-4), then

$$T\boldsymbol{a} = a_1, a_2, \cdots a_{N-1}, a_0 \tag{2.5.1-11}$$

It is easy to show, using the first equation in (2.5-10), that

$$\theta_{aT^k b}(l) = \theta_{ab}(l+k) \tag{2.5.1-12}$$

It is also clear that the following is true

$$\theta_{T^i a T^j b}(l) = \theta_{ab}(l + j - i) \tag{2.5.1-13}$$

and letting $\boldsymbol{a} = \boldsymbol{b}$ and $i = j$ that

$$\theta_{T^k a}(l) = \theta_a(l) \tag{2.5.1-14}$$

Two sequences $\boldsymbol{a}$ and $\boldsymbol{b}$ are said to be *uncorrelated* if

$$\theta_{ab}(l) = 0 \quad \text{for all} \quad l \tag{2.5.1-15}$$

A sequence is said to have a *two-valued autocorrelation function* if $\theta_a(l)$ equals some constant other than $\theta_a(0)$ for all $l \neq 0$ Mod N, with N the period.

2.5.2 Aperiodic Correlation Functions for Sequences

Now consider the case of aperiodic correlation functions for real valued sequences. These functions are very important in synchronization problems (e.g., where a short sequence is placed within a data stream and is used to obtain and maintain time synchronization). When the timing is not aligned, and a time error of say a few code chips exists, a periodic correlation will produce a value that has two parts. The first will depend on the portion of the local reference code that overlaps the received code, and the second part will depend on the correlation of the few chips of local code that correlates with the data. Thus, the only part of the correlation that is reliable (since the data pattern has an assumed random symbol sequence) is the overlap portion, which is an aperiodic correlation.

Again let the sequence values be εGF(2), with the understanding that when sums or correlations are used the values of the components are $\varepsilon\{\pm 1\}$. Thus for binary valued real sequences a and b we define the *discrete aperiodic cross-correlation function* [7] as

$$C_{ab}(l) = \begin{cases} \sum_{i=0}^{N-1-l} a_i b_{i+l} & 0 \leq l \leq N-1 \\ \sum_{i=0}^{N-1+l} a_{i-l} b_i = \sum_{k=-l}^{N-1} a_k b_{k+l} & 1\text{-}N \leq l \leq 0 \\ 0 & |l| \geq N \end{cases} \tag{2.5.2-1}$$

and we see that the number of symbols that are correlated in (2.5.2-1) is $N-|l|$. The discrete aperiodic autocorrelation function is obtained from (2.5.2-1) with $\boldsymbol{a} = \boldsymbol{b}$.

Now let us consider some properties of the aperiodic correlation function. First note that

$$C_{ab}(0) = \sum_{i=0}^{N-1} a_i b_i = \theta_{ab}(0) \tag{2.5.2-2}$$

Clearly if $\boldsymbol{b} = \boldsymbol{a}$ then one has

$$C_a(0) = \theta_a(0) \tag{2.5.2-3}$$

where $C_{aa}(l) = C_a(l)$. It can also be shown that

$$C_{ab}(-l) = C_{ab}(l) \qquad (2.5.2\text{-}4)$$

This is analogous to the periodic cross-correlation result given in (2.5.1-3). Now if $\boldsymbol{a} = \boldsymbol{b}$ then it follows from (2.5.2-4) that

$$C_a(-l) = C_a(l) \qquad (2.5.2\text{-}5)$$

Figure 2.5-1 illustrates the different correlation criteria and the resulting codes that were designed to satisfy that criteria [14]. More will be said about these criteria later in the chapter.

2.6 CODES FOR SPREAD SPECTRUM MULTIPLE ACCESS APPLICATIONS

In this section we will discuss some codes that are designed for multiple access usage in the sense that they have low periodic cross-correlation and periodic out-of-phase-autocorrelation values. In code division multiple access (CDMA) systems a common carrier frequency is used along with a unique code for each user. The user receiver is able to separate the desired code from all the others that are received by the antenna by a correlation process. This correlation process has the property that only the desired code has full correlation, and the other signals have a very low cross correlation and therefore appear as noise to the desired code receiver.

2.6.1 Binary Maximal Length Sequences

Before we present some properties of binary maximal length sequences (m-sequences), recall that the characteristic polynomial is obtained from a linear feedback shift register having n cells, by evaluating the equation

$$f(x) = \sum_{k=0}^{n} c_n x^n \quad \text{with } c_0 = c_n = 1 \qquad (2.6.1\text{-}1)$$

with c_i being 1 only if the connection is made at the i-th cell (and is zero otherwise) when counting from left to right. Now consider the conditions for a maximal length sequence. A *maximal length sequence* is one in which the sequence period is $N = 2^n-1$. Recall from Section 2.1.1 that the phrase “a polynomial over GF(2)” means a “polynomial with coefficients from GF(2).” A polynomial over GF(2) is called *irreducible* if it cannot be factored, that is, divided by another polynomial of degree less than n (other than the polynomial “1”).

An irreducible polynomial of degree n is called *primitive* if and only if it divides x^m+1 for no m less than 2^n-1. Theorem 1 establishes the fact that primitive polynomials generate maximal length codes.

Theorem 1 If 2^n-1 is prime, every irreducible polynomial of degree n corresponds to a shift register sequence of maximal length. However, if one requires a maximal length sequence for every n, we must restrict our characteristic polynomials to be primitive [6].

We now consider some properties of maximal length sequences.

Property I: The period of a maximal length sequence is $N = 2^n-1$, where n is the length of the shift register generating the sequence.

Proof Follows from Theorem 1. A table of primitive polynomials is discussed in Section 2.6.1.1 and given in Table 2A-1.

Property II: There is one more "one" in the sequence than "zeros."[3]

Proof: Since a maximal length sequence contains all possible sequences except the all zeros sequence, there is clearly one more "one" than "zeros." Note, translated to the symbols $\{1, -1\}$, the property reads: there is one more "–1" than "+1"s.

Property III: The modulo two sum of an m-sequence and a time shifted version of the m-sequence, is itself another time shift of the same m-sequence.

Proof: From (2.4-8) it follows that the modulo two sum of two versions of the same m-sequence is given by

$$G_1(x) + G_2(x) = G(x) \tag{2.6.1-2}$$

where

$$G_1(x) = \frac{\sum_{i=1}^{n} c_i x^i \left[a_{-i} x^{-i} + a_{-i+1} x^{-i+1} + \dots a_{-1} x^{-1} \right]}{\sum_{i=0}^{n} c_i x^i} \tag{2.6.1-3}$$

$$G_2(x) = \frac{\sum_{i=1}^{n} c_i x^i \left[b_{-i} x^{-i} + b_{-i+1} x^{-i+1} + \dots b_{-1} x^{-1} \right]}{\sum_{i=0}^{n} c_i x^i} \tag{2.6.1-4}$$

It follows that the sum of the two generating functions, which we can call $G(x)$ is given by

$$G(x) = \frac{\sum_{i=1}^{n} c_i x^i \left[(a_{-i} + b_{-i}) x^{-i} + (a_{-i+1} + b_{-i+1}) x^{-i+1} + \dots (a_{-1} + b_{-1}) x^{-1} \right]}{\sum_{i=0}^{n} c_i x^i} \tag{2.6.1-5}$$

Clearly the modulo two sum of the respective a and b sequences are either "1" or "0." This is equivalent to a new initial condition in the shift register, which implies that the output sequence is simply a shifted version of the same sequence. Note that the normalized autocorrelation function is defined in (2.5-10).

Property IV: The normalized discrete periodic autocorrelation function of an m-sequence has only two values, namely

$$\begin{aligned} \hat{\theta}_a(l) &= \frac{-1}{N} \qquad l \neq 0 \text{ modulo } N \\ \hat{\theta}_a(l) &= 1 \qquad l = 0 \text{ modulo } N \end{aligned} \tag{2.6.1-6}$$

[3] Recall when we speak of the sequence we will assume that the symbols are $\varepsilon\{0,1\}$, whereas when we speak of the correlation properties we will be dealing with symbols that are $\varepsilon\{1,-1\}$.

Proof It follows that when $l \neq 0$ (modulo N) the normalized autocorrelation function is the value of the symbol by symbol product of an m-sequence and a shifted version of the m-sequence. The symbol by symbol product of two m-sequences in the {1,–1} format is equivalent to the modulo two sum of the same two m-sequences in {0,1} format. Therefore it follows from property III that since there is one more "–1" than "+1" in the "new" m-sequence, the value of the normalized correlation function is given by $\hat{\theta}_a(l) = -1/N$. When $l = 0$ it is clear that the normalized correlation function is given by $\hat{\theta}_a(l) = 1$.

Denote a *run of length q,* a subsequence of the m-sequence of q consecutive symbols of the same type. A *run of q ones* denotes q consecutive "ones" in the subsequence, and no more. A *run of q zeros* is defined in a similar manner.

Property V: For every m-sequence generated from a shift register of length n there is one run of 1s of length n, one run of 0s of length n–1 with relative frequency 2^{n-1}. All the rest of the runs of length $r = n-1$ or less have relative frequency 2^{-r}. It is not possible to have run lengths greater than n, of course.

Proof For a shift register of length n there are 2^n-1 symbols in one period, which includes all the possibilities except the all zeros sequence, which is not included in the output sequence and has a period 1. Of all the 2^n binary vectors of length n, exactly half are even (have a zero in the right most bit) and half are odd (have a one in the right most bit). Since the all zeros vector is not included shift register output, the balance is offset by 1 out of 2^n. Thus there are 2^{n-1} odd sequences and $2^{n-1}-1$ even sequences.

Now consider $r + 2$ successive cells of the shift register generator where $r \leq n-2$. Consider the cell contents in the two forms

$$0, y_1, y_2, \cdots y_r, 0 \quad \text{and} \quad 1, y_1, y_2, \cdots y_r, 1$$

There are 2^r possibilities in each case and there will be just one that has a run length of exactly r; the all-ones of length r for the left sequence and the all-zeros of length r in the second sequence. Therefore for the $(2)2^r = 2^{r+1}$ possibilities for both cases considered together, exactly two have a run length of r. It follows that a fraction 2^{-r} have run length r for $1 \leq r \leq n-2$.

Consider now the case of run lengths of n–1. If this occurs then the contents of the cells in the shift register at some point must be either

$$\underbrace{0,0,\cdots 0}_{n-1 \text{ zeros}},1 \qquad \text{or} \qquad \underbrace{1,1,\cdots 1}_{n-1 \text{ ones}},0$$

In either case, since all n cells of the shift register are specified there can only be one possible input to the leftmost cell. For the sequence on the left it must be a one, since otherwise the register would hold all zeros, which can't happen. For the sequence on the right, the next bit that enters on the left must also be a one, since this is the only way the all-ones can occur. In addition the next bit after this one must be a zero, or else the all-ones state would occur indefinitely with a period of one.

See also [6, 11, 15] for more discussion of this proof. There are more properties of m-sequences and the interested reader should consult [6, 7, 15]. See also property VI in Section 2.6.3.

2.6.1.1 A List of *m*-Sequence Generators

In order to design a shift register sequence generator it is convenient to look up a primitive polynomial and not have to exhaustively search for it, since that can be quite time consuming. Fortunately a number of authors have compiled irreducible polynomials in which primitive polynomials are indicated (in list form) which makes for easy access to them. Peterson and Weldon [16] have a rather complete tabular list (see Table 2A-1). Before we discuss Table 2A-1 and its use, we first consider the reciprocal polynomial. Define the *reciprocal polynomial of degree n,* denoted by $f^R(x)$, of any polynomial of degree n ($f(x)$), by

$$f^R(x) = x^n f(1/x) \tag{2.6.1.1-1}$$

As an example consider

$$f(x) = x^4 + x + 1 \tag{2.6.1.1-2}$$

using (2.6.1.1-1) one obtains the result

$$f^R(x) = x^4(x^{-4} + x^{-1} + 1) = x^4 + x^3 + 1 \tag{2.6.1.1-3}$$

It is to be noted *that the reciprocal of a primitive polynomial is itself primitive*, and *the reciprocal of a irreducible polynomial is itself irreducible.* The reciprocal of a primitive polynomial generates the reverse sequence.

Table 2A-1 illustrates the octal representation of primitive characteristic polynomials, all of which are listed up to degree 7. Above degree 7 only a partial list is provided. The table entries that have an asterisk after the entry have only two feedback connections. This is advantageous for high-speed circuits. Consider the entry [1021]* for $n = 9$. Since the entry is in octal it is necessary to convert the octal numbers into binary. This translates as

$$\overbrace{001}^{1}\overbrace{000}^{0}\overbrace{010}^{2}\overbrace{001}^{1} \tag{2.6.1.1-4}$$

Reading from the right, with the first coefficient (in this case 1) being the coefficient of x^0, the second coefficient (in this case 0) the coefficient of x^1, the third coefficient the coefficient of x^2, and so on. Thus the characteristic polynomial is given by

$$f(x) = 1 + x^4 + x^9 \tag{2.6.1.1-5}$$

The corresponding shift register is shown in Figure 2.6-1.

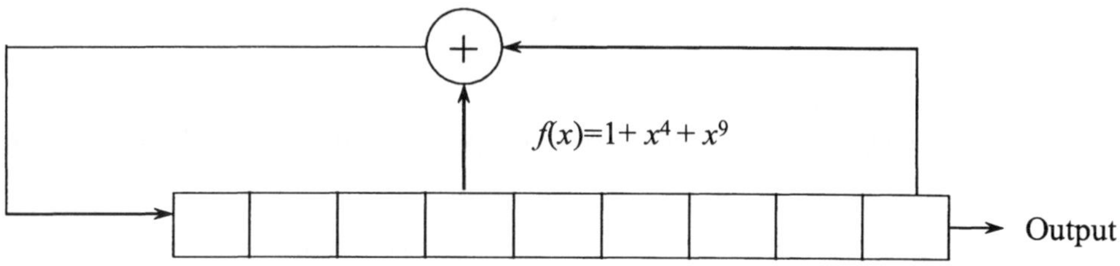

Figure 2.6-1 Shift register generator for $f(x) = 1+x^4+x^9$.

Note the x^4 term is represented by the fourth cell from the left, connected to the adder and the x^9 term is represented by the ninth cell from the left connected to the adder. It is seen that there are only two feedback taps, as indicated by the asterisk. It follows that the linear recursion relationship is given by (2.4-2).

$$a_m = a_{m-4} + a_{m-9} \quad m = 0,\ 1,\ 2,\ 3,... \tag{2.6.1.1-6}$$

can be used to generate a maximal length sequence.

2.6.1.2 Decimation of *m*-Sequences

Now consider decimating an *m*-sequence. Let **Z** denote the set of all integers and let q be a positive integer. The sequence b is said to be a *decimation of a* when b is formed by taking every q-th bit of a (i.e., $b_i = a_{qi}$ for all $i \in \mathbf{Z}$), and b will be denoted by $a[q]$. Denote *the greatest common divisor* of N and M by gcd(M,N). For example gcd(6,3) = 3. A proper decimation of b will result in a maximal length period sequence. It is shown in the paper by Sarwate and Pursley [7] that *a proper decimation of b* will occur and have period $N = 2^n - 1$ if and only if gcd(N,q) = 1, with N being the period of the sequence. In general the period of the decimation sequence will be N/gcd(N,q). It is to be noted that not all decimations will result in another *m*-sequence.

Example 10 As an example consider the case for a sequence of length $N = 63$. From [16] in Appendix C, the entry for the 6th degree ($2^6-1 = 63$) polynomials is as follows:

DEGREE 6 1 103 F 3 127 B 5 147 H 7 111 A
9 015 11 155 E 21 007

The letters E, F, G, H following the numbers indicate, among other things, that polynomials 103, 147, and 155 are primitive polynomials, and A, B, C, and D following the numbers indicate that the polynomials are nonprimitive. Table 2A-1, at the end of the chapter, also has the same primitive polynomials but no non-primitive polynomials. Let us assume that the polynomial 1 103 F represents our *m*-sequence under consideration. It is a primitive polynomial so that $N = 63$. The first integer of the codes indicates the decimation value. Now $a[3]$ (taking every third sample of the sequence) is generated by the polynomial 3 127 B, $a[5]$ is generated by the polynomial 5 147 H, and so forth.

First consider the second polynomial 3 127 B. We note that it is not primitive (it has the letter B following it) and has period 63/gcd(63,3) = 21, so that it will not be of maximal length. It is true, however, that the polynomial 5 147 H is primitive and therefore has a period of 63/gcd(63,5) =63, so that sampling every fifth bit out of the shift register will produce an *m*-sequence. Only primitive polynomials are indicated in Table 2A-1.

The *characteristic phase*[4] *of the m-sequence a*, denoted by $\tilde{a}$, is defined as the sequence that satisfies the condition

$$\tilde{a}_i = \tilde{a}_{2i} \quad \text{for all } i \in \mathbf{Z} \tag{2.6.1.2-1}$$

In other words the characteristic phase is generated when the initial conditions of the shift register are such that taking every second symbol in the output stream of the *m*-sequence produces the original sequence again.

Example 11 As an example of finding the characteristic phase of an *m*-sequence, consider the case that $n = 3$. From Table 2A-1 we see that the only primitive polynomial is given by [13]. Thus

$$f(x) = 1 + x + x^3 \tag{2.6.1.2-2}$$

Evaluating the recursion

$$a_k = a_{k-1} + a_{k-3} \tag{2.6.1.2-3}$$

produces the sequence for three periods (the period is $N=7$)

$$100111010011101001110 \tag{2.6.1.2-4}$$

[4] Sometimes this is called the characteristic *m*-sequence.

Trying the first sequence and comparing it to the sequence formed from every other chip produces the two sequences

$$\begin{matrix} 1001110 \\ 1010011 \end{matrix} \tag{2.6.1.2-5}$$

and they are obviously not the same. However trying the fourth chip as the initial chip, and recording the fourth through the tenth chip and comparing that sequence with the one obtained by starting at the fourth chip and recording every second one, produces the result

$$\begin{matrix} 1110100 \\ 1110100 \end{matrix} \tag{2.6.1.2-6}$$

which is indicated by the short vertical lines above the sequence in (2.6.1.2-4).

2.6.1.3 The Power Spectral Density of *m*-Sequences

Before discussing Gold codes and the other codes used for CDMA it is useful to develop the power spectral density of an *m*-sequence [13]. The baseband waveform of an *m*-sequence is given by (2.5-2)

$$u(t) = \sum_{m=-\infty}^{\infty} a_m p(t - mT_c) \tag{2.6.1.3-1}$$

where T_c is the chip duration, $p(t)$ is a rectangular pulse of unit amplitude starting at time zero and ending at time T_c, the period is assumed to be NT_c seconds long, and $a_m \in \{-1,1\}$ and is the value of the *m*-sequence chips. The time autocorrelation function is given by (2.5-3)

$$R_u(\tau) = \frac{1}{NT_c} \int_0^{NT_c} u(t)u(t+\tau)dt \tag{2.6.1.3-2}$$

For τ such that $-T_c < \tau < T_c$ it is clear that the value of the autocorrelation function is a linear combination of the autocorrelation for $\tau = 0$ and $\tau = T_c$. For example, if the time displacement is one half chip (see Figure 2.6-2) one half of the product will have a correlation value of 1 (shaded area) and one-half of the product will have a correlation value of $-1/N$.

Figure 2.6-2 illustrates an example of a length 7 *m*-sequence. Furthermore, when $|\tau| \geq T_c$ and $|\tau| \leq NT_c$, one half of the correlation value will be $-1/N$ and the other half will be $-1/N$. The resulting weighted sum is therefore $-1/N$. It is concluded that the autocorrelation function will be composed of straight-line segments as shown in Figure 2.6-3.

To obtain the power spectral density, the Fourier transform of the autocorrelation must be obtained. First define the triangle function $\Lambda(\tau/T_c)$ as

$$\Lambda(\tau / T_c) = \begin{cases} 1 - \dfrac{|\tau|}{T_c} & |\tau| \leq T_c \\ 0 & \text{elsewhere} \end{cases} \tag{2.6.3.1-3}$$

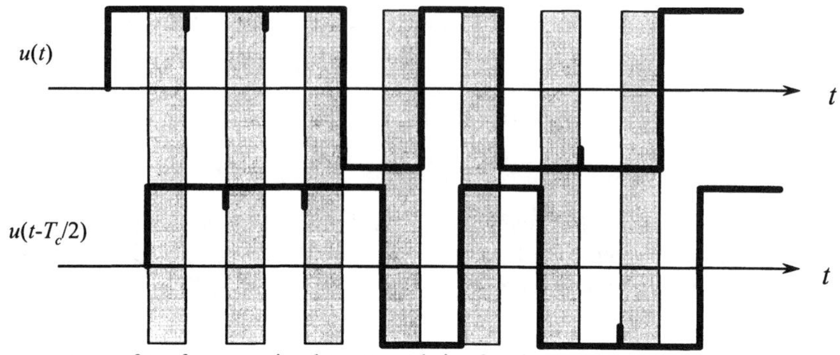

Figure 2.6-2 The *m*-sequence waveform for computing the autocorrelation function.

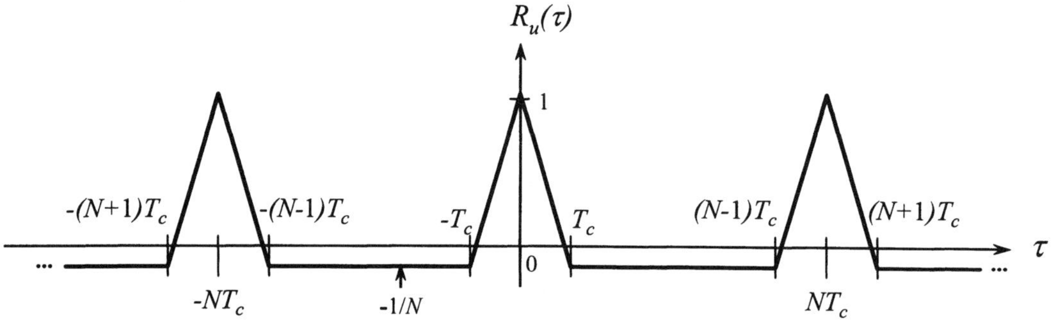

Figure 2.6-3 The *m*-sequence waveform autocorrelation function.

The Fourier transform of $\Lambda(\tau/T_c)$ is given by [17, 18]

$$F\left[\Lambda(\tau/T_c)\right]=T_c\text{sinc}(fT_c)^2 \qquad (2.6.1.3\text{-}4)$$

where

$$\text{sinc}(x)=\frac{\sin(\pi x)}{\pi x} \qquad (2.6.1.3\text{-}5)$$

Since $R_u(\tau)$ is periodic it can be expressed as the sum of a dc term $(-1/N)$ and an infinite series of triangle functions of amplitude $(N+1)/N$, separated by the period NT_c seconds. Thus

$$R_u(\tau)=-\frac{1}{N}+\frac{N+1}{N}\Lambda(\tau/T_c)*\sum_{m=-\infty}^{\infty}\delta(\tau+mNT_c) \qquad (2.6.1.3\text{-}6)$$

where the "*" denotes the convolution operation. Using the result [17]

$$F\left[\sum_{m=-\infty}^{\infty}\delta(\tau+mNT_c)\right]=\frac{1}{NT_c}\sum_{n=-\infty}^{\infty}\delta\left(f+\frac{n}{NT_c}\right) \qquad (2.6.1.3\text{-}7)$$

The power spectral density becomes

$$S_u(f)=-\frac{1}{N}\delta(f)+\frac{N+1}{N^2}\sum_{\substack{n=-\infty\\ n\neq 0}}^{\infty}\text{sinc}(fT_c)^2\delta\left(f+\frac{n}{NT_c}\right) \qquad (2.6.1.3\text{-}8)$$

Combining the terms at dc produces

$$S_u(f) = \frac{1}{N^2} + \frac{N+1}{N^2} \sum_{\substack{n=-\infty \\ n \neq 0}}^{\infty} \text{sinc}\left(\frac{n}{N}\right)^2 \delta\left(f + \frac{n}{NT_c}\right) \quad (2.6.1.3\text{-}9)$$

A sketch of this spectral density is shown in Figure 2.6-4 for nonnegative frequencies only.

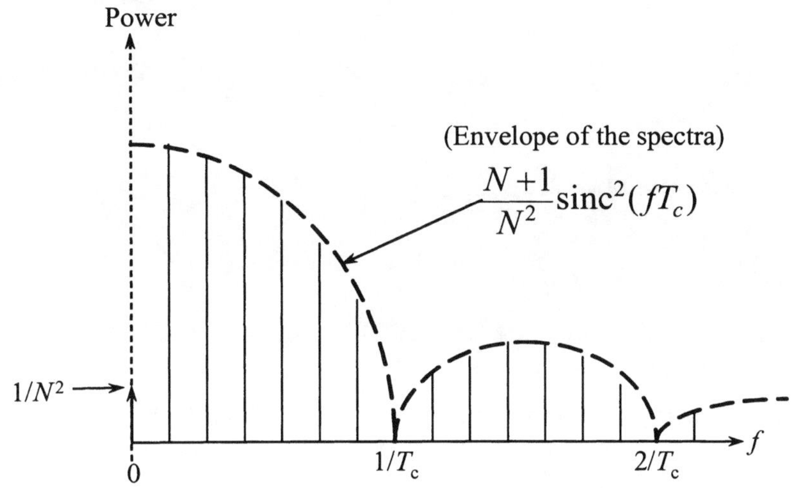

Figure 2.6-4 The m-sequence power spectral density with chip duration T_c and period NT_c (shown for $N=7$ and positive frequencies).

The spectral lines are symmetric about $f = 0$. Thus it is seen that the spectral lines are spaced $1/(NT_c)$ Hz apart, the nulls are spaced $1/T_c$ apart, and the envelope of the spectral lines varies with frequency as

$$\frac{N+1}{N^2}\text{sinc}^2(fT_c) \quad (2.6.1.3\text{-}10)$$

The spectral density of the baseband sequence is centered at $f = 0$. If an m-sequence $u(t)$ is biphase modulated onto a carrier and if the carrier and code are not coherent, the resulting signal can be modeled as

$$u_c(t) = \sqrt{2P}u(t)\cos(2\pi f_o t + \theta_o) \quad (2.6.1.3\text{-}11)$$

where P is the average signal power and θ_o is a random phase uniformly distributed over $(0,2\pi)$. The modulated power spectral density is given by [13]

$$S_{u_c}(f) = \frac{P}{2}S_u(f + f_o) + \frac{P}{2}S_u(f - f_o) \quad (2.6.1.3\text{-}12)$$

Therefore the biphase modulated m-sequence power spectral density is the sum of two baseband versions of the m-sequence spectra with one centered at $+f_o$ and one at $-f_o$, with each one weighted by $P/2$.

2.6.1.4 Power Spectral Density of a Despread *m*-Sequence

In the process of despreading a received spread spectrum signal, spread with an m-sequence, the decorrelated signal, when the receiver code and the received code are aligned, is just a constant. However when the two signals are not aligned, many spectral components are generated. This spectral density is useful in analyzing the performance of acquisition, tracking, and detection. Figure 2.6-5 illustrates a baseband model of a despreader which includes the product of the received code delayed by T seconds and the local code reference which is delayed by $\hat{T}$.

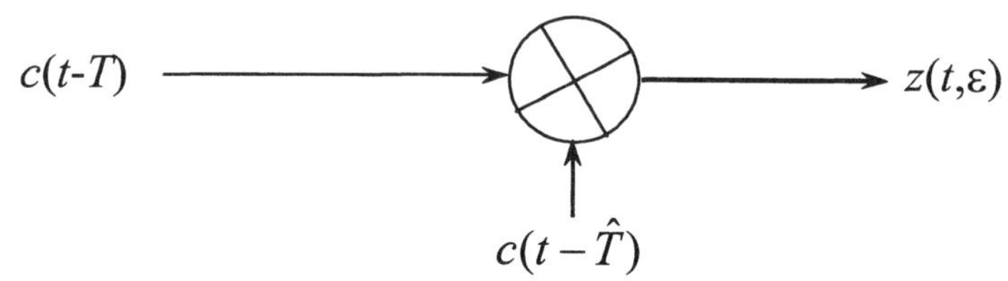

Figure 2.6-5 Baseband model of a despread signal.

The product is denoted by $z(t,\varepsilon)$, where

$$\varepsilon = T - \hat{T} \tag{2.6.1.4-1}$$

The spectral density of $z(t,\varepsilon)$ has been derived by Gill [19], Gill and Spilker [20], and Holmes [13]. We present the results here and refer the interested reader to [13] for the derivation.

$$\begin{aligned} S_z(f,\varepsilon) = {} & \left(1 - \frac{|\varepsilon|}{T_c}\frac{N+1}{N}\right)^2 \delta(f) + \frac{N+1}{N}\left(\frac{\varepsilon}{T_c}\right)^2 \sum_{\substack{m=-\infty \\ m\neq 0}}^{\infty} \operatorname{sinc}\left(\frac{m\varepsilon}{T_c}\right)\delta\left(f + \frac{m}{T_c}\right) + \\ & + \frac{N+1}{N^2}\left(\frac{\varepsilon}{T_c}\right)^2 \sum_{\substack{m=-\infty \\ m\neq 0}}^{\infty} \operatorname{sinc}\left(\frac{m\varepsilon}{NT_c}\right)^2 \delta\left(f + \frac{m}{T_c}\right) \qquad 0 \le |T_c| \end{aligned} \tag{2.6.1.4-2}$$

Thus it is seen that there is a dc term having decorrelation voltage (for reasonable values of N) of essentially $1-(|\varepsilon|/T_c)^2$. In addition to the dc term there are two spectra components, one with the line spectra spaced apart by $1/T_c$ Hz, and one spaced apart by $1/(NT_c)$ Hz. Figure 2.6-6 illustrates the power spectral densities for the case when $|\varepsilon| = 0$, $T_c/10$, $T_c/2$, and T_c and $N = 7$.

Note that the power spectral density shape changes according to the size of the timing error. When the timing error reaches T_c seconds then the spectral density is the same as the original power spectral density, since the product of an m-sequence and a shifted version of itself is itself a different phase of the original m-sequence with the same power spectral density.

2.6.1.5 Determination of the Generator Taps from the *m*-Sequence Output

Spread spectrum systems that use m-sequences for their spreading code should note that they are not very secure in the sense that they can be "read" and the shift register sequence can be determined by solving the resulting linear equations specified by the linear recursion of the sequence generator (2.2-3). Once the shift register taps of the transmitter are known, it becomes an easy task for the uninvited listener to detect the information modulated on the spread spectrum sequence. One of the main functions of a spread spectrum system is to minimize the vulnerability to being jammed and to preclude the uninvited listener from decoding the messages that the transmitter has sent. Consider the following result.

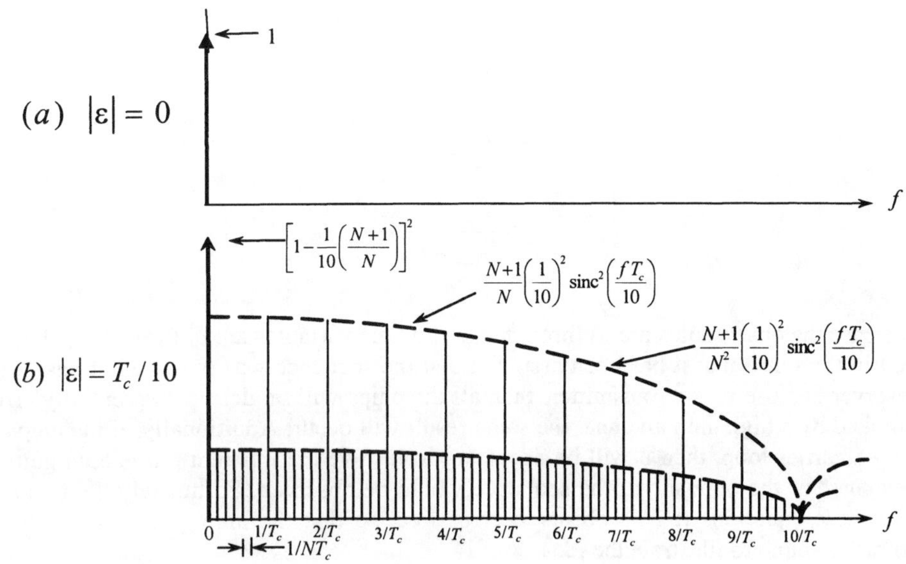

Figure 2.6-6 (a, b) Decorrelated spectra for one tenth chip error (not to scale).

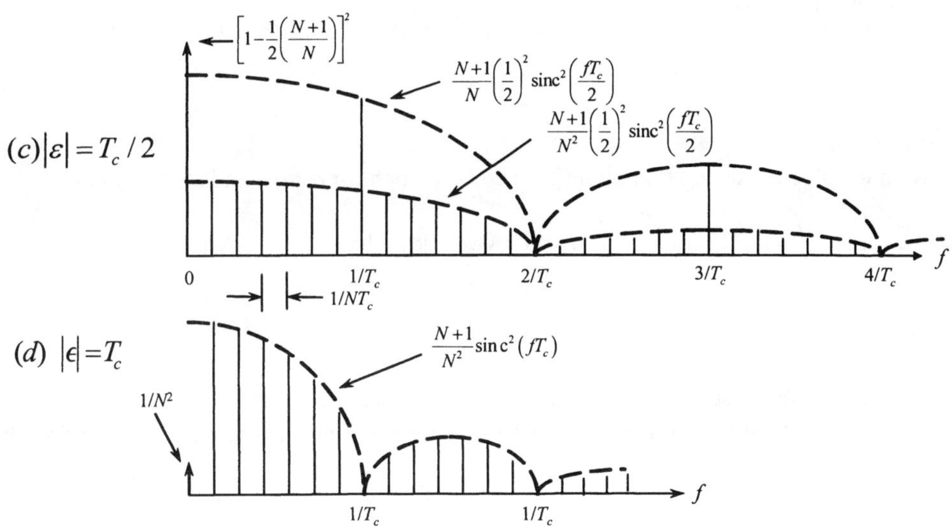

Figure 2.6-6 (c, d) Decorrelated spectra for one half and one chip error (not to amplitude scale).

Result: An m-sequence obtained from a shift register of known length n can be determined from the errorless observation of $2n$ chips.

The proof is based on the fact that from (2.2-3), one can write

$$a_k = \sum_{i=1}^{n} c_i a_{k-i} \qquad (2.6.1.5\text{-}1)$$

Setting $k = n, n+1, n+2, \ldots, 2n-1$, n equations in n variables can be displayed in the form

$$a_n = \sum_{i=1}^{n} c_i a_{n-i}$$
$$a_{n+1} = \sum_{i=1}^{n} c_i a_{n+1-i} \quad (2.6.1.5\text{-}2)$$
$$\vdots$$
$$a_{2n-1} = \sum_{i=1}^{n} c_i a_{2n-1-i}$$

In this case the observables are a_0 through a_{2n-1} and the variables are c_1 through c_n. This solution has been simplified in the sense that it has been assumed that the sequence was observed at baseband and noise free. If the observer is close to the transmitter, then all the chips will be detected essentially error free, or if the signal is tracked by a high gain antenna, the same result will occur. Additionally, if the chips are detected with a suppressed carrier loop, then it will be necessary to resolve the inherent phase ambiguity of the data. That is, the sequence or the opposite of the sequence would be observed. Fortunately the opposite sequence will not satisfy these equations if it is observed error free, so that the ambiguity can easily be resolved. Consider a simple example to illustrate the idea.

Example 12 Consider a simple length 3 SSRG illustrated in example 5. If 001 is the contents of the register initially, then the first seven output chips (one period) will be 1001011 with $a_0 = 1$, $a_1 = 0$, $a_2 = 0$, $a_3 = 1$, $a_4 = 0$, and $a_5 = 1$. From (2.4-3) one has

$$a_k = \sum_{i=1}^{3} c_i a_{k-i} \quad (2.6.1.5\text{-}3)$$

in which we wish to solve for the c_i. Equation (2.6.1.5-2), written for a_3, a_4, and a_5, becomes

$$a_3 = 1 = 0c_1 + 0c_2 + 1c_3$$
$$a_4 = 0 = 1c_1 + 0c_2 + 0c_3 \quad (2.6.1.5\text{-}4)$$
$$a_5 = 1 = 0c_1 + 1c_2 + 0c_3$$

Solving these equations, the first equation produces $c_3 = 1$, the second equation produces $c_1 = 0$, and the third equation produces $c_2 = 1$. If one would assume that $n = 4$ or $n = 2$ and try to obtain a solution, it would be determined that no solution exists.

Massey [18] has developed an efficient procedure for solving these general equations, and the reader is referred to that work for the details.

2.6.2 Gold Codes

Gold codes were developed by Dr. Robert Gold to provide the capability of code division multiple access (CDMA), which is an alternative to time division multiple access (TDMA), and frequency division multiple access (FDMA). Gold codes are one important class of periodic sequences, which provide reasonably large sets of codes with good periodic cross-correlation and autocorrelation properties. Gold codes have a code period of $N = 2^n - 1$ chips and have $N+2$ codes in the set. These codes are constructed from selected m-sequences in a manner that will be described next. This section follows [7, 21].

The set of cross-correlation values, together with the number of integers l $(0 \le l < N)$ for which $\theta_{ab}(l) = c$ for each c in this set, is called the *cross-correlation spectrum* for the pair of sequences $\boldsymbol{a}$ and $\boldsymbol{b}$. The cross-correlation of two m-sequences of length $N = 2^n - 1$ can have three, four, or possibly more values of c.

Certain pairs of m-sequences have cross-correlation values that are three valued, and those three values are given by [7]

$$\begin{aligned} &-t(n) \\ &-1 \\ &t(n)-2 \end{aligned} \tag{2.6.2-1}$$

where $t(n)=1+2^{(n+1)/2}$ n odd and $t(n)=1+2^{(n+2)/2}$ n even and $\neq 0 \bmod 4$. These pairs are called *preferred pairs* of m-sequences. It is to be pointed out that $\theta_c = \theta_a = t(n)$ for Gold codes [7].

It is necessary to find preferred pairs of m-sequences to construct Gold codes. The following conditions are sufficient to construct a preferred pair, a and b, of m-sequences of length $N = 2^n-1$:

$$\begin{aligned} &1.\ n \neq 0 \bmod 4;\ \text{that is to say, } n \text{ is odd or } n = 2 \bmod 4 \\ &2.\ b = a[q] \text{ where } q \text{ is odd and either has the value} \\ &q = 2^k+1 \quad \text{or} \quad q = 2^{2k}-2^k+1 \\ &3.\ \gcd(n,k) = \begin{cases} 1 & \text{for } n \text{ odd} \\ 2 & \text{for } n = 2 \bmod 4 \end{cases} \end{aligned} \tag{2.6.2-2}$$

Consider a preferred pair of m-sequences a and b of period $N = 2^n-1$ that are generated by the associated characteristic polynomials $f_1(x)$ and $f_2(x)$, respectively. If g denotes a nonzero sequence generated by the product $f_1(x)f_2(x)$, then it is true that

$$\begin{aligned} &g = T^i a \\ \text{or} \quad &g = T^j b \\ \text{or} \quad &g = T^i a + T^j b \end{aligned} \tag{2.6.2-3}$$

for $0 \le i, j \le N-1$, where $T^i a + T^j b$ denotes the modulo-2 sum, whose k-th element is given by $a_{i+k} + b_{j+k}$. It therefore follows that g is some phase of some sequence in the set $G(a,b)$ specified by

$$G(a,b) = \left\{a, b, a+b, a+Tb, a+T^2b, a+T^3b, \ldots a+T^{N-1}b\right\} \tag{2.6.2-4}$$

where, recall $T^i a$ denotes the operator that produces the sequence whose k-th element is given by a_{k+i}. We see that there are $N + 2 = 2^n + 1$ sequences in the code set, all with period N. This leads us to theorem 2.

Theorem 2 Let a and b be a preferred pair of m-sequences of period $N = 2^n - 1$ generated by primitive binary polynomials $f_1(x)$ and $f_2(x)$, which have no common factors where $n \neq 0 \bmod 4$ (i.e., n is odd or $n = 2 \bmod 4$). The set of sequences (or codes) defined in (2.6.2-4) is called a set of *Gold codes*. For any sequence in g_1 *and* $g_2 \in G(a,b)$,

$$\theta_{g_1 g_2}(l) \in \left\{-1, -t(n), t(n)-2\right\} \tag{2.6.2-5}$$

for all integers l, and

$$\theta_{g_1}(l) \in \{-1, -t(n), t(n)-2\} \tag{2.6.2-6}$$

for all integers $l \neq 0 \mod N$. Recall that $\theta_{g_1 g_1}(l) = \theta_{g_1}(l)$. Every sequence in $G(a,b)$ can be generated by $h(x)=f_1(x)f_2(x)$ [7].

To summarize the Gold code results, we conclude that the Gold codes have the following cross correlation values and approximate frequency of occurrence indicated in Table 2.6-1.

Table 2.6-1 Three-Level Gold Code Auto- and Cross-Correlation Values and Their Frequency

n Register Length	Code Length	Normalized Correlation Value	Frequency of Occurrence
n Odd	$N=2^n-1$	$-1/N$ $-(2^{(n+1)/2}+1)/N$ $(2^{(n+1)/2}-1)/N$	$\simeq 0.5$ $\simeq 0.25$ $\simeq 0.25$
n even and $n \neq 0 \mod 4$	$N=2^n-1$	$-1/N$ $-(2^{(n+2)/2}+1)/N$ $(2^{(n+2)/2}-1)/N$	$\simeq 0.75$ $\simeq 0.25$ $\simeq 0.25$

Here $N = 2^n-1$. The approximate frequency of occurrences is valid for $n > 4$. Note for Gold codes with $n = 0 \mod 4$, see [7]. The autocorrelation function at zero offset has a normalized correlation value of one.

Example 13 Now let us consider an example of forming a Gold code of length $2^5-1 = 31$ chips. The Gold sequence set of codes will total 33 (31+2). We need to find a preferred pair of m-sequences for $n = 5$. From [16] in Appendix C for irreducible polynomials, we see that the entry for $n = 5$ is given by

$$Degree\ 5:\ 1\ 45E,\quad 3\ 75G,\quad 5\ 67H$$

Alternatively, Table 2A-1 has the same three codes listed as primitive code generators. Hence if we pick the first polynomial 1 45E, call it a, we need to find $b = a[q]$. First, from (2.6.2-2), since $n = 5$ (an odd number), we need to find the decimation value q. The parameter q must satisfy the condition

$$q = 2^k + 1 \quad or \quad q = 2^{2k} - 2^k + 1 \tag{2.6.2-7}$$

Picking $k = 1$ produces $q = 3$ in either expression for q. Picking $k = 2$ produces $q = 5$ in the first expression for q, or 13 in the second expression for q. Clearly $a[q]$ is of maximal length if and only if gcd(N,q)=1. Now note that gcd(31,3) = 1, and gcd(31,5) = 1, and gcd(31,13) = 1. Therefore, one can select the first code a as 1 45 E and either 3 75G or 5 67H as the other preferred pair (i.e., b =a[3] or a[5]). Picking 1 45E and 5 67H as the two shift register polynomials and computing the product polynomial, we have

$$\begin{aligned} f_1(x) &= x^5 + x^2 + 1 \\ f_2(x) &= x^5 + x^4 + x^3 + x^2 + 1 \\ h(x) &= x^{10} + x^9 + x^8 + x^6 + x^5 + 1 \end{aligned} \tag{2.6.2-8}$$

with the length of the sequence being 31. The corresponding normalized cross-correlation values from Table 2.6-1 are given by

$$\hat{\theta}_{g_1 g_2}(l) = -\frac{1}{31}, -\frac{9}{31}, \frac{7}{31} \tag{2.6.2-9}$$

Figure 2.6-7 illustrates the double and single shift register representation.

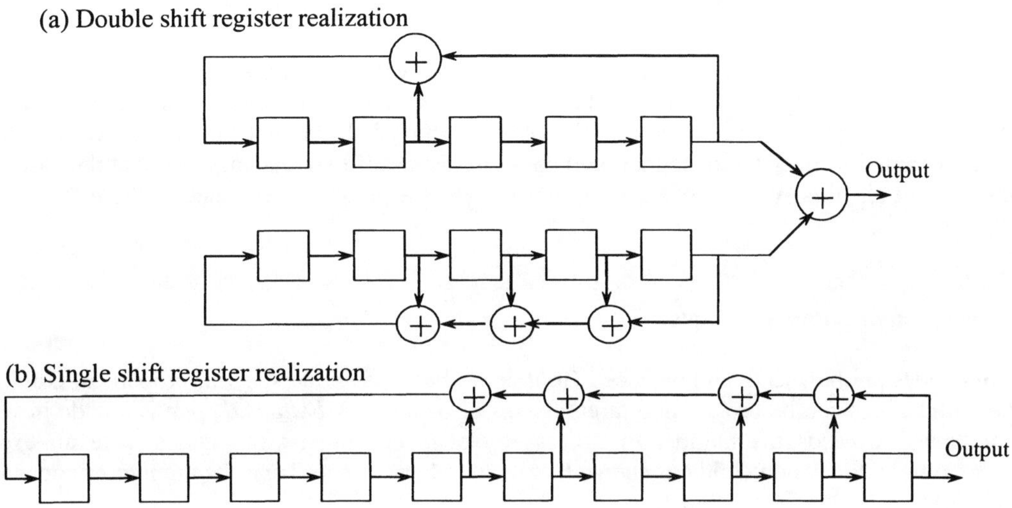

Figure 2.6-7 (a, b) Gold code generator of length 31.

Any nonzero initial conditions of the two preferred pairs will suffice. When the initial conditions 00001 are set in the first register and 00010 is set into the second one, the resulting autocorrelation function is shown in Figure 2.6-8.

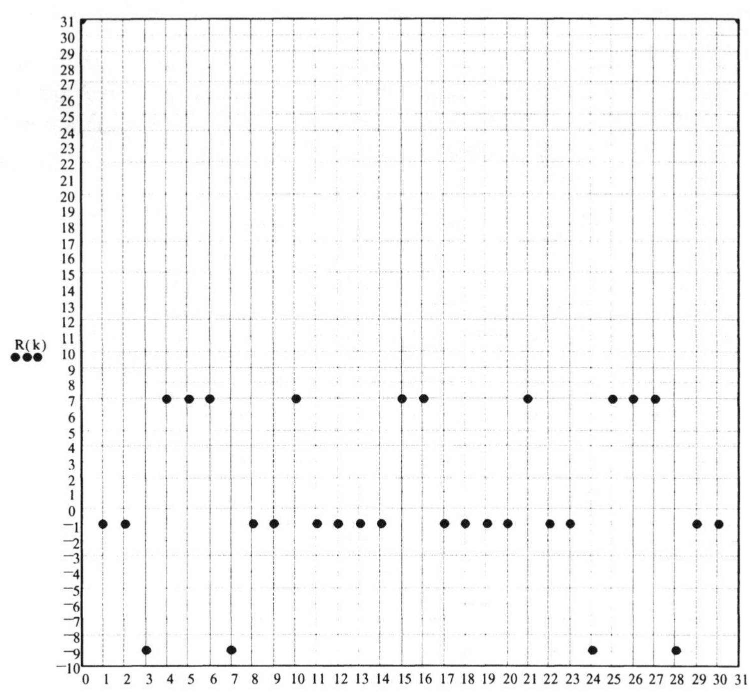

Figure 2.6-8 Autocorrelation function times 31, shown for the Gold code of example 13.

The autocorrelation function only has meaning on the integers so that only dots are shown in the figure. Note that most of the cross-correlation values are at the value of $-1/32$ in this case, whereas some correlation values take on the value of +7/32 and –9/32 as Table 2.6-1 indicates (the table relates the autocorrelation value).

It should be pointed out that all these autocorrelation values are based on the tacit assumption that all the code pairs are chip synchronous (i.e., the chip clock is exactly the same for any two received codes that are being cross correlated). In a typical CDMA environment this is not the case since oftentimes the signals are transmitted from different sources and therefore from different distances, so that the clocks are not synchronous; however, this synchronous assumption serves as a first-order estimate of the interference from other CDMA signals.

2.6.2.1 Generation of Balanced Codes

Not all Gold codes are balanced, and balanced Gold codes have desirable spectral properties. Gold [21] has shown that Gold codes can be broken into three classes of balance. A *balanced code* is a code in which the number of "ones" exceeds the number of "zeros" by one. The other two classes have an excess and deficiency of "ones." For n odd Gold has shown that the number of "ones" and the number of codes with that number of "ones" is as shown in Table 2.6-2.

Table 2.6-2 Number of Balanced Codes for n Odd

Class	Number of "Ones" in Code Sequence	Number of Codes with This Number of "Ones"
1	2^{n-1}	$2^{n-1}+1$
2	$2^{(n-1)}+2^{(n-1)/2}$	$2^{(n-2)}-2^{(n-3)/2}$
3	$2^{(n-1)}-2^{(n-1)/2}$	$2^{(n-2)}+2^{(n-3)/2}$

Note that in the first class there are 2^{n-1} "ones" in the sequence and therefore $2^{n-1}-1$ "zeros," and therefore set 1 is "balanced." It is clear from the expressions that classes 2 and 3 are not balanced. It should be noted that approximately half of all Gold codes are balanced. In order to generate balanced Gold codes, it is necessary to select the proper relative phases of the two original maximal length sequences (preferred pairs).

We must now determine the *characteristic phase of the sequence*. Recall that the property of the characteristic phase of the m-sequence is that when it is sampled at every other symbol, the same sequence results. Let $f(x)$ be the n-th degree characteristic polynomial corresponding to the m-sequence. It has been shown earlier (2.4-8) that any phase of the m-sequence can be represented by the ratio $G(x) = g(x)/f(x)$, where $g(x)$ is the numerator of the generating function and is of degree less than n.

As we have seen, long division[5] of these polynomials results in a formal binary power series whose binary coefficients are the symbols of the sequence generated by the shift register. The formula for the polynomial $g(x)$ that results in the characteristic phase of the m-sequence has been shown by Gold [22] to be given by

$$
\begin{aligned}
g(x) &= \frac{d\left[xf(x)\right]}{dx}, \quad f(x) \text{ of odd degree} \\
g(x) &= f(x) + \frac{d\left[xf(x)\right]}{dx}, \quad f(x) \text{ of even degree}
\end{aligned}
\qquad (2.6.2.1\text{-}1)
$$

Differentiation is carried out in the usual way, with the coefficients interpreted modulo 2.

[5] The division is based on writing both polynomials in ascending powers in the division operation.

Example 14 As a simple example of obtaining the characteristic phase of the sequence, let the characteristic polynomial be $f(x) = 1 + x^2 + x^3$. Since n is odd in this example, we obtain from the first equation in (2.6.2.1-1)

$$g(x) = \frac{d\left[x + x^3 + x^4\right]}{dx} = 1 + x^2$$

where the coefficients are taken modulo 2. The characteristic phase of this sequence is obtained from $G(x)$ when $g(x) = 1+x^2$. Therefore, evaluating $G(x)$ by long division produces

$$\begin{array}{r|l}
 & 1 + 0x + 0x^2 + x^3 + 0x^4 + x^5 + x^6 \\
\hline
1 + x^2 + x^3 & 1 + x^2 \\
 & \underline{1 + x^2 + x^3} \\
 & \quad x^3 \\
 & \quad \underline{x^3 + x^5 + x^6} \\
 & \qquad x^5 + x^6 \\
 & \qquad \underline{x^5 + x^7 + x^8} \\
 & \qquad\quad x^6 + x^7 + x^8 \\
 & \qquad\quad \underline{x^6 + x^8 + x^9} \\
 & \qquad\qquad x^7 + x^9
\end{array}$$

so the initial condition of the register is [001]; that is, $a_0 = 1$, $a_1 = 0$, $a_2 = 0$. If we observe the output, it will be

$$1\,0\,0\,1\,0\,1\,1, 1\,0\,0\,1\,0\,1\,1, \ldots$$

where the comma separates the length 7 m-sequence periods. Note that if every second symbol of the sequence is sampled then the sequence is repeated as shown here

$$\begin{array}{ll}
\text{sequence} & 1001011,1001011,1\cdots \\
\text{sampled sequence} & 1\ 0\ 0\ 1\ \ 0\ 1\ 1,\ 1\ 0\cdots
\end{array}$$

where the first, third, fifth, and so forth samples from the original sequence clearly produce the same sequence again, as seen in the sampled sequence output, and thus the characteristic phase for this sequence has been found.

2.6.2.2 Relative Phase Requirement for Balanced Gold Codes

We shall now describe the relative phase in which the preferred pair of m-sequences must be added in order to result in a balanced member of the family. Let $\boldsymbol{a}$ and $\boldsymbol{b}$ be the preferred pair of m-sequences in their characteristic phase. When the characteristic polynomial $f(x)$ is of odd degree, it is clear that the generator polynomial is of the form

$$G(x) = \frac{1 + c(x)}{1 + d(x)} \qquad (2.6.2.2\text{-}1)$$

where the degree of $d(x)$ is n and the degree of $c(x)$ is not greater than $n-1$. By long division it is clear that the quotient will be of the form $1+\ \ldots$ so that the initial symbol of the characteristic sequence will be a "one." Now we consider a theorem due to Gold [21] that describes how to generate a balanced Gold code.

Theorem 3 Any relative phase shifts of the sequence $\boldsymbol{a}$ and $\boldsymbol{b}$, in their characteristic phase, that are obtained by shifting the sequence $\boldsymbol{b}$ until its initial "one" corresponds to a "zero" in the sequence $\boldsymbol{a}$ will result in a balanced Gold code, when the two sequences are added together modulo 2.

Example 15 As an example of an application of this theorem, consider the following preferred pair of sequences for which $n = 3$

$$\begin{aligned} a &= \cdots 1110100,1110100,\cdots \\ b &= \cdots 1001011,1001011,\cdots \end{aligned} \tag{2.6.2.2-2}$$

We have already shown that sequence $\boldsymbol{b}$ is in its characteristic phase in example 14. Taking the first, third, fifth, and so forth samples of sequence $\boldsymbol{a}$ produces the sequence $\boldsymbol{a}$ again so that it too is in its characteristic phase. According to Theorem 3 we can shift the sequence $\boldsymbol{b}$ over to the right by three, five, or six positions in order that the initial "one" of sequence $\boldsymbol{b}$ lines up with a "zero" of sequence $\boldsymbol{a}$. To simplify the presentation, we show cyclically shifted sequences in which $\boldsymbol{a}$ is added to $\boldsymbol{b}$ to form the mod 2 sum.

$$\begin{aligned} \boldsymbol{a} &= 1110100 \\ \boldsymbol{T}^3\boldsymbol{b} &= \underline{0111001} \\ \boldsymbol{a}+\boldsymbol{T}^3\boldsymbol{b} &= 1001101 \end{aligned} \tag{2.6.2.2-3}$$

It is clear that the resulting sum sequence is "balanced" in the sense that it has one more "one" than "zero." However, if the first one of the sequence $\boldsymbol{b}$ is not lined up with a zero of $\boldsymbol{a}$, then the resultant sequence will not be balanced. That is

$$\begin{aligned} \boldsymbol{a} &= 1110100 \\ \boldsymbol{T}^1\boldsymbol{b} &= 1100101 \\ \boldsymbol{a}+\boldsymbol{T}^1\boldsymbol{b} &= \overline{0010001} \end{aligned} \tag{2.6.2.2-4}$$

This is clearly not balanced.

2.6.2.3 Procedure for Generating Balanced Gold Codes

The procedure to generate balanced Gold codes can now be specified. The generic Gold code generator is sketched in Figure 2.6-9.

(1) First select a preferred pair of m-sequences $\boldsymbol{a}$ and $\boldsymbol{b}$ of length $N=2^n-1$, via (2.6.2-2) and [16] (or Table 2A-1) with characteristic polynomials $f_1(x)$ and $f_2(x)$.

(2) The initial conditions for shift register 2 are those initial conditions of b that determine the characteristic phase as specified in (2.6.2.1-1). This is obtained by long division of the ratio $g(x)/f(x)$ ($g(x)$ and $f(x)$ are written in ascending order) and taking the first n binary coefficients of the first n terms of the quotient.

(3) The initial conditions for shift register 1 (sequence a) with corresponding characteristic polynomial $f_1(x)$, are only that the first stage (on the right in our shift register diagrams) of the number 1 shift register contain a "zero." This corresponds to the requirements of Theorem 3. And there should be at least one "one" in the register 1 to start with. Note in general there are 2^{n-1} choices for the remaining $n-1$ shift register cells, so that about half of all the Gold codes can be made balanced.

(4) The set of Gold codes is formed by modulo-2 addition of the two registers, 1 and 2, as determined from (2.6.2-4).

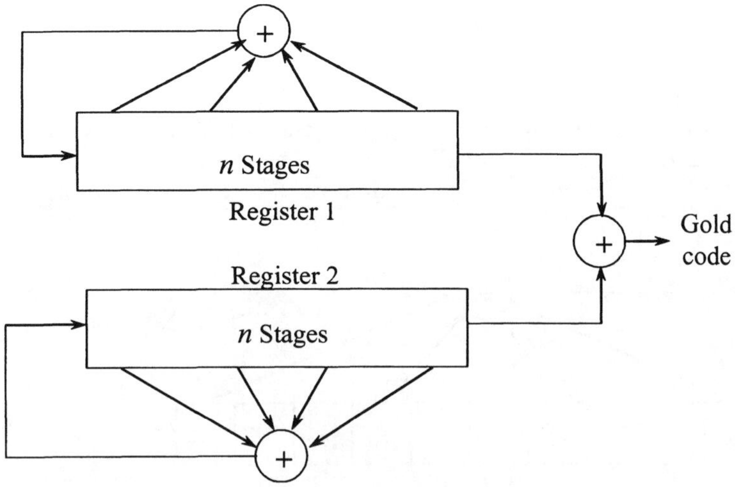

Figure 2.6-9 Generic Gold code generator configuration.

Example 16 Now consider an example construction of a balanced Gold code of register length $n = 11$, having period $2^{11}-1$. From Appendix C of [16] (or Table 2A-1) it is seen that "1 4005 E" is a primitive polynomial, which will be designated the b register (register 2). To find the polynomial for shift register a, use (2.6.2-2) and note that $q = 3$ ($k = 1$) satisfies all three conditions of (2.6.2-2) so that $a = b[3]$. The third decimation is represented by 3 4445 E, as determined from Appendix C of [16]. Thus the two characteristic polynomials are (recall the polynomials are determined by reading from right to left of the binary equivalent of the code, with the polynomial going from left to right)

$$\begin{aligned}&100{,}100{,}100{,}101\\&3\ 4445\ \mathrm{E}\quad f_1(x)=1+x^2+x^5+x^8+x^{11}\\&100{,}000{,}000{,}101\\&1\ 4005\ \mathrm{E}\quad f_2(x)=1+x^2+x^{11}\end{aligned} \tag{2.6.2.3-1}$$

The characteristic sequence generated by the second shift register corresponding to the polynomial 4005 is represented by the ratio

$$\frac{g(x)}{1+x^2+x^{11}} \tag{2.6.2.3-2}$$

where, since n is odd,

$$g(x)=\frac{d}{dx}\left(x+x^3+x^{12}\right)=1+x^2 \tag{2.6.2.3-3}$$

since the derivative of even powers of the variable x are zero. The initial conditions required for register 2, to be in its characteristic phase, are found from the quotient

$$G(x)=\frac{1+x^2}{1+x^2+x^{11}}=1+x^{11}+x^{13}\cdots \tag{2.6.2.3-4}$$

The initial conditions for the characteristic phase for register 2 becomes

$$[00000000001] \tag{2.6.2.3-5}$$

with a "1" in the rightmost cell and the rest being "zero."

The only constraint on the initial conditions of register 1 is that the entry in the first stage (the one on the right) must be a "zero." Therefore our balanced code generator of length $2^{11}-1$ is as indicated in Figure 2.6-10. There are about $2^{(10)}$ codes in this balanced Gold code set (since a zero must be located in the first register cell).

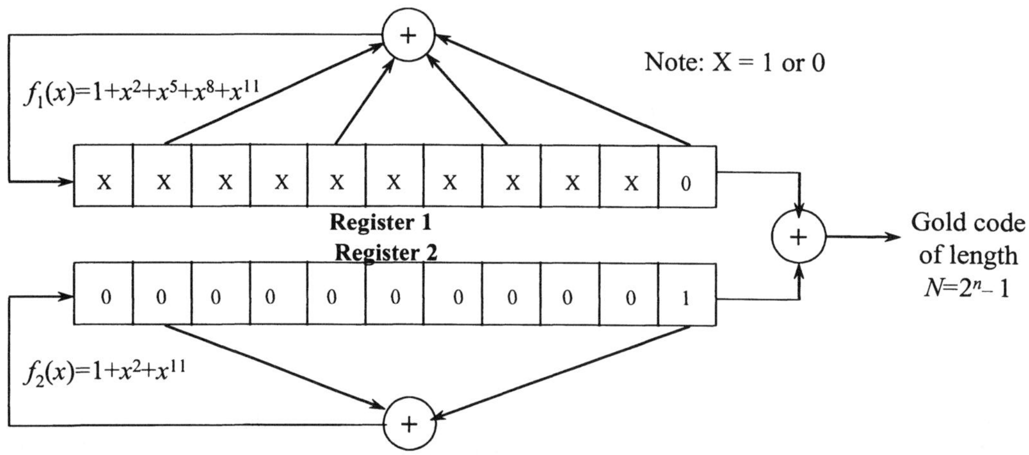

Figure 2.6-10 Balanced Gold code of length $2^{11}-1$.

2.6.2.4 Optimality of Gold Codes

It is of interest to point out the optimality of Gold codes in the following sense. Let ***a*** and ***b*** be two binary sequences. A bound due to Sidelnikov [23, 24] *(Sidelnikov's bound)* for a set of N or more binary sequences of period N satisfy the inequality

$$\theta_{\max} > \sqrt{2N-2} \tag{2.6.2.4-1}$$

where $\theta_{\max}$ is defined in (2.5-16). For all codes of length $N = 2^n-1$ it can be shown (2.6.2.4-1) implies that

$$\theta_{\max} > -1 + 2^{(n+1)/2} \tag{2.6.2.4-2}$$

When n is odd, the right hand side of (2.6.2.4-2) becomes $t(n)$–2. Since $\theta_{\max}$ must be and odd integer, and since (2.6.2.4-2) is a strict inequality, it follows that

$$\theta_{\max} \geq t(n) \tag{2.6.2.4-3}$$

It is noted that from (2.6.2-1) that *Gold codes for n odd form an optimal set* in the sense of minimizing the absolute magnitude of the periodic correlation parameter $\theta_{\max}$ in relationship to the Sidelnikov's bound. When n is even, the bound of (2.6.2.4-2) is not an integer, and is smaller than $t(n)$ by approximately the factor of $\sqrt{2}$ [7]. Hence *Gold sequences are not optimal when n is even.* Gold codes (sequences) have been used in a number of systems including the Global Positioning Satellite system (GPS) for the C/A code and the Tracking and Data Relay Satellite System (TDRSS), and in fact both GPS and TDRSS use balanced Gold codes.

2.6.3 Gold-Like Sequences and Dual BCH Sequences

Now two relatively unknown classes of sequences will be discussed that are related to Gold codes [7]. The following construction applies to both classes of sequences. Let n be an even integer and let q be an integer such that $\gcd(q,2^n-1)=3$. Let $\boldsymbol{a}$ denote an m-sequence of period $N=2^n-1$ with characteristic polynomial $f_1(x)$, and let the sequences $\boldsymbol{b}(k)$, $k=0,1,2$ denote the result of decimating $T^k\boldsymbol{a}$ by q. Consider now a property of maximal length sequences.

Property VI of Maximal Length Sequences Let $\boldsymbol{a}[q]$ be not be identically zero. Then $\boldsymbol{a}[q]$ has period $N/\gcd(N,q)$ and is generated by polynomial $f_2(x)$ which is the q-th decimation of the sequence of $\boldsymbol{a}$ [7].

From property VI it is clear that the $\boldsymbol{b}(k)$ are sequences having period $N'=N/3$ which are generated by the polynomial $f_2(x)$, which is the q-th decimation of the sequence $\boldsymbol{a}$. It is known that from linear shift register theory that if c denotes a (nonzero) sequence generated by $h(x)=f_1(x)f_2(x)$, then it is true that either

$$\boldsymbol{c}=T^i\boldsymbol{a} \tag{2.6.3-1}$$

or

$$\boldsymbol{c}=T^j\boldsymbol{b}^{(k)} \tag{2.6.3-2}$$

or

$$\boldsymbol{c}=T^i\boldsymbol{a}+T^j\boldsymbol{b}^{(k)} \tag{2.6.3-3}$$

where

$$0\le i\le N-1,\ 0\le j\le N'-1,\ and\ 0\le k\le 2$$

Noting that sequences of the form (2.6.3-2) have period $N/3$, it is seen that any sequence of period N generated by $h(x)$ is some phase of some sequence in the set $H_q(\boldsymbol{a})$ defined by [7]

$$\begin{aligned} H_q(\boldsymbol{a})=\{&\boldsymbol{a},\boldsymbol{a}+\boldsymbol{b}^{(0)},\boldsymbol{a}+T\boldsymbol{b}^{(0)},\cdots,\boldsymbol{a}+T^{N'-1}\boldsymbol{b}^{(0)},\\ &\boldsymbol{a}+\boldsymbol{b}^{(1)},\boldsymbol{a}+T\boldsymbol{b}^{(1)},\cdots,\boldsymbol{a}+T^{N'-1}\boldsymbol{b}^{(1)},\\ &\boldsymbol{a}+\boldsymbol{b}^{(2)},\boldsymbol{a}+T\boldsymbol{b}^{(2)},\cdots,\boldsymbol{a}+T^{N'-1}\boldsymbol{b}^{(2)}\} \end{aligned} \tag{2.6.3-4}$$

It is to be noted that $H_q(\boldsymbol{a})$ contains $N+1=2^n$ sequences of period N.

2.6.3.1 Gold-Like Sequences

Now consider the Gold-like sequences [7]. Let $\boldsymbol{a}$ be a m-sequence of period N of period 2^n-1 with characteristic function $f_1(x)$. According to property VI of m-sequences the $t(n)$-th decimation of $\boldsymbol{a}$, is generated by $f_2(x)$ where $\boldsymbol{b}=\boldsymbol{a}[t(n)]$. For $n\equiv 0 \bmod 4$, $\gcd(t(n),2^n-1)=3$ and (2.6.3-1) through (2.6.3-4) define the code set. When $q=t(n)$ the set $H_{t(n)}(\boldsymbol{a})$ is called the set of Gold-like sequences and contains 2^n code words. The correlation values belonging to the set of Gold-like sequences take on five values in the set $\{-1, -t(n), t(n)-2, -s(n), s(n)-2\}$, where for n even only, $s(n)$ is defined by

$$s(n)=1+2^{n/2}=\frac{1}{2}\left(t(n)+1\right) \tag{2.6.3.1-1}$$

2.6.3.2 Dual BCH Sequences

Dual BCH sequences, so named because the dual BCH sequence generator polynomial is the dual (see [16]) to a double error correcting BCH code [25] and is generated by the code generator polynomial $h(x) = f_1(x)f_2(x)$ where a is generated from $f_1(x)$ and $\boldsymbol{b} = \boldsymbol{a}[3]$ is generated from $f_2(x)$ [7]. When n is odd, $f_2(x)$ is a primitive polynomial and (2.6.2-3) and (2.6.2-4) define the code set. Therefore the set $G(\boldsymbol{a},\boldsymbol{a}[3])$ thus obtained is a set of Gold sequences because decimation by 3 is a preferred decimation when n is odd. In the case when n is even $\gcd(3,2^n-1) = 3$ and consequently (2.6.3-1) to (2.6.3-4) apply. For the set $H_3(\boldsymbol{a})$ the maximum correlation value is $\theta_{max} = t(n)$, and the correlation values take on the values contained in the following set [26]

$$\{-1, -t(n), t(n)-2, -s(n), s(n)-2\} \tag{2.6.3.2-1}$$

These sequences can be generalized to the dual of a t-error correcting BCH code, then a set of $(2^n-1)^{t-1}$ or more sequences can be constructed for which

$$\theta_{max} \le 1 + (t-1)2^{(n+2)/2} \tag{2.6.3.2-2}$$

Additional details can be found in [24, 27, 28]. The code set size is 2^n.

2.6.4 Kasami Sequences

Kasami sequences were discovered in the course of evaluating weight enumerators of linear cyclic codes in 1966 [24]. There are two types of Kasami sequences: *the small set of Kasami sequences* and *the large set of Kasami sequences*. Both will be considered briefly in what follows.

2.6.4.1 Small Set of Kasami Sequences

Let n be even and let $\boldsymbol{a}$ denote an m-sequence of period $N=2^n-1$, with characteristic polynomial $f_1(x)$. Let the second sequence $\boldsymbol{b}$ be related to $\boldsymbol{a}$ by a decimation of $\boldsymbol{a}$ of the form

$$\boldsymbol{b} = \boldsymbol{a}[s(n)] = \boldsymbol{a}\left[2^{(n/2)}+1\right] \tag{2.6.4.1-1}$$

It follows from Property VI of Section 2.6.3 that the period of sequence $\boldsymbol{b}$ is given by

$$N_b = \frac{2^n-1}{\gcd(2^n-1,\ 2^{n/2}+1)} = 2^{n/2}-1 \tag{2.6.4.1-2}$$

which is generated by a polynomial $f_2(x)$, which is the $s(n)$ decimation of $f_1(x)$. Furthermore since $f_2(x)$ can be shown to be a polynomial of degree $n/2$, $\boldsymbol{b}$ is an m-sequence of period $2^{(n/2)}-1$. The product polynomial $f_1(x)f_2(x)$ that generates the Kasami codes is of degree $3n/2$. Any sequence in this set must be one of the following forms

$$T^i\boldsymbol{a},\ T^j\boldsymbol{b},\ T^i\boldsymbol{a}+T^j\boldsymbol{b},\ 0 \le i < 2^n-1,\ 0 \le j < 2^{n/2}-1 \tag{2.6.4.1-3}$$

Thus any sequence of period 2^n-1 generated by $f_1(x)f_2(x)$ is some phase of some sequence in the set $K_s(\boldsymbol{a},\boldsymbol{b})$ defined by

$$K_s(\boldsymbol{a},\boldsymbol{b}) = \left\{\boldsymbol{a},\ \boldsymbol{a}+\boldsymbol{b},\ \boldsymbol{a}+T\boldsymbol{b},\cdots,\ \boldsymbol{a}+T^{2^{n/2}-2}\boldsymbol{b}\right\} \tag{2.6.4.1-4}$$

which is *the small set of Kasami sequences.* The code set size is $2^{n/2}$. Kasami showed that the correlation functions take on the following values, which are listed in Table 2.6-3 [7].

Table 2.6-3 Three-Level Small Set Kasami Normalized Correlation* Values

Register length n	Code Length N	Normalized Correlation Values
n even	$N=2^n-1$	$-1/N$ $(-1+2^{(n/2)})/N$ $(-1-2^{(n/2)})/N$

*Includes auto- and cross-correlation values, but of course the autocorrelation is one at zero offset.

It is to be noted that none of the compound sequences in the small set of Kasami codes have balanced sequences [29]. Clearly the (unnormalized) code correlation values have a value of $\theta_{max}=1+2^{(n/2)}$. Note that the value of θ_{max} for Kasami codes is approximately one half of the Gold codes; however, the number of codes in the set are $2^{(n/2)}=(N+1)^{1/2}$, or roughly $\sqrt{N}$, a much smaller quantity than N+2, which applies to Gold codes. Hence the small set of Kasami codes may be a better choice than Gold codes for a design that does not need many codes in the set ($<2^{(n/2)}$).

Welch's bound [24] is for K sequences of length N and is given by

$$\theta_{max} \geq N\left[\frac{K-1}{NK-1}\right]^{1/2} \tag{2.6.4.1-5}$$

so when $K=2^{(n/2)}$ and $N=2^n-1$ (2.6.4.1-5) can be shown to imply [7]

$$\theta_{max} > -1+2^{n/2} \tag{2.6.4.1-6}$$

Now for binary sequences and the fact that θ_{max} is odd, implies that

$$\theta_{max} \geq 1+2^{n/2} \tag{2.6.4.1-7}$$

By comparing the maximum correlation results in Table 2.6-3 and (2.6.4.1-7), it is clear that the small set of Kasami sequences are optimal in respect to minimizing the maximum absolute correlation of the bound of (2.1.6.4-7).

2.6.4.2 Large Set of Kasami Sequences

Let $f_1(x)$ and $f_2(x)$ be preferred pairs whose characteristic polynomials generate the Gold codes. That is, if $\boldsymbol{a}$ is generated from the characteristic polynomial $f_1(x)$, then $f_2(x)$ is the polynomial that generates the sequence $\boldsymbol{a}[t(n)]$ when $n \neq 0$ mod 4. If $n=0$ mod 4, then $f_1(x)f_2(x)$ generates the set of Gold-like sequences $H_{t(n)}(\boldsymbol{a})$. Consider all sequences generated by the product $h(x)=f_1(x)f_2(x)f_3(x)$, which can be realized by the modulo 2 sum

$$\boldsymbol{a}+\boldsymbol{b}+\boldsymbol{c} \tag{2.6.4.2-1}$$

where $f_1(x)$ generates sequence $\boldsymbol{a}$, $f_2(x)$ generates sequence $\boldsymbol{b}$, and $f_3(x)$ generates sequence $\boldsymbol{c}$.

Theorem 4 Let n be even and let $f_1(x)$ denote a primitive binary polynomial of degree n that generates the m-sequence $\boldsymbol{a}$. Let $\boldsymbol{b}=\boldsymbol{a}[s(n)]$ denote an m-sequence of period $2^{(n/2)}-1$ generated by the characteristic polynomial $f_2(x)$ of degree $n/2$, and let $f_3(x)$ denote the polynomial of degree n that generates $\boldsymbol{a}[t(n)]$. Then the set of sequences of period N generated by the characteristic polynomial $h(x) = f_1(x)f_2(x)f_3(x)$ is called *the large set of Kasami sequences* and is denoted by $K_L(\boldsymbol{a})$ where [7]

(1) If $n \equiv 2 \bmod 4$, then

$$K_L(a) = G(a,c) \cup \left[\bigcup_{i=0}^{2^{(n/2)}-2} \{T^i b + G(a,c)\} \right] \tag{2.6.4.2-2}$$

where $\boldsymbol{c} = \boldsymbol{a}[t(n)]$ and $\boldsymbol{b} = \boldsymbol{a}[s(n)]$, and $G(\boldsymbol{a},\boldsymbol{b})$ is defined in (2.6.2-4).

(2) If $n \equiv 0 \bmod 4$ then

$$K_L(\boldsymbol{a}) = H_{t(n)}(\boldsymbol{a}) \cup \left[\bigcup_{i=0}^{2^{(n/2)}-2} \{T^i \boldsymbol{b} + H_{t(n)}(\boldsymbol{a})\} \right] \cup \{\boldsymbol{c}^{(j)} + T^k \boldsymbol{b}:\ 0 \le j \le 2,\ 0 \le k < (2^{(n/2)} - 1)/3\} \tag{2.6.4.2-3}$$

where the code sequence $\boldsymbol{c}^{(j)}$ is the result of decimating $T^j\boldsymbol{a}$ by $t(n)$ and $H_{t(n)}(\boldsymbol{a})$ is given by (2.6.3-4) with $q = t(n)$. The correlation functions for $K_L(\boldsymbol{a})$ take on values in the set $\{-1, -t(n), t(n)-2, -s(n), s(n)-2\}$ from which it follows that $\theta_{max} = t(n)$. If $n \equiv 2 \bmod 4$, $K_L(\boldsymbol{a})$ contains $2^{(n/2)}(2^n + 1)$ sequences in the set whereas if $n \equiv 0 \bmod 4$, $K_L(a)$ contains $2^{(n/2)}(2^n + 1)-1$ sequences in the set [7].

It is to be noted that the large set of Kasami sequences contains both the set of Gold (or Gold-like) sequences and the small set of Kasami sequences as subsets.

2.6.5 Bent Sequences

Bent[6] sequences [29–31] are nonlinear binary sequences that achieve Welch's bound on cross- and autocorrelation. The period of the sequence is 2^n-1 with n the number of stages in the shift register. This value n must be a multiple of 4 ($n = 0$, mod 4). The number of sequences in the set is $2^{n/2}$ and the autocorrelation function has three values and $\theta_{max} = 2^{n/2} + 1$. The *linear span* is the minimum number of memory elements needed to build a Fibonacci generator to create the same sequence. The equivalent linear span (LS) of Bent sequences has the bound

$$LS \le \sum_{i=1}^{n/4} \binom{n}{i} \tag{2.6.5-1}$$

The actual span achieved is generally close to the bound [29]. These sequences are balanced. The high equivalent linear span makes the likelihood of analysis by a sophisticated enemy less likely than linear codes since the linear difference equation satisfied by the sequence can be orders of magnitude larger than the number of memory elements in the sequence generator. An example of a Bent sequence is shown in Figure 2.6-11. Details are contained in the previously cited references.

2.6.6 Comparison of CDMA Code Performance

In this section a brief summary of the code parameters for code division multiple access (CDMA) will be discussed. Table 2.6-4 [29] indicates the code parameters of interest for the various codes discussed.

---

[6] A bent function is a difference set in an elementary Abelian group.

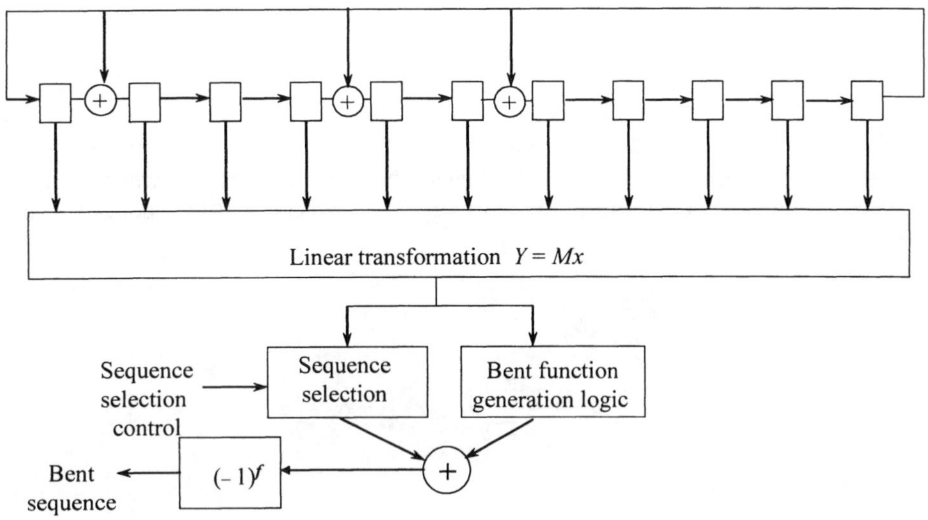

Figure 2.6-11 An example of a length 12 Bent sequence.

Gold sequences have lower correlation magnitude bounds when n is odd rather than even. They are approximately $\sqrt{2}$ larger when n is even. The Gold-like and dual BCH codes have essentially the same maximum correlation and code set size as the Gold codes that are even and equal to 2 mod 4. For even values of n, Gold codes have essentially twice the maximum correlation values as Bent and small set Kasami codes, but offer a code set size essentially the square of the latter set size. Bent codes have the edge over small set Kasami codes in the sense that they are balanced and since they are nonlinear, they have a much larger linear span.

Table 2.6-4 CDMA Code Properties with Period 2^n-1

Code Family	Shift Register Length n	Code Set Size	θ_{max}	Code Balance
Gold	Odd	2^n+1	$1+2^{(n+1)/2}$	Can be balanced
Gold	2 (mod 4)	2^n+1	$1+2^{(n+2)/2}$	Can be balanced
Gold-Like	Even	2^n	$1+2^{(n+2)/2}$	Cannot be balanced
Dual-BCH	Even	2^n	$1+2^{(n+2)/2}$	Cannot be balanced
Kasami (small set)	Even	$2^{n/2}$	$1+2^{n/2}$	Cannot be balanced
Kasami* (large set)	Even	$2^{n/2}(2^n+1)$	$1+2^{(n+2)/2}$	Cannot be balanced
Bent	0 mod 4	$2^{n/2}$	$1+2^{n/2}$	Balanced

* Actually when $n \equiv 0 \bmod 4$, the set contains $2^{n/2}(2^n+1)-1$ sequences in the set [7].

2.7 SEQUENCES WITH GOOD APERIODIC CORRELATION

Now consider sequences designed to have low aperiodic correlation values. These sequences are very important in synchronization problems, with one common application using a short sync sequence placed within a data stream and used to obtain and maintain time synchronization. Also it is possible to use the sync sequence for phase synchronization (in two pass systems). When the receiver timing is not aligned with the received transmitter timing and there is a time error of say a few code chips, a periodic correlation of the received signal will produce a correlation value that has two parts. The first will depend on the portion of the receiver sequence that overlaps the received sequence; however, the second part will contain a value that depends on the correlation with the adjacent data bits. Thus the only part of the correlation that is reliable is the overlap portion, which is an aperiodic correlation. This topic has been discussed in Section 2.5.2.

Again let the sequence chip values be $\varepsilon\{0,1\}$, with the understanding that when sums or correlations are used the values of the components are translated into $\{1,-1\}$. Recall that the aperiodic cross-correlation function for sequences is defined by

$$C_{ab}(l)=\begin{cases}\sum_{i=0}^{N-1-l} a_i b_{i+l} & 0 \le l \le N-1 \\ \sum_{i=0}^{N-1+l} a_{i-l} b_i = \sum_{k=-l}^{N-1} a_k b_{k+l} & 1-N \le l \le 0 \\ 0 & |l| \ge N \end{cases} \tag{2.7-1}$$

It is seen that $N-|l|$ symbols are used in this cross-correlation function. The aperiodic autocorrelation function for sequences is defined in (2.5.2-1) with $\boldsymbol{a} = \boldsymbol{b}$.

2.7.1 Barker and Williard Sequences

An important sequence designed to achieve the minimum aperiodic correlation values when $l \neq 0$ is known as a *Barker sequence*. These sequences are used primarily for synchronization of data streams, such as frame synchronization. A Barker sequence (denoted by a) is characterized by the fact that its aperiodic autocorrelation function satisfies

$$|C_a(l)| \le 1 \ \ for\ l \neq 0 \tag{2.7.1-1}$$

In other words as the code shifts by 1 or more symbols ($|l| \ge 1$) out-of-sync the magnitude of the autocorrelation function never exceeds one in magnitude, in its out of sync position. The only known Barker sequences [32] are for lengths of 1, 2, 3, 4, 5, 7, 11, 13. It is to be noted that the reverse and the inverse of a Barker sequence is a Barker sequence.

Williard sequences [33] minimize the probability of false synchronization with random data words in the overlap region for all values of overlap. Table 2.7-1 illustrates the Barker and Williard sequences up to length 13, with each symbol of the sequence being $\varepsilon\{+1,-1\}$.

Table 2.7-1 Barker and Williard Sequences to Length 13

Sequence Length	Barker Sequences	Williard Sequences
1	1	1
2	1,1 or 1,–1	1,–1
3	1,1,–1	1,1,–1
4	1,1,1,–1 or 1,1,–1,1	1,1,–1,–1
5	1,1,1,-1,1	1,1,–1,1,–1
7	1,1,1,–1,–1,1,–1	1,1,1,–1,1,–1,–1
11	1,1,1,–1,–1,–1,1,–1,–1,1,–1	1,1,1,–1,1,1,–1,1,–1,-1,–1
13	1,1,1,1,1,–1,–1,1,1,–1,1,–1,1	1,1,1,1,1,–1,–1,1,–1,1,–1,–1,–1

2.7.2 Neuman-Hofman Sequences

Many times it is necessary to use larger sequences to improve the performance of the correctly detected synchronization word. Neuman and Hofman [34] have utilized an extensive computer search to determine aperiodic synchronization sequences up to length 24. Table 2.7-2 illustrates N-H code sequences up to length 24. Almost all of the entries have two different sequences to choose from. For convenience of presentation the codes have been displayed with the code elements $\varepsilon\{0,1\}$. The elements can be converted to values of ± 1 via (2.5-1). It is to be noted that Massey [35] has determined the optimum frame synchronization algorithm, and the optimum statistic is based on a correlation term and a correction term. The magnitude of the maximum value of the aperiodic correlation value is listed as $|R_M|$, clearly the smaller the value the better the code in a certain sense.

2.7.3 Partial Period Correlation for *m*-Sequences

It is useful to observe what the aperiodic correlation function will be; in the case that code sequence period is large compared to the correlation time. In general, the aperiodic autocorrelation may not be nearly as good as the full period autocorrelation function. Let the number of code chips included in the correlation be M, and let the number of code chips in the sequence (period) be N ($M<N$). Results for the mean and the variance of the partial period autocorrelation value are presented next [13]. The *normalized partial period autocorrelation function* is given by

$$\tilde{C}_{\boldsymbol{a}}^{M}(k)=\frac{1}{M}\sum_{i=0}^{M-1}a_i a_{i+k} \tag{2.7.3-1}$$

Table 2.7-2 Neumann-Hoffman Sequences Up to Length 24

Length	Sync Sequence	$\lvert R_M\rvert$	Length	Sync Sequence	$\lvert R_M\rvert$
7	0001101	1	16	0000011001101011	2
8	00011101	2	16	0000111011101101	2
8	00001101	2	17	00001011001110101	4
9	000011101	2	17	00001111011011101	2
9	001111101	2	18	000010101101100111	3
10	0000011010	2	18	001100111110100101	2
10	000011010	2	19	0000111000100010010	2
11	00001011001	2	19	0001110111011011010	2
11	00011101101	1	20	00000100110101001110	2
12	000111101101	4	20	00010001111100101101	4
12	001100000101	2	21	000000101110100111001	2
13	0000001100101	2	21	001101100001000010101	2
13	0000010110011	2	22	0001000111110011011010	3
14	00001100110101	2	23	00000010101100110100111	5
14	00110011111010	2	23	00000011110011001001010	3
15	001111100110101	2	24	000111111001000011001010	9
15	000011001001010	2	24	000001110011101010110110	4

Theorem 5 Let $\boldsymbol{a}$ be an m-sequence of length N. Then the average value of the normalized aperiodic autocorrelation, averaged over all N starting positions, is just the normalized periodic autocorrelation function itself. That is, letting E denote the average value, that

$$E\left[\tilde{C}_{\boldsymbol{a}}^{M}(k)\right]=\frac{1}{N}\sum_{j=0}^{N-1}\left[\frac{1}{M}\sum_{i=0}^{M-1}a_{i+j}a_{i+j+k}\right]=\tilde{\theta}_{\boldsymbol{a}}(k) \quad M\leq N \tag{2.7.3-2}$$

where $\tilde{C}_a^M(k)$ is the normalized aperiodic autocorrelation function over M chips, and $\tilde{\theta}_a(k)$ is the normalized periodic autocorrelation function. Note that the wiggly line over the function $C(k)$ or $\theta(k)$, indicates the normalized version (i.e., $\tilde{C}(0) = 1$ and $\tilde{\theta}(0) = 1$).

Proof By definition the average over the code period produces

$$E\left[\tilde{C}_a^M(k)\right] = \frac{1}{N}\sum_{j=0}^{N-1}\left[\frac{1}{M}\sum_{i=0}^{M-1} a_{i+j}a_{i+j+k}\right] \tag{2.7.3-3}$$

Interchanging the order of the summations produces

$$E\left[\tilde{C}_a^M(k)\right] = \frac{1}{M}\sum_{i=0}^{M-1}\left[\frac{1}{N}\sum_{j=0}^{N-1} a_{i+j}a_{i+j+k}\right] = \frac{1}{\mathrm{M}}\sum_{i=0}^{M-1}\tilde{\theta}_a(k) = \tilde{\theta}_a(k) \tag{2.7.3-4}$$

as was to be proven. Therefore the average normalized aperiodic autocorrelation function, when averaged over each starting position, is just the normalized periodic autocorrelation function.

Theorem 6 Let $\boldsymbol{a}$ be an m-sequence of length N. Then the variance of the normalized aperiodic autocorrelation function for $k \neq 0$ and $M < N$ is given by

$$Var\left[\hat{C}_a^M(k)\right] = \frac{1}{M}\left(1+\frac{1}{N}\right)\left(1-\frac{M}{N}\right)$$

where the average is taken over all N starting positions.

Proof By definition

$$\mathrm{var}\left[\tilde{C}_a^M(k)\right] = E\left\{\left[\tilde{C}_a^M(k)\right]^2\right\} - \left(E\left\{\tilde{C}_a^M(k)\right\}\right)^2 \tag{2.7.3-5}$$

The second term has been computed in Theorem 5, so it is only necessary to consider the first term. The first term can be written as

$$E\left\{\left[\tilde{C}_a^M(k)\right]^2\right\} = \frac{1}{N}\sum_{j=0}^{N-1}\left[\frac{1}{M}\sum_{i=0}^{M-1} a_{i+j}a_{i+j+k}\right]^2 \tag{2.7.3-6}$$

so that

$$E\left\{\left[\tilde{C}_a^M(k)\right]^2\right\} = \frac{1}{NM^2}\sum_{j=0}^{N-1}\sum_{i=0}^{M-1}\sum_{l=0}^{M-1} a_{i+j}a_{i+j+k}a_{l+j}a_{l+j+k} \tag{2.7.3-7}$$

Now using the shift and add property this can be written as

$$E\left\{\left[\tilde{C}_a^M(k)\right]^2\right\} = \frac{1}{NM^2}\sum_{j=0}^{N-1}\left(\sum_{i=0}^{M-1}\sum_{l=0}^{M-1} a'_{i+j+k}a'_{l+j+k}\right) \tag{2.7.3-8}$$

where a'_m is the shifted resultant, based on Property III of Section 2.6.1 for m-sequences. The same property applies when the symbols are $\varepsilon\{1,-1\}$ for multiplication of the code symbols. Now express the inner double sum into the diagonal and off-diagonal terms as

$$E\left\{\left[\tilde{C}_a^M(k)\right]^2\right\} = \frac{1}{NM^2}\sum_{j=0}^{N-1}\left(\sum_{i=0}^{M-1} M + \sum_{i=0}^{M-1}\sum_{\substack{l=0\\ i\neq l}}^{M-1} a'_{i+j+k}\, a'_{l+j+k}\right) \tag{2.7.3-9}$$

or again using Property III for m-sequences one obtains

$$E\left\{\left[\tilde{C}_a^M(k)\right]^2\right\} = \frac{1}{M} + \sum_{i=0}^{M-1}\sum_{\substack{l=0\\ i\neq l}}^{M-1}\frac{1}{NM^2}\left[\sum_{j=0}^{N-1} a''_{i+j+l+k}\right] \tag{2.7.3-10}$$

where a''_k is the shifted resultant, based on Property III of Section 2.6.1. Since the last summation is just equal to -1, and since there are just $M(M-1)$ off-diagonal terms, one obtains the result

$$E\left\{\left[\tilde{C}_a^M(k)\right]^2\right\} = \frac{1}{M} - \frac{1}{NM}(M-1) \tag{2.7.3-11}$$

Now from Theorem 5, one has

$$E\left\{\left[\tilde{C}_a^M(k)\right]\right\} = \frac{-1}{N} \qquad k\neq 0, \quad |k|\leq N-1 \tag{2.7.3-12}$$

So that the variance becomes

$$\operatorname{var}\left[\tilde{C}_a^M(k)\right] = \frac{1}{M} - \frac{M-1}{NM} - \left(\frac{-1}{N}\right)^2 = \frac{1}{M}\left(1+\frac{1}{N}\right)\left(1-\frac{M}{N}\right) \tag{2.7.3-13}$$

which is what was to be shown. Notice that for large N and for $N >> M$, the variance is well approximated by the result

$$\operatorname{var}\left[\tilde{C}_a^M(k)\right] \cong \left(\frac{1}{M} - \frac{1}{N}\right) \cong \frac{1}{M} \tag{2.7.3-14}$$

Having the mean and variance of the partial period autocorrelation values allows one to determine the random interference that will be observed when the clock time is correct but the code position is not aligned with the received code sequence, and the correlation is over a time short compared to the code period.

2.7.4 Frequency-Hopping Multiple Access Code Generators

In this section asynchronous frequency-hopping (FH) patterns are considered for SS FHMA (or FH/SSMA) designs. The *frequency hopping synthesizer* generates the frequencies that range across the frequency band of interest, in a pseudorandom manner, at regular intervals, and with equal spacing between the contiguous hop frequencies.

Often data transmission in FH systems is based on either frequency shift keying (FSK) (or MFSK) or differential shift keying (DPSK). With FSK the frequency-hopping pattern hops to frequency f_0 and then the data adds Δf Hz or subtracts Δf Hz according to whether a mark or space is transmitted. In effect the signal occupies a *frequency slot* which is the FSK band in which the signal is contained within. It is also not uncommon in some designs to have more than one hop frequency per data bit, which is known as *fast frequency hopping (FFH)*. The complete set of slots defines the hopping bandwidth, assuming that each slot is contiguous. In some systems the frequency separation is much less than the data modulation bandwidth.

When DPSK is used, one or more reference bits are transmitted followed by the differentially encoded data, all on one frequency hop. Again a slot is the frequency band needed to pass the DPSK signal without much distortion.

The multiplexing aspect of FH systems is quite natural, since when one hopping signal is occupying one frequency slot, the other signals can occupy the other slots.

The receiver for an FH system consists of basically a frequency dehopper followed by a conventional FSK (or MFSK) or DPSK demodulator. It is of course necessary that the frequency hopper acquires and follows the receiver-hopping signal in frequency and time. The frequency dehopper functionally consists of a multiplier that multiplies the received hopping signal with the locally generated one, which produces an output that is fixed (except for the modulation) at an intermediate frequency when it is in synchronization.

The multiple access capability of an FH system comes about from the fact that a frequency hopped signal only occupies one slot at a given time and therefore other hopping signals can be located at other frequency slots without causing a problem to each other. If the hopping signals are all timed with the same source and are at the same range from the receiver, this frequency-hopped multiple access (FHMA) scheme would be feasible. It is possible to slave the ground terminals to a satellite, which is made the master clock, such as in the Milstar system.

In general such time coordination is not always feasible. In general for many systems the transmitters are asynchronous, and it is possible that two or more signals will hop to the same frequency slot at the same time causing what is known as a *hit* or *collision*. The result of this hit is degradation in the link performance. Because of this problem, error correction coding is commonly utilized in frequency hopped spread spectrum multiple access communication systems.

An asynchronous FH/SSMA system will be assumed in the modeling that follows. Since an asynchronous system is assumed, it is to be noted that when a hit occurs the hopping signal that causes the hit and arrives at the same frequency hop position as the desired hopping signal may overlap in time some fraction of a hop. This caused a partial hit. To simplify the analysis it will be assumed that the time delays between the various signals are integer multiples of the duration of a hop, and therefore the hits have total overlap in time and the hits are complete.

Let the number of frequency slots be denoted by $f_1, f_2, \ldots f_q$ with the total number frequency slots being q, and the hopping patterns having period N. Let

$$\boldsymbol{X} = (X_0, X_1, \ldots X_{N-1}) \text{ and } \boldsymbol{Y} = (Y_0, Y_1, \ldots Y_{N-1}) \tag{2.7.4-1}$$

denote two hopping patterns with

$$X_k, Y_k \in \{f_1, f_2, \ldots f_q\} \tag{2.7.4-2}$$

When the two frequency-hop patterns are being hopped according to these patterns, then the *Hamming correlation* [36] is defined by

$$H_{XY}(\tau) = \sum_{i=0}^{N-1} h(X_i, Y_{i+\tau}) \quad 0 \le \tau < N \tag{2.7.4-3}$$

where

$$h(x,y) = \begin{cases} 0 \text{ if } x \neq y \\ 1 \text{ if } x = y \end{cases} \tag{2.7.4-4}$$

where $H_{XY}(\tau)$ denotes the number of hits between the two patterns for a relative delay of τ (an integer) with the condition that the sum "$i+\tau$" is taken modulo N.

Let $\bar{H}(\boldsymbol{X},\boldsymbol{Y})$ denote the average

$$\bar{H}(\boldsymbol{X},\boldsymbol{Y}) = \frac{1}{N}\sum_{\tau=0}^{N-1} H_{XY}(\tau) \qquad (2.7.4\text{-}5)$$

where $\bar{H}(\boldsymbol{X},\boldsymbol{Y})$ is the average number of hits between the two hopping patterns $\boldsymbol{X}$ and $\boldsymbol{Y}$.

It can be shown that it is not possible to achieve the ideal of $H_{XY}(\tau) = 0$ for all values of (integer) τ unless the hop patterns $\boldsymbol{X}$ and $\boldsymbol{Y}$ are restricted to disjoint subsets of the available frequency slots [37].

2.7.4.1 *m*-Sequence Frequency-Hopping Multiple Access Code Generators

Now assume that there are q frequency slots that are to be used in the hop pattern design [37] and the sequence length $N = 2^n - 1$ is equal to the number of slots, q. One simple method of driving a frequency hopper with an m-sequence is by using the n bits of the register to control the frequency of a frequency synthesizer. Now each nonzero n-tuple occurs exactly once within the period of the m-sequence. Hence each of the q slots is used once during the period of the m-sequence.

Note that the longest number of runs of all 1s is n, for a length $N = 2^n - 1$ m-sequence. Further any m-sequence must go through the following sequence

$$011...111, \quad 111...111, \quad 111...110 \qquad (2.7.4.1\text{-}1)$$

where the second sequence has n 1s in direct succession. It is therefore clear that if two distinct m-sequences of the same length are used for the two hopping patterns, then as least as many as three hits can occur for some time shifts. Sarwate and Purlsey [37] have noted that things can be much worse; a length 31 sequence can have as many as five hits for some relative time delays. Let us consider an example to clarify these concepts.

Example 17 Assume that two m-sequence generators of length $N = 7$ are to be used for providing the frequencies of the two frequency hoppers. Obviously, the generator length is artificially low in order to illustrate the ideas involved. As seen in the figure that SRG1 generates the numeric sequence 4, 6, 7, 3, 5, 2, 1 by taking the three bits of the register and associating a numeric value equal to the binary value.

For example, 100 converts to $1\times2^2+0\times2^1+0\times2^0 = 4$. Let the output numeric sequence for both SRGs be written as a column vector as shown in the figure. X denotes the first generator numeric output (seven values). The other seven vectors $\{Y(k), k=0,1,\ldots 6\}$ are the numeric outputs of the second SRG with k cyclic shifts. The row of integers below the horizontal line denotes the number of hits that would occur for the k-th shift of the Y code generator output. Thus in comparing X row by row with $Y(4)$, it is seen that the second (6), third (7), and fourth (3) numeric values (rows) are the same, and thus three hits would occur for this cyclic phase shift. In this example it is seen that the largest number of hits is three and the smallest is zero ($Y(0)$, for example).

It can be seen from the example that the value of $\bar{H}(\boldsymbol{X},\boldsymbol{Y}) = 1$ since the sum of the bottom row divided by $N = 7$ is 1. Even though the average was 1, one delay ($Y(4)$) caused three hits and other delays caused no hits.

Clearly this concept can be extended to a higher order (more FH signals) by considering pair-wise comparisons.

Other methods of synthesizer sequence design have been discussed in the literature, including Reed-Solomon codes [37, 38] and higher order m-sequences [39].

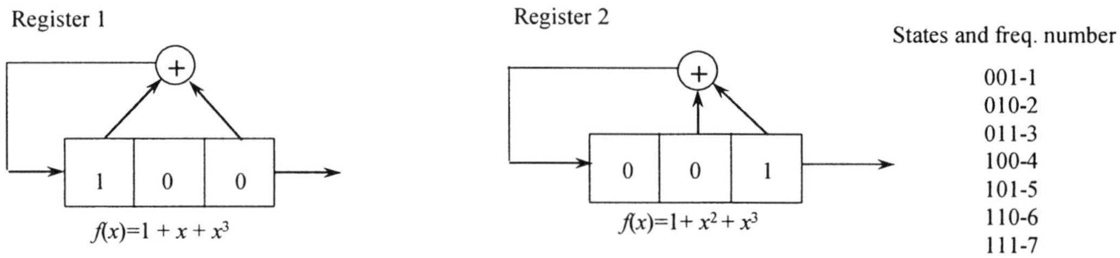

This register produces the following seven states: 4,6,7,3,5,2,1

This register produces the following seven states: 1,4,2,5,6,7,3

X	*Y*(0)	*Y*(1)	*Y*(2)	*Y*(3)	*Y*(4)	*Y*(5)	*Y*(6)
4	1	3	7	6	5	2	4
⑥	4	1	3	7	⑥	5	2
⑦	2	4	1	3	⑦	6	5
③	5	2	4	1	③	7	6
5	6	5	2	4	1	3	7
2	7	6	5	2	4	1	3
1	3	7	6	5	2	4	1
No. of hits:	**0**	**1**	**0**	**1**	**3**	**0**	**2**

Figure for Example 17. A simple example of a frequency-hopping generator.

2.8 SUMMARY

This chapter presented an introduction to the sequences used in CDMA spread spectrum communications. Matrix and polynomial characterizations of the SSRG were developed. Periodic and aperiodic correlation functions were presented. The *m*-sequence was introduced and many properties were presented. The power spectral density of an *m*-sequence was derived. Additionally the power spectral density of a despread *m*-sequence was developed.

Many commonly used codes that are employed in spread spectrum communications for periodic correlation properties, which include Gold codes, balanced Gold codes, Gold-like and dual BCH codes, both the small set and large set of Kasami codes, and Bent sequences were presented.

Codes used for aperiodic correlation applications, such as synchronization of data frames, are then presented. This was followed by a brief discussion of frequency-hopping sequences for asynchronous FH/SSMA applications.

This chapter did not address the area of nonlinear codes (other than Bent sequences) [29, 40]. In addition, polyphase sequences including the work on 4-phase codes have not been addressed here [41–44].

References

[1] E. N. Gilbert, "Quasi-Random Binary Sequences," Bell Telephone laboratories Memorandum MM-53-1400-42, November 27, 1953.

[2] S. W. Golomb, "Sequences with Randomness Properties," Terminal Progress Report under Contract R No. 639498, Martin Co., Baltimore, MD, June 1955.

[3] N. Zierler, "Several Binary-Sequence Generators," Tech. Report No. 95, Lincoln Lab., MIT, September 12, 1955.

[4] N. Zierler, "Linear Recurring Sequences," *J. Soc. Industrial and Applied Math*, Vol. 7, March 1959, pp. 31–48.

[5] T. G. Birdsall and M. P. Ristenbatt, "Introduction to Linear Shift-Register Generalized Sequences," Univ. of Michigan Research Institute, Dept. of Elect. Engr. Tech. Report, No. 90, October 1958.

[6] S. W. Golomb, *Shift Register Sequences*, San Francisco, CA: Holder-Day, 1967.

[7] D. V. Sarwate and M. B. Pursley, "Cross-Correlation Properties of Pseudorandom and Related Sequences," *Proceedings of the IEEE*, Vol. 68, No. 5, May 1980.

[8] G. Birkhoff and S. MacLane, *A Survey of Modern Algebra*, rev. ed., New York: Macmillan, 1953.

[9] B. L. Van Der Waerden, *Modern Algebra* (2 Vols.) Trans. by F. Blum and T. J. Benac, New York: Frederick Ungar, 1950 and 1953.

[10] S. Lin and D. J. Costello, *Error Control Coding: Fundamentals and Applications*, Englewood Cliffs, NJ: Prentice Hall, 1983.

[11] R. Ziemer and R. Peterson, *Digital Communications and Spread Spectrum Systems*, New York: Macmillan, 1985.

[12] D. J. Torrieri, *Principles of Secure Communication Systems,* 2nd ed., Norwood, MA: Artech House, 1992.

[13] J. K. Holmes, *Coherent Spread Spectrum Systems*, New York: Wiley-Interscience, 1982 (republished by Krieger Press, Malabar, FL 1990).

[14] T. Helleseth and P. V. Kumar in *The Mobile Communications Handbook*, J.D. Gibson editor, Chapter 8, Boca Raton, FL: CRC Press, 1999.

[15] A. J. Viterbi, *CDMA Principles of Spread Spectrum Communication*, Reading, MA: Addison-Wesley, 1995.

[16] W. W. Peterson and E. J. Weldon Jr., *Error Correcting Codes*, 2nd ed., Cambridge, MA: MIT Press, Appendix C, 1972.

[17] A. B. Carlson, *Communication Systems*, New York: McGraw-Hill, 1968, pp. 43, 52.

[18] J. L. Massey, "Shift-Register Synthesis and BCH Decoding," *IEEE Trans. on Information Theory*, Vol. IT-15, No. 1, January 1969.

[19] W. J. Gill, "Effect of Synchronization Error in Pseudorandom Carrier Communications," *First Annual IEEE Comm. Conv. Record*, Denver, CO, June 7–9, 1965, pp. 187–191.

[20] W. J. Gill and J. J. Spilker, "An Interesting Decomposition Property for the Self-Products of Random or Pseudorandom Binary Sequences," *IEEE Trans. on Communication Systems*, June 1966.

[21] R. Gold Associates, "TDRSS Telecommunication Systems PN Analysis Contract No. NAS 5-22546, Final Report," August 31, 1976.

[22] R. Gold, "Characteristic Linear Sequences and Their Coset Functions," *SIAM J. Appl. Math*., Vol. 14, 1966, pp. 980–985.

[23] V. M. Sidelnikov, "On Mutual Correlation of Sequences," *Soviet Math. Dokl*., Vol. 12, 1971, pp. 197–201.

[24] L. R. Welch, "Lower Bounds on the Maximum Cross Correlation of Signals," *IEEE Trans. on Information Theory*, Vol. IT-20, May 1974, pp. 397–399.

[25] E. R. Berlekamp, *Algebraic Coding Theory*, New York: McGraw-Hill, 1968.

[26] T. Kasami, "Weight Distribution of Bose-Chaudhuri-Hocquenghem Codes," in *Combinatorial Mathematics and Its Applications*, Chapel Hill, NC: Univ. of North Carolina Press, 1969.

[27] D. R. Anderson, "A New Class of Cyclic Codes," *SIAM J. Appl. Math*., Vol. 16, 1968, pp. 181–197.

[28] F. J. MacWilliams and N. J. A. Slone, *The Theory of Error-Correcting Codes*, the Netherlands: North-Holland, 1977.

[29] M. Simon, J. K. Omura, R. A. Scholtz, and B. K. Levitt, *Spread Spectrum Communications*, Vol. I, Chapter 5, Rockville, MD: Computer Science Press, 1985.

[30] J. D. Olsen, R. A. Scholtz, and L. R. Welch, "Bent-Function Sequences," *IEEE Trans. on Information Theory*, Vol. IT-28, No. 6, November 1982.

[31] A. Lempel and M. Cohn, "Maximal Families of Bent Sequences," *IEEE Trans. on Information Theory*, Vol. IT-28, No. 6, November 1982.

[32] R. H. Barker "Group Synchronization of Binary Digital Systems," in *Communication Theory*, W. Jackson, Editor, New York: Academic Press, 1953, pp. 273–287.

[33] M. W. Williard, "Optimum Code Patterns for PCM Synchronization," *1962 National Telemetry Conference*, paper 5-5, 1962.

[34] F. Neuman and L. Hofman, "New Pulse Sequences with Desirable Correlation Properties," *Proc. Nat. Telemetry Conf.*, 1971, pp. 272–282.

[35] J. L. Massey, "Optimum Frame Synchronization," *IEEE Trans. on Communications Technology*, April 1972, pp. 115–118.

[36] A. Lempel and H. Greenberger, "Families of Sequences with Optimal Correlation Properties," *IEEE Trans. on Information Theory*, Vol. 20, 1974, pp. 90–94.

[37] D. V. Sarwate and M. B. Pursley, "Hopping Patterns for Frequency-Hopped Multiple Access Communications," *IEEE ICC*, Vol. I, paper no. 7.4.1, 1978.

[38] R. M. Mersereau and T. S. Seay, Multiple Access Frequency Hopping Patterns with Low Ambiguity," *IEEE Trans. on Aero. and Elect. Sys.*, Vol. AES-17, No. 4, July 1981.

[39] J. J. Komo and S.-C. Liu, "Maximal Length Sequences for Frequency Hopping," *IEEE Journal on Selected Areas in Comm.*, Vol. 8, No. 5, June 1990.

[40] E. J. Groth, "Generation of Binary Sequences with Controllable Complexity," *IEEE Trans. on Information Theory*, Vol. IT-17, No. 3, May 1971.

[41] A. R. Hammons Jr. and P. V. Kumar, "On a Recent 4-Phase Design for CDMA," *IEICE Trans. Communications*, Vol. E76-B, No. 8, August 1993.

[42] A. R. Hammons et. al. "On the Apparent Duality of the Kerdock and Preparata Codes," *IEEE Int. Symposium on Information Theory*, San Antonio, TX, January 1993.

[43] I. Oppermann and B. S. Vucetic, "Complex Pseudo Random Spreading Sequences with a Wide Range of Correlation Properties," *Globecom 94*, San Francisco, CA, 1994.

[44] D. C. Chu, "Polyphase Codes with Good Periodic Correlation Properties," *IEEE Trans. on Information Theory*, July 1972.

Problems

1. (a) Show that A^{-1}, the inverse of A, is equivalent to running the SSRG backwards.
   (b) Show that A^{-1}, takes the form, in general, of

$$A^{-1} = \begin{bmatrix} 0 & 1 & 0 & 0 & \dots & 0 \\ 0 & 0 & 1 & 0 & \dots & 0 \\ 0 & 0 & 0 & 1 & \dots & 0 \\ 0 & 0 & 0 & 0 & \dots & 0 \\ \vdots & \vdots & \vdots & \vdots & \vdots & 1 \\ 1 & c_1 & c_2 & c_3 & \dots & c_{n-1} \end{bmatrix} \quad n \times n \tag{P2-1}$$

2. Consider the shift register shown in the figure. Using the results of Problem 1 determine the shift register state one and two shifts previous to the current state.

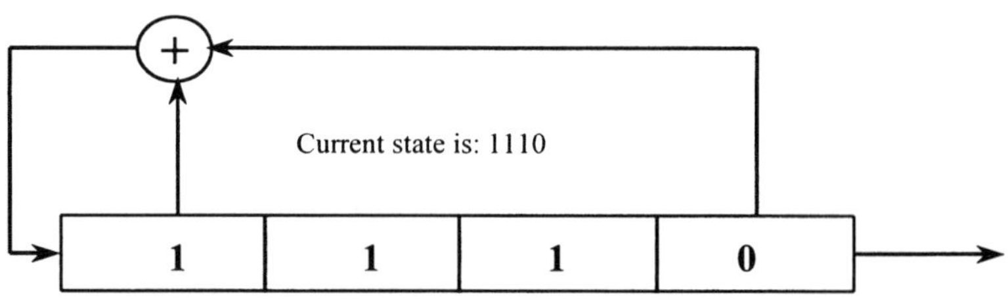

Figure for problem 2. Showing the shift register under study.

3. Determine the mathematical function that allows one to determine the k-th symbol in the output sequence. Hint: Consider the appropriate derivative of the generating function.

4. The SSRG shown in the figure has characteristic polynomial $f(x) = x^5+x^4+1$. Show that there are four different periods possible, with the actual periods of lengths 1, 3, 7, and 21. Hint: Consider direct enumeration of the sequence with different initial conditions.

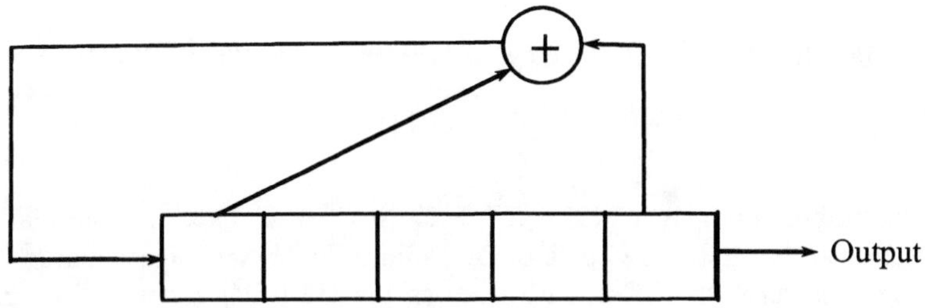

Characteristic polynomial: $f(x) = 1 + x + x^5$

Figure for Problem 4. Showing the shift register under study.

5. Show that the following is true for periodic sequences:

$$\sum_{l=0}^{N-1} \theta_{ab}(l) = \sum_{i=0}^{N-1} a_i \left(\sum_{k=0}^{N-1} b_k \right) \tag{P2-2}$$

6. Show that an irreducible polynomial must have an odd number of terms. Hint: First note that every polynomial of even order can be factored by at least $(x+1)$.

7. Consider the shift register sequence generator for Example 8, which has a characteristic polynomial given by

$$f(x) = 1 + x + x^3 \tag{P2-3}$$

(a) Find the reciprocal polynomial and (b) show that the shift register associated with the reciprocal polynomial generates the reverse sequence.

8. Show that the average of $H_{XY}(k)$, denoted by $\overline{H}(X,Y)$, satisfies the following

$$\overline{H}(X,Y) = \frac{1}{N}\sum_{k=0}^{N-1} H_{XY}(k) = 1 \tag{P2-4}$$

9. Show that for polynomials over GF(2) that for any i greater than 0 that

$$[f(x)]^{2^i} = f(x^{2^i}) \tag{P2-5}$$

10. Construct a balanced Gold code set of length $L=2^{10}-1$ that has 32 codes. What is the maximum cross correlation magnitude between the codes in this set?

11. Design 32 Gold codes of length 19. Use the following codes

$$\begin{aligned} &\boldsymbol{a}\ \ 2105575\ \ 1+x^2+x^3+x^4+x^5+x^6+x^8+x^9+x^{11}+x^{15}+x^{19} \\ &\boldsymbol{b}\ \ 2000605\ \ 1+x^2+x^7+x^8+x^{19} \end{aligned} \tag{P2-6}$$

Show that $G(x)$ is given by

$$G(x) = \frac{1 + x^2 + x^8}{1 + x^2 + x^7 + x^8 + x^{19}} \quad \text{(P2-7)}$$

By performing the appropriate division, show that the initial conditions for register ***b*** are given by

$$[1100000001010110011] \quad \text{(P2-8)}$$

corresponding to the polynomial $1 + x + x^9 + x^{11} + x^{13} + x^{14} + x^{17} + x^{18}$. Note that the only constraint on the "***a***" register is that the first entry be zero. One solution is to set stage 1 and k+2 through n to zero, while the remaining 5 be set with different ***b*** values so that 32 (=2^5) codes will be generated. The final generator is shown in the following figure.

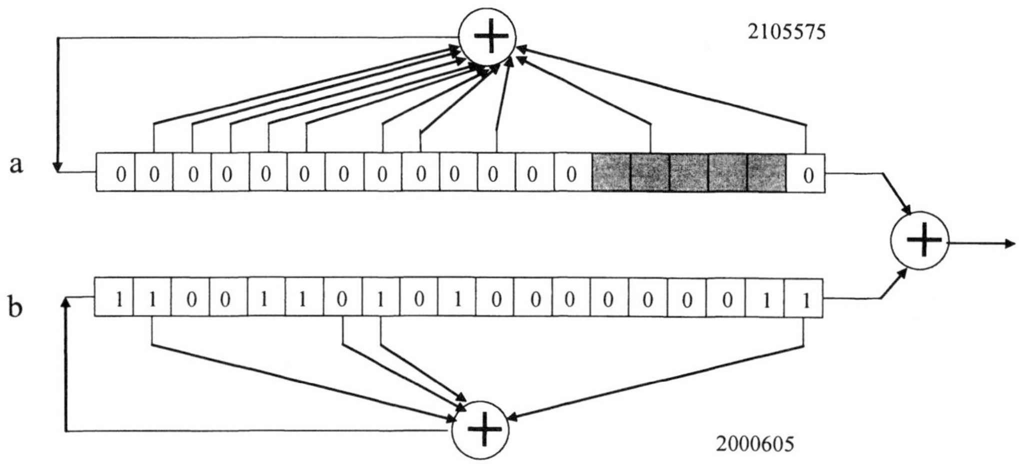

Figure for Problem 11. A generation of balanced Gold codes of period 2^{19}–1.

12. Construct a Kasami small set code generator of shift register length 6 (n = 6). Starting with the primitive polynomial 1 103F determine the proper decimation of the original shift register. Using the appropriate tables conclude that the polynomial 9 015 is the correct one. Sketch the code generator and state the code period of each component code. What are the correlation values?

Table 2A-1 Primitive Polynomials of Degree $n \leq 40$ or Less (and 61 and 89)

Degree	Octal Representation (f_0 on the Right and f_n on the Left)
2	[7]*
3	[13]*
4	[23]*
5	[45]*,[75],[67]
6	[103],[147],[155]
7	[211]*,[217],[235],[367],[277],[325],[203]*,[313],[345]
8	[435],[551],[747],[453],[545],[537],[703],[543]
9	[1021]*,[1131],[1461],[1423],[1055],[1167],[1541],[1333],[1605],[1751]
	[1743],[1617],[1553],[1157]
10	[2011]*,[2415],[3771],[2157],[3515],[2773],[2033],[2443],[2461],[3023]
	[3543],[2745],[2431],[3177],[3525],[2617]
11	[4005]*,[4445],[4215],[4055],[6015],[7413],[4143],[4563],[4053],[5023],[5623]
	[4577],[6233],[6673],[7237],[7335],[4505],[5337]
12	[10123],[15647],[16533],[16047],[11015],[14127],[17673],[13565],[15341]
	[15053],[15621],[15321],[11417],[13505],[13275],[11471],[16237]
13	[20033],[23261],[24623],[23517],[30741],[21643],[30171],[21277],[27777]
	[35051],[34723],[34047],[32535],[31425],[37505],[36515],[26077]
14	[42103],[43333],[51761],[40503],[77141],[62677],[44103],[45145],[76303]
	[64457],[57231],[64167],[60153],[55753],[70423],[47153],[53255],[41753]
15	[100003]*,[102043],[110013],[102067],[104307],[100317],[177775],[103451]
	[110075],[102061],[114725],[103251],[163005],[112611],[120265],[100021]*
16	[210013],[234313],[233303],[307107],[307527],[306357],[201735],[272201]
	[242413],[270155],[302157],[210205],[305667],[236107],[271055],[313371]
17	[400011]*,[400017],[400431],[525251],[410117],[40073],[411335],[600013]
	[403555],[525327],[411077],[404525],[401523],[466273][400041]*,[400101]*
18	[1000201]*,[1002241],[1703601],[1025711],[1115701],[1002241],[1002441]
	[1100045],[1000407],[1003011],[1020121],[1101005],[1000077],[1001361]
	[1001567],[1001727],[1002777]
19	[2000047],[2020471],[2013211],[2570103],[2561427],[2227023],[2001711]
	[2331067],[3146455],[3610353],[2766447],[2000641],[2001441],[2000107]
	[2000077],[2000157],[2000175]
20	[4000011]*,[4001051],[4004515],[6000031],[4442235]
21	[10000005]*,[10040205],[10020045],[10040315],[10000635],[10103075]
	[10050335],[10002135],[17000075]
22	[20000005]*,[20001043],[2222223],[25200127],[20401207],[20430607]
	[20070217]
23	[40000041]*,[40404041],[40000063],[40010061],[50000241],[40220151]
	[40006341],[40405463],[40103271],[41224445],[40435651]
24	[100000207],[125245661],[113763063]
25	[200000011]*,[200000017],[204000051],[200010031],[200402017]
	[252001251],[201014171],[200010031],[200402017],[252001251],[201014171]
	[204204057],[200005535],[200014731]
26	[400000107],[],[430216473],[402365755],[426225667]
	[510664323],[473167545],[411335571]
27	[1000000047],[1001007071],[1020024171],[1102210617]
	[1250025757],[1257242631],[1020560103],[1037530241]
28	[2000000011]*,[2104210431],[2000025051],[2020006031]
	[2002502115],[2001601071]
29	[4000000005]*,[4004004005],[4000010205],[4010000045]
	[4400000045],[4002200115],[4001040115],[4004204435]

	[4100060435],[4040003075],[4004064275]
30	[10040000007],[10104264207],[10115,131333],[11362212703]
	[10343244533]
31	[2000,0000,011]*,[2000,0000,017],[2000,0020,411],[2104,2104,211]
	[2001,0010,017],[2000,5000,251],[2000,4100,071],[2020,2040,217]
	[2000,0200,435],[2006,0140,231],[2104,2107,357]
32	[4002,0000,007],[4046,0216,667],[4003,5532,523],[42003247,143]
	[4176,0427,607]
33	[1000,0002,0001]*,[1000,2002,4001],[1040,0042,0001]
	[1000,2022,4401],[1111,0002,1111],[1000,0003,1463]
	[1040,2046,6001],[1005,0243,0041],[1006,0143,1001]
34	[2010,0000,0007],[2014,7202,4107],[3770,0000,7527]
	[2252,1343,3257],[2277,1224,0037],[2511,3251,6577]
	[2116,3622,0473]
35	[4000,0000,0005]*
36	[1000,0000,04001]*
37	[2000,0000,12005]
38	[4000,0000,00143]
39	[1000,0000,0000,21]*
40	[2000,0012,0000,05]
61	[2000,0000,0000,0000,0004,7]
89	[4000,0000,0000,0000,0000,0000,0001,51]

Entries that have an asterisk denote shift registers that have only two feedback connections. The table is based on [11, 13, 16].

CHAPTER 3

Jamming Performance of Uncoded Spread Spectrum Systems

3.0 INTRODUCTION

In this chapter the effects of jamming on bit error rate (BER) performance will be developed for various modulations and jammer types. A spread spectrum system is designed to minimize the effects of an intentional jammer, whose goal is to disrupt the communication between the spread spectrum transmitter and the spread spectrum receiver. A spread spectrum system is also designed to minimize the effects of unintentional interferences on the reception of the signal. The intent of this chapter is to provide results and a methodology for predicting the performance of spread spectrum (SS) systems with various data modulations, as well as various spread spectrum modulations in the face of various types of jammers or interferences.

All analyses in this chapter assume that no intermodulation products will develop and no harmonics are generated. In other words the receiver is assumed to be ideal. Carrier loop effects, code tracking effects, and bit synchronization effects are not considered in this chapter. These effects require additional refinements in the theory and are not treated here. This chapter does not address coding and interleaving, which improve the performance of the spread spectrum communication system under a jamming environment. Coding and interleaving will be addressed in Chapter 4.

It will be seen that the intelligent jammer will be able to cause serious degradation to the bit error rate performance. In fact, the intelligent jammer will change the bit error rate that is typically roughly negative exponential, in signal-to-noise ratio, to an inverse relationship in the signal-to-noise ratio, which requires much more power to overcome the jammer for a given bit error rate. When coding is considered in Chapter 4, it will be seen that the use of coding and interleaving recovers a good part of the lost performance compared to the case when coding and interleaving are not used.

Figure 3.0-1 illustrates the scenario that is applicable in a spread spectrum system. The spread spectrum transmitter sends a signal with transmit power P_t to the spread spectrum receiver with the intention of communicating, and the jammer, knowing the location of the receiver, tries to prevent the receiver from detecting the data by transmitting a jamming signal of power J_t to the receiver. At the receiver the jammer to signal power ratio is J/P, which may be different than J_t/P_t. This is true since the ranges will be different in general, and typically the jammer will be closer and therefore have a range advantage. To complicate things even further, fading may occur on one or both channels so that the fading losses may differ between links also.

It is useful to describe a general spread spectrum transmitter and receiver model before discussing the various types of jammer signals that may be employed by an adversary intent on jamming the communication link. Figure 3.0-2 illustrates a general transmitter and receiver, including coding, interleaving, and spread spectrum modulation, along with data modulation at the transmitter and the corresponding inverse functions at the receiver.

The data source is assumed to produce statistically independent, equally likely bits that take on the value of 0 or 1. The encoder maps the data bits into an Q-ary alphabet that takes on the values 0, 1, …, Q–1. In some applications these Q-ary symbols are mapped into another binary symbol stream taking on values 0 and 1. When the channel input $\boldsymbol{x}$ and channel output $\boldsymbol{y}$ (the input to the channel interleaver and the output from the channel deinterleaver) have a discrete number of states and there is no memory in the channel, then the channel is known as a *discrete memoryless channel* (DMC).

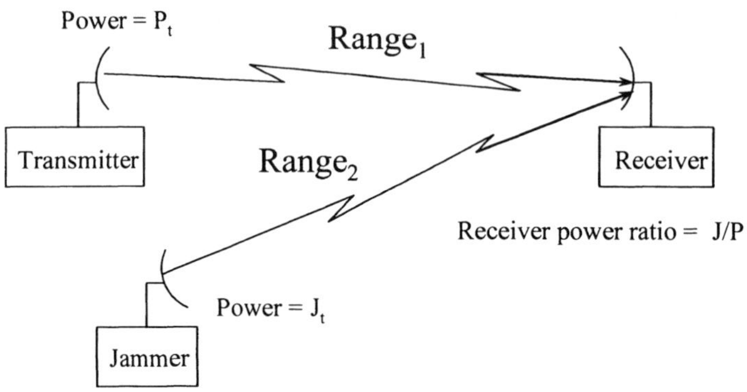

Figure 3.0-1 The transmitter, receiver, and jammer scenario.

The encoded symbols are sent through the interleaver, which randomizes the bit stream in such a way that at the receiver when a burst of errors enters the deinterleaver, the errors become randomized in time at the output. There is a delay due to the interleaving and deinterleaving process; however, the bit error rate performance in jamming is greatly improved by the deinterleaving process. After interleaving, the encoded symbols are carrier modulated, and then spread by the spreading modulator that is driven by a spreading code. Spreading the code creates the spread spectrum process. The signal is then transmitted to the receiver.

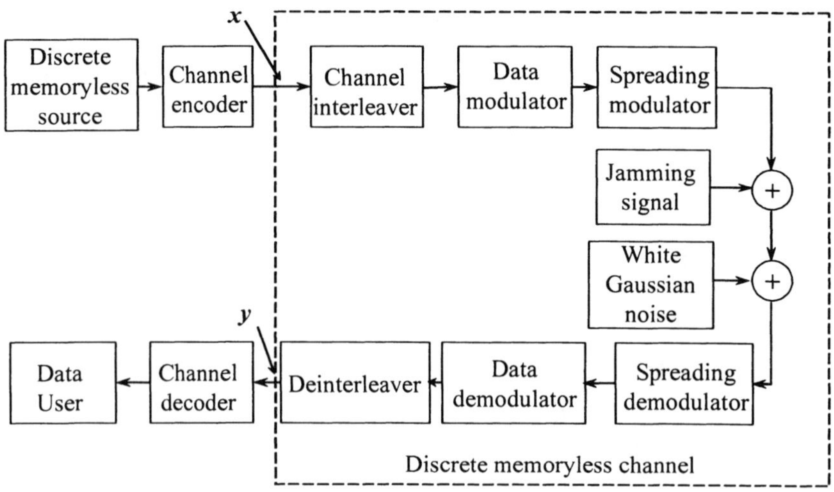

Figure 3.0-2 A model of a spread spectrum communications system.

At the receiver the signal is received along with a possible jamming signal and, of course, the ever-present receiver noise. The signal, jammer, and noise are despread and the data is demodulated. After demodulation the data is deinterleaved and then the data symbols are decoded. The actual performance depends on the actual coding scheme, the modulation, interleaver type and size, jammer type, and jammer to signal power ratio. Bit error rate performance will be presented in this chapter in the case that coding and interleaving are not employed.

3.1 JAMMER TYPES

Numerous jammer signals are postulated for performance evaluation in this chapter. Some of these jammers are realizable and some are optimal in the sense of causing the maximum degradation to the bit error rate performance.

The ever-present thermal noise competes with the signal, even without jamming. We model the thermal noise to have one-sided noise spectral density of N_0 W/Hz.

Perhaps the least efficient jammer in the sense of causing the least degradation to system performance is *wideband* or *barrage jamming*. This jammer transmits band-limited Gaussian noise, having one-sided noise spectral density of N_{0J} W/Hz over the spread spectrum bandwidth W, as shown in Figure 3.1-1(a). In addition to the jamming noise, thermal noise is also present, but often the jamming noise is much larger, so that in these cases the thermal noise can be neglected.

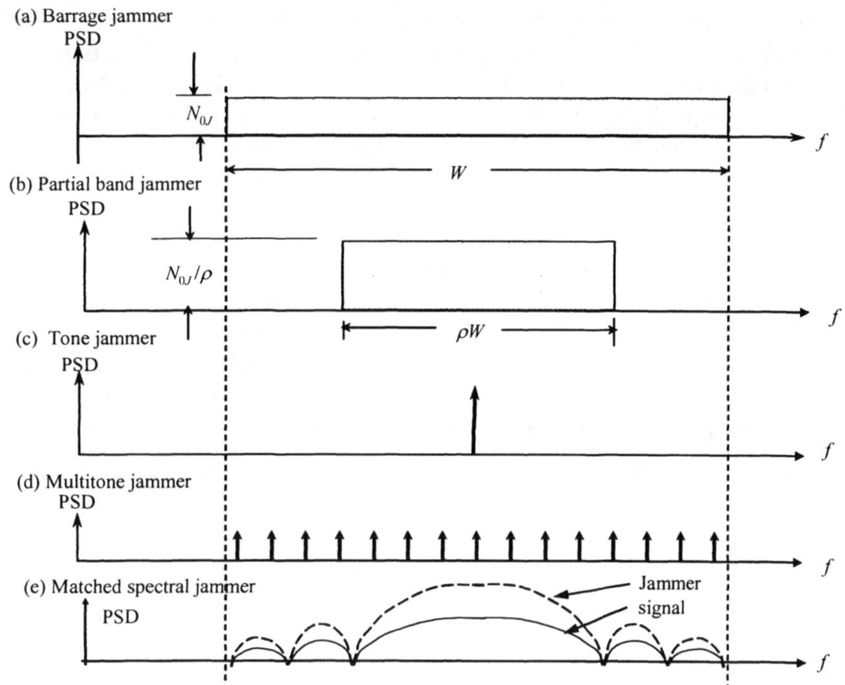

Figure 3.1-1 (a–e) Five jammer spectral densities types.

For a jammer power of J watts, spread over a bandwidth W Hz, the jammer spectral density is

$$N_{0J} = J / W \tag{3.1-1}$$

The second jammer shown in Figure 3.1-1(b) is the *partial band jammer*, which usually picks the bandwidth to maximize the bit error rate. The fractional portion of the bandwidth W that contains the partial band noise jammer is denoted by ρ. The jammer spectral density, for the same total power J, is given by

$$N_{0J}^{\rho} = J / (\rho W) = N_{0J} / \rho \tag{3.1-2}$$

over ρW Hz bandwidth, and is zero over the other $(1-\rho)W$ Hz bandwidth. Therefore the total spread bandwidth is W, and J is the total jammer power.

A third type of jammer is the *tone jammer* that has power J watts and it is illustrated in Figure 3.1-1(c). Usually a single tone jammer is most effective at the frequency (or frequencies) at which the direct sequence spectral density is largest and therefore is the best location for direct sequence systems for BER degradation. For frequency hopped systems the placement of the jammer frequency at the frequency (or frequencies) at which the spectral density is largest is also the best location; however, there are typically many frequencies where the maximum occurs in frequency-hopped systems.

The fourth jammer signal is illustrated in Figure 3.1-1(d) and is the *multitone jammer* with total power J watts. Typically the tones are equally spaced across the hopped bandwidth and the number of tones used in the jamming operation depends on the jammer-to-signal ratio. For the case that there are N jammers the power in each tone is

$$J_i = J / N \quad i = 1,\ 2,\ \ldots N \tag{3.1-3}$$

This multitone jammer is typically more effective in the jamming process than just one tone jammer, especially in light of the fact that as a countermeasure, a receiver could employ a narrowband filter that could remove a single tone jammer with only a small degradation in the received signal.

The fifth jammer type is called a *matched spectral jammer*. It is characterized by the fact that the jammer is commonly a Gaussian random process with the jammer spectral density "matched" to the signal power spectral density. This jammer is difficult to remove because the power spectral density is the same as the signal.

The sixth type of jammer that can occur by direct design or due to interference effects is a pulsed jammer. A *pulsed noise jammer* is a band-limited noise jammer signal that is pulsed on and off, normally in a periodic manner, with a duty factor of ρ. That is, the jammer is on ρ fraction of the time and is off $1-\rho$ fraction of the time. The corresponding one-sided jammer noise spectral density, when the jamming signal is on, is $J/(\rho W)$ and is zero otherwise. Normally, the jammer signal spectral density, when in the on-mode, is set to be equal to W Hz, which is the spread bandwidth. If the duty factor is ρ (the jammer is on ρ fraction of the time), then the jammer voltage, during the time that the pulsed noise jammer is on, is $1/\sqrt{\rho}$ and is zero when it is off. It is assumed that in this type of jammer, for convenience of the analysis, the average power is fixed but the peak power is not; however, in practice the peak power will have limits also.

In the analyses to follow, the peak power will not have a constraint, in order to simplify the calculations. In this analysis the action of the automatic gain control (AGC) will be ignored, even though it can have a significant effect on real systems. For example, in frequency-hopped systems if one hop is hit with interference the next hop may be partially affected due to the AGC response to the previous "hit." However, in the interest of using reasonably simplified models, the effect of the AGC or any other loop, such as a bit synchronizer or phased locked loop, will not be considered here.

A seventh type of jammer, not shown in Figure 3.1-1, is called the *repeater jammer*, or the *follower jammer*. A repeater jammer is a jammer that intercepts the signal, processes it (typically amplifying it), and then retransmits it toward the receiver at the same center frequency. In the case of a high chip rate direct sequence spread spectrum system, it is unlikely that the jammer can be close enough to receive, amplify, and retransmit the received signal within one code chip time. The jammer must be strong enough to capture the receiver. However, if the repeated jamming signal is sufficiently strong, it is possible to cause the receiver to lock on to an incorrect code phase of the jammer, then capture of the receiver is affected.

Once capture of the receiver has been accomplished, a slow frequency change, for example, of the jammer can cause the receiver to lose lock and force it to reacquire. For frequency hopped systems the repeater jammer attempts to have the jammer signal arrive with a delay that is small compared to the hop time. If the data modulation is FSK or MFSK then a carrier frequency offset can capture the receiver. The higher the hop rate, the closer the jammer has to be to the receiver to be effective in this manner. If the delay of the repeater frequency-hopped jammer is more than one hop, it would not be effective in a communication mode. However, in frequency-hopping acquisition, it could possibly be effective if the signal was missed, but the jammer occurring after the signal, and being stronger, was detected.

In addition to the seven intentional jammers described earlier there are three other types of interferences that the receiver has to tolerate. The first form of interference is due to other signals operating in the proximity of the desired signal. Many times in the crowded frequency spectrum most systems have to survive with some form of interference from other friendly signals that cause interference to the desired signal, whether it is a spread spectrum signal or not. There are international and national agreements as to what frequency band or bands a signal can operate in. And each new system must apply for approval in order not to cause significant degradation to already existing signals.

The second form of interference is what is sometimes called *CDMA noise*, and this is the interference caused to the desired signal by the other code division signals (CDMA) sharing the same frequency band. Normally they are designed to not cause significant degradation to the other users in the same CDMA channel. These systems go to great lengths to maintain the individual signal levels, so as to provide minimum interference to the other signals in the system.

A third type of interference to spread spectrum systems is *multipath interference fading*. Fading comes about from the fact that due to reflections there is more than one path between the transmitter and the receiver. These waves combine in the vector sense at the receiver antenna to produce a resultant signal that can vary widely in amplitude and phase, depending on the amplitude and phases of the individual component signals.

When the channel has a constant gain and linear phase over a bandwidth that is at least as great as the signal bandwidth, then *flat fading* occurs. When the channel has a constant gain and linear phase over a bandwidth that is much less than the signal bandwidth, then *frequency selective fading* occurs. Both types of interference can cause considerable degradation to the bit error rate performance when compared to the non-fading case.

3.2 BIT ERROR RATE PERFORMANCE IN BROADBAND NOISE JAMMING

In this section barrage jamming (broadband noise jamming) will be considered for various spread spectrum modulation schemes.

The following signal naming convention will be followed. The spread spectrum modulation type will be given first, followed by a forward slash, and then the data modulation type. For example, a direct sequence spread spectrum (SS) modulation used with binary phase shift keying data modulation would be denoted by DS/BPSK. The coherent direct sequence modulation DS can be (a) BPSK, (b) QPSK, (c) OQPSK, or (d) MSK. In what follows in this chapter DS will denote (a), (b), or (c). MSK will be denoted explicitly, by DS(MSK). Noncoherent frequency hopping can be either (a) slow frequency hopping, which will have an S prior to the FH designation (SFH) or (b) fast frequency hopping, which will have an F in front of FH (FFH). Hybrid modulations will be dashed between the SS modulation, with DS coming first. As an example, consider the SS modulation DS-SFH/MFSK. This would indicate a direct sequence using either (a), (b), or (c) SS modulation with a slow frequency-hopped SS modulation, used in conjunction with multiple frequency shift keying data modulation. The data modulation will be denoted by PSK to mean BPSK, QPSK, or OQPSK.

3.2.1 DS/PSK in Broadband Noise Jamming

In this section direct sequence (DS) phase modulation spread spectrum systems will be considered for BPSK, QPSK or OQPSK chip modulation and BPSK, QPSK, and OQPSK data modulation. Consider a phase modulated, direct sequence spread spectrum modulation system which, when received at the receiver, is of the form

$$y(t) = \sqrt{2P}\cos[\omega_0 t + \theta_d(t) + \theta_{PN}(t) + \theta] + n(t) + n_J(t) \qquad (3.2\text{-}1)$$

where P is the received signal power, ω_0 is the carrier angular frequency in radians/second, $\theta_d(t)$ is the binary data phase modulation in radians, $\theta_{PN}(t)$ is the chip phase modulation (two or four phase) in radians, and θ is the carrier phase in radians, which is assumed to be unknown but constant with time. The terms $n(t)$ and $n_J(t)$ are the thermal noise and the jammer noise processes, both which are assumed to be limited to W Hz bandwidth, centered at the carrier frequency f_0 (Hz) ($f_0 = \omega_o/(2\pi)$), as shown in Figure 3.2-1(a). The figure illustrates the sum of the noise and jammer spectral densities along with the simplified receiver model. The total two-sided noise spectral density level is represented by $(N_0'/2) = (N_{0J} + N_0)/2$ which includes the jammer and thermal noise power spectral densities, respectively.

The receiver model is shown in Figure 3.2-1(b). The signal is received at the antenna and multiplied by a local reference frequency that has the spreading modulation on it and is offset from the carrier by the IF frequency. After the despreading operation the data is left on the signal, which is denoted by $u(t)$, but is now centered at the IF frequency. There is also a sum term, which will be filtered out by the bandpass filter following $u(t)$. The bandpass filter removes the sum frequency and leaves just the IF frequency component, which is called $v(t)$. It is assumed that the bandpass filter negligibly affects the data. Finally the data is coherently detected and demodulated, with the output being the data estimate. The reference signal, assuming that the receiver code is synchronized with the received code, is

$$r(t) = 2\cos[(\omega_0 - \omega_{IF})t + \theta_{PN}(t)] \tag{3.2-2}$$

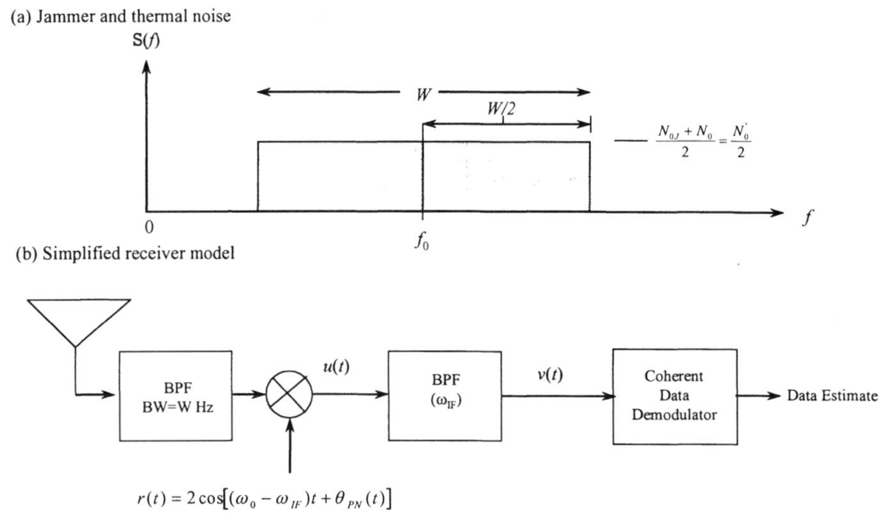

Figure 3.2-1 (a, b) Jammer and thermal noise interference and the receiver model.

After multiplying incoming signal by the reference signal, one obtains

$$\begin{aligned} u(t) = &\sqrt{2P}\cos[\omega_{IF}t + \theta_d(t) + \theta_{PN}(t) + \theta] \\ &+ 2(n_J(t) + n(t))\cos[(\omega_0 - \omega_{IF})t + \theta_{PN}(t)] + O(2\omega_0 - \omega_{IF}) \end{aligned} \tag{3.2-3}$$

where $O(x)$ denotes spectral terms that are centered at frequency x. The last term in (3.2-3) will be neglected from now on since the bandpass filter, which follows the despreader, will remove it. Since the reference signal and noise terms are statistically independent, the reference signal times the jammer plus thermal noise term can be written as a convolution of the spread reference signal power spectral density and the band-limited jammer plus thermal noise spectral density. For BPSK, QPSK, or OQPSK chip spreading, the two-sided spread reference signal spectral density of

$$2\cos[(\omega_0 - \omega_{IF})t + \theta_{PN}(t)] \tag{3.2-4}$$

is given by

$$S_r(f) = T_c \operatorname{sinc}^2\left[(f - f_0 + f_{IF})T_c\right] + T_c \operatorname{sinc}^2\left[(f + f_0 - f_{IF})T_c\right] \tag{3.2-5}$$

with T_c being the chip duration of the three types of chip modulations.

The jammer and thermal noise spectral density are shown in Figure 3.2-1 for the positive frequency and there is the corresponding term at $-f_0$. The interference term, which is the second term in (3.2-3), is the product of the spread signal and the noise plus jammer. The power spectral density of the product is given by the convolution of the two continuous spectral densities $S_r(f)$ (the reference signal spectral density) and $S_J(f)$ (the thermal noise plus jammer noise spectral density). The result is given by

$$S_{rJ}(f) = \int_{-\infty}^{\infty} S_r(f - f')S_J(f')df' \tag{3.2-6}$$

Therefore the power spectral density of the product of the spread reference signal and the jammer plus thermal noise, which has a two-sided spectral density of $N_0'/2$, is given by

$$\begin{aligned} S_{rJ}(f) = \frac{N_0'T_c}{2}\int_{-f_o-W/2}^{-f_o+W/2} &\{\operatorname{sinc}^2\left[(f - f' - f_0 + f_{IF})T_c\right] \\ &+\operatorname{sinc}^2\left[(f - f' + f_o - f_{IF})T_c\right]\}df' \\ +\frac{N_0'T_c}{2}\int_{f_o-W/2}^{f_o+W/2} &\{\operatorname{sinc}^2\left[(f - f' - f_o + f_{IF})T_c\right] \\ &+\operatorname{sinc}^2\left[(f - f' + f_o - f_{IF})T_c\right]\}df' \end{aligned} \tag{3.2-7}$$

In (3.2-7) the second and third terms are essentially zero, since those integrands are centered at around $2f_0-f_{IF}$ and $-2f_0+f_{IF}$, respectively. Thus we obtain

$$\begin{aligned} S_{rJ}(f) &= \frac{N_0'T_c}{2}\int_{-f_o-W/2}^{-f_o+W/2} \operatorname{sinc}^2\left[(f - f' - f_0 + f_{IF})T_c\right]df' \\ &+\frac{N_0'T_c}{2}\int_{f_o-W/2}^{f_o+W/2} \operatorname{sinc}^2\left[(f - f' + f_o - f_{IF})T_c\right]df' \end{aligned} \tag{3.2-8}$$

Letting $x = f - f' - f_0 + f_{IF}$ and $y = f - f' + f_0 - f_{IF}$ in the two integrals of (3.2-8) leads to

$$S_{rJ}(f) = \frac{N_0'T_c}{2}\int_{f+W/2+f_{IF}}^{f-W/2+f_{IF}} \operatorname{sinc}^2\left[xT_c\right](-dx) + \frac{N_0'T_c}{2}\int_{f+W/2-f_{IF}}^{f-W/2-f_{IF}} \operatorname{sinc}^2\left[yT_c\right](-dy) \tag{3.2-9}$$

Interchanging the upper and lower limits produces

$$S_{rJ}(f) = \frac{N_0'T_c}{2}\int_{f-W/2+f_{IF}}^{f+W/2+f_{IF}} \operatorname{sinc}^2\left[xT_c\right]dx + \frac{N_0'T_c}{2}\int_{f-W/2-f_{IF}}^{f+W/2-f_{IF}} \operatorname{sinc}^2\left[yT_c\right]dy \tag{3.2-10}$$

It follows from (3.2-10) that the two-sided power spectral density at f_{IF} and $-f_{IF}$ is given by

$$\begin{aligned} S_{rJ}(f_{IF}) &\cong \frac{N_0'T_c}{2}\int_{-W/2}^{W/2} \operatorname{sinc}^2\left[xT_c\right]dx \\ S_{rJ}(-f_{IF}) &\cong \frac{N_0'T_c}{2}\int_{-W/2}^{W/2} \operatorname{sinc}^2\left[yT_c\right]dy \end{aligned} \tag{3.2-11}$$

Letting $z = \pi x T_c$ produces the two-sided noise spectral density at both f_{IF} and $-f_{IF}$

$$S_{rJ}(f_{IF}) = S_{rJ}(-f_{IF}) \cong \frac{N_0'}{2\pi}\int_{-\pi WT_c/2}^{\pi WT_c/2}\frac{\sin^2[z]}{z^2}dz = \alpha\frac{N_0'}{2} \tag{3.2-12}$$

where α is the reduction in the two-sided noise spectral density from $N_0'/2$ due to the despreading operation. Evaluating the integral produces the values of α indicated in Table 3.2-1.

Table 3.2-1 Parameter α Versus the WT_c for BPSK, QPSK, and OQPSK Chip Modulation

WT_c	0.5	1	2	4	6	8	10
$\alpha(WT_c)$	0.467	0.773	0.903	0.95	0.966	0.974	0.979

This table is equivalent to the curve in Figure 1.4-10 in Chapter 1, with $B = W$. Note that when WT_c is 2, which corresponds to the null to null bandwidth, then α is 0.903 ($L = -0.44$ dB), and when WT_c is 4, which corresponds to the second null to second null bandwidth, then α is 0.95 ($L = -0.22$ dB). In the limit, as W becomes very large, α equals 1 ($L = 0$ dB). It should be noted that the effective two-sided noise spectral density has been determined at the IF; however, the spectral density is very flat around the IF, and therefore it is assumed to be the same value over the modulation bandwidth. Now the signal into the BPSK or QPSK data demodulator, from (3.2-6), is given by

$$v(t) = \sqrt{2P}\cos(\omega_{IF}t + \theta_d(t) + \theta) + n_1'(t) \tag{3.2-13}$$

in which $n_1'(t)$ is modeled as a band-limited white jammer plus thermal noise spectral density having two sided spectral density $\alpha N_0'/2$, where α is evaluated in Table 3.2-1. It follows that the bit error rate for BPSK or QPSK data modulation with BPSK or QPSK spread spectrum modulation is given by

$$PE_b = Q\left[\sqrt{\frac{2E_b}{\alpha(N_0 + N_{0J})}}\right] \tag{3.2-14}$$

where the denominator includes the thermal and jammer noise spectral densities. The function $Q(x)$ is defined by

$$Q(x) = \int_x^{\infty}\frac{1}{\sqrt{2\pi}}e^{-y^2/2}dy \tag{3.2-15}$$

Equation (3.2-14) can be rewritten as

$$PE_b = Q\left[\sqrt{\frac{2}{\dfrac{\alpha N_0 R_b}{P} + \dfrac{\alpha J}{P}\dfrac{R_b}{W}}}\right] \tag{3.2-16}$$

Equation (3.2-16) is plotted in Figure 3.2-2 for various values of E_b/N_0 for $\alpha = 1$, as a function of the power-bandwidth ratio (or jammer E_b/N_{0J}) of

$$\frac{P}{J}\frac{W}{R_b} = \frac{PR_b}{(J/W)} = \frac{E_b}{N_{0J}} \tag{3.2-17}$$

which is the common measure used in comparing spread spectrum systems with jamming, in terms of the signal to jammer ratio times the jammer bandwidth to data rate ratio. Note that the DS modulation could be either BPSK or QPSK. To allow for $\alpha < 1$, add $10\times\log(\alpha)$ to the abscissa value to obtain the BER performance for the case $\alpha < 1$ on the following BER plots.

Example 1 Assume a DS/PSK modulation system, in which broadband noise jamming jams the link. Let the jamming bandwidth be 20 MHz, the data rate be 7,943 bps, N_0 = –202 dBW/Hz, and the received signal power be given by $P = -150$ dBW. The question is what is the required jamming power needed to drive the detected bit error rate up to 10^{-2}? First compute the value of E_b/N_0. It follows that

$$E_b / N_0 = -150 + 202 - 10\log(7,943) = 13 \text{ dB}$$

Now compute the value of $(PW)/(JR_b)$ needed for the jammer to force the bit error rate to 10^{-2}. From Figure 3.2-2, it follows that this value must be 5.5 dB. Solving for J from $PW/(JR_b)$ produces

$$-150 - 39 + 73 - 5.5 = J = -121.5 \text{ dBW}$$

Therefore it follows that the required value of received jammer power needed to impose a bit error rate of 10^{-2} is –121.5 dBW. Of course the actual transmitted jammer power depends on the link losses, including the range and the antenna gains of the jammer transmitter and the receiver (victim) antenna gain.

In Figure 3.2-2 it is noted that the curves flatten out as the jammer E_b/N_{0J} ratio is increased, since eventually the thermal noise becomes the limiting factor. Thus, for the same chip rate ($1/T_c$), the performance for BPSK, QPSK, and OQPSK, for BPSK data modulation is the same.

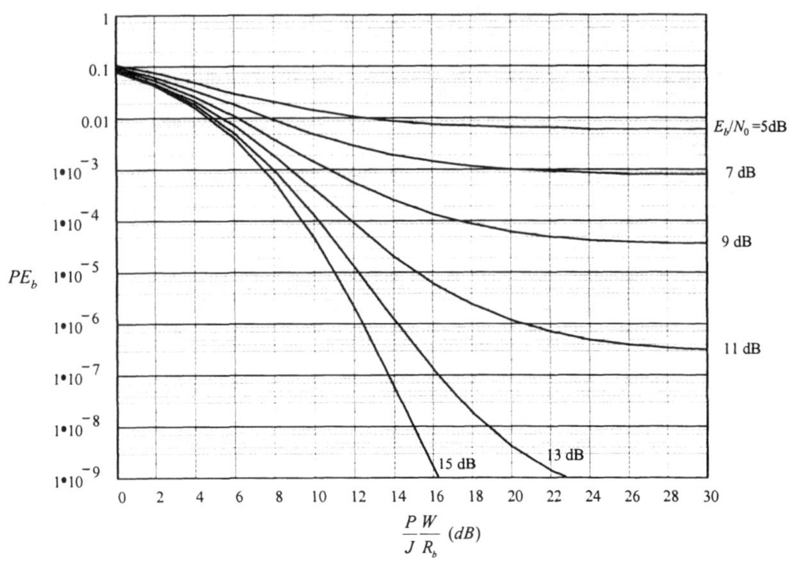

Figure 3.2-2 BER for DS/PSK, SFH/PSK, and DS-SFH/PSK with barrage jamming.

3.2.2 SFH/DPSK in Broadband Noise Jamming

Consider a slow frequency hopped (SFH) spread spectrum system with binary differential phase shift keyed data modulation. Figure 3.2-3 illustrates the receiver model in part (a) and a block diagram of the optimum DPSK demodulator used in part (b). The received signal is dehopped and bandpass filtered to remove the sum frequency term (the bandpass filter (bandwidth B) is assumed sufficiently wide to pass all the hopping frequencies) and fed to the DPSK demodulator. The optimum demodulator heterodynes the signal down to baseband components and integrates them over a bit time (symbol time if it is coded) and then processes the integrated values, which are hard limited to produce the data estimate.

The signal can be written from Chapter 1, Section 1.5.3, as

$$x(t)=\sqrt{2}A\sum_k p_{T_h}(t-kT_h)\sum_j d_j p_{T_d}(t-jT_d)\cos\left[\left(2\pi f_k+\omega_0\right)t+\theta_k\right] \tag{3.2-18}$$

in which A is the rms amplitude, T_h is the hop duration, and $p_{T_h}(t)$ is a unit amplitude pulse that starts at time 0 and lasts for T_h seconds and is zero elsewhere. Likewise d_j is a random variable representing the random differential data and $p_{T_d}(t)$ is a pulse of duration T_d seconds and takes the value of one when it is on (for T_d seconds) and the value of zero when it is off. The hopping frequencies are f_k, and ω_0 is the carrier angular frequency, and θ_k is the random phase for each hop that is constant over each hop. Since, during one hop, the binary DPSK signal is the same as the unhopped binary DPSK, the performance is well known [1–3] for a white Gaussian noise channel.

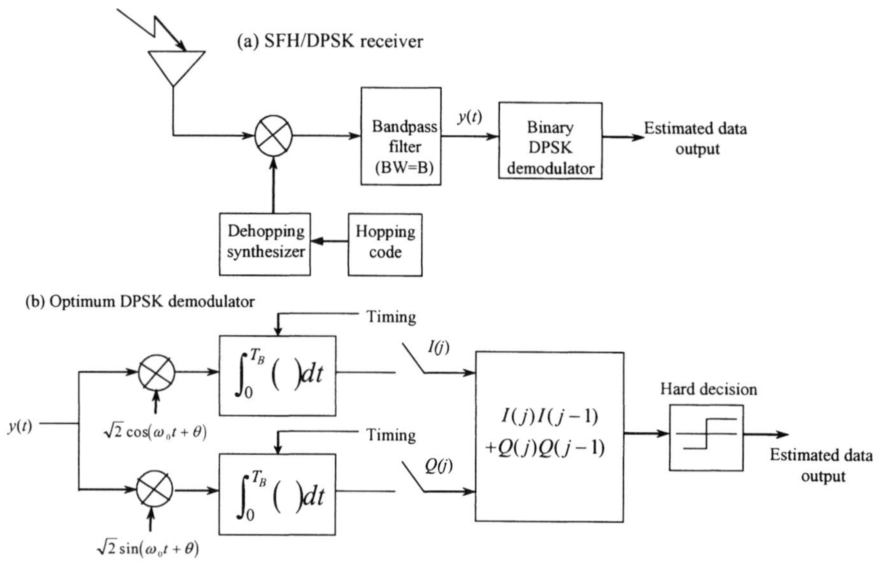

Figure 3.2-3 (a, b) Model for a SFH/DPSK receiver.

The bit error probability result is given by

$$PE_b=\frac{1}{2}\exp\left(-\frac{E_b}{N_0}\right) \tag{3.2-19}$$

When barrage noise is present, the thermal noise spectral density N_0 is replaced in (3.2-19) with the effective noise spectral density $N_0^{'}=N_0+N_{0J}$, which includes both the thermal and jammer noise. Following the argument that led to (3.2-16), in which E_b was divided by the sum $N_0^{'}=N_0+N_{0J}$, (3.2-19) can be written in the final form for the BER

$$PE_b = \frac{1}{2}\exp\left[\frac{-1}{\dfrac{N_0 R_b}{P} + \dfrac{J}{P}\dfrac{R_b}{W}}\right] \quad (3.2\text{-}20)$$

This result only applies to the optimal DPSK detector. The *suboptimal detector*, in which the dehopped signal is bandpass filtered and then the product of itself and a one bit delayed version is multiplied together and detected, produces a result that is degraded from (3.2-19) as discussed in Park [4].

The processing gain for frequency-hopped systems was defined in Section 1.5 to be

$$PG = \frac{W}{W_b} \cong \frac{W}{2R_b} \quad (3.2\text{-}21)$$

Thus it is seen for SFH/DPSK SS modulation for large jammer levels that the BER is exponentially dependent on negative product of the signal to jammer ratio and the twice the processing gain. Notice that the effect of the processing gain is to reduce the effect of the jammer on the BER performance. This results in (3.2-20) are for the optimal DPSK detector jammed by barrage noise. Figure 3.2-4 illustrates the bit error rate performance of the SFH/DPSK SS modulation scheme under consideration for various values of E_b/N_0, for the optimum DPSK demodulator.

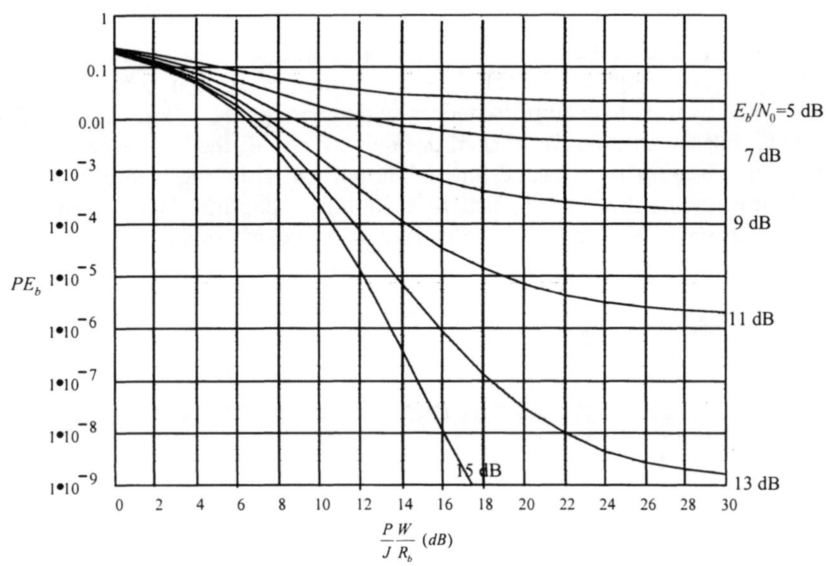

Figure 3.2-4 BER for SFH/DPSK and DS-SFH/DPSK with barrage jamming.

It is seen by comparing Figure 3.2-2 with Figure 3.2-4 that the performance of SFH/DPSK is not quite as good as DS/PSK, as would be expected, since carrier phase information is not utilized in DPSK demodulation.

Example 2 Consider the following spread spectrum link with SFH/DPSK modulation suffering barrage jamming. Suppose it is necessary to achieve a bit error rate of 10^{-7}. Determine what data rate will provide a bit error rate of 10^{-7}. The following conditions apply. The $P/N_0 = 35$ dB$-$Hz, $J/P = 20$ dB, and the jamming bandwidth is 300 kHz wide. Note that the following equations can be written for the signal-to-noise ratio and the effective signal-to-jamming ratio

$$E_b / N_0(\text{dB}) = 35 - 10\log(R_b)$$

and

$$\frac{PW}{JR_b}(\text{dB}) = -20 + 10\log(300{,}000) - 10\log(R_b)$$

By trial and error it is determined that about R_b = 100 bits per second yields a BER of 10^{-7}. Thus, under these jamming conditions and with this modulation type, the data rate must be restricted to 100 bps in order to achieve the low desired BER.

3.2.3 SFH/PSK in Broadband Noise Jamming

Another modulation scheme that has seen use in slow frequency hopped systems is SFH/PSK, in which a frequency-hopped SS signal has PSK data modulated on the hopped carrier. BPSK modulation, in conjunction with frequency hopping, is currently used on the crosslink of GPS, for example. Figure 3.2-5 illustrates the SFH/PSK receiver and the demodulator. The receiver shown in either figure first dehops the hopped signal and then bandpass filters the signal to remove the sum frequency. The bandwidth of the bandpass filter is assumed large enough to pass the hopping signal. This signal, denoted as $y(t)$, is fed into a BPSK, QPSK, or OQPSK demodulator to produce the estimated data. Since the phase after each hop is essentially random, carrier synchronization must be accomplished for each hop. Bit timing must also be maintained from hop to hop.

There are two basic approaches to obtain phase and time coherence with the received waveform. The first method is illustrated in Figure 3.2-5(a). The first portion of the hop, which contains only time and phase synchronization bits, is used to acquire frequency and carrier phase and bit timing. The remaining bits of the hop contain data. Phase and time are tracked through the rest of the hop. Since carrier phase changes randomly from hop to hop it must be estimated for each hop. Symbol timing is usually maintained from hop to hop since normally time is not affected significantly from hop to hop for moderate bit rates. This approach is illustrated in Figure 3.2-5(a). This is an example of an active tracking scheme. The current GPS cross-link utilizes this approach.

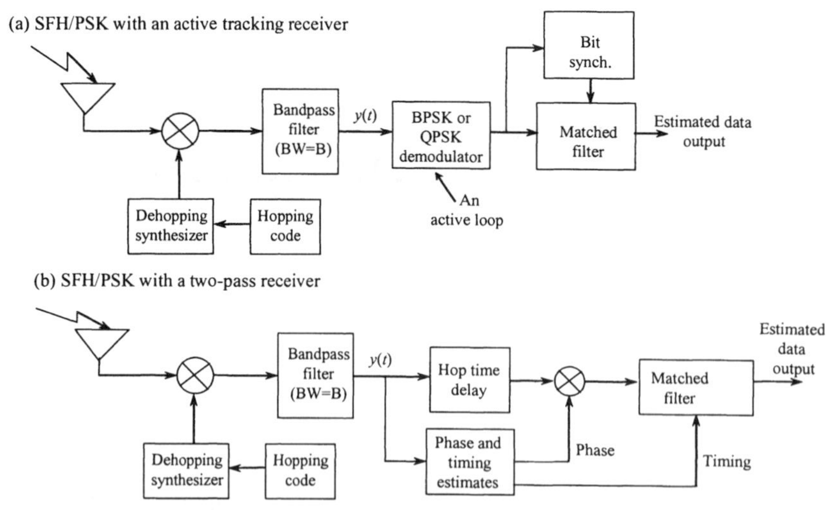

Figure 3.2-5 (a, b) SFH/PSK spread spectrum receiver model.

The second method shown in Figure 3.2-5(b) is to use one pass through the hop to estimate carrier phase and frequency and symbol timing from a sync word or words, and then using a one-hop delayed version of the signal with the just-obtained carrier phase estimate and timing, and demodulate the signal and provide the symbol timing for the symbol matched filter. The delay is needed since this is a two-pass scheme in which the first pass is used for phase and timing estimation, and the second pass is used for the demodulation.

Assuming perfect timing and phase estimation, the BER for coherent BPSK, QPSK, and OQPSK modulation is given by

$$PE_b = Q\left[\sqrt{\frac{2E_b}{N_o}}\right] \tag{3.2-22}$$

For broadband jamming, N_0 is replaced with $N_0^{'} = N_0 + N_{0J}$ in the form

$$\frac{E_b}{N_0^{'}} = \frac{1}{\dfrac{N_0 R_b}{P} + \dfrac{J}{P}\dfrac{R_b}{W}} \tag{3.2-23}$$

for the case that both thermal and jammer noise are present. Note that $\alpha = 1$ in this case (3.2-16) since it is assumed that the bandwidth W is large enough to pass both the signal modulation plus the hopping range. Hence using (3.2-23) for the effective signal to noise ratio in (3.2-22) produces the following expression for the bit error rate for SFH/PSK, assuming perfect phase and timing when thermal noise

$$PE_b = Q\left[\sqrt{\frac{2}{\dfrac{N_0 R_b}{P} + \dfrac{JR_b}{PW}}}\right] \tag{3.2-24}$$

and a wideband jammer are present at the input of the receiver. In (3.2-24) N_0 is the one-sided thermal noise spectral density, R_b is the bit rate in bits per second, J is the received jammer power in watts, P is the received signal power in watts, and W is the hopping bandwidth in Hz, which is assumed to be sufficiently wide to pass the hopped modulated signal. Figure 3.2-2 illustrates the results for SFH/PSK and also DS/PSK when both thermal and broadband jammer noise are present at the input of the receiver.

3.2.4 SFH/MFSK in Broadband Noise Jamming

In this section slow frequency (SFH) spread spectrum modulation, in conjunction with multiple frequency shift keyed information modulation (SFH/MFSK), will be considered with thermal noise and wideband jamming noise. It will be assumed that the jammer is spread over the hopping bandwidth W Hz. Slow frequency hopping will be considered in this section, and fast frequency hopping will be considered in the next section. Recall from Chapter 1 slow frequency hopping occurs when there are two or more symbols per frequency hop time. It will be assumed that orthogonal signaling is used for the MFSK frequency modulation; this means that the frequencies of the data modulation are spaced $1/T_d$ Hz apart, or integer multiples of $1/T_d$ Hz apart, with T_d being the data symbol duration.

The received SFH/MFSK signal, with slow hopping, is represented as

$$y(t) = \sqrt{2P}\sum_{i=-\infty}^{\infty} p_{T_h}(t - iT_h)\cos\left[\omega_i t + \theta_d(t) + \varphi_i\right] + n(t) + n_J(t) \tag{3.2-25}$$

where P is the received signal power, $p_{T_h}(t)$ is a unit amplitude rectangular pulse that starts at $t = 0$ and ends at $t = T_h$ seconds, corresponding to the hop time duration T_h seconds, ω_i is one of the hopping frequencies, which changes every T_h seconds, taken from the set of all the hopping frequencies, which covers the spread bandwidth W. This representation does not explicitly illustrate the MFSK frequency changes in contrast to (3.2-18). It is an alternative expression, however. The phase modulation $\theta_d(t)$ represents the MFSK data modulation, expressed as a phase process that changes every T_d seconds to one of M data modulation frequencies. It is assumed that T_h/T_d is an integer >1. The random phase φ_i represents the carrier phase over each hop, which changes randomly every hop time to a value between zero and 2π radians. The noise process $n(t)$ is modeled as a bandlimited white Gaussian noise process, and $n_J(t)$ is the barrage jammer noise process, which is also modeled as a band-limited white Gaussian noise process; both processes occupy W Hz bandwidth. Consequently, at any given time, the hopping frequency is assumed to be the sum of an unhopped carrier frequency, plus the particular hop frequency occurring at this particular time, plus the particular data modulation frequency occurring at this time.

The receiver and the demodulator are shown in Figure 3.2-6. This illustrates the dehopping synthesizer along with a bandpass filter, which has bandwidth B Hz and is sufficiently wide to pass all the hopped band and eliminate the sum frequency term, leaving only the dehopped MFSK signal. The MFSK signal is fed into the MFSK demodulator, which is illustrated in the part (b) of the figure. The demodulator shown in Figure 3.2-6(b) is based on an I-Q baseband approach to demodulation, although the same function can be accomplished at IF and yields essentially identical performance [3]. The output of the demodulator is one of the M possible choices, or $\log_2(M)$ bits. For example, if $M = 8$, then each data tone carries 3 bits of information.

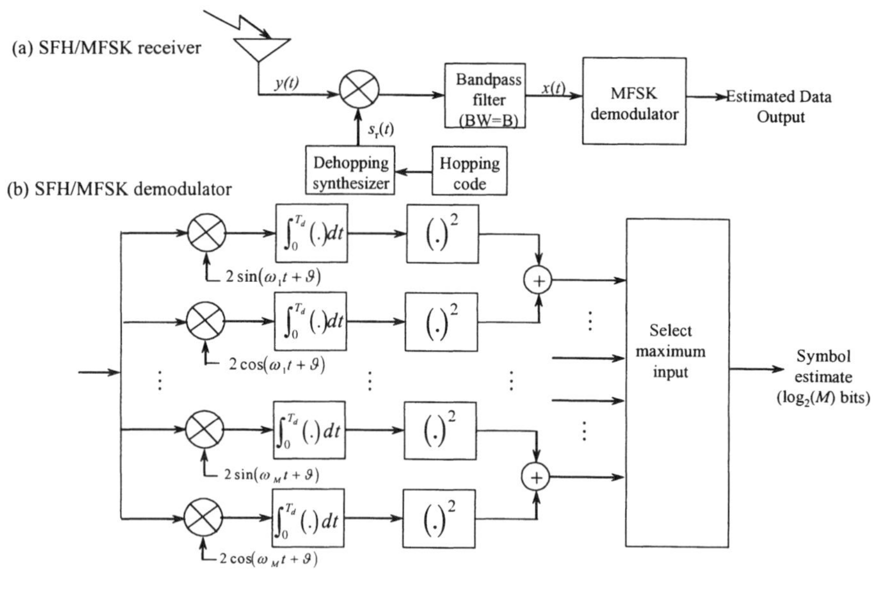

Figure 3.2-6 (a, b) Model of the SFH/MFSK spread spectrum receiver with slow hopping.

The reference signal $s_r(t)$, which provides the dehopping function, is modeled as

$$s_r(t) = 2\sum_{j=-\infty}^{\infty} p_{T_h}(t - jT_h)\cos[(\omega_j - \omega_{IF})t + \psi_j] \tag{3.2-26}$$

where ω_{IF} is the intermediate frequency and ψ_j is the random phase (within 0 to 2π radians) that occurs on every frequency hop of the receiver synthesizer. The result of the synchronized dehopping, after the broadband bandpass filter removes the sum frequency term, yields the following signal in the dehopped band

$$x(t)=\sqrt{2P}\sum_{i=-\infty}^{\infty}p_{T_h}(t-iT_h)\cos\left[\omega_{IF}t+(\varphi_i-\psi_i)+\theta_d(t)\right]+$$
$$(n(t)+n_J(t))\left\{2\sum_{i=-\infty}^{\infty}p_{T_h}(t-iT_h)\cos\left[(\omega_i-\omega_{IF})t+\psi_i\right]\right\} \tag{3.2-27}$$

The first term is the dehopped data modulation located at the IF with a random phase at each hop $(\varphi_i-\psi_i)$ and the second term is the thermal and jammer noise multiplied by the reference signal, which results in the jammer and thermal noise being located at the IF. Over each modulation frequency hop time, the jammer plus thermal noise is still white with effective one-sided noise power spectral density of $N_0' = N_0 + N_{0J}$ W/Hz. In other words, the band-limited signal and noise during one MFSK modulation tone time (after removing the sum frequency terms) is of the form

$$x(t)=\sqrt{2P}\cos\left[\omega_{IF}t+\theta_d(t)+\xi\right]+n'(t)+n_J'(t) \tag{3.2-28}$$

where ξ is the phase difference of φ and ψ at that hop, which is assumed to be a uniform random variable over 0 to 2π, and the primes on the two noise terms denote the frequency shifted versions of the unprimed noise terms. Arthurs and Dym [5] have solved the average MFSK tone symbol error rate, with the resulting expression[1]

$$PE_s=\frac{1}{M}\exp\left[\frac{-E_s}{2N_0'}\right]\sum_{n=2}^{M}\binom{M}{n}(-1)^n\exp\left[\frac{E_S(2-n)}{2nN_0'}\right] \tag{3.2-29}$$

where E_s is the tone (symbol) energy, orthogonal tone spacing is assumed as mentioned earlier, and N_0' is the effective noise plus jammer spectral density. Viterbi [6] has shown that when $k = \log_2(M)$ bits are encoded into M orthogonal signals, the mean number of bits in error when a k-bit coded word has been detected incorrectly is given by

$$\frac{\sum_{i=1}^{k}i\binom{k}{i}}{\sum_{i=1}^{k}\binom{k}{i}}=\frac{k2^{k-1}}{2^k-1}=\frac{kM}{2(M-1)} \tag{3.2-30}$$

Therefore, the mean bit error rate is related to the mean symbol error rate by

$$PE_b=\left(\frac{1}{k}\right)\frac{k2^{k-1}}{2^k-1}PE_s=\frac{M}{2(M-1)}PE_s \tag{3.2-31}$$

Consequently, the bit error rate can be expressed in terms of M and E_b/N_0' as

$$PE_b=\frac{1}{2(M-1)}\exp\left[\frac{-kE_b}{2N_0'}\right]\sum_{n=2}^{M}\binom{M}{n}(-1)^n\exp\left[\frac{kE_b(2-n)}{2nN_0'}\right] \tag{3.2-32}$$

where $E_s = kE_b$ has been used in (3.2-32), which is our final result for SFH/MFSK with slow hopping and barrage jamming. It should be noted that in the case of BFSK, (M=2), (3.2-32) becomes

[1] The authors use a two-sided noise spectral density of N_0 W/Hz, whereas we are using a two-sided noise spectral density of $N_0/2$. The conversion to make the results consistent with our notation has been made here.

$$PE_b = \frac{1}{2}\exp\left[\frac{-E_b}{2N_0'}\right] \tag{3.2-33}$$

All that is left to do now is relate the effective E_b / N_0' to the jamming and thermal noise components. As before (see (3.2-24)) one can write

$$\frac{E_b}{N_0'} = \frac{1}{\dfrac{N_0 R_b}{P} + \dfrac{J}{P}\dfrac{R_b}{W}} \tag{3.2-34}$$

Since thermal noise is usually small compared to the jamming noise, it will be neglected in the graph of bit error rate versus the effective E_b/N_0'. Here the effective E_b/N_0' is X where $X = PW/(JR_b)$. Figure 3.2-7 illustrates the final bit error rate result for five different values of M. As can be seen from the plots, the bit error rate decreases as the number of orthogonal signals increases.

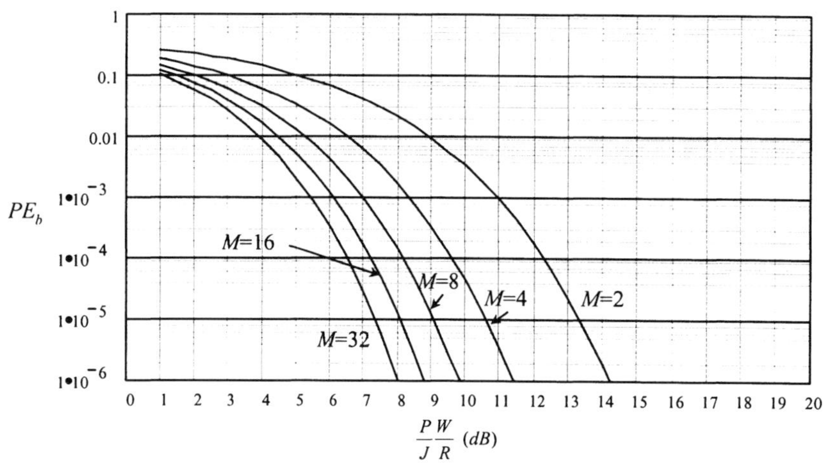

Figure 3.2-7 Bit error rate for SFH/MFSK and DS-SFH/MFSK for various values of M.

Example 3 (a) Consider a link in which slow frequency hop with BFSK (SFH/BFSK) modulation will be used. To achieve a probability of error of 10^{-5}, what effective bit signal-to-noise ratio is needed? (b) If thermal noise is neglected and the received broadband jammer of bandwidth 1 MHz has a received power of −120 dBW in the sidelobes of the receiver antenna, and a bit rate of 1 kbps is needed, what is the required signal power at the receiver input? Consider part (a) of the question first. From (3.2-33) based on numerical evaluation (or reading Figure 3.2-7), one finds that 13.352 dB provides a bit error rate of 10^{-5}. With regard to part (b) (3.2-34) yields

$$13.352 = P\text{dBW} + 10\log(10^6) + 120 - 10 * \log(1000)$$

Solving for P dBW, one obtains the value −136.65 dBW as the required signal power at the antenna output, neglecting the thermal noise.

3.2.5 FFH/BFSK in Broadband Noise Jamming

In this section fast frequency hopping (FFH) with binary frequency shift keying (FFH/BFSK) will be considered, with barrage jamming. The receiver is illustrated in Figure 3.2-8.

As indicated in Chapter 1, fast frequency hopping occurs when the hop rate is greater than or equal to the symbol rate. It will be assumed that there are L chips (or subintervals) per FSK tone. The receiver forms an estimate of the first chip matched filter output for both tone frequencies, adds the square of the matched filter output to the square of the next chip's matched filter output, and continues for each of the M tone frequencies until all L of the chips have been detected for both tone frequencies.

Proakis [2] has analyzed the FFH/BFSK case ($M = 2$) in white Gaussian noise. The bit error rate for the FFH/BFSK case is given by

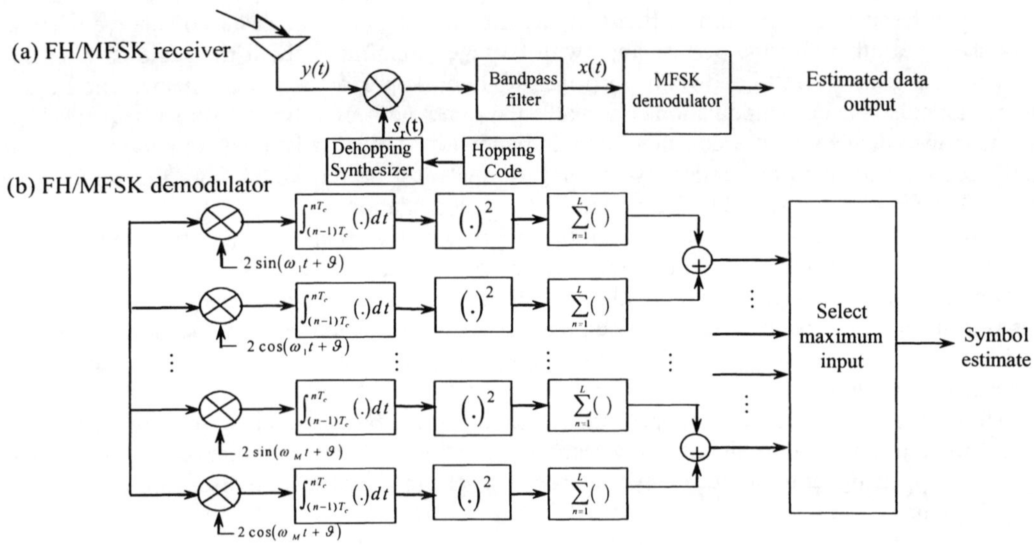

Figure 3.2-8 Model of the FFH/MFSK spread spectrum receiver.

$$PE_b = \frac{1}{2^{2L-1}} e^{-\frac{E_b}{2N_0'}} \sum_{i=0}^{L-1} K_i \left(\frac{E_b}{2N_0'} \right)^i \tag{3.2-35}$$

where the bit SNR is given by

$$\frac{E_b}{N_0'} = \frac{1}{\frac{N_0 R_b}{P} + \frac{J}{P} \frac{R_b}{W}} = L \frac{E_c}{N_0'} \tag{3.2-36}$$

in which E_b / N_0' is the effective bit SNR, E_c / N_0' is the effective chip SNR, and L times the effective chip SNR is equal to the bit SNR. In (3.2-35), K_i is defined by

$$K_i = \frac{1}{(i)! \sum_{j=0}^{L-1-i} \binom{2L-1}{j}} \tag{3.2-37}$$

Note that in the case $L = 1$, PE_b reverts to the simple expression found in (3.2-33)

$$PE_b = \frac{1}{2}\exp\left[\frac{-E_b}{2N_0^{'}}\right] \tag{3.2-38}$$

FFH/BFSK has poorer BER performance in thermal noise than SFH/BFSK but is more robust when jamming occurs.

3.2.6 Hybrid DS-SFH SS Modulation in Broadband Noise Jamming

With barrage jamming the advantage in using a hybrid spread spectrum modulation scheme is very minimal compared to just frequency hopping. It is true that larger bandwidths may be easier to achieve when a hybrid scheme is used, but otherwise the performance with barrage jamming is about the same as with frequency hopping, with the same overall bandwidth. In principle, hybrid spread spectrum schemes can be used with coherent or noncoherent data modulation. Normally the bandwidth of a typical direct sequence spreading process is narrower than a typical frequency-hopping bandwidth. One disadvantage in a hybrid system is that both the direct sequence and the frequency-hopping timing must be recovered from the signal in order to track the DS/FH SS modulation and to demodulate the data modulation.

Figure 3.2-9 illustrates a block diagram of a representative receiver, which first dehops the hopped signal and then utilizes a bandpass filter to remove the sum frequency.

Next the dehopped signal is despread by the direct sequence despreader and is followed by a second bandpass filter to limit the bandwidth of the output, if needed. For example, if the despreading process also contained a heterodyne operation as well as a despreading process, a bandpass filter would be needed to remove the sum frequency.

After the signal has been dehopped and despread, it can be demodulated according to the type of data modulation that was used. In the next few sections, the effect of having a direct sequence signal added to the frequency hopped signal to produce a hybrid spread spectrum signal will be discussed briefly in terms of system performance.

3.2.6.1 DS-SFH/PSK in Broadband Noise Jamming

When a direct sequence spreading code is used in addition to the frequency-hopping process, the resulting signal and noise are essentially statistically the same as in the frequency-hopped–only case. Actually, the direct sequence despreading process lowers the noise spectral density slightly when compared to the frequency-hopped–only case. Section 3.2.1 was developed for BPSK, QPSK, or OQPSK modulation with BPSK, QPSK, or OQPSK data modulation.

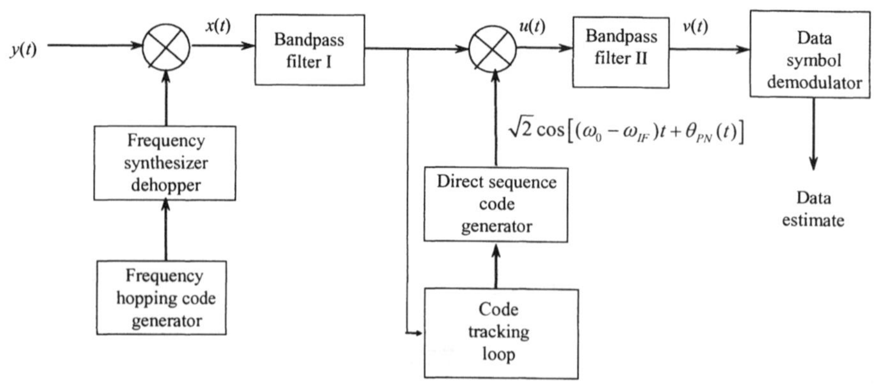

Figure 3.2-9 Block diagram model of a hybrid receiver with DS and FH.

Typically the bandwidth of the hybrid scheme would be considerably wider than just the direct sequence part of the hybrid scheme, so that filtering losses are normally negligibly small. Thus, Figure 3.2-2 applies to DS-SFH/PSK modulation. From (3.2.1-14), since it is assumed that the bandwidth is large enough to pass the FH signal plus the DS signal so that $\alpha = 1$, the BER performance is given by

$$PE_b = Q\left[\sqrt{\frac{2}{\frac{N_0 R_b}{P} + \frac{J}{P}\frac{R_b}{W}}}\right] \tag{3.2-39}$$

where P is the received signal power in watts, R_b is the data rate in bits per second, N_0 is the one-sided thermal noise spectral density in W/Hz, J is the received jammer power in watts, and W is the hybrid spread bandwidth in Hz.

3.2.6.2 DS-SFH/DPSK in Broadband Noise Jamming

When using direct sequence and frequency hopping with DPSK encoded data, the despreading process will decrease the effective noise power spectral density by the factor α that was derived in Section 3.2.1 for direct sequence BPSK spread spectrum signaling. However, since the bandwidth required for hopping is much greater than for the direct sequence signaling (by assumption), it follows that α will be essentially one. Therefore the bit error rate for the optimal detector is given by (3.2-20), which is reproduced in (3.2-40) and the results are plotted in Figure 3.2-4.

$$PE_b = \frac{1}{2}\exp\left[\frac{-1}{\frac{N_0 R_b}{P} + \frac{J}{P}\frac{R_b}{W}}\right] \tag{3.2-40}$$

Again, P is the received signal power in watts, R_b is the data rate in bits per second, N_0 is the one sided thermal noise spectral density in W/Hz, J is the received jammer power in watts, and W is the hybrid spread bandwidth in Hz.

3.2.6.3 DS-SFH/MFSK in Broadband Noise Jamming

When using MFSK and direct sequence spread spectrum modulation with slow frequency hopping, the performance is basically described in Section 3.2.4 for slow hopping, with the exception that the direct sequence despreading process will spread the noise somewhat depending on the hopping bandwidth and the spreading bandwidth.

Basically since it is assumed that the hopping bandwidth is much greater than the direct sequence bandwidth, the noise spreading will be negligible and the bit error rate performance will be given by (3.2-32) for slow hopping

$$PE_b = \frac{1}{2(M-1)}\exp\left[\frac{-kE_b}{2N_0'}\right]\sum_{n=2}^{M}\binom{M}{n}(-1)^n \exp\left[\frac{kE_b(2-n)}{2nN_0'}\right] \tag{3.2-41}$$

and

$$\frac{E_b}{N_0^{'}}=\frac{1}{\dfrac{N_0R_b}{P}+\dfrac{J}{P}\dfrac{R_b}{W}} \tag{3.2-42}$$

where P is the received signal power in watts, R_b is the data rate in bits per second, N_0 is the one sided thermal noise spectral density in W/Hz, J is the received jammer power in watts, and W is the hybrid spread bandwidth in Hz. Figure 3.2-7 illustrates the performance that is applicable to this modulation.

3.2.6.4 DS-FFH/FSK in Broadband Noise Jamming

When using FSK and direct sequence spread spectrum modulation with fast frequency hopping, the performance is basically described by Section 3.2.5 for fast hopping with the exception that the direct sequence despreading process will spread the noise somewhat depending on the hopping bandwidth and the spreading bandwidth. The results for fast hopping are discussed in Section 3.2.5. Basically since it is assumed that the hopping bandwidth is much greater than the direct sequence bandwidth, the noise spreading will be negligible and the bit error rate performance will be given from Section 3.2.5 for fast hopping as

$$PE_b=\frac{1}{2^{2L-1}}e^{-\frac{E_b}{2N_0^{'}}}\sum_{i=0}^{L-1}K_i\left(\frac{E_b}{2N_0^{'}}\right)^i \tag{3.2-43}$$

$$\frac{E_b}{N_0^{'}}=\frac{1}{\dfrac{N_0R_b}{P}+\dfrac{J}{P}\dfrac{R_b}{W}}=L\frac{E_c}{N_0^{'}} \tag{3.2-44}$$

and

$$K_i=\frac{1}{(i)!\sum_{j=0}^{L-1-i}\binom{2L-1}{j}} \tag{3.2-45}$$

In (3.2-44) P is the received signal power in watts, R_b is the data rate in bits per second, N_0 is the one-sided thermal noise spectral density in W/Hz, J is the received jammer power in watts, and W is the hybrid spread bandwidth in Hz.

3.3 BER PERFORMANCE IN PARTIAL BAND NOISE JAMMING

Partial band noise jamming is similar to full band noise jamming, except that the bandwidth is narrower for partial band noise jamming. Most jammer transmitters can produce more narrowband noise jammer power than they can wideband noise jammer power, and commonly partial band jammers are more effective than wideband jammers. Partial band jammers are more commonly used in electronic counter measures (ECM) operations.

3.3.1 DS/PSK in Partial Band Noise Jamming

Now consider a direct sequence spread spectrum communication system that uses DS spreading with either BPSK, QPSK, or OQPSK modulation with either BPSK, QPSK, or OQPSK data modulation in the presence of a partial band noise jammer (PBNJ). A block diagram of the system model is shown in Figure 3.2-1(b). It is assumed that the system-input bandwidth is W Hz, the partial band jammer has bandwidth

$W_\rho = \rho W$ Hz, and the data rate is R_b. In addition the received signal power is P, and the total received jammer power is J. The chip duration is T_c seconds for either BPSK or QPSK SS modulation. The received signal and the received partial band jammer plus thermal noise is given by

$$y(t) = \sqrt{2P}\cos[\omega_0 t + \theta_d(t) + \theta_{PN}(t) + \theta] + n(t) + n_J(t) \tag{3.3-1}$$

The reference signal is given by

$$r(t) = 2\cos[(\omega_0 - \omega_{IF})t + \theta_{PN}(t)] \tag{3.3-2}$$

After synchronously despreading, the resulting waveform is given by

$$\begin{aligned} u(t) = &\sqrt{2P}\cos[\omega_{IF}t + \theta_d(t) + \theta] \\ &+ 2(n_J(t) + n(t))\cos[(\omega_0 - \omega_{IF})t + \theta_{PN}(t)] + O(2\omega_0 - \omega_{IF}) \end{aligned} \tag{3.3-3}$$

After filtering by the bandpass filter one obtains the component at ω_{IF}

$$v(t) = \sqrt{2P}\cos[\omega_{IF}t + \theta_d(t) + \theta] + 2(n_J(t) + n(t))\cos[(\omega_0 - \omega_{IF})t + \theta_{PN}(t)] \tag{3.3-4}$$

The first term is the despread signal, the second term is the despread partial band jammer, and the last term is the despread thermal noise. The two-sided despread spectral density of this thermal noise term was already found in Section 3.2.1 to be given by

$$S(f_{IF}) = \alpha \frac{N_0}{2} \tag{3.3-5}$$

In this equation α turns out to be close to unity if WT_c is greater than three or four (see Table 3.2-1). Consider the despread partial band jammer term. From (3.2-5), one has for the power spectral density of the reference signal

$$S_r(f) = T_c \text{sinc}^2[(f - f_0 + f_{IF})T_c] + T_c \text{sinc}^2[(f + f_0 - f_{IF})T_c] \tag{3.3-6}$$

If the jammer and the despreading codes are uncorrelated, then the two-sided despread jammer spectral density is given by the following convolution

$$S_{JPN}(f) = \int_{-\infty}^{\infty} S_r(f - f')S_J(f')df' \tag{3.3-7}$$

Assume that the jammer bandwidth is equal to ρW Hz and the jammer spectral density is $N_{0J}^{\rho} = J/(\rho W)$ W/Hz, where J is the jammer power and the jammer is centered at frequency f_J, so that the jammer spectral density is centered at f_J and $-f_J$ and the jammer spectral density is given by

$$S_J(f) = \frac{N_{0J}^{\rho}}{2} \text{ for} |f + f_J| \le \frac{\rho W}{2} \text{ and for } |f - f_J| \le \frac{\rho W}{2} \tag{3.3-8}$$

Using (3.3-6) and (3.3-8) in (3.3-7) produces

$$\begin{aligned} S_{JPN}(f) &= \frac{N_{0J}^{\rho}T_c}{2}\int_{f_J-\frac{\rho W}{2}}^{f_J+\frac{\rho W}{2}}\left\{\begin{aligned}&\text{sinc}^2\left[(f-f_0+f_{IF}-f')T_c\right]\\&+\text{sinc}^2\left[(f+f_0-f_{IF}-f')T_c\right]\end{aligned}\right\}df' \\ &+\frac{N_{0J}^{\rho}T_c}{2}\int_{-f_J-\frac{\rho W}{2}}^{-f_J+\frac{\rho W}{2}}\left\{\begin{aligned}&\text{sinc}^2\left[(f-f_0+f_{IF}-f')T_c\right]\\&+\text{sinc}^2\left[(f+f_0-f_{IF}-f')T_c\right]\end{aligned}\right\}df' \end{aligned} \tag{3.3-9}$$

Assume that f_0 and f_J are not more than ρW Hz apart and $f_0 >> 2f_{IF}$. Then for $f = +f_{IF}$ only the second term contributes to the integral and for $f = -f_{IF}$ only the third term contributes to the integral, since the two integration regions correspond to the regions where the two-sided spectral density is nonnegligible. It follows that

$$\begin{aligned} S_{JPN}(f_{IF}) &= \frac{N_{0J}^{\rho}T_c}{2}\int_{f_J-\frac{\rho W}{2}}^{f_J+\frac{\rho W}{2}}\text{sinc}^2\left[(f_0-f')T_c\right]df' \\ S_{JPN}(-f_{IF}) &= \frac{N_{0J}^{\rho}T_c}{2}\int_{-f_J-\frac{\rho W}{2}}^{-f_J+\frac{\rho W}{2}}\text{sinc}^2\left[(-f_0-f')T_c\right]df' \end{aligned} \tag{3.3-10}$$

Clearly, the two integrals are the same so that the two-sided noise spectral density at $\pm f_{IF}$ is given by

$$S_{JPN}(f_{IF}) = S_{JPN}(-f_{IF}) = \frac{N_{0J}^{\rho}T_c}{2}\int_{f_J-\frac{\rho W}{2}}^{f_J+\frac{\rho W}{2}}\text{sinc}^2\left[(f_0-f')T_c\right]df' \tag{3.3-11}$$

This result applies for arbitrary jammer bandwidths ρW, as long as f_0 and f_J do not differ by more than ρW Hz and $f_0 >> 2f_{IF}$. Let us now consider two extreme cases. Consider first the case when the jammer bandwidth ρW is very narrowband compared to the chip rate $1/T_c$. Integrating under this assumption leads to the result that the interference spectral density given by

$$S_{JPN}(f_{IF}) = \frac{N_{0J}^{\rho}T_c\rho W}{2}\left[\text{sinc}(f_J-f_0)T_c\right]^2 = \frac{JT_c}{2}\left[\text{sinc}(f_J-f_0)T_c\right]^2 \tag{3.3-12}$$

Thus the jammer is most effective when the jammer frequency is equal to the signal center frequency, as would be expected, where the result becomes

$$S_{JPN}(f_{IF}) = \frac{JT_c}{2} \tag{3.3-13}$$

in the narrowband case. Note that if the jammer is offset in frequency by multiples of the chip rate R_c, the jammer spectral density is fully suppressed according to our theoretical model! Even a tone jammer will have a nonzero bandwidth, due to phase noise, so that complete suppression is not achieved in practice.

Now consider the case when the jammer is broadband, that is, we let $\rho = 1$. From (3.3-11) one has, after the change of variable $z = \pi T_c(f_0 - f')$ and $f_0 = f_J$, the despread jammer spectral density is given by

$$S_{JPN}(f_{IF}) = \frac{JT_c}{2W}\int_{-\pi(W/2)T_c}^{\pi(W/2)T_c}\frac{\sin^2(z)}{z^2}\frac{dz}{\pi T_c} \tag{3.3-14}$$

This integral can be evaluated exactly; however, a reasonable upper bound can be obtained with the result that

$$S_{JPN}(f_{IF}) \cong \frac{J}{2W} \tag{3.3-15}$$

Now if W is set to be the null-to-null bandwidth then ($W = 2/T_c$) then

$$S_{JPN}(f_{IF}) \cong \frac{JT_c}{4} = \frac{J}{4R_c} \tag{3.3-16}$$

with R_c being the chip rate $R_c = 1/T_c$. Thus, by comparing the narrowband case (3.3-13) with (3.3-16), it is seen that, when the narrowband jammer frequency is set equal to the signal carrier frequency f_0, then the narrow band jammer is approximately 3 dB more effective than a wideband jammer, which has an RF bandwidth of $W = 2/T_c$.

Now in order to easily compute the bit error rate, it is advantageous to assume that the jammer is a Gaussian noise jammer. In that case the bit error rate for the general partial band Gaussian noise jammer is given by

$$PE_b = Q\left[\sqrt{\frac{E_b}{\frac{N_0}{2} + S_{JPN}(f_{IF})}}\right] \tag{3.3-17}$$

where the two-sided thermal noise spectral density is $N_0/2$ and the two-sided jammer spectral density is $S_{JPN}(f_{IF})$. Under the assumption that the two random processes are statistically independent, the earlier result follows in Gaussian noise (see (3.2-14)).

For the vary narrowband Gaussian jammer with the jammer frequency at the signal frequency, one obtains the bit error rate of

$$PE_b = Q\left[\sqrt{\frac{2E_b}{N_0 + JT_c}}\right] = Q\left[\sqrt{\frac{2}{\left(\frac{E_b}{N_0}\right)^{-1} + \left(\frac{P}{J}\frac{R_c}{R_b}\right)^{-1}}}\right] = Q\left[\sqrt{\frac{2}{\left(\frac{E_b}{N_0}\right)^{-1} + \left(\frac{P}{J}PG\right)^{-1}}}\right] \tag{3.3-18}$$

where, for direct sequence spread spectrum systems with BPSK or QPSK chip modulation, PG is the processing gain and is given by

$$PG = \frac{R_c}{R_b} \tag{3.3-19}$$

The greater the processing gain the lower the bit error rate, and vice versa. Figure 3.3-1 illustrates the BER for various values of E_b/N_0 in dB versus the value of (PR_c / JR_b) in dB.

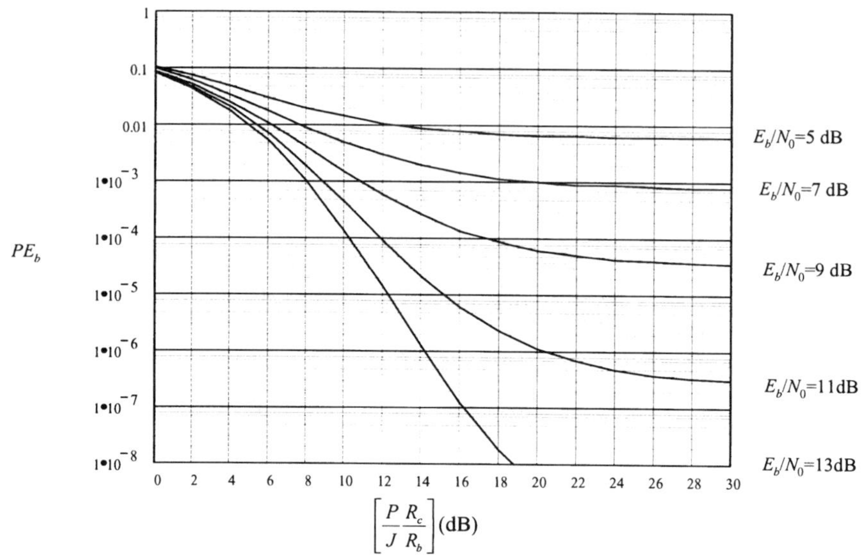

Figure 3.3-1 DS/PSK BER performance with a very narrowband Gaussian jammer.

Example 4 (a) Consider a DS/PSK spread spectrum communication link, which is being jammed by a partial band jammer. Assume the jammer utilizes the optimum bandwidth. Also assume that the data rate is 7,943 bps, N_0 = –202 dBW/Hz, the received signal power is P = –150 dBW, and the direct sequence chip rate is 10,000,000 chips per second. The questions are what jamming power is needed to cause a bit error rate of 10^{-2}? and (b) How does this compare with Example 1 with a broadband jammer? First consider (a). Compute the E_b/N_0 value via

$$E_b / N_0(\text{dB}) = 202 - 150 - 10\log(7943) = 13 \text{ dB}$$

From Figure 3.3-1 it is seen that a value of $PR_c/(JR_b)$ required to achieve a BER of 10^{-2} is about 5 dB. Expressing J in the equation $PR_c/(JR_b)$ in dB produces the relationship

$$5 \text{ dB} = -150 - J + 10 * \log(10{,}000{,}000 / 7{,}943) \text{ dB}$$

which yields the value of J = –124 dBW. Now consider part (b). Note that in this case, comparing to Example 1, that it requires a jammer that is 2.5 dB weaker compared to the wideband case to cause the same error rate.

3.3.2 SFH/DPSK Systems in Partial Band Noise Jamming

Consider now a frequency-hopped spread spectrum system with a hopped bandwidth of W Hz, in which binary differential encoding and decoding is used. Again Figure 3.2-3 illustrates the receiver and the optimum demodulator. Consider a one-sided spread bandwidth of ρW Hz, using J watts of jammer power, to produce a one-sided spectral density of $J/(\rho W)$ W/Hz. It is assumed that ρW is at least as large as $2R_b$, (tone jammers are not considered in this section). Under these conditions, the effective one-sided spectral density of the noise plus jammer is given by

$$S(f_{IF}) = N_0 + \frac{J}{\rho W} \tag{3.3-20}$$

The bit error rate for unjammed DPSK is given in (3.2-19). The hopping frequencies are assumed to be chosen (essentially) at random. If the hopping band is over the bandwidth ρW, then the approximate probability of falling in the jammer region is ρ and the approximate probability of falling outside of the jammer band is $(1-\rho)$. Therefore, from (3.2-19) the average bit error rate probability, in a frequency-hopped system, in which the jammer occupies ρW Hz (averaged over the jammed and unjammed environment), is approximately given by

$$\overline{PE}_b = (1-\rho)PE_b\left(\frac{E_b}{N_0}\right) + \rho PE_b\left(\frac{E_b}{S(f_{IF})}\right) \tag{3.3-21}$$

where $PE_b\left(\frac{E_b}{N_0}\right)$ is given by

$$PE_b\left(\frac{E_b}{N_0}\right) = \frac{1}{2}\exp\left(-\frac{E_b}{N_0}\right) \tag{3.3-22}$$

Hence the average BER can be written as

$$\overline{PE_b} = \frac{1}{2}(1-\rho)\exp\left(\frac{-P}{N_0 R_b}\right) + \frac{1}{2}\rho\exp\left(\frac{-1}{(N_0 R_b / P) + (J/P)(R_b/(\rho W)}\right) \tag{3.3-23}$$

Thus the average bit error rate for frequency-hopped DPSK modulation depends on the bandwidth occupancy ρ, as indicated in (3.3-23). Many times the thermal noise is negligible compared to the jammer in a typical system operation. Thus to reflect the typical case (and to simplify the calculations), it will be assumed that the first term in (3.3-23) is negligible so that the average bit error rate reduces to

$$\overline{PE_b} = \frac{1}{2}\rho\exp\left(\frac{-\rho}{(J/P)(R_b/(W)}\right) \tag{3.3-24}$$

when we neglect the thermal noise. Figure 3.3-2 illustrates the bit error rate for the case of frequency-hopped DPSK in partial band noise jamming.

Now consider the optimum value of ρ the fractional jammer bandwidth and the jammer bandwidth of the hopped bandwidth W. To do this we differentiate the expression for the average error rate (3.3-24) to obtain

$$\frac{\partial \overline{PE_b}}{\partial \rho} = \frac{1}{2}\exp\left[-\frac{\rho PW}{JR_b}\right] + \frac{\rho}{2}\left(-\frac{PW}{JR_b}\right)\exp\left[-\frac{\rho PW}{JR_b}\right] = 0 \tag{3.3-25}$$

Solving for the optimum fractional bandwidth ρ_0 produces

$$\rho_0 = \frac{1}{\left(\frac{PW}{JR_b}\right)} \tag{3.3-26}$$

as long as ρ_0 is not greater than 1. Note from (3.3-26) that ρ, being less than or equal to 1, implies that

$$\frac{PW}{JR_b} \geq 1 \tag{3.3-27}$$

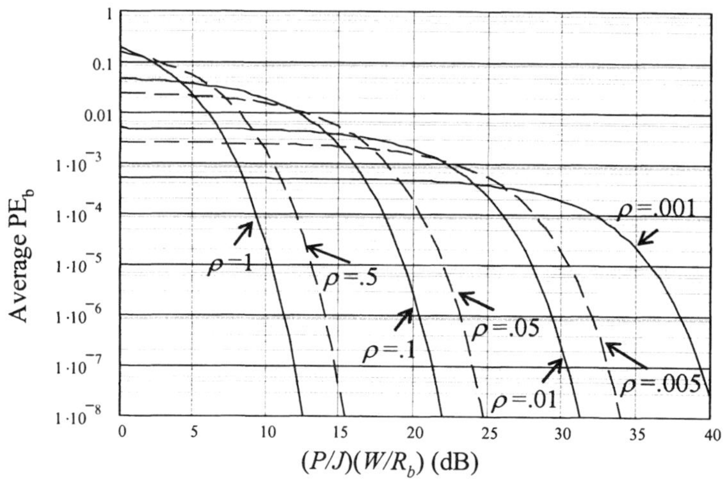

Figure 3.3-2 Average BER performance with a Gaussian partial band jammer and for a pulse jammer with SFH/DPSK. It also applies to DS-DPSK for both jammers.

Since the maximum value of ρ_0 is 1, it must be used whenever the optimum value of ρ_0 is equal to or greater than 1. Therefore, inserting the optimum value of ρ into (3.3-24) produces the average bit error rate of

$$\begin{aligned}
\left(\overline{PE}_b\right)_{\max} &= \frac{e^{-1}}{2\dfrac{P}{J}\dfrac{W}{R_b}} && \frac{P}{J}\frac{W}{R_b} \geq 1 \\
\left(\overline{PE}_b\right)_{\max} &= \frac{1}{2}\exp\left[-\frac{P}{J}\frac{W}{R_b}\right] && \frac{P}{J}\frac{W}{R_b} < 1
\end{aligned} \tag{3.3-28}$$

Hence in the region where the error rate is smallest $((P/J)(W/R_b)>1)$ the dependence on $(P/J)(W/R_b)$ is an inverse relationship and not exponential as one would expect. This behavior is typical of bit error rate performance with optimum jamming. Figure 3.3-2 illustrates the bit error rate for the worst-case (optimum) jammer for various values of ρ. Figure 3.3-3 illustrates the maximum bit error rate for the worst-case (optimum) jammer.

Due to the linear reduction in BER with (PW/JR_b) (rather than an exponential-like decrease), it takes a lot of signal power to overcome an optimal jammer if coding is not used.

3.3.3 SFH/PSK BER in Partial Band Noise Jamming

In this section, SFH/BPSK, SFH/QPSK, and SFH/OQPSK spread spectrum modulations are considered. Consider now a slow frequency-hopping spread spectrum system with hopped bandwidth of W Hz with one of the three data modulation indicated in the first line of this paragraph. Figure 3.2-5 illustrates both the active receiver and the two-pass receiver. Consider a one-sided spread bandwidth of ρW Hz, using J watts of jammer power, to produce a one-sided spectral density of $J/(\rho W)$ W/Hz. It is assumed that ρW is at least as large as $2R_b$; in other words, tone or narrowband jammers are not considered in this section. Under these conditions the effective one-sided spectral density of the noise plus jammer is given by

$$S(f_{IF}) = N_0 + \frac{J}{\rho W} \tag{3.3-29}$$

Since the hopping is assumed to be essentially random, if the hopping band is over the bandwidth ρW, then the probability of falling in the jammer region is ρ, and the probability of falling outside of the jammer band is $(1-\rho)$. Therefore, the average bit error rate [7] (averaged over the jammed and unjammed environment) is given by

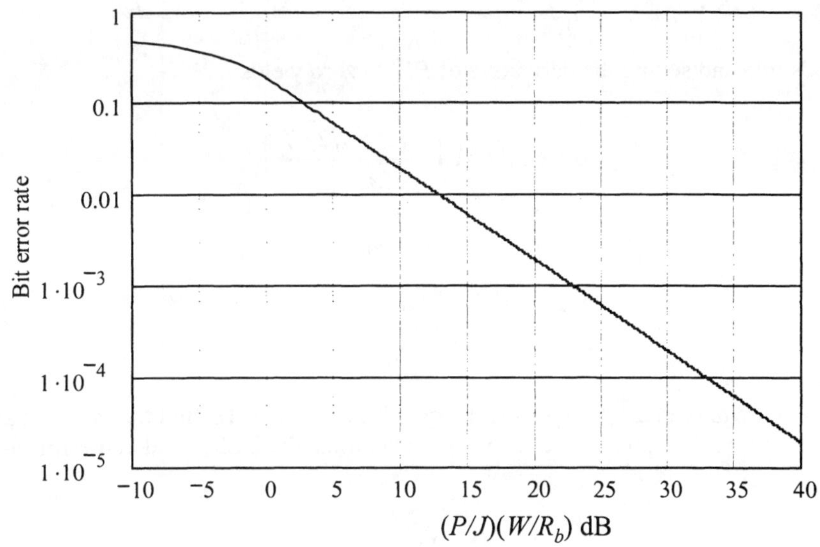

Figure 3.3-3 Average BER performance for the optimum Gaussian partial band jammer and for a pulse jammer with SFH/DPSK. It also applies to DS-SFH/DPSK for both jammers.

$$\overline{PE_b} = (1-\rho)PE_b\left(\frac{E_b}{N_0}\right) + \rho PE_b\left(\frac{E_b}{N_0 + J/(\rho W)}\right) \tag{3.3-30}$$

where $PE_b(E_b/N_0)$ is given by

$$PE_b(E_b/N_0) = Q\left[\sqrt{2E_b/N_0}\right] \tag{3.3-31}$$

Under the assumption that thermal noise is negligible, (3.3-30) can be written as

$$\overline{PE_b} = \rho Q\left(\sqrt{\frac{2\rho PW}{JR_b}}\right) \tag{3.3-32}$$

Hence it follows that

$$\frac{d(\overline{PE}_b)}{d\rho} = Q\left[\sqrt{2\rho X}\right] + \frac{\rho d}{d\rho}\left(Q\left[\sqrt{2\rho X}\right]\right) \tag{3.3-33}$$

where

$$X = \frac{PW}{JR_b} \tag{3.3-34}$$

Recall that Liebnitz's rule [8] states

$$\frac{d}{dx}\int_A^B f(x,t)dt = \int_A^B \frac{\partial f(x,t)}{\partial x}dt + f(x,B)\frac{dB}{dx} - f(x,A)\frac{dA}{dx} \tag{3.3-35}$$

Applying Liebnitz's rule and setting the derivative of PE_b to zero yields

$$0 = Q\left[\sqrt{2\rho X}\right] - \frac{e^{-\rho X}}{\sqrt{2\pi}}\frac{\sqrt{2\rho X}}{2} \tag{3.3-36}$$

From (3.3-36) it follows that the solution for ρ is of the form

$$\rho_0 = \frac{c_0}{X} \tag{3.3-37}$$

which corresponds to the maximum value of the average bit error rate. To find the value of ρ_0, fix X at some value, say 3, and determine the optimum value of c_0 to maximize the BER. Trial and error yields the value of c_0

$$c_0 = 0.709 \tag{3.3-38}$$

Since ρ can not be larger than 1, ρ_0 must be set equal to 1 whenever the optimum value of ρ_0 is 1 or greater. Thus the maximum average BER for SFH/PSK is given by

$$\begin{aligned} \left(\overline{PE}_b\right)_{\max} &= \frac{0.08286}{\left(\dfrac{PW}{JR_b}\right)} & \frac{PW}{JR_b} \geq 0.709 \\ \left(\overline{PE}_b\right)_{\max} &= Q\left[\sqrt{\frac{2PW}{JR_b}}\right] & \frac{PW}{JR_b} < 0.709 \end{aligned} \tag{3.3-39}$$

Figure 3.3-4 illustrates the BER plot for the range of X from 10 dB to 40 dB. Again it is seen that the BER drops off in a linear manner with (PW/JR_b). Notice that SFH/PSK has slightly better BER than SFH/DPSK for a given value of (PW/JR_b).

3.3.4 SFH/MFSK in Partial Band Noise Jamming

Consider now the frequency-hopped M-ary frequency shifted keyed modulation in partial band noise jamming, with the assumption that the hopping is slow (that is, there are one or more tones per hop). The analysis for partial band noise jamming is very similar to the analysis done in Section 3.2.4. Recall that there are k bits per tone symbol.

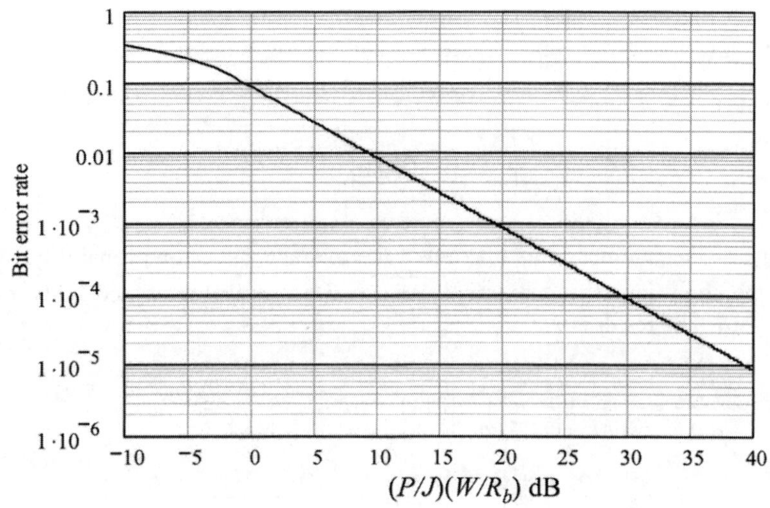

Figure 3.3-4 Average BER performance with a Gaussian optimum partial band jammer for SFH/BPSK.

Equation (3.2-28) is the starting point at one hop time and for one frequency-hop frequency, so that the signal plus noise can be written in the form at

$$x(t) = \sqrt{2P}\cos[\omega_{IF}t + \theta_d(t) + \xi] + n_{nJ}(t) \tag{3.3-40}$$

where the tone modulation is represented by $\theta_d(t)$, ξ is the random phase for that hop, and term $n_{nJ}(t)$ represents the thermal noise and jammer noise when the jammer noise is present.

As in Section 3.3.3, the effective one-sided spectral density of the noise plus jammer is given by

$$S(f_{IF}) = N_0 + \frac{J}{\rho W} \tag{3.3-41}$$

when the jammer hits the hop with probability ρ, it has to contend with the thermal plus jammer noise; however, the receiver has only thermal noise to contend with when the jammer does not hit the hop that the signal is at.

Since the hopping is assumed to be essentially random over W Hz, and the jammer has bandwidth ρW, then the probability of falling in the jammer region is ρ and the probability of falling outside of the jammer band is $(1-\rho)$. Therefore, the average bit error rate [7] (averaged over the jammed and unjammed environment) is given by

$$\overline{PE_b} = (1-\rho)PE_b\left(\frac{E_b}{N_0}\right) + \rho PE_b\left(\frac{E_b}{N_0 + J/(\rho W)}\right) \tag{3.3-42}$$

where $PE_b(E_b/N_0)$ is given by (3.2-32). Again making the assumption that thermal noise is negligible, and denoting the one-sided jammer spectral density J/W by N_{0J}, one has, from (3.2-32),

$$\overline{PE_b} = \frac{\rho}{2(M-1)} \exp\left[\frac{-\rho k E_b}{2N_{0J}}\right] \sum_{n=2}^{M} \binom{M}{n} (-1)^n \exp\left[\frac{k\rho E_b(2-n)}{2nN_{0J}}\right] \tag{3.3-43}$$

where $E_s = kE_b$. Bringing the exponential inside the summation allows us to write the average bit error probability as

$$\overline{PE_b} = \frac{\rho}{2(M-1)} \sum_{n=2}^{M} \binom{M}{n} (-1)^n \exp\left[\frac{k\rho E_b(1-n)}{nN_{0J}}\right] \tag{3.3-44}$$

The strategy of the jammer is to maximize the average bit error probability over the bandwidth fraction ρ. To find the conditions that maximize $\overline{PE_b}$, (3.3-44) is differentiated with respect to ρ and the result is set equal to zero. The result is

$$\begin{aligned} \frac{d\overline{PE_b}}{d\rho} &= \frac{1}{2(M-1)} \sum_{n=2}^{M} \binom{M}{n} (-1)^n \exp\left[\frac{k\rho E_b(1-n)}{nN_{0J}}\right] \\ &+ \frac{\rho}{2(M-1)} \sum_{n=2}^{M} \binom{M}{n} (-1)^n \exp\left[\frac{k\rho E_b(1-n)}{nN_{0J}}\right] \frac{kE_b(1-n)}{nN_{0J}} \end{aligned} \tag{3.3-45}$$

Making the substitutions $y = \rho kX$ and $X = PW/JR_b$, it is not difficult to show that (see also Houston [9]) (3.3-45) can be written as

$$\frac{\displaystyle\sum_{n=2}^{M} \binom{M}{n} \frac{(-1)^n}{n} \exp\left(\frac{y}{n}\right)}{\displaystyle\sum_{n=2}^{M} \binom{M}{n} (-1)^n \exp\left(\frac{y}{n}\right)} = \frac{y-1}{y} \tag{3.3-46}$$

where y and X have been introduced to facilitate the solution. Note again that X is the effective signal to jammer density ratio. Also recall that

$$k = \log_2(M) \tag{3.3-47}$$

The solution for y, call it y_0, can be solved from (3.3-46) to yield the value of ρ that maximizes the average bit error probability given by (3.3-44). It is clear from inspection that when $M = 2$ then $y_0 = 2$. Other values can be computed with the aid of a computer using an iterative solution. Hence when M and X are selected, the corresponding value of k is known, and from y_0 one can obtain the optimum value of ρ, call it ρ_0, and therefore the average bit error rate. Again, when the solution for ρ_0 is 1 or greater, the value of $\rho_0 = 1$ is used since the fractional bandwidth cannot exceed unity. When the optimum value of ρ is less than one, ρ is found from $\rho_0 = y_0/(kX)$. The value of $\rho = 1$ corresponds to the value of X_0 found when $\rho = 1$ from

$$X_0 = y_0/(k\rho_0) \tag{3.3-48}$$

Hence for $\rho < 1$ (or $E_b/N_{0J} > (E_b/N_{0J})_0$ ($(E_b/N_{0J})_0$ is the value where ρ_0 is 1) using (3.3-48)), one has from (3.3-43), the result

$$\overline{PE_b} = \frac{y_0}{2(M-1)(E_b/N_{0J})} \sum_{n=2}^{M} \binom{M}{n} (-1)^n \exp\left[\frac{y_0(1-n)}{n}\right] = \frac{c_0}{(E_b/N_{0J})} \tag{3.3-49}$$

where c_0 is a constant and is displayed in Table 3.3-1 for various values of M.

Table 3.3-1 Solution of (3.3-45) for Various Values of M [9]

M	y_0	c_0	$(E_b/N_{0J})_0$
2	2.00	0.3679	2.000
4	2.38	0.2329	1.170
8	2.78	0.1954	0.930
16	3.49	0.1803	0.872
32	3.62	0.1746	0.798

In the case when $\rho=1$, one has from (3.3-44) the value

$$\overline{PE_b} = \frac{1}{2(M-1)}\sum_{n=2}^{M}\binom{M}{n}(-1)^n \exp\left[\frac{kE_b(1-n)}{nN_{0J}}\right] \tag{3.3-50}$$

In summary then, the average bit error rate is given by

$$\begin{aligned} \overline{PE_b} &= \frac{c_0}{E_b/N_{0J}} && \frac{E_b}{N_{0J}} > \left(\frac{E_b}{N_{0J}}\right)_0 \\ \overline{PE_b} &= \frac{\rho}{2(M-1)}\sum_{n=2}^{M}\binom{M}{n}(-1)^n \exp\left[\frac{kE_b(1-n)}{nN_{0J}}\right] && \frac{E_b}{N_{0J}} \le \left(\frac{E_b}{N_{0J}}\right)_0 \end{aligned} \tag{3.3-51}$$

Notice that again the optimum jammer forces the bit error rate to drop off as the inverse SNR. The average bit error rate is shown in Figure 3.3-5 for four values of M. The value of $M = 32$ is almost the same value as the $M = 16$ case. For M greater than 8, there is almost no improvement in bit error rate performance. It can be shown that the optimum value of ρ is given by

$$\rho_{opt} = \frac{c_0}{(E_b/N_{0J})} \tag{3.3-52}$$

for ρ up to a value of 1 where it is limited to 1. Note that c_0 is obtained from Table 3.3-1.

3.3.5 FFH/MFSK in Partial Band Noise Jamming

Consider now the case of fast frequency hopping, in which it will be seen that diversity improves the performance of the MFSK signal. Refer to Figure 3.2-8 that illustrates a (fast) FFH/MFSK receiver. Again we will neglect thermal noise to simplify the analysis. Thus the jammer density over the bandwidth ρW is $J/(\rho W)$.

Each M-ary symbol is divided into L subsymbols (a different hop for each subsymbol) or chips with energy

$$E_c = \frac{kE_b}{L} \tag{3.3-53}$$

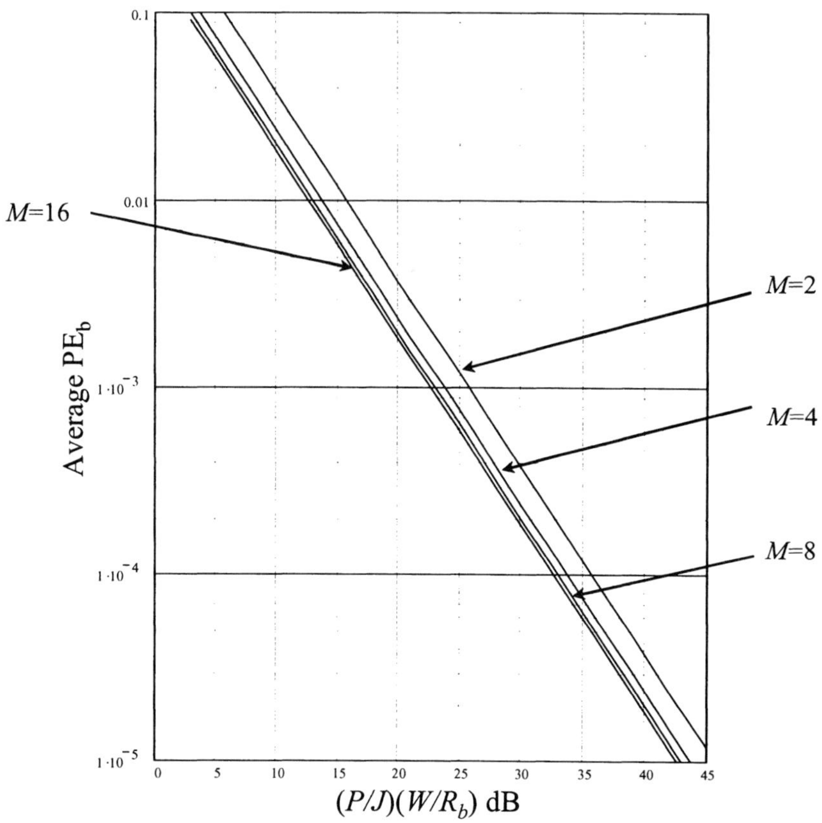

Figure 3.3-5 Average BER for SFH/MFSK and DS-SFH/MFSK for worst-case partial band jamming and worst-case pulse jamming.

where k is $\log_2(M)$. It is assumed that the receiver can detect perfectly when hops are jammed. A hop is jammed when *more than one* tone detector is high at the end of the symbol time [10, 11]. If any of the L chips are not jammed, error-free M-ary tone detection is made. If this does not occur, then choose the largest of the metrics

$$\lambda_i = \sum_{j=1}^{L} x_{ij}; \quad 1 \le i \le M \tag{3.3-54}$$

where x_{ij} is the energy detector output for the i-th M-ary tone on the j-th subsymbol or chip. Due to the fact that exact expressions for this metric generally do not yield closed form expressions, an exponentially tight upper bound [11] will be used for the average bit error rate. It is known that these bounds are close approximations. It has been shown [10, 11] that the Chernoff bound[2] for arbitrary L and $k = \log_2(M)$ is given by

$$\overline{PE_b} \le \frac{M}{4}\left[\frac{\rho}{1-\lambda^2}\exp\left(-\frac{2\lambda\,\beta}{1+\lambda}\right)\right]^L \tag{3.3-55}$$

[2] A technique to generate tight probability bounds that can then be used as an approximation. The exact solution would be very difficult to evaluate directly.

and maximizing over λ and ρ produces

$$\overline{PE}_b \le \frac{M}{4}\left[\frac{4Le^{-1}}{kE_b/N_0}\right]^L \qquad \rho_0 = \frac{3L}{kE_b/N_0}; \quad \frac{E_b}{N_0} \ge \frac{3L}{k} \tag{3.3-56}$$

where

$$\beta = \frac{kE_b/N_0}{2L}, \quad 2\lambda = \sqrt{(1+\beta)^2 + 4\beta} - (1+\beta) \tag{3.3-57}$$

Consider the bound when $E_b/N_0 \ge 3L/k$. When the jammer has optimized his bandwidth to maximize the average BER, the user at the receiver can optimize the number of chips L to minimize the average BER. Differentiating (3.3-56) with respect to L and utilizing the formula

$$\frac{dX(L)^L}{dL} = LX(L)^{L-1}\frac{dX(L)}{dL}\ln\left(\frac{4Le^1}{kE_b/N_0}\right) \tag{3.3-58}$$

yields the optimum value of L, call it L_0, which is given by

$$L_0 = \frac{kE_b}{4N_0} \tag{3.3-59}$$

The corresponding average BER is obtained by using (3.3-59) in (3.3-56), which produces

$$\overline{PE}_b \le \frac{M}{4}\exp\left[-\frac{kE_b}{4N_0}\right], \qquad \rho_0 = \frac{3}{4} \tag{3.3-60}$$

where the optimum value of ρ follows from using (3.3-59) in the expression for ρ_0 in (3.3-56).

Notice that the optimum diversity (L_0) restores the exponential reduction in the average bit error rate. Figure 3.3-6 illustrates the average bit error rate for fixed values of (a) $L = 4$ and (b) $L = 8$ (note that the optimized value is not used here).

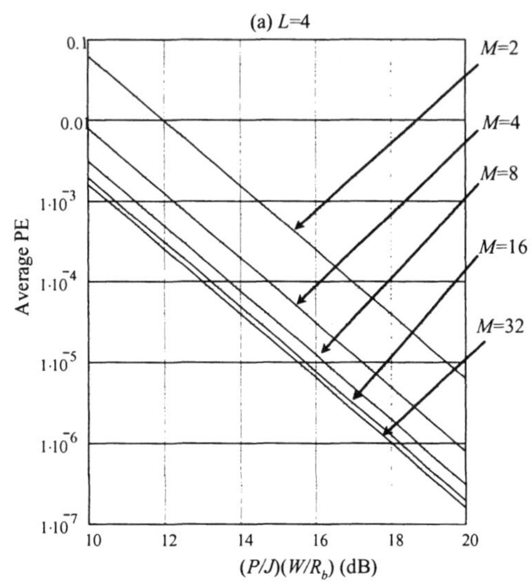

Figure 3.3-6 (a) Average BER for FFH/MFSK for worst-case partial band and pulse jamming and for DS-FFH/MFSK.

The curves start at a higher value of E_b/N_0 since $3L/k$ is larger in (b). Clearly fast frequency hopping with MFSK modulation is superior to slow frequency hopping in partial band jamming. Generally speaking, diversity greatly improves performance in partial band jamming.

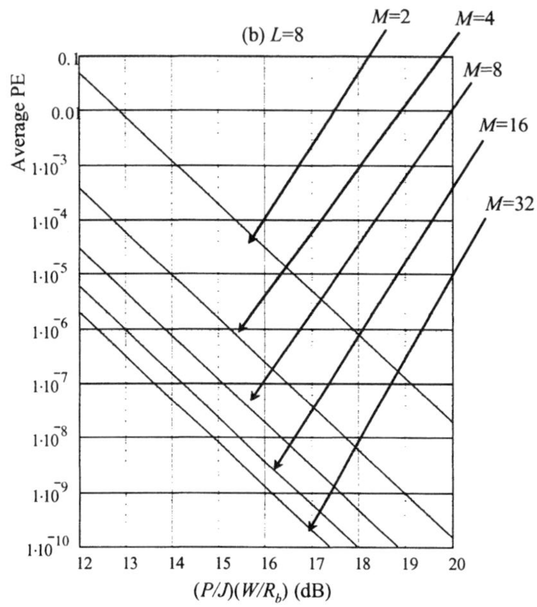

Figure 3.3-6 (b) Average BER for FFH/MFSK for worst-case partial band and pulse jamming and for DS-FFH/MFSK.

3.3.6 Hybrid DS-SFH/MFSK in Partial Band Noise Jamming

Figure 3.3-7 illustrates a block diagram of a BPSK-SFH/MFSK receiver. It will be assumed that the direct sequence signal is binary phase shift keyed (modulated) onto the carrier.

It will be assumed that slow frequency hopping is used. Figure 3.3-8 illustrates the two possible cases that can occur for the relative bandwidths of the jammer bandwidth compared to the pseudonoise (direct sequence) bandwidth (PNBW). In case 1 it is assumed that the jammer bandwidth is smaller than the direct sequence bandwidth. Note that the probability that the jammer falls into the PNBW bandwidth on a given hop is approximately given by (PNBW)/W. In case 2 it will be assumed that the jammer bandwidth is larger than PNBW. In this case the probability of the jammer hitting the signal is (quite) approximately ρ. Only the second case will be considered here.

From (3.2-28) the dehopped signal, when the thermal noise is neglected, is given by

$$x(t) = \sqrt{2P}PN(t)\cos\left[\omega_{IF}t + \theta_d(t) + \xi\right] + \sqrt{2}n_{Jc}(t)\cos(\omega_{IF}t) + \sqrt{2}n_{Js}(t)\sin(\omega_{IF}t) \qquad (3.3\text{-}61)$$

in which the two noise terms are statistically independent from one another, and each have a flat spectral density of $N_{0J}^{\rho}/2$ each. This jammer noise model represents the dehopped jammer noise. The spectral density $N_{0J}^{\rho} = J/(\rho W)$, with ρ the fractional jammer bandwidth.

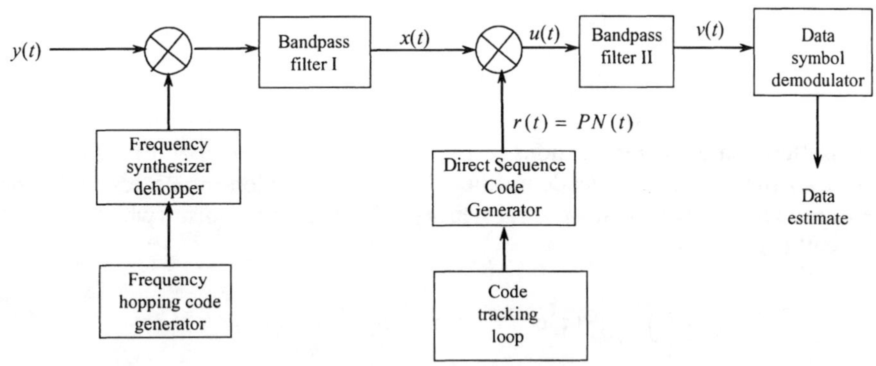

Figure 3.3-7 Block diagram of the BPSK-SFH/MFSK modulation receiver with BPSK direct sequence modulation.

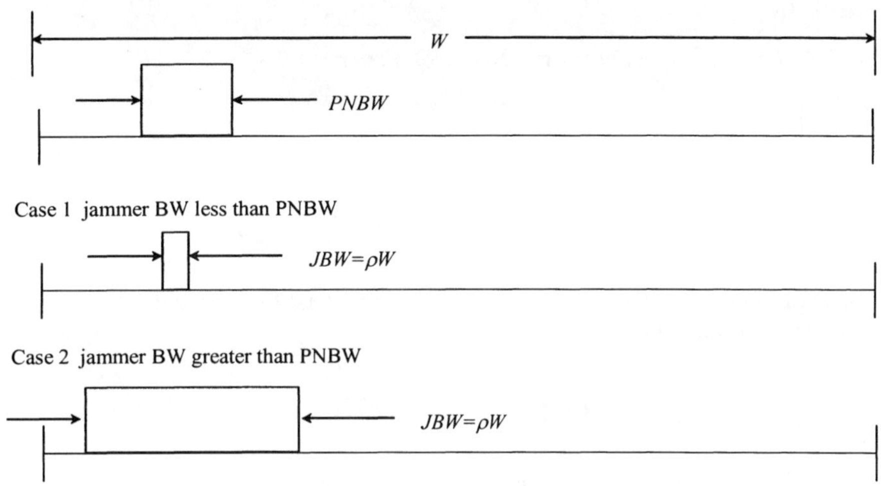

Figure 3.3-8 Model for determining the dehopped and despread jammer for DS-SFH/MFSK.

Since it has been assumed that the direct sequence spreading code is binary phase shift keyed onto the carrier, the BPSK DS despreading signal is of the form

$$r(t) = PN(t) \tag{3.3-62}$$

After despreading, the despread signal is of the form

$$\begin{aligned} u(t) &= \sqrt{2P}PN(t)\cos\left[\omega_{IF}t + \theta_d(t) + \xi\right] \\ &\quad + \sqrt{2}n'_{Jc}(t)\cos(\omega_{IF}t) + \sqrt{2}n'_{Js}(t)\sin(\omega_{IF}t) \end{aligned} \tag{3.3-63}$$

where the two noise terms are products of the jammer noise and the despreading code. The two jammer noise terms are given by

$$n'_{Jc}(t) = n_{Jc}(t)PN(t), \qquad n'_{Js}(t) = n_{Js}(t)PN(t) \tag{3.3-64}$$

After filtering by the second bandpass filter, the output can be represented by

$$v(t) = \sqrt{2P}PN(t)\cos\left[\omega_{IF}t + \theta_d(t) + \xi\right] + \sqrt{2}\tilde{n}_{Jc}^{'}(t)\cos(\omega_{IF}t) + \sqrt{2}\tilde{n}_{Js}^{'}(t)\sin(\omega_{IF}t) \quad (3.3\text{-}65)$$

where the two noise terms are filtered versions of (3.3-64) and are denoted as filtered terms by the "tilde" over the process. Two statistically independent, filtered, Gaussian random processes will approximate these two terms. The one-sided jammer noise spectral density of the jammer noise (which is the same for each baseband component) is given by

$$S_{\tilde{n}_{Jc}^{'}}(f) = \int_{-\infty}^{\infty} R_{\tilde{n}_{Jc}^{'}}(\tau)e^{-j2\pi f\tau}d\tau \cong \int_{-\infty}^{\infty} R_{\tilde{n}_{Jc}}(\tau)R_{PN}(\tau)e^{-j2\pi f\tau}d\tau \quad (3.3\text{-}66)$$

where $\tilde{n}_{jc}(t)$ and $\tilde{n}_{js}(t)$ are the bandpass equivalent filtered noise terms. The spectral density at $f = 0$ is given by

$$S_{\tilde{n}_{Jc}^{'}}(0) = \int_{-\infty}^{\infty} R_{\tilde{n}_{Jc}^{'}}(\tau)d\tau \cong \int_{-\infty}^{\infty} R_{\tilde{n}_{Jc}}(\tau)R_{PN}(\tau)d\tau = \int_{-\infty}^{\infty} S_{\tilde{n}_{Jc}}(f)S_{PN}(f)df \quad (3.3\text{-}67)$$

This can be evaluated as

$$S_{\tilde{n}_{Jc}^{'}}(0) = \int_{-\infty}^{\infty} \frac{J_0}{2}S_{PN}(f)\left|H_2(f)\right|^2 df = \alpha\frac{J_0}{2} \quad (3.3\text{-}68)$$

in which $S_{PN}(f)$ is a unit power PN code sequence, and α is the signal loss through the second bandpass filter. The filter loss through the filter is given by

$$\alpha = \int_{-\infty}^{\infty} S_{PN}(f)\left|H_2(f)\right|^2 df \quad (3.3\text{-}69)$$

Therefore the jammer spectral density, out of the bandpass filter, is given by

$$\alpha N_{0J}^{\rho} = \alpha\frac{J}{\rho W} \quad (3.3\text{-}70)$$

Thus the noise has been spread wider than in the slow hopped MFSK modulation case, due to the additional direct sequence despreading operation. Typically the value of α can be anywhere from a few dB to negligible, depending on the filter bandwidth and chip rate involved. The same results can be shown to hold for QPSK-SFH/MFSK and OQPSK-SFH/MFSK modulation.

It follows that DS-SFH/MFSK modulation is similar in performance to SFH/MFSK modulation except for the factor of α in the noise jammer density, under the assumption that the jammer bandwidth is larger than PNBW. Thus the results of Section 3.3.4 apply approximately with the understanding that

$$E_b / N_{0J} = \frac{PW}{\alpha JR_b} = \frac{X}{\alpha} \quad \text{with} \quad X = \frac{PW}{JR_b} \quad (3.3\text{-}71)$$

and α has to be evaluated according to the direct sequence modulation and the filter bandwidth according to (3.3-69). Figure 3.3-5 illustrates the BER performance.

3.3.7 Hybrid DS-SFH/DPSK in Partial Band Noise Jamming

It can be shown that DS-SFH/DPSK modulation has essentially the same performance as SFH/DPSK, as discussed in Section 3.3.2 and will not be discussed further here. Figure 3.3-2 illustrates the BER performance.

3.4 BIT ERROR RATE PERFORMANCE IN PULSED JAMMING

The analysis method for pulse jamming signals is conceptually very similar to the partial band jammer analysis. In this case the pulse jamming the parameter ρ is the fraction of time that the wideband noise jamming process is on and $(1-\rho)$ is the fraction of time that the pulsed jammer is off. The two-sided power spectral density of the jammer is $N_J/(2\rho)$ when the jammer is "pulsed on" and is $N_0/2$ when the jammer is "off." Here N_J is given

$$N_J = J/W \tag{3.4-1}$$

with W being the bandwidth of the filter preceding the multiplier that despreads or dehops, as shown in Figure 3.4-1.

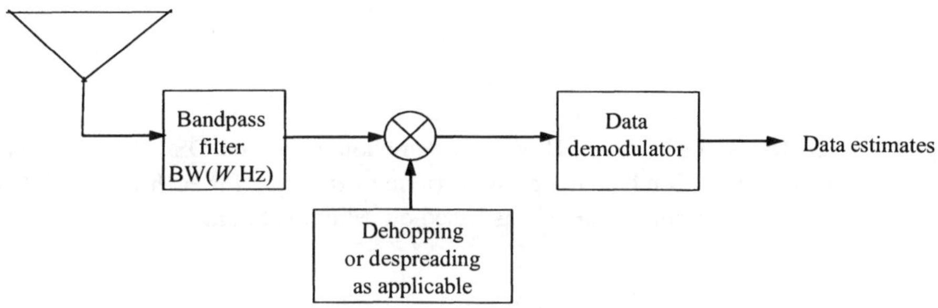

Figure 3.4-1 General model for pulsed jamming of spread spectrum signals.

We will see that a pulsed jammer can produce the same type of BER performance that occurs in a partial band jammer used in FH systems. One limitation with pulsed jamming is the fact that it is difficult to achieve very high pulse power levels.

The average bit error probability is given by

$$\overline{PE}_b(\rho) = (1-\rho)PE_b(E_b/\alpha N_0) + \rho PE_b(E_b/\alpha(N_0 + N_J(\rho))) \tag{3.4-2}$$

where α accounts for any despreading decrease in the noise or jammer spectral density level (see Section 3.2.1). The parameter ρ is the percentage of time the pulsed jammer is on. The object of the jammer is to optimize the value of ρ so as to maximize the value of the average error probability, $\overline{PE}_b$, subject to the allowable range of ρ.

3.4.1 Bit Error Rate Performance for DS/PSK in Pulsed Jamming

For all the coherent systems, the bit error rate is given by

$$PE_b = Q\left(\sqrt{\frac{2E_b}{\alpha N_0}}\right) \tag{3.4-3}$$

where N_0 is the one-sided noise spectral density. When jamming is present the density changes, and that fact must be reflected in the jamming calculations. The average *PE* is given by

$$\overline{PE}_b = (1-\rho)Q\left(\sqrt{\frac{2E_b}{\alpha N_0}}\right) + \rho Q\left(\sqrt{\frac{2E_b}{\alpha(N_0 + N_J(\rho))}}\right) \tag{3.4-4}$$

Normally the noise is negligible with respect to the jamming level so that (3.4-4) can be approximated by

$$\overline{PE}_b \cong \rho Q\left(\sqrt{\frac{2E_b}{\alpha N_J(\rho)}}\right) \tag{3.4-5}$$

Normally α is approximately unity, and that assumption will be made here to plot the curves versus ρ. Using (3.4-1) and the fact that $E_b = P/R_b$ and $N_J = J/W$ yields the expression

$$\overline{PE}_b \cong \rho Q\left(\sqrt{\frac{2\rho PW}{JR_b}}\right) \tag{3.4-6}$$

Figure 3.4-2 illustrates the average BER as a function of ρ for DS/PSK modulation in pulse jamming. From the figure it can be seen that an optimum value of ρ occurs for each value of $(P/J)(W/R)$. The easiest method of obtaining the optimum value of ρ is based on using the bound

$$\overline{PE}_b \le \frac{\rho}{\sqrt{4\pi\rho X}} \exp(-\rho X) \tag{3.4-7}$$

where $X = \left(\frac{P}{J}\right)\left(\frac{W}{R_b}\right)$. Consider the first derivative of (3.4-7) that yields

$$0 = \frac{e^{-\rho X}}{\sqrt{4\pi\rho X}} - \frac{1}{2}\frac{\rho e^{-\rho X}}{\sqrt{4\pi\rho X}\,\rho} - \frac{-X\rho e^{-\rho X}}{\sqrt{4\pi\rho X}} \tag{3.4-8}$$

Solving this equation yields $\rho X = 0.5$. Inserting the solution into (3.4-5) yields the following result

$$\begin{aligned} \overline{PE}_b &= \frac{1}{2X} Q\left(\sqrt{1}\right) \qquad 0.5 < \frac{PW}{JR_b} \\ \overline{PE}_b &= Q\left(\sqrt{\frac{2PW}{JR_b}}\right) \qquad \frac{PW}{JR_b} < 0.5 \end{aligned} \tag{3.4-9}$$

Equation (3.4-9) is plotted in Figure 3.4-3, which uses the optimum value or ρ at each $PW/(JR_b)$ value.

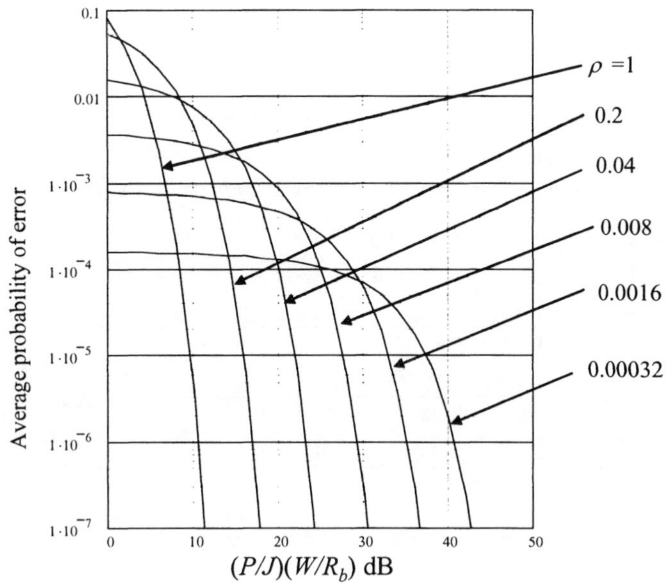

Figure 3.4-2 Average BER for various ρ versus the effective E_b/N_0 for DS/PSK in pulsed jamming.

Example 5 Consider a DS/BPSK link that is going to be jammed by an adversary. The received signal power is −145 dBW at the antenna output terminals, and the chip rate is 10^6 chips/sec. The data rate is 10 kbps, and the pulsed bandwidth is 10^6 Hz. Determine the optimum pulsed jammer power necessary to increase the bit error rate to 10^{-1}, neglecting thermal noise. Assume 0 dB gain for the receiver antenna. From Figure 3.4-3 the value of $PR_c/(JR_b)$ necessary to force the bit error rate to 10^{-1} is about 0 dB. The equation $PR_c/(JR_b)$ can be written in dB as

$$0 \text{ dB} = -145 - J + 10\log(10^6) - 10\log(10^4) \text{ dB}$$

This yields the value of $J = -125$ dBW received jammer power. Note that if the antenna gain were lower than 0 dB in the direction of the jammer the jammer power needed would be correspondingly larger. Also note that the jammer power is about 20 dB stronger than the signal power.

3.4.2 Performance of SFH/MFSK in Pulsed Jamming

In this section we consider a slow frequency-hopped system utilizing MFSK data modulation. The bit error probability is given by (3.2-32). Again the average bit error rate probability is given by

$$\overline{PE}_b(\rho) = (1-\rho)PE_b(E_b/\alpha N_0) + \rho PE_b(E_b/\alpha(N_0 + N_J(\rho))) \tag{3.4-10}$$

where ρ is the fraction of the time the pulse jammer is on. The analysis is identical to Section 3.3.4 so that the results are repeated here

$$\begin{aligned} \overline{PE}_b &= \frac{c_0}{E_b/N_{0J}} & \frac{E_b}{N_{0J}} &> \left(\frac{E_b}{N_{0J}}\right)_0 \\ \overline{PE}_b &= \frac{1}{2(M-1)}\sum_{n=2}^{M}\binom{M}{n}(-1)^n \exp\left[\frac{kE_b(1-n)}{nN_{0J}}\right] & \frac{E_b}{N_{0J}} &\le \left(\frac{E_b}{N_{0J}}\right)_0 \end{aligned} \tag{3.4-11}$$

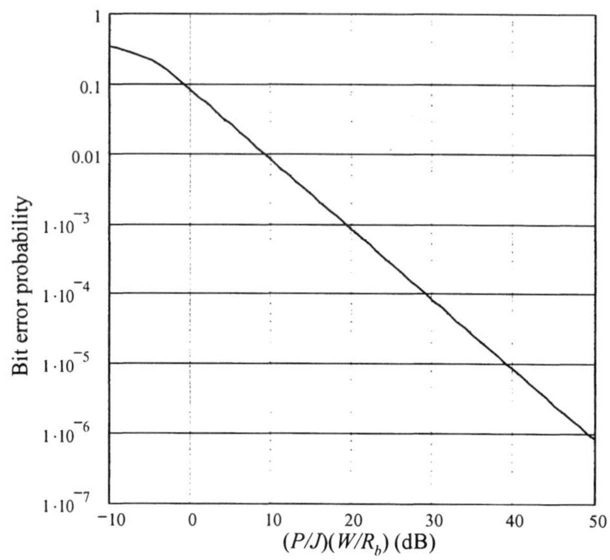

Figure 3.4-3 BER in optimum pulsed jamming for DS/PSK.

The bit error rate performance is plotted in Figure 3.3-5.

3.4.3 Performance of SFH/DPSK in Pulsed Jamming

The BER of DPSK is given by (3.2-19) and is of the form

$$\overline{PE_b} = \frac{1}{2}\exp\left(-\frac{E_b}{N_0}\right) \tag{3.4-12}$$

It follows that the average BER in pulse jamming is given by

$$\overline{PE_b} = \frac{1-\rho}{2}\exp\left(-\frac{E_b}{N_0}\right) + \frac{\rho}{2}\exp\left(-\frac{E_b}{N_0 + N_J/\rho}\right) \tag{3.4-13}$$

This expression is identical to the one in (3.3-23) so that it can be written as

$$\overline{PE_b} = \frac{1}{2}(1-\rho)\exp\left(\frac{-P}{N_0 R_b}\right) + \frac{1}{2}\rho\exp\left(\frac{-1}{(N_0 R_b / P) + (J/P)(R_b/(\rho W)}\right) \tag{3.4-14}$$

And the result is plotted in Figures 3.3-2 and 3.3-3.

3.4.4 Performance of Hybrid DS-SFH/MFSK in Pulsed Jamming

Again the approach used here is to evaluate the mean error probability based on (3.4-2) that is repeated here for convenience

$$\overline{PE}_b(\rho) = (1-\rho)PE_b(E_b/\alpha N_0) + \rho PE_b(E_b/\alpha(N_0 + N_J(\rho))) \tag{3.4-15}$$

where ρ is the fraction of the time that the jammer is on, $(1-\rho)$ is the fraction of the time it is off, and $N_J(\rho)$ is jammer spectral density that varies with ρ. When thermal noise is neglected and the worst-case jammer is used, the results of (3.3-51) apply. The results are plotted in Figure 3.3-5.

3.4.5 Performance of Hybrid DS-SFH/DPSK in Pulse Jamming

The approach is similar to the previous section except the data modulation is for DPSK. Following Section 3.3.2, the results are given in (3.3-28), and plotted in Figures 3.3-2 and 3.3-3.

3.5 BIT ERROR RATE PERFORMANCE IN TONE JAMMING

In this section the performance of direct sequence spread spectrum receivers will be addressed for a number of modulation types in the face of tone jamming.

3.5.1 Bit Error Rate Performance for DS(BPSK)/BPSK in Tone Jamming

In this section tone jamming will be investigated for its effect on BPSK data modulation with direct sequence DS BPSK spread spectrum signaling. The DS(BPSK)/BPSK signal is modeled as

$$s(t) = \sqrt{2P}d(t)PN(t)\cos(\omega_0 t + \theta_0) \tag{3.5-1}$$

where P is the received signal power, $d(t)$ ($\in\{-1,1\}$) is the binary valued data sequence, $PN(t)$ is the binary valued pseudonoise sequence, ω_0 is the angular carrier frequency, and θ is the carrier phase. The tone jammer is modeled as

$$n_J(t) = \sqrt{2J}\cos(\omega_0 t + \theta_J) \tag{3.5-2}$$

where J is the received jammer power, and θ_J is the jammer carrier phase. The received signal $y(t)$ is given by the sum of the signal, jammer, and thermal noise, which is denoted by $n(t)$. Thus

$$y(t) = s(t) + n_J(t) + n(t) \tag{3.5-3}$$

The receiver is illustrated in Figure 3.5-1.

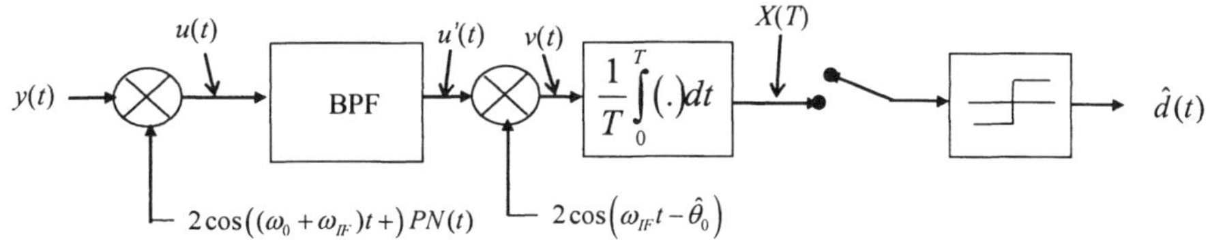

Figure 3.5-1 Model for a DS(BPSK)/BPSK receiver with tone jamming.

It is assumed that signal is despread perfectly (no time tracking error), and the signal is heterodyned to baseband coherently with no phase error (that is, $\hat{\theta}=\theta$). Consider the effect of the despreading operation on each component of the input signal. Consider $u'(t)$, which is the bandpass filtered despread response, which has three components

$$u'(t)=u_s'(t)+u_J'(t)+u_n'(t) \tag{3.5-4}$$

which includes the despread signal, jammer, and noise components.

Consider the signal component first. Taking the product of the signal and the first reference signal, after neglecting the sum frequency term, produces

$$u_s'(t)=\sqrt{2P}d(t)\cos\left(\omega_{IF}t-\theta_0\right) \tag{3.5-5}$$

After the second multiplier the following signal is obtained

$$v_s(t)=u_s'(t)2\cos(\omega_{IF}t-\hat{\theta}_0)=\sqrt{2P}+O(2\omega_{IF}) \tag{3.5-6}$$

Out of the integrate-and-dump matched filter one has

$$X_s(T)=\frac{d(T^-)}{T}\int_0^T\sqrt{2P}dt=d(T^-)\sqrt{2P} \tag{3.5-7}$$

where T^- is the time just before the end of the bit time.

Now consider the jammer. The despread jammer component is given by

$$u_J'(t)=\sqrt{2J}PN(t)\sin\left(\omega_{IF}t-\theta_J\right) \tag{3.5-8}$$

The jammer component, out of the second multiplier, is given by

$$v_J(t)=u_J'(t)2\cos\left(\omega_{IF}t-\hat{\theta}_0\right)=\sqrt{2J}PN(t)\cos\left(\theta_0-\theta_J\right)+O(2\omega_{IF}) \tag{3.5-9}$$

Again twice the IF frequency will be neglected in what follows since it will be filtered out by the integrator following the sine wave heterodyning.

Now consider the output of the integrate-and-dump filter. The jammer component is given by

$$X_J(T)=\frac{\sqrt{2J}}{T}\cos\left(\theta_0-\theta_J\right)\int_0^T PN(t)dt \tag{3.5-10}$$

This can be written as

$$X_J\left(T\right)=\frac{\sqrt{2J}}{T}\cos\left(\theta_0-\theta_J\right)\sum_{i=1}^{N}\int_{iT_c}^{(i+1)T_c}PN_i dt \tag{3.5-11}$$

where PN_i is the i-th term in the $PN(t)$ process that exists for T_c seconds. The PN_i will be modeled as random variables taking on the values of ±1 with each value having a probability of 1/2. Evaluating the integration produces

$$X_J(T) = \frac{\sqrt{2J}}{T}\cos(\theta_0 - \theta_J)\frac{1}{N}\sum_{i=1}^{N} PN_i(NT_c) \tag{3.5-12}$$

since $NT_c = T$, (3.5-12) can be rewritten as

$$X_J(T) = \sqrt{2J}\cos(\theta_0 - \theta_J)\frac{1}{N}\sum_{i=1}^{N} PN_i \tag{3.5-13}$$

The mean value of $X_J(T)$ is given by

$$E\{X_J(T)\} = 0 \tag{3.5-14}$$

The variance of $X_J(T)$ is given by

$$Var\{X_J(T)\} = \frac{2J}{N}\cos(\theta_0 - \theta_J)^2 \tag{3.5-15}$$

Now consider the thermal noise. The thermal noise can be written in the form

$$n(t) = \sqrt{2}n_c(t)\cos(\omega_{IF}t) + \sqrt{2}n_s(t)\sin(\omega_{IF}t) \tag{3.5-16}$$

The despread noise is given by, after a little trigonometry,

$$u_n^{'}(t) = \sqrt{2}n_c(t)PN(t)\sin(\omega_{IF}t) - \sqrt{2}n_s(t)PN(t)\cos(\omega_{IF}t) \tag{3.5-17}$$

Since the two baseband noise terms are approximately Gaussian with the spectral density of $N_0/2$, $u_n^{'}(t)$ can be approximated by

$$u_n^{'}(t) \cong \sqrt{2}n_c(t)\sin(\omega_{IF}t) + \sqrt{2}n_s(t)\cos(\omega_{IF}t) \tag{3.5-18}$$

Finally the noise component out of the second multiplier can be written as

$$v_n(t) = u_n^{'}(t)2\cos\left(\omega_{IF}t - \hat{\theta}_0\right) = \sqrt{2}n_c(t)\cos(\theta_0) + \sqrt{2}n_s(t)\sin(\theta_0) \tag{3.5-19}$$

This process is statistically (mean and variance) equivalent to

$$v_n(t) = \sqrt{2}n_c(t) \tag{3.5-20}$$

as is indicated by the fact that when $\theta_0 = 0$, $v_n(t)$ is exactly equal to (3.5-20), for example. The variance of the noise out of the integrate-and-dump filter is given by

$$Var\{X_n(T)\} = E\left\{\frac{1}{T^2}\int_0^T\int_0^T \sqrt{2}n_c(t)\sqrt{2}n_c(u)dtdu\right\} \tag{3.5-21}$$

Evaluating this expression and noting that $\sqrt{2}n_c^{'}(t)$ has two-sided noise spectral density of N_0 leads to

$$Var\{X_n(T)\} = \frac{N_0}{T} \tag{3.5-22}$$

Assuming that the sum of the thermal noise and the jammer noise can be modeled as Gaussian noise, the mean and standard deviation of the output of the integrate-and-dump is given by

$$\begin{aligned} \mu &= \sqrt{2P} \\ \sigma &= \sqrt{\frac{N_0}{T} + \frac{2J}{N}\cos\left(\theta_0 - \theta_J\right)^2} \end{aligned} \tag{3.5-23}$$

Given that a "plus one" bit was transmitted, the probability of error is given by

$$PE_b = \int_{\sqrt{2P}}^{\infty} \frac{1}{\sqrt{2\pi}} \exp\left(-\frac{x^2}{2\sigma^2}\right) dx \tag{3.5-24}$$

which can be evaluated to

$$PE_b = Q\left[\sqrt{\frac{2PT}{N_0 + 2JT_c \cos\left(\theta_0 - \theta_J\right)^2}}\right] \tag{3.5-25}$$

in which $Q(x)$ is defined as

$$Q(x) = \frac{1}{\sqrt{2\pi}} \int_x^{\infty} \exp\left(-\frac{u^2}{2}\right) du \tag{3.5-26}$$

When thermal noise is neglected compared to the jammer and the worst-case jammer phase angle is assumed ($\theta_0 = \theta_J$), then (3.5-25) can be written as

$$PE_b = Q\left[\sqrt{\frac{PT}{JT_c}}\right] = Q\left[\sqrt{\frac{P}{J}\frac{R_c}{R_b}}\right] \tag{3.5-27}$$

where the second factor in the right-hand side of (3.5-27) is the processing gain (*PG*) for BPSK systems; that is, $PG = R_c / R_b$. The loss would be less if $\Delta\omega$ is not zero [4]. Furthermore if $\theta_0 - \theta_J = (2n+1)\pi \quad n = 1, 2, 3...$, then the effect of the jammer is theoretically zero. It is to be noted, however, that it is essentially impossible to maintain the jammer coherent with the signal. In reality some small frequency error occurs with performance that is somewhat worse than predicted by the "perfect" jammer ($\theta_0 = \theta_J$). Figure 3.5-2 illustrates two curves relevant to DS(BPSK)/BPSK. The upper curve corresponds to the case (3.5-25) that the jammer takes on the worst-case phase causing the highest error rate ($\theta_0 = \theta_J$), and the second curve is based on averaging the phase difference of the jammer over 0 to 2π radians, assuming that the phase is uniformly distributed. Thus the second curve is based on the equation

$$PE_b = \frac{1}{2\pi} \int_0^{2\pi} Q\left[\sqrt{\frac{1}{\frac{1}{2E_b} + \frac{PT}{JT_c}\cos\left(\phi\right)^2}}\right] d\phi \tag{3.5-28}$$

Both curves of bit error rate are plotted versus the ratio $(P/J)(R_c/R_b)$ expressed in dB for the case that the worst-case jammer phase angle is assumed, and for $E_b/N_0 = 20$. Notice that when the jammer phase is randomized the jammer is about 1.5 dB less effective than the optimum jammer phase ($\theta_0 = \theta_J$).

3.5.2 Bit Error Rate Performance for DS(QPSK)/BPSK in Tone Jamming

In this section the case of BPSK data modulation with DS QPSK code spreading is considered. Figure 3.5-3 illustrates the receiver model that will be used for analysis purposes.

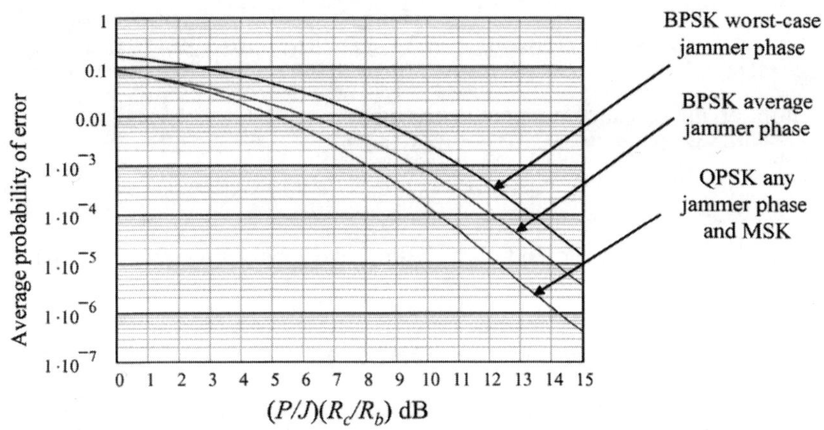

Figure 3.5-2 BER for a tone jammer at the worst-case phase.

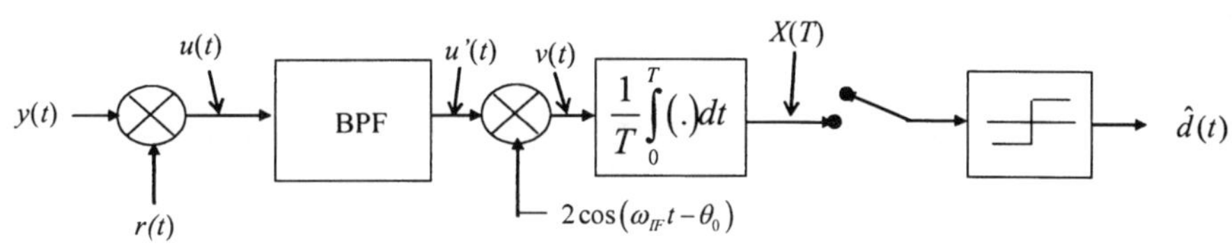

Figure 3.5-3 Model for a DS(PSK)/BPSK receiver with tone jamming.

The reference signal, which is the input to the first multiplier in Figure 3.5-3, is denoted by $r(t)$. The received signal plus jammer plus noise is described by

$$y(t) = s(t) + n_J(t) + n(t) \tag{3.5-29}$$

where the received signal is given by

$$s(t) = \sqrt{P}d(t)PN_1(t)\cos(\omega_0 t + \theta_0) + \sqrt{P}d(t)PN_2(t)\sin(\omega_0 t + \theta_0) \tag{3.5-30}$$

and the Gaussian jammer process is modeled as

$$n_J(t) = \sqrt{2J}\cos(\omega_0 t + \theta_J) \tag{3.5-31}$$

The thermal noise is described as

$$n(t) = \sqrt{2}n_c(t)\cos(\omega_0 t + \theta_0) + \sqrt{2}n_c(t)\sin(\omega_0 t + \theta_0) \tag{3.5-32}$$

Consider the signal component first. The signal component after the first multiplier is given by

$$\begin{aligned} u_s(t) = &\sqrt{P/2}d(t)\cos(\omega_{IF}t - \theta_0) + \sqrt{P/2}d(t)\cos(\omega_{IF}t - \theta_0) \\ &+ O(2\omega_0 + \omega_{IF}) \end{aligned} \tag{3.5-33}$$

where $O(x)$ denotes a signal at a frequency of x. After the bandpass filter, the signal component is given by

$$u_s^{'}(t) = \sqrt{P/2}d(t)\cos(\omega_{IF}t - \theta_0) + \sqrt{P/2}d(t)\cos(\omega_{IF}t - \theta_0) \tag{3.5-34}$$

After the second multiplier, the signal component is given by

$$v_s(t) = u_s^{'}(t)2\cos(\omega_{IF}t - \theta_0) = \sqrt{2P}d(t) + O(2\omega_{IF}) \tag{3.5-35}$$

After being integrated by the integrate-and-dump filter and neglecting the $O(2\omega_{IF})$ term, the signal component is given by

$$X_s(T) = \sqrt{2P}d(T^-) \tag{3.5-36}$$

where T^- is the value just prior to the end of a bit time. Now consider the jammer signal as it is processed from the input all the way to the output. After the first multiplier the jammer signal takes the form

$$\begin{aligned} u_J(t) = &\sqrt{J}\left[PN_1(t)\cos(\omega_{IF}t - \theta_J) + PN_2(t)\sin(\omega_{IF}t - \theta_J)\right] \\ &+ O(2\omega_0 + \omega_{IF}) \end{aligned} \tag{3.5-37}$$

The bandpass filter removes the second term in (3.5-37) so that after the second multiplier one obtains (using $\Delta\theta = \theta_J - \theta_0$)

$$v_J(t) = \sqrt{J}\left[PN_1(t)\cos(\Delta\theta) - PN_2(t)\sin(\Delta\theta)\right] + O(2\omega_{IF}) \tag{3.5-38}$$

The jammer signal out of the integrate-and-dump produces (neglecting the double frequency term since it will be filtered out by the integration)

$$X_J(T) = \frac{\sqrt{J}}{T}\int_0^T \left[PN_1(t)\cos(\Delta\theta) + PN_2(t)\sin(\Delta\theta)\right]dt \tag{3.5-39}$$

This can be written as

$$X_J(T) = \frac{\sqrt{J}\cos(\Delta\theta)}{T}\sum_{i=1}^{N}\int_{iT_c}^{(i+1)T_c} PN_{1i}dt + \frac{\sqrt{J}\sin(\Delta\theta)}{T}\sum_{i=1}^{N}\int_{iT_c}^{(i+1)T_c} PN_{2i}dt \tag{3.5-40}$$

where PN_{1i} and PN_{2i} are the i-th chip values of $PN_1(t)$ and $PN_2(t)$. Equation (3.5-40) can be rewritten as

$$X_J(T) = \sqrt{J}\cos(\Delta\theta)\frac{1}{N}\sum_{i=1}^{N} PN_{1i}NT_c + \sqrt{J}\sin(\Delta\theta)\frac{1}{N}\sum_{i=1}^{N} PN_{2i}NT_c \quad (3.5\text{-}41)$$

Note that the mean of $X_j(T)$ is zero, assuming that the PN_{ki} chips ($k = 1$ or 2) are random variables taking on the values of ± 1 with probability of 1/2. The variance can be evaluated by

$$\begin{aligned} Var\{X_J(T)\} &= J\cos(\Delta\theta)^2\frac{1}{N^2}\sum_{i=1}^{N}\sum_{j=1}^{N} E\{PN_{1i}PN_{1j}\} \\ &+ J\sin(\Delta\theta)^2\frac{1}{N^2}\sum_{i=1}^{N}\sum_{j=1}^{N} E\{PN_{2i}PN_{2j}\} \end{aligned} \quad (3.5\text{-}42)$$

This can be evaluated to the simple result

$$Var\{X_J(T)\} = J\frac{1}{N^2}\sum_{i=1}^{N} 1 = \frac{J}{N} \quad (3.5\text{-}43)$$

Now consider the despread thermal noise. The despread noise is given by

$$u_n(t) = r(t)\left[\sqrt{2}n_c(t)\cos(\omega_0 t + \theta_0) + \sqrt{2}n_s(t)\sin(\omega_0 t + \theta_0)\right] \quad (3.5\text{-}44)$$

After multiplying the terms out one has

$$\begin{aligned} u_n(t) &= n_c(t)PN_1(t)\cos(\omega_{IF}t) - n_s(t)PN_1(t)\sin(\omega_{IF}t) \\ &+ n_c(t)PN_1(t)\sin(\omega_{IF}t) - n_s(t)PN_2(t)\cos(\omega_{IF}t) + O(2\omega_0 + \omega_{IF}) \end{aligned} \quad (3.5\text{-}45)$$

Out of the second multiplier one has

$$v_n(t) = u_n^{'}(t)2\cos(\omega_{IF}t - \theta_0) \quad (3.5\text{-}46)$$

This can be simplified to the following expression

$$\begin{aligned} v_n(t) &= n_c(t)PN_1(t)\cos(\theta_0) + n_s(t)PN_2(t)\cos(\theta_0) \\ &+ n_c(t)PN_2(t)\sin(\theta_0) - n_s(t)PN_1(t)\sin(\theta_0) \end{aligned} \quad (3.5\text{-}47)$$

This sum is equivalent to the sum of two independent baseband white noise terms (multiplied by two independent codes) that is equivalent to one noise term multiplied by $\sqrt{2}$. That is to say

$$v_n(t) \cong \sqrt{2}n_c(t) \quad (3.5\text{-}48)$$

which has a two-sided density of $N_0/2$ W/Hz. The assumption that the despread jammer is Gaussian, since it is the sum of a large number of random variables having the values of ± 1, allows us to write the error probability as

$$PE = \int_{\sqrt{2P}}^{\infty} \frac{1}{\sqrt{2\pi}\sigma}\exp\left(-\frac{x^2}{2\sigma^2}\right)dx \quad (3.5\text{-}49)$$

This can be evaluated as

$$PE = Q\left[\sqrt{\frac{2PT}{N_0 + JT_c}}\right] \tag{3.5-50}$$

In comparing (3.5-50) with (3.5-25), it is seen that the worst-case tone jamming has 3 dB greater effect for DS-BPSK/BPSK than for DS-QPSK/BPSK! Figure 3.5-2 illustrates the worst-case DS-PSK/BPSK and the average jammer phase DS-BPSK/BPSK, along with the DS-QPSK/BPSK case.

3.5.3 Bit Error Rate Performance for DS(MSK)/BPSK in Tone Jamming

Now consider the bit error rate performance of uncoded direct sequence DS MSK SS signals with BPSK data modulation. The analysis in this section is similar to the two previous sections, so that the exposition will be brief. Figure 3.5-4 illustrates the receiver model for DS-MSK/BPSK.

The reference signal $r(t)$, letting $\omega_0' = \omega_0 + \omega_{IF}$, is given by

$$\begin{aligned} r(t) &= 2\sum_i p(t - iT_c)\sin\left[\frac{\pi}{T_c}(t - iT_c)\right]PN_1(t)\cos\left(\omega_0' t\right) \\ &+ 2\sum_i p(t - iT_c - T_c/2)\sin\left[\frac{\pi}{T_c}(t - iT_c - T_c/2)\right]PN_2(t)\sin\left(\omega_0' t\right) \end{aligned} \tag{3.5-51}$$

where $p(t)$ is a unit pulse of duration T_c seconds, starting at time zero. Note that the second set of terms is offset from the first set by one half of a chip, as in offset QPSK. Note also that only the positive half of the sinusoid is used in this representation of MSK modulation; see [4] for more details.

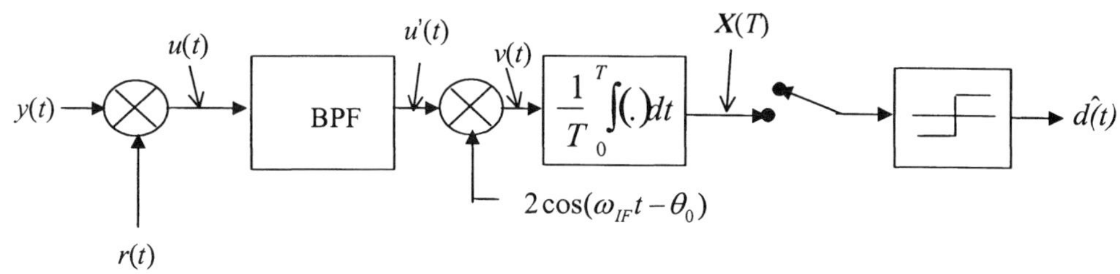

Figure 3.5-4 Model for a DS(MSK)/BPSK receiver with tone jamming.

The signal is represented by

$$\begin{aligned} s(t) &= \sqrt{2P}d(t)\sum_i p(t - iT_c)\sin\left[\frac{\pi}{T_c}(t - iT_c)\right]PN_1(t)\cos\left(\omega_0 t + \theta_0\right) \\ &+ \sqrt{2P}d(t)\sum_i p(t - iT_c - T_c/2)\sin\left[\frac{\pi}{T_c}(t - iT_c - T_c/2)\right]PN_2(t)\sin\left(\omega_0 t + \theta_0\right) \end{aligned} \tag{3.5-52}$$

and the tone jammer and the thermal noise are defined in (3.5-2) and (3.5-32), respectively. The reference signal for the first multiplier is given by

$$r(t)=2\sum_i p(t-iT_c)\sin\left[\frac{\pi}{T_c}(t-iT_c)\right]PN_1(t)\cos\left((\omega_0+\omega_{IF})t+\theta_0\right)$$
$$+2\sum_i p(t-iT_c-T_c/2)\sin\left[\frac{\pi}{T_c}(t-iT_c-T_c/2)\right]PN_2(t)\sin\left((\omega_0+\omega_{IF})t+\theta_0\right) \quad (3.5\text{-}53)$$

Define $u(t)$ as the product of the input and the reference signal, and $r(t)$ and $u_s(t)$ as the signal component of that product; hence

$$u_s(t)=s(t)r(t) \quad (3.5\text{-}54)$$

Evaluating produces the result

$$u_s(t)=\sqrt{2P}d(t)\sum_i p(t-iT_c)\sin\left(\frac{\pi}{T_c}(t-iT_c)\right)\sum_j p(t-jT_c)\sin\left(\frac{\pi}{T_c}(t-jT_c)\right)\cos\left(\omega_{IF}t-\theta\right)$$
$$+\sqrt{2P}d(t)\sum_i p(t-iT_c-\frac{T_c}{2})\sin\left(\frac{\pi}{T_c}(t-iT_c-\frac{T_c}{2})\right) \quad (3.5\text{-}55)$$
$$\times\sum_j p(t-jT_c-\frac{T_c}{2})\sin\left(\frac{\pi}{T_c}(t-jT_c-\frac{T_c}{2})\right)\cos\left(\omega_{IF}t-\theta\right)$$
$$+\text{cross terms}+O(2\omega_0+\omega_{IF})$$

Noting that the MSK shaping function can only exist at one time produces

$$u_s(t)=\sqrt{2P}d(t)\sum_i p(t-iT_c)\sin\left(\frac{\pi}{T_c}(t-iT_c)\right)^2\cos\left(\omega_{IF}t-\theta\right)$$
$$+\sqrt{2P}d(t)\sum_i p(t-iT_c-\frac{T_c}{2})\sin\left(\frac{\pi}{T_c}(t-iT_c-\frac{T_c}{2})\right)^2\cos\left(\omega_{IF}t-\theta\right) \quad (3.5\text{-}56)$$
$$+\text{cross terms}+O(2\omega_0+\omega_{IF})$$

Noting that

$$\sin\left(\frac{\pi}{T_c}(t-iT_c)\right)^2=\frac{1}{2}-\frac{1}{2}\cos\left(\frac{2\pi}{T_c}(t-iT_c)\right) \quad (3.5\text{-}57)$$

produces for $u_s(t)$ the signal term, neglecting the double frequency component

$$u_s'(t)=\sqrt{2P}d(t)\cos\left(\omega_{IF}t-\theta\right) \quad (3.5\text{-}58)$$

The second multiplier produces the value of the signal component into the integrator

$$v_s(t)=\sqrt{2P}d(t)+O(2\omega_{IF}) \quad (3.5\text{-}59)$$

Out of the integrator, the signal component is given by

$$X_s(T) = \sqrt{2P}d(T^-) \tag{3.5-60}$$

Now consider the jammer. The first component of the jammer signal is the product of the jammer and the reference signal, letting $\omega_0^{'} = \omega_0 + \omega_{IF}$, which yields

$$u_J(t) = \begin{pmatrix} 2\sum_i p(t-iT_c)\sin\left(\frac{\pi}{T_c}(t-iT_c)\right)PN_1(t)\cos\left(\omega_0^{'}t\right) \\ +2\sum_i p(t-iT_c-\frac{T_c}{2})\sin\left(\frac{\pi}{T_c}(t-iT_c-\frac{T_c}{2})\right)PN_2(t)\sin\left(\omega_0^{'}t\right) \end{pmatrix}(\sqrt{2J}\cos\left(\omega_0 t+\theta_J\right) \tag{3.5-61}$$

After the first multiplication with $r(t)$ produces

$$\begin{aligned} u_J^{'}(t) &= \sqrt{2J}\sum_i p(t-iT_c)\sin\left(\frac{\pi}{T_c}(t-iT_c)\right)PN_1(t)\cos\left(\omega_{IF}t-\theta_J\right) \\ &+\sqrt{2J}\sum_i p(t-iT_c)\sin\left(\frac{\pi}{T_c}(t-iT_c)\right)PN_1(t)\sin\left(\omega_{IF}t-\theta_J\right) \end{aligned} \tag{3.5-62}$$

The output of the second multiplier is given by the product of $u_J^{'}(t)$ and $2\cos\left(\omega_{IF}t-\theta\right)$ so that

$$v_J(t) = \sqrt{2J}\begin{Bmatrix} \sum_i p(t-iT_c)\sin\left(\frac{\pi}{T_c}(t-iT_c)\right)PN_1(t)\cos(\theta-\theta_J) \\ +\sum_i p\left(t-iT_c-\frac{T_c}{2}\right)\sin\left(\frac{\pi}{T_c}(t-iT_c-\frac{T_c}{2})\right)PN_2(t)\cos(\theta-\theta_J) \end{Bmatrix} + O(2\omega_{IF}) \tag{3.5-63}$$

Since the output component is the average over T seconds, one has

$$X_J(T) = \frac{\sqrt{2J}}{T}\int_0^T \begin{Bmatrix} \sum_i p(t-iT_c)\sin\left(\frac{\pi}{T_c}(t-iT_c)\right)PN_1(t)\cos(\Delta\theta) \\ +\sum_i p\left(t-iT_c-\frac{T_c}{2}\right)\sin\left(\frac{\pi}{T_c}(t-iT_c-\frac{T_c}{2})\right)PN_2(t)\cos(\Delta\theta) \end{Bmatrix} dt + O(2\omega_{IF}) \tag{3.5-64}$$

where $\Delta\theta = \theta - \theta_J$. Breaking up the integral over each chip time produces

$$\begin{aligned} X_J(T) &= \frac{\sqrt{2J}\cos(\Delta\theta)}{T}\sum_{i=1}^{N}\int_{iT_c}^{(i+1)T_c}\sin\left(\frac{\pi}{T_c}(t-iT_c)\right)PN_{1i}dt \\ &+\frac{\sqrt{2J}\sin(\Delta\theta)}{T}\sum_{i=1}^{N}\int_{iT_c}^{(i+1)T_c}\sin\left(\frac{\pi}{T_c}\left(t-iT-\frac{T_c}{2}\right)\right)PN_{2i}dt \end{aligned} \tag{3.5-65}$$

in which PN_{kj} denotes the j-th component of the k-th code. Since $X_J(T)$ is a random variable, consider the variance of the expression. The variance is given by

$$\operatorname{Var}\{X_J(T)\} =$$

$$\frac{2J\cos(\Delta\theta)^2}{T^2}\int_{iT_c}^{(i+1)T_c}\int_{jT_c}^{(j+1)T_c}\sum_{i=1}^{N}\sum_{j=1}^{N}\sin\left(\frac{\pi}{T_c}(t-iT_c)\right)\sin\left(\frac{\pi}{T_c}(t-jT_c)\right)PN_{1i}PN_{2j}dtds$$
$$+\frac{2J\sin(\Delta\theta)^2}{T^2}\int_{iT_c}^{(i+1)T_c}\int_{jT_c}^{(j+1)T_c}\sum_{i=1}^{N}\sum_{j=1}^{N}\sin\left(\frac{\pi}{T_c}\left(t-iT_c-\frac{T_c}{2}\right)\right)\sin\left(\frac{\pi}{T_c}\left(t-jT_c-\frac{T_c}{2}\right)\right)PN_{1i}PN_{2j}dtds \quad (3.5\text{-}66)$$
$$+\,cross\ terms$$

Utilizing (3.5-57), simplifying and neglecting the double frequency terms and the cross terms produces

$$Var\{X_J(T)\} = \frac{2J\cos(\Delta\theta)^2}{T^2}\left(\frac{1}{2}\right)\sum_{i=1}^{N}T_c + \frac{2J\sin(\Delta\theta)^2}{T^2}\left(\frac{1}{2}\right)\sum_{i=1}^{N}T_c \quad (3.5\text{-}67)$$

Simplifying even further produces

$$Var\{X_J(T)\} = \frac{J}{T^2}\sum_{i=1}^{N}T_c = \frac{J}{N} \quad (3.5\text{-}68)$$

Now consider the thermal noise component. The noise component is given by

$$u_n(t) = n(t)r(t) \quad (3.5\text{-}69)$$

which can be written as

$$u_n(t) = \left[\sqrt{2}n_c(t)\cos(\omega_0 t) + \sqrt{2}n_s(t)\sin(\omega_0 t)\right]\times$$
$$\left[2\sum_i p(t-iT_c)\sin\left(\frac{\pi}{T_c}(t-iT_c)\right)PN_1(t)\cos\left((\omega_0+\omega_{IF})t\right) + \right. \quad (3.5\text{-}70)$$
$$\left. +2\sum_i p\left(t-iT_c-\frac{T_c}{2}\right)\sin\left(\frac{\pi}{T_c}(t-iT_c)-\frac{T_c}{2}\right)PN_2(t)\sin\left((\omega_0+\omega_{IF})t\right)\right]$$

which, after assuming that the sum frequencies are removed by the BPF, can be expanded to the following form

$$u_n'(t) = \sqrt{2}n_c(t)\sum_i p(t-iT_c)\sin\left(\frac{\pi}{T_c}(t-iT_c)\right)PN_1(t)\cos(\omega_{IF}t)$$
$$+\sqrt{2}n_s(t)\sum_i p\left(t-iT_c-\frac{T_c}{2}\right)\sin\left(\frac{\pi}{T_c}\left(t-iT_c-\frac{T_c}{2}\right)\right)PN_2(t)\cos(\omega_{IF}t) \quad (3.5\text{-}71)$$
$$+\sqrt{2}n_s(t)\sum_i p(t-iT_c)\sin\left(\frac{\pi}{T_c}(t-iT_c)\right)PN_1(t)\sin(-\omega_{IF}t)$$
$$+\sqrt{2}n_c(t)\sum_i p\left(t-iT_c-\frac{T_c}{2}\right)\sin\left(\frac{\pi}{T_c}\left(t-iT_c-\frac{T_c}{2}\right)\right)PN_2(t)\sin(\omega_{IF}t)$$

The second multiplication produces the product

$$v_n(t) = 2\cos\left(\omega_{IF}t - \theta\right)u_n^{'}(t) \tag{3.5-72}$$

This can be written, after removing the sum frequencies (which are removed by the integrate-and-dump filter), by

$$\begin{aligned} v_n(t) &= \sqrt{2}n_c(t)\sum_i p(t-iT_c)\sin\left(\frac{\pi}{T_c}(t-iT_c)\right)PN_1(t)\cos(\theta) \\ &+\sqrt{2}n_s(t)\sum_i p\left(t-iT_c-\frac{T_c}{2}\right)\sin\left(\frac{\pi}{T_c}\left(t-iT_c-\frac{T_c}{2}\right)\right)PN_2(t)\cos(\theta) \\ &-\sqrt{2}n_s(t)\sum_i p(t-iT_c)\sin\left(\frac{\pi}{T_c}(t-iT_c)\right)PN_1(t)\sin(\theta) \\ &-\sqrt{2}n_c(t)\sum_i p\left(t-iT_c-\frac{T_c}{2}\right)\sin\left(\frac{\pi}{T_c}\left(t-iT_c-\frac{T_c}{2}\right)\right)PN_2(t)\sin(\theta) \end{aligned} \tag{3.5-73}$$

The output of the averaging integrate-and-dump filter, over the time from 0 to T seconds, is a random variable, and is given by

$$X_n(T) = \frac{1}{T}\int_0^T v_n(t)dt \tag{3.5-74}$$

Since it is a random variable, the variance will be computed. The variance of $X_n(T)$ is given by

$$\text{Var}\left(X_n(T)\right) = E\left\{\frac{1}{T^2}\int_0^T\int_0^T v_n(t)v_n(u)dtdu\right\} \tag{3.5-75}$$

where the $E\{x\}$ denotes the expectation of x. For convenience in the notation, an over bar will denote the expectation also. Evaluating (3.5-75) produces

$$\begin{aligned} &\text{Var}\left(X_n(T)\right) = \\ &\frac{\cos(\theta)^2}{T^2}\int_0^T\int_0^T \overline{2n_c(t)n_c(u)}\sum_i\sum_j p(t-iT_c)p(u-jT_c)\sin\left(\frac{\pi}{T_c}(t-iT_c)\right)\sin\left(\frac{\pi}{T_c}(u-jT_c)\right)\overline{PN_1(t)PN_1(u)}dudt \\ &+\frac{\cos(\theta)^2}{T^2}\int_0^T\int_0^T \overline{2n_s(t)n_s(u)}\sum_i\sum_j p\left(t-iT_c-\frac{T_c}{2}\right)p\left(u-jT_c-\frac{T_c}{2}\right)\sin\left(\frac{\pi}{T_c}\left(t-iT_c-\frac{T_c}{2}\right)\right)\sin\left(\frac{\pi}{T_c}\left(u-jT_c-\frac{T_c}{2}\right)\right) \\ &\qquad\times\overline{PN_2(t)PN_2(u)}dudt \\ &+\frac{\sin(\theta)^2}{T^2}\int_0^T\int_0^T \overline{2n_s(t)n_s(u)}\sum_i\sum_j p(t-iT_c)p(u-jT_c)\sin\left(\frac{\pi}{T_c}(t-iT_c)\right)\sin\left(\frac{\pi}{T_c}(u-jT_c)\right)\overline{PN_1(t)PN_1(u)}dudt \\ &+\frac{\sin(\theta)^2}{T^2}\int_0^T\int_0^T \overline{2n_c(t)n_c(u)}\sum_i\sum_j p\left(t-iT_c-\frac{T_c}{2}\right)p\left(u-jT_c-\frac{T_c}{2}\right)\sin\left(\frac{\pi}{T_c}\left(t-iT_c-\frac{T_c}{2}\right)\right)\sin\left(\frac{\pi}{T_c}\left(u-jT_c-\frac{T_c}{2}\right)\right) \\ &\qquad\times\overline{PN_2(t)PN_2(u)}dudt \end{aligned} \tag{3.5-76}$$

Equation (3.5-76) can be simplified by noting that the expected value of the noise pairs yield $(N_0/2)\delta(t-u)$. Thus

$$\begin{aligned}\mathrm{Var}(X_n(T)) &= \frac{2\cos(\theta)^2}{T^2}\int_0^T\left(\frac{N_0}{2}\right)\sum_i p(t-iT_c)^2\sin\left(\frac{\pi}{T_c}(t-iT_c)\right)^2 dt\\ &+\frac{2\cos(\theta)^2}{T^2}\int_0^T\left(\frac{N_0}{2}\right)\sum_i p\left(t-iT_c-\frac{T_c}{2}\right)^2\sin\left(\frac{\pi}{T_c}\left(t-iT_c-\frac{T_c}{2}\right)\right)^2 dt\\ &+\frac{2\sin(\theta)^2}{T^2}\int_0^T\left(\frac{N_0}{2}\right)\sum_i p(t-iT_c)^2\sin\left(\frac{\pi}{T_c}(t-iT_c)\right)^2 dt\\ &+\frac{2\sin(\theta)^2}{T^2}\int_0^T\left(\frac{N_0}{2}\right)\sum_i p\left(t-iT_c-\frac{T_c}{2}\right)^2\sin\left(\frac{\pi}{T_c}\left(t-iT_c-\frac{T_c}{2}\right)\right)^2 dt\end{aligned} \quad (3.5\text{-}77)$$

Using (3.5-57) and noting that the summation yields unity everywhere yields

$$\mathrm{Var}(X_n(T)) = \frac{2\cos(\theta)^2}{T^2}\int_0^T\frac{N_0}{T}dt + \frac{2\sin(\theta)^2}{T^2}\int_0^T\frac{N_0}{T}dt = \frac{N_0}{T} \quad (3.5\text{-}78)$$

Now approximate the sum of the jammer noise and the thermal noise as nonwhite Gaussian noise. Then use (3.5-60), (3.5-68), and (3.5-78), which leads to the final result for the bit error rate

$$PE = Q\left[\sqrt{\frac{2PT}{N_0 + JT_c}}\right] \quad (3.5\text{-}79)$$

It is to be observed that this result is the same as for the DS(QPSK)/BPSK of Figure 3.5-2! It should be pointed out that placing a notch filter at the jammer frequency could defeat a tone jammer. This will cause some degradation in the signal level but normally removes nearly all of the tone jammer so that it becomes quite ineffective.

3.6 MULTITONE JAMMING BIT ERROR RATE PERFORMANCE

Both SFH/MFSK and SFH/DPSK will be considered in this section. Noise will not be considered in the analysis to follow. This greatly simplifies the analysis, but provides only approximate answers when noise is present.

3.6.1 Multitone Jamming Bit Error Rate Performance for SFH/MFSK

In this section the bit error rate (BER) performance will be determined under idealized conditions for frequency-hopped systems only. Numerous analyses have been performed [9, 12–16] in the area of multitone jamming, including nonorthogonal frequency spacing as well as orthogonal frequency spacing. The analysis presented here follows [13] in which it is assumed that orthogonal frequency spacing of the modulated tones is used. Figure 3.2-6 illustrates a model of the FH/MFSK receiver.

It is assumed that the jammer has complete knowledge of the signal structure including the hopping bandwidth, and transmits only one jamming tone per MFSK frequency bin. Slow frequency hopping is assumed in this analysis. The MFSK symbol rate is given by $R_{sy} = R_b/\log_2(M)$, where R_b is the bit rate input to the MFSK modulator, and M is the number of tones used in the MFSK modulation. At each hop frequency, one of M possible MFSK tones can be transmitted. The tones are spaced R_{sy} Hz apart. Hence the bandwidth associated with each hop is given by $W_m = MR_{sy}/\log_2(M)$. The jammer transmits a total of n tones, each having a power of

$$J_n = J / n \tag{3.6-1}$$

and spaced apart by W_m Hz.

Neglect the thermal noise in the analysis to follow. Under this condition it follows that if $J_n < P$, where P is the received signal power, the correct MFSK tone will be detected. When $J_n > P$, then a symbol error will be made if the jamming tone is at the correct hop frequency, but not located at the correct signal MFSK frequency. It follows that the jammer should not use more power than necessary in order to maximize the number of frequency hops possible. In the limiting case set the jammer power to just more than the signal power so that the number n approximates

$$n = \left\lfloor \frac{J}{P} \right\rfloor \tag{3.6-2}$$

where $\lfloor x \rfloor$ denotes the largest integer less than or equal to x. If $J < P$, then $n = 0$. It will be assumed that $J > P$ in what follows. The maximum value of n, call it n_{max}, is bounded by the number of frequency hop bins, or

$$n_{max} \le \frac{W}{W_m} \tag{3.6-3}$$

The probability of any one FH band being jammed is given by the ratio

$$p_{Jam} = \left(\frac{n}{W/W_m} \right) \tag{3.6-4}$$

When all FH bands are jammed and n is chosen from (3.6-2), then the symbol error probability is just the probability that the jammer is not at the same MFSK frequency as the jammer so that

$$P_{sy} = 1 - \frac{1}{M} \tag{3.6-5}$$

when $W/W_m < \lfloor J/P \rfloor$. Now when all the FH bands are not jammed, then the probability of a symbol error is given by the probability that the correct hop band is chosen and the jammer does not fall at the frequency of the signal. Thus

$$P_{sy} = \left(1 - {1}/{M}\right) \frac{nW_m}{W} \tag{3.6-6}$$

under the condition that $1 \le \lfloor J/P \rfloor \le W/W_m$. As noted earlier when the jammer power is less than the signal power, then the symbol error probability is zero. Thus when can write the following

$$PE_{sy} = \begin{cases} \left(1 - {1}/{M}\right) & \left({W}/{W_m}\right) < \lfloor J/P \rfloor \\ \left(1 - {1}/{m}\right)\left({nW}/{W_m}\right) & 1 \le \lfloor J/P \rfloor \le W/W_m \\ 0 & \lfloor J/P \rfloor < 1 \end{cases} \tag{3.6-7}$$

In order to obtain the bit error rate probability, note that

$$PE_b = MPE_{sy}/(2(M-1)) \tag{3.6-8}$$

for orthogonal signals (MFSK) and that $n \simeq J/P$ and $W_m = MR/(\log_2 M)$. It is left to Problem 7 to show that

$$PE_b = \begin{cases} 0.5 & \left(\dfrac{PW}{JR_b}\right) < M\Big/(\log_2 M) \\ \left(M\Big/2\log_2 M\right)\left(\dfrac{JR_b}{PW}\right) & \dfrac{M}{\log_2 M} \le \dfrac{PW}{JW_m} \le W\Big/R_b \\ 0 & W\Big/R_b < \dfrac{PW}{JR_b} \end{cases} \tag{3.6-9}$$

This relationship is plotted in Figure 3.6-1 for M=2, 4, 8, and 16. This figure should be compared to Figure 3.3-5 for SFH/MFSK in partial band noise jamming, where it can be seen that multitone jamming is more efficient than partial band noise jamming in the sense that for the same $PW/(JR_b)$, the bit error rate is higher for multitone jamming.

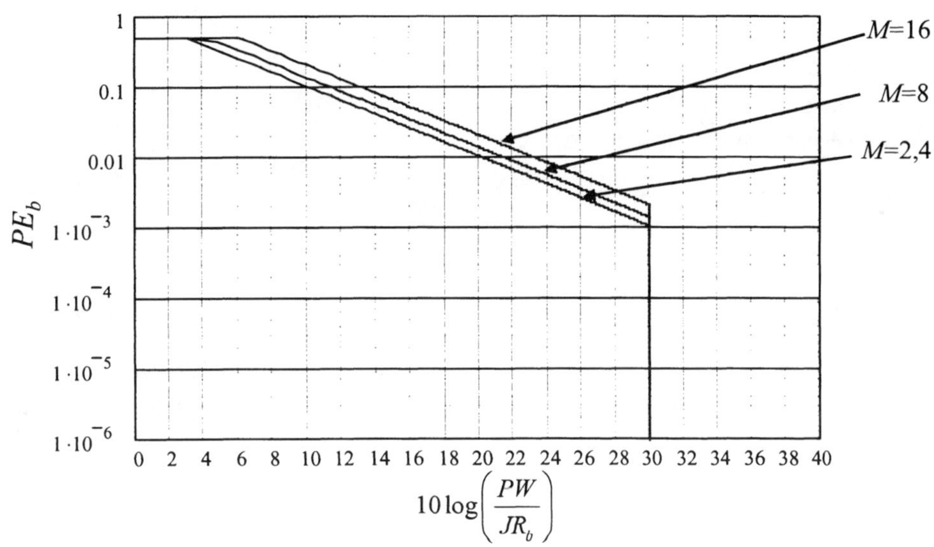

Figure 3.6-1 BER performance for multiple-tone jamming of SFH/MFSK.

This analysis assumed that only one tone per MFSK hop frequency was utilized. If the hop frequencies are known exactly, then this jamming procedure works well. If the hop frequencies are not known exactly then more tones than what was considered in the previous analysis may be a better choice; see Levitt [10, 14, 17]. Even with a more detailed analysis, it can be shown [13] that (3.6-9) is asymptotically accurate at high values of of $(PW_m)/(JR_b)$.

3.6.2 Multitone Jamming Bit Error Rate Performance for SFH/DPSK

In this section the performance of slow frequency hopping with binary differential phase shift keying is obtained. This work follows [9, 13]. Extensions to M-ary DPSK can be found in Simon [18].

Again the total received jammer power is assumed to be J W. Each jammer has power J_n where

$$J_n = \frac{J}{n} \tag{3.6-10}$$

The received signal power is assumed to be P W, and each frequency hop bandwidth is assumed to be R_s Hz, where R_s is the encoded symbol rate and $R_s \simeq R_b$, where R_b is the bit rate. The frequency hop bandwidth is W Hz. The number of frequency hop frequencies is given by

$$N_{FH} = \frac{W}{R_s} \tag{3.6-11}$$

It follows that the probability that a hop is jammed is given by

$$P_{Jam} = \frac{n}{N_{FH}} = \frac{n}{W/R_s} \tag{3.6-12}$$

It will be assumed that the noise is negligible. The DPSK demodulator functions by comparing the phase of the present symbol with the previous symbol. When the phase difference, call it $\Delta\theta$ $(\Delta\theta = \theta_2 - \theta_1)$, satisfies

$$-\frac{\pi}{2} < \Delta\theta \le \frac{\pi}{2} \tag{3.6-13}$$

then the receiver detects a "one." And when $\Delta\theta$ satisfies

$$\frac{\pi}{2} < \Delta\theta \le \frac{3\pi}{2} \tag{3.6-14}$$

the receiver detects a "zero." The received signal plus jammer is a vector sum of the two terms. The resulting angle of the sum of the signal plus jammer for the present symbol minus the resulting angle of the sum of the signal plus jammer for the past symbol is used to determine the phase difference, $\Delta\theta$, determines the detected data bit. Consider the case when a data bit "one" was transmitted, which is illustrated in Figure 3.6-2.

In Figure 3.6-2(a) the presently received data symbol and the previously received data symbol are shown on the right and left sides, respectively, for the case of a data bit "1" has been transmitted. Figure 3.6-2(b), the presently received data symbol and the previously received data symbol shows on the right and left sides, respectively, for the case of a data bit "0" that has been transmitted. Also the two diagrams have been joined at the point designated as "0" in Figure 3.6-2(b). Clearly in the case of the data bit "1" the phase angle difference is zero, and the detector correctly detects a data bit "1." In the second case of the transmitted data bit "0" the phase angle $\Delta\theta$ is not zero. Using the law of cosines on the triangle A0B yields the result

$$\cos(\Delta\theta) = \frac{R_1^2 + R_2^2 - \left(2\sqrt{2P}\right)^2}{2\sqrt{4J_n P}} \tag{3.6-15}$$

Solving for the angle ϕ from the triangle 0CA and 0BD produces the results

$$\cos(\phi)=\frac{2P+2J_n-R_2^2}{2\sqrt{4J_nP}} \tag{3.6-16}$$

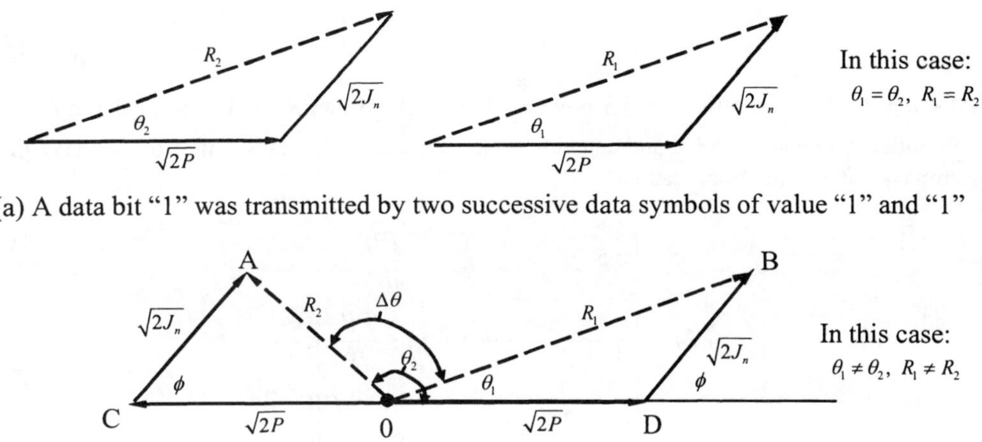

(a) A data bit "1" was transmitted by two successive data symbols of value "1" and "1"

(b) A data bit "0" was transmitted by two successive data symbols of value "1" and "0"

Figure 3.6-2 (a, b) Phase diagram showing the effect of a tone jammer on a data bit "1" and a "0" for SFH/DPSK modulation.

and

$$\cos(\pi-\phi)=-\cos(\phi)=\frac{2P+2J_n-R_1^2}{2\sqrt{4J_nP}} \tag{3.6-17}$$

Solving (3.6-16) and (3.6-17) for $R_1^2+R_2^2$ and substituting into (3.6-17) produces the result

$$\cos(\Delta\theta)=\frac{2(J_n-P)}{R_1R_2} \tag{3.6-18}$$

This result is established in Problem 8. When $\pi/2<\Delta\theta\leq 3\pi/2$, the demodulator produces the correct output, (i.e., a "0" data bit was transmitted). When $-\pi/2<\Delta\theta\leq\pi/2$, then an error is made on the data bit "0." In other words whenever $\cos(\Delta\theta)\geq 0$ then an error is made. This occurs whenever $J_n>P$. There are no errors when a data bit "1" is transmitted. Therefore the number of jammer tones is given by

$$n\simeq\frac{J}{P} \tag{3.6-19}$$

The optimum jamming power is just $J_n\simeq P$. The bit error rate probability is given by

$$\begin{aligned}PE_b&=0.5P(\text{error}|0\text{ was transmitted})\\&\quad+0.5P(\text{error}|1\text{ was transmitted})=0.5P(\text{error}|0\text{ was transmitted})\end{aligned} \tag{3.6-20}$$

The conditional bit error probability, given that a data bit "0" was transmitted, is given by the expression for P_{jam}, of (3.6-12). Combining (3.6-12) and (3.6-19) with (3.6-20) leads to the result

$$PE_b = \frac{(J/P)}{2(W/R_b)} = \frac{1}{2\left(\frac{PW}{JR_b}\right)} \tag{3.6-21}$$

Note that $1 \le n \le W/R_b$. When the jammer power is large, all FH bands can be jammed and $PE_b = 0.5$. Also when $J < P$, not even one single FH band can be jammed sufficiently to cause an error, so that $PE_b = 0$ in this case. In summary one has the final result

$$PE_b = \begin{cases} 0.5 & \frac{PW}{JR_b} < 1 \\ \frac{1}{2(PW)/(JR_b)} & 1 \le \frac{PW}{JR_b} < \frac{W}{R_b} \\ 0 & \frac{\mathrm{W}}{\mathrm{R_b}} < \frac{PW}{JR_b} \end{cases} \tag{3.6-22}$$

This result is plotted in Figure 3.6-3. Once again the inverse value of the parameter $X = (PW)/(JR_b)$ is seen in (3.6-22). If we compare Figure 3.6-1 with Figure 3.6-3, we see that SFH/DPSK is about 3 dB better than SFH/MFSK in the binary case.

3.7 DEGRADATION DUE TO INTERFERENCE OR JAMMING IN DS SYSTEMS

In this section the degradation due to interference or noise jamming to the bit error rate performance of DS/BPSK systems will be determined. Many times interference from other signals is a major problem in closing the link. This section will derive an expression useful for determining the degradation due to the effects of other signals, or jamming signals. This calculation is useful for code acquisition, bit error rate, and carrier loop tracking degradation; it is not directly applicable to code tracking degradation. This section follows [19].

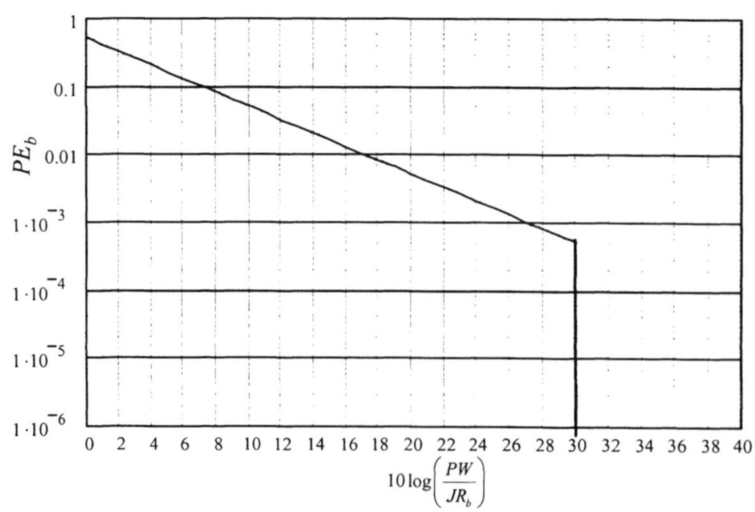

Figure 3.6-3 BER performance of SFH/DPSK spread spectrum signal in worst-case multiple-tone jamming.

3.7.1 Equivalent Noise Spectral Density for DS(BPSK)/BPSK Systems

This section will determine the equivalent noise spectral density of DS(BPSK)/BPSK signals. The calculation developed here is applicable to direct sequence systems, a model of which is illustrated in Figure 3.7-1. The simplified received signal plus interference model is given by

$$y(t)=\sqrt{2P_1}d_1(t)PN_1(t)\cos(\omega_0 t)+\sqrt{2}n_c(t)\cos(\omega_0 t)+\sqrt{2}n_s(t)\sin(\omega_0 t) \quad (3.7\text{-}1)$$

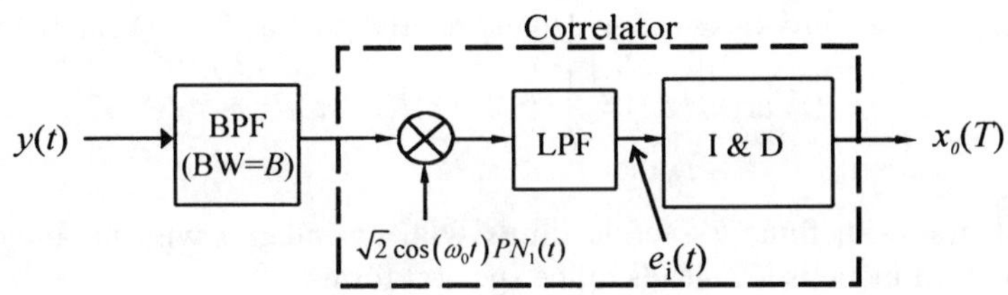

Figure 3.7-1 Simplified model for the correlation output for determining the interference and degradation due to the other signals in the band for DS(BPSK)/BPSK.

In (3.7-1) P_1 is the power of the received signal being tracked, $d_1(t)$ is the corresponding baseband data signal, and $PN_1(t)$ is the corresponding spread spectrum code used in signal 1; this is the one being tracked. Also the thermal noise and the jamming or interference signals are modeled with the last two terms in (3.7-1) as a Gaussian random process that can be decomposed as the sum of the white Gaussian thermal noise and the interfering signal Gaussian random process.

Demodulating the signal with the coherent reference signal and the input to the averaging integrator, out of the low-pass filter (which removes the double frequency term) is given by

$$e_i(t)=\sqrt{P}d(t)+PN_1(t)n_c(t) \quad (3.7\text{-}2)$$

At the output of the averaging integrator, one obtains the correlated signal plus noise

$$x_o(T)=\sqrt{P}\int_{-\infty}^{\infty}H_{BB}(f)S_{PN_2}(f)df+\frac{1}{T}\int_0^T n_c(t)PN_1(t)dt \quad (3.7\text{-}3)$$

where the correlated signal component [20] depends on the baseband equivalent low-pass filter $H_{BB}(f)$ of the bandpass filter at the input of the correlator system (bandpass filter of the receiver), with RF transfer function $H(f)$ and bandwidth B Hz. The output can be separated into the signal term and the noise term as suggested by (3.7-3)

$$x_0(T)=s_0(T)+n_0(T) \quad (3.7\text{-}4)$$

The variance of the noise term is given by

$$\text{Var}(n_o(T))=\frac{1}{T^2}E\left\{\int_0^T\int_0^T n_c(t)n_c(u)PN_1(t)PN_1(u)dtdu\right\} \quad (3.7\text{-}5)$$

Modeling the $PN_1(t)$ code as a stationary random process independent of the signal process and the thermal noise process yields

$$\mathrm{Var}(n_o(T)) = \frac{1}{T^2}\int_0^T\int_0^T R_{n_c}(t-u)R_{PN_1}(t-u)dtdu \tag{3.7-6}$$

This can be expressed as a single integration [21] with a change of variables to the form

$$\mathrm{Var}(n_0(T)) = \frac{1}{T^2}\int_{-T}^{T} R_{n_c}(\tau)R_{PN_1}(\tau)\left[T-|\tau|\right]d\tau \tag{3.7-7}$$

With the assumption that the two processes are wideband relative to $1/T$, (3.7-7) can be well approximated by

$$\mathrm{Var}(n_0(T)) = \frac{1}{T}\int_{-T}^{T} R_{n_c}(\tau)R_{PN_1}(\tau)d\tau \tag{3.7-8}$$

Approximating the integral with finite T second limits with an integral with unbounded limits and using Parseval's theorem for real functions [22] leads to the spectral form

$$\mathrm{Var}(n_0(T)) = \frac{1}{T}\int_{-\infty}^{\infty} S_{n_c}(f)S_{PN_1}(f)df \tag{3.7-9}$$

where $S_{n_c}(f)$ is the baseband equivalent power spectral density of the sum of white Gaussian (thermal) noise (WGN) and the jamming or interfering signals, which are modeled as Gaussian random processes, and $S_{PN_1}(f)$ is the unit power psd of the despreading signal. All code spectral densities denoted by the form $S_{PN}(f)$ will be assumed to have unit power in this section. This finite versus infinite time limit is justified if the correlation function becomes essentially zero in a small time compared to T seconds. Note that due to the front-end filter, the noise process spectral density $S_{n_c}(f)$ may have filtering on it. Consequently the limits of the integral may be finite for ideal filters.

Now consider the case that the other signals act as interference to signal 1. We can write the noise spectral density as the sum of thermal noise and interfering noise so that

$$S_{n_c}(f) = S_n(f) + S_{n_I}(f) \tag{3.7-10}$$

where $S_n(f)$ is the two-sided noise spectral density and $S_{n_I}(f)$ is the filtered, two-sided, interference spectral density.

Accounting for the band-limiting due to the front-end bandpass filter, one has

$$\mathrm{Var}(n_0(T)) = \frac{N_0}{2T}\int_{-\infty}^{\infty} |H_{BB}(f)|^2 S_{PN_1}(f)df + \frac{1}{T}\int_{-\infty}^{\infty} |H_{BB}(f)|^2 S_I(f)S_{PN_1}(f)df \tag{3.7-11}$$

where $S_I(f)$ is the interference or jamming spectral density, prior to being filtered by the baseband equivalent bandpass filter $H_{BB}(f)$, which is the baseband equivalent transfer function of the receiver front-end bandpass filter.

Note that the equivalent noise spectral density is obtained from (3.7-11) as

$$N_0' = 2T\mathrm{Var}(n_0(T)) \tag{3.7-12}$$

This can be written explicitly in terms of the code and interference spectral densities as

$$N_0^{'} = N_0 \int_{-\infty}^{\infty} |H_{BB}(f)|^2 S_{PN_1}(f)df + 2\int_{-\infty}^{\infty} |H_{BB}(f)|^2 S_I(f)S_{PN_1}(f)df \tag{3.7-13}$$

Noting that the baseband equivalent interference process $S_I(f)$ contains $I/2$ watts of interference power, (3.7-13) can be written as

$$N_0^{'} = N_0 \int_{-\infty}^{\infty} |H_{BB}(f)|^2 S_{PN_1}(f)df + I\int_{-\infty}^{\infty} |H_{BB}(f)|^2 S_I^{'}(f)S_{PN_1}(f)df \tag{3.7-14}$$

where $S_I^{'}(f)$ denotes the unit power, baseband equivalent, unfiltered, power spectral density (psd) of the interference or jammer and is related to the nonunit power spectral density by

$$S_I^{'}(f) = \frac{2}{I} S_I(f) \tag{3.7-15}$$

The integral in the second term in (3.7-14) is called the *spectral separation coefficient* (SSC) and is normally expressed in dB. It relates how much a given interferer interferes with the received signal. If the unit powered signal spectral densities don't overlap much, the SSC will be small and the interfering signal will have little effect. On the other hand if there is considerable overlap, the SSC will be larger and the degradation on the received signal will be more appreciable for a given interference power lever, I.

Equation (3.7-15) applies in the general case for arbitrary interference. In the special case that the RF filter is modeled as an ideal bandpass filter with a bandwidth of B Hz, the effective noise PSD due to an arbitrary interference simplifies for the equivalent noise spectral density to

$$N_0^{'} = N_0 \int_{-B/2}^{B/2} S_{PN_1}(f)df + I\int_{-B/2}^{B/2} S_{PN_1}(f)S_I^{'}(f)df \tag{3.7-16}$$

On the other hand, when the interference is due to another signal, say, signal 2, with baseband unit-power spectral density $S_{PN_2}(f)$, then (3.7-14) applies with $S_I^{'}(f) = S_{PN_2}(f)$.

3.7.2 Carrier to Equivalent Noise Spectral Density Ratio for DS(BPSK)/BPSK

It is now convenient to compute the signal to equivalent noise spectral density ratio after despreading a DS(BPSK)/BPSK signal. First define the output signal as the signal component out of the averaging filter. Thus the correlated signal voltage out of the averaging filter is given by [20]

$$s_o(T) = \sqrt{P}\,\mathrm{Re}\left(\int_{-\infty}^{\infty} H_{BB}(f)S_{PN_1}(f)df\right) \tag{3.7-17}$$

where P is the received signal power expressed in watts. When the baseband equivalent transfer function $H_{BB}(f)$ is complex conjugate symmetric, then (3.7-17) can be simplified [20] to

$$s_o(T) = \sqrt{P}\int_{-\infty}^{\infty} H_{BB}(f)S_{PN_1}(f)df \tag{3.7-18}$$

Note that complex conjugate symmetric means that $H_{BB}^{*}(j\omega)=H_{BB}(-j\omega)$. By definition, the output signal-to-noise ratio is given by

$$SNR_o(T)=\frac{s_o(T)^2}{Var(n_o(T))} \tag{3.7-19}$$

In WGN this expression can be evaluated, in the case of no filtering from (3.7-13) and (3.7-18), to

$$SNR_o(T)=\frac{P}{\left(\dfrac{N_0}{2T}\right)}=\frac{2PT}{N_0} \tag{3.7-20}$$

as is well known from matched filter theory.

When a front-end RF filter is present at the receiver, $s_0(T)$ is given by (3.7-17) and Var($n_0(T)$) is given by (3.7-14). From (3.7-14) it follows that

$$N_0^{'}=\nu N_0+I_0 \tag{3.7-21}$$

where ν is defined by

$$\nu=\int_{-\infty}^{\infty}\left|H_{BB}(f)\right|^2 S_{PN_1}(f)df \tag{3.7-22}$$

in the general case, where $H_{BB}(f)$ is the baseband equivalent receiver filter response.

The interference parameter I_0 is the SSC and is given by

$$I_0=I\int_{-\infty}^{\infty}\left|H_{BB}(f)\right|^2 S_I^{'}(f)S_{PN_1}(f)df \tag{3.7-23}$$

Note that the interference could be due to another signal or set of signals or an interference or jammer. The equivalent $P/N_0^{'}$ can be obtained from the relationship

$$\frac{P}{N_0^{'}}=(2T)SNR_0(T) \tag{3.7-24}$$

and $SNR_0(T)$ is defined in (3.7-19). When only thermal noise is present and there is no filtering, (3.7-24) reduces to

$$\frac{P}{N_0^{'}}=\frac{P}{N_0} \tag{3.7-25}$$

where P is the received signal power in watts. If the noise degradation is defined only relative to the increase in the effective noise level, then the effective one-sided noise spectral density in (3.7-21) can be written as

$$N_0^{'}=\nu N_0\left(1+\frac{I_o}{\nu N_0}\right) \tag{3.7-26}$$

where I_o is given by (3.7-23). When interference is present the equivalent value of carrier power to noise spectral density is reduced. Combining (3.7-14) and (3.7-18), the effective carrier ($C = P$) to one-sided noise spectral density ratio, in the general case, is given by

$$\frac{C}{N_0^{'}} = \frac{P\left(\operatorname{Re}\int_{-\infty}^{\infty} H_{BB}(f) S_{PN_1}(f)df\right)^2}{N_0 \int_{-\infty}^{\infty} \left|H_{BB}(f)\right|^2 S_{PN_1}(f)df + I\int_{-\infty}^{\infty} \left|H_{BB}(f)\right|^2 S_I^{'}(f) S_{PN_1}(f)df} \tag{3.7-27}$$

where I is the received interference power expressed in watts and all the power spectral densities have unit power. Equation (3.7-27) can be simplified in the case of an ideal BPF with bandwidth of B Hz to

$$\frac{C}{N_0^{'}} = \frac{P\left(\int_{-B/2}^{B/2} S_{PN_1}(f)df\right)^2}{N_0 \int_{-B/2}^{B/2} S_{PN_1}(f)df + I\int_{-B/2}^{B/2} S_I^{'}(f) S_{PN_1}(f)df} \tag{3.7-28}$$

We now determine the degradation for some particular cases in the following section.

3.7.3 Equivalent Noise Spectral Density Degradation for DS(BPSK)/BPSK Systems

Now consider the noise degradation due to an interference or jammer. Define the noise degradation as the additional noise due to interference. From (3.7-21) and (3.7-23) we have

$$N_0^{'} = \nu N_0 + I\int_{-\infty}^{\infty} \left|H_{BB}(f)\right|^2 S_I^{'}(f) S_{PN_1}(f)df \tag{3.7-29}$$

where ν is defined by (3.7-22). Factoring out the nominal noise yields

$$N_0^{'} = \nu N_0 \left(1 + \frac{I\int_{-\infty}^{\infty} \left|H_{BB}(f)\right|^2 S_I^{'}(f) S_{PN_1}(f)df}{\nu N_0}\right) \tag{3.7-30}$$

The noise degradation due to the interference-induced noise is therefore given by

$$Degr_n = \left(1 + \frac{I\int_{-\infty}^{\infty} \left|H_{BB}(f)\right|^2 S_I^{'}(f) S_{PN_1}(f)df}{\nu N_0}\right)^{-1} \tag{3.7-31}$$

where I is the interfering power (interfering signal) at the input to the receiver, expressed in watts. The carrier-to-noise density ratio, when the front-end filtering is accounted for, is given by (3.7-27), so that the total signal power to noise spectral density degradation, relative to the quantity $P/(\nu N_0)$, is given by

$$Degr_{total} = \frac{\gamma^2}{\left(1 + \frac{I\int_{-\infty}^{\infty}\left|H_{BB}(f)\right|^2 S_I'(f) S_{PN_1}(f)df}{\nu N_0}\right)} \tag{3.7-32}$$

By total it is meant the additional effective noise spectral density degradation as well as the correlated signal power loss, γ^2. The input signal level is taken to be P W, and the filtered thermal noise is taken as N_0. In (3.7-32), γ is the signal correlation loss component, where

$$\gamma = \mathrm{Re}\left\{\max_{\tau}\left[\int_{-\infty}^{\infty} H_{BB}(f) S_{PN}(f) e^{j2\pi f\tau} df\right]\right\} \tag{3.7-33}$$

which applies if the baseband equivalent filter $H_{BB}(f)$ is not complex conjugate symmetric [20]. However, if the baseband filter is complex conjugate symmetric [20], the following applies

$$\gamma = \max_{\tau}\left[\int_{-\infty}^{\infty} H_{BB}(f) S_{PN}(f) e^{j2\pi f\tau} df\right] \tag{3.7-34}$$

It is seen that $\gamma = \nu$ in the case of an ideal bandpass filter; however, the correlation loss in power is γ^2 whereas the power loss goes as γ, so that the correlation for an ideal bandpass filter is 20log(γ), whereas the power loss is 10log(γ).

Example 6 Assume that the unit-power signal spectral density is given by the following expression

$$S_s(f) = T_c \frac{\sin(\pi f T_c)^2}{(\pi f T_c)^2}$$

And furthermore assume that the interference is more of the same type of signal with the same power spectral density, but with distinct codes. The chip rate is 1.023×10^6. If the received interferer power is –145 dBW and the received signal power is –155 dBW and N_0 = –202 dBW/Hz, what is the noise degradation when the front-end bandwidth is large compared to the chip rate? Since the bandwidth is large, we can assume that $\nu = 1$. Furthermore, from (3.7-31), the following evaluation is made

$$Degr_n = 10\log\left(\left(1 + I\frac{\left(\int_{-\infty}^{\infty} S_s^2(f)df\right)}{N_0}\right)^{-1}\right) = -1.25 \text{ dB}$$

In addition if the interferer power was –140 dBW, then the degradation would increase to –3.14 dB.

Example 7 Determine the correlation loss associated with passing an NRZ signal having a chip rate of 5 Mcps, with a spectral density given by

$$S(f) = T\frac{\sin(\pi fT)^2}{(\pi fT)^2}$$

through an ideal bandpass filter, having bandwidth of 10 MHz. From (3.7-33) it follows that the integral produces

$$\int_{-5*10^6}^{5*10^6} S(f)df = 0.903$$

This is an equivalent loss of 20×log(0.903) = – 0.89 dB. In other words the correlation loss associated with using an ideal filter set to the first nulls of the signal spectra is about 0.9 dB.

3.7.4 Degradation to NRZ Signals Due to Narrowband Jammers for DS(BPSK)/BPSK Signals

Consider a narrowband Gaussian random process jammer at the center of the band, having a flat bandwidth of B_I Hz, degrading a nonreturn-to-zero (NRZ) DS(BPSK)/BPSK signal. Figure 3.7-2 illustrates the NRZ RF model of the narrowband interference, as well as the baseband version model.

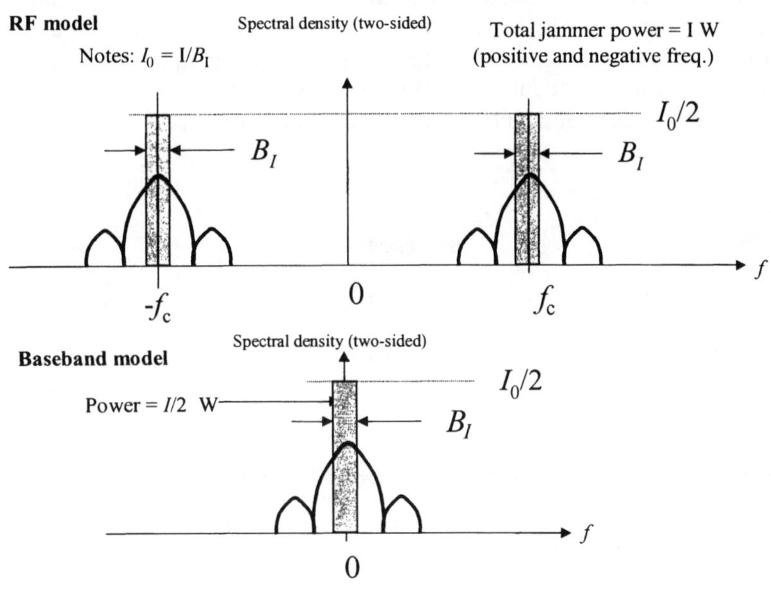

Figure 3.7-2 Radio frequency (RF) and baseband model for NRZ signals being jammed by narrowband jammers.

From (3.7-31) the noise degradation is given by

$$Degr_n = \left(1 + \frac{\frac{I}{B_I}\int_{-B_I/2}^{B_I/2} |H_{BB}(f)|^2 S_{PN_1}(f)df}{\nu N_0}\right)^{-1} \tag{3.7-35}$$

since the spectral density of $S_I^{'}(f)$ is $1/(B_I)$ for a unit-power interference, where ν is defined in (3.7-22), and $S_{PN_1}(f)$ is the received signal (unit) PSD.

This degradation expression applies to NRZ signals only. Note that when $B_I << 1/T_c$, then (3.7-35) can be well approximated by

$$Degr_n \cong \left(1+\frac{IH_{BB}(0)^2 S_{PN_1}(0)}{\nu N_0}\right)^{-1} \tag{3.7-36}$$

3.8 SUMMARY

This chapter dealt with the bit error rate performance with various types of jammers without coding or interleaving. The various types of jammers useful to jam the signal were presented. Some of the analyses were done without thermal noise present. Models were presented for each type of modulation. First broadband noise was the basis for a series of BER performance evaluations, followed by narrowband jamming BER performance. For coherent systems, when noise was considered, the evaluation was in terms of the parameter $PW/(JR_b)$ and E_b/N_0. Pulsed jamming for various SS and data modulation types were evaluated, and some similarities were found with partial band jamming. Single tone and multitone jamming followed the pulse jamming sections. Finally degradation due to interference or jamming a DS(BPSK)/BPSK system was developed. Both total degradation and noise degradation were presented.

It was observed that many modulation types, when subjected to optimum jamming led to a BER performance that goes as the inverse of the effective signal-to-noise ratio ($PW/(JR_b)$) in some regions of BER performance curves.

References

[1] M. K. Simon, S. M. Hinedi, and W. C. Lindsey, *Digital Communication Techniques*, Upper Saddle River, NJ: Prentice Hall, 1995.

[2] J. G. Proakis, *Digital Communications*, New York: McGraw-Hill, 1983.

[3] R. E. Ziemer and R. L. Peterson, *Digital Communications and Spread Spectrum Systems*, New York: Macmillan Publishing Company, 1985.

[4] J. H. Park, "On Binary Detection," *IEEE Trans. on Communications*, April 1978.

[5] E. Arthurs and H. Dym, "On the Optimum Detection of Digital Signals in the Presence of White Gaussian Noise—A Geometric Interpretation and a Study of Three Basic Data Transmission Systems," *IRE Transactions on Communications Systems*, December 1962.

[6] S. W. Golomb, L. D. Baumert, M. F. Esterling, J. J. Stiffler, and A. J. Viterbi, *Digital Communications With Space Applications*, Englewood Cliffs, NJ: Prentice-Hall, 1964.

[7] D. R. Martin and P. L. McAdam, "Convolutional Code Performance with Optimal Jamming," TRW Defense and Space Systems report, section 3, 1975.

[8] F. W. Hildebrand, *Advanced Calculus for Applications*, Englewood Cliffs, NJ: Prentice-Hall, Chapter 7, 1962.

[9] S. W. Houston, "Modulation Techniques for Communication, Part I: Tone and Noise Jamming Performance of Spread Spectrum M-ary FSK and 2, 4-ary DPSK Waveforms," *Conference Rec. of NAECON*, 1975.

[10] B. K. Levitt, "Use of Diversity to Improve FH/MFSK in Worst Case Partial Band Noise and Tone Jamming," *Proceedings of 1982 IEEE Military Communication Conference*, Vol. 2, October 17–20, 1982, pp. 28.2-1–8.2-5.

[11] A. J. Viterbi and I. M. Jacobs, "Advances in Coding and Modulation for Noncoherent Channels Affected by Fading, Partial Band, and Multiple Access Interference," in *Advances in Communication Systems*, Vol. 4, New York: Academic Press, 1975, pp. 279–308.

[12] D. J. Torrieri, "Frequency Hopping, Multiple-Frequency Shift Keying, Coding, and Optimal Partial-Band Jamming," Naval Air Systems Command, CM/CCM Center, Tech. Report CM/CCM-82-1, August 1982.

[13] R. E. Ziemer and R. L. Peterson, *Digital Communications and Spread Spectrum Systems,* New York: Macmillan Publishing Company, 1985.

[14] M. K. Simon, J. K. Omura, R. A. Scholtz, and B. K. Levitt, *Spread Spectrum Communications,* Vol. 2, Rockville, MD: Computer Science Press, 1985.

[15] L. B. Milstein, R. L. Pickholtz, and D. S. Schilling, "Optimizing of the Processing Gain of an FSK-FH System," *IEEE Trans. on Communications*, July 1980.

[16] M. P. Ristenbatt and J. L. Davis, Jr., "Performance Criteria for Spread Spectrum Communications," *IEEE Trans. on Communications,* August 1977.

[17] B. K. Levitt, "FH/MFSK Performance in Multitone Jamming," *IEEE J. Selected Areas in Communications,* Vol. SAC-3, September 1985, pp. 627–643.

[18] M. K. Simon, "The Performance of M-ary FH-DPSK in the Presence of Partial Band Multitone Jamming," *IEEE Trans. on Communications*, Vol. COM-30, May 1982, pp. 953–948.

[19] S. Raghavan and J. K. Holmes, "Frequency Band Sharing Between NRZ and Split Spectrum Signals—Analysis and Simulation Results," *AIAA ICSSC*, Montreal, 2002.

[20] J. K. Holmes, *Coherent Spread Spectrum Systems*, Malabar, FL: Kreiger Press, Chapter 8, 1990.

[21] A. Papoulis, *Probability, Random Variables, and Stochastic Processes*, 2nd ed., New York: McGraw-Hill, 1984.

[22] L. Couch II, *Digital and Analog Communication Systems*, 6th ed., Upper Saddle River, NJ: Prentice-Hall, Chapter 2, 2001.

[23] G. L. Lui and J. K. Holmes, "Power Spectral Density of General Offset Carrier Signals," *The Aerospace Corporation IOC*, November 27, 2001.

Problems

1. Show that for the binary case of MFSK, the bit error rate is given by

$$PE_b = \frac{1}{2}\exp\left[\frac{-E_b}{2N_0'}\right] \tag{P3-1}$$

2. Consider a narrowband and a wideband noise jammer. Show that when the wideband jammer bandwidth is set equal to $W = 2N/T_c$, then the on-frequency ($f_0 = f_J$) narrowband jammer is about 10 log ($2N$) dB more effective than the wideband jammer. More effective in this case means that it has a higher spectral density.
3. Show that the value of c_0 used in Section 3.3.3 for frequency-hopped BPSK and QPSK is given by $c_0 = 0.709$. In addition show that (3.3-28) is correct. Hint: Let $X = 3$ and show that the optimal value of ρ is given by $\rho = 0.2363$, so that $c_0 = 0.709$.
4. Show that (3.3-45), using the substitutions $y = \rho kX$ and $X = (PW)/(JR_b)$, can be written as (3.3-46) after canceling some common terms and rearranging the expressions. In addition show that when $M = 2$, the solution for y from (3.3-46) is $y_0 = 2$.

5. Show that the optimum value of L, L_0 is given by (3.3-58) for fast frequency hopping. Hint: Start with (3.3-56) and find the minimum value of L.

6. Show that (3.5-47) is statistically equivalent to (3.5-48). Do this by showing that the two equations have the same mean and same variance with the assumption that the expected value of $PN_1(t)PN_2(t)$ is zero.

7. Show that (3.6-9) follows from (3.6-7). Hint: Note that if $a > b > c$ is true, then it follows that $(1/a) < (1/b) < (1/c)$ is true.

8. Solving (3.6-16) and (3.6-17) for $R_1^2 + R_2^2$ and substituting into (3.6-15), show that the following result is obtained: $\cos(\Delta\theta) = \dfrac{2(J_n - P)}{R_1 R_2}$.

9. Show that when B_I is small compared to the chip rate $1/T_c$, the following expression for noise degradation applies to the case that a flat narrowband jammer of bandwidth B_I Hz can well be approximated by the expression

$$Degr_n \simeq \left(1 + \frac{I}{\nu N_0 (1/T_c)}\right)^{-1} \tag{P3-2}$$

when it is assumed that the low pass filter response at $f = 0$ is 1, and $S_{PN_1}(0) = T_c$, with T_c the chip duration.

10. Consider the effect of a pair of narrowband jammers or interference noise processes as shown in the following figure for jamming a signal spectra that has two equal-sized main lobe responses [23].

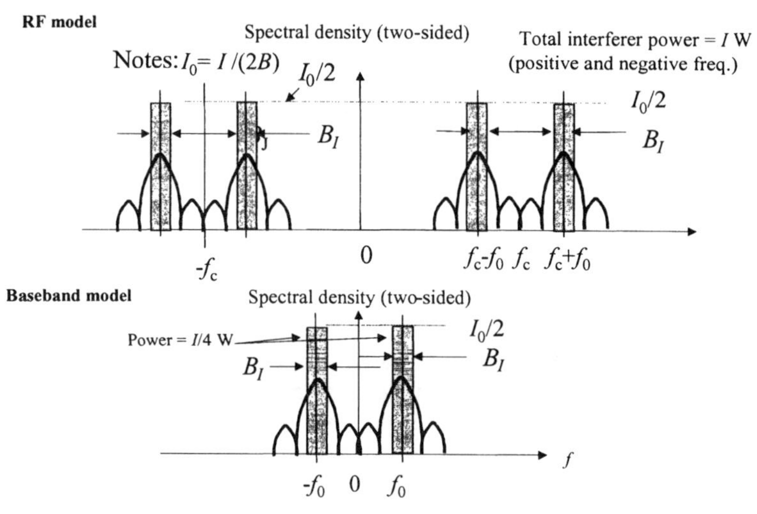

Figure for Problem 10 illustrating a pair of flat narrowband jammers.

(a) Show that the effective noise spectral density is given by

$$N_0' = N_0 \int_{-\infty}^{\infty} |H_{BB}(f)|^2 S_{PN_1}(f)df + I_0 \int_{-f_0 - B_I/2}^{-f_0 + B_I/2} |H_{BB}(f)|^2 S_{PN_1}(f)df + I_0 \int_{f_0 - B_I/2}^{f_0 + B_I/2} |H_{BB}(f)|^2 S_{PN_1}(f)df \tag{P3-3}$$

where f_0 is the positive frequency that maximizes the spectral density of the binary offset carrier (BOC) signal being interfered. Spectral symmetry implies

$$N_0^{'} = N_0 \int_{-\infty}^{\infty} |H_{BB}(f)|^2 S_{PN_1}(f)df + 2I_0 \int_{f_0-B_I/2}^{f_0+B_I/2} |H_{BB}(f)|^2 S_{PN_1}(f)df \qquad \text{(P3-4)}$$

which, due to the fact that $2I_0 = I/B_1$, becomes

$$N_0^{'} = N_0 \int_{-\infty}^{\infty} |H_{BB}(f)|^2 S_{PN_1}(f)df + \frac{I}{B_J} \int_{f_0-B_I/2}^{f_0+B_I/2} |H_{BB}(f)|^2 S_{PN_1}(f)df \qquad \text{(P3-5)}$$

(b) Further deduce that the noise degradation for a narrowband pair of interferers is given by

$$Degr_n = \left[1 + \frac{\dfrac{I}{B_I} \displaystyle\int_{f_0-B_I/2}^{f_0+B_I/2} |H_{BB}(f)|^2 S_{PN_1}(f)df}{\nu N_0} \right]^{-1} \qquad \text{(P3-6)}$$

where ν is given by (3.7-22). When the front-end filter is an ideal bandpass filter of bandwidth B Hz ($B>B_I$), where f_0 is the frequency offset that maximizes the dual peak spectral density, $S_{PN_2}(f_0)$ at $f=f_0$. If B_I is very small, show that (P3.4) can be approximated by

$$Degr_n = \left[1 + \frac{IS_{PN_1}(f_0)}{\nu N_0} \right]^{-1} \qquad \text{(P3-7)}$$

Note that in this case of an ideal front-end bandpass filter, the value of ν is given by

$$\nu = \int_{-B/2}^{B/2} S_{PN_1}(f)df \qquad \text{(P3-8)}$$

(c) Show that the total C/N_0 degradation is given by

$$Degr_{total} = \frac{\nu^2}{\left[1 + \frac{\dfrac{I}{B_J} \displaystyle\int_{f_0-B_I/2}^{f_0+B_I/2} S_{PN_1}(f)df}{\nu N_0} \right]} \qquad \text{(P3-9)}$$

where f_0 is 0 for the NRZ narrowband interferer signal case and is the frequency offset that maximizes the BOC spectral density in the BOC narrowband interferer pair case.

(d) Thus show in the general case for dual-spectral signals that the noise degradation can be described in the NRZ signal interferer case or the dual spectral signal case for dual-narrowband Gaussian noise interference and that the noise degradation is given by

$$Degr_n = \left[1 + \frac{\frac{I}{B_J} \int_{f_0 - B_J/2}^{f_0 + B_J/2} S_{PN_1}(f) df}{\nu N_0} \right]^{-1} \quad \text{(P3-10)}$$

CHAPTER 4

Jamming Performance of Coded Spread Spectrum Systems

4.0 INTRODUCTION

In this chapter the effects of coding and interleaving to the spread spectrum channel will be presented. The function of coding/interleaving is to improve upon the spread spectrum (SS) link performance in the face of jamming or interference. Recall in Chapter 3 the performance of SS links was better than non-SS links, due to the fact that the despreading function spread the interference over the spread bandwidth, thus diluting the effect of a narrowband jammer or interference. In fact the despread power that falls in the data bandwidth is the primary power causing a deleterious effect on the receiver performance. Since the power that falls in the data bandwidth is much less than the spread bandwidth, this ratio of spread bandwidth to data bandwidth yields the notion of processing gain.

Even with the employment of SS techniques, we found in Chapter 3 that worst-case jamming caused significant degradation to the bit error rate (BER) performance. The effect of worst-case jamming can be mitigated by the use of forward error correction (FEC) and the use of interleaving/deinterleaving. Jamming techniques that concentrate on a partial segment of the bandwidth, or on pulsing on a portion of the time that the jammer is on, tend to produce burst error patterns. However, most FEC techniques work best when the errors are distributed randomly; thus interleaving is assumed for all the FEC techniques that are used in this chapter. The effect of the interleaver/deinterleaver is to "randomize" the errors; the FEC methods have better performance with interleavers used in conjunction with deinterleavers. The model for the channel including the FEC and the interleaver/deinterleaver is shown in Figure 3.0-2 of Chapter 3.

The BER performance of the FEC schemes in this chapter is calculated as a function of the *discrete memoryless channel* (DMC) transition probabilities $p(y_j|x_j,z_j)$ that account for any jammer-state (side) information contained in z_j. This transition probability relates the output symbol y_j to the input symbol x_j and the jammer side information z_j at time j, which in its most rudimentary form indicates whether the jammer is on or off for channel use j. In some cases the input and output DMC alphabets are not the same. By having a different alphabet for the input and output, more information can be supplied to the decoder. As an example, the DMC can supply information indicating the reliability of the alphabet decision. When additional information is output from the DMC, the channel is called a *soft-decision channel*. The role of the interleaver/deinterleaver is to make the channel approximately memoryless. Consider the case when, in addition to outputting a binary one or zero, the DMC may also output information about the reliability of the one/zero decision. This occurs for the binary erasure channel for example. An example of a soft decision channel is illustrated in Figure 4.0-1. The input alphabet composed of a "1" and a "0." The output alphabet includes a "1," a "0," and an "erasure." An erasure indicates the channel is not sure what symbol was transmitted and decides not to guess what it was. The *binary symmetric channel* (BSC) is similar to the one shown in Figure 4.0-1 except that there is no "e" output. Thus a zero can become a zero or a one, and a one can become a one or a zero. The probabilities for the BSC are as follows:

$$p(1|0) = p(0|1) = p$$
$$p(1|1) = p(0|0) = 1 - p \qquad (4.0\text{-}1)$$

Thus a "one" will be received as a "1" with probability p and an error will occur with probability $1-p$. The same probabilities result when a "zero" is transmitted.

Jammer-state information (JSI) can include an indication of whether or not a jammer is detected. The use of JSI is critical for encoded systems that use soft-decision statistics, since without JSI a smart jammer can "break" the communication link with a small amount of jamming power. When considering hard-decision decoders, JSI is not a requirement; however, system performance can be improved with its use.

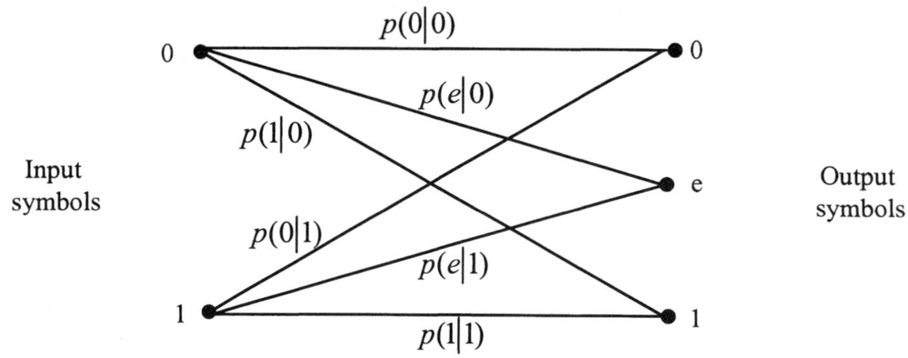

Figure 4.0-1 The transition diagram for soft-decision binary erasure channels.

The remainder of this chapter will address block, convolutional, and Turbo codes. Numerous references exist on the area of coding covering both convolutional and block coding. Some references that are relevant include, but are not limited to, Clark and Cain [1], Lin and Costello [2], Peterson and Weldon [3], Blahut [4], and Hanzo, Wong, and Yee [5].

4.1 INTERLEAVER STRUCTURES FOR CODED SYSTEMS

All of the systems analyzed in Chapter 3 did not necessarily have memoryless channels in the sense that a jamming burst could affect more than one coded bit in sequence. In the case of burst errors, an interleaver-deinterleaver combination renders the channel to be essentially memoryless. The approach is to interleave the coded data prior to transmission and deinterleave the coded data prior to decoding. Figure 4.1-1 illustrates the channel model.

An *interleaver* is a device that rearranges the ordering of the coded bits in a predetermined manner. Along with an interleaver is a compatible deinterleaver, which applies the inverse permutation to the coded bit stream. It is also used in turbo codes, as will be seen later in this chapter.

4.1.1 Block Periodic Interleaving

A typical block interleaver is implemented by writing the coded bits by columns into a rectangular array that has N rows and B columns.

Example 1 Figure 4.1-2 illustrates the interleaver for the case that $N = 8$ and $B = 10$. Coded symbols are read into the interleaver by columns and read out of the deinterleaver by rows, into the data modulator. This interleaver is referred to as a (B, N) block periodic interleaver.

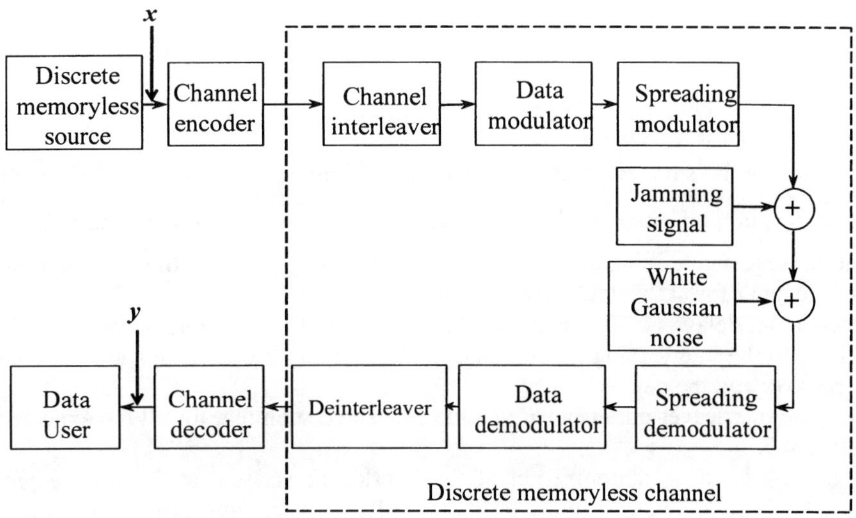

Figure 4.1-1 A model of a spread spectrum communication system.

B (columns), N (rows)

1	9	17	25	33	41	49	57	65	73
2	10	18	26	34	42	50	58	66	74
3	11	19	27	35	43	51	59	67	75
4	12	20	28	36	44	52	60	68	76
5	13	21	29	37	45	53	61	69	77
6	14	22	30	38	46	54	62	70	78
7	15	23	31	39	47	55	63	71	79
8	16	24	32	40	48	56	64	72	80

Figure 4.1-2 A block interleaver for the example when $N = 8$ and $B = 10$.

The deinterleaver does the inverse operation of the interleaver. The received coded bits are written into the deinterleaver by rows and read out by columns. It therefore follows that the coded bits in our example, which are fed into the data modulator, are as follows:

1,9,17,33,41,49,57,65,73,2,10,18,...

Suppose coded bits 27, 35, 43, 51, and 59 are corrupted by a burst event (they are shaded gray). Since they are transmitted in this sequence, and will be read out by columns in the deinterleaver, it is seen that each error will be spaced by $N = 8$ coded bits.

Good design dictates that for a block interleaver-deinterleaver the designer should make N large enough so that each error occurs in different code words, and thus the errors may be considered statistically independent. In addition B should be made so that all expected bursts of coded bits do not exceed length B. Clark and Cain [1] have shown that the following burst error characteristics of the block interleaver are as follows:

(1) Any burst of coded bit errors of length $b \leq B$ results in single errors at the deinterleaver output, each separated by at least N coded bits.

(2) Let $\lceil r \rceil$ denote the smallest integer greater than r. Any burst of length $b = rB$, for $r > 1$, results in bursts on the output of no more than $\lceil r \rceil$ coded bits, separated by no less than $N - \lceil r \rceil$ coded bits.

(3) A periodic sequence of single coded bit errors, spaced by B coded bits, results in a single burst of errors of length of N at the deinterleaver output.

(4) The end-to-end delay is $2NB$ coded bits, due to the interleaver-deinterleaver only. Channel delay is in addition to the interleaving process delay. The memory requirements are NB coded bits, in both the interleaver and the deinterleaver.

Typically, the interleaver parameter B would be selected such that for all the expected bursts lengths b, it would satisfy $b < B$.

It is to be noted that synchronization of the interleaver is associated with the problem of frame synchronization. The word boundaries must be known after deinterleaving in order to decode the coded words. One approach to deinterleaver synchronization is obtained by the use of a periodic addition of frame sync words, added to the data stream after interleaving and coding. Obtaining frame sync allows the receiver to synchronize to the decoder.

The hybrid interleaver has been described by Olsen [6] and is another type of interleaver, designed for use in packet switching applications on multiple access radio communication channels. This interleaving system has some of the characteristics of the block interleaver and some of the characteristics of the convolutional interleaving, which will be discussed next. Details are contained in the reference.

Another interleaver type discussed in the literature is the *uniform interleaver* [7], which is a probabilistic device that maps all words of weight w into all distinct $\binom{N}{w}$ permutations of it with equal probability given by $\Pr = 1/\binom{N}{w}$, where N is the size of the interleaver.

Additional interleaving types include the helical block interleaver system [8], and the pseudorandom interleavers [1].

4.1.2 Convolutional Interleaving

We have not addressed convolutional encoding; however, since we are addressing interleaving types, we will now address convolutional interleaving. Forney [9] has proposed the structure shown here. See also the work of Ramsey [10]. We will follow Forney in our presentation. It is assumed that all multiplexing functions are synchronous. Figure 4.1-3 illustrates a convolutional interleaver. It is to be noted that this interleaver will also be applicable to block codes.

There are B rows in both the interleaver and the deinterleaver. Each row has a multiple of M coded bit delays. Denote the parameter N as $N = BM$. This interleaver is referred to as a (B, N) interleaver. The multiplexer switches (shown with the curved double arrowheads) change position after each code bit time so that successive encoder coded bits enter different rows of the interleaver memory. Each interleaver and deinterleaver row contains a shift register memory of a certain number of units of delay of the coded bits of the form LM, where $L = 0, 1, 2, \ldots, B-1$, that are noted in the boxes.

The switching goes from top to bottom in a periodic fashion. First a coded bit is passed through the interleaver with no delay. It passes through the channel and is presented with a delay of $(B-1)M$ coded bits at the deinterleaver. The second coded bit is passed through M code bits of delay in the interleaver and sent through the burst error channel and presented with a delay of $(B-2)M$ coded bits in the deinterleaver. It follows that adjacent coded bits are M coded bits apart in the channel. Notice that the total interleaver/deinterleaver delay is $(B-1)M$ coded bits for all paths. It follows that the deinterleaved code bits are in the same order as they were at the input to the interleaver, delayed by $(B-1)M$ coded bits.

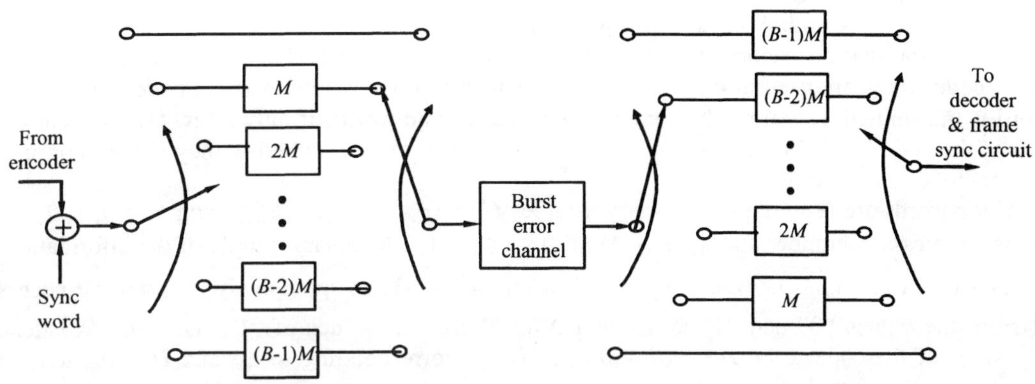

Figure 4.1-3 Convolutional interleaver/deinterleaver system for convolutional codes.

Clark and Cain [1] have summarized the properties and they are as follows:

(1) The minimum separation at the interleaver output is B symbols for any two coded bits that are separated by less than N coded bits at the interleaver input.
(2) Characteristic (1) implies that any burst of $b < B$ errors inserted by the channel results in single errors at the deinterleaver output, separated by at least N coded bits.
(3) A periodic pattern of single errors spaced $N + 1$ coded bits apart results in a burst of length of B at the deinterleaver output.
(4) The total memory required is $N(B-1)/2$ coded bits for both the interleaver and the deinterleaver. This is approximately half of what is required for a block interleaver/deinterleaver system. The total end to end delay is $N(B-1)$.

For the (B, N) interleaver, B is chosen to be larger than the length of any burst errors. N is chosen to be larger than the block length of block codes or the decoding constraint for convolutional codes. Under these conditions both interleaver/deinterleaver systems will have about the same burst error performance. It must be stressed that synchronization of the interleaver and the deinterleaver is necessary. The addition of a sync word at the input of the interleaver is one method of verifying that the deinterleaver is synchronous with the interleaver, by correlation at the receiver.

4.2 LINEAR BLOCK CODING

Block codes can be classified as either linear or nonlinear codes. Our discussion in this chapter will be limited to linear codes. A *linear block code* is a code for which any two code words can be added together, modulo-two, and the sum is another code word in the code set.

4.2.1 Linear Block Coding Concepts

A block code is composed of a set of vectors called *code words*. The length of the code word is the number of elements in the vector and is denoted by the letter n. The elements of the code word are selected from an alphabet of q symbols, or elements. When the code word is selected from a q symbol alphabet and q is a power of 2, then the code is nonbinary. When q is a power of 2 then it can be represented as a binary code word consisting of q-ary symbols that can be represented as b bits each, where $q = 2^b$. Thus each q-ary symbol can be represented as b bits each. Therefore a nonbinary code of block length N can be mapped into a binary code word of block length $n = bN$.

There are 2^n possible code words in a binary block of length n. From these code words it is possible to choose $M = 2^k$ code words where $k < n$ to form a code. In summary, a block of k information bits is mapped into a code word of length n, which is selected from the set of $M = 2^k$ code words. The resulting code block is denoted as an (n,k) code, and the *code rate* of the code is defined as the ratio $R_c = k/n$. For nonbinary codes having q symbols, there are q^n possible code words. In this case a subset of the $M = q^k$ code words may be selected to form the k-symbol blocks of information.

The reason that error correction capabilities of the code exist is due to the fact that not all of the possible 2^n code words are used in the code set. Consequently it is possible to select codes such that when a code word is transmitted, it will not be confused with other code words, if only a few errors occur upon the reception of the code word. More powerful codes will tolerate more errors before they will be confused with another member of the code set.

Code words are represented as binary n-tuples $\boldsymbol{C}_m = (c_{m0}, c_{m1}, \ldots c_{m(n-1)})$, with $m = 0, 1, 2, \ldots, 2^k-1$, where m is the message number and $c_{mi} \in (0,1)$ all $0 \le i \le n-1$. The binary k-tuple of the information bits is given by $\boldsymbol{X}_m=(x_{m0},x_{m1},\ldots x_{m(k-1)})$, where $x_{mi} \in (0,1)$ all $0 \le i \le k-1$. In some analysis it will be necessary to convert from the values "0" and "1" to the real AWGN channel values of "1" and "–1." We denote the channel values of the code to be $\boldsymbol{D}_m = (d_{m0}, d_{m1}, \ldots d_{mn-1})$, corresponding to $\boldsymbol{C}_m$ and $\boldsymbol{U}_m=(u_{m0},u_{m1},\ldots u_{m(k-1)})$ corresponding to $\boldsymbol{X}_m$. Note that

$$
\begin{aligned}
d_{mi} &= 1 - 2c_{mi} \\
u_{mi} &= 1 - 2x_{mi}
\end{aligned}
\qquad (4.2\text{-}1)
$$

Thus when $c_{mi} = 0$, $d_{mi} = 1$, and $c_{mi} = 1$, $d_{mi} = -1$, and similarly for u_{mi} and x_{mi}.

Only linear binary codes will be considered in what follows. An important property of linear codes is the fact that modulo-two sum of two code words is another code word in the same code word set, as was noted earlier.

Let $\boldsymbol{C}_i$ and $\boldsymbol{C}_j$ denote two code words in the code set of $M = 2^k$ code words. The *Hamming distance* between the two code words, having components c_{iq} and c_{jq} for $q = 0, 1, 2, \ldots, n-1$, is given by the number of elements in the two codes that differ and is denoted by d_{ij}.

Example 2 For example, consider the two code words $\boldsymbol{C}_1 = (1010)$ and $\boldsymbol{C}_2 = (0111)$. The Hamming distance is given by 3, since the two code words differ in three positions. It is clear that the code words agree in 4–3 = 1 position and disagree in 3 positions.

The smallest value of the set of all pairs of hamming distances $\{d_{ij}\}$, is called the *minimum distance* of the code and is denoted by d_{min}. Denote the *weight of a code word* as the sum of the nonzero elements in the code word. As an example, the weight of the word (0101) is two. In general, it can be shown that the distance d_{ij} between any code word pair $\boldsymbol{C}_i$ and $\boldsymbol{C}_j$ is simply the weight of the code word formed by taking the sum of the two code words, using modulo-2 arithmetic. The parameter d_{min} can be found from the expression

$$
d_{min} = \underset{r, r \neq 0}{Min}\{w_r\} \qquad (4.2\text{-}2)
$$

where w_r is the weight of each code word, and w_0 is the zero word, which is composed of all zeros, and has weight zero.

Let the number of code words that have Hamming distance d from the all-zeros code word by A_d. The *code weight distribution* of the code is the set of all A_d for $d = d_{min}, d_{min}+1, \ldots n$.

Example 3 Consider the following seven-symbol code indicated in Table 4.2-1. This code is rate R_c=4/7 and has $k = 4$ information bits and a code word length of 7 coded bits. The code weights can be determined from checking all but the all-zero code word. Thus there are seven of weight 3, so that $A_3 = 7$, seven of weight 4, so that $A_4 = 7$, and one of weight 7, so that $A_7 = 1$.

Another parameter is related to the minimum distance. It is the number of errors that can be corrected in a code word, is denoted by the letter "t," and is called the *error correction capability*. The error correction capability is given by

$$t = \left\lfloor \frac{d_{min} - 1}{2} \right\rfloor \tag{4.2-3}$$

where $\lfloor x \rfloor$ is the largest integer less than or equal to x. The value of t is the number of all combinations of t or fewer errors in any received code word that can be corrected. Note for error detection, $d_m - 1$ errors can be detected.

Table 4.2-1 Seven-Symbol (7,4) Code

Word number	Information bits					Code words						
0	0	0	0	0		0	0	0	0	0	0	0
1	0	0	0	1		0	0	0	1	1	0	1
2	0	0	1	0		0	0	1	1	0	1	0
3	0	0	1	1		0	0	1	0	1	1	1
4	0	1	0	0		0	1	1	0	1	0	0
5	0	1	0	1		0	1	1	1	0	0	1
6	0	1	1	0		0	1	0	1	1	1	0
7	0	1	1	1		0	1	0	0	0	1	1
8	1	0	0	0		1	1	0	1	0	0	0
9	1	0	0	1		1	1	0	0	1	0	1
10	1	0	1	0		1	1	1	0	0	1	0
11	1	0	1	1		1	1	1	1	1	1	1
12	1	1	0	0		1	0	1	1	1	0	0
13	1	1	0	1		1	0	1	0	0	0	1
14	1	1	1	0		1	0	0	0	1	1	0
15	1	1	1	1		1	0	0	1	0	1	1

Example 4 Consider the code set consisting of four code words: 00000, 11011, 11100, and 00111. Four messages can be transmitted with this code set. Note that 32 code words are possible. By limiting the number of possible code words, one increases the distance from the other code words. The minimum distance is 3, and therefore by (4.2-3) it can correct any single error in the received code word. It can detect at least two errors. Of the 32 possible code words, 4 are allowable and 28 are presumed to have at least one error in reception. One can construct a *decoding table* that is illustrated in Table 4.2-2.

Under each code word all the single position errors are listed, which are indicated as single errors in the table. Note that there are eight positions left over. In each of these eight words, the corresponding word differs from the correct code word by at least two errors. Note that there is not a unique way to assign them. For example, the word 10001 could be placed under the first column or the second column, since both have errors in two places.

This table is used in the decoding process by finding the column of the table that contains the received sequence and selecting the code word at the top of the column as the decoded output.

The reason that the table is constructed in this way is that single errors are more likely than double errors when the channel bit error rate (p) is less than 0.5. For a sequence of five channel bits the probability of k channel bit errors is given by $p^k(1-p)^{5-k}$. Clearly

$$(1-p)^5 > p(1-p)^4 > \ldots p^5 \tag{4.2-4}$$

which states that zero errors are more likely than a single error, followed by a double error, and so on, followed by five errors. No errors, a single error, and a double error are shown in Table 4.2-2. Therefore it follows that a decoder that decodes a particular received sequence into a code word that is nearest in the Hamming sense (distance) selects the most likely transmitted code word under the assumption that all code words are equally likely. A decoder that implements this decoding rule is called a *maximum-likelihood decoder*. In this sense it is an optimum decoder. This decoding approach via a decoding table is only feasible for short codes. This approach provides a benchmark to compare nonoptimum decoding approaches.

Table 4.2-2 Decoding Table for a Four-Word Code Set

Code words	0 0 0 0 0	1 1 0 1 1	1 1 1 0 0	0 0 1 1 1
Single errors	1 0 0 0 0	0 1 0 1 1	0 1 1 0 0	1 0 1 1 1
	0 1 0 0 0	1 0 0 1 1	1 0 1 0 0	0 1 1 1 1
	0 0 1 0 0	1 1 1 1 1	1 1 0 0 0	0 0 0 1 1
	0 0 0 1 0	1 1 0 0 1	1 1 1 1 0	0 0 1 0 1
	0 0 0 0 1	1 1 0 1 0	1 1 1 0 1	0 0 1 1 0
Double errors	1 0 0 0 1	0 1 0 1 0	0 1 1 0 1	1 0 1 1 0
	1 0 0 1 0	0 1 0 0 1	0 1 1 1 0	1 0 1 0 1

Each symbol is transmitted one at a time across the channel. Each symbol is corrupted by thermal noise or a jamming signal on the channel. The channel can be described by the probability of receiving a particular output n-tuple, given that the input was a particular input n-tuple, $\boldsymbol{C}_m$, and the jammer state vector is z. This probability is denoted as $p(y|C_m, z)$. The channel is assumed to be memoryless so that the joint probabilities is simply given as the product

$$p(y|C_m, z) = \prod_{i=1}^{n} p(y_i|c_{mi}, z_i) \tag{4.2-5}$$

4.2.1.1 Algebraic Elements of Linear Block Codes

Before we address the decoding rule for decoding, let us summarize some algebraic concepts that are used in dealing with linear block codes [11]. These operations are based on the convention of the algebraic field that has elements that are the symbols contained in the alphabet. A very common field is based on the two elements "0" and "1," which forms the binary alphabet. In general a field F consists of a set of elements that has two arithmetic operations defined on the elements: (1) addition and (2) multiplication that satisfy the following two axioms.

(A) Addition

1. The set F is closed under addition. That is, if $a,b \in F$, then $a+b \in F$.
2. Associative property. If $a, b,$ and $c \in F$, then $(a+b)+c = a+(b+c)$.
3. Addition is commutative. That is, $a+b=b+a$.
4. There is a "zero element" in the set that satisfies $a+0=a$.
5. For every element there exists its negative. That is, if b exists, then $-b$ exists. Subtraction is defined by $a-b=a+(-b)$

(B) Multiplication

1. The set F is closed under multiplication. That is, if $a,b \in F$ then, $ab \in F$.
2. The multiplication operation is associative. That is, $a(bc)=(ab)c$.
3. The multiplication operation is commutative. That is, $ab=ba$.
4. The multiplication operation is distributive over addition. That is, $(a+b)c = ac+bc$.
5. The set F contains the *identity element* that satisfied the condition $a(1)=a$.
6. Every element of F has an inverse except the element "0." If $b(\neq 0) \in F$, then its inverse is defined as b^{-1} and bb^{-1}=1.

The set of real numbers, the set of complex numbers, and the set of rational numbers are examples of fields. Codes are constructed from finite fields-fields that have a finite number of elements. Finite fields with q elements are called *Galois fields*, and are denoted by GF(q). The smallest field is the field GF(2). By these axioms every field must have a zero element and a one element. If q is a prime number then it is possible to construct GF(q) of the form $\{0,1,2...q-1\}$ with q an integer satisfying $q \geq 2$. The multiplication and addition operations on the elements in GF(q) are defined modulo q and are denoted as mod(q). The addition and multiplication tables for GF(2) are shown here.

+	0	1
0	0	1
1	1	0

Addition

•	0	1
0	0	0
1	0	1

Multiply

4.2.1.2 The Generator and Parity Check Matrices

Recall that the data is described as a row vector of the form $\boldsymbol{X}_m = (x_{m0}, x_{m1}, \ldots x_{m(k-1)})$ and the code word is of the form $\boldsymbol{C}_m = (c_{m0}, c_{m1}, \ldots c_{m(n-1)})$, with $m = 0, 1, 2, \ldots, 2^k-1$ and $\boldsymbol{C}_m$ is one of the 2^k code words. Each code word in a linear block encoder can be written as a set of n linear equations of the form

$$c_{mj} = x_{m1}g_{1j} + x_{m2}g_{2j} + x_{m3}g_{3j} + \ldots + x_{mk}g_{kj} \qquad j = 0,\ 1,\ 2, \ldots n-1 \tag{4.2-6}$$

where $x_{mi} \in (0,1)$ all $0 \leq i \leq k-1$ and $g_{ij} \in (0,1)$ and $c_{kj} \in (0,1)$ all $0 \leq j \leq n-1$. The linear equations of (4.2-6) can be written in matrix form as

$$C_m = X_m G \tag{4.2-7}$$

In (4.2-7) G is a $k \times n$ matrix called the *generator matrix* of the code and is represented by

$$G = \begin{bmatrix} g_0 \\ g_1 \\ g_2 \\ \vdots \\ g_{k-1} \end{bmatrix} = \begin{bmatrix} g_{00}, g_{01}, \dots g_{0(n-1)} \\ g_{10}, g_{11}, \dots g_{1(n-1)} \\ \vdots \\ g_{k0}, g_{k1}, \dots g_{k-1(n-1)} \end{bmatrix} \tag{4.2-8}$$

The rows of G generate or span the (n,k) code. The rows of G are known as the row vectors of G. A linear code with 2^k code words is a subspace of dimension k, and the row vectors g_j of the generator matrix G must be linearly independent (i.e., they must be a basis for the (n,k) code). Since the set of basis vectors g_j are not unique, the matrix G is also not unique. Since the subspace has dimension k, the rank of G is k. From (4.2-6) it can be seen that any code word can be written as the linear sum

$$C_m = \sum_{j=0}^{k-1} x_{mj} g_j \tag{4.2-9}$$

where the g_j are defined as the rows of the matrix in (4.2-8). To encode a code word the encoder only has to store the k rows of G and to form the linear combination of these k rows based on the input data message X_m, as indicated in (4.2-7).

Example 5 Consider the code indicated in Table 4.2-1. It has the following generator matrix

$$G = \begin{bmatrix} g_0 \\ g_1 \\ g_2 \\ g_3 \end{bmatrix} = \begin{bmatrix} 1\,1\,0\,1\,0\,0\,0 \\ 0\,1\,1\,0\,1\,0\,0 \\ 0\,0\,1\,1\,0\,1\,0 \\ 0\,0\,0\,1\,1\,0\,1 \end{bmatrix} \tag{4.2-10}$$

since the k row vectors are found by locating the unit vector data sequences. Thus the word number 9 of Table 4.2-1 is $\mathbf{g}_0$, the word number 4 is $\mathbf{g}_1$, and so on. If the data message to be coded is $X = (1\ 1\ 0\ 1)$, then its corresponding code word is given by, using modulo-2 addition

$$C = (1)g_0 + (1)g_1 + (1)g_3 \tag{4.2-11}$$

or

$$\begin{aligned} C &= (1)g_0 + (1)g_1 + (1)g_3 \\ C &= 1101000 \\ &\quad +0110100 \\ &\quad +0001101 \\ C &= 1010001 \end{aligned} \tag{4.2-12}$$

is the resulting code word as can be seen from Table 4.2-1 (code word 13).

Any generator matrix for a linear (n,k) code can be reduced by row operations and column permutations to the *systematic form of* G, which is a k by n matrix of the form:

$$G = [I_k | P] = \begin{bmatrix} 1\ 0\ 0 \ldots 0 & p_{11}\ p_{12} \cdots p_{1(n-k)} \\ 0\ 1\ 0 \ldots 0 & p_{21}\ p_{22} \cdots p_{2(n-k)} \\ \vdots\ \vdots\ \vdots\ \vdots & \vdots\ \vdots\ \vdots\ \vdots \\ 0\ 0\ 0 \ldots 1 & p_{k1}\ p_{k2} \cdots p_{k(n-k)} \end{bmatrix} \quad (4.2\text{-}13)$$

In (4.2-13) I_k is a $k \times k$ identity matrix and P is a $k \times (n–k)$ matrix that generates the $n–k$ redundant coded bits (parity check bits). The vertical line in the first matrix to the right of G denotes partitioning. The code generated by G of (4.2-13) is a *systematic code.* A linear block systematic code is a code in which the data bits are contained in the code word. The G of (4.2-13) defines the first k symbols of the code as the input data bits. If an (n,k) linear code generated by a generator matrix is not in systematic form, then the code is called *nonsystematic*. However a nonsystematic code can always be put into systematic form by elementary row and column operations.

It is possible to implement a linear systematic block code with the use of k bit shift registers and $n–k$ modulo-2 adders tied to the appropriate stages of the shift register. The sketch following illustrates a linear shift register for generating a (7,4) code, which has G given by

$$G = \begin{bmatrix} 1\,0\,0\,0\,1\,0\,1 \\ 0\,1\,0\,0\,1\,1\,1 \\ 0\,0\,1\,0\,1\,1\,0 \\ 0\,0\,0\,1\,0\,1\,1 \end{bmatrix} = [I_4 | P] \quad (4.2\text{-}14)$$

where I_4 is a 4×4 identity matrix and P is a 4×3 matrix.

It follows from (4.2-7) that the c_{mj} are given by:

$$\begin{aligned} c_{m4} &= x_{m0} + x_{m1} + x_{m2} \\ c_{m5} &= x_{m1} + c_{m2} + c_{m3} \\ c_{m6} &= x_{m0} + x_{m1} + x_{m3} \end{aligned} \quad (4.6\text{-}15)$$

The corresponding encoding scheme is shown in Figure 4.2-1. The general procedure (illustrated in Figure 4.2-1, for $k = 4$) is as follows. The k-bit block of data is shifted into the upper shift register in Figure 4.2-1. The $n–k$ parity check bits are generated in the lower shift register. Finally the data bits are shifted out followed by the parity check bits to form the n-bit coded word.

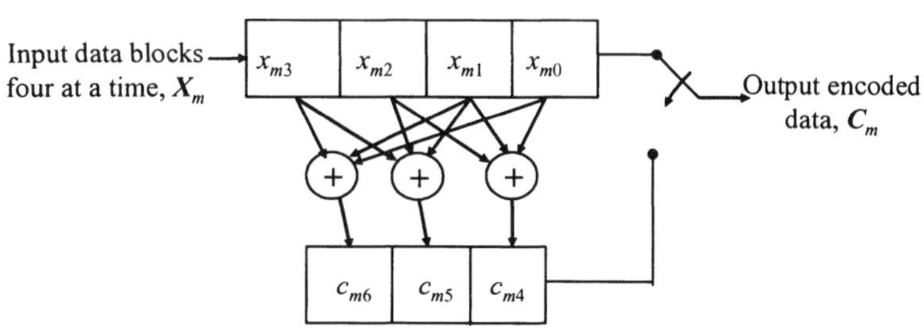

Figure 4.2-1 A linear shift register for generating a systematic (7,4) binary code.

It is to be noted that the resulting form of the code is as shown in Figure 4.2-2.

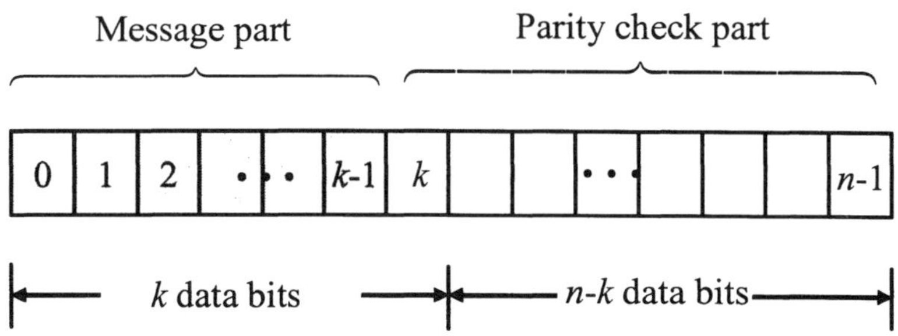

Figure 4.2-2 Systematic block code showing the data (shaded part) and parity bits.

The original data bits are at the front of the (block) word followed by the parity check bits as seen in Figure 4.2-2.

There is a *dual code* associated with any linear (n,k) code. The dual code is a linear $(n, n-k)$ code with 2^{n-k} code words, which is the null space of the original (n,k) code word. Let $\boldsymbol{H}$ denote the generator matrix for the dual code and note that it is an $n \times (n-k)$ matrix. It turns out that any code word of the (n,k) code word is orthogonal to any code word in the dual code [2]. Therefore any code word of the original (n,k) code is orthogonal to every row of the matrix[1] $\boldsymbol{H}$. That is

$$\boldsymbol{C}_m \boldsymbol{H}^T = \boldsymbol{0} \tag{4.2-16}$$

where the superscript "T" denotes the transpose of the given matrix, $\boldsymbol{0}$ is the all-zero row vector with $n-k$ elements, and $\boldsymbol{C}_m$ is one of the code words in the original (nondual) code. The matrix $\boldsymbol{H}$ is called the *parity check matrix* of the original code and has the property given in (4.2-16) and is an $(n-k) \times n$ matrix. Note that $\boldsymbol{H}^T$ is therefore an n by $n-k$ matrix. In addition to the relation indicated in (4.2-16), it can be shown that

$$\boldsymbol{G}\boldsymbol{H}^T = \boldsymbol{0} \tag{4.2-17}$$

where $\boldsymbol{0}$ is a $k \times n-k$ matrix with all-zero elements. Equivalently we have that since *the reversal law of transposed products of matrices* holds

$$([A][B])^T = [B]^T [A]^T \tag{4.2-18}$$

It follows that from (4.2-17) that

$$\boldsymbol{H}\boldsymbol{G}^T = \boldsymbol{0} \tag{4.2-19}$$

Consider now a linear block (n,k) code that is systematic and assume that its generator matrix $\boldsymbol{G}$ is given by (4.2-13), then since $\boldsymbol{G}\boldsymbol{H}^T = \boldsymbol{0}$, it follows [12] that for coefficients

$$\boldsymbol{H} = \left[-\boldsymbol{P}^T \middle| \boldsymbol{I}_{n-k}\right] = \left[\boldsymbol{P}^T \middle| \boldsymbol{I}_{n-k}\right] = \begin{bmatrix} p_{11}, p_{21} & \cdots & p_{k1}, 1, 0, 0, \ldots 0 \\ p_{12}, p_{22}, & \cdots & p_{k2}, 0, 1, 0, \ldots 0 \\ \vdots \quad \vdots & \vdots & \vdots \quad \vdots \\ p_{1(n-k)}, p_{2(n-k)}, & \cdots & p_{k(n-k)}, 0, 0, 0 \ldots 1 \end{bmatrix} \tag{4.2-20}$$

[1] It should be noted that not all authors define $\boldsymbol{G}$ and $\boldsymbol{H}$ as we do here. References [1, 4, 11, 12] have the same definitions as used here, but [2, 13] are different, for example.

since addition and subtraction modulo 2 is equivalent, the second term matrix form follows, in (4.2-20). Note that $\boldsymbol{H}$ is an $(n-k) \times n$ matrix. Clearly it is seen that

$$\boldsymbol{GH}^T = \left[\boldsymbol{I}_k \middle| \boldsymbol{P}\right]\left[\frac{\boldsymbol{P}}{\boldsymbol{I}_{n-k}}\right] \tag{4.2-21}$$

This is evaluated as

$$\boldsymbol{GH}^T = \boldsymbol{P} + \boldsymbol{P} = \boldsymbol{0} \tag{4.2-22}$$

which is a k by $n-k$ null matrix. (4.2-22) follows, since multiplication of a rectangular matrix by an identity matrix of compatible dimensions leaves the matrix unchanged [13].

Example 6 As an example of obtaining the generator matrix for the dual code $\boldsymbol{H}$, from $\boldsymbol{G}$, consider the $\boldsymbol{G}$ defined in (4.2-14). Repeating the $\boldsymbol{G}$ for the (7,4) code here

$$\boldsymbol{G} = \begin{bmatrix} 1\,0\,0\,0\,1\,0\,1 \\ 0\,1\,0\,0\,1\,1\,1 \\ 0\,0\,1\,0\,1\,1\,0 \\ 0\,0\,0\,1\,0\,1\,1 \end{bmatrix} \tag{4.2-23}$$

It follows that the $\boldsymbol{H}$ matrix is obtained from (4.2-20) to produce

$$\boldsymbol{H} = \begin{bmatrix} 1\,1\,1\,0\,1\,0\,0 \\ 0\,1\,1\,1\,0\,1\,0 \\ 1\,1\,0\,1\,0\,0\,1 \end{bmatrix} \tag{4.2-24}$$

Thus it is seen that the 3×3 identity matrix is located at the right side of $\boldsymbol{H}$, and the last three columns of $\boldsymbol{G}$ form the first four elements of the rows of H. Now let us check that (4.2-17) holds. Evaluating the indicated matrix product produces the 4×3 all-zero matrix.

$$\boldsymbol{GH}^T = \begin{bmatrix} 0\,0\,0 \\ 0\,0\,0 \\ 0\,0\,0 \\ 0\,0\,0 \end{bmatrix} \tag{4.2-25}$$

4.2.1.3 The Syndrome and Error Correction

Consider a linear (n,k) block code with a generator matrix $\boldsymbol{G}$ and a parity check matrix $\boldsymbol{H}$. Let $\boldsymbol{C}$ be a code word from this code set that was transmitted over a noise channel. Let $\boldsymbol{r} = (r_0, r_1 \ldots r_{n-1})$ be a received vector at the output of the channel. Due to the channel noise, $\boldsymbol{C}$ and $\boldsymbol{r}$ will not be the same. The n-component vector sum given by

$$\boldsymbol{e} = \boldsymbol{r} + \boldsymbol{C} = (e_0, e_1, \ldots e_{n-1}) \tag{4.2-26}$$

is called the *error pattern* or error vector. Each component of $\boldsymbol{e}$ satisfies $e_i = 1$ for $r_i \neq c_i$ and $e_i = 0$ for $r_i = c_i$. The "1s" in the error vector are the transmission errors caused by channel noise. It follows from (4.2-26) that

$$\boldsymbol{r} = \boldsymbol{C} + \boldsymbol{e} \tag{4.2-27}$$

The receiver does not know if errors exist in $\boldsymbol{r}$. When the presence of errors is detected then either action will be taken to correct the errors (forward error correction) or a retransmission will be requested.

The decoder computes the following $(n–k)$ tuple when $\boldsymbol{r}$ is received

$$\boldsymbol{s} = \boldsymbol{H}\boldsymbol{r}^T = \left(s_0, s_1, \ldots s_{n-k-1}\right)^T \tag{4.2-28}$$

which is called the *syndrome* of $\boldsymbol{r}$. Now $\boldsymbol{s} = \boldsymbol{0}$, if and only if $\boldsymbol{r}$ is a code word, and $\boldsymbol{s} \neq \boldsymbol{0}$ if and only if $\boldsymbol{r}$ is not a code word. Therefore if $\boldsymbol{s} \neq \boldsymbol{0}$, it is known that $\boldsymbol{r}$ is not a code word, and the presence of one or more errors has been detected. However if $\boldsymbol{s} = \boldsymbol{0}$, $\boldsymbol{r}$ is a code word, and the receiver assumes that $\boldsymbol{r}$ is the transmitted code word. It is to be noted that certain errors will not be detectable, (that is, if $\boldsymbol{r}$ has errors, but $\boldsymbol{s} = \boldsymbol{H}\boldsymbol{r}^T = \boldsymbol{0}$). This will occur when the error pattern $\boldsymbol{e}$ is identical to a nonzero code word. When this occurs, $\boldsymbol{r}$ is the sum of two code words, which is, of course, a code word and thus $\boldsymbol{s} = \boldsymbol{H}\boldsymbol{r}^T = \boldsymbol{0}$.

An error pattern of this kind is called an *undetectable error pattern*. It follows that since there are $2^k–1$ nonzero error patterns, there are $2^k–1$ undetectable error patterns. It is clear that when an undetectable error occurs, then a decoding error results. Note also that since the reversal law of transposed products of matrices holds, it follows that

$$\boldsymbol{s} = \left(s_0, s_1, \ldots s_{n-k-1}\right) = \boldsymbol{r}\boldsymbol{H}^T \tag{4.2-29}$$

It is to be noted that a circuit similar to the encoding circuit can be used to obtain the syndrome.

4.2.1.4 Polynomial Codes and Their Generation

Previously we defined a code word by the n-tuple

$$\boldsymbol{C}^{(i)} = (c_0, c_1, c_2, \ldots c_{n-1}) \tag{4.2-30}$$

An alternate method of expressing the same code word is to let the elements c_0, c_1, c_2,...$c_{n–1}$ be the coefficients of a polynomial in the variable x. Let this polynomial be of the form

$$p(x) = c_0 + c_1 x + c_2 x^2 + \ldots c_{n-1} x^{n-1} \tag{4.2-31}$$

With this notation a polynomial code is specified as the complete set of polynomials of degree $n–1$ or less that contains some specified polynomial $g(x)$ as a factor. In general $g(x)$ is of degree $n–k$. This polynomial $g(x)$ is called the *generator polynomial of the code.* It is necessary that the coefficients of these polynomials are elements of a finite field. The operations include adding, subtracting, multiplying, and dividing these coefficients, as indicated in Section 4.2.1.1.

Then the set of polynomials defined by

$$c(x) = g(x)\left(a_0 + a_1 x + a_2 x^2 + a_3 x^3 + a_{k-1} x^{k-1}\right) \tag{4.6-32}$$

forms an (n,k) code.

Example 7 Let $g(x)$ be given by

$$g(x) = 1 + x + x^3 \tag{4.2-33}$$

Then the set of polynomials described by the equation

$$c(x) = (1 + x + x^3)(a_0 + a_1 x + a_2 x^2) \tag{4.2-34}$$

form a (6,3) linear block code. Since there are three coefficients, there are eight distinct polynomials that correspond to the eight ways to choose the three coefficients (d_0, d_1, and d_2) from GF(2) (that is, $d_i = 0$ or 1). This will not lead to a systematic code.

Consider a method that will lead to a systematic code. If $(a_0 + a_1 x + a_2 x^2)$ is represented the input code words (i.e., a_0 the first, a_1 the second, and a_2 the third), then the code words would be represented by

$$c(x) = a_0 + (a_0 + a_1)x + (a_1 + a_2)x^2 + (a_0 + a_2)x^3 + a_1 x^4 + a_2 x^5 \tag{4.2-35}$$

A property of linear polynomial codes is that they may always be rendered in systematic form. For the above example, indicated in (4.2-35), let

$$\begin{aligned} b_0 &= a_0 \\ b_1 &= a_0 + a_1 \\ b_2 &= a_1 + a_2 \end{aligned} \tag{4.2-36}$$

By solving (4.2-36) for the a_i values and substituting into (4.2-35), the code polynomial becomes (see Problem 10 for the details)

$$c(x) = b_0 + b_1 x + b_2 x^2 + (b_0 + b_2)x^3 + (b_0 + b_1)x^4 + (b_0 + b_1 + b_2)x^5 \tag{4.2-37}$$

It is seen that this code polynomial is in the familiar (systematic) parity check form in which the first three bits are the input data and the last three bits are parity symbols.

Another option to obtaining a systematic form is to multiply the polynomial $(d_0 + d_1 x + d_2 x^2)$ by x^3 giving a result of $(d_0 x^3 + d_1 x^4 + d_2 x^5)$. Now, if this result is divided by the generator polynomial, it is shown in Problem 7 that one obtains a quotient of $((d_0 + d_2) + d_1 x + d_2 x^2)$ and a remainder of $(d_0 + d_2) + (d_0 + d_1 + d_2)x + (d_1 + d_2)x^2$. It is to be noted that in general for polynomials the coefficients are either "0" or "1" and the addition is modulo 2. Now notice that if the remainder is added to the dividend, the result is given by

$$p(x) = (d_0 + d_2) + (d_0 + d_1 + d_2)x + (d_1 + d_2)x^2 + d_0 x^3 + d_1 x^4 + d_2 x^5 \tag{4.2-38}$$

which represents a code word in systematic form. It is seen that the right three terms include the data and the left three terms are the parity check bits. This code polynomial procedure is an application of the Euclidean division algorithm, which states [1] that

$$\text{dividend} = (\text{quotient}) \times (\text{divisor}) + \text{remainder} \tag{4.2-39}$$

In the general case for this representation of a linear block (n, k) code, the pre-multiplication polynomial is $x^{(n-k)}$. It is true that the degree of the remainder will always be less than the degree of the divisor, which is $n-k$, since $g(x)$ is of degree $n-k$. The dividend will only contain terms higher than $n-k$, since the original sequence to be encoded was multiplied by x^{n-k}. It therefore follows that the adding the remainder to the dividend will not result in any cancellation since each group of terms have no polynomial terms in common.

A circuit model that can generate this code is shown in Figure 4.2-3. The first bit input is d_2. On the first shift it enters both stage 1 and stage 2 simultaneously. On the second shift d_1 enters stage 1, and d_1+d_2 is placed into stage 2 of the shift register. In stage 3 d_2 is placed. Finally on the last shift $d_0 + d_2$ ends up in stage 1, d_0+d_1+d_2 ends up in stage 2, and d_1+d_2 ends up in stage 3 of the shift register. All additions are modulo 2.

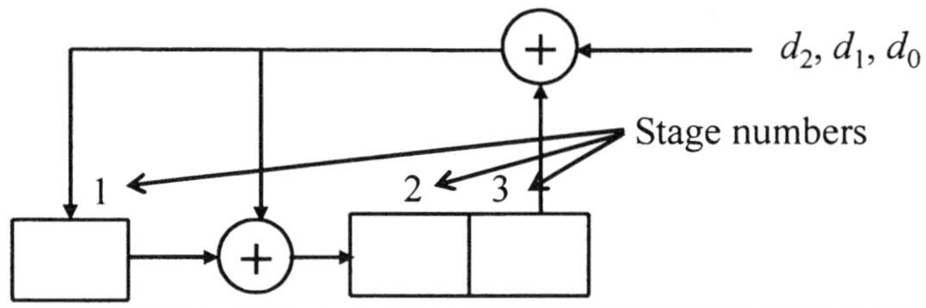

First stage	Second stage	Third stage	Shift number
0	0	0	0
d_2	d_2	0	1
d_1	$d_1 + d_2$	d_2	2
$d_0 + d_2$	$d_0 + d_1 + d_2$	$d_1 + d_2$	3

Figure 4.2-3 A circuit diagram that encodes the (6,3) code by simultaneously multiplying by x^3 and dividing by $g(x) = 1 = x + x^3$ [1].

Following Clark and Cain [1] the general encoder over GF(q) can be constructed as shown in Figure 4.2-4. Note that for binary codes defined over GF(2), all the minus signs can be removed with no effect on the encoder. The input is shown at the bottom of the figure, with k data bits that

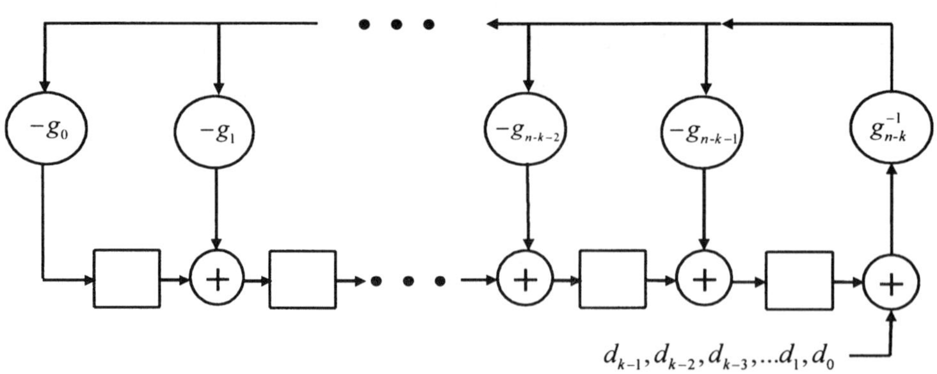

Figure 4.2-4 General circuit for encoding linear block codes in GF(q), which multiplies the input polynomial by x^{n-k} and divides by $g(x)$.

are input to the encoder. The inverse coefficient indicated on the right-hand side of the figure must be taken in GF(q). For more details see [1].

4.2.1.5 Cyclic Codes

In general for particular values of n the polynomial codes exhibit a *cyclic property*. That is when a code word $\boldsymbol{C} = (c_0, c_1, \ldots c_{n-1})$ is shifted one place to the right, one obtains another n-tuple code word

$$\boldsymbol{C}^{(i)} = (c_{n-1}, c_0, c_1, \ldots c_{n-2}) \tag{4.2-40}$$

which is called a *cyclic shift* of $\boldsymbol{C}$. That is to say, every cyclic shift generates a new code word.

Consider the code of Example 7 in the previous section. The generator polynomial for this code is given by $g(x) = 1 + x + x^3$. Consider a code word from the set of code words generated by $g(x)$. That is, if the code word is shifted one place to the right that is the same as multiplying the code word, expressed as a polynomial by x, and replacing the term x^7 with 1. This latter step is equivalent to reducing the polynomial mod(x^7–1). Now select a code word with a 1 in the right-hand position. We may write this code word as

$$c(x) = g(x)p(x) \tag{4.2-41}$$

where $g(x)$ was given earlier. Thus shifting this code word one place to the right yields the new code polynomial of the form

$$xc(x) = xc(x) - (x^7 - 1)\,\mathrm{mod}(x^7 - 1) \tag{4.2-42}$$

It is shown in Problem 8 that $g(x)$ divides (x^7–1). Since the right hand side clearly is divisible by $g(x)$ (see (4.2-41)) it is clear that the single shift is another code word. If our original code word did not have a 1 in the right-hand position, the result still holds [1].

This result can be extended to any (n,k) code in which $g(x)$ divides x^n–1. The least n in which $g(x)$ divides x^n–1 normally determines the largest useful code length n_{max} of the code. It can be shown [1] that the largest possible value of n_{max} is of the form (2^m–1) where m is the degree of $g(x)$. It is to be noted that for some polynomials the number of cycles can be much less than 2^m–1.

4.2.1.6 Cyclic Redundant Check Codes

Cyclic redundant check (CRC) codes are suited well for *error detection*, as opposed to error correction. In other words, they are suitable for the detection of one or more errors in a word, without the requirement to try to correct the word to its original uncorrupted form. The encoding and decoding process is practical, and they can be designed to detect many different types of errors. For these reasons CRC codes are almost universally used for all error detection applications in practice.

Define an *error burst of length B* in an n-bit received word as a contiguous sequence of B bits in which the first and the last bits of any number of intermediate bits are received in error. Binary (n,k) CRC codes are capable of detecting the following five error patterns:

(1) All error bursts of length n–k or less.
(2) A fraction of error bursts of length n–k+1; the fraction is $1-2^{-(n-k-1)}$.
(3) A fraction of error bursts of length greater than n–k+1; the fraction is $1-2^{-(n-k-1)}$.
(4) All combinations of $d_{min} - 1$ or fewer errors.
(5) All error patterns with an odd number of errors if the generator polynomial $g(x)$, for the code has an even number of nonzero coefficients.

Table 4.2-3 lists some CRC codes that are used in practice, and are international standards.

It is to be noted that the first three CRC codes listed in the table contain the factor (x + 1). By using the factor (x+1) in the generating polynomial ensures that all odd number of errors are detected. Often the CRC-12 is used for 6-bit characters, and the second and third codes are used for 8-bit characters. The last

CRC, CRC-32, is used in asynchronous transfer mode (ATM) applications. CRC-32 is difficult to compute because it is based on a polynomial of degree 32 that has many more terms (15) than any other CRC polynomial in common use.

Table 4.2-3 CRC Code Polynomials

CRC Code	Generator Polynomial	n–k
CRC-12	$x^{12}+x^{11}+x^3+x^2+x+1$	12
CRC-16	$x^{16}+x^{15}+x^2+1$	16
CRC-CCITT	$x^{16}+x^{12}+x^5+1$	16
CRC-32	$x^{32}+x^{26}+x^{23}+x^{22}+x^{16}+x^{12}+x^{11}+x^{10}+x^8+x^7+$ $x^5+x^4+x^2+x+1$	32

Now that we have finished the presentation of the basics of block codes, we will turn our attention to evaluating their performance.

4.2.2 Rule for Optimum Decoding with No Jammer Side Information

Now we will consider the optimum decoding rule when no side information is available for block codes.

It is assumed that the following is known to the decoder: (1) the received vector $\boldsymbol{y}$, (2) knowledge of the entire set of code words $\boldsymbol{C}_m$, (3) knowledge of the source probabilities of being transmitted, (4) and knowledge of the channel transition probabilities $p(\boldsymbol{y}|\boldsymbol{C}_m)$. Based on this information the decoding rule is designed so that the average number of bit errors outputted to the user is minimized. Denote the decoder output bit error probability as P_b.

First consider the problem of designing a decoding rule that results in the minimum probability of incorrectly estimating the transmitted code word without regard to the bit errors associated with a particular decoded code word. Denote the code word error probability as P_w. Denote the decoder estimate of the decoded code word as $\hat{\boldsymbol{C}}_m$, given that $\boldsymbol{y}$ was received. If $\boldsymbol{C}_m$ was transmitted and $\hat{\boldsymbol{C}}_m = \boldsymbol{C}_m$ when $\boldsymbol{y}$ was transmitted, then the decoder estimate is correct. The code word error probability P_w is minimized if the decoder chooses its estimate $\hat{\boldsymbol{C}}_m$ as the code word that was most likely to have been transmitted. In other words, choose $\hat{\boldsymbol{C}}_m$ to be $\boldsymbol{C}_m$ with the largest posterior probability $P(\boldsymbol{C}_m|\boldsymbol{y})$, and the decision based on this criterion is called the *maximum a posteriori probability criterion (MAP).*

Bayes' theorem [10] can be used to compute the probabilities $P(\boldsymbol{C}_m|\boldsymbol{y})$. Note that from Bayes' theorem one has

$$P(\boldsymbol{C}_m, \boldsymbol{y}) = P(\boldsymbol{C}_m|\boldsymbol{y})P(\boldsymbol{y}) = P(\boldsymbol{y}|\boldsymbol{C}_m)P(\boldsymbol{C}_m) \tag{4.2-43}$$

Hence one can solve for $P(\boldsymbol{C}_m|\boldsymbol{y})$ to be

$$P(\boldsymbol{C}_m|\boldsymbol{y}) = \frac{P(\boldsymbol{y}|\boldsymbol{C}_m)P(\boldsymbol{C}_m)}{P(\boldsymbol{y})} \tag{4.2-44}$$

From Papoulis [14] the denominator can be expressed as the following summation

$$P(\mathbf{y}) = \sum_{m'=0}^{2^k-1} P(\mathbf{C}_{m'})P(\mathbf{y}|\mathbf{C}_{m'}) \tag{4.2-45}$$

which is a positive number, independent of the signal message m. Therefore in maximizing (4.2-44) over $\mathbf{C}_m$ one can see that it is sufficient to maximize the following quantity

$$P(\mathbf{y}|\mathbf{C}_m)P(\mathbf{C}_m) \tag{4.2-46}$$

If all message probabilities are equal then

$$P(\mathbf{C}_m) = \frac{1}{2^k} \tag{4.2-47}$$

so when all messages are equally likely the decoding rule simplifies to the following rule: Choose $\hat{\mathbf{C}}_m$ to be $\mathbf{C}_m$ for which

$$P(\mathbf{y}|\mathbf{C}_m) \tag{4.2-48}$$

is maximized. A *maximum-likelihood decoder* is a decoder that maximizes (4.2-48) without regard for the message probabilities. This result is applicable [15] to all discrete memoryless channels, including hard-decision and soft-decision channels, which are memoryless and therefore characterized by the following expression

$$p(\mathbf{y}|\mathbf{C}_m) = \prod_{i=1}^{n} p(y_i|c_{mi}) \tag{4.2-49}$$

4.2.2.1 Hard Decisions

Consider now the case that hard decisions are made on the received coded symbols. This channel is modeled by the binary symmetric channel, so that

$$p(\mathbf{y}|\mathbf{C}_m) = p^d(1-p)^{n-d} \tag{4.2-50}$$

where d is the Hamming distance between the received $\mathbf{y}$ and $\mathbf{C}_m$, and p is the BSC error (transition) probability. Since the logarithm is a monotonic function of its argument, the log of $p(\mathbf{y}|\mathbf{C}_m)$ may be used as the maximum likelihood-decoding rule. Taking the log of (4.2-50) produces the result

$$\ln\{p(y|C_m) = d\ln(p) + (n-d)\ln(1-p) = d\ln\left(\frac{p}{(1-p)}\right) + n\ln(1-p) \tag{4.2-51}$$

The decoding rule is to choose $\hat{\mathbf{C}} = \mathbf{C}_m$ when the right-hand side of (4.2-51) is maximum. Only the first term of the right-hand side of (4.2-51) depends on d, and since $\ln(p/(1-p)) < 0$ for $p < 0.5$, maximizing (4.2-51) is equivalent to minimizing d. Hence the decoding rule is a *minimum-distance decoding rule* in the Hamming sense. In other words the decoding rule states the decoded word is the one with the smallest distance between the received code word and all of the possible code words.

The decoding rule just described achieves the minimum word (block) decoding error probability. Minimization of the word decoding error probability is also presumed to minimize the bit error probability. This is normally true in well-designed block codes. The messages are assigned such that the most likely word decoding errors induce the minimum number of bit errors in that decoded word. All reasonable codes have this property.

Example 8 Consider the $n = 7$, $k = 4$ code indicated in Table 4.2-1. Suppose that the demodulated output vector $\boldsymbol{y}$ is given by $\boldsymbol{y} = (0,1,1,0,0,1,0)$. Using Table 4.2-1, and using the minimum distance-decoding rule, determine the most likely transmitted word. By inspection the only code word that has a Hamming distance of one is code word ten, $\hat{\boldsymbol{C}} = (1,1,1,0,0,1,0)$.

4.2.2.2 Soft Decisions

Now consider a soft-decision memoryless, AWGN process with one-side noise power spectral density of N_0, where the coded bits take on the value of $d_k \in \pm 1$, where T is the coded bit duration in seconds, and the matched filter integration is weighted by the factor $1/T$. Since we are dealing with channel symbols that take on the values of ± 1 we will substitute $\boldsymbol{D}_m$ for $\boldsymbol{C}_m$. Assuming that the receiver uses this matched filter, the channel produces the conditional probability density

$$p(y_k | d_{mk}) = \frac{1}{\sqrt{\pi N_0 / T}} \exp\left[-\frac{(y_k - d_{mk})^2}{N_0 / T} \right] \tag{4.2-52}$$

for each bit number k. Since the channel is assumed to be memoryless, the joint probability density of all the coded bits in a word is given by (conditioning on $\boldsymbol{C}_m$ is equivalent to conditioning on $\boldsymbol{D}_m$)

$$p(\boldsymbol{y} | \boldsymbol{D}_m) = \prod_{k=1}^{n} p(y_k | d_{mk}) \tag{4.2-53}$$

Again it is more convenient to consider the natural logarithm of (4.2-53). Using (4.2-52) produces

$$\ln\left(p(\boldsymbol{y} | \boldsymbol{D}_m)\right) = -\frac{n}{2} \ln\left(\pi N_0 / T\right) - \frac{T}{N_0} \sum_{k=1}^{n} \left(y_k - d_{mk}\right)^2 \tag{4.2-54}$$

In order to maximize this result we note that the first term is independent of the code word $\boldsymbol{D}_m$ and the second term has a coefficient (T/N_0) that does not depend on the code choice. Hence it is sufficient to minimize the summation by choosing the code word l such that the minimum is obtained; that is,

$$\hat{\boldsymbol{D}}_l = \min_m \sum_{k=1}^{n} \left(y_k - d_{mk}\right)^2 \tag{4.2-55}$$

This quantity is the square of the Euclidean distance between the received code word plus noise and the code word. The optimum choice of the code word is the value of $\boldsymbol{D}_l$ that minimizes the Euclidean distance to the received data, the received vector $\boldsymbol{y}$ (n-tuple). It is to be noted that the values of the code in (4.2-55) are expressed in 1s and –1s, not 0s and 1s. Thus it is assumed that a binary "0" maps to a "1" and a "1" maps to a "–1." This fact is noted in Example 6.

In problem 2 it is shown that for the AWGN channel (4.2-55) can be simplified to the following

$$\hat{\boldsymbol{D}}_l = \max_m \sum_{k=1}^{n} \{y_k d_{mk}\} \tag{4.2-56}$$

In other words it is equivalent to maximize the correlation of the received code word over all possible code words and choose the one that yields the maximum.

Example 9 Reconsider Example 1 in which the 7-bit code is illustrated. Consider the all-zero code word C_0=(0,0,0,0,0,0,0) which corresponds to (0,0,0,0) as the input data. The additive real vector white Gaussian noise channel C_0 is represented by C_0'=(1,1,1,1,1,1,1). Hypothesize that the sampled output of the matched filter is given in ±1 notation by (–0.02, –0.03,1.0,1.0,1.0,1.0,1.0). In other words the first two values have been corrupted by noise, and the other five components have not been corrupted. When the hard-decision detector is used the threshold is set to zero volts, so that anything larger than zero is detected as a 1 and anything less as a –1. Thus, in this example the hard decision detector will detect (1,1,0,0,0,0,0) as the detected word. The closest code word from Table 4.2-1 is (1,1,0,1,0,0,0), which corresponds to the data sequence (1,0,0,0), producing an error in the first bit position.

The soft-decision decoder chooses the code word that is closest in Euclidean distance to the received word. Consider Table 4.2-4, in which the message number, the channel representation (±1) along with the Euclidean distance between the received vector and each word via (4.2-55) are shown. As can be from the table that the closest Euclidean fit is to message 0, which corresponds to the data sequence (0,0,0,0). Hence with soft decision with this assumed $\boldsymbol{y}$ vector data, no bit decoding error is made, unlike the hard-decision decoding which produced one bit error in the first bit position. An alternative calculation would be to maximize the correlation via (4.2-56), in lieu of minimizing the Euclidean distance.

4.2.3 Rule for Optimum Decoding with Jammer Side Information

When jammer side information (JSI) is available the jammer state vector $\boldsymbol{z}$ is available to the decoder to aid in forming decoding decisions. Equation (4.2-5) characterizes the channel when jammer side information is available.

4.2.3.1 Rule for Optimum Decoding with Jammer Side Information—Hard Decisions

Consider the case when jammer side information ($\boldsymbol{z}$) is available to the decoder to aid in making decisions in the decoding process and again assume a BSC.

Again consider the maximum-likelihood decoder, which is based on choosing the code word $\boldsymbol{C}_m$ for which

$$p(\boldsymbol{y}|\boldsymbol{C}_m, \boldsymbol{z}) \tag{4.2-57}$$

is maximum [11].

Assuming that the channel is memoryless, (4.2-5) applies. The analysis with JSI is similar to the case in which JSI is not available. Consider the natural logarithm of (4.2-5) to yield the quantity to maximize

$$\max_{C_m} \sum_{k=1}^{n} \ln\left\{p(y_k | c_{mk}, z_k\right\} \tag{4.2-58}$$

Table 4.2-4 Soft-Decision Decoding Euclidean Distance for Example 6

Message Number		Codewords (in Channel Format)								Euclidean Distance
0		1	1	1	1	1	1	1		2.101
1		1	1	1	− 1	-1	1	− 1		14.101
2		1	1	− 1	− 1	1	− 1	1		14.101
3		1	1	− 1	1	-1	− 1	− 1		18.101
4		1	− 1	− 1	1	-1	1	1		9.981
5		1	− 1	− 1	− 1	1	1	− 1		13.981
6		1	− 1	1	− 1	-1	− 1	1		13.981
7		1	− 1	1	1	1	− 1	− 1		9.981
8		−1	− 1	1	− 1	1	1	1		5.901
9		−1	− 1	1	1	-1	1	− 1		9.901
10		− 1	− 1	− 1	1	1	− 1	− 1		9.901
11		− 1	− 1	− 1	− 1	-1	− 1	− 1		21.901
12		− 1	1	− 1	− 1	-1	1	1		14.021
13		− 1	1	− 1	1	1	1	− 1		10.021
14		− 1	1	1	1	-1	− 1	1		10.021
15		− 1	1	1	− 1	1	− 1	− 1		14.021

Assume that the JSI is perfect in the sense that it knows what the jamming levels are exactly. Let **J** denote the set of all channel use indices j, for which $z_j = 1$ and denote $\overline{\mathbf{J}}$ the set of all channel indices j, for which $z_j = 0$, $j = 0,1,2 \ldots n{-}1$. Let d_1 and d_0 denote the Hamming distance between and $\boldsymbol{C}_m$ and $\boldsymbol{y}$ for the channel use indices (coded symbol number in a word) in **J** and $\overline{\mathbf{J}}$, respectively, and let p_1 and p_0 denote the BSC error probability for the channel use indices in **J** and $\overline{\mathbf{J}}$, respectively. In addition, let n_1 denote the number of elements in **J** and n_0 denote the number of elements in $\overline{\mathbf{J}}$, and note that $n_1 + n_0 = n$. Then $p(\boldsymbol{y}|\boldsymbol{C}_m, \boldsymbol{z})$ can be written as

$$\prod_{k\in \mathbf{J}} p(y_k|c_{mk},1) \prod_{k\in \overline{\mathbf{J}}} p(y_k|c_{mk},0) = p_1^{d_1}(1-p_1)^{n_1-d_1} p_0^{d_0}(1-p_0)^{n-n_1-d_0} \tag{4.2-59}$$

and recall that $n = n_1 + n_0$. Taking the natural logarithm of both sides produces

$$\ln\{p(\boldsymbol{y}|C_m, \boldsymbol{z})\} = d_1 \ln\left(\frac{p_1}{1-p_1}\right) + d_0 \ln\left(\frac{p_0}{1-p_0}\right) + n_1 \ln\left(\frac{p_1}{1-p_1}\right) + (n-n_1)\ln(1-p_0) \tag{4.2-60}$$

The decoder chooses as its estimate of $\boldsymbol{C}_m$, which is $\hat{\boldsymbol{C}}_m$, the code word in which (4.2-60) is the largest. The third and fourth terms do not depend on $\boldsymbol{C}_m$, and therefore can be disregarded in the search for the maximum. Thus the maximum of the following function can be used for the maximum-likelihood estimate of the transmitted code word $\hat{\boldsymbol{C}}_m$ of $\boldsymbol{C}_m$

$$\ln\{p(\boldsymbol{y}|C_m, \boldsymbol{z})\} = d_1 \ln\left(\frac{p_1}{1-p_1}\right) + d_0 \ln\left(\frac{p_0}{1-p_0}\right) \tag{4.2-61}$$

Note both terms are of the form $\ln(p_i/(1-p_i))$ and are negative for $p_i < 0.5$. It follows that when some bits are jammed, the corresponding probability of error satisfies, $p_1 \cong 0.5$, and for the other bits that are not jammed $p_0 << 0.5$, so that $\ln(p_0/(1-p_0))$ is very negative. In this case, basically only the unjammed bits are used by the decoder to determine the best choice of C_m. Minimizing d_0 does this. In the case that all bits are jammed the second term of (4.2-61) is gone ($d_0 = 0$) and then d_1 must be minimized [15].

4.2.3.2 Rule for Optimum Decoding with Jammer Side Information—Soft Decisions

Consider now the binary input soft-decision AWGN channel, which is characterized by (4.2-52). The function of the decoder is to choose its output code word that maximizes $p(\boldsymbol{y}|\boldsymbol{C}_m, \boldsymbol{z})$ or actually $p(\boldsymbol{y}|\boldsymbol{D}_m, \boldsymbol{z})$, where the $\boldsymbol{D}_m$ expresses each coded symbol as $\{\pm 1\}$, whereas $\boldsymbol{C}_m$ expresses each coded symbol as $\{0,1\}$. Defining $\mathbf{J}$ and $\bar{\mathbf{J}}$ as in Section 4.1.3.1, one can write

$$p(\boldsymbol{y}|\boldsymbol{D}_m, \boldsymbol{z}) = \prod_{k\in\mathbf{J}} p(y|d_{mk}, 1) \prod_{k\in\bar{\mathbf{J}}} p(y|d_{mk}, 0) \tag{4.2-62}$$

The one-sided thermal noise spectral density is given by N_0. When jamming is present the total spectral density is given by $N_0 + N_{0J}$. Using (4.2-52) in (4.2-62) produces

$$p(\boldsymbol{y}|\boldsymbol{D}_m, \boldsymbol{z}) = \prod_{k\in\mathbf{J}} \frac{1}{\sqrt{\pi(N_{0J}+N_0)/T}} \exp\left[-\frac{(y_k - d_{mk})^2}{(N_{0J}+N_0)/T}\right] \prod_{k\in\bar{\mathbf{J}}} \frac{1}{\sqrt{\pi(N_0)/T}} \exp\left[-\frac{(y_k - d_{mk})^2}{(N_0)/T}\right] \tag{4.2-63}$$

Again taking natural logarithms one obtains

$$\begin{aligned} ln\{p(\boldsymbol{y}|\boldsymbol{D}_m, \boldsymbol{z})\} = &-\frac{n_1}{2} ln\{\pi(N_{0J}+N_0)/T\} - \frac{T}{N_{0J}+N_0}\sum_{k\in\mathbf{J}} (y_k - d_{mk})^2 \\ &-\frac{n_0}{2}\ln\{\pi N_0/T\} - \frac{T}{N_0}\sum_{k\in\bar{\mathbf{J}}}(y_k - d_{mk})^2 \end{aligned} \tag{4.2-64}$$

As was observed before, the first and third terms on the right side of (4.2-64) are independent of the code word and can be ignored for determining the choice of the code word. The decoder, which determines the minimum error probability solution, chooses its output as the code word $\boldsymbol{D}_m$ in which

$$ln\{p(\boldsymbol{y}|\boldsymbol{D}_m, \boldsymbol{z})\} = -\frac{1}{N_{0J}+N_0}\sum_{k\in\mathbf{J}}(y_k - d_{mk})^2 - \frac{1}{N_0}\sum_{k\in\bar{\mathbf{J}}}(y_k - d_{mk})^2 \tag{4.2-65}$$

is the largest (we have removed the T since it doesn't affect the solution). The decoding metric is seen to be a weighted Euclidean distance metric that is influenced more by the received symbols that are not jammed than the ones that are jammed. This is true since it is assumed that N_{0J} is much larger than N_0 so that the second term (unjammed metric) is weighted more heavily.

4.2.4 Computation of the Block Coded Word and Bit Error Rate

In this section we shall compute the word error rate and the bit error rate for block codes.

4.2.4.1 Computation of the Block Coded Error Rates with No Jammer Side Information

The actual probability of word or bit error is dependent upon the code as well as the decoding rule used. In the following sections minimum distance or maximum likelihood decoding is assumed. Let us view the decoder as a partition or mapping of the space of all possible received n-tuples $\boldsymbol{y}$ into one of the 2^k possible code words. This mapping or partitioning results in $2^k = M$ disjoint subspaces [15, 16] Λ_m, where $m = 0, 1, 2,\ldots M-1$. Λ_m is defined by

$$\Lambda_m \equiv \left\{ \boldsymbol{y} : \frac{p(\boldsymbol{y}|\boldsymbol{C}_m)}{p(\boldsymbol{y}|\boldsymbol{C}_{m'})} \geq 1 \text{ for all } m' \neq m \right\} \tag{4.2-66}$$

Consider the definition $\boldsymbol{y} \in \bar{\Lambda}_m$. A word error (block of n bits) occurs when message m is transmitted and $\boldsymbol{y} \in \bar{\Lambda}_m$. A correct detection occurs when message m is transmitted and $\boldsymbol{y} \in \Lambda_m$. $\bar{\Lambda}_m$ is defined by

$$\bar{\Lambda}_m \equiv \left\{ \boldsymbol{y} : \frac{p(\boldsymbol{y}|\boldsymbol{C}_m)}{p(\boldsymbol{y}|\boldsymbol{C}_{m'})} < 1 \text{ for some } m' \neq m \right\} \tag{4.2-67}$$

Again assume that there are M code words, $\boldsymbol{C}_0, \boldsymbol{C}_1,\ldots \boldsymbol{C}_{M-1}$ with $M = 2^k$. The word error probability for the m-th code word is based on a result in [16] that states

$$P_w(m) = P\left(\boldsymbol{y} \in \bar{\Lambda}_m \middle| \boldsymbol{C}_m\right) = \sum_{\boldsymbol{y} \in \bar{\Lambda}_m} p(\boldsymbol{y}|\boldsymbol{C}_m) = 1 - P\left(\boldsymbol{y} \in \Lambda_m \middle| \boldsymbol{C}_m\right) \tag{4.2-68}$$

The overall average word error probability of all M words, given that $P(m)$ is the probability of selecting message m, is given by

$$P_w = \sum_{m=0}^{M-1} P(m) P_w(m) \tag{4.2-69}$$

Consider a bound on the word error probability. Let $\Lambda_{mm'}$ denote the decision region when there are only two messages m, and m'; that is,

$$\Lambda_{mm'} = \left\{ \boldsymbol{y} \middle| \frac{p(\boldsymbol{y}|\boldsymbol{C}_{m'})}{p(\boldsymbol{y}|\boldsymbol{C}_m)} \geq 1 \right\} \tag{4.2-70}$$

Note that $\Lambda_{mm'}$ separates message m and m'. Figure 4.2-5(a) illustrates both a four set space with Λ_1, Λ_2, Λ_3, and Λ_4 for partitioning four signals, shown as shaded areas. It is to be noted that the shaded areas are unbounded in reality, but are shown as finite areas for drawing purposes only. Also indicated in Figures 4.2-5(b–d) are Λ_{12}, Λ_{13}, and Λ_{14}, which partitions signal 1, signal 2, signal 3, and signal 4, as if there were only two signals total in the partitioning. For example, in Figure 4.2-5(b) the partition is a horizontal line that separates the two areas. Note that $\bar{\Lambda}_1 = \Lambda_{12} \cup \Lambda_{13} \cup \Lambda_{14}$ as can be seen from the figure. In general [16] it can be shown that

$$\bar{\Lambda}_m = \bigcup_{m \neq m'} \Lambda_{mm'} \tag{4.2-71}$$

One can write

$$P_w(m) = P(y \in \bar{\Lambda}_m | C_m) = P(y \in \bigcup_{m' \neq m} \Lambda_{mm'} | C_m) \leq \sum_{m' \neq m} P(y \in \Lambda_{mm'} | C_m) \quad (4.2\text{-}72)$$

since the probability of a union of events is less than or equal to the sum of probabilities of the component events. Restating (4.2-72), it follows that

$$P_w(m) \leq \sum_{m' \neq m} P\left(\frac{p(y|C_{m'})}{p(y|C_m)} \geq 1 | C_m \right) = \sum_{m' \neq m} P_w(m \to m') \quad (4.2\text{-}73)$$

where $P_w(m \to m')$ denotes the pairwise probability of error when C_m is transmitted and $C_{m'}$ is the only alternative. Equation (4.2-73) becomes an equality when there are only two messages (i.e., when $M = 2$). The bound in (4.2-73) is known as the *union bound* for word errors. Averaging over m one has the result

$$P_w \leq \sum_{m=0}^{M-1} P(m) \sum_{\substack{m'=0 \\ m' \neq m}}^{M-1} P_w(m \to m') \quad (4.2\text{-}74)$$

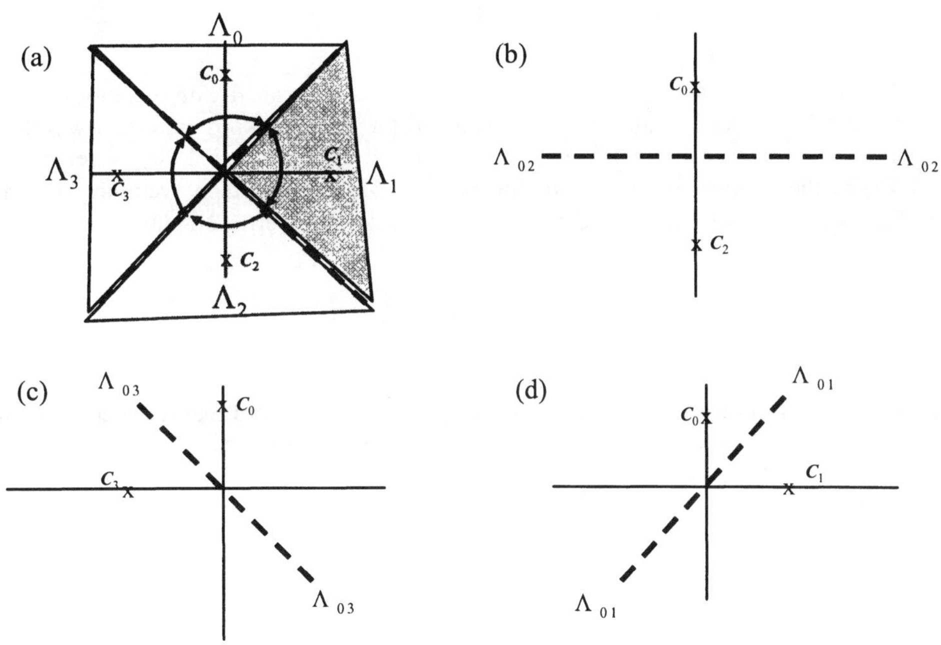

Figure 4.2-5(a–d) Λ_m regions and the Λ_{0m} regions for a four signal set.

A further simplification to (4.2-74) is dependent on the properties of linear block codes. A particular property of linear block codes is that the Hamming distance is the same between a code word C_m and code word $C_{m'}$ for which $m \neq m'$ and $m' = 0, 1, M-1$. In this case it is convenient to let $m = 0$, since the error probability is unchanged with this assumption, therefore (4.2-74) becomes

$$P_w \le \sum_{m=0}^{M-1} P(m) \sum_{m'=1}^{M-1} P_w(0 \to m') = \sum_{m'=1}^{M-1} P_w(0 \to m') \qquad (4.2\text{-}75)$$

and $P_w(m \to m')$ is determined by (4.2-79) or (4.2-80) and is the two-code word probability of error.

From (4.2-75) it is possible to obtain a useful expression for P_w. Assume the number of code words at Hamming weight d are known and are denoted A_d. The two-code word error probability, $P_w(0 \to m')$, is the same for all code words that have d errors, and is denoted by P_d. From (4.2-75) it follows [15] that the word error probability can be expressed in the alternative form

$$P_w \le \sum_{d=d_{min}}^{n} A_d P_d \qquad (4.2\text{-}76)$$

Given the word error probability, let us now determine the bit error probability. The bit error probability is obtained from the word error probability by considering the particular block decoding error events as well as the number of bit errors produced with each of the word error events. Let $P_w(m,m')$ denote the probability that message m was transmitted, and the decoded message was determined to be m' (i.e., $\boldsymbol{y} \in \Lambda_{m'}$). That is $P_w(m,m')$ is given by

$$P_w(m,m') = \sum_{y \in \Lambda_{m'}} P(\boldsymbol{y}|\boldsymbol{C}_m) \qquad (4.2\text{-}77)$$

Let $N(m,m')$ denote the number of information bits associated with this block error event. As an example in Table 4.2-1 if word number 4 was transmitted (m = 4) and word number 5 was decoded (m'= 5), then three bit errors would occur, and $N(4,5)$ = 3. The exact bit error probability is determined [15] by weighting (4.2-77) by the number of information bits in error $N(m,m')$ and then averaging over all possible transmitted messages m, and for each m value, over all possible decoding error outputs m' to produce

$$P_b = \frac{1}{k} \sum_{m=0}^{M-1} P(m) \sum_{\substack{m'=0 \\ m' \ne m}}^{M-1} N(m,m') \sum_{y \in \Lambda_{m'}} P(\boldsymbol{y}|\boldsymbol{C}_m) \qquad (4.2\text{-}78)$$

The factor of $(1/k)$ stems from the fact that k bits are decoded for each word that it decoded. In general this expression is quite difficult to evaluate, except for very simple cases. Fortunately bounds can be obtained that make the calculation much easier to accomplish.

Consider now the evaluation of the word error probability for the binary symmetric channel from (4.2-73) for $P_w(m \to m')$. Recall that $P_w(m \to m')$ is the probability that when there are only two signals, m is transmitted and m' is decoded. Let d denote the Hamming distance between these two codes. An error in the code word only matters when the errors occur where the code words bits differ in the two words. Consider first the case that d is an even number. An error only occurs when $(d/2)+1$ errors occur in the positions where $\boldsymbol{C}_m$ and $\boldsymbol{C}_{m'}$ differ. If only $d/2$ errors occur then a decoding error is assumed to occur only one half of the time [15]. Let p denote the BSC error probability. The probability of exactly ε errors in d coded bits is given by $\binom{d}{\varepsilon} p^{\varepsilon}(1-p)^{d-\varepsilon}$. Hence for d even [15] the probability $P_w(m \to m')$ is given by

$$P_w(m \to m') = \sum_{\varepsilon=(d/2)+1}^{d} \binom{d}{\varepsilon} p^{\varepsilon}(1-p)^{d-\varepsilon} + \frac{1}{2}\binom{d}{d/2} p^{d/2}(1-p)^{d/2} \qquad (4.2\text{-}79)$$

Errors greater than d have no effect on channel decoding, since the code words differ by only d channel symbols, and any errors where the code words agree increase the distance of the received word, $\boldsymbol{y}$, from both words $\boldsymbol{C}_m$ and $\boldsymbol{C}_{m'}$ equally.

Now consider the case that d is odd. The two-word word error probability, $P_w(m \to m')$, for odd d is given by

$$P_w(m \to m') = \sum_{\varepsilon=((d+1)/2)}^{d} \binom{d}{\varepsilon} p^{\varepsilon}(1-p)^{d-\varepsilon} \tag{4.2-80}$$

Using (4.2-74) along with either (4.2-79) or (4.2-80) produces an upper bound on the word error probability.

Consider a *bounded distance decoder*. This decoder has the property that it corrects up to E coded bit errors, and no more. Note that a maximum likelihood decoder is guaranteed to be able to correct at least t errors (t is defined in (4.2-3)). The bounded distance decoder will make an error whenever more than t errors are made on the channel. Thus the word error probability for a bounded distance decoder is given by

$$P_w = \sum_{k=E+1}^{n} \binom{n}{k} p^{k}(1-p)^{n-k} \tag{4.2-81}$$

For a maximum likelihood decoder this expression is an upper bound, since more than t errors may be corrected, and at least t error's will be corrected. Oldenwalder [17] has presented an argument that suggests that the bit error probability for a bounded distance decoder is given by

$$P_b = \frac{1}{n} \sum_{k=E+1}^{n} \beta_k \binom{n}{k} p^{k}(1-p)^{n-k} \tag{4.2-82}$$

where β_k is the average number of coded bit errors remaining in the corrected received code word, given that the channel caused k coded bit errors. Oldenwalder [17] claims that $\beta_k \simeq k$ is a good approximation. Thus the final approximation for the bit error probability of a bounded distance decoder is given by

$$P_b \simeq \frac{1}{n} \sum_{k=E+1}^{n} k \binom{n}{k} p^{k}(1-p)^{n-k} \tag{4.2-83}$$

Consider further the issue of estimating the bit error rate. Consider linear systematic block codes. We have already defined systematic block codes as a subset of linear block codes that have the k information bits used directly in the code word. In other words there are k information bits and n–k remaining bits, which all together form the coded bits of the code word. Recall for linear code words the number of code words that are Hamming distance d from any particular code word is the same for all code words. Using this fact, the word error probability calculation was simplified assuming that the "**0**" word was transmitted. This simplification occurs since (1) the number of code words at a Hamming distance d from a particular code word is the same for all code words, and (2) the number of bit errors caused by any Hamming distance d word decoding error is the same for all code words.

Example 10 Consider the (7,4) code of Table 4.2-1 in Example 1. Consider word number 14 and assume that there is an error in the fourth bit producing word number 15. Note that word 14 is three coded bits away from word 15 in distance. Now consider word "0" and let one bit error occur in the fourth data bit. That produces word number 1, which is also three code bits away from code word "0." The Hamming distance between the transmitted words and the decoded words is three (coded bits).

The probability of information bit error probability, P_b, is determined by assuming that message "0" is transmitted and weighting each block error event by the associated number of information bit errors. Denote the number of information bit errors that occur when $m = 0$ is transmitted and m' is decoded by $N(m')$. The average (information) bit error probability is then upper bounded by a modification of (4.2-75), which produces [15]

$$P_b \le \sum_{m'=1}^{M-1} \frac{1}{k} N(m') P_w(0 \to m') \tag{4.2-84}$$

Note that $P_w(0 \to m')$ is the same number for all code words $\boldsymbol{C}_m'$, with the same Hamming distance d from the "0" code word, having all zeros. Define B_d as the total number of information bit errors that occur in all code word error events involving code words that have a distance d from the all-zeros code word (the code word of all zeros). Then (4.2-84) can be rewritten as [15]

$$P_b \le \frac{1}{k} \sum_{d=d_{min}}^{n} B_d P_d \tag{4.2-85}$$

and d_{min} is, of course, the minimum distance of the code. In order to use this bound for the bit error probability, B_d must be determined by calculation for any particular code. In this equation P_d is the value of $P_w(0 \to m')$ with Hamming distance d and is the same for all code words $\boldsymbol{C}_{m'}$.

Example 11 Determine the upper bound for the bit error probability in terms of the BSC channel error probability parameter, p, for the code of Example 3, which is illustrated in Table 4.2-1. From Table 4.2-1 it is seen that, since $d_{min} = 3$, and the three values of d are 3, 4, and 7. It can be seen that $B_3 = 13$, $B_4 = 16$, and $B_7 = 3$. Note as a check, in this case the sum of all the data bits should equal the sum of the nonzero B_i: that is, $B_3 + B_4 + B_7 = 32$. From (4.2-79) and (4.2-80) we have

$$P_3 = \sum_{\varepsilon=2}^{3} \binom{3}{\varepsilon} p^{\varepsilon} (1-p)^{3-\varepsilon}$$

$$P_4 = \sum_{\varepsilon=2}^{4} \binom{4}{\varepsilon} p^{\varepsilon} (1-p)^{4-\varepsilon} + \frac{1}{2}\binom{4}{2} p^2 (1-p)^2$$

$$P_7 = \sum_{\varepsilon=4}^{7} \binom{7}{\varepsilon} p^{\varepsilon} (1-p)^{7-\varepsilon}$$

The bit error probability is given by (4.2-85); that is,

$$P_b \le \frac{1}{4}\left[13P_3 + 16P_4 + 3P_7\right]$$

Now consider another example to further illustrate the ideas presented so far.

Example 12 Consider an FH/BFSK spread spectrum system. Determine a bound on the bit error probability versus the E_b/N_0 value in dB for the case of worst-case partial band jamming. Assume that the code of Example 3, illustrated in Table 4.2-1 is used. Note that the bit energy per bit is related to the coded bit energy through the following relationship

$$\frac{E_s}{N_0} = \frac{k}{n} \frac{E_b}{N_0}$$

The coded bit (symbol) error probability has been determined in (3.3-51) of Chapter 3 and is included here with the 4/7 conversion factor relating the bit energy and the coded bit energy. Using Table 3.3-1 to determine the symbol error probability, which is the BSC crossover probability, one has

$$P_s = p = \begin{cases} \dfrac{0.3679}{(4/7)(E_J/N_0)} & \dfrac{E_b}{N_J} > \dfrac{7}{2} \\ \dfrac{1}{2}\exp\left(-\dfrac{4}{7}\dfrac{E_b}{N_J}\dfrac{1}{2}\right) & \dfrac{E_b}{N_J} \le \dfrac{7}{2} \end{cases}$$

Using p in this example along with the results in Example 11, the bit error probability is obtained as an upper bound. This example relates the bit error rate probability to the coded bit (symbol) error probability. The plot of both the uncoded and the coded cases are plotted in Figure 4.2-6.

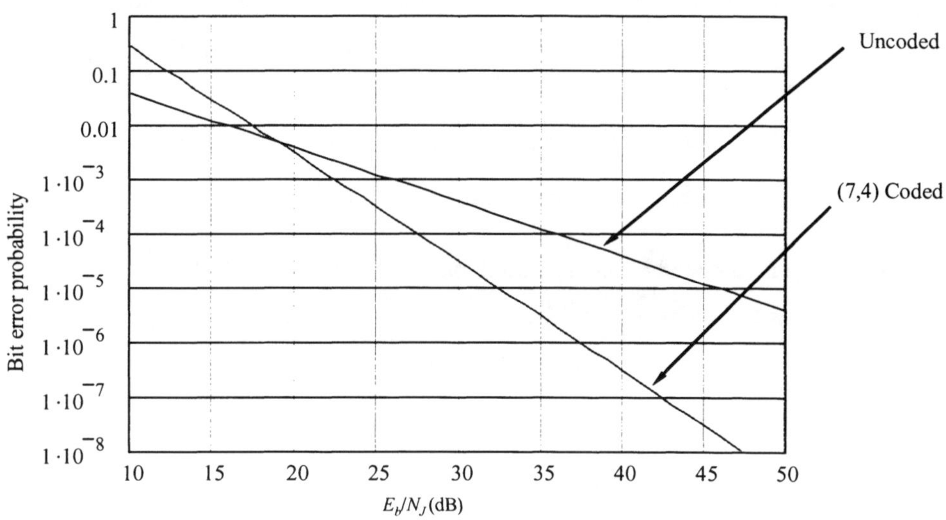

Figure 4.2-6 FH/BFSK comparison in worst-case partial band jamming of uncoded and (7,4) block coded bit error performance.

It is seen that the coded case has superior performance at higher values of E_b/N_J ratios. It is also to be noted that the bit error probability bound of (4.2-85) may produce bit error probabilities greater than unity. Clearly this is a poor bound when this occurs. These bounds tend to be more accurate at higher values of E_b/N_J ratios.

4.2.4.2 Computation of the Word Error Rates with Jammer Side Information

In this section we consider the computation of the word error rates when jammer side information is known. This section is based primarily on [15, 18]. It will be assumed that jammer side information is composed of two states, "1" corresponding to the jammer "on" condition, and "0" corresponding to the jammer "off" condition. Assume that the probability that a particular symbol in the coded word is jammed is "ρ." Assume that interleaving is present so that the channel has no memory, and each symbol may be jammed independently. Equation (4.2-57) is a discrete memoryless channel transition probability equation used for the maximum-likelihood decoder rule for choosing the code word $\boldsymbol{C}_m$, by searching for the maximum over $\boldsymbol{C}_m$.

For a code word of n coded bits with a known jammer state $z = (z_1, z_2, z_3 ..., z_n)$ regions for decoding, $\Lambda_m(z)$ is specified for each possible transmitted message m. $\Lambda_m(z)$ is given by modifying (4.2-66) to include the parameter z:

$$\Lambda_m(z) \equiv \left\{ \boldsymbol{y} : \frac{p(\boldsymbol{y}|\boldsymbol{C}_m, z)}{p(\boldsymbol{y}|\boldsymbol{C}_{m'}, z)} \geq 1 \text{ for all } m' \neq m \right\} \tag{4.2-86}$$

When message m is transmitted, a decoding error occurs when the received vector $\boldsymbol{y}$ does not fall within $\Lambda_m(z)$; in other words, an error occurs when $y \in \overline{\Lambda}_m(z)$. From (4.2-68) generalizing to the jammer side information case, it follows that

$$P_w(m,z) = P\left(\boldsymbol{y} \in \overline{\Lambda}_m | \boldsymbol{C}_m, z\right) = \sum_{\boldsymbol{y} \in \overline{\Lambda}_m} p(\boldsymbol{y}|\boldsymbol{C}_m, z) \tag{4.2-87}$$

which has the jammer side information contained within the expression. In a similar manner to the definition, with no jammer side information, one can write

$$\overline{\Lambda}_m(z) \equiv \left\{ \boldsymbol{y} : \frac{p(\boldsymbol{y}|\boldsymbol{C}_m, z)}{p(\boldsymbol{y}|\boldsymbol{C}_{m'}, z)} < 1 \text{ for some } m' \neq m \right\} \tag{4.2-88}$$

Now consider a union bound of the type used for the case of no JSI. Define

$$\Lambda_{mm'}(z) = \left\{ \boldsymbol{y} \left| \frac{p(\boldsymbol{y}|\boldsymbol{C}_{m'}, z)}{p(\boldsymbol{y}|\boldsymbol{C}_m, z)} \geq 1 \right. \right\} \tag{4.2-89}$$

Note that

$$\overline{\Lambda}_m(z) = \bigcup_{m \neq m'} \Lambda_{mm'}(z) \tag{4.2-90}$$

One can write

$$P_w(m,z) = P(\boldsymbol{y} \in \overline{\Lambda}_m |\boldsymbol{C}_m, z) = P(\boldsymbol{y} \in \bigcup_{m' \neq m} \Lambda_{mm'}(z)|\boldsymbol{C}_m, z) \leq \sum_{m' \neq m} P(\boldsymbol{y} \in \Lambda_{mm'}(z)|\boldsymbol{C}_m, z) \tag{4.2-91}$$

since the probability of a union of events is less than or equal to the sum of probabilities of the component events. Restating (4.2-91) it follows that

$$P_w(m,z) \leq \sum_{m' \neq m} P\left(\frac{p(\boldsymbol{y}|\boldsymbol{C}_{m'}, z)}{p(\boldsymbol{y}|\boldsymbol{C}_m, z)} \geq 1 |\boldsymbol{C}_m, z \right) = \sum_{\substack{m'=0 \\ m' \neq m}} P_w(m \to m', z) \tag{4.2-92}$$

where $P_w(m \to m', z)$ denotes the pair-wise probability of error, when $\boldsymbol{C}_m$ is transmitted and $\boldsymbol{C}_{m'}$ is the only alternative.

Consider the evaluation of the word error bound of (4.2-92) for the BSC in which the error probability of jammer state $z = 1$ corresponds to transition probability p_1, and jammer state $z = 0$ corresponds to transition probability p_0. From (4.2-61), it is seen that the maximum likelihood rule for this channel is used to decode when two jammer states exist, and is given by

$$d_1 \ln\left(\frac{p_1}{1-p_1}\right) + d_0 \ln\left(\frac{p_0}{1-p_0}\right) \tag{4.2-93}$$

which is a weighted Hamming distance. This metric is evaluated for both code words and the decoder output is the code word which yields the largest output.

In (4.2-93) d_i is the Hamming distance between $\boldsymbol{y}$ and the code word being tested against the jammed and unjammed channel coded bits, respectively. This rule is evaluated for each of the two code words, and the decoded output is determined by the code word for which (4.2-93) is largest.

Note that based on the analysis for jamming without jammer side information for BSC errors when $c_{mj} = c_{m'j}$ the effect on the Hamming decoding metric is the same for both code words. In a similar manner when JSI is available the value of the weighted Hamming distance for code symbols where $c_{mj} = c_{m'j}$ is affected identically. Hence again we can ignore those coded bits when they are the same. Define [15] $\boldsymbol{J}_1(\boldsymbol{C}_m, \boldsymbol{C}_{m'}, \boldsymbol{z})$ as the set of coded symbols for which $c_{mj} \neq c_{m'j}$ and $z_j = 1$, and in a similar manner define $\boldsymbol{J}_0(\boldsymbol{C}_m, \boldsymbol{C}_{m'}, \boldsymbol{z})$ as the set of coded symbols for which $c_{mj} \neq c_{m'j}$ and $z_j = 0$. Denote the Hamming distance between m and m' as $d(m,m')$. Denote by $n_J(\boldsymbol{C}_m, \boldsymbol{C}_{m'}, \boldsymbol{z})$ as the number of elements in $\boldsymbol{J}_1(\boldsymbol{C}_m, \boldsymbol{C}_{m'}, \boldsymbol{z})$, and therefore note that $d(m,m') - n_J(\boldsymbol{C}_m, \boldsymbol{C}_{m'}, \boldsymbol{z})$ is the number of elements in $\boldsymbol{J}_0(\boldsymbol{C}_m, \boldsymbol{C}_{m'}, \boldsymbol{z})$. Denote by $d_1(m)$ the Hamming distance between $\boldsymbol{y}$ and $\boldsymbol{C}_m$ for channel coded bits in $\boldsymbol{J}_1(\boldsymbol{C}_m, \boldsymbol{C}_{m'}, \boldsymbol{z})$, and denote by $d_0(m)$ the Hamming distance between $\boldsymbol{y}$ and $\boldsymbol{C}_m$ for channel coded bits in $\boldsymbol{J}_0(\boldsymbol{C}_m, \boldsymbol{C}_{m'}, \boldsymbol{z})$. Let $d_1(m') = n_J(\boldsymbol{C}_m, \boldsymbol{C}_{m'}, \boldsymbol{z}) - d_1(m)$ (this is true for the same coded bits in $\boldsymbol{J}_1(\boldsymbol{C}_m, \boldsymbol{C}_{m'}, \boldsymbol{z})$ only) and let $d_0(m') = d(m,m') - n_J(\boldsymbol{C}_m, \boldsymbol{C}_{m'}, \boldsymbol{z}) - d_0(m)$ denote the same distances for $\boldsymbol{C}_m'$. To avoid a cumbersome notation, the dependence on $\boldsymbol{C}_m$, $\boldsymbol{C}_m'$, $\boldsymbol{z}$ for $n_J(\boldsymbol{C}_m, \boldsymbol{C}_{m'}, \boldsymbol{z})$, $\boldsymbol{J}_1(\boldsymbol{C}_m, \boldsymbol{C}_{m'}, \boldsymbol{z})$, and $\boldsymbol{J}_0(\boldsymbol{C}_m, \boldsymbol{C}_{m'}, \boldsymbol{z})$ will be suppressed, and $d(m,m')$ will be denoted by d. Using the abbreviated notational form produces the decoder metrics

$$DM(m) \equiv d_1(m) \ln\left(\frac{p_1}{1-p_1}\right) + d_0(m) \ln\left(\frac{p_0}{1-p_0}\right) \tag{4.2-94}$$

and

$$DM(m') \equiv d_1(m') \ln\left(\frac{p_1}{1-p_1}\right) + d_0(m') \ln\left(\frac{p_0}{1-p_0}\right) \tag{4.2-95}$$

One can write

$$DM(m') \equiv \left(n_J - d_1(m)\right) \ln\left(\frac{p_1}{1-p_1}\right) + \left(d - n_J - d_0(m)\right) \ln\left(\frac{p_0}{1-p_0}\right) \tag{4.2-96}$$

The decoder chooses code word m if $D(m) > D(m')$ and m' if $D(m) < D(m')$. When $D(m) = D(m')$ the decoder chooses the decoded word randomly. Now initially ignore the case when the two metrics are equal. Consider the two-code word case that will be used in the word error probability of (4.2-92). The word error-probability $P_w(m \to m', z)$ is the probability that

$$d_1(m) \ln\left(\frac{p_1}{1-p_1}\right) + d_0(m) \ln\left(\frac{p_0}{1-p_0}\right) < \left(n_J - d_1(m)\right) \ln\left(\frac{p_1}{1-p_1}\right) + \left(d - n_J - d_0(m)\right) \ln\left(\frac{p_0}{1-p_0}\right) \tag{4.2-97}$$

under the condition that message m was transmitted and given z. Rearranging (4.2-97) produces the result

$$DM(m) = d_1(m)\ln\left(\frac{p_1}{1-p_1}\right) + d_0(m)\ln\left(\frac{p_0}{1-p_0}\right) < \frac{n_J}{2}\ln\left(\frac{p_1}{1-p_1}\right) + \frac{(d-n_J)}{2}\ln\left(\frac{p_0}{1-p_0}\right) \equiv \delta(d,n_J) \quad (4.2\text{-}98)$$

Following [15] the word error probability corresponding to the case that $DM(m) = DM(m')$ is given by

$$P_w(m \to m', z) = Pr\big(D(m) < \delta(d,n_J)\big) + \tfrac{1}{2} Pr\big(D(m) = \delta(d,n_J)\big) \quad (4.2\text{-}99)$$

where $Pr(a)$ denotes the probability event a. The Hamming distance between the received word $\boldsymbol{y}$ and the transmitted word $\boldsymbol{C}_m$ equals $d_1(m)$ when $\boldsymbol{C}_m$ is transmitted, and $d_1(m)$ transmission coded bit errors occur with n_J total coded bits that are affected by the jammer. Now the probability of $d_1(m)$ channel bit errors for n_J channel bits is given by

$$\binom{n_J}{d_1} p_1^{d_1}(1-p_1)^{n_J - d_1} \quad (4.2\text{-}100)$$

Note that $d_1(m)$ has been written as d_1, since there is no dependence on m' in the expression. In a similar manner the probability of $d_0(m)$ errors in the $d-n_J$ unjammed channel coded bits is given by

$$\binom{d-n_J}{d_0} p_0^{d_0}(1-p_0)^{d-n_J-d_0} \quad (4.2\text{-}101)$$

Now the two-code word error probability $P_w(m \to m', z)$, given jammer state condition z, is the sum of the product of the two probabilities in (4.2-100) and (4.2-101) over all d_1 and d_0, where $DM(m) < \delta(d,n_J)$ or $DM(m) = \delta(d,n_J)$. Therefore the two-word error probability is given by

$$P_w(m \to m', z) = \sum_{d_1=0}^{n_J}\sum_{d_0=0}^{d-n_J} \beta(d_1,d_0)\binom{n_J}{d_1} p_1^{d_1}(1-p_1)^{n_J-d_1}\binom{d-n_J}{d_0} p_0^{d_0}(1-p_0)^{d-n_J-d_0} \quad (4.2\text{-}102)$$

where $\beta(d_1,d_0)=1$ if $DM(m) < \delta(d,n_J)$, $\beta(d_1,d_0)=1/2$ if $DM(m) = \delta(d,n_J)$, and $\beta(d_1,d_0)=0$ otherwise. From (4.2-102) it is seen that the value of $P_w(m \to m', z)$ is a function of the Hamming distance $d(m,m')$ between messages m and m', and the number of jammed coded bits $n_J(\boldsymbol{C}_m, \boldsymbol{C}_{m'}, z)$ in which $c_{mj} \neq c_{m'j}$. From (4.2-92) the bound on the word error probability was derived as

$$P_w(m,z) \le \sum_{\substack{m'=0 \\ m'\neq m}} P_w(m \to m', z) \quad (4.2\text{-}103)$$

This expression depends on m as well as z. To remove the dependence on m and z, it is necessary to average over both the vector z and the message index m. Thus the word error probability is given by

$$P_w \le \sum_z \sum_{m=0}^{M-1} P_w(m,z)P(m)P(z) \quad (4.2\text{-}104)$$

From (4.2-103) it follows that

$$P_w \le \sum_z \sum_{m=0}^{M-1} \sum_{\substack{m'=0 \\ m' \ne m}}^{M-1} P_w(m \to m', z) P(m) P(z) \qquad (4.2\text{-}105)$$

Since the summations are finite, they may be rearranged to the following [15]

$$P_w \le \sum_{m=0}^{M-1} P(m) \left\{ \sum_{\substack{m'=0 \\ m' \ne m}} \left[\sum_z P(z) P_w(m \to m', z \right] \right\} \qquad (4.2\text{-}106)$$

Consider the summation inside the bracket. The function $P_w(m \to m', z)$ was noted in the paragraph following (4.2-102) to have the same value for all z for which the value of $n_J(\mathbf{C}_m, \mathbf{C}_{m'}, z)$ is the same. Consequently the summation may be reconstructed as a summation over all possible values of n_J. Therefore one can write

$$\sum_z P(z) P_w(m \to m', z) = \sum_{n_J}^{d(m,m')} \binom{d(m,m')}{n_J} \rho^{n_J} (1-\rho)^{d(m,m')-n_J} P_w(m \to m', z) \equiv PE_{Jd}(d(m,m'), \rho) \qquad (4.2\text{-}107)$$

where $\binom{d(m,m')}{n_J} \rho^{n_J} (1-\rho)^{d(m,m')-n_J}$ is the sum of the probabilities of all z for which $n_J(\mathbf{C}_m, \mathbf{C}_{m'}, z) = n_J$. The fraction ρ, which appears in (4.2-107), is the probability of a coded bit being jammed and applies to partial band jamming or partial pulse jamming. Insert (4.2-107) into (4.2-106) to yield the result

$$P_w(\rho) \le \sum_{m=0}^{M-1} P(m) \sum_{\substack{m'=0 \\ m' \ne m}}^{M-1} PE_{Jd}(d(m,m'), \rho) \qquad (4.2\text{-}108)$$

Now when linear codes are considered [15], the second summation is the same for all m, and therefore can be written as

$$P_w(\rho) \le \sum_{m'=1}^{M-1} PE_{Jd}(d(0,m'), \rho) = \sum_{d=d_{\min}}^{n} A_d PE_{Jd}(\rho) \qquad (4.2\text{-}109)$$

where $PE_{Jd}(\rho)$ denotes the value of $PE_J(d(0,m'),\rho)$ for $d(0,m') = d$ (the word error probability for a two-word code), and the summation over the variable m' has been changed to a summation over all possible Hamming distances d with each term of the summation weighted by the number of code words A_d, which are Hamming distance d from the all-zeros code word. Note that it is assumed that interleaving is present so that the channel has no memory. Each coded bit may be jammed independently with probability ρ and applies to partial band jamming or partial pulse jamming.

Thus an upper bound on the word error probability can be obtained for any specific block code that is subject to partial band jamming or pulse (partial time) jamming, as long as the number of code words A_d, at a distance d, is known. The bit error probability can be estimated from (4.2-109) by

$$P_b(\rho) \cong \frac{d}{n} \sum_{d=d_{\min}}^{n} A_d PE_{Jd}(\rho) \qquad (4.2\text{-}110)$$

which is applicable to all codes and for systematic codes one can modify (4.2-85) to write

$$P_b \le \frac{1}{k} \sum_{d=d_{min}}^{n} B_d PE_{Jd}(\rho) \tag{4.2-111}$$

where, as before, B_d is the total number of information bit errors that occurs in all code word error events involving code words that have a distance d from the all-zeros code word.

4.3 CONVOLUTIONAL CODES

In block coding the encoder accepts a k bit message and generates an n bit code word. Hence code words are generated on a block-by-block basis. This necessitates providing for a buffer to accumulate the k bit message before encoding it into a code word. Convolutional codes are different in that they are coded serially rather than in blocks. A *convolutional code* is a code generated by inputting the information bit stream into a linear finite state shift register.

4.3.1 Convolutional Code Encoder Characterization

The shift register contains K' k bit stages and there are n modulo-2 adders that connect to some of the individual shift register bits. The code rate is defined as

$$R_c = \frac{k}{n} \tag{4.3-1}$$

which is consistent with the definition of the code rate of a block code. The *constraint length* of the convolutional code is the value of $K = kK'$. When $k = 1$ the constraint length is simply $K=K'$. Inherently the constraint length of the encoder is the value of the memory of the encoder. Figure 4.3-1 illustrates the general convolutional encoder. For each k data bits entered into the encoder, n coded bits are sent to the modulator.

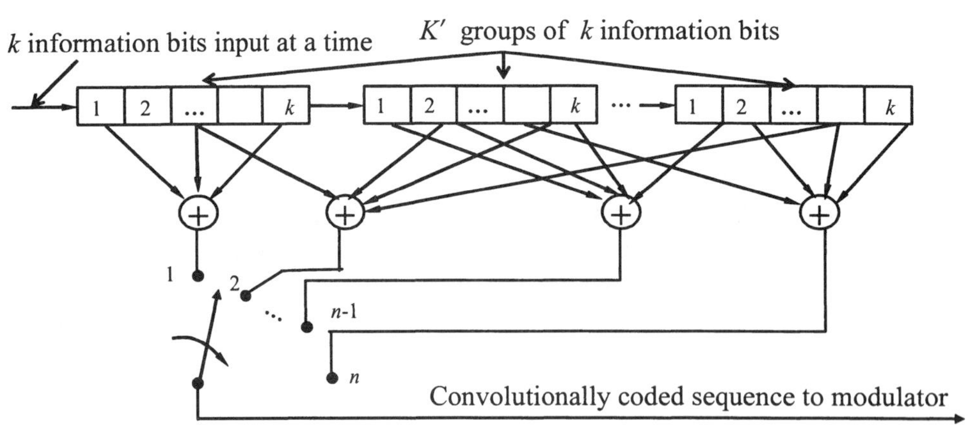

Figure 4.3-1 A general convolutional encoder of rate k/n.

Each modulo-2 adder, indicated by the symbol $\oplus$, has associated with it a generator that is denoted by a series of ones and zeros (contained in brackets) denoting connections (a 1) or nonconnections (a 0). Figure 4.3-2 [19] illustrates an example of a convolutional encoder that has $K = 3$ ($k = 1$) and $n = 2$, has a rate 1/2, and has two generators.

In the encoder the all-zeros bits are initially loaded into the registers. The upper code generator is denoted by g_1= [111] and the lower code generator is denoted by g_2 = [101]. In the first case all the connections to the modulo-2 adder are made, whereas for the lower code generator only the outer two are made. An alternative representation of the convolutional encoder is illustrated in Figure 4.3-3; it is an alternative form for Figure 4.3-2.

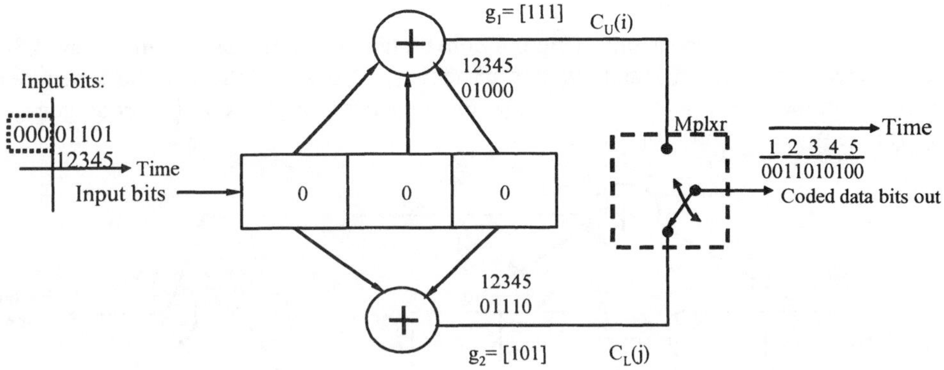

Figure 4.3-2 A $K = 3$, $n = 2$, rate ½ convolutional encoder.

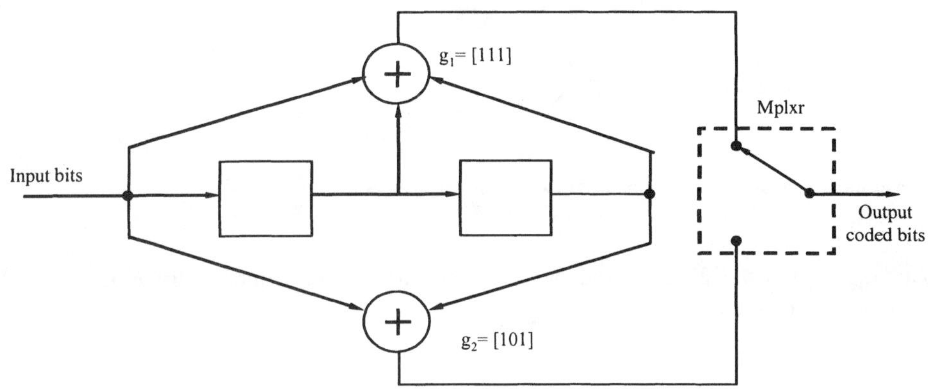

Figure 4.3-3 An alternative representation of a convolutional encoder with $R_c = ½$, $K = 3$, and $n = 2$.

In this alternative diagram the input bit and the two stored bits determine the coded output bits. In this alternative case the constraint length is one plus the number of memory elements. When a data bit is input into the encoder the coded output bits are based on the input bit and the two stored bits as is seen in the diagram. Both versions of the convolutional encoder are used in the literature. Figures 4.3-2 and 4.3-3 are alternative descriptions of the same encoder.

Again consider Figure 4.3-2 to determine the output coded bit sequence for a particular input bit sequence. Initially the three zero bits are loaded into the shift register. Let the input sequence be

$$\underbrace{10110\overbrace{000}^{initial}}_{Time\ to\ the\ left} \tag{4.3-2}$$

After the three initialization bits are loaded into the shift register (denoted by "initial" in the equation), it is seen that the first input bit (reading from right to left) is a "0." The upper mod 2 adder produces the modulo-2 sum: $0 \oplus 0 \oplus 0$, which of course yields zero. The lower modulo-2 adder also produces a zero. The next input is a "1." The upper generator produces a "1" and the lower generator produces a "1." Next, another "1" is input to the encoder. The upper generator produces a $1 \oplus 1 = 0$ and the lower generator produces $1 \oplus 0 = 1$. This continues in this manner, generating the sequence shown at the output based on having the input bits as shown in (4.3-2). Thus the sequence 01101 from first to last bit on the right produces the encoded output sequence of 0011010100 (first on the left last on the right). Again the initial sequence is 000.

It is to be noted that octal notation may be used to specify the code generators. For example, $g_1 = [111]$ can be represented as $g_1 = [7]$ and $g_2 = [1011]$ can be represented as $g_2 = [13]$.

Example 13 Consider a rate 2/3 constraint length 4 convolutional encoder shown in Figure 4.3-4. For this encoder there are three outputs ($n = 3$), and two bits are input per clock cycle so that the code rate is 2/3. The total number of bits in the shift register is 2×2=4, so the constraint length is 4. The code generators are as follows: g_1=[1011], g_2=[1101], and g_3=[1010].

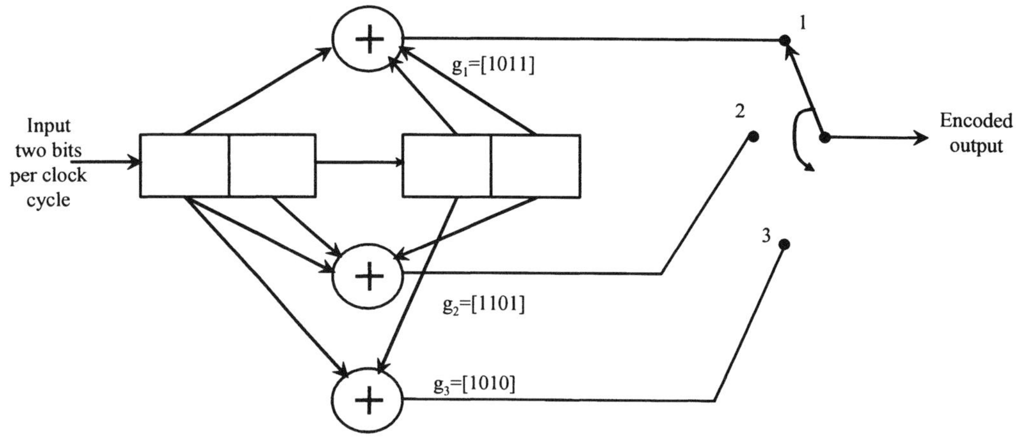

Figure 4.3-4 An example of a rate 2/3, constraint length 4, convolutional encoder.

Viewing Figure 4.3-2 again and denoting the input data sequence by $I(j)$, notice that the upper encoded sequence is given by

$$C_U(j) = I(j) + I(j-2) \tag{4.3-3}$$

and the lower encoded sequence is given by

$$C_U(j) = I(j) + I(j-1) + I(j-2) \tag{4.3-4}$$

Thus each coded data bit can be described by

$$C(j) = \sum_{k=0}^{2} W_k I(j-k) \tag{4.3-5}$$

where W_k is the tap weights, $I(k)$ in the input data sequence, and $C(j)$ is the output encoded data sequence. Thus the output sequence is the *convolution* of the input sequence and hence the name convolutional code.

There are three methods to represent a convolutional encoder. These are the (1) the tree diagram, (2), the trellis diagram, and (3) the state diagram.

Figure 4.3-5 illustrates an example of a rate 1/2, constraint length 3, convolutional code tree diagram for the convolutional encoder of Figure 4.3-2 (or 4.3-3). Starting at the left of the diagram a "0" is input into the encoder and "00" is output. Likewise, starting at the left and inputting a "1" produces an output of "11." Each upward movement on the tree implies a "0" was input to the encoder; likewise, each downward movement on the tree implies a "1" was input to the convolutional encoder. To determine the output of the encoder, it is observed that the first two bits (in a constraint length 3 encoder) is the state of the encoder, plus the input bit, determines the output encoded bit pair.

At the third stage (third set of output encoded bits) it is seen that there are two sets of the four states. It is observed from Figure 4.3-5 that all branches emanating from the same state are equivalent in the sense that they generate the identical sequences. Therefore the two nodes having the same label (state) can be *merged*, which results in a much simpler diagram. Merging of the nodes lead to the trellis diagram shown in Figure 4.3-6. In the trellis diagram shown in Figure 4.3-6 the input sequence 01101 (first bit on the left) is shown on the trellis diagram. Input "1" is denoted by a dashed line and an input of "0" is denoted by a solid line. Thus it can be seen that the input sequence 01101 (first bit on the left) produces the following output coded bit sequence: 00 11 01 01 00, which is the same as the result shown in Figure 4.3-2.

Now consider the trellis diagram, Figure 4.3-6, which is the second method of representing a convolutional encoder. Consider the level numbers at the bottom of the trellis diagram (Figure 4.3-6). If the message length is L, and the constraint length is K, then the trellis contains L+K levels, where a *level* is one stage in the trellis diagram. The levels of the trellis are labeled as 0, 1, 2, 3, ... L+K–1. In Figure 4.3-6, since K=3, L+K –1=L+2. Another term for level is the *trellis depth*. The first K–1 levels correspond to the encoders movement from state "a," and the last K–1 levels corresponds to the encoders return to the state "a." The trellis diagram shows the output symbols and the state of the encoder as the evolution through the encoder. In Figure 4.3-6 the initial state is "00." If a "1" is input (shown by the solid lines), the result is "00" being output and the encoder stays in state "00." However, if a "1" is input (shown by the dashed lines), an "11" is output and the changes to "10," as can be seen in the diagram. The dark arrows on the diagram illustrate the state progress and the output coded symbols as the trellis depth increases from depth zero to depth 5 for the input sequence 01101 (first bit on the left).

Now consider the third representation of the encoder, the state transition diagram. Figure 4.3-7 illustrates a state transition diagram for Figure 4.3-2. In this diagram all the states are shown, and the resulting coded outputs are shown with the "1" and "0" inputs.

The states are shown inside the circles, and when a "1" is input into the encoder, the solid arrow indicates the next state and the resulting output coded bits. As an example in state **11** a "1" bit input produces the state **11** and outputs 10 as the encoded bit pair. If a "0" is input to the encoder from state **11**, the state **01** results and 01 is the resulting encoded bit pair.

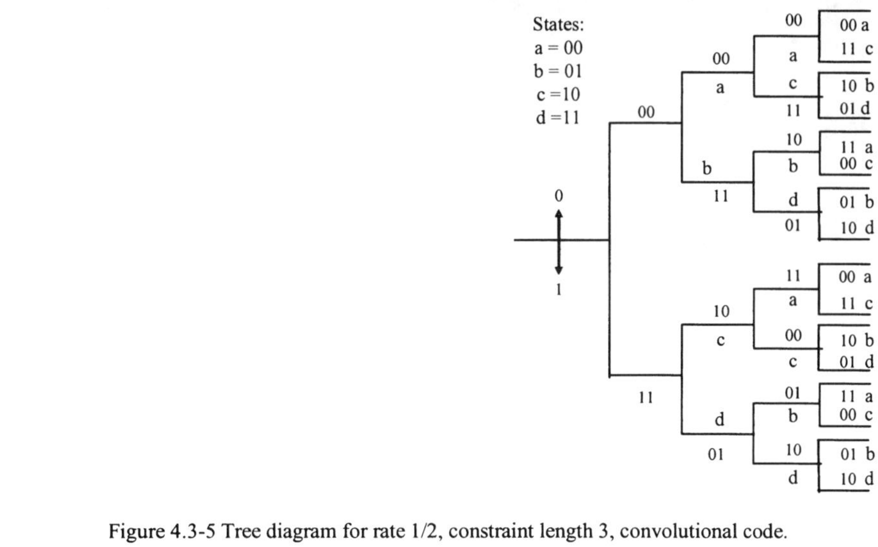

Figure 4.3-5 Tree diagram for rate 1/2, constraint length 3, convolutional code.

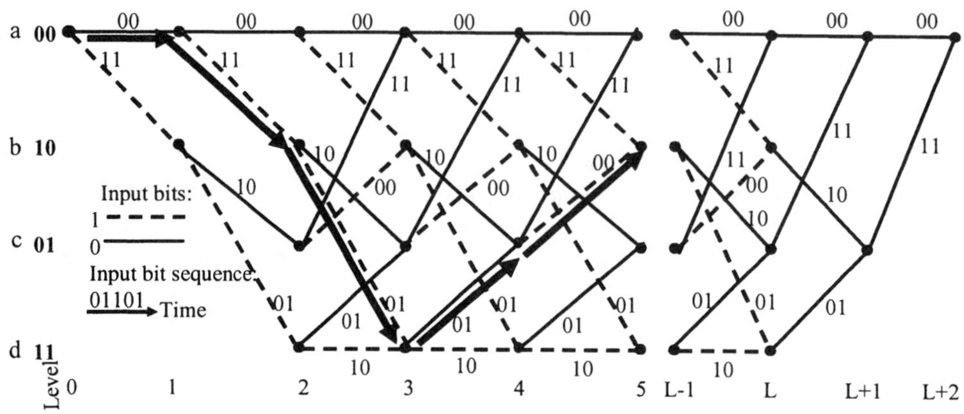

Figure 4.3-6 Trellis diagram for rate 1/2, constraint length 3, convolutional code.

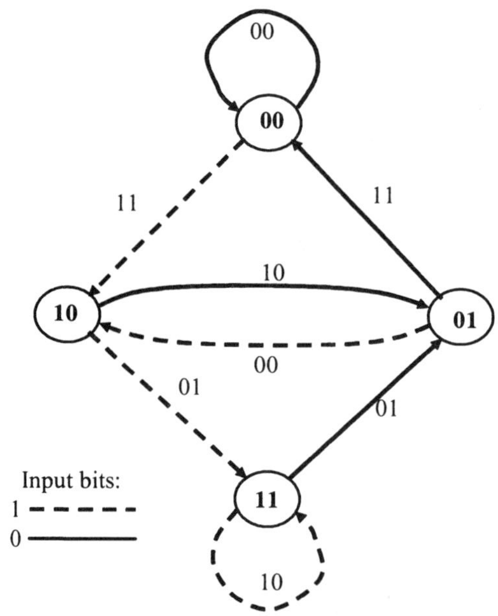

Figure 4.3-7 State diagram for rate 1/2, constraint length 3, $n = 2$ convolutional encoder.

A modification to convolutional codes is based on the concept of puncturing the output sequence. A punctured convolutional code is one in which some output coded bits are systematically deleted, or not transmitted, out of a low rate convolutional encoder. Since the trellis structure of the low rate code remains the same, the number of information bits per sequence does not change. As a consequence the output sequences belong to a higher rate *punctured convolutional code.* This process will not be discussed further here, but the reader can consult Clark and Cain [1] for more details.

4.3.2 The Transfer Function of a Convolutional Code and the Free Distance

The distance properties and the bit error rate performance of a convolutional code can be obtained from its associated state diagram. In addition the free distance can be obtained from the transfer function. Due to the fact that a convolutional code is linear, the set of Hamming distances of the code sequences generated up to some point in the tree, from the all-zero code sequence, is the same as the set of distances of the code sequences with respect to any other code sequence. Therefore we can state, without loss in generality, that the all-zero code sequence is the input to the encoder. The *free distance* of a convolutional code is the minimum Hamming distance between any two code words in the code set. A convolutional code with free distance d_f can correct t errors if and only if d_f is greater than $2t$.

The free distance of the convolutional code can be obtained from the transfer function of the code when the state diagram has been obtained. Consider the state diagram of Figure 4.3-7 for the encoder of Figure 4.3-2 (or Figure 4.3-3). First it is convenient to put the 00 state (state **a**) at the bottom of the diagram as shown in Figure 4.3-8(a). Since the only paths that are of interest are those that start and end in the state 00 and do not go through the state 00 in between, the self loop on the state 00 (zero state) can be removed and the state can be split into an input and an output. To do this one breaks apart state 00 (state a) and forms an input state ($\mathbf{a_0}$) and an output state ($\mathbf{a_1}$), as shown if Figure 4.3-8(b). It is convenient to label the branches with the associated weight of the output bits, with D raised to the power of the weight. The D parameter is a dummy parameter, which denotes a unit distance. In addition to the D parameter it is possible to place another factor N into the branch transition whenever a bit transition is caused by a input data bit "1." Hence, as each branch is traversed, the cumulative exponent on N increases by one only if that branch transition is due to an input data bit "1" [11].

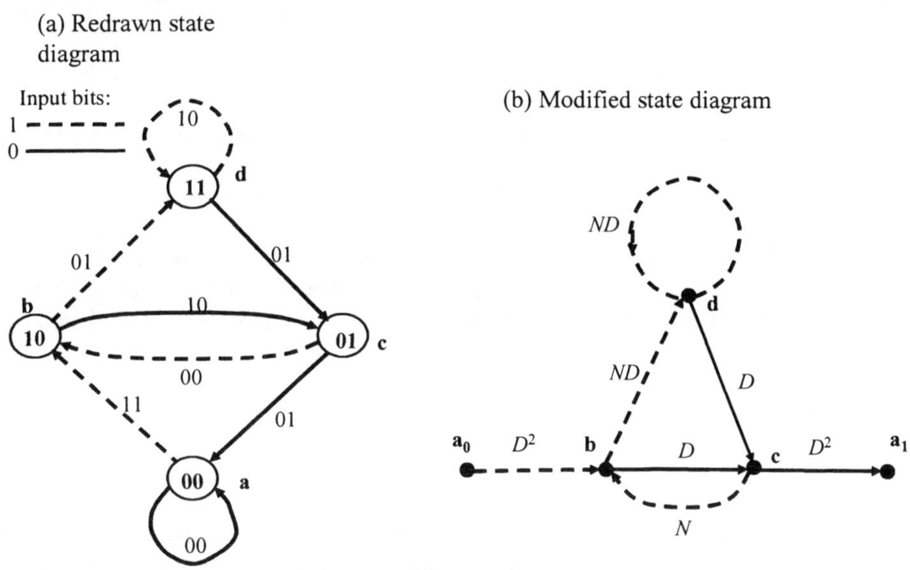

Figure 4.3-8 (a) Redrawn state diagram along with (b) the modified state diagram.

The modified state diagram can be viewed as a *signal flow graph* with one input and one output. Signal flow graphs are composed of *nodes* (large dots) and *directed branches* (lines connecting the nodes). The following rules [20] apply to flow graphs:

(1) A branch multiplies the signal at its input node by the *transmittance* characterizing that branch (for example, D).
(2) A node with incoming branches sums the signals produced by all of those branches.
(3) The signal at a node is applied equally to all the branches outgoing from that node.
(4) The transfer function of the graph is the ratio of the output signal to the input signal, expressed as a ratio.

The following four equations can be written from Figure 4.3-8(b):

$$\begin{aligned} b &= ND^2 a_0 + Nc \\ c &= Db + Dd \\ d &= NDd + NDb \\ a_1 &= D^2 c \end{aligned} \tag{4.3-6}$$

In Problem 4 it is shown that the ratio of the output to the input, as a function of D, is given by

$$T(D,N) = \frac{ND^5}{1-2ND} \tag{4.3-7}$$

Noting that $1/(1-x) = 1 + x + x^2 + x^3 + \ldots$ (4.3-7), evaluated at $N=1$, can be written as

$$T(D,1) = D^5 + 2D^6 + 4D^7 + 8D^8 + \ldots \tag{4.3-8}$$

The free distance is the minimum Hamming distance between any two code words in the code, and the distance transfer function $T(D)$ enumerates the number of code words that are a given distance apart. It is clear that the exponent of the first term in the expression for $T(D,1)$ is the free distance. Thus the convolutional code under study has a free distance of $d_f = 5$. In addition there are two paths of distance 6, four paths of distance 7, eight paths of distance 8 and so on. This is the weight structure of the code. Further information can be obtained from the state diagram; however, it will not be discussed here [1], except for the relationship to the bit error rate bounds.

The transfer function can be used to provide additional information besides the distance of the various paths [12]. One such augmentation is based on the introduction of a factor N into all branch transitions caused by a data "1" bit. Therefore, as each branch is traversed, the cumulative exponent on N increases by unity only if that branch transition is due to a data input of a "1." The resulting transfer function will be labeled $T(D,N)$ and will be used in the expression of the upper bound on Viterbi decoding error rate.

4.3.3 Decoding of Convolutional Codes

When we decoded a block code for a memoryless channel, we determined the Hamming distances (hard-decision channel) or the Euclidean distance (soft decision) between the received code word and the 2^k possible transmitted code words. Our job was to select the code word that was closest to the received code word. The computation of the 2^k metrics is the optimum decision rule, in the sense that it results in a minimum probability of error for the binary symmetric channel when the error probability is less than 0.5 in the white Gaussian noise channel.

A convolutional encoder, unlike a block code that has a fixed length of n channel bits, is essentially a finite state machine. It turns out that the optimum decoder is a search through the trellis for the most probable sequence.

The approach for the decoding rule for convolutional codes is based on the same basic approach used for block codes. It will be assumed that the data bits are equally likely and statistically independent from one another. In this case all paths through the trellis are equally probable, and a maximum-likelihood decoding rule is appropriate for this case. The decoder output is the path through the trellis whose data bit sequence was most likely to have been transmitted to the channel by the convolutional coder. Denote the convolutional encoder output infinite sequence, starting at time zero, of the message sequence (or path) m by the vector $\boldsymbol{C}_m$ where

$$\boldsymbol{C}_m = c_{m0} c_{m1} c_{m2} c_{m3} \ldots c_{mk} \ldots \tag{4.3-9}$$

The discrete memoryless channel sequence is denoted by the vector $\boldsymbol{y}$ where

$$\boldsymbol{y} = y_0 y_1 y_2 y_3 \cdots y_k \cdots \tag{4.3-10}$$

It follows that the probability of receiving the channel output $\boldsymbol{y}$, given that channel input $\boldsymbol{C}_m$ was transmitted, is given by

$$p(\boldsymbol{y}|\boldsymbol{C}_m) = \prod_{k=0}^{\infty} p(y_k|c_{mk}) \tag{4.3-11}$$

For a given received vector $\boldsymbol{y}$, the path that is most likely to be obtained through the trellis is the one that maximizes $p(\boldsymbol{y}|\boldsymbol{C}_m)$. The probability function $p(\boldsymbol{y}|\boldsymbol{C}_m)$ is the *metric* or measure used in working through the trellis that is used to compare all the possible code words $\boldsymbol{C}_m$ that were transmitted. In the case of a hard-decision binary symmetric channel the metric evolves to finding the path through the trellis diagram that minimizes the Hamming distance between a path and the received sequence. The most common method used to decode convolutional codes is the Viterbi decoder algorithm [21–23]. This algorithm will be discussed shortly.

Commonly the natural logarithm is taken of the likelihood function to form the log likelihood function. Since the natural log is a monotonic function of its argument, natural logarithms can be taken on both sides of (4.3-11) to produce

$$\ln\left[p(\boldsymbol{y}|\boldsymbol{C}_m)\right] = \sum_{k=0}^{\infty} \ln\left[p(y_k|c_{mk})\right] \tag{4.3-12}$$

since the log of a product yields the sum of the product components. The summation of the probabilities associated with a single code trellis branch is called the *branch metric* for that particular branch.

Up until now we have primarily discussed the hard-decision binary symmetric channel, however there is more information when the channel output is not heavily quantized. In principle the channel could be a continuous output value. However, in practice, the channel output is normally quantized to 2 or 3 bits. It has been determined that only about 0.2 to 0.25 dB can be gained by quantizing from 3 bits to infinite quantization [24].

Consider now the threshold levels for 1-, 2-, and 3-bit quantization. Figure 4.3-9 illustrates the three levels of quantization, assuming that the matched filter output takes on the values of $\pm\sqrt{E_s}$, where E_s is the symbol energy.

For 1-bit quantization the levels are divided between the negative voltage (0) as one region and the positive voltage for the other region (1). The 2-bit quantization produces four levels numbered 0, 1, 2, and 3, and the levels are smaller than the 1-bit case, with two levels below zero volts and two above. Finally the 3-bit quantization has eight levels, which can be represented by 3 bits. Figure 4.3-10 illustrates the channel models for the three cases noted earlier.

It can be seen from Figure 4.3-10 that the input is always binary, but the output can be 2, 4, or 8 values. The transition probabilities for any of the three channels in Figure 4.3-10 can be determined by computing the area under the Gaussian probability density function between the chosen thresholds that determine the output symbol. For example, for the case of the 3-bit quantization, the probability of having a "0" sent and an output of a "3" occurring is given by the area that has lines pointing up to the right, and when a "1" is transmitted the probability is given by the area that has the lines pointing up to the left (actually cross-hatched).

For both hard-decision and soft-decision channel quantization, the decoding rule is basically the same. Given the received sequence $\boldsymbol{y}$, the decoder output is the code path (sequence) in the trellis diagram such that $\boldsymbol{x}_m$ makes the probability $p(\boldsymbol{y}|\boldsymbol{x}_m)$ the maximum over all possible $\boldsymbol{x}_m$ values. The Viterbi algorithm is optimal in finding the correct path through the trellis diagram for either soft- or hard-decision processing.

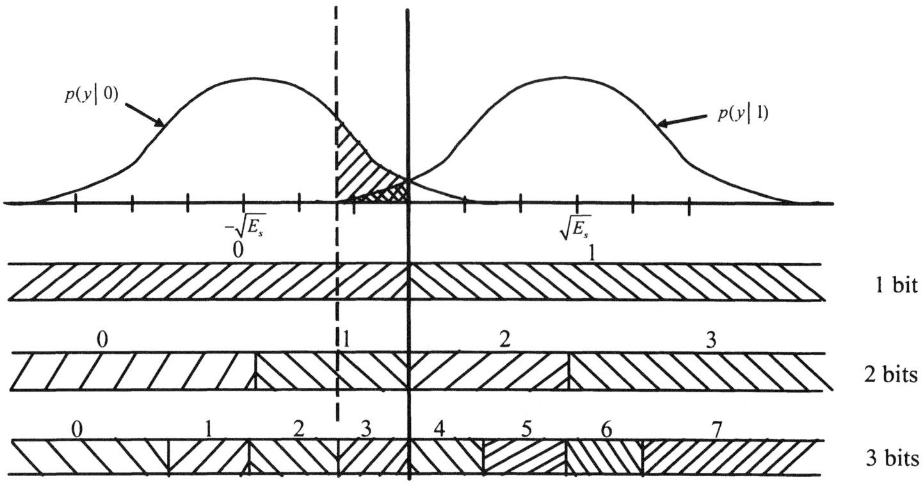

Figure 4.3-9 (a–c) Three levels of quantization for convolutional decoding.

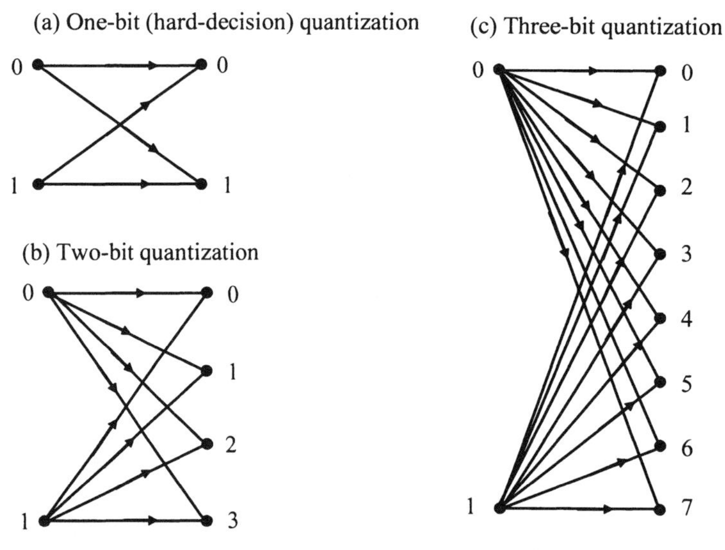

Figure 4.3-10 Three channel models for decoding convolutional codes.

4.3.4 The Viterbi Algorithm

The Viterbi algorithm is an efficient method for using maximum-likelihood methods to decode convolutional codes. Andrew Viterbi first described the Viterbi algorithm in 1967 [25]. However it was Forney [26] that proved that the algorithm was optimal and not just asymptotically optimal.

Note that the function of the maximum-likelihood decoder is to find the trellis path (sequence) that was most likely to have been transmitted, given the received channel output sequence. As was mentioned earlier, this corresponds to finding the path (sequence) through the trellis whose code sequence has the largest log-likelihood function as discussed in (4.3-12). For the binary symmetric channel, the maximizing of the log-likelihood function is equivalent to finding the path through the trellis whose code sequence is closest in Hamming distance to the received sequence.

4.3.4.1 The Viterbi Algorithm with Hard Decisions

We have shown in Section 4.1.2 for optimum decoding of the BSC that it is sufficient to find the code or path with the minimum Hamming distance from the received code. This is minimum distance decoding in the Hamming sense. Since a code tree is equivalent to the trellis diagram, we may limit our choice to the possible paths in the trellis representation of the convolutional code. The advantage in using the trellis over the tree diagram is the fact that the number of nodes, at any level of the trellis, does not continue to grow as the number of channel bits are imputed to the decoder. In fact the number of nodes remains at $2^{(K-1)}$, where K is the constraint length of the convolutional code. Consider the trellis diagram of Figure 4.3-6, which represents a $K = 3$, $R_c = 1/2$ convolutional encoder, as shown in Figure 4.3-2.

Consider for example Figure 4.3-6, which is the trellis diagram of the encoder shown in Figure 4.3-2. At depth (level) 3 there are two paths entering any of the four nodes in the trellis diagram. Furthermore these paths will be identical from this level onward until the convergence to node "a" at level L to L+2. It is clear that a minimum distance decoder may make a decision at that point in time on which path to retain, without suffering any loss of bit error rate performance.

Basically this is what the Viterbi algorithm for decoding does as it works its way through the trellis diagram. The *Viterbi algorithm* computes a metric for every possible path in the convolutional code trellis. The metric differs for soft and hard decisions. For the hard-decision channel the metric is the Hamming distance between the coded sequence for that path and the received sequence (which probably has errors on it). Thus for each state (node) in the trellis diagram of Figure 4.3-6, the algorithm compares the two paths entering the state, and the state with the lower Hamming distance is retained, and the other path is discarded. This comparison is repeated for every level j of the trellis in the range of $K-1 \leq j \leq L$, where K–1 is the encoder's memory, and L is the length of the data message sequence. The retained paths are called the *survivors* or *active paths*. For a length K convolutional code, there are no more than 2^{K-1} survivors' paths. The set of 2^{K-1} remaining paths is always guaranteed to contain the maximum-likelihood choice.

It is possible that when comparing paths entering a state, the metrics will be identical. In this case a choice is made by conceptually flipping a random coin, and letting the coin flip choose which path becomes a survivor.

To recap, the Viterbi algorithm is a maximum-likelihood decoder that is optimum in the sense of minimum sequence error probability estimate (or word error) for the case of white Gaussian channel noise. It does not necessarily minimize the bit error probability [27]. Even though Viterbi decoding is not optimal in the sense of minimization of the bit error probability, in most applications of interest, the performance of both the optimum and the Viterbi algorithm are essentially identical. The Viterbi algorithm decoding steps have been summarized by Lin and Costello [2] and are as follows:

Step 0 Initialization.
Label the all-zero state as state "0."

Step 1 Computational step.
Begin at step (depth) j = 1; compute the partial metric for the single path entering each state. Store the survivor path and its metric for each state.

Step 2 Additional computational steps.
Increase j by 1. Compute the partial metric for all the paths entering a state by adding the branch metric of the surviving connections at the preceding depth. For each state, store the path with the largest metric (the survivor path), together with its metric, and eliminate all the other paths. Note that for the hard-decision channel, one minimizes the Hamming distance.

Step 3 Test and computational step(s)
If $j < L+K-1$ repeat step 2; if not, stop the process.

Example 14 Assume that the encoder of Figure 4.3-2 or (4.3-3) is used as our encoder. Let our input coded sequence be <u>0100010000</u>..., where the underline indicates the encoded pair of coded bits.

It is also assumed that the all-zeros data bit sequence was sent. Following the algorithm rules determine that the 2-bit errors are not enough to prevent the Viterbi decoder from correctly detecting the all-zero data bit sequence.

Figure 4.3-6 is used to determine the Hamming distance for each path. Figure 4.3-11(a) and 4.3-11(b) illustrate the trellis development. At depth j=2 the algorithm first defines four paths, one for each of the four states of the encoder. The upper sequence is the input encoded data sequence. The values next to each branch, near the node at the right end of the branch, is the Hamming distance for that particular path. At j = 3, we see that the first decision must be made to eliminate one of the paths into each node. The figure labeled j = 3, survivors indicates what the survivors are at time j=3. Again we are down to four paths, one into each node. This process continues until j=5 occurs. At this point there are four survivor paths, but only one has the lowest Hamming distance, which corresponds correctly to the all-zeros data input sequence.

Thus the correctly decoded path is the all-zeros path shown horizontally from the state "00" in Figure 4.3-11(b) in the right-hand lower corner. Clearly it has the lowest Hamming distance value (2).

Now consider the Viterbi algorithm when soft decisions are employed.

4.3.4.2 The Viterbi Algorithm with Soft Decisions

Soft-decision decoding is quite similar to hard-decision decoding except that, instead of using the Hamming distance and minimizing it, one seeks the largest log likelihood along the path. For a soft decision memoryless channel with AWGN it was shown in (4.2-56) that one can maximize the correlation of the received code word over all possible code words, to obtain the path with the highest correlation. Consequently the discarded paths are the ones with the smaller correlation metric. The result is that the path that is most likely to have been followed by the encoder is chosen.

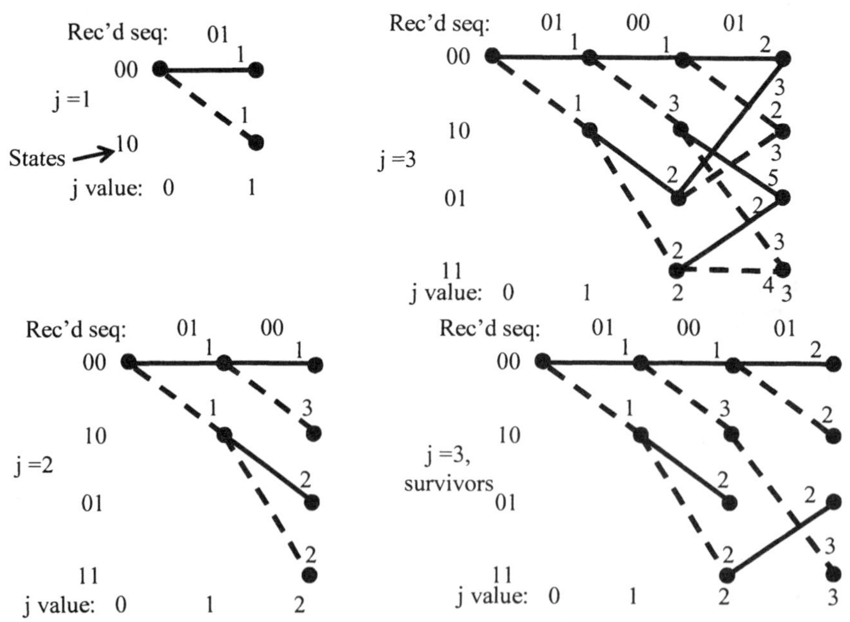

Figure 4.3-11 (a) An illustration of the decoding steps for Example 14. An input "1" is denoted by a dashed line and a "0" by a solid line.

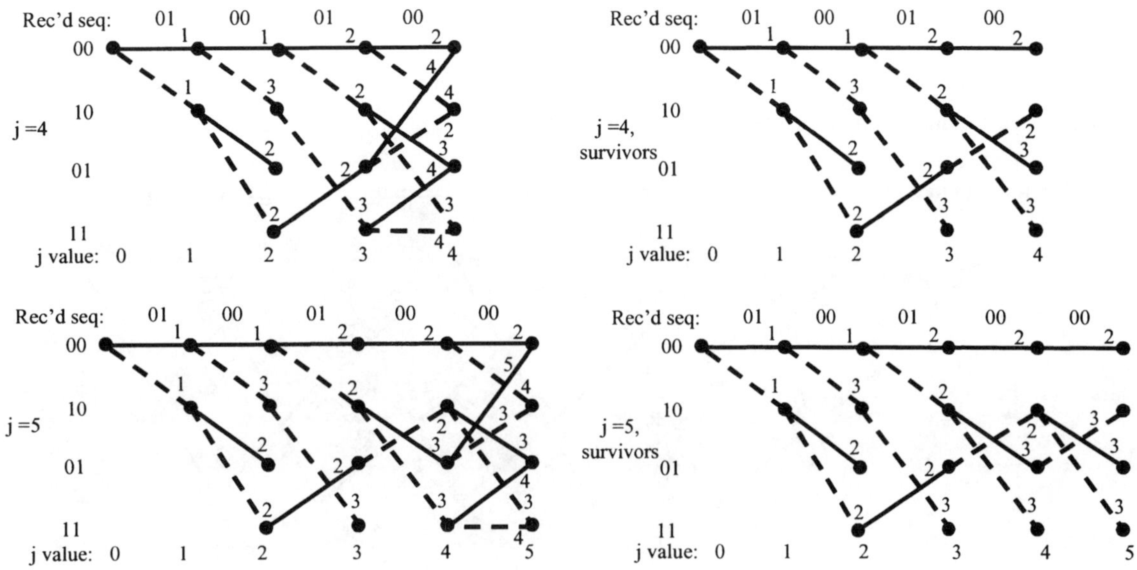

Figure 4.3-11 (b) An illustration of the decoding steps for Example 14.

Example 15 Consider an example to illustrate the soft-decision Viterbi decoder, for the encoder shown in Figures 4.3-2 or 4.3-3. The binary input with four levels of output is illustrated in Figure 4.3-12, along with the log likelihoods. For example the probability of receiving a "0" when a "0" is sent is 0.7. Conversely, the probability of receiving a "3" when a "0" was sent is 0.01. The natural log of $P(0|0)$ is –0.36.

Note that the natural log of the transition probabilities is the metric for the soft decision along the trellis. From (4.2-53) we see when taking natural logs that for the memoryless channel one has

$$\ln\left[p(\boldsymbol{y}|\boldsymbol{C}_m)\right] = \sum_{k=1}^{n} \ln\left[p(y_k|c_{mk})\right] \tag{4.3-13}$$

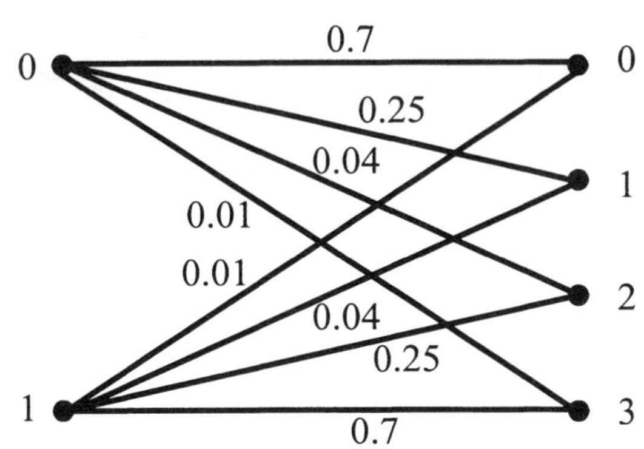

$$\ln P(0|0) = \ln P(3|1) = \ln(0.7) = -0.36$$
$$\ln P(1|0) = \ln P(2|1) = \ln(0.25) = -1.39$$
$$\ln P(2|0) = \ln P(1|1) = \ln(0.04) = -3.22$$
$$\ln P(3|0) = \ln P(0|1) = \ln(0.01) = -4.61$$

Figure 4.3-12 Binary input 2-bit soft-decision memoryless channel.

Therefore the sum of the log likelihoods is applicable to each trellis path. Figure 4.3-13(a) illustrates the trellis paths for each possibility, when five zeroes were transmitted along with two zeros to drive the state to "00." It was assumed that the initial state was "00."

The branch metrics were computed by adding the natural log likelihood. For example, consider the branch from state "00" at level 1 going to state "10" at level 2. Adding $\ln P(2|1)+\ln P(0|1) = |-1.39-4.61 = -6$. Now consider the cumulative branch metrics. Figure 4.3-13(b) illustrates the cumulative branch metrics.

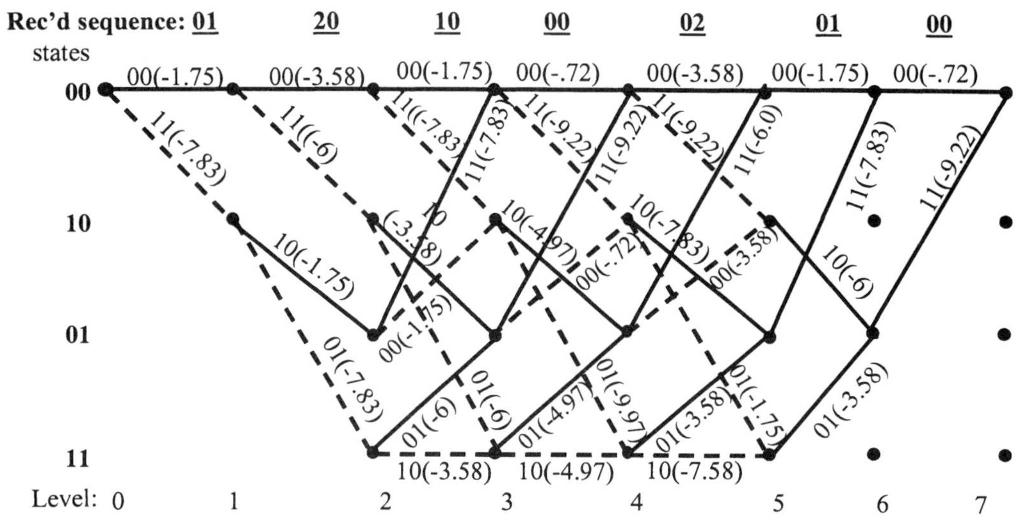

Figure 4.3-13 (a) Soft-decision Viterbi decoding branch metrics for the encoder of Figure 4.3-2.

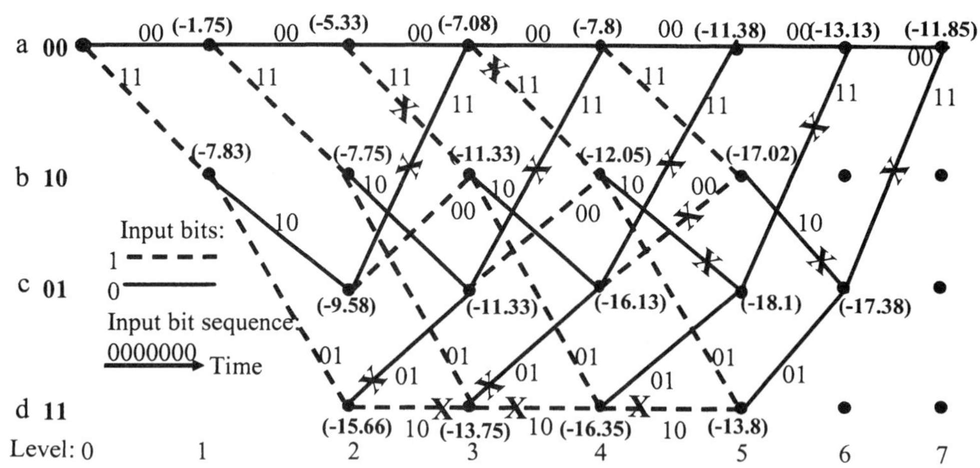

Figure 4.3-13 (b) Soft-decision Viterbi decoding cumulative branch metrics for the encoder of Figure 4.3-2. The "X" denotes that the path is cut.

The cumulative values are noted in parentheses above each node. An "X" on a branch in the figure indicates that that path has been eliminated, since the other path's cumulative branch metric was smaller into that node. For example, consider the node at depth 3 in the 00 state. The cumulative branch metric is –7.08 enclosed in a parenthesis.

From Figure 4.3-13(b) we see that moving from state 00 at depth 3 to state 00 at depth 4 yields a cumulative branch metric of –7.08 –0.72 = –7.8. However going from state 01 at level 3 to state 00 at level 4 yields a branch metric of –11.33 –9.22 = –20.55 that is smaller than –7.8. Thus this latter branch is severed, which is denoted by the "X" in its path. This process continues for all nodes until only one path is left and this is the Viterbi algorithm's choice.

Examples 14 and 15 have illustrated the Viterbi decoder for both the hard-decision and the soft-decision channels. We have seen that the decoding method is the identical; only the branch metric is different.

Now consider a list of optimum short constraint length convolutional codes that are shown in Table 4.3-1 [17]. The table lists the code rate, the constraint length, the free distance, and the associated tap connections for that code. Notice that the optimum code for rate 1/2, constraint length 3 has the tap connections of "111" and "101." Thus the convolutional code shown in Figure 4.3-2 (or alternatively 4.3-3) is optimum, based on the table. Note that the rate 1/2 codes are listed up to length 9 and codes with rate 1/3 are listed up to length 8.

4.3.4.3 Zero Filling and Tail Biting of the Viterbi Algorithm

Up until now all convolutional codes were assumed to start in the all-zero state. Furthermore at the end of a message, K–1 zeros, where K is the constraint length of the convolutional code, are added to the message to make sure the decoder converges to the all-zeros state. This is called *zero-filling*. *Tail biting* [12, 28] is an alternative means of ending in a known state. When tail biting is used, the code sequence starts from a state equal to the last K–1 bits of the information message. This insures that all code words start and end at the same state. This offers approximately 0.1 dB better bit error performance [29] than zero filling.

Table 4.3-1 Optimum Short Constraint Length R_c= 1/2 and 1/3 Convolutional Codes

Code Rate (R_c)	Constraint Length (K)	d_f	Code Generators
1/2	3	5	111, 101
1/2	4	6	1111, 1101
1/2	5	7	11101, 10011
1/2	6	8	111101, 101011
1/2	7	10	1111001, 1011011
1/2	8	10	11111001, 10100111
1/2	9	12	111101011, 101110001
1/3	3	8	111,111,101
1/3	4	10	1111,1101,1011
1/3	5	12	11111,11011,10101
1/3	6	13	111101,101011,100111
1/3	7	15	1111001,1110101,1011011
1/3	8	16	11110111,11011001,10010101

4.3.4.4 Branch Symbol Synchronization

The branch symbols must be synchronized with the received symbols. When misaligned, the performance deteriorates rapidly. Typically a misalignment can be detected by observing (a) path metric growth or (b) channel bit error estimates. In the first case (a) the path metrics are typically close to each other and grow at a large rate. For case (b) the channel bit error rate will approach 1/2.

Thus a synchronization stage must be included external to the decoder itself. The synchronization stage has the function of changing the symbol timing by one symbol for at most $n-1$ changes where the code rate is $1/n$. After each trial (there are a total of n trials at most) the metric is compared to what is expected for normal behavior. If normal behavior is not obtained, the symbol timing is shifted once and the metric is tried again. This continues until synchronization is achieved.

4.3.4.5 Required Memory for Viterbi Decoding

One of the problems in decoding with a Viterbi decoder is the fact that as the sequence length increases, the required storage of the branch metrics and the sequences increases. Moreover a delay in decoding is required, which depends directly on the length of the transmitted sequence. A solution to this problem is to modify the algorithm in such a way that it becomes simpler, but maintains about the same performance. The usual solution to this problem is to allow only a fixed delay and fixed memory in the trellis. Basically at any given time, the algorithm only needs the N most recent bits.

As each new information bit is received, a final decision is made on the bit received N branches (bits) back in the trellis, by comparing the branch metrics in the surviving sequences and deciding in favor of the bit in the sequence that has the largest path metric. If the parameter N is chosen to be large enough, all of the surviving sequences will contain the same decoded bit N branches back in time. In other words with a very high probability, all surviving sequences at a particular time t_1 will stem from the same node at t_1-N. It has been determined by simulation and hardware measurement that if $N \geq 5K$, then negligible degradation will occur compared to the case of infinite memory case (optimum Viterbi decoding).

4.3.5 Error Probabilities for Viterbi Decoding of Convolutional Codes

Perhaps the most useful techniques for estimating bit and word error performance of convolutional codes are union bounds and computer simulation [1]. Now consider the evaluation of the bit error probability. Following Clark and Cain [1], first event error probability per node denoted by P_w is evaluated. Since convolutional codes are linear, the decoding error probability is the same for all possible transmitted sequences. Therefore it is assumed that the all-zeros sequence corresponds to the transmitted sequence.

4.3.5.1 Error Probabilities for Viterbi Decoding for the BSC Via the Union Bound

Let p denote the channel symbol error probability for the binary symmetric channel (BSC). It is shown in [23] and [1] that for Viterbi decoding, the word or sequence error probability is overbounded by a union bound of the form

$$P_w < T(D,N)\big|_{N=1,D=2\sqrt{p(1-p)}} \tag{4.3-14}$$

where p is the BSC error probabilities. Again for Viterbi decoding, the union bound for the bit error probability is overbounded by [23] and [1]

$$P_b < \frac{1}{k}\frac{dT(D,N)}{dN}\bigg|_{N=1,D=2\sqrt{p(1-p)}} \tag{4.3-15}$$

where k is as before the number of bits input to the encoder at a time.

Example 16 Consider the upper bound on sequence and the bit error probability for the convolutional code of Figure 4.3-2. The transfer function for this code is repeated here.

$$T(D,N)=\frac{ND^5}{1-2ND} \tag{4.3-16}$$

Evaluating (4.3-14) produces the sequence error probability

$$P_w \le \frac{\left[2\sqrt{p(1-p}\right]^5}{1-4\left[\sqrt{p(1-p)}\right]^2} \tag{4.3-17}$$

Now consider the bit error probability upper bound. Taking the derivative with respect to the variable N produces the result

$$\left.\frac{dT(D<N)}{dN}\right|_{N=1}=\frac{D^5}{(1-2D)^2} \tag{4.3-18}$$

Combining (4.3-15) and (4.3-18) produces the result for the bit error probability

$$P_b \le \frac{\left[2\left(p(1-p)\right)^{1/2}\right]^5}{\left[1-4\{p(1-p)\}^{1/2}\right]^2} \tag{4.3-19}$$

since $k = 1$ in this case.

4.3.5.2 Error Probabilities for Viterbi Decoding for the Soft-Decision Decoding in the AWGN Channel

In a derivation similar to the derivation of (4.3-19) for the unquantized soft-decision decoding for the binary input AWGN channel, the following upper bound for the bit error probability is given by [16]

$$P_b \le \frac{1}{k}Q\left(\sqrt{\frac{2d_fE_s}{N_0}}\right)\exp\left(d_fE_s/N_0\right)\left.\frac{dT(D,N)}{dN}\right|_{N=1,D=e^{-E_s/N_0}} \tag{4.3-20}$$

The parameters in (4.3-20) are given by:

$E_c/N_0 = R_cE_b/N_0$ = the channel symbol energy to one-sided noise spectral density

R_c= the code rate of the code = k/n

N_0 is the one-sided noise spectral density in W/Hz

d_f is the free distance of the convolutional code

And for each k data bits entered into the encoder, n coded bits are sent to the modulator. This bound has been used very effectively to obtain tight bounds for the bit error rate probability on the AWGN channel for a variety of convolutional codes of constraint lengths (K) less than 10. It is pointed out in Viterbi's paper [23] that the transfer function bound is unmanageable for $K > 4$; however, a matrix inversion can be used to obtain the same generating function.

Example 17 Consider the convolutional code of Figure 4.2-3, which is a rate 1/2, $k = 1$, K=3 code. Utilize (4.3-20) to obtain a bound to the bit error probability. Since the derivative is given by

$$\left.\frac{dT(D,N)}{dN}\right|_{N=1} = \frac{D^5}{(1-2D)^2} \tag{4.3-21}$$

It follows that

$$P_b \le Q\left(\sqrt{\frac{5E_b}{N_0}}\right)\exp\left(\frac{5}{2}\frac{E_s}{N_0}\right)\frac{\exp\left(-\frac{5}{2}\frac{E_b}{N_0}\right)}{\left[1-2\exp\left(-\frac{E_b}{2N_0}\right)\right]^2} \tag{4.3-22}$$

Canceling the common terms produces

$$P_b \le \frac{Q\left(\sqrt{\frac{5E_b}{N_0}}\right)}{\left[1-2\exp\left(-\frac{E_b}{2N_0}\right)\right]^2} \tag{4.3-23}$$

which is our result for the bit error probability bound for soft decision in a WGN channel, for the rate ½, K = 3, k = 1, convolutional code of Figure 4.3-2. The reader should note that other bounds for these error probabilities exist and in some cases they may slightly tighter [30].

Figure 4.3-14 illustrates eight level (uniform spacing) simulation and transfer function bounds by Heller and Jacobs [24] of rate ½ convolutional codes of constraint lengths of K = 3 through K = 8 with k = 1.

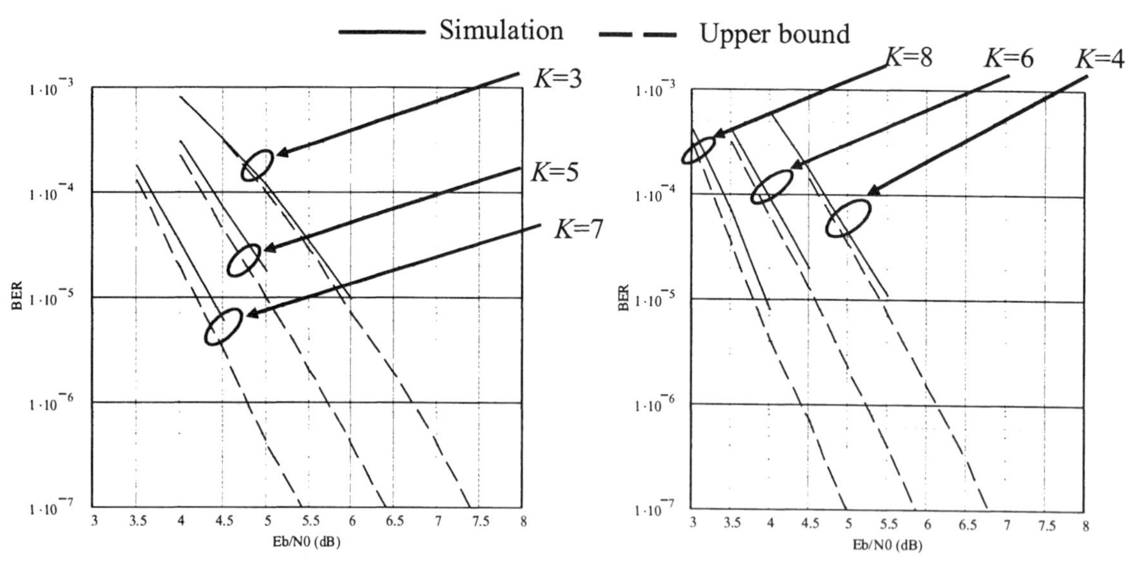

Figure 4.3-14 Bit error probability based on simulation and transfer function bounds.

One thing to notice immediately is the simulation results (solid lines) are above the upper bounds! This is due to the fact the bounds are for the infinitely quantized channel and the simulation was for 8-level quantization. Thus to convert the bounds to 8-level quantization also, one must add about 0.2–0.25 dB [24] to the bounds to bound the 8-level case. When this correction factor is added to the bounds, they agree quite well over the BER range of 10^{-4} to 10^{-5}. Furthermore, since the bounds become more accurate at lower bit error rates, it is believed that the bounds are a good estimate for the BER performance when the ¼ dB is added to correct for the infinite quantization used in the bounds.

Another method of estimating the bit error probability follows from the work of Viterbi [21]. Consider the all nonzero paths through the code trellis that first emerges with the all-zero path at depth j. Recall that the number of ones in a nonzero code word sequence is the Hamming distance between this sequence and the all-zero sequence. The free distance, as defined in Section 4.3.2, is just the minimum Hamming distance between any two distinct code sequences. The free distance is equal to the Hamming weight on the nonzero code sequence with the smallest Hamming weight, since the code is linear. For the error rate estimates, let a_d denote the number of nonzero code sequences that are Hamming distance d away from the all-zero sequence path. Therefore, in light of the definition of the free distance, for all $d < d_f$, $a_d = 0$. In addition define P_{Ed} as the probability that the convolutional decoder discards the all-zero code sequence at depth j in favor of a code having Hamming distance d. This probability is the same as the two-code word error probability for a block code with minimum distance d. Viterbi [21] has shown that P_{Ed} can be overbounded by

$$P_{Ed} < \sum_{d=d_f}^{\infty} a_d P_d \tag{4.3-24}$$

Equation (4.3-24) bounds the probability that the Viterbi decoder selects an incorrect path through the trellis. It should be noted that it is not a bound on the bit error probability. Viterbi [21] has shown that for the continuous output, soft-decision AWGN channel that

$$P_d = \int_{\sqrt{2dR_cE_b/N_0}}^{\infty} \frac{1}{\sqrt{2\pi}} \exp\left(\frac{-z^2}{2}\right) dz = Q\left(\sqrt{\frac{2dR_cE_b}{N_0}}\right) \tag{4.3-25}$$

where R_c is the code rate, E_b is energy per bit, d is the Hamming distance, and N_0 is the one-sided noise spectral density.

To bound the bit error rate probability, each decoding error event is weighted by the number of bit errors associated with that error event. Under the assumption that the all-zeros sequence is the correct path, the number of bit errors associated with a nonzero path is equal to the number of information bits that are "ones" associated with that path. In the general case there are a total of c_d "information bit" errors associated with all paths that are at Hamming distance d from the all-zeros sequence path. Viterbi [17] has shown that the bit error rate probability can be overbounded by

$$P_b < \sum_{d=d_f}^{\infty} c_d P_d \tag{4.3-26}$$

where P_d is given by (4.2-79) and (4.2-80) for the hard-decision binary symmetric channel and (4.3-25) for the soft-decision AWGN channel. When jammer state information (JSI) is available, Peterson et al. [15] have shown that (4.3-25) and (4.3-26) remain valid with the appropriate calculation of P_d. Peterson et al. [15] have shown that using analysis similar to that used for the binary symmetric channel, when JSI is available, P_d must be replaced with P_{jd} of (4.2-107).

The results that have been presented here are upper bounds and not exact results. Most upper bounds are reasonably accurate at lower BER values of say less than 10^{-4}. Simulation is another means of determining the BER values and of course works at all values of BER.

4.3.6 Sequential Decoding of Convolutional Codes

Before the discovery of Viterbi decoding of convolutional codes, convolutional codes were decoded using sequential decoding, as well as some of its variants. Sequential decoding is not maximum likelihood but can be made to be roughly of the same BER performance. The complexity of sequential decoders is roughly independent of the constraint length of the convolutional code. Typically sequential decoding is used when the constraint length is very large ($K > 10$).

Sequential decoding is a method that makes it feasible to decode large constraint length convolutional codes by checking only some of the branches, and the number of branches depends on the channel characteristics. When a period of reliable data is received, only a few paths need be checked, but when a batch of unreliable data is received, many more paths are checked.

A sequential decoder systematically searches through a code tree by extending, one branch at a time, a path that has already been examined and bases its decision on which path to extend only on the metrics of already-examined paths [17]. Sequential decoders work on paths of different lengths, and to accommodate these operations on different length paths, a fixed bias term is added to the symbol metric in order that the metric values that have been accumulated along the correct path tend to increase whereas those metric values along an incorrect path have a tendency to decrease. For a binary input soft decision the output has 2^q values and a code rate of R_c. The most commonly used sequential decoding symbol metrics add a bias of qR_c [17] to these maximum likelihood metrics. This metric bias is called the *Fano metric.*

The main sequential decoding algorithms are the Fano and Stack algorithms, or some variation of these. References [2] and [31] present good descriptions of these two algorithms. Figure 4.3-15 illustrates bit error rate performance for sequential decoding of a $K = 36$, $R_c = 1/2$ sequential decoder based on results presented in Oldenwalder [17].

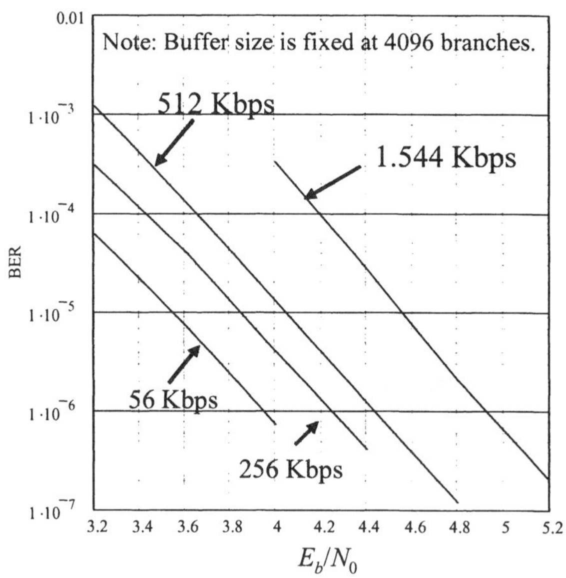

Figure 4.3-15 Measured BER of a $K = 36$, $R_c = 1/2$ 2-bit sequential decoded, convolutional code, on an AWGN channel with BPSK or QPSK modulation [17].

4.3.7 Threshold Decoding of Convolutional Codes

Feedback decoding of convolutional codes is a method of obtaining more modest coding gains than those that can be obtained with Viterbi or sequential decoding. The primary advantage of feedback decoding is that the decoder is simple to implement and that interleaving and deinterleaving can be included as part of the encoder and decoder [32]. Feedback decoded convolutional coding systems usually employ systematic codes along with hard-decision quantized receiver data.

4.3.8 Nonbinary Convolutional Codes

So far we have discussed binary convolutional codes. Binary convolutional codes are suitable for channels in which BPSK of QPSK modulation along with coherent demodulation is utilized. In some applications BPSK or QPSK modulation with coherent demodulation may not be appropriate; however, noncoherent modulations may be applicable, such as M-ary FSK. If M is picked to be divisible by 2, then nonbinary convolutional codes are directly applicable. We will briefly discuss *dual-k codes* for a rate 1/2 encoder. These codes can be decoded either by hard-decision or soft-decision methods. A rate 1/2 dual-k convolutional encoder can be represented as shown in Figure 4.3-16. The encoder is composed of two (K'=2) k-bit shift register stages and $2k$ function generators. The output is two k-bit symbols. The example shown is a dual-2 bit convolutional encoder (k = 2). We see that the code rate is ½ since there are k input bits and $2k$ output coded symbols. For more details see [17, 33].

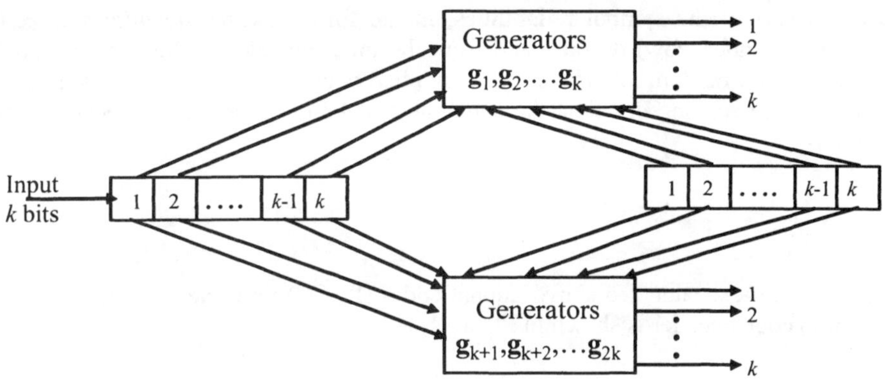

Figure 4.3-16 Convolutional encoder for rate 1/2 dual-2 codes.

4.4 ITERATIVELY DECODED CODES

Iterative decoding can be described as a decoding technique utilizing a soft-output decoding algorithm that is iterated several times to improve the bit error performance of a coding scheme, with the goal of obtaining true maximum-likelihood decoding, with less decoder complexity. These codes are also called *capacity approaching codes*, since their performance approaches channel capacity.

The early work of Elias [34] on iterated codes back in 1954 was the original work in the area of iterated codes. Iterated decoding was known as probabilistic decoding in those times. The basic concept was then, as it is currently, to maximize the a posteriori probability of a symbol being sent, given a noisy version of the coded sequence.

In this section a brief discussion of one branch of iterated codes, known as *product codes*, are presented. These codes include parallel-concatenated codes (turbo codes) and serial-concatenated codes. The codes can be either convolutional or block codes.

The other class of iterated codes, low-density parity check codes (LDPC), will be discussed briefly here also. Basic to turbo codes is the notion of *concatenation of codes*. Concatenation of two codes is a method of achieving large coding gains from two (or more) simple component or constituent codes. Concatenation was first proposed by Forney [35]. The most popular concatenated code consists of a Reed-Solomon outer block code and a convolutional inner code, usually utilizing Viterbi decoding of the convolutional code.

4.4.1 Turbo Codes

A *turbo code* can be viewed as generalization of a concatenated code in conjunction with the use of an iterative decoding algorithm for decoding the turbo code. The unique aspect of turbo codes is the iterative decoding algorithm. Turbo codes were discovered by Berrou, Glavieux, and Thitimajshima [36, 37]. They reported a coding and decoding scheme that offered phenomenal bit error probability results: A rate 1/2 code in AWGN, with BPSK modulation, produced a bit error probability of 10^{-5} at a E_b/N_0 of 0.7 dB!

4.4.1.1 Turbo Coding Structure

A turbo code has the following four attributes. (1) A turbo code system is a product code (as referred to in [36]) with a parallel concatenated code structure, with constituent recursive systematic convolutional encoders. (2) A turbo code system utilizes soft-input soft-output (SISO) maximum a posteriori (MAP) decoders for each component code in which to generate log-likelihood ratios. (3) A turbo code system utilizes feedback of part of the symbol reliabilities, in the form of *extrinsic information*, from an outer (column) decoder to an inner (row) decoder and from the inner decoder to the outer decoder. (4) A turbo code system utilizes a long random interleaver that is applied between the two encoders. Its primary function is to ensure that, at each decoding iteration, the component MAP decoders get independent estimates on the information symbols.

4.4.1.2 Turbo Encoder

First we consider parallel-concatenated convolutional codes (PCCC) with interleaving, commonly known as a turbo code. A turbo code encoder is shown in Figure 4.4-1.

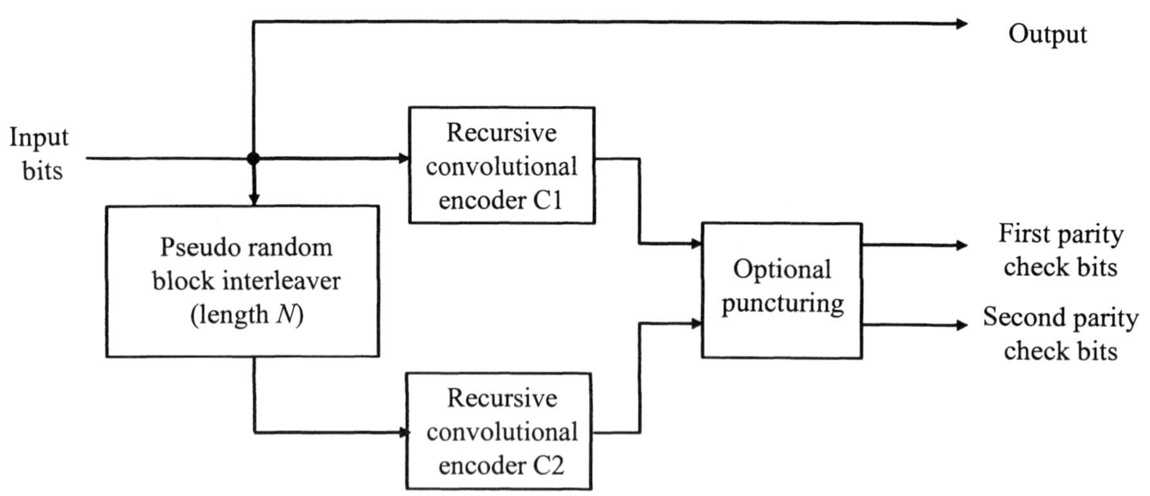

Figure 4.4-1 Turbo encoder for parallel concatenated code, a turbo code.

The encoder is a *recursive systematic convolutional (RSC) encoder* and is comprised of two RSC convolutional encoders operating in parallel, along with an interleaver that inputs the data in a scrambled manner. A recursive systematic convolutional encoder has the property that one of the outputs is the data input itself (systematic property). The two internal RSC encoders can be identical or different. The nominal code rate for a turbo encoder is 1/3, since for each input bit there are three output coded symbols. However by the act of puncturing, the code rate can be increased to 1/2, 2/3, or higher. The interleaver is designed to randomize the data bits prior to inputting them to the second RSC. The use of the interleaver in conjunction with the two encoders turns out to have relatively few nearest neighbors as shown by Benedetto and Montorsi [38]. They also have shown that the number of nearest neighbors, is reduced by N, with N being the interleaver length. Ryan [39] has shown that the bit error rate decreases as N is increased. Thus the interleaver length is an important parameter in the design of RSC codes.

Consider now the difference between a nonsystematic convolutional code (NSC) and a recursive systematic code (RSC). Figure 4.4-2 illustrates both an NSC and an RSC encoder.

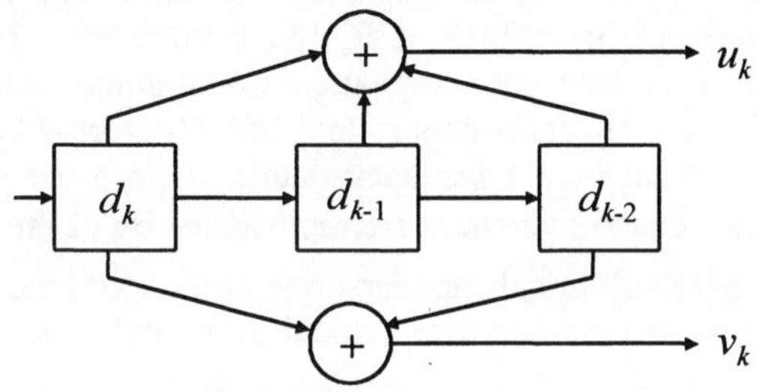

(a) Nonsystematic convolutional (NSC) example with $G_1 = \{1,1,1\}$ and $G_2 = \{1,0,1\}$.

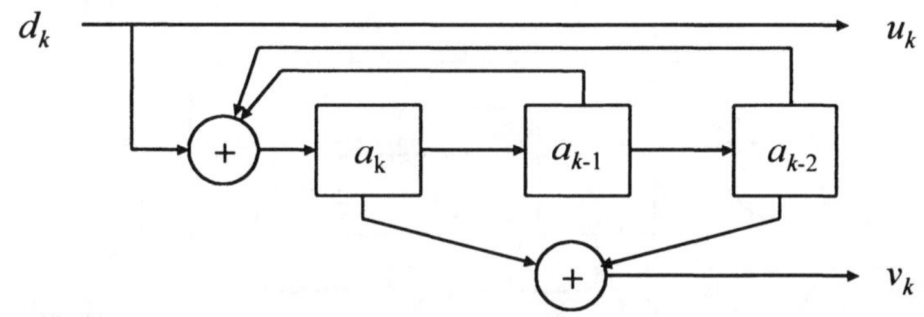

(b) Recursive systematic convolutional (RSC) example with $G_1 = \{1,1,1\}$ and $G_2 = \{1,0,1\}$.

Figure 4.4-2 (a, b) An example of NSC and RSC codes with $G_1 = \{1,1,1\}$ and $G_2 = \{1,0,1\}$.

In Figure 4.4-2(a) is an example of a nonsystematic convolutional encoder (NSC). It is a simple rate 1/2, constraint length $K = 3$ convolutional encoder. The input to the encoder at time k is the data bit d_k. The two outputs, in the general case, can be written as

$$\begin{aligned} u_k &= \sum_{i=0}^{K-1} g1_i d_{k-1} \quad \text{Mod}(2) \quad g1_i = 0,1 \\ v_k &= \sum_{i=0}^{K-1} g2_i d_{k-1} \quad \text{Mod}(2) \quad g2_i = 0,1 \end{aligned} \qquad (4.4\text{-}1)$$

where the code generators $G_1 = \{g1_i\}$ and $G_2 = \{g2_i\}$ are the connections for each coder. This encoder has a finite impulse response (FIR). Now consider the rate 1/2, K=3, recursive systematic convolutional encoder (RSC) of Figure 4.4-2(b), in generalized form. The variable a_k is defined by

$$\begin{aligned} a_k &= d_k + \sum_{i=0}^{K-1} g1_k a_{k-1} \quad \text{mod}(2) \\ v_k &= \sum_{i=0}^{K-1} g2_k a_{k-1} \quad \text{mod}(2) \\ u_k &= d_k \quad \text{mod}(2) \end{aligned} \qquad (4.4\text{-}2)$$

The trellises for both these encoders (NSC and RSC) have the same structure, and they have the same free distance [39]. As noted in the figure the two encoder generators are given by $G_1=\{1,1,1\}$ and $G_2=\{1,0,1\}$.

4.4.1.3 Parallel Concatenation of RSC Codes

It is well known that the BER of a classical nonsystematic convolutional code is lower than a systematic convolutional code with the same memory at high SNR, and at low SNR it is higher. For turbo codes at low SNR RSC are better in BER than the best nonsystematic convolutional codes [36]. An example of a turbo code based on two recursive convolutional encoders is depicted in Figure 4.4-3. Notice that the convolutional encoders are recursive in nature. That is, the feedback data adds into the input data. The data input is d_k ($d_k \in (0,1)$) at time k, and the two coded sequences are represented by $v1_k$ and $v2_k$ ($v1_k$ and $v2_k \in (0,1)$).

There are two recursive convolutional encoders; one of them has the data interleaved. There is also a third output that is composed of the unmodified data, which is output directly.

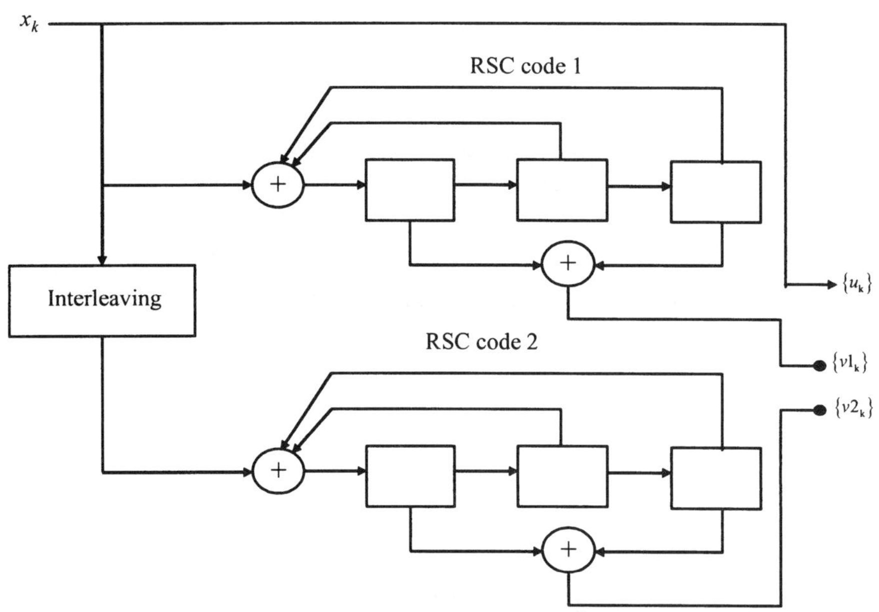

Figure 4.4-3 A turbo encoder composed of the parallel concatenation to RSC encoders.

4.4.1.4 A Parallel Concatenated Convolutional Decoder (Turbo Decoder)

The turbo decoder utilizes iterative decoding based on the maximum a posteriori (MAP) criterion. The Viterbi decoding algorithm is an optimal decoding process for minimizing the probability of a sequence error. It turns out that its output of a hard decision of a bit is not applicable to generate the a posteriori probability (APP) for each detected bit. The iteration-decoding algorithm that works with soft decision bits is oftentimes based on the Ball, Cocke, Jelinek, and Raviv [27] (BCJR) algorithm or a modification of the same. Figure 4.4-4 illustrates the iterative decoder for the parallel concatenated convolutional code (PCCC) (that is, a turbo code). The BCJR algorithm was modified by Berrou et al. [36] for the decoding of RSC codes. This section is based, in part, on Berrou et al. [37], Sklar [40], and Proakis [11].

Consider an RSC with constraint length K. Let vector $\boldsymbol{r}$ denote the outputs from the demodulator over a discrete, memoryless, AWGN channel.

The first decoder (Decoder 1) is fed inputs from the information bits and the first set of parity check bits $v1_k$ derived from the soft output demodulator. In a similar manner the second decoder (Decoder 2) is fed information bits and the second set of parity check bits from the soft output demodulator. Each soft input soft output (SISO) decoder attempts to decode the information sequence $\{u_i\}$ by computing the a posteriori probability for each information bit derived from its respective parity bits and then sending the soft output information to improve upon it the probability of correct decision. This soft output information that is exchanged between the two decoders is called the *extrinsic information*.

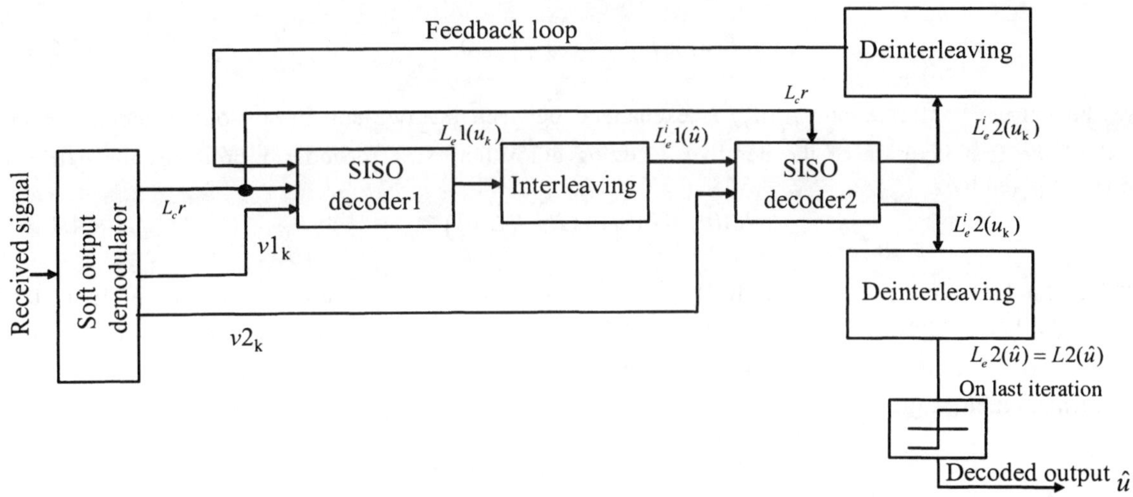

Figure 4.4-4 Block diagram representation of an iterative decoder for a PCCC (turbo code).

The output symbol-by-symbol MAP decoder is the a posteriori log-likelihood ratio for the condition that a transmitted information bit $u = 1$ and the transmitted bit $u = -1$. Thus one can write

$$L(\hat{u}) = \log\left(\frac{P(u=1|\boldsymbol{r})}{P(u=-1|\boldsymbol{r})}\right) \tag{4.4-3}$$

where the hat over the letter u denotes the estimate of the bit u. The decoder can make a decoded bit decision based on comparing the log-likelihood ratio (LLR) to a threshold of zero:

$$\begin{aligned} \hat{u}_k &= 1 \ \text{ if } L(\hat{u}_k) > 0 \\ \hat{u}_k &= -1 \ \text{ if } L(\hat{u}_k) < 0 \end{aligned} \tag{4.4-4}$$

The LLR (4.4-3) can be decomposed into the sum of three components [40]. The first component is the a priori likelihood and is defined as

$$L(u) = \log\left(\frac{P(u=1)}{P(u=-1)}\right) \tag{4.4-5}$$

which depends only on the a priori probability of $u = 1$ and $u = -1$ occurrence. The second component [12] is expressed as [11]

$$L_c r = \frac{4E_c}{N_0} r \tag{4.4-6}$$

where

$$L_c = \frac{4E_c}{N_0} \text{ for an AWGN channel} \tag{4.4-7}$$

and where $N_0/2$ is the two-sided noise spectral density, and E_c is the coded symbol energy.

Hagenauer et al. [40] have shown for systematic turbo codes that the soft output for the information bit u may be expressed as the sum of three additive terms for statistically independent transmission, in particular

$$L(\hat{u}) = L_c r + L(u) + L_e(\hat{u}) \tag{4.4-8}$$

where the extrinsic information, $L_e(\hat{u})$, is exchanged between the two soft-in soft-out decoders. In the process of the first iteration of the iterative decoding algorithm, SISO decoder 1 produces the extrinsic information given by

$$L_e 1(u) = L1(\hat{u}) - \left(L_c r + L(u)\right) \tag{4.4-9}$$

If a "1" and a "–1" are a priori equally likely, then $L(u) = 0$. This extrinsic information from the first SISO decoder, after being interleaved, is sent to the second SISO decoder, which uses $L_e 1(\hat{u})$ as the a priori value in place of $L(u)$ to compute $L2(\hat{u})$. Therefore, the extrinsic information value computed by decoder 2, for each information bit, is given by

$$L_e 2(\hat{u}) = L2(\hat{u}) - \left(L_c r + L_e 1(\hat{u})\right) \tag{4.4-10}$$

Next SISO decoder 1 will use the extrinsic information values $L_e 2(\hat{u})$ as prior information in the second SISO decoder iteration. This computation is repeated for each iteration.

The iterative process is commonly terminated after a predetermined number of iterations. This occurs typically when the successive soft output values $L2(\hat{u})$ stabilize, in the sense that they change very little between each successive iteration. The SISO decoder 2, in the final iteration combines the two extrinsic information values in determining the soft output values

$$L2(\hat{u}) = L_c r + L_e 1(\hat{u}) + L_e 2(\hat{u}) \tag{4.4-11}$$

The data bit estimate is obtained from a hard decision of $L2(\hat{u})$, as shown in Figure 4.4-3.

Berrou et al. [37] have simulated the performance of a rate ½ encoder with constraint length $K = 5$, having octal generators G_1=37 and G_2=21 and parallel concatenation. The parity bits are alternatively deleted producing a rate of 1/2. The interleaver is based on a 256×256 matrix, and the modified Bahl algorithm [27] is used. The block length is $N = 65,535$ bits. Figure 4.4-5 [37] illustrates the BER performance of a recursive parallel-concatenated turbo code with these parameters.

As can be seen from the figure, after one iteration, the BER versus E_b/N_0 ratio is considerably better than uncoded BPSK. Finally after 18 iterations the BER is 1×10^{-5} at $E_b/N_0 = 0.7$ dB.

The primary drawback with decoding turbo codes with large interleavers is the attendant decoding delay and the computational complexity resident in the iterative decoding algorithms. However, in communication applications some delay may be tolerable and the enhancement in bit error rate performance makes the option very appealing.

4.4.2 A Serial Concatenated Convolutional Code

An alternative to a PCCC is a serial concatenated convolutional code (SCCC). Benedetto, Divsalar, Montorsi, and Pollara [7] introduced serial concatenation of codes in 1998. Figure 4.4-6 illustrates a block diagram of a SCCC model. The encoder is composed of two constituent linear convolutional encoders and an interleaver.

This section follows Benedetto et al. [7] and their notation. The symbols $L(.;I)$ and $L(.;O)$ at the input and output ports of the SISO refer to the log-likelihood ratios (LLRs) unconstrained when the second argument is I and modified according to the code constraints when it is O. When the symbols are binary only one LLR is required. The first argument u refers to the information symbols of the encoder, and c refers to the code symbols. Finally, the superscript o refers to the outer encoder and the superscript i refers to the inner encoder.

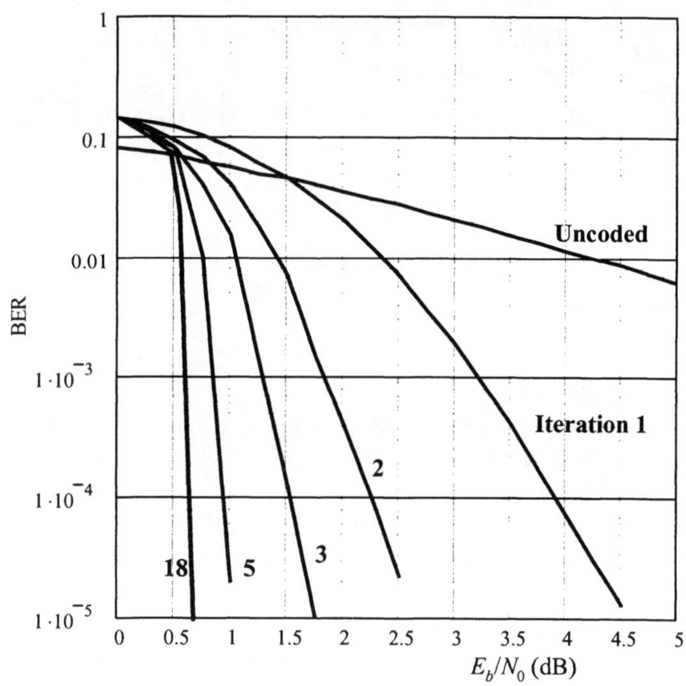

Figure 4.4-5 A plot of the BER for a PCCC (turbo code) against the SNR and iteration number.

The log-likelihood ratios are defined by [7]

$$L(u;.) = \log\left(\frac{P(u;.)}{P(u_{ref};.)}\right) \tag{4.4-12}$$

When u is a binary symbol, "1" or "0" u_{ref} is usually assumed to be "1." When u belongs to an M-ary alphabet, u_{ref} can be chosen as each one of the M symbols.

The outer code $C1$ has a code rate of $R_c^o = k/p$ and the inner code $C2$ has a code rate of $R_c^o = p/n$ connected by an interleaver of length N bits. This concatenation results in a code of length of $R_c^S = k/n$. It is assumed that N is a multiple of p.

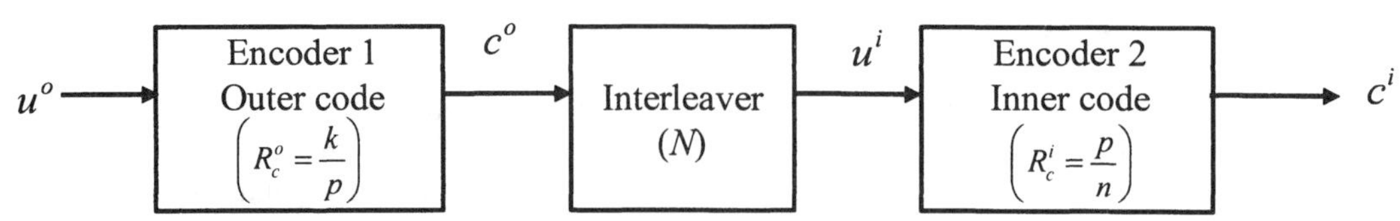

Figure 4.4-6 A block diagram of an encoder for an SCCC.

The iterative decoder for SCCC systems is illustrated in Figure 4.4-7 [41]. Again, soft input and soft output (SISO) devices are used in the decoding process. The SISO is a four-port device that accepts as inputs the probability likelihood ratios of the information and code symbols labeling the edges of the code trellis, and forms an update of these probabilities based on the code constraints.

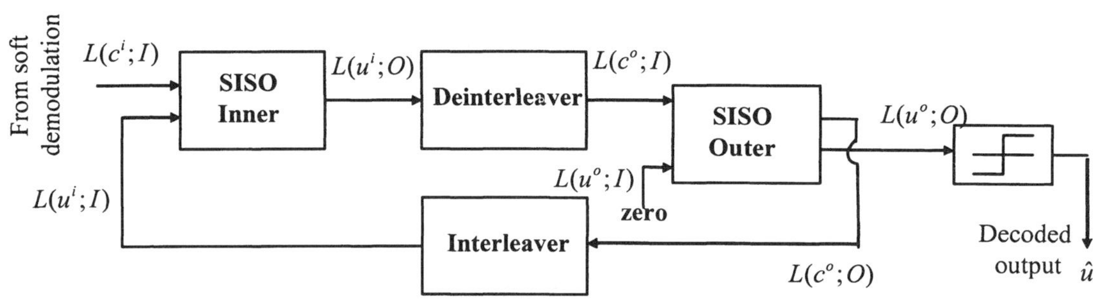

Figure 4.4-7 Block diagram of an iterative decoder for SCCC.

Unlike the turbo decoder, in which only the LLRs of information symbols are updated, for the SCCC both the LLRs of information and code symbols are updated, based on the code constraints.

During the first iteration of the SCCC algorithm, the SISO inner block is fed with the demodulator soft outputs, which consists of the LLRs of symbols received from the channels (that is, code symbols of the inner channel). If the a priori information of a "0" and a "1" are equally likely, then the log likelihood is set to zero on the first iteration. The second (lower) input $L(u^i;I)$ of the inner SISO is set to zero during the first iteration, because there is no a priori information available on the input symbols u^i of the inner encoder.

The inner SISO algorithm processes the LLRs $L(c^i;I)$ to produce the *extrinsic* LLRs of the inner encoder $L(u^i;O)$ conditioned on the inner code constraints. The extrinsic LLRs are transmitted through the deinterleaver (inverse interleaver) whose outputs correspond to the LLRs of the code symbols of the outer code—that is,

$$\pi^{-1}\left[L(u^i;O)\right] = L(c^o;I) \tag{4.4-13}$$

where π^{-1} denotes the inverse operation of the interleaver (i.e., the deinterleaver function).

These LLRs are then transmitted to the outer SISO (upper input), which corresponds to the code symbols. The outer SISO decoder processes the LLRs $L(c^o;I)$ of its unconstrained code symbols, and computes the LLRs of both the information and code symbols, based on the code constraints. The input $L(u^o;I)$ of the SISO outer decoder is always set equal to zero. This is based on the assumption that the information data bits are "0" and "1," and occur with an equal probability of 1/2.

The output LLRs of information symbols, which yields the a posteriori LLRs of the SCCC information symbols, is used in the final iteration to recover the data information bits. In addition, the LLRs of outer code symbols, after interleaving, are fed back to the lower input, which are the information symbols of the inner code of the inner SISO decoder to start the second iteration. It can be seen from the block diagram that

$$\pi\left[L(c^o;O)\right] = L(u^i;I) \tag{4.4-14}$$

That is to say, the interleaved value of $L(c^o;O)$ is equal to $L(u^i;I)$.

Benedetto et al. [7] have determined that PCCCs seem better suited to approaching capacity limits for bit error probabilities above the error floor, which is around 10^{-6} to 10^{-7}; on the other hand SCCCs seem to be better suited to provide near error-free performance. Further details are contained in [7].

4.4.3 Serial Concatenated Block Codes

Serial concatenated block codes (SCBC) are similar to SCCC codes in which an outer linear encoder feeds a block interleaver, which in turn drives another linear block encoder. Figure 4.4-8 is a block diagram of a serial concatenated block code system.

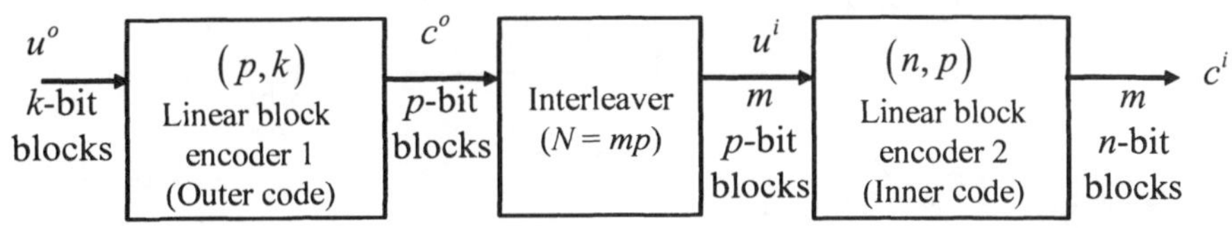

Figure 4.4-8 A block diagram of an encoder for an SCBC.

Both the inner and the outer codes are linear block codes. The outer code is a (p,k) linear code and the inner code is an (n,p) code. The block interleaver is chosen to be of size $N = mp$, in which m is the number of p bit blocks, or outer code words. The data processing is as follows: m blocks of k bits are encoded, one block at a time. This results in mp encoded bits. Each of the m blocks of interleaved and coded data is fed to the inner encoder, which produces m blocks of n coded bits. The resulting code rate is the product

$$R_c^{SCBC} = \frac{k}{p}\frac{p}{n} = \frac{k}{n} \tag{4.4-15}$$

The block length of the SCBC is nm coded bits, which could easily be significantly larger than the block length of a conventional serial concatenation of two block codes that do not utilize an interleaver between them.

The decoder for SCBC codes is the same as shown in Figure 4.4-7, and the decoding process description is the same for the SCBC as for the SCCC systems.

4.4.4 Parallel Concatenated Block Codes

Parallel-concatenated block codes (PCBC) can be constructed in parallel along with the use of an interleaver. Figure 4.4-9 illustrates a block diagram of an example of a PCBC, which is based on two constituent systematic linear block codes denoted as (n_1,k) and (n_2,k), respectively.

The pseudorandom block interleaver has length of $N = mk$ bits. The overall code has length $m(n_1+n_2-k)$ coded bits and has a code rate of $k/(n_1+n_2-k)$. Daneshgaran and Mondin [41] have investigated the design of block interleavers for parallel-concatenated codes. The effect of the interleaver when used with SCBC and PCBC is to produce code words that are both large in block length and relatively sparse (very few binary ones).

4.4.5 Low-Density Parity Check Codes

Gallagher, in 1962, introduced a class of codes known as low-density parity check (LDPC) codes and described two iterative probabilistic decoding algorithms [42]. Later Zyablov and Pinsker [43] showed that LDPC codes have a minimum distance that grows in a linear manner with the code length, and that errors up to the minimum distance could be corrected with a decoding algorithm with an almost linear complexity. Still later Tanner [44] extended Gallagher's probabilistic decoding algorithm to the more general case in which the parity checks are defined as subcodes rather than simply single parity check equations.

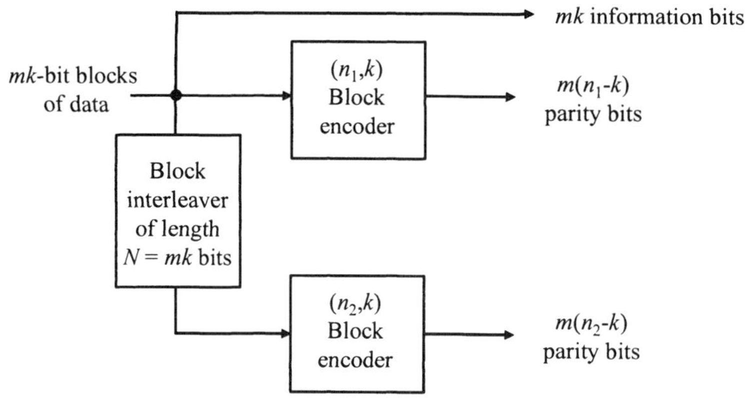

Figure 4.4-9 Parallel concatenated block code (PCBC) utilizing a block interleaver.

It has been shown that LDPC can get as close to the Shannon limit as turbo codes in papers by MacKay and Neal [45] as well as MacKay [46]. Recently it was shown that irregular LDPC codes were shown to outperform turbo codes when they are of the same approximate length and rate [47].

A *regular LDPC* is a linear (N,K) code with parity check matrix H having the Hamming weight J of the columns of H and weight K of the rows of H. Both J and K are much smaller than the code length N. As a consequence a LDPC has a very sparse (low number on ones) parity check matrix. When the Hamming weights are chosen in accordance with some nonuniform distribution, then irregular LDPC are obtained [47]. The reader may consult [12] for a more complete discussion of LDPC. An example of an LDPC parity check code is shown in Figure 4.4-10; indicated in Gallagher's paper [42].

$$H = \begin{bmatrix} 1111\ 0000\ 0000\ 0000\ 0000 \\ 0000\ 1111\ 0000\ 0000\ 0000 \\ 0000\ 0000\ 1111\ 0000\ 0000 \\ 0000\ 0000\ 0000\ 1111\ 0000 \\ 0000\ 0000\ 0000\ 00001111 \\ \\ 1000\ 1000\ 1000\ 1000\ 0000 \\ 0100\ 0100\ 0100\ 0000\ 1000 \\ 0010\ 0010\ 0000\ 0100\ 0100 \\ 0001\ 0000\ 0010\ 0010\ 0010 \\ 0000\ 0001\ 0001\ 0001\ 0001 \\ \\ 1000\ 0100\ 0001\ 0000\ 0100 \\ 0100\ 0010\ 0010\ 0001\ 0000 \\ 0010\ 0001\ 0000\ 1000\ 0010 \\ 0001\ 0000\ 1000\ 0100\ 1000 \\ 0000\ 1000\ 0100\ 0010\ 0001 \end{bmatrix}$$

Figure 4.4-10 LDPC parity check matrix based on Gallagher's paper [42].

It is seen that J=3 and K=4 in this code, since the weight of each column is 3 and the weight of each row is 4. This is a LDPC(20,5) code since from (4.2-20) it is clear that H is an $n \times (n-k)$ matrix and since $n-k = 15$, it follows that $k = 5$. Further details can be found in [12].

For every linear (n,k) code there exists a bipartite graph with incidence matrix H. The name of this graph is a *Tanner graph*. The Tanner graph of a linear (n, k) graph has n code nodes and at least $n-k$ parity nodes z, associated with the parity check equations. For a regular LDPC code the degrees of the code nodes are all equal to J and the degrees of the parity nodes are equal to K. The Tanner graph of the (20,5) Gallagher code is shown in Figure 4.4-11.

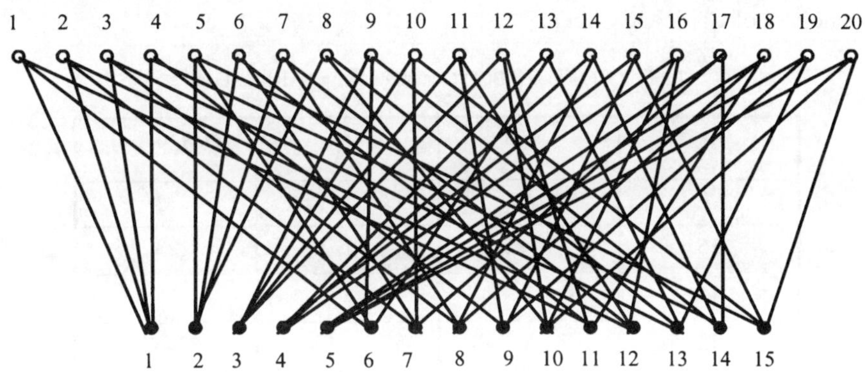

Figure 4.4-11 A Tanner graph for the Gallagher (20,5) LDPC code.

A few of the resulting parity check equations are given by

$$\begin{aligned} z_1 &= x_1 + x_2 + x_3 + x_4 \\ z_2 &= x_5 + x_6 + x_7 + x_8 \\ &\vdots \\ z_{15} &= x_5 + x_{10} + x_{15} + x_{20} \end{aligned} \tag{4.4-16}$$

Two iterative decoding algorithms are discussed by Gallagher [42] and Morelos-Zaragoza [12] for the interested reader.

4.5 SELECTED RESULTS FOR SOME ERROR CORRECTION CODES

Up until now the basic ideas of block, convolutional, and iterative decoding type codes have been discussed to present the general ideas and types of codes that are used in applications. In this section the goal is to present some specific codes. It is to be noted that only a small selection of the available codes are presented here.

4.5.1 Bose, Chaudhuri, and Hocquenghem Codes

The Bose, Chaudhuri, and Hocquenghem (BCH) linear codes are a clever generalization of the Hamming codes for multiple error correction. These codes were discovered independently by Hocquenghem in 1959 and Bose and Ray-Chaudhuri in 1960. This family of codes includes many code rates and a wide range of error correction capability for which excellent decoding algorithms exist. All error correction-coding books discuss BCH codes, oftentimes devoting a chapter to them. See, for example, Blahut [4] or Lin and Costello [2] for details on the decoding algorithms and the codes.

The relevant parameters for a BCH code that is capable of correcting t or fewer errors in a block of length n coded bits [2,17] are the following:

(1) Block length $n = 2^m - 1$, $m \geq 3$
(2) Number of parity check bits $n-k \leq mt$
(3) Minimum distance $d \geq 2t+1$
(4) Error correction capability $t < 2^{m-1}$

Table 4.5-1 lists a few selected generator polynomials for some BCH codes [20]. A larger table is provided in Appendix A of Clark and Cain [1].

Table 4.5-1 Selected BCH Generator Polynomials in Octal

n	*k*	*t*	**g(x) (octal)**
7	4	1	13
15	11	1	23
15	7	2	721
15	5	3	2467
31	26	1	45
31	21	2	3551
31	16	3	107657
31	11	5	5423325
31	6	7	313365047

Example 18 As an example of how to use the table, consider the case that $n = 31$, $k = 21$ BCH code. Using the octal 3551 produces the binary values of 011,101,101,001. Using the lowest order polynomial from the right side produces the polynomial sequence $x^{10} + x^9 + x^8 + x^6 + x^5 + x^3 + 1$. The leftmost "0" is ignored in the polynomial (actually it corresponds to $0x^{11}$). From the table $t = 2$ so that $d \geq 5$ and $m = 5$ and clearly $t < 2^4 = 16$ and $n-k = 10 \leq (5)(2) = 10$.

Example 19 Consider the (31,16) BCH code in which $t = 3$. The code rate is given by 16/30 = 0.53. Assume that this code is used in a SFH/DPSK spread spectrum communication link and also assume that $W/R_b = 1{,}000$. Determine the upper bound to the bit error rate under the condition that the worst-case tone jamming is employed. Using (3.6-22) for the channel symbol error rate and using the fact that $R_s = (30/16)R_b$, one obtains the following symbol error rate:

$$PE_s = \begin{cases} 0.5 & X < (30/16) \\ \dfrac{1}{2(P/J)(W/R_b)(30/16)} & (30/16) \leq X \leq (W/R_b) \\ 0 & (W/R_b) \leq X \end{cases} \tag{4.5-1}$$

where $X = (P/J)(W/R_b)$. Since most BCH decoder are bounded distance decoders one can use (4.2-82) to obtain the result shown in Figure 4.5-1.

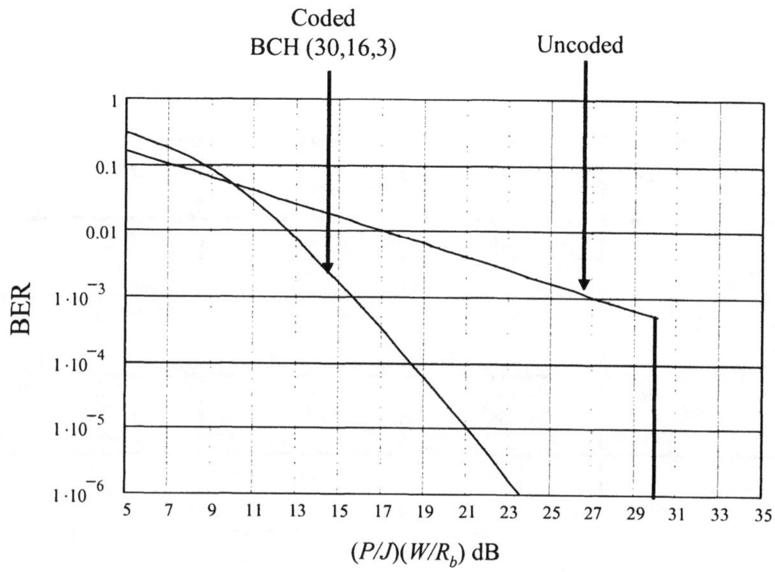

Figure 4.5-1 Bit error rate performance for SFH/DPSK with BCH (30,1,63) encoding for worst-case tone jamming with $W/R_b = 1{,}000$.

4.5.2 Reed-Solomon Codes

Reed-Solomon codes (RS) form a special and very important subclass of q-ary BCH codes. Here $q = 2^m$. The encoder for an RS (n,k) code on m-bit symbols ($m \geq 1$) groups the binary data stream into blocks, with each block containing k symbols (km bits). When m is an integer power of 2, the m bit symbols are called bytes. A common value of m is 8 and 8-bit RS codes are very powerful. The block length of a RS code is one less than 2^m. Efficient decoding techniques are available for RS codes.

The relevant parameters for an RS code, which is capable of correcting t or fewer errors in a block of length n symbols [2,17], are the following:

(1) Block length (symbols) $n = q-1 = 2^m-1$ ($m(2^m-1)$ bits)
(2) Number of message symbols k (mk bits)
(3) Number of redundancy symbols $n-k = 2t$ ($m(2t)$ bits)
(4) Minimum distance in symbols $d = 2t+1$ ($m(2t+1)$ bits)

The decoded bit error rate for RS codes has been bounded by Clark and Cain [1]. The result for the bound on the bit error probability is given by

$$P_b < \frac{2^{k-1}}{2^k - 1} \sum_{i=t+1}^{n} \frac{i+t}{n} \binom{n}{i} p_s^{\,i} (1-p_s)^{n-i} \tag{4.5-2}$$

in which p_s is the channel symbol error probability. It should be noted that the error probability may be for an actual nonbinary symbol such as in an FH/MFSK modulation, or may be the probability of one or more binary errors in an m-bit word on a binary channel. In the case the channel is binary [15] the following expression is used for the symbol error probability

$$p_s = \sum_{i=1}^{m} \binom{m}{i} p^i (1-p)^{m-i} \tag{4.5-3}$$

where p is the binary channel error probability. It should be noted that RS codes can be extended to have length $n = q$ and $n = q + 1$ [4]. A partial list of some RS codes used in practice in GF(8) are listed in Table 4.5-2.

Table 4.5-2 A Short List of RS Codes with $m = 6$ or 8

n	k	t	m
63	31	16	6
255	233	11	8
255	223	16	8
255	249	3	8

Example 20 Assume an RS code is being used in a SFH/DPSK spread spectrum communication system. Using the first entry (63,31,16) with $m = 6$ and estimate the bit error probability for worst-case multitone jamming. This code has a rate of 31/63 = 0.492. From (3.6-22) one has the channel symbol error probability is given by

$$p = \begin{cases} 0.5X < 63/31 & X < 63/31 \\ 1/2(31/63)X & 63/31 < X < 1000 \\ 0 & 1000 < X \end{cases} \tag{4.5-4}$$

where $X = (P/J)(W/R_b)$. Use of (4.5-3) specifies the symbol error probability, in terms of the binary channel error probability. Finally, using (4.5-2) provides the estimate of the bit error probability. The result is plotted in Figure 4.5-2.

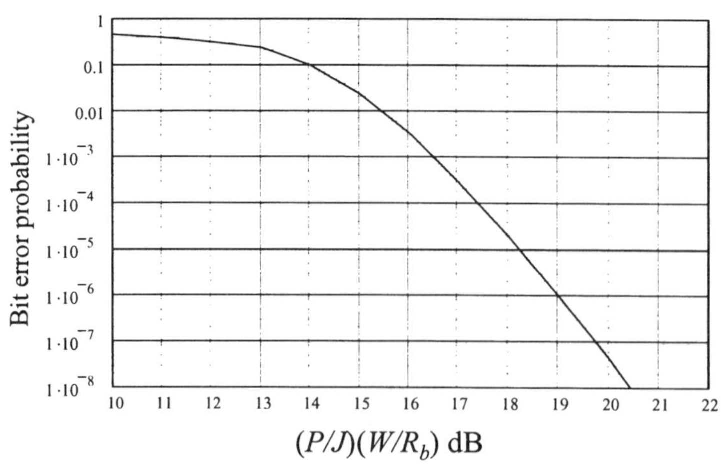

Figure 4.5-2 Bit error probability for an RS(63,31,16) code with $m = 6$ and $W/R_b = 1000$ for worst-case tone jamming.

4.5.2.1 Concatenated Reed-Solomon/Convolutional Codes

A very powerful coding scheme that has been used for some time is the concatenation of RS codes with Viterbi decoded convolutional codes [2]. Figure 4.5-3 illustrates the block diagram for the RS/convolutional code concatenation system. The information bits are grouped in blocks of m bits to form a 2^m-ary symbol alphabet used by the RS encoder. This first code, the RS code is called the *outer code*, in the sense it is out from the channel.

The RS encoded symbols are converted to the m-bit binary representation and input to the symbol-interleaving buffer. The interleaver along with the deinterleaver tends to spread symbol errors uniformly over the symbol sequence. The interleaved bits forming the interleaver are input to the convolutional encoder (the *inner encoder*). The encoded data bits are modulated and transmitted over the discrete memoryless channel (DMC). After the demodulation process the demodulated bits are input to the Viterbi decoder, which decodes the bit stream. The decoded bits are grouped into RS symbols and deinterleaved and fed into the RS decoder. Finally the RS decoder decodes the data in bit format and sends it to its destination. The portion of the block diagram between the interleaver input and the deinterleaver output constitutes a *super channel.* Measurements and simulations have yielded the fact that output errors from the Viterbi decoder tend to appear in bursts. Therefore it is necessary to utilize the interleaver and deinterleaver combination to randomize the error bursts, and make the channel to appear to be memoryless. In effect the RS decoder is able to correct some errors that the Viterbi decoder was not able to correct, thus improving the overall bit error rate.

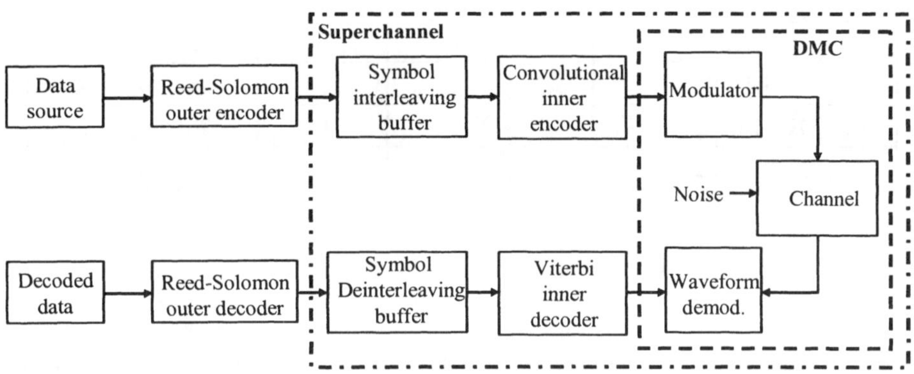

Figure 4.5-3 Block diagram of RS/convolutional concatenated coding system.

4.5.3 Convolutional Codes with Maximum Free Distance

An important measure of the goodness of a convolutional code is its weight structure. Typically this is found by an exhaustive search on a digital computer. For a maximum-likelihood approach the optimum weight structure is that which has the minimum number of bit errors in the paths through the code trellis that are closest to one another in the sense of the Hamming distance. Oldenwalder [48] first published the optimum rate 1/2 and 1/3 convolutional codes. Clark and Cain [1] and Peterson et al. [15] later published these results also. Tables 4.5-2(a) and 4.5-2(b) list the weights out to d_f+4. Note that for listing purposes d denotes d_f, the free distance in the table. The constraint length is denoted by K. The entries for each code at distance $d+i$, i = 0,1,...4, is the coefficient c_d needed to evaluate the upper bound on the bit error rate. Note (4.3-26) is the equation used to estimate the upper bound of the bit error rate probability.

Table 4.5-2 (a) Optimal Rate 1/2 Convolutional Codes and their Partial Weight Structure

R_c=1/2			c_d for d_f=				
K	Generators	d_f	d_f+0	d_f+1	d_f+2	d_f+3	d_f+4
3	7,5	5	1	4	12	32	80
4	17,15	6	2	7	18	49	130
5	35,23	7	4	12	20	72	225
6	75,53	8	2	36	32	62	332
7	171,133	10	36	0	211	0	1404
8	371,247	10	2	22	60	148	340
9	753,561	12	33	0	281	0	2179

Table 4.5-2 (b) Optimal Rate 1/3 Convolutional Codes and their Partial Weight Structure

R_c=1/3			c_d for d_f=				
K	Generators	d_f	d_f	d_f+1	d_f+2	d_f+3	d_f+4
3	7,7,5	8	3	0	15	0	58
4	17,15,13	10	6	0	6	0	58
5	37,33,25	12	12	0	12	0	56
6	75,53,47	13	1	8	26	20	19
7	171,165,133	14	1	0	20	0	53
8	367,331,225	16	1	0	24	0	113

To evaluate the BER, (4.3-26) is used with P_d given by (4.2-79) or (4.2-80) for the hard-decision binary symmetric channel and (4.3-25) for the soft-decision, AWGN channel. The value of c_d is given in the above table. When jammer-state information (JSI) is available Peterson et al. [15] have shown that (4.3-24) and (4.3-26) remain valid with the appropriate calculation of P_d. Peterson et al. [15] have also shown that using analysis similar to that used for the binary symmetric channel, when JSI is available, P_d must be replaced with P_{jd}.

4.5.3.1 SFH/DPSK BER Performance Convolutional Code in Optimal Tone Jamming

In this section the bit error rate performance of an SFH/DPSK modulation scheme with optimal tone jamming utilizing Viterbi decoding of a rate 1/2 K=7 convolutional code will be assessed for hard-decision decoding. The methodology applies to any convolutional code in which the c_d values are known.

The BER performance is based on the upper bound given in (4.3-26) with the c_d values given in Table 4.5-2 (or elsewhere) and the value of P_d given by (4.2-79) or (4.2-80) according to whether d_f is even or odd, and evaluation of the symbol error probability, p, from (3.6-22). Since the code rate is 1/2 the symbol error rate, the probability is given by

$$p \quad \begin{matrix} 0.5 & X<2 \\ \dfrac{1}{X} & 2 \le X < W/R_b \\ 0 & W/R_b \le X \end{matrix} \tag{4.5-5}$$

The convolutional code selected is a rate 1/2 K=7 code having the octal representation of the two code generators as (171,133), shown in Figure 4.5-4. The binary equivalent of the octal is given by (1111001,1011011). Figure 4.4-12 depicts the convolutional encoder, based on the tap connections to the decoder. The c_d values are the entries given in Table 4.5-2. Using (4.5-5) in (4.2-79), since d_f is even, produces the bit error rate performance result illustrated in Figure 4.5-5, for the case that W/R_b=1,000.

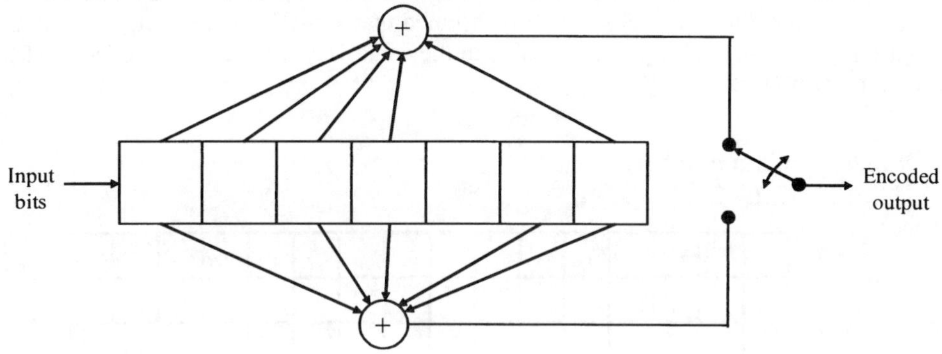

Figure 4.5-4 Rate 1/22 convolutional encoder with (171,133) code generators.

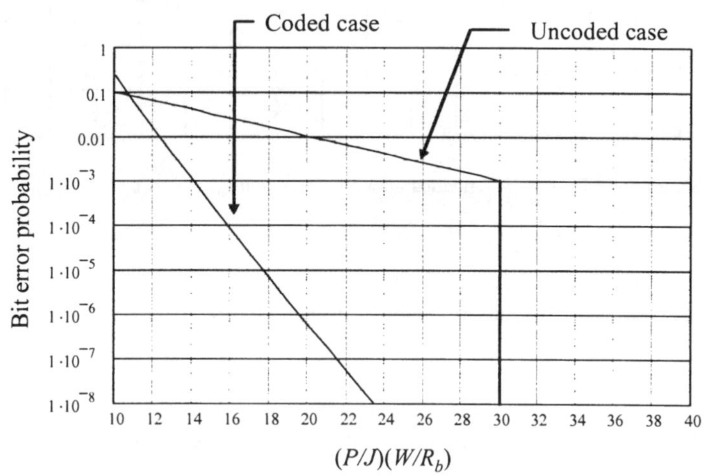

Figure 4.5-5 Bit error probability for SFH/DPSK modulation with (171,133) convolutional code with K = 7 and R_c = 1/2 and the uncoded case for comparison.

4.5.4 Hard- and Soft-Decision FFH/MFSK with Repeat Coding BER Performance

Repeat coding (RC), in which the same symbol is transmitted numerous times, is not very effective for the AWGN channel. However significant gains can be achieved with its use in a FFH/(orthogonal)MFSK modulation system. The area of repeat coding has been studied extensively in [15, 49–56]. It is to be noted that repeat coding and time diversity are the same. The fact that there are many references in this area reflects the fact that high-quality anti-jam performance, with reasonable hardware designs, can be obtained with this coding/modulation scheme. In this section worst-case two-state partial band jamming is assumed.

Consider a spread spectrum communication link that uses orthogonal MFSK modulation with fast frequency hopping. The approach in this section follows Peterson et al. [15]. As was discussed in Section 3.3.5 for FFH/MFSK, each MFSK tone is divided up into L chips, and each chip is transmitted with a different FH frequency. This is known as L state diversity. Thus the FFH rate is L times the MFSK symbol rate. Let the uncoded MFSK symbol rate be R_s and the chip rate be R_{ch} so that $R_{ch} = LR_s$. It is assumed that the jammer is intelligent in that it knows how much jamming power is needed to transmit over a particular bandwidth or over a particular time slot in order to provide the highest possible chip error rate probability. It is assumed that an interleaver is employed in the link and the assumption is made that all chip errors are considered to be statistically independent from each other.

Figure 4.5-6 illustrates the M different tones that are repeated over L times. The next MFSK symbol is also repeated L times, as are all the MFSK tones. What is not shown in Figure 4.5-6 is the fact that, at each chip time a different hop frequency is transmitted for the chip duration, another hop frequency is transmitted in the next chip duration, and so on. It is assumed that $M = 2k$, for k an integer (commonly $k = 3$). The jammer optimized its bandwidth (or duty factor) in the same way it would if uncoded FFH/MFSK were being utilized with a symbol rate of LR_s.

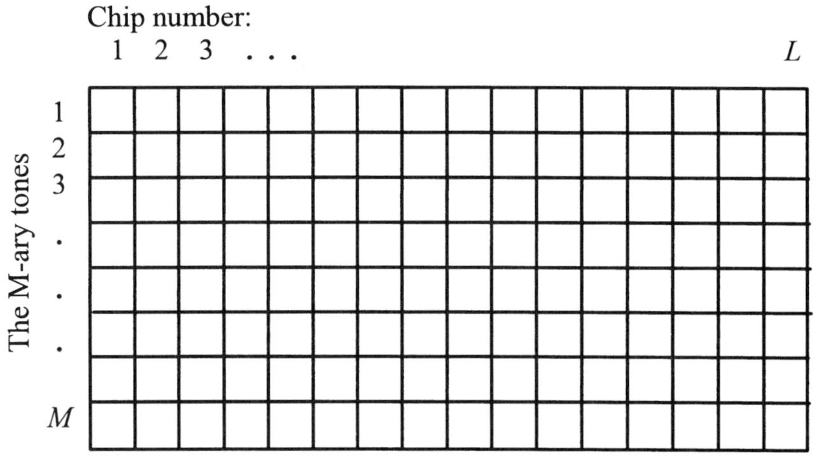

Figure 4.5-6 Transmitted MFSK tones for FFH/MFSK modulation with repeat coding ($M = 2k$).

The receiver dehops the frequency-hopping spread spectrum modulation, leaving the FSK modulation. The resulting dehopped signal plus interference is input to the MFSK demodulator, with the filters being matched to the chip duration. Figure 4.5-7 illustrates the demodulator.

4.5.4.1 Hard-Decision BER Performance of FFH/MFSK with RC Without JSI

The demodulator is composed of a bank of M noncoherent demodulators, with one detector pair for each of the M possible MFSK tones as shown in Figure 4.5-7. The frequency hopping is removed prior to the demodulation process.

For each chip time, j, (j =1, 2,... L), there is a noncoherent detector output denoted by e_{ij}. For the i-th symbol and the j-th chip, these outputs can be processed in a number of ways, a few of which will be discussed here. The simplest processing involves making a hard decision of the MFSK symbol for each demodulated chip. From Problem 9 note that the relationship between the MFSK symbol error rate and the bit error rate [57] is given by

$$PE_s = PE_b\left(\frac{M/2}{M-1}\right) \tag{4.5-6}$$

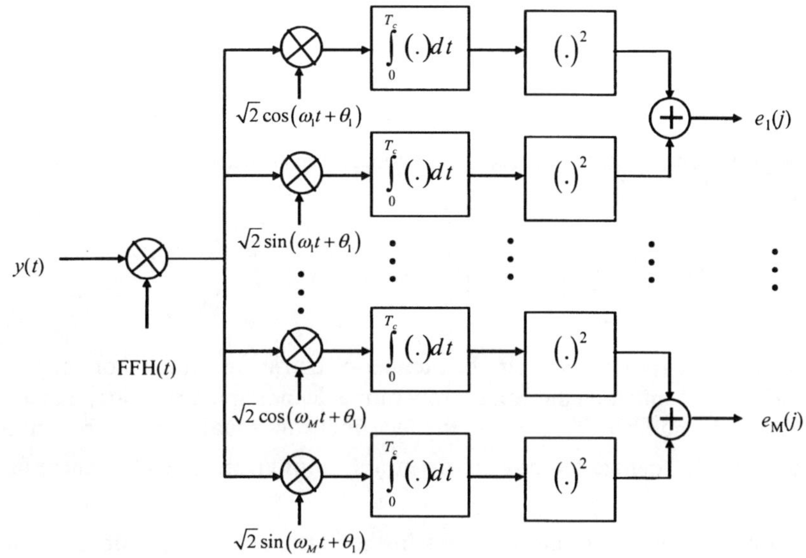

Figure 4.5-7 MFSK demodulation for the FFH/MFSK SS modulation ($j = 1, 2, \ldots L$).

Also noting that the bit energy is reduced by the factor L due to the repeat coding, and using (4.5-6) and (3.3-51), the chip error probability becomes

$$P_c = \begin{cases} \dfrac{2(M-1)}{M}\dfrac{Lc_0}{E_b/N_0} & E_b/N_{0J} > L(E_b/N_{0J})_0 \\ \dfrac{1}{M}\displaystyle\sum_{n=2}^{M}(-1)^n\binom{M}{n}\exp\left(\dfrac{kE_b}{LN_{0J}}\left(\dfrac{1-n}{n}\right)\right) & E_b/N_{0J} < L(E_b/N_{0J})_0 \end{cases} \tag{4.5-7}$$

In (4.5-7) c_0 and E_b/N_{0J} are obtained from Table 3.3-1.

The receiver generates an output sequence of L chips for each information symbol (1 thru M). One realization of the outcomes for L chips is illustrated in Figure 4.5-8. The dots indicate one possible detected sequence for each chip time and for every one of the eight frequencies. The decoding rule that is used by the receiver is to choose the decoder output symbol frequency (which frequency was sent) as the frequency with the greatest number of detects over the L (L = 10 in this case) chips within the M (eight in this case) frequencies. Note that in the example of Figure 4.5-8, frequency number 7 has the most detects. Note that P_c is the probability of received chip error probability. Denote by P_e the probability that there is no chip error, so that a correct entry occurs in the tone number versus chip number decoding matrix of Figure 4.5-8.

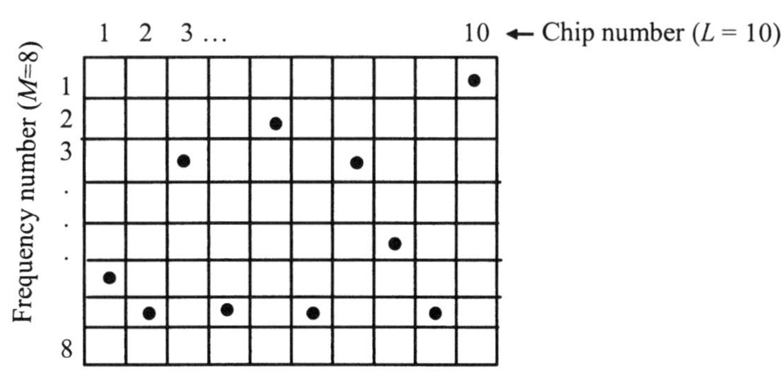

Figure 4.5-8 One possible realization of a chip detect.

Thus

$$p_e = 1 - P_c \tag{4.5-8}$$

Further the probability of n entries in the correct row and column of the decoding matrix is expressed by the following expression

$$PC(n) = \binom{L}{n} p_e^n (1-p_e)^{L-n} \tag{4.5-9}$$

Note that a symbol error occurs when the number of entries in some incorrect row of the decoding matrix exceeds the number of entries in the correct row of the decoding matrix. Note that a symbol error may be made in the case that the number of entries in the incorrect row equals the number in the correct row of the decoding matrix. Let $P_{m|n}$ denote the probability of a symbol error, given that there are n entries in the correct row of the decoding matrix.

Under the condition that there are n entries in the correct row, there are L–1 entries in all of the incorrect rows. Considering the decoding matrix, there are $(M-1)L$ incorrect positions in the decoding matrix, in which $(M-1)n$ do not contain entries since there are only n entries in the correct row. Consequently there are a total of $(M-1)(L-n)$ incorrect positions in the decoding matrix, which contains L–1 entries. It can be shown that the probability of n entries in a block that has $L-n$ elements, given that the total entries in $(M-1)(L-n)$ elements, is given by [15]

$$P(j|n) = \frac{\binom{L-n}{j}\binom{(M-2)(L-n)}{L-n-j}}{\binom{(M-1)(L-n)}{L-n}} \tag{4.5-10}$$

There are two distinct cases that must be dealt with when computing the total symbol error probability. The first one is $M = 2$ and the second one is $M \geq 4$. First consider the case $M = 2$, which corresponds to FSK modulation. In this case whenever a chip error is made, an entry in the decoding matrix appears in the only incorrect row of the decoding matrix. For L an even number the symbol error probability is given by

$$P_s = \sum_{n=0}^{(L/2)-1} PC(n) + \frac{1}{2} PC(L/2) \tag{4.5-11}$$

When L is an odd number, then the symbol error probability is given by

$$P_s = \sum_{n=0}^{(L-1)/2} PC(n) \tag{4.5-12}$$

Equation (4.5-11) is based on the fact that a decoding error always occurs when the correct row has anywhere from 0 to $(L/2)-1$ entries and occurs with probability 1/2 when there are $L/2$ entries in the correct row. Equation (4.5-12) is similar except that it acknowledges the fact that there cannot be $L/2$ entries in any row since L is an odd number.

In the case that $M \geq 4$, the error probability for L even is given by [15]

$$P_s = PC(0) + \sum_{n=1}^{(L/2)-1} (M-1)PC(n)\left[\frac{1}{2}P(n|n) + \sum_{j=n+1}^{L-1} P(j|n)\right] + (M-1)PC(L/2)\frac{1}{2}P\left(\frac{L}{2}\middle|\frac{L}{2}\right) \quad (4.5\text{-}13)$$

In the case that L is an odd number the symbol error rate is given by [15]

$$P_s = PC(0) + \sum_{n=1}^{(L-1)/2} (M-1)PC(n)\left[\frac{1}{2}P(n|n) + \sum_{j=n+1}^{L-1} P(j|n)\right] \quad (4.5\text{-}14)$$

The first term in (4.5-13) accounts for the case that there are no entries in the correct row, hence some other incorrect row must have more entries that the correct row. The last term in (4.5-13) accounts for the case that there are $L/2$ entries in the correct row. Note that an error occurs with probability 1/2 only when an incorrect row has $L/2$ entries. The probability of a particular incorrect row that has $L/2$ entries is $P(L/2|L/2)$ and the probability of any one of the $(M-1)$ incorrect rows having $L/2$ entries is $(M-1)P(L/2|L/2)$. The first summation in (4.5-13) is over all the remaining values of n for which errors can occur. The term in brackets of this equation is the probability of a particular incorrect row being decoded, given n entries in the correct row. The factor of $(M-1)$ in the summation term accounts for the $(M-1)$ incorrect rows.

Based on the bit error rate to symbol error rate relationship of (4.5-6) it follows that

$$P_b = \left(\frac{M}{2(M-1)}\right)P_s \quad (4.5\text{-}15)$$

It is to be noted that for M large $P_b \cong (1/2)P_s$.

4.5.4.2 Hard-Decision BER Performance of FFH/MFSK with RC with JSI

The analysis associated with having jammer side information (JSI) is very complicated. This section follows Peterson et al. [15] and Stark [18]. In order to render a solution only FSK ($M = 2$) will be considered here, so that the results of Section 4.2.4.2 can be utilized. Similar to the assumptions of Section 4.2.4 the jammer state $\boldsymbol{z} = (z_1, z_2, z_3,\ldots z_L)$ and the binary symmetric channel have error probabilities of p_0 and p_1 corresponding to the jammer states 0 and 1 which are known to the receiver. The time diversity, or L-repeat code, is composed of two code words (i.e., the all zero word $\boldsymbol{C}_0 = (0,0,0,\ldots 0)$ and $\boldsymbol{C}_1 = (1,1,1,\ldots 1)$). Clearly the Hamming distance between the two words is L. Recall Section 4.2.4 provides bounds on the average block code bit error probability for any linear binary block code. Those bounds can be applied to the L-repeat code discussed here. Denote the hard-decision output vector $\boldsymbol{y}$ by $\boldsymbol{y} = (y_1, y_2, y_3,\ldots y_L)$ in which the y_i satisfy $y_i \in \{0,1\}$. From (4.2-61) the block decoder determines the weighted Hamming distance

$$\ln\{p(\boldsymbol{y}|C_m, \boldsymbol{z}) = d_0 \ln\left(\frac{p_0}{1-p_0}\right) + d_1 \ln\left(\frac{p_1}{1-p_1}\right) + \quad (4.5\text{-}16)$$

between the hard-decision output vector $\boldsymbol{y}$ and each of the two code words. It estimates that the received code word is the code word in which the weighted Hamming distance (4.5-16) is the largest. Recall that in (4.5-16) d_0 is the Hamming distance between $\boldsymbol{y}$ and $\boldsymbol{C}_m$ for the symbols of the code that are not jammed, and d_1 is the Hamming distance between $\boldsymbol{y}$ and $\boldsymbol{C}_m$ for the symbols of the code that are jammed. Note that for this trivial code, the word error rate and the bit error rate are the same. The transition probabilities in (4.5-16) have the following meaning. The transition probability p_0 is based on assuming that only thermal noise is present, so that the BSC transition probability for a chip is given by

$$p_0 = Q\left(\sqrt{\frac{2E_b}{N_0 L}}\right) \tag{4.5-17}$$

where N_0 is the one-sided noise spectral density of thermal noise. The transition probability p_1 for a chip is based on assuming that thermal noise plus jamming is present, so that its value is given by

$$p_1 = Q\left(\sqrt{\frac{2E_b}{(N_0 + N_{0J})L}}\right) \tag{4.5-18}$$

where N_{0J} is the jammer spectral density. The bit error rate bound of (4.2-109) simplifies to

$$PE_b(\rho) = P_{Jd}(L, \rho) \tag{4.5-19}$$

where ρ is the fraction of the band that is jammed, and there is only one term in the series of (4.2-109). Note that $d(0,1) = L$ in this case of repeat coding, and there is only one term left in the series. Furthermore, since there is only one term in the union bound of (4.2-109), it then becomes equality, rather than a bound. The bit error probability can be evaluated from (4.2-107), which in this case becomes

$$P_{Jd}(L, \rho) = \sum_{n_{Jd}=0}^{L} \binom{L}{n_{Jd}} \rho^{n_{Jd}} (1-\rho)^{L-n_{Jd}} PE_w[L, n_{Jd}] \tag{4.5-20}$$

where $PE_w[L, n_{Jd}]$ is evaluated from (4.2-102), and using $PE_w(L, n_{Jd}) = P_w(m \to m', z)$ produces

$$PE_w(L, n_{Jd}) = \sum_{d_1=0}^{n_J} \sum_{d_0=0}^{L-n_J} \beta(d_1, d_0) \binom{n_J}{d_1} p_1^{d_1} (1-p_1)^{n_J - d_1} \binom{L-n_J}{d_0} p_0^{d_0} (1-p_0)^{L-n_J-d_0} \tag{4.5-21}$$

in which d_0 and d_1 have the same meaning as in (4.2-93). That is $\beta(d_1, d_0) = 1$ if $\delta < \alpha(L, n_{Jd})$, $\beta(d_1, d_0) = 1/2$ if $\delta = \alpha(L, n_{Jd})$, and $\beta(d_1, d_0) = 0$ if $\delta > \alpha(L, n_{Jd})$. Further δ and $\beta(L, n_{Jd})$ are defined by

$$\delta = d_0 \ln\left(\frac{p_0}{1-p_0}\right) + d_1 \ln\left(\frac{p_1}{1-p_1}\right) \tag{4.5-22}$$

and

$$\alpha(L, n_{Jd}) = \frac{L - n_{Jd}}{2} \ln\left(\frac{p_0}{1-p_0}\right) + \frac{n_{Jd}}{2} \ln\left(\frac{p_1}{1-p_1}\right) \tag{4.5-23}$$

The bit error rate probability may be determined from the use of (4.5-19)–(4.5-23) as a function of the jammer parameter ρ. The dependence on ρ simplifies if one assumes that ρ is optimized for each SNR, and recall that the optimum value of ρ is obtained from

$$\rho_{opt} = \frac{y_0}{(E_b / N_{0J})} \tag{4.5-24}$$

where y_0 is found in Table 3.3-1 of Chapter 3. In the case of the repeat code, with the assumption that the jammer operates on each chip, E_b is replaced with the chip energy (i.e., E_b/L for each repetition of the repeat code).

4.5.4.3 Soft-Decision BER Performance of FFH/MFSK with RC and with JSI

Now consider the L-repeat code for the case of FFH/MFSK modulation with soft-decision decoding and jammer state information $\boldsymbol{z} = (z_1, z_2, z_3, \ldots z_L)$ with worst-case partial band or a worst-case pulsed jammer. The soft decision SFH/MFSK demodulator of Figure 4.5-2 has an output of the vector $\boldsymbol{y}$, where

$$\boldsymbol{y} = (\boldsymbol{e}(1), \boldsymbol{e}(2), \boldsymbol{e}(3) \ldots \boldsymbol{e}(L)) \tag{4.5-25}$$

and $\boldsymbol{e}(k)$ is expressed by

$$\boldsymbol{e}(k) = (e_1(k), e_2(k), e_3(k) \ldots e_M(k)) \tag{4.5-26}$$

Hence $\boldsymbol{y}$ is a vector of L vectors, each one having M components, corresponding to the M FSK frequencies. Hence $e_j(k)$ is the j-th frequency output at the k-th chip time. Given the channel state information vector $\boldsymbol{z}$, and $\boldsymbol{y}$, the decoder's function is to estimate the transmitted code word. This problem has been treated by a number of authors [52, 58–64]. These analyses are beyond the scope of this book; however, the results will be summarized in what follows. The results indicated here are based on [15], [52], and [64]. This analysis assumes that the system is operating at very high SNR so that the thermal noise can be neglected. Thus if any of the L chips are detected without jamming, the decoder is able to make a perfect decision, in the sense that no error is committed. Additionally assume that the channel state information vector $\boldsymbol{z}$ is produced by the receiver for each received symbol $\boldsymbol{e}(k)$ and declare $z_k = 0$ if a single energy detector is high and $z_k = 1$ if two or more energy detectors are high. Note that the only way this detector can make an error is if all the L received symbols are jammed. If all the L received symbols are jammed, the decoder then examines the summation value

$$\Lambda_m = \sum_{k=1}^{L} e_m(k) \tag{4.5-27}$$

for all the possible m values $m = 1, 2, 3, \ldots M$, and outputs the value of m for which Λ_m is the largest. The "exact error probability" under these ideal conditions of no thermal noise has been shown to be given by [64]

$$P_b = \beta \left[\frac{L N_{0J}}{(\log_2(M)) E_b} \right]^L \tag{4.5-28}$$

whenever $E_b/N_{0J} > \gamma$, in which β and γ are listed in Table 4.5-3 for selected values of L from 2 to 10 and M from 2 to 32. It is to be noted that

$$\frac{E_b}{N_{0J}} = \frac{PW}{JR_b} \tag{4.5-29}$$

Figure 4.5-9 illustrates the results for the BSC with RC without JSI and soft decision with RC and with JSI for the FFH/FSK spread spectrum modulation scheme for worst-case partial-band or pulse noise jamming. The dotted curves represent the bit error probability for the case of Section 4.5.4.2 for BSC without

JSI, with repeat coding for various values of repeating parameter L. The solid lines represent the bit error probability for the case that JSI is available with soft decision and for the case that the thermal noise is negligible, and is based on Section 4.5.4.2. Clearly the soft-decision case with JSI is superior based on the curves which are valid for $E_b/N_{0J} > \gamma$, with γ being listed in Table 4.5-3.

Table 4.5-3 Parameters for the Exact Bit Error Performance of FFH/MFSK with L-Repeat Code in Worst-Case Partial Band Jamming

	$M = 2$		$M = 4$		$M = 8$		$M = 16$		$M = 32$	
L	β	Γ (dB)	β	Γ (dB)	β	Γ (dB)	β	Γ (dB)	β	Γ (dB)
2	0.4168	6.7	0.5959	4.2	0.8575	2.8	1.0796	2.0	1.3659	1.4
3	0.5210	8.8	0.8265	6.1	1.2565	4.6	1.8198	3.7	2.5937	3.0
4	0.6797	10.2	1.1401	7.4	1.8320	5.9	2.8251	4.8	4.1493	4.1
6	1.2392	12.1	2.2245	9.2	3.8464	7.6	6.4370	6.5	10.3502	5.7
8	2.3584	13.5	4.4101	10.5	8.0165	8.9	14.0997	7.7	24.0142	6.8
10	4.6110	14.5	8.8385	11.5	16.5774	9.8	30.2829	8.7	61.6169	7.8

Finally the dashed curve on the left of the figure illustrates the bit error probability performance for the case of broadband noise jamming spread over the hopping bandwidth W.

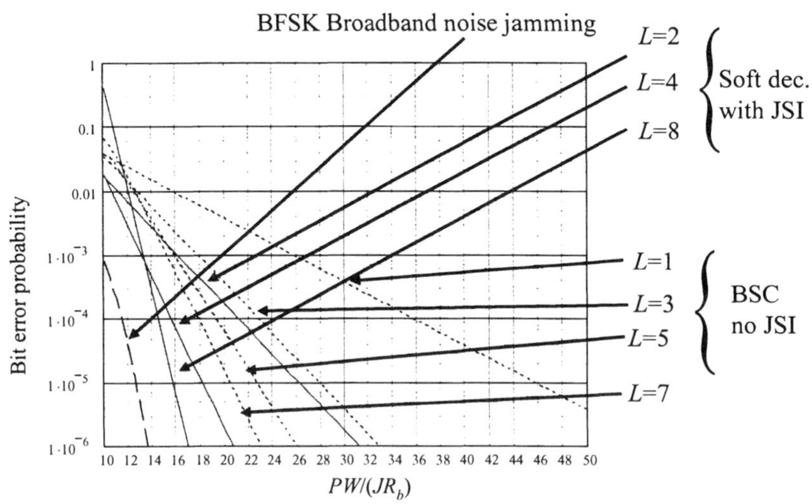

Figure 4.5-9 FFH/BFSK BER for the two-state partial band jamming or pulsed noise jamming versus the number of repeat coding values L for the worst-case partial band jamming or worst-case pulse noise jamming.

4.6 SHANNON'S CAPACITY THEOREM, THE CHANNEL CODING THEOREM, AND BANDWIDTH EFFICIENCY

We have discussed the performance enhancement that is obtainable with forward error correction coding and interleaving, and it has been dramatic in some cases, justifying the use of coding. In this section we will discuss the theorem that provides an existence statement on codes that exist to obtain very good performance, and we will discuss the bandwidth efficiency, or spectral efficiency, of the coded and uncoded digital transmission systems. This efficiency is useful in comparing combined coding and modulation schemes to see how power efficiency trades with bandwidth (BW) efficiency.

4.6.1 Shannon's Capacity Theorem

In 1948 Claude Shannon [65] derived the following channel capacity formula for the AWGN channel

$$C = W\log_2\left(1+S/N\right) \quad \text{bits/s} \tag{4.6-1}$$

where W is the channel bandwidth in Hz, S is the signal power in watts, and N is the total noise power of the channel in watts.

Example 21 Determine the channel capacity in bits/sec for the case that the channel bandwidth is 20 kHz and the signal-to-noise ratio is 30 dB. From (4.6-1) one has that $W = 20{,}000$ and $S/N = 1{,}000$ so that

$$C = 20000(9{,}967) = 200{,}200 \text{ Hz}$$

4.6.2 Channel Coding Theorem

The importance of the channel capacity is best illustrated by the *channel coding theorem*. This theorem, due to Shannon [66], can be broken down into the *direct part* and *the converse part*. The direct part states that for a bit rate of $R_b < C$ there exists a coding system with arbitrary low block and bit error rates as the code length $n \to \infty$. The *converse part* states that for $R_b \geq C$ the bit and block error rates are strictly bounded away from zero for any coding scheme. Thus the channel coding theorem established strict limits on the maximal supportable bit rate of an AWGN channel in terms of bandwidth and power.

4.6.3 Bandwidth Efficiency

Bandwidth efficiency, η, characterizes how efficiently the communication system uses its allocated bandwidth and is defined as

$$\eta = \frac{\text{data rate}}{\text{channel bandwidth}} = \frac{R_b}{W} \quad \left(\frac{\text{bits/s}}{\text{Hz}}\right) \tag{4.6-2}$$

From the bandwidth efficiency we compute the Shannon Limit as

$$\eta_{\max} = \log_2\left(1+S/N\right) \quad \left(\frac{\text{bits/s}}{\text{Hz}}\right) \tag{4.6-3}$$

In order to use this definition it is necessary to define what is meant by the bandwidth W. One common definition is the 99% bandwidth, although 95% has been used also. The average signal power is defined by

$$S = \frac{kE_b}{T_s} \tag{4.6-4}$$

where k is the number of bits per symbol, E_b is the bit energy, and T_s is the symbol time. Let N_0 denote the one-sided noise spectral density. Hence (4.6-3) can be written as

$$\eta_{\max} = \log_2\left(1+\frac{R_b E_b}{N_0 W}\right) \tag{4.6-5}$$

In Problem 10(a) it is shown that E_b/N_0 can be solved for from (4.6-5) to yield the *Shannon bound* on required signal-to-noise ratio

$$\frac{E_b}{N_0} \geq \frac{2^{\eta_{\max}}-1}{\eta_{\max}} \tag{4.6-6}$$

The *fundamental limit* is obtained for the case that the bandwidth increases without bound and therefore $\eta_{\max}$ approaches zero. From Problem 10(b) it is shown that

$$\frac{E_b}{N_0} \geq \lim_{\eta_{\max}\to 0}\left[\frac{2^{\eta_{\max}}-1}{\eta_{\max}}\right] = \ln(2) = -1.59 \text{ dB} \tag{4.6-7}$$

This is the absolute minimum value of signal energy to noise spectral density ratio that is required to reliably transmit information.

4.7 APPLICATIONS OF ERROR CONTROL CODING

In this section we will present some codes [67] that have been used in the past and some codes that are being considered in the future. One of the earliest attempts on improving on uncoded BPSK with applications of error control coding to satellite or space probe communication links includes the (32,6) Reed Muller bi-orthogonal codes that were used in the 1969 Mariner Mars missions developed at the Jet Propulsion Laboratory. It required an E_b/N_0 of 6.4 dB to achieve an error rate of 10^{-5}.

In 1972 and 1973 the Pioneer spacecraft utilized a constraint length 32, rate 1/2 convolutional encoder and achieved a BER of 10^{-5} at an E_b/N_0 of 2.7 dB! A modified sequential decoder of the Fano type was utilized in the decoding process.

Voyagers 1 and 2 were launched in 1977 and utilized maximum likelihood Viterbi decoding with nonsystematic convolutional codes of constraint length 7, with rate 1/2. The BER of 10^{-5} was obtained at a E_b/N_0 of 4.5 dB. Although this code was not as powerful as the Pioneer code, it did not suffer long decoding delays and was less complex in decoder implementation.

The concatenated scheme of the Consultative Committee on Planetary Standard Systems (CCSDS) consisted of a convolutional $K = 7$ rate 1/2 code concatenated with a (255,233) Reed-Solomon code that achieved a BER of 10^{-5} at E_b/N_0 of 2.5 dB.

Taking advantage of improved technology led to the Big Viterbi Decoder (BVD) which used a $K = 15$, rate 1/4 convolutional encoder with a Viterbi decoder. It achieved a BER of 10^{-5} at E_b/N_0 of 1.7 dB.

Globalstar and Iridium satellite networks utilize two convolutional codes, the first a $K = 9$, rate 1/2 code and the second a $K = 7$, rate 3/4 code.

A turbo code with rate 1/2 and information blocks of $N^{16} = 65{,}536$ bits with 18 iterations achieved a BER of 10^{-5} at an E_b/N_0 of 0.7 dB! It is to be noted that turbo codes cause a long delay in obtaining the decoded bits. However these delays are not always important except for voice transmission and operation at lower SNRs (below 0.7 dB) where the bit error probability is not very good.

High-speed memories utilize forward error correcting codes. Many of these codes are constructed from Reed-Solomon (RS) or BCH codes over GF(2^b). If each symbol of GF(2^b) is expressed as a b-bit byte, then RS codes or 2^b-ary BCH codes having minimum distance $d_{min} = 4$ are called SbEC/DbED codes.

Magnetic/optical discs utilize RS codes. The CD format uses a coding system called cross-interleaved RS codes (CIRC), which is a combination of two shortened RS codes of minimum distance $d_{min} = 5$ over GF(2^8). The information sequence is first encoded with a (28,24) code and then by a (32,28) code. In addition CIRC utilizes a CRC code.

The need for a powerful code arose in the use of FM multiplexed digital audio broadcasting when it was desired to transmit the Japanese language, which has ideograms called Kanji and the occurrence of even one error in the character may make the text incomprehensible. For the stated reasons a (273,191) code shortened to a (272,190) was selected to fit the data format of the teletext for digital audio/video transmission.

For mobile communications Global System for Mobile Communications (GSM) utilizes a $K = 5$, rate 1/2 nonsystematic code for speech coding. It is to be noted that GSM utilizes frequency hopping as part of its communication system. Other FEC schemes include rate 1/2, 1/3, and 1/6.

The FEC used in the CDMA mobile system (IS-95) is based on an outer convolutional code that has a $K = 9$ and rate 1/3 and an inner 64-ary orthogonal modulation scheme.

CRC codes are often used in applications including CRC-4, CRC-7, CRC-8, CRC-12, CRC-ANSI, CRC-CCITT, CRC-SDLC, CRC-24, and CRC-32 (three distinct types). In particular two important applications include, (1) the use of CRCs in disk drives: each block of data (commonly 512 bytes) is protected by a CRC code, so that bit errors in the block are protected, and (2) CRCs are used in data networks. In a data network each packet of data is terminated with check bits of the CRC. In both applications the CRCs were selected for their burst error detecting capabilities. In addition to single bit errors they can detect multiple bit errors that are clustered together within the disk or block. Burst errors are likely in disk drives due to defects in the drive, and burst errors are likely in networks due to noise bursts in communication links.

This sampling was meant to give a flavor of the type of codes used in practice but was not meant to be all encompassing. Further examples can be noted in studying most communication systems, communication textbooks, or coding textbooks.

4.8 SUMMARY

This chapter dealt with bit error rate performance with various types of jammers and various types of coding and interleaving. The various types of jammers useful to jam the signal were presented. First interleavers were discussed, including the various types that have been used in practice. Then linear block coding was presented along with some important properties of linear block codes. Following this decoding rules were discussed with and without jammer side information, including both hard and soft decisions. Word and bit error probability bounds were then discussed. Block codes were followed by convolutional codes. The trellis diagram for convolutional codes was presented along with a discussion of the Viterbi decoding algorithm. Iteratively decoded codes, such as the turbo code, were then discussed. Then a number of important codes used in applications, such as the Reed-Solomon, BCH, CRC, and convolutional codes, were presented. Next BER performance for various codes and jamming environments with and without JSI being available were analyzed with their performance plotted.

The hope is that the student will have gained some understanding of how important channel coding is and how it greatly improves the BER of properly designed codes.

References

[1] G. Clark and J. Cain, *Error-Correction Coding for Digital Communications,* New York and London: Plenum Press, 1981.

[2] S. Lin and D. Costello, *Error Control Coding Fundamentals and Applications,* Englewood Cliffs, NJ: Prentice-Hall: 1983.

[3] W. Peterson and E. Weldon, *Error-Correction Codes,* 2nd ed., Cambridge, MA: MIT Press, 1972.

[4] R. F. Blahut, *Theory and Practice of Error Control Codes*, Reading, MA: Addison-Wesley, 1983.

[5] L. Hanzo, C. H. Wong, and M. S. Yee, *Adaptive Wireless Transceivers Turbo-Coded, Turbo-Equalized and Space-Time Coded TDMA, CDMA, and OFDM Systems,* New York: IEEE Press and John Wiley & Sons, 2002.

[6] D. P. Olsen, "A Hybrid Interleaving Scheme That Enables Packet Switching on Multiple-Access Radio Communication Channels," *IEEE Trans. on Communications*, Vol. 47, No. 12, December 1999.

[7] S. Benedetto, D. Divsalar, G. Montorsi, and Pollara, "Serial Concatenation of Interleaved Codes: Performance Analysis, Design, and Iterative Decoding," *IEEE Trans. on Information Theory*, Vol. 44, No. 3, May 1998, pp. 909–926.

[8] A. S. Barbulescu and S. S. Pietrobon, "Terminating the Trellis of Turbo-Codes in the Same State," *Electronics Letters*, Vol. 31, issue 1, 1995, pp. 22–23.

[9] G. D. Forney, "Burst-Error Correction Codes for the Classic Bursty Channel," *IEEE Trans. Communications Technology*, Vol. COM-19, October 1971, pp. 772–781.

[10] J. L. Ramsey, "Realization of Optimum Interleavers," *IEEE Trans. on Information Theory*, IT-16, May 1970, pp. 338–345.

[11] J. G. Proakis, *Digital Communications*, 4th ed., New York: McGraw-Hill, 2001.

[12] R. H. Morelos-Zaragoza, *The Art of Error Correction Coding*, New York: John Wiley & Sons, Chapter 5, 2002.

[13] S. Haykin, *Communication Systems*, 3rd ed., New York: John Wiley & Sons, Chapter 11, 1994.

[14] A. Papoulis, *Probability, Random Variables, and Stochastic Processes*, New York: 2nd ed., McGraw-Hill, 1984.

[15] R. Peterson, R. Ziemer, and D. Borth, *Introduction to Spread Spectrum Communications,* Upper Saddle River, NJ: Prentice Hall, 1995.

[16] A. J. Viterbi and J. Omura, *Principles of Digital Communication and Coding,* New York: McGraw-Hill, 1979.

[17] J. P. Oldenwalder, *Data Communications, Networks, and Systems*, 2nd ed., Carmel, IN: Sams, Chapter 10, 1991.

[18] W. E. Stark, "Coding for Frequency-Hopped Spread-Spectrum Communication with Partial-Band Interference: Part II Coded Performance," *IEEE Trans. on Communications*, Vol. COM-33, October 1985, pp. 1045–1057.

[19] J. K. Holmes, *Coherent Spread Spectrum Systems*, New York: Wiley-Interscience, John-Wiley & Sons, 1982.

[20] S. Haykin, *Communication Systems*, 3rd ed., New York: John Wiley & Sons, Chapter 11, 1994.

[21] A. J. Viterbi, "Error Bounds for Convolutional Codes and an Asymptotically Optimum Decoding Algorithm," *IEEE Trans. on Information Theory*, Vol. IT-13, April 1967, pp. 260–269.

[22] G. D. Forney, Jr., "The Viterbi Algorithm," *Proc. IEEE* Vol. 61, April 1967, pp. 268–278.

[23] A. J. Viterbi, "Convolutional Codes and Their Performance in Communication Systems," *IEEE Trans. on Communications Tech.*, Vol. COM-19, October 1971, pp. 751–772.

[24] J. A. Heller and I. M. Jacobs, "Viterbi Decoding for Satellite and Space Communication," *IEEE Trans. on Communications, Technology*, October 1971.

[25] A. J. Viterbi, "Error Bounds for Convolutional Codes and an Asymptotically Optimum Decoding Algorithm," *IEEE Trans. on Information Theory*, Vol. IT-13, 1967, pp. 260–269.

[26] G. D. Forney, Jr., "The Viterbi Algorithm," *Proc. of the IEEE*, Vol. 61, 1973 pp. 268–278.

[27] L. R. Bahl, J. Cocke, F. Jelinek, and J. Raviv, "Optimal Decoding of Linear Codes for Minimizing Symbol Error Rate," *IEEE Trans. on Information Theory*, March 1974, pp. 284–287.

[28] H. Ma and J. K. Wolf, "On Tail Biting Convolutional Codes," *IEEE Trans. on Communications*, Vol. COM-34, February 1986.

[29] S. Crozier, A. Hunt, K. Gracie, and J. Lodge, "Performance and Complexity Comparison of Block Turbo-Codes, Hyper-Codes, and Tail-Biting Convolutional Codes," *19th Symposium on Communications*, Kingston, Canada, June 1998, pp. 84–88.

[30] S. Benedetto, E. Bigleri, and V. Castellani, *Digital Transmission Theory*, Upper Saddle River, NJ: Prentice-Hall, 1987.

[31] K. S. Zigangirov, "Some Sequential Decoding Procedures," *Probl. Peredach Inform.* (in Russian), Vol. 2, No. 4, 1966, pp. 13–25.

[32] J. A. Heller, "Feedback Decoding of Convolutional Codes," *Advances in Communication Systems*, Vol. 4, A. J. Viterbi, editor, New York: Academic Press, 1975, pp. 261–278.

[33] A. J. Viterbi and I. M. Jacobs, "Advances in Coding and Modulation for Noncoherent Channels Affected by Fading, Partial Band, and Multiple-Access Interference," in *Advances in Communication Systems*, Vol. 4, A. J. Viterbi, editor, New York: Academic Press, 1975.

[34] P. Elias, "Error-Free Coding," *IRE Trans.*, Vol. PGIT-4, 1954, pp. 29–37.

[35] G. D. Forney, Jr., *Concatenated Codes*, Cambridge, MA: MIT Press, 1966.

[36] C. Berrou, A. Glavieux, and P. Thitimashima, "Near Shannon Limit Error-Correcting Coding and Decoding: Turbo Codes," *IEEE Proc. of the International Conf. on Communications*, Geneva, Switzerland, May 1993, pp. 1064–1070.

[37] C. Berrou and A. Glavieux, "Near Optimum Error Correcting Coding and Decoding: Turbo Codes," *IEEE Trans. on Communications*, Vol. 44, No. 10, October 1996, pp. 1261–1271.

[38] S. Benedetto and G. Montorsi, "Unveiling Turbo Codes: Some Results on Parallel Concatenated Coding Schemes," *IEEE Trans. on Information Theory*, Vol. 42, 1996, pp. 409–428.

[39] W. E. Ryan, "A Turbo Code Tutorial," unpublished paper located on the New Mexico State University Web site, circa 1998.

[40] J. Hagenauer, E. Offer, and L. Papke, "Iterative Decoding of Binary Block and Convolutional Codes," *IEEE Trans. on Information Theory*, Vol. 42, March 1996.

[41] F. Daneshgaran and M. Mondin, "Design of Interleavers for Turbo Codes: Iterative Interleaver Growth Algorithms of Polynomial Complexity," *IEEE Trans. on Information Theory*, Vol. IT-28, 1999, pp. 793–799.

[42] R. G. Gallager, "Low Density Parity-Check Codes," *IRE Trans. Info. Theory*, Vol. 8, No. 1, 1962, pp. 21–28.

[43] V. V. Zyablov and M. S. Pinsker, "Estimation of the Error-Correction Complexity for Gallagher Low-Density Codes," *Prob. Pered. Inform.* Vol. 11, No. 1, 1975, pp. 26–36 (in Russian).

[44] R. M. Tanner, "A Recursive Approach to Low Complexity Codes," *IEEE Trans. on Information Theory*, Vol. IT-27, No. 5, September 1981, pp. 533–547.

[45] D. J. C. MacKay and R. M. Neal, "Near Shannon Limit Performance of Low-Density Parity Check Codes," *Electronics Letters*, Vol. 32, 1966, pp. 1645–1646.

[46] D. J. C. MacKay, "Good Error-Correcting Codes Based on Very Sparse Matrices," *IEEE Trans. on Information Theory*, Vol. 45, No. 2, 1996, pp. 399–432.

[47] T. J. Richardson, M. A. Shokrollahi and R. L. Urbanke, "Design of Capacity-Approaching Irregular Low-Density Parity-Check Codes," *IEEE Trans. on Information Theory*, Vol. 47, No. 2, February 2001, pp. 619–637.

[48] J. P. Oldenwalder, "Optimal Decoding of Convolutional Codes," Ph.D. Dissertation, University of California, 1970.

[49] W. E. Stark, "Coding for Frequency-Hopped Spread Spectrum Communications with Partial-Band Interference: Part II Coded Performance," *IEEE Trans. on Communications*, Vol. COM-33, October 1985, pp. 1045–1057.

[50] H. H. Ma and M. A. Poole, "Error-Correcting Codes Against Worst-Case Partial Band Jammer," *IEEE Trans. on Communications*, Vol. COM-32, February 1984, pp. 124–133.

[51] J. S. Lee, R. H. French, and L. E. Miller, "Error-Correcting Codes and Nonlinear Diversity Combining Against Worst Case Partial-Band Noise Jamming of Frequency-Hopping MFSK Systems," *IEEE Trans. on Communications*, Vol. COM-36, April 1988, pp. 471–478.

[52] M. K. Simon, J. K. Omura, R. A. Scholtz, and B. K. Levitt, *Spread Spectrum Communications*, Vols. I, II, and III, Rockville, MD: Computer Science Press, 1985.

[53] M. B. Pursley, and W. E. Stark, "Performance of Reed-Solomon Coded Frequency-Hop Spread-Spectrum in Partial Band Interference," *IEEE Trans. on Communications*, Vol. COM-33, August 1985, pp. 858–862.

[54] D. J. Torrieri, "Frequency Hopping with Multiple Frequency Shift Keying and Hard Decisions," *IEEE Trans. on Communications*, Vol. COM-32, May 1984, pp. 574–582.

[55] B. K. Levitt, "Use of Diversity to Improve FH/MFSK Performance in Worst Case Partial Band Noise and Multitone Interference," *Conf. Rec. 1982, IEEE Military Communications Conference*, October 1982, pp. 28.2-1 to 28.2-5.

[56] W. E. Stark, "Coding for Frequency-Hopped Spread-Spectrum Communications with Partial-Band Interference: Part I. Capacity and Cutoff Rate," *IEEE Trans. on Communications*, Vol. COM-33, October 1985, pp. 1036–1044.

[57] S. W. Golomb, L. D. Baumert, M. F. Easterling, J. J. Stiffler, and A. Viterbi, *Digital Communications with Space Applications*, Englewood Cliffs, NJ: Prentice-Hall, Chapter 7, 1964.

[58] A. J. Viterbi,, "A Robust Ratio-Threshold Technique to Mitigate Tone and Partial Band Jamming in Coded MFSK Systems," *Conf. Rec., 1982 IEEE Military Communications Conference*, October 1982, pp. 22.4-1 to 22.4-5.

[59] J. S. Lee, R. H. French, and L. E. Miller, "Probability of Error Analysis of BFSK Frequency-Hopping System with Diversity Under Partial-Band Jamming Interference: Part I. Performance of Square-Law Linear Combining Soft Decision Receiver," *IEEE Trans. on Communications*, Vol. COM-32, June 1984, pp. 645–653.

[60] J. S. Lee, L. E. Miller, and Y. K. Kim, "Probability of Error Analysis of BFSK Frequency-Hopping system with Diversity under Partial-Band Jamming Interference: Part I. Performance of Square-Law Nonlinear Combining Soft Decision Receiver," *IEEE Trans. on Communications*, Vol. COM-32, December 1984, pp. 1243–1250.

[61] B. K. Levitt, "FH/MFSK Signaling with Diversity in Worst-Case Partial-Band Noise," *IEEE Journal of Selected Areas of Communications*, Vol. SAC-3, September 1985, pp. 622–626.

[62] B. K. Levitt, "FH/MFSK Performance in Multitone Jamming," *IEEE Journal of Selected Areas of Communications*, Vol. SAC-3, September 1985, pp. 627–643.

[63] C. M. Keller and M. B. Pursley, "Clipped Diversity Combining for Channels with Partial-Band Interference," *IEEE Journal of Selected Areas of Communications*, Vol. SAC-5, February 1987, pp. 248–260.

[64] B. Trumpis, "On the Optimum Detection of Fast Frequency Hopped MFSK Signals in Worst Case Jamming," TRW Internal Memorandum, June 1981.

[65] C. E. Shannon, "A Mathematical Theory of Communication," *Bell System Technical Journal*, Vol. 27, Part I, pp. 379–423, Part II, pp. 623–656.

[66] J. M. Wozencraft and I. M. Jacobs, *Principles of Communication Engineering*, New York: Wiley, 1965.

[67] D. J. Costello Jr., J. Hagenauer, H. Imai, and S. B. Wicker, "Applications of Error-Control Coding," *IEEE Trans. on Information Theory*, Vol. 44, No. 6, October 1998.

Selected Bibliography

J. P. Oldenwalder, *Error Control Coding Handbook*, Final report, Linkabit Corporation, for the U.S. Air Force, July 15, 1976.

C. Schlegel, Lecture notes, Winter term 2003, EE 780 on "Error Control Coding," University of Alberta, Edmonton Canada, AB, Canada.

B. Sklar, *Digital Communications Fundamentals and Applications*, 2nd ed., Englewood Cliffs, NJ: Prentice-Hall, 2000.

J. K. Wolf, A. M. Michelson, and A. H. Levesque, "On the Probability of Undetected Error for Linear Block Codes," *IEEE Trans. on Communications*, Vol. COM-30, No. 2, February 1982.

J. K. Wolf and R. D. Blackeney II, "An Exact Evaluation of the Probability of Undetected Error for Certain Shortened Binary CRC Codes," *MILCOM 88*, 1988, pp. 15.2.1–15.2.6.

M. D. Yacoub, *Foundations of Mobile Radio Engineering*, Boca Ratan, FL: CRC Press, Chapter 6, 1993.

Problems

1. Using the property that the sum of two linear codes produces a word, show that the minimum distance is just the smallest nonzero code word weight.

2. Starting with (4.2-55), which is based on the white Gaussian noise channel, show that an equivalent calculation requires that the correlation be maximized; that is, choose the decoded code word based on $\hat{C}_l = \max_m \sum_{k=1}^{n} \{y_k c_{mk}\}$ where the c_{mk} take on the values of ± 1.

3. Consider the encoder shown here. It has parameters $K = 3$, $k = 1$, and $n = 3$ and is a rate 1/3 convolutional code. Determine the tree diagram, the trellis diagram, and the state diagram, for this encoder. What is the output coded bits when an all-ones input occurs?

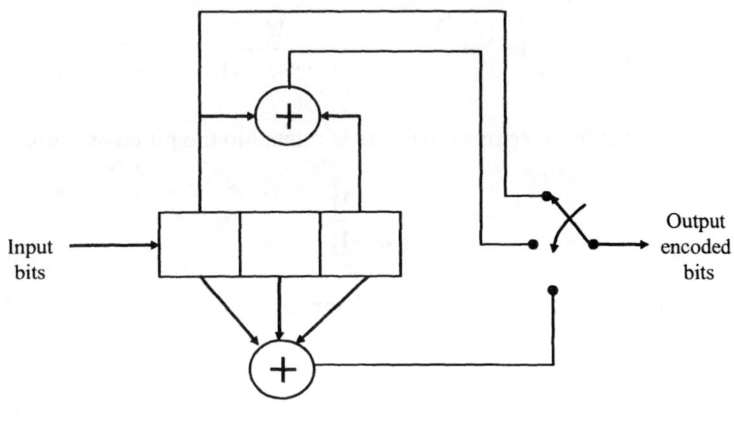

Figure for Problem 5 illustrating a rate 1/3, constraint length 3, convolutional code.

4. Show that the transfer function for the encoder of Figure 4.3-8 is given by

$$\mathrm{T}(D,N) = \frac{ND^5}{1-2ND} \tag{P4-1}$$

5. Consider the encoder of Example 14, shown in Figure 4.3-2, and consider the input coded sequence given by 1100010000.... Show that the Viterbi algorithm, for the hard-decision channel, does not correctly decode the correct path in this case.

6. Construct the addition and multiplication tables for the GF(5) field using the elements $\{0,1,2,3,4\}$.

7. Show that when one divides x^3+x+1 into $b_2x^5+b_1x^4+b_0x^3$ that the quotient is given by $b_2x^2 + b_1x + (b_0 + b_2)$ and the remainder is given by $(b_1 + b_2)x^2 + (b_0 + b_1 + b_2)x + (b_0 + b_2)$. Hint: Order the polynomials starting on the left with the highest order in x and perform the long division.

8. Show that $1 + x + x^3$ divides x^7-1.

9. Establish the relationship for orthogonal MFSK modulation between the bit error rate and the symbol error rate, (4.5-6), by showing the following [57]. Show that for orthogonal coding, since all the errors are equally probable, the expected number of bits in error when a k-bit ($M = 2^k$) code word has been detected incorrectly is given by

$$\sum_{i=1}^{k} i \frac{\binom{k}{i}}{\sum_{j=1}^{k}\binom{k}{j}} = \frac{k(2^{k-1})}{2^k - 1} \tag{P4-2}$$

Thus conclude that the conditional probability that a given bit is in error when the k-bit word within which it was encoded is incorrect is given by

$$\frac{1}{k}\left(\frac{k(2^{k-1})}{2^k - 1}\right) = \frac{2^{k-1}}{2^k - 1} = \frac{(M/2)}{(M-1)} \tag{P4-3}$$

Finally conclude that the word error probability is related to the bit error probability by

$$PE_b = \frac{(M/2)}{(M-1)} PE_s \tag{P4-4}$$

10. Solve for the values of a_i for $i = 1$, 2, and 3, from the values of the b_i in (4.2-37) by combining the three equations in three unknowns

$$\begin{aligned} b_0 &= a_0 \\ b_1 &= a_0 + a_1 \\ b_2 &= a_1 + a_2 \end{aligned} \tag{P4-5}$$

CHAPTER 5

Carrier Tracking Loops and Frequency Synthesizers

5.0 INTRODUCTION

This chapter will address carrier-tracking loops for residual carrier signals, as well as for the coherent reception of MPSK signals, and it will also address frequency synthesis. The starting point in our discussion will be the phase locked loop (PLL). Later, useful forms, such as the Costas loop and the Squaring loop, that are applicable to tracking suppressed carrier signals, such as for BPSK signaling, will be discussed. After the concept of coherent tracking loops has been presented, a discussion of frequency synthesis will be developed. A brief presentation of the following types of synthesizers will then be discussed: (1) digital frequency synthesis, (2) direct frequency synthesis, and (3) indirect frequency synthesis. The last portion of this chapter will address tracking loops for $N > 2$ for N-phase modulation systems.

5.1 TRACKING OF RESIDUAL CARRIER SIGNALS

In many phase modulation schemes the modulation approach is such that there remains a residual line spectral component after modulating the signal onto the carrier. For example consider a simple phase modulation case in which the phase modulation index is θ. It follows that the transmitted signal is of the form

$$s(t) = \sqrt{2P}\sin(\omega_0 t + \theta d(t)) \tag{5.1-1}$$

Expanding the sine wave term in terms of both of its components produces the result

$$s(t) = \sqrt{2P}\left[\sin(\omega_0 t)\cos(\theta d(t)) + \cos(\omega_0 t)\sin(\theta d(t))\right] \tag{5.1-2}$$

Simplifying leads to the following form

$$s(t) = \sqrt{2P}\left[\cos(\theta)\sin(\omega_0 t) + d(t)\sin(\theta)\cos(\omega_0 t)\right] \tag{5.1-3}$$

From (5.1-3) it is clear that the transmitted signal is composed of the sum of a BPSK signal, having power $P\sin(\theta)^2$, and an unmodulated tone (spectral line) having the power of $P\cos(\theta)^2$. Thus this latter signal could be tracked with a PLL, since it is an unmodulated tone. When $\theta = 90°$, the unmodulated component disappears, and the only component would be a BPSK signal (suppressed carrier). Later we will show that it is necessary to use a suppressed carrier type of loop carrier tracking loop to track a fully suppressed BPSK signal.

5.2 PLL FOR TRACKING A RESIDUAL CARRIER COMPONENT

In this section we will derive the mathematical basis for the device known as a phase locked loop (PLL). The PLL is a device that is capable of tracking the phase of a sinusoidal signal, such as the first signal term in (5.1-3).

5.2.1 The Likelihood Function for Phase Estimation

The two basic criteria that are used in signal parameter estimation is (1) maximum a posteriori probability (MAP) estimation approach and (2) the maximum-likelihood (ML) approach [1]. The signal parameter θ to be estimated is modeled as a random variable and is assumed to have an a priori probability density function $p(\theta)$. Let the received signal be described as

$$y(t) = s(t,\theta) + n(t) = \sqrt{2P}\cos\left(\omega_0 t + \theta\right) + n(t) \tag{5.2-1}$$

where $n(t)$ is a white Gaussian noise process, P is the received signal power, and ω_0 is the carrier radian frequency.

Using an orthonormal series expansion of N orthonormal functions $\{f_n(t)\}$, $y(t)$ may be represented by the N component vector $\boldsymbol{y} = (y_1, y_2, \ldots y_N)$. The ML estimate of θ is that value of θ that maximizes $p(\boldsymbol{y}|\theta)$. However the MAP estimate is the estimate of θ that maximizes the value of the a posteriori probability density function

$$p(\theta|\boldsymbol{y}) = \frac{p(\boldsymbol{y}|\theta)p(\theta)}{p(\boldsymbol{y})} \tag{5.2-2}$$

Now note that if the variable θ is assumed to be uniformly distributed over its range of operation, and since $p(\boldsymbol{y})$ does not depend on θ, then maximizing $p(\boldsymbol{y}|\theta)$ and $p(\theta|\boldsymbol{y})$ lead to the same estimate. In our derivation of the optimum estimator it will be assumed that θ is an unknown constant and the ML criterion will be used to obtain the optimum estimate. The time duration of T seconds will be assumed as the estimation observation interval.

Now let us consider a time continuous representation of the maximization of $p(\boldsymbol{y}|\theta)$. Since $n(t)$ is assumed to be a white Gaussian random process, having a zero mean value, the joint probability density function can be written as

$$p(\boldsymbol{y}|\theta) = \frac{1}{\left(\sqrt{2\pi}\sigma\right)^N}\exp\left[-\sum_{i=1}^{N}\frac{(y_n - s_n(\theta))^2}{2\sigma^2}\right] \tag{5.2-3}$$

where y_n and $s_n(\theta)$ are defined by

$$\begin{aligned} y_n &= \int_T y(t) f_n(t)dt \\ s_n(\theta) &= \int_T s(t,\theta) f_n(t)dt \end{aligned} \tag{5.2-4}$$

and T is the integration interval used in the expansion of $y(t)$ and $s(t,\theta)$. The functions $f_n(t)$ are *orthonormal functions* having the property

$$\int_T f_n(t) f_m(t) dt = \begin{cases} 1 & \text{for } m = n \\ 0 & \text{for } m \neq n \end{cases} \tag{5.2-5}$$

In the limit, as N increases without bound, the series in (5.2-3) can be shown to converge to a functional of the form

$$\lim_{N\to\infty} \left(\frac{1}{2\sigma^2}\right) \sum_{i=1}^{N} [y_n - s_n(\theta)]^2 = \frac{1}{N_0} \int_T [y(t) - s(t,\theta)]^2 dt \tag{5.2-6}$$

where N_0 is the one-sided noise spectral density of the noise. From (5.2-3) it is clear that maximization of $p(\mathbf{y}|\theta)$ is equivalent to maximization of the likelihood function given by

$$\Lambda(\theta) = \exp\left[-\frac{1}{N_0} \int_T \{y(t) - s(t,\theta)\}^2 dt \right] \tag{5.2-7}$$

5.2.2 The Maximum-Likelihood Estimation of Carrier Phase

Now we are in a position to determine the optimum estimator of the received signal carrier phase, θ. To proceed, expand (5.2-7) to produce

$$\Lambda(\theta) = \exp\left[-\frac{1}{N_0} \int_T \left\{ y(t)^2 - 2y(t)s(t,\theta) + s(t,\theta)^2 \right\} dt \right] \tag{5.2-8}$$

Both the first and third terms do not depend on the value of θ. The third integral of the squared signal is constant, independent of θ. Consequently these terms may be neglected in the search for the optimum value. It follows that the likelihood function can be expressed as

$$\Lambda(\theta) = C \exp\left[\frac{2}{N_0} \int_T y(t)s(t,\theta) dt \right] \tag{5.2-9}$$

where C is a fixed constant.

The maximum-likelihood estimate of θ, which we denote by $\hat{\theta}$, can be obtained from the log-likelihood function, since the logarithm is a monotonic function of its argument. Thus it is sufficient to consider the log-likelihood function

$$\Lambda_{\log}(\theta) = \frac{2}{N_0} \int_T y(t)s(t,\theta) dt \tag{5.2-10}$$

Using the definition of $s(t,\theta)$ from (5.2-1) yields, for the log-likelihood, the functional

$$\Lambda_{\log}(\theta) = \frac{2}{N_0} \int_T y(t) \sqrt{2P} \cos(\omega_0 t + \theta) dt \tag{5.2-11}$$

A necessary condition to find the maximum is given by the condition

$$\left.\frac{d\Lambda_{\log}(\theta)}{d\theta}\right|_{\theta=\hat{\theta}}=0 \tag{5.2-12}$$

Differentiating (5.2-11) produces

$$\int_T y(t)\sin(\omega_0 t+\hat{\theta})dt=0 \tag{5.2-13}$$

Expanding (5.2-13) produces

$$\cos(\hat{\theta})\int_T y(t)\sin(\omega_0 t)dt=-\sin(\hat{\theta})\int_T y(t)\cos(\omega_0 t)dt \tag{5.2-14}$$

Solving for the estimate, $\hat{\theta}$, produces

$$\hat{\theta}=-\tan^{-1}\left[\frac{\int_T y(t)\sin(\omega_0 t)dt}{\int_T y(t)\cos(\omega_0 t)dt}\right] \tag{5.2-15}$$

Thus (5.2-15) is an estimate that requires a pair of quadrature carriers to obtain the received carrier phase estimate. However, (5.2-13) suggests that a closed-loop form can be utilized to estimate the carrier phase. Figure 5.2-1 illustrates the PLL based on the ML estimate.

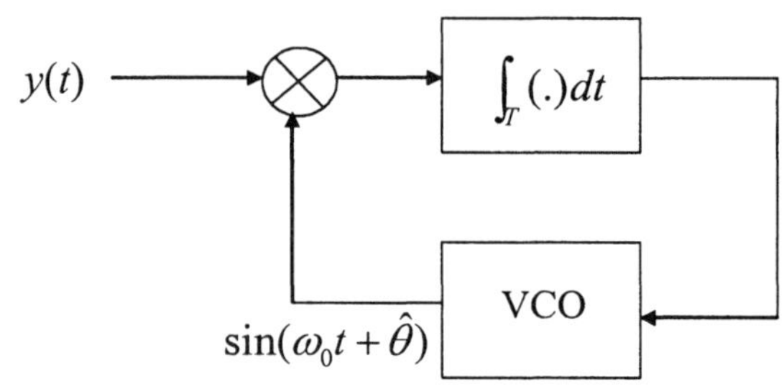

Figure 5.2-1 The estimate of the phase based on the ML estimate of an unmodulated carrier signal.

where VCO stands for a voltage controlled oscillator and the circle with two crossed lines denotes multiplication of the two signals that are input to it.

The upper block denotes a filter that has a bandwidth proportional to the reciprocal of the integration time *T*. In practice the loop filter contains a term that represents integration plus other components. Loop filter types will be discussed later in this section.

5.2.3 Long Loops and Short Loops

Before we delve into the details of the PLL operation, we shall first define what we mean by a short and a long loop structure. The term *short loop* denotes that there is only one multiplication of the signals, such as is shown in Figure 5.2-1. A *long loop* is a loop that has more than one multiplier. Since all phase locked loops can be modeled as short loops, we will confine our study to short loops in what follows. Figure 5.2-2 illustrates an example of a long loop.

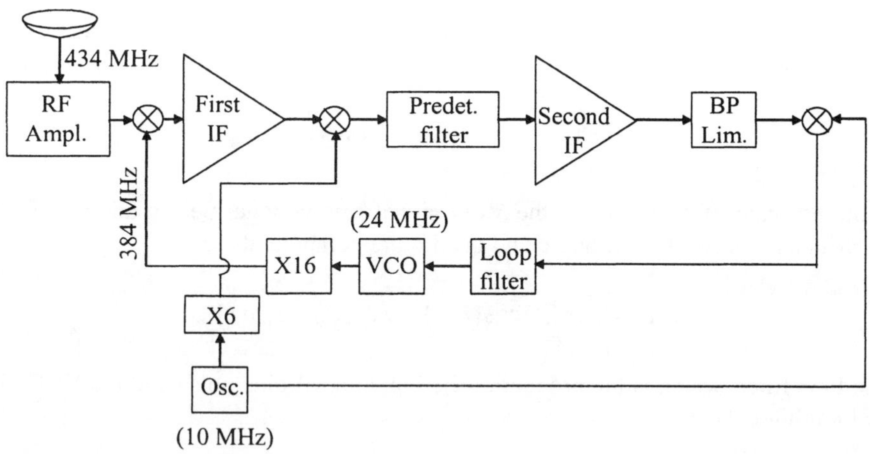

Figure 5.2-2 Long loop model of a PLL showing two IFs.

As can be seen in Figure 5.2-2 there are three multiplications, and thus it is a long loop. Almost all practical receivers that employ PLLs utilize long loops. Implementation advantages occur with a long loop. It is to be noted that it is always possible to find an equivalent short PLL of a long PLL, for the purposes of analysis.

5.2.4 The Stochastic Differential Equation of Operation

As before, we model the input to the PLL as a sinusoid signal plus thermal noise of the form

$$y(t) = \sqrt{2P}\sin(\omega_0 t + \theta) + n(t) \tag{5.2-16}$$

in which the received power is P (W), the carrier frequency is $f_0=\omega_0/(2\pi)$ Hz, the carrier phase is θ radians, and the white Gaussian noise process with power spectral density $N_0/2$ W/Hz is represented by $n(t)$. The short loop version of a PLL, which we will use as our model, is shown in Figure 5.2-3.

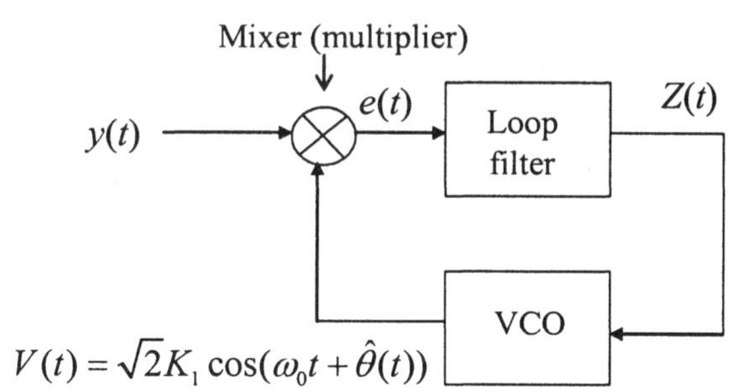

Figure 5.2-3 The basic PLL in its short form as the model to ascertain its equation of operation.

In all of the work to follow on tracking loops, only the analog loop model will be discussed. However most modern phase lock loops are implemented digitally. The basic concepts can be presented in analog form without loss of generality. Because digital PLLs operate much like analog loops, if designed properly, they will not be treated separately here. In fact the analog PLL can be viewed as an approximate upper bound on the performance of a digital PLL. Of course quantization, aliasing, and other issues must be dealt with in a digital phase locked loop design.

It is assumed that the VCO output includes the loop estimate of the carrier phase, $\hat{\theta}$, which is the estimate of the unknown carrier phase, θ, at the receiver. The VCO output is modeled as [2, 3]

$$V(t) = \sqrt{2}K_1 \cos(\omega_0 t + \hat{\theta}(t)) \tag{5.2-17}$$

with $\hat{\theta}$ being the estimate of θ, and K_1 is the VCO signal rms voltage. See in addition [4–6]. The WGN process can be written in terms of its in-phase and quadrature components as

$$n(t) = \sqrt{2}n_c(t)\cos(\omega_0 t) + \sqrt{2}n_s(t)\sin(\omega_0 t) \tag{5.2-18}$$

The input to the loop filter, which is the product of the input signal plus noise and the VCO signal, provides the input to the loop filter. Thus

$$\begin{aligned} e(t) = K_m V(t) y(t) = \sqrt{P}K_m K_1 &\left\{\sin(\theta - \hat{\theta}(t)) + \sin(2\omega_0 t + \theta + \hat{\theta}(t)\right\} \\ &+ K_1' n_c(t)\cos(\hat{\theta}(t)) - K_1' n_s(t)\sin(\hat{\theta}(t)) \\ &+ K_1' n_s(t)\sin(2\omega_0 t + \hat{\theta}(t)) \\ &+ K_1' n_c(t)\cos(2\omega_0 t + \hat{\theta}(t)) \end{aligned} \tag{5.2-19}$$

where $K_1' = K_m K_1$ and K_m is the multiplier gain. The product $\sqrt{P}K_1 K_m = K_\theta$ is the *phase detector gain* and has dimensions of V/rad. Typically the mixer and the loop filter cannot respond significantly to the sum $2\omega_0$ frequency terms in (5.2-19) so that (5.2-19) can be well approximated by

$$\begin{aligned} e(t) &= \sqrt{P}K_m K_1 \sin(\theta - \hat{\theta}(t)) \\ &+ K_m K_1 n_c(t)\cos(\hat{\theta}(t)) - K_m K_1 n_s(t)\sin(\hat{\theta}(t)) \end{aligned} \tag{5.2-20}$$

Oftentimes in actual loop design a low-pass filter, which is wide compared to the loop noise bandwidth, is used to suppress the sum frequency. Normally it is chosen to have negligible effect on the loop response. Denoting the noise term as $n'(t)$ as the sum of the two noise terms in (5.2-20) allows us to write

$$e(t) = \sqrt{P}K_m K_1 \sin(\phi(t)) + K_m K_1 n'(t) \tag{5.2-21}$$

where we have denoted the *phase error* as $\phi(t) = \theta - \hat{\theta}(t)$. It can be shown [7] that $n'(t)$ has the same statistics as $n_c(t)$ or $n_s(t)$, and therefore has two-sided noise spectral density of $N_0/2$, as long as $\phi(t)$ is slowly varying compared to the noise process bandwidth.

The VCO has as an output phase estimate of the form

$$\hat{\theta}(t) = \int_0^t K_{vco} Z(u)du \tag{5.2-22}$$

where $Z(t)$ is the output of the loop filter, and K_{vco} is the VCO gain constant which has dimensions of radian frequency/V. Equation (5.2-22) can be conveniently represented in operator notation as

$$\hat{\theta}(t) = \frac{K_{vco}}{s} Z(t) \tag{5.2-23}$$

where s is the La Place transform variable, and by comparing (5.2-22) with (5.2-23) it can be seen that $(1/s)(g(t))$ denotes integration of $g(t)$ and $sg(t)$ denotes the derivative of $g(t)$; that is, $\frac{dg(t)}{dt}$. Using operator notation the signal out of the loop filter is given by

$$Z(t) = F(s)e(t) = F(s)K_m K_1\left[\sqrt{P}\sin(\phi(t)) + n'(t)\right] \tag{5.2-24}$$

From (5.2-23) it follows that

$$\hat{\theta}(t) = \frac{K_{vco}}{s}Z(t) = \frac{K_{vco}}{s}F(s)K_m K_1\left[\sqrt{P}\sin(\phi(t)) + n'(t)\right] \tag{5.2-25}$$

Now since $\hat{\theta}(t) = \theta - \phi(t)$ it follows that (5.2-25) can be written as

$$\theta = \phi(t) + \frac{K}{s}F(s)\left[\sqrt{P}\sin(\phi(t)) + n'(t)\right] \tag{5.2-26}$$

where $K = K_m K_1 K_{vco}$ is the *open loop gain* and has dimensions of radian frequency (rad/s). Equation (5.2-26) is the basic stochastic differential equation of operation. Multiplying both sides of (5.2-26) by s produces the equivalent equation

$$\dot{\theta}(t) = \dot{\phi}(t) + KF(s)\left[\sqrt{P}\sin(\phi(t)) + n'(t)\right] \tag{5.2-27}$$

The equivalent baseband nonlinear model for this equation is shown in Figure 5.2-4. In this model and the previous equation we have allowed θ to be slowly dependent on time and have denoted it by $\theta(t)$ to account for transient errors.

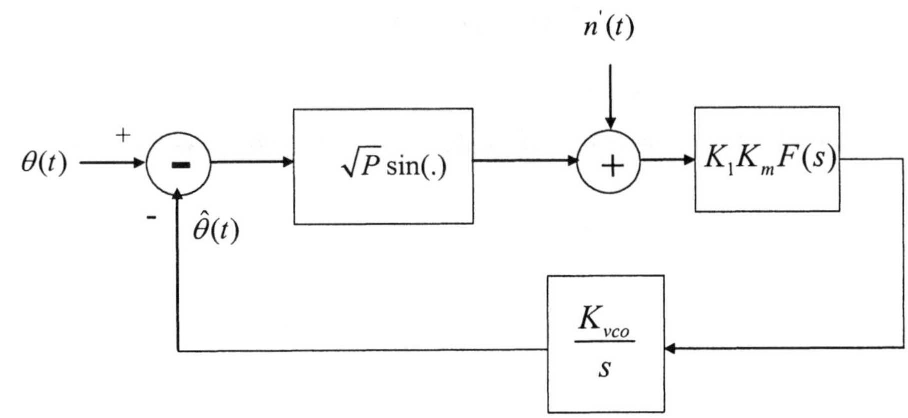

Figure 5.2-4 The baseband PLL nonlinear model.

5.2.5 The Linear Model of the PLL with Noise

Now we turn our attention to the linear model of the PLL. In normal operation it is desirable to make sure the phase error is small so that demodulation will not induce much degradation to the demodulated bit error rate

(BER). If we assume that the error does not exceed 0.8 rad then the relative error between ϕ and $\sin(\phi)$ is no greater than 10%. Under this assumption one can write

$$\sin(\phi(t)) \cong \phi(t) \tag{5.2-28}$$

From the basic loop equation, (5.2-26), one has

$$\theta(t) = \phi(t) + \frac{K}{s}F(s)\left[\sqrt{P}\phi(t) + n'(t)\right] \tag{5.2-29}$$

This can be rewritten as

$$\theta(t) = \frac{s + K\sqrt{P}F(s)}{s}\phi(t) + \frac{K\sqrt{P}F(s)}{s}\frac{n'(t)}{\sqrt{P}} \tag{5.2-30}$$

Rearranging this equation produces

$$\phi(t) = \frac{s}{s + K\sqrt{P}F(s)}\theta(t) - \frac{K\sqrt{P}F(s)}{s + K\sqrt{P}F(s)}\frac{n'(t)}{\sqrt{P}} \tag{5.2-31}$$

Now using the fact that $\phi(t) = \theta(t) - \hat{\theta}(t)$ allows us to write (5.2-31) as

$$\hat{\theta}(t) = \frac{\sqrt{P}KF(s)/s}{1 + \sqrt{P}KF(s)/s}\theta(t) \tag{5.2-32}$$

Thus we see that we can write the output phase estimate in terms of the input phase process via

$$\hat{\theta}(t) = H(s)\theta(t) \tag{5.2-33}$$

where $H(s)$ is known as the *closed-loop transfer function*. Let $\varphi(s)$ denote the La Place transform of $\theta(t)$ and let $\hat{\varphi}(s)$ denote the La Place transform of $\hat{\theta}(t)$. The closed-loop transfer function can be written as

$$H(s) = \frac{\sqrt{P}KF(s)/s}{1 + \sqrt{P}KF(s)/s} = \frac{\hat{\varphi}(s)}{\varphi(s)} \tag{5.2-34}$$

Equation (5.2-31) can be written in terms of $H(s)$ and $1 - H(s)$ using the results of Problem 1, as follows

$$\phi(t) = [1 - H(s)]\theta(t) - H(s)\frac{n'(t)}{\sqrt{P}} \tag{5.2-35}$$

If the derivation included transmitter oscillator phase noise $\theta_{VCO}(t)$ then it is easy to show that (5.2-35) generalizes to the form

$$\phi(t) = [1 - H(s)]\theta(t) - H(s)\frac{n'(t)}{\sqrt{P}} - [1 - H(s)]\theta_{VCO}(t) \text{ rad} \tag{5.2-36}$$

which is the error process based on a linear model of the PLL. The first term yields the error due to the phase process due to relative movement between the receiver and the transmitter, the second term is the effects of thermal noise at the input of the PLL, and the last term is the effect of oscillator phase noise of the transmitter on the loop error. The linearized PLL baseband model is shown in Figure 5.2-5.

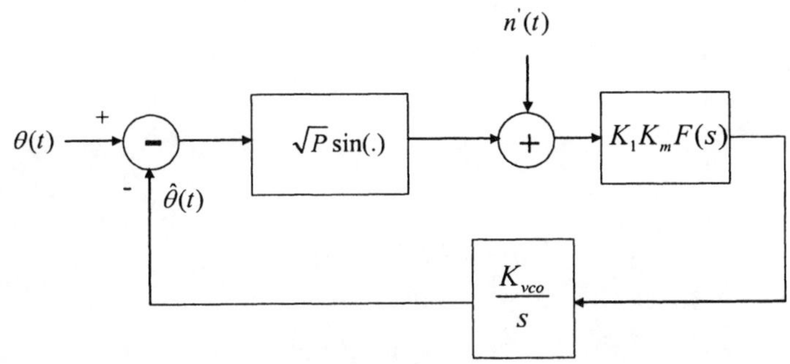

Figure 5.2-5 The baseband PLL linear model.

5.2.6 The Various Loop Filter Types

So far we have not discussed the loop filter. There are many types of loop filters, designed for many different requirements. The *first-order phase locked loop* occurs when the loop filter is of the form

$$F(s) = 1 \tag{5.2-37}$$

which actually is no filter at all, just a constant. The corresponding closed-loop transfer function (CLTF) is given by

$$H(s) = \frac{\sqrt{P}K}{s + \sqrt{P}K} \tag{5.2-38}$$

The reason that this CLTF is called a first-order loop is because the denominator of the closed-loop transfer function is first-order.

There are two basic types of second-order filters that are used as models. The first type of loop filter leads to a passive second-order loop and is of the form

$$F(s) = \frac{1 + \tau_2 s}{1 + \tau_1 s} \tag{5.2-39}$$

The circuit shown in Figure 5.2-6 can implement this loop filter.

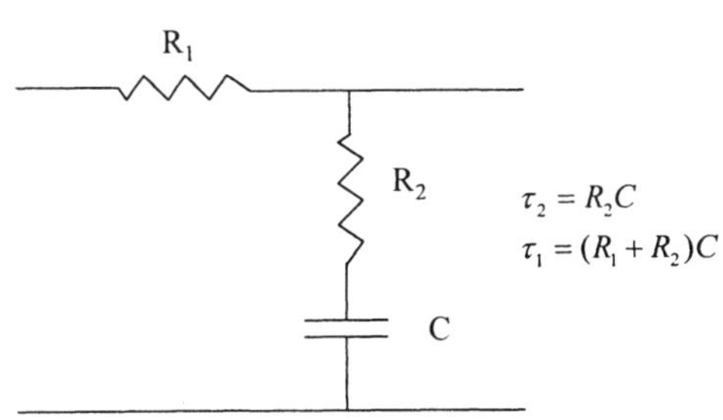

Figure 5.2-6 Loop filter for the passive second-order PLL.

The corresponding CLTF of the passive second-order loop is given by inserting (5.2-39) in (5.2-34) to produce

$$H(s)=\frac{1+\tau_2 s}{1+(\tau_2+1/\sqrt{PK})s+(\tau_1/\sqrt{PK})s^2} \tag{5.2-40}$$

To simplify this notation [3] let

$$r=\frac{\sqrt{PK}\tau_2^2}{\tau_1}$$
$$\omega_n^2=\frac{\sqrt{PK}}{\tau_1} \tag{5.2-41}$$
$$\frac{2\varsigma}{\omega_n}=\left(\tau_2+\frac{1}{\sqrt{PK}}\right)$$

Then it is possible to express the CLTF in standard control theory format of the form

$$H(s)=\frac{s\left(2\omega_n\zeta-(\omega_n^2/\sqrt{PK})\right)+\omega_n^2}{s^2+2\omega_n\zeta s+\omega_n^2} \tag{5.2-42}$$

where ω_n is the natural frequency (rad/s) and ς is the damping factor (no dimensions).

The second type of loop filter, the (ideal) active loop filter, has a transfer function of the form

$$F(s)=\frac{1+\tau_2 s}{\tau_1 s} \tag{5.2-43}$$

The corresponding CLTF in this case is given by inserting (5.2-43) in (5.2-34) and yields

$$H(s)=\frac{1+\tau_2 s}{1+\tau_2 s+(\tau_1/\sqrt{PK})s^2} \tag{5.2-44}$$

An active amplifier of the form [2] shown in Figure 5.2-7 can implement this loop filter.

The transfer function, expressed in the damping factor, ς, and the natural frequency, ω_n, is expressed by

$$H(s)=\frac{2\omega_n\zeta s+\omega_n^2}{s^2+2\omega_n\zeta s+\omega_n^2} \tag{5.2-45}$$

where

$$\omega_n^2=\frac{\sqrt{PK}}{\tau_1}$$
$$\varsigma=\frac{\tau_2}{2}\sqrt{\frac{\sqrt{PK}}{\tau_1}} \tag{5.2-46}$$

Now consider a third-order PLL. Based on the optimum Jaffe-Rechtin optimization for a third-order PLL,[1] to a frequency ramp (quadratic in phase), yields a loop filter which is of the idealized form

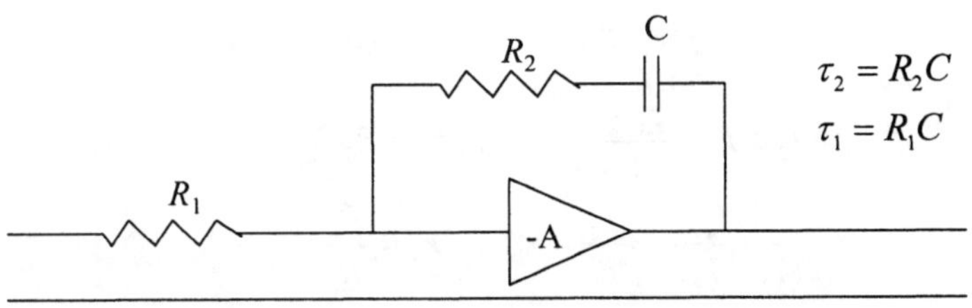

Figure 5.2-7 The filter for a second-order (active) PLL.

$$\sqrt{P}KF(s) = \frac{\omega_n^3 + 2\omega_n^2 s + 2\omega_n s^2}{s^2} \tag{5.2-47}$$

From Problem 2, using (5.2-34) for the definition of $H(s)$ produces the closed-loop transfer function

$$H(s) = \frac{\omega_n^3 + 2\omega_n^2 s + 2\omega_n s^2}{\omega_n^3 + 2\omega_n^2 s + 2\omega_n s^2 + s^3} \tag{5.2-48}$$

For this loop $B_L = (5/6)\omega_n = 5.236f_n$. This is very similar functionally to the filter discussed in (5.2-47). Problem 2 also shows that the response to phase noise is given by

$$|1 - H(f)|^2 = \frac{f^6}{f_n^6 + f^6} \tag{5.2-49}$$

An equivalent form of the loop filter, utilizing another set of parameters, is of the form [8]

$$F(s) = \frac{1 + \tau_2 s}{\tau_1 s} + \frac{1}{2\tau_1\tau_2 s^2} \tag{5.2-50}$$

Since it is not possible to synthesize perfect integrators (all have finite gain), the loop filter of (5.2-50) was modified to

$$F(s) = \frac{1 + \tau_2 s}{1 + \tau_1 s} + \frac{1}{(1 + \tau_1 s)(\delta + \tau_3 s)} \tag{5.2-51}$$

This loop filter approximates (5.2-50) except at $f = 0$ Hz. Note that $1/\delta$ is assumed to be large number. Tausworthe and Crow [8] have used the following loop filter, which is essentially equivalent to (5.2-51)

$$F(s) = \frac{\kappa_1\kappa_2(1 + T_2 s)(1 + T_4 s)}{(1 + T_1 s)(1 + T_3 s)} \tag{5.2-52}$$

where [8]

[1] This method minimizes the mean squared tracking error plus the time integral (0 to ∞) of the transient error squared.

$$
\begin{aligned}
&T_1 = \tau_1 && T_1 = (R_2 + R_3)C_1 \\
&T_3 = \tau_3 / \delta && T_2 = R_3 C_1 \\
&T_2 T_4 = \frac{\tau_2 \tau_3}{1+\delta} \cong \tau_2 \tau_3 && T_3 = (R_5 + R_6)C_2 \\
&T_2 + T_4 = \frac{\tau_3 + \tau_2 \delta}{1+\delta} \cong \tau_3 && T_4 = R_6 C_2 \\
&\frac{T_2 T_4}{(T_2 + T_4)^2} = \frac{\tau_2}{\tau_3} = k && \kappa_1 = \frac{R_2}{R_1} \\
&\kappa_1 \kappa_2 = \frac{1+\delta}{\delta} \cong \frac{1}{\delta} && \kappa_2 = \frac{R_5}{R_4}
\end{aligned}
\tag{5.2-53}
$$

Figure 5.2-8 illustrates the loop filter model for this third-order PLL. Abramovitch [9] and Cahn [10] have discussed other forms of third-order loop filters.

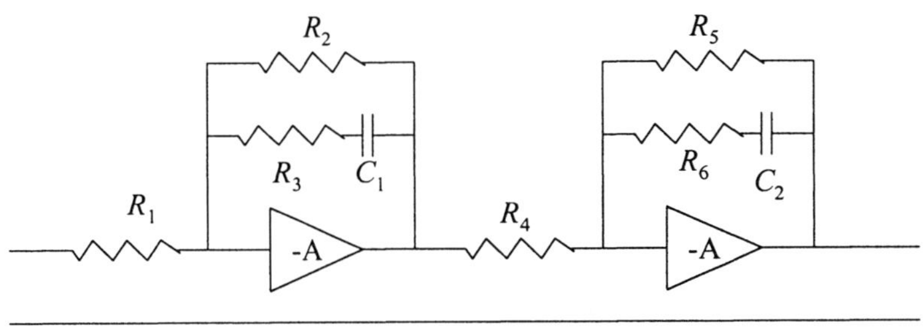

Figure 5.2-8 One implementation of the third-order loop filter $F(s)$.

5.2.7 Transient Response of a Second-Order Loop

Oftentimes it is useful to determine the transient response of a PLL when subject to various dynamical Doppler excitations. Let us now consider the often-utilized perfect second-order PLL. Based on (5.2-45) the error transfer function, $1 - H(s)$, for this loop can be written as

$$1 - H(s) = \frac{s^2}{s^2 + 2\omega_n \zeta s + \omega_n^2} \tag{5.2-54}$$

This function characterizes the phase error function within the linear region. Equation (5.2-35) characterizes the transient phase error response versus time, in terms of the natural frequency (ω_n) and the damping factor (ζ), when noise is neglected. Equation (5.2-35) repeated here is given by

$$\phi(t) = [1 - H(s)]\theta(t) \tag{5.2-55}$$

Now consider the phase error for the perfect second-order PLL with (5.2-54) defining the $1 - H(s)$ function. The input phase function is given by

$$\theta(t) = \theta_0 u(t) + \Omega_0 t u(t) + (1/2)\Lambda_0 t^2 u(t) \tag{5.2-56}$$

where the first term ($u(t)$) denotes a phase step starting at time $t = 0$ and the phase step has magnitude θ_0 in rad, the second term ($\Omega_0 tu(t)$) denotes a frequency step of magnitude of Ω_0 in rad/s, and the third term

($(1/2)\Lambda_0 t^2 u(t)$) denotes a frequency ramp Λ_0 in rad/s^2. Table 5.2-1 illustrates the transient responses for each input for a perfect second-order loop with natural frequency ω_n, and damping factor ζ [4].

Table 5.2-1 Transient Phase Error for Various Dynamics for the Perfect Second-Order PLL

Damping factor	Phase Step θ_0 (rad)	Frequency Step Ω_0 (rad/s)	Frequency Ramp Λ_0 (rad/s^2)
$\zeta<1$	$\theta_0\cos\left(\sqrt{1-\zeta^2}\,\omega_n t\right)e^{-\zeta\omega_n t}-\frac{\theta_0\zeta}{\sqrt{1-\zeta^2}}\sin\left(\sqrt{1-\zeta^2}\,\omega_n t\right)e^{-\zeta\omega_n t}$	$\frac{\Omega_0}{\omega_n}\left(\frac{1}{\sqrt{1-\zeta^2}}\sin\left(\sqrt{1-\zeta^2}\,\omega_n t\right)\right)e^{-\zeta\omega_n t}$	$\frac{\Lambda_0}{\omega_n^2}-\frac{\Lambda_0}{\omega_n^2}\cos\left(\sqrt{1-\zeta^2}\,\omega_n t\right)e^{-\zeta\omega_n t}-\frac{\Lambda_0\zeta}{\sqrt{1-\zeta^2}\,\omega_n^2}\sin\left(\sqrt{1-\zeta^2}\,\omega_n t\right)e^{-\zeta\omega_n t}$
$\zeta=1$	$\theta_0(1-\omega_n t)e^{-\omega_n t}$	$\Omega_0(\omega_n t)e^{-\omega_n t}$	$\frac{\Lambda_0}{\omega_n^2}-\frac{\Lambda_0}{\omega_n^2}(1+\omega_n t)e^{-\omega_n t}$
$\zeta>1$	$\theta_0\cosh\left(\sqrt{\zeta^2-1}\,\omega_n t\right)e^{-\zeta\omega_n t}-\frac{\theta_0\zeta}{\sqrt{\zeta^2-1}}\sinh\left(\sqrt{\zeta^2-1}\,\omega_n t\right)e^{-\zeta\omega_n t}$	$\frac{\Omega_0}{\omega_n}\left(\frac{1}{\sqrt{\zeta^2-1}}\sin\left(\sqrt{\zeta^2-1}\,\omega_n t\right)\right)e^{-\zeta\omega_n t}$	$\frac{\Lambda_0}{\omega_n^2}-\frac{\Lambda_0}{\omega_n^2}\cosh\left(\sqrt{\zeta^2-1}\,\omega_n t\right)e^{-\zeta\omega_n t}-\frac{\Lambda_0\zeta}{\sqrt{\zeta^2-1}\,\omega_n^2}\sinh\left(\sqrt{\zeta^2-1}\,\omega_n t\right)e^{-\zeta\omega_n t}$

As can be seen in the table three cases are delineated: (1) the damping factor less than unity, (2) the damping factor equal to unity, and (3) the damping factor greater than unity. The results only apply for the case that the phase errors remain in the linear region.

5.2.8 Steady State Tracking Error When the Phase Error Is Small

Now consider the tracking error that manifests itself when relative motion (dynamics) between the transmitter and the receiver occur. Under the assumption that $\phi(t)$ is small, the error can be written as indicated in (5.2-35), with the noise set equal to zero

$$\phi(t)=\left[1-H(s)\right]\theta(t)=\frac{s}{s+\sqrt{P}KF(s)}\theta(t) \tag{5.2-57}$$

where $\phi(t)$ is the error due to the dynamically induced phase modulation $\theta(t)$. In order to determine the long-term ($t\to\infty$) tracking error in steady state assume that the phase modulation can be written as, starting at time zero, the form

$$\theta(t)=\theta_0 u(t)+\Omega_0 t u(t)+(1/2)\Lambda_0 t^2 u(t) \tag{5.2-58}$$

where, as noted before, $u(t)$ denotes the unit step function starting at time $t = 0$. Based on this model the instantaneous angular frequency at $t > 0$, is given by

$$\dot{\theta}(t)=\Omega_0+\Lambda_0 t \tag{5.2-59}$$

where the dots over θ denote the time derivatives with one dot denoting the first derivative, two dots the second derivative, and so on. Clearly the instantaneous angular frequency rate is given by

$$\ddot{\theta}(t) = \Lambda_0 \tag{5.2-60}$$

The final value theorem of La Place transform theory allows us to compute the steady state phase error. Let $\Phi(s)$ denote the La Place transform of $\phi(t)$. The *final value theorem of La Place transform theory* states that the steady state value of the phase error is given by

$$\phi_{ss} = \lim_{s\to 0}\left[s\Phi(s)\right] \tag{5.2-61}$$

Therefore, in our case, using (5.2-57) in (5.2-61) produces the result

$$\phi_{ss} = \lim_{s\to 0}\left[\left(\frac{s^2}{s+\sqrt{P}KF(s)}\right)\Phi(s)\right] \tag{5.2-62}$$

Our three-term polynomial of (5.2-58) is

$$\Phi(s) = \frac{\theta_0}{s} + \frac{\Omega_0}{s^2} + \frac{\Lambda_0}{s^3} \tag{5.2-63}$$

Now consider the first-order loop for which the loop filter is of the form $F(s) = 1$. For this loop we must let $\Lambda_0 = 0$. We have from (5.2-62) and (5.2-63), and using the fact that $F(s) = 1$ produces

$$\phi_{ss} = \lim_{s\to 0}\left[\left(\frac{s^2}{s+\sqrt{P}K}\right)\left(\frac{\theta_0}{s} + \frac{\Omega_0}{s^2}\right)\right] = \frac{\Omega_0}{\sqrt{P}K} \tag{5.2-64}$$

when Λ_0 is zero and the steady state error is unbounded when Λ_0 is not zero.

Now consider the passive second-order PLL case, for which $F(s) = (1+\tau_2 s)/(1+\tau_1 s)$. Again let $\Lambda_0 = 0$. The steady state phase error in this case is given by

$$\phi_{ss} = \lim_{s\to 0}\left[\left(\frac{s^2}{s+\left(\dfrac{1+\tau_2 s}{1+\tau_1 s}\right)\sqrt{P}K}\right)\left(\frac{\theta_0}{s} + \frac{\Omega_0}{s^2}\right)\right] = \frac{\Omega_0}{\sqrt{P}K} \tag{5.2-65}$$

which is the same result as for the first-order loop.

Now consider the active second-order PLL case. In this case the loop filter transfer function is given by $F(s) = (1+\tau_2 s)/(\tau_1 s)$. It therefore follows from (5.2-62) that the steady state phase error is given by

$$\phi_{ss} = \lim_{s\to 0}\left[\left(\frac{s^2}{s+\left(\dfrac{1+\tau_2 s}{\tau_1 s}\right)\sqrt{P}K}\right)\left(\frac{\theta_0}{s} + \frac{\Omega_0}{s^2} + \frac{\Lambda_0}{s^3}\right)\right] = \frac{\Lambda_0\tau_1}{\sqrt{P}K} = \frac{\Lambda_0}{\omega_n^2} \tag{5.2-66}$$

where ω_n is the loop natural frequency and will be discussed shortly.

Now consider the Jaffe-Rechtin optimum third-order loop which has a loop filter transfer function of the form $\sqrt{P}KF(s) = (\omega_n^3 + 2\omega_n^2 s + 2\omega_n s^2)/s^2$. Consider an extension to the definition of the dynamic phase input process of (5.2-58) to the form

$$\theta(t) = \theta_0 u(t) + \omega_0 t u(t) + (1/2)\Lambda_0 t^2 u(t) + (1/6)\Psi_0 t^3 u(t) \tag{5.2-67}$$

where a rate of change of the rate of change term has been added. The corresponding third derivative of the phase input function is given by

$$\dddot{\theta}(t) = \Psi_0 \tag{5.2-68}$$

The corresponding La Place transform of the phase process is therefore given by

$$\Phi(s) = \frac{\theta_0}{s} + \frac{\Omega_0}{s^2} + \frac{\Lambda_0}{s^3} + \frac{\Psi_0}{s^4} \tag{5.2-69}$$

Therefore the steady state phase error follows from (5.2-62) so that

$$\phi_{ss} = \lim_{s\to 0}\left[\left(\frac{s^2}{s + \left(\dfrac{\omega_n^3 + 2\omega_n^2 s + 2\omega_n s^2}{s^2}\right)}\right)\left(\frac{\theta_0}{s} + \frac{\Omega_0}{s^2} + \frac{\Lambda_0}{s^3} + \frac{\Psi_0}{s^4}\right)\right] = \frac{\Psi_0}{\omega_n^3} \tag{5.2-70}$$

Thus the ideal third-order loop can track dynamics that have a constant third derivative.

The results for the steady state phase error for the types of loops we have discussed are indicated in Table 5.2-2.

Table 5.2-2 Steady State Phase Error for Various Loop Types and Dynamics

Type \ Dynamics	Step θ_0	Step Ω_0	Step Λ_0	Step Ψ_0
First order	0	$\Omega_0/(\sqrt{P}K)$	Unbounded	Unbounded
Passive second order	0	$\Omega_0/(\sqrt{P}K)$	Unbounded	Unbounded
Active second order	0	0	Λ_0/ω_n^2	Unbounded
Jaffee-Rechtin third order	0	0	0	Ψ_0/ω_n^3

Thus it is seen that the higher order loops will tolerate higher order dynamics, as would be expected.

5.2.9 The Variance of the Linearized PLL Phase Error Due to Thermal Noise

Now consider the effects of thermal noise on the PLL tracking error. As we shall see both relative dynamics and thermal noise affect the loop tracking error. From (5.2-36) one has that the phase error due to thermal noise is given by

$$\phi(t) = -H(s)\frac{n'(t)}{\sqrt{P}} \quad \text{rad} \tag{5.2-71}$$

Since we have modeled the thermal noise as white and Gaussian, if we assume that the error is small, we can determine that the variance of the phase error is given by

$$\sigma_\phi^2 = \int_{-\infty}^{\infty} |H(i2\pi f)|^2 \frac{N_0}{2P} df \quad \text{rad}^2 \tag{5.2-72}$$

This can be simplified to

$$\sigma_\phi^2 = \frac{N_0 B_L}{P} = \frac{1}{\rho} \quad \text{rad}^2 \tag{5.2-73}$$

where ρ is the *PLL SNR* and where B_L is the one-sided loop noise bandwidth, and is defined by

$$2B_L = \int_{-\infty}^{\infty} |H(i2\pi f)|^2 df \quad \text{Hz} \tag{5.2-74}$$

Equation (5.2-73) is a very basic result that states for small errors that the PLL phase error variance in rad^2 due to thermal noise is equal to the inverse signal-to-noise ratio in the one-sided loop noise bandwidth (B_L).

Example 1 Suppose that a first-order PLL has a one-sided loop noise bandwidth of 5 Hz, and that the signal it is tracking has a received power of –180 dBW, and assume that the white Gaussian noise has spectral density of –202 dBW/Hz. Using linear theory estimate the one sigma tracking error in degrees, due to thermal noise.

First compute the loop signal-to-noise ratio (SNR) to be

$$\rho(\text{dB}) = 202 - 180 - 10 \times \log(5) = 15 \text{ dB} \tag{5.2-75}$$

Obtaining ρ in arithmetic terms yields the value $\rho = 31.6$. Since the inverse of ρ is the variance of the tracking error, compute

$$\sigma_\phi^2 = 1/31.6 \quad \text{rad}^2 \tag{5.2-76}$$

Taking the square root and converting to degrees yields $\sigma_\phi = 10.9\,°$.

In order to tabulate the noise bandwidth for various cases, define the closed-loop transfer function $H(s)$ by

$$H(s) = \frac{C_0 + C_1 s + C_2 s^2 + \ldots + C_{n-1} s^{n-1}}{d_0 + d_1 s + d_2 s^2 + \ldots + d_n s^n} \tag{5.2-77}$$

Lindsey and Simon [11] have compiled closed-loop noise bandwidths up to order 3, and we have added the fourth-order case to produce Table 5.2-3.

Now consider the first-order PLL, which has a loop filter transfer function of $F(s) = 1$, so that

$$H(s) = \frac{\sqrt{PK}}{s + \sqrt{PK}} \tag{5.2-78}$$

Table 5.2-3 Loop Noise Bandwidths for First to Fourth-Order Loops

$2B_{L1} = \dfrac{C_0^2}{2d_0 d_1}$	$2B_{L2} = \dfrac{C_0^2 d_2 + C_1^2 d_0}{2d_0 d_1 d_2}$
$2B_{L3} = \dfrac{C_2^2 d_0 d_1 + (C_1^2 - 2C_0 C_2) d_0 d_3 + C_0^2 d_2 d_3}{2d_0 d_3 (d_1 d_2 - d_0 d_3)}$	
$2B_{L4} = \dfrac{C_3^2 (d_0 d_1 d_2 - d_0^2 d_3) + (C_2^2 - C_1 C_3) d_0 d_1 d_4 + (C_1^2 - 2C_0 C_2) d_0 d_3 d_4 + C_0^2 (d_2 d_3 d_4 - d_1 d_4^2)}{2d_0 d_4 (d_1 d_2 d_3 - d_1^2 d_4 - d_0 d_3^2)}$	

From Table 5.2-3 it follows that

$$2B_{L1} = \frac{\sqrt{PK}}{2} \tag{5.2-79}$$

or

$$B_{L1} = \frac{\sqrt{PK}}{4} \tag{5.2-80}$$

From Problem 3, for the second-order passive loop filter, we have the result

$$B_{L2P} = \frac{r+1}{4\tau_2\left(1 + \dfrac{\tau_2}{r\tau_1}\right)} \cong \frac{r+1}{4\tau_2} \quad \textit{for} \quad \tau_1 \gg \tau_2 \tag{5.2-81}$$

Additionally, for the second-order passive loop filter, from (5.2-40) if we set

$$\omega_n^2 = \frac{\sqrt{PK}}{\tau_1} \quad \text{and} \quad \left(\tau_2 + \frac{1}{\sqrt{PK}}\right) = \frac{2\varsigma}{\omega_n} \tag{5.2-82}$$

then the closed-loop response $H(s)$ can be written as

$$H(s) = \frac{s\left(2\varsigma\omega_n - \omega_n^2/(\sqrt{PK})\right)}{s^2 + 2\varsigma\omega_n + \omega_n^2} \tag{5.2-83}$$

which is in standard control theory notation. Note that the natural frequency, ω_n, is given by

$$\omega_n = \sqrt{\frac{\sqrt{PK}}{\tau_1}} \tag{5.2-84}$$

and the loop damping factor is given by

$$\varsigma = \frac{\sqrt{\sqrt{PK}}}{2\sqrt{\tau_1}}\left(\tau_2 + \frac{1}{\sqrt{PK}}\right) \tag{5.2-85}$$

Now consider the active second-order PLL. Using Table 5.2-3, one obtains the loop noise bandwidth of

$$B_{L2A} = \frac{r+1}{4\tau_2} \tag{5.2-86}$$

which is almost the same as the passive filter second-order PLL closed-loop noise bandwidth.

The optimum Jaffe-Rechtin third-order loop noise bandwidth which can be obtained via Table 5.2-3, using (5.2-48), from Problem 4 yields

$$B_{L3} = \frac{5\omega_n}{6} \tag{5.2-87}$$

5.2.10 Frequency Response of the Active Filter Second-Order PLL

Now consider the frequency response of the second-order PLL with an active filter. From (5.2-44) and making the following substitutions

$$\begin{aligned} \omega_n &= \sqrt{\frac{\sqrt{PK}}{\tau_1}} \\ \varsigma &= \frac{\tau_2}{2}\sqrt{\frac{\sqrt{PK}}{\tau_1}} \end{aligned} \tag{5.2-88}$$

where ω_n is the loop natural frequency in radians/sec and ς is the loop damping factor (a number without units). It is shown in Problem 5 that the closed-loop transfer function, $H(s)$, can be written as

$$H(s) = \frac{2\omega_n\varsigma s + \omega_n^2}{s^2 + 2\omega_n\varsigma s + \omega_n^2} \tag{5.2-89}$$

and it follows that

$$1 - H(s) = \frac{s^2}{s^2 + 2\omega_n\varsigma s + \omega_n^2} \tag{5.2-90}$$

Letting $s = i2\pi f$ in (5.2-89) produces the responses for the case $\varsigma = 0.707$

$$\left|H(f)\right|^2 = \frac{1+2(f/f_n)^2}{1+(f/f_n)^4}$$
$$\left|1-H(s)\right|^2 = \frac{f^4}{f_n^4+f^4} \tag{5.2-91}$$

Both $H(f)$ and $1-H(f)$, for the second-order ideal filter PLL, with $\zeta = 0.707$, are plotted in Figure 5.2-9. It can be that $\left|H(f)\right|$ is a low-pass response, whereas $\left|1-H(f)\right|$ is a high-pass response.

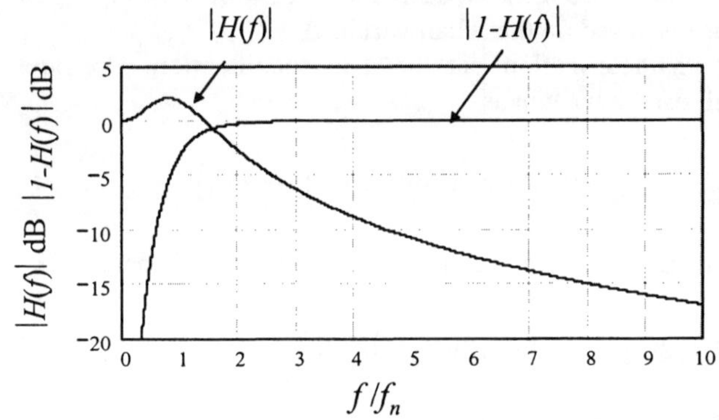

Figure 5.2-9 Magnitude response of $H(f)$ and $1-H(f)$ for ς =0.707.

Example 2 Determine the value of f for the second-order ideal loop response with a damping factor of 0.707 and a B_L= 5 Hz that is 3 dB down for (a) $\left|H(f)\right|$ and (b) $\left|1-H(f)\right|$.

From the results of Problem 5 we have that $B_L = \pi f_n\left(\zeta + 1/(4\varsigma)\right)$. For $\zeta = 0.707$ one has that $B_L = 3.33 f_n$, so that when B_L = 5 Hz, f_n =1.5 Hz. From (5.2-91) one has that (a) f/f_n = 2.0558 or f = 3 Hz for the low-pass response and (b) f/f_n = 1.0012 or f = 1.5 Hz for the high-pass response.

5.2.11 Phase Noise Effects of the Total Phase Error in the PLL

Consider the effects of phase noise on the received carrier and the PLL VCO. From (5.2-36) it is seen that the phase noise the PLL VCO contributes to the phase error through the filtered response

$$\sigma_{VCO}^2 = \int_{-\infty}^{\infty} \left|1-H(f)\right|^2 S_{VCO}(f)df \ \text{rad}^2 \tag{5.2-92}$$

and the effect of the carrier phase noise on the phase error is given by

$$\sigma_\theta^2 = \int_{-\infty}^{\infty} \left|1-H(f)\right|^2 S_\theta(f)df \ \text{rad}^2 \tag{5.2-93}$$

As can be seen from Figure 5.2-9 $|1-H(f)|$ is a high pass filter function that filters out the low-frequency components of the phase noise due to both the VCO and carrier phase noise components. The total phase noise due to thermal noise, carrier, and VCO oscillator noise is given by the sum of the variances

$$\sigma_T^2 = \sigma_\theta^2 + \sigma_{VCO}^2 + \sigma_\phi^2 \ \text{rad}^2 \tag{5.2-94}$$

where σ_ϕ^2 is given by (5.2-73). Designing the PLL loop closed-loop noise bandwidth to minimize the overall tracking error variance depends on the actual phase noise spectral density characteristics of both the carrier and the VCO.

Phase noise is normally characterized in two different ways. The first characterization is based on a spectral plot in dBc. The letters dBc denote decibels below the carrier signal. The second method of characterizing the phase noise is based on the Allan variance.

Consider the first characterization via the dBc specification. Consider the carrier signal only, without thermal noise, which can be written as

$$y(t) = \sqrt{2P}\sin\left[\omega_0 t + \theta(t) + \theta_0\right] \tag{5.2-95}$$

The following heuristic argument describes the meaning of term dBc. Expand (5.2-95) to yield

$$y(t) = \sqrt{2P}\sin\left(\omega_0 t + \theta_0\right)\cos(\theta(t)) + \sqrt{2P}\cos\left(\omega_0 t + \theta_0\right)\sin(\theta(t)) \tag{5.2-96}$$

If the variance of $\theta(t)$ is small then one can write the approximation

$$y(t) \cong \sqrt{2P}\sin\left(\omega_0 t + \theta_0\right) + \sqrt{2P}\cos\left(\omega_0 t + \theta_0\right)\theta(t) \tag{5.2-97}$$

If the phase process is stationary then one can write

$$R_y(\tau) = P\cos(\omega_0\tau) + R_\theta(\tau)\cos(\omega_0\tau) \tag{5.2-98}$$

It follows that the power spectral density is given by

$$S_y(f) = \frac{P}{2}\left[\delta(f+f_0) + \delta(f-f_0)\right] + \frac{P}{2}\left[S_\theta(f+f_0) + S_\theta(f-f_0)\right] \tag{5.2-99}$$

Thus if one specifies the one-sided phase noise power spectral density relative to the one-sided carrier power, it is clear that the phase noise spectral density is going to be "down" from the carrier power by $10\log(S_\theta(f))$ in dB or dB below the carrier (i.e., dBc).

Example 3 Determine the phase error variance seen by an ideal second-order phase locked loop with natural frequency f_n and damping factor $\varsigma = 0.707$ to the following single-sided phase noise spectral density model [12]

$$S_\theta(f)=\begin{cases}\dfrac{k_4}{f^4} & 0\le f\le f_1\\ \dfrac{k_3}{f^3} & f_1\le f\le f_2\\ \dfrac{k_2}{f^2} & f_2\le f\le f_3 \quad \text{rad}^2/\text{Hz}\\ \dfrac{k_1}{f} & f_3\le f\le f_4\\ k_0 & f_4\le f\le f_5\end{cases} \qquad (5.2\text{-}100)$$

where the phase noise spectral density is continuous at the ends of the piecewise linear segments, so that it follows that

$$\frac{k_4}{f_1^4}=\frac{k_3}{f_1^3},\ \frac{k_3}{f_2^3}=\frac{k_2}{f_2^2},\ \frac{k_2}{f_3^2}=\frac{k_1}{f_3^1},\ \frac{k_1}{f_4^1}=k_0 \qquad (5.2\text{-}101)$$

The figure for the example is shown next

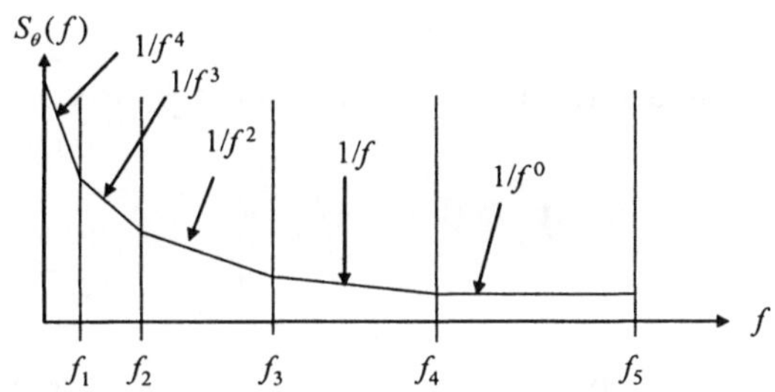

Figure for Example 3 showing the one-sided phase noise spectral density as a function of f.

To obtain the solution we must integrate the phase noise one-sided spectral density to obtain

$$\sigma_\theta^2=\int_0^\infty |1-H(f)|^2 S_\theta(f)df=2\int_0^\infty |1-H(f)|^2 L(f)df \quad \text{rad}^2 \qquad (5.2\text{-}102)$$

where $L(f)$ is the single sideband phase noise power spectral density. Commonly, the single sideband phase noise power spectral density is specified; however, $S_\theta(f)$ can easily be obtained from it via $S_\theta(f)=2L(f)$. From (5.2-91) we have that

$$|1-H(s)|^2=\frac{f^4}{f_n^4+f^4} \qquad (5.2\text{-}103)$$

Hence the phase error variance due to carrier phase noise is given by

$$\sigma_{\theta}^2 = \int_0^{f_1} \frac{f^4}{f_n^4 + f^4} \frac{k_4}{f^4} df + \int_{f_1}^{f_2} \frac{f^4}{f_n^4 + f^4} \frac{k_3}{f^3} df + \int_{f_2}^{f_3} \frac{f^4}{f_n^4 + f^4} \frac{k_2}{f^2} df + \int_{f_3}^{f_4} \frac{f^4}{f_n^4 + f^4} \frac{k_1}{f} df + \int_{f_4}^{f_5} \frac{f^4}{f_n^4 + f^4} k_0 df \quad (5.2\text{-}104)$$

This expression can be evaluated numerically by commercial software computational packages or with the use of integration tables.

The *two-sample Allan variance* is a measure of short-term oscillator stability. Let the signal be specified by

$$V(t) = \sqrt{2} \sin\left(2\pi f_0 t + \theta(t)\right) \quad (5.2\text{-}105)$$

where f_0 is the center frequency and $\theta(t)$ is the phase noise on the carrier. The two sample Allan variance is defined by

$$\sigma_y^2(\tau) = \left\langle \frac{\left(\overline{y}_{k+1} - \overline{y}_k\right)^2}{2} \right\rangle \quad (5.2\text{-}106)$$

in which $\langle \ \rangle$ denotes the time average, and where $y(t)$ is given by

$$y(t) = \frac{1}{2\pi f_0} \frac{d\theta(t)}{dt} \quad (5.2\text{-}107)$$

and is the fractional instantaneous frequency deviation. In addition $\overline{y}_k$ is the τ second average frequency and is defined by

$$\overline{y}_k = \frac{1}{\tau} \int_{t_k}^{t_k+\tau} y(t) dt \quad (5.2\text{-}108)$$

This model assumes that all the y_k samples are contiguous. A more general N sample variance with $T-\tau$ seconds between samples is given by

$$\sigma_y^2(N, T, \tau) = \left\langle \frac{1}{N-1} \sum_{n=1}^{N} \left(\overline{y}_n - \frac{1}{N} \sum_{k=1}^{N} \overline{y}_k \right)^2 \right\rangle \quad (5.2\text{-}109)$$

Cutler [13] has shown that the two sample Allan variance with adjacent samples can be evaluated in terms of the two-sided phase noise spectral density as

$$\sigma_y^2(\tau) = \frac{2}{(\pi f_0 \tau)^2} \int_0^{f_{max}} S_{\theta}(f) \sin(\pi f \tau)^4 df \quad (5.2\text{-}110)$$

where f_0 is the center frequency expressed in Hz, τ is the averaging time, f_{max} is the upper frequency limit determined by receiver filters, and $S_{\theta}(f)$ is the two-sided noise spectral density. Note that the Allan variance is a normalized variance, with the normalization being the actual carrier frequency.

5.2.12 Nonlinear PLL Results

Now consider the performance of PLLs in the region where the nonlinear behavior is important.

5.2.12.1 PLL Mean Time to Loss of Lock

One important parameter is the mean time to lose lock. By losing lock we mean reaching either +2π or –2π radians phase error. Viterbi [7] has shown that the mean time to reach $\pm 2\pi$ for the first-order PLL is given by

$$2B_L\overline{T} = \frac{\pi^2}{\sigma^2}\int_0^{2\pi}\int_{\phi}^{2\pi}\exp\left(\frac{-G(\phi)}{\sigma^2}\right)\exp\left(\frac{G(x)}{\sigma^2}\right)d\phi dx \quad (5.2\text{-}111)$$

where $G(x) = \cos(x)$ is the integral of the phase error S curve, normalized to unit slope at $x = 0$. Note that this expression is unitless. This double integral expression can be evaluated to produce the result

$$2B_L\overline{T} = \frac{\pi^2}{\sigma^2}I_0^2\left(\frac{1}{\sigma^2}\right) \quad (5.2\text{-}112)$$

where I_0 is the modified Bessel function of zeroth order. This result only strictly applies to the first-order loop; however, it would be an approximation to both types of second-order loops also, but with a degradation of 1 dB to the loop SNR. If one uses the asymptotic approximation for $I_0(x)$ of the form $I_0(\rho) \cong e^{\rho}/\sqrt{2\pi\rho}$, one can approximate (5.2-112) by $(\rho = 1/\sigma^2)$

$$2B_L\overline{T} \cong \frac{\pi}{2}\exp\left(\frac{2}{\sigma^2}\right) \quad (5.2\text{-}113)$$

Viterbi [7] has shown that the cycle slipping rate is the inverse of the mean time to slip, so that the mean cycle slipping rate is given by

$$\overline{R}_{slip} = 1/\overline{T} \ \ \text{s}^{-1} \quad (5.2\text{-}114)$$

where $\overline{T}$ is given by (5.2-112). In addition Smith [14] has shown that the slip process is approximately Poisson distributed, so that the probability of not slipping in t seconds is given by

$$P_{not\ slipping} = \exp\left(-R_{slip}t\right) \quad (5.2\text{-}115)$$

And the probability of K slips in t seconds is given by

$$P_{slip}(K,t) = \frac{\left(\overline{R}_{slip}t\right)^K \exp\left(-\overline{R}_{slip}t\right)}{K!} \quad (5.2\text{-}116)$$

Example 4 Determine the mean time to lose lock for a first-order PLL in which $\varsigma = 0.707$, B_L=10 Hz, and C/N_0 = 25 dB/Hz. Utilize the asymptotic form for simplicity. To use (5.2-113) it is necessary to determine σ^2 first. From (5.2-73) we compute σ^2 = 10/($10^{2.5}$) = 0.0317 rad$^{2.}$. From (5.2-113) it follows that $\overline{T} = 1.98*10^{26}$ s, or 5.5*10^{22} h!

5.2.12.2 PLL Acquisition by Sweeping the VCO

Loop acquisition can be obtained by sweeping at a constant rate the VCO frequency from one end of the expected frequency range to the other. Although there is not much analysis in the way of acquisition performance, there is an approximate model that is based on the Frazier-Page simulation study [15]. Their result states that the sweep rate in Hz/s, for a 90% chance of being successful in acquisition, for a second-order loop, is given by

$$R_{90} = \frac{\left(1-\sqrt{2}\sigma_{\phi}\right)\omega_n^2}{2\pi\left(1+d\right)} \qquad \text{Hz}/s \tag{5.2-117}$$

where d is given by

$$\begin{aligned} d &= \exp\left(-\pi\frac{\zeta}{\sqrt{1-\zeta^2}}\right) \qquad && \zeta < 1 \\ d &= 1 && \zeta \geq 1 \end{aligned} \tag{5.2-118}$$

and where σ_{ϕ} is the one sigma value of phase error from the linear model. Faster rates will produce lower probabilities of success and slower sweep rates a higher probability of success.

5.2.12.3 Steady State PLL Phase Error Distribution

Based on the Fokker-Planck method of obtaining the probability density function from the stochastic differential equation of PLL operation in steady state, Tikhonov [16] and Viterbi [7] have derived the phase error probability density function for the first-order PLL, with the phase error reduced modulo 2π. The result, assuming no dynamically induced phase error, is given by

$$p(\phi) = \frac{\exp\left(\rho\cos(\phi)\right)}{2\pi I_0(\rho)} \qquad \left|\phi \leq \pi\right| \tag{5.2-119}$$

It can be shown that the phase error density can be approximated as a Gaussian random variable, by

$$p(\phi) \cong \frac{\exp\left(-\dfrac{\phi^2}{2\sigma_{\phi}^2}\right)}{\sqrt{2\pi\sigma_{\phi}^2}} \tag{5.2-120}$$

when σ_{ϕ}^2 is small or when $\rho = 1/\sigma_{\phi}^2$ is large.

5.2.12.4 Phase Error Performance with a BPF Limiter in Front of the PLL

The question arises of what is the effect when a bandpass filter and limiter is placed in front of the phase detector of a PLL. Figure 5.2-10 illustrates a PLL with a bandpass filter prior to the limiter, which is placed in front of the phase detector (mixer).

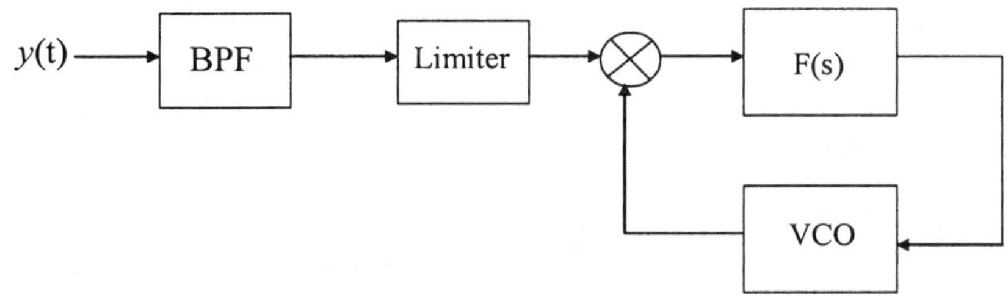

Figure 5.2-10 PLL showing the insertion of a limiter before the phase detector.

For one thing, the amplitude variation is limited to a small range (about 3 dB) if the input signal to noise ratio is at least unity. This is useful since it minimizes the need for a coherent automatic gain control (AGC) circuit, which is used to maintain a constant output signal level, and therefore the damping factor and the loop noise bandwidth will be constant. Forney, in an unpublished report, has shown that the phase error variance is unaffected by the limiter at high input signal-to-noise ratios in the input bandpass filter (BPF) bandwidth. In other words no degradation in tracking error performance occurs when a high input SNR is chosen in the design, when a BPF-limiter system is inserted in front of the PLL. In addition, he showed that there is no enhancement in tracking performance when a limiter is inserted in front of a PLL.

5.3 FREQUENCY SYNTHESIZERS

A frequency synthesizer is a device capable of generating many precise frequencies from a single or few very stable oscillator source(s) [17–19]. System applications include frequency hopping and dehopping, frequency division multiple access in satellite communications, test equipment, and HF radio.

There are three methods of frequency synthesis: (1) direct, (2) indirect (PLL based), and (3) digital (table lookup). This section follows Egan [6] in part.

5.3.1 Digital Frequency Synthesis

The digital frequency synthesizer is based on utilizing a lookup table to construct the signal segment by segment [6, 17]. A simplified block diagram model is illustrated in Figure 5.3-1. The description follows Egan [6]. Every clock cycle (f_{clock}) $\Delta\theta$ is shifted into the accumulator, which adds $\Delta\theta$ radians to the previous accumulation to obtain the total phase θ. The output of the accumulator is a step-wise approximation to a linear phase ramp. The output phase consists of changes by $\Delta\theta$ radians every ($T_{clock}=1/f_{clock}$) seconds, as seen in Figure 5.3-2. The value of θ serves as the address for the cos(θ) memory, which is output in discrete approximation form to the digital to analog (D/A) converter. The signal, which is a discrete approximation to a cosine wave, is filtered by the low-pass filter, rendering it an analog cosine wave.

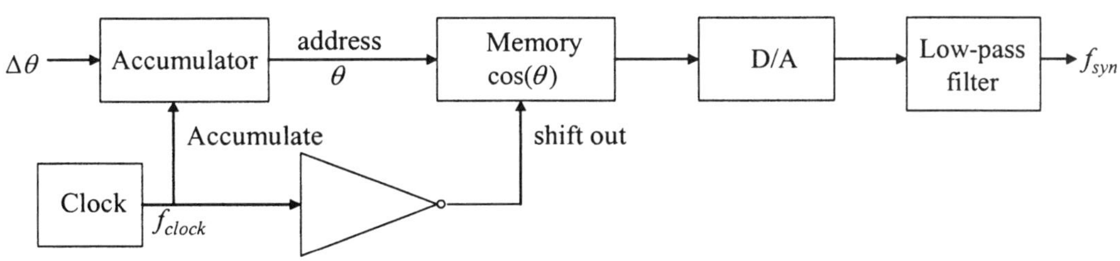

Figure 5.3-1 Block diagram model of a digital synthesizer.

The accumulator contains N bits of resolution, so that there are $n_C = 2^N$ levels of the phase output. The capacity of the accumulator is 2π radians. Let n_i be the input number that corresponds to the desired input phase change per clock period. Since the accumulator rolls over (counts modulo 2π) it is clear that the output frequency is just the ratio

$$f_{syn} = \frac{n_i}{n_c} f_{clock} = \frac{n_i}{2^N} f_{clock} \tag{5.3-1}$$

Thus, controlling the value of the input number (n_i) controls the output frequency (f_{syn}). The frequency resolution is given by the minimum size of n_i which is one; thus

$$f_{res} = \frac{1}{n_c} f_{clock} = \frac{1}{2^N} f_{clock} \tag{5.3-2}$$

The highest practical synthesized frequency, to allow reasonable low pass filtering is about

$$|f_{\max}| = \frac{f_{clock}}{4} \tag{5.3-3}$$

even though the theoretical limit is twice that of (5.3-3), since $N = 1$ corresponds to the case of two phase values per clock period. Figure 5.3-2 illustrates an example of the type of output obtained from a digital frequency synthesizer. The step-wise function shown at the bottom of the figure illustrates the approximate linear increase in phase with time, until the value of 2π is reached, and then it is reset to zero. Each step is one step of the clock. After passing the signal through the memory, a step-wise approximation to the cosine of the phase function is constructed. Finally the D/A converter signal is fed into the low-pass filter and the approximate cosine function is generated, as shown in the top of the figure.

Example 5 Assume that a digital synthesizer will be used to generate a 500-Hz frequency resolution and that 10 bits are used in the accumulator. Determine the clock frequency needed to produce a 500-Hz resolution. Also, based on this clock frequency, what input value must be added to the input of the accumulator to produce an output frequency of 50 kHz?

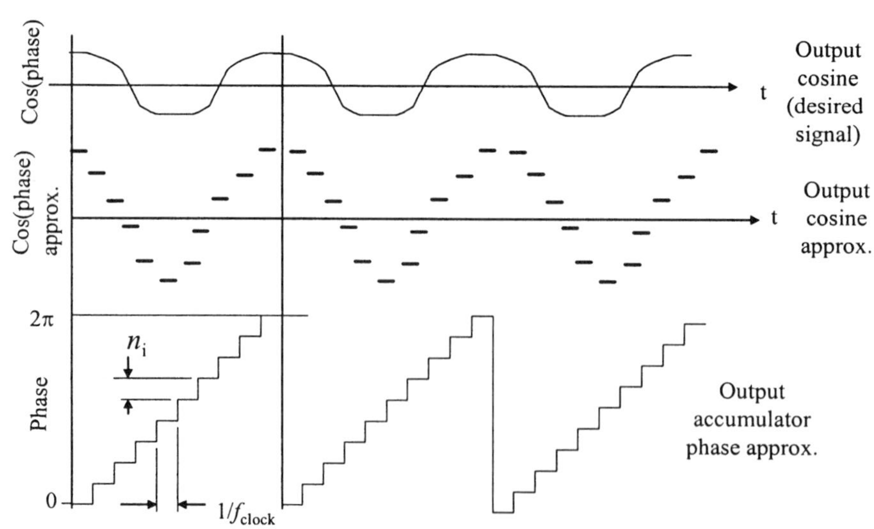

Figure 5.3-2 Evolution of the cosine function from the accumulated phase of a digital synthesizer.

First note that (5.3-2) relates the frequency resolution so that

$$f_{clock} = n_c f_{res} = 2^{10}(500) = 512{,}000\ Hz$$

Now the input level needed follows from (5.3-1) so that the input needed is

$$n_i = \frac{f_{syn}}{f_{clock} / n_c} = \frac{50{,}000}{512{,}000 / 2^{10}} = 100$$

5.3.2 Direct Frequency Synthesis

A direct frequency synthesizer (DFS) is a synthesizer that utilizes mixing, multiplication, and division to generate the desired output frequency from a single or limited number of frequency sources. Figure 5.3-3 illustrates how a frequency can be synthesized. Switches denoted "SW" may connect one of the ten possible frequencies for all "a" through "g" connections (seven in total). One of the frequencies is switched on in the first mixer (multiplier) and is added to the 3-MHz signal. The added frequency "a" has a value of 0–9 MHz. The resulting frequency, which has value 3a MHz, is divided by 10 to yield a frequency of 3.a MHz. Next the second switch is set to produce 3b.a MHz. Note that we are neglecting the sum frequencies in this discussion and retaining only the difference frequencies. This process continues until the next to last (ninth) mixer is used to create the frequency 3g.fedcba MHz. The last mixer converts the final output frequency to the value of g.fedcba MHz. Greater frequency precision can be obtained by adding more switches and mixers. Note that in this process "a" through "g" can take on the values of "0" to "9."

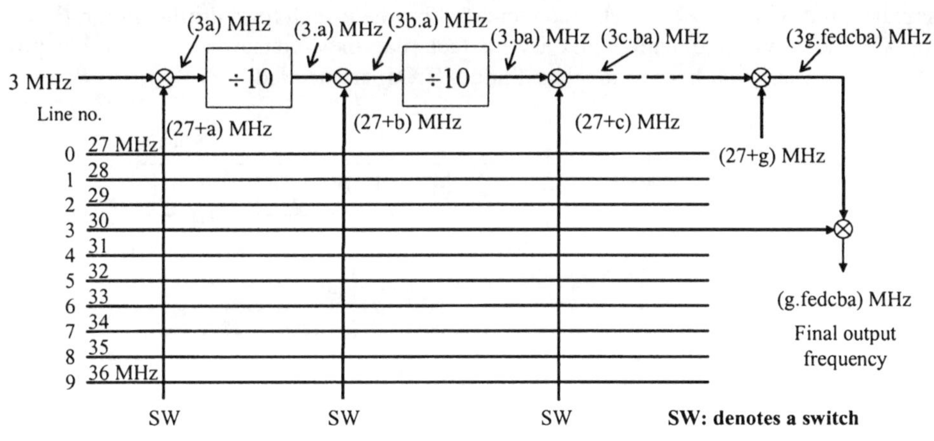

Figure 5.3-3 Simplified direct synthesis synthesizer concept.

Example 6 Determine the necessary switch closures and mathematical process required to produce the frequency of 6.353 891 MHz. First we notice that the following is true: g = 6, f = 3, e = 5, d = 3, c = 8, b = 9, and a = 1. Figure 5.3-4 illustrates the connections that are needed to make the desired frequency.

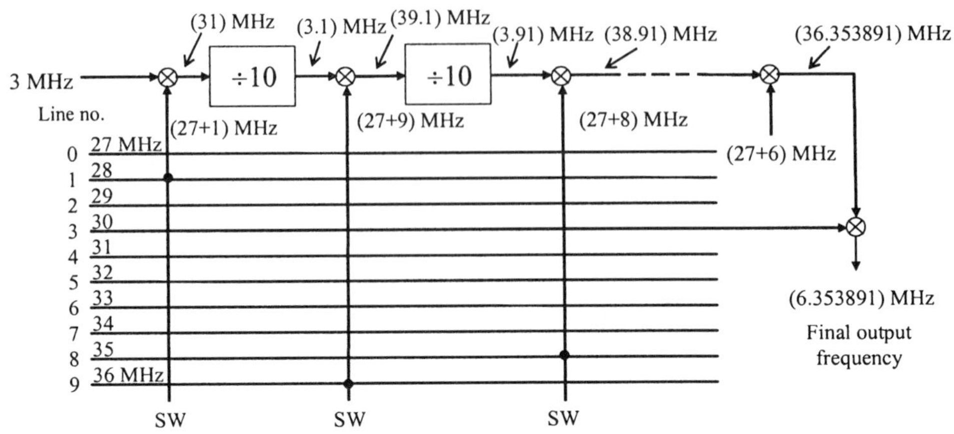

Figure 5.3-4 Example of generating via direct synthesis the frequency 6.63489 MHz.

The systems shown in Figures 5.3-4 and 5.3-5 are not very practical due to the fact that it is difficult to remove the mixer input in the 27–36-MHz range from the synthesized output frequency, since the input frequency overlaps the output frequency. Figure 5.3-5 illustrates a more practical implementation. Note that filters are normally utilized in the design, but are not illustrated in the figure.

All of these systems are based on a decimal scheme, in which one of ten frequencies are chosen and the divisions are by 10, however other schemes such as binary [19] and BCD [20] can be used. One advantage of a direct synthesizer is the fact that it can switch frequencies at the speed of the switches and any associated filters used in the design. Switching in less than a few microseconds is usually feasible. Another advantage of DFS is the fact that the output spectral density can be very clean, basically a replica of the reference oscillator, but with the FM sidebands increased by the effective multiplication ratio from the input to the output. One disadvantage with DFSs is the fact that they consume considerable power and are relatively bulky, since a large number of local oscillators (LOs) are needed.

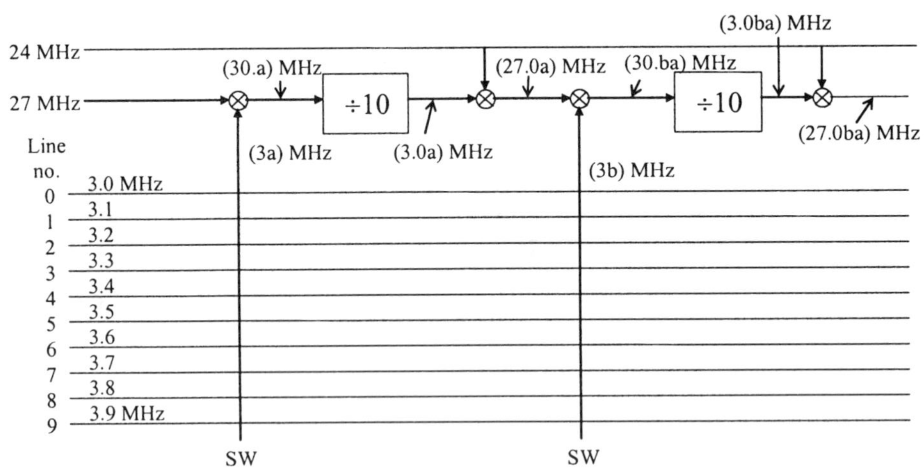

Figure 5.3-5 First portion of a direct synthesis synthesizer that avoids the mixer feed-through problem.

It is to be noted that ideal multipliers do not exist. Thus when two signals are "multiplied" they produce the desired multiplication product plus a multitude of other frequencies. If the reference frequencies are labeled "*r*" and the mixer frequencies are labeled "*m*" then it is true that

$$f_{out} = nf_r + mf_m \tag{5.3-4}$$

where f_{out} is the output frequency and n and m are integers. Commonly only the frequency given by $f_{out} = f_r + f_m$ is desired.

5.3.3 Indirect Frequency Synthesis

Indirect frequency synthesis is also known as phase locked loop synthesis, because it is based on using a PLL to generate the desired frequencies. Figure 5.3-6 illustrates a PLL synthesizer (PLLS) that uses a divide ratio to multiply up the reference frequency. It is clear that for the loop to be locked the output frequency must be related to the input frequency by

$$f_{out} / N_1 = f_{ref} / N_2 \tag{5.3-5}$$

where N_1 is the divide ratio in the feedback loop and N_2 is the divide ratio in the input; both numbers are integers. The VCO is represented by the VCO parameter K_v and the integration by 1/s. The loop divider has a ratio of N_1. The loop filter has the function of suppressing the undesirable FM on the VCO. From (5.3-5) it follows that

$$f_{out} = \frac{N_1}{N_2} f_{ref} \tag{5.3-6}$$

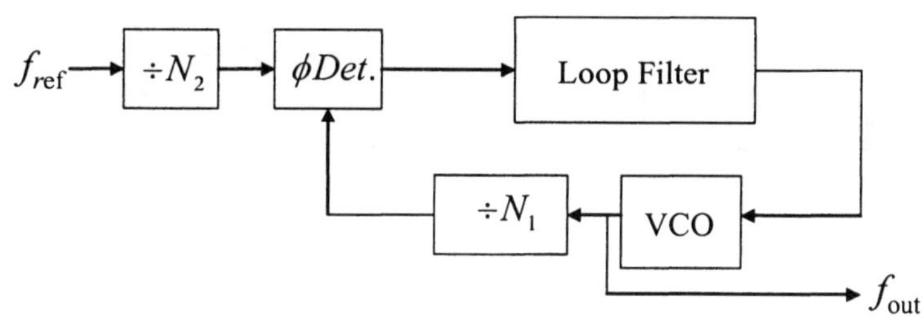

Figure 5.3-6 Synthesizer model of a PLLS.

The minimum frequency increment is given by

$$f_{out} = \frac{1}{N_2} f_{ref} \tag{5.3-7}$$

A practical synthesizer may have a number of additional features, such as a voltage source based on a D/A converter to drive the VCO to the nominal correct frequency, so as to increase the speed of the PLL synthesizer. Another feature that is commonly used in the implementation is a phase/frequency detector to speed up the switch from the last frequency to the next frequency [21]. The PLLS is slow compared to both the DFS and the direct frequency synthesizers.

Clearly combinations of these three techniques can be utilized in a synthesizer design.

5.3.4 Indirect Frequency Synthesis Transfer Functions

Now consider the various transfer functions that apply for different sources of noise and phase noise. Figure 5.3-7 illustrates a second-order indirect PLL frequency synthesizer.

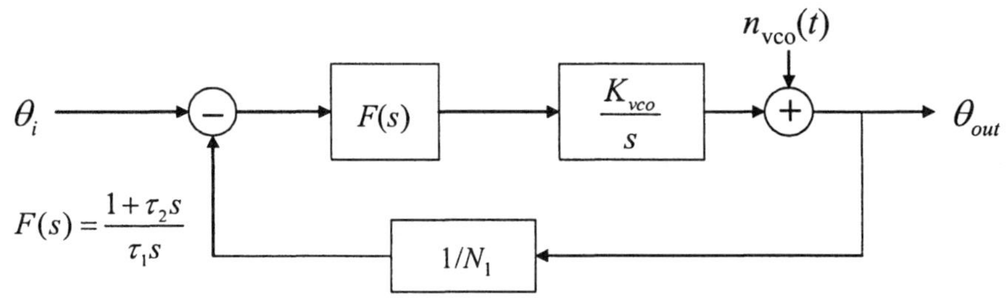

Figure 5.3-7 Baseband phase model of a second-order PLL synthesizer.

Let us determine the transfer function of Figure 5.3-7 from the VCO noise to the output, which is θ_{out} in the figure. Redraw the figure in terms of a forward and a feedback term as shown in Figure 5.3-8.

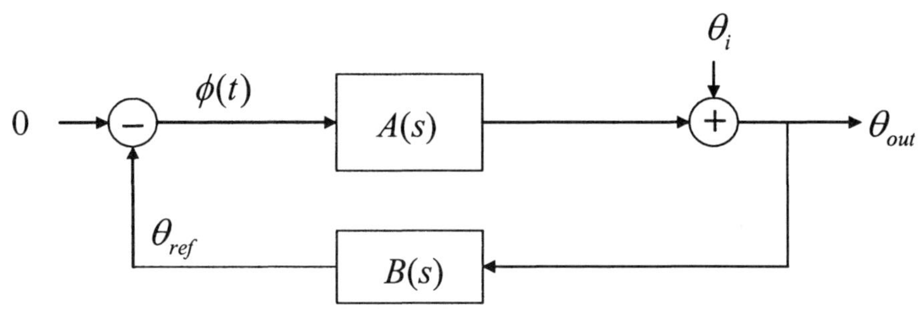

Figure 5.3-8 Redrawn baseband model of a second-order PLL synthesizer in terms of phase.

In Figure 5.3-8 the forward path, which includes the loop filter and the VCO, is denoted by $A(s)$, and the division by N_1 is denoted by $B(s)$. The actual input is set to "0" whereas the VCO noise is labeled as θ_i. Now let us determine the transfer function from the VCO noise (θ_i) to the output θ_{out}. Note all the phase and phase error terms are a function of time. The following four equations can be obtained from Figure 5.3-8.

$$\begin{aligned}
&(1)\quad \phi(t) = -\theta_{ref}(t) \\
&(2)\quad \theta_{vco}(t) = A(s)\phi(t) \\
&(3)\quad \theta_0(t) = \theta_{vco}(t) + \theta_i(t) \\
&(4)\quad \theta_{ref}(t) = B(s)\theta_0(t)
\end{aligned} \tag{5.3-8}$$

From (1) and (2) one obtains

$$(5)\quad \theta_{vco}(t) = -A(s)\theta_{ref}(t) \tag{5.3-9}$$

From (5) and (3) one obtains

$$(6)\quad \theta_0(t) = -A(s)\theta_{ref}(t) + \theta_i(t) \tag{5.3-10}$$

Combining (4) and (6) yields

$$(7)\ \ \theta_0(t) = -A(s)B(s)\theta_i(t) + \theta_i(t) \qquad (5.3\text{-}11)$$

Let $\varphi_0(s)$ denote the La Place transform $\theta_0(t)$ and $\varphi_i(s)$ denote the La Place transform of $\theta_i(t)$. From (7) one can solve for the ratio of $\varphi_0(s)$ to $\varphi_i(s)$ to yield

$$\frac{\varphi_0(s)}{\varphi_i(s)} = \frac{1}{1 + A(s)B(s)} \qquad (5.3\text{-}12)$$

which can be rewritten as

$$\frac{\varphi_0(s)}{\varphi_i(s)} = \frac{s^2}{s^2 + 2\omega_n \varsigma s + \omega_n^2} \qquad (5.3\text{-}13)$$

Note that if a linearizer is used in front of the VCO, then the linearizer noise, since the VCO is modeled as K_{vco}/s, has transfer function given by

$$\frac{\varphi_0(s)}{\varphi_i(s)} = \frac{K_{vco} s}{s^2 + 2\omega_n \varsigma s + \omega_n^2} \qquad (5.3\text{-}14)$$

because passing through to the output of the VCO one multiplies the output by K_{vco}/s. In fact it is possible to extend this analysis to cover other sources of noise in a PLL, and this is listed in Table 5.3-1 [22, 23]. Note that a perfect second-order PLL has been assumed. The parameter K_m is the multiplier gain in the phase detector.

Table 5.3-1 Summary of the Transfer Function for Various Noise Sources

Noise Source	Resultant Transfer Function	Filter Response
VCO	$\frac{s^2}{s^2 + 2\omega_n \varsigma s + \omega_n^2}$	High-pass
Linearizer Coarse tuning DAC	$\frac{K_{vco} s}{s^2 + 2\omega_n \varsigma s + \omega_n^2}$	Bandpass
Phase detector	$\frac{N_1}{K_m \sqrt{P}} \frac{2\omega_n \varsigma s + \omega_n^2}{s^2 + 2\omega_n \varsigma s + \omega_n^2}$	Low-pass
Reference OSC	$N_1 \frac{2\omega_n \varsigma s + \omega_n^2}{s^2 + 2\omega_n \varsigma s + \omega_n^2}$	Low-pass

Thus it is clear that the transfer functions varies from bandpass to low-pass to high-pass filter response depending on where the noise source is located in the loop.

5.4 TRACKING OF BPSK SIGNALS

Now consider the types of tracking loops needed to track suppressed carrier signals. In particular we consider BPSK modulated signals, and then later we consider multiphase signals.

5.4.1 Tracking a BPSK Signal with a Squaring Loop

The simplest way to track a BPSK, suppressed carrier signal is to square the BPSK signal and track the resulting carrier, which occurs at twice the input frequency. This is the basis of the squaring loop. Numerous authors have analyzed the squaring loop [2, 21–28]. Now consider a model of a squaring loop illustrated in Figure 5.4-1.

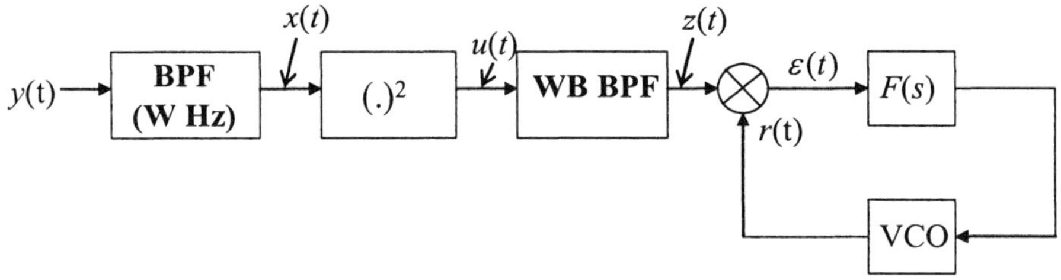

Figure 5.4-1 Model of a squaring loop in the short loop form.

The basic elements of a squaring loop include the bandpass filter (BPF), followed by a squaring device, which mathematically squares the input signal plus noise, followed by a wideband zonal filter to remove the dc component out of the squaring device. The squaring system is then fed into a PLL, which is tuned to twice the input frequency.

First we develop the stochastic differential equation (SDE) governing the operation of the phase error process of the squaring process. Let the input signal be described by

$$y(t) = \sqrt{2P}d(t)\sin(\omega_0 t + \theta_0) + n(t) \tag{5.4-1}$$

where P is the signal power in watts, $n(t)$ is modeled as a white Gaussian noise (WGN) process, ω_0 is the angular carrier frequency of the input BPSK signal, θ_0 is the carrier phase at the input of the receiver, and $d(t)$ is the baseband data sequence that is modeled by

$$d(t) = \sum_{k=-\infty}^{\infty} d_k p(t - kT_b) \tag{5.4-2}$$

In (5.4-2) it is assumed that d_k is a random bit sequence that takes on the values of ± 1 with probability 1/2 and the sequence is assumed to be statistically independent from bit to bit. The function $p(t)$ is a unit amplitude pulse that exists for $t \in (0, T_b)$ and is zero otherwise. The time period T_b is the bit duration, in seconds. For coded systems, the symbol period would be substituted for the bit period. The WGN process is modeled as

$$n(t) = \sqrt{2}n_c(t)\cos(\omega_0 t + \theta_0) + \sqrt{2}n_s(t)\sin(\omega_0 t + \theta_0) \tag{5.4-3}$$

where $n_c(t)$ and $n_s(t)$ and statistically independent WGN processes, spectrally located at baseband, and each has a two-sided noise spectral density of $N_0/2$.

After filtering by the bandpass filter (BPF) the received signal plus noise is given by

$$y(t)=\sqrt{2P}\tilde{d}(t)\sin(\omega_0 t+\theta_0)+n(t)+\sqrt{2}\tilde{n}_c(t)\cos(\omega_0 t+\theta_0)+\sqrt{2}\tilde{n}_s(t)\sin(\omega_0 t+\theta_0) \tag{5.4-4}$$

where $\tilde{d}(t)$ is the filtered version of $d(t)$, and the two baseband noise terms are the filtered versions of the respective unfiltered noise terms. That is,

$$\tilde{d}(t)=\int_{-\infty}^{\infty} h_{LP}(\xi)d(t-\xi)d\xi \tag{5.4-5}$$

and for each noise component (c or s)

$$\tilde{n}_{c,s}(t)=\int_{-\infty}^{\infty} h_{LP}(\xi)n_{c,s}(t-\xi)d\xi \tag{5.4-6}$$

where $h_{LP}(u)$ is the equivalent low-pass impulse response function [29]. After squaring the output, the following results

$$\begin{aligned} x(t)=&-\left[P\tilde{d}(t)\right]^2\cos(2\omega_0 t+2\theta_0)+2\sqrt{P}\tilde{d}(t)\tilde{n}_c(t)\sin(2\omega_0 t+2\theta_0)\\ &-2\sqrt{P}\tilde{d}(t)\tilde{n}_s(t)\cos(2\omega_0 t+2\theta_0)+\tilde{n}_c^2(t)\cos(2\omega_0 t+2\theta_0)\\ &-\tilde{n}_s^2(t)\cos(2\omega_0 t+2\theta_0)+2\tilde{n}_s(t)\tilde{n}_c(t)\sin(2\omega_0 t+2\theta_0)\\ &+\textit{baseband signals}\end{aligned} \tag{5.4-7}$$

It is seen that if the squared term in front of the $\cos(2\omega_0 t+2\theta_0)$ term has a mean value (which, by definition, it does), a line component exists at $2\omega_0$. The wideband (zonal) bandpass filter removes the baseband terms leaving

$$\begin{aligned} z(t)=&-\left[P\tilde{d}(t)\right]^2\cos(2\omega_0 t+2\theta_0)+2\sqrt{P}\tilde{d}(t)\tilde{n}_c(t)\sin(2\omega_0 t+2\theta_0)\\ &-2\sqrt{P}\tilde{d}(t)\tilde{n}_s(t)\cos(2\omega_0 t+2\theta_0)+\tilde{n}_c^2(t)\cos(2\omega_0 t+2\theta_0)\\ &-\tilde{n}_s^2(t)\cos(2\omega_0 t+2\theta_0)+2\tilde{n}_s(t)\tilde{n}_c(t)\sin(2\omega_0 t+2\theta_0)\end{aligned} \tag{5.4-8}$$

Now denote $\omega_0 t+\theta_0$ by $\Phi(t)$ and $\omega_1 t+\hat{\theta}$ by $\hat{\Phi}(t)$. The reference signal can be modeled by

$$r(t)=2K_1\sin\left(2\hat{\Phi}(t)\right) \tag{5.4-9}$$

and $\sqrt{2}K_1$ is the rms voltage of the reference signal. The loop error control signal out of the phase detector (multiplier) is given by

$$\varepsilon(t)=K_m z(t)r(t) \tag{5.4-10}$$

Therefore the error signal becomes

$$\varepsilon(t)=K_mK_1\left\{P\left(\tilde{d}(t)\right)^2+2\sqrt{P}\tilde{d}(t)\tilde{n}_s(t)-\tilde{n}_c^2(t)+\tilde{n}_s^2(t)\right\}\sin(2\phi(t)) \\ +\left\{2\sqrt{P}\tilde{d}(t)\tilde{n}_c(t)+2\tilde{n}_c(t)\tilde{n}_s(t)\right\}\cos(2\phi(t)) \tag{5.4-11}$$

where $\phi(t)=\Phi(t)-\Phi(t)$ is the instantaneous phase error at time t. Note that by viewing the signal component of the error signal it is seen that any dynamics present on the input phase process (due to dynamics) is effectively doubled in the loop ($\sin(2\phi(t))$, and this must be accounted for in determining the loop noise bandwidth. The PLL portion of the squaring loop operates at $2\omega_0$, so that the instantaneous frequency of the VCO is given by

$$2\frac{d\tilde{\Phi}(t)}{dt}-2\frac{d\Phi(t)}{dt}=K_{vco}F(s)\varepsilon(t) \tag{5.4-12}$$

When the local VCO and the received frequencies are the same, $\omega_0=\omega_1$, then (5.4-12) becomes

$$2\dot{\phi}(t)=-K_{vco}F(s)\varepsilon(t) \tag{5.4-13}$$

Assuming that $\omega_0=\omega_1$, then using (5.4-11) in (5.4-13) produces

$$2\dot{\phi}(t)=-KF(s)\left\{P\left(\tilde{d}(t)\right)^2\sin(2\phi(t))+N(t,2\phi(t))\right\} \tag{5.4-14}$$

where $K=K_1K_mK_{vco}$ and where the noise process is defined by

$$N(t,\phi(t))=\left\{2\sqrt{P}\tilde{d}(t)\tilde{n}_s(t)-\tilde{n}_c^2(t)+\tilde{n}_s^2(t)\right\}\sin(2\phi(t)) \\ +\left\{2\sqrt{P}\tilde{d}(t)\tilde{n}_c(t)+2\tilde{n}_c(t)\tilde{n}_s(t)\right\}\cos(2\phi(t)) \tag{5.4-15}$$

If we let $\Phi_e(t)=2\phi(t)$ be *effective phase error* in the loop, we can write

$$\dot{\Phi}_e(t)+KPF(s)\left(\tilde{d}(t)\right)^2\sin(\Phi_e(t))=-KF(s)N(t,\Phi_e(t)) \tag{5.4-16}$$

To simplify the analysis it is expeditious to replace $\left(\tilde{d}(t)\right)^2$ with its average (assumes that all the components besides the dc component are filtered out by the loop)

$$E\left\{(\tilde{d}(t))^2\right\}=\int_{-\infty}^{\infty}S_d(f)\left|H_{LP}(f)\right|^2df=\alpha \tag{5.4-17}$$

That is simply the amount of power that passes through the BPF equivalent low-pass filter response $H_{LP}(f)$. This approximation neglects self-noise that could be an issue at very high signal-to-noise ratios (SNR), but can be neglected at thermal noise dominated levels. Thus (5.4-16) can be approximated at low SNRs by

$$\dot{\Phi}_e(t)+\alpha KPF(s)^2\sin(\Phi_e(t))=-KF(s)N(t,\Phi_e(t)) \tag{5.4-18}$$

In order to analyze the linearized model, we must linearize (5.4-18) to obtain

$$\Phi_e(t) = \frac{\alpha KPF(s)}{s + \alpha KPF(s)} \left[\frac{N(t, \Phi_e(t))}{\alpha P} \right] \tag{5.4-19}$$

Note that we have used the fact that $\dot{x}(t) = sx(t)$ in (5.4-19). We can view the operator in front of the noise term in brackets as the closed-loop response that has a noise bandwidth of $2B_L$ Hz. Consequently if we obtain the noise spectral density, normalized by αP, we can determine the tracking error variance as

$$\sigma^2_{\Phi_e} = \frac{2B_L S_N(0)}{\alpha^2 P^2} \tag{5.4-20}$$

Thus, to continue, the spectral density of the noise process at $f = 0$ Hz in (5.4-19) must be obtained, under the assumption that $W >> B_L$. Now let us determine the autocorrelation function of $N(t, \Phi_e(t))$. Consider $N(t, \Phi_e(t))$ near $\Phi_e(t) = 0$. In this case $N(t, 0)$ becomes

$$N(t,0) = N(t) = \left\{ 2\sqrt{P}\tilde{d}(t)\tilde{n}_c(t) + 2\tilde{n}_c(t)\tilde{n}_s(t) \right\} \tag{5.4-21}$$

The autocorrelation function becomes

$$R_N(\tau) = E\left\{ N(t)N(t+\tau) \right\} \tag{5.4-22}$$

Since the expectation of an odd number of zero mean Gaussian random variables is zero one has

$$R_N(\tau) = 4PR_{\tilde{d}}(\tau)R_{\tilde{n}_c}(\tau) + 4R^2_{\tilde{n}_c}(\tau) \tag{5.4-23}$$

since $R_{\tilde{n}_c}(\tau) = R_{\tilde{n}_s}(\tau)$. It follows that the spectral density at $f = 0$ is given by

$$S_N(0) = 4P \int_{-\infty}^{\infty} R_{\tilde{d}}(\tau)R_{\tilde{n}_c}(\tau)d\tau + 4\int_{-\infty}^{\infty} R^2_{\tilde{n}_c}(\tau)d\tau \tag{5.4-24}$$

Using Parseval's theorem one obtains

$$S_N(0) = 2N_0P \int_{-\infty}^{\infty} T\frac{\sin(\pi fT)^2}{(\pi fT)^2} \left| H_{LP}(f) \right|^4 df + N_0^2 \int_{-\infty}^{\infty} \left| H_{LP}(f) \right|^4 df \tag{5.4-25}$$

which can be written as

$$S_N(0) = 2N_0P\alpha' + 2N_0^2 B'_{LP} \tag{5.4-26}$$

With

$$\alpha' = \int_{-\infty}^{\infty} T\frac{\sin(\pi fT)^2}{(\pi fT)^2} \left| H_{LP}(f) \right|^4 df \tag{5.4-27}$$

and

$$2B_{LP}^{'} = \int_{-\infty}^{\infty} \left|H_{LP}(f)\right|^4 df \tag{5.4-28}$$

From (5.4-20) it follows that the *effective*[2] *linearized squaring loop tracking error variance* is given by

$$\sigma_{\Phi_e}^2 = \frac{\left(4N_0 P\alpha' + 4N_0^2 B_{LP}^{'}\right) B_L}{\alpha^2 P^2} \tag{5.4-29}$$

which can be simplified to

$$\sigma_{\Phi_e}^2 = \frac{4N_0 B_L}{\alpha P}\left(\frac{\alpha'}{\alpha} + \frac{N_0 B_{LP}^{'}}{\alpha P}\right) \tag{5.4-30}$$

When comparing this expression to the PLL result, (5.2-73), it is seen that the effective linearized squaring loop tracking error variance is at least four times as large. However, if the term in the parenthesis is larger, then the tracking error variance could be more than four times as large. The term in parentheses together with the factor of 4 is termed the *squaring loss.* It is clear that if threshold occurs at a particular error variance and if the term in parentheses is near unity, then the squaring loop has an approximate 6 dB poorer threshold. It is clear from (5.4-30) that to obtain the *actual linearized tracking error,* we let $\phi(t) = \Phi_e(t)/2$. Therefore

$$\sigma_\phi^2 = \frac{N_0 B_L}{\alpha P}\left(\frac{\alpha'}{\alpha} + \frac{N_0 B_{LP}^{'}}{\alpha P}\right) \tag{5.4-31}$$

The effective tracking error variance, not the actual linear tracking error variance, is the important parameter with regard to threshold loss of lock conditions. However, the actual linearized tracking error is applicable to BER degradation, for example.

5.4.2 Tracking a BPSK Signal with an Integrate-and Dump Costas Loop

Another tracking loop that can track BPSK signals is the *Costas loop*. The basic Costas loop is shown in Figure 5.4-2. The VCO feeds the 90° phase splitter, which produces two signals, 90° apart in phase. Both of the signals are fed into the integrate-and-dump (I & D) filters and held in the sample and hold (S & H) devices. They are multiplied together to drive the loop filter and in turn the VCO. When I & D filters are used it is necessary to have timing information fed back from the bit synchronizer. If this is not feasible then

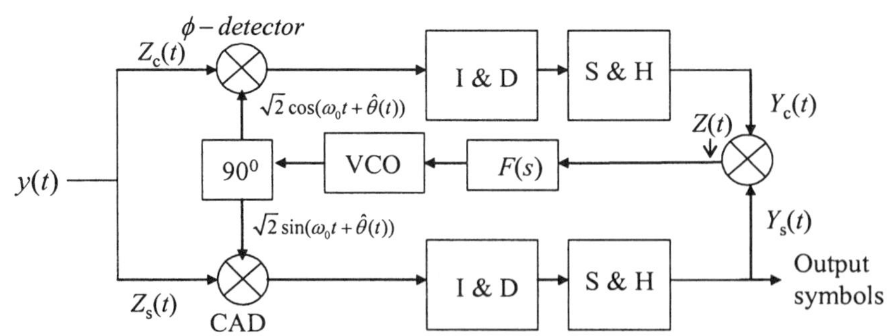

Figure 5.4-2 Costas loop with I & D filters in the arms.

[2] By effective we mean in relationship to threshold conditions such as loss of lock.

the passive filter Costas loop is the best choice (see Section 5.4.3). The phase detector is the upper multiplier and the coherent amplitude detector (CAD) the lower.

Various versions of the Costas loop exist. Riter [30] has shown that the optimum device for tracking suppressed carrier signals at low SNRs is either a Costas loop or a squaring loop. He showed that the inphase arm (with the CAD) should be weighted by tanh(cI(t)), where c is a constant dependent on the SNR and I(t) is the inphase arm output. It can be shown that when the SNR in the I & D filter output is low the hyperbolic tangent becomes linear, and the weighting becomes unimportant. Also at high SNRs the hyperbolic tangent becomes a limiter function and again the weighting function becomes unimportant. In this book only the linear inphase arm case will be considered in detail. It will be assumed that the signals are NRZ-L symbols (i.e., rectangular symbols) [2].

Let the received BPSK modulated signal plus noise be modeled as

$$y(t)=\sqrt{2P}d(t)\sin\left(\omega_0 t+\theta_0\right)+n(t) \tag{5.4-32}$$

where P is the received signal power in watts, and $d(t)$ is the baseband ± 1 valued NRZ-L data stream. Again the noise can be written in terms of its in-phase and quadrature phase noise components if the form

$$n(t)=\sqrt{2}n_c(t)\cos\left(\omega_0 t+\theta\right)+\sqrt{2}n_s(t)\sin\left(\omega_0 t+\theta\right) \tag{5.4-33}$$

in which both $n_c(t)$ and $n_s(t)$ are modeled as statistically independent processes having two-sided power spectral density of $N_0/2$. At the phase detector output one has

$$\begin{aligned} Z_c(t) &= \sqrt{P}d(t)\sin(\phi(t))+\sqrt{P}d(t)\sin\left(2\omega_0 t+\theta+\hat{\theta}(t)\right) \\ &+n_c(t)\cos(\phi(t))+n_c(t)\cos\left(2\omega_0 t+\theta+\hat{\theta}(t)\right) \\ &+n_s(t)\sin(\phi(t))+n_s(t)\sin\left(2\omega_0 t+\theta+\hat{\theta}(t)\right) \end{aligned} \tag{5.4-34}$$

in which $\phi(t)=\theta-\hat{\theta}(t)$ is the instantaneous phase error. For notational simplicity the loop will be analyzed during the time between t = 0 and t = T seconds. At the output of the I & D filter, neglecting the double frequency terms since they will be filtered out produces

$$Y_c(t)=\sqrt{P}d(T)T\sin(\phi(T))+N_c(T)\cos\left(\phi(T)\right)+N_s(T)\sin(\phi(T)) \tag{5.4-35}$$

where $N_c(t)$ and $N_s(t)$ are independent Gaussian random variables that change their value every T seconds, and both have zero means and variances of $N_0T/2$. Now the output of the CAD is given by

$$\begin{aligned} Z_s(t) &= \sqrt{P}d(t)\cos(\phi(t))-\sqrt{P}d(t)\cos\left(2\omega_0 t+\theta+\hat{\theta}(t)\right) \\ &-n_c(t)\sin(\phi(t))+n_c(t)\sin\left(2\omega_0 t+\theta+\hat{\theta}(t)\right) \\ &+n_s(t)\cos(\phi(t))-n_s(t)\cos\left(2\omega_0 t+\theta+\hat{\theta}(t)\right) \end{aligned} \tag{5.4-36}$$

After the I & D filter, and neglecting the double frequency terms, yields

$$Y_s(t)=\sqrt{P}d(T)T\cos(\phi(t))-N_c(T)\cos\left(\phi(t)\right)+N_s(T)\sin(\phi(t)) \tag{5.4-37}$$

with the understanding that $d(T)$ denotes the value of d at $t=T^-$, that is, just before T seconds. Now $Z(t)$ is the product of $Y_s(t)$ and $Y_c(t)$, so that

$$\begin{aligned}Z(t) = {} & \frac{PT^2}{2}\sin(2\phi(t)) + \sqrt{P}Td(T)N_s(t)\sin(2\phi(t)) + \frac{(N_s(T))^2}{2}\sin(2\phi(t)) \\ & - \frac{(N_c(t))^2}{2}\sin(2\phi(t)) - \left(\sqrt{P}Td(T) + N_s(T)\right)N_c(T)\cos(2\phi(t))\end{aligned} \tag{5.4-38}$$

Now the output of the VCO is given by

$$\hat{\theta}(t) = \frac{K_{vco}K_m}{s}Z(t) \tag{5.4-39}$$

where K_m is the multiplier coefficient of the multiplier that outputs $Z(t)$. Letting K denote the product $K = K_m K_{vco}$ (we have assumed that $K_1 = 1$ here) and assuming that $\dot{\theta}(t) = 0$, one has

$$\dot{\phi}(t) + \frac{PT^2KF(s)}{2} = KF(s)\left\{\begin{aligned}&\left(\frac{N_c^2(T) - N_s^2(T)}{2} - \sqrt{P}Td(T)N_s(T)\right)\sin(2\phi(t)) \\ &-\left[\left(\sqrt{P}Td(T)N_c(T) + N_c(T)N_s(T)\right)\cos\left(2\phi(t)\right)\right]\end{aligned}\right\} \tag{5.4-40}$$

Again letting $\Phi(t) = 2\phi(t)$ we have

$$s\Phi(t) + PT^2KF(s)\sin(\Phi(t)) = KF(s)\left\{\begin{aligned}&\left(N_c^2(T) - N_s^2(T) - 2\sqrt{P}Td(T)N_s(T)\right)\sin(\Phi(t)) \\ &-2\left(\sqrt{P}Td(T)N_c(T) + N_c(T)N_s(T)\right)\cos(\Phi(t))\end{aligned}\right\} \tag{5.4-41}$$

Denoting the term inside the brackets as $N(T)$, then letting $|\Phi(t)| << 1$ produces the linearized equation of operation

$$\Phi(t) = \left[\frac{PT^2KF(s)}{s + PT^2KF(s)}\right]\left[\frac{N(T)}{PT^2}\right] \tag{5.4-42}$$

where it is understood that the S & H function only allows the noise process to change every T seconds. It follows that the equivalent tracking error variance is given by

$$\sigma_\Phi^2 = \frac{N_0^{'}B_L}{\left(P^2T^4\right)} \tag{5.4-43}$$

in which B_L is defined by

$$B_L = \int_{-\infty}^{\infty}\left|\frac{PT^2KF(i2\pi f)}{i2\pi f + PT^2KF(i2\pi f)}\right|^2 df \quad Hz \tag{5.4-44}$$

and $N_0^{'}$ is the one-sided noise spectral density of $N(T)$, assuming that $N(T)$ changes every T seconds and therefore is a step-wise constant function with time. As long as $TB_L << 1$ then the delay in the loop, due to the bit decision, will not affect loop performance. To finish the calculation of the effective phase error variance, we must determine the spectral density of the noise term $N(T)$. In Problem 8 it is shown that the autocorrelation function of a random amplitude, fixed duration, rectangular signal pulse (that is, statistically independent from pulse to pulse) is given by

$$R_N(\tau) = \sigma_N^2\left(1-\frac{|\tau|}{T}\right) \quad |\tau|\le T$$
$$0 \quad |\tau|>T \tag{5.4-45}$$

where σ_N is the standard deviation of the noise process. The Fourier transform of (5.4-45) yields

$$S_N(f) = \sigma_N^2 T\frac{\sin(\pi fT)^2}{(\pi fT)^2} \tag{5.4-46}$$

At $f = 0$ the spectral density becomes

$$S_N(0) = \sigma_N^2 T \tag{5.4-47}$$

The only parameter needed to evaluate the spectral density of the noise process at $f = 0$ is σ_N^2. Consider σ_N^2. Therefore

$$\sigma_N^2 = E\left[N(T)^2\right] \tag{5.4-48}$$

Evaluating from (5.4-41), and assuming $\Phi(T)$ is near zero, and is approximated as zero, one obtains

$$\sigma_N^2 = 4\left(PT^2\overline{N}_c^2 + \overline{N}_s^2\overline{N}_c^2\right) = 4\left(PT^2\overline{N}_c^2 + \left(\overline{N}_s^2\right)^2\right) \tag{5.4-49}$$

Noting that for Gaussian random variables that

$$\overline{N}_c^2 = \overline{N}_s^2 = \frac{N_0T}{2} \tag{5.4-50}$$

and that

$$\sigma_N^2 = 4\left(\frac{PN_0T^3}{2} + \frac{N_0^2T^2}{4}\right) \tag{5.4-51}$$

Hence from (5.4-47) it follows that

$$S_N(0) = \frac{N_0^{'}}{2} = 2PN_0T^4 + N_0^2T^3 \tag{5.4-52}$$

Using (5.4-52) in (5.4-43) produces our result for the I & D Costas loop linear effective phase error variance

$$\sigma_\Phi^2 = \frac{4N_0B_L}{P}\left(1+\frac{N_0}{2PT}\right) \quad \text{rad}^2 \tag{5.4-53}$$

It is to be noted that the Costas loop also doubles the input dynamics, and this must be accounted for when determining the Costas loop noise bandwidth B_L. The actual phase error variance is given by

$$\sigma_\phi^2 = \frac{N_0 B_L}{P}\left(1 + \frac{N_0}{2PT}\right) \text{ rad}^2 \tag{5.4-54}$$

When comparing the Costas loop performance of (5.4-54) to the squaring loop tracking error performance, (5.4-30), it can be shown that the Costas loop has slightly lower tracking error. Moreover the Costas loop is much more commonly implemented in practice, since the processing is at baseband and is more easily implemented digitally.

5.4.3 Tracking a BPSK Signal with a Passive Arm Filter Costas Loop

One disadvantage of the I & D Costas loop is the fact that bit or symbol timing must be provided to the loop to provide the timing needed for the filters. The passive filter Costas loop utilizes a filter that does not require timing. There is a small loss in this case, but it simplifies the Costas loop acquisition process since during carrier acquisition the bit or symbol timing is not known, and therefore acquisition of the I & D Costas loop is more difficult. Figure 5.4-3 illustrates a passive filter Costas loop.

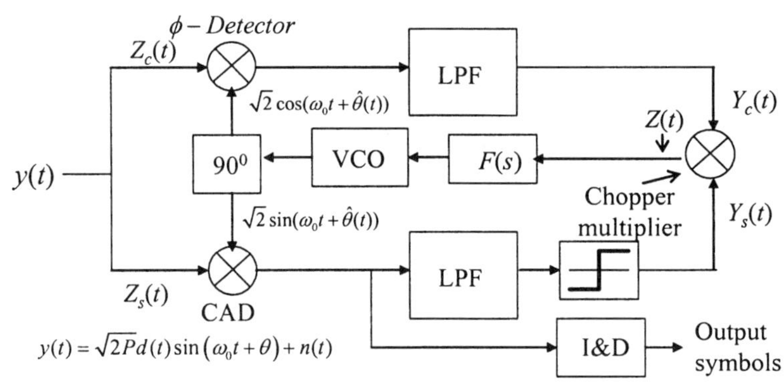

Figure 5.4-3 Costas loop with low pass filters in the arms.

One solution is to use passive filters for the Costas loop acquisition and then switch to I & D filters in the tracking mode, when bit timing in available. Figure 5.3-11 illustrates the passive filter Costas loop model.

The solution to the linearized tracking error variance with passive filters, in the Costas loop, is very similar to the analysis of the squaring loop model and will be summarized here [2]. The effective tracking error variance of the passive filter Costas loop is given by

$$\sigma_\Phi^2 = \frac{4N_0 B_L}{\alpha P}\left[\frac{\alpha'}{\alpha} + \frac{N_0 B_{LP}'}{\alpha P}\right] \text{ rad}^2 \tag{5.4-55}$$

where α, α', and B_L' are defined in (5.4-17), (5.4-27), and (5.4-28). The actual tracking error variance is given by

$$\sigma_\phi^2 = \frac{N_0 B_L}{\alpha P}\left[\frac{\alpha'}{\alpha} + \frac{N_0 B_{LP}'}{\alpha P}\right] \text{ rad}^2 \tag{5.4-56}$$

Thus if the low pass equivalent filter of the squaring loop is the same as the low pass filter of the Costas loop the performance is identical.

5.4.4 Steady State Tracking Error for the Costas and Squaring Loops

We have mentioned earlier that the effective dynamics are doubled relative to a PLL. Thus, for either the squaring loop or the Costas loop, the fact that the error signal is proportional to $\sin(2\phi)$ means that any perturbation in ϕ is doubled, so that any offset due to dynamics is doubled, as far as loop threshold is concerned. This applies to noise performance and noise threshold also, as was determined in the expressions for the effective tracking error variance. Thus from Table 5.2-1 if the steady state tracking error is required for either a Costas or a squaring loop, the entries in the table must be doubled when compared to a PLL.

5.4.5 Costas Loop with Hard-Limited In-Phase Arm Processing

Often Costas loops have the in-phase arm hard limited to improve the performance and also to allow the use of a chopper multiplier to be used in lieu of an analog multiplier, as seen in Figure 5.4-4.

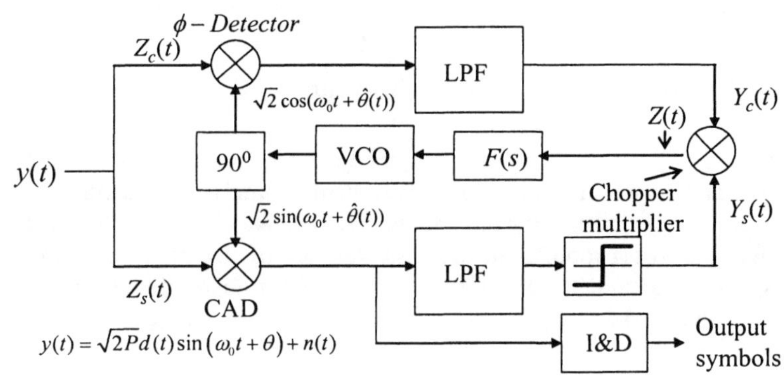

Figure 5.4-4 Costas loop with a hard-limited in-phase arm.

Simon [31] has shown that for a one-pole arm filter Costas loop, the tracking error variance is at most 1.08 dB greater than the linear (limiter removed) Costas loop. At nominal E_b/N_0 values of 6 dB, the tracking error variance of the hard-limited version is about 0.5 dB smaller. In conclusion the hard limiter in the inphase arm of a Costas loop can improve or diminish the performance of the tracking loop, depending on the arm filter SNR. However, the hardware implementation is simplified with the use of a chopper multiplier (hard limiter) as the third multiplier in the loop.

5.4.6 Improved Frequency Acquisition of a Passive Filter Costas Loop

Cahn [32] has observed that removing one of the low pass filters of an analog Costas loop improves the frequency acquisition performance and diminishes the likelihood of false lock (this will be discussed shortly). In effect Cahn has shown that by removing the quadrature arm filter, the loop acts like a hybrid frequency and phase tracking loop. This loop has been called the *modified Costas loop*. Figure 5.4-5 illustrates the hard-limited Costas loop (modified Costas loop) with no quadrature arm filter. Another method of obtaining frequency acquisition is to add an additional delay in one arm relative to the other. This alternate approach will not be pursued here, however.

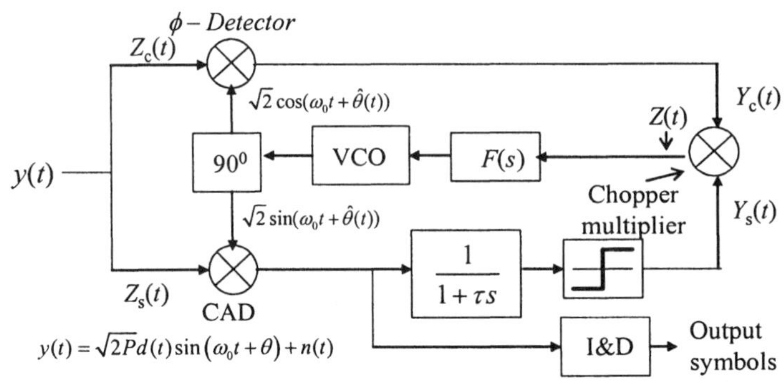

Figure 5.4-5 Hard-limited Costas loop with no quadrature arm filter.

As can be seen in the figure the arm filter in the inphase channel is a one-pole low pass filter, and the quadrature arm has no filter. If the input frequency ω is not equal to the loop frequency estimate $\hat{\omega}$, so that a frequency error $\Delta\omega$ exists, then Cahn has shown that the time average of the error signal, $Z(t)$, is given by

$$\langle Z(t)\rangle = \frac{2}{\pi}\frac{\Delta\omega\tau}{\sqrt{1+(\Delta\omega\tau)^2}} \tag{5.4-57}$$

which exhibits a restoring force related to the frequency error. However, as noted by Cahn, when the frequency error is small and approaches zero, the phase error dominates, and tracking is controlled by the phase error rather than the frequency error. Thus the modified Costas loop has the advantage of combining a frequency lock loop with a Costas loop, with the disadvantage of not being able to control the automatic frequency control (AFC) independently from the Costas loop transfer function. Based on (5.4-57) and assuming a second-order Costas loop, the linearized AFC loop, assuming that $|\Delta\omega\tau| << 1$, has the transfer function

$$\frac{\hat{\omega}}{\omega} = \frac{2\tau}{\pi}\frac{\sqrt{2}\omega_n s + \omega_n^2}{\left(1+2\sqrt{2}\tau\omega_n/\pi\right)s + 2\tau\omega_n^2/\pi} \tag{5.4-58}$$

To a rough approximation this loop appears to be a first-order AFC loop (for small frequency errors) with the AFC natural frequency given by

$$\omega_{nAFC} = \left(2\tau/\pi\right)\omega_n^2 \tag{5.4-59}$$

Thus the AFC natural frequency is proportional to the square of the Costas natural frequency. One might ask what the price is of leaving off one arm filter. The answer is given in Problem 11 where it is shown that the effective tracking phase error variance is given by

$$\sigma_\Phi^2 = \frac{4N_0B_L}{\alpha P}\left[1+\frac{N_0B_{LP}}{\alpha P}\right] \tag{5.4-60}$$

where B_{LP} is defined by

$$2B_{LP} = \int_{-\infty}^{\infty}|H_{LP}(f)|^2 df \tag{5.4-61}$$

This analysis is for the nonhard-limited in-phase channel processing assumption with a one-pole Butterworth filter. When comparing (5.4-60) with (5.4-55) it is seen that at low SNR the effective tracking error variance is increased by the ratio

$$\frac{\sigma^2_{\Phi 1 fil}}{\sigma^2_{\Phi 2 fil}} = \frac{B_{LP}}{B'_{LP}} \geq 1 \tag{5.4-62}$$

which is the ratio of the effective tracking error variance of a linear Costas loop with one arm filter to a linear Costas loop with both LPFs in place. In Problem 12 B'_{LP} has been evaluated for an n pole Butterworth filter and it was shown that

$$B'_{LP} = B_{LP}\left(1 - \frac{1}{2n}\right) \tag{5.4-63}$$

where n is the order of the Butterworth arm filter. Thus for the case on a one-pole arm filter it is seen that

$$B'_{LP} = B_{LP}/2 \tag{5.4-64}$$

or

$$\frac{\sigma^2_{\Phi 1 fil}}{\sigma^2_{\Phi 2 fil}} = 2 \tag{5.4-65}$$

at low loop SNR. At high SNR values it can be shown that by comparing (5.4-60) with (5.4-55) it is seen that

$$\frac{\sigma^2_{\Phi 1 fil}}{\sigma^2_{\Phi 2 fil}} = \frac{\alpha}{\alpha'} \geq 1 \tag{5.4-66}$$

When f_0/R varies between 1 and 3 the ratio of (5.4-66) varies from 1.72 dB to 0.43 dB for a one-pole Butterworth arm filter [33].

5.4.7 Lock Detectors for Costas and Squaring Loops

The reality of communication system design is that every loop must have some type of monitor, or lock detector, to indicate whether or not synchronization lock has been achieved and maintained. Normally the bit synchronizer should not attempt to obtain time synchronization until the carrier loop has been successful in acquisition. Often the signal generated for use in the lock detector can be used to drive a coherent automatic gain control (AGC) circuit. Consider the squaring loop lock detector of Figure 5.4-6.

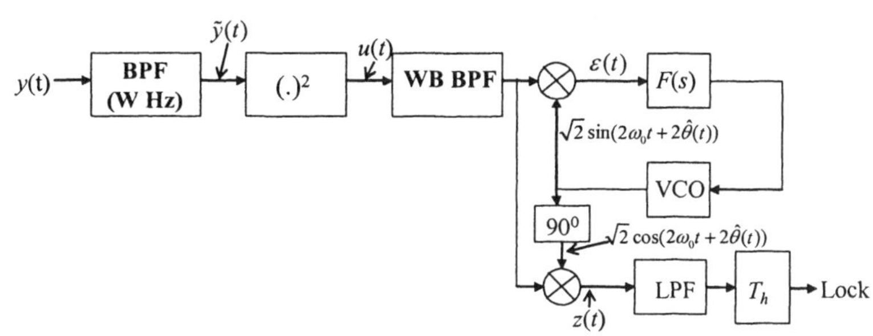

Figure 5.4-6 Model of a squaring loop with a lock detector.

As mentioned earlier the squaring loop basically squares the input signal to yield a tone at $2\omega_0$ from the input signal operating at ω_0, which can be tracked by a PLL operating at $2\omega_0$, as shown in Figure 5.4-6. Assume that the input is BPSK modulation so that out of the bandpass filter is the received signal plus noise

$$\tilde{y}(t)=\sqrt{2P}\tilde{d}(t)\cos(\omega_0 t+\theta)+\sqrt{2}\tilde{n}_c(t)\cos(\omega_0 t+\theta)+\sqrt{2}\tilde{n}_s(t)\sin(\omega_0 t+\theta) \tag{5.4-67}$$

where the terms with tildes imply the low pass equivalent version of the original signal or noise term. Upon squaring $\tilde{y}(t)$ the following is obtained

$$\begin{aligned}\tilde{y}(t)^2 &= P\tilde{d}(t)^2\cos(2\omega_0 t+2\theta)+\tilde{n}_c^2(t)\cos(2\omega_0 t+2\theta)-\tilde{n}_s^2(t)\cos(2\omega_0 t+2\theta)\\ &+2\sqrt{P}\tilde{d}(t)\tilde{n}_c(t)\cos(2\omega_0 t+2\theta)+2\sqrt{P}\tilde{d}(t)\tilde{n}_s(t)\sin(2\omega_0 t+2\theta)\\ &+2\tilde{n}_s(t)\tilde{n}_c(t)\,\text{in}(2\omega_0 t+2\theta)+\text{dc terms}\end{aligned} \tag{5.4-68}$$

After multiplying by the in-phase reference at $2\omega_0$, into the low-pass filter (LPF) denoted as $z(t)$ in the figure, one obtains the result

$$z(t)=\frac{P\tilde{d}(t)^2+\tilde{n}_c^2(t)-\tilde{n}_s^2(t)+2\sqrt{P}\tilde{d}(t)\tilde{n}_c(t)}{\sqrt{2}} \tag{5.4-69}$$

Note that $z(t)$ has a mean signal component when it is present, but the noise terms have zero mean value. Thus when the signal appears and is locked, the output will have mean value proportional to the signal power. This process can be analyzed to determine the loss of performance in terms of false alarm probability and detection probability.

Now consider two types of lock detectors for a Costas loop. Figures 5.4-7(a) and 5.4-7(b) illustrate two different lock detector block diagrams. In Figure 5.4-7(a), a Costas loop lock detector has zero mean value output, only when noise is present; when signal plus noise is present the output is nonzero. When the signal (and noise) is present and synchronized by the Costas loop, the output becomes positive. However, the lock detector of Figure 5.4-7(b) has an output that is positive even when the signal is not present at the input. When the signal is present at the input, the mean output is proportional to the sum of the signal and the noise out of the arm low-pass filter, and therefore the threshold must be placed between the noise only and the signal plus noise levels.

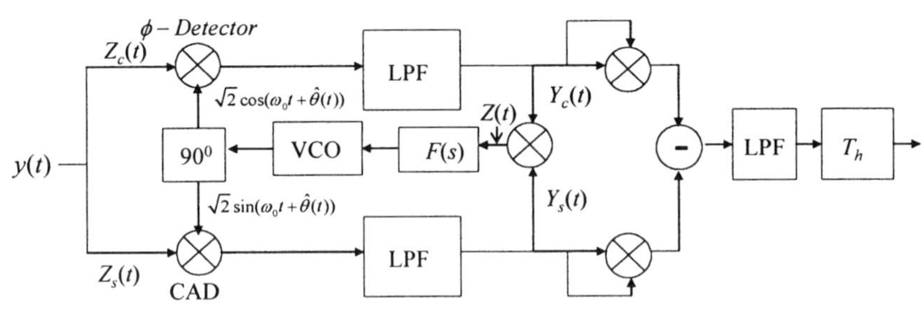

Figure 5.4-7 (a) Costas loops with first lock detector model.

5.4.8 False Lock in Costas Loops

False (frequency) lock is a phenomenon that allows a loop to stably "track" a received signal at the wrong carrier frequency. The wrong carrier frequency means not tracking the center frequency of the signal. In Costas loops there are at least four mechanisms that can be identified and can lead to false lock. One mechanism is due to spurious oscillations (resulting from mixers, for example), which can generate sufficient voltage to provide a stable lock point to that spurious oscillation. Another mechanism is due to the accumulated delay in a longloop implementation. A method of analyzing this has been reported in [4, 34–37] and others. A third mechanism is due to periodic motion of the receive antenna (or transmit antenna). The fourth mechanism is data related and is the subject of this section. The principal reason for data-related false lock is distortion due to filtering of the frequency-offset signal.

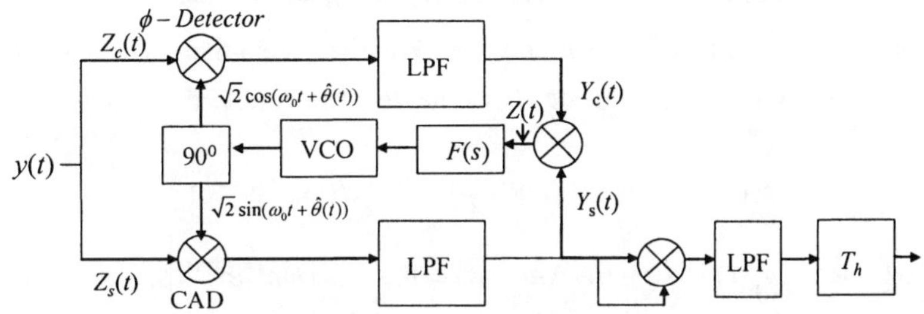

Figure 5.4-7 (b) Costas loop with second lock detector model.

It will be seen that widening the arm filter bandwidth will weaken the false lock tracking capability, however, it has been found that keeping the arm filter bandwidth near the data rate bandwidth optimizes the tracking performance due to thermal noise. Thus a compromise may be required in the selection of the arm filter bandwidth when false lock can occur. It is to be noted that reducing the frequency uncertainty also reduces the possibility of false lock.

The results given here closely follow an original study by Hedin, Holmes, Lindsey, and Woo [38] that was later published as a journal paper [39].

5.4.8.1 Costas Loop Error Signal when in False Lock

Consider the low-pass filter model of a Costas loop with linear in-phase arm processing (no limiter) shown in Figure 5.4-8.

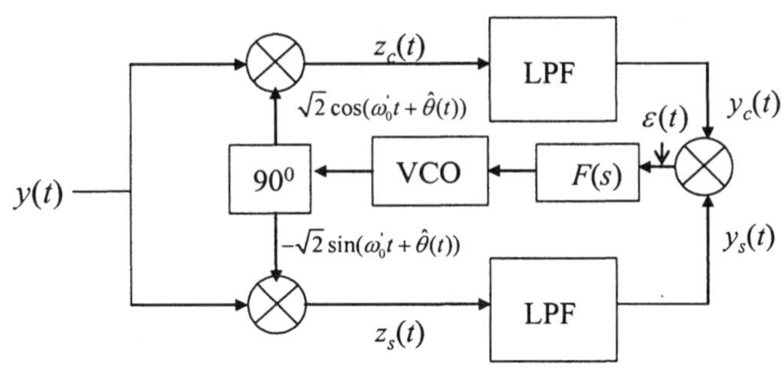

Figure 5.4-8 Costas loop model for false lock analysis.

In this analysis noise will not be considered, so that only the signal will be modeled. Let the received signal be represented by

$$s(t) = \sqrt{2P}d(t)\cos\left(\omega_0 t + \theta\right) \tag{5.4-70}$$

in which ω_0 is the input frequency in radians/second and θ is the input carrier phase in radians, P is the received power in $s(t)$ (W), and $d(t)$ is the ± 1 data bit stream which has data rate of R_b bps. Let the local oscillator operate at $\omega_0^{'}$ radians/second. The two reference signals are represented by

$$r_c(t) = \sqrt{2}\cos\left(\omega_0' t + \hat{\theta}\right)$$
$$r_s(t) = -\sqrt{2}\sin\left(\omega_0' t + \hat{\theta}\right) \quad (5.4\text{-}71)$$

Let $\Delta\omega = \omega_c - \omega_c'$ be the frequency error between the input signal and the frequency of the receiver VCO frequency, and let $\phi(t)$ be the phase difference between θ and $\hat{\theta}(t)$ so that $\phi(t) = \theta - \hat{\theta}(t)$. We will assume that $\phi(t)$ is essentially a constant, with value ϕ. If we neglect the double frequency terms then we can write

$$z_c(t) = d(t)\cos\left(\Delta\omega t + \phi\right) = \text{Re}\left[d(t)e^{j\phi}e^{j\Delta\omega t}\right]$$
$$z_s(t) = d(t)\sin\left(\Delta\omega t + \phi\right) = \text{Re}\left[\frac{1}{j}d(t)e^{j\phi}e^{j\Delta\omega t}\right] \quad (5.4\text{-}72)$$

where of course Re[x] is the real part of x. Note that the signals written in (5.4-72) are bandpass signals with complex envelopes $d(t)$ and $(1/j)d(t)$, respectively. Let $D(\omega)$ be the Fourier transform of the baseband signal $d(t)$:

$$D(\omega) = \int_{-\infty}^{\infty} d(t)e^{-j\omega t}dt \quad (5.4\text{-}73)$$

It follows that

$$Z_c(\omega) = (1/2)D(\omega - \Delta\omega)e^{j\phi} + (1/2)D^*(-\omega - \Delta\omega)e^{-j\phi}$$
$$Z_s(\omega) = (1/2)D(\omega - \Delta\omega)e^{j\phi} + (1/2j)D^*(-\omega - \Delta\omega)e^{-j\phi} \quad (5.4\text{-}74)$$

where $Z_c(\omega)$ and $Z_s(\omega)$ are the Fourier transforms of $z_c(t)$ and $z_s(t)$, respectively. Now let $H(\omega)$ be the transfer function of the each arm filter. Let $y_c(t)$ and $y_s(t)$ have the following spectra:

$$Y_c(\omega) = (1/2)D(\omega - \Delta\omega)e^{j\phi}H(\omega) + (1/2)D^*(-\omega - \Delta\omega)e^{-j\phi}H(\omega)$$
$$Y_s(\omega) = (1/2j)D(\omega - \Delta\omega)e^{j\phi}H(\omega) + (1/2j)D^*(-\omega - \Delta\omega)e^{-j\phi}H(\omega) \quad (5.4\text{-}75)$$

Now it is true that for a real filter $H(\omega) = H^*(-\omega)$, where the asterisk denotes the complex conjugate. Therefore, (5.4-75) can be written as

$$Y_c(\omega) = (1/2)\tilde{D}(\omega - \Delta\omega)e^{j\phi} + (1/2)\tilde{D}^*(-\omega - \Delta\omega)e^{-j\phi}$$
$$Y_s(\omega) = (1/2j)\tilde{D}(\omega - \Delta\omega)e^{j\phi} - (1/2j)\tilde{D}^*(-\omega - \Delta\omega)e^{-j\phi} \quad (5.4\text{-}76)$$

in which the following defines $\tilde{D}(\omega)$

$$\tilde{D}(\omega) = D(\omega)H(\omega + \Delta\omega) \quad (5.4\text{-}77)$$

Note that $\tilde{D}(\omega) \neq \tilde{D}^*(-\omega)$ since $H(\omega + \Delta\omega) \neq H^*(-\omega + \Delta\omega)$ if the frequency error, $\Delta\omega$, is not zero. However, if we define the complex envelope $\tilde{d}(t)$ to be the inverse transform of $\tilde{D}(\omega)$ where

$$\tilde{d}(t) = \frac{1}{2\pi}\int_{-\infty}^{\infty} \tilde{D}(\omega)e^{j\omega t}d\omega \quad (5.4\text{-}78)$$

then since $Y_c(\omega)$ and $Y_s(\omega)$ indicated in (5.4-76) have the same form as $z_c(t)$ and $z_s(t)$ defined in (5.4-72), they are also bandpass signals and can be represented analogously as

$$y_c(t) = \mathrm{Re}\left[\tilde{d}(t)e^{j(\Delta\omega+\phi)}\right]$$
$$y_s(t) = \mathrm{Re}\left[\frac{1}{j}\tilde{d}(t)e^{j(\Delta\omega+\phi)}\right] \tag{5.4-79}$$

Now using the complex identity

$$\mathrm{Re}[Z_1]\mathrm{Re}[Z_2] = \frac{1}{2}\left(\mathrm{Re}[Z_1Z_2] + \mathrm{Re}\left[Z_1Z_2^*\right]\right) \tag{5.4-80}$$

the Costas loop error signal, $\varepsilon(t)$, can be determined from the product of $y_c(t)$ and $y_s(t)$ from (5.4-79) as

$$\varepsilon(t) = y_c(t)y_s(t) = \frac{1}{2}\mathrm{Re}\left[\frac{1}{j}\left(\tilde{d}(t)\right)^2 e^{j2(\Delta\omega t+\phi)}\right] \tag{5.4-81}$$

Equation (5.4-81) can be rewritten in the form

$$\varepsilon(t) = \frac{1}{2}\mathrm{Im}\left[\left(\tilde{d}(t)\right)^2 e^{j2(\Delta\omega t+\phi)}\right] \tag{5.4-82}$$

where Im[x] denotes the imaginary part of x. Since $\tilde{d}(t)^2$ is in general complex (except when $\Delta\omega = 0$), the Costas error signal can be written as

$$\varepsilon(t) = \frac{1}{2}\left|\left(\tilde{d}(t)\right)\right|^2 \sin\left(2(\Delta\omega t + \phi) + \varsigma(t)\right) \tag{5.4-83}$$

where $\tilde{d}(t)^2$ has been written in its magnitude and phase angle form

$$\tilde{d}(t)^2 = \left|\tilde{d}(t)\right|^2 e^{j\varsigma(t)} \tag{5.4-84}$$

Clearly, when $\Delta\omega = 0$ so that true lock can be achieved, $\tilde{d}(t)$ is real, so that the error signal has the well-known form $(1/2)\alpha\sin(2\phi)$, and α is the average power loss through the LPF. However when $\Delta\omega \neq 0$ the error signal has phase variation $\varsigma(t)$ as well as amplitude variation, so that false lock is a very complex process.

During sweeping for carrier acquisition the main effect that controls the loop to complete the phase acquisition process is dc component in $\varepsilon(t)$. During false lock the data term $\tilde{d}(t)^2$ has a dc value and harmonics, which can "beat" with the sine term in (5.4-83) to yield a dc component that can sustain false lock.

5.4.8.2 Costas Loop False Lock Levels for Random Data with One-Pole Arm Filters

Now consider an application of the derived equations to understand what the false lock levels are. It will be assumed that nonreturn to zero (NRZ) symbols are being transmitted, and each symbol (or bit if not coded) takes on the value of ±1 with equal probability. From (5.4-82) we have shown that

$$\varepsilon(t) = \frac{1}{2}\mathrm{Im}\left[\left(\tilde{d}(t)\right)^2 e^{j2(\Delta\omega t+\phi)}\right] \tag{5.4-85}$$

Now write the data explicitly as

$$d(t) = \sum_{k=-\infty}^{\infty} d_k p(t-kT) \tag{5.4-86}$$

where d_k takes on the values ± 1 with equal probability, and $p(t)$ is an NRZ pulse of amplitude 1 and duration T seconds. Since $D(\omega)$ is the Fourier transform of $d(t)$ we have

$$D(\omega) = \sum_{k=-\infty}^{\infty} d_k P(\omega) e^{-jk\omega T} \tag{5.4-87}$$

It follows that $\tilde{D}(\omega)$ is given by

$$\tilde{D}(\omega) = \sum_{k=-\infty}^{\infty} d_k P(\omega) H(\omega+\Delta\omega) e^{-jk\omega T} \tag{5.4-88}$$

From (5.4-88) it is clear that the Fourier transform of $\tilde{D}(\omega)$ is given by

$$\tilde{d}(t) = \sum_{k=-\infty}^{\infty} d_k q(t-kT) \tag{5.4-89}$$

where $q(t)$ is defined from

$$q(t) = \frac{1}{2\pi}\int_{-\infty}^{\infty} H(\omega+\Delta\omega) P(\omega) e^{j\omega t} d\omega \tag{5.4-90}$$

since $H(\omega+\Delta\omega)$ is not symmetric with respect to zero frequency when $\Delta\omega \neq 0$, $q(t)$ is not real in the general case. From (5.4-89) it follows that

$$\left[\tilde{d}(t)\right]^2 = \sum_{k=-\infty}^{\infty} q^2(t-kT) + \sum_{k=-\infty}^{\infty}\sum_{\substack{n=-\infty \\ k\neq n}}^{\infty} d_k d_n q(t-kT) q(t-nT) \tag{5.4-91}$$

The first term in (5.4-91) is a complex periodic function with period T. This periodicity will give rise to line spectra with harmonics at the symbol rate. For the case of random data, the second term of (5.4-91) only gives rise to continuous spectra. In the case under study, in which the d_k take on the values of ± 1 with equal probability and are statistically independent of each other, the second term has zero mean and does not affect false lock. Thus in the case of random data the loop error signal is equal to

$$\varepsilon(t) = \frac{1}{2}\mathrm{Im}\left[\left(\sum_{k=-\infty}^{\infty} q(t-kT)^2\right) e^{j2(\Delta\omega t+\phi)}\right] \tag{5.4-92}$$

Equation (5.4-92) is our general result for false lock in a Costas loop with passive arm filters.

Since the function $\sum_{k=-\infty}^{\infty} q(t-kT)^2$ is periodic in period T, it can be represented by a Fourier series of the form

$$\sum_{k=-\infty}^{\infty} q(t-kT)^2 = \sum_{n=-\infty}^{\infty} C_n \, e^{j(2n\pi t/T)} \tag{5.4-93}$$

where the coefficients C_n are given by

$$C_n = \frac{1}{T} F\left[q(t)^2\right]\Big|_{\omega=2\pi n/T} \tag{5.4-94}$$

where $F[q(t)^2]$ denotes the Fourier transform of $q(t)^2$. It turns out that it is easier to compute $F[q(t)^2]$ by first computing the time domain function $q(t)^2$. The calculation will only be carried out for the case of a one-pole low pass arm filter, of the form

$$H(\omega) = \frac{1}{1 + j\left(\omega / \omega_0\right)} \tag{5.4-95}$$

in which $\omega_0 = 1/RC$ is the 3-dB cutoff of the filter in radians/second. We will evaluate false lock in a Costas loop for the case that NRZ[3] symbols are used in the symbol stream, and Manchester[4] symbols will be considered in a problem. Now

$$Q(\omega) = F\left[q(t)\right] = P(\omega)H(\omega + \Delta\omega) \tag{5.4-96}$$

or

$$Q(\omega) = \left[1 - e^{-j\omega T}\right]\left(\frac{1}{j\omega}\right)\frac{1}{1 + j\left(\dfrac{\omega + \Delta\omega}{\omega_0}\right)} \tag{5.4-97}$$

Taking the inverse Fourier transform, after a little algebra

$$q(t) = \frac{1}{1 + j\dfrac{\Delta\omega}{\omega_0}}\left[p(t) - F^{-1}\left\{\left(1 - e^{-j\omega T}\right)\left(\frac{1}{\left(\omega_0 + j\Delta\omega\right) + j\omega}\right)\right\}\right] \tag{5.4-98}$$

where $F^{-1}(x)$ is the inverse Fourier transform of x. Evaluating the inverse transform yields

$$q(t) = \frac{1}{1 + j\left(\dfrac{\Delta\omega}{\omega_0}\right)}\left[p(t) - e^{-(\omega_0 + j\Delta\omega)t}U(t) + e^{-(\omega_0 + j\Delta\omega)(t-T)}U(t-T)\right] \tag{5.4-99}$$

in which $U(t)$ is the unit step function, starting at $t = 0$. Letting $\gamma = \omega_0 + j\Delta\omega$ allows us to write (5.4-99) as

---

[3] A nonreturn to zero (NRZ) symbol is simply a pulse that is either +1 or –1 over the symbol time.
[4] A Manchester symbol is a symbol that for the first half of the symbol is +1 and the second half –1, or just the reverse (i.e., +1 to –1).

$$q(t)=\frac{1}{1+\left(\dfrac{j\Delta\omega}{\omega_0}\right)}\left[p(t)-e^{-\gamma t}U(t)+e^{-\gamma(t-T)}U(t-T)\right] \tag{5.4-100}$$

which is clearly complex. Forming

$$\frac{1}{T}F\left\{q(t)^2\right\}=\frac{1}{\left[1+\dfrac{j\Delta\omega}{\omega_0}\right]^2}\left\{\begin{array}{l}\left[\dfrac{1-e^{-j\omega T}}{j\omega T}\right]+\dfrac{1}{(\gamma+j\omega)T}\left[-2+2e^{-\gamma T}e^{-j\omega T}\right]\\ +\dfrac{1}{(2\gamma+j\omega)T}\left[1+e^{-j\omega T}-2e^{-\gamma T}e^{-j\omega T}\right]\end{array}\right\} \tag{5.4-101}$$

Therefore C_n, after some algebra, is given by

$$C_n=\frac{1}{\left[1+\dfrac{j\Delta\omega}{\omega_0}\right]^2}\left\{\begin{array}{l}(-1)^n\operatorname{sinc}(n\pi)+\dfrac{1}{(\gamma T+j2\pi n)}\left[2e^{-\gamma T}e^{-j2n\pi}-2\right]\\ +\dfrac{1}{(2\gamma T+j2\pi n)}\left[1+e^{-j2\pi n}-2e^{-\gamma T}e^{-j2\pi n}\right]\end{array}\right\} \tag{5.4-102}$$

Let

$$R=\frac{\omega_0 T}{2\pi}=f_0 T \tag{5.4-103}$$

where $\omega_0/(2\pi)=f_0$ is the 3-dB frequency of the LPF. Note that since the harmonics of $\{C_n\}$ are separated by $1/T$, the possible values of $\Delta\omega$, which give rise to a dc component in the error signal and thus create an error voltage that drives the loop into lock, are at *multiples of half the symbol rate*. That is

$$\Delta\omega=\frac{k\pi}{T}\quad k=0,\ \pm1,\ \pm2,\ldots \tag{5.4-104}$$

or equivalently $\Delta f=k/(2T)$. That is to say that the false lock points are located at multiples of one-half of the bit rate. Clearly $\Delta\omega=0$ is true lock, and all the other values of k correspond to the false lock phenomena. After some algebra one can write

$$C_n^k=\frac{1}{\left[1+\dfrac{jk}{2R}\right]^2}\left\{\begin{array}{l}(-1)^n\operatorname{sinc}(n\pi)+\dfrac{1}{(2\pi R+j\pi(2n+k))}\left[2e^{-2\pi R}(-1)^k-2\right]\\ +\dfrac{1}{(4\pi R+j2\pi(n+k))}\left[2-2e^{-2\pi R}(-1)^k\right]\end{array}\right\} \tag{5.4-105}$$

Now let $k=-n$ to get the false lock points, hence

$$C_{-k}=\frac{1}{\left[1+\dfrac{jk}{2R}\right]^2}\left\{\begin{array}{l}(-1)^k\operatorname{sinc}(k\pi)-\dfrac{1}{(2\pi R-jk\pi)}\left[-2e^{-2\pi R}(-1)^k+2\right]\\ +\dfrac{1}{(4\pi R)}\left[2-2e^{-2\pi R}(-1)^k\right]\end{array}\right\} \tag{5.4-106}$$

Note that when $\Delta\omega = k\pi / T$ the false lock point is always real. Hence C_{-k} is actually the magnitude of the loop error signal; that is,

$$\varepsilon(t)\big|_{dc} = \frac{1}{2} C_{-k} \sin(2\phi) \tag{5.4-107}$$

Evaluating C_{-k} for $k = 0$, produces the value

$$C_0 = 1 - \frac{1}{2\pi R}\left[1 - e^{-2\pi R}\right] = \alpha \tag{5.4-108}$$

where α is the power loss of the baseband signal through a one-pole low-pass filter.

For odd values of k, it can be shown [39] that

$$C_{-k} = \frac{-\left(1 + e^{-2\pi R}\right)}{2\pi R\left[1 + \left(\frac{k}{2R}\right)\right]^2} \quad k \text{ odd}, \quad \Delta f_{FL} = k/(2T) \tag{5.4-109}$$

where $\Delta\omega_{FL} = 2\pi \; \Delta f_{FL}$ is the angular false lock frequency, and $k = \pm 1, \; \pm 3, \; \pm 5, \ldots$ Now for k even and not equal to zero we have

$$C_{-k} = \frac{-\left(1 - e^{-2\pi R}\right)}{2\pi R\left[1 + \left(\frac{k}{2R}\right)^2\right]} \quad k \text{ even and} \neq 0, \quad \Delta f_{FL} = k/(2T) \tag{5.4-110}$$

where $k = \pm 2, \; \pm 4, \; \pm 6, \ldots$. Note that the false lock level is the same for $\pm k$ values. Figure 5.4-9 illustrates the levels, expressed in dB, for $k = 0$ through 5.

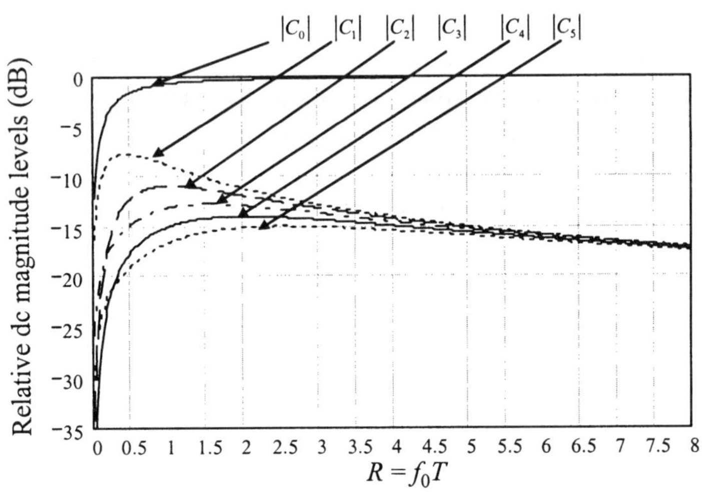

Figure 5.4-9 Relative false lock levels for different lock frequencies, NRZ symbols, and a one-pole LPF.

Clearly the largest false lock point occurs at the offset frequency of $1/(2T)$ and is about 6 dB weaker than true lock ($k = 0$). As the 3-dB bandwidth increases, the false lock signal level diminishes, at least when $R = f_0T$ is greater than some number between 1 to 2. In fact it is possible to plot the protection as the difference to true lock and false lock. Define the *protection level* of the k-th false lock level as the ratio

$$PL_k = 10\log\left(\left|\frac{C_0}{C_k}\right|\right) \tag{5.4-111}$$

These values are plotted for the five levels in Figure 5.4-10.

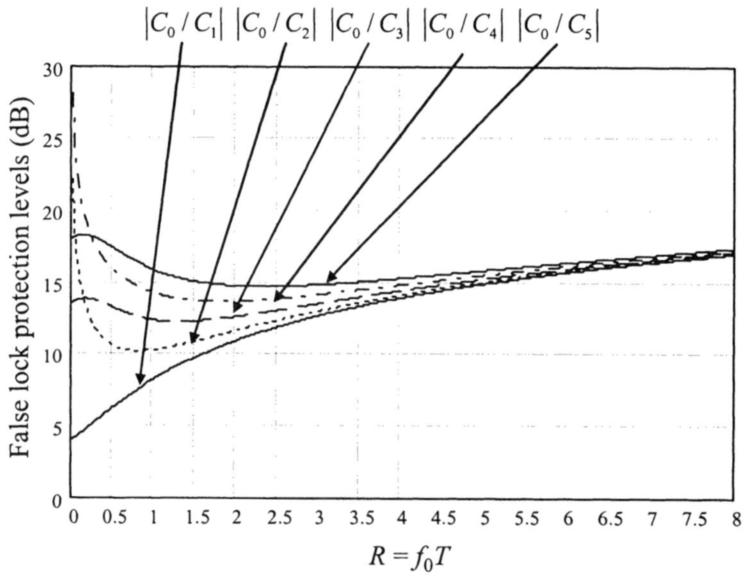

Figure 5.4-10 Relative false lock protection levels for different lock frequencies, NRZ symbols, and a one-pole LPF.

It can be seen that for $R \geq 0.5$ the protection level is at least 6 dB for the strongest line component at $\Delta f_{FL} = 1/(2T)$, and increases for the higher order false lock points. Shown in Table 5.4-1 are the false lock levels for Manchester (Bi-ϕ-L) symbols in a Costas loop with one-pole arm filters, based on the results of Problem 13.

It should be noted that periodic data can also cause false lock and has been analyzed in [39] and [2].

5.4.9 Decision-Directed Feedback Loops

Consider the decision-directed feedback loop (DDFL), which also tracks a BPSK signal. The loop operates by estimating the modulation and then multiplies it by a delayed version of the input to remove it. The loop model is shown in Figure 5.4-11.

Model the received signal plus thermal noise, expressed as in-phase and quadrature components in the form

$$y(t) = \sqrt{2P}d(t)\sin(\omega_0 t) + \sqrt{2}n_c(t)\cos(\omega_0 t) + \sqrt{2}n_s(t)\sin(\omega_0 t) \tag{5.4-112}$$

Table 5.4-1 Table of False Lock Levels at the Given Lock Frequencies for Manchester Symbols and a One-Pole LPF

False Lock Offset (Hz)	$\lvert C_k \rvert$
$\Delta f = 0$ (true lock)	$1-\frac{1}{2\pi R}\left[3-4e^{-\pi R}+e^{-2\pi R}\right]$
$\Delta f = \pm k/(2T)$ k odd	$\frac{1+e^{-2\pi R}}{2\pi R\left[1+\left(\frac{k}{2R}\right)^2\right]}$
$\Delta f = \pm k/(2T)$ k even, $\neq 0$	$\frac{3-(-1)^{k/2}4e^{-\pi R}+e^{-2\pi R}}{2\pi R\left[1+\left(\frac{k}{2R}\right)^2\right]}$

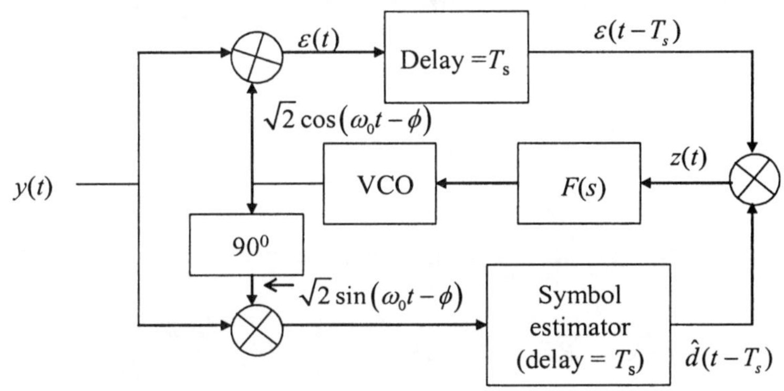

Figure 5.4-11 Decision-directed feedback loop with delay.

in which P is the received signal power in s, $d(t)$ is the ± 1 valued baseband data stream, ω_0 is the carrier angular frequency in radians/second, and $n_s(t)$ and $n_c(t)$ are the quadrature noise terms in the white Gaussian noise model. The symbol duration is assumed to be T_s seconds. The term ϕ is the slowly varying phase error of the loop. The error signal, $\varepsilon(t)$, generated from the upper arm (neglecting the double frequency) is given

$$\varepsilon(t) = \sqrt{P}d(t)\sin(\phi) + n_c(t) + n_s(t)\cos(\phi) \tag{5.4-113}$$

The symbol estimator, derived from the lower arm, which is commonly a matched filter, produces the delayed estimate $\hat{d}(t-T_s)$, which is multiplied by $\varepsilon(t-T_s)$ to form the error control signal $z(t)$

$$\begin{aligned} z(t) &= \sqrt{P}\hat{d}(t-T_s)d(t-T_s)\sin(\phi) \\ &\quad + \hat{d}(t-T_s)\left[n_c(t-T_s)\cos(\phi) + n_s(t-T_s)\sin(\phi)\right] \end{aligned} \tag{5.4-114}$$

Under the normal condition that the symbol rate is much greater than the loop noise bandwidth, the delay can be neglected as far as tracking performance is considered, so that one can write

$$z(t) = \sqrt{P}\hat{d}(t-T_s)d(t-T_s)\sin(\phi) \\ +\hat{d}(t-T_s)n_c(t-T_s) \tag{5.4-115}$$

in which it has been noted that the noise properties are independent of ϕ so that the noise term can be written as one baseband noise term ($n_c(t)$). Now consider the product $\hat{d}(t-T_s)d(t-T_s)$. Since the loop has many symbols within the loop memory, we can approximate the product with the time and ensemble average:

$$\left\langle E\left[\hat{d}(t-T_s)d(t-T_s)\middle|\phi\right]\right\rangle \cong 1-2PE(\phi) \tag{5.4-116}$$

where $\langle x(t)\rangle$ denotes the time average of $x(t)$, and PE(ϕ) is the symbol error rate of a BPSK signal in white Gaussian noise. With this approximation, (5.4-115) becomes

$$z(t) = \sqrt{P}\left[1-2PE(\phi)\right]\sin(\phi) \\ +\hat{d}(t-T_s)n_c(t-T_s) \tag{5.4-117}$$

Now to close the loop we write

$$\hat{\theta}(t) = \frac{K_{VCO}}{s} \tag{5.4-118}$$

Noting that in this case since we have defined the input phase as zero, that $\phi(t) = -\hat{\theta}(t)$ and

$$\dot{\phi}(t) = -K_{VCO}\sqrt{P}F(s)\left\{\left[1-2PE(\phi(t))\right]\sin(\phi(t)) + \frac{n_c(t-T_s)}{\sqrt{P}}\right\} \tag{5.4-119}$$

where the product of the baseband delayed noise and data estimate has been approximated as the noise alone, since the input noise is assumed to be very broadband compared to the data process, and $F(s)$ is the loop filter transfer function, expressed in the La Place variable s. It is to be noted that the quantity $\left[1-2PE(\phi(t))\right]\sin(\phi(t))$ is periodic in π radians, and therefore the DDFL has the same π radian ambiguity as the Costas and squaring loops.

In order to analyze the control loop that is nonlinear in the error signal, it is necessary to compute the linear equivalent gain of the loop. This is obtained from

$$A_e = \frac{d}{d\phi}\left\{\left[1-2PE(\phi)\right]\sin(\phi)\right\}\Big|_{\phi=0} \tag{5.4-120}$$

In (5.4-120) $\phi(t)$ is viewed as a variable that depends negligibly on time. For matched filter detection of PSK data in white Gaussian noise (WGN), the symbol probability of error is given by

$$PE(\phi) = \frac{1}{\sqrt{2\pi}}\int_{\sqrt{2R}\cos(\phi)}^{\infty}\exp\left(-t^2/2\right)dt = Q\left[\sqrt{2R}\cos(\phi)\right] \tag{5.4-121}$$

where $R = E_s/N_0$ is the symbol energy to one-sided noise spectral density ratio. Evaluation of A_e leads to

$$A_e = \left[1 - 2Q\left(\sqrt{R}\right)\right] \tag{5.4-122}$$

Hence the linearized loop equation from (5.4-119) becomes

$$s\phi(t) + K_{vco} A_e \sqrt{P} F(s)\phi(t) = \frac{n_c(t - T_s)}{\sqrt{P}} \tag{5.4-123}$$

Solving for $\phi(t)$ produces

$$\phi(t) = \frac{A_e \sqrt{P} K_{vco} F(s)}{s + A_e \sqrt{P} K_{vco} F(s)} \frac{n(t - T_s)}{A_e \sqrt{P}} \tag{5.4-124}$$

The phase error variance is obtained from (5.4-124) as the product of the one-sided noise spectral density and the one-sided noise loop noise bandwidth, and is given by

$$\sigma_\phi^2 = N_0 B_L \left(\frac{1}{A_e \sqrt{P}}\right)^2 = \frac{N_0 B_L}{P} \left[\frac{1}{\left(1 - 2Q(\sqrt{2R}\right)}\right]^2 \text{ rad}^2 \tag{5.4-125}$$

From Problem 13 one has for $R \geq 2$ that the following approximation is applicable

$$\sigma_\phi^2 \cong \frac{N_0 B_L}{P} \left[1 + \frac{1}{\sqrt{2\pi R}} \exp(-R)\right]^2 \text{ rad}^2 \tag{5.4-126}$$

Example 7 It is of interest to plot the variance of the tracking error of the DDFL and the I & D Costas loop, assuming the following parameters, R_s = 50 symbols/sec, B_L=5 Hz, and variable C/N_0 values. The results are shown in Figure 5.4-12.

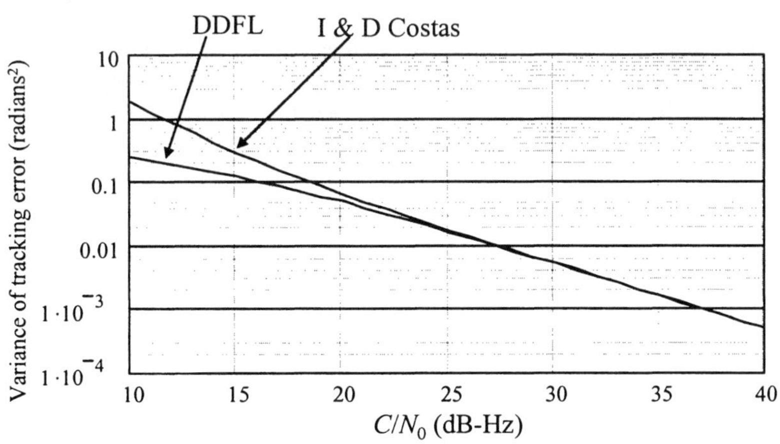

Figure 5.4-12 Comparison of the DDFL and I & D Costas loops for tracking error variance as a function of C/N_0.

It can be seen that the DDFL provides lower (better) tracking error performance below about 25 dB-Hz, for the parameters assumed in this example.

5.5 MULTIPHASE TRACKING LOOPS

Now we consider multiphase tracking loops. Since multiphase modulation is quite commonly used in communication systems, carrier tracking for this signal is required for coherent reception.

5.5.1 The *N*-th Power Loop

Consider an N (= M) phase signal (MPSK) of the form

$$s(t) = \sqrt{2P}\sin\left(\omega_0 t + (n-1)\frac{2\pi}{N} + \theta_0\right) \quad n = 1,\ 2,\ 3...N \tag{5.5-1}$$

where N is normally a power of 2 (i.e., $N = 2^n$) where n is an integer. If it is desired to track an N phase signal, it is sufficient to form the N-th power of the signal, which produces a trackable (unmodulated tone) signal. For the N phase case there will be N possible lock points that must be resolved with some additional information, such as a phase coded synchronization word. One of the problems in forming high powers of the received signal plus noise is the fact that there are high losses associated with large values of N. Figure 5.5-1 illustrates a N-th power loop.

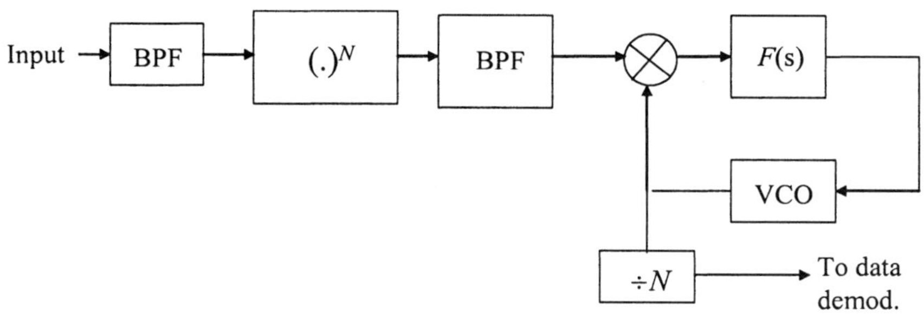

Figure 5.5-1 *N*-th power loop and the reference for data demodulation.

It is seen that the N-th power of the signal plus noise is formed and then filtered to $N\omega_0$, where ω_0 is the original angular carrier frequency.

As a concrete example consider a quadriphase ($N = 4$) encoded signal of the form

$$s(t) = \sqrt{2P}\sin\left(\omega_0 t + (n-1)\frac{\pi}{2} + \theta_0\right) \quad n = 1,\ 2,\ 3,\ 4 \tag{5.5-2}$$

Taking the fourth power of s(t) produces

$$s^4(t) = \frac{P^2}{2}\cos\left(4\omega_0 t + (n-1)2\pi + 4\theta_0\right) + \text{non } 4\omega_0 \text{ terms} \tag{5.5-3}$$

Notice that the $2\pi(n-1)$ term can be removed from the expression, since for any value of n (1, 2, 3, 4) the angle equals zero modulo 2π. Thus (5.5-3) can be simplified for terms near $4\omega_0$ by

$$s^4(t)\Big|_{near\,4\omega_0} = \frac{P^2}{2}\cos\left(4\omega_0 t + 4\theta_0\right) \tag{5.5-4}$$

which is an unmodulated tone operating at $4\omega_0$, and thus can be tracked by a PLL tuned to $4\omega_0$. Of course taking the fourth power of the input signal plus noise causes a very significant fourth power loss similar to, but more severe than, the squaring loop squaring loss. In addition a more severe threshold will occur due to the fact that the error signal will contain a term of the form $\sin(4\phi)$, with ϕ the phase error. Thus one can expect at least a 12 dB loss in threshold for the fourth power tracking loop, when compared to a PLL.

5.5.2 The *N*-Phase Costas Loop

The *N*-phase Costas loop is a generalization of the biphase Costas loop. Figure 5.5-2 illustrates the quadriphase Costas loop.

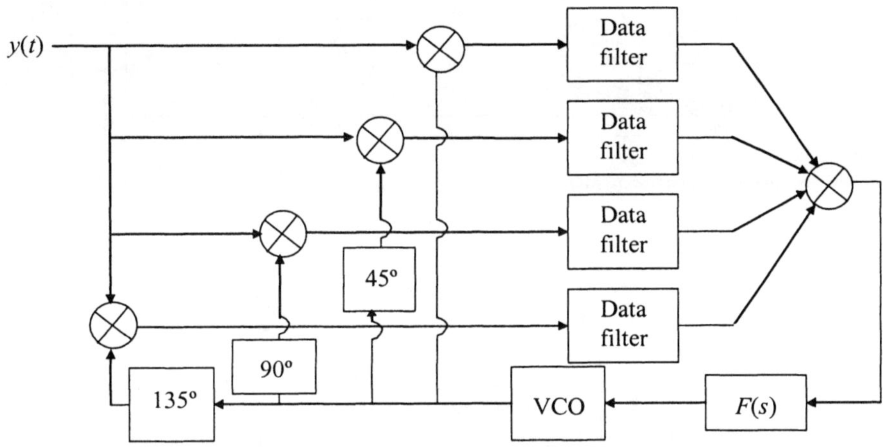

Figure 5.5-2 The quadriphase Costas loop for tracking quadriphase data signals.

It is seen that the quadriphase loop needs 4 multipliers plus the final multiplier, for a total of 5. The error signal is proportionate to $\sin(N\phi)$, where ϕ is the phase error in the loop and $N = 4$ in the figure. This loop can be generalized to an N phase loop in which N reference signals of the form $(k/N)180°$ are the angles used in the loop, and $k = 1, 2, \ldots N-1$.

5.5.3 Demod-Remod Quadriphase Tracking Loop

The demod-remod quadriphase tracking loop is another type of loop that will successfully track a quadriphase signal. Figure 5.5-3 illustrates the demod-remod quadriphase loop. The loop works by estimating the data polarity on both I and Q arms, in the demodulation section, and then remodulating the data in the remodulation section [40] onto a locally generated carrier reference signal. This reference signal, when multiplied by a delayed version of the input, produces an appropriate signal for phase tracking.

5.5.4 Modified Four-Phase Costas Loop-SQPSK Modulation

Now consider a tracking loop that was used on the TDRSS multiple access ground receiver for tracking both a balanced and an unbalanced quadriphase signal. The signal structure is staggered quadriphase (SQPSK) modulation. Braun and Lindsey [41] have shown that the closed-loop approximation to the maximum a posteriori probability (MAP) estimate leads to a carrier demodulator, at high E_b/N_0, as shown in Figure 5.5-4, for both balanced and unbalanced quadriphase signals.

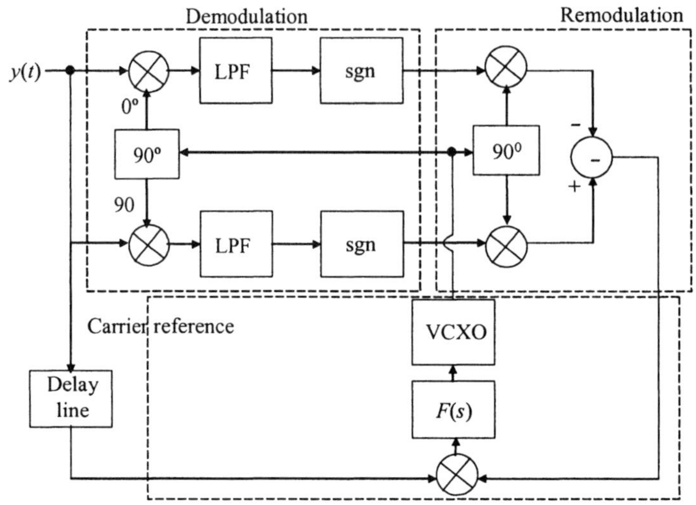

Figure 5.5-3 Demod-remod quadriphase tracking loop.

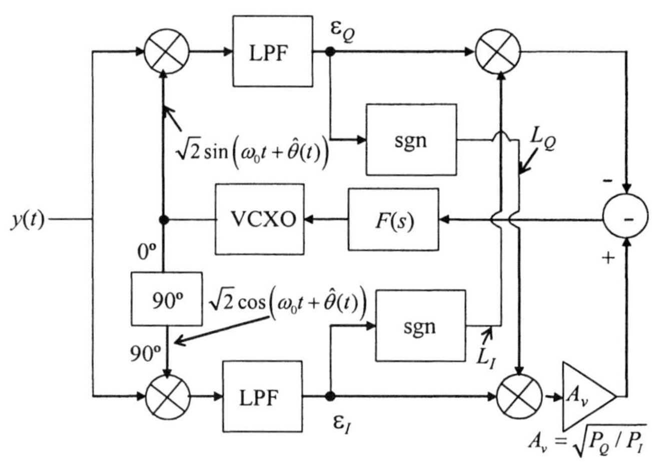

Figure 5.5-4 Modified four-phase Costas loop with arbitrary P_Q/P_I ratio.

It is assumed that the power in the Q channel is at least as large as the I channel (that is, $P_Q \ge P_I$). It is assumed that the noise bandwidth of the low-pass filters (LPFs) is B Hz. The input quadriphase signal is assumed to be of the form

$$y(t)=\sqrt{2(1-\alpha)P}d_Q(t)\sin\left(\omega_0 t+\theta\right)+\sqrt{2\alpha P}d_I(t)\cos\left(\omega_0 t+\theta\right)+n(t) \tag{5.5-5}$$

and the thermal noise is assumed to be expressed in terms of its in-phase and quadrature components of the form

$$n(t)=\sqrt{2}n_c(t)\cos(\omega_0 t+\theta)+\sqrt{2}n_s(t)\sin(\omega_0 t+\theta) \tag{5.5-6}$$

in which $n_c(t)$ and $n_s(t)$ are white Gaussian noise processes, statistically independent of each other, and each has a two-sided power spectral density of $N_0/2$. In (5.5-5) the following applies

αP is the power in the I channel ($\alpha \leq 1/2$ to insure that $P_Q \geq P_I$)
$(1-\alpha)P$ is the power in the Q channel
$d_Q(t)$ is the Q channel data stream ± 1, symbol duration T_Q seconds
$d_I(t)$ is the I channel data stream ± 1, symbol duration T_I seconds
$\mathcal{R}_Q(t)$ is the Q channel data rate ($\mathcal{R}_Q = 1/T_Q$)
$\mathcal{R}_I(t)$ is the I channel data rate ($\mathcal{R}_I = 1/T_I$)

Notice that in Figure 5.5-4 as P_Q/P_I increases the gain in the lower arm increases. The reference signals are given by

$$r_s(t) = \sqrt{2}\sin(\omega_0 t + \hat{\theta}(t)) \tag{5.5-7}$$

and

$$r_c(t) = \sqrt{2}\cos(\omega_0 t + \hat{\theta}(t)) \tag{5.5-8}$$

The output of the LPFs, which are assumed to be one-pole RC filters, are given by

$$\varepsilon_Q(t) = \sqrt{(1-\alpha)P}\tilde{d}_Q(t)\cos(\phi) - \sqrt{\alpha P}\tilde{d}_I(t)\sin(\phi) - \tilde{n}_c(t)\sin(\phi) + \tilde{n}_s(t)\cos(\phi) \tag{5.5-9}$$

$$\varepsilon_I(t) = \sqrt{(1-\alpha)P}\tilde{d}_Q(t)\sin(\phi) + \sqrt{\alpha P}\tilde{d}_I(t)\cos(\phi) + \tilde{n}_c(t)\cos(\phi) + \tilde{n}_s(t)\sin(\phi) \tag{5.5-10}$$

where

$\tilde{n}_c(t)$ is the filtered version of $n_c(t)$, ($\sigma^2_{\tilde{n}_c} = N_0 B$)
$\tilde{n}_s(t)$ is the filtered version of $n_s(t)$, ($\sigma^2_{\tilde{n}_s} = N_0 B$)
$\tilde{d}_Q(t)$ is the filtered version of $d_Q(t)$
$\tilde{d}_I(t)$ is the filtered version of $d_I(t)$
B is the arm filter noise bandwidth in Hz
$\phi = \theta - \hat{\theta}$ is the phase error in radians and assumed to be a slow function of time (the dependence on time has been suppressed)

Out of the hard limiters we have

$$L_Q(t) = \text{sgn}\left(\tilde{d}_Q(t)\right) \tag{5.5-11}$$

and

$$L_I(t) = \text{sgn}\left(\tilde{d}_I(t)\right) \tag{5.5-12}$$

The error control signal, which is used to control the VCO, is given by

$$C(\phi) = A_v \varepsilon_I(t) L_Q(t) - \varepsilon_Q(t) L_I(t) \tag{5.5-13}$$

Using (5.5-9), (5.5-10), (5.5-11), and (5.5-12) produces

$$\begin{aligned} C(\phi) = {} & A_v\sqrt{(1-\alpha)P}\tilde{d}_Q(t)\text{sgn}(\varepsilon_Q(t))\sin(\phi) + A_v\sqrt{\alpha P}\tilde{d}_I(t)\text{sgn}(\varepsilon_Q(t))\cos(\phi) \\ & - \sqrt{(1-\alpha)P}\tilde{d}_Q(t)\text{sgn}(\varepsilon_I(t))\cos(\phi) + \sqrt{\alpha P}\tilde{d}_I(t)\text{sgn}(\varepsilon_I(t))\sin(\phi) + n'(t) \end{aligned} \tag{5.5-14}$$

where

$$\begin{aligned} n'(t) = \tilde{n}_c(t)\left[A_v \operatorname{sgn}(\varepsilon_Q(t))\cos(\phi) + \operatorname{sgn}(\varepsilon_I(t)\sin(\phi)\right] \\ +\tilde{n}_s(t)\left[A_v \operatorname{sgn}(\varepsilon_Q(t))\sin(\phi) + \operatorname{sgn}(\varepsilon_I(t)\cos(\phi)\right] \end{aligned} \quad (5.5\text{-}15)$$

Having the equations for the error signal, it is possible to obtain the closed-loop tracking error curve $S(\phi)$ so that the gain—the slope of $S(\phi)$ that is at $\phi = 0$—can be obtained. Denote the *S*-curve (tracking curve) as

$$\bar{S}(\phi) = E\left\{S(\phi|\phi)\right\} \quad (5.5\text{-}16)$$

where $E\{x\}$ denotes the expectation of x. In order to process with the solution, it is necessary to make the following approximations

$$\tilde{d}_Q(t) = \sqrt{\gamma_Q}\, d_Q(t) \quad (5.5\text{-}17)$$

and

$$\tilde{d}_I(t) = \sqrt{\gamma_I}\, d_I(t) \quad (5.5\text{-}18)$$

in which the filtered symbols are approximated by the unfiltered symbols, but the voltage attenuation of the signals are accounted for. In other words, the distortion in the filtered pulsed symbols are neglected. The parameter γ is the relative power loss; that is,

$$\gamma_{I,Q} = \int_{-\infty}^{\infty} S_{d_{I,Q}}(f)\left|H(f)\right|^2 df \quad (5.5\text{-}19)$$

in which $H(f)$ is the low-pass filter transfer function. This approximation allows us to use conditional expectations on random variables, rather than on random processes. With this assumption it can be shown [42, 43] that

$$\begin{aligned} \bar{S}(\phi) &= \left[A_v\sqrt{(1-\alpha)\gamma_Q P} + \sqrt{\alpha\,\gamma_I P}\right]\sin(\phi) \\ &+ A_v\left[-\sqrt{(1-\alpha)\gamma_Q P}\sin(\phi) + \sqrt{\alpha\,\gamma_I P}\cos(\phi)\right]Q(F) \\ &- A_v\left[\sqrt{(1-\gamma)\gamma_Q P}\sin(\phi) + \sqrt{\alpha\,\gamma_I P}\cos(\phi)\right]Q(G) \\ &+ \left[\sqrt{(1-\gamma)\gamma_Q P}\cos(\phi) - \sqrt{\alpha\,\gamma_I P}\sin(\phi)\right]Q(A) \\ &- \left[\sqrt{(1-\gamma)\gamma_Q P}\cos(\phi) + \sqrt{\alpha\,\gamma_I P}\sin(\phi)\right]Q(B) \end{aligned} \quad (5.5\text{-}20)$$

where

$$\begin{aligned} F &= \sqrt{2R_Q}\sqrt{1-\alpha}\cos(\phi) + \sqrt{2R_I}\sqrt{\alpha}\sin(\phi) \\ G &= \sqrt{2R_Q}\sqrt{1-\alpha}\cos(\phi) - \sqrt{2R_I}\sqrt{\alpha}\sin(\phi) \end{aligned} \quad (5.5\text{-}21)$$

and

$$\begin{aligned} A &= \sqrt{2R_I}\sqrt{\alpha}\cos(\phi) + \sqrt{2R_Q}\sqrt{(1-\alpha)}\sin(\phi) \\ B &= \sqrt{2R_I}\sqrt{\alpha}\cos(\phi) - \sqrt{2R_Q}\sqrt{(1-\alpha)}\sin(\phi) \end{aligned} \quad (5.5\text{-}22)$$

with the definition

$$Q(x) = \int_{x}^{\infty} \frac{\exp(-t^2/2)}{\sqrt{2\pi}} dt \tag{5.5-23}$$

For the two-arm filter SNR parameters, we obtain

$$\begin{aligned} R_Q &= \frac{\gamma_Q P}{2N_0 B} \\ R_I &= \frac{\gamma_I P}{2N_0 B} \end{aligned} \tag{5.5-24}$$

in which B is the one-sided noise bandwidth of the one-pole arm filters. The S-curves (discriminator curves) are plotted in Figures 5.5-5 through 5.5-8 for both cases.

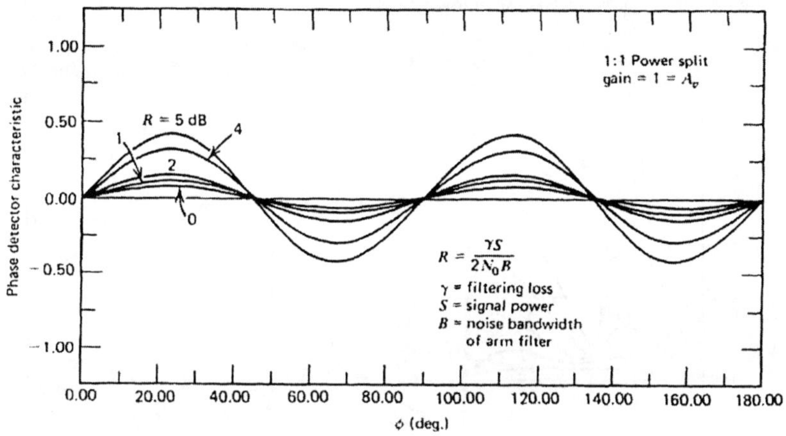

Figure 5.5-5 Normalized S-curve with balanced data rates at low SNR $\gamma_I = \gamma_Q$.

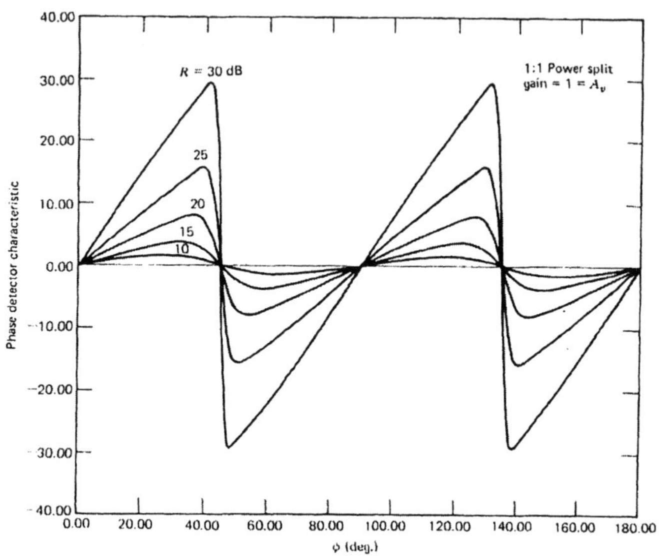

Figure 5.5-6 Normalized S-curve for the balanced case at high SNR.

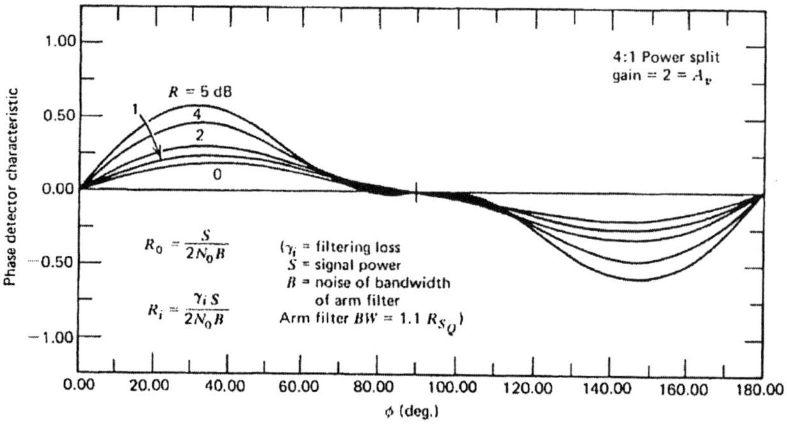

Figure 5.5-7 Normalized S-curve for the unbalanced case at low SNR.

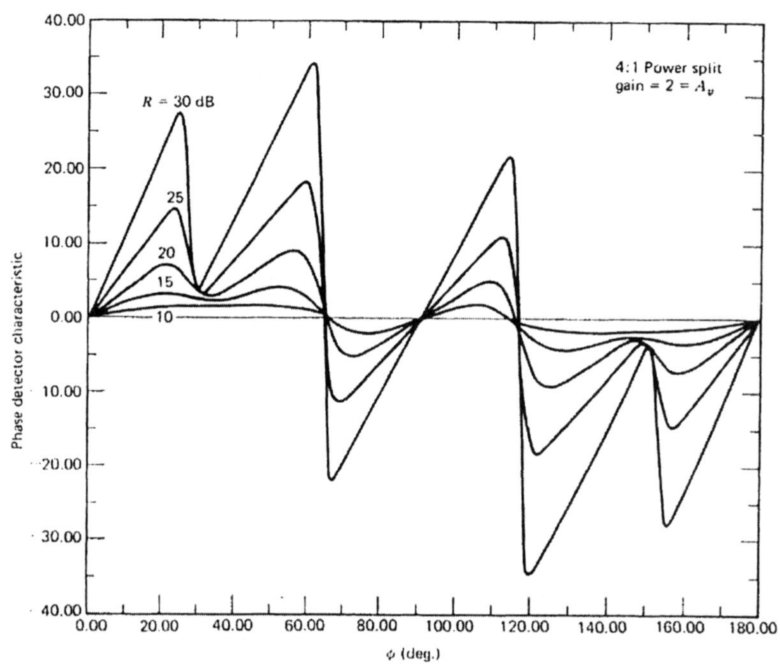

Figure 5.5-8 Normalized S-curve for the unbalanced case at high SNR.

The range of R in the balanced data and power case is 0–5 dB on Figure 5.5-5 and Figure 5.5-6, whereas it is 10–30 dB for Figures 5.5-7 and 5.5-8. The value of R in the figures is actually the range of $R_Q(\geq R_I)$. Notice that stable lock points for the balanced case occur at 0°, 90°, 180°, and 270°. Thus the correct carrier phase must be determined from the four possible, unless four-phase differential coding is used.

In the case when the data and the power are in the ratio of four to one, for the unbalanced case, only two stable lock points exist at 0° and 180°. Thus the receiver only has to resolve the two stable lock points via a correlation word, for example.

Now consider the linearized tracking error variance performance. To do so it is necessary to obtain the slope of the *S*-curve at the origin, which we shall call $\bar{S}'(0)$. It has been shown in [42] for the balanced case that

$$\bar{S}'(0) = \sqrt{2\gamma P}\left[1 - 2Q\left(\sqrt{R}\right) - \sqrt{\frac{2}{\pi}}\sqrt{R}\exp(-R/2)\right] \tag{5.5-25}$$

where γ is defined in (5.5-19) with the understanding that $\gamma_I = \gamma_Q$. To compute the tracking error variance it is convenient to obtain the stochastic differential equation of operation. Note that the control signal can be written as

$$S(\phi) = \bar{S}(\phi) + \left[S(\phi) - \bar{S}(\phi)\right] \tag{5.5-26}$$

that is in terms of the mean value and the variation around the mean. The second term can be viewed as the noise on the error signal and will be denoted by $N(t, \phi)$. It is convenient to express the error signal in terms of the unit gain at $\phi = 0$. Define the unit slope *S*-curve by $g_n(\phi)$, where it is defined through

$$\bar{S}(\phi) = \bar{S}'(0) g_n(\phi) \tag{5.5-27}$$

so that

$$g_n'(0) = 1 \tag{5.5-28}$$

has unit slope. Let K_v denote the VCO gain in Hz/second. Then the loop phase estimate can be expressed as

$$\hat{\theta} = \frac{S'(0) K_v F(s)}{s}\left[g_n(\phi) + \frac{N(t,\phi)}{S'(0)}\right] \tag{5.5-29}$$

Now since the phase error is defined by

$$\phi = \theta - \hat{\theta} \tag{5.5-30}$$

We can write (5.5-29) as

$$\theta - \phi = \frac{S'(0) K_v F(s)}{s}\left[g_n(\phi) + \frac{N(t,\phi)}{S'(0)}\right] \tag{5.5-31}$$

In order to proceed it is convenient to linearize the loop and assume that $\theta = 0$. Hence (5.5-31) can be written in linearized form as

$$-\phi = \frac{S'(0) K_v F(s)}{s}\left[\phi + \frac{N(t,\phi)}{S'(0)}\right] \tag{5.5-32}$$

Solving for ϕ produces the result

$$\phi = -H(s)\left[\frac{N(t,\phi)}{S'(0)}\right] \tag{5.5-33}$$

where $H(s)$ is defined as

$$H(s) = \frac{K_v S'(0)F(s)}{s + K_v S'(0)F(s)} \tag{5.5-34}$$

the closed-loop transfer function. It follows that the variance of the phase error is given by

$$\sigma_\phi^2 = \int_{-\infty}^{\infty} \frac{|H(j2\pi f)|^2 S_N(f)df}{[S'(0)]^2} \quad \text{rad}^2 \tag{5.5-35}$$

where $j = \sqrt{-1}$, and $S_N(f)$ is the power spectral density of the equivalent noise process. Under the assumption that $S_N(f)$ is constant near $f = 0$, one can accurately approximate the variance of the phase error as

$$\sigma_\phi^2 = \frac{S_N(0)(2B_L)}{[S'(0)]^2} \quad \text{rad}^2 \tag{5.5-36}$$

where B_L is the closed-loop noise bandwidth and is defined by

$$2B_L = \int_{-\infty}^{\infty} |H(j2\pi f)|^2 df \qquad \text{Hz} \tag{5.5-37}$$

In order to compute the tracking error variance, it is necessary to evaluate the autocorrelation function of the effective noise process. First consider the balanced case with NRZ symbols. The autocorrelation function of the equivalent noise is defined by

$$R_N(\phi,\tau) = E\left\{\left[S(\phi,t) - \bar{S}(\phi,t)\right]\left[S(\phi,t+\tau) - \bar{S}(\phi,t+\tau)\right]\right\} \tag{5.5-38}$$

Weber [40] has shown to a good approximation that

$$R_N(\tau,\phi) \cong R_N(\tau) \tag{5.5-39}$$

that is at $\phi = 0$.

Using the approximation of (5.5-39) and others with a considerable amount of algebra, it can be shown that

$$\begin{aligned} R_N(\tau) = {} & \frac{4}{\pi}\Lambda(\tau)e^{-R}R_{\tilde{n}}(\tau) + \frac{2}{\pi}R^2 e^{-R}\Lambda(\tau)^2 \frac{R_{\tilde{n}}(\tau)^2}{\sigma^2} \\ & + \frac{2}{\pi}R\mathrm{e}^{-R}\Lambda(\tau)\frac{R_{\tilde{n}}(\tau)^3}{3\sigma^4}\left(R^2 - 2R + 1\right) \\ & - \frac{4\sqrt{2R}}{\sqrt{\pi}}\Lambda(\tau)R_{\tilde{n}}(\tau)e^{-R/2}erf\left(\sqrt{R/2}\right) \\ & + 2R_{\tilde{n}}(\tau)\Lambda(\tau)\left(erf\left(\sqrt{R/2}\right)\right)^2 + \frac{2R_{\tilde{n}}(\tau)}{\pi}e^{-R}\rho(\tau)^2\Lambda(\tau) \\ & + \frac{2R_{\tilde{n}}(\tau)}{\pi}e^{-R}\frac{\rho(\tau)^3}{3}\left(R^2 - 2R + 1\right) \end{aligned} \tag{5.5-40}$$

where the following definitions have been used:

$$\Lambda(\tau)=1-|\tau|/T \quad ,|\tau|\le T \qquad (5.5\text{-}41)$$
$$= 0 \quad ,|\tau|>T$$

$$\sigma^2 = N_0 B \qquad (5.5\text{-}42)$$

$$erf\left(x\right)=\frac{2}{\sqrt{\pi}}\int_0^x e^{-t^2}dt \qquad (5.5\text{-}43)$$

$$R_{\tilde{n}}(\tau)=\sigma^2\rho(\tau)=N_0 B e^{-4B|\tau|} \qquad (5.5\text{-}44)$$

In (5.5-44) the arm filters were assumed to be one-pole RC (Butterworth) filters. Now the power spectral density of the noise process is given by

$$S_N(0)=F\left\{R_N(\tau)\right\}\Big|_{f=0} \qquad (5.5\text{-}45)$$

Evaluating (5.5-40), after some algebra yields

$$\begin{aligned}S_N(0)=&\left[\frac{4}{\pi}\mathrm{Re}^{-R}\,2\left(erf\left(\sqrt{R/2}\right)^2\right)-\frac{4\sqrt{2R}}{\sqrt{\pi}}e^{-R/2}erf\left(\sqrt{R/2}\right)\right]\frac{\gamma N_0}{2}\\&+\left[\frac{2}{3\pi}R\,\mathrm{e}^{-R}\left(R^2-2R+1\right)+\frac{2R}{\pi}e^{-R}\right]\left\{\frac{N_0}{6}\left(1-e^{-12BT}\right)-\frac{N_0}{72BT}+e^{-12BT}\left[\frac{N_0}{6}+\frac{N_0}{72BT}\right]\right\}\\&+\left[\frac{2}{\pi}R^2e^{-R}\right]\left[\frac{N_0}{4}-\frac{N_0}{16BT}+\frac{N_0}{128B^2T^2}-\frac{N_0e^{-8BT}}{128B^2T^2}\right]\\&+\left[\frac{2}{3\pi}e^{-R}\left(R^2-2R+1\right)\right]\frac{N_0}{8}\end{aligned} \qquad (5.5\text{-}46)$$

Finally using (5.5-36), (5.5-25), and (5.5-46) yields our result for the variance of the phase tracking error for the balanced case, when NRZ symbols are used

$$\sigma_\phi^2=\frac{N_0B_L}{\gamma P\left[1-2Q(\sqrt{R})-\sqrt{2R/\pi}e^{-R/2}\right]^2}\left\{\begin{aligned}&\gamma\left[\frac{2}{\pi}R\,\mathrm{e}^{-R}+\left(erf(\sqrt{R/2})\right)^2-\frac{\sqrt{8R}}{\sqrt{\pi}}e^{-R/2}erf(\sqrt{R/2})\right]\\&+\left[\frac{1}{3\pi}R\,\mathrm{e}^{-R}\left(R^2-2R+1\right)+\frac{R}{\pi}e^{-R}\right]\\&\cdot\left[\frac{1}{3}(1-e^{-12BT})-\frac{1}{36BT}+e^{-12BT}\left(\frac{1}{3}+\frac{1}{36BT}\right)\right]\\&+\left(\frac{1}{\pi}R^2e^{-R}\right)\left(\frac{1}{2}-\frac{1}{8BT}+\frac{1}{64B^2T^2}-\frac{e^{-8BT}}{64B^2T^2}\right)\\&+\left[\frac{1}{12\pi}e^{-R}\left(R^2-2R+1\right)\right]\end{aligned}\right\} \qquad (5.5\text{-}47)$$

It is not hard to show that when R becomes large,

$$\sigma_\phi^2 \to \frac{N_0 B_L}{P} \tag{5.5-48}$$

In a similar manner to the balanced case, consider the 4 to 1 unbalanced case, for which $\mathcal{R}_Q = 4\mathcal{R}_I$ (α=1/5 in this case). The loop gain is given by

$$\begin{aligned} S'(0) &= \left[A_v\sqrt{(1-\alpha)\gamma_Q P} + \sqrt{\alpha\gamma_I P} \right] \\ &+ A_v \begin{bmatrix} -2\sqrt{(1-\alpha)\gamma_Q P}Q\left(\sqrt{2R_Q(1-\alpha)}\right) \\ +2\sqrt{\alpha\gamma_Q P}Q\left(\sqrt{4R_Q(1-\alpha)\alpha R_I}\right) \end{bmatrix} \\ &+ \left[-2\sqrt{\alpha\gamma_I P}Q\left(\sqrt{2R_I\alpha}\right) + 2\sqrt{(1-\alpha)\gamma_Q P}Q\left(\sqrt{4R_I R_Q \alpha(1-\alpha)}\right) \right] \end{aligned} \tag{5.5-49}$$

It can be shown that the phase error variance is given by (5.5-50) for unbalanced NRZ data in which $\mathcal{R}_Q(t)$ is the Q channel data rate, ($\mathcal{R}_Q$ =1/T_Q), and $\mathcal{R}_I$ is the I channel data rate, ($\mathcal{R}_I$ =1/T_I), and $\mathcal{R}_Q$= 4$\mathcal{R}_I$.

This expression is obviously very messy, but is an expression of the tracking error variance of the modified Costas loop for NRZ signals when the power ratio of the Q channel to I channel is four and the data rate satisfies $\mathcal{R}_Q$= 4$\mathcal{R}_I$. Further results for the case of Manchester symbols are contained in [2, 42, 43].

$$\sigma_\phi^2 = \frac{N_0 B_L}{S'(0)^2} \left\{ \begin{aligned} &\frac{16}{5}\frac{R_Q}{\gamma_Q}\left[\sqrt{\gamma_I}\,erf\left(\sqrt{4/5\,R_Q}\right) - \sqrt{\gamma_Q}\,erf\left(R_I/5\right)\right]^2 BT_Q\left(1-\frac{1}{3}\frac{T_Q}{T_I}\right) \\ &+\gamma_Q\left[\frac{16R_Q}{5\pi}e^{-2R_I/5} - \frac{16\sqrt{R_Q}}{\sqrt{5}\pi}erf\left(\sqrt{4R_Q/5}\right)e^{-R_I/5} + 4\left(erf\left(\sqrt{4R_Q/5}\right)\right)^2\right] \\ &+\gamma_Q\left[\frac{16R_I}{5\pi}e^{-8R_Q/5} - \frac{8\sqrt{R_Q}}{\sqrt{5}\pi}erf\left(\sqrt{R_I/5}\right)e^{-4R_Q/5} + \left(erf\left(\sqrt{R_I/5}\right)\right)^2\right] \\ &+\frac{64}{25\pi}R_I R_Q\left(e^{-2R_I/5} + 4e^{-8R_Q/5}\right)\left[\frac{1}{8} - \left(\frac{1}{T_Q}+\frac{1}{T_I}\right)\frac{1}{64B} + \frac{1}{256B^2 T_I T_Q}\right] \\ &\left\{\frac{4}{\pi}e^{-8R_Q/5} + \frac{1}{\pi}e^{-2R_I/5} - \frac{4}{\pi}e^{-\left[\frac{R_I+4R_Q}{5}\right]}\right\} \\ &\left[\frac{32R_Q}{5\pi}e^{-8R_Q/5} + \frac{16R_Q}{30\pi}e^{-2R_I/5}\left(\frac{2}{5}R_I - 1\right)^2\right]\left(\frac{1}{3} - \frac{1}{36BT_Q}\right) \\ &\left[\frac{2R_I}{5\pi}e^{-2R_I/5} + \frac{16R_Q}{30\pi}e^{-8R_Q/5}\left(\frac{8}{5}R_Q - 1\right)^2\right]\left(\frac{1}{3} - \frac{1}{36BT_I}\right) \\ &\left[\frac{1}{12\pi}e^{-2R_I/5}\left(\frac{2}{5}R_I - 1\right)^2 + \frac{1}{3\pi}e^{-8R_Q/5}\left(\frac{8}{5}R_Q - 1\right)^2\right] \end{aligned} \right\} \tag{5.5-50}$$

5.6 FREQUENCY LOCKED LOOPS

Commonly a phase locked loop cannot handle the frequency acquisition problem readily unless augmented by some acquisition aid such as a frequency sweep across the frequency band of interest. Normally, only when the carrier loop (PLL, squaring, or Costas, for example) is within the frequency error of the one-sided loop noise bandwidth does frequency and phase acquisition occur quickly. Consequently frequency locked loops (FLL) are often employed for initial (frequency) acquisition, and then the loop configuration is switched to a carrier tracking loop (phase tracking). Whenever the carrier tracking loop is detected to have lost track, the FLL will be configured to aid in the frequency reacquisition. Now we will consider a FLL configuration.

5.6.1 The Cross Product FLL

The *cross product FLL* (CPFLL) is a useful discriminator model for tracking BPSK signals. Let the input signal plus white Gaussian noise be modeled as

$$y(t) = \sqrt{2P}d(t)\cos\left(\omega_0 t + \theta\right) + n(t) \tag{5.6-1}$$

where P is the received signal power in W, and $d(t)$ is the NRZ data bits with data bit duration T_b. In addition θ is the fixed unknown phase of the carrier, and $n(t)$ is the white Gaussian noise process with two-sided noise spectral density $N_0/2$. Assume that the arm integrators are T seconds long and exactly KT integration comprises one bit duration, so that

$$T_b = KT \tag{5.6-2}$$

The model for the CPFLL is shown in Figure 5.6-1 [44–46]. For spread spectrum systems the timing of the data will be known very accurately after code acquisition, since the data is related to the direct sequence code. Therefore integrators can be used in the arm filters for frequency tracking applications. Bit timing is assumed to be known in what follows. Notice that each output is multiplied by the delayed version of the other branch, which is then differenced to form the error control signal. This is enhanced by summing for K' samples where K' can be equal to K or some smaller value depending on the frequency range of the discriminator that is needed.

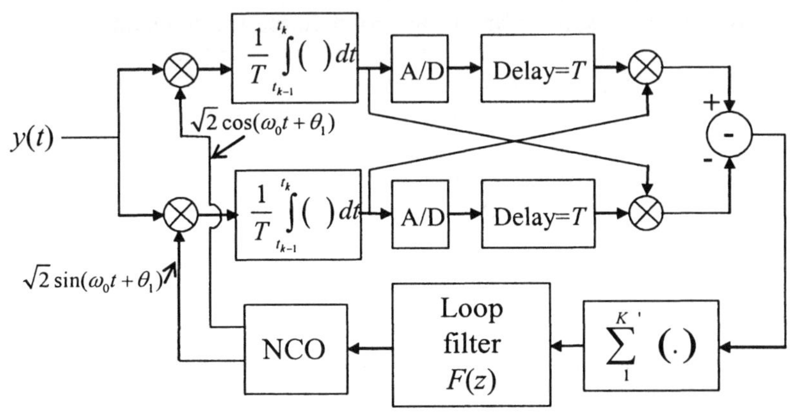

Figure 5.6-1 The cross product FLL (CPFLL) block diagram.

Natali [46] has shown that the discriminator function for this FLL is given by

$$D(\omega)=(P/2)\text{sinc}\left(\frac{\Delta\omega T}{2}\right)^2\sin(\Delta\omega T) \quad (5.6\text{-}3)$$

where $\Delta\omega$ is the angular frequency error (rad./s) and T is the integration time of the integrate-and-dump filters. The discriminator function when $P = 1$ and $T = 0.02$ second is plotted in Figure 5.6-2.

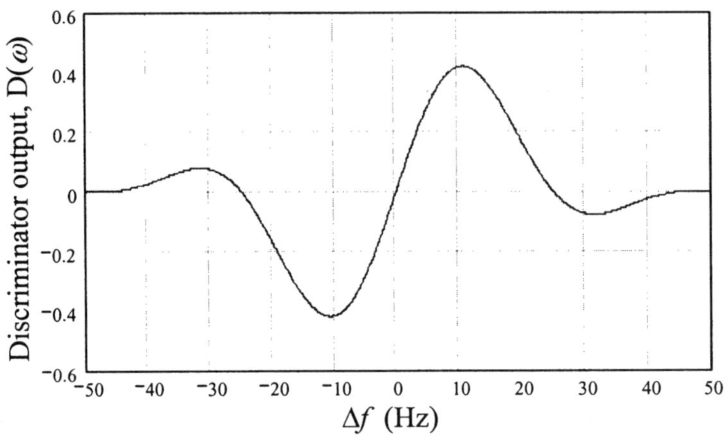

Figure 5.6-2 The discriminator function for the CPFLL with $T = 20$ ms.

Natali has also shown that the variance of the tracking error is given by the expression

$$\sigma_{\Delta f}^2=\frac{1}{2\pi^2}\frac{1}{SNR_T\left[1-2KB_LT\right]}\left(\frac{K}{K-1}\right)^2\left[\frac{1}{K}+\left(\frac{K-1}{K}\right)\frac{1}{2SNR_T}\right] \quad \text{Hz}^2,\ N\geq 2 \quad (5.6\text{-}4)$$

where B_L is the one-sided loop noise bandwidth of the FLL and $SNR_T=(PT/N_0)$ is the signal-to-noise ratio in the T second integration time. Other configurations are possible but will not be discussed here. Refer to [45, 47] for other configurations. Sometimes the loop configuration is modified to avoid computing the cross product over two adjacent bits. If the bit changes sign during the computation, it will introduce a small error; usually this term is dropped to avoid this problem. An arc-tangent detector is discussed in [47].

Normally the carrier loop will be fed back to mixer located before the code tracking loop. This provides for the carrier to be where it should be and prevents correlation loss due to carrier frequency errors.

In Section 5.4.6 a modified Costas loop was described that has better frequency tracking capabilities and diminishes the propensity for false lock, but has somewhat poorer phase tracking capabilities when compared to a similar unmodified Costas loop.

5.7 SUMMARY

This chapter dealt with carrier tracking, including residual carrier signal tracking, and suppressed carrier signal tracking, and a brief summary of frequency synthesizers. First, the structure of a phase locked loop (PLL) was derived from maximum-likelihood theory. Then the stochastic differential equation was derived for the PLL. This was followed by linear and nonlinear results for the PLL. A discussion of frequency synthesizers was presented discussing: (1) digital frequency synthesis, (2) direct frequency synthesis, and (3) indirect frequency synthesis.

Various loop configurations for tracking suppressed carrier signals were presented, including various forms of a Costas loop, along with a discussion of lock detectors for these loops. A theory of false (frequency) lock in a Costas loop was presented. Finally multiphase tracking loops including the N-phase Costas loop and the demod-remod loop, followed by a detailed analysis of the tracking performance of the modified four phase loop, were presented.

We concluded our discussion of tracking loops by presenting a cross product frequency lock loop indicating the variance of its frequency error.

References

[1] J. G. Proakis, *Digital Communications,* 4th ed., New York: McGraw-Hill, 2001, Chapter 6.

[2] J. K. Holmes, *Coherent Spread Spectrum Systems*, New York: Wiley-Interscience, 1982.

[3] R. C. Tausworthe, "Theory and Practical Design of Phase-Locked Receivers" Jet Propulsion Lab. Tech., Vol. 1, Report No. 32-819, February 15, 1966.

[4] F. M. Gardner, *Phaselock Techniques*, 2nd ed., New York: Wiley-Interscience, 1979.

[5] A. Blanchard, *Phase-Locked Loops Application to Coherent Receiver Design,* New York: Wiley-Interscience Publication, 1976.

[6] W. F. Egan, *Frequency Synthesis by Phase Lock*, New York: Wiley-Interscience Publication, John Wiley & Sons, 1981.

[7] A. J. Viterbi, *Principles of Coherent Communication*, New York: McGraw-Hill, 1966.

[8] R. C. Tausworthe and R. B. Crow, "Practical Design of Third-Order Tracking Loops," *JPL Interim Report 900-450*, April 27, 1971.

[9] D. Y. Abramovitch, "Analysis and Design of a Third Order Phase-Lock Loop," *1988 IEEE Military Communications Conference*, October 1988, pp. 23–26.

[10] C. Cahn, "Spread Spectrum Applications and State-of-the-Art Equipments," Paper no. 5, AGARD-NATO Lecture series, No. 58, May 29–June 6, 1973.

[11] W. C. Lindsey and M. Simon, *Telecommunication Systems Engineering*, Englewood Cliffs, NJ: Prentice-Hall, 1973.

[12] E. Drucker, "Model PLL Dynamics and Phase-Noise Performance, Model PLL Dynamics, Part 2," *Microwaves & RF*, February 2000.

[13] L. S. Cutler and C. L. Searle, "Some Aspects of the Theory and Measurement of Frequency Standards," *Proc. IEEE*, Vol. 54, 1966.

[14] B. M. Smith, "The Phase Lock Loop with Filter: Frequency of Skipping Cycles-Theoretic Considerations," Letters to the Editor, *Proc. of IEEE,* 1966, pp. 295.

[15] J. P. Frazier and J. Page, "Phase-Lock Loop Frequency Acquisition Study," *IRE Trans. on Space Electronics and Telemetry*, SET-8, September 1962, pp. 210–227.

[16] V. I. Tikhonov, "The Operation of Phase and Automatic Frequency Control in the Presence of Noise," *Automation and Remote Control*, Vol. 21, No. 3, 1960, pp. 209–214.

[17] R. Peterson, R. Ziemer, and D. Borth, *Introduction to Spread Spectrum Communications,* Englewood Cliffs, NJ: Prentice-Hall, 1995.

[18] V. Manassewitsch, *Frequency Synthesizers: Theory and Practice*, 2nd ed., New York: John Wiley & Sons, 1976.

[19] F. Oropeza and J. P. Schoenberg, "Binary Frequency Synthesis," *Frequency*, September/October 1976, pp. 14–17.

[20] R. Papaiech and R. Coe, "New Technique Yields Superior Frequency Synthesis at Lower Cost," *Electronic Design News*, October 20, 1975, pp. 73–79.

[21] U. L. Rhode, *Microwave and Wireless Synthesizer Theory and Design*, New York: Wiley-Interscience, 1997.

[22] C. Diorni and T. Humes, TRW notes posted on the Internet.

[23] TRW notes on indirect synthesizer phase noise, circa 1970.

[24] R. L. Didday and W. C. Lindsey, "Subcarrier Tracking Methods and Communication System Design," *IEEE Trans. on Communications Technology*, Vol. 16, August 1968.

[25] H. L. Van Trees, "Optimum Power Division in Coherent Communication Systems," Lincoln Lab., Lexington, MA, Technical Report 301, 1963.

[26] A. J. Viterbi, *Principles of Coherent Communications,* New York: McGraw-Hill, 1966.

[27] J. W. Layland, "An Optimum Squaring Loop Prefilter," *IEEE Trans. on Communications Technology*, October 1970.

[28] J. J. Stiffler, *Theory of Synchronous Communications*, Chapter 8, Englewood Cliffs, NJ: Prentice-Hall, 1971.

[29] J. G. Proakis, *Digital Communications*, 2nd ed., New York: McGraw-Hill, 1989.

[30] S. Riter, "An Optimum Phase Reference Detector for Fully Modulated Phase Shift Keyed Signals," *IEEE AES-5*, No. 4, July 1969.

[31] M. K. Simon, "Tracking Performance of Costas Loops with Hard-Limited In-Phase Channels," *IEEE Trans. on Communications*, April 1978.

[32] C. Cahn, "Improving Frequency Acquisition of a Costas Loop," *IEEE Trans. on Communications*, Vol. COM-25, No. 12, December 1977.

[33] J. K. Holmes, "Unpublished Notes on the Effective Tracking Error Variance Effect of Removing One Arm Filter for the Costas Loop," August 23, 1977.

[34] R. Leonhardt and H. H. Fleischmann, "Pull-In Range of Phase-Lock Circuits with Arbitrary Feedback Filter," *The Radio and Electronic Engineer*, August 1978.

[35] W. C. Lindsey, *Synchronization Systems in Communication and Control,* Englewood Cliffs, NJ: Prentice-Hall, 1972.

[36] J. A. Develet, Jr., "The Influence of Time Delay on Second-Order Phase-Locked Loop Acquisition Range," STL Report No. 9332.6-9, September 1962.

[37] R. C. Tausworthe, "Acquisition and False-Lock Behavior of Phase Locked Loops with Noisy Inputs," JPL SPS 37-46, Vol. IV.

[38] G. Hedin, J. K. Holmes, W. C. Lindsey, and K. T. Woo, "Analysis of False Lock Phenomena in Costas Loops During Acquisition," TRW Systems Memo SCT-50-76-249/KTW, July 15, 1976.

[39] G. Hedin, J. K. Holmes, W. C. Lindsey and K. T. Woo, Theory of False Lock in Costas Loops, *IEEE Trans. on Communications* Vol. COM-26, No. 1, January 1978, pp. 1–12.

[40] C. Weber, D. Vandervoet, R. Salcedo, and D. Jarett, "Spacecraft Switched TDMA Analyses and Laboratory Demonstration," TRW Final report, Vol. II, No. 25588-6001-RV-00, February 1976.

[41] W. R. Braun and W. C. Lindsey, "Carrier Synchronization Techniques for Unbalanced QPSK Signals, Parts I & II," *IEEE Trans. on Communications*, September 1978.

[42] J. K. Holmes and H. C. Osborne, "DG-2 Modified Costas Loop Tracking Performance, Part I, Balanced QPSK," TRW Memo No. LRD-78-200-035, November 1978.

[43] J. K. Holmes and H. C. Osborne, "DG-2 Modified Costas Loop Tracking Performance, Part II, Unbalanced QPSK," TRW Memo No. LRD-78-200-036, November 1978.

[44] F. D. Natalie, "AFC Tracking Algorithms," *IEEE Trans. on Communications*, Vol. COM-32, No. 8, August 1984.

[45] L. H. Enloe, "Decreasing the Threshold in FM by Frequency Feedback," *Proc. IRE*, Vol. 50, January 1962.

[46] F. D. Natali, "Noise Performance of a Cross-Product AFC with Decision Feedback for DPSK Signals," *IEEE Trans. on Communications,* Vol. COM-34, No. 3, March 1986, pp. 303–307.

[47] G. McGraw and B. Schnaufer, "A Modified Frequency-Locked Loop for Improved WAAS Carrier Tracking," *Proceedings of the ION GPS-95, 8th International Meeting*, Palm Springs CA, Part 2 of 2, September 12–15, 1995.

Problems

1. Starting with (5.2-31), (5.2-34), and the fact that $\phi = \theta - \hat{\theta}$, show that (5.2-35) follows.

2. Using (5.2-34) and letting

$$\sqrt{P}KF(s) = \frac{\omega_n^3 + 2\omega_n^2 s + 2\omega_n s^2}{s^2} \tag{P5-1}$$

show that the Jaffe-Rechtin closed-loop transfer function is given by

$$H(s) = \frac{\omega_n^3 + 2\omega_n^2 s + 2\omega_n s^2}{\omega_n^3 + 2\omega_n^2 s + 2\omega_n s^2 + s^3} \tag{P5-2}$$

In addition, show that $|1 - H(f)|^2 = f^6 / (f_n^6 + f^6)$ for this third-order loop.

3. For the passive second-order PLL show that

$$2B_{L2P} = \frac{r+1}{2\tau_2\left(1 + \dfrac{\tau_2}{r\tau_1}\right)} \cong \frac{r+1}{2\tau_2} \quad \mathit{for} \quad \tau_1 \gg \tau_2 \tag{P5-3}$$

where

$$r = \frac{\sqrt{P}K\tau_2^2}{\tau_1} \quad \tau_1 = (R_1 + R_2)C, \quad \tau_2 = R_2 C \tag{P5-4}$$

4. Starting from (5.2-48) show that the two-sided loop noise bandwidth for the Jaffe-Rechtin third-order PLL is given by

$$2B_{L3} = \frac{5\omega_n}{3} \quad \text{or} \quad B_{L3} = \frac{5\omega_n}{6} \tag{P5-5}$$

5. Show that for the ideal second-order PLL of (5.2-44) the loop noise bandwidth can be written as

$$B_{L2I} = \frac{\omega_n}{2}\left(\varsigma + \frac{1}{4\varsigma}\right) \tag{P5-6}$$

where

$$\omega_n = \sqrt{\frac{\sqrt{P}K}{\tau_1}} \qquad \varsigma = \frac{\tau_2}{2}\sqrt{\frac{\sqrt{P}K}{\tau_1}} \tag{P5-7}$$

Hint: Use the same definition of r as in Problem 3.

6. Show that the steady state probability density for the first-order PLL can be approximated by a Gaussian probability density for small phase error variance.

7. Show that for when the low-pass equivalent of the n-pole Butterworth filter bandpass filter has magnitude squared response of

$$|H_{LP}(f)|^2 = \frac{1}{1+\left(\frac{f}{f_0}\right)^{2n}} \tag{P5-8}$$

where f_0 is the 3-dB bandwidth, then $B_{LP}^{'}$ is given by

$$B_{LP}^{'} = B_{LP}\left(1-\frac{1}{2n}\right) \tag{P5-9}$$

and the equivalent linearized tracking error variance of the squaring loop is given by

$$\sigma_{\Phi_e}^2 = \frac{N_0 B_L}{\alpha P}\left[\frac{\alpha^{'}}{\alpha}+\frac{N_0 W}{2\alpha P}\left(1-\frac{1}{2n}\right)\right] \quad \text{rad}^2 \tag{P5-10}$$

8. Show that the autocorrelation function of a random amplitude, fixed duration, rectangular signal pulse that is statistically independent from pulse to pulse is given by

$$R_N(\tau) = \begin{cases} \sigma_N^2\left(1-\frac{|\tau|}{T}\right) & |\tau| \le T \\ 0 & |\tau| > T \end{cases} \tag{P5-11}$$

9. Determine the linearized tracking performance of the modified Costas loop, with a one-pole Butterworth arm filter. Assume that no frequency error exists. Model the received signal plus WGN as

$$y(t) = \sqrt{2P}d(t)\sin\left(\omega_0 t+\theta\right)+n(t)$$

10. Using Figure 5.4-5, show that the error signal is given by (P5-12)

$$Z(t) = Pd(t)\tilde{d}(t)\frac{\sin(2\phi)}{2}+\sqrt{P}d(t)\tilde{n}_s(t)\sin(\phi)+\sqrt{P}\tilde{d}(t)n_c(t)\cos(\phi)+n_c(t)\tilde{n}_s(t) \tag{P5-13}$$

where k_m^2 has been assumed to be unity for convenience, and the tildes denote filtering by the low pass filters. Letting $2\phi = \Phi$ show that the differential equation of operation is given by

$$\dot{\Phi}(t)+KF(s)Pd(t)\tilde{d}(t)\sin(\Phi(t)) = \left(-\sqrt{P}n_c(t)\tilde{d}(t)-n_c(t)\tilde{n}_s(t)\right) \tag{P5-14}$$

Now show that the time average of $d(t)\tilde{d}(t)$ has the value of

$$\left\langle d(t)\tilde{d}(t)\right\rangle = \frac{1}{T}\int_{-\infty}^{\infty} P(f)H^{*}(f)P^{*}(f)df = \int_{-\infty}^{\infty} S_d(f)H^{*}(f)df \tag{P5-15}$$

Evaluate the above average by noting that

$$\mathrm{Re}\left\{H^{*}(f)\right\} = \mathrm{Re}\left\{\left(\left(\frac{1}{1+j\dfrac{f}{f_0}}\right)\right)^{*}\right\} = \mathrm{Re}\left\{\frac{\left(1-j\dfrac{f}{f_0}\right)^{*}}{1+\left(\dfrac{f}{f_0}\right)^{2}}\right\} = \frac{1}{1+\left(\dfrac{f}{f_0}\right)^{2}} = \left|H(f)\right|^{2} \tag{P5-16}$$

Based on this show that

$$\left\langle d(t)\tilde{d}(t)\right\rangle = \int_{-\infty}^{\infty} S_d(f)\left|H(f)\right|^{2} df = \alpha \tag{P5-17}$$

Now show that

$$\Phi(t) = \frac{\alpha PKF(s)}{s+\alpha PKF(s)}\left[\frac{-\sqrt{P}n_c(t)\tilde{d}(t) - n_c(t)\tilde{n}_s(t)}{\alpha P}\right] \tag{P5-18}$$

Finally, show that the linearized effective tracking error variance is given by

$$\sigma_{\Phi}^{2} = \frac{4N_0B_L}{\alpha P}\left[1+\frac{N_0B_{LP}}{\alpha P}\right] \tag{P5-19}$$

11. Show that the relationship between the two bandwidths used in the linearized tracking error variance B_{LP} and $B_{LP}^{'}$ (see (5.4-28)) for a Butterworth arm filter is given by

$$B_{LP}^{'} = B_{LP}\left(1-\frac{1}{2n}\right) \tag{P5-20}$$

where the low pass Butterworth filter has a frequency response given by

$$\left|H_{LP}(f)\right|^{2} = \frac{1}{1+\left(\dfrac{f}{f_0}\right)^{2n}} \tag{P5-21}$$

where f_0 is the 3 dB bandwidth of the low pass filter.

12. Consider the false lock of a Costas loop when Manchester symbols are used. Following the approach in the book for the NRZ symbols show that

$$C_{-k} = \left(\frac{1}{1+j\left(\dfrac{k}{2R}\right)}\right)^{2}\left\{(-1)^{k}\frac{\sin(k\pi)}{k\pi} - \frac{1}{2\pi R}\left(\frac{1+j\dfrac{k}{2R}}{1-j\dfrac{k}{2R}}\right)\left[1+2(-1)^{k}+4e^{-\pi R}\cos\left(\frac{k\pi}{2}\right)+(-1)^{k}e^{-2\pi R}\right]\right. \tag{P5-22}$$

From this result show that the false lock levels of Table 5.4-1 follow.

13. Consider the approximate performance for the DDFL (Section 5.4-8) when $R \geq 2$. Note that

$$Q(x) \cong \frac{1}{\sqrt{2\pi R}} \exp(-x^2/2) \tag{P5-23}$$

is a good approximation when $R \geq 2$. Also note that

$$\frac{1}{1-x} \cong 1+x \tag{P5-24}$$

for small x. Using these two approximations, show that the tracking error variance is given by

$$\sigma_\phi^2 \cong \frac{N_0 B_L}{P}\left[1+\frac{1}{\sqrt{2\pi R}}\exp(-R)\right]^2 \tag{P5-25}$$

when $R \geq 2$.

CHAPTER 6

Code Acquisition in Direct Sequence Receivers

6.0 INTRODUCTION

This chapter deals with pseudonoise (PN) code acquisition techniques based primarily on both active search methods and passive matched filter search methods; however, other approaches to achieving acquisition will also be discussed. By code acquisition we mean the aligning of the local receiver PN code with the received PN code to within a fraction of a code chip (of duration T_c seconds). Complete synchronization is effected when both code acquisition has been accomplished and the code loop acquisition transient has settled out into steady state tracking.

Within the realm of active search methods for obtaining acquisition, both fixed dwell time methods, including single and double dwell time schemes, as well as variable dwell time methods will be discussed. Parallel processing will be analyzed, which is the method of choice for modern code acquisition systems, when acquisition time is important. Although most of the development in this chapter is for PN code acquisition, there is also a brief section on acquisition of frequency-hopped systems.

There are a number of parameters to consider when designing an acquisition system for a spread spectrum receiver. The frequency uncertainty and time uncertainty are two parameters that are of major importance in acquisition performance. Another design parameter, for example, is the receiver front-end bandwidth. This bandwidth is a function of both the signal spread spectrum bandwidth and the difference between the received signal carrier frequency and the receiver oscillator heterodyning frequency. The time uncertainty that is an important parameter includes both the actual clock uncertainties between the transmitted clock and the receiver clock and the range uncertainty. The larger the effective time uncertainty, the greater the time it will take to achieve acquisition, or the greater parallel searching is required for a given acquisition time requirement. The data rate also affects the performance of the link, since the greater the data rate the greater the required signal power to maintain a reasonable bit error rate, and in some situations the shorter the coherent correlation time available.

In addition to the parameters described earlier, the type of acquisition circuit that is used also affects the acquisition time performance. Of course the greater the number of acquisition circuits that are used in parallel, the faster the acquisition of the PN signal is accomplished.

One common method to obtain code acquisition that only utilizes a single correlator is to first advance the local code relative to the received code the maximum expected amount (minimum delay). Then the product of the local receiver code and the received code is correlated and filtered to the data bandwidth and detected with a square law detector. For baseband implementations this is done at baseband with a two channel I-Q arrangement, and for bandpass implementations only one RF channel is needed. The resultant signal is then integrated for a fixed time duration that is known as the coherent dwell time (or correlation time). In many designs a number of these coherent integrations are added noncoherently to form a noncoherent integration period. At the end of the integration period the output of the noncoherent sum is compared to a fixed, known threshold. If the output is greater than the threshold the system declares, the signal is synchronized. If the output is not greater than the threshold, it is presumed that the signal is not located at that time delay and the system retards the code phase by some fraction of a code chip (usually one half of a code chip) and the process starts over at this code phase.

This process continues until either the code is detected or the whole time uncertainty is searched, and the process is started over again by advancing the local code the maximum expected amount repeating the process. This method of acquisition is known as an active search method and the system that obtains acquisition is known as a *conventional sliding correlator*. Typically in modern receivers sliding correlators are implemented digitally, with many operating in parallel.

Another method that is used in code acquisition is *passive matched filtering*. Here the local code (or a segment of it) is stored in a register and a chip-by-chip comparison is made (by summing the chip-by-chip comparison up). The result is that when the received code and the local code are lined up in time, a large pulse occurs at the output of the correlator, which signifies the occurrence of code acquisition (code alignment). This method is typically much faster than a single active search method but may not be applicable to all systems depending on the interference environment which may include jamming. Passive matched filters are commonly implemented digitally and commonly at baseband so that both an I and Q correlator are needed.

A simplified pseudonoise binary phase shift keyed (PN BPSK) acquisition and tracking system, along with the transmitter and receiver, is shown in Figure 6.0-1. In Figure 6.0-1(a) a bandpass version is shown.

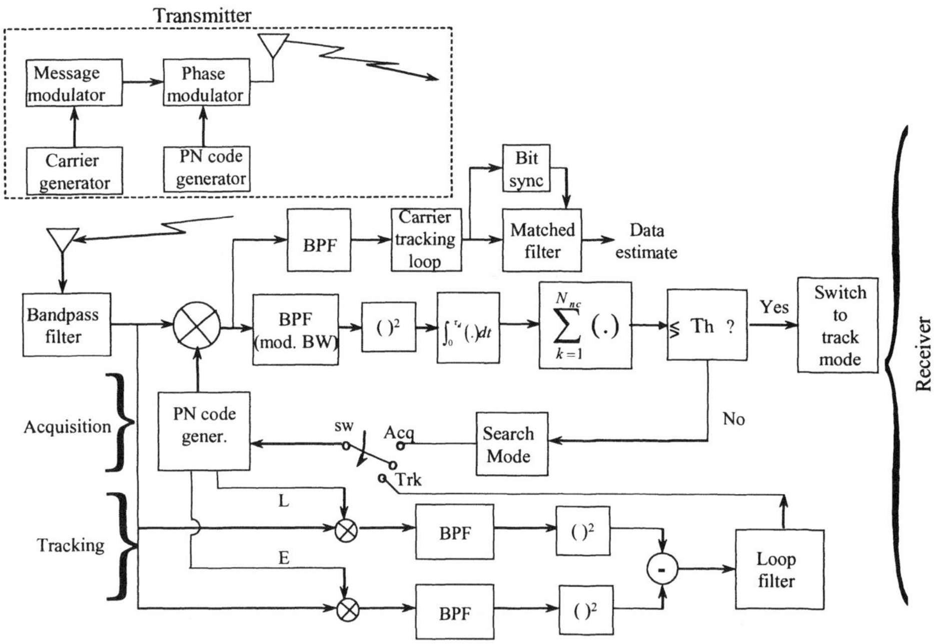

Figure 6.0-1 (a) BPSK transmitter and receiver with acquisition and tracking functions using a bandpass acquisition system.

The bandpass version of the acquisition and tracking model processes the signal at the carrier (or IF). First consider the acquisition circuit. The filtered signal plus noise and possible interference and jamming is multiplied by the local code set to the minimum delay possible; it then filters it to a bandwidth that depends on the modulation bandwidth plus the Doppler frequency offset. In this introduction it is assumed that there is no carrier frequency error. The resulting signal plus noise (or noise plus interference plus jamming if they are present) is integrated for τ_d seconds, which is called the *correlation time*. N_{nc} noncoherent correlations are summed (thus, $\tau_D^{'} = N_{nc} \times \tau_D$ is the total noncoherent correlation time), and this summation is compared to a predetermined threshold. If the noncoherent sum exceeds the threshold, a detection is assumed, and the code tracking function is initiated. Otherwise it is assumed that the local code was not sufficiently close to the received code, and the received code delay is advanced a small fraction of a code symbol (commonly 1/2 of a code symbol) and the process is tried again on the new code timing

hypothesis. If the noncoherent correlation signal is above the threshold the code loop is turned on and a longer correlation is tried to hopefully verify the first "hit." If the signal-to-noise ratio is sufficient, then normally the code loop will lock on to the received code and the second noncoherent correlation will verify the true "hit." Code-tracking loops will be discussed in the next chapter; however, the lower portion of the block diagram shows the elements of the code tracking loop. Once code loop synchronism is obtained the carrier loop will be initiated, and when it has stabilized the data can be demodulated. Normally bit timing can be obtained from the code itself. When the carrier frequency error between the received signal and the local receiver frequency is not negligible, both time uncertainty and frequency uncertainty must be searched.
In Figure 6.0-1(b) a baseband version is illustrated. The description of its operation is very similar to the bandpass version, except the processing is done at baseband with I and Q baseband processes. Modern day receivers usually process the signals at baseband.

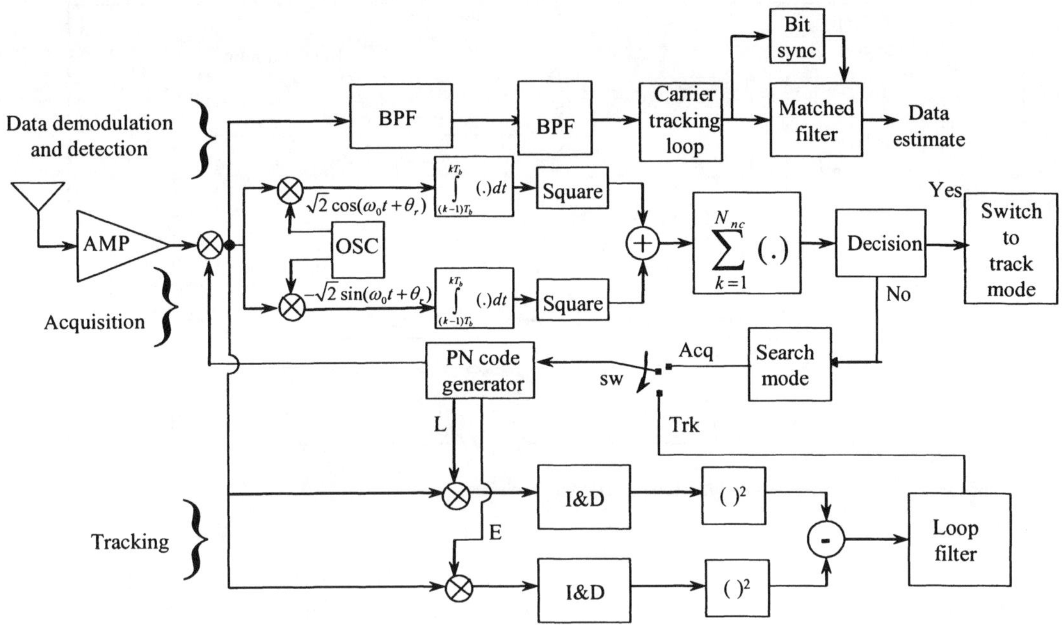

Figure 6.0-1 (b) Baseband acquisition system of the receiver for BPSK PN code modulation.

For both figures the modulator modulates both data and the PN signal onto the transmitted BPSK carrier by means of phase modulation. The received signal is of the form

$$y(t)=\sqrt{2P}d(t)PN(t)\cos(\omega_0 t+\theta)+n(t) \tag{6.0-1}$$

where P is the received signal power, $PN(t)$ is the unit amplitude PN code process, $d(t)$ is the unit amplitude (± 1) data process, ω_0 is the carrier angular frequency in radians/second, θ is carrier phase, and $n(t)$ is the noise process corrupting the reception of the signal.

Before embarking on the theory of active and passive acquisition, some other PN code acquisition techniques will be discussed that are neither active nor passive correlation methods, but offer the potential for fast acquisition. Basically these approaches attempt to determine the state of the PN encoder in order to obtain quick acquisition. An approach that was introduced by Ward [1] obtains synchronization by loading n samples of the received chip sequence signal plus noise (assumed to be at baseband) into the n-stage linear PN code shift register, which defines the initial state of the shift register. Figure 6.0-2 illustrates the acquisition block diagram. Then, after the first n chips are loaded into the local shift register, the clock starts the local shift register that is correlated against the incoming signal. The basic idea is that knowing the initial sequence of the PN code generator is all that is needed to continue to generate the sequence. After a period of time the lock detector decides if the local code is matching the incoming code; if not, a new set of samples of the chip sequence are input to the local chip generator. This process continues until the lock detector indicates that the signal is present. When this occurs, the chip sequence is then used as the code sequence for the code-tracking loop.

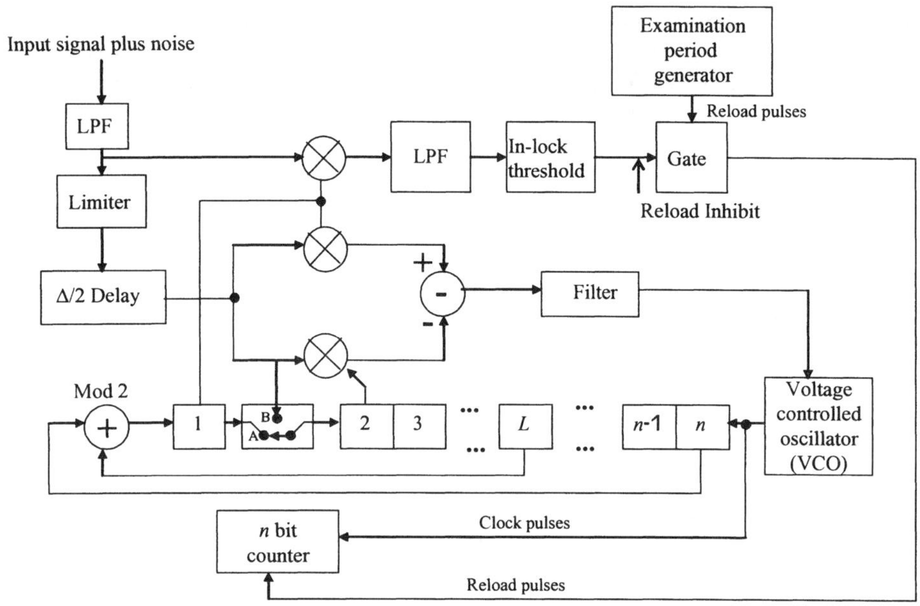

Figure 6.0-2 The rapid acquisition by sequential estimation (RASE) scheme.

This acquisition method is known as the rapid acquisition by sequential estimation (RASE) method. This scheme only works for an unmodulated baseband system, since with BPSK the actual sequence or its inverse could be received and the inverted code initial state will not generate the inverted sequence. The RASE system makes its best estimate of the first n received code chips, loads the receiver sequence generator with that estimate, and starts the operation of the code generator and the code tracking circuits. If the correct estimate is loaded, tracking will occur.

At the same time a cross-correlation is performed between the input signal plus noise and the local code generator. If the cross-correlation indicates the receiver has the correct sequence and tracking is occurring, no further action is taken. If the cross-correlation indicates that an incorrect state is loaded, a new estimate is made, loaded, and tracked. This process continues until true lock occurs. The low pass filter on the input and the limiter form the estimate of the individual chips. The n cycle shift register is tried for the examination period (T_e seconds).

When the estimate is finished, the loading switch is changed to disconnect the local code generator from the limiter and close the feedback loop. The local code generator then produces the sequence starting from the initial state with which it was loaded. The delay lock loop consists of a VCO, a loop filter, and a differential correlator, which uses reference signals from taps of adjacent storage elements of the local code generator. As noted earlier code-tracking loops will be discussed in the next chapter.

In summary the RASE system makes estimates of the present state of the input sequence, starts the local sequence code generator at that present state, and attempts tracking from that condition. New estimates are made at a rate controlled by an examination period code generator until an in-lock detector determines that the estimate was made. Further discussion can be found in [1].

A modification of this technique known as RARASE or recursion-aided RASE involves a preliminary determination of the shift register contents to determine if it is a likely candidate or not. If not, it is rejected quickly and a new initial condition is loaded into the chip sequence register. This method is described in Ward and Yiu [2]. This method also is dependent on having a baseband signal present at the input. The RARASE method provides better performance than RASE at high SNRs, but requires a significant amount of logic hardware.

Kilgus [3] has improved on these two acquisition techniques by using a portion of the 2^n–1 chips in a maximal length sequence and obtaining a number of independent estimates of the n chips of the initial state. The chip estimates are 1-bit estimates. The estimates for the initial state are majority voted and the resultant initial state is then loaded into the local shift register generator. When the number of estimates is equal to one, it reduces to the Ward acquisition system. Kilgus notes that the probability of acquisition can be made close to unity for moderate channel SNRs. Kilgus did not compare his approach with the sliding correlator method.

Another modification of the RASE technique is due to Chiu and Lee [4]. Their scheme works with modulated signals, which is a big advantage over the three schemes, which don't work with data modulation. Basically Chiu and Lee developed an extended characteristic polynomial that will generate the PN sequence or the inverted PN sequence, depending upon the initial condition or its inverse. As with the other techniques discussed, the scheme works best when the chip SNR (PT_c/N_0) is above –15 dB. Thus this scheme will work for BPSK signals and is not restricted to baseband use, as are the other three methods [1–3] described earlier.

Another offshoot of the Ward approach is the one due to Rawlins, Koller, and Belkerdid [5], which is based on digital technologies. This method does work with BPSK signaling. Again a chip detector is used to estimate the chip sequence off of the in-phase arm of a Costas loop, which is in turn fed into an "algebraic despreader." This algebraic despreader controls a chip clock phase locked loop. The code acquisition is nestled inside of a Costas loop. The chip phase lock loop has as an input, the estimated chip sequence and has a clock signal that is fed back to the algebraic despreader. The algebraic despreader is composed of two local PN generators, two dot product calculators, and a smoothing filter. The authors have built this acquisition device and have shown that it works. The fact that this technique is contained inside a Costas loop suggests that it will not work well at low SNR. Based on the data in their paper it appears that it works best above –10 dB. More details are given in their paper.

A set of three direct sequence spread spectrum code phase acquisition schemes have been proposed by Dodds and Zhang [6] that use cyclic accumulation and a matched filter to provide a rapid estimate of the correct code phase. The authors claim that the proposed schemes will achieve much better acquisition time performance than the other methods. More details are contained in [6].

Stiffler [7] has suggested another approach to fast acquisition, while at the Jet Propulsion Laboratory (JPL). This method was used for ranging code acquisition, which allowed acquisition of the code by steps in which $n = \log_2(N)$ binary decisions need be made to search a code of length of $N = 2^n$. When N is large and the time uncertainty is on the order of N chips, the improvement in the Stiffler method can be considerable. This approach may not be useful for code multiple access without major modification; it was designed for a single code stream acquisition problem.

Curry, Schwartz, and Collins [8] have developed an approach related to the Ward approach, with the added feature of searching the code states with the aid of a trellis search using an extension of the Fano search algorithm. Their method is called the state transition assisted receiver/synchronizer (STARS).

Figure 6.0-3 illustrates the relationships for the various types of noncoherent acquisition detectors. Code acquisition detectors can be subdivided into four groups: active searches, passive searches (matched filter searches), code state searches, and transform searches.

Active code searches can be further subdivided into fixed dwell times (fixed correlation times) and variable dwell times. Variable dwell time schemes are easier to implement with a single correlator. However fixed dwell time methods are commonly used in highly parallel correlator designs.

Matched filter acquisition can be done with or without the aid of an FFT (fast Fourier transform) to search frequency.

Code state estimators are acquisition techniques that attempt to determine the shift register state of the PN sequence as opposed to the standard sliding correlator that searches over the actual sequence in a serial manner. We have discussed a few in the above discussions.

Finally transform acquisition procedures are based on taking the complex conjugate of the FFT[1] of the input sequence and then storing it in memory. Then an FFT is performed on the code to be searched. Then the two FFTs are multiplied point by point and the inverse FFT is taken. Searching this product of FFTs for the maximum value yields the time and frequency of the received signal.

The remainder of this chapter will be concerned with the active and passive code searches and the transform method. Typically active searches are done in a highly parallel manner, with as many as thousands or hundreds of thousands of correlators (effective correlators) used in parallel.

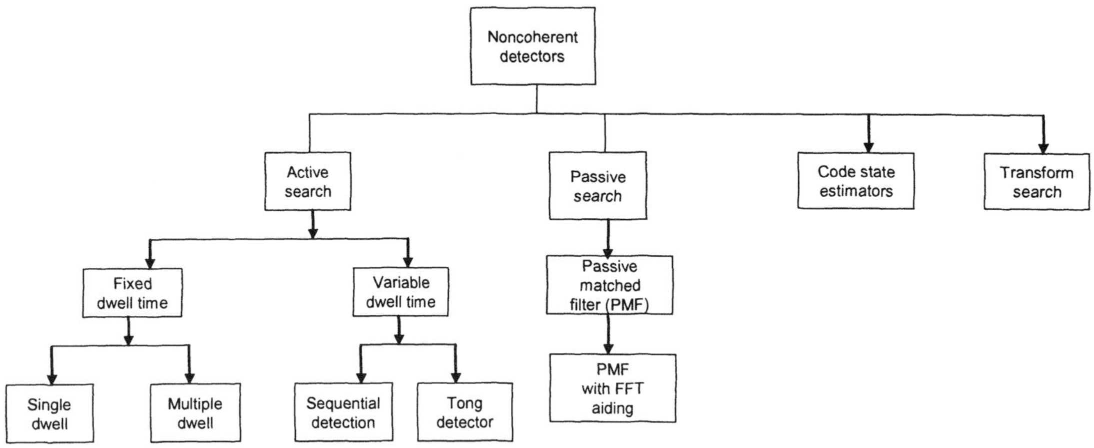

Figure 6.0-3 Pseudonoise noncoherent code acquisition detector types.

6.1 THE ACQUISITION PROBLEM

PN code acquisition can be specified in various different ways. For example, "the mean time to acquire is 10 seconds" is a valid statistical specification for acquisition time performance. On the other hand the requirements of "acquisition in 18 seconds with a probability of 0.9" is an equally reasonable specification. What is not mathematically acceptable is to state that the acquisition time is 10 seconds. Mathematically or statistically this is nonsense. Either some statistic associated the acquisition time (mean or mode time and so forth), or else a time along with an associated probability, must be specified. It is not uncommon to see specifications for system performance that state the acquisition time in so many seconds, under some initial time and frequency uncertainty, without a mean or probability specified.

The first topic to be addressed will be the mean time to acquisition for serial search acquisition schemes (conventional sliding correlator) with a single channel system. It is to be noted that this theory is not limited to BPSK signals; the only requirement is that the acquisition process be characterized as a discrete time Markov process. It will be assumed that the detection probability (P_D) and the false alarm probability (P_{FA}) or the effective false alarm probability[2] is known and is constant in what follows.

6.2 ACTIVE SEARCH ACQUISITION (SLIDING CORRELATOR)

The first problem that we now consider is how long it takes to align the receiver code timing to some accuracy with the received code; this time is defined as the acquisition time. Normally it is desirable to align the timing of the local code to within 1/2 to 1/4 of a chip (code symbol)[3] error for an active correlation system. We assume an additive white Gaussian noise (WGN) channel, the actual acquisition time is a random

[1] FFT denotes the fast Fourier transform.

[2] The effective false alarm probability will be discussed later in the chapter.

[3] For BPSK, by chip we mean the smallest duration for which the baseband waveform is constant. Manchester symbols are an example in which the symbol has two chips. Thus symbol duration is a more general term than chip duration. Of course, for NRZ symbols the symbol duration is equal to the chip duration.

variable and therefore it must be characterized by statistical techniques. This analysis follows the work of Holmes and Chen [9] for the single correlator model for BPSK code acquisition, without noncoherent combining. Before delving into the theory, let us consider a simple example to clarify the concepts.

Example 1 Assume that the time and range uncertainty corresponds to N PN code chips or $NT_c = \Delta T$ seconds of time uncertainty, where T_c is the code chip duration and there is no carrier frequency uncertainty. Assume that the probability of detection at the correct hypotheses is one (P_D=1), the probability of false alarm at any incorrect hypotheses is zero (P_{FA} = 0), and the dwell time (correlation or the integration time) is τ_D seconds (coherent correlation only). Figures 6.0-1(a) and 6.0-1(b) indicate representative bandpass and baseband code acquisition systems. It is also assumed that there is no transmitter or receiver oscillator jitter or time drift. In addition it will be assumed that the timing search update is in one half-chip increments, so that $2N$ timing positions (timing hypotheses) have to be searched. The time to search all N cells ($2N$ timing positions) is given by

$$T_{acq} = 2N\tau_D \quad \text{(s)} \tag{6.2-1}$$

Thus, in this case of the example, the probability of acquisition in T_{acq} seconds is one. The mean acquisition time is therefore given by one half of T_{acq}, or

$$\overline{T}_{acq} \cong N\tau_D \quad \text{(s)} \tag{6.2-2}$$

Unfortunately, in general, the detection probability P_D is not one, the false alarm probability is not zero, and the calculation for the mean acquisition time is not nearly as simple for a nonideal system.

It should be noted that the N chips that are to be searched over (correlated) in our model are physically determined from the timing uncertainty between the transmitter and the receiver clocks and the position uncertainty between the transmitter and the receiver. Since position uncertainty can also be expressed as a time uncertainty in chips, the value of N chips reflects the totality of position and timing uncertainty, expressed in chips.

We now develop the expression for the mean time to acquire the PN code with a single correlator and without noncoherent combining. Readers not familiar with flow graph manipulations and construction should consider reading the appendix to this chapter or some other source for an introduction to flow graph theory.

6.2.1 Mean Acquisition Time Model for an Active Search System

We initially assume that the signal, be it BPSK, QPSK, OQPSK, or MSK, is immersed in additive white Gaussian noise (AWGN), and is centered at some frequency ω_0 with some signal power P. In the case of BPSK code modulation, the filter, square, and integrate detector acquisition system is shown in Figure 6.2-1 for the baseband version.

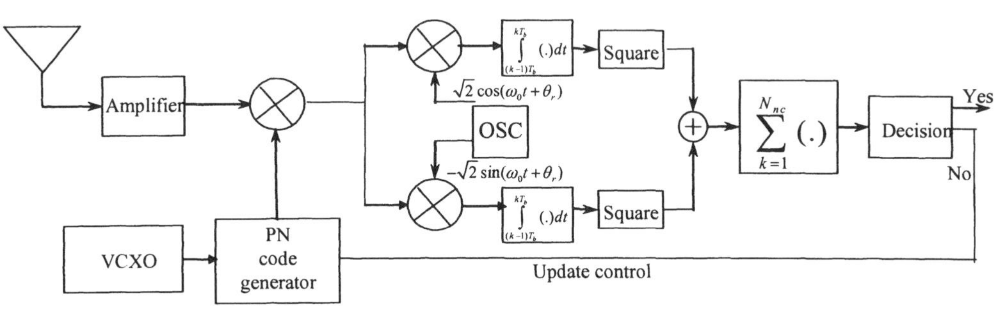

Figure 6.2-1 A baseband version of a PN code acquisition system for BPSK modulation.

The effective noncoherent dwell time is taken as τ_D' [4] ($\tau_D' = N_{nc}T_b$) seconds for the code acquisition system. We follow the method of [9–11] in the analysis to follow. The baseband and a bandpass version will be roughly similar if the baseband filters are approximate translates of the bandpass filter.

If QPSK, OQPSK, or MSK were used for the code modulation the detector system would differ, but the algorithm and model and the results that will be derived in the following still apply.

Our model is as follows. Assume that due to timing and range uncertainty there are N code chips of time uncertainty to be searched. It is assumed that there is no frequency error between the receiver and the received signal. In addition assume that the N code chips result in q code phases to be searched, where q may be equal to the number of PN code chips to be searched, or some multiple of it. For example if the code is searched in one-half chip steps then $q = 2N$ and in one-third chip steps $q = 3N$, and so forth. Further, it is assumed that if a "hit" occurs (the output is above the threshold), the system goes into a verification mode that may include both an extended duration noncoherent dwell time and an initiation of the code loop-tracking mode. In any event, we model the associated "penalty of time" of verifying a false alarm as $K\tau_D'$ seconds (the time needed to determine that it was a false alarm and not a true detection), where K is a number typically larger than one. If a true "hit" (signal is present) is observed, the system has acquired the signal, and the search is completed.

Assume that the false alarm probability is P_{FA} and the probability of detection is P_D, and they are both given. Clearly the time to obtain acquisition (a true hit) is a random variable. Although the problem of obtaining the distribution function of the acquisition time is obtainable in principle, it is very difficult to obtain in practice, at least in closed form. Consequently we shall be content with the mean and variance of the acquisition time at this time.

Initially we shall assume that Doppler effects (carrier frequency error) are not present; later we will relax this assumption. Also we will assume that the correlation error in the acquisition search will be fixed at one-half the update size, and the update size is often chosen to be one-half of a code chip, so that the correlation error would then be one-fourth of a chip. This assumption implies that P_D is a constant (time invariant) and as a consequence, the analysis of the model can be carried out. A basic assumption is that the detection and false alarm probabilities are constant with time.

6.2.2 Analysis of the Active Search System

In this section the mean time to acquire the PN code, using a flow graph method to determine the generating function of the acquisition process, will be used. Let each cell to be searched be numbered from left to right (from 1 to q) so that the k-th cell has an a priori probability of having the signal being located in that cell, given that it was not present in cells 1 through k–1, of

$$P_k = \frac{1}{q+1-k}, \quad 1 \le k \le q \tag{6.2-3}$$

The generating function flow graph diagram is given in Figure 6.2-2 using the rule that at each node the sum of the probabilities emanating from the node equals unity. The **S** denotes the start position, and the **F** denotes the final (acquisition) position.

Consider node 1. The a priori probability of having the signal actually located in state one is $P_1=1/q$, and the probability of it not being present in the node 1 is $1–P_1$ or $(q–1)/q$. Suppose the signal were not present. Then we advance to the next node (node 1a); since it corresponds to a probabilistic decision and not a unit time delay, no z multiplies the branch going to it. A z in any path denotes one unit delay (one dwell time, τ_D' seconds). At node 1a either a false alarm occurs with probability of $P_{FA} = \alpha$, which requires one unit of time to determine (τ_D' seconds) and then K units of time ($K\tau_D'$ seconds) are needed to verify that

[4] In general, we will use τ_D for the coherent dwell time and τ_D' for the noncoherent dwell time, that is, for multiple dwell times.

there was only a false alarm, or alternatively there is no false alarm with probability $(1-\alpha)$, which takes one dwell time to determine, and which requires the $(1-\alpha)z$ branch going to node 2.

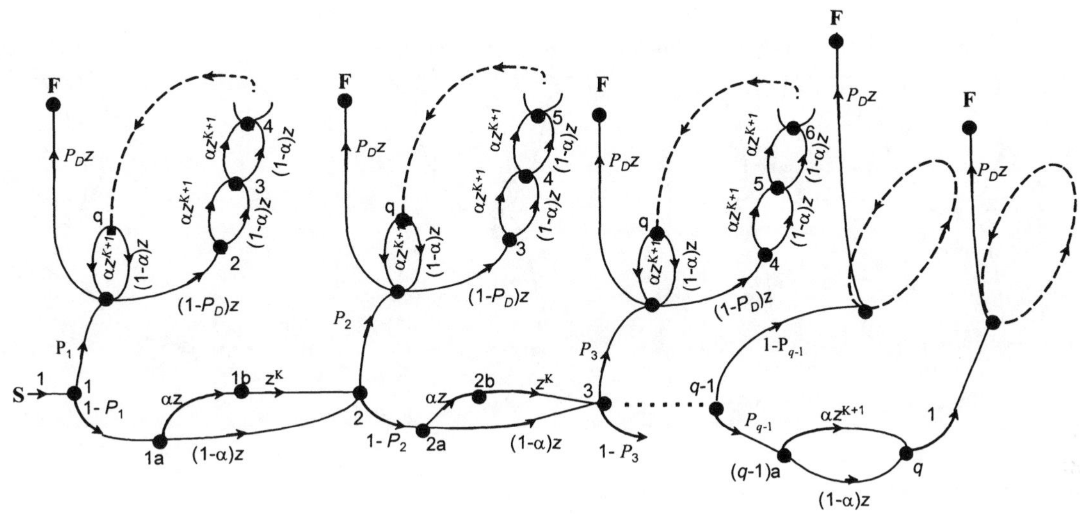

Figure 6.2-2 Flow graph diagram for determining acquisition time via the generating function.

Now consider the situation at node 1 when the signal does occur there. If the signal is detected then acquisition, as we have defined it, occurs and the process is terminated, hence the node **F** denoting "finish." If the signal was not detected at node 1 (the integrator output was below the threshold), which occurs with probability $1-P_D$, one unit of time would be consumed. This is represented by the branch $(1-P_D)z$, leading to node 2.

At node 2 (assuming that the signal was present in cell 1) in the upper left part of the diagram, either a false alarm occurs with a probability α and a delay of $(K+1)$ units occurs, or a false alarm does not occur with probability of $1-\alpha$ and a delay of one unit occurs. The remaining portion of the generating function flow graph is a repetition of the portion just discussed with the appropriate node number changes.

In order to obtain the transfer function of the flow graph we shall reduce it by combining sections at a time. Let

$$H(z) = \alpha z^{K+1} + (1-\alpha)z \tag{6.2-4}$$

Then the flow graph can be drawn as shown in Figure 6.2-3. Letting $\beta = 1-P_D$ and letting

$$Q(z) = \frac{z}{1-\beta(H(z))^{q-1}z} \tag{6.2-5}$$

we can reduce the flow graph to that of Figure 6.2-4, since $Q(z)$ takes into account the feedback loops in the path to the finish (**F**, or acquisition).

Define

$$B_l(z) = P_l(1-\beta)Q(z) = \frac{P_l(1-\beta)z}{1-\beta z(H(z))^{q-1}} \tag{6.2-6}$$

and

$$A_l(z) = (1 - P_l)H(z) \tag{6.2-7}$$

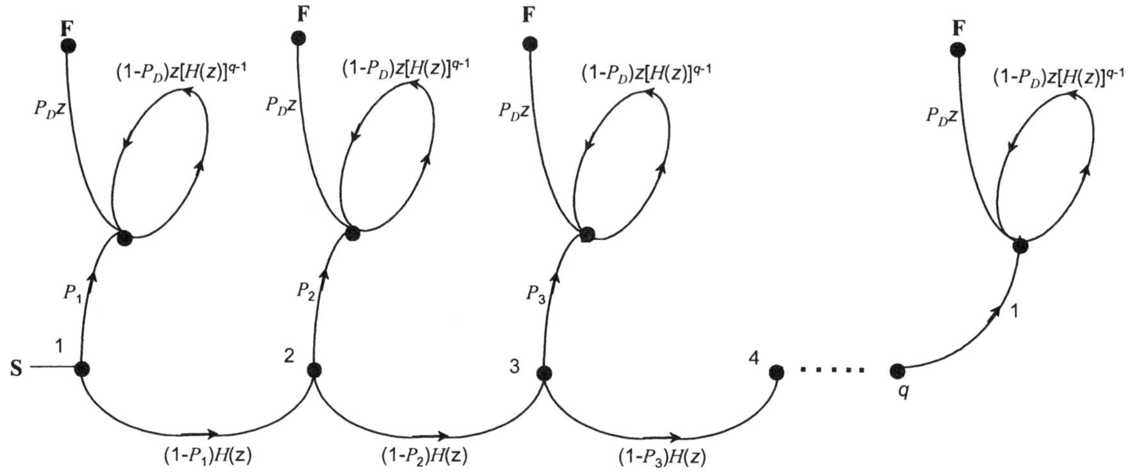

Figure 6.2-3 Partially reduced flow graph for the determination of acquisition time.

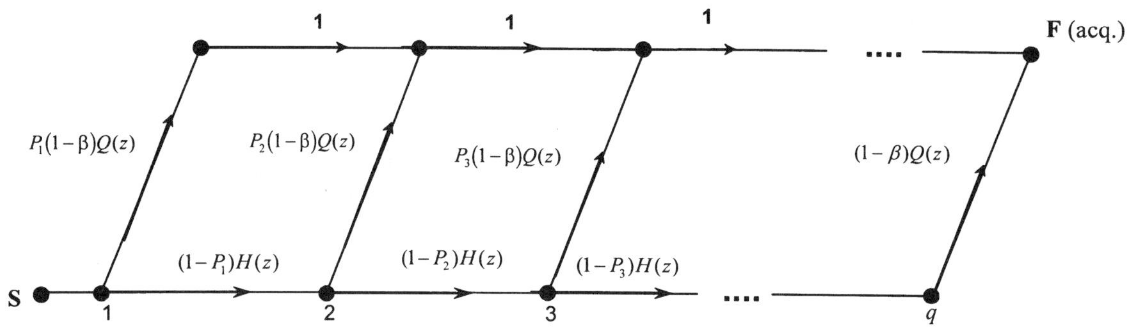

Figure 6.2-4 A further reduction of the signal flow graph for acquisition time.

By inspection the flow graph, from the start to finish, is given by

$$U(z) = B_1 + A_1B_2 + A_1A_2B_3 + A_1A_2A_3B_4 + ... + A_1...A_{q-1}B_q \tag{6.2-8}$$

or

$$U(z) = \frac{(1-\beta)z}{1-\beta zH(z)^{q-1}}[P_1 + (1-P_1)H(z)P_2 + (1-P_1)(1-P_2)P_3H(z)^2 \\ +...+(1-P_1)(1-P_2)...(1-P_{q-1})H(z)^{q-1}] \tag{6.2-9}$$

Writing P_i and $1-P_i$ in terms of q, from (6.2-3), we have

$$U(z) = \frac{(1-\beta)z}{1-z\beta H(z)^{q-1}}\left(\frac{1}{q} + \frac{H(z)}{q} + \frac{1}{q}H(z)^2 + ... + \frac{H(z)^{q-1}}{q}\right) \tag{6.2-10}$$

Summing the terms yields

$$U(z)=\frac{(1-\beta)z}{1-\beta zH(z)^{q-1}}\frac{1}{q}\left[\sum_{l=0}^{q-1}H^{l}(z)\right] \tag{6.2-11}$$

As a check $U(1)$ should be unity, since the probability of acquiring in infinite time is one; therefore,

$$U(1)=\frac{(1-\beta)/q}{1-\beta}\left[\sum_{l=0}^{q-1}1\right]=1 \tag{6.2-12}$$

The mean acquisition time is given by (after some algebra)

$$\bar{T}=\left.\frac{d\ln U(z)}{dz}\right|_{z=1}\tau_D^{'}=\frac{2+(2-P_D)(q-1)(1+KP_{FA})}{2P_D}\tau_D^{'} \tag{6.2-13}$$

with $\tau_D^{'}$ included in the formula to relate back from the unit time flow graph calculations to the actual $\tau_D^{'}$ second time period. To reiterate the meaning of the terms in (6.2-13) $\tau_D^{'}$ is the (noncoherent) dwell time, P_D is the detection probability, P_{FA} is the false alarm probability, and K is the false alarm penalty (in units of $\tau_D^{'}$) and is the time it takes to determine that a false alarm is not a true detection. As a partial check on (6.2-13) let $P_D = 1$ and $P_{FA} = 0$. Then we have from (6.2-13) that

$$\bar{T}=\frac{1+q}{2}\tau_D^{'} \tag{6.2-14}$$

This result can be obtained by direct calculation by noting that the mean time to acquire is given by (the a priori probability is $1/q$)

$$\bar{T}=\frac{1}{q}\tau_D^{'}+\frac{2}{q}\tau_D^{'}+\ldots+\frac{q}{q}\tau_D^{'} \tag{6.2-15}$$

Summing this finite series produces the result

$$\bar{T}=\frac{\tau_D^{'}}{q}\frac{q(q+1)}{2}=\frac{q+1}{2}\tau_D^{'} \tag{6.2-16}$$

Now looking at (6.2-13) for the case when $q \gg 1$ (the usual case), the mean acquisition time is well approximated by[5]

$$\bar{T}\cong\frac{(2-P_D)(1+KP_{FA})}{2P_D}(q\tau_D^{'}) \tag{6.2-17}$$

The variance of the acquisition time is given by

$$\sigma^2=\left[\frac{d^2U}{dz^2}+\frac{dU}{dz}-(\frac{dU}{dz})^2\right]_{z=1} \tag{6.2-18}$$

[5] For the result with coherent correlation, replace $\tau_D^{'}$ with τ_D.

or alternately by the equivalent, but usually simpler formula (when $U(1) = 1$)

$$\sigma^2 = \left[\frac{d^2 \ln U}{dz^2} + \frac{d \ln U}{dz} \right] \Bigg|_{z=1} \tag{6.2.19}$$

Using the simpler expression (6.2.19) it can be shown that the expression for the variance of the acquisition time, σ^2, is given by ($q >>1$)

$$\sigma^2 = (\tau_D^{'})^2 \{ (1 + KP_{FA})^2 q^2 (\frac{1}{12} - \frac{1}{P_D} + \frac{1}{P_D^{\,2}}) + 6q[K(K+1)P_{FA}(2P_D - P_D^{\,2}) \\ +(1+\text{P}_{\text{FA}}K)(4 - 2P_D - P_D^{\,2})] + \frac{(1-P_D)}{P_D^{\,2}} \} \tag{6.2-20}$$

When $1<<K<< q$, and $KP_{FA}<<1$, then the variance of the acquisition time can be simplified to

$$\sigma^2 \cong (\tau_D^{'})^2 (1 + KP_{FA})^2 q^2 \left(\frac{1}{12} - \frac{1}{P_D} + \frac{1}{P_D^{\,2}} \right) \tag{6.2-21}$$

As a partial check on the results for the variance of the acquisition time, let $P_{FA} = 0$ and $P_D = 1$. Then we have the result

$$\sigma^2 \cong \frac{(q\tau_D^{'})^2}{12} \tag{6.2-22}$$

which is the variance of the uniformly distributed random variable, as one would expect for the limiting case. Equations (6.2-13) and (6.2-17) are the main results for the mean time to acquire and (6.2-20) and (6.2-21) are the main results for the variance of the acquisition time, again for a single correlator model.

These results provide a useful estimate of the mean acquisition time for all systems that use an active search PN-type acquisition system with a fixed dwell time with a single correlator and which can be characterized as a discrete time Markov process. In addition, it is necessary that P_D and P_{FA} be assumed to be constant during the code search. Some basic modifications can be made to adjust the acquisition time estimates to reflect some nonidealized conditions that real systems operate in.

The first effect that will be considered is Doppler. The result of code Doppler is to "smear" the relative code phase during the acquisition dwell time, which increases or reduces the probability of detection depending on the code phase and the algebraic sign of the code Doppler rate. The Doppler also affects the effective sweep rate, which in the extreme case can reduce the code sweep rate to zero, to cause the search time to increase without bound. This topic will be discussed in the next section.

The second refinement to the model concerns the handover process between acquisition and tracking. Typically after a "hit" the code-tracking loop is turned on to attempt to pull the code into tight code lock. In low SNR systems, where both an acquisition (pull-in) bandwidth and a tracking bandwidth are often used, multiple code loop tracking bandwidths may be utilized in order to soften the transition from the acquisition to the tracking modes. Consequently, the probability of going from the acquisition bandwidth to the final tracking bandwidth may occur with a probability less than one.

The estimation of the probability of making the bandwidth transition is a difficult task. At high SNRs the probability quickly converges to unity so this is not an issue. In any case the probability of detection P_D should be replaced with the product P_D and P_{HO} of the form

$$P_D^{'} = P_D P_{HO} \tag{6.2-23}$$

where P_{HO} is the probability of making the bandwidth transition (hand over) and maintaining code lock with the code-tracking loop, given that the signal has been detected. P_D is the probability of detection of the signal as before. At moderate SNRs P_{HO} is essentially one, but under high jamming this may not be the case and will affect acquisition, since acquisition is not finished until the code loop is tracking in a stable manner.

6.2.3 Single Dwell Mean Acquisition Time Formula with Doppler

It is possible to account for code Doppler effects, as far as sweep rate is concerned, in the following way [9]. Define

N = number of chips to be searched

$\Delta T_c/T_c$ = step size of search in fractions of a chip (typically 1/2)

Δf_c = code Doppler in chips per second

$\Delta f_c \tau'_D$ = PN code phase timing shift due to code Doppler during the noncoherent dwell the time

$\Delta f_c K \tau'_D$ = code phase shift during hit verification

P_{FA} = probability of a false alarm

P_D = probability of detection

The mean (search phase) update timing change, accounting for code Doppler is given by

$$\mu = \frac{\Delta T_c}{T_c} + \tau'_D \Delta f_c + K\tau'_D \Delta f_c P_{FA} \tag{6.2-24}$$

We have, therefore, as an approximation to the mean acquisition time, the expression

$$\overline{T} \cong \frac{(2-P_D)(1+KP_{FA})(N\tau'_D)}{2P_D\left|\dfrac{\Delta T_c}{T_c} + \tau'_D \Delta f_c + K\tau'_D \Delta f_c P_{FA}\right|} \tag{6.2-25}$$

And using the same type of argument for the variance produces the expression

$$\sigma^2 \cong \frac{(1+KP_{FA})N^2\left(\dfrac{1}{12} - \dfrac{1}{P_D} + \dfrac{1}{P_D^{\,2}}\right)}{\left(\dfrac{\Delta T_c}{T_c} + \tau'_D \Delta f_c + K\tau'_D \Delta f_c P_{FA}\right)^2}(\tau'_D)^2 \tag{6.2-26}$$

In this expression N is the number of chips to be searched, and not the timing positions (q) to be searched. The denominator accounts for this conversion.

In (6.2-25) and (6.2-26), the algebraic sign of the code Doppler must be assigned so that the search is (apparently) speeded up by the code Doppler (Δf_c) when it is positive, and the search is slowed down if it is negative. Note that when $\Delta f_c = 0$ then $q = N/(\Delta T_c/T_c)$, and the result collapses to the previous results of (6.2-17) and (6.2-21). Actually, the two Doppler effects tend to counteract each other to a certain extent, since when the search is speeded up (due to Doppler) the correlation values tend to be reduced.

If the code change per update is larger than the search step size, then acquisition may not occur. Furthermore if large code Doppler is expected, it is possible to predict out a fair amount of the code Doppler and thereby reduce the effect of the code Doppler on the search rate.

6.2.4 Mean Acquisition Time for the Double Dwell Time Search

A simple conceptual generalization of the single dwell time acquisition system produces the double dwell time system. It has two integration periods, one to search the code phase quickly (τ'_{D1} is small), and the second (τ'_{D2}) to provide a "better" estimate of whether the "in-sync" code phase has been found. The basic idea is to apportion some of the false alarm protection in the first integration and place the remaining false alarm protection in the second integration, with $\tau'_{D1} < \tau'_{D2}$.

Generally speaking this double dwell time procedure reduces acquisition time. The amount of improvement depends upon the parameters involved. The system model is shown in Figure 6.2-5 for a BPSK modulated code. In the figure the δ_i $i=1, 2, 3$ are the three thresholds used in this approach. This approach can be used for QPSK, OQPSK, MSK, and any other code modulation format with the appropriate system model changes and the appropriate values of P_D and P_{FA}.

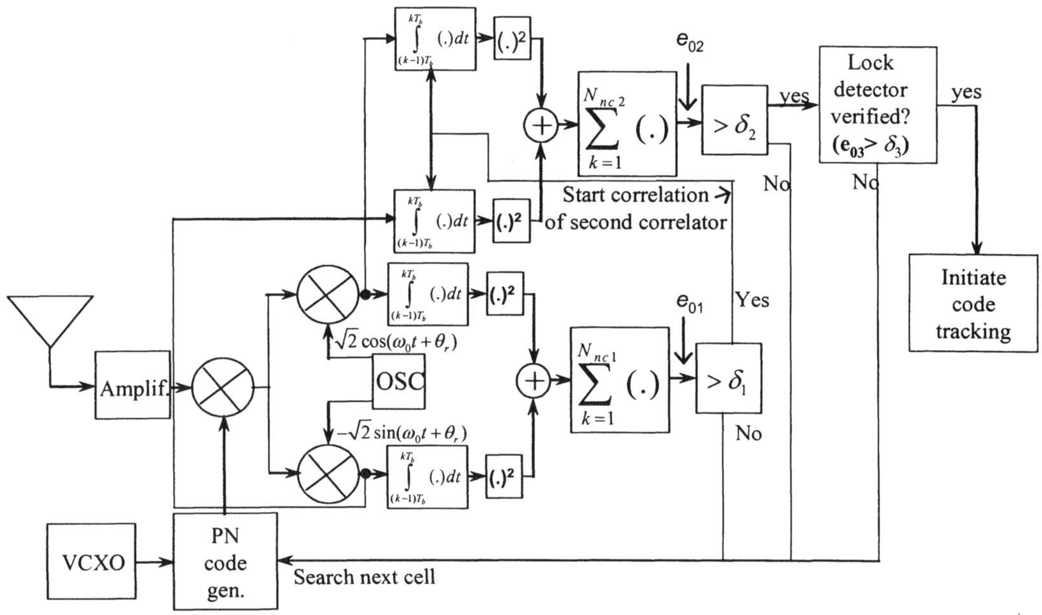

Figure 6.2-5 Double dwell acquisition time system model.

The system can be best understood by looking at the acquisition track algorithm shown in Figure 6.2-6. Typically the search starts by advancing the reference code phase to the minimum time delay end of the extremity of the range ambiguity region; then a trial integration of τ'_{D1} seconds is made on the received signal plus noise.

If the threshold is not exceeded, then the reference (system) code is delayed, for example, by one half chip, and again the dwell (integration) time is τ'_{D1} seconds. The process continues in this manner until a hit (threshold is exceeded) occurs. Then without changing the code phase, the integration time is increased to τ'_{D2}. This dwell time provides both a higher probability of detection and a lower probability of false alarm. If the second threshold is exceeded, typically the code loop is activated and a third integration (the lock detector function) of the input signal and noise is performed (in practice it could be the same duration as τ'_{D2} or longer). If the threshold were not exceeded at this point, the search would continue. This final integration period forms the basis of a lock detector. Note that this example lock detector algorithm requires three consecutive misses before it declares loss of lock.

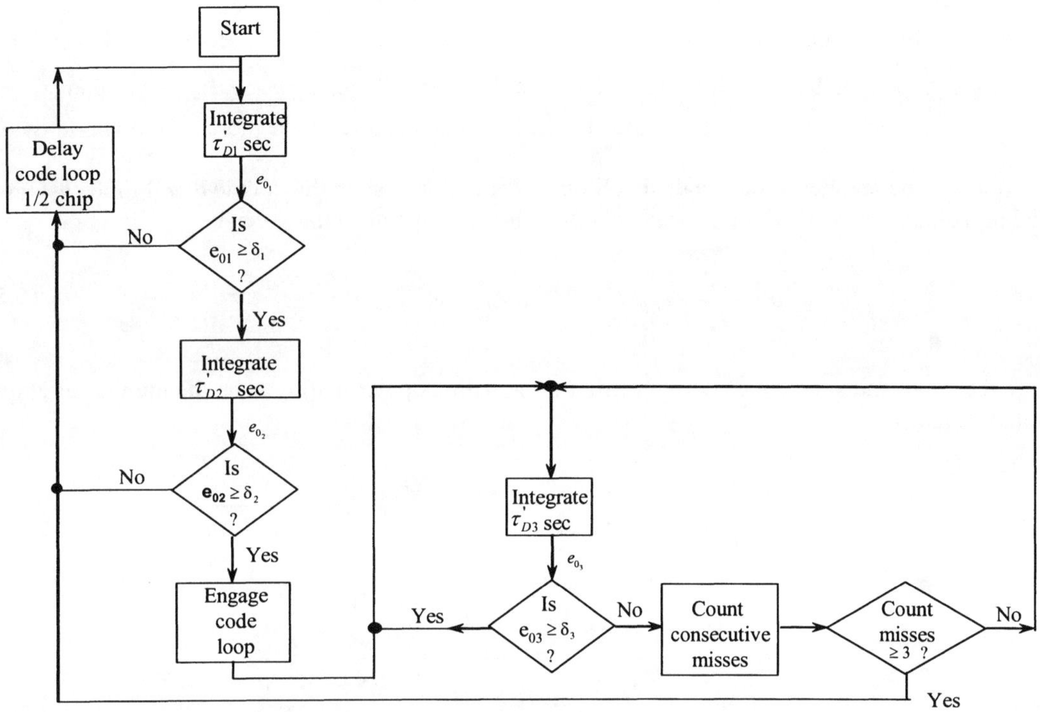

Figure 6.2-6 Double integration flow graph.

A slight modification of DiCarlo's [12] results produce the following results for mean acquisition and variance of this double dwell time scheme:

$$\bar{T} = \frac{2-P_D}{2P_D}[\tau'_{D1} + \tau'_{D2} P_{FA1}(1+KP_{FA2})]q, \quad q >> 1 \tag{6.2-27}$$

and

$$\sigma^2 = (\tau'_{D1})^2 + (\tau'_{D2})^2 P_{FA}^{\,2}(1+KP_{FA2})^2 q^2 \left(\frac{1}{12} - \frac{1}{P_D} + \frac{1}{P_D^{\,2}}\right), \quad q >> P_{FA2}K(K+1) \tag{6.2-28}$$

This result is an approximation that is accurate when the conditions on q are satisfied. In the previous two equations

τ'_{D1} = first noncoherent dwell time

τ'_{D2} = second noncoherent dwell time

$P_D = P_{D1} P_{D2}$ the overall detection probability

P_{D1} = the detection probability of the first dwell

P_{D2} = the detection probability of the second dwell

P_{FA1} = false alarm probability of the first dwell

P_{FA2} = false alarm probability of the second dwell
q = number of cells to be searched
K = penalty for a false alarm at the second detector (number of τ'_{D2} units of time)

In the previous equations the conditions $q >> 1$ and $q >> P_{FA2}K(K+1)$ are normally met. The more exact results can be found in DiCarlo [12]. Notice, as a check, if $\tau'_{D1} = 0$ and $P_{FA1} = 1$, then the system collapses to the single dwell time system and the analytic results agree with those of the single dwell time system.

In the same manner as the single dwell time acquisition system the double dwell time system can be modified to include Doppler, with the mean code search phase update being given by

$$\mu = \frac{\Delta T_c}{T_c} + \tau'_{D1}\Delta f_c + P_{FA1}\tau'_{D2}\Delta f_c + P_{FA1}P_{FA2}K\tau'_{D2}\Delta f_c \tag{6.2-29}$$

Therefore the mean and variance of the acquisition time with Doppler, letting N be the number of chips to be searched, become

$$\bar{T} \cong \frac{2-P_D}{2P_D}\frac{[\tau'_{D1} + \tau'_{D2}P_{FA1}(1+KP_{FA2})]N}{\left|(\Delta T_c/T_c) + \tau'_{D1}\Delta f_c + P_{FA1}\tau'_{D2}\Delta f_c + K\tau'_{D2}\Delta f_c P_{FA1}P_{FA2}\right|} \tag{6.2-30}$$

$$\sigma^2 \cong \frac{N^2[\tau'_{D1} + \tau'_{D2}P_{FA1}(1+KP_{FA2})]^2\left(\frac{1}{12} - \frac{1}{P_D} + \frac{1}{P_D^2}\right)}{\left[\frac{\Delta T_c}{T_c} + \tau'_{D1}\Delta f_c + \tau'_{D2}\Delta f_c P_{FA1} + K\tau'_{D2}\Delta f_c P_{FA1}P_{FA2}\right]^2} \tag{6.2-31}$$

Again with the algebraic sign of the code Doppler assigned such that when the search is speeded up by the code Doppler, Δf_c is positive and vice versa.

Further work in the area of acquisition time calculations, can be found in [13–15] and the appendix of this chapter.

6.2.5 Active Acquisition System Structures Used for Acquisition for BPSK, QPSK, OQPSK, and MSK

In this section active acquisition system models for BPSK, QPSK, OQPSK, and MSK chip modulations will be discussed.

6.2.5.1 Active Acquisition System Structures for BPSK

RF and baseband I-Q system acquisition models have already been presented in Section 6.0 for BPSK spread spectrum modulated signals with BPSK data modulation and will be discussed further. The signal model for BPSK is given by

$$s(t) = \sqrt{2P}d(t)PN(t-T)\cos(\omega_0 t + \theta) \tag{6.2-32}$$

Figures 6.0-1(a) and 6.0-1(b) are bandpass and I-Q versions of the same basic acquisition detection system along with code tracking and data demodulation for BPSK.

BPSK code modulation is quite commonly used in direct sequence spread spectrum (DSSS) systems. Our discussion is centered on the acquisition portion of the system (for the baseband version that is much more common in practice) and the acquisition system is indicated in the middle of Figure 6.0-1(b).

Typically the SS code minimum delay expected, based on both lack of precise transmitter and receiver clock time and precise range, is set into the acquisition system. The integrate-and-dump filters (T_b seconds duration) are set to integrate over the *coherent integration (or correlation) time* which is upper bounded by the bit duration (it is possible to correlate for less than but not more than the bit duration). In general either the carrier Doppler or the bit duration will set the upper limit of the coherent correlation time. The I and Q components are then squared and N_{nc} integrations are added in the noncoherent adder. The N_{nc} sum of the noncoherent integrations is called the *noncoherent integration time*. The noncoherent integration time is limited by code Doppler rates and transmitter and receiver oscillator stability. Basically the change in code time slippage should be limited by some fraction of a chip, say, about 0.4 of a chip. Thus if the code slippage rate (between the received code rate and the receiver code rate) is, for example, 0.1 chip/sec, then the total noncoherent summation time cannot exceed 4 seconds.

The noncoherent sum is compared to the *search threshold* and a decision is made on whether the local code and the received code are in approximate alignment. If the threshold is exceeded then the code tracking loop and a verification dwell time are turned on, and if the *verification threshold* is not exceeded the local code is delayed one half chip and the process is continued until the search threshold is exceeded. As we have seen in this discussion for BPSK SS code modulation, only one I-Q pair is needed to for code acquisition.

6.2.5.2 Active Acquisition System Structures for QPSK

Quadriphase modulation with equal inphase and quadrature phase power is sometimes used, because they are harder to detect when used in low probability of detection applications and are more resistant to some types of jammers. When QPSK SS PN code and QPSK data are used in a balanced quadriphase SS signaling format, the received signal can be written in the form

$$s(t) = \sqrt{P}d_1(t)PN_1(t-T)\cos(\omega_0 t+\theta) + \sqrt{P}d_2(t)PN_2(t-T)\sin(\omega_0 t+\theta) \qquad (6.2\text{-}33)$$

where P is the power of the signal in watts, ω_0 is the angular frequency in radians/sec, θ is the fixed RF phase of the carrier, $d_i(t)$ (i is 1 or 2) are the two ± 1 valued data signals and $PN_i(t)$ are the two ± 1 valued pseudonoise coded, baseband spread spectrum signals, which are normally selected to be approximately orthogonal over a correlation time. Sometimes the same data is placed on both phases of the signal. However, it will be assumed that they are different in this discussion. It should be noted that in general, the power in either carrier phase does not have to be the same value, for simplicity here it will be assumed that they are. The generalization to two different power levels is easy to accomplish. This quadriphase signal can also be viewed as two, essentially independent BPSK signals that are in phase quadrature.

Figure 6.2-7 illustrates an acquisition system for QPSK DSSS signals. The acquisition system consists of two BPSK acquisition detectors that have their $I^2 + Q^2$ sum for the first SS code added to the $I^2 + Q^2$ sum for the second SS code to drive the noncoherent integration (the summer in the figure). It is assumed that the cross correlation over a bit time of the two codes is low. This arrangement has better signal to noise performance than the more simply implemented system shown in Figure 6.2-8. In the simpler system of Figure 6.2-8 it is shown in Problem 6 that the performance is inferior to that of Figure 6.2-7. In fact it is shown that the signal power into the coherent summer is the same for both configurations, but the system of Figure 6.2-8 has the same noise terms as the system in Figure 6.2-7 plus additional noise terms, and therefore has inferior acquisition performance. If a "hit" occurs the tracking loop and lock detector are initiated and otherwise the code delay is increased by, say, 1/2 of a code chip.[6]

If offset QPSK (OQPSK) is used, then the Figure 6.2-7 has to be slightly modified. The signal structure for offset QPSK is the same as indicated in (6.2-33) except that the second (for example) code is delayed one half of a code chip relative to the first code (normally the second channel data is offset one half of a code chip also).

---

[6] When we refer to a chip, we are assuming NRZ formatting; in a more general case the term symbol should be used (for example, with Manchester coding).

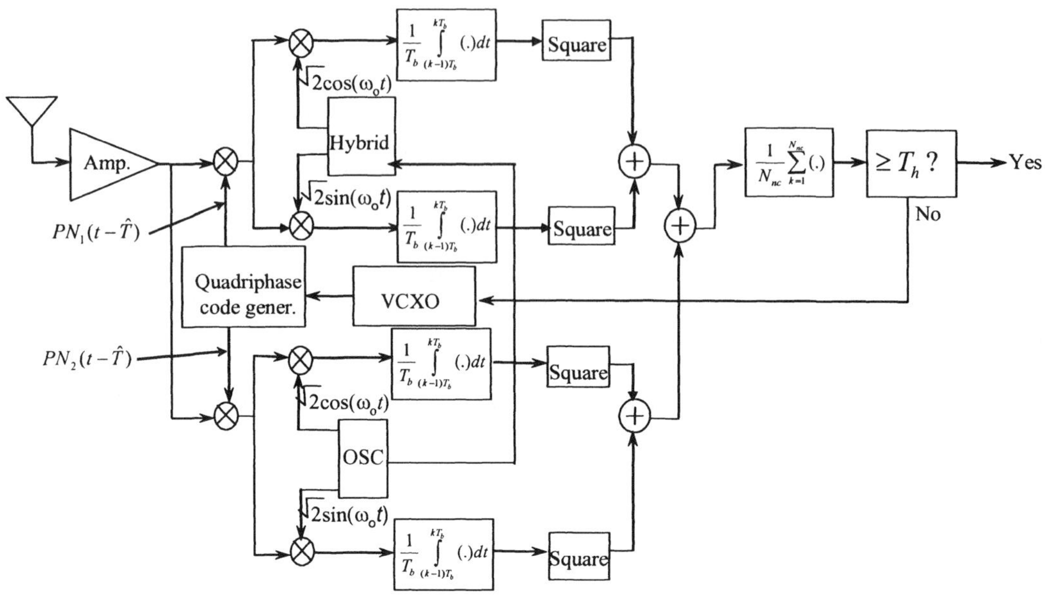

Figure 6.2-7 An acquisition system for balanced QPSK direct sequence modulation.

When OQPSK is received, it can be written in the form

$$s(t)=\sqrt{P}d_1(t)PN_1(t-T)\cos(\omega_0 t+\theta)+\sqrt{P}d_2(t-T_c/2)PN_2(t-T-T_c/2)\sin(\omega_0 t+\theta) \quad (6.2\text{-}34)$$

Normally when using OQPSK DSSS modulation, the power in each phase of the carrier is the same and the data rates are the same. If the powers are not equal, spectral regrowth[7] will be more pronounced. An alternative acquisition system for QPSK modulation is illustrated in Figure 6.2-8.

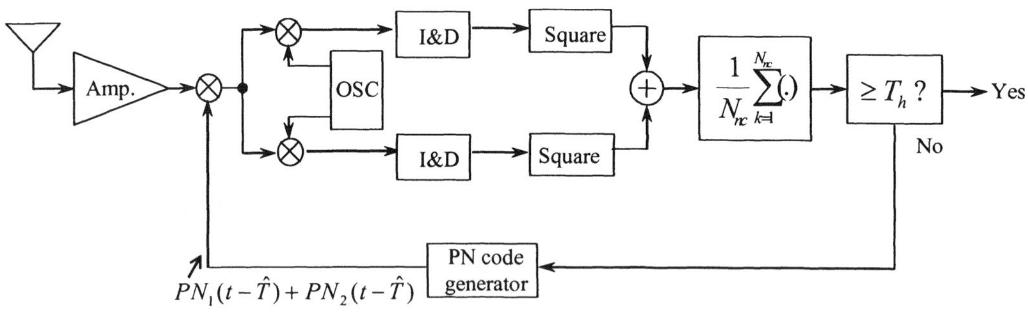

Figure 6.2-8 Alternative code acquisition system for balanced QPSK.

The acquisition system for OQPSK is shown in Figure 6.2-9. Note that the delays in the integrators are set to $T_c/2$ seconds and the upper code branch is delayed by $T_c/2$ seconds.

[7] Spectral regrowth refers to the fact that when transmitting a filtered, nonconstant amplitude signal, through a nonlinear power amplifier, the spectrum "regrows" or tends to revert back to the unfiltered version.

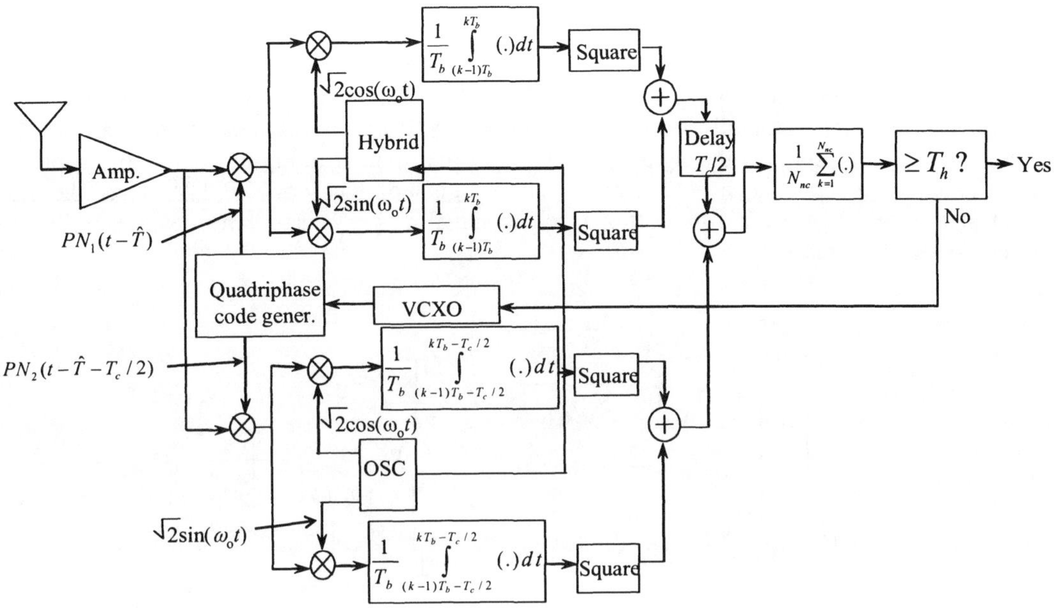

Figure 6.2-9 Acquisition system for offset balanced OQPSK.

6.2.5.3 Active Acquisition System Structures for MSK

When minimum shift keying is used for the code modulation, the received signal structure is given by

$$s(t) = \sqrt{2P}d(t)\left[PN_1(t)\cos(\pi t/T_c)\cos(\omega_0 t+\theta) + PN_2(t)\sin(\pi t/T_c)\sin(\omega_0 t+\theta)\right] \quad (6.2\text{-}35)$$

where P is the total power in the MSK DSSS signal, T_c is the chip duration, ω_0 is the carrier frequency in radians per second, and $PN_1(t)$ and $PN_2(t)$ are the two direct sequence codes used for spreading. First it will be shown that the power in the received signal is in fact P watts. Consider the mean square power

$$P_{rms} = \frac{1}{T_c}\int_0^{T_c} s(t)^2 dt \quad (6.2\text{-}36)$$

Applying (6.2-35) to (6.2-36) produces the result

$$P_{rms} = 2P\frac{1}{T_c}\int_0^{T_c}\left[\cos(\pi t/T_c)^2\cos(\omega_0 t)^2 + \sin(\pi t/T_c)^2\sin(\omega_0 t)^2\right]dt \quad (6.2\text{-}37)$$

where we have neglected the cross term, since it will be assumed that the two codes are nearly orthogonal. In addition the fact that the square of $PN(t)$ is unity for either code and the square of $d(t)$ is also unity also was used. Expanding the cosine and sine terms at the carrier frequency, one obtains

$$P_{rms} \cong 2P\frac{1}{T_c}\int_0^{T_c}\left[\cos(\pi t/T_c)^2(1/2) + \sin(\pi t/T_c)^2(1/2)\right]dt \quad (6.2\text{-}38)$$

which can be evaluated to

$$P_{rms} \cong P \tag{6.2-39}$$

as was claimed. Some approximations were used in this calculation; however it is an accurate estimate as long as the two codes are essentially orthogonal. Figure 6.2-10 can be used to acquire an MSK direct sequence SS system. Consider now the signal coming out of the summer in Figure 6.2-10. In Problem 7 it is found that the input to the noncoherent summer has a signal of the form $2P(T_d)^2$. Thus the signal is independent of the carrier phase and proportional to the signal power P, and can be successfully used for acquisition of MSK SS signals.

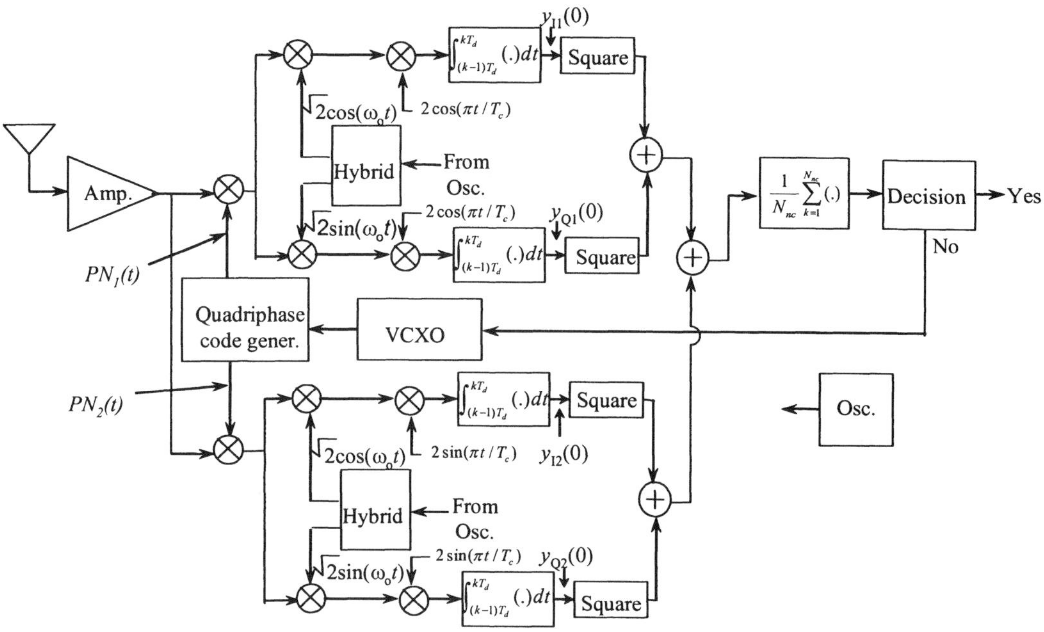

Figure 6.2-10 Acquisition system for MSK modulation.

6.3 ACQUISITION PROBABILITY VERSUS TIME FOR ACTIVE CORRELATION

We have determined how to compute the mean time to acquire for active correlation single dwell time systems in Section 6.2 with no carrier frequency error. In this section we will approximate the probability of acquisition versus time for the single dwell time, active acquisition system. Often, in practice, the probability of acquisition is specified, rather than the mean time to obtain code acquisition. So this analysis is an approximation that will allow the acquisition probability to be approximated.

It is assumed that P_D and P_{FA} are given and constant. As in Section 6.2 we will assume that the search starts from the minimum possible delay point in time and the timing is retarded in code phase, to reduce the chance of acquiring a multipath signal. After q cell checks the code phase uncertainty will have been searched one time, assuming that there are q cells in the search uncertainty. As the search passes through the code phase uncertainty more times the probability of acquisition accumulates.

To clarify the difference between P_D and the acquisition probability we consider the following. The symbol P_D denotes the probability of detection given that the signal is present and it is checked at that code phase position. Let $P_{ACQ}(n\overline{T}_{SW})$ denote the probability of acquisition after n searches or $n\overline{T}_{SW}$ seconds through the uncertainty region, in terms of the mean sweep time $\overline{T}_{SW}$.[8] $P_{ACQ}(n\overline{T}_{SW})$ depends on P_D and will be larger than P_D if more than one complete search of the code uncertainty is made.

If n complete searches of the time uncertainty have been made, the probability of acquisition, at the end of the n-th search, is given by

$$P_{ACQ}(n\overline{T}_{SW}) = P_D + P_D(1-P_D) + P_D(1-P_D)^2 + ...P_D(1-P_D)^{n-1} = 1-(1-P_D)^n \quad (6.3\text{-}1)$$

where n is the number of complete searches through the code phase uncertainty (or timing uncertainty) and P_D is the probability of detection, given that the signal is present at that tested code phase. Another way to derive this equation is to note that $(1-P_D)$ is the probability of not detecting the signal; $(1-P_D)^n$ is the probability of not detecting the signal in n sweeps (searches through the code phase uncertainty). Thus $1-(1-P_D)^n$ is the probability of detecting the signal in n sweeps of the uncertainty. Figure 6.3-1 illustrates how the acquisition probability "builds up" as the number of complete sweeps is completed, under the assumption of uniform a priori timing uncertainty. It is to be noted that if the a priori probability density of the signal timing location is assumed to be uniform in the timing uncertainty range, then the probability of acquisition increases linearly between the full sweeps as shown in Figure 6.3-1.

The acquisition time for a given probability of acquisition ($P_{ACQ}(n\overline{T}_{SW})$) can be written in terms of the mean sweep time ($\overline{T}_{SW}$) and n by the expression

$$T_{ACQ}(n) = n\overline{T}_{SW} \quad (6.3\text{-}2)$$

where

n is the number of sweeps (integer or not)

$\overline{T}_{SW}$ is the mean time to search the search one sweep

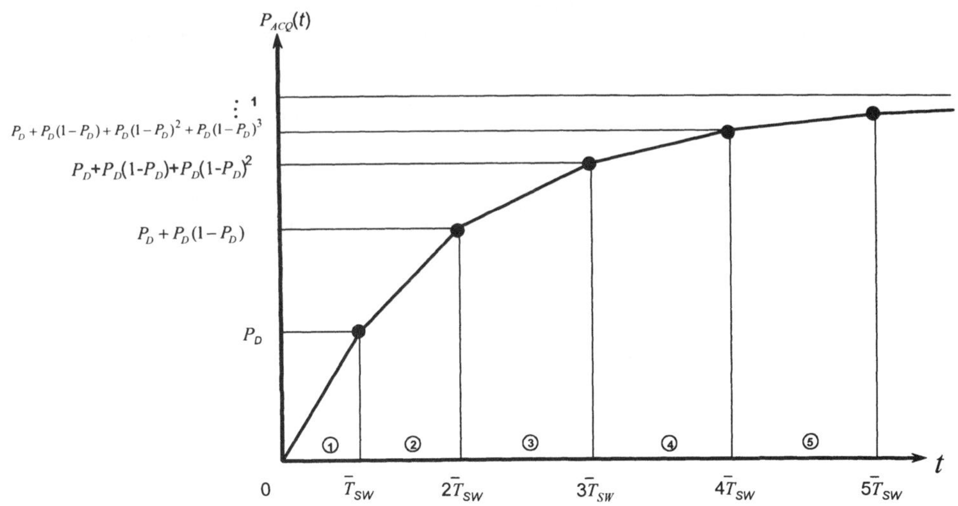

Figure 6.3-1 Illustration of how the acquisition probability accumulates with time.

[8] Note that $\overline{T}_{SW} = (1+KP_{FA})\tau_D' q$ for the single active search approach, where τ_D' is the noncoherent dwell time, K is the false alarm penality in dwell times for a false alarm, and q is the number of hypotheses to be searched.

When the desired acquisition probability does not correspond to (6.3-1) for some integer n, then it is necessary to interpolate between multiple sweep times to find the actual n needed to achieve that probability of acquisition.

We will develop the equations for the case in which the desired acquisition probability ($P_{ACQ}(n\overline{T}_{SW})$) is not satisfied by (6.3-1) for an integer value of n. In region 1 of Figure 6.3-1 we have

$$T_{ACQ}(n)=\frac{P_{ACQ}^D}{P_D}\overline{T_{SW}} \quad \text{when} \quad P_{ACQ}^D \le P_D \quad \text{and when } n \le 1 \tag{6.3-3}$$

where

T_{ACQ} (n) is the acquisition time at the desired probability of acquisition P_{ACQ}^D

$P_{ACQ}^D = P_{ACQ}(n\,\overline{T}_{SW})$ is the desired acquisition time probability

P_D is the probability of detection when the local code and the received code are aligned at a single correlation

The regions are indicated with numbers with circles around them in Figure 6.3-1. For region 2 of Figure 6.3-1 we have

$$T_{ACQ}(n)=(1+\frac{(P_{ACQ}^D-P_D)}{P_D(1-P_D)})\overline{T_{SW}} \quad \text{when } P_D \le P_{ACQ}^D \le P_D+P_D(1-P_D) \ and\ 1\le n\le 2 \tag{6.3-4}$$

For region 3 we have

$$T_{ACQ}(n)=(2+\frac{P_{ACQ}^D-(P_D+P_D(1-P_D))}{P_D(1-P_D)^2})\overline{T_{SW}} \tag{6.3-5}$$

when $P_D+P_D(1-P_D)\le P_{ACQ}^D \le P_D+P_D(1-P_D)+(1-P_D)^2$ and $2\le n\le 3$. For region 4 we have

$$T_{ACQ}(n)=(3+\frac{P_{ACQ}^D-(P_D+P_D(1-P_D)+P_D(1-P_D)^2}{P_D(1-P_D)^3})\overline{T_{SW}} \tag{6.3-6}$$

when $P_D+P_D(1-P_D)+P_D(1-P_D)^2 \le P_{ACQ}^D \le P_D+P_D(1-P_D)+P_D(1-P_D)^2+P_D(1-P_D)^3$ and $3\le n\le 4$.

For region 5 we have

$$T_{ACQ}(n)=(4+\frac{P_{ACQ}^D-(P_D+P_D(1-P_D)+P_D(1-P_D)^2+P_D(1-P_D)^3}{P_D(1-P_D)^4})\overline{T_{SW}} \tag{6.3-7}$$

when $P_D+P_D(1-P_D)+P_D(1-P_D)^2+P_D(1-P_D)^3 \le P_{ACQ}^D \le P_D+...+P_D(1-P_D)^4$ and $4\le n\le 5$.

From Section 6.2 it can be deduced that the mean sweep time (assuming no acquisitions) is given by

$$\overline{T}_{sw}=(1+kP_{FA})q\tau_D^{'} \tag{6.3-8}$$

If $P_D \le P_{ACQ}^D$, the desired acquisition probability can be achieved by employing multiple sweeps to build up to P_{ACQ}^D. If $P_D \ge P_{ACQ}^D$ then only one sweep or less is required.

Let n be the solution to the number of sweeps needed to achieve the desired acquisition probability P_{ACQ}^{D}. Then the acquisition time is given by

$$T_{ACQ}(n) = n\overline{T}_{SW} \tag{6.3-9}$$

which produces an acquisition probability of $P_{ACQ}^{D} = P_{ACQ}(n)$.

Example 2 As an example of this theory suppose that $\overline{T_{SW}} = 1$ second and $P_D = 0.9$ and it is desired to achieve an acquisition probability of 0.99. After one complete sweep of the uncertainty region $P_{ACQ}(1) = 0.9$. After two complete sweeps $P_{ACQ}(2) = 0.99$ according to (6.3-1). Hence the solution, in this case, is that it takes 2 seconds to achieve a probability of acquisition of 0.99. In general noninteger values of sweeps will be required to achieve a specific value of P_{ACQ}^{D}. In this case one of the (6.3-3) to (6.3-7) must be used to determine n, then T_{ACQ} is given by (6.3-7).

An alternative to evaluating (6.3-3) to (6.3-7) is to ("solve" (6.3-1)) directly for n to obtain

$$n \cong \frac{\log(1 - P_{ACQ}^{D})}{\log(1 - P_D)} \tag{6.3-10}$$

where n should be greater than or equal to 2 to be a satisfactory approximation.

From our Example 2 above we can solve for n from (6.3-10) to find

$$n \cong \frac{\log(1 - .99)}{\log(1 - .9)} = 2 \tag{6.3-11}$$

which fortuitously agrees exactly with our previous calculation. The larger the value of n, the better the accuracy of (6.3-9) along with (6.3-10). A more complete approach to the time to first fix (TTFF) which includes code acquisition of at least four codes and reading the data necessary for acquisition has been presented recently in Holmes et al. [16].

6.4 PARALLEL METHODS OF ACTIVE CODE ACQUISITION

Typically, modern SS acquisition methods are based on digitizing the incoming signal and performing the correlation, or correlations, digitally. Usually the sample rate is taken to be a little more than twice the chip rate. From hardware considerations it is desirable to minimize the sampling rate and the number of bits per sample, which minimizes the number of bits to be processed for the correlation (or correlations). Baseband filtering of the received signal normally limits the spectral content to the first nulls (with NRZ formatting) in order to sample the baseband chip sequence at a little over twice the chip rate for BPSK, to avoid aliasing problems.

If the approach is done digitally at slightly over the Nyquist rate of two samples per chip, then the baseband signal has to be filtered appropriately (to the first null of a BPSK signal, for example). Then the correlation search can be done in time increments of about 1/2 chip. This poses no problem since most active searches are done in 1/2-chip increments anyway. Oftentimes, these code acquisition systems will utilize a fast Fourier transform (FFT) to search frequency, so that both time and frequency can be searched simultaneously.

Figure 6.4-1 illustrates a parallel digital correlator model in which parallel processing is used to speed up the acquisition time.

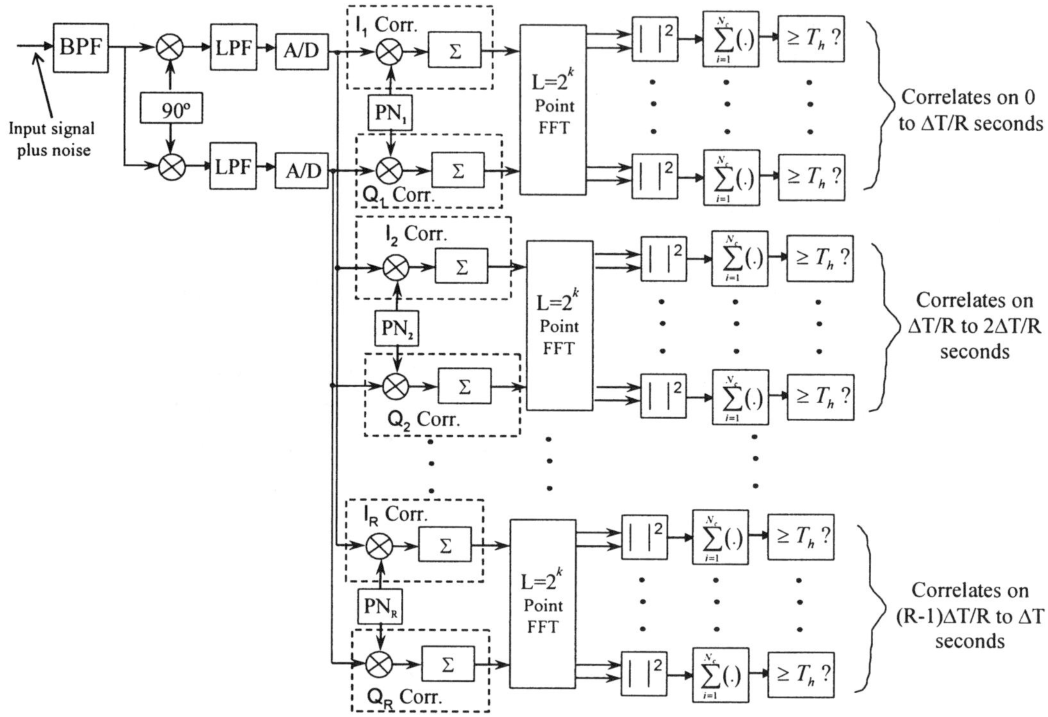

Figure 6.4-1 Parallel correlator model.

In this model R correlators are arranged so as to assign the first correlator to search the first (ΔT/R) seconds of the time uncertainty. The second correlator searches the $\Delta T/R$ to $2\Delta T/R$ seconds of time uncertainty, and so on. until the R-th correlator searches the last segment of time uncertainty, which is $(R-1)$ $\Delta T/R$ seconds to ΔT seconds. Thus, the time uncertainty is divided by R, which speeds up the time search by a factor of about R, for R parallel correlators. In reality a slight overlap in time of the time uncertainty intervals is necessary to compensate for any code drift during the search time. Let the correlation time of each correlator be T_{i0}. When the bit time is T_i, seconds the number of points a nonzero padded FFT (the FFT adds the T_{i0} segments to form a T_i second correlation) satisfies

$$N_{FFT}T_{i0} = T_i \tag{6.4-1}$$

where typically $T_i = T_b$, the bit duration (or coded symbol when a coded system is used). A frequency search can also be accomplished with the FFTs, as will be discussed in the Section 6.5. The correlators could also be utilized to search frequency in parallel or a combination of frequency and time in parallel.

It must be noted that with multiple correlators, especially with each correlator having an FFT, the likelihood of having a false alarm is much higher than with a single correlator. Therefore the threshold must be set somewhat higher to reduce the individual false alarm probability so that the aggregate false alarm probability is kept to the desired level.

6.4.1 Active Search Mean Acquisition Time with Parallel Processing

In this section the single active correlator will be generalized to the important case when multiple frequency and time hypotheses are tested simultaneously [17]. Section 6.2.3 considered the case in which only one time hypothesis was tested at a time. It will be assumed that both time and frequency will be searched in parallel in this section. Parallel processing is used, for example, in Global Positioning System (GPS) receivers to speed acquisition time.

Consider Figure 6.4-2 which illustrates the time and frequency uncertainty that the receiver has to contend with at initial acquisition. Let ΔT denote the total uncertainty in seconds, which is based on the clock time uncertainty plus range equivalent time uncertainty. Let $\Delta T = 2\Delta t$, so that the time uncertainty can be described as $\pm\Delta t$ seconds. Let δt denote the separation in time between consecutive time hypotheses in seconds (in our last example it was 1/2 of a chip, which can be related to seconds) and N the number of hypothesis per PN code chip. N is typically 2 in many receivers. It follows that the total number of time cells to be searched is

$$N_{ct} = \frac{\Delta T}{\delta t} = \frac{\Delta T}{T_c / N} = NR_c \Delta T \tag{6.4-2}$$

where $R_c = 1/T_c$ is the code chip rate, in chips per second.[9] Now consider the frequency search in which ΔF is the total frequency uncertainty in Hz, and δf is the separation between hypotheses in Hz. Let $\Delta F = 2\Delta f_m$. where Δf_m is the maximum frequency error to be searched. The total number of frequency cells to be searched is given by

$$N_{cf} = \frac{\Delta F}{\delta f} \tag{6.4-3}$$

Therefore the total number of time and frequency cells to be searched is given by

$$N_c = N_{ct} N_{cf} = NR_c \Delta T \frac{\Delta F}{\delta f} \tag{6.4-4}$$

Now assume that M_c cells can be searched in parallel, in other words on every correlation M_c cells are searched (in both frequency and time) as shown in Figure 6.4-2.

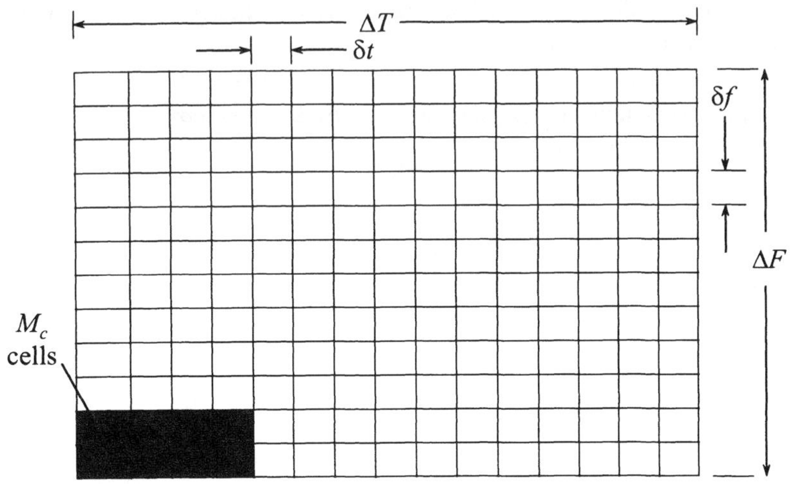

Figure 6.4-2 Time and frequency uncertainties showing a search of M_c cells.

[9] When the code format is not NRZ, then the chip is the smallest time period for which the symbol is constant. If the code is based on one squarewave then a chip will be one half of the squarewave period.

Further assume that M_c divides[10] N_c. It follows that a total of

$$N_c^{'} = \frac{N_c}{M_c} \quad (6.4\text{-}5)$$

blocks of cells must be tested in order to search the whole time-frequency uncertainty.

When M_c cells are searched in parallel, each cell can have a false alarm. It will be assumed here that verifying[11] one or more false alarms requires $K\ \tau_D$ seconds, where K is a number that is typically greater than one and τ_D is the coherent dwell time. M_c separate verification channels must verify false alarms in order to determine whether the correct signal or a false alarm has occurred. It is assumed that the time in seconds to determine that the signal is present or not is $K\ \tau_D$ seconds. Assuming that all the channels are statistically independent, the false alarm probability of the group of M_c cells is given by

$$P_{FA}^{'} = 1 - \left(1 - P_{FA}\right)^{M_c} \quad (6.4\text{-}6)$$

since the false alarm probability of one or more individual cells is one minus the probability that there are no individual cell false alarms. It is assumed that lock verification circuits are available for up to M_c false alarms.

When $N_c^{'}$ groups of cells have to be searched, the mean time to acquire is well approximated from (6.2-17)

$$\overline{T} \cong \frac{(2 - P_D)(1 + KP_{FA}^{'})}{2P_D} N_c^{'} \tau_D \quad \text{(s)} \quad (6.4.7)$$

where P_D is the probability of detection, K is the false alarm penalty in units of τ_D, τ_D is the coherent dwell time in seconds (here noncoherent combining is not included), and $N_c^{'}$ ($N_c^{'}$ is assumed to be >>1) is the number of blocks of M_c cells to be searched over the total time and frequency uncertainty.

Commonly additional detectability is derived from noncoherently combining an integer number of the coherent dwells (this is called noncoherent integration or correlation), so that the total dwell time is a multiple of the coherent dwell time, N_{nc}. The detection decision is made after multiples (N_{nc}) of τ_D seconds. Thus the total dwell time, when noncoherent integration is used, is given by

$$\tau_D^{'} = N_{nc} \tau_D \quad (6.4\text{-}8)$$

In practice N_{nc} can be considerably larger than one to provide the desired acquisition time performance in the face of a jammer or significant interference. Using (6.4-8) and (6.4-6) in (6.4.7) produces

$$\overline{T} \cong \frac{(2 - P_D)(1 + K[1 - (1 - P_{FA})^{M_c}])}{2P_D} \frac{N_c^{'}}{M_c} \tau_D^{'} \quad (6.4\text{-}9)$$

This is the main result for the mean time to acquire when there are $N_c^{'}$ hypotheses to be searched and M_c hypotheses are searched at one correlation time period and $\tau_D^{'}$ the noncoherent dwell time. If M_c does not divide $N_c^{'}$, then $N_c^{'}/M_c$ in (6.4-9) is replaced with $\lceil N_c^{'}/M_c \rceil$, where $\lceil x \rceil$ denotes the least integer greater than x. Hence using this notation in (6.4-9) yields our main result

[10] By M divides N we mean that $N = qM$ where q is a positive integer.

[11] This may entail observing a lock detector for a time equivalent to multiples of the dwell time.

$$\bar{T} \cong \frac{(2-P_D)(1+K[1-(1-P_{FA})^{M_c}])}{2P_D}\left\lceil \frac{N_c'}{M_c} \right\rceil \tau_D' \tag{6.4-10}$$

for the mean acquisition time when N_c' blocks of cells are correlated, M_c cells at a time with noncoherent combining. As noted earlier the noncoherent dwell time is given by $\tau_D' = N_{nc}\tau_D$.

When $M_c P_{FA} << 1$, then (6.4-10) reduces to

$$\bar{T} \cong \frac{(2-P_D)(1+KM_c P_{FA})}{2P_D}\left\lceil \frac{N_c'}{M_c} \right\rceil \tau_D' \tag{6.4-11}$$

Thus, the effective false alarm probability, uner these conditions, is $M_c P_{FA}$, which is M_c times larger than for the single hypothesis test case, as one would expect. Also, from (6.4-5) $N_c'=N_c/M_c$, therefore the search is about M_c time faster than a nonparallel search.

6.5 ACTIVE CODE SEARCH UTILIZING THE FFT

In this section active code acquisition, utilizing the FFT, will be analyzed for parallel frequency searching and coherent integration.

6.5.1 Signal Modeling for BPSK Code Acquisition Utilizing the FFT

In this section the use of an FFT, in an active search acquisition scheme, will be discussed in terms of the signal modeling and performance. Typically the frequency of the carrier is unknown to within some accuracy, so that multiple frequency hypotheses must be searched. One way to accomplish the frequency search is to utilize an FFT, when the signal processing is done digitally. Another option is to use highly parallel equipment to search many frequencies at the same time.

Figure 6.5-1 illustrates an *I-Q* baseband system derived from the quadrature carrier references. Although this is an analog implementation to simplify the analysis, a digital receiver would utilize analog to digital (A/D) converters following each low-pass filter. The function of the A/D is to sample the signal and noise and to convert the samples to a digital word. The low pass filters in the figure are there to do two things: (1) to remove the sum frequency term and (2) for a digital model to prevent aliasing. It will be assumed that the low pass filters have a bandwidth of R_c Hz, where R_c is the chip rate, and that two samples per second is used in the digital implementation.

It will also be assumed that the filters are ideal in their amplitude and phase response (that is, a sharp cutoff filter). To simplify the analysis, analog processing will be assumed. Later, the losses associated with digitizing the signal will be discussed.

Assume that the input signal is a BPSK direct sequence DSSS signal of the form

$$y(t) = \sqrt{2}APN(t-T)d(t)\cos(\omega_o t+\theta) + \sqrt{2}n_c(t)\cos(\omega_o t+\hat{\theta}) + \sqrt{2}n_s(t)\sin(\omega_o t+\hat{\theta}) \tag{6.5-1}$$

where A is the rms signal voltage, $d(t)$ is the data sequence, and $PN(t\text{-}T)$ is the direct sequence code with delay T seconds from the transmitter to the receiver. The inphase and quadrature noise components are Gaussian random processes and assumed to be in phase with each of the reference signals for mathematical convenience.

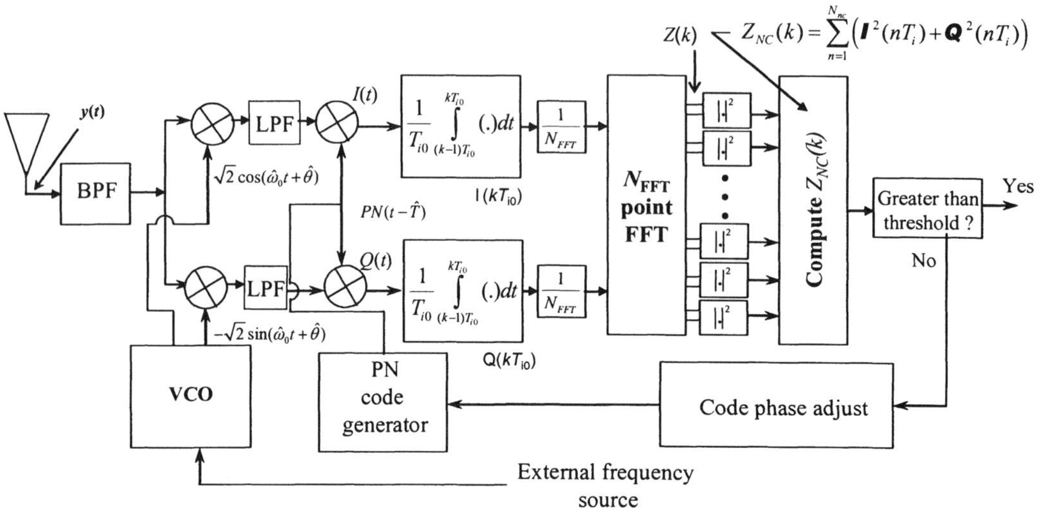

Figure 6.5-1 Baseband version of a PN code acquisition system for BPSK.

This assumption doesn't affect the results. The reference signals are composed of the local code at delay estimate $\hat{T}$ and the sine and cosine references, which are represented by

$$r_c(t) = \sqrt{2}\cos(\hat{\omega}_0 t + \hat{\theta}) \quad r_s(t) = -\sqrt{2}\sin(\hat{\omega}_0 t + \hat{\theta}) \tag{6.5-2}$$

where $\hat{\omega}_0, \hat{\theta}$, and $\hat{T}$ are the receiver's estimated received signal angular frequency, carrier phase, and estimated code time delay from the transmitter to the receiver, respectively. After demodulating the received signal plus noise with the cosine reference signal, one obtains

$$I(t) = \sqrt{P}PN(t-T)PN(t-\hat{T})d(t)\cos(\Delta\omega' t + \phi) + n_c(t)PN(t-\hat{T}) + O(\omega_0 + \hat{\omega}_0) \tag{6.5-3}$$

where $O(\omega_0 + \hat{\omega}_0)$ denotes a term at $\omega_0 + \hat{\omega}_0$, the angular frequency error is $2\pi\Delta f = \Delta\omega = \omega_0 - \hat{\omega}_0$, where Δf is the frequency error and the phase difference is $\phi = \theta - \hat{\theta}$. Similarly the product of the input and the Q channel reference signal gives the Q channel output

$$Q(t) = \sqrt{P}PN(t-T)PN(t-\hat{T})d(t)\sin(\Delta\omega' t + \phi) - n_s(t)PN(t-\hat{T}) + O(\omega_0 + \hat{\omega}_0) \tag{6.5-4}$$

In Figure 6.5-1, it is assumed that the low pass filter essentially removes the sum frequency term. In addition, it will be assumed that the ideal BPF filters the signal such that there are no aliasing errors. It will be assumed for computational convenience that the correlation is formed from an analog integration, and not a summation as in a digital receiver, to simplify the calculations. Amplitude quantization is not modeled in the analysis. Later, losses for the amplitude quantization, assuming a digital correlation was used, will be estimated. Time quantization will be accounted for by assuming that there are two time hypotheses per chip.

The correlation on the I and the Q channels, over the coherent correlation time of T_{i0} seconds, will be modeled as a T_{i0} second average. This is based on the fact that an analog matched filter for the code sequence is equivalent to an analog correlation. It will also be assumed that the N_{FFT} summed T_{i0} segments ($N_{FFT}T_{i0}$ seconds = T_i) is equal to T_i seconds. Normally T_i is the bit duration, $T_b = 1/R_b$, since the data bit boundaries upper limit the coherent integration time. The n-th in-phase T_{i0} second correlation is modeled as

$$\mathsf{I}(n)=\frac{1}{T_{i0}}\int_{(n-1)T_{i0}}^{nT_{i0}}\left[\sqrt{P}d(t)PN(t-T)PN(t-\hat{T})\cos(\Delta\omega t+\phi)+n_c(t)PN(t-\hat{T})\right]dt \tag{6.5-5}$$

It is to be noted that the signal component of $\mathsf{I}(n)$ is dependent on the difference of the actual and estimated delays $T-\hat{T}$. This expression can be evaluated as

$$\mathsf{I}(n)\cong\frac{1}{T_{i0}}d(n)\sqrt{P}R_{PN}(T-\hat{T})\int_{(n-1)T_{i0}}^{nT_{i0}}\cos(\Delta\omega t+\phi)dt+\frac{1}{T_{i0}}\int_{(n-1)T_{i0}}^{nT_{i0}}n_c(t)PN(t-\hat{T})dt \tag{6.5-6}$$

where $R_{PN}(T-\hat{T})$ is the autocorrelation function of the spreading code and has been factored out of the integral as an approximation assuming that Δf is small compared to the chip rate, and $d(n)$ is the value of the data sequence of the bit[12] during the n-th T_{i0} seconds of time. Let $\tau=T-\hat{T}$ be the timing error between the received code and the local matched filter code. The integral in (6.5-6) can be evaluated by writing the cosine as the real part of a complex exponential, integrating and taking the real part. This results in

$$\mathsf{I}(n)=d(n)\sqrt{P}R_{PN}(\tau)\mathrm{sinc}\left(\Delta fT_{i0}\right)\cos\left[n\Delta\omega T_{i0}+\phi'\right]+\frac{1}{T_{i0}}\int_{(n-1)T_{i0}}^{nT_{i0}}n_c(t)PN(t-\hat{T})dt \tag{6.5-7}$$

where $\phi'=\phi-\Delta\omega T_{i0}/2$ is the constant phase that results from each integration. The term $R_{PN}(\tau)$ is the autocorrelation function of the spread spectrum code, without filtering effects included. In (6.5-7) sinc(x) is defined as $\mathrm{sinc}(x)=\sin(\pi x)/(\pi x)$.

In the same manner as in the I channel, but using the imaginary part of the exponential, $\mathsf{Q}(n)$ can be shown to be given by

$$\mathsf{Q}(n)\cong\sqrt{P}R(\tau)d(n)\mathrm{sinc}\left(\Delta fT_{i0}\right)\sin\left[n\Delta\omega T_{i0}+\phi'\right]-\frac{1}{T_{i0}}\int_{(n-1)T_{i0}}^{nT_{i0}}n_s(t)PN(t-\hat{T})dt \tag{6.5-8}$$

Thus we see that the correlation loss, expressed in power out of the correlator when considering both the I and Q channels, is given by $\mathrm{sinc}(\Delta fT_{i0})^2$ times the autocorrelation function squared evaluated at the timing error τ. In (6.5-8) $\Delta f=\Delta\omega/(2\pi)$.

The maximum frequency uncertainty is given by Δf_m Hz. Then the integration time, T_{i0}, should not be too large in order to limit the frequency error losses in the T_{i0} second correlation. If the loss is to be limited to 0.91 dB, for example, then it is necessary that

$$L=10\log\left[\frac{\sin(\pi\Delta f_mT_{i0})^2}{(\pi\Delta f_mT_{i0})^2}\right]=0.91 \tag{6.5-9}$$

Solving (6.5-9) yields the result that $\Delta f_mT_{i0}=1/4$, or

$$T_{i0}=1/(4\Delta f_m) \tag{6.5-10}$$

This correlation loss at the extremes of the frequency error is known as the *outer bin loss*. It occurs at the maximum frequency error between the received signal and the reference signal. If the product of frequency error and correlation time is doubled to Δf_mT_{i0} = 1/2, the outer bin loss is increased to 3.91 dB.

[12] It is assumed that the bit time is a multiple of T_{i0}. That is, $N_{FFT}T_{i0}=T_b$. At a bit time transition $d(n)$ assumes the value of the bit just before the transition.

Usually the gain of about 3 dB in signal level is worthwhile, so that the condition of (6.5-10) is often used in practice. The frequency search range is doubled when $\Delta f_m T_{i0} = 1/2$ is used compared to $\Delta f_m T_{i0} = 1/4$.

Now let us determine the frequency response of the FFT. Let the output of the I and Q correlators, with T_{i0} seconds correlation time, be represented as the complex number $z(n)$ (which is the input to the FFT)

$$z(n) = \mathrm{I}(n) + j\mathrm{Q}(n) \tag{6.5-11}$$

The N_{FFT}-point DFT[13] [18] of (6.5-11) is given by[14]

$$Z(k) = \frac{1}{N_{FFT}} \sum_{n=0}^{N_{FFT}-1} z(n) e^{-j(2\pi/N_{FFT})nk} \qquad k = 0,1,\ldots N_{FFT}\text{-}1 \tag{6.5-12}$$

Letting

$$W = e^{-j(2\pi/N_{FFT})} \tag{6.5-13}$$

produces the following

$$N_{FFT} Z(k) = \sum_{n=0}^{N_{FFT}-1} z(n)(W)^{nk} \tag{6.5-14}$$

The N_{FFT} outputs of the N_{FFT} point FFT shown at the output of the FFT in Figure 6.5-1, which correspond to filtered outputs that are offset at a set of frequencies, as will be shown shortly. Normally not all of the outputs will be checked for code acquisition, since they will be out of the frequency range of the signal. The effect of the FFT is to act as a filter bank, albeit with crude filters, with rather high sidelobes and a fair amount of overlap between the filters. An alternate way to form similar statistics[15] is to have N_{FFT} different I-Q pair of reference frequencies to offset the carrier and then form the magnitude. Typically using one FFT requires less hardware than using N_{FFT} I-Q pairs for a digital system implementation.

To determine the form of the filter response of the filters in the FFT filter bank, it is to be noted that the impulse response of a N_{FFT} point FFT is given by [18]

$$h(n) = e^{-j(2\pi kn/N)}, \quad n = 0,\, 1,\, N_{FFT}\text{-}1 \tag{6.5-15}$$

Assume that the initial correlation time is one (that is, $T_{i0} = 1$) so that the z transform of the filter response is given by

$$H(z) = \sum_{n=0}^{N_{FFT}-1} e^{-j(2\pi kn/N_{FFT})} z^{-n} = \frac{1 - z^{-n}}{1 - z^{-1} e^{-j(2\pi k/N_{FFT})}} \tag{6.5-16}$$

Evaluating $H(z)$ on the unit circle produces the result

$$H(e^{j\omega}) = e^{-j\omega((N_{FFT}-1)/2)} e^{j\pi k/N} \frac{\sin(N_{FFT}\omega/2)}{\sin(\omega/2 + \pi k/N_{FFT})} \tag{6.5-17}$$

where ω is the discrete time frequency variable [18]. To relate the discrete time frequency variable (ω) to the analog frequency variable (f_k) note that

[13] This would be implemented by an FFT in most systems.
[14] We initially neglect the division by N_{FFT}, and in the next section we will account for it.
[15] This method does not incur the outer bin loss, whereas the FFT method does incur the outer bin loss.

$$f_k = \frac{k}{N_{FFT}T_{i0}} \quad \text{and} \quad \Omega_k = 2\pi f_k \tag{6.5-18}$$

so that the relationship is

$$\omega = \Omega_k T_{i0} \tag{6.5-19}$$

Thus, the normalized amplitude magnitude response is given by (letting $f_k = f$)

$$\frac{1}{N_{FFT}}\left|\left(\frac{\sin(\pi N_{FFT} f T_{i0})}{\sin(\pi f T_{i0} + \pi k / N_{FFT})}\right)\right| \quad \text{for } 0 < f < 1/T_{i0} \tag{6.5-20}$$

$$\text{and for } k = 0,1,2,...N_{FFT} - 1$$

The frequency response of $|A(f)|$ is illustrated in Figure 6.5-2. Artistic liberty has been taken in the drawing, since the shape of each magnitude curve is actually a $|\sin(x)/x|$ like shape that extends for all frequency. However, to avoid cluttering the figure only inverted U shapes were used to represent the main magnitude response at each bin frequency. More details can be found in [18]. Note that the frequency range is $1/T_{i0}$, which is just the FFT input precorrelation rate, and the spacing between the filter centers is $1/(N_{FFT}\,T_{i0})$, which is $1/T_i$, or the inverse of the coherent correlation time. There are N_{FFT} bin frequencies for the FFT filter. Other filter bin arrangements are possible and are discussed in [19]. Typically N_{FFT} is a power of 2, so assume $N_{FFT} = 2^{n_f}$, where n_f is an integer. As was noted earlier, with the use of an FFT the total coherent correlation time is increased from T_{i0} seconds to $N_{FFT}T_{i0} = T_i$ seconds. Thus, each time correlation can also be searched in frequency over the applicable portion of the N_{FFT} frequency bins, and consequently multiple frequency hypotheses can be searched in parallel.

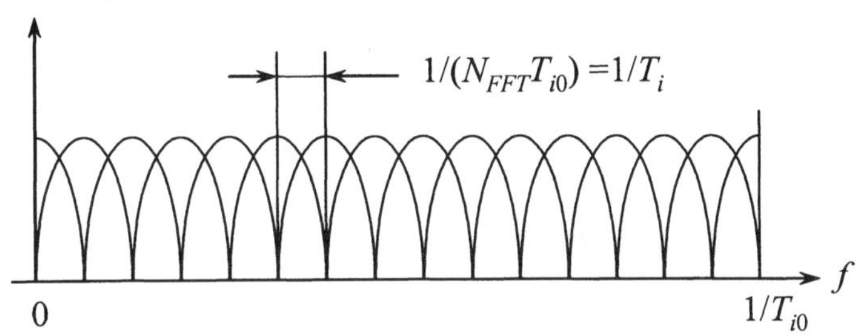

Figure 6.5-2 FFT frequency response for the case of $N_{FFT} = 16$.

Now it will be established that the upper half range of the FFT is in fact the negative frequency part of the FFT filter bank. Again let $Z(k)$ denote the complex FFT output of the I-Q input pair. Denote $\mathrm{I}(n)$ as the real FFT input and denote $\mathrm{Q}(n)$ as the imaginary part of FFT input. Thus each complex output FFT value is of the form

$$N_{FFT}Z(k) = \sum_{n=0}^{N_{FFT}-1} \left(\mathrm{I}(n) + j\mathrm{Q}(n)\right) e^{-j(2\pi/N_{FFT})kn} \quad k = 0,1,...N_{FFT} - 1 \tag{6.5-21}$$

where N_{FFT} is the size of the FFT. Let $z(n) = \mathrm{I}(n) + j\mathrm{Q}(n)$ represent the complex FFT input. Note that nonbold, nonitalic, Arial font is used to represent the I and Q components of the complex input of the FFT. Then it follows that the output of the FFT is given by

$$N_{FFT}Z(k)=\sum_{n=0}^{N_{FFT}-1} z(n)e^{-j(2\pi/N_{FFT})kn} \tag{6.5-22}$$

With the index k (k = 0, 1, N_{FFT} −1) denoting the N_{FFT} frequency bins, produced by the FFT. Let the exponential be represented by

$$W=e^{-j(2\pi/N_{FFT})} \tag{6.5-23}$$

so that (6.5-22) can be written as

$$N_{FFT}Z(k)=\sum_{n=0}^{N_{FFT}-1} z(n)W^{kn} \tag{6.5-24}$$

Now consider negative values of k. Therefore let $k = -m$. It follows that

$$N_{FFT}Z(-m)=\sum_{n=0}^{N_{FFT}-1} z(n)W^{-nm} \tag{6.5-25}$$

However it is true that

$$\left(W^n\right)^{-m}=\left(W^n\right)^{-m+N_{FFT}} \tag{6.5-26}$$

Since, for any n

$$\left(W^n\right)^{N_{FFT}}=1 \tag{6.5-27}$$

Therefore, it follows that

$$N_{FFT}Z(-m)=\sum_{n=0}^{N_{FFT}-1} z(n)\left(W^n\right)^{-m}=\sum_{n=0}^{N_{FFT}-1} z(n)\left(W^n\right)^{-m+N_{FFT}}=N_{FFT}Z(N_{FFT}-m) \tag{6.5-28}$$

as was to be demonstrated. Thus the negative frequency errors are located in the upper half of the FFT frequency range of the FFT output. For example, when N_{FFT} = 16, $Z(-1) = Z(15)$.

6.5.2 Model for the Correlator Output Out of the FFT

Now consider the effect that the FFT has on the signal and the noise of the inphase and quadrature terms out of the FFT. Note that the effect of the FFT is to increase the correlation time to T_i seconds from T_{i0} seconds. In the previous section the scale factor of $1/N_{FFT}$ has been neglected. Now it will be accounted for. See Figure 6.5-1 for the BPSK model incorporating the FFT. Following this section more details on the losses and other effects of the FFT will be discussed. After the FFT summation and the normalization by N_{FFT}, the inphase and quadrature channels are given by

$$Z(k)=\mathbf{I}(k)+j\mathbf{Q}(k)=\frac{1}{N_{FFT}}\sum_{n=0}^{N_{FFT}-1}\left[\mathrm{I}(n)+j\mathrm{Q}(n)\right]e^{-jkn(2\pi/N_{FFT})}\ ,\quad k=0,1,2,N_{FFT}-1 \tag{6.5-29}$$

in which for each time hypothesis and for some portion of each of the N_{FFT} frequencies, these correlations are generated for the acquisition search process. Note that the prime indicates the normalization by N_{FFT}.

From (6.5-7) and (6.5-8), it can be seen that, neglecting the frequency error effect which will be addressed later in the chapter, the signal component will not be affected by the normalized FFT. Denote the FFT frequency offset and the k-th bin effect on the signal by $L(\Delta f, T_i, k)$, and the signal components out of the FFT are given by

$$\boldsymbol{I}_S(k) = d(T_i)\sqrt{P}R_{PN}(\tau)\text{sinc}\left(\Delta f T_{i0}\right)L\left(\Delta f, T_i, k\right)\cos(n\Delta\omega + \phi') \tag{6.5-30}$$

$$\boldsymbol{Q}_S(k) = d(T_i)\sqrt{P}R_{PN}(\tau)\text{sinc}\left(\Delta f T_{i0}\right)L\left(\Delta f, T_i, k\right)\sin(n\Delta\omega + \phi') \tag{6.5-31}$$

we have used Arial Black italic font for the components out of the FFT. The noise term will be affected by the FFT. If we denote real and imaginary outputs in terms of signal and noise then we can write

$$\begin{aligned} Z(k) &= \boldsymbol{I}(k) + j\boldsymbol{Q}(k) = \boldsymbol{I}_S(k) + j\boldsymbol{Q}_S(k) + \boldsymbol{I}_N(k) + j\boldsymbol{Q}_N(k) \\ &= Z_S(k) + Z_N(k), \quad k = 0, 1, 2, N_{\text{FFT}} - 1 \end{aligned} \tag{6.5-32}$$

where the subscript N denotes noise and the subscript S denotes the signal.

The FFT input noise terms $\mathrm{I}_N(k)$ and $\mathrm{Q}_N(k)$ are assumed to be Gaussian, independent, random variables with zero mean value. The FFT input variables can be broken into signal and noise terms as

$$\begin{aligned} \mathrm{I}(n) &= \mathrm{I}_S(n) + \mathrm{I}_N(n) \\ \mathrm{Q}(n) &= \mathrm{Q}_S(n) + \mathrm{Q}_N(n) \end{aligned} \tag{6.5-33}$$

The normalized, discrete Fourier transform (DFT) of the noise components for complex inputs is defined as

$$\boldsymbol{I}_N(k) + j\boldsymbol{Q}_N(k) = Z_N(k) = \frac{1}{N_{FFT}}\sum_{n=0}^{N_{FFT}-1}\left[\mathrm{I}_N(n) + j\mathrm{Q}_N(n)\right]e^{-jkn(2\pi/N_{FFT})} \tag{6.5-34}$$

where $Z_N(k)$ is the noise component of the complex FFT output ($Z(k) = Z_S(k) + Z_N(k)$). Note that italic, Arial Black font is used for the FFT output variables $\boldsymbol{I}$ and $\boldsymbol{Q}$, and Arial font, without italics and not bold, is used for the input variables I and Q. The subscript N, in (6.5-34), denotes noise only.

Note that $Z_N(k)$ is a linear sum of zero mean Gaussian random variables and therefore is also a zero mean Gaussian random variable. Since $\mathrm{I}_n(k)$ and $\mathrm{Q}_n(k)$ are zero mean Gaussian input random variables, it follows that the real and imaginary part of $Z_N(k)$ represents the output Gaussian random variables.

Therefore the variance of the real and imaginary parts represent the FFT output inphase and quadrature noise terms. Consider Re($Z_N(k)$) which can be written as

$$\boldsymbol{I}_N(k) = \text{Re}\left(Z_N(k)\right) = \frac{1}{N_{FFT}}\sum_{n=0}^{N_{FFT}-1}\left[\mathrm{I}_N(n)\cos\left(nk\frac{2\pi}{N_{FFT}}\right) + \mathrm{Q}_N(n)\sin\left(nk\frac{2\pi}{N_{FFT}}\right)\right] \tag{6.5-35}$$

The variance can be evaluated to

$$Var(\mathbf{I}_N(k)) = \frac{1}{N_{FFT}^2} E\left\{ \begin{matrix} \sum_{m=0}^{N_{FFT}-1} \sum_{n=0}^{N_{FFT}-1} \left[\mathsf{I}_N(n)\cos\left(\frac{kn2\pi}{N_{FFT}}\right) + \mathsf{Q}_N(n)\sin\left(\frac{kn2\pi}{N_{FFT}}\right) \right] \times \\ \left[\mathsf{I}_N(m)\cos\left(\frac{mk2\pi}{N_{FFT}}\right) + \mathsf{Q}_N(m)\sin\left(\frac{mk2\pi}{N_{FFT}}\right) \right] \end{matrix} \right\} \quad (6.5\text{-}36)$$

where $E\{x\}$ denotes the statistical expectation of x. Since the input $\mathsf{I}_N(n)$ and $\mathsf{Q}_N(n)$ noise terms are assumed to be statistically independent, each random variable is independent from sample to sample, so that one can write

$$Var(\mathbf{I}_N(k)) = \frac{1}{N_{FFT}^2} \sum_{n=0}^{N_{FFT}-1} \left[E\{\mathsf{I}_N^2(n)\}\cos\left(\frac{kn2\pi}{N_{FFT}}\right)^2 + E\{\mathsf{Q}_N^2(n)\}\sin\left(\frac{kn2\pi}{N_{FFT}}\right)^2 \right] \quad (6.5\text{-}37)$$

Since the expected value of both terms is the same, and the summation produces the value N_{FFT}, it follows that

$$Var(\mathbf{I}_N(k)) = \frac{1}{N_{FFT}} Var(\mathsf{I}_N(i)) \quad (6.5\text{-}38)$$

Thus the output correlation from the FFT has a normalized variance that is $1/(N_{FFT})$ times as large as the input correlation, as would be expected. It follows that the output noise out of the FFT is given at the m-th bit time (T_i) by

$$\mathbf{I}_N(m) = \frac{1}{T_i} \int_{(m-1)T_i}^{mT_i} n_c(t) PN(t-\hat{T}) dt \quad (6.5\text{-}39)$$

$$\mathbf{Q}_N(m) = \frac{-1}{T_i} \int_{(m-1)T_i}^{mT_i} n_s(t) PN(t-\hat{T}) dt \quad (6.5\text{-}40)$$

where $N_{FFT}T_{i0} = T_i$. Now let us consider the evaluation of the variance of the two noise terms.

6.5.3 Evaluation of the FFT Enhanced Acquisition System Output Variance for an Arbitrary Gaussian Noise Process

Now consider the evaluation of the variance of the noise random variables, $\mathbf{I}_N(m)$ and $\mathbf{Q}_N(m_i)$. The variance is needed in order to obtain the false alarm and detection probabilities. Consider $\mathbf{I}_N(m)$ first, which is defined by in (6.5-39).

Taking the expected value of the square of $\mathbf{I}_n^{'}(m)$ produces

$$Var(\mathbf{I}_N(m)) = \frac{1}{T_i^2} \int_{(m-1)T_i}^{mT_i} \int_{(m-1)T_i}^{mT_i} E\left[n_c(t)n_c(u)\right] E\left[PN(t-\hat{T})PN(u-\hat{T})\right] dt du \quad (6.5\text{-}41)$$

where the statistical independence of the noise and code processes has been assumed in (6.5-41). Denoting the autocorrelation function of the PN code process by $R_{PN}(\tau)$, one obtains

$$Var(\boldsymbol{I}_N(m)) = \frac{1}{T_i^2} \int_{(m-1)T_i}^{mT_i} \int_{(m-1)T_i}^{mT_i} R_{n_c}(t-u) R_{PN}(t-u) dt du \tag{6.5-42}$$

Now let $m = 1$ for notational convenience, and make the change of variables

$$t' = t - T_i/2 \quad \text{and} \quad u' = u - T_i/2 \tag{6.5-43}$$

to produce

$$Var(\boldsymbol{I}_N(1)) = \frac{1}{T_i^2} \int_{-T_i/2}^{T_i/2} \int_{-T_i/2}^{T_i/2} R_{n_c}(t-u) R_{PN}(t-u) dt du \tag{6.5-44}$$

Using a result in Papoulis [20] one can convert the double integral to a single integral

$$Var(\boldsymbol{I}_N(1)) = \frac{1}{T_i^2} \int_{-T_i}^{T_i} R_{n_c}(\tau) R_{PN}(\tau) \left[T_i - |\tau| \right] d\tau \tag{6.5-45}$$

Making the reasonable assumption that T_i is much larger than T_c, one has

$$Var(\boldsymbol{I}_N(1)) \cong \frac{1}{T_i} \int_{-\infty}^{\infty} R_{n_c}(\tau) R_{PN}(\tau) d\tau \tag{6.5-46}$$

Now with (6.5-46) it is easy to express the variance in terms of the respective power spectral densities, via Parseval's equality, to yield

$$\sigma^2 = Var(\boldsymbol{I}_N(1)) \cong \frac{1}{T_i} \int_{-\infty}^{\infty} S_{n_c}(f) S_{PN}(f) df \tag{6.5-47}$$

where $S_{PN}(f)$ is the power spectral density of the unit power signal code, assuming that the chips are randomly generated with a +1 or −1, and each value occurring with a probability of ½, and σ^2 is the variance of $\boldsymbol{I}_n'(m)$. The spectral density $S_{n_c}(f)$ is the baseband version of the input noise, due to interference, thermal noise, and jamming. The same result can be shown to be true for $\boldsymbol{Q}_N(1)$ also. Thus

$$\sigma^2 = Var(\boldsymbol{Q}_N(m)) \cong \frac{1}{T_i} \int_{-\infty}^{\infty} S_{n_c}(f) S_{PN}(f) df \tag{6.5-48}$$

Thus (6.5-47) or (6.5-48) is the general result for σ^2. For the case of white Gaussian thermal noise using (6.5-47), one obtains

$$\sigma_{WGN}^2 = Var(\boldsymbol{I}_N(m)) = Var(\boldsymbol{Q}_N(m)) \cong \frac{1}{T_i} \int_{-\infty}^{\infty} \frac{N_0}{2} \delta(\tau) R_{PN}(\tau) d\tau = \frac{N_0}{2T_i} \tag{6.5-49}$$

6.5.4 BPSK Code Modulation Evaluation of P_D and P_{FA} for Arbitrary Noise

In this section the detection and false alarm probabilities for code acquisition are computed for the typical modern BPSK receiver. Most modern receivers are digitally implemented and utilize baseband I-Q designs, and therefore the baseband version will be analyzed here. However, the analog baseband version will be evaluated here due to mathematical convenience, since there are many different digital implementations, and since the analog model normally provides the best one can hope to achieve in performance. Digital receiver performance can be estimated from the analog performance when digitization losses are accounted for.

The test statistic is given by the noncoherent sum (N_h **I-Q** pairs) of the squared real and imaginary components, as shown in Figure 6.5-1. The sum is compared to a threshold T_h. The noncoherent sum is given by

$$Z_{NC}(k)=\sum_{p=1}^{N_{nc}}\left[\left(\mathbf{I}(pT_i)\right)^2+\left(\mathbf{Q}(pT_i)\right)^2\right] \tag{6.5-50}$$

Consider first the noise-only case, in order to determine the probability of false alarm. When the signal is not present, one can write

$$Z_{NC}(k)=\sum_{p=1}^{N_{nc}}\left(\mathbf{I}_N(pT_i)^2+\mathbf{Q}_N(pT_i)^2\right)=\sum_{p=1}^{2N_{nc}}\mathsf{n}^2(pT_i) \tag{6.5-51}$$

where $\mathsf{n}(pT_i)$ is a zero mean, Gaussian, random variable independent from sample to sample having the same statistics as $\mathbf{I}_N(pT_i)$ and $\mathbf{Q}_N(pT_b)$ when the signal is absent. From [20, 21] it follows that $Z_{NC}(k)$ is a central chi-square distribution with $2N_{nc}$ degrees of freedom. Since it is a noise term its statistics do not depend on k. Denote by the random variable y the density function of $Z_{NC}(k)$. It has the density function

$$p_Y(y)=\frac{1}{(2\sigma^2)^{N_{nc}}(N_{nc}-1)!}y^{(N_{nc}-1)}e^{-y/(2\sigma^2)} \tag{6.5-52}$$

where $\sigma^2=\mathrm{Var}(\mathbf{I}_n^{'}(1))$. Making the change of variable $z=y/(2\sigma^2)$ and noting that the Jacobean of the transformation is just $2\sigma^2$, one obtains the normalized noise-only probability density function that determines the false alarm probability

$$p_n(z)=\frac{(z)^{N_{nc}-1}e^{-z}}{(N_{nc}-1)!}\quad ,z\geq 0 \tag{6.5-53}$$

and is zero otherwise.

Now consider the probability density of the case when the signal present. Let

$$y=\sum_{k=1}^{2N_{nc}}x_k^2 \; and \; s^2=\sum_{k=1}^{2N_{nc}}m_k^2 \tag{6.5-54}$$

where s^2 is the *noncentrality parameter* of the distribution, with m_k being the mean value of both the $\mathbf{I}(mT_i)$ and the $\mathbf{Q}(mT_i)$. Clearly s^2 is the N sum of the $\mathbf{I}_S(mT_i)^2$ plus the N_{nc} sum of $\mathbf{Q}_S(pT_i)^2$. From (6.5-30) and (6.5-31) one has

$$s^2=N_{nc}PR_{PN}^2(\tau)\mathrm{sinc}^2(\Delta fT_{i0})L_{FFT}^2(\Delta f,T_i,k) \tag{6.5-55}$$

Let L^2 be the square of the loss[16] so that

$$L^2 = R_{PN}^2(\tau)\,\text{sinc}^2(\Delta f T_{i0}) L_{FFT}^2(\Delta f, T_i, k) \tag{6.5-56}$$

where the functional dependence of the loss has been dropped. Therefore s^2 can be written as

$$s^2 = PL^2 N_{nc} \tag{6.5-57}$$

and note that it only applies to a particular T_{i0}, k, Δf, etc. Again, using [21] one has for the probability density of the signal plus noise

$$p_Y(y) = \frac{1}{2\sigma^2}\left(\frac{y}{s^2}\right)^{\frac{N_{nc}-1}{2}} e^{\left(-\frac{s^2+y}{2\sigma^2}\right)} I_{N_{nc}-1}\left(\sqrt{y}\frac{s}{\sigma^2}\right) \quad y \geq 0 \tag{6.5-58}$$

To obtain the probability density of the normalized random variable $z = y/(2\sigma^2)$ one makes the same transformation of variables as for the noise-only case, with the Jacobean of the transformation again being just $(2\sigma^2)$. Thus one obtains for the probability density function of the signal plus noise, the result

$$p_{S+N}(z) = \left(\frac{z}{\gamma}\right)^{\frac{N_{nc}-1}{2}} e^{-\gamma} e^{-z} I_{N_{nc}-1}\left(2\sqrt{\gamma z}\right) \tag{6.5-59}$$

where the signal-to-noise parameter, γ, is defined in the general case by

$$\gamma = \frac{s^2}{2\sigma^2} = \frac{PL^2 N_{nc}}{\frac{2}{T_i}\int_{-\infty}^{\infty} S_{n_c}(f) S_{PN}(f)\,df} \tag{6.5-60}$$

The particular value of σ^2 depends on the type of noise interference, including thermal noise, unintentional interference, and jamming that can be determined from (6.5-60). When the interference is the sum of thermal noise with one-sided density N_0, then (6.5-60) becomes

$$\gamma = \frac{s^2}{2\sigma^2} = \frac{PL^2 N_{nc}}{\frac{2}{T_i}\int_{-\infty}^{\infty} S_{n_c}(f) S_{PN}(f)\,df + \frac{N_0}{T_i}} \tag{6.5-61}$$

with $S_{n_c}(f)$ now denoting all interference minus white Gaussian noise.

Now with the appropriate density functions it is possible to compute the false alarm and detection probabilities. First consider the false alarm probability. A probability of false alarm occurs when the noise exceeds the threshold after the summation of N_{nc} pairs of terms. Thus

$$P_{FA} = \Pr(z > T_h/(2\sigma^2)) = \int_{T_h/(2\sigma^2)}^{\infty} p_N(z)\,dz \tag{6.5-62}$$

so that the false alarm, letting $\eta = T_h/(2\sigma^2)$, can be written as

[16] This loss does not include filtering, A/D sampling, or other miscellaneous losses.

$$P_{FA} = \int_{\eta}^{\infty} \frac{1}{(N_{nc}-1)!} z^{N_{nc}-1} e^{-z} dz = e^{-\eta} \sum_{k=0}^{N_{nc}-1} \frac{\eta^k}{k!} \tag{6.5-63}$$

Thus this result is expressed entirely in terms of the normalized threshold and the *noncoherent combining number* N_{nc}.

Consider now the detection probability. The probability of detection is the probability that the normalized threshold is exceeded when the signal is present. Thus, letting $\eta = T_h/(2\sigma^2)$, one can write

$$P_D = \Pr(z > \eta)) = \int_{\eta}^{\infty} p_{S+N}(z) dz \tag{6.5-64}$$

Evaluating, one obtains the result

$$P_D = \int_{\eta}^{\infty} \left(\frac{z}{\gamma}\right)^{\frac{N_{nc}-1}{2}} e^{-\gamma} e^{-z} I_{N_{nc}-1}\left(2\sqrt{\gamma z}\right) dz \tag{6.5-65}$$

where γ and L^2 are defined in (6.5-60) and (6.5-56), respectively. This result can be expressed as a generalized Q function, although we will obtain a more computationally useful result. If we define the normalized signal-to-noise ratio parameter γ_0 as

$$\gamma_0 = \frac{\gamma}{N_{nc}} = \frac{s^2}{N_{nc}(2\sigma^2)} = \frac{PL^2}{2\sigma^2} \tag{6.5-66}$$

then the detection probability can be written in a form

$$P_D(\gamma_0, \eta) = \int_{\eta}^{\infty} \left(\frac{z}{N_{nc}\gamma_0}\right)^{\frac{N_{nc}-1}{2}} e^{-N_{nc}\gamma_0} e^{-z} I_{N_{nc}-1}\left(2\sqrt{N_{nc}\gamma_0 z}\right) dz \tag{6.5-67}$$

It is to be noted that evaluation of the expression for P_D is not straightforward. The most accurate method to date is to employ a series expansion suggested by Shnidman [22]. A discussion of the series used to evaluate the detection probability is presented in a later section for the matched filter acquisition case. The results are applicable for the present case also. The results for the Shnidman series to evaluate the detection probability are given by

$$P_D(\gamma_0, \eta) = \sum_{m=0}^{N_{nc}-1} e^{-\eta} \frac{\eta^m}{(m!)} + \sum_{m=N_{nc}}^{\infty} e^{-\eta} \frac{\eta^m}{(m!)} \left(1 - \sum_{k=0}^{m-N_{nc}} e^{-N_{nc}\gamma_0} \frac{(N_{nc}\gamma_0)^k}{(k!)}\right) \tag{6.5-68}$$

where γ_0 is defined by (6.5-66). This series converges reasonably rapidly and is a useful expression for evaluating the detection probability. Thus with P_D and P_{FA} evaluated the mean acquisition time can be evaluated via (6.4-10).

6.5.5 Gaussian Approximations of the Detection Probability for BPSK

In addition to this series for the detection probability, a Gaussian approximation to the expression for P_D can be easily obtained and provides a simple approximation to P_D that avoids dealing with a large number of terms needed for the detection probability computation. Since the false alarm probability can be easily evaluated exactly, only an approximation for the detection probability will be obtained here. The un-normalized probability density function in (6.5-58) has a mean and variance given by [21]

$$E\{y\} = 2N_{nc}\sigma^2 + s^2$$
$$Var\{y\} = 4N_{nc}\sigma^4 + 4\sigma^2 s^2 \tag{6.5-69}$$

Now let $z = y/(2\sigma^2)$ so that the mean and variance of z is given by

$$E\{z\} = N_{nc} + \gamma = N_{nc}(1+\gamma_0)$$
$$Var\{z\} = N_{nc} + 2\gamma = N_{nc}(1+2\gamma_0) \tag{6.5-70}$$

where γ_0 is defined in (6.5-66). It follows that the probability of detection is given by

$$P_D = Q\left(\frac{\eta - N_{nc}(1+\gamma_0)}{\sqrt{N_{nc}(1+2\gamma_0)}}\right) \tag{6.5-71}$$

where the Q function is defined by

$$Q(x) = \int_x^{\infty} \frac{1}{\sqrt{2\pi}} e^{-\frac{z^2}{2}} dz \tag{6.5-72}$$

Thus (6.5-71) is an alternative approximation to the exact series representation given in (6.5-68).

It should be pointed out that the noncoherent combining number N_{nc} cannot be increased arbitrarily to improve detection performance. The limitations are caused by the clock drift between the transmitter and the receiver and receiver to transmitter dynamics that cause the receiver code timing to drift away from the received signal. That is to say, as the number of noncoherent combinings increase the dwell time increases to $N_{nc}T_i$ seconds. If the clock drift changes on the order of 1/2 to one code chip during the dwell time (correlation time at one time hypothesis) increasing the noncoherent combining number beyond this point will lead to decreased performance.

6.5.6 Losses Between Bins in a Zero Padded FFT

Besides the outer bin loss discussed in Section 6.5.1 there is another loss that occurs in active FFT aided acquisition and is known as the *between bin losses*. This can be reduced with an augmentation of the FFT for acquisition is based on *zero padding* the FFT. Zero padding is composed of typically adding an equal number of zeros after the actual data, so that the number of FFT points is doubled.

First consider the losses between bins when zero padding is not used. The amplitude response of (6.5-20) can be well approximated by

$$|A(f)|^2 = \frac{\sin(\pi fT)^2}{(\pi fT)^2} \tag{6.5-73}$$

near bin number zero, without zero padding and is of the same form centered around the kth bin. Each bin has a response around the k-th bin of the form of (6.5-73). The integration due to the FFT produces output samples at the rate of $1/T_i$, where $T_i = N_{FFT}T_{i0}$. Consider Figure 6.5-2 which illustrates the unpadded FFT frequency response. At the center of any two adjacent bins (i.e., at $f = 1/(2T_i)$ from the number zero bin) the response is down by $\text{sinc}(\pi/2)^2$ or 3.92 dB. Hence, using a standard FFT to search the frequency cells involves a loss of about 3.9 dB, at the worst-case frequency position (i.e., midpoint between two frequency bins).

Now consider adding N_{FFT} zeros to the N_{FFT} data points for a total of $2N_{FFT}$ points, which requires an FFT with $2N_{FFT}$ taps. Let N_{FFT}^{z} denote the number of points in the zero padded FFT so that $N_{FFT}^{z} = 2N_{FFT}$. Figure 6.5-3 illustrates the zero padded multiple filter response. The result is that the number of bins is doubled, and each one is now half as far apart, but the shape is the same [19].

Note: $N_{FFT}^{z} = 2N_{FFT}$, $N_{FFT}^{z} = 32$

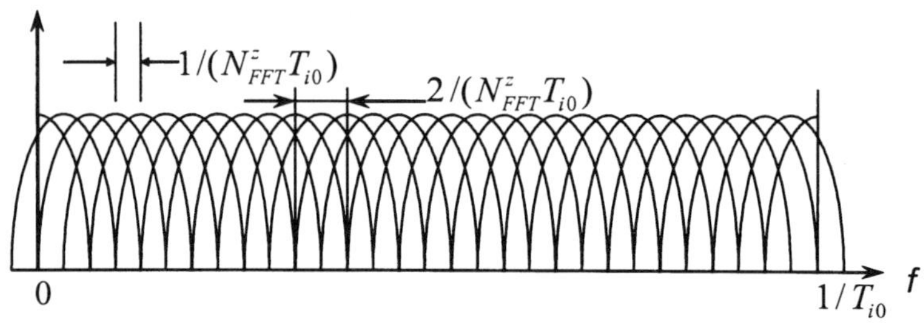

Figure 6.5-3 FFT response with zero padding for the case that $N_{FFT} = 16$ and with 16 zeros added.

Hence the worst-case loss with zero padding is given by $\text{sinc}(\pi/4)^2$ or 0.91 dB at the midpoint between two adjacent bins when zero padding is used. The disadvantage of zero padding is the fact that the number of frequency bins to be checked is doubled. Hence, doubling the number of frequency cells to search on the one hand, and losing about 3 dB less loss in the worst-case frequency position, is the tradeoff. Most times the increase in the 3-dB signal level will more than compensate for the doubling in the number of frequency cells to be searched.

6.5.7 The Frequency Search Range and Total Frequency Losses Using an FFT

In some applications a correlation of T_{i0} seconds is performed subject to the frequency error limitation of $\Delta f_m^s T_{i0} = 1/4$ or $\Delta f_m^s T_{i0} = 1/2$, where Δf_m^s is the maximum frequency error allowed in the search process between the received and local carrier frequencies. These T_{i0} second correlations are then fed into the FFT that increases the correlation time to $N_{FFT}\ T_{i0} = T_i$. When, for example, $\Delta f_m^s T_{i0} = 1/2$ is chosen for the relationship between the precorrelation time T_{i0} and the allowable frequency error Δf_m^s, the frequency search range out of the FFT includes the lower half of the FFT bins for the positive frequency errors and the upper half for the negative frequency errors. Thus, in this case, all the FFT frequency bins are searched. If $\Delta f_m^s T_{i0} = 1/4$ is selected then the frequencies from 0 Hz to $1/(4T_{i0})$ are searched for positive frequency errors, and the frequency range from $3/(4T_{i0})$ to $1/T_{i0}$ are searched for negative frequency errors. Therefore in this case only half of the frequency bins need to be searched. Figure 6.5-4 illustrates this case.

Note that in the first case the value of T_{i0} is twice the value of T_{i0} in the latter case, so that both search ranges are the same in actual frequency.

It is to be noted that this loss associated with a frequency error on the initial correlation is in addition to the loss associated with signal frequency being at the center of two bins in the FFT. In other words, without zero padding, the initial frequency error could cause as much as 0.91 dB loss if T_{i0} was subject to the equality in (6.5-10). In addition, if the signal frequency fell in between two adjacent bins the loss would be an additional 3.92 dB, for a total of 4.83 dB. This is of course a double worst case. If zero padding of the FFT were used, the total loss would be no larger than 0.91 + 0.91 or 1.82 dB.

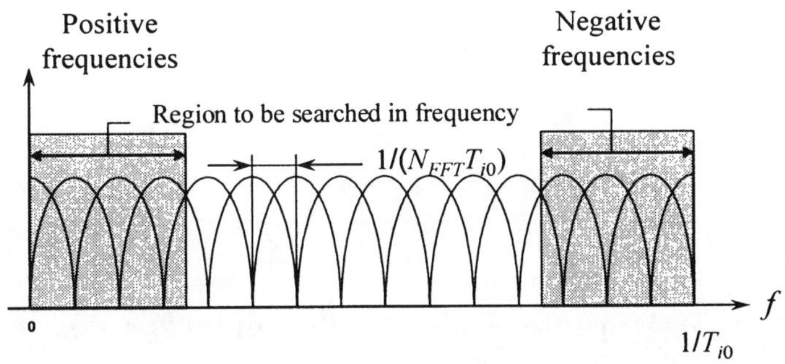

Figure 6.5-4 FFT frequency search region with precorrelation when $\Delta f_m^s T_{i0} = 1/4$.

6.5.8 The Frequency Bins of the FFT Are Uncorrelated

It will now be demonstrated that the correlator outputs of each filter bin are uncorrelated when the input is composed of signal plus thermal (white Gaussian) noise and the signal is centered at one of the frequency bins. When the signal is centered at one of the bin frequencies, all of the other bin frequency outputs signal components are zero.

All the bin frequencies have Gaussian random variable outputs. The k-th frequency bin (assumed to be a noise-only output) will have an unnormalized output given by (we are considering $Z_n(k)$ and not $Z_n'(k)$ since the orthogonality doesn't depend on the factor $1/N_{FFT}$)

$$Z_N(k) = \sum_{n=0}^{N_{FFT}-1} (\mathsf{I}_N(n) + j\mathsf{Q}_N(n))e^{-j(2\pi/N)kn} \tag{6.5-74}$$

where $\mathsf{I}_N(n)$ and $\mathsf{Q}_N(n)$ are independent Gaussian random variables that are statistically independent. Clearly $Z_N(k)$ is a linear sum of Gaussian random variables. Let

$$E(\mathsf{I}_n^{\,2}) = E(\mathsf{Q}_n^{\,2}) = \sigma_1^2 \tag{6.5-75}$$

and also assume that the following is true

$$E(\mathsf{I}_n(n)\mathsf{Q}_n(n+p)) = 0 \qquad \text{for all } p \tag{6.5-76}$$

$$E(\mathsf{I}_n(n)\mathsf{I}_n(n+p)) = 0 \qquad \text{for all } p \neq 0 \tag{6.5-77}$$

$$E(\mathsf{Q}_n(n)\mathsf{Q}_n(n+p)) = 0 \qquad \text{for all } p \neq 0 \tag{6.5-78}$$

Then the following is also true for $k \neq l$

$$E(Z_N(k)Z_N^*(l)) = E\sum_{n=0}^{N_{FFT}-1} (\mathsf{I}_N(n) + j\mathsf{Q}_N(n))e^{-j(2\pi/N_{FFT})kn} \sum_{m=0}^{N_{FFT}-1} (\mathsf{I}_N(m) - j\mathsf{Q}_N(m))e^{+(j2\pi/N_{FFT})lm} \tag{6.5-79}$$

This can be rewritten as

$$E(Z_N(k)Z_N^*(l)) = E\sum_{n=0}^{N_{FFT}-1} \mathsf{I}_N(n)\mathsf{I}_N(m)e^{-j(2\pi/N_{FFT})kn}e^{+j(2\pi/N_{FFT})lm} + E\sum_{n=0}^{N_{FFT}-1} \mathsf{Q}_N(n)\mathsf{Q}_N(m)e^{-j(2\pi/N_{FFT})kn}e^{+j(2\pi/N_{FFT})lm} \quad (6.5\text{-}80)$$

This can be simplified to

$$E(Z_N(k)Z_N^*(l)) = 2\sigma_1^2\sum_{n=0}^{N_{FFT}-1} e^{-j(2\pi/N_{FFT})(k-l)n} = 0 \quad \text{if } k-l \neq rN_{FFT} \quad (6.5\text{-}81)$$

where r is an integer including zero [20]. So that for $0 \leq k \leq N_{FFT}-1$ and $k \neq l$, the expectation is zero. Thus any distinct pair of the N_{FFT} bin frequencies outputs is uncorrelated, and since they are Gaussian, they are statistically independent. It can be shown that this is true for all the outputs including the signal. This fact is useful in determining the detection and false alarm probabilities.

6.5.9 BPSK Code Modulation γ in a Matched Spectral Jammer

The results developed so far are for arbitrary Gaussian noise interference. These results can be specialized to two jamming cases of interest. The first one is a narrowband random Gaussian process jammer that has a spectral density that is rectangular in shape and is very narrow compared to the chip rate. The second one is the matched spectral jammer. The matched spectral jammer (MSJ) is a Gaussian random jammer process that has the same power spectral density as the signal.

The input noise process is now modeled to be the sum of white Gaussian noise and the jammer noise. It is easy to show the RF noise process $n(t)$ has power that is twice the power of each individual baseband component $n_c(t)$ and $n_s(t)$ in (6.5-1). Therefore a jammer power of J watts received will have half the jammer power in each baseband component. Now consider BPSK signals that have a chip duration of T_c seconds and that have not been filtered. The spectral density of the baseband version of the in-phase noise component has a spectral density of the form

$$S_{n_c}(f) = \frac{N_0}{2} + \frac{J}{2}T_c\frac{\sin(\pi fT_c)^2}{(\pi fT_c)^2} \quad \forall f \quad (6.5\text{-}82)$$

where "$\forall$" denotes the word "all" and J is the jammer power (matched spectral jammer) at the receiver. Equation (6.5-47) can be used to evaluate the variance of the $\boldsymbol{I}_N(k)$ component, which was denoted by σ^2 for simplicity. Therefore

$$\sigma^2 = Var(\boldsymbol{I}_N(k)) = \frac{1}{T_i}\int_{-\infty}^{\infty}\left[\frac{N_0}{2} + \frac{J}{2}T_c\frac{\sin(\pi fT_c)^2}{(\pi fT_c)^2}\right]\left[T_c\frac{\sin(\pi fT_c)^2}{(\pi fT_c)^2}\right]df \quad (6.5\text{-}83)$$

when the bandpass filter effects are neglected. If the bandpass filter (BPF) is modeled as a ideal BPF with radio frequency bandwidth B, then (6.5-83) is modified to

$$\sigma^2 = Var(\boldsymbol{I}_N(k)) = \frac{1}{T_i}\int_{-B/2}^{B/2}\left[\frac{N_0}{2} + \frac{J}{2}T_c\frac{\sin(\pi fT_c)^2}{(\pi fT_c)^2}\right]\left[T_c\frac{\sin(\pi fT_c)^2}{(\pi fT_c)^2}\right]df \quad (6.5\text{-}84)$$

Neglecting the BPF effects, (6.5-84) can be evaluated to

$$\sigma^2 = Var(I_N(k)) = \frac{N_0}{2T_i} + \frac{JT_c}{3T_i} \tag{6.5-85}$$

since the integral of the flat noise term yields $N_0/(2T_i)$ and the second term can be evaluated utilizing the fact that

$$\int_{-\infty}^{\infty} \frac{\sin(\pi x)^4}{(\pi x)^4} dx = \frac{2}{3} \tag{6.5-86}$$

Now let us evaluate the γ parameter for the MSJ. Using the definition of γ (the left side of (6.5-60)).

$$\gamma = \frac{PL^2 N_{nc}}{2\sigma^2} \tag{6.5-87}$$

leads to the result

$$\gamma_{MSJ} = \frac{PL^2 N_{nc} T_i}{N_0 + \frac{2}{3} JT_c} = N_{nc} \gamma_{0MSJ} \tag{6.5-88}$$

When the jammer level is much stronger than the thermal noise, one obtains the asymptotic result

$$\gamma_{MSJ} \rightarrow \frac{3}{2} \frac{L^2 N_{nc}}{\frac{J}{P}\left(\frac{T_c}{T_i}\right)} = \frac{3}{2} \frac{L^2 N_{nc} PG}{\left(\frac{J}{P}\right)} \tag{6.5-89}$$

Where PG is the *processing gain* and is defined for direct sequence BPSK systems by

$$PG = \frac{T_i}{T_c} = \frac{R_c}{R_i} \tag{6.5.90}$$

which is the chip rate to coherent integration time rate (typically the bit or coded bit rate). The false alarm and detection probabilities are given by (6.5-63) and (6.5-68) (or (6.5-71)) using (6.5-88) for the signal-to-noise ratio parameter.

6.5.10 BPSK Code Modulation γ for a Narrowband Jammer

A Gaussian process narrowband jammer (NBJ) is a Gaussian random process that is narrowband compared to the chip rate and has bandwidth B_J. The input noise process is again assumed to be the sum of white Gaussian noise and narrowband jammer noise. As was shown in the previous section the jammer power in each baseband component is $J/2$ watts received. Again consider a BPSK signal that has a chip duration of T_c seconds. The spectral density of the baseband version of the in-phase noise component has a spectral density of the form

$$S_{n_c}(f)=\frac{N_0}{2}+\frac{J}{2B_J} \qquad -B_J/2<f<B_J/2$$
$$S_{n_c}(f)=\frac{N_0}{2} \qquad |f|\geq B_J/2 \tag{6.5-91}$$

where J is the jammer power at the receiver and B_J is the bandwidth of the jammer. The value of B_J in (6.5.90) is not important as long as it is small compared to the chip rate as will be seen shortly. Equation (6.5-47) can be used to evaluate the variance of the $\mathbf{I}_N(k)$ component. Therefore, again neglecting the effects of the BPF

$$\sigma^2=Var(\mathbf{I}_N(k))=\frac{N_0}{2T_i}+\frac{J}{2B_J}\frac{1}{T_i}\int_{-B_J/2}^{B_J/2}T_c\frac{\sin(\pi fT_c)^2}{(\pi fT_c)^2}df\cong\frac{N_0}{2T_i}+\frac{JT_c}{2T_i} \tag{6.5-92}$$

where is has been assumed that $B_JT_c<<1$. If the BPF is not neglected then the following equation applies

$$\sigma^2=Var(\mathbf{I}_N(k))=\frac{N_0}{2T_i}\int_{-B/2}^{B/2}T_c\frac{\sin(\pi fT_c)^2}{(\pi fT_c)^2}df+\frac{J}{2B_J}\frac{1}{T_i}\int_{-B_J/2}^{B_J/2}T_c\frac{\sin(\pi fT_c)^2}{(\pi fT_c)^2}df\cong\frac{N_0}{2T_i}+\frac{JT_c}{2T_i} \tag{6.5-93}$$

Notice that the narrowband jammer is about 10 log(3/2) = 1.8 dB more efficient in jamming when compared to the matched spectral jammer, but the matched spectral jammer is more difficult to remove at the receiver without also removing the signal, since it completely overlays the signal spectra.

Now let us evaluate the γ parameter for the NBJ. Using the definition of γ one has

$$\gamma_{NBJ}=\frac{PT_iL^2N_{nc}}{N_0+JT_c}=N_{nc}\gamma_{0NBJ} \tag{6.5-94}$$

When the jammer level is much stronger than the thermal noise, one obtains the asymptotic result

$$\gamma_{NBJ}\rightarrow\frac{L^2N_{nc}}{\frac{J}{P}\left(\frac{T_c}{T_i}\right)}=\frac{L^2N_{nc}PG}{\left(\frac{J}{P}\right)} \tag{6.5-95}$$

Notice that the narrowband jammer causes the signal-to-noise ratio parameter to be smaller than in the MSJ case, implying that an NBJ is more effective than the MSJ. The false alarm and detection probabilities for the narrowband jammer are given by (6.5-63) and (6.5-68) (or (6.5-71)) using (6.5-94) for the signal-to-noise ratio parameter.

Example 3 Consider an example to illustrate the results for a mean acquisition time calculation with a parallel search in time and in frequency. The model for the system is shown in Figure 6.5-1 for BPSK code acquisition where the bandpass filter (modeled as an ideal filter) bandwidth is assumed to be 24 MHz. It will be assumed that the circuit is implemented digitally (not shown in the figure). Assume a direct sequence chip rate of 10.23 megachips per second for the code. In addition, assume a total frequency uncertainty of ΔF = 1,600 Hz, and a total time uncertainty of ΔT = 20 milliseconds. Assume an equivalent thermal noise density of N_0 = –201.5 dB/Hz, a padded FFT size of 128 (corresponding to an unpadded FFT size of 64) along with a precorrelation time of $T_{i0}=3.125*10^{-6}$ seconds. Assume a bit duration of 20 ms. The assumed losses are indicated in Table 6.5-1.

Table 6.5-1 Example 3 Losses for the Acquisition Time Calculation

Loss Type	Loss in dB
Filtering (includes signal correlation loss)	0.89
Miscellaneous	0.8
Quantization	1.2
Time error	2.5
Outer bin frequency	0.91
Frequency error with zero padding	0.91
Total	7.21

The losses include a filtering loss from the assumed ideal bandpass filter with a 24-MHz bandwidth, miscellaneous losses, a quantization loss for an assumed 2-bit quantization, and a time error loss assuming that the samples are taken with a chip time error of 1/4 of a chip. The outer bin loss has been included, along with the worst-case FFT bin loss with zero padding. The total losses total 7.2 dB. The fact that two possible acquisitions could occur with a timing error of 0.25 chips has been neglected in this calculation (this would increase the P_D if accounted for). The mean acquisition time curves for the case of $N_{nc}=1$, $N_{nc}=2$, $N_{nc}=4$, $N_{nc}=8$, and $N_{nc}=16$ are shown in Figure 6.5-5. Notice that as the number of noncoherent combinings (N_{nc}) is increased, the mean acquisition time increases at low J/S ratios, but is smaller at some larger J/S values.

However, at high J/S ratios increasing N_{nc} improves the mean acquisition time. For example $\overline{T}1$ ($N_{nc}=1$) has a mean acquisition time of about 100 seconds at $J/S = 37$ dB, but $\overline{T}8$ ($N_{nc}=8$) has a mean acquisition time of about 34 seconds at $J/S = 40$ dB, which has a faster mean acquisition time at a higher J/S ratio! Thus at a given J/S ratio an optimum value of N_{nc} exists. The value of N_{nc} is limited by the difference in the transmitter clock and receiver clock times, and uncompensated motion between the transmitter and receiver. Since the dwell time is equal to $N_{nc}T_i$, when dwell time is large enough to cause the received code to slip about a chip relative to the local code, the maximum value of N_{nc} has been reached. Greater values will yield decreased performance.

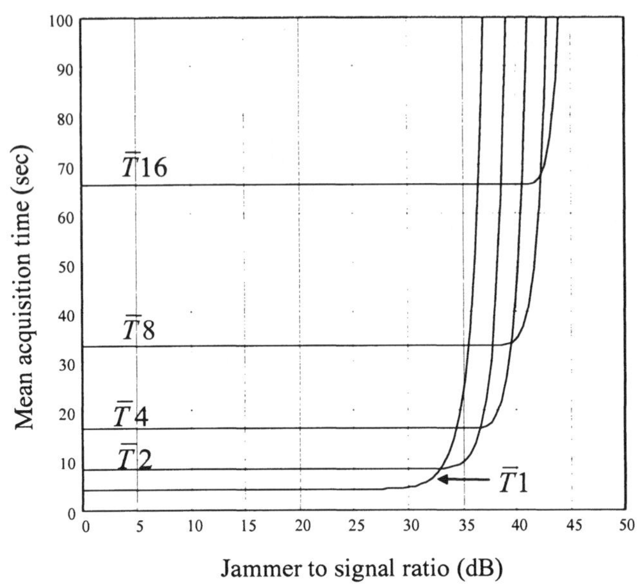

Figure 6.5-5 Mean acquisition time for a BPSK parallel active search.

6.5.11 Balanced QPSK and Balanced OQPSK Acquisition Performance

In this section balanced (equal power in both signal phases) QPSK and balanced OQPSK SS modulation with BPSK data modulation will be considered with regard to the probability of detection and the probability of false alarm performance, which can be used in mean time to acquisition calculations. It will be assumed that the data is common to both direct sequence codes, which are distinct from one another, and which operate at the same code rate. Consider Figure 6.5-6 first as the model for a QPSK acquisition detector in which the two codes operate at the same code rate and have the same data on each code. It can be viewed as two BPSK acquisition detectors in parallel, one for the $PN_1(t)$ code and one for the $PN_2(t)$ code. Each code component is similar to the BPSK code acquisition circuit shown in Figure 6.5-1. The received balanced QPSK input signal is modeled as

$$\begin{aligned} y(t) &= \sqrt{P}d(t)PN_1(t)\cos(\omega_0 t+\theta)+\sqrt{P}d(t)PN_2(t)\sin(\omega_0 t+\theta) \\ &\quad + \sqrt{2}n_c(t)\cos(\omega_0 t)+ \sqrt{2}n_s(t)\sin(\omega_0 t) \end{aligned} \qquad (6.5\text{-}96)$$

where P is the total power in the QPSK signal waveform and $PN_i(t)$ (i =1 and 2) are the two quasi-orthogonal (have a small correlation over T_i seconds) direct sequence codes on the I and Q arms of the quadriphase signal, which has carrier frequency of ω_0, and phase θ. In offset QPSK (OQPSK) $PN_2(t)$ is assumed to be delayed one half of a chip from $PN_1(t)$. The two noise terms are statistically independent, quadrature noise processes $n_c(t)$ and $n_s(t)$, each of which is modeled as independent Gaussian random processes. It is assumed that the bandwidth at the front end of the receiver is an ideal filter with bandwidth B. The power in each code is $P/2$, since balanced QPSK SS modulation was assumed.

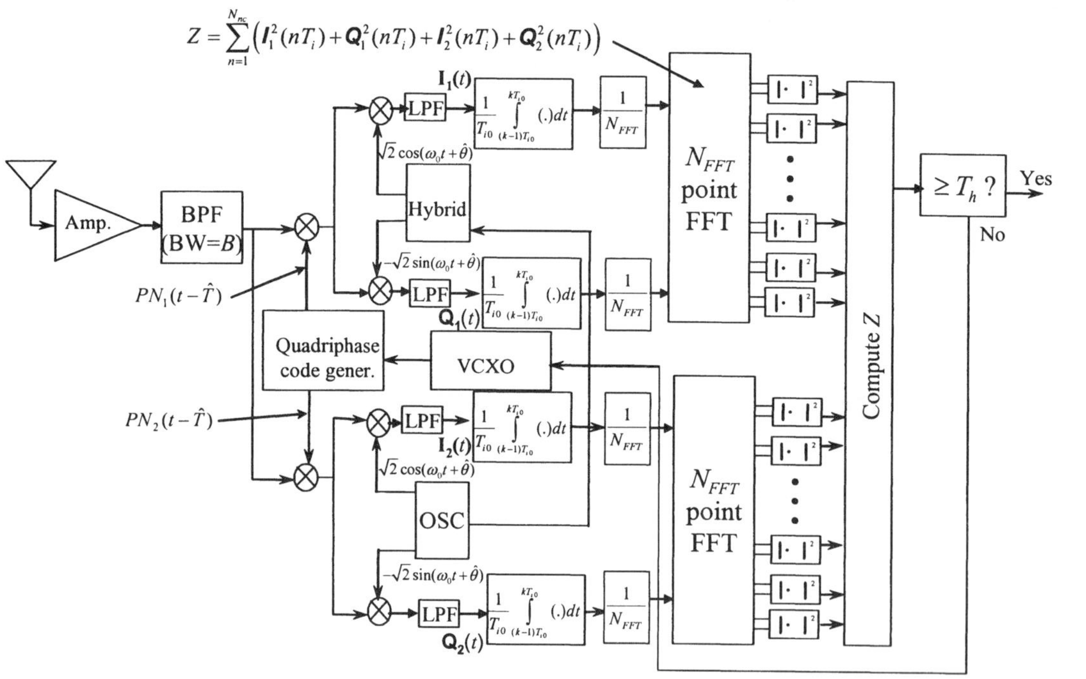

Figure 6.5-6 Model for QPSK acquisition performance (baseband version).

The signals out of the four I-Q multipliers shown in Figure 6.5-6, after heterodyning to baseband and synchronously despreading, are given by

$$I_1(t)=\sqrt{P/2}PN_1(t-T)PN_1(t-\hat{T})d(t)\cos(\Delta\omega' t+\phi)+n_c(t)PN_1(t-\hat{T})+O(\omega_0+\hat{\omega}_0) \qquad (6.5\text{-}97)$$

$$Q_1(t) = \sqrt{P/2}PN_1(t-T)PN_1(t-\hat{T})d(t)\sin(\Delta\omega' t+\phi) - n_s(t)PN_1(t-\hat{T}) + O(\omega_0+\hat{\omega}_0) \quad (6.5\text{-}98)$$

$$I_2(t) = \sqrt{P/2}PN_2(t-T)PN_2(t-\hat{T})d(t)\cos(\Delta\omega' t+\phi) + n_c(t)PN_2(t-\hat{T}) + O(\omega_0+\hat{\omega}_0) \quad (6.5\text{-}99)$$

$$Q_2(t) = \sqrt{P/2}PN_2(t-T)PN_2(t-\hat{T})d(t)\sin(\Delta\omega' t+\phi) - n_s(t)PN_2(t-\hat{T}) + O(\omega_0+\hat{\omega}_0) \quad (6.5\text{-}100)$$

which are obtained in the same manner as the BPSK case. It is to be noted that each signal component has half the power as the BPSK case, but there are twice as many.

Following the approach in the BPSK section, the output of the T_{i0} second correlators is given by

$$\mathsf{I}_1(kT_{i0}) = d(kT_{i0})\sqrt{P/2}L\cos\left[(k-1/2)\Delta\omega T_{i0}+\phi\right] + \frac{1}{T_{i0}}\int_{(k-1)T_{i0}}^{kT_{i0}} n_c(t)PN_1(t-\hat{T})dt \quad (6.5\text{-}101)$$

$$\mathsf{Q}_1(kT_{i0}) = d(kT_{i0})\sqrt{P/2}L\sin\left[(k-1/2)\Delta\omega T_{i0}+\phi\right] - \frac{1}{T_{i0}}\int_{(k-1)T_{i0}}^{kT_{i0}} n_s(t)PN_1(t-\hat{T})dt \quad (6.5\text{-}102)$$

$$\mathsf{I}_2(kT_{i0}) = d(kT_{i0})\sqrt{P/2}L\cos\left[(k-1/2)\Delta\omega T_{i0}+\phi\right] + \frac{1}{T_{i0}}\int_{(k-1)T_{i0}}^{kT_{i0}} n_c(t)PN_2(t-\hat{T})dt \quad (6.5\text{-}103)$$

$$\mathsf{Q}_2(kT_{i0}) = d(kT_{i0})\sqrt{P/2}L\sin\left[(k-1/2)\Delta\omega T_{i0}+\phi\right] - \frac{1}{T_{i0}}\int_{(k-1)T_{i0}}^{kT_{i0}} n_s(t)PN_2(t-\hat{T})dt \quad (6.5\text{-}104)$$

Using the same definition as in the BPSK case for L, one can write

$$L = R_{PN}(\tau)\text{sinc}(\Delta f T_{i0})L_{FFT}(\Delta f, T_i, k) \quad (6.5\text{-}105)$$

which accounts for the correlation, outer bin, and between bin FFT losses. Following the derivation in the BPSK section, the outputs of the FFTs in Figure 6.5-6 are given by

$$\boldsymbol{I}_1(k) = d(kT_i^-)\sqrt{P/2}L\cos\left[k\Delta\omega T_{i0}+\phi'\right] + \frac{1}{T_i}\int_{(k-1)T_i}^{kT_i} n_c(t)PN_1(t-\hat{T})dt \quad (6.5\text{-}106)$$

$$\boldsymbol{Q}_1(kT_i) = d(kT_i^-)\sqrt{P/2}L\sin\left[k\Delta\omega T_{i0}+\phi'\right] - \frac{1}{T_i}\int_{(k-1)T_i}^{kT_i} n_s(t)PN_1(t-\hat{T})dt \quad (6.5\text{-}107)$$

$$\boldsymbol{I}_2(kT_i) = d(kT_i^-)\sqrt{P/2}L\cos\left[k\Delta\omega T_{i0}+\phi\right] + \frac{1}{T_i}\int_{(k-1)T_i}^{kT_i} n_c(t)PN_2(t-\hat{T})dt \quad (6.5\text{-}108)$$

$$\boldsymbol{Q}_2(kT_i) = d(kT_i^-)\sqrt{P/2}L\sin\left[k\Delta\omega T_{i0}+\phi\right] - \frac{1}{T_i}\int_{(k-1)T_i}^{kT_i} n_s(t)PN_2(t-\hat{T})dt \quad (6.5\text{-}109)$$

Typically the coherent correlation time T_i is equal to the bit time, T_b. All the noise terms in these four equations can be shown to be statistically independent, and all have a spectral density that is the convolution of the noise spectral density and the PN code spectral density. The variance of each noise term above is denoted by σ^2.

The output of each averaging integrator is squared and averaged for N_{nc} bits. At this point it is to be noted that the four squared outputs are very similar to the BSPK case, except that there are four terms instead of two terms. Hence generalizing (6.5-50) produces

$$Z_{NC}(k)=\sum_{p=1}^{N_{nc}}\left[\boldsymbol{I}_1(pT_b)^2+\boldsymbol{Q}_1(pT_b)^2+\boldsymbol{I}_2(pT_b)^2+\boldsymbol{Q}_2(pT_b)^2\right] \quad (6.5\text{-}110)$$

which can be written in a simpler form as

$$y=\sum_{p=1}^{n}x_p^2 \quad (6.5\text{-}111)$$

where $n = 4N_{nc}$ in this QPSK case, and the x_k are Gaussian random variables that have nonzero mean values. An important component in this equation is the QPSK noncentrality parameter s_Q^2, which in this case is given by

$$s_Q^2=\sum_{p=1}^{4N_{nc}}E\left\{x_p^2\right\} \quad (6.5\text{-}112)$$

with the $E\{.\}$ denoting the ensemble average of the quantity in braces. Evaluating s_Q^2 for the QPSK case yields

$$s_Q^2=N_{nc}\frac{PL^2}{2}+N_{nc}\frac{PL^2}{2}=N_{nc}PL^2 \quad (6.5\text{-}113)$$

with both code detectors contributing equally to the value of s_Q^2. Note that the mean value is the same as in the BPSK case. Recall that L is defined in (6.5-105). Again letting the signal-to-noise ratio parameter γ_Q be defined by

$$\gamma_Q=s_Q^2/(2\sigma^2) \quad (6.5\text{-}114)$$

and using a result in [21], one has the density function of the signal plus noise

$$p_y(y)=\frac{1}{2\sigma^2}\left(\frac{y}{s_Q^2}\right)^{N_{nc}-1/2}e^{-\frac{s_Q^2+y}{2\sigma^2}}I_{2N_{nc}-1}\left(\sqrt{y}\frac{s_Q}{\sigma^2}\right) \quad ,y\geq 0 \quad (6.5\text{-}115)$$

Again making the transformation $z = y/(2\sigma^2)$ produces the density function

$$p_z(z)=\left(\frac{z}{\gamma_Q}\right)^{N_{nc}-1/2}e^{-\gamma_Q}e^{-z}I_{2N_{nc}-1}\left(2\sqrt{\gamma_Q z}\right) \quad ,z\geq 0 \quad (6.5\text{-}116)$$

The probability of detection is given by (since the distribution of the random variable z is that of the signal plus noise)

$$P_D = \Pr(z > T_h^Q/(2\sigma^2)) = \int_{T_h^Q/(2\sigma^2)}^{\infty} p_{s+n}(z)dz \tag{6.5-117}$$

where T_h^Q is the threshold for the QPSK acquisition system. Again using the normalized threshold $\eta_Q = T_h^Q/(2\sigma^2)$ produces the result

$$P_D = \int_{\eta_Q}^{\infty} \left(\frac{z}{\gamma_Q}\right)^{\frac{2N_{nc}-1}{2}} e^{-\gamma_Q} e^{-z} I_{2N_{nc}-1}\left(2\sqrt{\gamma_Q z}\right)dz \tag{6.5-118}$$

for the detection probability.

Defining a new normalized signal-to-noise ratio parameter by

$$\gamma_{Q0} = \frac{\gamma_Q}{2N_{nc}} \tag{6.5-119}$$

one can write (6.5-118) as

$$P_D = \int_{\eta_Q}^{\infty} \left(\frac{z}{2N_{nc}\gamma_{Q0}}\right)^{\frac{2N_{nc}-1}{2}} e^{-2N_{nc}\gamma_{Q0}} e^{-z} I_{2N_{nc}-1}\left(2\sqrt{2N_{nc}\gamma_{Q0} z}\right)dz \tag{6.5-120}$$

From (6.5-120) one can write the Shnidman series for the probability of detection as

$$P_D = \sum_{m=0}^{2N_{nc}-1} e^{-\eta_Q} \frac{\eta_Q^m}{(m!)} + \sum_{m=2N_{nc}}^{\infty} e^{-\eta_Q} \frac{\eta_Q^m}{(m!)} \left(1 - \sum_{k=0}^{m-2N_{nc}} e^{-2N_{nc}\gamma_{Q0}} \frac{(2N_{nc}\gamma_{Q0})^k}{(k!)}\right) \tag{6.5-121}$$

It is seen that (6.5-121) for P_D for QPSK is similar to (6.5-68) for P_D for BPSK except that the value of N_{nc} is effectively doubled and γ_{Q0} is one half of the values in the case of BPSK. Note that in the expansion of (6.5-121) γ_{Q0} is defined in (6.5-119).

The false alarm probability is given by

$$P_{FA} = \Pr(z > T_h^Q/(2\sigma^2)) = \int_{T_h^Q/(2\sigma^2)}^{\infty} p_n(z)dz \tag{6.5-122}$$

Again this can be evaluated in the same manner as before, to yield

$$P_{FA} = e^{-\eta_Q} \sum_{k=0}^{2N_{nc}-1} \frac{\eta_Q^k}{k!} \tag{6.5-123}$$

Comparing the value of P_{FA} for BPSK and QPSK we see that the value of N_{nc} is effectively doubled.

The analysis has been done in the general case. Borrowing from the BPSK analysis for a matched spectral jammer plus thermal noise, the signal-to-noise parameter is given by

$$\gamma_Q^{MSJ} = \frac{PL^2 N_{nc} T_i}{N_0 + \frac{2}{3} J T_c} = 2\gamma_{Q0}^{MSJ} N_{nc} \tag{6.5-124}$$

For a narrowband jammer plus thermal noise, the signal-to-noise ratio parameter is given by

$$\gamma_Q^{NB} = \frac{PT_i L^2 N_{nc}}{N_0 + JT_c} = 2\gamma_{Q0}^{NB} N_{nc} \tag{6.5-125}$$

The general case for the QPSK signal-to-noise ratio parameter, based on (6.5-113) and (6.5-48), is given by

$$\gamma_Q = \frac{PL^2 N_{nc} T_i}{2\int_{-\infty}^{\infty} S_{n_c}(f) S_{PN}(f) df} = 2N_{nc}\gamma_{Q0} \tag{6.5-126}$$

where $2\sigma^2$ is given by

$$2\sigma^2 = \frac{2}{T_i}\int_{-\infty}^{\infty} S_{n_c}(f) S_{PN}(f) df \tag{6.5-127}$$

and where $S_{n_c}(f)$ is the input spectral density and $S_{PN}(f)$ is the local code power spectral density normalized to unit power, since the variance of each random variable is the same for BPSK and QPSK. The main difference between BPSK and QPSK is the fact that the number of variables is doubled for QPSK. Now consider a Gaussian approximation for the detection probability for QPSK modulation.

6.5.12 A Gaussian Approximation for P_D for Balanced QPSK and Balanced OQPSK

In this section an approximation for the detection probability will be developed. Again using a result in [21] yields the mean of the unnormalized random variable y

$$E(y) = n\sigma^2 + s_Q^2 = 4N_{nc}\sigma^2 + s_Q^2 \tag{6.5-128}$$

The variance of y is given by

$$Var(y) = 2n\sigma^4 + 4\sigma^2 s_Q^2 = 8N_{nc}\sigma^4 + 4\sigma^2 s_Q^2 \tag{6.5-129}$$

Now normalize the random variable y by $2\sigma^2$ to yield the normalized variable z, which has a mean given by

$$E(z) = \frac{4N_{nc}\sigma^2 + s_Q^2}{2\sigma^2} = 2N_{nc} + \gamma_Q = 2N_{nc}(1+\gamma_{Q0}) \tag{6.5-130}$$

where $\gamma_Q = 2N_{nc}\gamma_{Q0}$ and the variance of z is given by

$$Var(z) = \frac{8N_{nc}\sigma^4 + 4\sigma^2 s_Q^2}{4\sigma^4} = 2N_{nc}(1+2\gamma_{Q0}) \tag{6.5-131}$$

where γ_{Q0} is the signal-to-noise ratio parameter

$$\gamma_{Q0} = \frac{PL^2}{4\sigma^2} \text{ in general, } \quad \gamma_{Q0} = \frac{PL^2 T_i}{2N_0} \text{ in white Gaussian noise} \tag{6.5-132}$$

and as noted earlier

$$\gamma_Q = \frac{PL^2 N_{nc}}{2\sigma^2} = 2N_{nc}\gamma_{Q0} \text{ in general, } \quad \gamma_Q = \frac{N_{nc} PL^2 T_b}{N_0} \text{ in white Gaussian noise} \tag{6.5-133}$$

Equations (6.5-130) and (6.5-131) are the mean and variance of z. The case when the signal is not present can be obtained by letting $\gamma_{Q0} = 0$ in (6.5-130) and (6.5-131). Thus with the Gaussian assumption for the random variable z, it follows that the probability of the detection is given by

$$P_D = \int_{\eta}^{\infty} \frac{1}{\sqrt{2\pi(2N_{nc}(1+2\gamma_{Q0}))}} e^{-\frac{(z-2N_{nc}(1+\gamma_{Q0}))^2}{2(2N_{nc}(1+2\gamma_{Q0}))}} dz \tag{6.5-134}$$

This can be written as

$$P_D = Q\left(\frac{\eta - 2N_{nc}(1+\gamma_{Q0})}{\sqrt{2N_{nc}(1+2\gamma_{Q0})}}\right) \tag{6.5-135}$$

which can be compared to the BPSK case in (6.5-71). γ_{Q0} is one half the BPSK value and the number of random variables is twice as large. Figure 6.5-7 illustrates the detection performance for QPSK SS and BPSK SS modulation for the case that $P_{FA}=10^{-5}$ for the cases when N_{nc} = 1 and 10. It is seen that QPSK SS modulation is about 0.5 to 1.0 dB poorer in detection probability compared to BPSK SS modulation, when comparing the required signal-to-noise ratio at a false alarm probability of 10^{-5}. Thus QPSK detection is slightly less efficient as BPSK as far as acquisition detection is concerned. It should be noted that the exact expressions for the false alarm probability were used, but the Gaussian approximation was used for the detection probabilities.

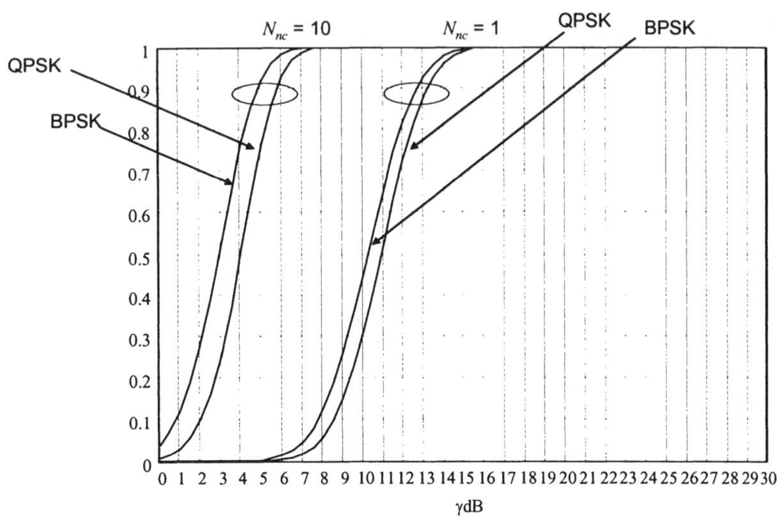

Figure 6.5-7 Comparison of detection for BPSK SS and QPSK SS for acquisition.

Offset QPSK SS modulation is quite similar to QPSK SS modulation. Figure 6.5-8 illustrates the OQPSK SS acquisition system. Note that an additional delay is needed in the upper code channel ($PN_1(t)$), since it was assumed that $PN_1(t)$ was early by one half chip, compared to $PN_2(t)$. The detection and false alarm performance will be the same for OQPSK SS and QPSK SS modulation, which as noted above is somewhat worse than BPSK SS modulation.

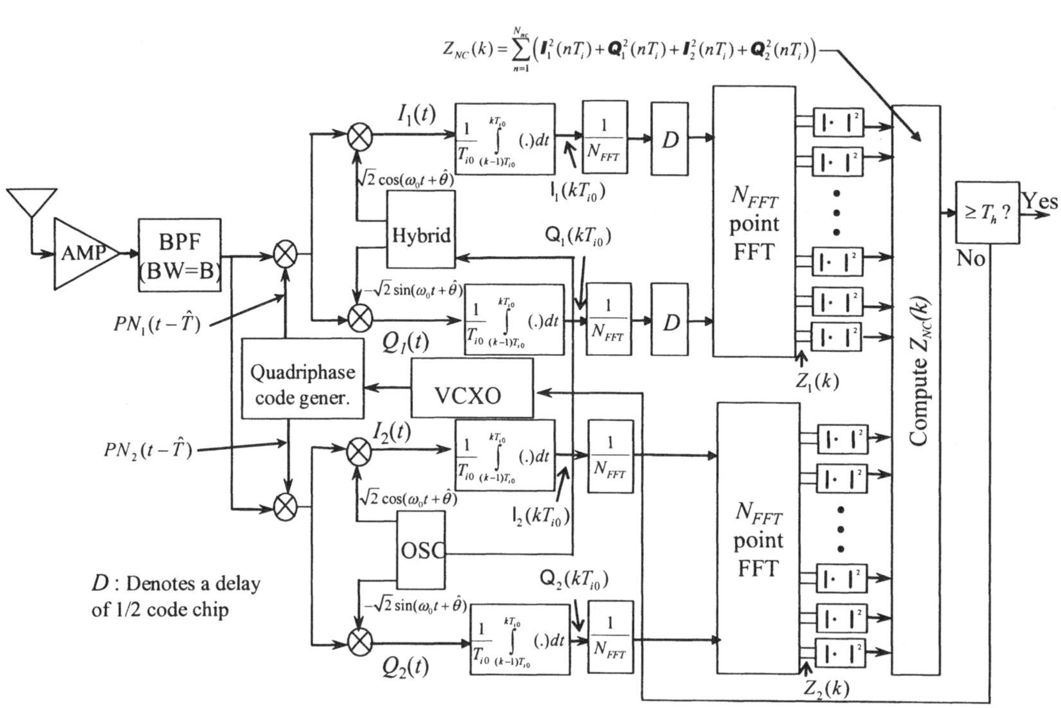

Figure 6.5-8 Model for OQPSK acquisition performance (baseband version).

6.6 AN OPTIMUM SWEEP SEARCH TECHNIQUE FOR ACTIVE ACQUISITION

In this section a modified search acquisition scheme is discussed that will improve the acquisition time for a single correlator channel, serial, search scheme if the time uncertainty is not uniformly distributed. This work follows [23]; other work in this area includes [24–27].

This section treats the optimization of the sweep path in order to minimize the acquisition time by accounting for the a priori signal location probability density function. If the a priori probability density function is uniformly distributed, then this method will not improve acquisition time performance. The idea of this approach can be illustrated by a simple example. For example, if it was known that the absolute timing uncertainty of the code epoch was $\pm$ 10 ms, but that with a probability of 90% the code timing uncertainty was within $\pm$ 2 ms, it would seem reasonable to search in the $\pm$ 2 ms region first, for one or two sweeps, and then start expanding the search to the outer regions of $\pm$ 10 ms.

Consider the a priori probability density function that is illustrated in Figure 6.6-1. Figure 6.6-1 illustrates the probability density function that is modeled to extend only over the length of L_4. The number four is taken as an example; any number of sweeps could be considered.

Figure 6.6-1(b) illustrates the code sweep procedure. The sweep starts at the starting point (the most advanced timing position) and then delays the timing (sweeps right) until it comes to the end of the line of length L_1. At the end of the line of length L_1 it is assumed that the local code is retarded to the position at the left end of the second line (of length L_2). The sweep continues to the right (delaying the code timing) until the end of line L_2 is reached at the right end. Again the local code is retarded to the position of minimum delay at the left of line 3 (code timing is advanced).

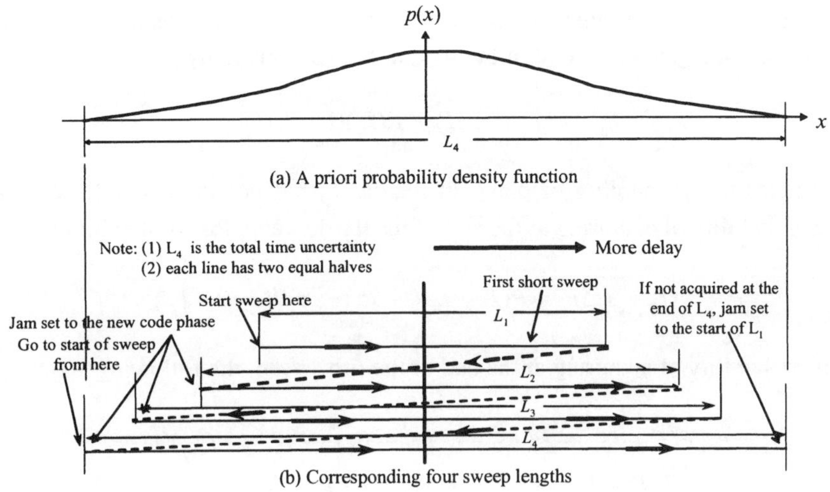

Figure 6.6-1 (a, b) A priori density function and resulting four sweep sizes.

The sweep continues to the right (retarding the code) until the end of the line of length L_3 is reached. At the right end of line L_3 the local code is retarded to the position at the left end of the position of line L_4 (least delay). The code is delayed until the right end of the fourth line is reached. If the signal has not been acquired the process starts over at the starting point by jam setting the local code to the timing position associated with the start of the process at the left end of line L_1. It should be understood that the searches along the lines of length L_i are affected in discrete correlation searches, typically 1/2 chip at a time. The large arrows on the diagram indicate the search path. The lengths are broken into halves as seen in figure 6.6-1(b). All searches are from the most advanced timing to the most delayed, to minimize the probability of detecting delayed replicas (multipath).

The length of each line in Figure 6.6-1(b) is proportional to the number of code chips searched in that sweep. Each sweep has L_i chips (i = 1, 2, 3, 4) and is symmetric about the most probable timing error location point. It is assumed that the a priori density function is symmetric about the maximum point. The search is discrete, in the sense that a given code chip timing is correlated for τ_d seconds (the dwell time), and either accepted as the correct signal timing, or the next code phase (for example, 1/2 chip away) is then correlated against for τ_d seconds, and so on.

The optimization problem is as follows. Given N sweeps, determine the optimum lengths (in chips), L_i (number of chips), to achieve a cumulative probability Q_N, at the end of the N-th sweep, so that the total acquisition time, T_N (proportional to the number of chips to be searched),

$$T_N = \alpha L_T = \alpha \sum_{i=1}^{N} L_i \tag{6.6-1}$$

is minimized. The parameter α is a proportionality constant that relates the number of chips to be searched to the acquisition time needed to search that number of chips. This analysis makes the reasonable assumption that the acquisition time is proportional to the number of chips to be searched.

First, the cumulative probability of acquisition will be formulated, assuming that the probability of signal detection is P_D (given that the signal is present). At the end of the first sweep the probability of acquisition is given by

$$P_1 = P_D P(L_1) \tag{6.6-2}$$

where $P(L_1)$ is the probability that the signal is located in region L_1. Thus the probability of acquisition is the product of the probability that the signal is in the region L_1, times the probability of detection, P_D. The probability of acquiring the signal at the end of the second sweep is given by

$$P_2 = P(L_2 - L_1)P_D + P(L_1)(1-P_D)P_D \tag{6.6-3}$$

where $P(L_2-L_1)$ denotes the probability of being in region L_2, but not in region L_1. In the same fashion, denoting P_3 as the probability of acquiring at the end of the third sweep, that probability is given by

$$P_3 = P(L_3 - L_2)P_D + P(L_2 - L_1)(1-P_D)P_D + P(L_1)(1-P_D)^2 P_D \tag{6.6-4}$$

Denoting P_i as the probability of acquiring at the end of the i-th sweep, the following is obtained

$$\begin{aligned} P_i = P_D[P(L_i - L_{i-1}) + P(L_{i-1} - L_{i-2})(1-P_D) + P(L_{i-2} - L_{i-3})(1-P_D)^2 \\ +...+P(L_2 - L_1)(1-P_D)^{i-2} + P(L_1)(1-P_D)^{i-1}] \end{aligned} \tag{6.6-5}$$

with the understanding that any term in (6.6-5) that has a nonpositive integer subscript is zero, and any term that has a factor of $(1-P_D)$ that has a negative exponent is zero. As before $P(L_i-L_{i-1})$ is the probability of being in region L_i, but not in region L_{i-1}. Now denote Q_N as the probability of acquiring in N sweeps in time T_N. By definition

$$Q_N = \sum_{i=1}^{N} P_i \quad \text{and} \quad T_N = \sum_{i=1}^{N} T_i = \alpha \sum_{i=1}^{N} L_i \tag{6.6-6}$$

In Problem 9 it is shown that Q_N can be written as

$$Q_N = P_D \sum_{k=1}^{N} P(L_k)(1-P_D)^{N-k} \tag{6.6-7}$$

In order to find the optimum lengths L_i for each portion of the N sweep approach, the Lagrangian multiplier technique can be used. Thus the function $f(\lambda, L_1, L_2, L_3,...L_N)$ is considered, where

$$f(\lambda, L_1, L_2,...L_N) = \alpha \sum_{i=1}^{N} L_i + \lambda P_D (1-P_D)^N \sum_{k=1}^{N} P(L_k)\left(\frac{1}{1-P_D}\right)^k \tag{6.6-8}$$

where λ is the Lagrangian multiplier. To proceed further it is necessary to specify an a priori density function of the code phase. Therefore it will be assumed that the a priori density function is a zero mean Gaussian random variable with a variance of σ^2, since it seems reasonable to assume that the a priori distribution function would be Gaussian in some cases. This approach will work for other density functions also. Thus, the values of $P(L_i)$ are given by

$$P(L_i) = \frac{2}{\sqrt{2\pi}\sigma} \int_0^{L_i/2} e^{-t^2/2\sigma^2} dt \tag{6.6-9}$$

and symmetry has been used in (6.6-9). Define the two parameters

$$\lambda' = \frac{-\lambda P_D(1-P_D)^N}{\alpha} \quad \text{and} \quad q = \frac{1}{1-P_D} \tag{6.6-10}$$

so that

$$\frac{1}{\alpha}\frac{\partial f(\lambda', L_1, L_2, \ldots L_N)}{\partial L_i} = 1 - \lambda' q^i \frac{1}{\sqrt{2\pi}} e^{-L_i^2/8\sigma^2} \tag{6.6-11}$$

Denoting L_i/σ as l_i and setting the left side of (6.6-11) to zero leads to the result

$$L_i = \sigma\, l_i = 2\sqrt{2}\sigma \sqrt{C + i\ln\left(\frac{1}{1-P_D}\right)} \tag{6.6-12}$$

where C is a constant that depends on α, σ, and λ'. Thus the optimum value of L_i, is given by (6.6-12), such that C satisfies the equation

$$Q_N = P_D \sum_{k=1}^{N} P(L_k)(1-P_D)^{N-k} \tag{6.6.13}$$

for the specified value of Q_N. While explicit solutions are not known, this system can be solved by simple iteration. For example, one assumes a value of C, evaluates (6.6-12) to obtain the value of the L_i, and then checks the value of Q_N to see if it agrees with the specified value of Q_N. If it does not agree another value of C has to be used and then it is necessary to check the corresponding value of Q_N. This procedure continues until the specified value of Q_N, and the calculated value of Q_N, agree for a specified value of N. The actual minimum occurs when all values of N are checked. In practice only a few values of N must be checked. The resulting values of L_i will produce the minimum acquisition time for the given a priori density function of code phase.

When the search range is $\pm 3\sigma$, based on a uniform search, it has been determined that a reduction of about 60% in acquisition time has been obtained when $P_D = 0.25$ for a Gaussian density a priori signal location, using an optimized sweep. The actual improvement depends on the value of P_D and how "peaked" the a priori probability density function is. It was found that with the Gaussian a priori distribution assumption the improvement decreases as P_D increases. Also as the a priori density function becomes less peaked the improvement decreases. The theory is quite general, the only requirement being that the a priori density function be unimodal and symmetric and that the $P(L_k)$ be differentable. More details are contained in [23, 24].

6.7 SEQUENTIAL DETECTION

Up to this point in this chapter, active detection schemes that have fixed integration times have been considered. However, when a variable integration time is considered it is possible to decrease the acquisition time for the same value of P_D and P_{FA}. The genesis of the theory of sequential detection is due to Wald [28]. Further work has been contributed by Bussgang and Middleton [29], Wald and Wolfowitz [30], and others [31–33]. In addition, DiFranco and Rubin [34] and Hancock and Wintz [35] have chapters on sequential detection.

The motivation in this section is to consider an acquisition scheme that can decrease the acquisition time on the average, for a single correlator.

6.7.1 Sequential Probability Ratio Test

Consider a process $x(t)$ for which successive data samples are labeled x_1, x_2, x_3, ...x_m. The conditional probability density function of the m samples is denoted by $p_m(\boldsymbol{x}|\boldsymbol{\theta})$, where $\boldsymbol{\theta}$ is a k-dimensional vector having components θ_1, θ_2, ...θ_k and $\boldsymbol{x}$ is an m dimensional vector of samples x_i, i =1, 2, 3...m. Let $p_m(\boldsymbol{x}|\boldsymbol{0})$ denote the conditional probability density function when the signal is not present.

The likelihood ratio is then given by

$$\Lambda_m(\boldsymbol{x}) = \frac{p_m(\boldsymbol{x} \mid \boldsymbol{\theta})}{p_m(\boldsymbol{x} \mid \boldsymbol{0})} \tag{6.7-1}$$

defines a conditional probability ratio, or likelihood ratio, for testing the null hypothesis H_0 against the hypothesis that the signal is present, H_1. Note that the likelihood ratio is a function of the sample size m.

The sequential test of Hypothesis H_1 against H_0 is as follows. Denote A and B as two positive constants such that $A > B$. Then at the m-th sample accept H_0 if

$$\Lambda_m(\boldsymbol{x}) \leq B \tag{6.7-2}$$

and accept H_1 if

$$\Lambda_m(\boldsymbol{x}) \geq A \tag{6.7-3}$$

and continue observing samples if

$$B < \Lambda_m(\boldsymbol{x}) < A \tag{6.7-4}$$

Based on this algorithm one realization of two possible outputs is shown in Figure 6.7-1. In Figure 6.7-1 one particular realization of the samples crosses the value of A, when the signal is present. In the other case when the signal is absent, the ratio crosses the lower boundary. Assuming that Hypothesis H_1 was signal plus noise, and H_0 was noise only, one would conclude that the signal was present on this particular trial with circles and was absent on the trial with squares. Once the probability density functions are known it is possible to obtain the structure of the detector. It is to be noted that Wald and Wolfowitz [30] have shown that the sequential probability ratio test is optimum in the sense that no other test, on the average, can achieve the same error probabilities with fewer samples. Let us now consider the noncoherent detector for an important case.

6.7.2 Sequential Detection for DS Acquisition with BPSK Data Modulation

Consider the noncoherent detection of a sine wave signal of amplitude A, corrupted by WGN of two-sided noise spectral density of $N_0/2$. It will be assumed that an ideal bandpass filter and an envelope detector are used to generate the output statistics.

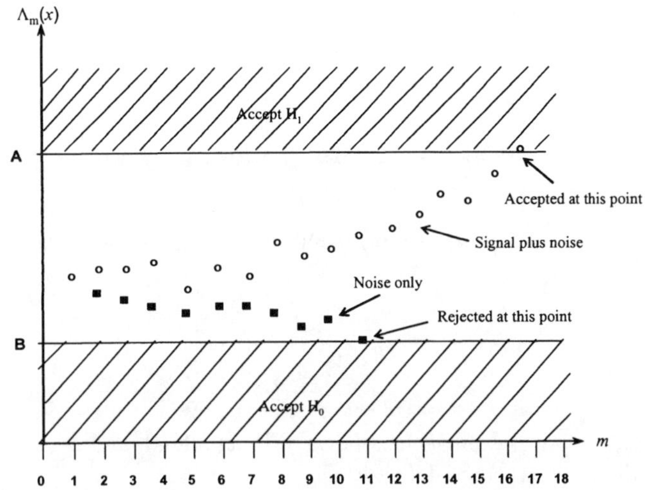

Figure 6.7-1 Two realizations of a sequential test: one with signal plus noise, and one with noise only.

Figure 6.7-2(a) illustrates the model that has a bandpass implementation.

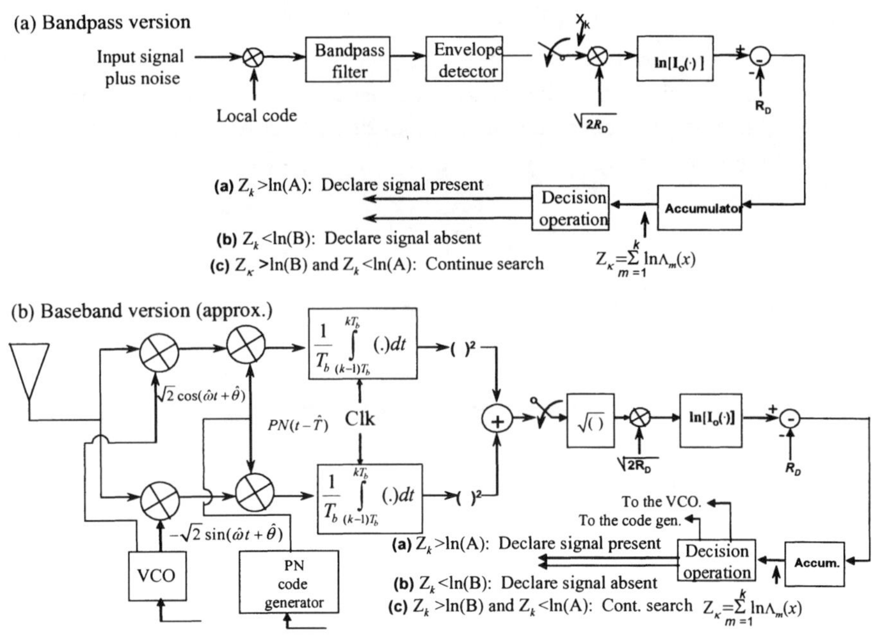

Figure 6.7-2 (a, b) Sequential detector, based on the likelihood ratio.

Figure 6.7-2(b) illustrates an approximate baseband equivalent version of the same detector. If the baseband version utilized ideal low-pass filters rather than integrate-and-dump filters it would be equivalent. It is desirable to have independent samples to simplify the analysis; therefore the assumption of an ideal bandpass filter preceding the envelope detector for the bandpass version yields statistically independent samples. The baseband version also yields independent samples since the integrator outputs are statistically independent every bit time, or T_b seconds, with white Gaussian noise on the input. In order to obtain the form of the optimum detector it is necessary to form the likelihood ratio.

Now consider the bandpass model. The probability density function of the envelope of a sinusoidal tone plus noise out of a noncoherent envelope detector of the type shown in Figure 6.7-2(a) is given by DiFranco and Rubin [34] as

$$p(r_i \mid R) = r_i \exp\left(-\frac{r_i^2 + 2R}{2}\right) I_0(\sqrt{2R}r_i) \quad , r_i \geq 0 \tag{6.7-5}$$

In this equation R is the input signal-to-noise ratio (SNR) to the envelope detector assumed to be constant and is given by

$$R = \frac{P}{N_0 B} \tag{6.7-6}$$

In (6.7-6) $P = A^2/2$ and is the input power of the sinewave input signal, and B is the noise bandwidth of the ideal bandpass filter preceding the envelope detector.

Now the natural log of the likelihood ratio is given by

$$\ln(\Lambda_i) = \ln\left[\frac{p(r_i \mid R_D)}{p(r_i \mid 0)}\right] \tag{6.7-7}$$

with R_D the design value of R. The density functions for (6.7-7) are obtained from (6.7-5) when R is the specified value of the input SNR for the numerator probability density term (R_D) and R is set equal to zero for the denominator probability density term. The logarithm is used to simplify the computations used, since the natural log of a product becomes the sum of the natural logs of each term. Let z_i denote the log of Λ_i. Then z_i, using (6.7-5) and (6.7-7), becomes

$$z_i = -R_D + \ln\left[I_0(\sqrt{2R_D}r_i)\right] \tag{6.7-8}$$

With the assumption of independent samples, the optimum detector then compares the quantity

$$Z_k = \sum_{i=1}^{k} \ln\Lambda_i = \sum_{i=1}^{k} z_i \tag{6.7-9}$$

to the two thresholds. Thus Figure 6.7-2(a) illustrates the form of the optimum detector implemented in bandpass form, and Figure 6.7-2(b) illustrates the form of the approximate optimum detector implemented in baseband form.

Consider the bandpass version of the optimum detector. The input code modulated carrier plus noise is despread by the local code, which may or may not be in code phase with the input code. This resulting signal plus noise is (ideal) bandpass filtered and envelope detected. Then the detected signal plus noise is sampled at the Nyquist rate to generate the independent samples, which are scaled by $\sqrt{2R_D}$, and then $\ln[I_0(\ )]$ of the quantity is taken. The SNR is subtracted from the $\ln[I_0(\ )]$ processed value and accumulated and the result is compared at each sample to the two boundaries of ln A and ln B. If Z_k is not less than ln B or more than ln A, the process continues until one of the boundaries is exceeded. At this point when the processed signal goes below ln B it indicates that the signal is not present on that trial, and exceeding ln A indicates that the signal was present.

It is clear that theoretically this scheme can run forever, since there is no time-out in the design. A practical implementation of a sequential detector will be discussed in the next section which incorporates a fixed time-out duration.

6.7.3 A Sequential Detection Implementation

In this section an implementation of a practical sequential detector system for BPSK SS modulated signals with BPSK data modulation that utilize the essence of Wald's sequential detector will be discussed. Consider Figure 6.7-3(a) which is a practical implementation of a bandpass version of the sequential detector. In the figure it is assumed that it is set to operate at start time $t = 0$. A baseband version is shown in Figure 6.7-2(b). The center channel of Figure 6.7-3(a) (with a subtractor and an adder) is the sequential detector, similar to what we have discussed in the last section. The upper channel is the reference channel.

First consider the center channel, which is the acquisition channel. The timing of the local code generator is set to the minimum time delay expected to minimize the chance of acquiring a multipath signal. The input signal plus noise, after being multiplied by the local code generator signal, is bandpass filtered to the data modulation plus carrier uncertainty bandwidth (B) and square law detected. The reference channel (the upper branch) provides an estimate of N_0W+P_{SP}, which is the noise power plus the spread signal power via the BPF, which has a bandwidth of W Hz and is wide enough to pass the spread signal plus any Doppler.

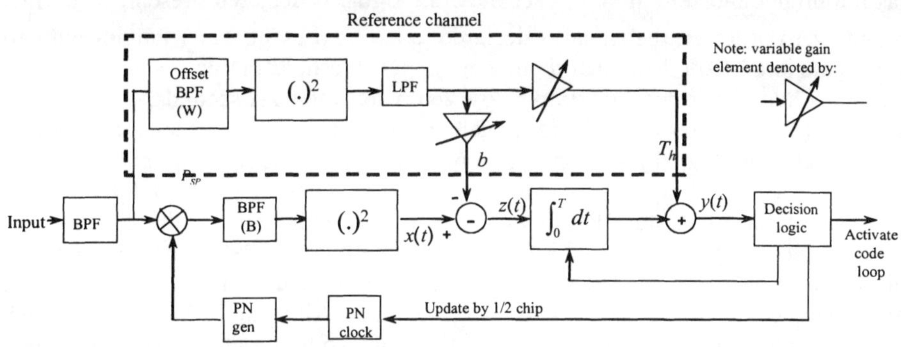

Figure 6.7-3 (a) A practical RF version of a sequential detector for BPSK signals, starting at $t = 0$.

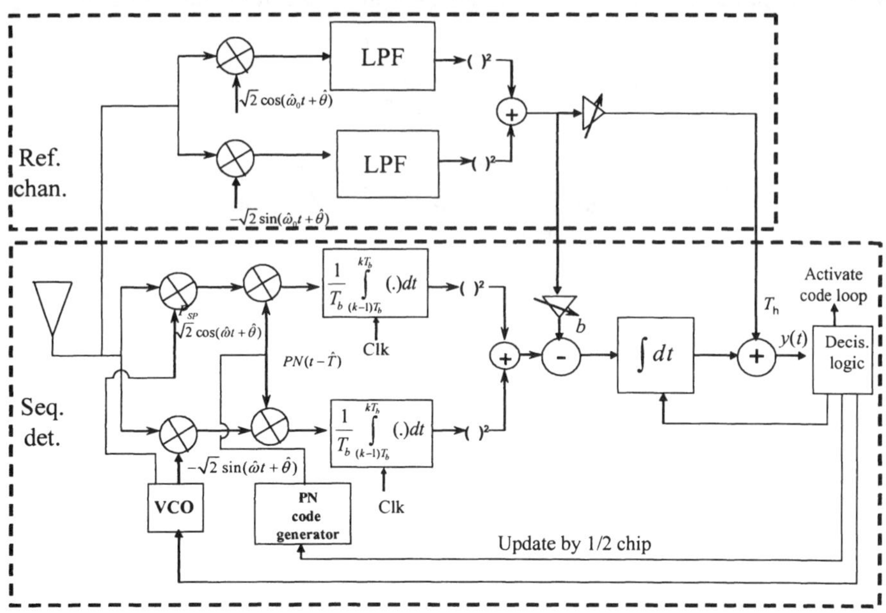

Figure 6.7-3 (b) A practical baseband sequential detector for BPSK signals.

The center frequency of the offset BPF is offset in frequency, away from the data spectrum, to "see" primarily noise. The parameter N_0 is the one-sided noise spectral density, W is the noise bandwidth of the modulation, and P_{SP} is the spread signal power that is passed through the BPF with bandwidth W Hz. Typically the spread signal power (P_{SP}) detected out of the reference channel is small compared to the power of the total despread signal P in the bandwidth W, and the spread signal power is also small compared to the noise power. Thus the quantity N_0W+P_{SP} is just the total mean power that exists in the reference channel, which is primarily noise. For typical direct sequence acquisition conditions $N_0W >> P_{SP}$, so that the estimate out of the low pass filter (LPF) in the reference channel is proportional to N_0, since it can be obtained from the total noise power, after dividing by the noise bandwidth (W) of the bandpass filter.

The reference channel output, after heavy channel low-pass filtering and a scale change, is subtracted as a bias (b) from the direct channel output $x(t)$ to yield an integrator input of $z(t) = x(t)–b$. The threshold, T_h, which is derived from the reference channel signal, is added to the integrator output. This final signal, $y(t)$, is compared in the decision logic to the value of zero, until a truncation time T_{TR} occurs. If the signal, $y(t)$, has not dropped to zero after T_{TR} seconds, the signal is declared present. If, during this time the output $y(t)$ drops to zero or lower, the signal is declared to not be present. The triangles with arrows through them are variable gain amplifiers. Note that there is no upper threshold in this system and detection is noted when the signal under hypothesis does not drop below zero value after T_{TR} seconds.

Figure 6.7-3(b) illustrates a baseband implementation of a practical sequential detector. The upper branch provides the reference signal just as in Figure 6.7-3(a). However, it is implemented at baseband rather than at a carrier or intermediate frequency. Again the input to the integrator is b, and the threshold value T_h is added to the output of the integrator. The lower section of the diagram is the baseband version of the acquisition circuitry. Normally the code and bit timing are in a known relationship, so that when the code is aligned, the bit times are also aligned. The signal is beat down to I and Q and then despread, and then both I and Q are integrated and dumped, squared, and added together. After the bias b is subtracted off from the processed signal, it is integrated and then the threshold T_h is added to it and compared to zero. The threshold and processing are identical to the bandpass version.

Figure 6.7-4 illustrates three sample functions, one with no signal present and two with different levels of signal power. All three sample functions start from the value of the threshold, T_h.

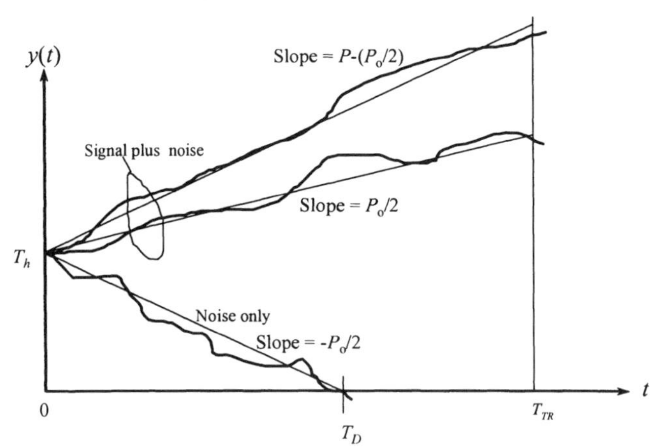

Figure 6.7-4 A practical sequential detector output with time.

Let P_0 denote the minimum power level at threshold. The nominal slope (slope of the straight line approximation) of the threshold signal is $+P_0/2$ and the slope of the noise only case is $–P_0/2$. The slope of an arbitrary signal is $P–P_0/2$ when the received power is P watts. The nominally optimum bias value for parameter b is $N_0B+P_0/2$, where P_0 is the threshold value of signal power. Note that the approximate optimum bias can be written as $b = N_0B(1+SNR_0/2)$, where $SNR_0 = P_0/(N_0B)$ with P_0 being the threshold power.

In Figure 6.7-4, the sample function that corresponds to noise only (the lowest curve) decreases to the zero from the T_h level, and the signal hypothesis is rejected at that time. The middle plot increases at a slope of $P_0/2$ and is for a signal that is at the level of P_0 watts. Finally the upper graph is for a signal at power P $(P > P_0)$ which increases with a slope of $P–(P_0/2)$ watts.

In Figure 6.7-5 the probability of false alarm and probability of detection are plotted versus truncation time for three different settings, based on a sequential detector simulation, due to Ricker et al. [36]. The results are illustrative of the type of performance that one can expect with sequential detection. As can be seen from Figure 6.7-5, P_D saturates with truncation time but P_{FA} decreases as truncation time increases. Both P_{FA} and P_D are strongly dependent on the bias setting, however. The three biases used in the simulation are b_{opt}, $0.94b_{\text{opt}}$, and $1.05b_{\text{opt}}$, where b_{opt} is the optimum bias setting.

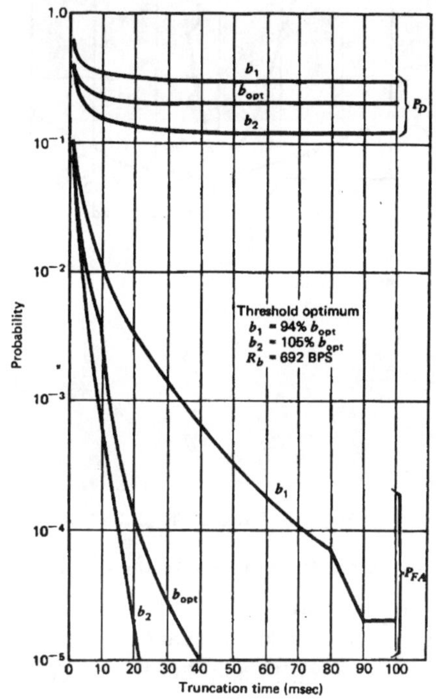

Figure 6.7-5 P_{FA} and P_D versus truncation time.

As can be seen in Figure 6.7-6 acquisition time is not as nearly as sensitive to threshold as it is to bias, since the bias is the input to the integrator. The variation with bias and threshold depend on signal to noise ratio of course, but the above curves give the general type of performance expected. The curves in Figures 6.7-5 and 6.7-6 are representative but not scalable to other bandwidths. The best performance estimates of sequential detection are made via simulation, since the analysis appears to be intractable.

6.7.4 Acquisition Time of a Sequential Detector

In this section an approximate expression for the probability of acquisition with probability P_A will be developed for the sequential detector. The results are based on computer simulation results of Ricker et al. [36] for the false alarm and the probability of detection. The acquisition time, for a given probability of acquisition, is obtained assuming that a uniform a priori probability distribution represents the a priori statistical environment. Let T_{PA} denote the time to acquire with probability P_A.

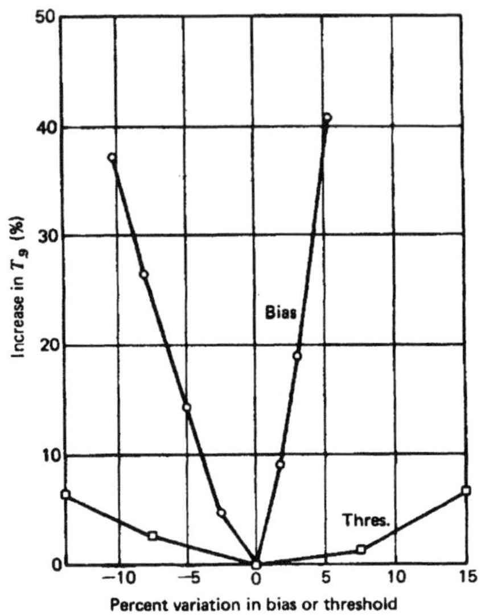

Figure 6.7-6 Acquisition time sensitivity to bias and threshold.

Then it follows that T_{PA} is given by

$$T_{PA} = T_{P\bar{S}} + T_{PS} \qquad (6.7\text{-}10)$$

where $T_{P\bar{S}}$ is the time required to search that part of the code phase uncertainty region where the signal is not present and T_{PS} is the time required to search the cell where the signal is located. Let $\overline{T_{DIS}}$ denote the mean dismissal time per cell required for the sequential detector to dismiss the hypothesis when the signal is not present (noise only). Let T_{TR} denote the time it takes to reach truncation, and therefore declare initially that the signal is present, when it was a false alarm. In addition let T_{VR} denote the time required (in addition to T_{TR}) to verify that a false alarm occurred and not a true detection, based on turning on the code tracking loop and a lock detector. It follows that

$$T_{P\bar{S}} \cong [\overline{T_{DIS}} + P_{FA}(T_{TR} + T_{VR})]N(q-1) \qquad (6.7\text{-}11)$$

where N is the total number of passes through the uncertainty region needed to achieve the acquisition probability P_A, and q is the total number of code cells to be searched in the code uncertainty region. There are $(q-1)$ cells to be searched on a single pass in which the signal is not present.

Let the probability of one cell (or the total equivalent probability of the four partial correlation values if the step size is one half chip) be denoted by P_D. It is shown in Problem 10 that the number of N (N is an integer) sweeps through the uncertainty region to achieve probability $P_{acq}(N)$ satisfies the expression

$$P_{acq}(N) = P_D + (1-P_D)P_D + (1-P_D)^2 P_D + ...(1-P_D)^{N-1} P_D \qquad (6.7\text{-}12)$$

When N is not an integer, with the assumption that the a priori probability density function is uniformly distributed, the probability of acquisition increases linearly (between integers) so that

$$N = \left\{ (K-1) + \frac{P - \hat{P}(K-1)}{\hat{P}(K) - \hat{P}(K-1)} \right\} \tag{6.7-13}$$

where $\hat{P}_A(K)$ is the probability of acquisition after the K-th sweep of the uncertainty region, and K is given by

$$K = \min[m \ni \hat{P}_A(m) > P_A] \tag{6.7-14}$$

A good approximation to the number of sweeps N for $N \geq 2$, is given by the equation

$$N \cong \frac{\ln(1-P_A)}{\ln(1-P_D)} \tag{6.7-15}$$

When the signal is present, it is not always detected in the first pass because $P_D < 1$ in general. Denote $\overline{T_{miss}}$ as the average time required rejecting the desired signal at the correct signal cell. Then the time to search the code cell, where the signal is located, is given by

$$T_{PS} \cong (N-1)\overline{T_{miss}} + T_{TR} \tag{6.7-16}$$

where T_{TR} is the truncation time.

Finally we have that the acquisition time for an acquisition probability of P_A is given by the sum in (6.7-11), which yields

$$T_{PA} = N\overline{T_{DIS}}\left[1 + \frac{P_{FA}(T_{TR} + T_{VR})}{\overline{T_{DIS}}}\right](q-1) + (N-1)\overline{T_{miss}} + T_{TR} \tag{6.7.17}$$

where N is given by (6.7.17). Note that if N is equal to 2 or larger then, to a good approximation, it follows that

$$T_P \cong \overline{T_{DIS}}\left[1 + \frac{P_{FA}(T_{TR} + T_{VR})}{\overline{T_{DIS}}}\right] q \frac{\ln(1-P_A)}{\ln(1-P_D)} + \left(\frac{\ln(1-P_A)}{\ln(1-P_D)} - 1\right)\overline{T_{DIS}} + T_{TR} \tag{6.7.18}$$

For this approximation the design parameters P_A (the probability of the acquisition), T_{VR}, T_{TR}, and the simulation values of $\overline{T_{miss}}$, $\overline{T_{DIS}}$, P_{FA}, and P_D are needed to estimate the acquisition time with probability P_A. Note also that if c chips are to be searched and 1/2 chip steps are to be used, then $2c$ cells have to be searched. Also $4c$ cells have to be searched for 1/4-chip steps, and so on. When $q >> 1$ the following approximation is accurate:

$$T_P \cong \overline{T_{DIS}}\left[1 + \frac{P_{FA}(T_{TR} + T_{VR})}{\overline{T_{DIS}}}\right] q \frac{\ln(1-P_A)}{\ln(1-P_D)} \tag{6.7-19}$$

Example 4 In Figure 6.7-7 an example is illustrated having the parameters of a 0.9 probability of acquisition, with q = 8,860 cells, a TVR value of 50 ms, a BPF noise bandwidth of 8 kHz, and a data rate of 1,400 (NRZ) symbols per second. The design point SNR of –3.5 dB (in the 8-kHz bandwidth) was used to optimize the threshold and bias. This value of SNR supports a BER of 10^{-5} using a rate 1/3 Viterbi decoder at an E_b/N_0 of 4.6 dB. Note that the acquisition time drops rapidly with predetection SNR.

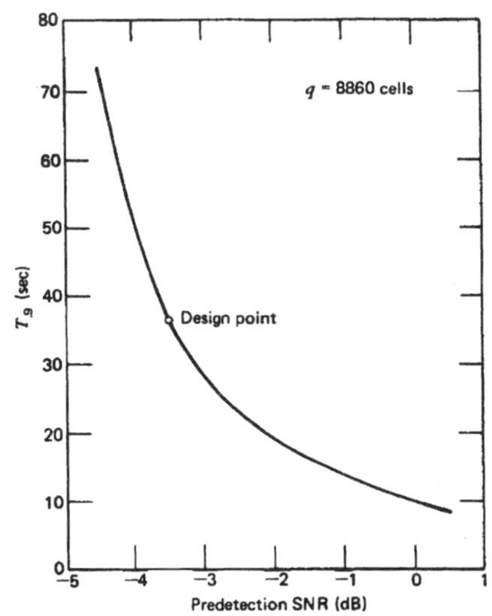

Figure 6.7-7 $T_{0.9}$ time versus the predetection signal-to-noise ratio.

Sequential detection can considerably reduce the acquisition times compared to a fixed dwell system. It should be noted that sequential detectors are not readily compatible with highly parallel correlator acquisition systems since the decision time is not fixed. Thus, the use in modern highly parallel systems is not typical.

6.7.5 The Tong Detector

An acquisition scheme that is closely related to the sequential detector is the *Tong detector*. The Tong detector has a variable detection time just like the sequential detector. Tong [37], Scott [38], and Kaplan [39] have described the Tong detector. This detector is suboptimum but more efficient than a fixed dwell time scheme. The Tong detector described here will include a lock detector function to monitor true lock and reject false alarms that passed the Tong detector. It is assumed that the time uncertainty is divided up into cells that are (typically) 1/2 chip apart. It is also assumed that the code and the data are synchronous. Figure 6.7-8 illustrates a baseband realization of the Tong detector. The front end of the receiver heterodynes the signal down to baseband with I and Q components. The I and Q channels are square law detected (some implementations are based on magnitude detectors) and compared to a threshold. The search starts at the minimum delay over the time uncertainty and is searched cell by cell, typically in increments of 1/2 a chip, until the end of the time uncertainty is reached and then it is started over again. At time 0, which corresponds to k = 1 in the figure, the correlator correlates for T seconds.

If at the end of the T second correlation the threshold is exceeded, a tentative positive decision is made and one is added to the up-down counter. If, on the other hand, the threshold is not exceeded, a one is subtracted from the up-down counter. At the initiation of the cell test the up-down counter is set to state B (B > 0). If the counter reaches the value zero, the hypothesis is rejected and the next cell is checked. If the count reaches state A (A > B), synchronization is declared. If neither state A nor B is reached, the test continues until a decision is made.

In addition to this logic it will be assumed that a lock detector is employed to monitor true lock conditions (for loss of lock) and flush out false alarms that made it through the up-down counter. It will be assumed that KT seconds are needed to perform this verification, with KT being a multiple of T.

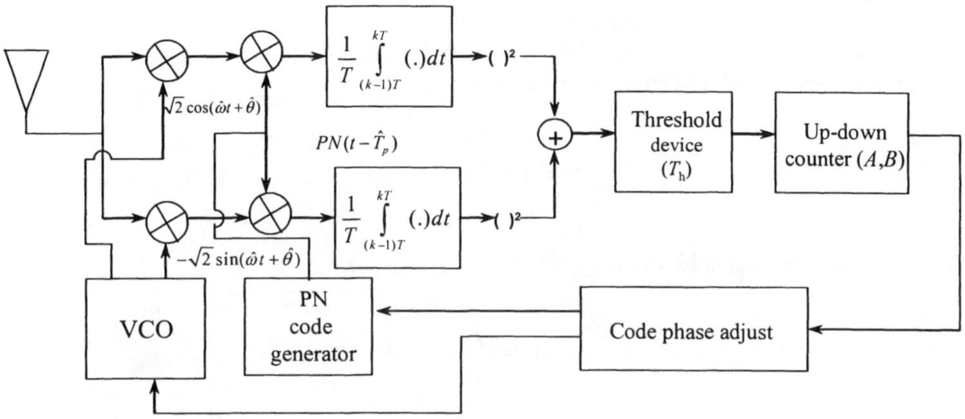

Figure 6.7-8 Baseband diagram of the Tong detector.

Figure 6.7-9 illustrates the Tong search detector algorithm with a lock detector.

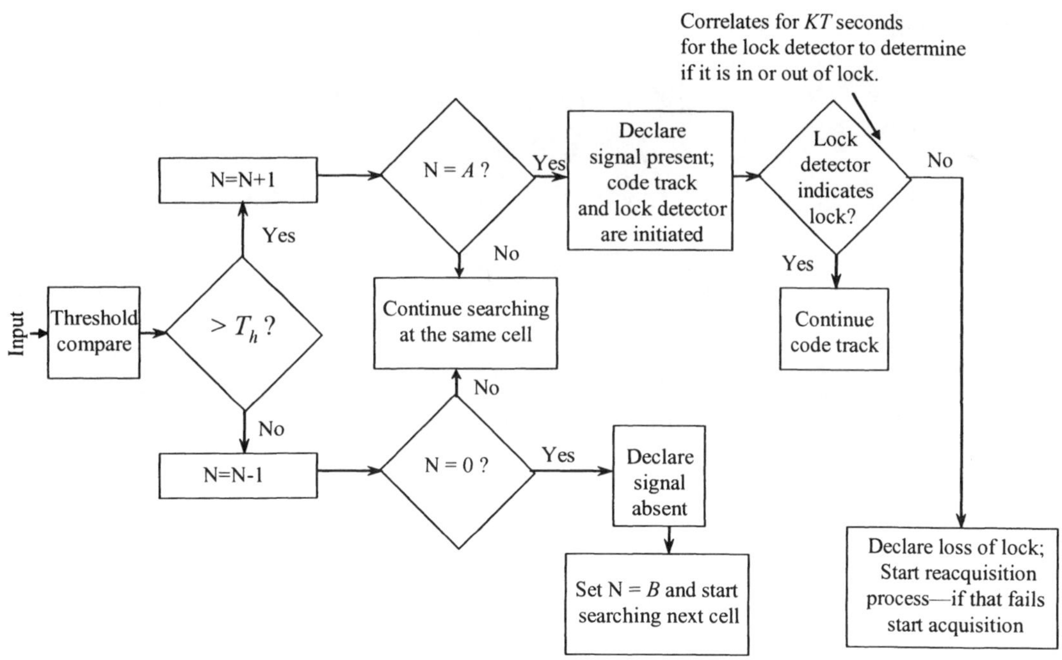

Figure 6.7-9 The Tong detector search algorithm used with a lock detector.

As seen in Figure 6.7-9, the lock detector algorithm is integrated into the acquisition algorithm. The lock detector monitors the tracking mode in general and initiates reacquisition when necessary. It also checks for false alarms by looking at the first KT second integration. If this first integration indicates that code lock is not present, it rejects this false alarm and sets the counter to B and starts searching the next cell.

In summary it is evident that there are five important parameters that are used in the Tong detector: A, B, T_h (the threshold), T, and K. Scott [38] has determined that the results in the original paper by Tong were somewhat in error. The corrected results will be presented here without derivation. Let P_d denote the probability of exceeding the threshold (T_h) after correlating for T seconds, and P_{fa} denote the probability of exceeding the threshold (T_h) after correlating for T seconds.

Now consider the evaluation of the detection probability P_d and the false alarm probability P_{fa}. From Section 6.5.4 the false alarm and detection probabilities are given by

$$P_{fa} = \int_{\eta}^{\infty} e^{-z} dz = e^{-\eta} \tag{6.7-20}$$

where η is the normalized threshold and is given by

$$\eta = \frac{T_h}{2\sigma^2} \tag{6.7-21}$$

and the noise variance is given by

$$\sigma^2 = \frac{1}{T_i} \int_{-\infty}^{\infty} S_{n_c}(f) S_{PN}(f) df \tag{6.7-22}$$

The detection probability is given by

$$p_d(\gamma_0, \eta) = \int_{\eta}^{\infty} e^{-\gamma_0} e^{-z} I_0(2\sqrt{\gamma_0 z}) dz \tag{6.7-23}$$

where γ_0 is given by

$$\gamma_0 = \frac{s^2}{2\sigma^2} = \frac{PL^2}{\frac{2}{T_i} \int_{-\infty}^{\infty} S_{n_c}(f) S_{PN}(f) df} \tag{6.7-24}$$

Again utilizing the Shnidman series expansion [22] produces

$$P_d(\gamma_0, \eta) = P_d \triangleq e^{-\eta} + \sum_{m=1}^{\infty} e^{-\eta} \frac{\eta^m}{(m!)} \left(1 - \sum_{k=0}^{m-1} e^{-\gamma_0} \frac{(\gamma_0)^k}{(k!)}\right) \tag{6.7-25}$$

Then the probability of having a true detection P_D for the cell under test is given by

$$P_D = \frac{\left(\dfrac{1-P_d}{P_d}\right)^{B} - 1}{\left(\dfrac{1-P_d}{P_d}\right)^{A+B-1} - 1} \tag{6.7-26}$$

Correspondingly the probability of having a false alarm P_{FA} for the cell under test is given by

$$P_{FA} = \frac{\left(\dfrac{1-P_{fa}}{P_{fa}}\right)^{B} - 1}{\left(\dfrac{1-P_{fa}}{P_{fa}}\right)^{A+B-1} - 1} \tag{6.7-27}$$

Once T, T_h, and C/N_0 (the carrier to noise spectral density ratio) are determined, P_{fa} and P_d can be determined. Then when A and B are chosen, P_{FA} and P_D can be determined. The mean number of correlations for a cell (that does not have the signal in it) of duration T seconds can be determined from [38]

$$E\{N_N\} = \overline{N_N} \cong \frac{1}{1-2P_{fa}} \tag{6.7-28}$$

for $P_{fa} < 0.2$ and $A > 5$. Commonly B is set equal to 1 or 2 and A is set between 5 and 100. Hence the mean time to search q cells can be approximated by

$$\overline{T} \cong \frac{2-P_D}{2P_D} qT\overline{N_N}\left(1+KP_{FA}\right) \tag{6.7-29}$$

where q is the number of cells to be searched (twice the number of chips to search if ½ chip is the separation between cells), T is the integration time to form a decision against the threshold, $\overline{N_n}$ is the mean number of (T second) correlations needed per chip to dismiss it given that the signal is not present, and K is the lock verification parameter.

6.8 TRANSFORM METHODS USED IN CODE ACQUISITION

Now consider Fourier transforms as an aid to code acquisition. We will consider only periodic sequences in this discussion. Consider the cross-correlation of two periodic sequences $x(m)$ and $y(m)$ having period $N/2$, assuming that there are two samples per code chip. Thus the cross-correlation is given by

$$R_{xy}(n) = \sum_{m=0}^{N-1} x(m)y(m+n) \tag{6.8-1}$$

If one sample per chip were used then the period of the code would be N chips. It will be shown that the cross-correlation is given by [40]

$$DFT^{-1}\left(X(k)Y^{*}(k)\right) = \sum_{m=0}^{N-1} x(m)y(m+n) \tag{6.8-2}$$

In other words the inverse discrete Fourier transform (DFT) of the product of the DFT of the x sequence and the DFT of the complex conjugate of the y sequence is the cross-correlation of the x and y sequences. Note that this relationship (6.8-2) only holds for periodic sequences.

Now let us demonstrate (6.8-2). Define the transforms of the two sequences as

$$Y(k) = \sum_{r=0}^{N-1} y(r)e^{-j\frac{2\pi rk}{N}} \tag{6.8-3}$$

and

$$X(k)=\sum_{s=0}^{N-1}x(s)e^{-j\frac{2\pi sk}{N}} \tag{6.8-4}$$

Then

$$DFT^{-1}\left(Y(k)X(k)^{*}\right)=\frac{1}{N}\sum_{k=0}^{N-1}Y(k)X^{*}(k)e^{j\frac{2\pi mk}{N}} \tag{6.8-5}$$

which can be written as

$$DFT^{-1}\left(Y(k)X(k)^{*}\right)=\frac{1}{N}\sum_{k=0}^{N-1}\left[\sum_{r=0}^{N-1}y(r)e^{-j\frac{2\pi rk}{N}}\right]\left[\sum_{r=0}^{N-1}x(s)e^{j\frac{2\pi sk}{N}}\right]e^{j\frac{2\pi mk}{N}} \tag{6.8-6}$$

which can be written as

$$DFT^{-1}\left(Y(k)X(k)^{*}\right)=\sum_{r=0}^{N-1}\left[\sum_{s=0}^{N-1}y(r)x(s)\right]\left[\frac{1}{N}\sum_{r=0}^{N-1}e^{j\frac{2\pi k(m-r+s)}{N}}\right] \tag{6.8-7}$$

Since it is true that

$$\left[\frac{1}{N}\sum_{r=0}^{N-1}e^{j\frac{2\pi k(m-r+s)}{N}}\right]=\delta(m-r+s) \tag{6.8-8}$$

Let *CC(m)* denote the cross-correlation function at the *m*-th offset, so that

$$DFT^{-1}\left(Y(k)X(k)^{*}\right)=\sum_{s=0}^{N-1}x(s)y(s+m)=CC(m) \tag{6.8-9}$$

and thus the inverse DFT of two periodic sequences yields the cross-correlation function.

Figure 6.8-1 illustrates the receiver structure to utilize DFTs (or FFTs) for code acquisition for periodic codes. The input signal plus noise is sampled to produce

$$y(rT_s)=\sqrt{2P}d(rT_s)PN(rT_s)\cos(\omega_0 rT_s+\theta)+n(rT_s) \tag{6.8-10}$$

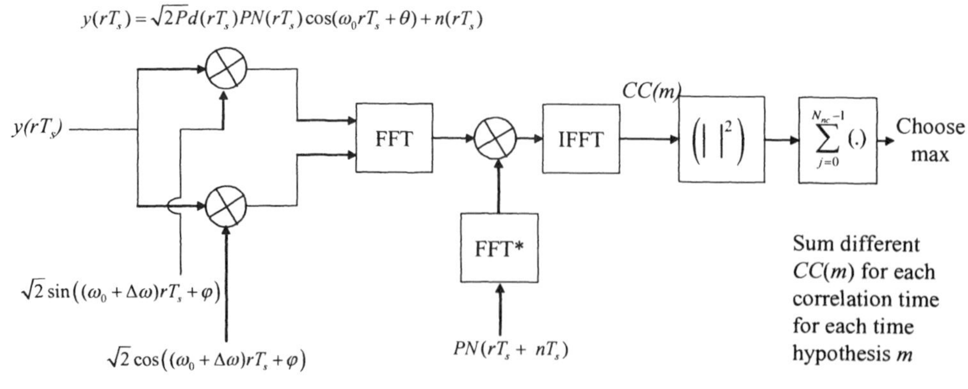

Figure 6.8-1 Model used to obtain code acquisition with a periodic code using FFTs.

It is assumed that there are two samples per chip in this discussion. The local code has the complex conjugate of the FFT taken, and the signal plus noise samples have a complex FFT taken. The product of the complex FFT and the conjugated FFT of the local code are multiplied and the inverse FFT is taken. At this point the N correlation values are available for checking for the location of the signal.

Figure 6.8-1 illustrates the use of noncoherent combining [41] to increase the SNR by adding successive correlations for each correlation delay hypothesis. Other related papers based on FFT methods of code acquisition include [42–44]. One application of this transform approach is with the code acquisition of the C/A code in the GPS. Basically the C/A code is a direct sequence spread spectrum system utilizing BPSK modulation, for both the data and the PN sequence (a set of Gold codes). The C/A code repeats every millisecond. The chip rate is 1.023 mega chips/sec. The period is 1,023 chips. The data bit rate is 50 bps. Thus every 20 ms or every 20 periods of the C/A code, the data may cause the sequence to be inverted. Hence, except for every 20 code cycles the code is periodic, and at the bit transition it may or may not invert depending upon the current and last data bits. In fact, if the contiguous bits don't change signs, 40 C/A code cycles would occur before the next possible bit transition. Thus this would be a possible application of the transform technique.

6.9 CODE ACQUISITION USING A PASSIVE MATCHED FILTER

An alternative method of obtaining code acquisition to the active correlator is to utilize a matched filter. This also called a passive code search, as opposed to the active code search already discussed in the previous sections. This section describes and determines the performance of noncoherent (passive) matched filters used for obtaining code acquisition. All the previous sections, except for the FFT section, were based on active correlation methods. We start with a discussion of a matched filter and then consider the optimum noncoherent time estimator and a practical realization of it.

6.9.1 The Matched Filter

A matched filter for a known signal [34] is a filter that maximizes the filter output SNR when the signal is immersed in white Gaussian noise (WGN). The impulse response of a matched filter for a signal $s(t)$ that exists for T seconds is well known and is of the form

$$h_{MF}(t) = ks(T-t) \quad 0 \le t \le T \tag{6.9-1}$$

where k is a constant. Thus the matched filter impulse response is proportional to a time-reversed version of the original signal. The Fourier transform of the matched filter is given by

$$H_{MF}(\omega) = kS^*(\omega)e^{-j\omega T} \tag{6.9-2}$$

where $S^*(\omega)$ is complex conjugate of the Fourier transform of $s(t)$.

Now consider a matched filter for a finite length segment of a direct sequence PSK SS signal [34]. The N chip PSK SS waveform is modeled by

$$s(t) = \sum_{n=1}^{N} a_n p[t-(n-1)T_c] \tag{6.9-3}$$

where a_n takes on the values ± 1 according to the particular sequence involved and $p(t)$ denotes an NRZ pulse of a cosine wave signal, of duration T_c seconds, starting at $-T_c/2$ seconds. Thus the complete signal is a series of pulses of cosine waves. The Fourier transform of (6.9-3) produces

$$S(\omega) = P(\omega)\sum_{n=1}^{N} a_n e^{-i\omega(n-1)T_c} \tag{6.9-4}$$

where $P(\omega)$ is Fourier transform of the basic pulse $p(t)$. However, a PSK SS signal would normally be produced by using a code generator phase shift keying a modulator. The corresponding matched filter is of the form

$$H_{MF}(\omega) = kS^*(\omega)e^{-i\omega(N-1)T_c} \tag{6.9-5}$$

where the delay term $H_{MF}(\omega) = kS^*(\omega)e^{-i\omega(N-1)T_c}$. ($T$ = $(N-1)T_C$) has been included to provide physical realizability [34]. Using (6.9-2) and (6.9-4) in (6.9-5), it is concluded that the matched filter for a BPSK SS signal is given by

$$H_{MF}(\omega) = kP^*(\omega)\sum_{n=1}^{N} a_n e^{-i\omega(N-n)T_c} \tag{6.9-6}$$

The Fourier transform of the cosine wave pulse $p(t)$, with an rms amplitude of A volts, is given by

$$P(\omega) = \int_{-T_c/2}^{T_c/2} \sqrt{2}A\cos(\omega_0 t)e^{-i\omega t}dt \tag{6.9-7}$$

This result can be evaluated to produce

$$P(\omega) = \frac{AT_c}{2}\left\{\frac{\sin[(\omega-\omega_0)T_c/2]}{[(\omega-\omega_0)T_c/2]} + \frac{\sin[(\omega+\omega_0)T_c/2]}{[(\omega+\omega_0)T_c/2]}\right\} \tag{6.9-8}$$

which is clearly real so that $P^*(\omega) = P(\omega)$. Therefore (6.9-6) can be rewritten as

$$H_{MF}(\omega) = kP(\omega)\sum_{n=1}^{N} a_n e^{-i\omega(N-n)T_c} \tag{6.9-9}$$

where $P(\omega)$ is given by (6.9-8). Thus the matched filter is the sum of chip delayed segments modified by the chip sequence $\{a_n\}$ and multiplied by the Fourier transform of the pulse signal.

Figure 6.9-1 illustrates a practical digital matched filter for the direct sequence BPSK SS signal.

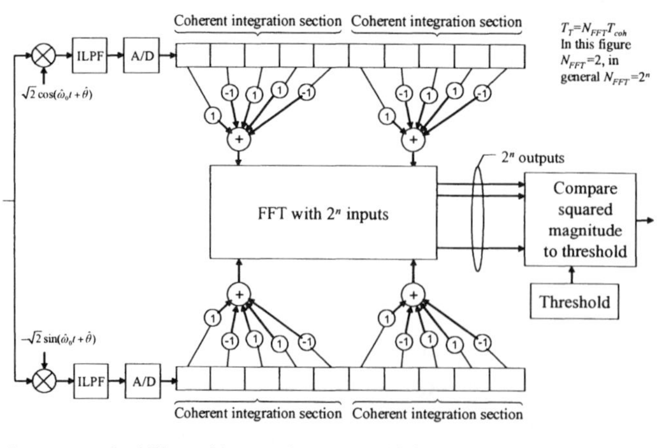

Figure 6.9-1 Model for a noncoherent matched filter with noncoherent combining.

The input signal plus noise is heterodyned down to inphase and quadrature baseband components, filtered to be compatible with the sampling, and shifted into the respective shift registers. The FFT coherently combines the individual correlation segments to increase the decision signal-to-noise ratio. A more detailed look at the digital matched filter will follow. The coherent integration is accomplished by summation.

6.9.2 Optimum Time of Arrival Estimator

In this section we consider the optimum time of arrival estimator for the case of noncoherent signals, in which the phase is unknown. The carrier frequency will be used to estimate the received carrier signal and the resulting signal will be heterodyned down to an I and Q output at baseband, where baseband matched filtering will be employed for both the I and the Q channels. A BPSK SS signal is assumed in the following. The signal is modeled over T_d seconds duration as

$$s(t) = \sqrt{2P}PN(t-T)\cos\left[\omega_0(t-T)+\theta(t-T)+\varphi\right] \tag{6.9-10}$$

The received signal plus noise is modeled as the sum

$$y(t) = s(t) + n(t) \tag{6.9-11}$$

where $n(t)$ is modeled as white Gaussian noise (thermal noise). The Bayes [34] solution to the detection of the PN coded signal is found from the generalized likelihood ratio [34] that has been averaged over phase, and is given by

$$\Lambda(\boldsymbol{r}) = \frac{p}{q}\frac{E\{p_s(\boldsymbol{r}/\varphi)\}}{p(\boldsymbol{r}/0)} \tag{6.9-12}$$

where $\Lambda(\boldsymbol{r})$ is the generalized likelihood ratio [34], and $\boldsymbol{r}$ denotes the sampled waveform with the samples taken at the Nyquist rate with the assumption that the signal plus noise passes through an ideal bandpass filter, which has a bandwidth that is large compared to the signal bandwidth. The parameter φ is the unknown phase in the received signal, with the assumption that the phase is constant (but unknown) during the receiver processing time. The subscript s on the conditional probability density function denotes that the signal is present, and the conditional density function without the s denotes the fact that the signal is not present. The $E\{\ \}$ denotes the ensemble average over the phase variable φ. The letter p denotes the a priori probability that the signal is present and q denotes the a priori probability that the signal is not present.

DiFranco and Rubin [34] have derived the optimum detection structure, based on evaluation of (6.9-12) for the signal of the form (6.9-10), and is shown in Figure 6.9-2. In the derivation the phase was assumed to be uniformly distributed over 0 to 2π radians, but it assumed that the frequency was known. The receiver heterodynes the signal down to inphase and quadrature components, then the envelope squared is obtained. The parameter $k = 2/(N_0\sqrt{R})$ and R is defined by

$$R = \frac{E_T}{N_0} \tag{6.9-13}$$

where E_T is the energy (watts-seconds) in the code burst and N_0 is the one-sided noise spectral density (watts/Hz). This is followed by a square root operation and then it is sampled at the end of the burst. Finally the function $I_0(x)$ is formed from the input x out of the sampler, and this output value is compared to a threshold. The function $I_0(x)$ is the modified Bessel function of the first kind and zeroth order. $I_0(x)$ is a monotonic function of x. The optimum detector is difficult to implement since it requires the exact signal level and noise level and the $I_0(x)$ function.

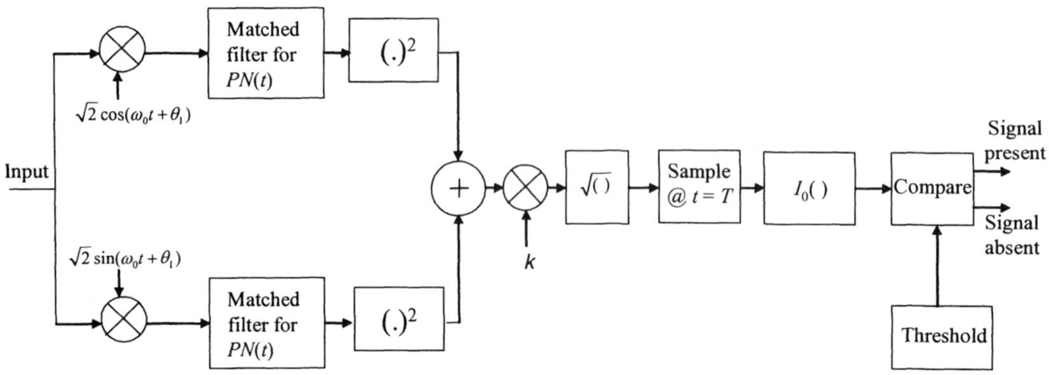

Figure 6.9-2 The optimum detector for noncoherent reception of BPSK SS signals.

However, a reasonable approximation can be made to the receiver structure, which removes the need for signal and noise strength by recognizing that the optimum receiver deals with a monotonic function of the envelope of the signal. Figure 6.9-3 illustrates the approximate optimum structure. Notice that the square root and the $I_0(x)$ function are not present in the block diagram and this structure does not require signal-to-noise ratio values.

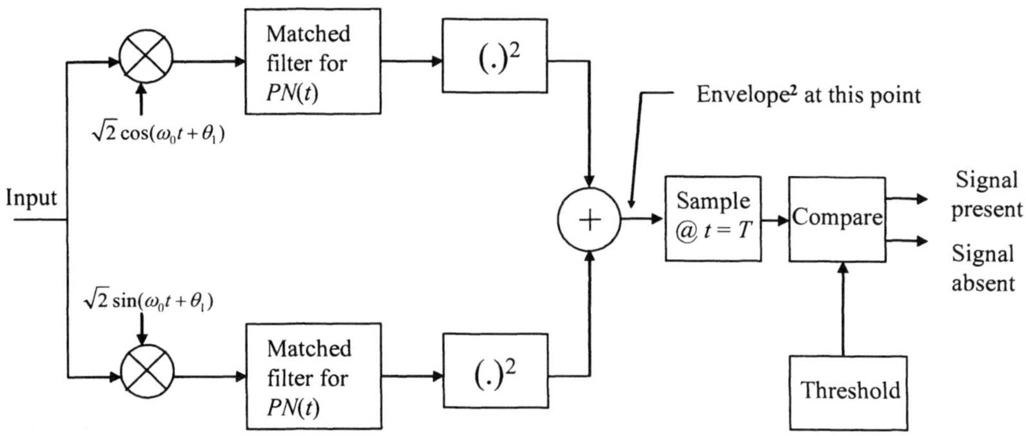

Figure 6.9-3 An approximately optimum detector for noncoherent reception of BPSK SS signals.

6.9.3 Digital Passive Matched Filter Introduction

Models of the passive and active acquisition search implementations have been discussed in numerous prior publications [34, 45–47]. These models, however, do not treat the more general case, where a receiver must search both an appreciable time uncertainty and a frequency uncertainty in the presence of noise and interference. While noncoherent combining of successive coherent integrations has been used to enable higher jammer power to signal power ratio (J/S) tolerance for active parallel search implementations, recent passive matched filter (PMF) search descriptions have not used this technique to extend the PMF dwell time beyond the bit duration [46]. Similarly, FFT-based frequency search implementations used in active parallel searches [48, 49] are not usually applied to the PMF approach [46]. The following sections will present the system model and performance of a digital passive matched filter (DPMF).

6.9.4 DPMF Acquisition Model

Consider a noncoherent digital passive matched filter (DPMF) for a random-like code with BPSK signal modulation that has a chip rate of $R_c = 1/T_c$ chip/sec, where T_c is the chip duration in seconds. The signal is assumed to be a BPSK SS modulated signal of the form

$$y(t) = \sqrt{2}Ad(t)PN(t-T)\cos(\omega_o t+\theta) \quad (6.9\text{-}14)$$

where P is the received signal power in watts, $d(t)$ is the data sequence that has nonreturn to zero (NRZ) symbols and bit a rate of R_b bps, $PN(t)$ is the spreading sequence with a chip rate of R_c cps, ω_o is the carrier frequency in radians/sec, θ is the unknown carrier phase in radians, T is unknown delay from the transmitter to the receiver, and $n(t)$ represents the bandpass noise process that can be represented in inphase and quadrature baseband noise components.

It is assumed that the clock on the receiver is accurate to $\pm\Delta T/2$ seconds, or equivalently the code chip uncertainty is a total of $M_u = R_c\Delta T$ chips. In addition to the time uncertainty, there is an assumed total frequency uncertainty of ΔF Hz. The maximum frequency error is assumed to be Δf, where $\Delta f = \Delta F/2$. The acquisition system consists of a T_{i0} second pre-FFT correlation, followed by an FFT-aided summation to increase the correlation time to 1 bit at all the FFT frequencies. Noncoherent combining extends the correlation time to τ_d seconds, which will also be referred to here as the dwell time, in multiples of bit times. Before the digital matched filter is discussed, a brief introduction of a matched filter will be presented.

Figure 6.9-4 illustrates a DPMF model for an NRZ code. This architecture is an extension of the real time implementation described in [45]. The noncoherent matched filter is a digital approximation to an analog matched filter. It is assumed that the input BPSK signal plus noise and interferer are filtered by the B_{RF} Hz ideal bandpass filter, at the input. The filtered signal plus noise is heterodyned down to baseband by two tones that are 90 degrees apart in phase and are within Δf_m Hz of the carrier frequency. Thus in-phase (I) and quadrature (Q) signals are obtained from the bandpass signal. The I and Q signals are next filtered by Nyquist filters with bandwidth B_{NQ} that are compatible with the sampling rate to remove the sum frequency terms. Thus the bandwidth of the Nyquist filters is equal to the chip rate. These filtered signals plus noise are sampled with an analog-to-digital converter (A/D).

Consider the noncoherent matched filter illustrated in Figure 6.9-4. It will be assumed that each matched filter will use two samples per chip, as is standard practice for many applications, including many GPS receivers. Hence the number of actual time cells (time hypotheses) to be searched in time is given by

$$M_{ct} = 2M_u = 2\Delta TR_c \quad (6.9\text{-}15)$$

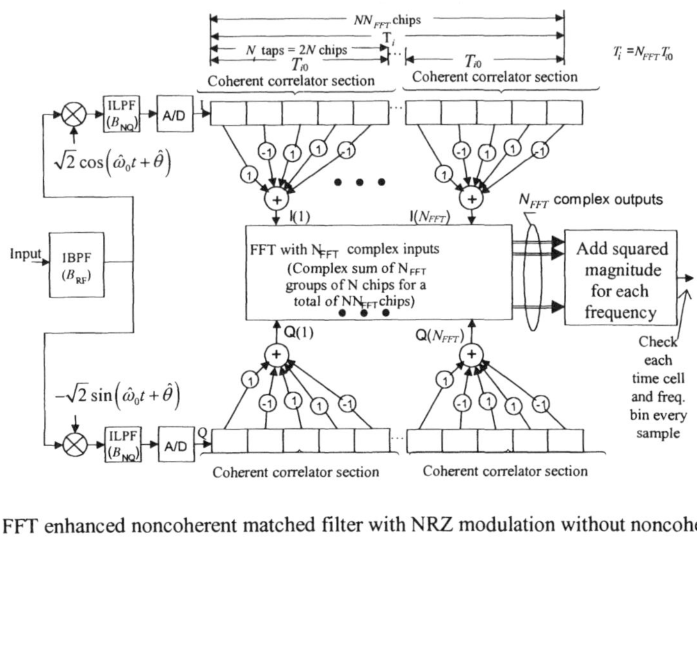

Figure 6.9-4 Model for the FFT enhanced noncoherent matched filter with NRZ modulation without noncoherent combining.

For illustrative purposes Figure 6.9-4 illustrates one sample per shift register cell; however, it is to be understood that the actual implementation would have two samples per chip.

For simplicity, it will be assumed in this analysis that the A/D process has the effect of causing only degradation to the detected signal and does not cause distortion. It follows that the bandpass filter and the anti-aliasing filter (Nyquist ideal filter with bandwidth B_{NQP}) cause some filter distortion on the received signal.

It is assumed that N ($N = T_{i0}R_c$) chips or $N_t = 2N = 2T_{i0}R_c$ taps are used for the pre-FFT coherent correlation portion of the matched filter of T_{i0} seconds, as indicated in Figure 6.9-4. The actual correlation would be accomplished digitally, however analog correlations will be analyzed for mathematical convenience. It will be assumed that the FFT forms the N_{FFT} sums of T_{i0} segments ($T_i = N_{FFT}T_{i0}$) to produce a coherent correlation time T_i seconds corresponding to the bit duration. Therefore, the total coherent correlation time is $T_i = N_{FFT}NT_c$ seconds, for each channel correlation of the I-Q pair.

The N_{FFT} correlations are input to the FFT as complex inputs and summed by the FFT to provide complex outputs at all the N_{FFT} frequencies of the FFT. The input sequence is complex since both the I and the Q phases are input to the FFT. In addition to the coherent integration (correlation), this system utilizes noncoherent combining by noncoherently combining N_{nc} coherent correlations for a total noncoherent correlation time of $\tau_D^{'}$ $=N_{nc}N_{FFT}T_{i0}$ seconds, or $N_{nc}N_{FFT}N$ chips. Noncoherent combining is not shown in Figure 6.9-4, but will be discussed shortly and is shown in Figure 6.9-5 for the complete acquisition system model.

Now consider noncoherent combining. Noncoherent combining of the coherent sections is employed to increase the detection probability at the expense of increased dwell time. Contiguous segments of code that are T_i seconds in duration are noncoherently combined (the squares of the ***I***s and ***Q***s are added together) as shown in Figure 6.9-5. The ***I*** and ***Q*** values are the output of the FFT. Thus more SNR is obtained, since more energy is collected with the N_{nc} matched filter outputs being combined noncoherently at each frequency of the FFTs for each time hypothesis. Therefore there is an added hardware complexity associated with the N_{nc} FFTs and N_{nc} noncoherent matched filters that are required.

6.9.5 Digital Matched Filter Acquisition Time Model

Consider Figure 6.9-6 to understand the acquisition algorithm. As noted earlier, the time uncertainty in the signal arrival time is assumed to be $\pm \Delta T/2$ seconds relative to the local (receiver) code (L.C.) time, for a total time uncertainty of ΔT seconds. The code segment that is to be detected has a dwell time of τ_d seconds, or equivalently $N_{nc}R_cT_i$ chips, as seen in Figure 6.9-5.

The algorithm is as follows. Based on the receiver time t_0, the matched filter (MF) receiver (local) code is instantly delayed by $\Delta T/2$ seconds relative to the receiver time t_0 and loaded into the local code register, which guarantees that even the earliest arrival of the signal will be detected. In Figure 6.9-6(a) the earliest, latest, and nominal arrival times of the segment of code that is to be detected is shown. These three cases are shown in Figures 6.9-6(a) through 6.9-6(f). We have used τ_d = $\tau_D^{'}$ in Figure 6.9-6(b) for visual simplification. Figure 6.9-6(b) shows that when the code segment of τ_d seconds arrives at the nominal time, it could possibly be detected by the matched filter after waiting $\tau_d + \Delta T/2$ seconds.

This is the first opportunity for detection of the nominal arrival time of the code segment. Figure 6.9-6(c) illustrates that $\tau_d + \Delta T$ seconds after t_0, the latest arrival time of the received code will just align itself with the local code and could be detected. At this point in time a new segment of code, of time duration τ_d seconds, is jam-set to correspond to the time associated with just being equal to the start of the earliest arrival code segment.

Figure 6.9-6(d) illustrates the situation when time has progressed to $2\tau_d+\Delta T$ seconds past the initial time t_0. At this point in time the earliest possible arrival time has just aligned with the local code in its second position. It can be seen that $\Delta T/2$ seconds later, the nominal code segment will be aligned, and ΔT seconds later the latest possible arrival of the code segment will be aligned.

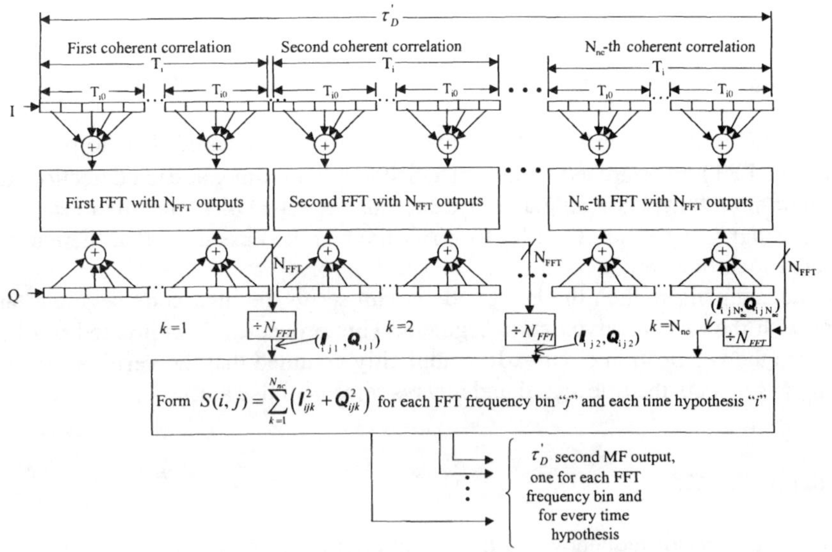

Figure 6.9-5 Code MF acquisition circuit illustrating noncoherent combining of coherent sections.

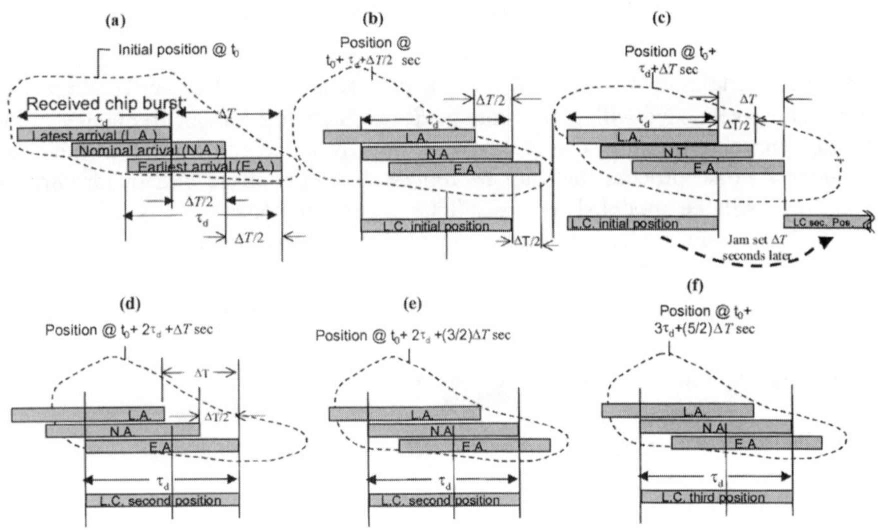

Figure 6.9-6 (a–f) Time diagram for matched filter acquisition time calculations.

After $2\tau_d + (3/2)\Delta T$ seconds after t_0, the nominal arrival code will be aligned, as shown in Figure 6.9-6(e). At this point in time the nominal code could again be detected, since the received and local codes are aligned in time. At $2\tau_d + 2\Delta T$ seconds after t_0, the third local code segment is jam-set to be ready for the earliest arrival time for the third acquisition try. Figure 6.9–6(f) illustrates the situation when the next attempt can be made for the nominal arrival code time, which occurs at $3(\tau_d + \Delta T)$–$\Delta T/2$ seconds after t_0. Based on the previous discussion, the possible acquisition times of the nominal time arriving signal segment is given by

$$T_{acq}(n) = n(\Delta T + \tau_d) - \Delta T/2 \tag{6.9-16}$$

for $n=1,2,\ldots,\infty$. It is clear that the earliest arriving code signal burst would be detected $\Delta T/2$ seconds earlier, and the latest arriving code burst would be detected $\Delta T/2$ seconds later than the nominal arrival time segment.

This procedure continues in this manner until detection is declared. It is to be noted that both time and frequency (via the FFT) are searched with each τ_d second dwell time, at each detection attempt.

After one or more "hits" (signal plus noise exceeds the threshold), the hits must be verified as either false alarms or a true signal detection. In order to verify the hits, it is assumed that there are N_v code-tracking loops and associated lock detectors available for verification of up to N_v hits. If none of the N_v verification circuits detect a true acquisition, then the local code is jam-set to the next code segment, and the correlator waits to correlate with the next received code segment. This procedure is continued until a successful true lock is obtained from the verification circuits. It is implicitly assumed that the verification circuit is perfect in the sense that it always detects the true signal and always rejects a false alarm.

6.9.6 Signal Model for DPMF

In this section the losses due to frequency and time errors (Doppler, clock error, and so on) and the variance of the noise over the initial coherent correlation time for the NRZ signal will be obtained. The calculations are very similar to Section 6.5.1. The correlation time for this correlation is T_{i0} seconds. The received BPSK SS signal into the acquisition circuit is modeled as

$$y(t)=\sqrt{2}Ad(t)PN(t-T)\cos(\omega_0 t+\theta)+n(t) \tag{6.9-17}$$

where P is the received signal power in watts, $d(t)$ is the data sequence that has NRZ symbols and bit rate of R_b, $PN(t)$ is the spreading sequence with a chip rate of R_c cps, ω_0 is the carrier frequency in radians/sec, θ is the unknown carrier phase in radians, T is unknown delay from the transmitter to the receiver, and $n(t)$ represents the bandpass noise process that can be represented in inphase and quadrature baseband noise components. The noise will be modeled as a bandpass Gaussian noise process with a symmetric power spectral density. The noise term can be represented in terms of the inphase and quadrature components by[17]

$$n(t)=\sqrt{2}n_c(t)\cos(\hat{\omega}_0 t+\hat{\theta})+\sqrt{2}n_c(t)\sin(\hat{\omega}_0 t+\hat{\theta}) \tag{6.9-18}$$

The cosine and sine reference signals are assumed to be of the form

$$r_c(t)=\sqrt{2}\cos(\hat{\omega}_0 t+\hat{\theta}) \quad r_s(t)=-\sqrt{2}\sin(\hat{\omega}_0 t+\hat{\theta}) \tag{6.9-19}$$

where $\hat{\omega}_0$, $\hat{\theta}$, and $\hat{T}$ are the receiver's estimated received signal angular frequency, carrier phase, and estimated code time delay from the transmitter to the receiver, respectively. Following the analysis in Section 6.5.1 it is shown in Problem 12 that the inphase T_{i0} second correlation can be modeled as

$$\mathsf{I}(k)=\frac{1}{T_{i0}}\int_{(k-1)T_{i0}}^{kT_{i0}}\left[\sqrt{P}d(t)PN(t-T)PN(t-\hat{T})\cos(\Delta\omega t+\phi)+n_c(t)PN(t-\hat{T})\right]dt \tag{6.9-20}$$

with the assumption that the correlation starts at $t=0$ ($k=1$). It is to be noted that the signal component of $\mathsf{I}(k)$ is dependent on the difference of the actual and estimated delays $T-\hat{T}$. Let $\tau=T-\hat{T}$ be the timing error between the received code and the local matched filter code.

The signal component of $\mathsf{I}(k)$ is denoted by $\mathsf{I}_s(k)$ and can be shown to be equal to

[17] Note that we have assumed that the phase of the reference is aligned with the noise process to simplify the noise analysis.

$$\mathsf{I}_S(k) \cong d(k)\sqrt{P}R_{PN}(\tau)\text{sinc}\left(\Delta f T_{i0}\right)\cos\left[k\Delta\omega T_{i0} + \phi_k\right] \tag{6.9-21}$$

where ϕ_k is the phase that results from the k-th integration including a constant phase term that is unimportant in this analysis, $R_{PN}(\tau)$ is the autocorrelation function of the spreading code, and $\text{sinc}(x) = \sin(\pi x)/(\pi x)$.

Note that $d(k)$ denotes the data value in the k-th T_{i0} second integration time, which will be within a bit when synchronized. The term $d(k)$ is equal to 1 or −1 during the k-th bit time.

In the same manner the Q channel signal component can be described by

$$\mathsf{Q}_S(k) \cong d(k)\sqrt{P}R_{PN}(\tau)\text{sinc}\left(\Delta f T_{i0}\right)\sin\left[\Delta\omega T_{i0}/2 + \phi_k\right] \tag{6.9-22}$$

Thus the correlation loss out of the first coherent combining matched filter output, when considering both the I and Q channels, is given by

$$L = R_{PN}^2(\tau)\left[\frac{\sin(\pi\Delta f T_{i0})^2}{(\pi\Delta f T_{i0})^2}\right] \tag{6.9-23}$$

Filtering has been neglected in this expression but will be accounted for later.

In order to keep the carrier frequency error losses small, the maximum allowable coherent integration time must satisfy the outer bin frequency loss (carrier frequency error) equation

$$L_{outerbin} = 10\log\left[\frac{\sin(\pi\Delta f T_{i0})^2}{(\pi\Delta f T_{i0})^2}\right] \text{ dB} \tag{6.9-24}$$

which is the same for the I and Q channels and is maximum when $\Delta f = \Delta f_m$. In order to limit the power loss to 0.91 dB, for example, it is necessary to limit the coherent correlation time to

$$T_{i0} \le \frac{1}{4\Delta f_m} \tag{6.9-25}$$

If we denote the integrated noise terms on the I channel and the Q channel by $\mathsf{I}_N(k)$ and $\mathsf{Q}_N(k)$ then the I and Q channel outputs from the averaging integrate-and-dump filters (T_{i0} correlators) is given by

$$\mathsf{I}(k) = \mathsf{I}_S(k) + \mathsf{I}_N(k) \tag{6.9-26}$$

and

$$\mathsf{Q}(k) = \mathsf{Q}_S(k) + \mathsf{Q}_N(k) \tag{6.9-27}$$

where

$$\mathsf{I}_N(k) = \frac{1}{T_{i0}}\int_{(k-1)T_{i0}}^{kT_{i0}} n_c(t)PN(t-\hat{T})dt \tag{6.9-28}$$

and

$$\mathsf{Q}_N(k) = \frac{-1}{T_{i0}} \int_{(k-1)T_{i0}}^{kT_{i0}} n_s(t) PN(t - \hat{T}) dt \tag{6.9-29}$$

6.9.7 Noise Variance Equation for a Gaussian Random Process Model of a Jammer

In this section the variance of a Gaussian noise matched spectral jammer (MSJ) will be determined. Consider now the evaluation of the noise terms in (6.9-28) and (6.9-29). Since $\mathsf{I}_N(k)$ is zero mean valued, the variance is given by

$$\mathrm{Var}(\mathsf{I}_N(k)) = \frac{1}{T_{i0}^2} \int_{(k-1)T_{i0}}^{kT_{i0}} \int_{(k-1)T_{i0}}^{kT_{i0}} R_{n_c}(t-u) R_{PN}(t-u) dt du \tag{6.9-30}$$

where the statistical independence of the noise and code processes has been assumed. Making a change of variables and using a result from Papoulis [20], it can be shown that

$$\mathrm{Var}(\mathsf{I}_N(k)) = \frac{1}{T_{i0}^2} \int_{-T_{i0}}^{T_{i0}} R_{n_c}(\tau) R_{PN}(\tau) \left[T_{i0} - |\tau| \right] d\tau \tag{6.9-31}$$

Making the assumption that T_{i0} is much larger than the chip time, T_c, it follows that

$$\mathrm{Var}(\mathsf{I}_N(k)) \cong \frac{1}{T_{i0}} \int_{-\infty}^{\infty} R_{n_c}(\tau) R_{PN}(\tau) d\tau \tag{6.9-32}$$

Now (6.9-32) can be written in terms of the respective power spectral densities via Parseval's equality in the form

$$\mathrm{Var}(\mathsf{I}_N(k)) \cong \frac{1}{T_{i0}} \int_{-\infty}^{\infty} S_{n_c}(f) S_{PN}(f) df \tag{6.9-33}$$

where $S_{PN}(f)$ is the power spectral density of the unit power code, assuming that the chips are randomly generated with a +1 or −1, and each value occurs with a probability of 1/2.

It can be shown that the variance out of the FFT summed correlations (T_i sec) is simply modified to

$$\mathrm{Var}(\boldsymbol{I}_N(k)) \cong \frac{1}{T_i} \int_{-\infty}^{\infty} S_{n_c}(f) S_{PN}(f) df = \mathrm{Var}(\boldsymbol{Q}_n(k)) \tag{6.9-34}$$

where the bold italic Arial font denotes the FFT output terms. In other words the time T_{i0} is replaced by T_i.in (6.9-34). It is easy to show that $\mathrm{Var}(\boldsymbol{I}_N(k)) = \mathrm{Var}(\boldsymbol{Q}_N(k))$.

6.9.8 Variance Evaluation of an MSJ

Consider now a matched spectral jammer having a received power of J watts that is centered at the carrier frequency. A matched spectral jammer (MSJ) is a Gaussian random process that has the power spectral density of the signal. The Nyquist baseband filter has an ideal low-pass filter bandwidth equal to twice the sample rate (assuming two samples per chip). The variance of the sum of the thermal noise and the matched spectral interference has a variance defined in (6.9-34) that can be written as

$$\sigma_{MSI}^2 \cong \frac{1}{T_i}\int_{-B_{\min}}^{B_{\min}} \frac{N_0}{2}\left|H_{BB}(f)\right|^2 \left|H_{NQ}(f)\right|^2 T_c \frac{\sin(\pi f T_c)^2}{(\pi f T_c)^2}df + \frac{1}{T_i}\int_{-B_{\min}}^{B_{\min}} \frac{J}{2}(T_c)^2 \left|H_{BB}(f)\right|^2 \left|H_{NQ}(f)\right|^2 \frac{\sin(\pi f T_c)^4}{(\pi f T_c)^4}df \quad (6.9\text{-}35)$$

where $B_{\min}$ is the minimum of $(B/2, B_{NQ})$. It will be assumed that $B = 2B_{NQ}$ in what follows. Thus the variance can be written as

$$\sigma_{MSI}^2 \cong \frac{N_0}{2T_i}\int_{-B_{NQ}}^{B_{NQ}} T_c \frac{\sin(\pi f T_c)^2}{(\pi f T_c)^2}df + \frac{1}{T_i}\int_{-B_{NQ}}^{B_{NQ}} \frac{J}{2}(T_c)^2 \frac{\sin(\pi f T_c)^4}{(\pi f T_c)^4}df \quad (6.9\text{-}36)$$

since the interference power in contained in $n_c(t)$ is $J/2$ when the total jammer power is J watts. With the assumptions that low-pass filter bandwidths is equal to the chip rate, (6.9-36) can be evaluated to

$$\sigma_{MSJ}^2 = 0.903\frac{N_0}{2T_i} + 0.665\frac{JT_c}{2T_i} = Var(\boldsymbol{I}_n(k)) = Var(\mathbf{Q}_n(k)) \quad (6.9\text{-}37)$$

where

$$\int_{-1}^{1} \frac{\sin(\pi x)^2}{(\pi x)^2}dx = 0.903 \text{ and } \int_{-1}^{1} \frac{\sin(\pi x)^4}{(\pi x)^4}dx = 0.665 \quad (6.9\text{-}38)$$

Clearly other jammer types can be evaluated with the use of the basic expression (6.9-34).

6.9.9 Correlation Signal Voltage Loss

It is to be noted that not only the noise but also the correlated signal voltage is reduced when it is passed through the bandpass filter. The correlated signal filtering voltage loss, for a complex symmetric baseband signal, can be described by [11]

$$\int_{-1}^{1} \frac{\sin(\pi x)^2}{(\pi x)^2}dx = 0.903 \text{ and } \int_{-1}^{1} \frac{\sin(\pi x)^4}{(\pi x)^4}dx = 0.665 \quad (6.9\text{-}39)$$

where $H_{BB}(f)$ is the baseband equivalent filter associated with the bandpass filter. It is to be noted that the transfer function itself appears in the expression, and not the magnitude squared, which would specify the power passed through the filter. Evaluation of (6.9-39) with the Nyquist filter set equal to the chip rate produces the correlation voltage given by

$$(L_s)^{1/2} = \int_{-1}^{1} \frac{\sin(\pi x)^2}{(\pi x)^2}dx = 0.903 \quad (6.9\text{-}40)$$

Since the signal filtering voltage loss is reduced by 0.903, the signal power loss is $(0.903)^2$. This signal correlation loss will be accounted for in the total losses.

Based on the assumption of two samples per chip, it can be shown that adjacent samples are correlated. In this paper the correlation will be ignored to simplify the analysis.

Now consider the detection of the signal. Since the samples are spaced one-half chip apart, and are asynchronous with the received signal, there will be at least two points where the signal can be detected. For simplicity it will be assumed that the two sample points adjacent to the signal are spaced ± 1/4 of a code chip away from the peak of the correlation curve. Each sample will suffer a timing error-induced power loss given by approximately $(3/4)^2$ or 2.5 dB, neglecting the antialiasing, transmitter, and receiver filters. In reality the relative loss will be less than 2.5 dB due to filtering effects.

In addition to the time error–induced losses, there will be frequency errors also. It will be shown in the next section that frequency errors at the midpoint of any two cells (an error of $\pm 1/(2T_i)$ Hz) out of the FFT will produce a loss of 3.92 dB (without zero padding.) When zero padding is employed the losses for the adjacent frequency cells will be 0.91 dB. However, there will be two different possible frequencies that the signal can be detected in this case with this loss.

All four possible detection points (for each of the two frequencies, there will be two time points that could yield detection) will be considered in the analysis to follow.

6.9.10 Combining Coherent Segments of the DMF with the FFT

In this section the FFT will be used to coherently add the coherent segments of the correlations, for an NRZ modulated code, when the correct frequency output of the FFT is selected. Figure 6.9-4 illustrates the block diagram showing how the T_{i0} segments are combined by the FFT to produce a correlation of T_i seconds. The FFT output at FFT frequency bin "i" is given by the following N_{FFT} rotated sum, described mathematically by

$$Z(i) = \sum_{k=0}^{N_{FFT}-1} \left(\mathrm{I}(k) + j\mathrm{Q}(k)\right) e^{-jki(2\pi/N_{FFT})} = \boldsymbol{I}(i) + j\boldsymbol{Q}(i) \tag{6.9-41}$$

where $j = \sqrt{-1}$ and $i = 0, 1, 2, \ldots N_{FFT}-1$.

The effect of the FFT is to rotate the complex input I-Q signal at each of the N_{FFT} frequency rates so that one of these FFT frequencies will approximately align itself properly (be equal and opposite) with the frequency of the I-Q pair frequency offset at baseband. In effect the FFT forms parallel additional correlators at the N_{FFT} different frequencies. The different frequencies out of the FFT are given by

$$f_i = \frac{i}{N_{FFT}T_{i0}} \tag{6.9-42}$$

and have power responses around each FFT frequency given by [50]

$$\left|A^2(f)\right| = \frac{1}{N_{FFT}^2}\frac{\sin(\pi N_{FFT} f T_{i0})^2}{\sin(\pi f T_{i0})^2} \tag{6.9-43}$$

This can be approximated by the $\text{sinc}^2(x)$ function, in the following way

$$A^2(f) \cong \frac{1}{N_{FFT}^2}\frac{\sin(\pi N_{FFT} f T_{i0})^2}{\sin(\pi f N_{FFT} T_{i0}/N_{FFT})^2} \cong \frac{\sin\left(\pi f T_i\right)^2}{\left(\pi f T_i\right)^2} \tag{6.9-44}$$

for $T_i = N_{FFT}T_{i0}$ and when $f \le 1/(N_{FFT}T_{i0})$. The main lobe response of the FFT is seen in Figure 6.9-7.

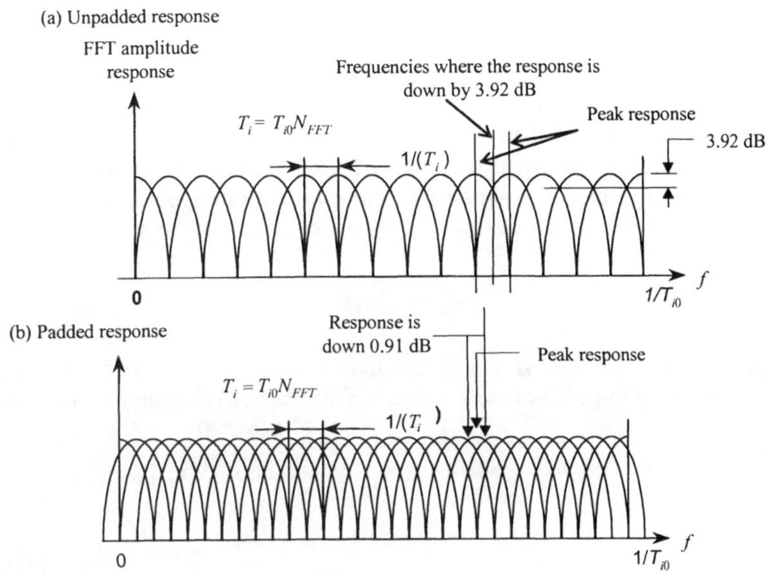

Figure 6.9-7 (a, b) Main lobe response of the FFT with and without padding.

It can be seen that bin peaks are spaced apart by $1/(T_i)$ Hz, as are the bin crossover points where any two adjacent curves are equal in amplitude at the same point and the response is down 3.92 dB for the nonzero padded FFT. Hence when the frequency is offset by $1/(2T_i)$ Hz from a peak response, the response for either curve is down by $\sin(\pi/2)^2/(\pi/2)^2$, or 3.92 db. The total frequency range of the N_{FFT} frequency cells is $1/T_{i0}$. Based on (6.9-25) at equality and the fact that both positive and negative frequency errors must be checked, it is clear that the total frequency search range is $\Delta F_{\text{srch}} = 1/(2T_{i0}) = 2\Delta f_m$.

It is also clear that only $N_{FFT}/2$ of the frequency bins of the nonzero padded FFT are used. In fact, the first ¼ and the last ¼ of the frequencies are used, since the first ¼ of the frequencies correspond to the positive frequency error range and the last ¼ provides the negative frequencies. In addition to the outer bin losses, there is another possible loss that can occur when using the FFT to enhance the coherent integration range. This loss stems from the fact that the actual frequency error may not fall at one of the N_{FFT} frequencies illustrated in Figure 6.9-7. In reality if the frequency error falls exactly between two FFT response peaks the two adjacent correlators will have a chance to detect the signal; however, each output will have a loss of 3.92 dB, as discussed previously.

To summarize, then, the coherent correlations are designed so that the outer bin loss does not exceed 0.91 dB, by obeying (6.9-25). The total coherent correlation time is given by

$$N_{FFT}T_{i0} = T_i \tag{6.9-45}$$

The total frequency error induced losses are the sum of 0.91-dB outer bin loss and the loss from falling in between the FFT bins. In this work it will be assumed that the signal frequency falls in between two FFT cells (when zero padding is not used) as discussed earlier. Also it will be assumed that two of the bins containing the signal are down by 3.92 dB.

The number of frequency cells that are to be searched is $(N_{FFT}/2)$. Typically N_{FFT} is a power of 2, so let $N_{FFT} = 2^{n_f}$, where n_f is an integer. Thus the total coherent integration time is given by (6.9-45). The number of frequency cells to be searched when zero padding is not used is given by

$$N_{cf} = (N_{FFT}/2) \tag{6.9-46}$$

It can be shown that the upper range of the FFT is in fact the negative frequency part of the FFT filter bank for detection [50].

The previous discussion did not consider zero padding the FFT in order to reduce the loss between FFT frequency bins. Zero padding will now be considered to see how the results are affected. A zero-padded FFT is an FFT that is double the size of the corresponding unpadded FFT. In a zero-padded FFT, the input is composed of the signal followed by an equal number of zeros to double the N_{FFT} input length. The result of doing this is to provide FFT bins that are one-half as far apart as they would be without zero padding, and there are twice as many bins. Thus the number of cells to search, using zero padding, is given by

$$N_{cf}^{zp} = N_{FFT} \tag{6.9-47}$$

in terms of the nonzero padded FFT size N_{FFT}. The loss between bins with zero padding is 0.91 dB at the cross over locations. Comparing to (6.9-46) we see that the number of frequency cells to search is double that of the unpadded case, so that

$$N_{FFT}^{zp} = 2N_{FFT} \tag{6.9-48}$$

One small disadvantage of zero padding is that the number of frequency bins to be checked is doubled, however the reduction of about 3 dB in correlation losses normally more than compensates for the doubled frequency search.

In order to simplify the calculation of the signal and the noise variance, it will be assumed that the FFT is normalized by the parameter N_{FFT}. Then the FFT-aided correlations can be modeled as $N_{FFT}\,T_{i0}$ second integrations that are normalized by $N_{FFT}T_{i0}$. Zero padding will be assumed in all the calculations to follow.

6.9.11 Detection and False Alarm Probability Densities for the NRZ Code Case

The probability density function of the noncoherently summed statistic will now be obtained for the DMF. The square of the real part of each FFT output (N_{FFT} samples) $\boldsymbol{Q}_i$ is added to the square of the imaginary part of each FFT output $\boldsymbol{I}_i$, and these are summed over the N_{nc} values as shown in Figure 6.9-5. This final sum is compared to the threshold of T_h.

The output measure after N_{nc} noncoherent combinings is given by

$$S(i,j) = \sum_{k=1}^{N_{nc}} \left(\boldsymbol{I}^2(i,j,k) + \boldsymbol{Q}^2(i,j,k) \right) \tag{6.9-49}$$

where i is the index on the time hypotheses, j is the index on the N_{FFT} FFT frequency hypotheses, and k is the index on the noncoherent combinings ($k = 1, 2, \ldots N_{nc}$), for a total correlation duration of $N_{nc}T_i$ seconds. The two terms $\boldsymbol{I}(i,j,k)$ and $\boldsymbol{Q}(i,j,k)$ are the real and imaginary components out of the k-th FFT output at the i th timing position and the j-th frequency bin.

In order to determine the probability of false alarm, consider first the noise-only case for the N_{nc} sum on the index k of the metric given in (6.9-49).

This decision variable is then compared to a threshold after the summation. When the signal is not located in any of the time-frequency bins (hypotheses) under investigation, one can write the N_{nc} noncoherent sum of the $\boldsymbol{I}^2$ and $\boldsymbol{Q}^2$ terms, in the simplifying notational form for any i and any j,

$$Y = \sum_{m=1}^{2N_{nc}} n_F^2(m) \tag{6.9-50}$$

where $n_F(m)$ is a zero mean, Gaussian random variable, which is essentially independent from integration to integration, having statistics that are given in (6.9-37) for the matched spectral jammer.

In Figure 6.9-5 the division by N_{FFT} at the input to the FFT converts the divisor from T_{i0} to T_i seconds, and the FFT summation adds N_{FFT} T_{i0} seconds to produce T_i. The dependence on i and j has been dropped for convenience. The statistics for each i and each j are the same. The index m in (6.9-50) is over each noise term, which is twice the number of N_{nc} summations since both the $\boldsymbol{I}$ and $\boldsymbol{Q}$ noise terms are added. When these noise terms are added N_{nc} times, the metric S is formed. For this case it is assumed that either the time or the frequency is not aligned with the signal.

In Section 6.5.4 it was determined that the normalized random variable $Z = Y/(2\sigma^2)$ has a false alarm probability that is given by

$$P_{FA} = \Pr(z > Th/(2\sigma^2)) = \int_{T_h/(2\sigma^2)}^{\infty} p_n(z)dz \tag{6.9-51}$$

where the normalized threshold for the MSJ be given by

$$\eta = \frac{T_h}{2\sigma^2} \tag{6.9-52}$$

The normalized random variable has a false alarm probability density given by

$$p_n(z) = \frac{(z)^{N_{nc}-1} e^{-z}}{(N_{nc}-1)!}, \quad z \geq 0 \tag{6.9-53}$$

From (6.9-51), (6.9-52), and (6.9-53), it follows that the false alarm probability is given by

$$P_{FA} = e^{-\eta} \sum_{k=0}^{N_{nc}-1} \frac{(\eta)^k}{k!} \tag{6.9-54}$$

Again following Section 6.5.4 the random variable associated with the decision statistic, given in (6.9-50), has a noncentrality parameter s^2 given by

$$s^2 = \sum_{m=1}^{N_{nc}} \left[\left(\overline{n_{IS}(m)}\right)^2 + \left(\overline{n_{QS}(m)}\right)^2 \right] \tag{6.9-55}$$

with $\overline{n_{IS}(m)}$ and $\overline{n_{QS}(m)}$ denoting the mean values of the $\boldsymbol{I}_S(i_0, j_0, m)$ and $\boldsymbol{Q}_S(i_0, j_0, m)$, respectively. Clearly, s^2 is the sum of the signal-squared terms. From (6.9-54), one has, for s^2,

$$s^2 = N_{nc} P R_{PN}^2(\tau) \frac{\sin(\Delta\omega T_{i0}/2)^2}{(\Delta\omega T_{i0}/2)^2} = N_{nc} P R_{PN}^2(\tau) \text{sinc}^2(\Delta f T_{i0}) \tag{6.9-56}$$

which accounts for the timing error loss and the outer bin frequency loss only.

Let L_T be the total power loss for the signal, which will account for the losses indicated in (6.9-56), as well as all other applicable losses. Then

$$L_T = L_s L_{quan} L_{freq} L_{outerbin} L_{time} L_{misc} \tag{6.9-57}$$

where L_s is the signal correlation filtering loss, including the Nyquist filter and the front-end filter. L_{quan} is the quantization loss that depends on how many bits of quantization are used. L_{freq} is the FFT frequency loss (–0.91 dB) at the output of the FFT as discussed in Section 6.5.6 for the zero padded FFT when the signal falls between bins. $L_{outerbin}$ is the outer-bin loss and described by $\text{sinc}^2(\Delta f_m T_{i0})$ (-0.91 dB), which has been determined earlier, and L_{time} is the time error loss, which is given by $R^2_{PN}(\tau)$ and assumed to be –2.5 dB, based on an unfiltered correlation curve and a chip error of ¼ chip. L_{misc} accounts for any other miscellaneous losses not specifically addressed previously. Therefore the noncentrality parameter s^2 for the NRZ code signal, (accounting for all the losses), can be written as

$$s^2 = N_{nc} P L_T \tag{6.9-58}$$

Again using a result indicated in [21], assuming all the random variables are statistically independent and normalizing by $Z = Y/(2\sigma^2)$, one has the following expression for the normalized probability density function of the signal plus noise decision metric

$$p_{S+N}(z) = \left(\frac{z}{\gamma}\right)^{\frac{N_{nc}-1}{2}} e^{-\gamma} e^{-z} I_{N_{nc}-1}\left(2\sqrt{\gamma z}\right) \tag{6.9-59}$$

where the signal-to-noise parameter γ (in $N_{nc}T_i$ seconds) is defined by

$$\gamma = \frac{s^2}{2\sigma^2} = N_{nc}\gamma_0 \tag{6.9-60}$$

The signal-to-noise ratio (SNR) γ_0 is the SNR in one bit time in T_i, seconds ($T_i = N_{FFT}T_{i0}$).

The SNR out of the FFT for the matched spectral jammer is given by

$$\gamma_{MSJ} = \frac{L_T N_{nc}}{\left[0.903\dfrac{N_0}{PT_i} + 0.665\dfrac{J}{P}\dfrac{T_c}{T_i}\right]} = N_{nc}\gamma_{0MSJ} \tag{6.9-61}$$

where (6.9-37), (6.9-58), and (6.9-60) have been used in (6.9-61). The ratio J/P is the jammer to signal ratio, and the ratio T_i/T_c is the processing gain (PG).

Again using a result indicated in [21], assuming all the random variables are statistically independent and normalizing by $Z = Y/(2\sigma^2)$, one has the following expression for the normalized probability density function of the signal plus noise decision metric

$$p_{S+N}(z) = \left(\frac{z}{\gamma}\right)^{\frac{N_{nc}-1}{2}} e^{-\gamma} e^{-z} I_{N_{nc}-1}\left(2\sqrt{\gamma z}\right) \tag{6.9-62}$$

Now consider the detection probability. The probability of detection is the probability that the normalized threshold is exceeded when the signal is present. Thus

$$P_D = \Pr(z > T_h/(2\sigma^2)) = \int_{T_h/(2\sigma^2)}^{\infty} p_{s+n}(z)dz \tag{6.9-63}$$

From (6.9-62) it follows that the detection probability is given by

$$P_D = \int_{\eta}^{\infty} \left(\frac{z}{\gamma}\right)^{\frac{N_{nc}-1}{2}} e^{-\gamma} e^{-z} I_{N_{nc}-1}\left(2\sqrt{\gamma z}\right) dz \tag{6.9-64}$$

where γ is defined in (6.9-61). This result can be expressed as a generalized Q function [21]. An alternative and more computationally useful form, using a power series for an exact expression, is given in Shnidman [22]. Starting with

$$P_D(\gamma_0, \eta) = \int_{\eta}^{\infty} \left(\frac{z}{N_{nc}\gamma_0}\right)^{\frac{N_{nc}-1}{2}} e^{-N_{nc}\gamma_0} e^{-z} I_{N_{nc}-1}\left(2\sqrt{N_{nc}\gamma_0 z}\right) dz \tag{6.9-65}$$

and using form 2 in [22] it can be shown that the detection probability is given by

$$P_D(\gamma_0, \eta) = \sum_{m=0}^{N_{nc}-1} e^{-\eta} \frac{\eta^m}{(m!)} + \sum_{m=N_{nc}}^{\infty} e^{-\eta} \frac{\eta^m}{(m!)} \left(1 - \sum_{k=0}^{m-N_{nc}} e^{-N_{nc}\gamma_0} \frac{(N_{nc}\gamma_0)^k}{(k!)}\right) \tag{6.9-66}$$

In addition a Gaussian approximation to the expression for P_D can be easily obtained and provides a simple approximation to P_D that avoids dealing with large summations. It can be shown that with the normalization by $2\sigma^2$, the mean and variance of Z is given by

$$\begin{aligned} E\{Z\} &= N_{nc} + \gamma = N_{nc}(1+\gamma_0) \\ Var\{Z\} &= N_{nc} + 2\gamma = N_{nc}(1+2\gamma_0) \end{aligned} \tag{6.9-67}$$

where γ_0 is defined in (6.9-61) for the MSJ. It follows that the probability of detection is given by

$$P_D = Q\left(\frac{\eta - N_{nc}(1+\gamma_0)}{\sqrt{N_{nc}(1+2\gamma_0)}}\right) \tag{6.9-68}$$

where the Q function is defined by

$$Q(x) = \int_{x}^{\infty} \frac{1}{\sqrt{2\pi}} e^{-\frac{z^2}{2}} du \tag{6.9-69}$$

Now that the detection and false alarm probabilities have been developed, the derivation of the acquisition probability will be presented.

6.9.12 The Acquisition Probability

The acquisition probability is the probability that the true signal has been identified and all false alarms have been rejected. To evaluate the mean time to obtain code acquisition, one must specify the search space and the detection probability as well as the effective false alarm probability. Our model shall be as follows. Assume that there are M_{ct} (see (6.9-15)) half-chip cells to be checked for the signal for each frequency bin. In addition to the time search, there is also a frequency search, which is given by (6.9-47) for the case of a zero-padded FFT. Therefore the total effective number of time and frequency cells (bins) to be searched, for the zero-padded FFT case, is

$$M_c^z = M_{ct} N_{cf}^z = 2R_c \Delta T N_{FFT}, \quad N_{FFT} \geq 2 \tag{6.9-70}$$

where M_c^z denotes the totality of time and frequency cells to be searched in the zero padded case. In this model it is assumed that four cells contain the signal at equal level with the losses indicated in (6.9-57) and M_{cf}^z denotes the number of frequency cells to be searched with zero padding.

It will be further assumed that there are five verification circuits available in parallel to the matched filter detector. This assumption is only justifiable when the false alarm is very small, such that the effective false alarm probability, given by $M_c P_{FA}$, is very small. The verification circuits are used to determine if the "hit" that exceeds the threshold is a real detection or a false alarm. Thus, up to five simultaneous verifications can be made at one time. Clearly, more verification circuits would improve the performance negligibly under the assumption that the effective false alarm probability is small. The results provided here can be modified to include any any number of verification circuits.

The verification circuits would typically be composed of a code-tracking loop dedicated to that code timing and frequency offset and a code lock detector used to verify true lock. Since there are only four possible detection points for the signal, there is no mean acquisition time advantage in having more than four when the false alarm probability is very small, which it has to be when the dimensionality is large. However, in this analysis, $N_v = 5$ verification circuits will be assumed.

It will also be assumed that the verifications will require less than ΔT seconds,[18] so that they will be ready for the next possible verification on the next segment (ΔT seconds later) of code chips to be checked.

Consider the probability of acquisition, which will be denoted by P_{acq}, which is the probability that at least one signal plus noise correlation exceeds the threshold. Since the verification circuits are assumed to be perfect, a hit at a signal location will result in successful acquisition. To evaluate the acquisition probability, let a "hit" denote the fact that the threshold has been exceeded either due to noise (false alarm) or the signal plus noise. The probability of acquisition is given by

$$P_{acq} = Pr(\text{of acquisition with } j = 1 \text{ or } 2 \text{ or } 3 \text{ or ... hits}) \tag{6.9-71}$$

Since these events of the number of hits will be assumed to be mutually exclusive, one can write

$$P_{acq} = Pr(\text{acq. with } j = 1) + Pr(\text{acq. with } j = 2) + Pr(\text{acq. with } j = 3) + \ldots \tag{6.9-72}$$

Let the probability of acquisition with k hits be denoted by $P_{acq}(k)$. Then it follows that in the case of no-zero padding, $P_{acq}(1)$ is given by

$$P_{acq}(1) = \binom{4}{1} P_D (1 - P_D)^3 \left[\binom{M_c'}{0} (1 - P_{FA})^{M_c'} \right] \tag{6.9-73}$$

where $M_c' = M_c - 4$ is the number of possible cells that do not have the signal present. It is also assumed that four cells have the signal present in our model. The first probability in (6.9-73) is the probability that one of the possible signals is detected and the second is the probability that no false alarm occurs.

Consider the case of two hits. In order for acquisition to occur, there must be at least one true hit (signal plus noise exceeds the threshold). Hence it follows that

---

[18] If this is not true, a multiple of four verification circuits will be needed.

$$P_{acq}(2)=\binom{4}{1}P_D(1-P_D)^3\left[\binom{M_c'}{1}P_{FA}(1-P_{FA})^{M_c'-1}\right]+\binom{4}{2}P_D^2(1-P_D)^2\left[\binom{M_c'}{0}(1-P_{FA})^{M_c'}\right] \quad (6.9\text{-}74)$$

where the two products in (6.9-74) reflect the fact that the first terms in each product account for the case of one true detection and one false alarm in the former case and two signal detections with no false alarms in the second case. It is to be noted that the last combinatorial term in (6.9-74) is equal to one.

When there are three hits and acquisition occurs, there must be at least one true hit. This can occur with one true hit and two false alarms, or two true hits and one false alarm, or three true hits. Therefore one can write

$$\begin{aligned}P_{acq}(3)=&\binom{4}{1}P_D(1-P_D)^3\left[\binom{M_c'}{2}P_{FA}^2(1-P_{FA})^{M_c'-2}\right]\\&+\binom{4}{2}P_D^2(1-P_D)^2\left[\binom{M_c'}{1}P_{FA}(1-P_{FA})^{M_c'-1}\right]\\&+\binom{4}{3}P_D^3(1-P_D)(1-P_{FA})^{M_c'}\end{aligned} \quad (6.9\text{-}75)$$

Continuing in the same manner $P_{acq}(4)$ is given by

$$\begin{aligned}P_{acq}(4)=&\binom{4}{1}P_D(1-P_D)^3\left[\binom{M_c'}{3}P_{FA}^3(1-P_{FA})^{M_c'-3}\right]\\&+\binom{4}{2}P_D^2(1-P_D)^2\left[\binom{M_c'}{2}P_{FA}^2(1-P_{FA})^{M_c'-2}\right]\\&+\binom{4}{3}P_D^3(1-P_D)\left[\binom{M_c'}{1}P_{FA}(1-P_{FA})^{M_c'-1}\right]+\binom{4}{4}P_D^4(1-P_{FA})^{M_c'}\end{aligned} \quad (6.9\text{-}76)$$

Now consider the fifth term, which can be written as

$$\begin{aligned}P_{acq}(5)=&\binom{4}{1}P_D(1-P_D)^3\left[\binom{M_c'}{4}P_{FA}^4(1-P_{FA})^{M_c'-4}\right]\\&+\binom{4}{2}P_D^2(1-P_D)^2\left[\binom{M_c'}{3}P_{FA}^3(1-P_{FA})^{M_c'-3}\right]\\&+\binom{4}{3}P_D^3(1-P_D)\left[\binom{M_c'}{2}P_{FA}^2(1-P_{FA})^{M_c'-2}\right]+\binom{4}{4}P_D^4\left[\binom{M_c'}{1}P_{FA}(1-P_{FA})^{M_c'-1}\right]\end{aligned} \quad (6.9.77)$$

Now for the model under study, the false alarm probability is assumed to be small such that

$$M_c'P_{FA}<<1 \quad (6.9\text{-}78)$$

If we keep only terms of order zeroth power and the first power of P_{FA}, we obtain the acquisition probability

$$P_{acq} \cong \left[(1-P_{FA})^{M_c'} + M_c' P_{FA}(1-P_{FA})^{M_c'-1}\right] \times \left[\binom{4}{1} P_D(1-P_D)^3 + \binom{4}{2} P_D^2(1-P_D)^2 + \binom{4}{3} P_D^3(1-P_D) + \binom{4}{4} P_D^4\right] \tag{6.9-79}$$

This equation can be simplified to the following form

$$P_{acq} \cong \left[(1-P_{FA})^{M_c'} + M_c' P_{FA}(1-P_{FA})^{M_c'-1}\right]\left[1-(1-P_D)^4\right] \tag{6.9-80}$$

and is our final expression for the acquisition probability under the assumption of large M_c' and subject to the constraint of (6.9-78). It is clear from (6.9-80) when P_D approaches one and P_{FA} approaches zero, P_{acq} approaches 1 also.

In the case that zero padding of the form discussed earlier is employed, the acquisition probability (6.9-80) is modified in the sense that M_c is replaced with M_c^{zp}. Thus $M_c' = M_c - 4$ when zero padding is not used and $M_c' = M_c^{zp} - 4$ when zero padding is being used. It is assumed in the analysis presented here that the false alarm probability is fixed, independent of how the interference varies.

6.9.13 Mean Acquisition Time Calculation

The operation of the matched filter was discussed in the first section. It was noted that the possible acquisition times for the nominal arrival time of the code segment occurred as indicated in (6.9-16).

In the calculation for the mean acquisition time, it will be assumed that the true code timing can occur anywhere over the ΔT second duration with uniform likelihood, going from $+\Delta T/2$ to $-\Delta T/2$ seconds about the receiver time t_0. It will be assumed that the initial wait time, from the receiver time t_0, is the mean wait time of $\Delta T/2$ seconds.

The mean time to acquire will be defined as the mean time to correctly detect the signal for the code arriving at nominal time. The mean time to acquire expression is given by an infinite sum of terms. The first term is the time to fill (load) the receiver shift register (τ_d seconds) plus the mean time to wait for the nominal signal to enter the receiver correlator ($\Delta T/2$ seconds), multiplied by the probability of acquisition on the first code burst. The second term can be determined from Figure 6.9-6(d). It is composed of the time it takes on the second code burst to wait the mean time for the signal to arrive, plus the time it takes to fill the correlator, all multiplied by the probability of not acquiring on the first code burst times the probability of acquiring on the second code burst.

This sequence of contributing terms continues in that manner with the appropriately increased times, weighted by the probability of missing the first "(j–1)-th" code bursts, and acquiring the j-th code burst. Thus, the mean time to acquire is given by the infinite sum

$$\begin{aligned}\overline{T}_{acq} &= P_{acq}\left(\tau_d + \Delta T - \Delta T/2\right) + P_{acq}(1-P_{acq})\left(2(\tau_d+\Delta T) - \Delta T/2\right) \\ &+ \ldots + P_{acq}(1-P_{acq})^{j-1}\left(j(\tau_d+\Delta T) - \Delta T/2\right) + \ldots\end{aligned} \tag{6.9-81}$$

where P_{acq} is given in (6.9-80). This result, for the mean time to obtain acquisition, can be written more succinctly as

$$\overline{T}_{acq} = P_{acq}\sum_{j=1}^{\infty}\left(1-P_{acq}\right)^{j-1}\left[j\left(\tau_d+\Delta T\right) - \Delta T/2\right] \tag{6.9-82}$$

Equation (6.9-82) can be written in the following form

$$\overline{T}_{acq} = P_{acq}\sum_{j=1}^{\infty}\left(1-P_{acq}\right)^{j-1} j\left(\tau_d+\Delta T\right) - P_{acq}\sum_{j=1}^{\infty}\left(1-P_{acq}\right)^{j-1}(\Delta T/2) \tag{6.9-83}$$

The following identities can be used to evaluate the expression for the mean acquisition time, when $|x|<1$:

$$\left(\frac{1}{1-x}\right) = \sum_{j=1}^{\infty} x^{j-1}$$
$$\left(\frac{1}{1-x}\right)^2 = \sum_{j=1}^{\infty} jx^{j-1} \tag{6.9-84}$$

Letting $x = 1-P_{acq}$ in (6.9-83) and (6.9-84) produces the following results for the mean acquisition time

$$\overline{T}_{acq} = P_{acq}(\tau_d+\Delta T)\frac{1}{\left(1-(1-P_{acq})\right)^2} - P_{acq}\frac{\Delta T}{2}\frac{1}{\left(1-(1-P_{acq})\right)} \tag{6.9-85}$$

which simplifies to

$$\overline{T}_{acq} = \frac{\tau_d+\Delta T}{P_{acq}} - \frac{\Delta T}{2} \tag{6.9-86}$$

Equation (6.9-86) can also be written in the following convenient equivalent form

$$\overline{T}_{acq} = \frac{\tau_d}{P_{acq}} + \left[\frac{2-P_{acq}}{2P_{acq}}\right]\Delta T \tag{6.9-87}$$

It is easy to see from (6.9-87) that when P_{acq} approaches one, the mean acquisition time approaches

$$\overline{T}_{acq} \to \tau_d + \Delta T/2 \tag{6.9-88}$$

which is often assumed for P_D almost one and P_{FA} almost zero. Furthermore, when P_{acq} becomes very small, the mean acquisition time approaches

$$\overline{T}_{acq} \to \frac{\tau_d+\Delta T}{P_{acq}} \tag{6.9-89}$$

In the calculation for the mean acquisition time it will be assumed that the true code timing is uniformly distributed during the ΔT second duration. Therefore the mean wait time to have the true correlation occur will be $\Delta T/2$ seconds.

Example 5 As an example of the DPMF theory consider the code acquisition of the GPS P(Y) code in which $\Delta T = 2$ sec, and $\Delta F = 3{,}200$ Hz. The P(Y) code can be considered a random code (very long period) that runs at 10.23 Mcps. It is assumed that the sample rate is 20.46 Msps. The Nyquist filter (low-pass filter-located just before the A/D converters) bandwidths were chosen to be 10.23 MHz. Table 6.9-1 illustrates the signal losses for the P(Y) code.

Table 6.9-1 Summary of P(Y) Signal Losses

Loss Type	Loss in dB
Filtering (includes signal correlation loss)	0.89
Miscellaneous	0.8
Quantization	1.2
Time error	2.5
Outer bin frequency	0.91
Frequency error with zero padding	0.91
Total	7.21

The filtering loss is the signal correlation loss. The quantization loss is based on the work of Chang [51]. The time error is based on the assumption of a quarter of a chip timing error or $20\times\log(3/4)$ = 2.5 dB. The outer bin loss was based on setting $\Delta T = 1/(4\Delta f_m)$. The final loss was the loss due to frequency error when zero padding was used (it would be 3.92 dB without zero padding). The power was assumed to be −159.6 dBW. The jammer was assumed to be a spectrally flat Gaussian noise process that had a 24-MHz bandwidth, so that the J/S ratio is the ration of the jammer power in 24 MHz, due to a spectrally flat noise jammer, to the signal power.

Figure 6.9-8 illustrates the exact (6.9-66) and the Gaussian approximation (6.9-68) for the mean acquisition time for three values of N_{nc}.

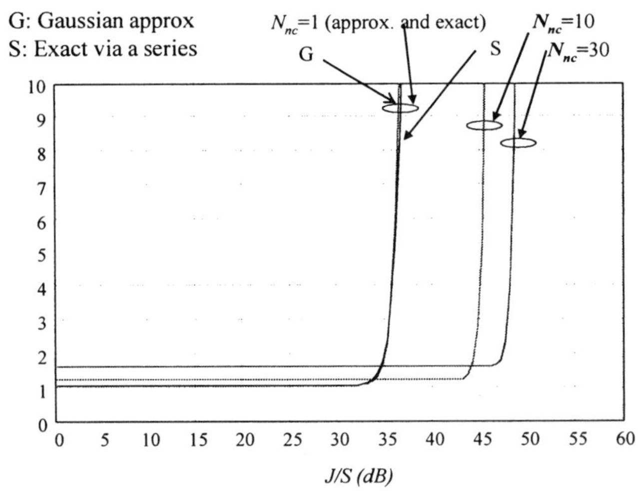

Figure 6.9-8 Comparison of the Gaussian and the exact values of mean acquisition time versus the J/S ratio for the GPS P(Y) code.

It is clear that the Gaussian approximation is very good for N_{nc} values of 10 or greater (both lines lay on top of each other), in fact it is not that bad for $N_{nc} = 1$. All acquisition time calculations are based on the series representation for P_D for $N_{nc} \leq 8$, and the Gaussian approximation for $N_{nc} > 8$. In Figure 6.9-9 the mean time to code acquisition results are illustrated for many cases of N_{nc}.

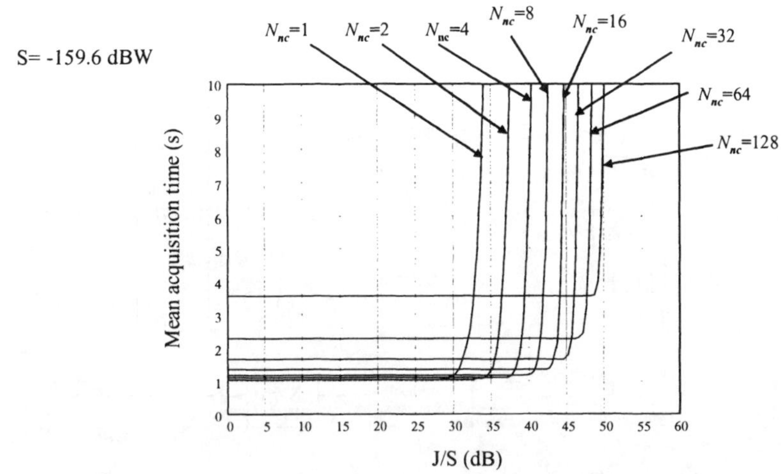

Figure 6.9-9 Comparison of the mean acquisition time performance with a matched spectral jammer.

Note that as the number of noncoherent combinings increases, the point where the mean acquisition time rapidly increases also increases. However the mean time to acquire is higher at low J/S ratios due to the fact that when the P_{acq} is near 1, the mean acquisition time is approximately $\tau_d + \Delta T/2$, and increasing N_{nc} increases τ_d. Consider the case for N_{nc} = 128 and denote the mean acquisition time at T. At low J/S ratios we have $\overline{T} \cong \tau_d + \Delta T/2$. Consider the case that N_{nc} = 128 then $\tau_d \cong (0.02)(128) + 1 = 3.56$ seconds, as is seen in Figure 6.9-9 for N_{nc} = 128. More details can be found in [52].

6.10 SERIAL ACTIVE SEARCH FOR ACQUISITION OF FH/MFSK SIGNALS

In this section the acquisition of fast frequency-hopped signals and slow frequency-hopped signals will be considered via an active serial acquisition approach. MFSK modulation changes the carrier frequency to one of M tones, so that there are M possible frequencies around the carrier frequency. For fast frequency hopping (FFH) the M-ary frequency is the same over the hop duration, and there are "N" hops per modulation symbol. For slow frequency hopping (SFH) it will be assumed that there will be one MFSK tone over the hop duration.

6.10.1 Serial Active Search for Acquisition of FFH/MFSK Signals

Figure 6.10-1 illustrates the acquisition model for the acquisition of FFH/MFSK signals. Figure 6.10-1(a) illustrates the overall system for acquisition and Figure 6.10-1(b) illustrates the demodulator that is used for fast frequency or slow frequency hopping with the appropriate modifications. Each MFSK frequency is correlated over the N hops and the maximum is selected. If more SNR is needed for acquisition then multiple MFSK symbols can be combined over, say, N_S MFSK symbols.

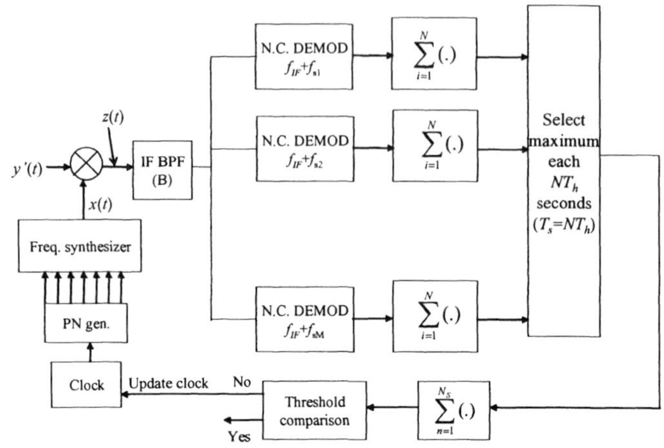

Figure 6.10-1 (a) FFH/MFSK model for fast FH acquisition.

Since it is assumed that there are N frequency hops per MFSK tone modulation for fast frequency hopping, then a symbol period has duration $T_s = NT_h$. For fast frequency hopping each MFSK symbol frequency is the same over each frequency hop; however, the phase will not be the same due to the action of the hopping and dehopping synthesizers. It will be assumed in the following analysis that the phases are random after each hop.

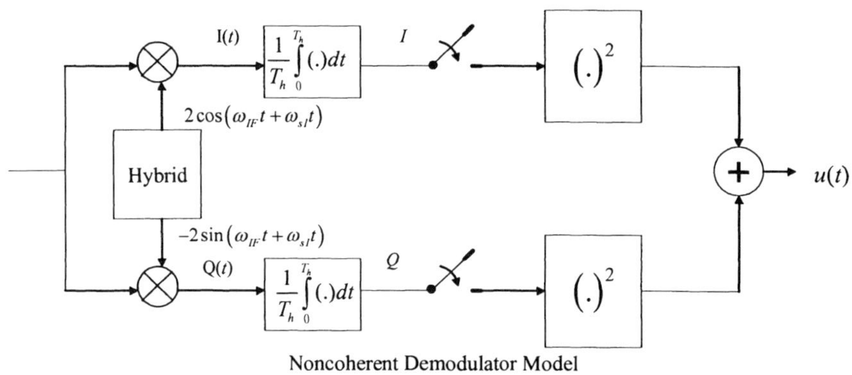

Figure 6.10-1 (b) FH/MFSK noncoherent demodulator model for fast FH acquisition.

Figure 6.10-1(b) also applies to slow frequency hopping when the hop period T_h is changed at the symbol period T_s in the figure.

Figure 6.10-2 illustrates the timing assumptions and the relationship of the frequency hops for FFH/MFSK and the MFSK symbols. It will also be assumed that when the receiver dehopping synthesizer is offset by ε seconds from the received hopping waveform (when ε is less than T_h) that the previous and the following hop frequencies are sufficiently different from the current hop frequency and that no energy is detected in the detectors of Figure 6.10-1. With this assumption the "detectable portion" of the signal will be of duration $(T_h-|\varepsilon|)$. It may be necessary to accumulate noncoherently N_S symbols to enhance the signal-to-noise ratio.

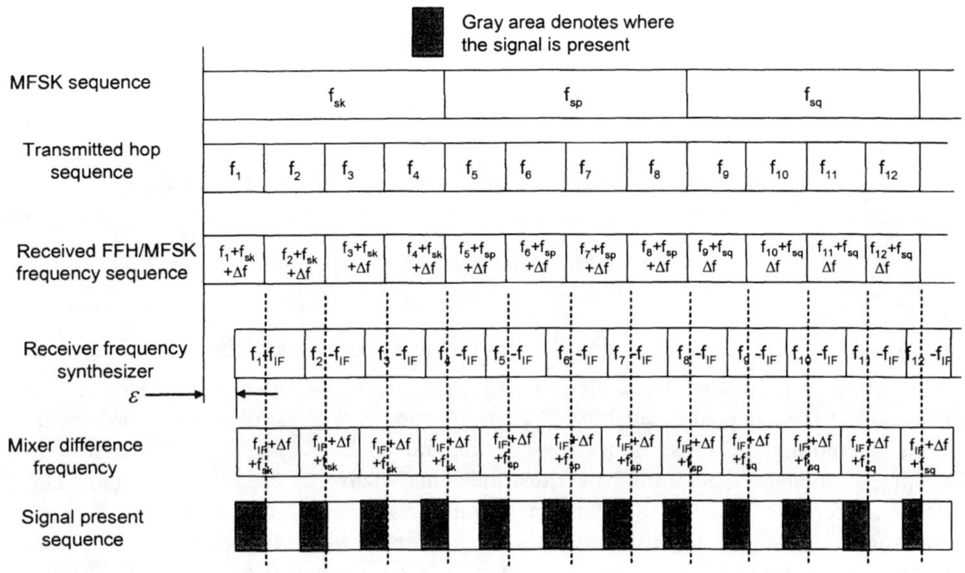

Figure 6.10-2 Timing and frequency diagram for the FFH/MFSK received and local signals.

Let the transmitted signal defined over one symbol time be given by

$$y(t) = \sqrt{2P_T}\sum_{i=1}^{N} p(t - iT_h)\cos\left[(\omega_i + \omega_{sj})t + \theta_i\right] \tag{6.10-1}$$

where P_T is the received signal power, T_h is the fast frequency-hopping duration, and ω_i and ω_{sj} are the FH angular frequency and the MFSK modulation angular frequencies, respectively, both expressed in radians/second. Also the index i is on time whereas the index j is on the signals in the modulation signal set. Note that $\omega_i = 2\pi f_i$ and $\omega_{sj} = 2\pi f_{sj}$. The phase θ_i is a phase value for the i-th transmitted frequency-hopped signal and is expressed in radians. The pulse function $p(t)$ is defined by

$$\begin{aligned} p(t) &= 1 \qquad 0 \le t \le T_h \\ &= 0 \qquad \text{elsewhere} \end{aligned} \tag{6.10-2}$$

The received signal, assuming that it is only modified by the Doppler shift Δf, and neglecting the time delay, is given by

$$\begin{aligned} y'(t) &= \sqrt{2P}\sum_{i=1}^{n} p(t - iT_h)\cos\left[(\omega_i + \omega_{sj} + \Delta\omega)t + \theta_i\right] \\ &+ \sqrt{2}n_c(t)\cos(\omega_i t) + \sqrt{2}n_s(t)\sin(\omega_i t) \end{aligned} \tag{6.10-3}$$

where $\Delta\omega = 2\pi\Delta f$, and the last term, $n(t)$, is the receiver noise and is assumed to be white Gaussian noise. The dehopping signal is of the form

$$x(t) = \sqrt{2}\sum_{k=1}^{n} p(t - kT_h - \varepsilon)\cos\left[(\omega_k - \omega_{IF})(t - \varepsilon) + \theta_r\right] \tag{6.10-4}$$

and is delayed by ε seconds ($|\varepsilon| \leq T_h$) relative to the received hopping signal. It is assumed that the error is not as large as the hop duration. When the error is larger than the hop duration it is assumed that no signal correlation will occur. The difference frequency term out of the mixer is given by

$$\begin{aligned} z(t) = \sqrt{P}\sum_{k=1}^{n} p(t-kT_h)p(t-kT_h-\varepsilon)\cos\left[\left(\omega_{sj}+\omega_{IF}+\Delta\omega\right)t+\phi_{ir}\right]+O(2\omega_1) \\ +\sum_{i=1}^{n} p(t-iT_h-\varepsilon)\left[n_c(t)\cos(\omega_{IF}t-\omega_{IF}\varepsilon-\theta_r)+n_s(t)\sin(\omega_{IF}t-\omega_{IF}\varepsilon-\theta_r)\right] \end{aligned} \tag{6.10-5}$$

where $\phi_{ir} = \theta_i - \theta_r + (\omega_i - \omega_{IF})\varepsilon$, since the double sum collapses to a single sum when the error is less than one hop duration, and $O(x)$ denotes terms of order of x. Figure 6.10-2 illustrates the timing and frequency diagram for the FFH/MFSK signal acquisition system. The top row illustrates the data frequency, which in this example is operating at 1/4 the frequency-hopping rate. The second row illustrates the transmitted frequency-hopping sequence. The third row illustrates the received FFH/MFSK signal. The fourth row illustrates the local receiver frequency synthesizer frequency and time error. The mixer frequency and time delay are shown in row five. The last row indicates by gray areas where the received signal and the receiver frequency synthesizer overlap in time, and therefore the difference signal is present in the receiver.

Consider the signal term out of the noncoherent demodulator. Figure 6.10-1(b) illustrates the noncoherent demodulator for FFH or SFH demodulation. The upper term after the multiplication $I(t)$ is given by

$$\begin{aligned} I(t) = \sqrt{P}\sum_{k=1}^{n} p(t-kT_h)p(t-kT_h-\varepsilon)\cos\left[(\omega_{sj}+\Delta\omega-\omega_{sl})t+\phi_{ir}\right] \\ +\sum_{k=1}^{n} p(t-kT_h-\varepsilon)\left\{n_c(t)\cos(\omega_{sl}t+\omega_{IF}\varepsilon+\theta_r)-n_s(t)\sin(\omega_{sl}t+\omega_{IF}\varepsilon+\theta_r)\right\} \end{aligned} \tag{6.10-6}$$

and the sum frequencies have been neglected, since they will be filtered out in the correlation process. In a similar manner the quadrature term is given by

$$\begin{aligned} Q(t) = \sqrt{P}\sum_{i=1}^{n} p(t-kT_h-\varepsilon)p(t-kT_h)\sin\left[(\omega_{sj}-\omega_{sl}+\Delta\omega)t+\phi_{ir}\right] \\ -\sum_{i=1}^{n} p(t-kT_h-\varepsilon)\left[n_c(t)\sin(\omega_{sl}t+\omega_{IF}\varepsilon+\theta_r)+n_s(t)\cos(\omega_{sl}t+\omega_{IF}\varepsilon+\theta_r)\right] \end{aligned} \tag{6.10-7}$$

and again the sum frequency terms have been neglected. The second terms in both (6.10-6) and (6.10-7) can be simplified as follows. The summation multiplying the noise terms forms a constant, so that the white Gaussian noise terms are scaled by $1/\sqrt{2}$. Therefore $I(t)$ and $Q(t)$ can be written as equivalent to

$$I(t) = \sqrt{P}\sum_{k=1}^{n} p(t-kT_h)p(t-kT_h-\varepsilon)\cos\left[(\omega_{sj}+\Delta\omega-\omega_{sl})t+\phi_{ir}\right]+n_I(t) \tag{6.10-8}$$

$$Q(t) = \sqrt{P}\sum_{i=1}^{n} p(t-kT_h-\varepsilon)p(t-kT_h)\sin\left[(\omega_{sj}-\omega_{sl}+\Delta\omega)t+\phi_{ir}\right]-n_Q(t) \tag{6.10-9}$$

where both $n_I(t)$ and $n_Q(t)$ are white Gaussian statistically independent random processes. Let the I and Q correlator outputs be denoted by

$$I = I_S + I_N \tag{6.10-10}$$

and

$$Q = Q_S + Q_N \tag{6.10-11}$$

Hence the two noise terms out of the correlators are given by

$$I_N = \frac{1}{T_h}\int_0^{T_h} n_I(t)dt \tag{6.10.12}$$

$$Q_N = \frac{1}{T_h}\int_0^{T_h} n_Q(t)dt \tag{6.10-13}$$

It is easy to show that the variance of each noise term is given by

$$\sigma^2 = Var(I_N) = Var(Q_N) = \frac{N_0}{2T_h} \tag{6.10-14}$$

It follows that the output of the signal component out of the demodulator is given by the sum of the squares of the inphase and quadrature components, so that

$$I_S^2 + Q_S^2 = \left(\frac{1}{T_h}\int_0^{T_h} I(t)dt\right)^2 + \left(\frac{1}{T_h}\int_0^{T_h} Q(t)dt\right)^2 = \frac{P\sin\left(\pi\left(\Delta f + f_{sj} - f_{sl}\right)(T_h - |\varepsilon|)\right)^2}{\left(\pi\left(\Delta f + f_{sj} - f_{sl}\right)T_h\right)^2} \tag{6.10-15}$$

The correlation of the FH signal (the ambiguity function) for a frequency error Δf and a time error of ε seconds is given by [52] and is the same as $\sqrt{I_S^2 + Q_S^2}$ when $P = 1$ and $f_{sj} = f_{sl}$.

$$R(\Delta f, \varepsilon) = \left|\frac{\sin\left(\pi\Delta f(T_h - |\varepsilon|)\right)}{\pi\Delta f T_h}\right| \qquad |\varepsilon| \le T_h \tag{6.10-16}$$
$$= 0 \qquad \text{otherwise}$$

Note that when the transmitter modulation frequency is the same as the receiver symbol frequency then $f_{sj} = f_{sl}$ so that $\sqrt{I_S^2 + Q_S^2}$ becomes

$$\sqrt{I_S^2 + Q_S^2} = \frac{P\sin\left(\pi\left(\Delta f\right)(T_h - |\varepsilon|)\right)^2}{\left(\pi\left(\Delta f\right)T_h\right)^2} \tag{6.10-17}$$

Consider the case when the frequency error times hop duration product is small compared to unity and the time error is small compared to one hop time. We assume that the noncoherent FSK modulation is selected such that each nearest tone is separated by $1/T_h$ Hz to ensure orthogonality. Thus from (6.10-15) when signal f_{sl} is being detected by the demodulator for f_{sj} one has for the signal component

$$I_S^2 + Q_S^2 = \frac{P\sin\left(\pi\left(f_{sj} - f_{sl}\right)T_h\right)^2}{\left(\pi\left(f_{sj} - f_{sl}\right)T_h\right)^2} = \frac{P\sin\left(\pi(k/T_h)(T_h)\right)^2}{\left(\pi(k/T_h)T_h\right)^2} = 0 \tag{6.10-18}$$

since the frequency separation is given by $k/(T_h)$ for k = 1, 2,... Thus only the desired signal will output a nonzero response when the time and frequency errors are small. With this assumption consider the probability distribution for each noncoherent detector.

6.10.2 Detection and False Alarm Probabilities for FFH/MFSK Serial Active Search

First consider the case when the signal is not present. Consider the variable Y where

$$Y = \sum_{i=1}^{N} \left(I_i^2 + Q_i^2 \right) = \sum_{i=1}^{2N} X_i^2 \tag{6.10-19}$$

where I_i and Q_i are I and Q components for the i-th hop, in one modulation tone period, and X_i represents either I_i or Q_i. The probability density function of Y is given by [21]

$$p_Y(y) = \frac{1}{\sigma^{2N} 2^N \Gamma(N)} y^{N-1} e^{-y/2\sigma^2}, \; y \ge 0 \tag{6.10-20}$$

Consider the change of variable

$$Z = Y/(2\sigma^2) \tag{6.10-21}$$

Hence noting that the Jacobian of the transformation is given by $2\sigma^2$, the new probability density function is given by

$$p_N(z) = \frac{z^{N-1} e^{-z}}{(N-1)!} \tag{6.10-22}$$

which has the distribution function

$$F_n(x) = \int_0^x p_N(z) dz \tag{6.10-23}$$

Now consider the case when the signal is present. From Proakis [21] the probability density function of the signal present case is given by

$$p_Y(y) = \frac{1}{2\sigma^2} \left(\frac{y}{s^2} \right)^{(N-1)/2} e^{-\left(\frac{s^2 + y}{2\sigma^2} \right)} I_{N-1}\left(\sqrt{y} \frac{s}{\sigma^2} \right) \tag{6.10-24}$$

where s^2 is given by

$$s^2 = \sum_{i=1}^{N} \left(I_s^2 + Q_s^2 \right) = NP \left(\frac{\sin\left(\pi \Delta f (T_h - |\varepsilon|) \right)}{\pi \Delta f T_h} \right)^2 \tag{6.10-25}$$

Let $Z = Y/(2\sigma^2)$ so that

$$p_{S+N}(z) = \left(\frac{z}{\gamma} \right)^{(N-1)/2} e^{-\gamma} e^{-z} I_{N-1}\left(2\sqrt{\gamma z} \right), \quad z \ge 0 \tag{6.10-26}$$

where γ is given by

$$\gamma = \frac{s^2}{2\sigma^2} = \frac{PNT_h L\left(\dfrac{\sin(\pi\Delta f(T_h - |\varepsilon|))}{(\pi\Delta f(T_h - |\varepsilon|))}\right)^2}{N_0} \tag{6.10-27}$$

where L was added to the signal term as a signal power loss and accounts for the miscellaneous losses due to filtering, quantization, and so on. When the timing is more than a hop in error, the signal is absent and a false alarm occurs when one of the M noise-only correlator outputs exceeds the threshold. Let z_i denote the i-th normalized correlator output. The probability of not having a false alarm is given by

$$1 - P_{FA} = P(z_1 < \eta,\ z_2 < \eta, ...z_M < \eta) = P(z_1 < \eta)P(z_2 < \eta)...P(z_M < \eta) \tag{6.10-28}$$

From (6.5-63) it follows that

$$P_{FA} = 1 - \left[1 - e^{-\eta}\sum_{k=0}^{N-1}\frac{\eta^k}{k!}\right]^M \tag{6.10-29}$$

Note that when $M = 1$ it agrees with (6.5-63). The probability of detection occurs when the correlator with the signal exceeds the threshold and all other (noise-only) correlator outputs. Accounting for the assumption that all the correlator variables are statistically independent, one has

$$P_D = P(z_2 < z_1,\ z_3 < z_1, ...,\ z_M < z_1 \text{ and } z_1 > \eta) \tag{6.10-30}$$

This can be written as

$$P_D = \int_\eta^\infty P_n(z_2 < z_1)^{M-1} p_s(z_1)dz_1 \tag{6.10-31}$$

where the distribution function of the noise is given by

$$P_N(z_2 < z_1) = 1 - e^{-z_1}\sum_{k=0}^{N-1}\frac{z_1^k}{k!} \tag{6.10-32}$$

and the density function of the signal is given by (6.10-26) and (6.10-27). Thus the detection probability is given by

$$P_D = \int_\eta^\infty \left[1 - e^{-z_1}\sum_{k=0}^{N-1}\frac{z_1^k}{k!}\right]^{M-1}\left(\frac{z_1}{\gamma}\right)^{(N-1)/2} e^{-\gamma}e^{-z_1}I_{N-1}\left(2\sqrt{\gamma z_1}\right)dz_1 \tag{6.10-33}$$

This expression is difficult to evaluate but it can be upper bounded by the following expression

$$P_D \le \int_\eta^\infty \left(\frac{z_1}{\gamma}\right)^{(N-1)/2} e^{-\gamma}e^{-z_1}I_{N-1}\left(2\sqrt{\gamma z_1}\right)dz_1 \tag{6.10-34}$$

The probability of detection can also be lower bounded by the following expression

$$P_D \geq \left[1-e^{-\eta}\sum_{k=0}^{N-1}\frac{\eta^k}{k!}\right]^{M-1}\int_{\eta}^{\infty}\left(\frac{z_1}{\gamma}\right)^{(N-1)/2} e^{-\gamma}e^{-z_1}I_{N-1}\left(2\sqrt{\gamma z_1}\right)dz_1 \tag{6.10-35}$$

From (6.5-68) it is possible to write the upper bound of the detection probability (6.10-34) in terms of an expression that is readily computable in the form

$$P_D \leq \sum_{m=0}^{N-1} e^{-\eta}\frac{\eta^m}{(m!)} + \sum_{m=N}^{\infty} e^{-\eta}\frac{\eta^m}{(m!)}\left(1-\sum_{k=0}^{m-N} e^{-N\gamma_0}\frac{(N\gamma_0)^k}{(k!)}\right) \tag{6.10-36}$$

where

$$\gamma_0 = \gamma / N = \frac{PT_hL\left(\dfrac{\sin(\pi\Delta f(T_h-|\varepsilon|))}{(\pi\Delta f(T_h-|\varepsilon|))}\right)^2}{N_0} \tag{6.10-37}$$

The upper bound for the detection probability (6.10-35) can be expressed as

$$P_D \geq \left[1-e^{-\eta}\sum_{k=0}^{N-1}\frac{\eta^k}{k!}\right]^{M-1}\sum_{m=0}^{N-1} e^{-\eta}\frac{\eta^m}{(m!)} + \sum_{m=N}^{\infty} e^{-\eta}\frac{\eta^m}{(m!)}\left(1-\sum_{k=0}^{m-N} e^{-N\gamma_0}\frac{(N\gamma_0)^k}{(k!)}\right) \tag{6.10-38}$$

Figure 6.10-3 illustrates the detection and false alarm probabilities for FFH/MFSK with 8-ary modulation ($M = 8$) for N hops per MFSK symbol running from 1 to 16 for the case in which P_{FA} was set equal to 10^{-6}. In the curves it was assumed that $|\varepsilon| = 0.25\ Th$ and $\Delta fT_h = 0.2$. Actually both the upper and lower bounds for P_D are illustrated; however, the upper and lower bounds are indistinguishable numerically so only one curve is visible for each value of N. The reason this is true is the fact that the probability that a nonsignal correlator output will exceed the threshold is very small (10^{-6}). If the false alarm probability were selected to be larger than the curves for the upper and lower bounds, the detection probability would diverge when plotted.

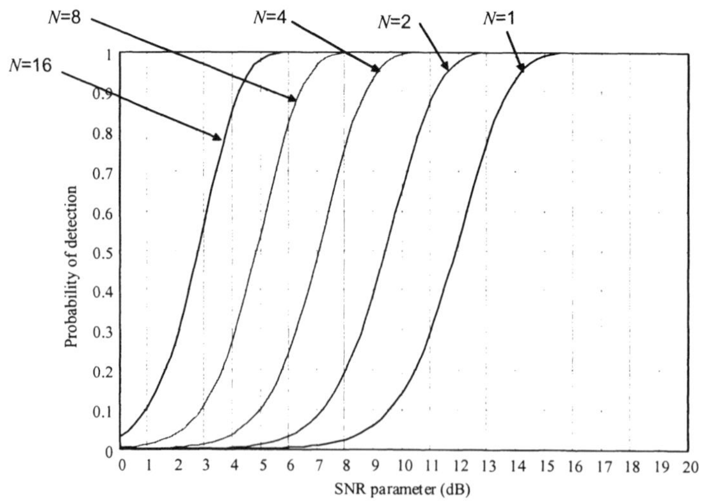

Figure 6.10-3 Probability of detection and false alarm for the FFH/8FSK acquisition system.

Note that to achieve a detection probability of 0.9 with $N = 1$ requires a value of γ_0 (6.10-37) or about 13.5 dB. Whereas with $N = 16$, only a value of γ_0 of about 4 dB is needed. It should be noted that since the hops have a fixed SNR, the more hops, the higher the SNR per MFSK symbol, and thus the lower per hop SNR is needed, as is expected.

6.10.3 Detection and False Alarm Probabilities for SFH/MFSK Serial Active Search

Now consider the slow frequency-hopping case in which there are more than one MFSK symbol per FH tone. In the case of slow frequency-hopped code acquisition with MFSK modulation, a selection of the maximum correlator output is obtained per modulation symbol and those maximums are summed. Figure 6.10-4 illustrates the SFH/MFSK acquisition model. The received signal is multiplied by the local time estimate of the hopping frequency and heterodyned down to an IF frequency. The signal is fed into the M noncoherent demodulators. The largest of the correlator outputs is summed over each MFSK symbol time (sum of N correlator outputs), which is compared to a threshold. If additional SNR is needed it is possible to sum N_h values of the N sum of the largest correlator values. The calculations in this section will not account for N_h; in other words, it will be assumed that $N_h = 1$. The timing and frequency diagram is illustrated in Figure 6.10-5.

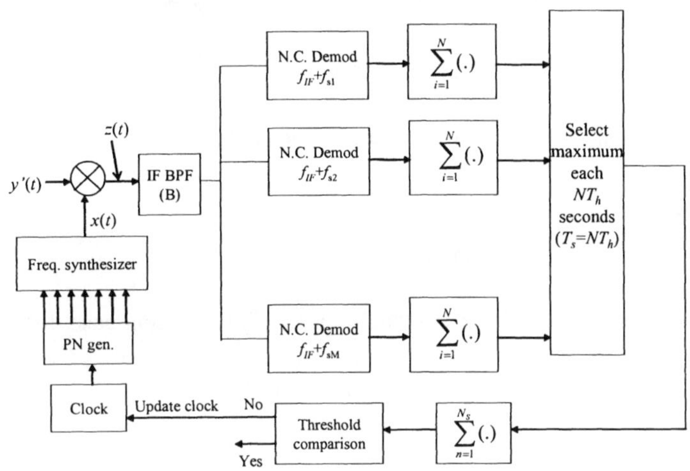

Figure 6.10-4 SHF/MFSK model for slow FH acquisition.

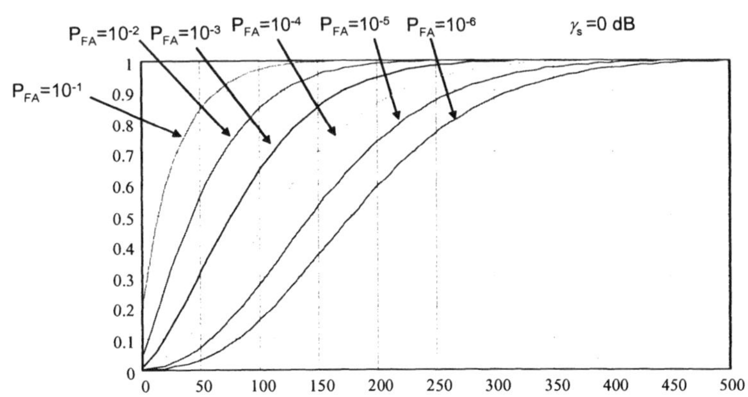

Figure 6.10-5 SFH/MFSK timing and frequency diagram for SFH/MFSK acquisition.

Basically we have the same model as in FFH/MFSK except the symbols operate at a higher rate than the hop rate. In order to the compute the probability of a false alarm or the detection probability, it is necessary to determine the probability that the sum of N maximums of M variates exceeds a threshold. When the signal is not present all the variates have the same distribution function; in the case when the signal is present M–1 variates are due to noise, and one is due to signal plus noise. Based on Lindgren [53] the distribution function of the maximum of M variates is given by

$$F_{\max}(x) = (F_N(x))^{M-1} F_{S+N}(x) \tag{6.10-39}$$

where $F_N(x)$ is the distribution function of the noise variates $F_{S+N}(x)$ and is the distribution function of the signal plus noise. It follows that the density function of the corresponding to this distribution function is given by

$$p_{\max}(x) = \frac{d}{dx}\left[F_{S+N}(x)\left(F_N(x)\right)^{M-1}\right] \tag{6.10-40}$$

or

$$p_{\max}(x) = p_{S+N}(x)\left[F_N(x)\right]^{M-1} + (M-1)F_{S+N}(x)p_N(x)\left[F_N(x)\right]^{M-2} \tag{6.10-41}$$

When the signal is not present $p_{S+N}(x) = p_N(x)$ so that the probability density function of the maximum is obtained from (6.10-41) as

$$p_{\max}(x) = Mp_N(x)\left[F_N(x)\right]^{M-1} \tag{6.10-42}$$

From (6.10-26), (6.10-22), and (6.10-23) one obtains, for the signal present case, the result

$$\begin{aligned} p_{S+N}(x) &= e^{-(\gamma+x)}I_0\left(2\sqrt{\gamma_s x}\right)\left[1-e^{-x}\right]^{M-1} \\ &+ (M-1)e^{-x}\left[1-e^{-x}\right]^{M-2}\int_0^x e^{-(y+\gamma_s)}I_0(2\sqrt{\gamma y_s})dy, \quad x \geq 0 \end{aligned} \tag{6.10-43}$$

where γ_s is the symbol SNR parameter, that is,

$$\gamma_s = \frac{s^2}{2\sigma^2} = \frac{PNT_sL\left(\dfrac{\sin(\pi\Delta f(T_s-|\varepsilon|))}{(\pi\Delta f(T_s-|\varepsilon|))}\right)^2}{N_0} \tag{6.10-44}$$

When the signal is not present, that is under false alarm conditions, the density function of the maximum of M variates is given by (6.10-43) when $\gamma_s = 0$, which yields

$$p_N(x) = Me^{-x}\left[1-e^{-x}\right]^{M-1}, \quad x \geq 0 \tag{6.10-45}$$

An approximate method of obtaining the probability density function of the sum of M maximum of M variates is to approximate the density function with a Gaussian approximation. To do this it is necessary to obtain the mean and variance of the maximum of M variates. Following Simon [15], the mean, when the signal is present, becomes

$$\mu_{S+N} = \int_{-\infty}^{\infty} x p_{S+N}(x) dx \tag{6.10-46}$$

or

$$\mu_{S+N} = 1 + \gamma_s + (M-1)\sum_{k=0}^{M-2} \frac{(-1)^k}{(k+2)(k+1)^2} \binom{M-2}{k} e^{\left[-\left(\frac{k+1}{k+2}\right)\gamma_s\right]} \tag{6.10-47}$$

When the signal is not present, one can let $\gamma_s = 0$ to obtain

$$\mu_N = 1 + (M-1)\sum_{k=0}^{M-2} \frac{(-1)^k}{(k+2)(k+1)^2} \binom{M-2}{k} \tag{6.10-48}$$

The mean squared value of x when the signal is present is given by

$$\overline{(x^2)}_{S+N} = \int_{-\infty}^{\infty} x^2 p(x) dx \tag{6.10-49}$$

or

$$\begin{aligned} \overline{(x^2)}_{S+N} = {} & 2 + 4\gamma_s + \gamma_s^2 + 2(M-1)\sum_{k=0}^{M-2} \frac{(-1)^k}{(k+2)(k+1)^2} \binom{M-2}{k} \\ & \times \left[\left(\frac{1}{k+2}\right)\left(\frac{\gamma_s}{k+2} + 1\right) + \frac{1}{k+1} \right] \exp\left[-\left(\frac{k+1}{k+2}\right)\gamma_s \right] \end{aligned} \tag{6.10-50}$$

When the signal is not present, one can let $\gamma_s = 0$ to obtain

$$\overline{(x^2)}_N = 2 + 2(M-1)\sum_{k=0}^{M-2} \frac{(-1)^k}{(k+2)(k+1)^2} \binom{M-2}{k} \left[\left(\frac{1}{k+2}\right) + \frac{1}{k+1} \right] \tag{6.10-51}$$

Thus the in-sync (signal present) and the out-of-sync (signal absent) variances can be written as

$$\begin{aligned} \sigma_{S+N}^2 &= \overline{\left(x^2\right)}_{S+N} - \mu_{S+N}^2 \\ \sigma_N^2 &= \overline{\left(x^2\right)}_N - \mu_N^2 \end{aligned} \tag{6.10-52}$$

where the values of the second moments and the mean values are given by (6.10-47), (6.10-48), (6.10-50), and (6.10-51).

The decision statistic is given by the sum of the N largest variates so that

$$y = \sum_{i=1}^{N} x_i^2 \tag{6.10-53}$$

which is compared to the normalized threshold η to determine if synchronization is detected. We note that the mean and variance for the sum of N independent decision random variables has a mean that is N times the individual random variable mean and a variance that is N times as large as the single random variable variance. The false alarm probability is given by

$$P_{FA} = \int_{\eta}^{\infty} \frac{1}{\sqrt{2\pi N\sigma^2}} \exp\left[-\frac{(y - N\mu_n)^2}{2N\sigma^2}\right] dy = Q\left(\frac{(\eta - N\mu_n)}{\sqrt{N\sigma_n^2}}\right) = Q(\beta) \tag{6.10-54}$$

where β is defined via (6.10-54). The detection probability is given by the following value

$$P_D = \int_{\eta}^{\infty} \frac{1}{\sqrt{2\pi N\sigma_s^2}} \exp\left[-\frac{(y - N\mu_s)^2}{2N\sigma_s^2}\right] dy = Q\left(\frac{(\eta - N\mu_s)}{\sqrt{N\sigma_s^2}}\right) \tag{6.10-55}$$

The false alarm probability can be written as a function of β in the following way

$$P_D = Q\left(\frac{(\eta - N\mu_N - N(\mu_{S+N} - \mu_N))}{\sqrt{N\sigma_N^2}(\sigma_{S+N}/\sigma_N}\right) = Q\left(\frac{\left(\beta - \sqrt{\frac{N}{\sigma_N^2}}(\mu_{S+N} - \mu_N\right)}{\sigma_{S+N}/\sigma_N}\right) \tag{6.10-56}$$

Using (6.10-54) to solve for β produces the final result of the relationship between P_D and P_{FA}

$$P_D = Q\left(\frac{\left(Q^{-1}(P_{FA}) - \sqrt{\frac{N}{\sigma_N^2}}(\mu_{S+N} - \mu_N\right)}{\sigma_{S+N}/\sigma_N}\right) \tag{6.10-57}$$

where β and P_{FA} are linked through (6.10-54). Recall that the Q function is defined by

$$Q(x) = \int_{x}^{\infty} \frac{1}{\sqrt{2\pi\sigma^2}} e^{-\left[\frac{z^2}{2\sigma^2}\right]} dz \tag{6.10-58}$$

Now consider the evaluation of the acquisition time for FFH/MFSK and SFH/MFSK.

6.10.4 Acquisition Time Calculations for FFH/MFSK and SFH/MFSK

Both the mean and the variance of the acquisition time will be obtained for SFH/MFSK and FFH/MFSK. From Section 6.2.1 one has for the mean time to acquire, for a single active correlator,

$$\overline{T} = \frac{2 + (2 - P_D)(q - 1)(1 + KP_{FA})}{2P_D} \tau_D' \tag{6.10-59}$$

where $\tau_D^{'}$ is given by

$$\begin{aligned} \tau_D^{'} &= NT_h \text{ for FFH/MFSK} \\ \tau_D^{'} &= NT_s \text{ for SFH/MFSK} \end{aligned} \tag{6.10-60}$$

and K is the number of $\tau_D^{'}$ seconds that are needed to verify a true detection from a false alarm. The variance of the acquisition time is also given in Section 6.2.1 for an active serial search and is given by

$$\sigma^2 \cong (\tau_D^{'})^2(1+KP_{FA})^2 q^2\left(\frac{1}{12}-\frac{1}{P_D}+\frac{1}{P_D^{\;2}}\right) \tag{6.10-61}$$

Now consider an example calculation for the acquisition time results.

Example 6 Consider an example for acquisition of SFH/MFSK. Assume that the time is unknown to 1000 symbols and 8FSK modulation is used. Let $K = 3$, and assume that P_{FA} is equal to 10^{-4}, 10^{-5}, 10^{-6}, and 10^{-7}. Also let $\gamma = 1$ (0 dB). Determine the mean acquisition time in this case as a function of the N, the number of modulation symbols that are combined per hop. Figure 6.10-6 illustrates the results. It can be seen that for each assumed value of the false alarm probability, there is an optimum value of N, which minimizes the mean acquisition time. For example for $P_{FA} = 10^{-4}$ the optimum value of N is about 190.

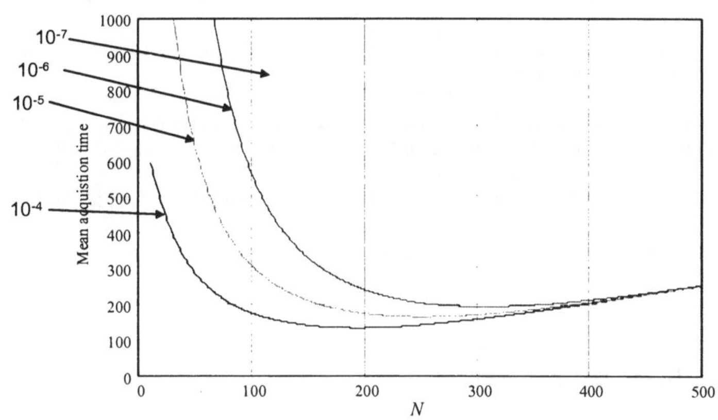

Figure 6.10-6 SFH/MFSK mean acquisition time for SFH/MFSK acquisition.

6.11 SUMMARY

In this chapter many types of acquisition schemes have been discussed; these include the single active correlator, including adjustments for Doppler, and an active search with parallel processing. System structures for BPSK, QPSK, OQPSK, and MSK have been presented. The use of the FFT to enhance the frequency search of the active search approach also has been presented. Acquisition time performance for BPSK acquisition for the case of the general jammer, the matched spectral jammer, and the narrowband jammer was developed. Acquisition performance for balanced QPSK and balanced OQPSK was also developed. Along with "exact" expressions for detection probability a Gaussian approximation was developed and presented. A discussion of sequential detection and the Tong detector were also presented, as well as the use of transform techniques for code acquisition.

Following the active search schemes, passive code acquisition was presented, including the general jammer and the matched spectral jammer. The use of the FFT for passive code searches was developed along with mean acquisition time performance.

The chapter concluded by addressing fast and slow frequency hopped spread spectrum acquisition with MFSK modulation. Mean acquisition time performance was developed for both cases. The appendix includes an introduction to signal flow graphs and discrete time invariant Markov processes.

References

[1] R. B. Ward, "Acquisition of Pseudorandom Signals by Sequential Estimation," *IEEE Trans. on Communication Technology*, Vol. Com-13, No. 4, December 1965, pp. 475–483.

[2] R. B. Ward and K. P. Yiu, "Acquisition of Pseudonoise Signals by Recursion Aided Sequential Estimation," *IEEE Trans. on Communication Technology*, Vol. COM-25, No. 8, August 1977, pp. 784–794.

[3] C. C. Kilgus, "Pseudonoise Code Acquisition Using Majority Logic Decoding," *IEEE Trans. on Communication Technology*, Vol. COM-21, No. 6, June 1973, pp. 772–774.

[4] J. H. Chiu and L. S. Lee, "An Improved Sequential Estimation Scheme for PN Acquisition," *IEEE Trans. on Communications*, Vol. 36, No. 10, October 1988, pp. 1182–1184.

[5] G. S. Rawlings, G. G. Koller, and M. A. Belkerdid, "Rapid Acquisition of Spread Spectrum Signals Utilizing Integrated Technologies," *RF Expo East Proceedings*, Tampa, FL, September 22–24, 1992.

[6] D. Dodds and X. Zhang, "Spread Spectrum Code Phase Acquisition Using a Cyclic Accumulation Matched Filter," *Globe Com 94*, San Francisco, CA, October 1994.

[7] J. J. Stiffler, "Rapid Acquisition Sequences," *IEEE Trans. on Information Theory*, March 1968, pp. 221–225.

[8] S. Curry, D. Schwartz and J. Collins, "State Transition Assisted DSSS Code Synchronization," *MILCOM 99*, Atlantic City, NJ, Unclassified proceeding, October 31–Novovember 3, 1999.

[9] J. K. Holmes and C. C. Chen, "Acquisition Time Performance of PN Spread Spectrum Systems," *IEEE Trans. on Communication Special Issue on Spread Spectrum Communications*, Vol. COM-25, August 1977.

[10] J. K. Holmes, "Predicted Mean SSP Acquisition Times with the New Algorithm," TRW IOC SCTE-50-76-251/JKH, July 21, 1976.

[11] K. Holmes, *Coherent Spread Spectrum Systems*, New York: Wiley-Interscience, 1982.

[12] D. M. DiCarlo, "Mean Time and Variance for the Two-Step Acquisition System," TRW IOC 75-7131.67-04, May 29, 1975.

[13] A. Polydoros and C. L. Weber, "A Unified Approach to Serial Search Spread-Spectrum Code Acquisition," *IEEE Trans. on Communication*, Vol. COM-32, No. 5, May 1984, pp. 542–560.

[14] A. Polydoros, "Generalized Serial Code Search Acquisition: The Equivalent Circular State Diagram Approach," *Milcom 83*, Conference Record, Washington, D.C., October 1983.

[15] M. K. Simon et al., *Spread Spectrum Communications*, Vol. III, Chapter 1, Rockville, MD: Computer Science Press, 1985.

[16] J. K. Holmes, N. Morgan, and P. A. Dafesh, "A Theoretical Approach to Determining the 95% Probability of TTFF for the P(Y) Code Utilizing Active Code Acquisition," *AIAA-ICSSC Conf.*, San Diego, CA, June 11–14, 2006.

[17] J. B. Lozow, "Analysis of Direct P(Y) Code Acquisition," *Journal of the Institute of Navigation*, Vol. 44, No. 1, Spring 1997.

[18] L. Rabiner and B. Gold, *Theory and Application of Digital Signal Processing*, Chapter 6, Upper Saddle River, NJ: Prentice-Hall, 1975.

[19] A. V. Oppenheim and R.W. Schafer, *Discrete-Time Signal Processing*, Upper Saddle River, NJ: Prentice-Hall, 1989.

[20] A. Papoulis, *Probability, Random Variables and Stochastic Processes*, 2nd ed., New York: McGraw-Hill, 1984.

[21] J. Proakis, *Digital Communications*, Chapter 1, 2nd ed., New York: Mc Graw-Hill, 1989.

[22] D. A. Shnidman, "The Calculation of the Probability of Detection and the Generalized Marcum Q-Function," *IEEE Trans. on Information Theory*, Vol. 35, No. 2, March 1989.

[23] J. K. Holmes, and K. T. Woo, "An Optimum PN Code Search Technique for a Given A Priori Signal Location Density," Section 18.6, *NTC 78 Conference Record*, Birmingham, AL, December 3–6, 1978.

[24] J. K. Holmes, and K. T. Woo, "An Optimum Asymmetric PN Code Search Strategy," *1979 International Telemetering Conference*, San Diego, CA, November 19–21, 1979, pp. 319–330.

[25] W. Braun, "Performance Analysis for the Expanding Search PN Acquisition Algorithm," *IEEE Trans. on Communications*, COM-30, March 1982, pp. 424–435.

[26] A. Weinberg, "Search Strategy Effects on PN Acquisition Performance," *NTC'81 Conf. Record*, November 29–December 3, 1981, New Orleans, LA, pp. F1.5.1–F1.5.5.

[27] A. Weinberg, "Generalized Analysis For the Evaluation of Search Strategy Effects on PN Acquisition Performance," *IEEE Trans. on Communications*, COM-31, January 1983, pp. 37–39.

[28] A. Wald, *Sequential Analysis*, New York: Wiley, 1947.

[29] J. J. Bussgang and D. Middleton, "Optimum Sequential Detection of Signals and Noise," *IRE Trans.*, Vol IT-1, December 1955, pp. 5–14.

[30] A. Wald, and J. J. Wolfowitz, "Optimum Character of the Sequential Probability Ratio Test," *Ann. Math. Statis.*, Vol. 19, 1948, p. 326.

[31] H. Blasburg, "Experimental Results in Sequential Detection," *IRE Trans.*, Vol IT-5, No. 2, June 1959, p. 41.

[32] W. B. Kendall, "Performance of the Biased Square-Law Sequential Detector in the Absence of Signal," *IEEE Trans. Information Theory*, January 1964.

[33] G. E. Albert, *Ann. Math. Statist.*, Vol. 25, September 1954.

[34] J. Y. DiFranco and W. L. Rubin, *Radar Detection*, Chapter 16, Englewood Cliffs, NJ: Prentice-Hall, 1968.

[35] J. C. Hancock and P. A. Wintz, *Signal Detection Theory*, Chapter 4, New York: McGraw-Hill, 1966.

[36] F. Ricker, P. McAdam, and B. Trumpis, "Sequential Acquisition of PN Codes for TDRSS," paper submitted for publication, circa 1970.

[37] P. S. Tong, "A Suboptimum Synchronization Procedure for Pseudo Noise Communication Systems," *National Telemetry Conference*, November 26–28, 1973.

[38] H. L. Scott, "Code Search Analysis," Texas Instruments Inc. Report, prepared for the Charles Stark Draper Lab., Contract no. DL-H-169021, March 28, 1980.

[39] E. Kaplan, (ed.), *Understanding GPS Principles and Applications*, Norwood, MA: Artech House, 1996.

[40] L. Rabiner and B. Gold, *Theory and Application of Digital Signal Processing*, Chapter 6, Englewood Cliffs, NJ: Prentice-Hall, 1975.

[41] C. Yang, "Fast Code Acquisition with FFT and Its Sampling Schemes," *Proc. of the ION*, September 1996.

[42] R. C. Borgiolio, "Fourier Transform Correlation Versus Direct Discrete Time Correlation," *Proc. IEEE*, Vol. 56, No. 9, September 26, 1968.

[43] R. Davenport, "FFT Processing of Direct Sequence Spreading Codes Using Modern DSP Microprocessors," *Proc. of the 1991 IEEE National Aerospace and Electronics Conf.*, 1991.

[44] D. Van Nee and A. Coenen, "New Fast GPS Code Acquisition Technique Using FFT," *Electronic Letters*, No. 27, January 1991.

[45] M. K. Simon, J. K. Omura, R. A. Scholtz, and B. K. Levitt, *Spread Spectrum Communications,* Rockville, MD: Computer Science Press, 1985.

[46] P. M. Fishman, and J. Betz, "Predicting Performance of Direct Acquisition for the M-Code Signal," *Proc. ION National Technical Meeting,* January 2000.

[47] A. Polydoros and C. Weber, "A Unified Approach to Serial Search Spread-Spectrum Code Acquisition-Part II: A Matched-Filter Receiver," *IEEE Trans. on Communications*, Vol. COM-32, May 1984, pp. 550–560.

[48] R. Wolfert, S. Chen, S. Kohli, D. Leimer, and J. Lascody, "Direct P(Y)-Code Acquisition Under a Jamming Environment," *Proc. of IEEE PLANS*, April 1998.

[49] S. F. Rounds, "A Low-Cost Direct-Y Code Fast Acquisition SAASM," *Proceedings of the ION GPS Conference*, 1998, p. 215.

[50] R. A. Roberts and C. T. Mullis, *Digital Signal Processing*, Chapter 11, Reading, MA: Addison-Wesley, 1987.

[51] H. Chang, "Presampling Filtering, Sampling and Quantization Effects on the Digital Matched Filter Performance," *Proc. of Int. Tel. Conf.*, 1982.

[52] C. R. Cahn, "Spread Spectrum Applications and State-of-the-Art Equipments," Lecture No. 5 in the Agard-NATO Lecture Series, January 31, 1973.

[53] B. W. Lindgren, *Statistical Theory*, New York: Macmillan, 1962.

Selected Bibliography

F. Amoroso, "Adaptive A/D Converter to Suppress CW Interference in DSPN Spread Spectrum Communications," *IEEE Trans. on Communication*, Vol. COM-31, No. 10, October 1983.

F. Amoroso, "Performance of the Adaptive A/D Converter in Combined CW and Gaussian Interference," *IEEE Milcom 84*, October 21–24, 1984.

E. O. Brigham, *The Fast Fourier Transform and Its Applications*, Chapter 10, Englewood Cliffs, NJ: Prentice-Hall, 1988.

P. A. Dafesh and J. K. Holmes, "Practical and Theoretical Tradeoffs of Active Parallel Correlator and Passive Matched Filter Acquisition implementations," *U.S. ION Annual Meeting*, San Diego, CA, June 26–28, 2000.

P. A. Dafesh and J. K. Holmes, "A Trade Study for the Selection of an Acquisition Signal for GPS Modernization," TOR 2000(1590)-1, The Aerospace Corporation, January 2000.

J. K. Holmes, S. Raghavan, and S. Lazar, "Acquisition and Tracking Performance of NRZ and Square Wave Modulated Symbols for Use in GPS," *Institute of Navigation 54th Annual Meeting,* June 1–3, 1998.

J. K. Holmes and P. A. Dafesh, "Matched Filter Mean Acquisition Time Performance For P(Y), BOC(10,5), and MAN(10) Codes with FFT Aiding and Noncoherent Combining," *ION Conference*, Albuquerque, NM, June 1–14, 2001.

L. B. Milstein and P.K. Das, "Spread Spectrum Receiver Using Surface Acoustic Wave Technology," *IEEE Trans. on Communications*, COM-25, No. 8, August 1977, pp. 841–847.

G. D. O Clock, "Matched Filters Boost Receiver Gain," *Electronic Design*, Vol. II, May 24, 1977.

M.W. Thompson and S. Dianat "Non-Coherent Code Acquisition in Direct Spread Spectrum Systems Using a Neural Network," *IEEE Milcom 93*, Boston, MA, October 11–14, 1993.

D. Torrieri, *Principles of Secure Communication Systems*, Chapter 4, Dedham, MA: Artech House, 1985.

TRW *LSI Products Inc. Data Handbook*, Digital output correlator section, 1991.

H. Urkowitz, "Energy Detection of Unknown Deterministic Signals," *Proc. of IEEE*, Vol. 55, No. 4, April 1967.

R. Ziemer and R. Peterson, *Digital Communications and Spread Spectrum Systems*, New York: Macmillan, 1985.

Problems

1. Show that for Example 1, described in Section 6.2, if $P_D \leq 1$ and $P_{FA} = 0$, then

$$\overline{T}_{acq} \cong \frac{2 - P_D}{P_D} N\tau_D \tag{P6-1}$$

by summing the appropriate series. Note that the series can be summed in closed form. Notice that when $P_D = 1$ the result agrees with (6.2-2).

2. Derive the expression for the mean and the variance of the acquisition time for the double dwell system as described in the text. Note that this is a problem of very advanced difficulty.

3. Consider the code correlation curve shown here. Denoting $P_D^{(i)}$ as the probability of detection at the i-th point on the correlation curve (i = 1, 2, 3, 4), show that the total probability of detection, including all four points, is given by

$$P_D^{\ T} = P_{D1}^{(1)} + P_D^{(2)}(1 - P_D^{(1)}) + P_D^{(3)}(1 - P_D^{(1)})(1 - P_D^{(2)}) \\ + P_D^{(4)}(1 - P_D^{(1)})(1 - P_D^{(2)})(1 - P_D^{(3)})$$

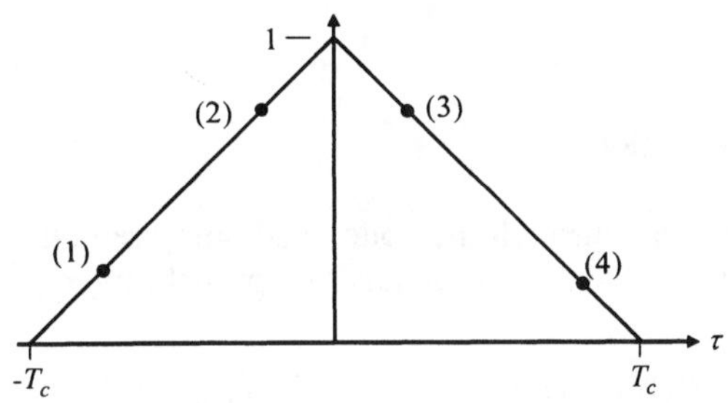

Figure for Problem 3.

4. Show, based on Section 6.5.6, that the minimum response between two FFT filters responses is about 3.9 dB down from the maximum response in the nonzero padded case and is down about 0.91 dB in the zero padded case. These values of the responses occur at the midpoint frequency between any two contiguous filter responses.

5. Based on Section 6.5.4, note that the false alarm probability, expressed as an integral, is given by

$$P_{FA} = \int_{\eta}^{\infty} \frac{1}{(N_{nc} - 1)!} u^{N_{nc}-1} e^{-u} du \tag{P6-2}$$

Show that this can be evaluated to the following finite series form

$$P_{FA} = e^{-\eta} \sum_{k=0}^{N_{nc}-1} \frac{\eta^k}{k!} \tag{P6-3}$$

6. Consider Figures 6.2-7 and 6.2-8. Let the received QPSK signal plus thermal noise be modeled as

$$y(t) = \sqrt{P}PN_1(t-T)\cos(\omega_o t) + \sqrt{P}PN_2(t-T)\sin(\omega_o t) + \sqrt{2}n_c(t)\cos(\omega_0 t) + \sqrt{2}n_s(t)\sin(\omega_0 t) \tag{P6-4}$$

Assume that the carrier references are given by $\sqrt{2}\cos(\omega_0 t)$ and $\sqrt{2}\sin(\omega_0 t)$. Assume that $T = \hat{T}$. Show that the acquisition system in Figure 6.2-8 is inferior to Figure 6.2-7 by considering the signal and noise terms at the input to the noncoherent summer. Observe that the system of Figure 6.2-8 has the same signal and has the noise terms of Figure 6.2-7 plus additional noise terms, and is therefore uniformly worse.

7. Show that the signal into the noncoherent summer of the MSK acquisition system of Figure 6.2-10 is given by $2PT_d^2$.

8. Starting with Section 6.5.5 and letting the signal to noise ratio γ_0 be zero, show that the false alarm probability is given by $P_{FA} = Q(\beta)$, where $\beta = \frac{\eta - N_{nc}}{\sqrt{N_{nc}}}$. Then show that using the definition of P_{FA} in terms of Q(β) that P_D is given by $P_D = Q\left(\frac{\beta - \sqrt{N_{nc}}\gamma_0}{\sqrt{1+2\gamma_0}}\right)$ for the case of BPSK SS modulation, when using the Gaussian approximations.

9. Referring to Section 6.6, show that the total probability of acquisition is given by

$$Q_N = P_D \sum_{k=1}^{N} P(L_k)(1-P_D)^{N-k} \tag{P6-5}$$

for the problem under consideration.

10. Determine the number of sweeps through the code uncertainty region that must occur to obtain the probability of acquisition of P. Note that the acquisition probability $P_{acq}(N)$ is given by

$$P_{acq}(N) = P_D + (1-P_D)P_D + (1-P_D)^2 P_D + ...(1-P_D)^{N-1} P_D = P \tag{P6-6}$$

where N is an integer. Further note that when N is not an integer, with the assumption that the a priori probability density function is uniformly distributed, the probability of acquisition increases linearly (between integers) so that

$$N = \left\{ (K-1) + \frac{P - \hat{P}(K-1)}{\hat{P}(K) - \hat{P}(K-1)} \right\} \tag{P6-7}$$

with

$$K = \min[m \ni \hat{P}(m) > P] \quad k = 1, 2, 3, 4, ... \tag{P6-8}$$

Also show that for $N > 1$ that

$$N \cong \frac{\ln(1-P)}{\ln(1-P_D)} \tag{P6-9}$$

11. Assume that an NRZ baseband SS process is ideally filtered to R_c Hz. Show that the samples taken at twice the chip rate ($2R_c$) for this case are statistically independent if the noise is modeled as flat white Gaussian noise with two-sided spectral density of $N_0/2$. Hint: Consider the autocorrelation function and show that it is zero when the time variable is any multiple of $1/(2R_c)$. Conclude therefore that samples taken at twice the chip rate are statistically independent.

12. Consider the digital matched filters of Figures 6.9-4 and 6.9-5. Compute the variance of a narrowband Gaussian jammer plus thermal noise over the correlator coherent correlation time, T_i. Extend the results of Section 6.9.8. It is to be assumed that the jammer is narrow compared to the chip rate. Let the bandwidth of the narrowband interferer be B_J. If the spectral density is J_0 over the bandwidth B_J then it follows that $J = J_0B_J$. Since the interferer noise is expressed in the form (6.9-18) each baseband component has only $J/2$ watts received. Show that the thermal plus jammer noise variance for both the real and imaginary components out of the FFT is given by

$$\sigma_{NBI}^2 = 0.903\frac{N_0}{2T_i} + \frac{JT_c}{2T_i} \tag{P6-10}$$

where T_i is the coherent correlation time, T_c is the chip duration, N_0 is the one-sided noise spectral density, J is the received jammer power in watts, and the Nyquist filters have ideal bandwidth of the chip rate.

13. Starting with the two I and Q signals

$$I(t) = \sqrt{P}PN(t-T)d(t)\cos(\Delta\omega' t + \phi) + n_c(t) \tag{P6-11}$$

$$Q(t) = \sqrt{P}PN(t-T)d(t)\sin(\Delta\omega' t + \phi) - n_s(t) \tag{P6-12}$$

where the sum terms has been neglected since they will be filtered out by the low-pass filters following the multipliers in Figure 6.9-4, and neglecting the effect of the analog-to-digital (A/D) converter, show that the inphase T_{i0} second correlation can be modeled as

$$\mathsf{I}(k) = \frac{1}{T_{i0}}\int_{(k-1)T_{i0}}^{kT_{i0}} \left[\sqrt{P}d(t)PN(t-T)PN(t-\hat{T})\cos(\Delta\omega t + \phi) + n_c(t)PN(t-\hat{T})\right]dt \tag{P6-13}$$

with the assumption that the correlation starts at $t = 0$ ($k = 1$). It is to be noted that the signal component of $\mathsf{I}(k)$ is dependent on the difference of the actual and estimated delays $T-\hat{T}$. Let $\tau = T - \hat{T}$ be the timing error between the received code and the local matched filter code, and show that

$$\mathsf{I}_s(k) \cong d(k)\sqrt{P}R_{PN}(\tau)\text{sinc}\left(\Delta f T_{i0}\right)\cos\left[k\Delta\omega T_{i0} + \phi_k\right] \tag{P6-14}$$

where ϕ_k is the phase that results from the k-th integration including a constant phase term that is unimportant in this analysis, $R_{PN}(\tau)$ is the autocorrelation function of the spreading code, and $\text{sinc}(x) = \sin(\pi x)/(\pi x)$.
In the same manner show that the Q channel signal component can be described by

$$\mathsf{Q}_s(k) \cong d(k)\sqrt{P}R_{PN}(\tau)\text{sinc}\left(\Delta f T_{i0}\right)\sin\left[\Delta\omega T_{i0}/2 + \phi_k\right] \tag{P6-15}$$

Thus the correlation loss out of the first coherent combining matched filter output, when considering both the I and Q channels, is given by

$$L = R_{PN}^2(\tau)\left[\frac{\sin(\pi\Delta f T_{i0})^2}{(\pi\Delta f T_{i0})^2}\right] \tag{P6-16}$$

14. Consider the DMF. Starting with (6.9-59) and letting $z = y \times N_{nc}/(2\sigma^2)$ show that (6.9-69) results for the detection probability where $\gamma = N_{nc}\gamma_0$.

APPENDIX 6A

SIGNAL FLOW GRAPHS AND DISCRETE TIME INVARIANT MARKOV PROCESSES

In this appendix an introduction to signal flow graphs and their application to discrete time invariant Markov processes will be given. First signal flow graphs will be presented and then an introduction to discrete time invariant Markov processes will be discussed along with its relationship to flow graphs.

6A1.0 SIGNAL FLOW GRAPHS

A signal flow graph is a topological representation of the simultaneous equations describing a system function. Signal flow graphs were developed by Mason [6A1, 6A2] based on some preliminary work of Tustin [6A3]. Once a system is represented by a flow graph it becomes a relatively simple matter to determine the closed-loop transfer function. One natural application of flow graphs is to control theory. However many other applications exist also, as long as a transfer function is defined in the application. The material that follows is based on [6A1–6A7].

Denote variables by *nodes* (small dots) and consider the following equation:

$$x_i = T_{ij} x_j \tag{6A1.0-1}$$

where the variables x_i and x_j can be functions of a variable such as time or states, for example, in a state diagram. The nodes are connected by *directed branches* that represent the operator T_{ji} mapping x_i into x_j. These operators are called *transmission functions* (or branch transmittances). Figure 6A1.0-1 illustrates this equation.

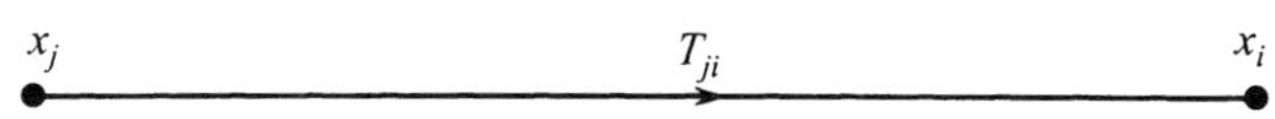

Figure 6A1.0-1 A simple signal flow graph with one branch.

Branches are always unidirectional, with the direction of the flow denoted by the arrowhead. Every variable in a flow graph is represented by a node and every transmission function by a branch. The value of the variable represented by a node is equal to the sum of all the signals entering the node. Let

$$x_i = \sum_{i=1}^{n} T_{ji} x_j \tag{6A1.0-2}$$

Consider the representation given by Figure 6A1.0-2.

As an example consider the two equations

$$y = 5x \quad w = 2y \tag{6A1.0-3}$$

The associated flow graph is shown in Figure 6A1.0-3.

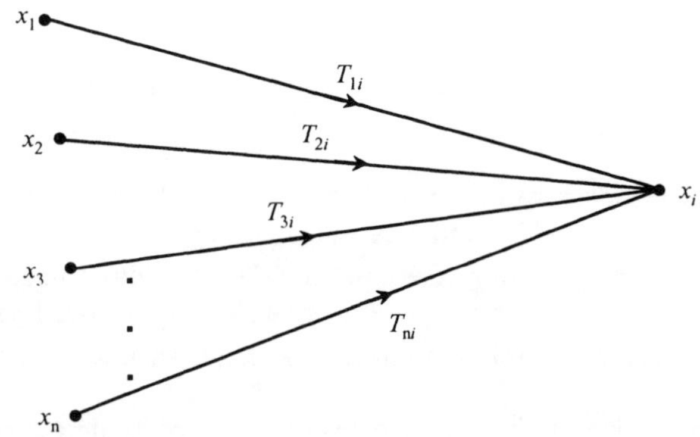

Figure 6A1.0-2 Pictorial representation of addition.

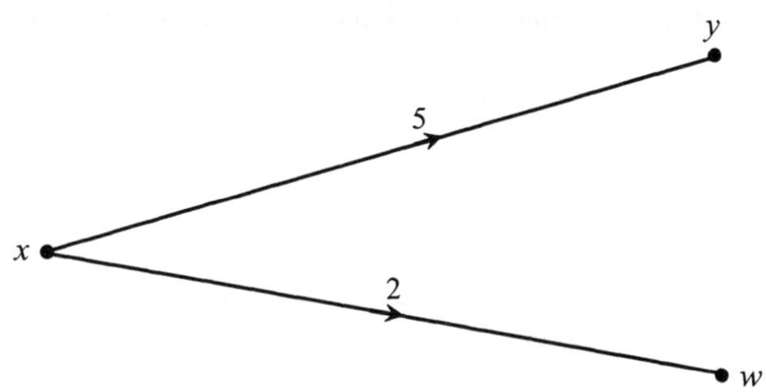

Figure 6A1.0-3 Signal flow graph for $y = 5x$ and $w = 2x$.

A series connection on n–1 branches and n nodes can be replaced by a single branch with a new transmission function equal to the product of the old original transmission functions. The proof is obvious.

6A1.1 SIGNAL FLOW GRAPH DEFINITIONS

Nodes and branches have already been defined as well as some elementary operations. Now we consider some definitions useful in signal flow graph theory and signal flow graph reduction.

An *open path* is a continuous, unidirectional succession of branches along which no node is passed more than once. In Figure 6A1.1-1 traversing from node x_2 to node x_3 and back to node x_2 is an example of an open path. Also x_1 to x_2 to x_4 to x_5 is another example, and in addition so is x_1 to x_2 to x_3 to x_4 to x_5. An *input node* (or source node) is a node with only outgoing branches. In Figure 6A1.1-1 x_1 is the only input node. An *output node* is a node with only incoming branches. x_5 is the only output node in Figure 6A1.1-1.

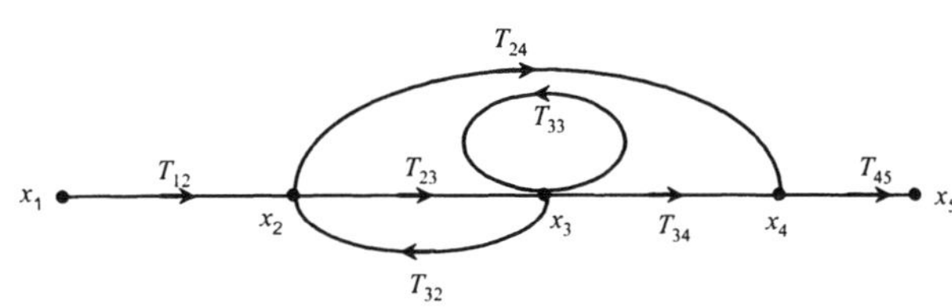

Figure 6A1.1-1 A signal flow graph with an open path, an input node and an output node.

Sometimes a variable in the system is actually the output variable. For example, in the classical feedback system the output is also the variable that is fed back to the feedback transfer function. In the flow graph for this example the output node would have an outgoing branch contrary to the definition of an output node. This dilemma can be solved by adding a branch with a unit transmission function entering a dummy node.

A *feedback loop* (or feedback path) is a path that starts and terminates on the very same node. In Figure 6A1.1-4 the path from x_2 to x_3 and back to x_2 is a feedback path.

A *forward path* is an open path from the input node to the output node. In our example shown in Figure 6A1.1-4 x_1 to x_2 to x_3 to x_4 to x_5 and x_1 to x_2 to x_4 to x_5 are the two forward paths of Figure 6A1.1-1.

The *gain of a branch* is the transmission function of that branch when the transmission function is a multiplicative operator.

The *loop gain* is the product of the branch gains of the loop. Again referring to Figure 6A1.1-1, we mention that the loop gain of the feedback from x_2 to x_3 and back to x_2 is $T_{23}\,T_{32}$.

The *path gain* is the product of the branch gains found in traversing a path. In Figure 6A1.1-1 the path gain of the forward path from x_1 to x_2 to x_3 to x_4 to x_5 is $T_{12}T_{23}T_{34}T_{45}$.

A *self-loop* is a feedback loop composed of a single branch. For example, in Figure 6A1.1-1 T_{33} is a self-loop.

One more definition is included in Section 6A1.3.

6A1.2 FLOW GRAPHS FROM BLOCK DIAGRAMS

Given a block diagram, a flow graph can easily be constructed by observing the following rule: Construct every variable of the block diagram as a node and each block diagram as a branch. For example, consider the basic control loop diagram shown in Figure 6A1.2-1. Notice that since the output variable C' was constructed using the branch gain of 1, C' is a proper output node, since it only has inputs and no outputs. More general block diagrams or systems of equations can be handled in this manner.

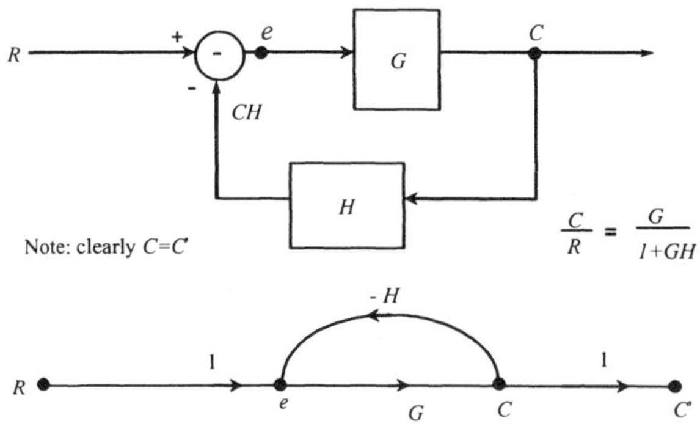

Figure 6A1.2-1 Block diagram and the associated signal flow diagram.

6A1.3 SIGNAL FLOW GRAPH REDUCTION AND MASON'S GAIN FORMULA

In the study of flow graphs it is desirable to reduce the original signal flow graph to two nodes and one branch—in particular, to the input and output nodes and the overall gain. There are two ways to achieve this goal. First, one can use reduction methods; second, one can use Mason's gain formula. Although reduction techniques will not be described in detail here, Table 6A1.3-1 illustrates some equivalents that are useful in flow graph reduction.

Table 6A1.3-1 Some Equivalent Flow Graphs

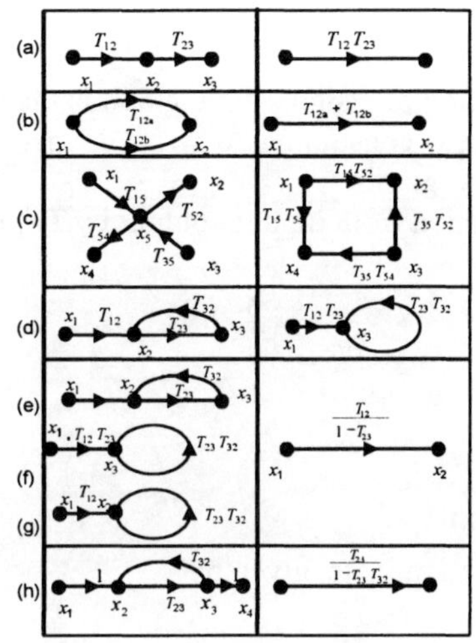

An example of a *nontouching loop* is a set of loops that have no nodes in common. For an example, see Figure 6A1.3-1.

In Table 6A1.3-1, part (a), for example, we have

$$x_2 = T_{12}x_1$$
$$x_3 = T_{23}x_2 = T_{12}x_1 \tag{6A1.3-1}$$

For the parallel paths of Table 6A1.3-1 part (b), we have

$$x_2 = (T_{12a} + T_{12b})x_1 \tag{6A1.3-2}$$

In part (c) of the table, node x_5 is eliminated by the following equations:

$$x_2 = T_{52}x_5$$
$$x_4 = T_{54}x_5 \tag{6A1.3-3}$$
$$x_5 = T_{15}x_1 + T_{35}x_3$$

Eliminating the x_5 variable lead to the relationships:

$$x_2 = T_{52}T_{15}x_1 + T_{35}T_{52}x_3$$
$$x_4 = T_{15}T_{54}x_1 + T_{35}T_{54}x_3 \tag{6A1.3-4}$$

which is shown as the equivalent in the table, part (c). This is an example of node elimination. The equivalence of Table 6A1.3-1, parts (d), (f), and (h), are left as an exercise for the reader. To obtain the equivalence of (g), write from (e)

$$x_2 = T_{12}x_1 + T_{32}x_3$$
$$x_3 = T_{23}x_2 \tag{6A1.3-5}$$

Now eliminate x_3 to get

$$x_2 = T_{12}x_1 + T_{32}T_{23}x_2 \tag{6A1.3-6}$$

or

$$x_2 = \frac{T_{12}}{1 - T_{23}T_{32}} x_1 \quad (6A1.3\text{-}7)$$

The second method of reduction of a signal flow graph is the use of Mason's gain formula. In many cases it is possible and much less time consuming to write down the input-output relationship by a few simple rules applied directly to the flow graph.

If we denote the transfer function from input to output by TF, then *Mason's gain formula* states that

$$TF = \frac{\sum_i T_i \Delta_i}{\Delta_g} \quad (6A1.3\text{-}8)$$

where

T_i is the i-th forward path gain.

Δ_g is the signal flow graph determinate, given by

$$\Delta_g = 1 - \sum_i T_i^1 + \sum_i T_i^2 - \sum_i T_i^3 + ... \quad (6A1.3\text{-}9)$$

$\sum_i T_i^1$ is the sum of all feedback loop gains.

$\sum_i T_i^2$ is the sum of all gain products of two nontouching feedback loops. Nontouching loops have neither branch nor node in common. In general

$\sum_i T_i^n$ is the sum of all gain products of n nontouching feedback loops.

Δ_i is Δ_g evaluated with all feedback loops touching the i-th forward path eliminated (or delete all branches of the i-th forward path).

Although the formula appears formidable, it is actually quite simple to use. Consider the example shown in Figure 6A1.3-1. There is only one path from input to output, hence

$$T_1 = T_{12}T_{23}T_{34}T_{45} \quad (6A1.3\text{-}10)$$

There are three feedback loops:

$$\begin{aligned} &x_3 \to x_2 \to x_3 \qquad T_1^1 = T_{23}T_{32} \\ &x_4 \to x_3 \to x_4 \qquad T_2^1 = T_{34}T_{43} \\ &x_4 \to x_5 \to x_4 \qquad T_3^1 = T_{45}T_{54} \end{aligned} \quad (6A1.3\text{-}11)$$

Hence $\sum_{i=1}^{3} T_1^i = T_{23}T_{32} + T_{34}T_{43} + T_{45}T_{54}$.

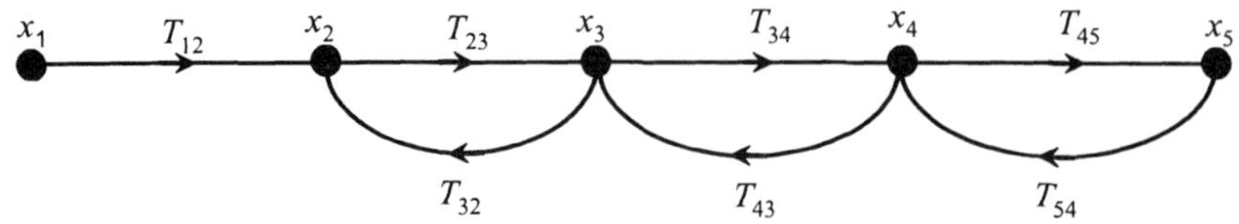

Figure 6A1.3-1 Signal flow graph example.

There are two nontouching feedback loops:

$$x_2 \to x_3 \to x_2 \qquad x_4 \to x_5 \to x_4 \tag{6A1.3-12}$$

since all three feedback loops touch the path from input to output. Therefore it follows that $\sum_{i=1}^{1} T_i^2 = T_{23}T_{32}T_{45}T_{54}$ and $T_2 = 0$ and $T_3 = 0$. Now the signal flow graph determinate is given from (6A1.3-9) by

$$\Delta_g = 1 - (T_{23}T_{32} + T_{34}T_{43} + T_{45}T_{54}) + T_{23}T_{32}T_{45}T_{54} \tag{6A1.3-13}$$

Therefore, since $\Delta_1 = 1$, one obtains the result for the formula for the transfer function

$$TF = \frac{T_{12}T_{23}T_{34}T_{45}}{1 - T_{23}T_{32} - T_{34}T_{43} - T_{45}T_{54} + T_{23}T_{32}T_{45}T_{54}} \tag{6A1.3-14}$$

6A2.0 DISCRETE TIME INVARIANT MARKOV PROCESSES AND FLOW GRAPHS

Although discrete Markov processes have been characterized by many techniques such as differential equations or matrix equations, it will be convenient in this section to describe them by state transition diagrams. From these diagrams, what we shall call *generating function flow graphs* will be established and will lead to the generating function of the process [6A5–6A7] under consideration. This section is based primarily on [6A5–6A7].

Markov processes with a finite (or countable) number of states and time invariant transition probabilities can be described with the aid of a state transition diagram. The state transition diagram is characterized by states indicated by dots that are connected by directed lines that indicate the probability of going from the originating dot to the terminating dot.

Consider as an example a two-state time invariant discrete Markov process that is characterized by two states:

$$\begin{aligned} \text{in state 1}: p(S_2|S_1) &= 3/8 \\ p(S_1|S_1) &= 5/8 \\ \text{in state 2}: p(S_1|S_2) &= 1/2 \\ p(S_2|S_2) &= 1/2 \end{aligned} \tag{6A2.0-1}$$

where S_2 denotes state 2 and S_1 denotes state 1. The state transition diagram is shown in Figure 6A2.0-1.

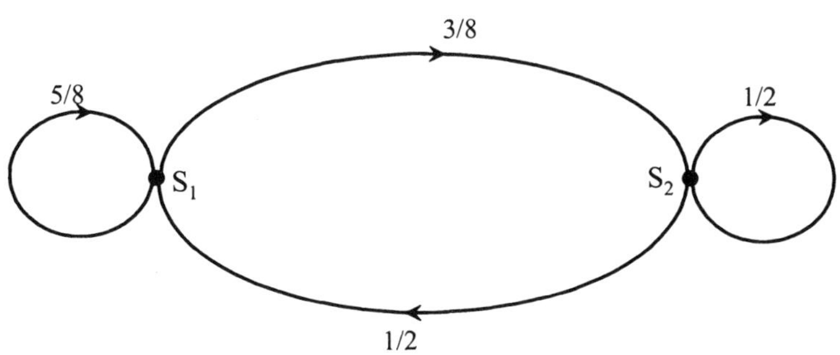

Figure 6A2.0-1 State transition diagram for the example.

Suppose that at time $t = n$ the process is in state 1; then at time $t = n+1$ the process will be in state 1 with probability 5/8 and in state 2 with probability 3/8. A similar relationship holds if we start at state 2, except the probabilities are 1/2 and 1/2.

It is to be noted that we may write down the difference equations that relate the probabilities of being in each state. Let $p_1(n)$ and $p_2(n)$ be the probabilities of being in states 1 and 2, respectively, at $t = n$. Then one time unit later the probabilities are

$$\begin{aligned} p_1(n+1) &= \frac{5}{8}p_1(n) + \frac{1}{2}p_2(n) \\ p_2(n+1) &= \frac{3}{8}p_1(n) + \frac{1}{2}p_2(n) \end{aligned} \tag{6A2.0-2}$$

Since the process is determined by a set of linear, constant coefficient, difference equations, one wonders whether a linear system technique such as flow graphs could be used to obtain the desired probabilities. The answer is yes, as we shall see shortly, and they are used to great advantage.

It is convenient to view the probabilities $p_1(n)$ and $p_2(n)$ as electrical signals and note that the difference equation shows that these signals are devised as linear combinations of themselves after a unit time delay. Hence we can associate with Figure 6A2.0-1 an associated linear system flow graph shown in Figure 6A2.0-2.

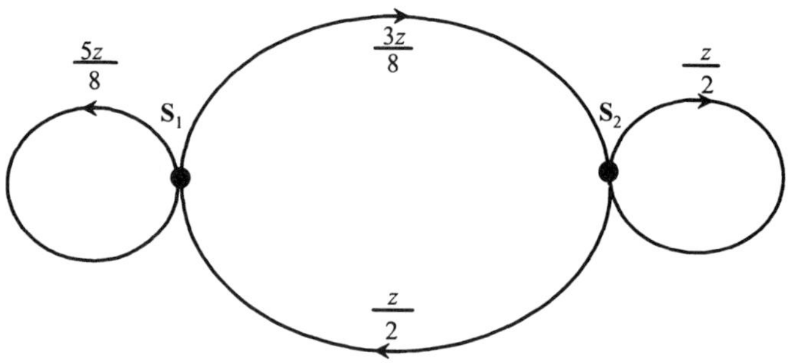

Figure 6A2.0-2 Associated linear system (generating function) flow graph 6A2.0-1.

The signal at each node is the probability $p_1(n)$ and $p_2(n)$, respectively. The presence of the z variable on the branches, with the transition probability gains, denote the delay of one unit time in going from one node to the next. The reason this diagram leads to the generating function will be demonstrated shortly. Define $p_{ij}(n)$ as the (time invariant) probability of going from state i to state j in n units of time; this is the n-step transition probability. For convenience usually $p_{ij}(1)$ is written as p_{ij}.

As is usual with linear systems, operations in the transform domain are most convenient. We introduce the z transform, also called the generating function, or the geometric transform, of the discrete probability function $p_{ij}(n)$ as

$$P_{ij}(z) = \sum_{n=0}^{\infty} z^n p_{ij}(n) \tag{6A2.0-3}$$

This series converges for all values of z inside the unit circle. Note that

$$\frac{dP_{ij}(1)}{dz} = \sum_{n=0}^{\infty} np_{ij}(n) = \overline{T}_{ij} \tag{6A2.0-4}$$

where $\overline{T}_{ij}$ is the mean time to go from state i to state j. It is left to the reader to show that the variance of the time to go from state i to state j is given by

$$Var(T_{ij}) = \frac{d^2P_{ij}(1)}{dz^2} + \frac{dP_{ij}(1)}{dz} - \left[\frac{dP_{ij}(1)}{dz}\right]^2 \tag{6A2.0-5}$$

Hence it is seen that $P_{ij}(z)$ is the generating function of the transition process. In addition to the moments, it is possible to find the probability $p_{ij}(n)$ by the following formula:

$$p_{ij}(n) = \frac{1}{n!}\frac{d^nP_{ij}(0)}{dz^n} \tag{6A2.0-6}$$

Consider now using the generating function flow graph of Figure 6A2.0-2 in the previous example. Suppose that we are given that the system is in state 1 at $t = 0$, and we desire to obtain the probability of starting in state 1 at $t = 0$ and arriving in state 1 at $t = n$. The approach to the problem in this section is to view it as a "system," with the input composed of a unit sample at $t = 0$ to node 1 and the output a tap off of node 1. We can accomplish our end by determining the transfer function $P_{11}(z)$, which is the z transform of the required probability signal. Given $P_{11}(z)$ it is a simple matter to use (6A2.0-6) to determine the individual transition probabilities.

We note in general that with arbitrary initial values we apply to each node, i, of the system at $t = 0$, samples equal to the prescribed initial values $p_i(0)$. Using the superposition property of linear systems, such a problem can be solved by adding linear sums of simple node to node responses. Hence if $P_{ij}(z)$ is the transfer function from node i to node j, then the signal at node j, in the general case, is given by the linear sum

$$P_j(z) = \sum_i p_i(0)P_{ij}(z) \tag{6A2.0-7}$$

In our example let $p_1(0) = 1$, $p_2(0) = 0$, and $P_1(z) = P_{11}(z)$. Many problems have the characteristic that, say, $p_1(0) = 1$ and $p_i(0) = 0$ for $i \neq 1$. Continuing with the example, we obtain the generating function flow graph of Figure 6A2.0-3.

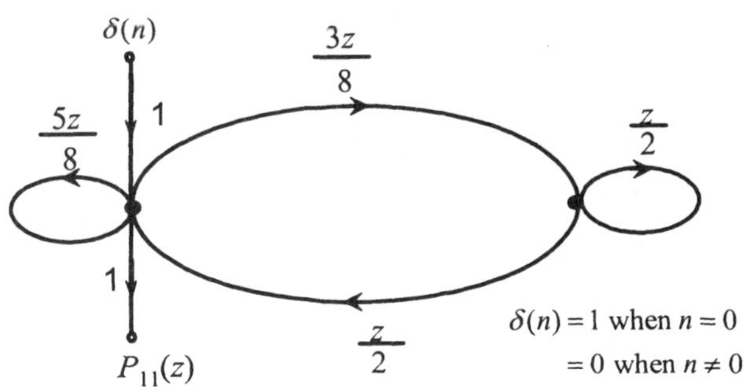

Figure 6A2.0-3 Generating function flow graph for the example following Problem 6A.3.

Using Mason's gain formula, we find that

$$T_1 = 1 \quad \Delta_1 = 1 - \frac{z}{2} \quad \Delta_g = 1 - \frac{5z}{8} - \frac{z}{2} - \frac{3z^2}{16} + \frac{5z^2}{16} \tag{6A2.0-8}$$

so that Mason's gain formula produces the transfer function from state 1 to state 1

$$P_{11}(z) = \frac{1 - \frac{z}{2}}{1 - \frac{9}{8}z + \frac{1}{8}z^2} \tag{6A2.0-9}$$

Note that $P_{11}(0)$ is 1 as it should be, since the probability of going from state 1 to state 1 at time 0 is one. It is to be pointed out that if the Markov process has a finite number of states, then the resulting transfer function will be a rational function.

To obtain $p_{ij}(n)$ we may (1) differentiate $P_{11}(z)$, (2) use long division to obtain the series, or (3) use partial fractions to obtain a separate series. We shall utilize the latter technique in what follows. Factoring (6A2.0-9) produces

$$P_{11}(z) = \frac{1 - \frac{z}{2}}{(1 - z)(1 - \frac{z}{8})} \tag{6A2.0-10}$$

Expanding in partial fractions produces

$$P_{11}(z) = \frac{4/7}{1 - z} + \frac{3/7}{1 - z/8} \tag{6A2.0-11}$$

so that

$$P_{11}(z) = \frac{4}{7}(1 + z + z^2 + z^3 + \ldots) + \frac{3}{7}(1 + \frac{z}{8} + \frac{z^2}{8^2} + \frac{z^3}{8^3} + \ldots) \tag{6A2.0-12}$$

Hence we see by using (6A2.0-6) that

$$p_{11}(n) = \frac{4}{7} + \frac{3}{7}\left(\frac{1}{8}\right)^n \quad n \geq 0 \tag{6A2.0-13}$$

As another example illustrating the techniques, consider a simple random walk problem with absorbing boundaries. Suppose a particle has probability of 1/2 of moving to the right or the left. The process ends when the particle reaches either boundary located two steps away. One possible problem is to find the probability that the process ends in n steps, and another possible problem is to find the mean time to end the process. We assume that we start in state 0. The generating function flow graph is shown in Figure 6A2.0-5. Since we start at state (node) zero, we apply a unit sample at $t = 0$ to our "system." Node 3 corresponds to the event the process terminates and hence is our required output point to determine $P_{03}(z)$, the generating function for this absorbing random walk problem.

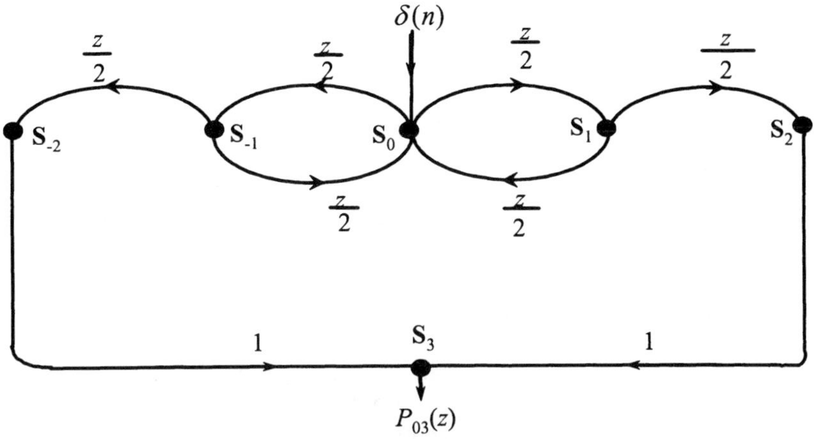

Figure 6A2.0-5 Generating function flow graph for particle absorption.

Using Mason's gain formula we find that

$$T_1 = \frac{1}{4}z^2 \qquad \Delta_1 = 1 \qquad T_2 = \frac{1}{4}z^2$$
$$\Delta_2 = 1 \qquad \Delta_g = 1 - \frac{1}{4}z^2 - \frac{1}{4}z^2 \tag{6A2.0-14}$$

so that

$$P_{03}(z) = \frac{\frac{1}{2}z^2}{1 - \frac{1}{2}z^2} \tag{6A2.0-15}$$

writing the denominator as a power series, we have

$$P_{03}(z) = \frac{1}{2}z^2 - (\frac{1}{2}z^2)^2 + (\frac{1}{2}z^2)^3 - \ldots \tag{6A2.0-16}$$

Hence the probability of ending on the n-th toss is given by (using (6A2.0-6))

$$p_{03}(n) = \left(\frac{1}{2}\right)^{n/2} \quad n = 2,4,6\ldots$$
$$0 \quad \text{otherwise} \tag{6A2.0-17}$$

Note that $p_{03}(n)$ is the probability of being in the final state, starting at state 0, at time n. Now the mean time for the process to terminate is given by

$$\overline{T_{03}} = \frac{dP_{03}(z)}{dz}\Big|_{z=1} = \frac{1}{(1 - \frac{1}{2}z^2)^2}\Big|_{z=1} = 4 \tag{6A2.0-18}$$

Therefore the process terminates in 4 steps on average. The standard deviation of the time to terminate the process is, from (6A2.0-5)

$$\sigma_{T_{03}} = \sqrt{\frac{d^2 P_{03}(1)}{dz^2} + \overline{T_{03}} - (\overline{T_{03}})^2} = 2 \tag{6A2.0-19}$$

For further examples and applications of the flow graph technique, [6A6-6A8] should be consulted.

6A3.0 Mathematical Basis for Generating Function Flow Graph Techniques

We have seen many examples of the generating function flow graph technique for solving time invariant, finite state Markov chain problems. Here we present a proof establishing the transition from a state transition diagram to a generating function flow graph. This section is based on [6A5].

Recall that we defined $p_{ij}(n)$ as the probability of going from state i to state j in n steps. It is shown in any text [6A9] on Markov chains that the following recursion formula

$$p_{ij}(n+1) = \sum_{k=1}^{\infty} p_{ik}(n) p_{kj} \quad 1 \le i, j \le N, \quad n = 0,1,2... \tag{6A3.0-1}$$

is satisfied where N is the number of states. Denote $p_{kj}(1) = p_{kj}$ for convenience. Multiplying this equation by z^n and summing over n, produces the result

$$\frac{1}{z}\left[P_{ij}(z) - p_{ij}(0)\right] = \sum_{k=1}^{N} p_{kj} P_{ik}(z) \tag{6A3.0-2}$$

where $P_{ij}(z)$ is the generating function and is defined by (6A2.0-6) since $p_{ij}(0) = \delta_{ij}$ (it allows the recursion relation to hold at $n = 0$), where δ_{ij} is the Kronecker δ function and is defined by

$$\begin{aligned} \delta_{ij} &= 1, \quad i = j \\ &\ 0 \quad i \neq j \end{aligned} \tag{6A3.0-3}$$

Solving for $P_{ij}(z)$ from (6A3.0-2), we obtain

$$P_{ij}(z) = \sum_{k=1}^{\infty} P_{ik} p_{kj} z + \delta_{ij} \qquad 1 \le i, j \le N \tag{6A3.0-4}$$

This relationship is the set of equations that describes the N-node flow graph of Figure 6A3.0-1. This flow graph is just the state transition diagram of the Markov chain with each of its branches labeled with $p_{ij}(n)z$ instead of $p_{ij}(n)$. The generating function $P_{ij}(z)$ (z transform) is the transformed "signal" at node j when a unit input is applied only to node i; hence the δ_{ij} term applied at time $n = 0$. Since the desired transform $P_{ij}(z)$ requires an input at node i and the output taken at node j, we see that the flow graph for the N-state Markov chain and our generating function flow graph are one and the same. Therefore the generating flow graph technique represents precisely the same Markov chain as the state transition diagram.

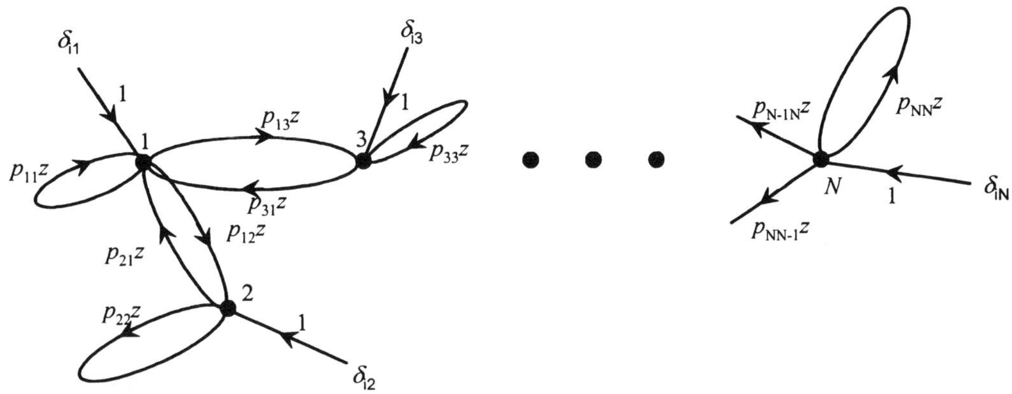

Figure 6A3.0-1 Generating function flow graph of an n-state Markov chain.

Appendix 6A Problems

6A.1 Show the equivalence in the two signal flow graphs in parts (d), (e), and (f) of Table 6A1.3-1.

6A.2 Using Mason's gain formula show that the transfer function for the flow graph shown here is given by

$$TF = \frac{G_1G_2}{1 - G_1H_1 - G_1G_2H_2 - G_2G_3} \tag{6AP.1}$$

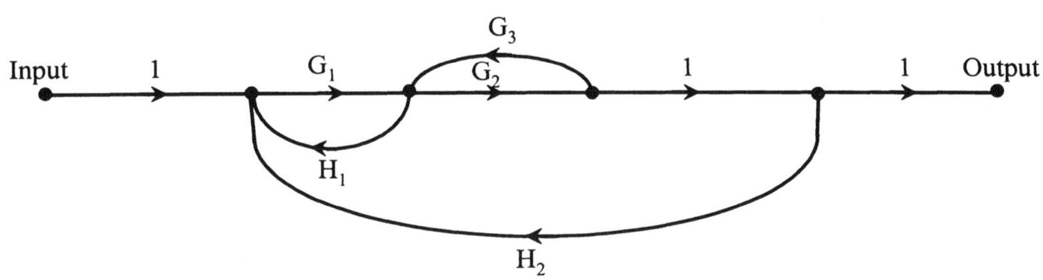

Figure 6A6.1 The figure for Problem 6A.2.

6A.3 Show that (6A2.0-5) is true; that is,

$$Var(T_{ij}) = \frac{d^2P_{ij}(1)}{dz^2} + \frac{dP_{ij}(1)}{dz} - \left[\frac{dP_{ij}(1)}{dz}\right]^2 \tag{6AP.2}$$

6A.4 Find $p_{ij}(n)$ for the model described by the state transition diagram here using the techniques of this section. Show that

$$p_{11}(n) = \frac{2}{3} + \frac{1}{3}\left(\frac{1}{4}\right)^n \quad n \geq 0 \tag{6AP.3}$$

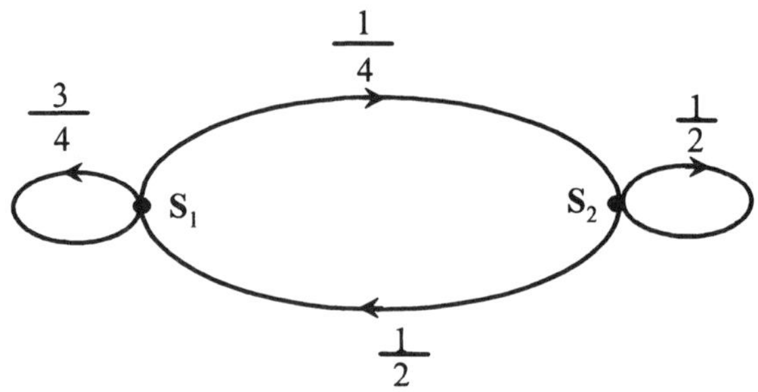

Figure 6A6.2 The figure for Problem 6A.4.

6A.5 Consider the Markov process to define the state transition diagram shown here. State 5 is an absorption state. Show that the mean time to be absorbed is nine units of time.

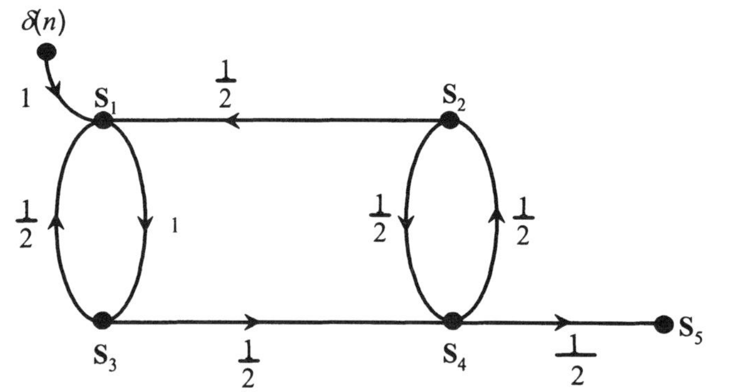

Figure 6A6.3 The figure for Problem 6A.5.

6A.6 Consider a fair coin having a head (H) and a tail (T), each having probability of 1/2 of occurrence.
(a) Show that generating function flow graph for the event, "obtaining a head (H) for the first time on the n-th toss," is given by

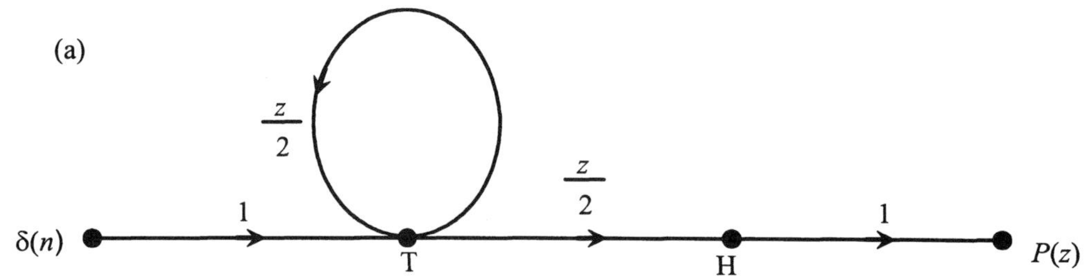

Figure 6A6.4 The figure for Problem 6A.6 (a)

so that the desired probability is $(1/2)^n$.

(b) Show that the generating function flow graph for the event "obtaining the third head on the n-th toss of a coin," is given by

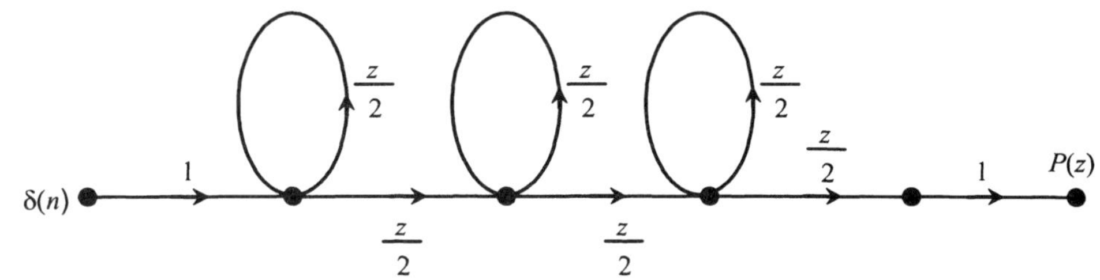

Figure 6A6.5 The figure for Problem 6A.6 (b)

so that

$$P(z) = \left(\frac{z/2}{1 - z/2} \right)^3 \quad (6AP.6)$$

and therefore the desired probability is given by $(n-1)(n-2)(1/2)^{n+1}$, $(n \geq 3)$.

6A.7 Show that (6A3.0-2) follows from (6A3.0-1).

Appendix 6A References

[6A1] S. J. Mason, "Feedback Theory-Some Properties of Signal Flow Graphs," *Proc. IEEE*, Vol. 41, No. 9, September 1953.

[6A2] S. J. Mason, "Feedback Theory-Further Properties of Signal Flow Graphs," *Proc. IEEE*, Vol. 44, No. 7, July 1956.

[6A3] A. Tustin, *Direct Current Machines for Control Systems*, New York: Macmillan, 1955.

[6A4] J. K. Holmes, *Coherent Spread Spectrum Systems*, Malabar, FL: Robert E. Krieger Publishing Co., 1990.

[6A5] A. Di Stefano, A. Stubberud, and S. Williams, *Feedback and Control Systems,* Schaums Outline Series, New York: McGraw-Hill, 1967.

[6A6] R. Howard, Class Notes for Industrial Engineering, MIT, 1963.

[6A7] R. W. Sittler, "Systems Analysis for Discrete Markov Processes," *IRE Trans. Circuit Theory*, Vol. CT-3, December 1956.

[6A8] W. H. Huggins, "Signal Flow Graphs and Random Signals," *Proc. IRE*, January 1957.

[6A9] R. Howard, *Dynamic Programming and Markov Processes,* New York: Technology Press and Wiley, 1960.

[6A10] W. Feller, *An Introduction to Probability Theory and Its Applications,* 2nd ed., Vol. 1, New York: Wiley, 1950.

Selected Bibliography

A. Di Stefano, A. Stubberud, and S. Williams, *Feedback and Control Systems,* Schaums Outline Series, New York: McGraw-Hill, 1967.

CHAPTER 7

Direct Sequence Code-Tracking Loops

7.0 INTRODUCTION

The intent of this chapter is to introduce the concept of a spread spectrum code-tracking loop. Its function is to track the pseudonoise (PN) code of the received signal by the receiver. Our goal in this chapter is to present the performance for the various types of code-tracking loops. There are four basic types of performance that are addressed in this chapter: (1) steady state code-tracking variance of the tracking error, (2) transient pull-in performance, (3) multipath range of tracking errors, and (4) loss of lock performance.

The first parameter, the tracking error variance, is only meaningful for the linearized model, since the concept of steady state does not exist for real models. That is to say that "real loops" will eventually lose lock, if one waits long enough. However, it is convenient to consider the linearized model in the case that the loop SNR is sufficiently high so that practically it never loses lock.

The second type of performance is related to the process after code acquisition has been completed and code-tracking ensues. At the initial code acquisition point the code error may be a substantial fraction of a code symbol (perhaps ½ symbol); however, when the code loop is turned on it will tend to drive the mean code-tracking error to near zero. Thus a transient code-tracking error typically follows the initial code acquisition. After the mean code error returns to zero or some other value, if there are relative dynamics between the transmitter and the receiver, the code loop operates in steady state.

The third facet is related to the range of steady state errors the code loop will endure when receiving both the desired signal and a multipath version of the desired signal, which commonly has smaller amplitude and a different carrier phase.

The fourth parameter deals with the fact that code-tracking loops, if one waits long enough, will eventually lose lock. If eventually means 50 years, no one probably cares. The parameter that is often used is the mean time to lose lock. Obviously this depends on the signal-to-noise density ratio, as well as the loop bandwidth and the loop configuration. The larger the mean time to lose lock, the better.

Figure 7.0-1 illustrates the sequence of events that must occur in order for a coherent BPSK receiver to be synchronized. This includes code acquisition, code pull-in, carrier acquisition bit synchronization, data demodulation, and data decoding (if used).

In this chapter we will consider multiple symbol format types, and will not be restricted to nonreturn to zero formatting for the code symbols. Code symbol formatting will be discussed in more detail later in this chapter.

7.1 BASIS FOR THE EARLY-LATE GATE CODE-TRACKING LOOP

Consider now the optimum estimate [1] of the timing of a baseband code sequence with data on it. It will be assumed that the received signal is immersed in thermal noise (white Gaussian noise (WGN)) so that the received signal plus noise is given by

$$r(t) = s(t,\tau) + n(t) \tag{7.1-1}$$

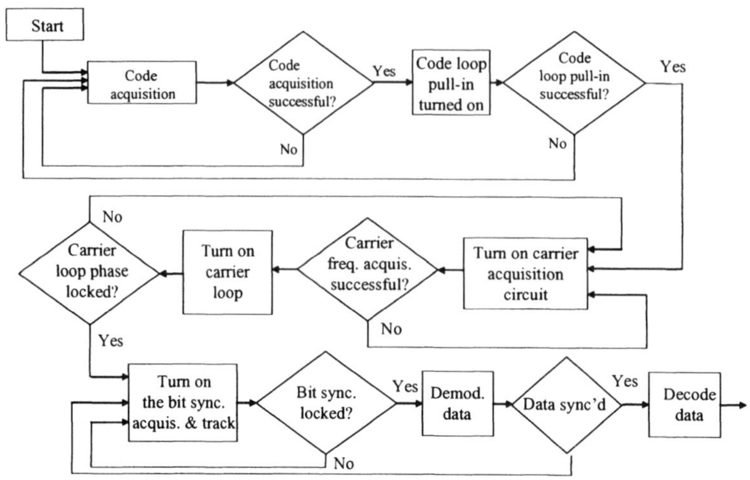

Figure 7.0-1 Logic flow for direct sequence acquisition, tracking, and data acquisition.

where $s(t, \tau)$ is the PN code signal and τ is the timing to be estimated. The maximum-likelihood criterion assumes that the timing parameter, τ, is unknown but deterministic.

7.1.1 Maximum-Likelihood Estimate Formulation

Consider an orthonormal expansion of $r(t)$ by the orthonormal functions $f_n(t)$ that have the property

$$\int_{T_0} f_n(t) f_m(t) dt = \delta_{nm} = \begin{cases} 1 & n = m \\ 0 & n \neq m \end{cases} \tag{7.1-2}$$

Expanding $r(t)$ in this orthonormal expansion produces

$$r(t) = \sum_{i=1}^{\infty} r_n f_n(t) \tag{7.1-3}$$

where

$$r_n = \int_{T_0} r(t) f_n(t) dt \tag{7.1-4}$$

The vector $\boldsymbol{r}$ is composed of an infinite number of components r_n. The signal can also be expanded in terms of orthogonal components in the form

$$s(t,\tau) = \sum_{i=1}^{\infty} s_i(\tau) f_i(t) \tag{7.1-5}$$

where

$$s_i(\tau) = \int_{T_0} s(t,\tau) f_i(t) dt \tag{7.1-6}$$

It is assumed that the noise is white Gaussian noise, and it can be shown [1] that the noise components are statistically independent from component to component. Thus one can write for the likelihood function, for the first N components

$$\Lambda(\tau) = p(\mathbf{r}|\tau) = \left(\frac{1}{\sqrt{2\pi}\sigma}\right)^N \exp\left(-\sum_{i=1}^{N}\frac{[r_i - s_i(\tau)]^2}{2\sigma^2}\right) \tag{7.1-7}$$

In the limit it can be shown [1] that

$$\Lambda(\tau) = \exp\left[\frac{1}{N_0 T_0}\int_{T_0}[r(t) - s(t,\tau)]^2 dt\right] \tag{7.1-8}$$

7.1.2 Maximum-Likelihood Estimate of the PN Code Timing

We are now in a position to formulate the optimum pseudonoise (PN) direct sequence code timing estimate operating at baseband having a data rate of $R_b = 1/T_b$ bits per second in which the code repeats every T_b seconds. Thus the signal can be written as

$$s(t,\tau) = PN(t-\tau) \tag{7.1-9}$$

where $PN(t)$ is a sequence of M PN code chips over the time T_b seconds. Hence $PN(t)$ is given by

$$PN(t) = \sum_{j=1}^{M} a_j p(t-(j-1)T_c) \tag{7.1-10}$$

such that $T_b = MT_c$ and $a_j = \pm 1$, $p(t)$, random from chip to chip, is a unit amplitude pulse that is the unit value for 0 to T_c seconds and zero otherwise. Equation (7.1-8) can be written more simply as

$$\Lambda(\tau) = \exp\left[\frac{C}{2}\int_{T_0}[r(t) - s(t,\tau)]^2 dt\right] \tag{7.1-11}$$

Expanding (7.1-11) produces

$$\Lambda(\tau) = \exp\left[-\frac{C}{2}\int_{T_b} r(t)^2 dt + C\int_{T_b} r(t)PN(t-\tau)dt - \frac{C}{2}\int_{T_b} PN(t-\tau)^2 dt\right] \tag{7.1-12}$$

We note that the first term does not depend on τ, and the third term is also independent of τ. Thus it is sufficient to maximize the second term since it is the only one dependent on the parameter τ. Since the "a_n" are random variables, it is necessary to average over the distribution of a_n. Hence the modified likelihood function is proportional to

$$\Lambda'(\tau) = K\exp\left[C\int_{T_b} r(t)PN(t-\tau)dt\right] \tag{7.1-13}$$

The probability function of $p(a_j)$ is given by

$$q(a_j) = \frac{1}{2}\delta(a_j - 1) + \frac{1}{2}\delta(a_j + 1) \tag{7.1-14}$$

Hence averaging over the log-likelihood $p(a)$ yields the average likelihood function

$$\bar{\Lambda}'(\tau)=\exp\left[C\int_{T_b}\sum_{j=1}^{M}r(t)a_j p(t-(j-1)T_c-\tau)\right]q(\bar{a})d\bar{a} \tag{7.1-15}$$

where $q(\bar{a})=q(a_1)q(a_2)...q(a_M)$ and $d\bar{a}=da_1 da_2...da_M$. Let

$$PN_j(\tau)=\int_{T_b}r(t)p(t-(j-1)T_c-\tau)dt \tag{7.1-16}$$

Then factor (7.1-15) as a product and use (7.1-16) to achieve

$$\bar{\Lambda}'(\tau)=\int_{T_b}\exp\left[C\sum_{j=1}^{M}a_j PN_j(\tau)\right]q(\bar{a})d\bar{a}=\prod_{j=1}^{M}\int_{-1.1}^{1.1}\exp\left[Ca_j PN_j(\tau)\right]q(a_j)da_j \tag{7.1-17}$$

Evaluating (7.1-17) yields

$$\bar{\Lambda}'(\tau)=\prod_{j=1}^{M}\left\{\frac{\left(\exp\left[CPN_j\right]+\exp\left[-CPN_j\right]\right)}{2}\right\}=\prod_{j=1}^{M}\cosh\left[CPN_j\right] \tag{7.1-18}$$

Now consider the log-likelihood function (actually the natural log, denoted by "ln") denoted here by $\bar{\Lambda}_L(\tau)$, where

$$\bar{\Lambda}_L(\tau)=\ln\left\{\bar{\Lambda}'(\tau)\right\} \tag{7.1-19}$$

Thus taking the natural log of (7.1-18) produces

$$\bar{\Lambda}_L(\tau)=\sum_{j=1}^{M}\ln\cosh\left[CPN_j(\tau)\right] \tag{7.1-20}$$

Now at small values of the argument we may approximate the lncosh(x) function by $x^2/2$, so that

$$\bar{\Lambda}_L(\tau)=\frac{1}{2}C^2\sum_{j=1}^{M}\left(PN_j\right)^2 \tag{7.1-21}$$

where

$$PN_j(\tau)=\int_{T_b}r(t)p(t-(j-1)T_c-\tau)dt \tag{7.1-22}$$

To obtain the estimate of τ, consider taking the approximate derivative of $\bar{\Lambda}_L(\tau)$ via

$$\frac{d\bar{\Lambda}_L(\tau)}{d\tau}\cong\frac{\bar{\Lambda}_L(\tau+\delta T_c)-\bar{\Lambda}_L(\tau-\delta T_c)}{2\delta T_c} \tag{7.1-23}$$

Hence using (7.1-21) in (7.1-23) produces the result (letting $C^2/2=1$)

$$\frac{d\bar{\Lambda}_L(\tau)}{d\tau} \cong \frac{1}{2\delta T_c}\left[\sum_n \left(PN_n(\tau+\delta T_c)\right)^2 - \sum_n \left(PN_n(\tau-\delta T_c)\right)^2\right] \tag{7.1-24}$$

Or using (7.1-22) produces

$$\frac{d\bar{\Lambda}_L(\tau)}{d\tau} \cong \frac{1}{2\delta T_c}\left[\begin{array}{l}\sum_n \left(\int_{T_b} r(t)PN(t-nT_b-\tau-\delta T_c)dt\right)^2 \\ -\sum_n \left(\int_{T_b} r(t)PN(t-nT_b-\tau+\delta T_c)dt\right)^2\end{array}\right] \tag{7.1-25}$$

This equation can be implemented by the following closed-loop structure illustrated in Figure 7.1-1, ignoring the multiplicative constant in front of the integrals.

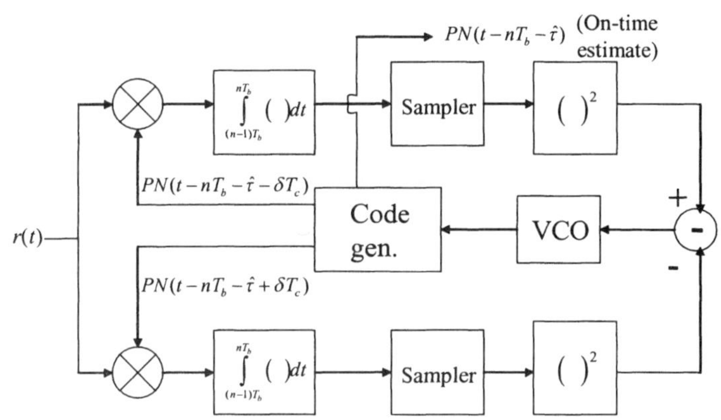

Figure 7.1-1 Noncoherent code-tracking loop configuration derived from the maximum-likelihood estimate of the timing delay τ.

7.2 FULL-TIME CODE-TRACKING LOOPS

First we will treat full-time code-tracking loops in which both the early and late correlators are processed on a full time basis. Then we will deal with *time-shared correlator* code-tracking loops in which the early and late correlators are processed sequentially (i.e., time-shared). When the carrier frequency and phase are known completely, the code loops are known as *coherent code-tracking loops.* When the frequency is known but the carrier phase is unknown *noncoherent code-tracking loops* are required, since they do not need to have knowledge of the carrier phase to function. All these models assume that no filtering effects have been accounted for. Following the unfiltered models we will address the case on a noncoherent code-tracking loop in which filtering and interference is addressed. All analysis is based on the assumption that $B_L T << 0$, where B_L is the one-sided closed-loop noise bandwidth in Hz (defined later) and the symbol or bit time, if uncoded (T seconds), of the data bearing signals. In other words the loop update time is T seconds. If this condition is not upheld the tracking error variance will be larger than predicted here, and in the extreme case the loop could become unstable.

7.2.1 Baseband Early-Late Gate Code-Tracking Loop with NRZ Symbols

The baseband code-tracking loop is designed to track baseband direct sequence PN nonreturn to zero (NRZ) signals. Applications would include high-speed wireless applications and direct sequence CDMA systems. Baseband code-tracking loops are not useful when the signal is placed on a carrier. Since the baseband code-tracking loop is based on an unmodulated PN code signal, it may be viewed as a degenerate case of a coherent code-tracking loop, since the carrier frequency and carrier phase are zero (and therefore known).

Consider the received waveform, which is composed of a baseband PN code, corrupted by additive white Gaussian noise (WGN).

$$y(t) = \sqrt{P}PN(t-T) + n(t) \tag{7.2-1}$$

Note that no data is assumed in this model. If data were used, a noncoherent code-tracking loop would be required. Figure 7.2-1 illustrates a baseband code-tracking loop.

Thus the maximum-likelihood estimate can motivate the closed-loop code-tracking model. Note that this is a baseband loop that has data; thus, noncoherent processing was required. A modification of this structure will accommodate carrier modulated BPSK signals, as we will see later in this chapter.

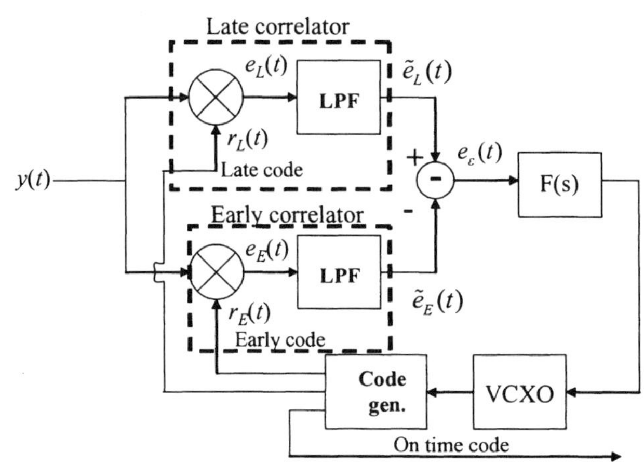

Figure 7.2-1 Baseband early-late gate code-tracking loop block diagram.

It is also called a delay lock loop that utilizes an early-late gate discriminator function. In (7.2-1) P is the received signal power, T is the unknown propagation time delay from the transmitter to the receiver, and $n(t)$ is the WGN process having two-sided noise spectral density of $N_0/2$. The chip duration of the PN code is T_c seconds, and the corresponding symbol rate[1] of the PN code is $R_s = 1/T_s$.

The two correlators are separated by dT_s, or $2\delta T_s$ seconds. Normally d or 2δ is less than or equal to one. Historically, the code early-late separation d was chosen to be 1 (one chip); however, it will be seen shortly, that this is not optimum. The late and early correlators are described by

$$\begin{aligned} &r_L(t-\hat{T}-\delta T_s) \\ &r_E(t-\hat{T}+\delta T_s) \end{aligned} \tag{7.2-2}$$

in which $\hat{T}$ is the estimate of T. It is assumed that the low pass filters (LPF) are selected to encompass the uncertainty in carrier frequency and are sufficiently narrow to form an accurate estimate of the correlation function between the received code and the locally generated one.

[1] We consider symbols rather than chips to be more general, since chips normally imply NRZ symbols.

The signal out of the late correlator is given by

$$\tilde{e}_L(t) = K_m\sqrt{P}R_{PN}(T-\hat{T}-\delta T_s) + K_m L(s)\left[n(t)PN(t-\hat{T}-\delta T_s\right] \tag{7.2-3}$$

where L(s)[$g(t)$] denotes the Heaviside operator $L(s)$ representing the arm filters, operating on the time function $g(t)$, and K_m is the multiplier gain constant of the multipliers (assumed to be both the same). In this expression the mean output has been used and the self-noise[2] has been neglected, which is reasonable at lower signal levels, so that the thermal noise is dominant. The meaning of the operator in (7.2-3) is as follows:

$$L(s)[g(t)] = \int_{-\infty}^{\infty} l(u)g(t-u)du = l(t)*g(t) \tag{7.2-4}$$

and $L(s)$ is defined by

$$L(s) = \mathcal{L}\{l(t)\} \tag{7.2-5}$$

In (7.2-5) $\mathcal{L}\{l(t)\}$ denotes the Laplace transform of $l(t)$, and the symbol $*$ denotes the convolution. Out of the early correlator we obtain

$$\tilde{e}_E(t) = K_m\sqrt{P}R_{PN}(T-\hat{T}+\delta T_s) + K_m L(s)\left[n(t)PN(t-\hat{T}+\delta T_s\right] \tag{7.2-6}$$

It will be assumed that $L(0) = 1$, that is the filter produces unit gain at $f = 0$.

The error signal is the difference between the early and late correlator signals

$$e_\varepsilon(t) = K_m\sqrt{P}\left[R_{PN}(T-\hat{T}-\delta T_s) - R_{PN}(T-\hat{T}+\delta T_s)\right] + K_m\left(n_L(t) - n_E(t)\right) \tag{7.2-7}$$

The bracketed term in (7.2-7) forms the S-curve for tracking. If we define the code-tracking error as

$$\varepsilon = T - \hat{T} \tag{7.2-8}$$

then (7.2-7) can be written as

$$e_\varepsilon(t) = K_m\sqrt{P}\left[R_{PN}(\varepsilon-\delta T_s) - R_{PN}(\varepsilon+\delta T_s)\right] + K_m\left(n_L(t) - n_E(t)\right) \tag{7.2-9}$$

The S-curve for NRZ symbols can be written seen in Figure 7.2-2. The S-curve for the baseband code-tracking loop can be seen in Figure 7.2-2 with four values of δ in the case of NRZ symbols.

As the value of δ decreases, the S-curve is limited in amplitude first to 0.5 at δ=1/4, then 0.25 at δ = 1/8, and finally to 0.125 at δ = 1/16. However, surprisingly, the slope of the S-curve (discriminator curve), at the origin, remains at 2 for all cases.

The linearized error signal can be written as

$$e_\varepsilon(t) = 2\sqrt{P}K_m\frac{\varepsilon}{T_s} + K_m\left(n_E(t) - n_L(t)\right) \tag{7.2-10}$$

[2] The self-noise is the difference between the mean output and the actual output of the correlated signal.

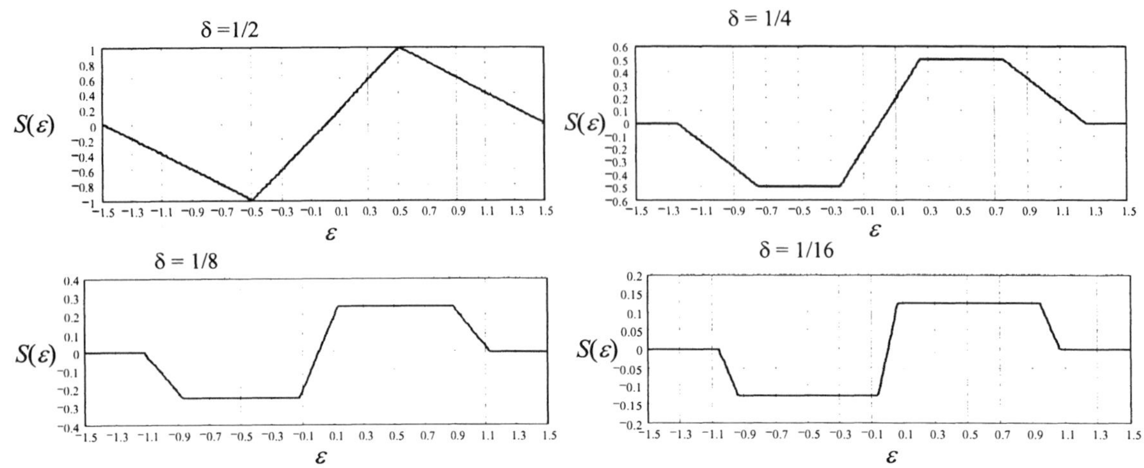

Figure 7.2-2 *S*-curve for the baseband code-tracking loop with four values of δ for NRZ symbols.

where the units of ε are in seconds. It follows that the code loop timing estimate can be written as

$$\frac{\hat{T}(t)}{T_s}=\frac{-2\sqrt{P}KF(s)}{s}\left[\frac{\varepsilon}{T_c}+\frac{n_E(t)-n_L(t)}{2\sqrt{P}}\right] \tag{7.2-11}$$

where

$$K=K_m K_{VCO} \tag{7.2-12}$$

with K_{VCO} being the voltage controlled oscillator gain in symbols/sec/volt (it is assumed that the VCO frequency is equal to the symbol rate). Using (7.2-8) in (7.2-11) produces

$$\frac{\varepsilon}{T_s}=\frac{-2\sqrt{P}KF(s)/s}{1+\left(2\sqrt{P}KF(s)/s\right)}\left(\frac{n_E(t)-n_L(t}{2\sqrt{P}}\right)+\frac{1}{1+\left(2\sqrt{P}KF(s)/s\right)}\left(\frac{T}{T_s}\right) \tag{7.2-13}$$

Note that we have implicitly not shown the time dependence of the error ε. Also note that T is the input timing phase process due to relative dynamics between the transmitter and the receiver.

Now define the closed-loop transfer function by

$$H(s)=\frac{2\sqrt{P}KF(s)}{s+2\sqrt{P}KF(s)} \tag{7.2-14}$$

Hence (7.2-13) can be written as

$$\frac{\varepsilon}{T_s}=-H(s)\left(\frac{n_E(t)-n_L(t)}{2\sqrt{P}}\right)+\left[1-H(s)\right]\frac{T}{T_s} \tag{7.2-15}$$

Since we are initially concerned about the influence of thermal noise we can let $T = C$, a constant, and since $H(s)$ is a high-pass function, the error will eventually approach zero. For simplicity we can let $C = 0$. Thus with this assumption the error signal depends only on the two noise terms so that

$$\frac{\varepsilon}{T_s} = -H(s)\left(\frac{n_E(t) - n_L(t)}{2\sqrt{P}}\right) \quad (7.2\text{-}16)$$

It follows from (7.2-16) that the variance of the tracking error for the baseband code-tracking loop is given by

$$E\left\{\left(\frac{\varepsilon}{T_c}\right)^2\right\} = \frac{\sigma_\varepsilon^2}{T_c^2} = \int_{-\infty}^{\infty} |H(j2\pi f)|^2 \frac{S_N(f)}{4P} df \quad (7.2\text{-}17)$$

where $S_N(f)$ is the noise spectral density of the difference of the early and late noise terms $(N(t) = n_E(t) - n_L(t))$.

Under the assumption that the two-sided noise spectral density is large compared to the loop bandwidth, one can write

$$\frac{\sigma_\varepsilon^2}{T_s^2} \cong \frac{S_N(f)B_L}{2P} \quad (7.2\text{-}18)$$

with little error, where B_L is the one-sided closed-loop noise bandwidth of the code-tracking loop. Twice this value ($2B_L$) is defined by

$$2B_L = \int_{-\infty}^{\infty} |H(2\pi f)|^2 df \quad (7.2\text{-}19)$$

where $H(2\pi f)$ is the closed-loop transfer function of the baseband code loop.

In order to continue it is necessary to compute first the autocorrelation function of the noise followed by the power spectral density at $f = 0$. We write the noise as

$$\begin{aligned} N(t) = n_E(t) - n_L(t) = &\int_{-\infty}^{\infty} h(t-u)n(u)PN(u - \hat{T} + \delta T_s)du \\ &- \int_{-\infty}^{\infty} h(t-u)n(u)PN(u - \hat{T} - \delta T_s)du \end{aligned} \quad (7.2\text{-}20)$$

By definition the autocorrelation function is given by

$$R_N(\tau) = E\left\{\left(n_L(t) - n_E(t)\right)\left(n_L(t+\tau) - n_E(t+\tau)\right)\right\} \quad (7.2\text{-}21)$$

Utilizing (7.2-20) produces

$$
\begin{aligned}
R_N(\tau) = & E\int_{-\infty}^{\infty}\int_{-\infty}^{\infty} h(t-u)h(t+\tau-u')PN(u-\hat{T}+\delta T_s)PN(u'-\hat{T}+\delta T_s)n(u)n(u')dudu' \\
& + E\int_{-\infty}^{\infty}\int_{-\infty}^{\infty} h(t-v)h(t+\tau-v')PN(v-\hat{T}-\delta T_s)PN(v'-\hat{T}-\delta T_s)n(v)n(v')dvdv' \\
& - E\int_{-\infty}^{\infty}\int_{-\infty}^{\infty} h(t-u)h(t+\tau-v)PN(u-\hat{T}+\delta T_s)PN(v-\hat{T}-\delta T_s)n(u)n(v)dudv \\
& - E\int_{-\infty}^{\infty}\int_{-\infty}^{\infty} h(t-u)h(t+\tau-v)PN(u-\hat{T}-\delta T_s)PN(v-\hat{T}+\delta T_s)n(u)n(v)dudv
\end{aligned}
\tag{7.2-22}
$$

We model the PN sequence as a statistically independent random sequence taking on the values of ± 1, each with probability of ½. We can write

$$
\begin{aligned}
R_N(\tau) &= N_0\int_{-\infty}^{\infty} h(t-u)h(t-u+\tau)du - N_0 R_{PN}(2\delta T_s)\int_{-\infty}^{\infty} h(t-u)h(t-u+\tau)du \\
&= N_0\left[1-R_{PN}(2\delta T_s)\right]\int_{-\infty}^{\infty} h(t-u)h(t-u+\tau)du
\end{aligned}
\tag{7.2-23}
$$

It follows that the power spectral density at $f = 0$ is the integral over τ, so that

$$
S_N(0) = N_0\left[1-R_{PN}(2\delta T_s)\right]\int_{-\infty}^{\infty}\int_{-\infty}^{\infty} h(t-u)h(t-u+\tau)dud\tau \tag{7.2-24}
$$

After some algebra this can be evaluated to

$$
S_N(0) = N_0\left[1-R_{PN}(2\delta T_s)\right]\int_{-\infty}^{\infty}\int_{-\infty}^{\infty} \left|H(\omega)^2\right| e^{j\omega\tau} d\omega d\tau/(2\pi) \tag{7.2-25}
$$

where $H(s)$ is the same as $H(j\omega)$ and we have dropped the j for convenience. Noting that

$$
\frac{1}{2\pi}\int_{-\infty}^{\infty} e^{j\omega\tau}d\tau = \delta(\omega) \tag{7.2-26}
$$

the *Dirac delta function*. Using (7.2-26) in (7.2-25) produces

$$
S_N(0) = N_0\left[1-R_{PN}(2\delta T_s)\right]\left|H(0)\right|^2 \tag{7.2-27}
$$

By assumption of the low pass filter $\left|H(0)\right| = 1$, so that

$$
S_N(0) = N_0\left[1-R_{PN}(2\delta T_s)\right] \tag{7.2-28}
$$

Hence the general expression for the tracking error variance is given by

$$
\sigma^2_{sym} = \frac{N_0 B_L\left[1-R_{PN}(2\delta T)_s\right]}{2P} \quad \text{symbols}^2 \tag{7.2-29}
$$

as our general result. Now let us specialize the case to NRZ symbols (therefore chips). We note that $[1 - R_{PN}(2\delta T_s)] = d$ (for $d \leq 1$), for NRZ chips, where $2\delta = d$, and d is the early-late correlator spacing in chips. It follows from (7.2-29) that the linearized tracking error variance for the baseband code-tracking loop is given by

$$\sigma_{sym}^2 = \frac{N_0 B_L d}{2P} \quad \text{chips}^2 \qquad (7.2\text{-}30)$$

It is to be noted that the linearized tracking performance only holds for small errors, that is, for errors that are small compared to the linear tracking region.

Thus we see that the smaller the value of "*d*," the lower the tracking error variance. Note that this result is based on the linearized tracking error variance, and as such there are some limitations. There is a practical limit based on the sample rate, since the time between samples is the minimum early-late spacing available. In addition as d is decreased the maximum restoring force is diminished. This restoring force is related to the mean time to lose lock in a complex way, but basically the higher and the broader the restoring voltage, the better. This phenomenon will be discussed in Section 7.9, later in this chapter.

7.2.2 Noncoherent Early-Late Gate I-Q Code-Tracking Loop

In this section we will treat [2] the case of a noncoherent I-Q code-tracking loop with an early-late gate error discriminator function, implemented at baseband using I and Q correlators. In all further discussions about I-Q code-tracking we will mean the translation to baseband from RF or IF with the subsequent processing.

7.2.2.1 Noncoherent I-Q Early-Late Gate Code-Tracking Loop Noise and Signal Components

A block diagram of a noncoherent I-Q version of a code-tracking loop (to be contrasted with the RF version, which is implemented at IF or RF) with an *early-late gate discriminator function* is illustrated in Figure 7.2-3. Notice that since carrier phase is assumed to be unknown, both the sine and cosine components are processed with early and late processing. Normally it is not possible to obtain phase estimates before code timing is obtained. Hence it is assumed that nominally, the frequency is known but the carrier phase is not. Certainly during code acquisition and code pull-in the carrier phase is not known; however, after code pull-in carrier tracking would be feasible.

The loop is a baseband version of a noncoherent code-tracking loop in which the early and late I and Q correlation arms are used in the error control of the loop. The upper arms with both the inphase and quadrature phase signals are used for the late correlation, and the lower arms with both the inphase and quadrature phase signals are used for the early correlation. The late minus the early correlations squared fed into the loop filter, represented by $F(s)$, and then into the voltage controlled crystal oscillator (VCXO), which in turn drives the code generator to produce the late and early versions of the receiver estimated codes.

This analysis will be general, in the sense that the symbols do not have to be NRZ symbols they can be any unit amplitude symbol shape that has a duration of T_s seconds. As an example, Figure 7.2-4 illustrates both NRZ and a binary offset carrier (BOC(2*M*,*M*)) example. The BOC (2*M*,*M*) symbol shape will be used on the new modernized GPS military signal (*M* code) in which $M = 5$. The argument (2*M*) denotes the multiplier of the underlying squarewave signal operating at 10(1.023 MHz) or 10.23 MHz, and the second argument denotes the rate of the underlying code at 5(1.023 MHz) or 5.115 MHz.

In Figure 7.2-4(a) seven NRZ code symbols are shown, the first, fourth, sixth, and seventh are +1s and the others are −1s. Likewise in the (b) part of the figure, three BOC(10,5) code symbols are shown, the first two are +1s and the last a −1 hence the symbol is inverted.

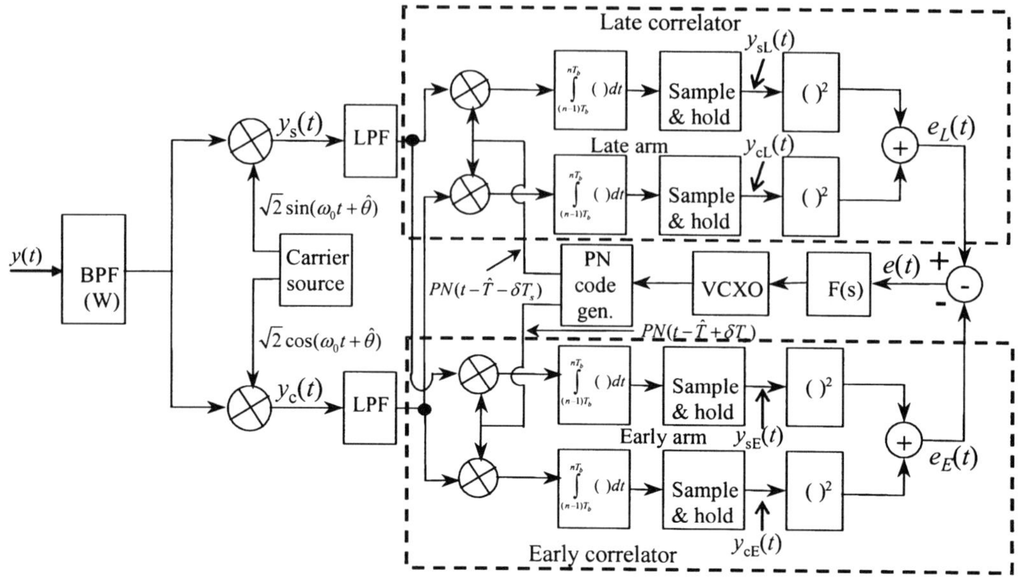

Figure 7.2-3 I-Q version of a noncoherent code-tracking loop with early-late gate and early-late separation of $d = 2\delta$.

The received BPSK signal plus thermal noise is modeled by

$$y(t) = \sqrt{2P}d(t)PN(t-T)\cos(\omega_0 t+\theta) + \sqrt{2}n_c(t)\cos(\omega_0 t+\theta) + \sqrt{2}n_s(t)\sin(\omega_0 t+\theta) \quad (7.2\text{-}31)$$

in which P is the received power in watts, $d(t)$ (±1) is the data modulation with a data rate of $R_d = 1/T_d$, $PN(t-T)$ is the unit amplitude (±1) pseudonoise code process, and the last two terms of (7.2-31) are the in-phase and quadrature noise processes in which $n_c(t)$ and $n_s(t)$ are independent Gaussian random processes with one-sided noise spectral density N_0. The object of this noncoherent code-tracking loop is to track the delay, T, which is the time delay between the transmitter and the receiver.

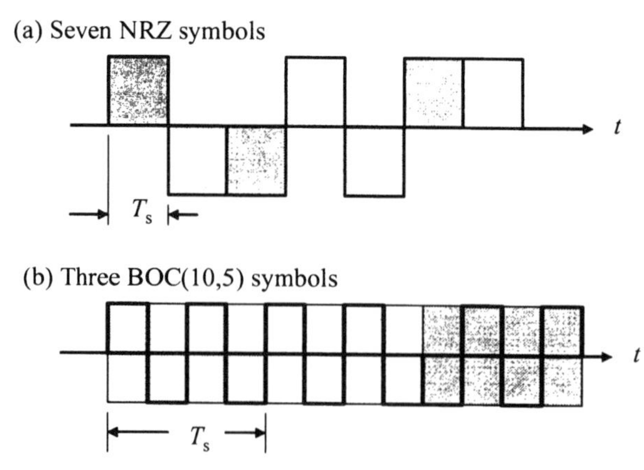

Figure 7.2-4 Plot of (a) NRZ and (b) BOC(10,5) code symbols.

The sine and cosine components indicated in Figure 7.2-3 are given by

$$y_s(t) = \sqrt{P}d(t)PN(t-T)\sin(\hat{\theta}-\theta) + n_c(t)\sin(\hat{\theta}-\theta) + n_s(t)\cos(\theta-\hat{\theta}) + O(2\omega_c) \quad (7.2\text{-}32)$$

and

$$y_c(t) = \sqrt{P}d(t)PN(t-T)\cos(\hat{\theta}-\theta) + n_c(t)\cos(\hat{\theta}-\theta) + n_s(t)\sin(\theta-\hat{\theta}) + O(2\omega_c) \quad (7.2\text{-}33)$$

in which $O(2\omega_c)$ denotes a term at a angular frequency of $2\omega_c$. During code-tracking the carrier loop would normally be in lock and $\hat{\theta}$ would roughly the same as θ, modulo 2π. For notational simplicity we have not explicitly shown the time dependence for the estimate $\hat{\theta}$. Now define the *carrier phase error* as

$$\phi(t) = \theta - \theta(t) \quad (7.2\text{-}34)$$

where the time dependence has been shown explicitly. Again suppressing time dependence on $\phi(t)$ for notational simplicity and neglecting the double frequency term we have

$$y_s(t) = -\sqrt{P}d(t)PN(t-T)\sin(\phi) - n_c(t)\sin(\phi) + n_s(t)\cos(\phi) \quad (7.2\text{-}35)$$

and

$$y_c(t) = \sqrt{P}d(t)PN(t-T)\cos(\phi) + n_c(t)\cos(\phi) + n_s(t)\sin(\phi) \quad (7.2\text{-}36)$$

Denote the two noise terms in (7.2-35) by $n_s^{'}(t)$ and let the two noise terms in (7.2-36) be denoted by $n_c^{'}(t)$. That is,

$$\begin{aligned} n_s^{'}(t) &= -n_c(t)\sin\phi + n_s(t)\cos\phi \\ n_c^{'}(t) &= n_c(t)\cos\phi + n_s(t)\sin\phi \end{aligned} \quad (7.2\text{-}37)$$

the two noise terms are mutually statistically independent of each other and have the same statistical properties as the input band-limited noise processes $n_s(t)$ and $n_c(t)$. That is they are band-limited to bandwidth B and have two-sided noise spectral density of $N_0/2$ over the bandwidth B. Hence it is convenient to write the two equations in (7.2-37) in the form

$$\begin{aligned} y_s(t) &= -\sqrt{P}d(t)PN(t-T)\sin(\phi) + n_s^{'}(t) \\ y_c(t) &= \sqrt{P}d(t)PN(t-T)\cos(\phi) + n_c^{'}(t) \end{aligned} \quad (7.2\text{-}38)$$

Carrying out some algebra, one can show out of the sample and hold circuits (see Figure 7.2-3) that

$$y_{sL} = -\sqrt{P}d(t)\int_0^{T_d} PN(t-T)PN(t-\hat{T}-\delta T_s)dt\sin\phi + n_{sLI} \quad (7.2\text{-}39)$$

$$y_{cL} = \sqrt{P}d(t)\int_0^{T_d} PN(t-T)PN(t-\hat{T}-\delta T_s)dt\cos\phi + n_{cLI} \tag{7.2-40}$$

$$y_{sE} = -\sqrt{P}d(t)\int_0^{T_d} PN(t-T)PN(t-\hat{T}+\delta T_s)dt\sin\phi + n_{sEI} \tag{7.2-41}$$

$$y_{cE} = \sqrt{P}d(t)\int_0^{T_d} PN(t-T)PN(t-\hat{T}+\delta T_s)dt\cos\phi + n_{cEI} \tag{7.2-42}$$

where it has been assumed that the bit under consideration occurs from 0 to T_d seconds for notational simplicity and where

$$n_{sL} = \int_0^{T_d} n_s'(t)PN(t-\hat{T}-\delta T_s)dt \tag{7.2-43}$$

$$n_{cL} = \int_0^{T_d} n_c'(t)PN(t-\hat{T}-\delta T_s)dt \tag{7.2-44}$$

$$n_{sE} = \int_0^{T_d} n_s'(t)PN(t-\hat{T}+\delta T_s)dt \tag{7.2-45}$$

$$n_{cE} = \int_0^{T_d} n_c'(t)PN(t-\hat{T}+\delta T_s)dt \tag{7.2-46}$$

where it has been assumed that the integration over the bit time (T_d) is sufficient to form the autocorrelation function out of the integrator for the signal terms in (7.2-39) through (7.2-42), so that

$$y_{sL} = -\sqrt{P}d(t)T_d R_{PN}(T-\hat{T}-\delta T_s)\sin\phi + n_{sL} \tag{7.2-47}$$

$$y_{cL} = \sqrt{P}d(t)T_d R_{PN}(T-\hat{T}-\delta T_s)\cos\phi + n_{cL} \tag{7.2-48}$$

$$y_{sE} = -\sqrt{P}d(t)T_d R_{PN}(T-\hat{T}+\delta T_s)\sin\phi + n_{sE} \tag{7.2-49}$$

$$y_{cE} = \sqrt{P}d(t)T_d R_{PN}(T-\hat{T}+\delta T_s)\cos\phi + n_{cE} \tag{7.2-50}$$

Now consider that the correlation properties of the four random variables defined in (7.2-43)–(7.2-46) note that

$$\begin{aligned}\text{Var}(n_{sL}) &= E\left\{\int_0^{T_d}\int_0^{T_d} n_s'(t)n_s'(u)PN(t-\hat{T}-\delta T_s)PN(u-\hat{T}-\delta T_s)dtdu\right\} \\ &= \int_0^{T_d}\frac{N_0}{2}dt = \frac{N_0}{2}T_d\end{aligned} \tag{7.2-51}$$

In a similar manner one can show that

$$\text{Var}(n_{sL}) = \text{Var}(n_{cL}) = \text{Var}(n_{sE}) = \text{Var}(n_{cE}) = \frac{N_0}{2}T_d \tag{7.2-52}$$

Cross-correlations of these noise terms when they have independent noise process are zero, so that

$$E\{n_{sL}n_{cL}\}=E\{n_{sE}n_{cE}\}=E\{n_{sL}n_{cE}\}=E\{n_{cL}n_{sE}\}=0 \tag{7.2-53}$$

The following two terms are not statistically independent

$$\begin{aligned} E\{n_{cL}n_{cE}\} &= \int_0^{T_d}\int_0^{T_d} E(n_c'(t)n_c'(u))PN(t-\hat{T}+\delta T_s)PN(u-\hat{T}-\delta T_s)dtdu \\ &= \int_0^{T_d}\int_0^{T_d} \frac{N_0}{2}\delta(t-u)PN(t-\hat{T}+\delta T_s)PN(u-\hat{T}-\delta T_s)dtdu \\ &= \frac{N_0}{2}T_d R_{PN}(2\delta T_s) = E\{n_{sL}n_{sE}\} \end{aligned} \tag{7.2-54}$$

where again it has been assumed that the integration over the bit time (T_d) is sufficient to form the autocorrelation function out of the integrator. Now the late and early signals can be written during the time 0 to T_d as

$$\begin{aligned} e_L(t) &= y_{sL}^2 + y_{cL}^2 \\ e_E(t) &= y_{sE}^2 + y_{cE}^2 \end{aligned} \tag{7.2-55}$$

Noting that the error signal $e(t)$ is given by

$$e(t) = e_L(t) - e_E(t) \tag{7.2-56}$$

One can compute the error signal (which is constant over T_d seconds) to be

$$\begin{aligned} e(t) = {} & T_d^2 P\left[R_{PN}^2(T-\hat{T}-\delta T_s) - R_{PN}^2(T-\hat{T}+\delta T_s)\right] + \left(n_{sL}^2+n_{cL}^2\right) - \left(n_{sE}^2+n_{cE}^2\right) \\ & -2\sqrt{P}d(t)\left[R_{PN}(T-\hat{T}-\delta T_s)n_{sL} - R_{PN}(T-\hat{T}+\delta T_s)n_{sE}\right]T_d\sin\phi \\ & +2\sqrt{P}d(t)\left[R_{PN}(T-\hat{T}-\delta T_s)n_{cL} - R_{PN}(T-\hat{T}+\delta T_s)n_{cE}\right]T_d\cos\phi \end{aligned} \tag{7.2-57}$$

After every T_d seconds the value of $e(t)$ changes to a new one, with the primary change coming from the noise terms in (7.2-57). Now let the code symbol timing error $\varepsilon(t)$ be defined as

$$\varepsilon(t) = T(t) - \hat{T}(t) \tag{7.2-58}$$

where we have shown the time dependence of the input and estimated time delays. Since the sum of the second and third noise terms have a mean and variance invariant with ϕ, the error signal of (7.2-57) can be simplified, with no loss in generality, to ($\phi = 0$)

$$e_i = T_d^2 P\left[R_{PN}^2(\varepsilon-\delta T_s) - R_{PN}^2(\varepsilon+\delta T_s)\right] + n_{1_i} + n_{2_i} \tag{7.2-59}$$

with the subscript "i" denoting the i-th bit sample. The two noise terms in (7.2-59) are given by

$$\begin{aligned} n_{1_i} &= 2\sqrt{P}d(T_{d_i})\left[R_{PN}(\varepsilon-\delta T_s)n_{cE_i} - R_{PN}(\varepsilon+\delta T_s)n_{cL_i}\right]T_d \\ n_{2_i} &= \left(n_{sL_i}^2 + n_{cL_i}^2\right) - \left(n_{sE_i}^2 + n_{cE_i}^2\right) \end{aligned} \tag{7.2-60}$$

The first term in (7.2-59) is the S-curve, which provides the error control of the code-tracking loop. Figure 7.2-2 illustrates that control curve for the case that δ is 1/2. Note that the error S-curve is linear over one symbol and is very nonlinear outside the linear region. The fact that the *S*-curve is linear around the zero error ($\varepsilon(t) = 0$) makes the linearization analysis feasible. Note that the error signal is $e(t)$ and the actual timing error is $\varepsilon(t)$.

Since the noise terms are constant over one bit time and vary from bit to bit, the control signal can be written as

$$e(t) = PT_d^2\left[R_{PN}^2(\varepsilon(t)-\delta T_s) - R_{PN}^2(\varepsilon(t)+\delta T_s)\right] + \sum_{i=-\infty}^{\infty} N_i p(t-iT_d) \tag{7.2-61}$$

in which $p(t)$ is a unit pulse of duration T_d seconds and where the noise terms are given by

$$N_i = n_{1_i} + n_{2_i} \tag{7.2-62}$$

At this point it will be assumed that $\varepsilon(t)$ is very slowly varying and is small in magnitude. For convenience we will suppress the time dependence of $\varepsilon(t)$ and call it ε.

7.2.2.2 Noncoherent I-Q Early-Late Gate Code-Tracking Loop Performance

Before we analyze the noise properties let us consider the *S*-curve shape for NRZ symbols. Figure 7.2-5 illustrates the *S*-curve ($R_{PN}^2(\varepsilon+\delta T_s) - R_{PN}^2(\varepsilon-\delta T_s)$) for four values of δ equal to 1/2, 1/4, 1/8, and 1/16 for the early-late gate error detector (L^2–E^2). Thus we see that as δ is reduced the linear region reduces, but the slope increases.

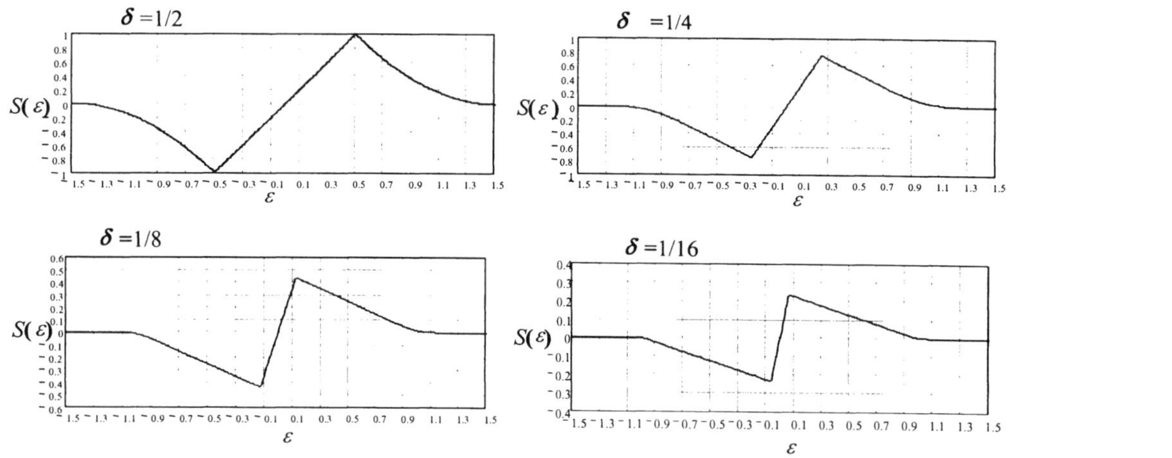

Figure 7.2-5 The *S*-curve and associated slope (gain) for NRZ symbols for the noncoherent early-late gate code-tracking loop.

Assuming for the moment the noise can be characterized as stationary noise, it is desirable to be able to model the code-tracking loop as a control loop with feedback. The analysis can most easily be done when the loop is linearized. As noted in Figure 7.2-5 the region of the error control is linear around the $\varepsilon = 0$ point. Hence it is assumed that the error ε is small and stays in the linear region for all time. Let the slope or "gain" of the error curve (often called the *S*-curve in phase locked loop literature) be denoted by *K*, where *K* is defined as

$$K = \frac{d}{d\varepsilon}\left[R_{PN}^2(\varepsilon - \delta T_s) - R_{PN}^2(\varepsilon + \delta T_s)\right]\Big|_{\varepsilon=0} \tag{7.2-63}$$

Denote the second (noise) term as $N(t)$. Then the error signal can be represented by

$$e(t) = KPT_d^2\varepsilon + N(t) \tag{7.2-64}$$

where

$$N(t) = \sum_{i=-\infty}^{\infty} N_i p(t - iT_d) \tag{7.2-65}$$

where the random variables N_i are statistically independent from sample to sample since they are composed of integrations of time disjoint segments of WGN. The estimate of the timing from the oscillator can be put in Heaviside operator notation [3]. Therefore one can write the code-timing estimate as

$$\hat{T} = \left\{\frac{K_v}{s}F(s)\right\}e(t) \tag{7.2-66}$$

where K_v is the VCXO scale parameter in chips per volt, $F(s)/s$ is a Heaviside operator operating on the signal $e(t)$. From (7.2-58), (7.2-64), and (7.2-66), one obtains

$$\frac{K_v}{s}F(s)\left[KPT_d^2\varepsilon + N(t)\right] = T(t) - \varepsilon \tag{7.2-67}$$

where $\varepsilon(t)$ has been written as ε to simplify the equations. It is to be noted that $e(t)$ and $\varepsilon(t)$ are defined by (7.2-57) and (7.2-58), respectively. Equation (7.2-64) can be used with $\varepsilon(t)$ to produce

$$\varepsilon(t) = \frac{T(t)}{\left[1 + \frac{K_v}{s}F(s)KPT_d^2\right]} - \frac{\frac{K_v F(s)}{s}KPT_d^2 N(t)}{\left[1 + \frac{K_v}{s}F(s)KPT_d^2\right]\left(KPT_d^2\right)} \tag{7.2-68}$$

with $T(t)$ being the input time delay function, which is assumed to be constant with time in the analysis to follow. Now define the *closed-loop transfer function* of the code-tracking loop, $H(s)$, by

$$H(s) = \frac{KPT_d^2 \frac{K_v F(s)}{s}}{\left[1 + KPT_d^2 \frac{K_v F(s)}{s}\right]} \tag{7.2-69}$$

From (7.2-68) and (7.2-69) it follows that

$$\varepsilon(t) = \left(1 - H(s)\right)T(t) - H(s)\frac{N(t)}{KPT_d^2} \tag{7.2-70}$$

which is the stochastic differential equation in Heaviside notation that specifies the code-tracking loop. We will assume that the input timing is constant so that the only term that produces an error is the loop noise term that is the second term in (7.2-70) since the first term converges to zero. Therefore the error under consideration is given by

$$\varepsilon(t) = -H(s)\left\{\frac{N(t)}{KPT_d^2}\right\} \tag{7.2-71}$$

The linearized tracking loop timing error variance, due to thermal noise, is given from (7.2-71) by the integral[3]

$$\sigma_\varepsilon^2 = \int_{-\infty}^{\infty} |H(f)|^2 \frac{S_N(f)}{K^2P^2T_d^4} df \cong \frac{2B_L S_N(0)}{K^2P^2T_d^4} \tag{7.2-72}$$

where K is the slope of the S-curve, P is the power of the signal, T_d is the data bit duration in seconds, $S_N(0)$ is the noise power spectral density of the noise at zero frequency in watts/Hz, and B_L is the one-sided closed-loop noise bandwidth in Hz.

It was assumed that the code-tracking loop is very narrowband relative to a chip rate. It follows that if the noise spectral density of $N(t)$ can be evaluated at $f = 0$, then the code-tracking error variance can be evaluated from (7.2-72).

7.2.2.3 Evaluation of the Sample and Hold Noise Sequence for the Noncoherent I-Q Early-Late Loop

It is to be noted from (7.2-60) that

$$E\{n_{1_i} n_{2_i}\} = 0 \tag{7.2-73}$$

since moments of odd numbers of zero mean Gaussian random variables are zero. Recall that $N(t)$ is given by (7.2-65). From (7.2-62) and (7.2-65) it follows that

$$Var(N_i) = Var(n_{1_i}) + Var(n_{2_i}) \tag{7.2-74}$$

Consider the autocorrelation function of $N(t)$ where

$$R_N(\tau) = \langle E\{N(t)N(t+\tau)\}\rangle = E\{N_i^2\}\left[1 - \frac{|\tau|}{T_d}\right] \qquad 0 \le |\tau| \le T_d \tag{7.2-75}$$

where $\langle x \rangle$ is the time average value of x. It therefore follows that the power spectral density is given by

$$S_N(f) = \left[Var(n_{1_i}) + Var(n_{2_i})\right] T_d \left(\frac{\sin(\pi f T_d)}{\pi f T_d}\right)^2 \qquad \forall f \tag{7.2-76}$$

With the assumption that the code loop bandwidth is small compared to the input bandwidth B, the power spectral density at $f = 0$ is given by

$$S_N(f) \cong \left[Var(n_{1_i}) + Var(n_{2_i})\right] T_d \quad \text{for} \quad |f| << 1/T_d \tag{7.2-77}$$

[3] It is to be noted that the delay in the loop was assumed to satisfy $B_L T_b << 1$, since the major part of the delay is the bit time integrators. This assumption will be made on all the remaining analyses for code-tracking error in this chapter.

In (7.2-77) the two constituent noise terms are indicated. Consider the first noise term n_{1_i} first, so that it follows that

$$Var(n_{1_i}) = 4PE\left\{d_i^2\right\}R_{PN}^2(\delta T_s)E\left\{(n_{cL} - n_{cE})^2\right\}T_d^2 \tag{7.2-78}$$

From (7.2-52) and (7.2-54) one has the result

$$Var(n_{1_i}) = 4PT_d^2 R_{PN}^2(\delta T_s)N_0 T_d\left[1 - R_{PN}(2\delta T_s)\right] \tag{7.2-79}$$

Now consider the variance of the second term n_{2_i}, which is given by

$$Var\left(n_{2_i}\right) = E\left\{\left(\left[n_{sL}^2 + n_{cL}^2\right] - \left[n_{sE}^2 + n_{cE}^2\right]\right)\left(\left[n_{sL}^2 + n_{cL}^2\right] - \left[n_{sE}^2 + n_{cE}^2\right]\right)\right\} \tag{7.2-80}$$

Sixteen terms are obtained by direct expansion. They can be broken down into four terms of four terms each. The four terms are

$$\begin{aligned}
T_1 &= E\left\{n_{sL}^4\right\} + E\left\{n_{sL}^2 n_{cL}^2\right\} - E\left\{n_{sL}^2 n_{sE}^2\right\} - E\left\{n_{sL}^2 n_{cE}^2\right\} \\
T_2 &= E\left\{n_{cL}^2 n_{sL}^2\right\} + E\left\{n_{cL}^4\right\} - E\left\{n_{cL}^2 n_{sE}^2\right\} - E\left\{n_{cL}^2 n_{cE}^2\right\} \\
T_3 &= -E\left\{n_{sE}^2 n_{sL}^2\right\} - E\left\{n_{sE}^2 n_{cL}^2\right\} + E\left\{n_{sE}^4\right\} + E\left\{n_{sE}^2 n_{cE}^2\right\} \\
T_3 &= -E\left\{n_{cE}^2 n_{sL}^2\right\} - E\left\{n_{cE}^2 n_{cL}^2\right\} + E\left\{n_{cE}^2 n_{sE}^2\right\} + E\left\{n_{cE}^4\right\}
\end{aligned} \tag{7.2-81}$$

For Gaussian, zero mean, random variables it is true that [4],

$$E\{uvxy\} = E\{uv\}E\{xy\} + E\{ux\}E\{vy\} + E\{uy\}E\{vx\} \tag{7.2-82}$$

Since it turns out that all four terms in (7.2-81) yield the same result, consider the first term T_1

$$T_1 = \frac{N_0^2 T_d^2}{4}\left[2 - 2R_{PN}^2(2\delta T_s)\right] \tag{7.2-83}$$

Since all the other three terms evaluate to the same quantity the result is

$$Var(n_{2_i}) = 2N_0^2 T_d^2\left[1 - R_{PN}^2(2\delta T_s)\right] \tag{7.2-84}$$

So that $T_d Var(N_i)$ is given by

$$\begin{aligned}
S_N(0) = T_d Var(N_i) &= 4N_0 P T_d^4 R_{PN}^2(\delta T_s)\left[1 - R_{PN}(2\delta T_s)\right] \\
&+ 2N_0^2 T_d^3\left[1 - R_{PN}^2(2\delta T_s)\right]
\end{aligned} \tag{7.2-85}$$

7.2.2.4 Evaluation of the Noncoherent I-Q Early-Late Gate Code Loop Tracking Error Performance

At this point in the analysis we are in a position to evaluate the noncoherent baseband code loop tracking error variance, which from (7.2-72) and (7.2-85) yields

$$\sigma_{sym}^2 = \frac{2N_0B_L}{K^2P}\left[\left[4R_{PN}^2(\delta T_s)\left(1-R_{PN}(2\delta T_s)\right)\right]+\frac{2N_0\left[1-R_{PN}^2(2\delta T_s)\right]}{PT_d}\right], \quad \text{symbols}^2 \qquad (7.2\text{-}86)$$

for values of δ such that $\delta < 0.5$, and the units of the variance is in symbols^2. This equation is our final general result for unit amplitude code symbols, which includes NRZ, Manchester, and BOC symbol type plus any others that are binary valued (± 1) over a symbol. Note that to convert to the actual time error from the symbol error, the following relationship applies

$$\sigma_t = T_s \sigma_{sym} \ \text{(s)} \qquad (7.2\text{-}87)$$

It is to be recalled that $R_{PN}(\tau)$ is the autocorrelation function of the symbol waveform evaluated at time offset τ from zero. In addition it is to be noted that channel filtering is not accounted for in this expression.

7.2.2.5 Noncoherent I-Q Early-Late Gate Code Loop Tracking Error Performance for NRZ Symbols

Now consider using (7.2-86) to obtain the tracking error performance for the case of NRZ symbols as seen in Figure 7.2-4. In Problem 2 it is shown that the slope of the *S*-curve at $\varepsilon = 0$ is given by

$$K = 4(1-\delta) \ \text{for} \ 0 < \delta \le 0.5 \qquad (7.2\text{-}88)$$

It is clear that gain (K) in volts/symbol has the range of 4 to 2 as δ goes from 0 to 0.5. Now the two correlation function values can be obtained in closed form in Problem 2 and are given by

$$R_{PN}(\delta T_s) = 1-\delta \ \text{and} \ R_{PN}(2\delta T_s) = 1-2\delta \qquad (7.2\text{-}89)$$

In Problem 2 it is shown that using (7.2-88) and (7.2-89) in (7.2-86) yields the following result, after some algebra,

$$\sigma_{sym}^2 = \frac{N_0B_Ld}{2P}\left[1+\frac{2}{(2-d)(E_b/N_0)}\right], \quad \text{symbols}^2 \qquad (7.2\text{-}90)$$

which is our general result for NRZ symbols with arbitrary values of $d = 2\delta$, with the condition that $\delta \le 0.5$ ($d \le 1$), for the early-late gate, noncoherent, baseband implemented, code-tracking loop. It can be seen that the smaller the value of d is, the smaller the variance of the tracking error, since $d/(2-d)$ is monotonically decreasing in d. This equation agrees with Van Dierendonck, Fenton, and Ford [5] based on an independent derivation. As noted earlier, this derivation assumes that there is no bandwidth limitation on the front-end filter of the tracking loop (no channel filtering). It will be shown later in this chapter that the bandwidth can affect the variance result significantly [6].

7.2.3 Noncoherent Early-Late Gate RF Implemented Code-Tracking Loop

An alternative model for code-tracking implemented at radio frequency (RF) is shown in Figure 7.2-6. This loop has been analyzed by Holmes and Raghavan [7] for binary values symbol types (± 1 amplitude), and it has been determined that the mean squared tracking error variance is given by

$$\sigma_{sym}^2 = \frac{2N_0 B_L}{K^2 \alpha P}\left[\left[4R_{PN}^2(\delta T_s)(1-R_{PN}(2\delta T_s))\right] + \frac{2N_0 B\left[1-R_{PN}^2(2\delta T_s)\right]}{\alpha P}\right], \quad \text{symbols}^2 \qquad (7.2\text{-}91)$$

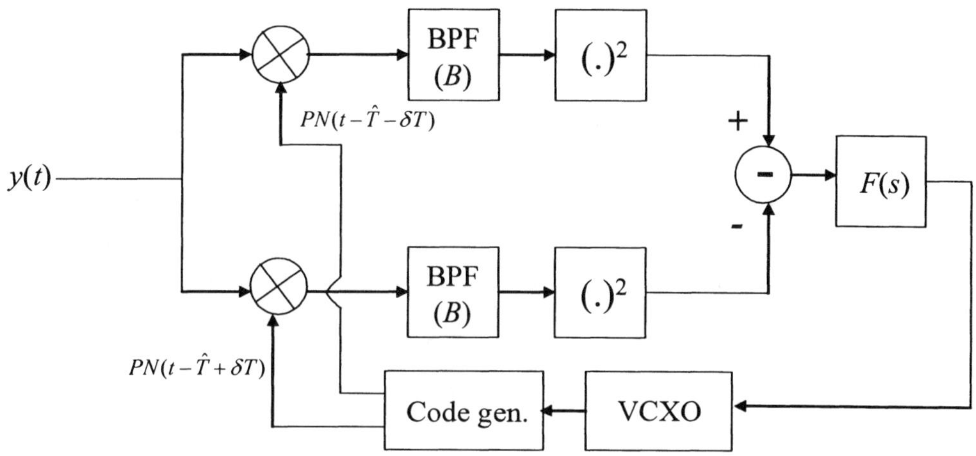

Figure 7.2-6 Early-late gate (L^2-E^2) noncoherent code-tracking loop implemented at RF.

In (7.2-91) N_0 is the one-sided noise spectral density, B is the noise bandwidth of the arm filters, and $R_{PN}(2\delta T_s)$ is code autocorrelation function evaluated at $2\delta T_s$ symbols for any ± 1 valued PN code sequence. The term α in the expression for the variance of the tracking error is the filter loss of the signal from the arm filters, so that

$$\alpha = \int_{-\infty}^{\infty} S(f)\left|H_{BB}(2\pi f)\right|^2 df \qquad (7.2\text{-}92)$$

and is the filtering loss due to the bandpass filters in each arm of the spread spectrum signal. In this equation $H_{BB}(2\pi f)$ is the baseband equivalent of the bandpass filters (BPFs) of Figure 7.2-6. It should be noted that the expression (7.2-91) is very similar to the baseband implemented noncoherent loop shown in (7.2-86), as one's intuition might suggest. One advantage with this loop implementation is the fact that only two correlators are needed, as opposed to four needed in the baseband I-Q loop.

7.2.3.1 Noncoherent Early-Late Gate RF Implemented Code-Tracking Loop with NRZ Symbols

When the code symbols are NRZ, then (7.2-91) can be shown to collapse to

$$\left(\frac{\sigma_c}{T_s}\right)^2 = \frac{N_0 B_L d}{2\alpha P}\left[1 + \frac{2N_0 B}{(2-d)(\alpha P)}\right] \; chips^2 \qquad (7.2\text{-}93)$$

with the signal loss α defined in (7.2-92).

7.2.4 Noncoherent I-Q Dot Product Code-Tracking Loop with Passive Arm Filters

In this section we will treat the *noncoherent dot product* code-tracking loop with passive arm filters (PAF). We will first consider the case in which the arm filters are passive; later we will consider the case in which the arm filters are integrate-and-dump types.

7.2.4.1 Noncoherent I-Q Dot Product Code Loop Model with PAF Model

The dot product code-tracking loop is shown in Figure 7.2-7. Again we consider the following model to describe the input signal plus noise

$$y(t) = \sqrt{2P}d(t)PN(t-T)\cos(\omega_0 t + \theta) + \sqrt{2}n_c(t)\cos(\omega_0 t + \theta) + \sqrt{2}n_s(t)\sin(\omega_0 t + \theta) \tag{7.2-94}$$

where, as before, P is the received power in watts, $d(t)$ (having values of ± 1) is the data modulation with a data rate of $R_d = 1/T_d$, $PN(t-T)$ is the unit amplitude (± 1) pseudonoise code process, and the last two terms of (7.2-94) are the inphase and quadrature noise processes in which $n_c(t)$and $n_s(t)$ are independent Gaussian random processes with one-sided noise spectral density N_0.

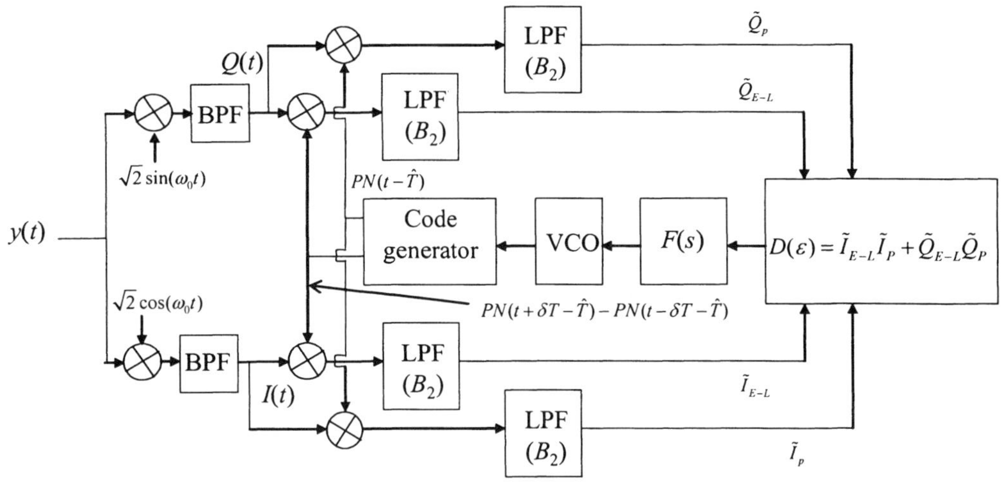

Figure 7.2-7 Dot product I-Q noncoherent code-tracking loop model.

Notice that the dot product noncoherent code-tracking loop utilizes the late minus early difference times the on-time correlation for both the I and Q channels to obtain the S-curve. The analysis is quite cumbersome so the results will only be outlined here.

7.2.4.2 Noise Analysis of the Noncoherent I-Q Dot Product Code Loop with PAFs

After the two bandpass filters (BPF) remove the sum frequency[4] at $2\omega_0$, the signals $Q(t)$ and $I(t)$ are given by

$$Q(t) = -\sqrt{P}d(t)PN(t-T)\sin(\theta) - n_c(t)\sin(\theta) + n_s(t)\cos(\theta) \tag{7.2-95}$$

and

$$I(t) = \sqrt{P}d(t)PN(t-T)\cos(\theta) + n_c(t)\cos(\theta) + n_s(t)\sin(\theta) \tag{7.2-96}$$

[4] It is assumed that the bandpass filter does not affect the signal quality but removes only sum frequency term.

It is easy to show that the Q channel noise terms are uncorrelated with the I channel noise terms and since they are Gaussian they are statistically independent; that is,

$$E\left[\left(-n_c(t)\sin(\theta)+n_s(t)\cos(\theta)\right)\left(-n_c(t)\sin(\theta)+n_s(t)\cos(\theta)\right)\right]=0 \tag{7.2-97}$$

Consequently let the I channel and Q channel noise terms be represented by one term with the same $N_0/2$ two-sided noise spectral density so that

$$\begin{aligned} Q(t) &= -\sqrt{P}d(t)PN(t-T)\sin(\theta)+n_Q(t) \\ I(t) &= \sqrt{P}d(t)PN(t-T)\cos(\theta)+n_I(t) \end{aligned} \tag{7.2-98}$$

and we have already observed that $n_Q(t)$ and $n_I(t)$ are statistically independent. Now consider the punctual and on time despread terms. We have

$$\begin{aligned} I_{L-E}(t) &= \left[\sqrt{P}d(t)PN(t-T)\cos(\theta)+n_I(t)\right]\left[PN(t-\delta T_s-\hat{T})-PN(t+\delta T_s-\hat{T})\right] \\ I_P(t) &= \left[\sqrt{P}d(t)PN(t-T)\cos(\theta)+n_I(t)\right]\left[PN(t-\hat{T})\right] \\ Q_{L-E}(t) &= \left[-\sqrt{P}d(t)PN(t-T)\sin(\theta)+n_Q(t)\right]\left[PN(t-\delta T_s-\hat{T})-PN(t+\delta T_s-\hat{T})\right] \\ I_P(t) &= \left[-\sqrt{P}d(t)PN(t-T)\sin(\theta)+n_Q(t)\right]\left[PN(t-\hat{T})\right] \end{aligned} \tag{7.2-99}$$

Denote the low-pass arm filter Laplace transform, with bandwidth B_2, as $F_2(s)$. Now it is assumed that the filter is narrow enough to approximate the correlation function of the code. The arm filtered four outputs become

$$\tilde{I}_{L-E}(t)=\left[\sqrt{P}d(t)\cos(\theta)\right]\left[R_{PN}(\varepsilon-\delta T_s)-R_{PN}(\varepsilon+\delta T_s)\right]+\tilde{n}_{I_{L-E}}(t) \tag{7.2-100}$$

where

$$\tilde{n}_{I_{E-L}}(t)=F_2(s)\left\{n_I(t)\left[PN(t-\delta T_s-\hat{T})-PN(t+\delta T_s-\hat{T})\right]\right\} \tag{7.2-101}$$

In a similar manner

$$\tilde{I}_P(t)=\sqrt{P}d(t)\cos(\theta)R_{PN}(\varepsilon)+\tilde{n}_{I_P}(t) \tag{7.2-102}$$

where

$$\tilde{n}_{I_P}(t)=F_2(s)\left\{n_I(t)PN(t-\hat{T})\right\} \tag{7.2-103}$$

Also

$$\tilde{Q}_{L-E}(t)=\left[-\sqrt{P}d(t)\sin(\theta)\right]\left[R_{PN}(\varepsilon+\delta T_s)-R_{PN}(\varepsilon-\delta T_s)\right]+\tilde{n}_{Q_{L-E}}(t) \tag{7.2-104}$$

where

$$\tilde{n}_{Q_{L-E}}(t)=F_2(s)\left\{n_Q(t)\left[PN(t-\delta T_s-\hat{T})-PN(t+\delta T_s-\hat{T})\right]\right\} \tag{7.2-105}$$

and

$$\tilde{Q}_P(t) = -\sqrt{P}d(t)\sin(\theta)R_{PN}(\varepsilon) + \tilde{n}_{Q_P}(t) \tag{7.2-106}$$

where

$$\tilde{n}_{Q_P}(t) = F_2(s)\left\{n_Q(t)PN(t-\hat{T})\right\} \tag{7.2-107}$$

In Problem 5 it is shown that the error signal can be written as

$$\begin{aligned} D(\varepsilon,t) \cong{} & P\left[R_{PN}(\varepsilon-\delta T_s) - R_{PN}(\varepsilon+\delta T_s)\right]R_{PN}(\varepsilon) \\ & +\left(\sqrt{P}\cos(\theta)\tilde{n}_{I_P}(t) - \sqrt{P}\sin(\theta)\tilde{n}_{Q_P}(t)\right)\left[R_{PN}(\varepsilon-\delta T_s) - R_{PN}(\varepsilon+\delta T_s)\right] \\ & +\left(\sqrt{P}\tilde{n}_{I_{L-E}}(t)\cos(\theta) - \sqrt{P}n_{Q_{L-E}}(t)\sin(\theta)\right)R_{PN}(\varepsilon) \\ & +\left(\tilde{n}_{I_{L-E}}(t)\tilde{n}_{I_P}(t) + \tilde{n}_{Q_{L-E}}(t)\tilde{n}_{Q_P}(t)\right) \end{aligned} \tag{7.2-108}$$

The first term is the error control term and the rest can be viewed as noise. Thus the error control is given by

$$D(\varepsilon) \cong P\left[R_{PN}(\varepsilon-\delta T_s) - R_{PN}(\varepsilon+\delta T_s)\right]R_{PN}(\varepsilon) \tag{7.2-109}$$

The *S*-curve is plotted in Figure 7.2-8 for various values of δ.

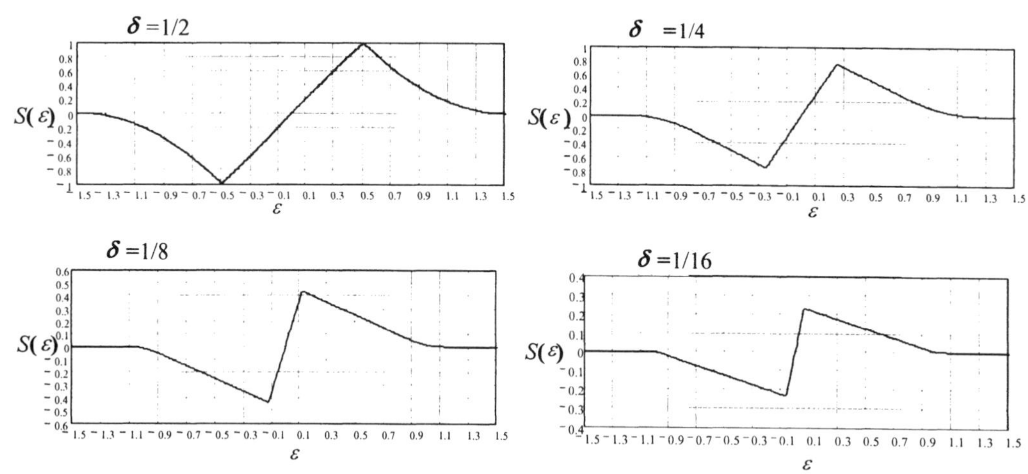

Figure 7.2-8 The *S*-curve for the dot product code-tracking loop.

It can be shown that the gain slope is always 2, independent of δ, as was observed in constructing these *S*-curves. Therefore at small errors we may linearize the error signal to be written as

$$D(\varepsilon,t) \cong 2P\frac{\varepsilon}{T_s}$$
$$+\left(\sqrt{P}\cos(\theta)\tilde{n}_{I_P}(t)-\sqrt{P}\sin(\theta)\tilde{n}_{Q_P}(t)\right)\left[R_{PN}(\varepsilon-\delta T_s)-R_{PN}(\varepsilon+\delta T_s)\right] \quad (7.2\text{-}110)$$
$$+\left(\sqrt{P}\tilde{n}_{I_{L-E}}(t)\cos(\theta)-\sqrt{P}n_{Q_{L-E}}(t)\sin(\theta)\right)R_{PN}(\varepsilon)$$
$$+\left(\tilde{n}_{I_{L-E}}(t)\tilde{n}_{I_P}(t)+\tilde{n}_{Q_{L-E}}(t)\tilde{n}_{Q_P}(t)\right)$$

Now define the two noise terms as

$$n_1(t)=\left(\sqrt{P}\cos(\theta)\tilde{n}_{I_P}(t)-\sqrt{P}\sin(\theta)\tilde{n}_{Q_P}(t)\right)\left[R_{PN}(\varepsilon-\delta T_s)-R_{PN}(\varepsilon+\delta T_s)\right]$$
$$+\left(\sqrt{P}\tilde{n}_{I_{L-E}}(t)\cos(\theta)-\sqrt{P}\tilde{n}_{Q_{L-E}}(t)\sin(\theta)\right)R_{PN}(\varepsilon) \quad (7.2\text{-}111)$$

and

$$n_2(t)=\tilde{n}_{I_{L-E}}(t)\tilde{n}_{I_P}(t)+\tilde{n}_{Q_{L-E}}(t)\tilde{n}_{Q_P}(t) \quad (7.2\text{-}112)$$

Therefore $D(\varepsilon)$ can be written succinctly as (neglecting the sign of the gain slope)

$$D(\varepsilon,t) \cong 2P\frac{\varepsilon}{T_s}+n_1(t)+n_2(t) \quad (7.2\text{-}113)$$

If we denote the multiplier gain of each multiplier as K_m (v/v) and the VCO gain as K_{VCO} in Hz/v, then the code loop delay estimate is given by

$$\frac{\hat{T}}{T_s}=\frac{K_m K_{VCO}F(s)2P}{s}\left(\frac{\varepsilon}{T_s}+\frac{n_1(t)+n_2(t)}{2P}\right) \quad (7.2\text{-}114)$$

and the loop filter has $F(s)$ as its transfer function and K_{VCO}/s is the transfer function of the VCO.

Letting $H(s)$ denote the closed-loop code loop transfer which can be written as

$$H(s)=\frac{2PK_{VCO}K_m F(s)/s}{1+2PK_{VCO}K_m F(s)/s} \quad (7.2\text{-}115)$$

it is not hard to show that the linearized delay error can be written as

$$\frac{\varepsilon}{T_s}=-H(s)\left[\frac{n_1(t)+n_2(t)}{2P}\right]+\left[1-H(s)\right]\frac{T}{T_s} \quad (7.2\text{-}116)$$

with T the input delay function (properly written as a function of time). Now the mean squared delay tracking error, assuming that the input delay variation is zero, can be written as

$$\sigma_{sym}^2=E\left\{\left(\frac{\varepsilon}{T_s}\right)^2\right\}=\int_{-\infty}^{\infty}\left|H(j2\pi f)\right|^2\frac{S_{n_1+n_2}(f)}{(2P)^2}\cong\frac{(2B_L)S_{n_1+n_2}(f)}{(2P)^2} \quad (7.2\text{-}117)$$

Now in order to continue it is necessary to evaluate the spectral density of the two noise terms. First consider the first noise term. Since we are assuming that the error is small we shall let it be zero so that we greatly simplify $n_1(t)$ as

$$n_1(t) = \left(\sqrt{P}\tilde{n}_{I_{L-E}}(t)\cos(\theta) - \sqrt{P}\tilde{n}_{Q_{L-E}}(t)\sin(\theta)\right)R_{PN}(0) \tag{7.2-118}$$

where $R_{PN}(0) = 1$. Since both $\tilde{n}_{I_{L-E}}(t)$ and $\tilde{n}_{Q_{L-E}}(t)$ are statistically independent the power spectral density of $n_1(t)$ at $f = 0$ can be written as (let $\theta = 0$)

$$S_{n_1}(0) = PS_{\tilde{n}_{I_{L-E}}}(0) \tag{7.2-119}$$

Thus it is necessary to evaluate $S_{\tilde{n}_{I_{L-E}}}(0)$. In Problem 6 it is shown that $S_{n_1}(0)$ can be evaluated as

$$S_{n_1}(0) = PN_0\left[1 - R_{PN}(2\delta T_s)\right] \tag{7.2-120}$$

Now consider obtaining the power spectral density of $n_2(t)$. Recall that it is defined by (7.2-112)

$$n_2(t) = \tilde{n}_{I_{L-E}}(t)\tilde{n}_{I_P}(t) + \tilde{n}_{Q_{L-E}}(t)\tilde{n}_{Q_P}(t) \tag{7.2-121}$$

with the four terms in (7.2-121) defined by (7.2-96), (7.2-98), (7.2-100), and (7.2-102).

In order to proceed, it is convenient to approximate a Gaussian noise term times a code term by a Gaussian noise term. For example, from (7.2-103) we can write

$$\tilde{n}_{I_P}(t) = \int_{-\infty}^{\infty} h_2(v)n_I(t-v)PN(t-v)dv \cong \int_{-\infty}^{\infty} h_2(v)n_I''(t-v)dv \tag{7.2-122}$$

where

$$n_I''(t) = n_I(t)PN(t) \tag{7.2-123}$$

In a similar manner one can write

$$\tilde{n}_{Q_P}(t) = \int_{-\infty}^{\infty} h_2(v)n_Q(t-v)PN(t-v)dv \cong \int_{-\infty}^{\infty} h_2(v)n_Q''(t-v)dv \tag{7.2-124}$$

where

$$n_Q''(t) = n_Q(t)PN(t) \tag{7.2-125}$$

The other two terms are approximated as

$$\begin{aligned}\tilde{n}_{I_{L-E}}(t) &= \int_{-\infty}^{\infty} h_2(v)n_I(t-v)\left[PN(t-\delta T_s - \hat{T} - v) - PN(t+\delta T_s - \hat{T} - v)\right]dv \\ &\cong \int_{-\infty}^{\infty} h_2(v)n_I'(t-v)dv\end{aligned} \tag{7.2-126}$$

and

$$\tilde{n}_{Q_{L-E}}(t) = \int_{-\infty}^{\infty} h_2(v) n_Q(t-v)\left[PN(t-\delta T_s - \hat{T} - v) - PN(t+\delta T_s - \hat{T} - v)\right]dv \\ \cong \int_{-\infty}^{\infty} h_2(v) n_Q^{'}(t-v)dv \tag{7.2-127}$$

where the primed noise terms are defined by (7.2-126) and (7.2-127); that is, by

$$n_I^{'}(t) \cong n_I(t)\left[PN(t-\delta T_s - \hat{T}) - PN(t+\delta T_s - \hat{T})\right] \tag{7.2-128}$$

and

$$n_Q^{'}(t) \cong n_Q(t)\left[PN(t-\delta T_s - \hat{T}) - PN(t+\delta T_s - \hat{T})\right] \tag{7.2-129}$$

The primed and double primed noise terms are modeled as Gaussian random processes with zero mean values variances that are to be determined. Now consider the autocorrelation function of $n_I^{'}(t)$. By definition

$$E\left\{n_I^{'}(t)n_I^{'}(t-\tau)\right\} = E\left\{n_I(t)n_I(t-\tau)\right\} \times \\ E\left\{\begin{matrix}\left[PN(t-\delta T_s - \hat{T}) - PN(t+\delta T_s - \hat{T})\right] \times \\ \left[PN(t-\delta T_s - \hat{T} - \tau) - PN(t+\delta T_s - \hat{T} - \tau)\right]\end{matrix}\right\} \tag{7.2-130}$$

This can be evaluated as

$$R_{n_I^{'}}(\tau) = E\left\{n_I^{'}(t)n_I^{'}(t-\tau)\right\} = \frac{N_0}{2}\delta_D(\tau)\left[2\left(1 - R_{PN}(2\delta T_s)\right]\right. \tag{7.2-131}$$

in which $\delta_D(\tau)$ is the Dirac delta function—a function that integrates to 1 if the origin is covered in the integration and integrates to zero otherwise. Furthermore it is zero everywhere except $\tau = 0$ and is unbounded at $\tau = 0$. Now consider $R_{n_I^{\cdot}}(\tau)$

$$R_{n_I^{\cdot}}(\tau) = E\left\{PN(t-\hat{T})PN(t-\hat{T}-\tau)\right\} E\left\{n_I(t)n_I(t-\tau)\right\} \tag{7.2-132}$$

which can be evaluated to

$$R_{n_I^{\cdot}}(\tau) = \frac{N_0}{2}\delta_D(\tau) \tag{7.2-133}$$

In Problem 7 it is shown that power spectral density of $n_2(t)$ is given by

$$S_{n_2}(0)/2 = F\left\{R_{n_2}(\tau)\right\} = E\left\{\int_{-\infty}^{\infty} \tilde{n}_{I_{L-E}}(t)\tilde{n}_{I_{L-E}}(t-\tau)\tilde{n}_{I_P}(t)\tilde{n}_{I_P}(t-\tau)d\tau\right\} \tag{7.2-134}$$

To evaluate (7.2-134), write each noise term in terms of its filter function, so that

$$S_{n_2}(0)/2=\int_{-\infty}^{\infty}\int_{-\infty}^{\infty}\int_{-\infty}^{\infty}\int_{-\infty}^{\infty}h_2(u)h_2(v)h_2(x)h_2(y)E\{n_I^{'}(t-u)n_I^{'}(t-v-\tau)\}E\{n_I^{''}(t-x)n_I^{''}(t-y-\tau)\}dudvdxdy \quad (7.2\text{-}135)$$

and $n_I^{'}(t)$ and $n_I^{''}(t)$ are defined via (7.2-128) and (7.2-123). Problem 8 shows that (7.2-135) can be evaluated as

$$S_{n_2}(0)=2N_0^2B_2^{'}\left[1-R_{PN}(2\delta T_s)\right] \quad (7.2\text{-}136)$$

where $B_2^{'}$ is given by

$$B_2^{'}=\int_{-\infty}^{\infty}\left|H_2(f)\right|^4df \quad (7.2\text{-}137)$$

in which $H_2(f)$ is the transfer function of the low pass arm filters that have impulse response of $h_2(t)$. Note that the bandwidths indicated in Figure 7.2-5 denote the noise bandwidth, which is given by

$$B_2=\int_{-\infty}^{\infty}\left|H_2(f)\right|^2df \quad (7.2\text{-}138)$$

We are now in a position to evaluate the tracking error variance of the dot product code-tracking loop. From (7.2-117) we have

$$\sigma_{sym}^2=E\left\{\left(\frac{\varepsilon}{T_s}\right)^2\right\}=\int_{-\infty}^{\infty}\left|H(j2\pi f)\right|^2\frac{S_{n_1+n_2}(f)}{(2P)^2}\cong\frac{(2B_L)S_{n_1+n_2}(f)}{(2P)^2} \quad (7.2\text{-}139)$$

and since $n_1(t)$ and $n_2(t)$ are statistically independent, we have

$$S_{n_1+n_2}(f)=S_{n_1}(f)+S_{n_2}(f) \quad (7.2\text{-}140)$$

One has for the mean squared tracking error

$$\sigma_{sym}^2=\frac{2B_L}{4P^2}\left[PN_0\left(1-R_{PN}(2\delta T_s)\right)+2N_0^2B_2^{'}\left[1-R_{PN}(2\delta T_s)\right]\right] \quad (7.2\text{-}141)$$

Or in a little simpler form, our result for the noncoherent dot product with PMFs is

$$\sigma_{sym}^2=\frac{N_0B_L\left(1-R_{PN}(2\delta T_s)\right)}{2P}\left[1+\frac{2N_0B_2^{'}}{P}\right],\ \text{symbols}^2 \quad (7.2\text{-}142)$$

which is the general result. From [8] the relationship between B_2 and $B_2^{'}$ for an n-pole Butterworth filter

$$B_2^{'}=\left(1-\frac{1}{2n}\right)B_2 \quad (7.2\text{-}143)$$

Clearly $B_2^{'}$ is smaller than B_2. In the limit for large n, it is seen that $B_2^{'} \cong B_2$. This general result can yield the common NRZ case result, noting that $(1-R_{PN}(2\delta T_s)) = d$. Therefore

$$\sigma_{sym}^2 = \frac{N_0 B_L d}{2P}\left[1+\frac{2N_0 B_2^{'}}{P}\right] \quad \text{symbols}^2 \qquad (7.2\text{-}144)$$

Conversion to the error variance in $(s)^2$ is accomplished by multiplying the right side of (7.2-144) by T_s^2. Furthermore conversion from time to distance for pseudonoise error estimates is accomplished by multiplying the variance of the time error by c^2 so that

$$\sigma_r^2 = T_s^2 c^2 \frac{N_0 B_L (1-R_{PN}(2\delta T_s))}{2P}\left[1+\frac{2N_0 B_2^{'}}{P}\right] \quad \text{meters}^2 \qquad (7.2\text{-}145)$$

where c is the speed of light. In the same manner for NRZ symbols we have

$$\sigma_r^2 = T_s^2 c^2 \frac{N_0 B_L d}{2P}\left[1+\frac{2N_0 B_2^{'}}{P}\right] \quad \text{meters}^2 \qquad (7.2\text{-}146)$$

7.2.5 Noncoherent I-Q Dot Product Code-Tracking Loop with Active Arm Filters

Another important code-tracking loop is the dot product loop with active arm filters. Its model is shown in Figure 7.2-9 which illustrates the loop is very similar to the previous loop model, except that the arm filters are now active, rather than passive.

In an analysis similar to the previous one [9], it has been shown than that the code-tracking error performance, in the general case, is given by

$$\sigma_{sym}^2 = \frac{N_0 B_L (1-R_{PN}(2\delta T_s))}{2P}\left[1+\frac{N_0}{PT_d}\right] \quad \text{symbols}^2 \qquad (7.2\text{-}147)$$

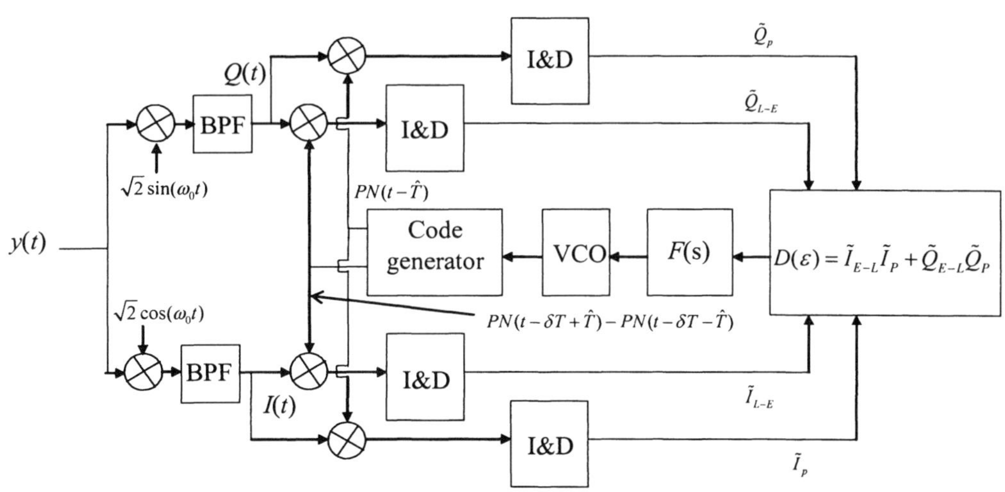

Figure 7.2-9 Dot product noncoherent I-Q code-tracking loop model with integrate-and-dump arm filters.

in which $R_d = 1/T_d$ is the symbol or bit rate depending on whether the signal is coded or not. When NRZ symbols are used the general result of (7.2-147) simplifies to

$$\sigma_{sym}^2 = \frac{N_0 B_L d}{2P}\left[1 + \frac{N_0}{PT_d}\right] \quad symbols^2 \qquad (7.2\text{-}148)$$

This result agrees with [10] for NRZ symbols.

When the dot product loop (7.2-148) is compared to the early-late gate noncoherent loop performance (7.2-90), we see that the dot product loop has slightly lower tracking error variance at the lower P/N_0 values, since $(2/(2-d)) \geq 1$.

Example 1 Let us compare the dot product and the late minus early squared code-tracking loops for tracking error performance. Assume that the BPSK signal is uncoded, and has a 50 bps data rate. Compare the two cases in tracking error squared. A comparison is illustrated in Figure 7.2-10.

As can be seen in the figure the performance is very close with the dot product being slightly better (smaller sigma) at the lower values of C/N_0.

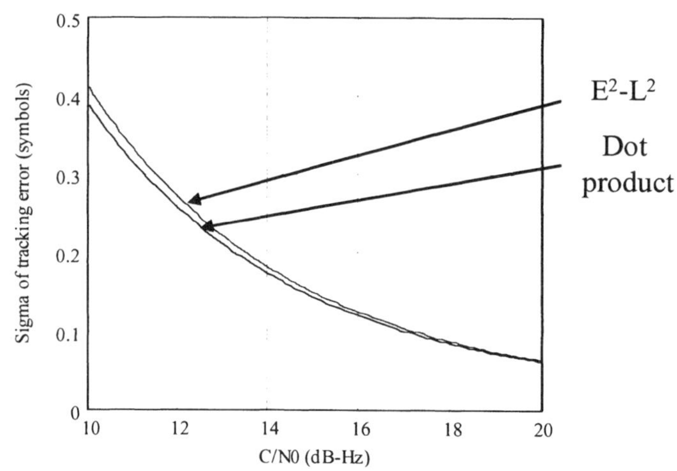

Figure 7.2-10 Comparison of the sigma for the dot product and early-late gate noncoherent code-tracking loops with $B_L = 2$ Hz, 50 bps data rate, and $d = ¼$.

7.3 EARLY-LATE GATE NONCOHERENT I-Q CODE-TRACKING WITH FILTERING AND INTERFERENCE

In this section we will treat the effects of channel filtering on the performance of the early-late gate code-tracking loop when corrupted by white Gaussian noise and Gaussian interference [11]. The analysis is quite lengthy, so it will be summarized here and the curious reader can consult [11].

7.3.1 Signal Model for the Noncoherent I-Q Early-Late Gate Code-Tracking Loop with Channel Filtering

The received signal is composed of the sum of a band-limited BPSK direct sequence signal with BPSK data modulation and a band-limited Gaussian random noise interference, which could be just thermal noise or the sum of thermal and Gaussian interference noise (including jammer noise). It will be assumed that the noise, which includes the thermal noise and the interfering noise-like signal, has a spectral density that is even in the frequency variable f.

The received signal, at the input of the receiver, is described by

$$y(t)=\sqrt{2P}d(t)PN(t-T)\cos(\omega_0 t)+\sqrt{2}n_1(t)\cos(\omega_0 t)+\sqrt{2}n_2(t)\sin(\omega_0 t) \tag{7.3-1}$$

where P is the received signal power in watts, $PN(t)$ is the binary valued (± 1) spread spectrum direct sequence spreading code with code symbol duration T_s, $d(t)$ is the nonreturn to zero (NRZ) data signal, and T is the (initially) unknown delay from the transmitter to the receiver. It will be assumed that the data and the code sequence are connected to the same clock. Hence knowing the code-state implies knowing the data transition points in time. The last two terms in (7.3-1) represent the inphase and quadrature phases of the sum of two filtered, statistically independent random noise processes, representing the sum of thermal noise and a Gaussian interference process. Thus $y(t)$ can be written as $y(t)=x(t)+n(t)$, where $x(t)$ is the signal and $n(t)$ is the sum of thermal noise and interference. The channel filtering is due to the presence of a bandpass filter (BPF) at the front end of the receiver.

The code-tracking loop under consideration is shown in Figure 7.3-1, which we have studied earlier. It will be assumed that $B_L T_b \ll 1$, where B_L is the closed-loop noise bandwidth of the loop and T_b is the bit duration. The input is filtered with a B Hz (3 dB) bandwidth filter, which produces the filtered input $\tilde{y}(t)$. The transfer function of the bandpass filter is $H(f)$, and the baseband equivalent transfer function is $H_{BB}(f)$. The baseband equivalent transfer function $H_{BB}(f)$ can be related to the bandpass filter transfer function $H(f)$, by the following [12]:

$$\begin{aligned} H_{BB}(f-f_c)&=H(f) \quad f>0 \\ &= 0 \quad\quad\;\; f<0 \end{aligned} \tag{7.3-2}$$

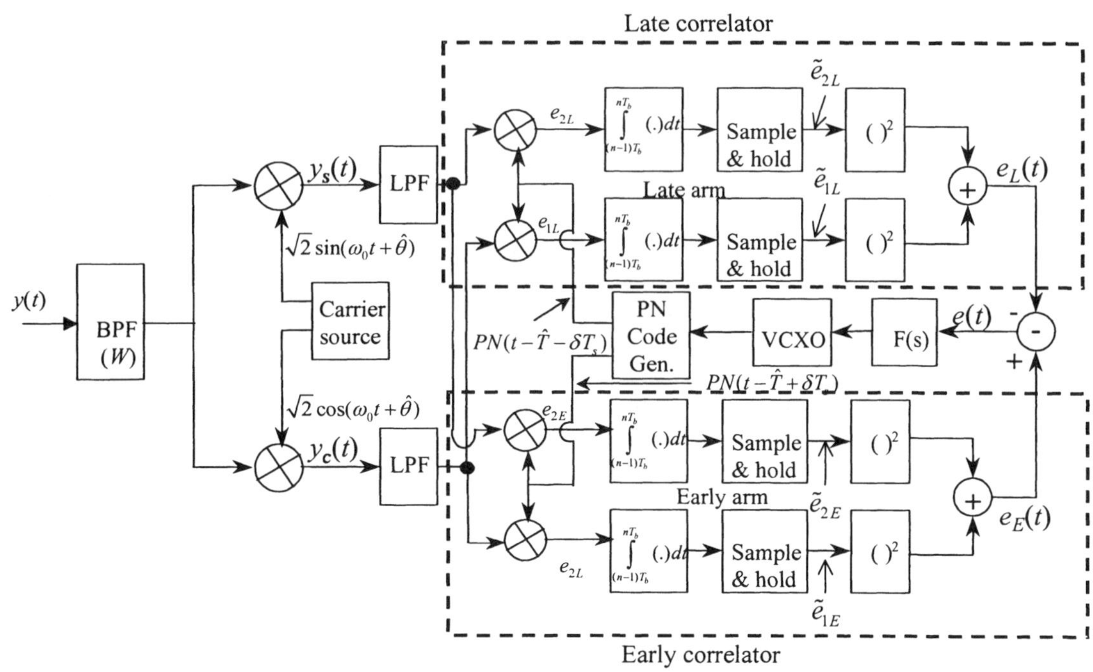

Figure 7.3-1 Noncoherent early-late gate code-tracking loop implemented at baseband with channel filtering.

Let $h(t)$ denote the radio frequency filter impulse response and the Fourier transform of $H(f)$; then since $h(t)$ is real, it follows that

$$H^*(-f)=H(f) \tag{7.3-3}$$

Using (7.3-3) in (7.3-2) produces

$$\begin{aligned} H^*_{BB}(-f-f_c)&=0 \qquad f>0 \\ &= H^*(-f) \quad f<0 \end{aligned} \tag{7.3-4}$$

It follows that the RF filter transfer function can be written in terms of the baseband transfer function as

$$H(f)=H_{BB}(f-f_c)+H^*{}_{BB}(-f-f_c) \tag{7.3-5}$$

In addition, the impulse function response can be written [12] as

$$h(t)=2\,\mathrm{Re}\left[h_{BB}(t)e^{j2\pi f_c t}\right] \tag{7.3-6}$$

where $h_{BB}(t)$ is the Fourier transform of $H_{BB}(f)$. Thus, the baseband impulse response can be written as

$$h_{BB}(t)=F\{H_{BB}(f)\}=\int_{-\infty}^{\infty} H_{BB}(f)e^{j2\pi ft}df \tag{7.3-7}$$

where $F\{H_{BB}(f)\}$ denotes the Fourier transform of $h_{BB}(t)$. It will be assumed that the baseband equivalent filter transfer function is *complex conjugate symmetric*; that is,

$$H_{BB}(f)=H^*_{BB}(-f) \tag{7.3-8}$$

The received signal, after being filtered by the front-end bandpass filter, can be expressed as

$$\begin{aligned} \tilde{y}(t)&=\sqrt{2P}d(t)\widetilde{PN}(t-T)\cos(\omega_0 t) \\ &+\sqrt{2}\tilde{n}_1(t)\cos(\omega_0 t)+\sqrt{2}\tilde{n}_2(t)\sin(\omega_0 t) \end{aligned} \tag{7.3-9}$$

where the tilde over a term in an equation implies that the front-end filter filters it. It was assumed that the data rate is very low compared to the PN code rate, and thus the filtered data and filtered code sequence can be factored with very small error. In fact since the data rate is so low, the filtering effect on the data will be negligible and is ignored in the analysis to follow.

In this analysis the filter on the input models the product of the transmit filter and the receive filter(s). The baseband equivalent process can be described in terms of the baseband equivalent channel impulse response function as

$$\begin{aligned} \widetilde{PN}(t-T)&=\int_{-\infty}^{t-T} h_{BB}(t-T-u)PN(u)du \\ &=\int_{-\infty}^{t} h_{BB}(t-v)PN(v-T)dv \end{aligned} \tag{7.3-10}$$

Consider now the outputs at the $y_s(t)$ and $y_c(t)$ points in the code-tracking loop. The upper output labeled as $y_s(t)$ in Figure 7.3-1, neglecting the sum frequency terms, is given by

$$y_s(t) = \sqrt{2P}d(t)\tilde{P}N(t-T)\cos(\omega_0 t)\sqrt{2}\sin(\omega_0 t - \phi) \\ -\tilde{n}_1(t)\sin\phi + \tilde{n}_2(t)\cos\phi \tag{7.3-11}$$

It is assumed that the carrier reference is at some random phase ϕ, relative to the input carrier phase. In the same manner, the lower signal $y_c(t)$ is given by

$$y_c(t) = \sqrt{2P}d(t)\tilde{P}N(t-T)\cos(\omega_0 t)\sqrt{2}\cos(\omega_0 t - \phi) \\ +\tilde{n}_1(t)\cos\phi + \tilde{n}_2(t)\sin\phi \tag{7.3-12}$$

where the filtered noise terms are given by the baseband equivalent impulse responses as

$$\tilde{n}_i(t) = \int_{-\infty}^{t} h_{BB}(t-u)n_i(u)du \quad i = 1 \text{ or } 2 \tag{7.3-13}$$

7.3.2 Signal and Noise Terms in the Noncoherent I-Q Early-Late Gate Code Loop with Channel Filtering

With the signal and the interference modeled, the internal signals and interference can be modeled. Rewrite $y_c(t)$ as

$$y_c(t) = e_{cc}(t) + \tilde{n}_1(t)\cos\phi + \tilde{n}_2(t)\sin\phi \tag{7.3-14}$$

Where

$$e_{cc}(t) = \sqrt{2P}d(t)\widetilde{PN}(t-T)\cos(\omega_0 t)\sqrt{2}\cos(\omega_0 t - \phi) \tag{7.3-15}$$

In a similar manner $y_s(t)$ can be written as

$$y_s(t) = e_{cs(t)} - \tilde{n}_1(t)\sin\phi + \tilde{n}_2\cos\phi \tag{7.3-16}$$

where

$$e_{cs(t)} = \sqrt{2P}d(t)\widetilde{PN}(t-T)\cos(\omega_0 t)\sqrt{2}\sin(\omega_0 t - \phi) \tag{7.3-17}$$

Now consider the four despread terms e_{2L}, e_{2E}, e_{1L}, and e_{1E}, which can be written as

$$e_{2L} = e_{cs}(t)\cdot PN(t-\hat{T}-\delta T_s) - \tilde{n}_1(t)\sin\phi \ PN(t-\hat{T}-\delta T_s) \\ +\tilde{n}_2(t) \ \cos\phi \ PN(t-\hat{T}-\delta T_s) \tag{7.3-18}$$

$$e_{2E} = e_{cs}(t)\cdot PN\left(t-\hat{T}+\delta T_s\right) - \tilde{n}_1(t)\sin\phi \;\; PN\left(t-\hat{T}+\delta T_s\right) + \tilde{n}_2(t)\;\cos\phi\;\; PN\left(t-\hat{T}+\delta T_s\right) \tag{7.3-19}$$

$$e_{1L} = e_{cc}(t)\cdot PN\left(t-\hat{T}-\delta T_s\right) + \tilde{n}_1(t)\cos\phi \;\; PN\left(t-\hat{T}-\delta T_s\right) + \tilde{n}_2(t)\;\sin\phi\;\; PN\left(t-\hat{T}-\delta T_s\right) \tag{7.3-20}$$

$$e_{1E} = e_{cc}(t)\cdot PN\left(t-\hat{T}+\delta T_s\right) + \tilde{n}_1(t)\cos\phi \;\; PN\left(t-\hat{T}+\delta T_s\right) + \tilde{n}_2(t)\;\sin\phi\;\; PN\left(t-\hat{T}+\delta T_s\right) \tag{7.3-21}$$

Now to simplify the analysis let $\phi = 0$ on the noise terms only since it can be shown that this assumption does not affect the result (it is a noncoherent loop), so that one obtains

$$e_{2L} = e_{cs}(t)\cdot PN\left(t-\hat{T}-\delta T_s\right) + \tilde{n}_2(t)\;\; PN\left(t-\hat{T}-\delta T_s\right) \tag{7.3-22}$$

$$e_{2E} = e_{cs}(t)\cdot PN\left(t-\hat{T}+\delta T_s\right) + \tilde{n}_2(t)\;\; PN\left(t-\hat{T}+\delta T_s\right) \tag{7.3-23}$$

$$e_{1L} = e_{cc}(t)\cdot PN\left(t-\hat{T}-\delta T_s\right) + \tilde{n}_1(t)\;\; PN\left(t-\hat{T}-\delta T_s\right) \tag{7.3-24}$$

$$e_{1E} = e_{cc}(t)\cdot PN\left(t-\hat{T}+\delta T_s\right) + \tilde{n}_1(t)\;\; PN\left(t-\hat{T}+\delta T_s\right) \tag{7.3-25}$$

The respective filtered signals with a "tilde" over the filtered variable, at the output of the averaging integrate-and-dump filters, are given by

$$\tilde{e}_{2L} = \overbrace{\frac{1}{T_b}\int_0^{T_b} e_{cs}(t)\;PN\;\left(t-\hat{T}-\delta T_s\right)dt}^{e_{CSL}} + \overbrace{\frac{1}{T_b}\int_0^{T_b} \tilde{n}_2(t)PN\;\left(t-\hat{T}-\delta T_s\right)dt}^{N_{2L}} \tag{7.3-26}$$

$$\tilde{e}_{2E} = \overbrace{\frac{1}{T_b}\int_0^{T_b} e_{cs}(t)\;PN\;\left(t-\hat{T}+\delta T_s\right)dt}^{e_{CEL}} + \overbrace{\frac{1}{T_b}\int_0^{T_b} \tilde{n}_2(t)PN\;\left(t-\hat{T}+\delta T_s\right)dt}^{N_{2E}} \tag{7.3-27}$$

$$\tilde{e}_{1L} = \overbrace{\frac{1}{T_b}\int_0^{T_b} e_{cc}(t)\;PN\;\left(t-\hat{T}-\delta T_s\right)dt}^{e_{CEL}} + \overbrace{\frac{1}{T_b}\int_0^{T_b} \tilde{n}_1(t)PN\;\left(t-\hat{T}-\delta T_s\right)dt}^{N_{1L}} \tag{7.3-28}$$

$$\tilde{e}_{1E} = \overbrace{\frac{1}{T_b}\int_0^{T_b} e_{cc}(t)\;PN\;\left(t-\hat{T}+\delta T_s\right)dt}^{e_{CSL}} + \overbrace{\frac{1}{T_b}\int_0^{T_b} \tilde{n}_1(t)PN\;\left(t-\hat{T}+\delta T_s\right)dt}^{N_{1E}} \tag{7.3-29}$$

7.3.3 Signal Terms in the Noncoherent I-Q Early-Late Gate Code Loop with Channel Filtering

In this section the signal terms out of the early and late correlators will be considered and complex notation will be utilized for analytical convenience. Figure 7.3-2 illustrates the signal model that will be used in the analysis to follow.

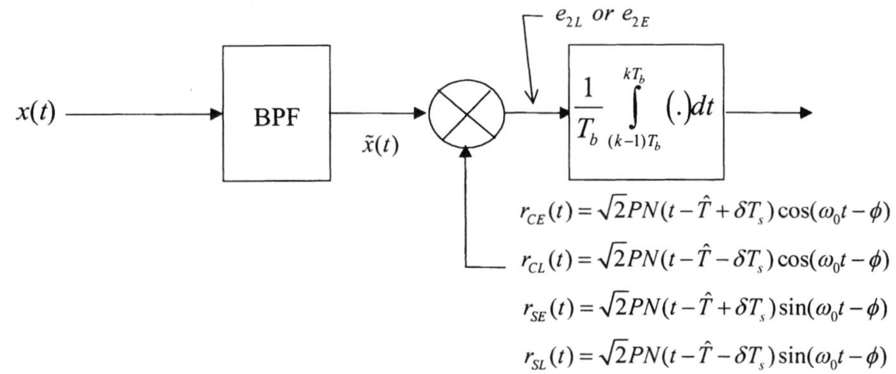

Figure 7.3-2 Model for the signal components.

The input is given by $x(t)$, as indicated in Figure 7.3-2. The filtered signal, out of the bandpass filter (BPF), is denoted by $\tilde{x}(t)$. The two reference signals used for $e_{1E}(t)$ and $e_{1L}(t)$ in the code-tracking loop of Figure 7.3-1, $r_{CE}(t)$ and $r_{CL}(t)$, are shown in Figure 7.3-2.

Repeating (7.3-26)–(7.3-29) and explicitly writing out $e_{CS}(t)$ and $e_{CC}(t)$, one has, in slightly different order,

$$\tilde{e}_{1E}=\frac{1}{T_b}\int_0^{T_b}\sqrt{2P}d(t)\tilde{P}N(t-T)\cos(\omega_0 t)\overbrace{\sqrt{2}\;PN(t-\hat{T}+\delta T_s)\cos(\omega_0 t-\phi)}^{r_{CE}(t)}dt+N_{1E} \tag{7.3-30}$$

$$\tilde{e}_{1L}=\frac{1}{T_b}\int_0^{T_b}\sqrt{2P}d(t)\tilde{P}N(t-T)\cos(\omega_0 t)\overbrace{\sqrt{2}PN(t-\hat{T}-\delta T_s)\cos(\omega_0 t-\phi)}^{r_{CL}(t)}dt+N_{1L} \tag{7.3-31}$$

$$\tilde{e}_{2E}=\frac{1}{T_b}\int_0^{T_b}\sqrt{2P}d(t)\tilde{P}N(t-T)\cos(\omega_0 t)\overbrace{\sqrt{2}PN(t-\hat{T}+\delta T_s)\sin(\omega_0 t-\phi)}^{r_{SL}(t)}dt+N_{2E} \tag{7.3-32}$$

$$\tilde{e}_{2L}=\frac{1}{T_b}\int_0^{T_b}\sqrt{2P}d(t)\tilde{P}N(t-T)\cos(\omega_0 t)\overbrace{\sqrt{2}\;PN(t-\hat{T}-\delta T_s)\sin(\omega_0 t-\phi)}^{r_{SE}(t)}dt+N_{2L} \tag{7.3-33}$$

Now consider evaluating $\tilde{e}_{1E}$ and $\tilde{e}_{1L}$. To do this we shall neglect the data term, since in the actual model the correlation will only last for the data integration time and thus will not affect the noncoherent loop performance. To do this write the two terms in complex form. Hence[5]

$$\tilde{x}(t)=\sqrt{2P}\,\mathrm{Re}\left[e^{j\omega_0 t}\widetilde{PN}(t-T)\right]=\sqrt{2P}\,\mathrm{Re}\left[e^{j\omega_0 t}\frac{1}{2\pi}\int_{-\infty}^{\infty}A_{T_1}(\omega)\;H_{BB}(\omega)\;e^{j\omega(t-T)}d\omega\right] \tag{7.3-34}$$

where the integral in brackets represents the filtered version of the Fourier transform of the voltage waveform *PN*(*t*) over T_1 seconds. Later the limit, as T_1 becomes unbounded,[6] will be obtained. The two reference signals can be written in complex form, as

$$r_{CE}(t)=\sqrt{2}\;\mathrm{Re}\left[e^{j(\omega_0 t-\phi)}PN(t-T+\delta T_s)\right]=\sqrt{2}\;\mathrm{Re}\left[e^{j\omega_0 t}e^{-j\phi}\frac{1}{2\pi}\int_{-\infty}^{\infty}A_{T_1}(\omega')\;e^{j\omega'(t-\hat{T}+\delta T_s)}d\omega'\right] \tag{7.3-35}$$

$$r_{CL}(t)=\sqrt{2}\;\mathrm{Re}\left[e^{j(\omega_0 t-\phi)}PN(t-T-\delta T_s)\right]=\sqrt{2}\;\mathrm{Re}\left[e^{j\omega_0 t}e^{-j\phi}\frac{1}{2\pi}\int_{-\infty}^{\infty}A_{T_1}(\omega')\;e^{j\omega'(t-\hat{T}-\delta T_s)}d\omega'\right] \tag{7.3-36}$$

where A_{T_1} is the $2T_1$ second Fourier transform of the PN code, that is

$$PN(t)=\frac{1}{2\pi}\int_{-\infty}^{-\infty}A_{T_1}(\omega)e^{j\omega t}d\omega \tag{7.3-37}$$

where $A_{T_1}(j\omega)$ is the Fourier transform and is given by

$$A_{T_1}(\omega)=\int_{-T_1}^{T_1}PN(t)e^{-j\omega t}dt \tag{7.3-38}$$

Now consider the signal part of $\tilde{e}_{1E}$. Denote the signal part of $\tilde{e}_{1E}$ as $\tilde{s}_{1E}$, and let T_1 become unbounded, so that

$$\tilde{s}_{1E}=\tilde{e}_{1E}-N_{1E}\cong\lim_{T_1\to\infty}\frac{1}{2T_1}E\left\{\int_{-T_1}^{T_1}\tilde{x}(t)r_{CE}(t)dt\right\} \tag{7.3-39}$$

where $E\{x\}$ denotes ensemble averaging, with the PN code sequence regarded as a random, equally likely (uncorrelated from chip to chip) sequence.

Note that for complex variables Z_1 and Z_2, one has

$$\mathrm{Re}(Z_1)\mathrm{Re}(Z_2)=\frac{1}{2}\mathrm{Re}(Z_1Z_2)+\frac{1}{2}\mathrm{Re}\left[Z_1Z_2^*\right] \tag{7.3-40}$$

---

[5] We have used $H_{BB}(f)$ previously, but for convenience we now use $H_{BB}(\omega)$ to mean the same thing, although technically it should be $H_{BB}(\omega/(2\pi))$.

[6] In the analysis the correlation is taken over all time, but in the actual loop operation the correlation is only for the bit duration. Actually the infinite time correlation used in the analysis simplifies some calculations.

and the first term is at the frequency of order $2\omega_0$, and the second term is at baseband and is the term of interest in what follows. Using (7.3-39) in (7.3-40) yields (letting the $E\{.\}$ denote the ensemble average)

$$\tilde{s}_{1E} = \sqrt{P} \lim_{T_1 \to \infty} \frac{1}{2T_1} \int_{-T_1}^{T_1} \operatorname{Re} E \left\{ \begin{matrix} \left\{ \int_{-\infty}^{\infty} A_{T_1}(\omega) H_{BB}(j\omega) e^{j\omega(t-T)} \frac{d\omega}{2\pi} \right\} \times \\ \left\{ e^{j\phi} \int_{-\infty}^{\infty} A_{T_1}^{*}(\omega') e^{-j\omega'(t-\hat{T}+\delta T_s)} \frac{d\omega'}{2\pi} \right\} dt \end{matrix} \right\} \quad (7.3\text{-}41)$$

Or collecting terms and noting that

$$\frac{1}{2\pi} \int_{-\infty}^{\infty} e^{j(\omega-\omega')t}\, dt = \delta(\omega - \omega') \quad (7.3\text{-}42)$$

and that

$$S_{PN}(\omega) = \lim_{T_1 \to \infty} \left(\frac{E\left\{A_{T_1}(\omega) A_{T_1}^{*}(\omega')\right\}}{2T_1} \right) \quad (7.3\text{-}43)$$

it can be shown [8] that

$$\tilde{s}_{1E} = \sqrt{P} \operatorname{Re}\left[\int_{-\infty}^{\infty} H_{BB}(\omega) S_{PN}(\omega) e^{-j\omega(T-\hat{T})} e^{-j\omega\delta T_s} e^{j\phi} \frac{d\omega}{2\pi} \right] \quad (7.3\text{-}44)$$

Now define $T - \hat{T} = \varepsilon$ as the code timing error from input code to the reference code at the output of the bandpass filter. Then one can write (7.3-44) as

$$\tilde{s}_{1E} = \sqrt{P} \operatorname{Re}\left[\int_{-\infty}^{\infty} H_{BB}(\omega)\ S_{PN}(\omega) e^{-j\omega(\varepsilon+\delta T_s)} e^{j\phi} \frac{d\omega}{2\pi} \right] \quad (7.3\text{-}45)$$

Changing to the variable f from the variable ω and suppressing[7] the factors of 2π in the arguments of $H_{BB}(j\omega)$, and reinserting the data terms (which were left out of the infinite time correlation) yield

$$\tilde{e}_{1E} = \sqrt{P} d(t) \overbrace{\operatorname{Re}\left[\int_{-\infty}^{\infty} H_{BB}(f) S_{PN}(f) e^{-j2\pi f(\varepsilon+\delta T_s)} e^{j\phi} df \right]}^{I_{1E}(\varepsilon)} + N_{1E} \quad (7.3\text{-}46)$$

In a similar manner, one can show that

[7] Strictly speaking, we should use a different symbol for $H_{BB}(jf)$ and $H_{BB}(j\omega)$ as well as $S_{PN}(f)$ and $S_{PN}(\omega)$; however, to simplify the notation we will not. If the baseband response is written in a normalized form—that is, $H_{BB}(jf/f_0)$ or $H_{BB}(j\omega/\omega_0)$—there will be no confusion, and the same is true for the power spectral density.

$$\tilde{e}_{1L} = \sqrt{P}d(t)\text{Re}\left[\overbrace{\int_{-\infty}^{\infty} H_{BB}(f)S_{PN}(f)e^{-j2\pi f(\varepsilon-\delta T_s)}e^{j\phi}df}^{I_{1L}(\varepsilon)}\right] + N_{1L} \tag{7.3-47}$$

where $I_{1E}(\varepsilon)$ and $I_{1L}(\varepsilon)$ are defined from (7.3-46) and (7.3-47). The two expressions, (3.15) and (3.16), neglect the self-noise terms (we have used the mean value only), and they also neglect the double-frequency terms. These effects would normally be swamped out by thermal noise anyway except at very high signal-to-noise ratios.

Now a simplification will be introduced that will reduce the number of terms in the analysis without affecting the result. First we will show that the value of ϕ in (7.3-46) and (7.3-47) that maximizes the output of the in-phase correlated channel is $\phi = 0 \mod 2\pi$! As a consequence, assuming $\phi = 0$, $e_{cs}(t) = 0$ (neglecting the sum frequency term which will have no effect at baseband) at baseband so that at the output of the sample and hold

$$\tilde{e}_{2E} = N_{2E} \tag{7.3-48}$$

$$\tilde{e}_{2L} = N_{2L} \tag{7.3-49}$$

$$\tilde{e}_{1E} = \sqrt{P}d(t)\text{Re}\left[\int_{-\infty}^{\infty} H_{BB}(f)S_{PN}(f)e^{-j2\pi f(\varepsilon+\delta T_s)}df\right] + N_{1E} \tag{7.3-50}$$

$$\tilde{e}_{1L} = \sqrt{P}d(t)\text{Re}\left[\int_{-\infty}^{\infty} H_{BB}(f)S_{PN}(f)e^{-j2\pi f(\varepsilon-\delta T_s)}df\right] + N_{1L} \tag{7.3-51}$$

Now let us see why $\phi = 0$ maximizes the signal component of the in-phase channel and at the same time zeros the signal component of the quadrature channel (see (7.3-48) and (7.3-49)) which indicate that the signal components of $\tilde{e}_{2E}$ and $\tilde{e}_{2L}$ are zero. Also, we will show that the "real part of" the operator can be removed when $\phi = 0$. From (7.3-46) and (7.3-47), it follows that

$$\tilde{e}_{1E} = \sqrt{P}d(t)I_{1E}(\varepsilon) + N_{1E} \tag{7.3-52}$$

where

$$I_{1E}(\varepsilon) = \text{Re}\left[\int_{-\infty}^{\infty} H_{BB}(f)S_{PN}(f)e^{-j2\pi f(\varepsilon+\delta T_s)}e^{j\phi}df\right] \tag{7.3-53}$$

For complex symmetric baseband equivalent filter transfer functions one has

$$H_{BB}^{*}(f) = H_{BB}(-f) \tag{7.3-54}$$

and since Re{Z}= (Z+Z*)/2, one has

$$I_{1E}(\varepsilon) = \frac{1}{2}\int_{-\infty}^{\infty} S_{PN}(f)\left[H_{BB}(f)e^{-j2\pi f(\varepsilon+\delta T_s)}e^{j\phi} + H_{BB}(-f)e^{j2\pi f(\varepsilon+\delta T_s)}e^{-j\phi}\right]df \tag{7.3-55}$$

Let $-f \to f'$ in the last integral to obtain

$$I_{1E}(\varepsilon)=\frac{1}{2}\int_{-\infty}^{\infty} S_{PN}(f)\ H_{BB}(f)e^{-j2\pi f(\varepsilon+\delta T_s)}e^{j\phi}\,df + \frac{1}{2}\int_{+\infty}^{-\infty} H_{BB}^{*}(f')e^{-j2\pi f'(\varepsilon+\delta T_s)}e^{-j\phi}(-df') \tag{7.3-56}$$

or simplifying the second term, one has

$$I_{1E}(\varepsilon)=\int_{-\infty}^{\infty} S_{PN}(f)\ H_{BB}(f)e^{-j2\pi f(\varepsilon+\delta T_s)}\left[\frac{e^{j\phi}+e^{-j\phi}}{2}\right]df \tag{7.3-57}$$

or

$$I_{1E}(\varepsilon)=\int_{-\infty}^{\infty} S_{PN}(f)\ H_{BB}(f)e^{-j2\pi f(\varepsilon+\delta T_s)}\cos\phi df \tag{7.3-58}$$

The maximum occurs at $\phi = 0 \bmod 2\pi$, so that one has

$$I_{1E}(\varepsilon)=\int_{-\infty}^{\infty} S_{PN}(f)\ H_{BB}(f)e^{-j2\pi f(\varepsilon+\delta T_s)}df \tag{7.3-59}$$

In a similar manner one has

$$I_{1L}(\varepsilon)=\int_{-\infty}^{\infty} S_{PN}(f)\ H_{BB}(f)e^{-j2\pi f(\varepsilon-\delta T_s)}df \tag{7.3-60}$$

From Figure 7.2-3 and (7.3-48) through (7.3-51), one obtains

$$e_E(\varepsilon)=\left(\tilde{e}_{1E}(\varepsilon)\right)^2+\left(\tilde{e}_{2E}(\varepsilon)\right)^2 \quad \text{and} \quad e_L(\varepsilon)=\left(\tilde{e}_{1L}(\varepsilon)\right)^2+\left(\tilde{e}_{2L}(\varepsilon)\right)^2 \tag{7.3-61}$$

$$e_E(\varepsilon)=\left(N_{2E}\right)^2+\left(\sqrt{P}d(t)I_{1E}(\varepsilon)+N_{1E}\right)^2 \tag{7.3-62}$$

Expanding $e_E(\varepsilon)$, one has

$$e_E(\varepsilon)=N_{2E}^2+P\left(I_{1E}(\varepsilon)\right)^2+2\sqrt{P}d(t)I_{1E}(\varepsilon)N_{1E}+N_{1E}^2 \tag{7.3-63}$$

and

$$e(\varepsilon)=e_E(\varepsilon)-e_L(\varepsilon) \tag{7.3-64}$$

Expanding $e_L(\varepsilon)$ one has

$$e_L(\varepsilon)=\left(N_{2L}\right)^2+P\left(I_{1L}(\varepsilon)\right)^2+2\sqrt{P}d(t)I_{1L}(\varepsilon)N_{1L}+\left(N_{1L}\right)^2 \tag{7.3-65}$$

Summarizing, the signal terms are given by (7.3-59) and (7.3-60), and the noise terms by (7.3-26)–(7.3-29). It follows that the error control signal of the code-tracking loop, from (7.3-64), (7.3-63), and (7.3-65), is given by

$$e(\varepsilon)=\overbrace{\left[\left(\sqrt{P}I_{1L}(\varepsilon)\right)^2-\left(\sqrt{P}I_{1E}(\varepsilon)\right)^2\right]}^{T_1(\varepsilon)}+2\left(\sqrt{P}\,d(t)I_{1L}(\varepsilon)N_{1L}-\sqrt{P}d(t)I_{1E}(\varepsilon)N_{1E}\right) \quad (7.3\text{-}66)$$
$$\left[(N_{2L})^2+(N_{1L})^2\right]-\left[(N_{2E})^2+(N_{1E})^2\right]$$

We are now in a position to derive the closed-loop response of the code-tracking loop. However before we do let us observe the S-curves for the noncoherent early-late gate code-tracking loop. Figure 7.3-3 illustrates the S-curves for d = 0.2, 0.4, 0.6, 0.8, and 1 for an NRZ random code running at 1.023 Mcps, with a 4-and 14-MHz bandwidth ideal bandpass filter.

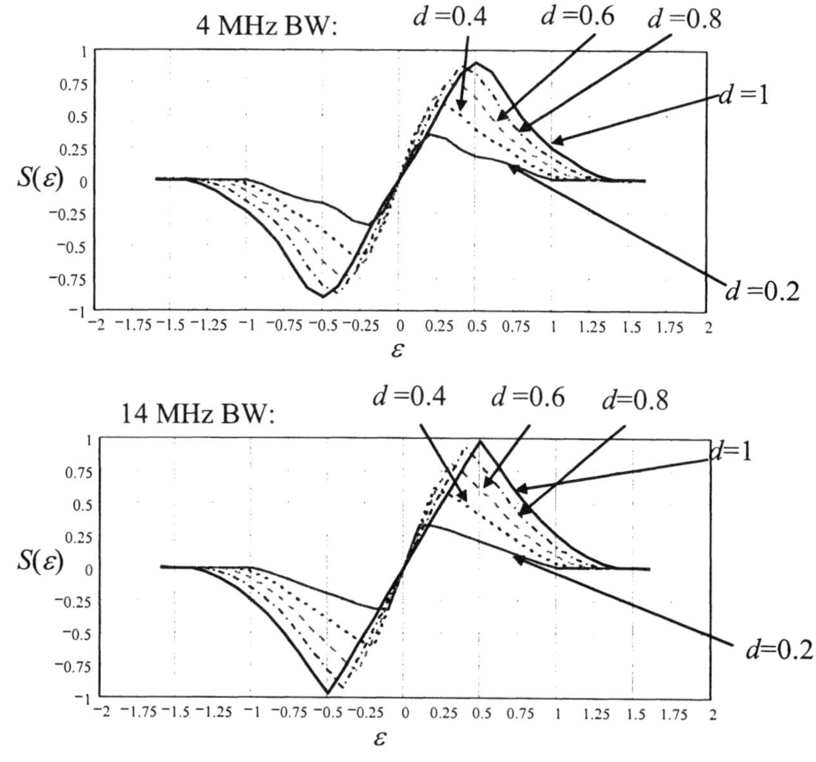

Figure 7.3-3 S-curve with 4-and 14-MHz ideal bandpass filter for the noncoherent early-late gate code-tracking loop.

Notice that the peaks are rounded off in the 4-MHz case, and the nearly linear error curve peak value is reduced for smaller values of d. In the 14-MHz bandwidth case the S-curve is has very little rounding off.

7.3.4 Closed-Loop Operation of the Noncoherent I-Q Early-Late Gate Code Loop with Channel Filtering

In this section the linearized code-tracking loop model will be developed. To do this, a linear error model must be obtained from the signal component of the error signal of the code-tracking loop. To this end, consider the gain (slope) around the point $\varepsilon = \varepsilon_0$, the lock point,[8] by analyzing $T_1(\varepsilon)$ in(7.3-66) for small values of error, one has

$$T_1(\varepsilon) = PKT_c\varepsilon \quad \text{(for small } \varepsilon\text{)} \tag{7.3-67}$$

where P is the received signal power, K is the gain which has units of seconds^{-1}, $K_1 = KT_s$ is the gain per PN symbol, and the code timing error ε is in units of seconds. From (7.3-66), $T_1(\varepsilon)$ is the signal-dependent error control signal (also called the S-curve[9] (or discriminator function)). Consider the determination of K.

$$T_1(\varepsilon) = P\left(I_{1E}^2(\varepsilon) - I_{1L}^2(\varepsilon)\right) \tag{7.3-68}$$

The (power-normalized) slope K is defined by

$$K = \frac{\partial T_1(\varepsilon)}{P\partial\varepsilon}\Big|_{\varepsilon=\varepsilon_0} \tag{7.3-69}$$

and has units of seconds^{-1}. Consider

$$\frac{\partial}{\partial\varepsilon}\left[T_1(\varepsilon)\right] = \left[\overbrace{P\frac{\partial}{\partial\varepsilon}\left[I_{1E}^2(\varepsilon)\right]}^{T_{11}(\varepsilon)} - \overbrace{P\frac{\partial}{\partial\varepsilon}\left[I_{1L}^2(\varepsilon)\right]}^{T_{12}(\varepsilon)}\right] \tag{7.3-70}$$

So that one can write

$$\frac{\partial T_1}{\partial\varepsilon}(\varepsilon) = T_{11}(\varepsilon) - T_{12}(\varepsilon) \tag{7.3-71}$$

where

$$T_{11}(\varepsilon) = \frac{\partial}{\partial\varepsilon}P\left\{\left[\int_{-\infty}^{\infty} H_{BB}(f)S_{PN}(f)e^{-j2\pi f\varepsilon}e^{-j2\pi f\delta T_s}df\right]\right\}^2 \tag{7.3-72}$$

Taking the derivative

[8] Note that the lock point will not be zero error except when the filtering is negligible for real filters. The lock point has to be determined from the S-curve. ε_0 is obtained when the error control signal equals zero in the linear region of the S-curve.

[9] Note in this section we are using $E^2 - L^2$ rather than the appropriate L^2-E^2; however, the final results for tracking error are the same.

$$T_{11}(\varepsilon)=2P\left\{\left[\int_{-\infty}^{\infty} H_{BB}(f)S_{PN}(f)e^{-j2\pi f(\varepsilon+\delta T_s)}df\right]\right\} \times\left\{\left[\int_{-\infty}^{\infty} H_{BB}(f)S_{PN}(f)e^{-j2\pi f(\varepsilon+\delta T_s)}(-j2\pi f)df\right]\right\} \tag{7.3-73}$$

Now evaluate this at the stable lock point, $\varepsilon=\varepsilon_0$

$$T_{11}(\varepsilon_0)=2P\left\{\left[\int_{-\infty}^{\infty} H_{BB}(f)S_{PN}(f)e^{-j2\pi f(\varepsilon_0+\delta T_s)}df\right]\right\} \times\left\{\left[\int_{-\infty}^{\infty} H_{BB}(f)S_{PN}(f)e^{-j2\pi f(\varepsilon_0+\delta T_s)}(-j2\pi f)df\right]\right\} \tag{7.3-74}$$

Consider $T_{12}(\varepsilon)$, from (7.3-70) and (7.3-53) one has

$$T_{12}(\varepsilon)=\frac{\partial}{\partial\varepsilon}P\left\{\left[\int_{-\infty}^{\infty} H_{BB}(f)S_{PN}(f)e^{-j2\pi f(\varepsilon-\delta T_s)}df\right]\right\}^2 \tag{7.3-75}$$

so that

$$T_{12}(\varepsilon_0)=2P\left\{\left[\int_{-\infty}^{\infty} H_{BB}(f)S_{PN}(f)e^{-j2\pi f(\varepsilon_0-\delta T_s)}df\right]\right\} \times\left\{\left[\int_{-\infty}^{\infty} H_{BB}(f)S_{PN}(f)e^{-j2\pi f(\varepsilon_0-\delta T_s)_s}(-j2\pi f)df\right]\right\} \tag{7.3-76}$$

Now the gain (K, the slope) can be written as

$$K=\left[\frac{\partial}{\partial\varepsilon}\left(I_{1E}^2(\varepsilon)\right)-\frac{\partial}{\partial\varepsilon}\left(I_{1L}^2(\varepsilon)\right)\right]\Bigg|_{\varepsilon=\varepsilon_0} \tag{7.3-77}$$

which can be also written in the per chip form[10] as

$$KT_s=2\left\{\begin{matrix}\left[\int_{-\infty}^{\infty} H_{BB}(f)S_{PN}(f)e^{-j2\pi f(\varepsilon_0+\delta T_s)}df\right]\cdot\left[\int_{-\infty}^{\infty} H_{BB}(f)S_{PN}(f)(-j2\pi f)e^{-j2\pi f(\varepsilon_0+\delta T_s)}df\right]\\ -\left[\int_{-\infty}^{\infty} H_{BB}(f)S_{PN}(f)e^{+j2\pi f(\varepsilon_0-\delta T_s)}df\right]\cdot\left[\int_{-\infty}^{\infty} H_{BB}(f)S_{PN}(f)(-j2\pi f)e^{+j2\pi f(\varepsilon_0-\delta T_s)}df\right]\end{matrix}\right\}T_c \tag{7.3-78}$$

So, for small ϵ, (7.3-66), one can write

$$e(\varepsilon)=KP\varepsilon+\overbrace{2\sqrt{P}d(t)\left[I_{1L}(0)N_{1L}-I_{1E}(0)N_{1E}\right]}^{n_A(t)}+\overbrace{\left[N_{2L}^2+N_{1L}^2\right]-\left[N_{2E}^2+N_{1E}^2\right]}^{n_B(t)} \tag{7.3-79}$$

---

[10] Note that since we assumed E^2-L^2 we will generally have a negative gain. If we had used L2-E2 we would obtain the negative of (7.3-78).

where the assumption that small ε was used for the signal term, but $\varepsilon = 0$ was used in the noise terms, under the understanding that only small errors are considered. Therefore, in terms of the error signal and the two noise terms, one can write for small errors

$$e(\varepsilon) = KT_sP\varepsilon + n_A(t) + n_B(t) = K_1P\varepsilon + n_A(t) + n_B(t) \tag{7.3-80}$$

where $n_A(t)$ and $n_B(t)$ are defined in (7.3-79) and $K_1 = KT_s$. The delay estimate of the loop takes the error signal, filters it by the loop filter $F(s)$ and the VCO model (K_v/s), and using the Heaviside operator notation, produces the code loop delay estimate $\hat{T}$ given by

$$\hat{T}(s) = \frac{K_vF(s)}{s}\left[K_1P\varepsilon + \overbrace{n_A(t) + n_B(t)}^{N(t)}\right] \tag{7.3-81}$$

where $N(t)$ is the sum of the two noise terms

$$N(t) = n_A(t) + n_B(t) \tag{7.3-82}$$

Since the code loop timing error is given by $\varepsilon = T_i - \hat{T}(s)$, one can write

$$\hat{T}(s) = T_i - \varepsilon \tag{7.3-83}$$

where T_i is the input time delay process, which is assumed to be constant in this analysis. It follows that

$$T_i - \varepsilon = \left(\frac{(K_vK)PF(s)}{s}\right)\left[\varepsilon + \frac{N(t)}{K_1P}\right] \tag{7.3-84}$$

Now let $K' = K \cdot K_v$, then

$$T_i - K'PF(s)\frac{1}{s}\cdot\frac{N(t)}{K_1P} = \frac{K'PF(s)}{s}\varepsilon + \varepsilon \tag{7.3-85}$$

Therefore the code timing error can be written as

$$\varepsilon = \frac{T_i - \left(K'PF(s)\dfrac{(n_A(t) + n_B(t))}{K_1P}\right)}{\left[1 + \dfrac{K'F(s)}{s}\right]} \tag{7.3-86}$$

Define the closed-loop code transfer function by

$$H(s) = \frac{K'PF(s)/s}{1 + K'PF(s)/s} \quad \text{and therefore } 1 - H(s) = \frac{1}{1 + K'PF(s)/s} \tag{7.3-87}$$

Then, from (7.3-86) and (7.3-87) and dividing by the symbol duration T_s, one can solve for

$$\frac{\varepsilon(t)}{T_s} = (1 - H(s))\left(\frac{T_i(t)}{T_s}\right) - H(s)\left[\frac{N(t)}{K_1 P T_s}\right] \tag{7.3-88}$$

Now the code-tracking loop variance, in symbols², is given by

$$E\left[\frac{\varepsilon^2}{T_s^2}\right] = \sigma_\varepsilon^2 = \int_{-\infty}^{\infty} |H(f)|^2 \frac{S_N(f)}{(K_1 P T_s)^2} df, \text{ symbols}^2 \tag{7.3-89}$$

where $E(x)$ denotes the ensemble average of x, and KT_s is the normalized gain. Now approximate $S_N(f)$ with $S_N(0) = S_{n_A+n_B}(0)$; in other words, assume that the noise spectral density is essentially flat across the loop bandwidth, so that

$$\left[\frac{\sigma_\varepsilon^2}{T_s^2}\right] \cong \frac{2B_L \cdot S_N(0)}{(K_1 P T_s)^2}, \text{ symbols}^2 \tag{7.3-90}$$

where B_L is the one-sided closed-loop noise bandwidth of the code-tracking loop, in Hz. Thus the closed-loop tracking error variance has been reduced to evaluating the noise spectral density at $f = 0$ and evaluating the gain of the discriminator function for a specified loop bandwidth B_L.

7.3.5 Noncoherent I-Q Early-Late Gate Code-Tracking Loop with Channel Filtering of $N(t)$ at $f = 0$

We now obtain spectral density at $f = 0$ of the $N(t)$ process in order to evaluate (7.3-82). The noise process is based on the outputs of the integrate-and-dump arm filters, which change every T_b seconds. The noise process can be described in the following form [13], with the random variable θ being uniform on $(0, T_b)$, independent of N_i

$$N(t) = \sum_{i=-\infty}^{\infty} N_i p(t - iT_b - \theta) \tag{7.3-91}$$

where $p(t)$ is a unit amplitude pulse being unity at $t = 0$ and having a duration of T_b seconds. Now the random variable N_i is given by the output of the error control point in the loop from (7.3-79); hence

$$N_i = \left(N_{2L_i}^2 + N_{1L_i}^2\right) - \left(N_{2E_i}^2 + N_{1E_i}^2\right) + 2\sqrt{P} d_i \left(I_{1L_i}(0) N_{1L_i} - I_{1E_i}(0) N_{1E_i}\right) \tag{7.3-92}$$

where the explicit dependence on the sequence at the i-th time has been used and d_i denotes $d(iT_b)$. In order to determine the spectral density it is necessary to determine the autocorrelation function of the noise process N_i. The autocorrelation function is defined by

$$R_N(t + \tau, t) = E\left[N(t) N(t + \tau)\right] \tag{7.3-93}$$

Therefore, from (7.3-91) one has

$$R_N(t + \tau, t) = \sum_{i=-\infty}^{\infty} \sum_{j=-\infty}^{\infty} R_N(i - j) \frac{1}{T_b} \int_0^{T_b} p(t + \tau - iT_b - \theta) p(t - jT_b - \theta) d\theta \tag{7.3-94}$$

Now let

$$i - j = n \quad \text{and} \quad t - jT_b - \theta = \phi$$
$$\text{and} \quad \theta = -\phi - jT_b + t \tag{7.3-95}$$

Then

$$R_N(t+\tau, t) = \sum_{n=-\infty}^{\infty} R_N(n) \sum_{j=-\infty}^{\infty} \frac{1}{T_b} \int_{t-jT_b}^{t-jT_b+T_b} p(\phi)p(\tau - nT_b + \phi)\,d\phi \tag{7.3-96}$$

or

$$R_N(\tau) = \sum_{n=-\infty}^{\infty} R_N(n) R_b(\tau - nT_b) \tag{7.3-97}$$

where

$$R_b(\tau) = \frac{1}{T_b}\int_{-\infty}^{\infty} p(\tau+\phi)p(\phi)\,d\phi \tag{7.3-98}$$

and is the finite autocorrelation function of the bit process (pulse) [13]. The power spectral density is therefore given by

$$S_N(f) = \sum_{n=-\infty}^{\infty} R_N(n) \int_{-\infty}^{\infty} R_b(\tau - nT_b) e^{-j2\pi f\tau} d\tau \tag{7.3-99}$$

Let $\tau - nT_b = x$ to yield

$$S_N(f) = \sum_{n=-\infty}^{\infty} R_N(n) \int_{-\infty}^{\infty} R_b(x) e^{-j2\pi f(x+nT_b)} dx \tag{7.3-100}$$

or

$$S_N(f) = \sum_{n=-\infty}^{\infty} R_N(n) e^{-j2\pi fnT_b} \int_{-\infty}^{\infty} R_b(x) e^{-j2\pi fx} dx \tag{7.3-101}$$

the integral is just the power spectral density of the PN code process, so that one can write

$$S_N(f) = \sum_{n=-\infty}^{\infty} R_N(n) e^{-j2\pi fnT_b} S_b(f) \tag{7.3-102}$$

At $f = 0$, the power spectral density becomes

$$S_N(0) = \sum_{n=-\infty}^{\infty} R_N(n) S_b(0) \tag{7.3-103}$$

Now define

$$\Lambda(t/T_b)=\begin{cases}1-\dfrac{|t|}{T_b} & |t|\le T_b\\ 0 & |t|>T_b\end{cases} \tag{7.3-104}$$

which is the autocorrelation function, normalized to unit power, for the random, unfiltered, bit sequence (NRZ shape assumed, and not related to the code symbol shape). Then the Fourier transform of the autocorrelation function is given by

$$F\{\Lambda(t/T_b)\}=T_b\frac{\sin^2(\pi fT_b)}{(\pi fT_b)^2} \tag{7.3-105}$$

so that

$$S_b(f)=T_b\frac{\sin^2(\pi fT_b)}{(\pi fT_b)^2} \tag{7.3-106}$$

So that (7.3-103) becomes

$$S_N(0)=\sum_{n=-\infty}^{\infty}R_N(n)T_b \tag{7.3-107}$$

Recall from (7.3-90) that

$$\left[\frac{\sigma_\varepsilon^2}{T_s^2}\right]\cong\frac{2B_L\cdot S_N(0)}{(K_1PT_s)^2},\ \text{symbols}^2 \tag{7.3-108}$$

Now the noise process $N(t)$ power spectral density, can be written as

$$S_N(0)=S_{n_A+n_B}(0)=\sum_{n=-\infty}^{\infty}R_{n_A+n_B}(n)S_b(0)=\sum_{n=-\infty}^{\infty}\left[R_{n_A}(n)+R_{n_B}(n)\right]T_b \tag{7.3-109}$$

since $n_A(t)$ and $n_B(t)$ are statistically independent and $S_b(0)=T_b$. One has

$$S_N(0)\cong T_b\left[R_{n_A}(0)+R_{n_B}(0)\right] \tag{7.3-110}$$

It can be shown that $R_{n_A}(n)\cong 0$ and $R_{n_B}(n)\cong 0,\ n\neq 0$. Rewrite N_i from (7.3-92) as

$$\begin{aligned}N_i=&\overbrace{2\sqrt{P}\,d_i\left[I_{1E}(\varepsilon,i)\cdot N_{1E}(i)-I_{1L}(\varepsilon,i)\ N_{1L}(i)\right]}^{n_A(i)}\\&\overbrace{+\left[N_{2E}^2(i)+N_{1E}^2(i)\right]-\left[N_{2L}^2(i)+N_{1L}^2(i)\right]}^{n_B(i)}\end{aligned} \tag{7.3-111}$$

and clearly N_i can be written as the sum of two noise processes

$$N_i = n_A(i) + n_B(i) \tag{7.3-112}$$

Consider the autocorrelation of the process N_i

$$R_N(n) = E[N_i N_{i+n}] \tag{7.3-113}$$

The calculations for this autocorrelation function calculation are quite detailed, so that the results will be (7.3-110) summarized here. The interested reader should consult [11] for more details. Starting from the fact that

$$S_N(0) \cong T_b R_N(0) \tag{7.3-114}$$

one can show that $R_M(0)$ can be written as the sum of two terms of the form

$$R_N(0) = R_A(0) + R_B(0) \tag{7.3-115}$$

The first autocorrelation function can be written as

$$\begin{aligned} R_A(0) = &\frac{4P}{T_b}\left[I_{1E}^2(0) + I_{1L}^2(0)\right] \cdot \int_{-\infty}^{\infty} S_{PN}(f)\left|H_{BB}(f)\right|^2 S_n(f)\,df \\ &-\frac{8P}{T_b} I_{1L}(0) I_{1E}(0) \int_{-\infty}^{\infty} S_{PN}(f)\left|H_{BB}(f)\right|^2 S_n(f)\cos\left[2\pi f dT_s\right]df \end{aligned} \tag{7.3-116}$$

The second autocorrelation function can be written as

$$\begin{aligned} R_B(0) = &\frac{8}{T_b^2}\left[\int_{-\infty}^{\infty} S_{n_1}(f)\left|H_{BB}(f)\right|^2 S_{PN}(f)\,df\right]^2 \\ &-\frac{8}{T_b^2}\left[\int_{-\infty}^{\infty} S_{PN}(f)\left|H_{BB}(f)\right|^2 S_{n_1}(f) e^{-i2\pi f dT_s}\,df\right]^2 \end{aligned} \tag{7.3-117}$$

Hence, summing (7.3-116) and (7.3-117) and multiplying by T_b yields the total noise spectral density at $f = 0$. Now the closed-loop tracking error variance can be obtained.

7.3.6 Noncoherent Early-Late Gate I-Q Code Loop Tracking Error Variance with Channel Filtering

In this section the code-tracking error variance will be determined using the results obtained in Section 7.3.5. From (7.3-116) and our results for $R_A(0)$ and $R_B(0)$, one has our general result for the code loop tracking error variance expressed in symbols^2.

$$\left(\frac{\sigma^2}{T_s^2}\right) = \frac{2B_L \cdot S_N(0)}{(K_1 P T_s)^2}, \quad \text{symbols}^2 \tag{7.3-118}$$

Using (7.3-118), (7.3-117), (7.3-110), (7.3-115), and (7.3-114) and utilizing the assumptions that $S_{n_1}(f)$ is even in f, $S_{PN}(f)$ is even in f, and the magnitude of $H_{BB}(f)$ is even in f, we have our general result

$$\left(\frac{\sigma^2}{T_s^2}\right)=\frac{2B_L}{\left(K_1PT_s\right)^2}\left\{\begin{array}{l}\overbrace{4\mathrm{P}\left(I_{1E}^2(0)+I_{1L}^2(0)\right)\left(\int\limits_{-\infty}^{\infty} S_{n_1}(f)\left|H_{BB}(f)\right|^2 S_{PN}(f)\right)df}^{T_1}\\ \overbrace{-8\mathrm{P}\left(I_{1E}(0)I_{1L}(0)\right)\left(\int\limits_{-\infty}^{\infty} S_{n_1}(f)\left|H_{BB}(f)\right|^2 S_{PN}(f)\cos\left(2\pi f\Delta T_s\right)\right)df}^{T_2}\\ \overbrace{+\frac{8}{T_b}\left(\int\limits_{-\infty}^{\infty} S_{n_1}(f)\left|H_{BB}(f)\right|^2 S_{PN}(f)\,df\right)^2}^{T_3}\\ \overbrace{-\frac{8}{T_b}\left(\int\limits_{-\infty}^{\infty} S_{n_1}(f)\left|H_{BB}(f)\right|^2 S_{PN}(f)\cos\left(2\pi f\Delta T_s\right)df\right)^2}^{T_4}\ \text{chips}^2\end{array}\right\} \tag{7.3-119}$$

where we have used $\Delta = d = 2\delta$ and where K_1 is defined in (7.3-78), that is,

$$K_1=KT_s=2\left\{\begin{array}{l}\left[\int\limits_{-\infty}^{\infty} H_{BB}(f)S_{PN}(f)e^{-j2\pi f(\varepsilon_0-\delta T_s)}df\right]\times\\ \left[\int\limits_{-\infty}^{\infty} H_{BB}(f)S_{PN}(f)(-j2\pi f)e^{-j2\pi f(\varepsilon_0-\delta T_s)}df\right]\\ -\left[\int\limits_{-\infty}^{\infty} H_{BB}(f)S_{PN}(f)e^{-j2\pi f(\varepsilon_0+\delta T_s)}df\right]\times\\ \left[\int\limits_{-\infty}^{\infty} H_{BB}(f)S_{PN}(f)(-j2\pi f)e^{-j2\pi f(\varepsilon_0+\delta T_s)}df\right]\end{array}\right\}T_s \tag{7.3-120}$$

and note that KT_s is the normalized slope in arithmetic units (per symbol). Recall that from (7.3-59) and (7.3-60) for $\varepsilon = 0$ one has

$$I_{1E}(0)=\int\limits_{-\infty}^{\infty} S_{PN}(f)\left(H_{BB}(f)\right)e^{-j2\pi f\delta T_s}df \tag{7.3-121}$$

$$I_{1L}(0)=\int\limits_{-\infty}^{\infty} S_{PN}(f)\left(H_{BB}(f)\right)e^{+j2\pi f\delta T_s}df \tag{7.3-122}$$

It is to be noted that this is the general case including both thermal noise plus interference with channel filtering for binary valued code formatting (± 1).

The important case for which there is no interference other than thermal noise and negligible transmitter and receiver filtering is considered in the next section as a check against our original unfiltered results.

7.3.7 Noncoherent Early-Late Gate Code I-Q Tracking Error Variance with Thermal Noise and Without Channel Filtering

In this section the receiver tracking performance without channel filtering is considered for binary valued code symbols. The only interference is the naturally occurring thermal noise. This can be accomplished by utilizing a baseband equivalent filter of the form $H_{BB}(f)=1$, for all f. Furthermore, the thermal noise is modeled as a white Gaussian noise process with two-sided noise spectral density of $N_0/2$ W/Hz; therefore the baseband equivalent process is also white Gaussian noise. From (7.3-121) and (7.3-122) one has (in the case that $H_{BB}(f) = 1$)

$$I_{1E}(\varepsilon) = R_{PN}(\varepsilon + \delta T_s) \tag{7.3-123}$$

and

$$I_{1L}(\varepsilon) = R_{PN}(\varepsilon - \delta T_s) \tag{7.3-124}$$

Also note that

$$\frac{N_0}{2}\int_{-\infty}^{\infty} S_{PN}(f)df = \frac{N_0}{2} \tag{7.3-125}$$

$$S_{n_1}(f) = N_0/2 \tag{7.3-126}$$

From (7.3-119), using (7.3-123)–(7.3-126) yields

$$\frac{\sigma^2}{T_s^2} = \frac{2B_L}{(K_1T_sP)^2}\left\{\begin{array}{l} \overbrace{(4P)\left[\left(R_{PN}(\delta T_s)\right)^2 + \left(R_{PN}(-\delta T_s)\right)^2\right]\left(\frac{No}{2}\right)}^{T_1} \\ \overbrace{-8P\ R_{PN}(\delta T_s)R_{PN}(-\delta T_s)\left[\frac{No}{2}\int_{-\infty}^{\infty} S_{PN}(f)e^{-i2\pi f\Delta T_s}df\right]}^{T_2} \\ \overbrace{+\frac{8}{T_b}\left[\left(\frac{No}{2}\right)\right]^2}^{T_3} \overbrace{-\frac{8}{T_b}\left[\frac{No}{2}\int_{-\infty}^{\infty} S_{PN}(f)e^{-i2\pi f\Delta T_s}df\right]^2}^{T_4} \end{array}\right\} \tag{7.3-127}$$

and

$$K = \frac{\partial}{\partial\varepsilon}\left[R_{PN}^2(\varepsilon + \delta T_s) - R_{PN}^2(\varepsilon - \delta T_s)\right]\Big|_{\varepsilon=0} \tag{7.3-128}$$

Therefore the general results for the unfiltered signal with binary valued code symbols (± 1), in white Gaussian noise (WGN), is given by:

$$\left(\frac{\sigma}{T_s}\right)^2 = \frac{2N_0B_L}{(K_1T_s)^2 P}\left[4R_{PN}^2(\delta T_s)\left\{1 - R_{PN}(2\delta T_s)\right\} + \frac{2N_0\left\{1 - R_{PN}^2(2\delta T_s)\right\}}{PT_b}\right], \tag{7.3-129}$$

which is in units of symbols^2, and the normalized gain is given by

$$KT_s = T_s \frac{\partial}{\partial \varepsilon}\left[R_{PN}^2(\varepsilon+\delta T_s)-R_{PN}^2(\varepsilon-\delta T_s)\right]\Big|_{\varepsilon=0}, \quad \text{symbols} \tag{7.3-130}$$

Equations (7.3-129) and (7.3-130) form the general solution for unfiltered, binary valued, code-tracking error variance in white Gaussian thermal noise with arbitrary early-late correlator spacing and is the same as (7.2-86). Thus the model for the noncoherent early-late gate code-tracking loop tracking error with filtering collapses to the known unfiltered case, as it should.

We have already established the result the variance of the tracking error performance for NRZ symbols

$$\left(\frac{\sigma}{T_s}\right)^2 = \frac{N_0 B_L d}{2P}\left[1+\frac{2}{(2-d)(E_b/N_0)}\right], \quad \text{symbols}^2 \tag{7.3-131}$$

which is the same as chips^2 for the NRZ code formatting case.

To summarize the parameters, N_0 is the one-sided noise spectral density in watts/Hz, B_L is the code loop closed-loop noise equivalent bandwidth in Hz, d is the early minus late correlator spacing ($d = 2\delta$) in chips, P is the received signal power in watts, and E_b/N_0 is the received signal to noise ratio in the data bandwidth (symbol bandwidth for the coded case)

$$E_b/N_0 = \frac{PT_b}{N_0} \tag{7.3-132}$$

where $R_b = 1/T_b$ is the bit rate and T_b is the bit duration. It is to be noted that the tracking error in terms of distance (global navigational satellite systems (GNSS) applications) can be found from

$$\sigma_m = \sqrt{\frac{N_0 B_L d}{2P}\left(1+\frac{2}{(2-d)(E_b/N_0)}\right)}\, cT_s, \quad \text{m} \tag{7.3-133}$$

where c is the speed of light (3×10^8 m/s) and T_s is the symbol duration in seconds.

7.3.8 Noncoherent Early-Late Gate I-Q Code-Tracking Error Variance with Channel Filtering in White Gaussian Noise with NRZ Symbols

In this section we will specialize the general equations for the tracking error variance to a case in which filtering effects are included in the face of WGN. From (7.3-119) the equation for the tracking error variance can be simplified for the case of white Gaussian noise, by letting $S_{n_1}(f) = N_0/2$ so that one obtains for WGN

$$\left(\frac{\sigma}{T_s}\right)^2 = \frac{2B_L N_0}{K_1^2 T_s^2 P}\left[\begin{array}{l} 2\left(I_{1E}^2(0)+I_{1L}^2(0)\right)\left(\int\limits_{-\infty}^{\infty} \left|H_{BB}(f)\right|^2 S_{PN}(f)df\right) \\ -4I_{1E}(0)I_{1L}(0)\left(\int\limits_{-\infty}^{\infty}\left|H_{BB}(f)\right|^2 S_{PN}(f)\cos(2\pi f\Delta T_s)df\right) \\ +\frac{2N_0}{PT_b}\left[\int\limits_{-\infty}^{\infty} \left|H_{BB}(f)\right|^2 S_{PN}(f)df\right]^2 \\ -\frac{2N_0}{PT_b}\left[\int\limits_{-\infty}^{\infty} \left|H_{BB}(f)\right|^2 S_{PN}(f)\cos(2\pi f\Delta T_s)df\right]^2 \end{array}\right] \quad (7.3\text{-}134)$$

Where again $\Delta = d = 2\delta$. This equation holds for all symbols of the type studied here (that is, with amplitudes of ± 1) with only WGN present at the input. In order to visualize the tracking performance for NRZ symbols (also chips in this case) consider a model of an ideal bandpass filter that has 3-dB channel bandwidths of $W = 2$, 4, and 6, and ∞ MHz, as seen in Figure 7.3-4.

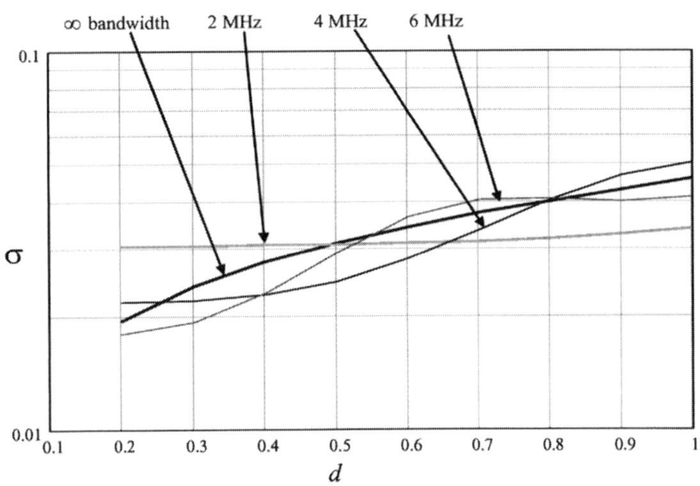

Figure 7.3-4 Early-late gate (L^2-E^2) I-Q code-tracking loop one sigma error for a 1.023 Mcps code symbol rate for various RF bandwidths and early-late separation.

In the figure the chip rate is 1.023 Mcps, $C/N_0 = 25$ dB-Hz, $B_L = 1$ Hz, and the data rate is 50 bps (uncoded).

As can be seen from the plots, as the correlator spacing (d) is reduced, the tracking error tends to be reduced. When the bandwidth is very wide compared to the chip rate, the result for no filtering ($W = \infty$) is a good approximation (see (7.3-131) for the $W = \infty$ case). Furthermore, using the infinite bandwidth case as an estimate with filtering is not unreasonable as a quick approximation to the tracking performance.

7.3.9 Noncoherent Early-Late Gate I-Q Code-Tracking Performance with Narrowband Gaussian Interference Plus White Gaussian Noise with NRZ Symbols and No Channel Filtering

In this section, we consider the code-tracking error variance for an unfiltered narrowband interference process. The objective of this calculation is to determine that the response to narrowband interference is quite different from wideband interference.

Let the radio frequency interference be a pair of narrowband Gaussian random process in which each has a rectangular power spectral density as shown in Figure 7.3-5(a). In this figure the positive frequency bandwidth of the process is $2B_1$ Hz and the interference is centered at $f_c + f_I$ and $f_c - f_I$ Hz. Each of the narrowband pair has a bandwidth of B_1 Hz. It is assumed that f_I is variable; in other words, the interference spectra are not necessarily centered at the carrier frequency.

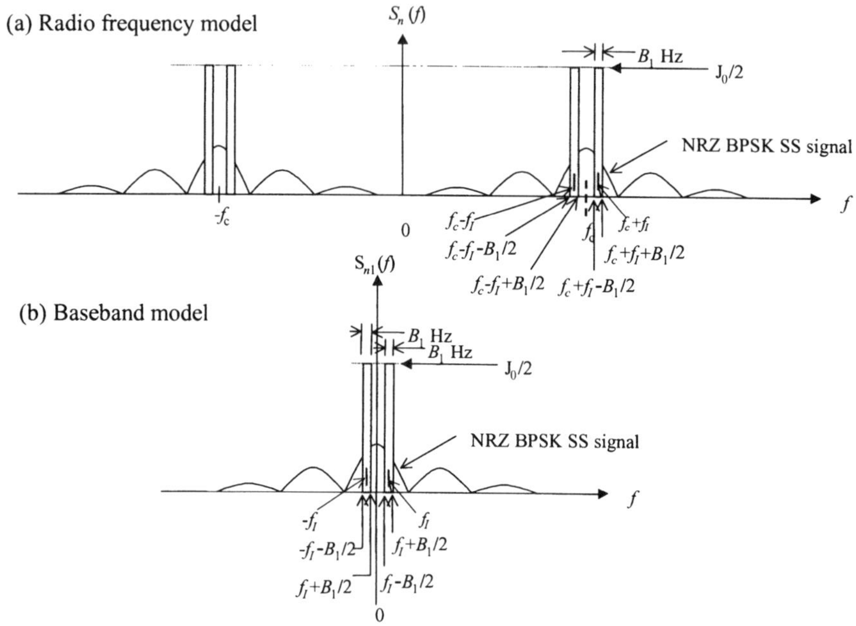

Figure 7.3-5 (a, b) Narrowband interference model spectral characteristics.

Figure 7.3-5(b) illustrates the baseband equivalent power spectral density. Basically one pair of interferers has been translated down to zero center frequency. The signal spectrum illustrated in Figure 7.3-4 is for an NRZ encoded BPSK SS signal. The signal could well have been any other ± 1 valued symbol shape. We will see that the optimum value of the interferer frequency offset f_I is not zero for NRZ waveforms.

Thus the interference model for the power spectral density is given by

$$S_{I_1}(f) = \frac{J_0}{2}\begin{Bmatrix} f_I - B_1/2 \le f \le f_I + B_1/2 \\ -f_I - B_1/2 \le f \le -f_I + B_1/2 \end{Bmatrix} \tag{7.3-135}$$

where f_I is the interferer frequency and B_1 is the positive frequency bandwidth of the interference.

From (7.3-121) and (7.3-122), it follows that

$$I_{1E}(0) = \int_{-\infty}^{\infty} e^{-2\pi f \delta T_s} S_{PN}(f)df = R_{PN}(-\delta T_s) \tag{7.3-136}$$

and

$$I_{1L}(0)=\int_{-\infty}^{\infty} e^{+2\pi f\delta T_s} S_{PN}(f)df = R_{PN}(\delta T_s) \tag{7.3-137}$$

So, (7.3-119), for the tracking error variance, can be written in the form

$$\left(\frac{\sigma}{T_s}\right)^2 = \frac{2B_L}{(K_1T_s)^2P}\left\{\begin{array}{l} \overbrace{4\left[R_{PN}^2(\delta T_s)+R_{PN}^2(-\delta T_s)\right]}^{A}\overbrace{\left[\int_{-\infty}^{\infty} S_{n_1}(f)S_{PN}(f)df\right]}^{B1} \\ -\overbrace{8R_{PN}(\delta T_s)R_{PN}(-\delta T_s)}^{C}\cdot\overbrace{\left(\int_{-\infty}^{\infty} S_{n_1}(f)S_{PN}(f)\cos(2\pi f\Delta T_s)df\right)}^{D} \\ +\dfrac{8}{PT_b}\left(\overbrace{\int_{-\infty}^{\infty} S_{n_1}(f)S_{PN}(f)df}^{B1}\right)^2 \\ -\dfrac{8}{PT_b}\left(\overbrace{\int_{-\infty}^{\infty} S_{n_1}(f)S_{PN}(f)\cos(2\pi f\Delta T_s)df}^{D}\right)^2 \end{array}\right\} \tag{7.3-138}$$

where $R_{PN}(x)$ is the autocorrelation function evaluated at the time difference x. Equation (7.3-138) applies to any ± 1 valued PN code, without filtering effects. Now consider the case when the noise is the sum of thermal noise and narrowband noise of the type just discussed.

We will assume that the PN code has random NRZ symbols, with the probability of being flipped or not flipped is ½. The noise process in (7.3-138) will be modeled as the sum of WGN noise and narrowband Gaussian noise, which are mutually statistically independent of each other. Thus the power spectral density of the total noise is given by, for all f

$$S_{n_1}(f)=\frac{N_0}{2}+S_{J_1}(f), \quad \forall f \tag{7.3-139}$$

and $S_{J_1}(f)$ is the narrowband Gaussian noise process (see Figure 7.3-5) whose spectral density is modeled in (7.3-135), and B_1 is assumed to be very small.

Now consider the evaluation of the terms A, $B1$, C, and D in (7.3-138). For the NRZ symbols the power spectral density is given by

$$S_{PN}(f)=T_s\frac{\sin^2(\pi fT_s)}{(\pi fT_s)} \tag{7.3-140}$$

Consider first the evaluation of the term A in (7.3-138). A can be evaluated as

$$A=4\left[(1-\delta)^2+(1-\delta)^2\right]=8(1-\delta)^2 \tag{7.3-141}$$

Consider next the evaluation of the term $B1$ in (7.3-138). Recall that the noise power spectral density and the PN waveform are the baseband versions of the respective processes. From Figure 7.3-5 one has

$$J = \frac{J_0}{2}(4B) = 2J_0 B \tag{7.3-142}$$

Hence

$$B1 = \overbrace{\int_{-\infty}^{\infty} \frac{N_0}{2} T_s \frac{\sin^2(\pi f T_s)}{(\pi f T_s)^2} df}^{B1_1} + \overbrace{\int_{-\infty}^{\infty} S_{J_1}(f) T_s \frac{\sin^2(\pi f T_s)}{(\pi f T_s)^2} df}^{B1_2} \tag{7.3-143}$$

The value of $B1$ can be evaluated as

$$B1 = \frac{N_0}{2} + \frac{JT_s}{2} \frac{\sin(\pi f_J T_s)^2}{(\pi f_J T_s)^2} \tag{7.3-144}$$

Now consider the term denoted as D.

$$D = \overbrace{\int_{-\infty}^{\infty} \frac{N_0}{2} S_{PN}(f) \cos(\pi f \Delta T_s) df}^{D_1} + \overbrace{\int_{-\infty}^{\infty} S_{J_1}(f) S_{PN}(f) \cos(\pi f \Delta T_s) df}^{D_2} \tag{7.3-145}$$

The term D can be shown to be given by

$$D = \frac{N_0}{2} R_{PN}(\Delta T_s) + \frac{JT_c}{2} \frac{\sin^2(\pi f_J T_s)}{(\pi f_J T_s)^2} \cos(2\pi f_J \Delta T_s) \tag{7.3-146}$$

Using A, $B1$, C, and D in (7.3-138) yields (for $\delta \le 1/2$) hereh

$$\left(\frac{\sigma}{T_s}\right)^2 = \frac{2B_L}{(K_1 T_s)^2 P} \left\{ \begin{array}{l} 8(1-\delta)^2 \left(N_0 \delta + \frac{JT_s}{2} \frac{\sin^2(\pi f_J T_s)}{(\pi f_J T_s)^2} \left[1 - \cos(2\pi f_J \Delta T_s)\right] \right) \\ + \frac{8}{PT_b} \left(\begin{array}{l} \left[\frac{N_0}{2} + \frac{JT_s}{2} \frac{\sin^2(\pi f_J T_s)}{(\pi f_J T_s)^2} \right]^2 \\ - \left[\frac{N_0}{2}(1-2\delta) + \frac{JT_s}{2} \frac{\sin^2(\pi f_J T_s)}{(\pi f_J T_s)^2} \cos(2\pi f_J T_s) \right]^2 \end{array} \right) \end{array} \right\} \tag{7.3-147}$$

Now consider the effect of thermal noise along with a narrowband interferer of the type considered in (7.3-135). From (7.3-147) one can simplify the expression somewhat to

$$\left(\frac{\sigma}{T_s}\right)^2 = \frac{N_0 B_L}{2P}\left\{ \begin{aligned} &\left(d + \frac{JT_s}{N_0}\frac{\sin^2(\pi f_J T_s)}{(\pi f_J T_c)^2}\left[1-\cos(2\pi f_J \Delta T_s)\right]\right) \\ &+ \frac{1}{(1-\left(\frac{d}{2}\right)^2 PT_b}\frac{N_0}{2}\left(\begin{aligned} &\left[1+\frac{JT_s}{N_0}\frac{\sin^2(\pi f_J T_s)}{(\pi f_J T_s)^2}\right]^2 \\ &-\left[(1-d)+\frac{JT_c}{N_0}\frac{\sin^2(\pi f_J T_s)}{(\pi f_J T_s)^2}\cos(2\pi f_J T_s)\right]^2 \end{aligned} \right) \end{aligned} \right\} \quad (7.3\text{-}148)$$

and again the units are symbols^2 (or chips^2 when NRZ formatting is used). This result, for tracking error variance, is applicable for NRZ symbols with both thermal noise and narrowband interference, and $\Delta = d$.

Figure 7.3-6 illustrates the effect of interference plus thermal noise affecting the code-tracking loop for NRZ symbol formatting. In this figure the following parameters have been used: I/P = 40 dB, P/N_0 = 40 dB (I/N = 20 dB), $T_c = 1/(1.023*10^6)$, B_L = 1 Hz, and R_b = 50 Bps. It is seen that the interference has an optimum frequency offset from the carrier frequency that depends on the early-late spacing, d.

The tracking error varies with the carrier frequency offset and the correlator spacing (d). The error is greatly suppressed when d is made very small.

In [14] the authors have measured a somewhat similar type of frequency dependence on the location of the narrowband interference for NRZ symbols, with a very low response at the carrier frequency for a coherent baseband code-tracking loop.

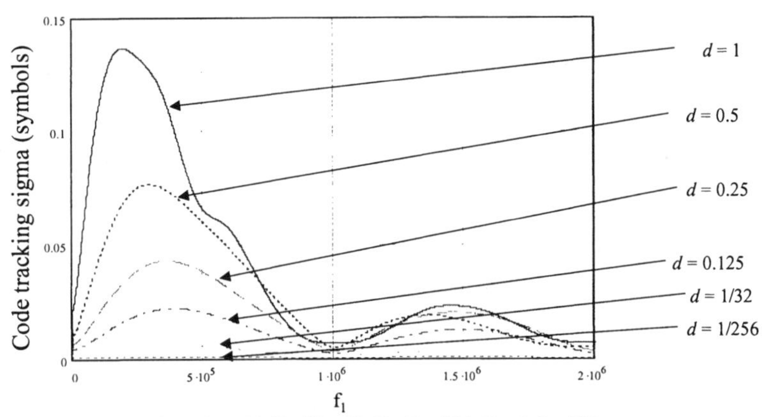

Notes: I/P = 40 dB, P/N_0 = 40 dB, I/N = 20 dB, $T_c = 10^{-6}$, B_L =1, R_b= 50 bps

Figure 7.3-6 Noncoherent code-tracking sigma with thermal noise with a 1.023 Mcps NRZ code versus the offset frequency of the interference from the carrier frequency.

An alternate viewpoint to the analysis of code-tracking in the presence of Gaussian random interference and filtering can be found in [15]. Another analysis for the filtered code-tracking of the dot product code-tracking loop has been presented in [16].

7.4 TIME-SHARED NONCOHERENT CODE-TRACKING LOOPS

Now we consider an efficient method of sharing the early and late correlators by time-sharing. This work follows Holmes [8]. Numerous other authors have considered the time-shared loop, including Simon [17], Stone [18], Huang [19], and Hartmann [20].

Consider a code-tracking loop that time-shares the early and late correlators as shown in Figure 7.4-1.

This loop operates by time-sharing the early and late code reference signals $PN(t-\hat{T}+T_c/2)$ and $PN(t-\hat{T}-T_c/2)$ in a periodic manner controlled by the periodic switch function $g(\,)$. Thus only one of the correlators is correlating against the input $y(t)$ at any one time. In this loop only two correlators are needed, one for the early and late correlations and one for the on-time dispreading channel.

We model the received signal plus noise by

$$y(t)=\sqrt{2P}d(t)PN(t-T)\cos(\omega_0 t+\theta)+n(t) \tag{7.4-1}$$

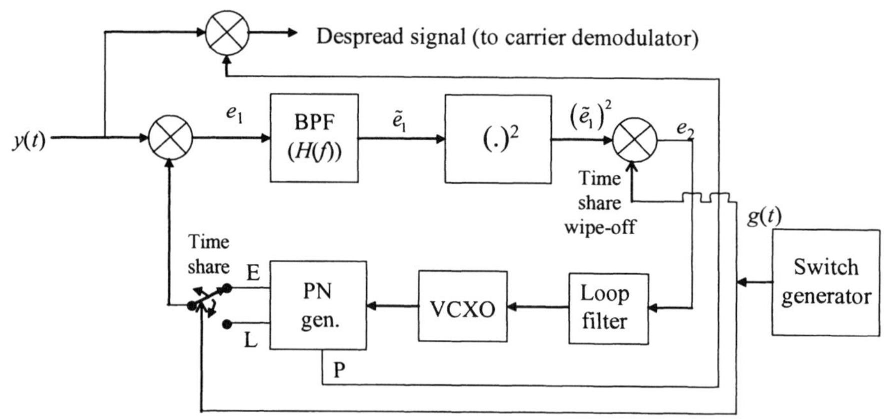

Figure 7.4-1 Noncoherent time-shared RF implemented code-tracking loop with the early (E), late (L), and punctual channels (P).

where P is the received signal power, T is the delay in the propagation time from the transmitter to the receiver, ω_0 is the carrier radian frequency, and θ is the carrier phase. The noise is assumed to be white Gaussian noise. We model the gating function along with two multiplexing functions in Figure 7.4-2.

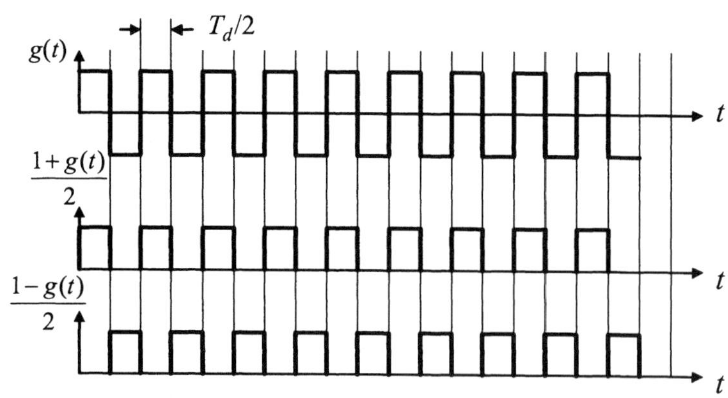

Figure 7.4-2 Gating and multiplexing functions for the time-shared, code-tracking loop.

The waveform at the point e_1 in Figure 7.4-1 is expressed by

$$e_1(t) = \left[\sqrt{2P}d(t)PN(t-T)\cos(\omega_0 t+\theta)+n(t)\right] \times\left[\left(\frac{1-g(t)}{2}\right)PN(t-\hat{T}+T_s/2)+\left(\frac{1+g(t)}{2}\right)PN(t-\hat{T}-T_s/2)\right] \tag{7.4-2}$$

Out of the bandpass filter the signal is modeled, neglecting the self-noise term due to the variation about the mean output, as

$$\tilde{e}_1 = \sqrt{2P}H_{BB}(s)\left[\begin{array}{l}\left(\frac{1-g(t)}{2}\right)d(t)PN(t-T)PN(t-\hat{T}+T_s/2)\cos\left(\omega_0 t+\theta\right)\\ +\left(\frac{1+g(t)}{2}\right)d(t)PN(t-T)PN(t-\hat{T}-T_s/2)\cos\left(\omega_0 t+\theta\right)\end{array}\right] + H_{BB}(s)\left[\left(\frac{1-g(t)}{2}\right)n(t)PN(t-\hat{T}+T_s/2)+\left(\frac{1+g(t)}{2}\right)n(t)PN(t-\hat{T}-T_s/2)\right] \tag{7.4-3}$$

where $H_{BB}(s)$ is the baseband equivalent filter of the bandpass filter $H(s)$.

Now we neglect the effects of the filter on the data and the multiplexing function since they normally run at much lower speed. We also assume that the filter produces the correlation of the PN codes, out of the filter process. Thus we have

$$\tilde{e}_1 \cong \sqrt{2P}\left[\begin{array}{l}\left(\frac{1-g(t)}{2}\right)d(t)R_{PN}(T-\hat{T}+T_s/2)\cos\left(\omega_0 t+\theta\right)\\ +\left(\frac{1+g(t)}{2}\right)d(t)R_{PN}(T-\hat{T}-T_s/2)\cos\left(\omega_0 t+\theta\right)\end{array}\right] + \sqrt{2P}\left[\left(\frac{1-g(t)}{2}\right)n_E(t)+\left(\frac{1+g(t)}{2}\right)n_L(t)\right] \tag{7.4-4}$$

in which the noise terms are given by

$$\begin{aligned} n_E(t) &= H_{BB}(s)\left[n(t)PN(t-T+\delta T_s)\right] \\ n_L(t) &= H_{BB}(s)\left[n(t)PN(t-T-\delta T_s)\right] \end{aligned} \tag{7.4-5}$$

With the timing error $\varepsilon = T-\hat{T}$, within the linear region we can rewrite (7.4-4) as

$$\tilde{e}_1(t) = \sqrt{2P}d(t)\cos(\omega_0 t+\theta)\left[\left(\frac{1-g(t)}{2}\right)R_{PN}(\varepsilon+T_s/2)+\left(\frac{1+g(t)}{2}\right)R_{PN}(\varepsilon-T_s/2)\right] + \left[\left(\frac{1-g(t)}{2}\right)n_E(t)+\left(\frac{1+g(t)}{2}\right)n_L(t)\right] \tag{7.4-6}$$

Now consider the square of $\tilde{e}_1(t)$

$$\begin{aligned}\tilde{e}_1(t)^2 &= P\left[\left(\frac{1-g(t)}{2}\right)^2 R_{PN}^2(\varepsilon+T_s/2)+\left(\frac{1+g(t)}{2}\right)^2 R_{PN}^2(\varepsilon-T_s/2)\right]\\ &+\left[\left(\frac{1-g(t)}{2}\right)n_E(t)+\left(\frac{1+g(t)}{2}\right)n_L(t)\right]^2\\ &+2\sqrt{2P}\left[\left(\frac{1-g(t)}{2}\right)R_{PN}(\varepsilon+T_s/2)+\left(\frac{1+g(t)}{2}\right)R_{PN}(\varepsilon-T_s/2)\right]\\ &\times\left[\left(\frac{1-g(t)}{2}\right)n_E(t)+\left(\frac{1+g(t)}{2}\right)n_L(t)\right]\cos(\omega_0 t+\theta)\end{aligned} \tag{7.4-7}$$

Simplifying (7.4-7) produces

$$\begin{aligned}\tilde{e}_1(t)^2 &= P\left[\left(\frac{1-g(t)}{2}\right)R_{PN}^2(\varepsilon+T_s/2)+\left(\frac{1+g(t)}{2}\right)R_{PN}^2(\varepsilon-T_s/2)\right]\\ &+\left[\left(\frac{1-g(t)}{2}\right)n_E^2(t)+\left(\frac{1+g(t)}{2}\right)n_L^2(t)\right]\\ &+2\sqrt{2P}\left[\begin{matrix}\left(\frac{1-g(t)}{2}\right)R_{PN}(\varepsilon+T_s/2)n_E(t)\\ +\left(\frac{1+g(t)}{2}\right)R_{PN}(\varepsilon-T_s/2)n_L(t)\end{matrix}\right]\cos(\omega_0 t+\theta)\end{aligned} \tag{7.4-8}$$

since it is true that

$$\left(\frac{1+g(t)}{2}\right)^2=\left(\frac{1+g(t)}{2}\right),\quad \left(\frac{1-g(t)}{2}\right)^2=\left(\frac{1-g(t)}{2}\right) \tag{7.4-9}$$

and is also true that

$$\left(\frac{1+g(t)}{2}\right)\left(\frac{1-g(t)}{2}\right)=0 \tag{7.4-10}$$

Now $e_2(t)$ is obtained when $\left(\tilde{e}_1(t)\right)^2$ is multiplied by $g(t)$. Hence letting $\varepsilon=0$ for the correlation function weights of the noise terms, we obtain

$$\begin{aligned}e_2(t) &= P\left[\left(-\frac{1-g(t)}{2}\right)R_{PN}(\varepsilon+T_s/2)^2+\left(\frac{1+g(t)}{2}\right)R_{PN}(\varepsilon-T_s/2)^2\right]\\ &+\left[\left(-\frac{1-g(t)}{2}\right)n_E^2(t)+\left(\frac{1+g(t)}{2}\right)n_L^2(t)\right]\\ &+2\sqrt{2P}d(t)\cos(\omega_0 t+\theta)\left[\begin{matrix}\left(-\frac{1-g(t)}{2}\right)R_{PN}(T_s/2)n_E(t)\\ +\left(\frac{1+g(t)}{2}\right)R_{PN}(T_s/2)n_L(t)\end{matrix}\right]\end{aligned} \tag{7.4-11}$$

Since it is true that

$$g(t)\left(1+g(t)\right)=\left(1+g(t)\right)$$
$$g(t)\left(1-g(t)\right)=-\left(1-g(t)\right) \tag{7.4-12}$$

Equation (7.4-11) can be rewritten as

$$\begin{aligned} e_2(t) &= P\left[\begin{array}{l}\left(\frac{1}{2}\right)R_{PN}(\varepsilon-T_s/2)^2-\left(\frac{1}{2}\right)R_{PN}(\varepsilon+T_s/2)^2\\ +\frac{g(t)}{2}\left(R_{PN}(T_s/2)+R_{PN}(-T_s/2)\right)\end{array}\right]\\ &+\left[\left(-\frac{1-g(t)}{2}\right)n_E^2(t)+\left(\frac{1+g(t)}{2}\right)n_L^2(t)\right]\\ &+2\sqrt{2P}d(t)\cos(\omega_0 t+\theta)\left[\begin{array}{l}\left(-\frac{1-g(t)}{2}\right)R_{PN}(T_s/2)n_E(t)\\ +\left(\frac{1+g(t)}{2}\right)R_{PN}(T_s/2)n_L(t)\end{array}\right]\end{aligned} \tag{7.4-13}$$

The first two terms in (7.4-13) represent the correction signal to maintain tracking and the third and fourth terms represent a high frequency dither that can be assumed to be negligible if $2/T_d \ll B_L$. In this inequality B_L is the closed-loop noise bandwidth of the time-shared code-tracking loop. Therefore we can rewrite (7.4-13) as

$$e_2(t)\cong P\left[\left(\frac{1}{2}\right)R_{PN}(\varepsilon-T_s/2)^2-\left(\frac{1}{2}\right)R_{PN}(\varepsilon+T_s/2)^2\right]+n_1(t)+n_2(t) \tag{7.4-14}$$

where

$$\begin{aligned} n_1(t) &= \left[\left(\frac{1+g(t)}{2}\right)n_L^2(t)-\left(\frac{1-g(t)}{2}\right)n_E^2(t)\right]\\ n_2(t) &= 2\sqrt{2P}d(t)\cos(\omega_0 t+\theta)\left[\begin{array}{l}\left(-\frac{1-g(t)}{2}\right)R_{PN}(T_s/2)n_E(t)\\ +\left(\frac{1+g(t)}{2}\right)R_{PN}(T_s/2)n_L(t)\end{array}\right]\end{aligned} \tag{7.4-15}$$

If we linearize the *S*-curve for NRZ code symbol shapes, when the early-late spacing is one symbol, and let $\varepsilon = T-\hat{T}$, (7.4-14) can be simplified to

$$e_2(t)\cong P\left[\left(\frac{1}{2}\right)P\left(\frac{2\varepsilon}{T_s}\right)\right]+n_1(t)+n_2(t) \tag{7.4-16}$$

Now to close the loop, write the code loop estimate of the code phase as

$$\frac{\hat{T}(t)}{T_s}=\frac{K_{VCO}}{s}F(s)\left[P\frac{\varepsilon}{T_s}+n_1(t)+n_2(t)\right] \tag{7.4-17}$$

Using (7.4-17), it can be shown that the following relationship is obtained from (7.4-17)

$$\frac{\varepsilon(t)}{T_s}=\frac{-K'PF(s)\left[\dfrac{n_1(t)+n_2(t)}{P}\right]}{s+K'PF(s)}=-H(s)\left[\frac{n_1(t)+n_2(t)}{P}\right] \tag{7.4-18}$$

where $H(s)$ is the closed-loop response of the time-shared code-tracking loop, and K' is equal to PK_{VCO}. The explicit dependence on t for $\varepsilon(t)$ and $\hat{T}(t)$ has been used in the previous equations, that is, $\varepsilon(t)=T(t)-\hat{T}(t)$. The variance of the tracking error can be described by

$$\frac{\sigma_\varepsilon^2}{T_s^2}=\int_{-\infty}^{\infty}\left|H(j2\pi f)\right|^2\frac{\left(S_{n_1}(f)+S_{n_2}(f)\right)}{P^2}df \quad \text{chips}^2 \tag{7.4-19}$$

where $S_{n_i}(f)$ are the two spectral densities of the two noise terms. Using the reasonable assumption that the two noise terms are wide compared to the closed-loop noise bandwidth (B_L) produces

$$\frac{\sigma_\varepsilon^2}{T_s^2}\cong 2B_L\frac{\left(S_{n_1}(0)+S_{n_2}(0)\right)}{P^2} \quad \text{chips}^2 \tag{7.4-20}$$

Therefore to complete our analysis it is necessary to evaluate the two noise spectral densities at $f=0$. Consider the first noise term $n_1(t)$ defined in (7.4-15) and repeated here

$$R_{n_1}(\tau)=\frac{1}{4}\left\langle E\left\{\begin{matrix}\left(-(1-g(t))n_L^2(t)+(1+g(t))n_E^2(t)\right)\\ \times\left(-(1-g(t+\tau))n_L^2(t+\tau)+(1+g(t+\tau))n_E^2(t+\tau)\right)\end{matrix}\right\}\right\rangle \tag{7.4-21}$$

where $\langle x\rangle$ denotes the time average of x. This can be written as

$$\begin{aligned}R_{n_1}(\tau)=&\frac{1}{4}\left\langle(1+g(t))(1+g(t+\tau))\right\rangle E\left\{n_E^2(t)n_E^2(t+\tau)\right\}\\&-\frac{1}{4}\left\langle(1+g(t))(1-g(t+\tau))\right\rangle E\left\{n_E^2(t)n_L^2(t+\tau)\right\}\\&+\frac{1}{4}\left\langle(1-g(t))(1-g(t+\tau))\right\rangle E\left\{n_L^2(t)n_L^2(t+\tau)\right\}\\&-\frac{1}{4}\left\langle(1-g(t))(1-g(t+\tau))\right\rangle E\left\{n_L^2(t)n_L^2(t+\tau)\right\}\end{aligned} \tag{7.4-22}$$

Recall that $\Lambda_d(2\tau/T_d)$ is defined by (7.3-104) and can be written in the form

$$\Lambda_d(2\tau/T_d)=\begin{cases}\left[1-\dfrac{2|\tau|}{T_d}\right] & \text{when}\,|\tau|>T_d/2\\ 0 & \text{when}\,|\tau|>T_d/2\end{cases} \tag{7.4-23}$$

Denote the two autocorrelation functions

$$R_{1+g}(\tau) = \langle (1+g(t))(1+g(t+\tau)) \rangle = R_{1-g}(\tau) = \langle (1-g(t))(1-g(t+\tau)) \rangle \tag{7.4-24}$$

and the two cross-correlation functions by

$$R_{(1+g)(1-g)}(\tau) = \langle (1+g(t))(1-g(t+\tau)) \rangle = R_{(1-g)(1+g)}(\tau) = \langle (1-g(t))(1+g(t+\tau)) \rangle \tag{7.4-25}$$

These auto- and cross-correlation functions are plotted in Figure 7.4-3.

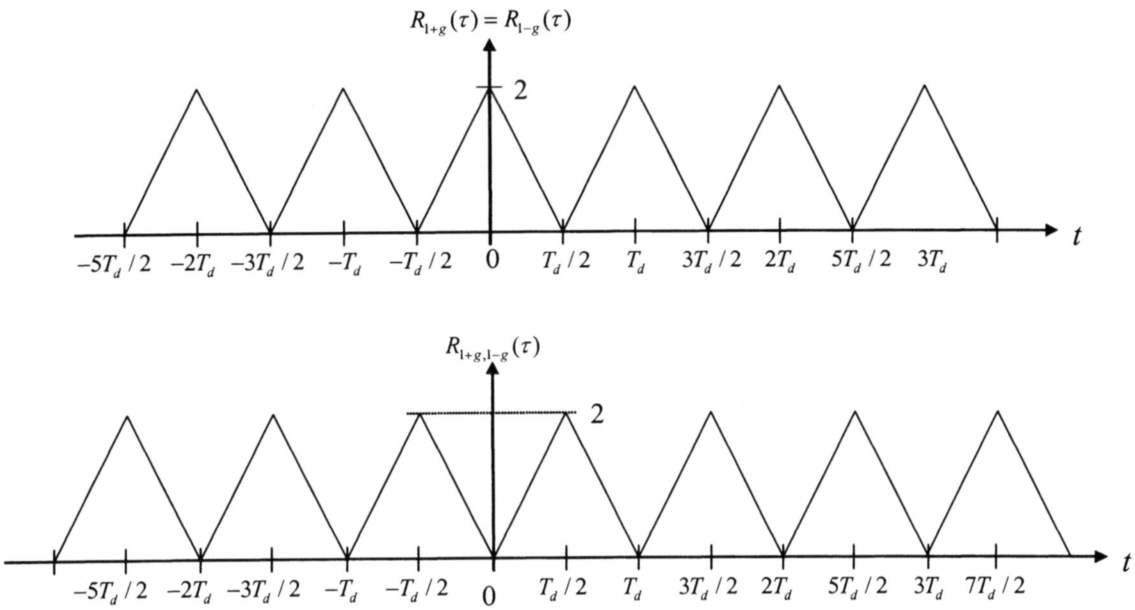

Figure 7.4-3 The auto- and cross-correlation functions related to the multiplexing function.

Now the autocorrelation function $R_{1+g}(\tau)$ can be written as

$$R_{1-g}(\tau) = R_{1+g}(\tau) = 2\Lambda_d(\tau) \otimes \sum_{n=-\infty}^{\infty} \delta(\tau + nT_d) \tag{7.4-26}$$

where the $\otimes$ symbol denotes the convolution of the functions before and after it. In addition, it is clear that the autocorrelation and the cross-correlation functions are related by

$$R_{(1+g)(1-g)}(\tau) = 2 - R_{1+g}(\tau) = 2 - 2\Lambda_d(\tau) \otimes \sum_{n=-\infty}^{\infty} \delta(\tau + nT_d) \tag{7.4-27}$$

Noting that

$$E\left\{ n_E^2(t) n_E^2(t+\tau) \right\} = \left(N_0 B \right)^2 + 2R_{n_E}^2(\tau) \tag{7.4-28}$$

Using (7.4-26)–(7.4-28) in (7.4-22) produces

$$R_{n_1}(\tau) = \left(2R_{n_E}^2(\tau) + 2(N_0 B)^2\right)\left(\Lambda_d(\tau) \otimes \sum_{n=-\infty}^{\infty} \delta(\tau + nT_d)\right) - (N_0 B)^2 \tag{7.4-29}$$

In order to obtain the spectral density at $f = 0$ note that

$$\begin{gathered} F\left\{\sum_{n=-\infty}^{\infty} \delta(\tau + nT_d)\right\} = \frac{1}{T_d}\sum_{n=-\infty}^{\infty} \delta(f + n/T_d) \\ F\{\Lambda_d(\tau)\} = \frac{T_d}{2}\operatorname{sinc}\left(\frac{fT_d}{2}\right) \end{gathered} \tag{7.4-30}$$

Now the spectral density at $f = 0$ can be obtained by integrating (7.4-29) over all τ. Thus

$$\begin{aligned} S_{n_1}(0) = & -(N_0 B)^2 \delta(f) + 2(N_0 B)^2 \frac{T_d}{2}\operatorname{sinc}\left(\frac{fT_d}{2}\right)\frac{1}{T_d} \\ & + 2\int_{-\infty}^{\infty} R_{n_E}^2(\tau)\left[\Lambda_d(\tau) \otimes \sum_{n=-\infty}^{\infty} \delta(\tau + nT_d)\right] d\tau \Bigg|_{f=0} \end{aligned} \tag{7.4-31}$$

We have neglected the higher order terms in the second term in (7.4-31), since it is assumed that it will not be passed by the loop. Note that the first two terms in (7.4-31) cancel. Thus (7.4-31) can be rewritten as

$$S_{n_1}(0) = 2\int_{-\infty}^{\infty} \left(S_{n_E}(f) \otimes S_{n_E}(f)\right)\left[\frac{T_d}{2}\operatorname{sinc}^2\left(\frac{fT_d}{2}\right)\frac{1}{T_d}\sum_{n=-\infty}^{\infty} \delta(f + n/T_d)\right] \tag{7.4-32}$$

Evaluating (7.4-32) produces

$$S_{n_1}(0) = 2\int_{-\infty}^{\infty} \sum_{n=-\infty}^{\infty} S_{n_E}(f')S_{n_E}(n/T_d - f')\operatorname{sinc}^2\left(\frac{n}{2}\right) df' \tag{7.4-33}$$

With the assumption that the bandpass filter bandwidth satisfies $B >> 1/T_d$, (7.4-33) can be rewritten as

$$S_{n_1}(0) = \frac{N_0^2}{2} B' \left[1 + 2\left(\frac{2}{\pi}\right)^2 \left[1 + \frac{1}{3^2} + \frac{1}{5^2} + \ldots\right]\right] \tag{7.4-34}$$

where B' is defined by

$$2B' = \int_{-\infty}^{\infty} \left|H_{BB}(f)\right|^4 df \tag{7.4-35}$$

where $H_{BB}(f)$ is the baseband equivalent filter of the RF filter. From Jollcy [21] we have

$$1 + \frac{1}{3^2} + \frac{1}{5^2} + \ldots = \frac{\pi^2}{8} \tag{7.4-36}$$

so that (7.4-34) can be written as

$$S_{n_1}(0) = N_0^2 B' \tag{7.4-37}$$

Now consider the calculation of $S_{n_2}(0)$. The noise terms $n_E(t)$ and $n_L(t)$ can be written in the form

$$\begin{aligned} n_E(t) &= \sqrt{2}n_{E_c}(t)\cos(\omega_0 t+\theta)+\sqrt{2}n_{E_s}(t)\sin(\omega_0 t+\theta) \\ n_L(t) &= \sqrt{2}n_{L_c}(t)\cos(\omega_0 t+\theta)+\sqrt{2}n_{L_s}(t)\sin(\omega_0 t+\theta) \end{aligned} \tag{7.4-38}$$

Using (7.4-38) in $n_2(t)$ of (7.4-15) produces

$$n_2(t)=\sqrt{P}d(t)\left[-\left(\frac{1-g(t)}{2}\right)n_{E_c}(t)+\left(\frac{1+g(t)}{2}\right)n_{L_c}(t)\right]+O(2\omega_0) \tag{7.4-39}$$

where $O(2\omega_0)$ denotes on the order of $2\omega_0$. The autocorrelation function of $n_2(t)$ is given by

$$R_{n_2}(\tau)=PR_d(\tau)\left\langle E\left\{\begin{aligned}&\left(-\left(\frac{1-g(t)}{2}\right)n_{E_c}(t)+\left(\frac{1+g(t)}{2}\right)n_{L_c}(t)\right)\\ &\times\left(-\left(\frac{1-g(t+\tau)}{2}\right)n_{E_c}(t+\tau)+\left(\frac{1+g(t+\tau)}{2}\right)n_{L_c}(t+\tau)\right)\end{aligned}\right\}\right\rangle \tag{7.4-40}$$

Equation (7.4-40) can be simplified to

$$R_{n_2}(\tau)=\frac{PR_d(\tau)}{4}\langle(1+g(t))(1-g(t))\rangle\left(2R_{n_{E_c}}(\tau)\right) \tag{7.4-41}$$

where the "2" on the right side comes from the fact that both $R_{n_{E_c}}(\tau)$ and $R_{n_{L_c}}(\tau)$ have the same autocorrelation functions. From (7.4-26) we have

$$R_{n_2}(\tau)=PR_{n_{E_c}}(\tau)R_d(\tau)\Lambda_d(\tau)\otimes\sum_{n=-\infty}^{\infty}\delta(\tau+nT_d) \tag{7.4-42}$$

It follows that the power spectral density of $n_2(t)$ is given by

$$S_{n_2}(0)=P\int_{-\infty}^{\infty}R_d(\tau)R_{n_{E_c}}(\tau)\left\{\Lambda_d(\tau)\sum_{n=-\infty}^{\infty}\delta(\tau+nT_d)\right\}d\tau \tag{7.4-43}$$

Since the Fourier transform of a convolution is the product of the Fourier transforms we have

$$S_{n_2}(0)=\int_{-\infty}^{\infty}S_{n_{E_c}}(f')S_d(f-f')\frac{T_d}{2}\text{sinc}^2\left(\frac{fT_d}{2}\right)\frac{1}{T_d}\sum_{n=-\infty}^{\infty}\delta(f+n/T_d)df' \tag{7.4-44}$$

Evaluating the delta functions produces

$$S_{n_2}(0)=\frac{P}{2}\int_{-\infty}^{\infty}\sum_{n=-\infty}^{\infty}S_{n_{E_c}}(f')S_d\left(\frac{n}{T_d}-f'\right)\text{sinc}^2\left(\frac{n}{2}\right)df' \tag{7.4-45}$$

If $T_d >> T_s$ then one can write

$$S_{n_2}(0) = \frac{PN_0}{4} \int_{-\infty}^{\infty} S_d(f')|H_{BB}(f')|^2 df' \sum_{n=-\infty}^{\infty} \text{sinc}^2\left(\frac{n}{2}\right) \tag{7.4-46}$$

Since

$$\sum_{n=-\infty}^{\infty} \text{sinc}^2\left(\frac{n}{2}\right) = \sum_{\substack{n=-\infty \\ n\neq 0}}^{\infty} \left(\frac{2}{n\pi}\right)^2 + 1 = 1 + \left(\frac{2}{\pi}\right)^2 \sum_{n=-\infty}^{\infty} \frac{1}{n^2} = 2 \tag{7.4-47}$$

So $S_{n_2}(0)$ is well approximated by

$$S_{n_2}(0) = \frac{PN_0}{2}\alpha \tag{7.4-48}$$

where α is given by

$$\alpha = \int_{-\infty}^{\infty} S_d(f)|H_{BB}(f)|^2 df \cong 1 \tag{7.4-49}$$

since the data rate is small compared to the RF filter bandwidth, B. It therefore follows that the tracking error variance of the time-shared code-tracking loop is given by

$$\frac{\sigma_\varepsilon^2}{T_s^2} = \frac{N_0 B_L}{P}\left(1 + \frac{2N_0 B}{P}\right), \quad \text{chips}^2 \tag{7.4-50}$$

In comparing (7.4-50) with (7.2-93) it is seen that the variance of the time-shared loop is exactly twice that of the full-time bandpass version loop, when α is essentially one, which is the normal case. If fact, if α was not approximated as one, (7.4-50) would contain α as a product with P in the expression, and this would be exactly twice that of (7.2-93).

7.5 PERFORMANCE OF A NONCOHERENT RF IMPLEMENTED TIME GATED EARLY-LATE GATE BANDPASS CODE-TRACKING LOOP

Up until this point we have been concerned with tracking constant envelope signals. Now we will turn our attention to time gated or pulsed envelope signals such as is used in time division multiple access (TDMA) systems. The approach presented here differs from that of Spilker [22] and Huff and Reinhard [23]. The gating function, which we denote by $g(t)$, is illustrated in Figure 7.5-1. It is assumed that the receiver is synchronized with the received pulse signal so that the noise is shut off when the signal is in the "off" state, and it will be further assumed that the repetition (gating) rate ($1/T_0$) is much greater than the code loop bandwidth B_L (that is, it is assumed that $B_L T_0 \ll 1$). It will also be assumed that the data symbol rate is much greater than the gating rate (i.e., $(T_0/T_s) \gg 1$).

The gating function is on for $d_f T_0$ seconds every T_0 second. Thus the duty factor is d_f. It therefore follows that the gating function is off $(1-d_f)$ T_0 seconds every T_0 seconds, and T_0 is the period of the gating function. The received, gated signal is gated synchronously by the receiver and is modeled by

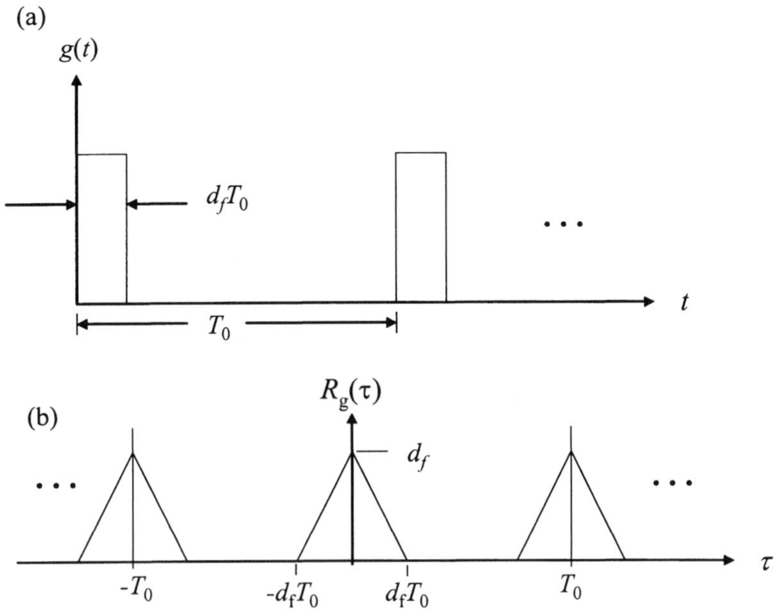

Figure 7.5-1 Gating function and its autocorrelation function: (a) the gating function and (b) the gating autocorrelation function.

$$y(t) = \sqrt{2P}d(t)PN(t)g(t)\cos\left(\omega_0 t + \theta\right) + g(t)n(t) \tag{7.5-1}$$

where it is clear that the input noise is gated also. In (7.5-1) P is the received power, $d(t)$ is the baseband data stream, $PN(t)$ is the direct sequence code, ω_0 is the received angular frequency in radians/second, θ is the received carrier phase in radians, and $n(t)$ is the received white noise process with dimensions of volts.

The late and early signals are given

$$\begin{aligned} L(t) &= PN(t - \hat{T} - T_s/2) \\ E(t) &= PN(t - \hat{T} + T_s/2) \end{aligned} \tag{7.5-2}$$

Now consider the calculation of the code-tracking error variance. Figure 7.5-2 illustrates the code-tracking loop model with gating.

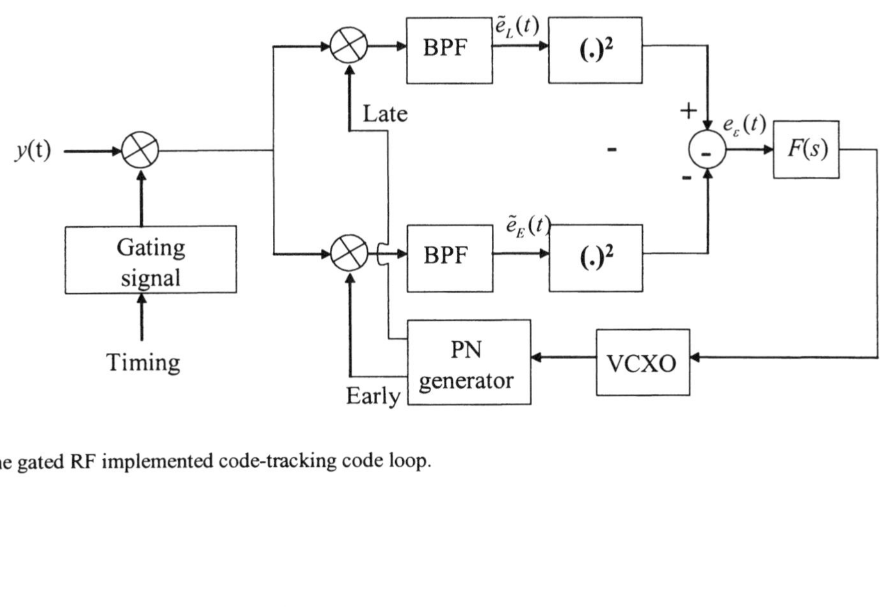

Figure 7.5-2 Time gated RF implemented code-tracking code loop.

The two correlator outputs are expressed as follows

$$\tilde{e}_L(t) = H(s)\left[d(t)g(t)R_{PN}(T-\hat{T}-T_s/2)\cos(\omega_0 t+\theta)\right]+n_L(t) \tag{7.5-3}$$

and

$$\tilde{e}_E(t) = H(s)\left[d(t)g(t)R_{PN}(T-\hat{T}+T_s/2)\cos(\omega_0 t+\theta)\right]+n_E(t) \tag{7.5-4}$$

in which $H(s)$ is the Heaviside operator representing the bandpass filters, and the two noise terms are defined by

$$\begin{aligned} n_L(t) &= H(s)\left[n(t)PN(t-\hat{T}-T_s/2)\right] \\ n_E(t) &= H(s)\left[n(t)PN(t-\hat{T}+T_s/2)\right] \end{aligned} \tag{7.5-5}$$

After squaring and forming the difference one obtains

$$\begin{aligned} e_\varepsilon(t) \cong\ & PH_{BB}(s)\left[d(t)^2 g(t)\right]\left[R_{PN}^2(T-\hat{T}-T_s/2)-R_{PN}^2(T-\hat{T}+T_s/2)\right] \\ & +2\sqrt{2P}H(s)\left[d(t)g(t)\cos(\omega_0 t+\theta)\right]\begin{bmatrix} n_L(t)g(t)R_{PN}(T-\hat{T}-T_s/2) \\ -n_E(t)g(t)R_{PN}(T-\hat{T}+T_s/2) \end{bmatrix} \\ & +n_L^2(t)g(t)-n_E^2(t)g(t) \end{aligned} \tag{7.5-6}$$

where $H_{BB}(s)$ is the Heaviside operator representing the baseband equivalent filter of the RF filter. With the assumption of an early-late spacing of one symbol and taking the average dc signal as the error control component, we can write the error signal as

$$e_\varepsilon(t) = \frac{2P}{T_s}d_f + n_1(t)+n_2(t)+n_3(t) \tag{7.5-7}$$

where

$$n_1(t) = 2\sqrt{2P}H(s)\left[d(t)g(t)\cos(\omega_0 t+\theta)\right]\begin{bmatrix} n_L(t)g(t)R_{PN}(T-\hat{T}-T_s/2) \\ -n_E(t)g(t)R_{PN}(T-\hat{T}+T_s/2) \end{bmatrix} \tag{7.5-8}$$

and

$$n_2(t) = n_L^2(t)g(t)-n_E^2(t)g(t) \tag{7.5-9}$$

and $n_3(t)$ is the non-dc component of self-noise due to the time gating process. We will neglect this latter term since it will only be significant at high loop SNRs, where it normally is not a concern.

Now to close the loop, write the code loop estimate of the code phase as

$$\frac{\hat{T}}{T_s} = \frac{K_{VCO}}{s}F(s)\left[2d_f P\frac{\varepsilon}{T_s}+n_1(t)+n_2(t)\right] \tag{7.5-10}$$

Using $\varepsilon = T-\hat{T}$ it can be shown that the following relationship is obtained

$$\frac{\varepsilon}{T_s} = \frac{-K'PF(s)\left[\dfrac{n_1(t)+n_2(t)}{2d_f P}\right]}{s+K'PF(s)} = -H(s)\left[\frac{n_1(t)+n_2(t)}{2d_f P}\right] \quad (7.5\text{-}11)$$

where $H(s)$ is the closed-loop response of the time-shared code-tracking loop and $K' = 2K_{VCO}d_f$. It is to be noted that both ε and $\hat{T}$ are functions of time, even though they have not been indicated as functions of time explicitly in the loop equations. The variance of the tracking error can be described by

$$\frac{\sigma_\varepsilon^2}{T_s^2} = \int_{-\infty}^{\infty} |H(j2\pi f)|^2 \frac{\left(S_{n_1}(f)+S_{n_2}(f)\right)}{4d_f^2 P^2} df \quad \text{chips}^2 \quad (7.5\text{-}12)$$

where $S_{n_i}(f)$ are the two spectral densities of the two noise terms. Using the reasonable assumption that the two noise terms are wide compared to the closed-loop noise bandwidth (B_L) produces

$$\frac{\sigma_\varepsilon^2}{T_s^2} \cong 2B_L \frac{\left(S_{n_1}(0)+S_{n_2}(0)\right)}{4d_f^2 P^2} \quad \text{chips}^2 \quad (7.5\text{-}13)$$

It is reasonably assumed that the gating process is statistically independent of the receiver noise, so that the autocorrelation function of $n_1(t)$ is given by

$$R_{n_1}(\tau) = 2PR_d(\tau)R_g(\tau)R_{n_{Lc}}(\tau)\left[R_{PN}(\varepsilon - T_s/2) + R_{PN}(\varepsilon + T_s/2)\right] \quad (7.5\text{-}14)$$

For NRZ code symbols the sum of the last two terms can be approximated by 1, when the error, ε, is small (during strong signal tracking for example). Thus (7.5-14) can be approximated by

$$R_{n_1}(\tau) = 2PR_d(\tau)R_g(\tau)R_{n_{Lc}}(\tau) \quad (7.5\text{-}15)$$

In order to determine the power spectral density at $f = 0$, it is necessary to obtain the autocorrelation function of the gating function. Figure 7.5-1 illustrates the autocorrelation function. We therefore have from (7.3-104)

$$\Lambda_{d_f}(\tau) = \Lambda\left(\tau/(d_f T_0)\right) = \begin{cases} \left(1 - \dfrac{|\tau|}{d_f T_0}\right) & |\tau| \le d_f T_0 \\ 0 & \text{elsewhere} \end{cases} \quad (7.5\text{-}16)$$

It follows that the autocorrelation function can be written as

$$R_g(\tau) = d_f \Lambda_{d_f}(\tau) \otimes \sum_{n=-\infty}^{\infty} \delta(\tau + nT_0) \quad (7.5\text{-}17)$$

Since the noise spectral density at $f = 0$ is obtained by obtaining the Fourier transform of the autocorrelation function we can write for $n_1(t)$ that

$$S_{n_1}(0) = 2P\int_{-\infty}^{\infty} R_{n_{Lc}}(\tau)R_d(\tau)\left\{d_f \Lambda_{d_f}(\tau) \otimes \left(\sum_{n=-\infty}^{\infty} \delta(\tau + nT_0)\right)\right\} d\tau \quad (7.5\text{-}18)$$

However since it is true that

$$F\{f_1(t)\otimes f_2(t)\}=F_1(f)F_2(f) \tag{7.5-19}$$

where $F_i(f)$ is the Fourier transform of $f_i(t)$. Further noting from Section 7.4 that

$$\begin{aligned}&F\left\{\sum_{n=-\infty}^{\infty}\delta(\tau+nT_d)\right\}=\frac{1}{T_d}\sum_{n=-\infty}^{\infty}\delta(f+n/T_d)\\&F\{\Lambda_d(\tau)\}=d_fT_d\text{sinc}\left(fd_fT_d\right)\end{aligned} \tag{7.5-20}$$

Using (7.5-19) and (7.5-20) in (7.5-18) produces

$$S_{n_1}(0)=2P\int_{-\infty}^{\infty}\int_{-\infty}^{\infty}S_{n_{Lc}}(f')S_d(f-f')df'(d_f)\left[d_fT_0\text{sinc}^2(fd_fT_0)\right]\frac{1}{T_0}\sum_{n=-\infty}^{\infty}\delta\left(f+\frac{n}{T_0}\right)df \tag{7.5-21}$$

Evaluating the above expression produces

$$S_{n_1}(0)=2d_f^2P\int_{-\infty}^{\infty}S_{n_{Lc}}(f')S_d\left(\frac{n}{T_0}-f'\right)\left[\sum_{n=-\infty}^{\infty}\text{sinc}^2(nd_f)\right]df' \tag{7.5-22}$$

By assumption the symbol rate was much larger than the gating rate, so that we obtain

$$S_{n_{Lc}}(0)=N_0d_f^2P\int_{-\infty}^{\infty}|H_{BB}(f)|^2S_d(f)df\left[\sum_{n=-\infty}^{\infty}\text{sinc}^2\left(nd_f\right)\right] \tag{7.5-23}$$

where $H_{BB}(f)$ is the baseband equivalent low-pass filter corresponding to the bandpass filter arm filters.
Consider the summation in (7.5-23). It can be written as

$$S=\sum_{n=-\infty}^{\infty}\frac{\sin\left(n\pi d_f\right)^2}{\left(n\pi d_f\right)^2}=\left(\frac{1}{(\pi d_f)^2}\right)\sum_{n=-\infty}^{\infty}\frac{\sin\left(n\theta\right)^2}{\left(n\right)^2} \tag{7.5-24}$$

where we have substituted $\theta=\pi d_f$. Breaking the summation into the $n=0$ term and all the rest of the terms we have

$$S=\frac{1}{(\pi d_f)^2}\left[\theta^2+2\sum_{n=1}^{\infty}\frac{\sin\left(n\theta\right)^2}{\left(n\right)^2}\right] \tag{7.5-25}$$

From Jolley [21] it is shown that

$$\sum_{n=1}^{\infty}\frac{\sin(n\theta)^2}{(n\theta)^2}=\frac{1}{2}\theta(\pi-\theta)\qquad 0\leq\theta\leq\pi \tag{7.5-26}$$

In Problem 10 it is shown that

$$S = \frac{1}{d_f} \tag{7.5-27}$$

Therefore it follows that

$$S_{n_1}(0) = d_f P N_0 \alpha \tag{7.5-28}$$

where α is the filter loss of the data passing through the bandpass filter. Since this loss is very small we approximate it as 1. Thus we have

$$S_{n_1}(0) \cong d_f P N_0 \tag{7.5-29}$$

Now we are in a position to evaluate the second noise term $n_2(t)$. Recall that

$$n_2(t) = n_L^2(t)g(t) - n_E^2(t)g(t) \tag{7.5-30}$$

From Section 7.4 it can be deduced that the autocorrelation function can be written as

$$R_{n_2}(\tau) = N_0^2 \left(\int_{-\infty}^{\infty} |H(f)|^2 e^{-i2\pi f\tau} df \right)^2 R_g(\tau) \tag{7.5-31}$$

in which $H(f)$ is the RF frequency response of the bandpass filters in the arms of the time gated code-tracking loop. Evaluating the Fourier transform at $f = 0$ yields

$$S_{n_2}(0) = N_0^2 \int_{-\infty}^{\infty}\int_{-\infty}^{\infty}\int_{-\infty}^{\infty} |H(f'')|^2 |H(f')|^2 e^{-2\pi f'\tau} e^{2\pi f''\tau} R_g(\tau) df' df'' d\tau \tag{7.5-32}$$

Using (7.5-17) to represent $R_g(\tau)$, and noting that $F\{\Lambda_d(\tau)\} = d_f T_d \text{sinc}\left(f d_f T_d\right)$, produces

$$S_{n_2}(0) = N_0^2 d_f^2 \int_{-\infty}^{\infty}\int_{-\infty}^{\infty}\int_{-\infty}^{\infty} |H(f'')|^2 |H(f')|^2 \int_{-\infty}^{\infty} \left\{ \int_{-\infty}^{\infty} e^{-i2\pi(f''-f'+f)\tau} d\tau \right\}$$
$$\times \text{sinc}^2(f d_f T_d) \sum_{n=-\infty}^{\infty} \delta\left(f + \frac{n}{T_d}\right) df df' df'' \tag{7.5-33}$$

This expression is evaluated in Problem 11 with the result that

$$S_{n_2}(0) = N_0^2 d_f^2 \sum_{n=-\infty}^{\infty} \int_{-\infty}^{\infty} |H(f')|^2 \left| H\left(f' + \frac{n}{T_d}\right)\right|^2 \text{sinc}^2(n d_f) df' \tag{7.5-34}$$

Evaluating (7.5-34), assuming that $R_s/R_d >> 1$ produces

$$S_{n_2}(0) = N_0^2 d_f (2B') \tag{7.5-35}$$

where

$$2B' = \int_{-\infty}^{\infty} |H(f)|^4 df \tag{7.5-36}$$

Using (7.5-28) and (7.5-35) in (7.5-13) produces the result

$$\left(\frac{\sigma_\varepsilon^2}{T_s^2}\right) = \frac{N_0 B_L}{2d_f P}\left[1 + \frac{2N_0 B'}{P}\right] \tag{7.5-37}$$

Thus we see that by pulsing the signal with a power of P watts, the increase in the tracking error variance is just $1/d_f$ of what it would have been if the duty factor was 100%.

7.6 STEADY STATE ERROR OF CODE-TRACKING LOOPS WITHOUT NOISE

Up to this point in this chapter we have considered various code loop configurations with noise or with noise and Gaussian interference. The code-tracking error is determined by the noise or interference. However, many times the noise or interference is not the dominant source of the code loop error. This can occur when the noise or interference is small and the dynamics on the line of sight between the transmitter and the receiver has relative movement. Starting with (7.3-86), it is not hard to show that in general for a gate late noncoherent code-tracking loop without noise the code error is given by

$$\frac{\varepsilon}{T_s} = \frac{1}{1 + KPF(s)/s}\left(\frac{T_i}{T_s}\right) = [1 - H(s)]\left(\frac{T_i}{T_s}\right) \tag{7.6-1}$$

where K is an appropriate gain, ε is the timing error in seconds, $F(s)$ is the loop filter transfer function, P is the received signal power, $H(s)$ is the closed-loop transfer function, and T_i is the input timing delay, which is a function of time. Again for convenience, we have left off the dependence on time on both ε and T_i. If the code loop is a baseband code loop the received power P would be replaced with $\sqrt{P}$ in (7.6-1). In this expression the closed-loop transfer function (CLTF) is given by

$$H(s) = \frac{PKF(s)/s}{1 + PKF(s)/s} \tag{7.6-2}$$

where K in this equation includes the multiplier and VCO gains, and the loop gain is KP.

7.6.1 First-Order Noncoherent I-Q Early-Late Gate Code-Tracking Loop

First we consider the first-order code-tracking loop. In this case $F(s) = 1$, so that

$$H(s) = \frac{\sqrt{P}K}{s + \sqrt{P}K} \tag{7.6-3}$$

has one pole. It is to be noted that most GPS receivers utilize first-order code-tracking loops with carrier aiding. That is, the carrier loop transfers Doppler frequency information to the code loop, weighted by the ratio of the code symbol rate to carrier frequency. The steady state error can be computed in a similar manner as was used in Chapter 5 for carrier loops. The steady state code timing error is given by

$$\frac{\varepsilon_{ss}}{T_s} = \lim_{s \to 0} \left(sT_i(s) \right) \tag{7.6-4}$$

where $T_i(s)$ is the Laplace transform of $T_i(t)$. Now it is true that

$$1 - H(s) = \frac{s}{s + \sqrt{PK}} \tag{7.6-5}$$

From Chapter 5 it is known that the noise bandwidth, B_L, is ¼ of the loop gain, so that the loop transfer function and one minus the transfer function can be written as

$$\begin{aligned} H(s) &= \frac{4B_L}{s + 4B_L} \\ 1 - H(s) &= \frac{s}{s + 4B_L} \end{aligned} \tag{7.6-6}$$

The final value theorem of Laplace transform theory allows us to compute the steady state code-tracking error just as we did in Chapter 5. Write the input line of sight time delay as a function of time as

$$\frac{T_i}{T_s} = \left(T_0 + \dot{T}t + \frac{1}{2}\ddot{T}t^2 \right) u(t) \text{ symbols} \tag{7.6-7}$$

where $u(t)$ is the unit step function that is zero before zero and unity after $t = 0$. Then the corresponding Laplace transform of the input function is given by

$$\frac{T_i(s)}{T_s} = \frac{T_0}{s} + \frac{\dot{T}}{s^2} + \frac{\ddot{T}}{s^3} \tag{7.6-8}$$

In general the resulting error signal in Laplace transform variables is given by

$$\frac{\varepsilon(t)}{T_s} = \left[1 - H(s) \right] \frac{T_i(s)}{T_s} \tag{7.6-9}$$

Now apply the Laplace transform final value theorem to our problem

$$\lim_{t \to \infty} \left(\frac{\varepsilon(t)}{T_s} \right) = \lim_{t \to \infty} \left(s\left[1 - H(s) \right] \frac{T_i(s)}{T_s} \right) \tag{7.6-10}$$

For the first-order loop one has with $\ddot{T} = 0$, the result[11]

$$\lim_{t \to \infty} \left(\frac{\varepsilon(t)}{T_s} \right) = \frac{\dot{T}}{4B_L} \tag{7.6-11}$$

Therefore if the signal time delay is at a constant rate of $\dot{T}t$, the long-term (asymptotic) time tracking error is $\dot{T}/(4B_L)$ as long as this error lies within the linear region.

---

[11] If $\ddot{T}$ is not set to zero, the limit is unbounded.

7.6.2 Second-Order Ideal Noncoherent I-Q Early-Late Gate Code-Tracking Loop

Now consider the ideal second-order code-tracking loop. The loop filter transfer function is given by

$$F(s) = \frac{1+\tau_2 s}{\tau_1 s} \tag{7.6-12}$$

It follows from (7.6-12) that the CLTF is given by

$$H(s) = \frac{1+\tau_2 s}{1+\tau_2 s + (\tau_1 / KP) s^2} \tag{7.6-13}$$

This can be written as

$$\begin{aligned} H(s) &= \frac{2\varsigma\omega_n s + \omega_n^2}{s^2 + 2\varsigma\omega_n s + \omega_n^2} \\ 1 - H(s) &= \frac{s^2}{s^2 + 2\varsigma\omega_n s + \omega_n^2} \end{aligned} \tag{7.6-14}$$

where ς is the *damping factor* and ω_n is the *natural frequency*. These parameters can be related to the loop parameters, just as was done for the carrier loop results, in the form

$$\begin{aligned} \omega_n^2 &= \frac{PK}{\tau_1} \\ \varsigma &= \frac{\tau_2}{2}\sqrt{\frac{PK}{\tau_1}} \end{aligned} \tag{7.6-15}$$

In addition the loop natural frequency and the damping factor can be related to the closed-loop noise bandwidth B_L via

$$\begin{aligned} B_L &= \frac{r+1}{4\tau_2} \\ r &= \frac{KP\tau_2^2}{\tau_1} \end{aligned} \tag{7.6-16}$$

or alternatively

$$B_L = \frac{\omega_n}{2}\left(\varsigma + \frac{1}{4\varsigma}\right) \tag{7.6-17}$$

When $\varsigma = 1/\sqrt{2}$, a commonly selected value in practice, it can be seen that $B_L = 0.53\omega_n$.

Now consider the steady state response to a varying input delay process. Using (7.6-14) and assuming a delay model of the form of (7.6-8), the steady state error becomes

$$\lim_{t\to\infty}\left(\frac{\varepsilon(t)}{T_s}\right) = \lim_{t\to\infty}\left(s\left[\frac{s^2}{s^2 + 2\varsigma\omega_n s + \omega_n^2}\right]\left(\frac{T_0}{s} + \frac{\dot{T}}{s^2} + \frac{\ddot{T}}{s^3}\right)\right) = \frac{\ddot{T}}{\omega_n^2} \tag{7.6-18}$$

This can also be written, for an active second-order code-tracking loop with $\zeta = 0.707$ as

$$\lim_{t\to\infty}\left(\frac{\varepsilon(t)}{T_s}\right)=\frac{\ddot{T}}{\omega_n^2}=\frac{\ddot{T}}{3.56B_L^2} \tag{7.6-19}$$

Thus we see that a first-order code-tracking loop can track a delay rate $(\dot{T})$ if the loop bandwidth is large enough so that the error does not exceed the tracking range, and similarly an ideal second-order code-tracking loop can track a delay acceleration if the bandwidth is large enough. It is clear that the third-order code-tracking loop will track a constant $\dddot{T}$, if the loop bandwidth is sufficiently large enough.

7.7 EARLY-LATE GATE NONCOHERENT I-Q CODE-TRACKING LOOP PULL-IN WITHOUT NOISE

Now consider the pull-in behavior of a noncoherent I-Q early-late gate code-tracking loop when no noise is present. The tool to characterize the behavior is the *phase plane*, which is a plot of the timing error, in symbols, versus the timing error rate in symbols per second. The presentation given here follows [22, 24] with modifications. Consider now a second-order active code-tracking loop with loop transfer function

$$F_p(s)=\frac{1+\tau_2 s}{\tau_1 s} \tag{7.7-1}$$

for the active loop filter. In general the second-order code-tracking loop has a timing estimate of the form

$$\hat{T}=\frac{GF(s)}{s}D(\varepsilon) \tag{7.7-2}$$

where $D(\varepsilon)$ is defined by

$$D(\varepsilon)=\frac{S(\varepsilon)}{\dot{S}(0)} \tag{7.7-3}$$

is the normalized to unity slope *S*-curve characteristic. Now use the definitions in (7.7-4) and divide

$$\begin{aligned} x&=\frac{\varepsilon}{T_c}\\ y&=\frac{T_i}{T_c}\\ D(\varepsilon)&=\frac{D(\varepsilon)}{T_c} \end{aligned} \tag{7.7-4}$$

Equation (7.7-2) by T_c produces

$$s(y-x)=GF(s)D(x) \tag{7.7-5}$$

Let $g=G/\omega_n$ in (7.7-5) to yield

$$\frac{s}{\omega_n}(y-x) = gGF(s)D(x) \tag{7.7-6}$$

Consider the active loop filter of the form

$$F(s) = \frac{1+\tau_2 s}{\tau_1 s} \tag{7.7-7}$$

For the passive loop filter we can write the following

$$\tau_2 = \frac{2\varsigma}{\omega_n} \tag{7.7-8}$$

and

$$\tau_1 = \frac{G}{\omega_n^2} \tag{7.7-9}$$

to produce for the filter response in terms of ς and ω_n

$$gF(s) = \frac{1+2\varsigma s/\omega_n}{s/\omega_n} \tag{7.7-10}$$

Define

$$\begin{aligned} \dot{y} &= \frac{1}{\omega_n}\frac{dy}{dt} \\ \dot{x} &= \frac{1}{\omega_n}\frac{dx}{dt} \end{aligned} \tag{7.7-11}$$

Using (7.7-6) in terms of the (7.7-11) yields

$$\ddot{y} - \ddot{x} = D(x) + 2\varsigma\frac{dD(x)}{dx}x \tag{7.7-12}$$

Noting that

$$\frac{\ddot{x}}{\dot{x}} = \frac{d\dot{x}}{dx} \tag{7.7-13}$$

along with (7.7-12) produces the *phase plane equation*

$$\frac{d\dot{x}}{dx} = \frac{-D(x) - 2\varsigma\left(\dfrac{dD(x)}{dx}\right)\dot{x} + \ddot{y}}{\dot{x}} \tag{7.7-14}$$

Nielson [25] observed that the two previously cited authors [22, 24] did not use the required normalization

$$\left|\frac{dD(x)}{dx}\right\|_{x=0} = 1 \qquad (7.7\text{-}15)$$

The results used here are based on this normalization. The phase plane is symmetric about $\dot{x} = 0$. The phase plane is generated by picking a value of $\omega_n \Delta t$ and $\dot{x}(0)$ that determines $\Delta x(0)$ via

$$\Delta x(0) = \dot{x}(0)\left(\omega_n \Delta t\right) \qquad (7.7\text{-}16)$$

Therefore after Δt seconds one has

$$\begin{aligned} x(\Delta t) &= x(0) + \Delta x(0) \\ \dot{x}(\Delta t) &= \dot{x}(0) + \Delta \dot{x}(0) \end{aligned} \qquad (7.7\text{-}17)$$

This procedure continues with an appropriately small value of $\omega_n \Delta t$ until the code loop moves off the restricted phase plane or converges to the lock point. This yields one curve on the phase plane. This process continues with a new initial point $(x(0), \dot{x}(0))$ that also constructs a curve on the phase plane. This process continues until the phase plane is sufficiently populated with curves. Figures for the baseband loop and with the second-order active filter loop are plotted in Figure 7.7-1. Both Figures 7.7-1(a) and 7.7-1(b) are for the second order active code loop with $\varsigma = 0.707$ and $\ddot{y} = 0$. In the figures the ordinate is $(1/\omega_n)(dx/dt)$ = $(1/\omega_n)(d(\varepsilon/dt)(1/T_c))$. The abscissa is $x = \varepsilon/T_c$. Note that the trajectories on the phase plane rotate clockwise; when $\dot{x}$ is positive, then x must increase, whereas when $\dot{x}$ is negative, x must decrease. The plots are taken from Holmes [8].

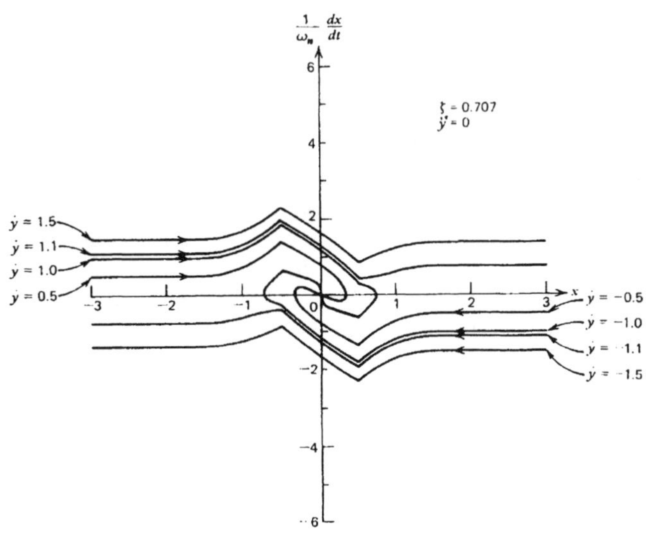

(a)

Figure 7.7-1 (a) Acquisition trajectories for the noncoherent loop with $\Delta = T_c$.

Consider Figure 7.7-1(a), which is the phase plane for the early-late spacing of one chip (T_c seconds) for the noncoherent early-late gate active filter, code-tracking loop. In both Figures 7.7-1(a) and

7.7-1(b), it is assumed that $\ddot{y} = 0$, and both x and $(\dot{x}/\omega_n)$ are assumed to be some nonzero value. As an example in Figure 7.7-1(a) if $\dot{y}(0) = 1.1$ and $x = -3$, then with time the error and normalized error rate converge to zero, making a clockwise convergence on the diagram. If, however, $\dot{y}(0) = 1.5$ and $x = -3$, then it is seen that the error reduces, but then slowly increases and x increases. Figure 7.7-1(b) is similar to Figure 7.7-1(a), except that the allowable error rates and errors are smaller to be able to achieve lock, that is—when $x = 0$ and $\dot{x} = 0$.

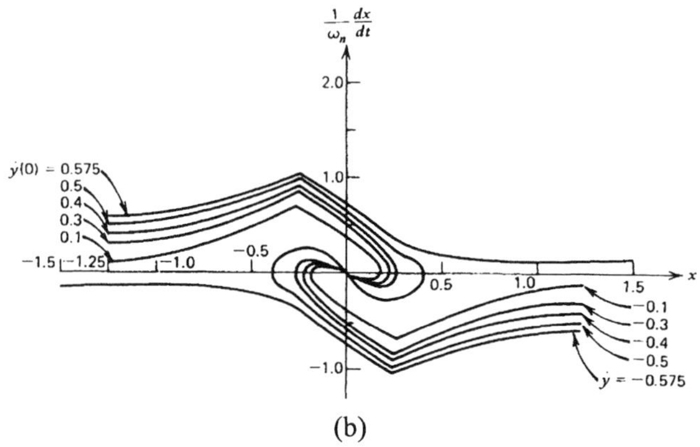

Figure 7.7-1 (b) Acquisition trajectories for the noncoherent loop with $\Delta = T_c/2$.

7.8 MULTIPATH EFFECTS

For the clear access (C/A) code in the GPS multipath errors are perhaps the largest of any location error contributor, including thermal noise.[12] Thus multipath cannot normally be ignored when discussing the performance of a code-tracking loop in an Earth involved propagation environment.

Multipath relates to the phenomena of a signal reaching an antenna by two or more paths. Commonly, the direct path signal arrives first and subsequent versions of the signal arrive slightly later in time, due to the fact that they are reflections from (normally) nearby objects. Typically the reflected versions of the signal are delayed in time and are normally weaker in signal strength and have a different carrier phase. The multipath phenomenon affects both the code-tracking process and the carrier-tracking process. The early-late *S*-curve only has a range of 1.5 symbols to as small as about 1 symbol, depending on the early-late spacing. Delays larger than these values (depending on the early-late spacing) of multipath delay of the signal have a negligible effect due to the much lower correlation value and can normally be neglected.

Ideally the receiver should be located in a location in which the receiver antenna is far away from items that can cause reflections, such as buildings, metal structures, water bodies, and the ground. However, in practice, reflections normally cannot be avoided. We will only consider the effect of multipath on code-tracking loops here. Figure 7.8-1 illustrates a direct ray along with two reflected rays.

7.8.1 Multipath Effects on Filtered Noncoherent Code-Tracking Loops

Now consider the characterization of the effects of multipath on a early-late gate noncoherent code-tracking loop. Figure 7.8-2 illustrates an I-Q baseband version of a noncoherent code-tracking loop. This loop has been discussed previously. The method that is commonly used to assess the impact of a single reflected component is to plot the code-tracking loop bias error against the multipath delay for a given relative level of the multipath component. We will refer to this as the *multipath diagram* (or more appropriately the multipath induced code-tracking error diagram).

[12] This excludes intentional or unintentional jamming or interference, of course.

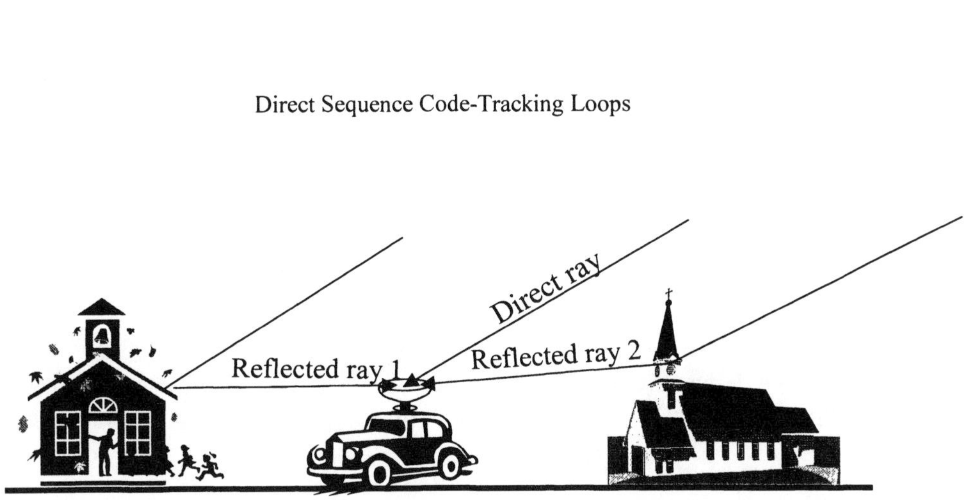

Figure 7.8-1 An example of multipath with two reflected rays and the direct ray.

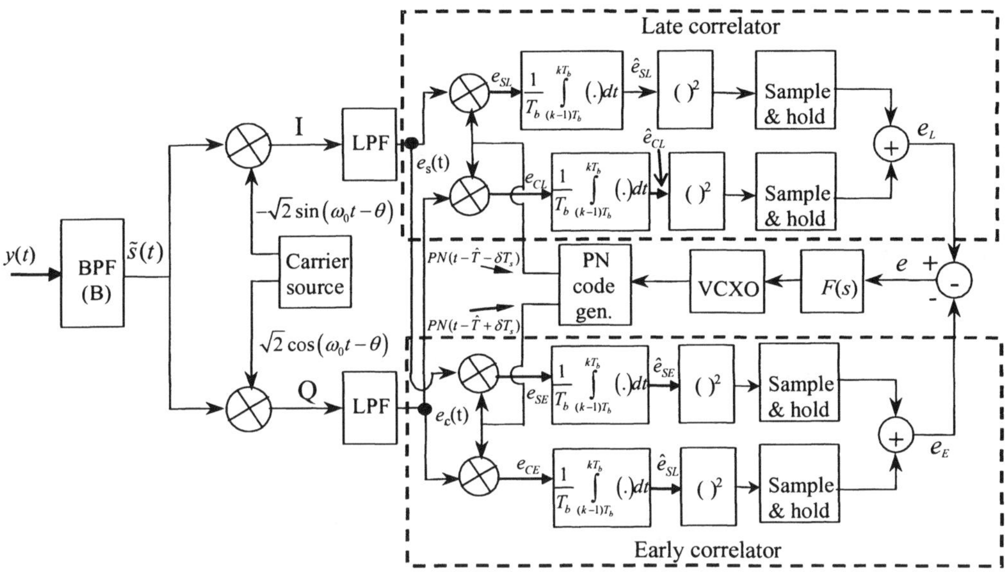

Figure 7.8-2 Early-late gate (L^2-E^2) noncoherent I-Q loop for code-tracking.

Model the received, channel filtered signal as

$$\tilde{s}(t)=\sqrt{2P}\widetilde{PN}(t-T)\cos\left(\omega_0 t+\theta\right)+\sqrt{2\alpha P}\widetilde{PN}(t-T-\Delta t)\cos\left(\omega_0 t+\theta+\Delta\theta\right) \qquad (7.8\text{-}1)$$

where A is the signal amplitude, α is the multipath relative power, Δt is the time delay (in seconds) of the multipath line component from the direct component, θ is the direct component signal carrier phase, $\Delta\theta$ is the carrier phase difference between the multipath and the direct components, and $\widetilde{PN}(t-T)$ denotes the baseband equivalent filtered (by the RF filter with bandwidth B Hz) pseudonoise code process that is delayed by T seconds and that has a symbol time of T_s seconds.[13] The early-late spacing is $2\delta T_s$, where δ is a number between 0 and 0.5. The effect of a channel filter on the multipath error is determined by the modification of the correlation function and therefore the S-curve (discriminator function). The filtering effect on the correlation function is determined by (see Section 1.4.1.2 of Chapter 1)

[13] Note that as before when the symbol shape is a simple rectangular pulse (i.e., NRZ), the chip time equals the symbol time.

$$\tilde{R}_{PN}(\tau) = \text{Re}\left\{ \int_{-\infty}^{\infty} H_{BB}(f) S(f) \exp(j2\pi f\tau) df \right\} \tag{7.8-2}$$

and is the resulting cross-correlation function between unfiltered and filtered PN codes.

Let the output $e_s(t)$ be derived from the product of the signal and the sine wave reference signal, with the sum frequency suppressed by the low-pass filter, and assuming that the carrier estimate is perfect. Then we have

$$e_s(t) = \tilde{s}(t)\left(-\sin(\omega_0 t + \theta)\right) \tag{7.8-3}$$

Doing the indicated multiplication yields (leaving the sum frequency off, since it will be removed by the low-pass filter)

$$e_s(t) = -\sqrt{2}\sin(\omega_0 t + \theta)\tilde{s}(t) \tag{7.8-4}$$

or evaluating we have

$$e_s(t) = \sqrt{2\alpha P}\,\widetilde{PN}(t - T - \Delta t)\sin(\Delta\theta) \tag{7.8-5}$$

In a like manner $e_c(t)$ is defined by

$$e_c(t) = \sqrt{2}\cos(\omega_0 t + \theta)\tilde{s}(t) \tag{7.8-6}$$

And after the multiplication, and leaving off the sum term, we are left with

$$e_c(t) = \sqrt{2P}\,\widetilde{PN}(t - T) + \sqrt{2\alpha P}\,\widetilde{PN}(t - T - \Delta t)\cos(\Delta\theta) \tag{7.8-7}$$

Now determine the four components: $\hat{e}_{SL}(T_b)$, $\hat{e}_{CL}(T_b)$, $\hat{e}_{SE}(T_b)$, and $\hat{e}_{CL}(T_b)$. Consider first the definition of $\hat{e}_{CL}(T_b)$ and consider the approximation suggested on the right-hand side of (7.8-8)

$$\hat{e}_{CL}(T_b) = \frac{1}{T_b}\int_0^{T_b} e_c(t) PN(t - \hat{T} - \delta T_s) dt \tag{7.8-8}$$

where $\hat{T}$ is the receiver code timing delay estimate. The integration time is T_b seconds, which could be the bit time, the coded symbol time, or some multiple divisor of the coded symbol time. Evaluating (7.8-8) leads to the result

$$\hat{e}_{CL}(T_b) \cong \sqrt{P}\tilde{R}_{PN}(\varepsilon - \delta T_s) + \sqrt{\alpha P}\tilde{R}_{PN}(\varepsilon - \delta T_s + \Delta t)\cos(\Delta\theta) \tag{7.8-9}$$

where $\tilde{R}(\varepsilon)$ denotes the cross-correlation of the filtered baseband code $\widetilde{PN}(t - T)$ and the unfiltered receiver code estimate $PN(t - \hat{T})$ and $\varepsilon = T - \hat{T}$. In the same manner $\hat{e}_{CE}(T_b)$ is evaluated as

$$\hat{e}_{CE}(T_b) = \frac{1}{T_b}\int_0^{T_b} e_c(t) PN(t - \hat{T} + \delta T_s) dt \tag{7.8-10}$$

This can be evaluated as

$$\hat{e}_{CE}(T_b) \cong \sqrt{P}\tilde{R}_{PN}(\varepsilon+\delta T_s)+\sqrt{\alpha P}\tilde{R}_{PN}(\varepsilon+\delta T_s+\Delta t)\cos(\Delta\theta) \tag{7.8-11}$$

Furthermore $\hat{e}_{SL}$ is defined as

$$\hat{e}_{SL}(T_b)=\frac{1}{T_b}\int_0^{T_b} e_s(t)PN(t-\hat{T}-\delta T_s)dt \tag{7.8-12}$$

which can be evaluated as

$$\hat{e}_{SL}(T_b) \cong \sqrt{\alpha P}\tilde{R}_{PN}(\varepsilon-\delta T_s+\Delta t)\sin(\Delta\theta) \tag{7.8-13}$$

In a similar manner $\hat{e}_{SE}$ is defined by

$$\hat{e}_{SE}(T_b)=\frac{1}{T_b}\int_0^{T_b} e_s(t)PN(t-\hat{T}+\delta T_s)dt \tag{7.8-14}$$

Which can be simplified to

$$\hat{e}_{SE}(T_b) \cong \sqrt{\alpha P}\tilde{R}_{PN}(\varepsilon+\delta T_s+\Delta t)\sin(\Delta\theta) \tag{7.8-15}$$

The discriminator function is defined from Figure 7.8-2 as

$$D=\hat{e}_{SL}^2(T_b)+\hat{e}_{CL}^2(T_b)-\hat{e}_{SE}^2(T_b)-\hat{e}_{CE}^2(T_b) \tag{7.8-16}$$

Using (7.8-9), (7.8-11), (7.8-13), and (7.8-15) in (7.8-16) produces the result

$$\begin{aligned}D_{nc}(\varepsilon) &= \alpha P\left[\tilde{R}_{PN}(\varepsilon-\delta T_s+\Delta t)\right]^2\sin^2(\Delta\theta)-\alpha P\left[\tilde{R}_{PN}(\varepsilon+\delta T_s+\Delta t)\right]^2\sin^2(\Delta\theta)\\ &+P\left[\tilde{R}_{PN}(\varepsilon-\delta T_s)\right]^2+2\sqrt{\alpha}P\tilde{R}_{PN}(\varepsilon-\delta T_s+\Delta t)\tilde{R}_{PN}(\varepsilon-\delta T_s)\cos(\Delta\theta)\\ &+\alpha P\left[\tilde{R}_{PN}(\varepsilon-\delta T_s+\Delta t)\right]^2\cos^2(\Delta\theta)\\ &-P\left[\tilde{R}_{PN}(\varepsilon+\delta T_s)\right]^2-2\sqrt{\alpha}P\tilde{R}_{PN}(\varepsilon+\delta T_s+\Delta t)\tilde{R}_{PN}(\varepsilon+\delta T_s)\cos(\Delta\theta)\\ &-\alpha P\left[\tilde{R}_{PN}(\varepsilon+\delta T_s+\Delta t)\right]^2\cos^2(\Delta\theta)\end{aligned} \tag{7.8-17}$$

Equation (7.8-17) can be simplified to

$$\begin{aligned}D_{nc}(\varepsilon) &= P\left\{\left[\tilde{R}_{PN}(\varepsilon-\delta T_s)\right]^2-\left[\tilde{R}_{PN}(\varepsilon+\delta T_s)\right]^2\right\}\\ &+\alpha P\left\{\left[\tilde{R}_{PN}(\varepsilon-\delta T_s+\Delta t)\right]^2-\left[\tilde{R}_{PN}(\varepsilon+\delta T_s+\Delta t)\right]^2\right\}\\ &-2\sqrt{\alpha}P\cos(\Delta\theta)\left[\tilde{R}_{PN}(\varepsilon-\delta T_s+\Delta t)\tilde{R}_{PN}(\varepsilon-\delta T_s)-\tilde{R}_{PN}(\varepsilon+\delta T_s+\Delta t)\tilde{R}_{PN}(\varepsilon+\delta T_s)\right]\end{aligned} \tag{7.8-18}$$

This result is similar to as M. S. Braasch's result, in Chapter 14 of [26] with some algebraic sign differences. The solution of the problem at hand involves determining the range of ε values that causes $D(\varepsilon)$ to be zero. This will yield the envelope of errors (ε values) that can provide a stable lock point of the code-tracking loop ($D_{nc}(\varepsilon) = 0$) with the appropriate assumptions of the value of $\Delta\theta$. This will be discussed in the following example.

Example 2 In (7.8-18) the following parameters will be used in order to plot the error envelope due to multipath as a function of the multipath delay f_0 = 1,575.42 MHz ($f_0 = \omega_0/(2\pi)$), $A = 1$, $\alpha = 0.5$ (the power of the multipath is one-fourth the power of the signal, or 6 dB down), $\delta = 0.5$ (an early-late separation of one chip), Δt a variable delay time plotted in chips (representing the multipath delay), and $T_s = 0.009775$ µs. The value of $\Delta\theta$ will be set to the two extreme values of 0 and π radians. The reader might notice that these parameters are for the L1 P(Y) code in the GPS navigation system. When $\Delta\theta = \pi$ the lower portion of the curve is generated. When $\Delta\theta = 0$ the upper portion of the curve is generated. Figure 7.8-3 illustrates the resulting multipath induced error diagram in this case. The ordinate shows the error in chips (ε) versus the value of the multipath component delay Δt on the abscissa. Basically the interior area of the allowable code-tracking errors has a diamond-like shape and is the allowable set of all tracking errors that can occur for all values of the multipath delay, for that particular relative multipath amplitude (value of α). At, for example, a delay of 0.25 chip, the error can range from –0.25 to +0.08 chips. This plot is nearly identical to the dot product multipath plot reported by Van Dierendonck et al. [27]. The line defines the possible value of tracking error as a function of the delay of the multipath component relative to the desired signal. It is to be noted that even though the multipath component is always delayed with respect to the direct component, the multipath induced error can be either positive or negative!

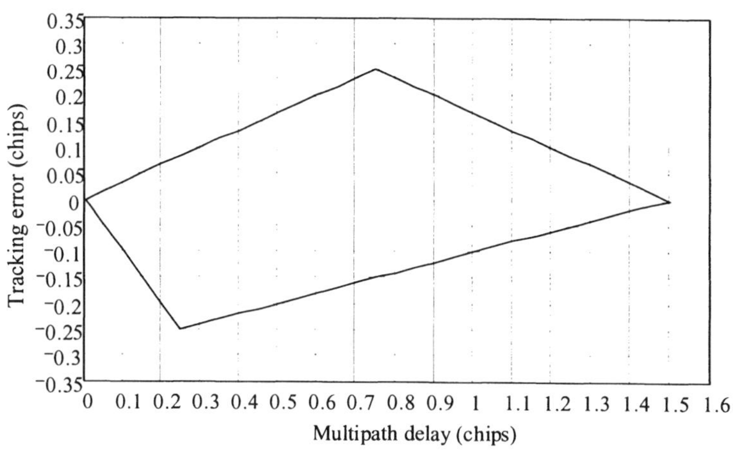

Figure 7.8-3 Multipath diagram (envelope of the timing error in chips as a function of the multipath delay in chips) for the early-late gate I-Q discriminator function, with δ=1/2, infinite bandwidth.

Now consider what happens when the early-late amplitude is reduced. Consider the case when δ = 0.05 chip, with the other parameters the same. Figure 7.8-4 illustrates the new multipath induced error diagram.

The first thing that is noticed about the plot with δ reduced to one-tenth the value is the fact that the peak error is reduced to 0.025 chip! Thus it is clear that reducing the early-late correlator spacing reduces the peak multipath induced tracking error (when no filtering is accounted for). However it is seen that the peak error occurs over a larger range of multipath errors. It is to be noted that the range of delay in which the errors are not negligible is greater. Another aspect of the multipath problem is the fact that both the error and the multipath delay are plotted in chips. Therefore, if the chip rate is high, the actual error, in time, is small and the converse is true also. In addition with a high chip rate, the induced tracking error occurs only over a small multipath time delay window. The converse is true also.

The relationship between chip errors and actual timing errors is given by

$$1 \text{ chip error} = T_s \text{ seconds error} \tag{7.8-19}$$

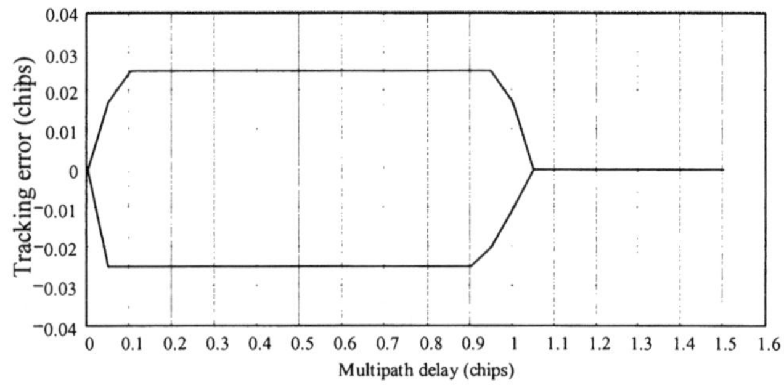

Figure 7.8-4 Multipath diagram (envelope of the timing error in chips as a function of the multipath delay in chips) for the early-late gate I-Q discriminator function, with δ=0.05, infinite bandwidth.

Let us now consider an example in which filtering is accounted for. It will be seen that the filtering has a major effect on the range of the code-tracking error.

Example 3 Consider the same parameters as in Example 2, but include an ideal bandpass filter (IBP) having a RF bandwidth of 24 MHz. Evaluation of (7.8-18) produces the results indicated in Figures 7.8-5 and 7.8-6.

It is seen that the reduction in multipath induced code-tracking error is not nearly as much at a narrower correlator spacing, when significant filtering is present in the channel. In fact when the 24 MHz ideal bandpass filter is present the maximum errors are about 0.18 chip at $\delta = 0.5$, and about 0.16 chip at $\delta = 0.05$. Thus narrow correlator spacing tends to not yield significantly lower multipath errors when the bandwidth of the channel is commensurate with the signal bandwidth and the errors are not as small as when the bandwidth is very wide. In addition, it is noticed that channel filtering causes the sharp corners of the diagram to be rounded off and the shapes are smoother in general. In summary the higher the chip rate, the smaller the effect of the multipath induced code-tracking errors.

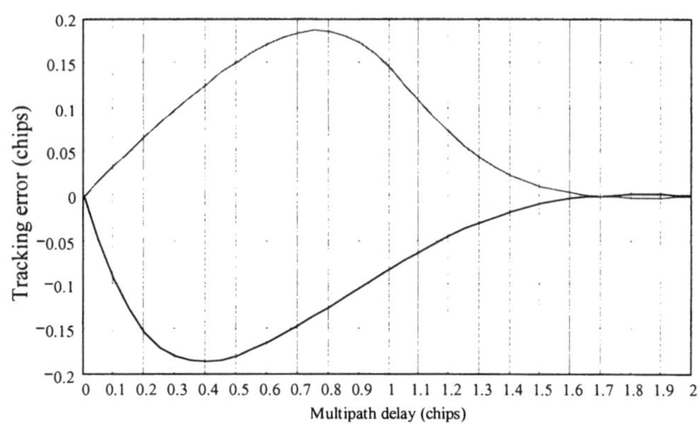

Figure 7.8-5 Multipath diagram (envelope of the timing error in chips as a function of the multipath delay in chips) for the early-late gate I-Q discriminator function, with δ=0.5 with 24-MHz IBP filtering.

It is to be noted that a great amount of work has been done in the effort to minimize the effect of multipath on the receiver solution. The additional references list a few pertinent articles to aid the reader in understanding the techniques used to minimize the effects of multipath.

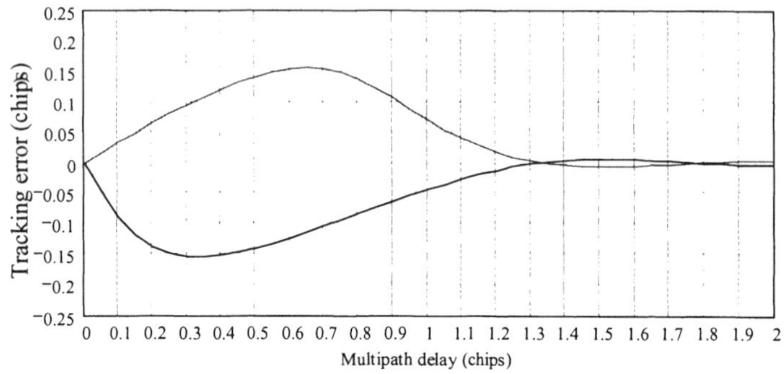

Figure 7.8-6 Multipath diagram (envelope of the timing error in chips as a function of the multipath delay in chips), for the early-late gate discriminator function, with δ=0.05 with 24-MHz IBP filtering.

7.8.2 Multipath Effects on Baseband Coherent I-Q Code-Tracking Loops

In this section we will establish the equation that defines the multipath error for a single delayed multipath component. Again the input is modeled as (again including the channel filtering on the received code)

$$\tilde{s}(t)=\sqrt{2P}\widetilde{PN}(t-T)\cos\left(\omega_0 t+\theta\right)+\sqrt{2\alpha P}\widetilde{PN}(t-T-\Delta t)\cos\left(\omega_0 t+\theta+\Delta\theta\right) \qquad (7.8\text{-}20)$$

Figure 7.8-7 illustrates the coherent code tracking model.

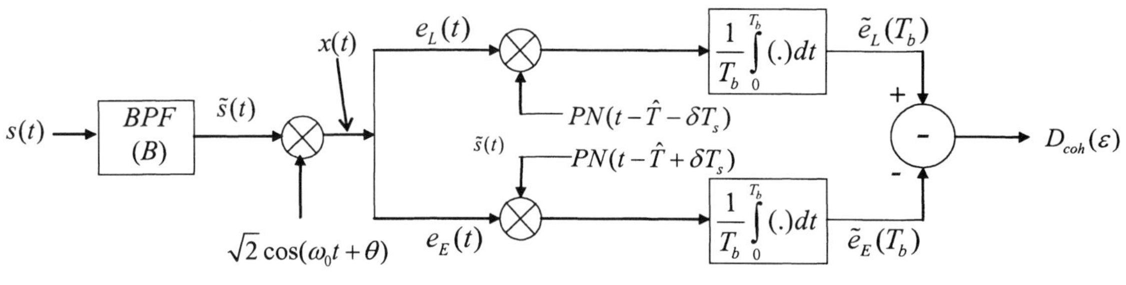

Figure 7.8-7 The baseband coherent code-tracking loop with integrators used in the correlation processes.

It is assumed that the coherent carrier reference is exactly in phase with the received carrier signal. Thus the signal at the point called $x(t)$ is given by

$$x(t)=\sqrt{P}\widetilde{PN}(t-T)+\sqrt{\alpha P}\widetilde{PN}(t-T-\Delta t)\cos(\Delta\theta) \qquad (7.8\text{-}21)$$

where the double carrier frequency term has been dropped since it will be filtered out by the integrate-and-dump filter. The late and early signals ($e_L(t)$ and $e_E(t)$) are then given by

$$e_E(t)=\sqrt{P}\widetilde{PN}(t-T)PN(t-\hat{T}+\delta T_s)+\sqrt{\alpha P}\widetilde{PN}(t-T-\Delta t)PN(t-\hat{T}+\delta T_s)\cos(\Delta\theta) \qquad (7.8\text{-}22)$$

and

$$e_L(t) = \sqrt{P}\widetilde{PN}(t-T)PN(t-\hat{T}-\delta T_s) + \sqrt{\alpha P}\widetilde{PN}(t-T-\Delta t)PN(t-\hat{T}-\delta T_s)\cos(\Delta\theta) \tag{7.8-23}$$

Again letting $\varepsilon = T - \hat{T}$ we can obtain the outputs of the integrators

$$e_L(T_b) = \sqrt{P}\tilde{R}_{PN}(\varepsilon - \delta T_s) + \sqrt{\alpha P}\tilde{R}_{PN}(\varepsilon + \Delta t - \delta T_s)\cos(\Delta\theta) \tag{7.8-24}$$

and

$$e_E(T_b) = \sqrt{P}\tilde{R}_{PN}(\varepsilon + \delta T_s) + \sqrt{\alpha P}\tilde{R}_{PN}(\varepsilon + \Delta t + \delta T_s)\cos(\Delta\theta) \tag{7.8-25}$$

The *S*-curve is given by

$$D_{coh}(\varepsilon) = \tilde{e}_L(T_b) - \tilde{e}_E(T_b) \tag{7.8-26}$$

After rearranging, the *S*-curve can be written as

$$D_{coh}(\varepsilon) = \sqrt{P}\left[\tilde{R}_{PN}(\varepsilon - \delta T_s) - \tilde{R}_{PN}(\varepsilon + \delta T_s) + \sqrt{\alpha}\left\{\tilde{R}_{PN}(\varepsilon + \Delta t - \delta T_s) - \tilde{R}_{PN}(\varepsilon + \Delta t + \delta T_s)\right\}\cos(\Delta\theta)\right] \tag{7.8-27}$$

which is our final result for the discriminator function in the coherent case to determine the envelope of timing errors when one delayed multipath signal is present. The envelope of multipath induced errors involves determining the range of ε values that causes $D_{coh}(\varepsilon)$ to be zero. This will yield the envelope of errors (ε values) that can provide a stable lock point of the code-tracking loop ($D_{coh}(\varepsilon) = 0$) with the appropriate assumptions of the value of $\Delta\theta$. Now we will show that the multipath error envelope is the same for the coherent and the noncoherent code-tracking loops.

7.8.3 The Multipath Error Plots Are the Same for Coherent and Noncoherent Code-Tracking Loops

In this section we will prove that the error envelope plots of the coherent baseband code-tracking loop are the same as for the noncoherent I Q code-tracking loop for a single delayed multipath component. To do this we will show that the values of ε are the same in the case that $D_{coh}(\varepsilon) = 0$ and $D_{nc}(\varepsilon) = 0$.[14]

The normalized value of the coherent discriminator function is given by

$$\frac{D_{coh}(\varepsilon)}{\sqrt{P}} = \left[\tilde{R}_{PN}(\varepsilon - \delta T_s) - \tilde{R}_{PN}(\varepsilon + \delta T_s) + \sqrt{\alpha}\left\{\tilde{R}_{PN}(\varepsilon + \Delta t - \delta T_s) - \tilde{R}_{PN}(\varepsilon + \Delta t + \delta T_s)\right\}\cos(\Delta\theta)\right] \tag{7.8-28}$$

We will consider two cases that define the limits of the multipath error: (1) $\Delta\theta = 0$ and (2) $\Delta\theta = \pi$. First consider case (1): for $D_{coh}(\varepsilon) = 0$ we require from (7.8-28) that

$$\tilde{R}_{PN}(\varepsilon - \delta T_s) + \sqrt{\alpha}\tilde{R}_{PN}(\varepsilon + \Delta t - \delta T_s) = \tilde{R}_{PN}(\varepsilon + \delta T_s) + \sqrt{\alpha}\tilde{R}_{PN}(\varepsilon + \Delta t + \delta T_s) \tag{7.8-29}$$

and for the second case when $\Delta\theta = \pi$ in order that $D_{coh}(\varepsilon) = 0$, we must have

$$\tilde{R}_{PN}(\varepsilon - \delta T_s) - \sqrt{\alpha}\tilde{R}_{PN}(\varepsilon + \Delta t - \delta T_s) = \tilde{R}_{PN}(\varepsilon + \delta T_s) - \sqrt{\alpha}\tilde{R}_{PN}(\varepsilon + \Delta t + \delta T_s) \tag{7.8-30}$$

[14] This approach on equivalence was based on an unpublished paper written by G. Fan of The Aerospace Corporation.

Now we consider the noncoherent code-tracking loop case for the late^2–early^2 discriminator function. From (7.8-18) we have for the normalized discriminator function

$$\begin{aligned}\frac{D_{nc}(\varepsilon)}{P} &= \left\{\left[\tilde{R}_{PN}(\varepsilon-\delta T_s)\right]^2 - \left[\tilde{R}_{PN}(\varepsilon+\delta T_s)\right]^2\right\} \\ &+\alpha\left\{\left[\tilde{R}_{PN}(\varepsilon-\delta T_s+\Delta t)\right]^2 - \left[\tilde{R}_{PN}(\varepsilon+\delta T_s+\Delta t)\right]^2\right\} \\ &-2\sqrt{\alpha}\cos(\Delta\theta)\left[\tilde{R}_{PN}(\varepsilon-\delta T_s+\Delta t)\tilde{R}_{PN}(\varepsilon-\delta T_s)-\tilde{R}_{PN}(\varepsilon+\delta T_s+\Delta t)\tilde{R}_{PN}(\varepsilon+\delta T_s)\right]\end{aligned} \tag{7.8-31}$$

First consider the case when $\Delta\theta = 0$. Then it is not difficult to show that

$$\left[\tilde{R}_{PN}(\varepsilon-\delta T_s)+\sqrt{\alpha}\tilde{R}_{PN}(\varepsilon-\delta T_s+\Delta t)\right]^2 = \left[\tilde{R}_{PN}(\varepsilon+\delta T_s)+\sqrt{\alpha}\tilde{R}_{PN}(\varepsilon+\delta T_s+\Delta t)\right]^2 \tag{7.8-32}$$

It follows from (7.8-32) by taking the square root of both sides that the following must be true

$$\left[\tilde{R}_{PN}(\varepsilon-\delta T_s)+\sqrt{\alpha}\tilde{R}_{PN}(\varepsilon-\delta T_s+\Delta t)\right] = \left[\tilde{R}_{PN}(\varepsilon+\delta T_s)+\sqrt{\alpha}\tilde{R}_{PN}(\varepsilon+\delta T_s+\Delta t)\right] \tag{7.8-33}$$

It is noticed that (7.8-33) is identical to the coherent case in (7.8-29). Now consider the second case when $\Delta\theta = \pi$. In this case the normalized discriminator function is given by

$$\frac{D_{nc}(\varepsilon)}{P} = \left[\tilde{R}_{PN}(\varepsilon-\delta T_s)-\sqrt{\alpha}\tilde{R}_{PN}(\varepsilon-\delta T_s+\Delta t)\right]^2 - \left[\tilde{R}_{PN}(\varepsilon+\delta T_s)-\sqrt{\alpha}\tilde{R}_{PN}(\varepsilon+\delta T_s+\Delta t)\right]^2 \tag{7.8-34}$$

In order for the noncoherent discriminator function to be zero we see that the square of the first term must equal the square of the second term. So taking the square root of both sides produces the requirement

$$\left[\tilde{R}_{PN}(\varepsilon-\delta T_s)-\sqrt{\alpha}\tilde{R}_{PN}(\varepsilon-\delta T_s+\Delta t)\right] = \left[\tilde{R}_{PN}(\varepsilon+\delta T_s)-\sqrt{\alpha}\tilde{R}_{PN}(\varepsilon+\delta T_s+\Delta t)\right] \tag{7.8-35}$$

Notice that (7.8-35) is the same as (7.8-30). Thus we have proven that the same conditions apply for the coherent and noncoherent code-tracking loops when only one multipath component exists for late^2–early^2 discriminator functions. Now since the multipath diagrams are the same for both the noncoherent and coherent baseband I-Q code-tracking loops (7.8-27) may be plotted rather than the more cumbersome (7.8-18).

7.9 MEAN TIME TO LOSE LOCK FOR A FIRST-ORDER EARLY-LATE GATE RF CODE-TRACKING LOOP

The threshold to losing lock in a code-tracking loop is difficult to define by arbitrarily specifying a specific code timing error or the code loop signal-to-noise ratio. As a consequence a more refined notion of threshold will be proposed here; although also arbitrary, it is hopefully more satisfying in a conceptual sense. We will use the mean time to lose lock in the sense of exceeding the discriminator function lock range as the criteria. This work follows Holmes and Biederman [28] and was generalized to arbitrary, binary valued (± 1) code symbols in [7].

The analysis that is used here is based on an analog, first-order, noncoherent, RF code-tracking loop. The reason for the analog model was to ameliorate the analysis to some extent, using known Markov mean first passage time techniques. This model is shown in Figure 7.9-1. It is based on an early-late gate discriminator function model.

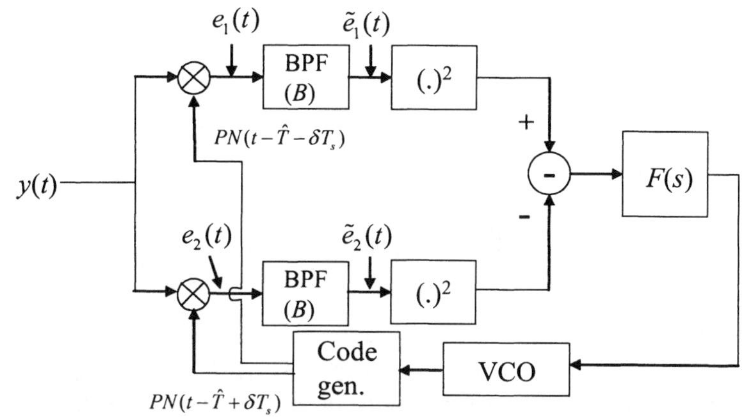

Figure 7.9-1 First-order BPF implemented code-tracking loop model.

The discriminator function controls the loop filter. The loop filter, a constant in this case for first order loop, drives a voltage-controlled oscillator (VCO) that provides the clock frequency for the early and late local estimates of the receiver codes. However, the baseband implementation shown in Figure 7.9-2 is more common.

We have already seen both loops earlier in this chapter. The I^2+Q^2 of the early correlators is subtracted from the I^2+Q^2 of the late correlators to achieve an error control signal to control the code-tracking loop. This error signal drives the loop filter, which in turns controls the voltage-controlled crystal oscillator (VCXO), which drives the timing of the early and late receiver estimates of the received pseudonoise (PN) code.

The approach used here is to formulate the mean slip time with the IF code loop model which depends on the linearized code loop error variance and then substitute the known code loop error variance for the integrate-and-dump (I&D) code loop model. The basic approximation is based on replacing a discrete time process (I&D loop) with a continuous code loop model (IF loop). The goal is to obtain an analysis to allow prediction of the I&D model.

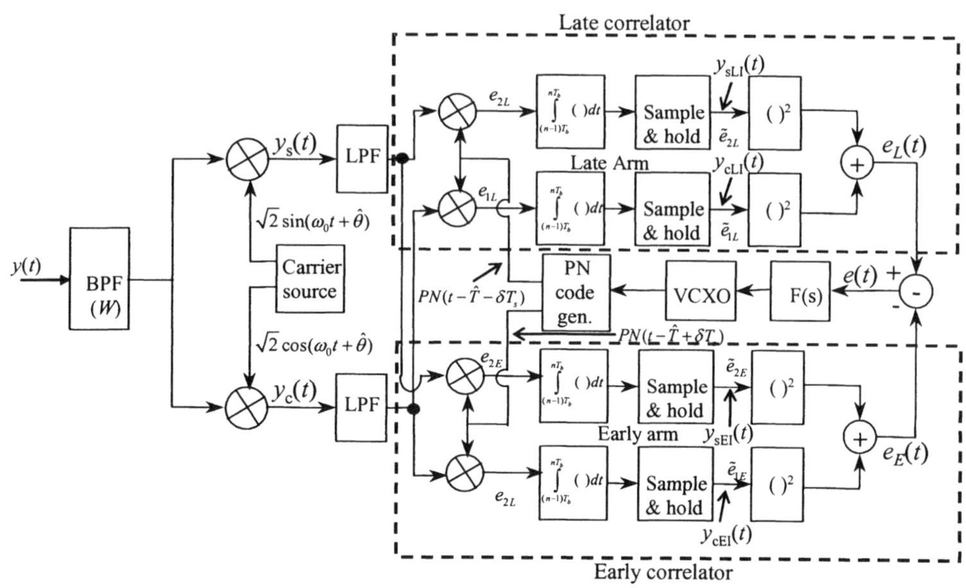

Figure 7.9-2 Baseband I-Q version of a noncoherent code-tracking loop with early-late gate code separation of $d = 2\delta$.

7.9.1 Model for the Analysis of the Mean Slip Time Performance of the Early-Late Gate Code-Tracking Loop with RF Implementation

It is shown in Appendix 7A that the stochastic nonlinear differential equation governing the operation of the IF code loop model is given by

$$\varepsilon = T - \frac{k_v}{s} F(s)\left[\alpha P K g_n(\varepsilon) + n_1(t) + n_2(t)\right] \tag{7.9-1}$$

where ε is the code timing error ($\varepsilon = T - \hat{T}$) and is a function of time, T is the delay between the transmitter and the receiver, $\hat{T}$ is the code estimate of the delay, $F(s)$ is the loop filter transfer function in the Laplace variable s, P is the received power, K is the discriminator function gain, K_v is the VCO frequency per voltage scale parameter, $g_n(\varepsilon)$ is the normalized unit gain discriminator function for the symbol shape under discussion, and $n_1(t)$ and $n_2(t)$ are the signal times noise and noise times noise terms that result and are described in the Appendix.

After considerable analysis, the Fokker-Planck Equation [29–31], for the density function as a function of the code phase error ε at time t, as long as the phase error stays inside the code error boundaries ($\forall|\varepsilon| < \varepsilon_{LL}$), for the problem at hand is given by

$$\frac{\partial q(\varepsilon,t)}{\partial t} = \alpha P K K_v \frac{\partial}{\partial \varepsilon}\left[g_n(\varepsilon) q(\varepsilon,t)\right] + \frac{N_0' K_v^2}{4} \frac{\partial^2}{\partial t^2}\left[q(\varepsilon,t)\right] \tag{7.9-2}$$

where N_0' is the one-sided noise spectral density, P is the received signal power, the Ks are gains (see the appendix), and α is a signal power loss. This equation is not valid after a boundary is reached. The partial differential equation is subject to the initial conditions

$$q(\varepsilon,0) = \delta(\varepsilon) \tag{7.9-3}$$

and the boundary conditions

$$\begin{aligned} q(\varepsilon,t) &= 0 \qquad && \forall|\varepsilon| \geq \varepsilon_{LL} \\ g(\varepsilon,t) &= q(-\varepsilon,t) && \forall t \end{aligned} \tag{7.9-4}$$

where $\forall x$ denotes "all values of x."

From Appendix 7A the solution to the mean slip time performance, for the IF model (continuous time), leads to the following general formulation

$$\overline{T} = \frac{1}{4B_L\sigma^2}\int_{-\varepsilon_{LL}}^{\varepsilon_{LL}} e^{-\left[\frac{G(\varepsilon)}{\sigma^2}\right]} \int_{-\varepsilon_{LL}}^{\varepsilon} e^{\left[\frac{G(\varepsilon')}{\sigma^2}\right]} \left[\frac{1}{2} - u(\varepsilon')\right] d\varepsilon' d\varepsilon \tag{7.9-5}$$

In this expression $\overline{T}$ denotes the mean time to lose lock in seconds, ε_{LL} is the magnitude of the upper and lower code phase error in symbols where loss of lock is defined, B_L is the code one-sided closed-loop noise bandwidth in Hz, σ^2 is the linearized code-tracking error expressed in symbols squared for either the RF or baseband versions of the code-tracking loops, $u(x)$ is the unit step function that is one for $x \geq 0$ and zero otherwise, and $G(\varepsilon)$ is an even function of ε and is the indefinite integral of $g_n(\varepsilon)$. Figure 7.9-3 illustrates an example of the $G(\varepsilon)$ function.

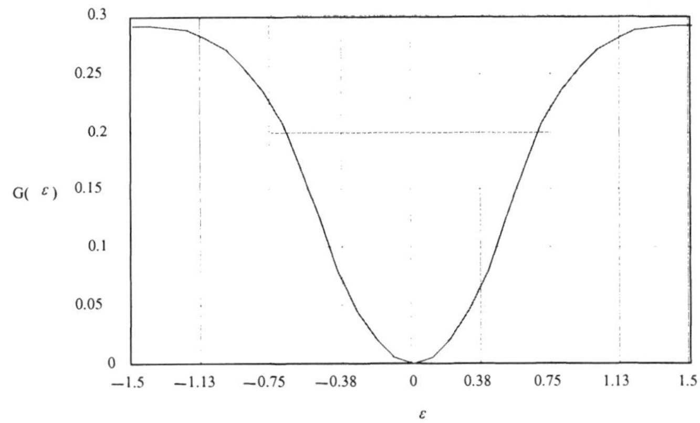

Figure 7.9-3 $G(\varepsilon)$ for BPSK(10) with $d = 1$ symbol correlation separation.

The variance of the bandpass filter (BPF) code loop model is given from Appendix 7A, as

$$\sigma^2 = \frac{N_0 B_L}{K^2(\alpha P)}\left[8R_{PN}^2(\delta T_s)\left[1 - R_{PN}(2\delta T_s)\right] + \frac{4N_0 W}{\alpha P}\left[1 - R_{PN}^2(2\delta T_s)\right]\right] \quad (7.9\text{-}6)$$

in units of symbols squared. However the model of interest is the baseband model, illustrated in Figure 7.9-2, which has [2] the linearized tracking variance (this does not include quantization errors) given by

$$\sigma_{I\&D}^2 = \frac{N_0 B_L}{K^2 P}\left[8R_{PN}^2(\delta T_s)\left(1 - R_{PN}(2\delta T_s)\right) + \frac{4N_0\left[1 - R_{PN}^2(2\delta T_s)\right]}{PT_b}\right] \quad (7.9\text{-}7)$$

in units of symbols squared. This expression for the tracking error variance was used in the actual mean slip time results. Note that (7.9-6) and (7.9-7) are very similar in form. It is to be noted that it has not been proved that it is legitimate to use (7.9-7) in (7.9-5), but it is conjectured that it is correct.

7.9.2 Mean Slip Time Comparison of Theory and Simulation for the Early-Late Gate Code-Tracking Loop with RF Implementation

Figure 7.9-4 compares the theoretical and simulated values of the mean slip time for the BOC(10,5)[15] and the NRZ(10) modulations. The simulation was developed to model the loss of code lock phenomena with two samples per code chip.

As can be seen from Figure 7.9-4, although there is agreement between the theoretical and the simulation results in terms of the trend in which mean slip time depends on the C/N_0 value, the absolute values of loss of lock (LOL) do not agree exactly. The shapes are similar but the simulation performance prediction was better than the theory predicted in the range of comparison. The theory suggests that there is a 1-2-dB improvement in threshold for NRZ waveforms at $d = 1$, compared to the optimized spacing ($d = 1/7$) BOC(10,5) signal. However the simulation shows about a 3-dB improvement in the threshold performance of NRZ (at $d = 1$) compared to the optimum BOC modulation. It is expected that narrower spacing ($d < 1$) will additionally improve (raise the LOL time) the NRZ slip time performance. The main point here is the fact that increasing the loop SNR greatly improves the mean time to lose lock. In fairness, it should be stated that

[15] Note the BOC(10,5) code is based on the synchronous product of a squarewave operating at 10.23 MHz and a 5.115 Msps underlying NRZ code. It is used in at least one navigational system.

the BOC(10,5) signal has much better code-tracking performance above threshold than the NRZ(10) signal has.

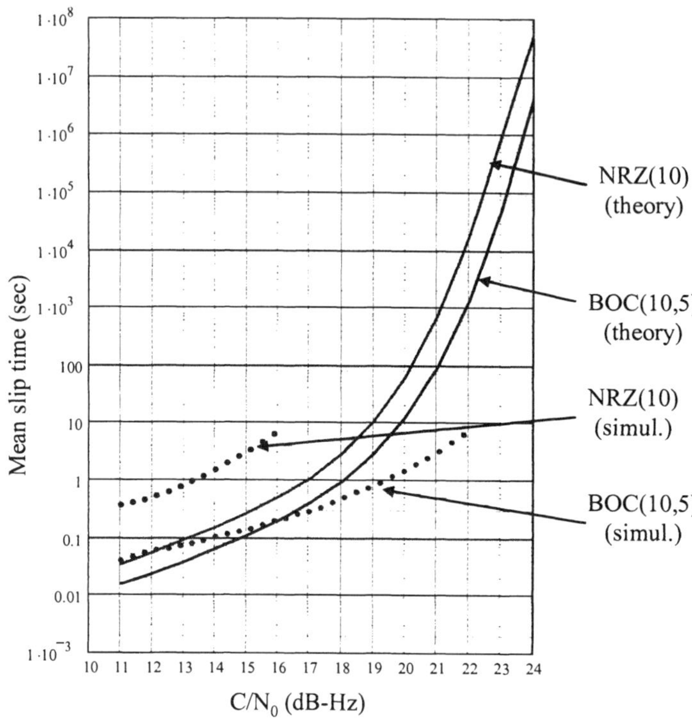

Figure 7.9-4 BOC(10,5) and NRZ(10) mean slip time in seconds at B_L = 5 Hz for the early-late gate code-tracking loop with RF implementation.

The reason for the difference in the absolute values between theory and simulation is not fully understood at the time of this writing, although the simulation did include the discrete time model of the loop. The IF model analyzed here is based on a continuous error control model, whereas the baseband loop is piecewise continuous (every 20 ms the correction is updated). This is not an essential difference as long as the $B_L\times$(Update period) <<1. Quantization was not accounted for in the theoretical model.

7.10 WIDEBAND JAMMING EFFECTS ON TRACKING AND MEAN TIME TO LOSE LOCK FOR THE EARLY-LATE GATE CODE-TRACKING LOOP WITH RF IMPLEMENTATION

In this section we desire to show the effect of broadband jamming on the performance of code-tracking and the mean time to lose lock with unfiltered signals. Assume that the input signal plus interference is given

$$y(t)=\sqrt{2P}d(t)PN(t)\cos(\omega_0 t+\theta)+n(t)+J(t) \tag{7.10-1}$$

in which the first term denotes the received BPSK signal, the second term denotes white Gaussian noise, and the third term is a broadband flat jammer, assumed to be flat over the band of $f_0+B_J/2$ to $f_0-B_J/2$, with a one-sided jammer spectral density given by

$$J_0 = J / B_J \text{ for } f_0 - B_J / 2 \text{ to } f_0 + B_J / 2 \tag{7.10-2}$$

It is assumed that the bandwidth B_J is wide compared to the signal in the following discussion. Consequently the total jammer plus noise spectral density over the bandwidth B_J (Hz) is given by

$$I_0 = N_0 + J_0 \tag{7.10-3}$$

The degradation relative to the only thermal noise is given by the factor multiplying the noise spectral density; thus from

$$I_0 = N_0\left(1 + \left(\frac{J_0}{N_0}\right)\right) = N_0\left(1 + \left(\frac{J}{N_0 B_J}\right)\right) \tag{7.10-4}$$

where I_0 is the effective one-sided noise spectral density, we obtain the degradation

$$L_D = \frac{1}{\left(1 + \left(\dfrac{J}{N_0 B_J}\right)\right)} \tag{7.10-5}$$

Therefore the effective E_b/N_0 performance of, say, a code-tracking loop, under broadband flat jamming, can be written as

$$\frac{E_b}{I_0} = \frac{PT_b}{N_0\left(1 + \left(\dfrac{J}{N_0 B_J}\right)\right)} = \frac{1}{\left(\dfrac{1}{E_b / N_0} + \dfrac{J}{P(B_J T_b)}\right)} \tag{7.10-6}$$

where I_0 is the total effective noise-like spectral density and the interference is expressed as a function of J/P, the jammer-to-signal power ratio. Note that it is common to call degradation due to jamming the *jamming degradation*

$$Degr = \left(1 + \left(\frac{J}{N_0 B_J}\right)\right) \tag{7.10-7}$$

The metric is $P/(N_0 B_{sys})$, where B_{sys} is the appropriate system bandwidth, such as the arm filter bandwidth in a bandpass implemented code-tracking loop. In this case the following expression can be used

$$\frac{P}{I_0 B_{sys}} = \frac{1}{\left(\dfrac{1}{P / N_0 B_{sys}} + \dfrac{JB_{sys}}{PB_J}\right)} \tag{7.10-8}$$

and the interference is again expressed as a function of J/P.

For code acquisition N_0 can be replaced by I_0, just as in the code-tracking case. To see how these equations apply in the code-tracking case, consider an example.

Example 4 Determine the expression for the variance of the code loop tracking error as a function of J/P when a broadband noise jammer is utilized for NRZ symbols and there is no channel filtering. Consider (7.2-90), which is repeated here

$$\sigma_{sym}^2 = \frac{N_0 B_L d}{2P}\left[1+\frac{2}{(2-d)(E_b/N_0)}\right] \quad \text{symbols}^2$$

Replacing N_0 everywhere with $I_0 = N_0\left(1+\left(\frac{J}{N_0 B_J}\right)\right)$ produces the result

$$\sigma_{sym}^2 = \frac{\left(N_0 + J/B_J\right)B_L d}{2P}\left[1+\frac{2\left(N_0+J/B_J\right)}{(2-d)E_b}\right] \quad \text{symbols}^2$$

as our result as a function of the J/P ratio.

7.11 CRAMER-RAO BOUND ON CODE-TRACKING ERROR

The *Cramer-Rao bound* (CRB) on code-tracking error is a useful lower bound on the performance one can expect from an estimate of code tracking. In other words, a bound on code-tracking error variance is useful so as to compare it to a practical estimator such as a code-tracking loop. The CRB is a general lower bound on the minimum mean square estimate of a random parameter [32]. This bound has been applied to code-tracking [33], symbol synchronizers [34], Carrier tracking [35], and joint carrier and code-tracking [36], as well as other applications.

Let the set of N scalar measurements at time $i\Delta t$ be denoted as

$$x(\Delta t) = s(i\Delta t, \tau) + n(i\Delta t) \tag{7.11-1}$$

for $i = 0, 1, 2, \ldots N-1$. The $n(i\Delta t)$ are white Gaussian noise terms of zero mean and covariance given by

$$E\left\{\boldsymbol{n}^T\boldsymbol{n}\right\} = \sigma^2\boldsymbol{I} \tag{7.11-2}$$

where $\boldsymbol{I}$ is the unit matrix of dimension $N \times N$ and $\boldsymbol{n}$ is the row vector described by

$$\boldsymbol{n} = \left(n(0), n(\Delta t), n(2\Delta t), \ldots n((N-1)\Delta t)\right) \tag{7.11-3}$$

and τ is the parameter to be estimated. Assume that the probability density function (pdf) for the samples is given by $p\left(\boldsymbol{n}\middle|\tau\right)$. Let $\hat{\tau}$ be the maximum-likelihood (ML) estimator of the parameter τ, then let the bias of the estimator be $b(\tau)$ so that

$$E\left\{\hat{\tau}\right\} = \tau + b(\tau) \tag{7.11-4}$$

Then the Cramer-Rao bound states in the general case that

$$Var\left(\hat{\tau}-\tau\right) \geq \frac{-\left[1+\dfrac{db(\tau)}{d\tau}\right]}{E\left\{\dfrac{\partial^2 \ln p(x|\tau)}{\partial\tau^2}\right\}} \tag{7.11-5}$$

where $p(\boldsymbol{x}|\tau)$ is the likelihood function viewed as function of τ, where $\boldsymbol{x}=\boldsymbol{s}+\boldsymbol{n}$ and $\boldsymbol{n}$ is defined previously and $\boldsymbol{s}$ is defined in a similar fashion; that is,

$$\boldsymbol{s}=\left(s(0),s(\Delta t),s(2\Delta t),...s((N-1)\Delta t)\right) \tag{7.11-6}$$

In the case that the estimate is unbiased ($b(\tau) = 0$) then the CRB reduces to

$$Var\left(\hat{\tau}-\tau\right)\geq\frac{-1}{E\left\{\dfrac{\partial^2 \ln p(\boldsymbol{x}|\tau)}{\partial\tau^2}\right\}} \tag{7.11-7}$$

An alternate expression for the bound on the estimator error is obtained from Problem 12 as

$$Var\left(\hat{\tau}-\tau\right)\geq\frac{\left[1+\dfrac{db(\tau)}{d\tau}\right]}{E\left\{\left(\dfrac{\partial \ln p(\boldsymbol{x}|\tau)}{\partial\tau}\right)^2\right\}} \tag{7.11-8}$$

where the negative second derivative is replaced with the first derivative squared. The denominator is called the *Fisher information matrix*, $J(\tau)$ [33]; that is,

$$J(\tau)=E\left\{\left(\frac{\partial \ln p(\boldsymbol{x}|\tau)}{\partial\tau}\right)^2\right\} \tag{7.11-9}$$

The Fisher information matrix is roughly a measure of the sharpness of the peak of the likelihood function. For a Gaussian noise example, one has for the likelihood function

$$p(\boldsymbol{x}|\tau)=\left(\frac{1}{\sigma\sqrt{2\pi}}\right)^N \exp\left[-\frac{1}{2\sigma^2}\sum_{i=0}^{N-1}\left(x(i\Delta t)-s(i\Delta t,\tau)\right)^2\right] \tag{7.11-10}$$

The Fisher information matrix is evaluated as follows

$$J(\tau)=-\frac{1}{\sigma^2}\sum_{i=0}^{N-1}\left[\frac{\partial s(i\Delta t|\tau)}{\partial\tau}\right] \tag{7.11-11}$$

More generally [33] if $\boldsymbol{\tau}$ is a vector parameter with components τ_i, then the Cramer-Rao bound states that the following is true

$$\sigma_{ij}^2\geq\left[\boldsymbol{I}^{-1}(\tau)\right]_{ij} \tag{7.11-12}$$

where the rectangular matrix $\boldsymbol{I}(\boldsymbol{\tau})$ has components given by

$$\left[\boldsymbol{I}(\boldsymbol{\tau})\right]_{ij}=-E\left\{\frac{\partial^2[\ln(p(\boldsymbol{x}|\boldsymbol{\tau}))]}{\partial\tau_i\partial\tau_j}\right\} \tag{7.11-13}$$

Assume that the signal is received through an ideal low-pass filter of bandwidth W Hz, the signal samples are taken at the Nyquist rate (i.e., Δt = $1/(2W)$ seconds), and the delay is given by $s(i\Delta t,\tau)=s(i\Delta t-\tau)$; then the CRB is given by [33]

$$\sigma_\tau^2 \geq \frac{\sigma^2}{\sum_{i=0}^{N-1}\left(\frac{\partial s(i\Delta t-\tau)}{\partial \tau}\right)^2} = \frac{2}{\frac{E_s}{N_0}\overline{\Delta\omega^2}} \tag{7.11-14}$$

where

$$E_s = \int_0^{N\Delta t} s^2(t)dt \tag{7.11-15}$$

and the mean squared bandwidth, $\overline{\Delta\omega^2}$, is given by

$$\overline{\Delta\omega^2} = (2\pi)^2 \frac{\int_{-\infty}^{\infty} f^2 |S(f)|^2 df}{\int_{-\infty}^{\infty} |S(f)|^2 df} \tag{7.11-16}$$

where $S(f)$ is the Fourier transform of $s(t)$. The square root of $\overline{\Delta\omega^2}$ is the *Gabor bandwidth* and is defined by $\sqrt{\overline{\Delta\omega^2}}$. Note that if the numerator and the denominator are divided by T_s, the symbol duration of the signal $s(t)$, then we may write for random code sequences

$$\overline{\Delta\omega^2} = (2\pi)^2 \frac{\int_{-\infty}^{\infty} f^2 P_{PN}(f)df}{\int_{-\infty}^{\infty} P_{PN}(f)df} \tag{7.11-17}$$

where $P_{PN}(f)$ is the power spectral density of the baseband code signal. Alberty [35] has shown that when a filter limits the PSD, (7.11-17) is modified to

$$\overline{\Delta\omega^2} = (2\pi)^2 \frac{\int_{-\infty}^{\infty} f^2 P_{PN}(f)|H_{BB}(f)|^2 df}{\int_{-\infty}^{\infty} P_{PN}(f)|H_{BB}(f)|^2 df} \tag{7.11-18}$$

where $|H_{BB}(f)|^2$ is the magnitude squared of the low-pass equivalent combined transmitter and receiver filter.

Example 5 Consider an NRZ ranging code operates at R chip/sec. Assume that the code is essentially a random code. Determine the CRB under the assumption that the equivalent baseband filter is an ideal low-pass filter with a bandwidth of $4R$ Hz. Evaluating produces

$$\overline{\Delta\omega^2} = (2\pi)^2 \int_{-4R}^{4R} f^2 \frac{\sin(\pi f T_c)^2}{(\pi f T_c)^2} df = 16R^2$$

Thus we see via (7.11-14) that the mean squared error is proportional to the inverse chip rate^2 under these conditions. Clearly a higher code rate improves the performance in the sense that the minimum mean squared tracking error bound is reduced. For additional insight see Problem 13.

7.12 PHASE ROTATION AND HETERODYNING FOR RECEIVER USE

In this section we discuss the difference between a heterodyning operation and a phase rotation translation (also known as single sideband rotation). In a receiver it is often necessary to translate the signal from an IF frequency to baseband in order to process the signal when the IF frequency is smaller than the bandwidth of the signal. Typically navigational receivers utilize phase rotation.

7.12.1 Heterodyning the Signal to Near Baseband

Now we will consider a convenient method of changing the frequency of a signal by in effect phase rotating it up or down in frequency. The need to phase rotate the signal down in frequency (or up in frequency) occurs, for example, when an approximate frequency is known a priori and the signal then is to be heterodyned down to essentially zero frequency. This approximate frequency estimate may be obtained, for example, from known dynamics of the transmitter and receiver. When a carrier-tracking loop is employed to track the carrier signal, the final shift to baseband is normally accomplished with a phase rotation.

Let the received BPSK signal (or one phase of a quadriphase signal) be the received signal of the form

$$y(t) = \sqrt{2P} d(t) PN(t) \cos(\omega_0 t + \theta) \tag{7.12-1}$$

Consider a heterodyne operation for generating the I and Q baseband components for processing, such as code-tracking, code acquisition or bit error rate detection, that is shown in Figure 7.12-1. Figure 7.12-2 illustrates the heterodyning process in which one output resides at the sum frequency and the other one at the difference frequency for the signals only (noise is not considered).

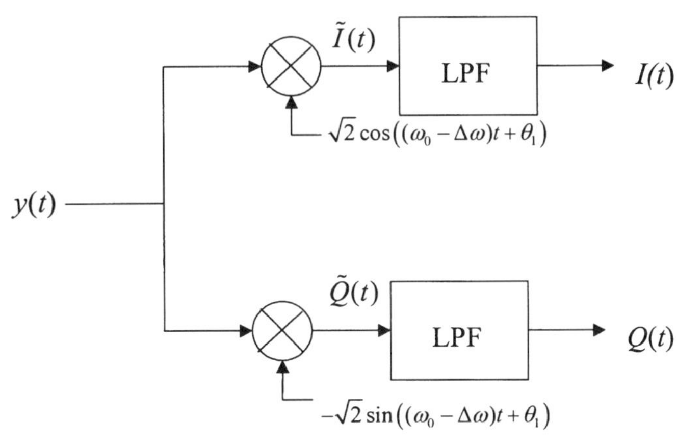

Figure 7.12-1 Baseband heterodyned output for a BPSK signal where $f_0 = \omega_0/(2\pi)$.

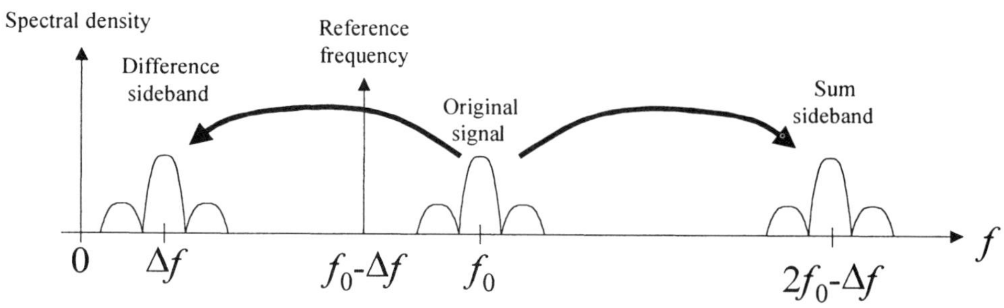

Figure 7.12-2 The results of heterodyning a BPSK carrier to near zero frequency.

It is seen that the heterodyne process has produced two signals (the sum and difference terms) per channel (I and Q). There is one at near baseband and one at near double the carrier frequency.

It is easy to show that the two components out of the I-Q demodulator, before the low-pass filters (LPF), is of the form

$$\tilde{I}(t) = \sqrt{P}d(t)\cos\left(\Delta\omega t + \theta - \theta_1\right) + \sqrt{P}d(t)\cos\left((2\omega_0 - \Delta\omega)t + \theta + \theta_1\right) \quad (7.12\text{-}2)$$

and

$$\tilde{Q}(t) = \sqrt{P}d(t)\sin\left(\Delta\omega t + \theta - \theta_1\right) - \sqrt{P}d(t)\sin\left((2\omega_0 - \Delta\omega)t + \theta + \theta_1\right) \quad (7.12\text{-}3)$$

when the I and Q references are lower by $\Delta f = \Delta\omega/2\pi$ Hz.

After filtering with a low-pass filter that cuts off at around ω_0, the two baseband components become

$$I(t) = \sqrt{P}d(t)\cos\left(\Delta\omega t + \theta - \theta_1\right) \quad (7.12\text{-}4)$$

and

$$Q(t) = \sqrt{P}d(t)\sin\left(\Delta\omega t + \theta - \theta_1\right) \quad (7.12\text{-}5)$$

Thus, after the low-pass filtering we see that the effect of the heterodyning is to shift the signal to near baseband. If the frequency uncertainty $\Delta\omega$ were zero, then the signals would be right at baseband. It is clear that the low-pass filter was important to remove the double frequency or sum term. It should be noted that as long as the difference term (near baseband signal) and the sum term (signal near $2\omega_0$) do not overlap, the frequency translation to baseband is easily accomplished via heterodyning.

Now let us consider a method to translate the heterodyned signal directly to baseband, without generating the other sideband. The alternative method is known as a phase rotation (translation).

7.12.2 Phase Rotation or Single Sideband Translation

Assume that the I and Q signals have already been generated, as shown in Figure 7.12-1, from an initial heterodyne process. Also assume that $\Delta\omega$ is the IF frequency and it is desired to do an additional frequency correction via the control element of a Costas loop, for example, for carrier tracking. One means to do this and to avoid overlapping of the signal spectra is to use a phase rotation translation. This process only produces one signal (sideband) at a lower frequency and does not produce two sidebands, and thus there is no concern about overlapping signal spectra.

Now consider the single sideband translation,[16] to lower the frequency by, say, $\Delta\hat{\omega}$, which is also known as a phase rotation downward. The translation (rotational transformation) from I' and Q' from I and Q is given by

$$\begin{aligned} I'(t) &= I(t)\cos\left(\hat{\Phi}(t)\right) + Q(t)\sin\left(\hat{\Phi}(t)\right) \\ Q'(t) &= Q(t)\cos\left(\hat{\Phi}(t)\right) - I(t)\sin\left(\hat{\Phi}(t)\right) \end{aligned} \tag{7.12-6}$$

where $\hat{\Phi}(t)$ is given by

$$\hat{\Phi}(t) = \Delta\hat{\omega}t + \theta - \theta_1 = \Delta\hat{\omega}t + \Delta\theta \tag{7.12-7}$$

and is the estimate of $\Delta\omega$ and the phase difference. Let $\Phi(t) = \Delta\omega t + \theta - \theta_1 = \Delta\omega t + \Delta\theta$; then the transformation (phase rotation) can be written as

$$I'(t) = \sqrt{P}d(t)\cos(\Phi)\cos\left(\hat{\Phi}\right) + \sqrt{P}d(t)\sin(\Phi)\sin\left(\hat{\Phi}\right) \tag{7.12-8}$$

which can be simplified to

$$I'(t) = \sqrt{P}d(t)\cos\left[\Phi - \hat{\Phi}\right] \tag{7.12-9}$$

Now consider the Q' term. It can be written as

$$Q'(t) = \sqrt{P}d(t)\sin(\Phi)\cos(\hat{\Phi}) - \sqrt{P}d(t)\cos(\Phi)\sin(\hat{\Phi}) \tag{7.12-10}$$

Simplifying one obtains

$$Q'(t) = \sqrt{P}d(t)\sin(\Phi - \hat{\Phi}) \tag{7.12-11}$$

Thus from (7.12-9) and (7.12-11) it is seen that the effect of the transformation is to shift the frequencies of the signal down. If $\Phi = \hat{\Phi}$ then the signal I and Q components have been "rotated" to baseband, and furthermore since this is a true translation in frequency there is only a difference frequency term coming out of the transformation. Figure 7.12-3 illustrates the phase rotation pictorially.

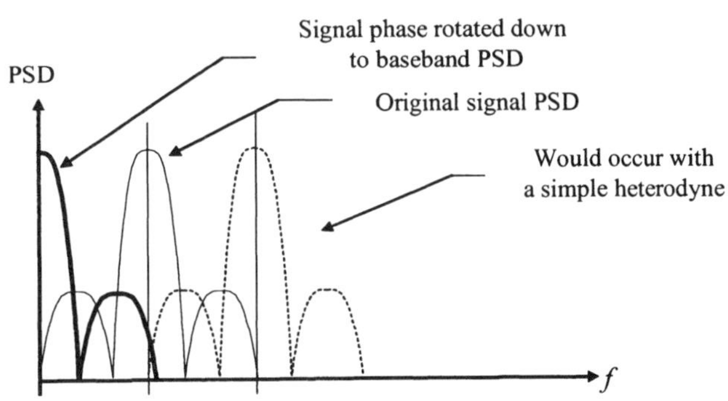

Figure 7.12-3 An example of how phase rotation shifts the signal spectra down but does not produce a sum term like heterodyning does.

[16] In this case the phase is assumed to be known, so it is included in the rotation.

One application of this rotation is after an approximate heterodyne to near baseband is completed second and final heterodyne exactly to baseband may well produce sum terms that are overlapping the desired terms at baseband. Thus a single sided translation has the advantage of not only producing the desired translation to the difference frequency located at baseband; it does not produce the sum frequency term. It can be shown that the signal-to-noise performance is the same in either case.

This translation subtracted the estimate frequency from the residual frequency on the I and Q terms. In order to add to the residual frequency on the I and Q signals the following transformation may be used

$$\begin{aligned} I'(t) &= I(t)\cos\left(\hat{\Phi}(t)\right) - Q(t)\sin\left(\hat{\Phi}(t)\right) \\ Q'(t) &= Q(t)\cos\left(\hat{\Phi}(t)\right) + I(t)\sin\left(\hat{\Phi}(t)\right) \end{aligned} \quad (7.12\text{-}12)$$

By comparing (7.12-12) with (7.12-6) it is seen that they are similar but the signs are changed in the second terms. This transformation can alternately be viewed as a complex rotation transforming $I(t)+jQ(t)$ to $I'(t)+jQ'(t)$.

It is to be noted that another way of looking at single sideband rotation is to view it as a vector rotation. That is, consider (7.12-6) written as a vector relationship, let $Z' = I' + jQ'$ and let $Z = I + jQ$, then

$$Z' = Z\exp(j\omega t) \quad (7.12\text{-}13)$$

This leads to

$$\begin{aligned} I' &= I\cos(\omega t) - Q\sin(\omega t) \\ Q' &= Q\cos(\omega t) + I\sin(\omega t) \end{aligned} \quad (7.12\text{-}14)$$

which is equivalent to (7.12-6), in that the signal is rotated down by ω radians/sec. Alternatively to rotate the signal up by ω radians/second we just change the sign of the rotating vector; that is,

$$Z'' = Z\exp(-j\omega t) \quad (7.12\text{-}15)$$

This leads to

$$\begin{aligned} I'' &= I\cos(\omega t) + Q\sin(\omega t) \\ Q'' &= Q\cos(\omega t) - I\sin(\omega t) \end{aligned} \quad (7.12\text{-}16)$$

A model of an actual direct sequence receiver in which the phase rotation is used is shown in Figure 7.12-4. It is based on the Rockwell Collins miniature airborne GPS receiver (MAGR) described in Parkinson and Spilker [10].

The receiver converts the signal down to an IF frequency and is sampled at IF and then the phase rotation removes the residual Doppler frequency. The I and Q baseband components are correlated and accumulated and sent to the microprocessor, which controls the acquisition functions (carrier and code), the tracking functions (carrier and code), as well as the lock detection monitoring and data detection. Finally the navigational solution (user's location) is provided based on the pseudorange estimates and other inputs.

7.13 PULSING AND BLANKING IN A BASEBAND EARLY-LATE CODE-TRACKING LOOP

Some special GPS-like transmitters, known as pseudolites (PL), are considered augmentations to the normally present GPS signals. They are commonly used in local area augmentation systems (LAAS). Many authors view pulsing as a necessary method to overcome the near-far problem that can occur with PLs. The

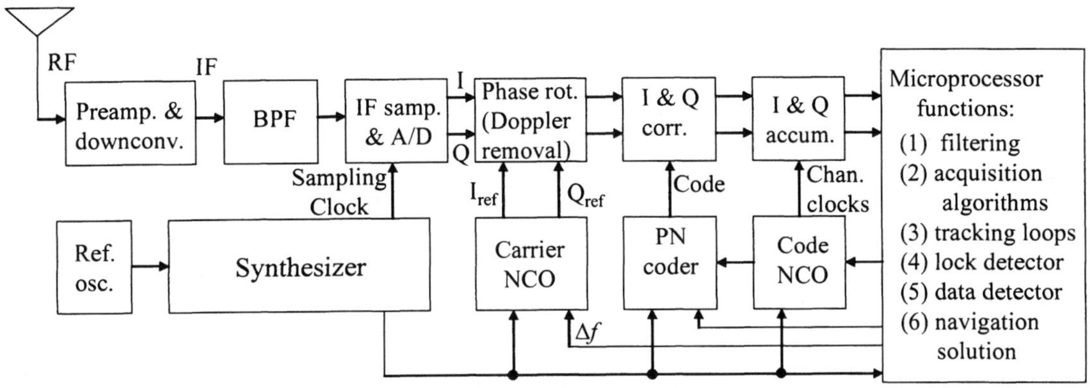

Figure 7.12-4 MAGR functional block diagram showing the phase rotation [10].

near-far problem occurs when a receiver is located very close to one PL and some distance from another PL. If the closer PL is much closer, its received power will be much greater than the one farther away. In this environment, the weaker PL signal might not be properly acquired and tracked. This is the essence of the near-far problem. One way to overcome this problem is to use pulsing for each PL signal at high power for a short duration burst at nonoverlapping times. One important application for PLs is for airports, where high location accuracy is required [37].

This section will only deal with receiver code-tracking performance with pulsing when (1) the receiver is designed to blank out the signal plus noise when the signal pulse occurs and (2) the case when the receiver is not designed to blank out the signal plus noise when the signal pulse occurs. This work is based on unpublished notes of the author. Thus we only consider receivers that are not designed to process the PL signals, but may or may not be able to blank out the PL signals.

7.13.1 Baseband Signal and Code Loop Model for a Baseband Early-Late Gate I-Q Code-Tracking Loop with Pulsing

Consider a baseband direct sequence signal of the form

$$y(t) = \sqrt{P}PN(t-T)g(t) + n(t) \tag{7.13-1}$$

in which P is the signal power, T is the unknown delay from the transmitter, $PN(t-T)$ is the pseudonoise sequence with NRZ code chips (or symbols), $g(t)$ is the transmitter pulse gating function that gates on the transmitted signal, and $n(t)$ is the assumed continuous input white Gaussian noise process. The gating function is shown in Figure 7.13-1.

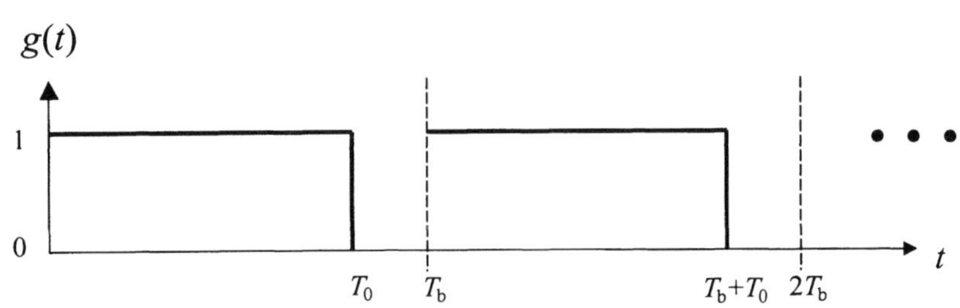

Figure 7.13-1 Model of the gating function that turns the signal on and off.

We see that the gating function takes the value of "1" (is on) for $d_f = T_0/T_b$ fractions of time and is "0" (off) for $(1-d_f)$ fractions of time. The fraction d_f is called the duty factor. It should be noted that the time that the gating function is off could correspond to when the code symbols are periodically a higher bandwidth symbol type with a low duty factor, such as when BOC(6,1) may be inserted in place of BOC(1,1) on GPS L1C for example, as well as when the signal is blanked off by the transmitter. It is convenient to model the gating function as shown for simplicity, however. The goal of this section is to show that the tracking error variance of the code-tracking loop varies as $1/(d_f)^2$ if no receiver noise blanking is employed in the receiver, and the tracking error variance of the code-tracking loop varies as $1/(d_f)$ when receiver blanking is employed. Thus synchronous blanking reduces tracking error performance compared to no blanking when the signal is pulsed. Clearly the best performance (minimum tracking error) occurs when $d_f = 1$, of course, but this only occurs when signal pulsing is not employed.

7.13.2 Full Correlation in the Coherent Baseband I-Q Code-Tracking Loop When the Signal Is Pulsed

First we consider the case that the receiver does not blank or mask the portion of the signal when it is pulsed off. Figure 7.13-2 shows the model for the code-tracking loop when blanking is not being utilized. It is also assumed that the receiver correlates for the full correlation time which is a bit time (T_b seconds) in this analysis.

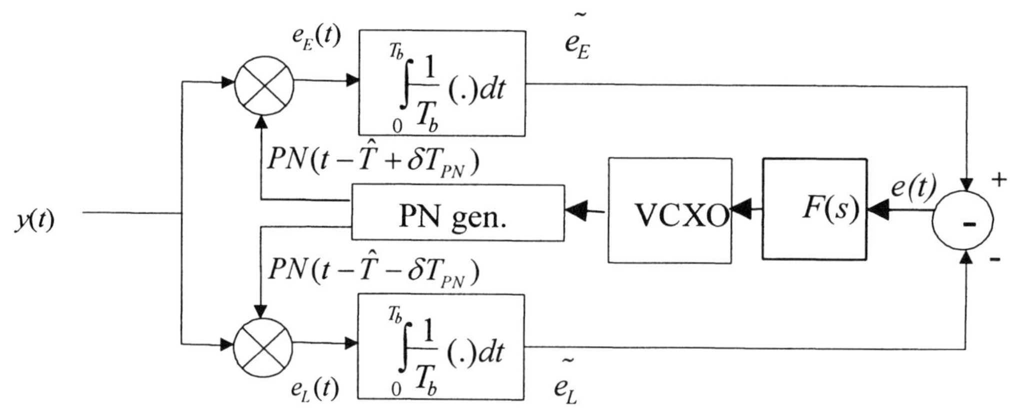

Figure 7.13-2 Baseband coherent early-late gate code-tracking loop model that does not employ receiver blanking.

Let $d = 2\delta$ where d is the early minus late gate code separation, and δ is the one-sided early or late offset, in code symbols. The output out of the upper multiplier is given by

$$e_E(t) = \sqrt{P}g(t)PN(t-T)PN(t-\hat{T}+\delta T_{PN}) + n(t)PN(t-\hat{T}+\delta T_{PN}) \tag{7.13-2}$$

where $\hat{T}$ is the code loop estimate of T. Now let $\varepsilon = T - \hat{T}$ and consider the signal out of integrator (part of the correlator) so that

$$\tilde{e}_E = \sqrt{P}d_f R_{PN}(\varepsilon + \delta T_{PN}) + \frac{1}{T_b}\int_0^{T_b} n(t)PN(t-\hat{T}+\delta T_{PN})dt \tag{7.13-3}$$

and d_f is the duty factor and was defined as T_0/T_b. In a similar manner the late correlation value, $\tilde{e}_L$ can be evaluated as

$$\tilde{e}_L = \sqrt{P}d_f R_{PN}(\varepsilon - \delta T_{PN}) + \frac{1}{T_b}\int_0^{T_b} n(t)PN(t - \hat{T} - \delta T_{PN})dt \quad (7.13\text{-}4)$$

The loop error control signal is defined as e is defined as

$$e = \tilde{e}_E - \tilde{e}_L \quad (7.13\text{-}5)$$

Therefore, combining (7.13-2), (7.13-3), and (7.13-4) produces over one bit time

$$\begin{aligned} e &= \sqrt{P}d_f\left[R_{PN}(\varepsilon + \delta T_{PN}) - R_{PN}(\varepsilon - \delta T_{PN})\right] \\ &+ \frac{1}{T_b}\int_0^{T_b} n(t)\left[PN(t - \hat{T} + \delta T_{PN}) - PN(t - \hat{T} - \delta T_{PN})\right]dt \end{aligned} \quad (7.13\text{-}6)$$

Let the noise term be defined over one bit time as

$$N(t) = \frac{1}{T_b}\int_0^{T_b} n(t)\left[PN(t - \hat{T} + \delta T_{PN}) - PN(t - \hat{T} - \delta T_{PN})\right]dt(p(t)) \quad (7.13\text{-}7)$$

where $p(t)$ is a T_b second unit amplitude pulse that repeats every T_b seconds. Note that the integral that is in front of $p(t)$ is a random variable that varies from bit to bit. Equation (7.13-6) can be rewritten for all time as

$$e = \sqrt{P}d_f K\varepsilon + N(t) \quad (7.13\text{-}8)$$

in which it is assumed that the code loop error is small (within the linear region), and K is the discriminator function slope at zero tracking error, and the noise has been extended to all bits. Now the timing estimate out of the receiver code generator is given by

$$\frac{\hat{T}}{T_{PN}} = \frac{K\sqrt{P}d_f F(s)}{s}\left[\frac{\varepsilon}{T_{PN}} + \frac{N(t)}{KT_{PN}}\right] \quad (7.13\text{-}9)$$

in which $F(s)$ is the loop filter expressed in the Laplace variable, $1/s$ denotes the integration operation in Laplace notation due to the VCO, and T_{PN} is the PN code symbol duration in seconds. Since

$$\frac{\varepsilon}{T_{PN}} = \frac{T}{T_{PN}} - \frac{\hat{T}}{T_{PN}} \quad (7.13\text{-}10)$$

Using (7.13-10) in (7.13-9) produces

$$\frac{\varepsilon}{T_{PN}} = -\frac{K\sqrt{P}d_f F(s)}{s}\frac{\varepsilon}{T_{PN}} - \frac{K\sqrt{P}d_f F(s)}{s}\left[\frac{N(t)}{Kd_f\sqrt{P}}\right] + \frac{T(t)}{T_{PN}} \quad (7.13\text{-}11)$$

Solving for the error (ε/T_{PN}), one has

$$\left[1 + K\sqrt{P}d_f F(s)/s\right]\frac{\varepsilon}{T_{PN}} = -K\sqrt{P}d_f F(s)/s\left[\frac{N(t)}{Kd_f\sqrt{P}}\right] + \frac{T(t)}{T_{PN}} \quad (7.13\text{-}12)$$

or

$$\frac{\varepsilon}{T_{PN}} = -H(s)\left[\frac{N(t)}{Kd_f\sqrt{P}}\right] + [1 - H(s)]\frac{T(t)}{T_{PN}} \tag{7.13-13}$$

where $H(s)$ is the closed-loop response of the code-tracking loop and $1-H(s)$ are given by

$$H(s) = \frac{K\sqrt{P}d_f F(s)/s}{1 + K\sqrt{P}d_f F(s)/s}$$
$$1 - H(s) = \frac{1}{1 + K\sqrt{P}d_f F(s)/s} \tag{7.13-14}$$

We have let the input time delay $T(t)$ be written as a function of t to reflect the case that it is not necessarily constant. However in our case, we are only interested in the noise induced error so we will let $T(t)$ be zero.

Under the zero input conditions ($T(t) = 0$), we have

$$\frac{\varepsilon}{T_{PN}} = -H(s)\left[\frac{N(t)}{d_f\sqrt{PK}}\right] \tag{7.13-15}$$

Now $N(t)$ can be written for one bit time as

$$N_i\big|_{i=0} = \frac{1}{T_b}\int_0^{T_0} n(t)\left[PN(t-\hat{T}+\delta T_{PN}) - PN(t-\hat{T}-\delta T_{PN})\right]dt \tag{7.13-16}$$

over the first T_b seconds, where $N_i\big|_{i=0}$ is the value of $N(t)$ for time between 0 and T_b seconds. Since the integrate-and-dump has a constant output over T_b seconds and changes every T_b seconds we can model $N(t)$ over all time as

$$N(t) = \sum_{i=-\infty}^{\infty} N_i p(t - iT_b) \tag{7.13-17}$$

Evaluation of this noise representation yields

$$R_N(\tau) = \langle E[N(t)N(t+\tau)]\rangle = \langle E(N_i^2)\rangle\left[1 - \frac{|\tau|}{T_{PN}}\right], \quad |\tau| \le T_{PN}$$
$$R_N(\tau) = 0 \ \text{ otherwise} \tag{7.13-18}$$

where $\langle x\rangle$ denotes the time average of x.

Thus it is only necessary to evaluate the variance of the correlation of the noise over one bit time. The variance can be written as

$$E\{N^2\} = \frac{1}{T_b^2}\int_0^{T_b}\int_0^{T_b} n(t)n(u)\times\left\langle \begin{matrix}\left[PN(t-\hat{T}+\delta T_{PN}) - PN(t-\hat{T}-\delta T_{PN})\right] \\ \times\left[PN(u-\hat{T}+\delta T_{PN}) - PN(u-\hat{T}-\delta T_{PN})\right]\end{matrix}\right\rangle dtdu \tag{7.13-19}$$

Noting that

$$E\{n(t)n(u)\} = \frac{N_0}{2}\delta(t-u) \tag{7.13-20}$$

and $\delta(u)$ is the Dirac delta function of the variable u. Using this property it can be shown that

$$E\{N^2\} = \frac{N_0}{2T_{PN}}[2R_{PN}(0) - 2R_{PN}(2\delta T_{PN})] \tag{7.13-21}$$

Therefore the autocorrelation function can be written as

$$R_N(\tau) = \frac{N_0}{2T_{PN}}[2R_{PN}(0) - 2R_{PN}(2\delta T_{PN})]\left[1 - \frac{|\tau|}{T_{PN}}\right] \tag{7.13-22}$$

Since the closed-loop noise bandwidth is very narrow, we can simply evaluate the noise spectral density (NSD) at $f = 0$ as a simplifying assumption. The NSD is given by the Fourier transform of the autocorrelation function, so that at $f = 0$ we have

$$S_N(0) = \int_{-\infty}^{\infty} R_N(\tau)d\tau \tag{7.13-23}$$

or

$$S_N(0) = \frac{N_0 d}{T_{PN}}(T_{PN}) = N_0 d \tag{7.13-24}$$

where is can be shown [1] that

$$R_{PN}(0) - R_{PN}(2\delta T_{PN}) = d = 2\delta, \quad d \le 1 \tag{7.13-25}$$

in symbols. From (7.13-11), the loop gain is given by

$$\text{loop gain} = Kd_f\sqrt{P} \tag{7.13-26}$$

It follows that the code loop tracking error is given by

$$\sigma^2 = \frac{S_N(0)(2B_L)}{\left(2\sqrt{P}d_f\right)^2} = \frac{dN_0B_L}{2P(d_f)^2} \tag{7.13-27}$$

where the fact that $K = 2$, regardless of the early-late spacing, for a baseband code loop that has an E-L discriminator function, was established earlier in this chapter. Thus, from (7.13-27) we see that the variance is $(1/d_f)^2$ times larger than the case that the signal lasts for the full correlation time (T_b is this analysis). Now consider the case when the receiver noise is synchronously blanked off when the signal pulse is off, thus eliminating some noise. This requires that the receiver have timing knowledge of signal pulsing. This timing could be derived from the tracking of the PN code.

7.13.3 Synchronous Blanking of the Noise When the Signal Is Pulsed Off

In this section we consider the effect of blanking the receiver noise when the transmitter signal has been pulsed off. Since the code-tracking loop knows time down to a small fraction of a symbol it will be assumed that the receiver code loop knows time perfectly, as a first-order assumption for the analysis. The advantage in blanking the noise is that in this case a smaller signal component has to contend with less noise when the noise is blanked. We model the baseband-received signal plus noise as

$$y(t) = \sqrt{P}PN(t-T)g(t) + n(t) \tag{7.13-28}$$

with $g(t)$ defined in Figure 7.13-1. It is assumed for convenience that the "on time" of the signal pulse is the first T_0 seconds and is off for the remaining T_b–T_0 seconds. Figure 7.13-3 illustrates the receiver model with signal and noise blanking.

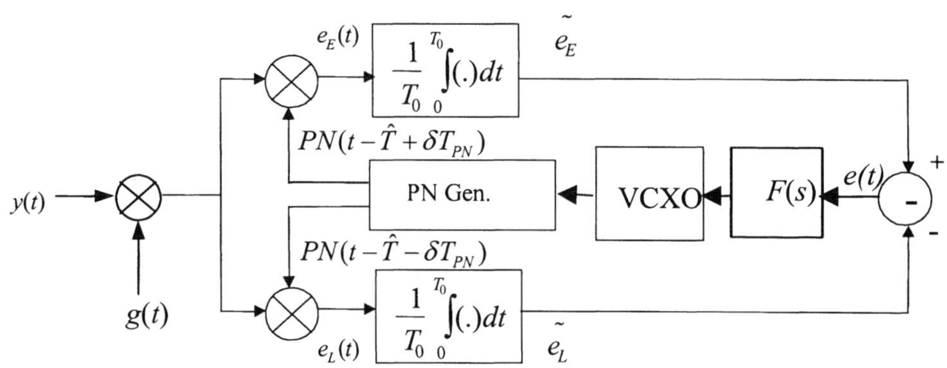

Figure 7.13-3 Model of the baseband early-late gate I-Q code-tracking loop with signal and noise blanking.

This loop has been modified from that of Figure 7.13-1, in that the input is assumed to be synchronous to the pulse signal so that only the signal plus noise is admitted to the loop when the signal is present and is off otherwise. The signal at the upper multiplier is given by

$$\varepsilon_E(t) = \sqrt{P}g(t)^2 PN(t-T)PN(t-\hat{T}+\delta T_{PN}) + n(t)g(t)PN(t-\hat{T}+\delta T_{PN}) \tag{7.13-29}$$

For the time period of 0 to T_0 seconds the output of the upper correlator is given by

$$\tilde{\varepsilon}_E = \sqrt{P}d_f R(\varepsilon+\delta T_{PN}) + \frac{1}{T_b}\int_0^{T_0} n(t)PN(t-\hat{T}+\delta T_{PN})dt \tag{7.13-30}$$

And in a similar manner, the output of the lower correlator is given by

$$\tilde{\varepsilon}_L = \sqrt{P}d_f R(\varepsilon+\delta T_{PN}) + \frac{1}{T_b}\int_0^{T_0} n(t)PN(t-\hat{T}-\delta T_{PN})dt \tag{7.13-31}$$

Over the time period of 0 to T_0 seconds, the error signal that controls the loop is given by

$$e = \tilde{\varepsilon}_E - \tilde{\varepsilon}_L \tag{7.13-32}$$

This can be written as

$$e = \sqrt{P}d_f\left[R(\varepsilon + \delta T_{PN}) - R(\varepsilon - \delta T_{PN})\right] + \frac{1}{T_b}\int_0^{T_0} n(t)\left[PN(t - \hat{T} + \delta T_{PN}) - PN(t - \hat{T} - \delta T_{PN})\right]dt \tag{7.13-33}$$

For small errors (ε small) we have

$$e = \sqrt{P}d_f\left[K\varepsilon\right] + n_E - n_L \tag{7.13-34}$$

where K is the gain slope of the early-late gate discriminator function, which is equal to 2, as noted earlier. In addition the two noise terms are given by

$$n_E = \frac{1}{T_b}\int_0^{T_0} n(t)PN(t - \hat{T} + \delta T_{PN})dt$$
$$n_L = \frac{1}{T_b}\int_0^{T_0} n(t)PN(t - \hat{T} - \delta T_{PN})dt \tag{7.13-35}$$

The code loop estimate of the time delay is $\hat{T}$, so that the estimate, in symbols, is given by

$$\frac{\hat{T}}{T_{PN}} = K\sqrt{P}d_f\frac{F(s)}{s}\left[\frac{\varepsilon}{T_{PN}} + \frac{n_E - n_L}{K\sqrt{P}d_f}\right] \tag{7.13-36}$$

Now the normalized code-tracking error is given by

$$\frac{\varepsilon}{T_{PN}} = \frac{T_i}{T_{PN}} - \frac{\hat{T}}{T_{PN}} \tag{7.13-37}$$

in which T_i is the input time delay, which will be neglected in this analysis. Therefore the code-tracking error is given by (see (7.13-13))

$$\frac{\varepsilon}{T_{PN}} = -H(s)\left[\frac{N(t)}{Kd_f\sqrt{P}}\right] + \left[1 - H(s)\right]\frac{T(t)}{T_{PN}} \tag{7.13-38}$$

We will assume that $T(t)$ is zero since we are only interested in the noise performance.

The variation with time has been suppressed; however, the functions are assumed to vary slowly with time, changing from bit time to bit time. The function $H(s)$ is the closed-loop transfer function of the code loop. The variance of the code-tracking loop due to the noise is given by

$$\sigma_\varepsilon^2 = \int_{-\infty}^{\infty} |H(f)|^2 S_N(f)df \cong 2B_L S_N(0)\text{ , symbols}^2 \tag{7.13-39}$$

The noise term $N(t)$ is given by

$$\frac{N(t)}{K\sqrt{Pd_f}} = \left(\frac{n_E(t) - n_L(t)}{K\sqrt{Pd_f}}\right) \tag{7.13-40}$$

with the understanding that the noise only changes every bit time (T_b). Therefore $N(t)$ can be written as

$$N_i\big|_{i=0} = \frac{1}{T_b}\int_0^{T_0} n(t)\left[PN(t-\hat{T}+\delta T_{PN}) - PN(t-\hat{T}-\delta T_{PN})\right]dt \tag{7.13-41}$$

over the first T_b seconds, where $N_i\big|_{i=0}$ is the value of $N(t)$ for time between 0 and T_b seconds. Since the integrate-and-dump has a constant output over T_b seconds and changes every T_b seconds we can model $N(t)$ over all time as

$$N(t) = \sum_{i=-\infty}^{\infty} N_i p(t - iT_b) \tag{7.13-42}$$

Following the same approach as in Section 7.3.2, we have

$$R_N(\tau) = \left\langle E\left[N(t)N(t+\tau)\right]\right\rangle = \left\langle E\left(N_i^2\right)\right\rangle\left[1 - \frac{|\tau|}{T_{PN}}\right], \quad |\tau| \le T_{PN} \tag{7.13-43}$$

$$R_N(\tau) = 0 \text{ otherwise}$$

where $\langle x \rangle$ denotes the time average of x. This can be evaluated as

$$\left\langle E\left(N_i^2\right)\right\rangle = \frac{1}{T_b^2}\int_0^{T_0}\int_0^{T_0} n(t)n(u)\left\langle \begin{array}{l}\left[\left(PN(t-\hat{T}+\delta T_{PN}) - PN(t-\hat{T}-\delta T_{PN})\right)\right]\cdot \\ \left[\left(PN(u-\hat{T}+\delta T_{PN}) - PN(u-\hat{T}-\delta T_{PN})\right)\right]\end{array}\right\rangle dtdu \tag{7.13-44}$$

This can be evaluated to

$$\left\langle E\left(N_i^2\right)\right\rangle = \frac{1}{T_b^2}\int_0^{T_0}\int_0^{T_0} \frac{N_0}{2}\delta(t-u)\left[2R_{PN}(0) - 2R_{PN}(2\delta T_{PN})\right]dtdu \tag{7.13-45}$$

Finally this can be written as

$$\left\langle E\left(N_i^2\right)\right\rangle = \frac{1}{T_b}\int_0^{T_0}\left[R_{PN}(0) - R_{PN}(2\delta T_{PN})\right]dt = d_f N_0\left[R_{PN}(0) - R_{PN}(2\delta T_{PN})\right] \tag{7.13-46}$$

Using the result from (7.13-20) it follows that

$$R_N(\tau) = d_f N_0 d\left[1 - \frac{|\tau|}{T_{PN}}\right], \quad |\tau| \le T_{PN} \tag{7.13-47}$$

$$R_N(\tau) = 0 \text{ otherwise}$$

The power spectral density of the noise process is given by

$$S_N(f)=\int_{-\infty}^{\infty} R_N(\tau)e^{-j\omega\tau}d\tau=\frac{d_f dN_0}{T_b}T_b\frac{\sin(\pi f T_b)^2}{(\pi f T_b)^2} \tag{7.13-48}$$

Approximating the noise spectral density by the value at $f = 0$ yields

$$S_N(0)=d_f dN_0 \tag{7.13-49}$$

From (7.13-39) and (7.13-49) the tracking error variance is given by

$$\sigma^2=\frac{S_N(0)(2B_L)}{d_f^2 4P}=\frac{dN_0 B_L}{2d_f P} \tag{7.13-50}$$

Upon comparing (7.13-50) with (7.13-27) we see that receiver blanking has reduced the tracking error variance of a baseband early minus late, first-order code-tracking loop by d_f. In other words, the tracking error variance of the unblanked code loop tracking error variance depends on $1/(d_f)^2$, whereas the blanked code loop tracking error variance depends on only $1/(d_f)$! Since $d_f < 1$ it clearly pays to blank the noise when the signal is not present. However, it is necessary to know the timing of the PL signal pulse in order to blank the signal and noise when the PL signal is pulsed on. For a direct sequence system this information could normally be obtained from the code itself. However, for nonspread spectrum systems, the timing would have to be obtained from the receiver system clock or by other means. This result is compatible with the result of Section 7.5 for blanked bandpass implemented early-late gate code-tracking loops.

This work was done for a baseband code-tracking loop. However, it is expected that a similar degradation would occur for noncoherent (nonbaseband) I-Q code-tracking loops.

7.14 SUMMARY

This chapter has attempted to summarize many known or recently discovered results for code-tracking loops that have the function of tracking the codes modulated onto the carrier. Accurate code-tracking is very important (for example, in global navigational satellite systems (GNSS) such as GPS, the code-tracking accuracy directly affects the pseudorange error used in computing the navigational solution).

First we determined the maximum-likelihood estimate for the code delay and showed that it provides a closed-loop model of a code-tracking loop. Both a coherent and three noncoherent code-tracking loop error performances were developed. For the noncoherent code loops: (a) the late^2–early^2 code loop base on an I-Q architecture was analyzed, (b) the radio frequency (RF) version of the same loop implemented at RF, and (c) the I-Q dot product code-tracking loop was analyzed. Next a new result for the effects of channel filtering and interference was developed for the I-Q late^2–early^2 type discriminator function showing how channel filtering affects performance. Various cases were obtained and the very wide bandwidth case was shown to agree with a known result. In addition, steady state errors in a code-tracking loop, time-shared early-late gate implementations were investigated for tracking error performance. Additional topics include the following: (1) the effects of a single component of multipath was analyzed for its effect on both coherent and noncoherent code-tracking loops and the fact that they both have the same multipath diagram, (2) the mean time to lose lock of a code-tracking loop was examined, (3) phase rotation (single sideband frequency shifting) was also considered, and (4) the effect of PL pulsing and receiver blanking on coherent code-tracking loop tracking error performance was explored.

References

[1] J. G. Proakis, *Digital Communications*, 4th ed., New York: McGraw-Hill, 2001.

[2] J. K. Holmes, S. Raghavan, and S. Lazar, "Acquisition and Tracking Performance of NRZ and Square Wave Modulated Symbols for Use in GPS," *Institute of Navigation 54th Annual Meeting*, June 1–3, 1998.

[3] W. C. Lindsey and M. K. Simon, *Telecommunication Systems Engineering*, Upper Saddle River, NJ: Prentice-Hall, 1973.

[4] W. B. Davenport and W. L. Root, *An Introduction to Random Signals and Noise*, New York: McGraw-Hill, 1958.

[5] A. J. Van Dierendonck, P. Fenton, and T. Ford, "Theory and Performance of Narrow Correlator Spacing in a GPS Receiver," *Navigation,* Vol. 39, No. 3, Fall 1992.

[6] J. K. Holmes, "Code-tracking Loop Performance Including the Effects of Channel Filtering and Gaussian Interference," *Proceedings of the 56th Annual Meeting of the ION*, San Diego, CA, June 2000.

[7] J. K. Holmes and S. Raghavan, "Code Loop Mean Slip Time Comparison of NRZ and BOC Signals," *IEEE Aerospace Conference*, Big Sky, Montana, August 2003.

[8] J. K. Holmes, *Coherent Spread Spectrum Systems*, New York: Wiley-Interscience, 1982, p. 127.

[9] J. K. Holmes, Unpublished notes on the I&D arm filter dot product code-tracking loop tracking performance.

[10] B. W. Parkinson and J. J. Spilker (eds.), *Global Positioning System: Theory and Application*, Vol. 1, Washington, D.C.: Progress in Astronautics and Aeronautics, Chapter 8, 1996.

[11] J. K. Holmes, "Code-tracking Performance Including the Effects of Channel Filtering and Gaussian Interference," *Proceedings of the 56th Annual Meeting of the Institute of Navigation*, San Diego, CA, June 2000.

[12] J. Proakis, *Digital Communications*, 2nd ed., New York: McGraw-Hill, 1989.

[13] L. E. Franks, *Signal Theory*, Englewood Cliffs, NJ: Prentice-Hall, 1969.

[14] J.-M. Lopez-Almansa and P. A. Pablos, "Derivation of an Analytical Model for Estimation of Measurement Error Induced by an Arbitrary Disturbance on a Coherent Delay Lock Loop," *ION 98*, June 1998.

[15] J. Betz, "Effect of Jamming on GPS M Code Signal SNIR and Code-Tracking Accuracy," *Proceedings of the Institute of Navigation, 2000 National Technical Meeting*, January 2000.

[16] C. J. Hegarty, "Noise Performance of Narrow Correlator GPS Receivers with the Inmarsat Transponder Package," MITRE CAASD study report done for the FAA contracts office, June 1994.

[17] M. K. Simon, "Noncoherent Pseudonoise Code-tracking Performances of Spread Spectrum Receivers," *IEEE Trans. on Communications.*, Vol. COM 25, No. 3, March 1977.

[18] M. Stone, "Noncoherent τ-Dither Code-tracking Loop Performance Analysis," TRW IOC 7131.50-2, June 13, 1972.

[19] T. C. Huang, "Analysis of a Time Shared Early/Late PN Code-Tracking Loop and Signal Dropout Considerations," TRW IOC SCTE-50-75-063/TCH, September 1975.

[20] H. P. Hartmann, "Analysis of a Dithering Loop for PN Code-Tracking," *IEEE Trans. on Aerospace and Electronic Systems*, January 1974.

[21] L. B. W. Jolley, *Summation of Series*, 2nd ed. (rev.), New York: Dover Publications, 1961.

[22] J. J. Spilker, *Digital Communications by Satellite*, Englewood Cliffs, NJ: Prentice-Hall, Chapter 18, 1977.

[23] R. J. Huff and K. L. Reinhard, "A Delay Lock Loop for Tracking Pulsed-Envelope Signals," *IEEE Trans. on Aerospace and Electronics Systems*, May 1971.

[24] W. J. Gill, "A Comparison of Binary Delay-Lock Tracking-Loop Implementations," *IEEE Trans. on Aerospace and Electronics Systems*, Vol. AES-2. No. 4, July 1966.

[25] P. T. Nielsen, "On the Acquisition of Binary Delay-Lock Loops," *IEEE Trans. on Aerospace and Electronics Systems*, Vol. AES-11, No. 3, May 1975.

[26] B. W. Parkinson and J. J. Spilker (eds.), *Global Positioning System: Theory and Application*, Vol. 1, Washington, D.C.: Progress in Astronautics and Aeronautics, Chapter 14, 1996.

[27] A. J. Van Dierendonck, P. Fenton and P. Ford, "Theory and Performance of Narrow Correlator Spacing in a GPS Receiver," *Navigation: Journal of The Institute of Navigation*, Vol. 39, No. 3, Fall 1992.

[28] J. K. Holmes and L. Biederman, "Delay-Lock Loop Mean Time to Lose Lock," *IEEE Trans. on Communications*, November 1978.

[29] A. J. Viterbi, *Principles of Coherent Communication*, Upper Saddle River, NJ: Prentice-Hall, 1972.

[30] R. L. Stratonovich, *Topics in the Theory of Random Noise Vol. I*, Chapter 4, New York: Gordon and Breach, 1966.

[31] T. C. Huang and J. K. Holmes, "Performance of Noncoherent Time Shared PN Code-Tracking Loops," *National Telemetry Conference (NTC)*, November 29–December 1, 1976.

[32] A. P. Sage and J. C. Melsa, *Estimation Theory with Applications to Communications and Control,* New York: McGraw-Hill, 1971.

[33] B. W. Parkinson and J. J. Spilker, editors, *Global Positioning System: Theory and Application,* Vol. 1, Washington, D.C.: Progress in Astronautics and Aeronautics, Chapter 3, 1996.

[34] M. Moeneclaey, "Simple Lower Bound on the Linearized Performance of Practical Symbol Synchronizers," *IEEE Trans. on Communications*, Vol. COM-31, No. 9, September 1983.

[35] T. Alberty, "Frequency Domain Interpretation of the Cramer-Rao Bound for Carrier and Clock Synchronization," *IEEE Trans. on Communications*, Vol. 43, No. 2/3/4, February/March/April 1995.

[36] M. Moeneclaey, "Fundamental Lower Bound on the Performance of Practical Joint Carrier and Bit Synchronizers," *Trans. on Communications*, Vol. COM-32, No. 9, September 1984.

[37] A. J. Van Dierendonck, P. Fenton, and C. Hegarty, "Proposed Airport Pseudolite Signal Specification," NovAtel report TP0046, circa 1999.

Selected Bibliography

R. Schwiekert and T. Worz, "Technical Note on Cramer-Rao Bound for Chip Tracking," Doc. No. SDS-TN-DLR/NT-01/97, Issue No. 3, January 10, 1997.

L. Weill, "Achieving Theoretical Accuracy Limits for Pseudoranging in The Presence of Multipath," *8th International Technical Meeting of the Satellite Division of The Institute of Navigation*, Part 2 of 2, September 12–15, 1995.

L. Weill, "Conquering Multipath: The GPS Accuracy Battle," *GPS World*, April 1977.

L. Weill, "GPS Multipath Mitigation by Means of Correlator Reference Waveform Design," *National Technical Meeting Satellite Navigation of the Institute of Navigation*, January 14–16, 1997.

Problems

1. Assume that one is designing an NRZ baseband code-tracking loop in which $B_L = 2$ Hz, $P/N_0 = 30$ dB-Hz, and the early-late correlator spacing is ½ of a code chip. Determine the standard deviation of the code-tracking error in chips.

2. Using the general expression for unit amplitude NRZ symbols, (7.2-86) and (7.2-88), show that the variance of the tracking error in symbols^2 for NRZ symbol types is given by

$$\left(\frac{\sigma_{sym}}{T_s}\right)^2 = \frac{N_0 B_L d}{2P}\left[1+\frac{2}{(2-d)(E_b/N_0)}\right] \text{ symbols}^2 \qquad \text{(P7-1)}$$

3. Assume that one is designing a noncoherent code-tracking loop in which $B_L = 2$ Hz, $P/N_0 = 30$ dB-Hz, $R_d = 50$ bps, and the early-late correlator spacing satisfies $d = ½$ of a code symbol (chip in this case). Determine the standard deviation of the code-tracking error in chips. Compare to the result of Problem 1.

4. Show that the $I(t)$ and $Q(t)$ noise terms for the PAF dot product code-tracking loop are statistically independent; that is, show that

$$E\left[\left(-n_c(t)\sin(\theta)+n_s(t)\cos(\theta)\right)\left(-n_c(t)\sin(\theta)+n_s(t)\cos(\theta)\right)\right]=0 \quad \text{(P7-2)}$$

5. Starting with the error signal $D(\varepsilon,t)=\tilde{I}_{L-E}\tilde{I}_P+\tilde{Q}_{L-E}\tilde{Q}_P$ show that the error signal can be written as

$$\begin{aligned}D(\varepsilon,t)&=P\left[R_{PN}(\varepsilon-\delta T)-R_{PN}(\varepsilon+\delta T)\right]R_{PN}(\varepsilon)\\&+\left(\sqrt{P}\cos(\theta)\tilde{n}_{I_P}(t)-\sqrt{P}\sin(\theta)\tilde{n}_{Q_P}(t)\right)\left[R_{PN}(\varepsilon-\delta T)-R_{PN}(\varepsilon+\delta T)\right]\\&+\left(\sqrt{P}\tilde{n}_{I_{L-E}}(t)\cos(\theta)-\sqrt{P}n_{Q_{L-E}}(t)\sin(\theta)\right)R_{PN}(\varepsilon)\\&+\left(\tilde{n}_{I_{L-E}}(t)\tilde{n}_{I_P}(t)+\tilde{n}_{Q_{L-E}}(t)\tilde{n}_{Q_P}(t)\right)\end{aligned} \quad \text{(P7-3)}$$

where the noise terms are defined in (7.2-101) to (7.2-107), and $d(t)$ times a noise term has been approximated by the noise term since $d(t)$ is very narrowband compared to the noise term, by assumption.

6. Evaluate the first noise term in $n_1(t)$ for the dot product code-tracking loop with PAFs. Starting with (7.2-119) show that $\tilde{n}_{I_{L-E}}(t)$ can be written as

7.

$$\tilde{n}_{I_{L-E}}(t)=\int_{-\infty}^{\infty}n_I(t-u)\left[PN(t-u-\delta T_s-\hat{T})-PN(t-u+\delta T_s-\hat{T})\right]h_2(u)du \quad \text{(P7-4)}$$

Furthermore, show that the autocorrelation function can be written as

$$R_{\tilde{n}_{I_{L-E}}}(\tau)=\int_{-\infty}^{\infty}\frac{N_0}{2}\left[1-R_{PN}(2\delta T_s)\right]h_2(v+\tau)h_2(v)dv \quad \text{(P7-5)}$$

By taking the Fourier transform of the autocorrelation function obtain the spectral density at $f=0$ in the form

$$S_{\tilde{n}_{I_{L-E}}}(0)=N_0\left[1-R_{PN}(2\delta T_s)\right]\int_{-\infty}^{\infty}\int_{-\infty}^{\infty}h_2(v+\tau)h_2(v)dvd\tau \quad \text{(P7-6)}$$

finally show that $S_{n_1}(0)$ can be written as

$$S_{n_1}(0)=PN_0\left[1-R_{PN}(2\delta T_s)\right] \quad \text{(P7-8)}$$

8. Starting with (7.2-121), show that $S_{n_2}(0)$ can be written as

$$S_{n_2}(0)=\int_{-\infty}^{\infty}2R_{n_{I_{L-E}}}(\tau)R_{n_{I_P}}(\tau)d\tau \quad \text{(P7-9)}$$

by evaluating the autocorrelation function $R_{n_2}(\tau)$ and noticing the autocorrelation functions of the first product and the second product of (7.2-121) are the same.

9. Show that (7.2-135) can be written as (7.2-136) by direct evaluation. Represent the noise terms as white Gaussian noise with the appropriate variances, that is

$$E\left\{n_I^{'}(t)n_I^{'}(t-\tau)\right\}N_0\left[1-R_{PN}(2\delta T_s)\right]\delta_D(\tau) \quad \text{(P7-10)}$$

and

$$E\left\{n_I^{''}(t)n_I^{''}(t-\tau)\right\}=\frac{N_0}{2}\delta_D(\tau) \quad \text{(P7-11)}$$

10. Derive the equation for the phase plane for the passive loop filter code-tracking loop. Start with

$$\frac{s}{\omega_n}(y-x)=GF(s)D(x) \quad \text{(P7-12)}$$

and use

$$\begin{aligned}F(s)&=\frac{1+\tau_2 s}{1+\tau_1 s}\\ \frac{1}{G}+\tau_2&=\frac{2\varsigma}{\omega_n}\\ \tau_1&=\frac{G}{\omega_n^2}\end{aligned} \quad \text{(P7-13)}$$

so that

$$gF(s)=\frac{1+2\varsigma s/\omega_n}{1/g+s/\omega_n} \quad \text{(P7-14)}$$

and finally show that the phase plane for the passive second-order code-tracking loop is given by

$$\frac{d\dot{x}}{dx}=\frac{-D(x)-2\varsigma\left(\dfrac{dD(x)}{dx}\right)\dot{x}+\ddot{y}-(1/g)(\dot{x}-\dot{y})}{\dot{x}} \quad \text{(P7-15)}$$

11. Show, for the time gated code-tracking loop, using (7.5-25) and (7.5-26), that S can be evaluated as $S=1/d_f$.

12. By evaluating the delta functions on the variable t' show that (7.5-33) can be written as (7.5-34).

13. Show that (7.11-5) and (7.11-8) are equivalent, by noting that

$$\frac{\partial^2}{\partial\theta^2}\left[\frac{f'}{f}\right]=\frac{f''f-(f')^2}{f^2} \quad \text{(P7-16)}$$

and the fact that (establish this fact)

$$E\left\{\frac{f''}{f}\right\}=0 \quad \text{(P7-17)}$$

14. Reconsider Example 5 by allowing the bandwidth to be increased from 4R to 40R. Show that the standard deviation of the mean squared error is increased by $\sqrt{10}$. Thus increased chip rate or bandwidth yields improvement in the minimum mean squared error possible in code-tracking.

15. Show that (7.12-12) does in fact shift the center frequency up by $\Delta\omega$.

APPENDIX 7A

MEAN TIME TO LOSE LOCK FOR A FIRST-ORDER EARLY-LATE GATE CODE-TRACKING LOOP WITH EITHER BANDPASS ARM FILTERS OR BASEBAND ARM FILTERS

7A1.0 SUMMARY OF THE TRACKING ERROR VARIANCE

In this appendix the mean slip time expression is derived for an analog first-order code-tracking loop. The derivation will turn out to be in terms of the linearized code-tracking loop variance, the loop bandwidth, and the integral of the discriminator function curve. The evaluations will be based on replacing the variance expression for the bandpass loop model expression with the baseband code-tracking loop variance expression. The received signal plus thermal noise is of the form

$$y(t) = \sqrt{2P}d(t)PN(t-T)\cos(\omega_0 t+\theta) + \sqrt{2}n_c(t)\cos(\omega_0 t+\theta) + \sqrt{2}n_s(t)\sin(\omega_0 t+\theta) \qquad (7A1.0\text{-}1)$$

where P is the received power, $PN(t)$ has values of ± 1, $d(t)$ is the baseband data process, ω_0 is the carrier angular frequency (rad/sec), and θ is the carrier phase in radians. The second and third terms together comprise the white Gaussian noise process expressed in terms of the two baseband Gaussian white noise processes $n_c(t)$ and $n_s(t)$. Figure 7.9-2 illustrates the bandpass filter analog first-order code-tracking loop model. From Section 7.2.2 it was determined that the variance of the tracking error was given by

$$\sigma_\varepsilon^2 = \frac{N_0 B_L}{(\alpha P)^2 K^2}\left[8R_{PN}^2(\delta T_s)\left[1-R_{PN}(2\delta T_s)\right] + \frac{4N_0 W}{P}\left(1-R_{PN}^2(2\delta T_s)\right)\right] \qquad (7A1.0\text{-}2)$$

where the bandpass RF filter at the input is assumed to be W Hz (positive frequency) wide. For NRZ symbols it is known that $K = 2$ if $\delta T_S = T_s/2$, so that this reduces to the known result

$$\sigma_\varepsilon^2 = \frac{N_0 B_L}{2\alpha P}\left[1 + \frac{2N_0^2 W}{\alpha P}\right] \qquad (7A1.0\text{-}3)$$

7A2.0 MEAN SLIP TIME DERIVATION

In this section the derivation of the mean time to lose lock for arbitrary binary valued symbol shapes will be obtained. From Section 7.2.2 one has that the code loop estimate of the signal timing is given by

$$\hat{T} = \frac{K_v}{s} F(s)\left[\alpha PK g_n(\varepsilon) + n_1(t) + n_2(t)\right] \quad \text{(7A2.0-1)}$$

Since $\hat{T} = T - \varepsilon$ so that (7A2.0-1) can be written as

$$T - \varepsilon = \alpha PKK_v \frac{F(s)}{s}\left[g_n(\varepsilon) + \frac{n_1(t) + n_2(t)}{\alpha KP}\right] \quad \text{(7A2.0-2)}$$

Noting that multiplication by s denotes the derivative, for a first order loop $F(s) = 1$, and assuming that dT/dt is zero, one has

$$\frac{d\varepsilon}{dt} = -\alpha PKK_v\left[g_n(\varepsilon) + \frac{n_1(t) + n_2(t)}{\alpha KP}\right] \quad \text{(7A2.0-3)}$$

Stratonovich [7A1] shows that the probability density satisfies the Fokker-Planck equation. Let $\varepsilon = \pm\varepsilon_{LL}$ denote the limit when loss of lock is declared (the value when ε (the code phase error) causes the error signal to go to zero). When the code timing error ε, satisfies $|\varepsilon| < \varepsilon_{LL}$, then

$$\frac{\partial P(\varepsilon,t)}{\partial t} = \frac{\partial}{\partial \varepsilon}\left[A_1(\varepsilon)P(\varepsilon,t)\right] + \frac{1}{2}\frac{\partial^2}{\partial \varepsilon^2}\left[A_2(\varepsilon)P(\varepsilon,t)\right] \quad \text{(7A2.0-4)}$$

where the two normalized moments are defined by

$$A_1(\varepsilon) = \lim_{\Delta t \to 0}\left\{\frac{E\left[(\Delta\varepsilon)|\varepsilon\right]}{\Delta t}\right\} \qquad A_2(\varepsilon) = \lim_{\Delta t \to 0}\left\{\frac{E\left[(\Delta\varepsilon)^2|\varepsilon\right]}{\Delta t}\right\} \quad \text{(7A2.0-5)}$$

From (7A2.0-3) integrate from t to $t + \Delta t$ to yield

$$\varepsilon(t+\Delta t) - \varepsilon = \Delta\varepsilon = -\alpha PKK_v g_n(\varepsilon)\Delta t - \alpha PKK_v \int_t^{t+\Delta t} \frac{n_1(t) + n_2(t)}{\alpha KP} dt \quad \text{(7A2.0-6)}$$

From the definition of $A_1(\varepsilon)$, (7A2.0-5), one has

$$A_1(\varepsilon) = \lim_{\Delta t \to 0}\left[\frac{E(\Delta\varepsilon|\varepsilon)}{\Delta t}\right] = -\frac{\alpha PKK_v g_n(\varepsilon)\Delta t}{\Delta t} \quad \text{(7A2.0-7)}$$

Now consider $A_2(\varepsilon)$, where

$$A_2(\varepsilon) = \lim_{\Delta t \to 0}\left[\frac{E\left[(\Delta\varepsilon)^2|\varepsilon\right]}{\Delta t}\right] \quad \text{(7A2.0-8)}$$

Therefore

$$A_2(\varepsilon)=\lim_{\Delta t\to 0}\frac{1}{\Delta t}E\left\{\alpha^2P^2K^2K_v^2\left(g_n(\varepsilon)\right)^2\Delta t^2+\alpha^2P^2K^2K_v^2g_n(\varepsilon)\Delta t\int_t^{t+\Delta t}\frac{n_1(t)+n_2(t)}{\alpha KP}dt\right\}$$
$$+\frac{\alpha^2P^2K^2K_v^2}{d^2K^2P^2}\lim_{\Delta t\to 0}\frac{1}{\Delta t}E\left\{\int_t^{t+\Delta t}\int_t^{t+\Delta t}\left(n_1(u)+n_2(u)\right)\left(n_1(v)+n_2(v)\right)dudv\right\} \quad (7A2.0\text{-}9)$$

The first and second terms are zero since the first one goes to zero as $\Delta t\to 0$ and the second one has zero mean value; hence

$$A_2(\varepsilon)=\lim_{\Delta t\to 0}\frac{K^2E\left\{\int_t^{t+\Delta t}\int_t^{t+\Delta t}\left(n_1(u)+n_2(u)\right)\left(n_1(v)+n_2(v)\right)dudv\right\}}{\Delta t} \quad (7A2.0\text{-}10)$$

Model the sum of the two new processes as essentially white as far as the effect on the loop is concerned. So

$$A_2(\varepsilon)=K_v^2\frac{N_0^{'}}{2} \quad (7A2.0\text{-}11)$$

where $N_0^{'}$ is the effective one-sided noise spectral density. From Section 7.2.2 we have

$$\frac{N_0^{'}}{2}=4N_0PR_{PN}^2(\delta T_s)\left[1-R_{PN}(2\delta T_s)\right]\alpha+2N_0^2W\left[1-R_{PN}^2(2\delta T_s)\right] \quad (7A2.0\text{-}12)$$

For as long as $|\varepsilon|<\varepsilon_{LL}$ the density function of ε, which we call $q(\varepsilon, t)$, satisfies the Fokker-Planck equation so that

$$\frac{\partial q(\varepsilon,t)}{\partial t}=-\frac{\partial}{\partial\varepsilon}\left[A_1(\varepsilon)q(\varepsilon,t)\right]+\frac{1}{2}\frac{\partial^2}{\partial t^2}\left[A_2(\varepsilon)q(\varepsilon,t)\right] \quad (7A2.0\text{-}13)$$

after inserting the values of $A_1(\varepsilon)$ and $A_2(\varepsilon)$ one obtains

$$\frac{\partial q(\varepsilon,t)}{\partial t}=\alpha PKK_v\frac{\partial}{\partial\varepsilon}\left[g_n(t)q(\varepsilon,t)\right]+\frac{N_0^{'}K_v^2}{4}\frac{\partial^2}{\partial t^2}\left[q(\varepsilon,t)\right] \quad (7A2.0\text{-}14)$$

with the initial conditions

$$q(\varepsilon,0)=\delta(\varepsilon) \quad (7A2.0\text{-}15)$$

and with the boundary conditions

$$q(\varepsilon,t)=0 \qquad \forall|\varepsilon|\geq\varepsilon_{LL}$$
$$g(\varepsilon,t)=q(-\varepsilon,t) \qquad \forall t \tag{7A2.0-16}$$

A comment on the first equation in (7A2.0-16) is worthwhile. The density function exists as long as the error does not exceed the boundaries ($|\varepsilon| < \varepsilon_{LL}$). The integral over the interval ($-\varepsilon_{LL}$, ε_{LL}), $\Phi(t)$, gives the probability that ε has not exceeded $|\varepsilon_{LL}|$, so

$$\Phi(t)=\int_{-\varepsilon_{LL}}^{\varepsilon_{LL}} q(\varepsilon,t)d\varepsilon\leq 1 \tag{7A2.0-17}$$

Let $\psi(t)=1-\Phi(t)$ to denote the probability that that the boundary has been reached at time t; note that

$$\frac{\partial\psi(t)}{\partial t}=-\frac{\partial\Phi(t)}{\partial t} \tag{7A2.0-18}$$

and $\psi(0)=\psi(\infty)=0$. The probability density function is given by this derivative so that the mean time to lose lock is given by

$$\overline{T}=\int_0^\infty t\frac{\partial\psi}{\partial t}dt=\int_0^\infty(-t)\frac{\partial\Phi}{\partial t}dt \tag{7A2.0-19}$$

Integrate by parts, and let $u = -t$ and $v=\dfrac{\partial\Phi}{\partial t}$ and recall that

$$\int udv=uv=\int vdu \tag{7A2.0-20}$$

Now the mean time to lose lock is given by

$$\overline{T}=-t\Phi(t)\Big|_0^\infty+\int_0^\infty\Phi(t)dt=\int_0^\infty\Phi(t)dt \tag{7A2.0-21}$$

since $\Phi(t)$ must decrease with t faster than t to exist and clearly the lower limit is zero. Using the definition of $\Phi(t)$ produces

$$\overline{T}=\int_0^\infty\int_{-\varepsilon_{LL}}^{\varepsilon_{LL}} q(\varepsilon,t)dt \tag{7A2.0-22}$$

To evaluate (7A2.0-22) it is necessary to obtain $q(t)$. To do this integrate the Fokker-Planck equation, (7A2.0-14) from 0 to ∞ to produce

$$\int_0^\infty\frac{\partial}{\partial t}q(\varepsilon,t)dt=K\alpha PK_v\frac{\partial}{\partial\varepsilon}\left[q_n(\varepsilon)\int_0^\infty q(\varepsilon,t)dt\right]+\frac{N_0'}{4}K_v^2\frac{\delta^2}{\delta\varepsilon^2}\left[\int_0^\infty q(\varepsilon,t)dt\right] \tag{7A2.0-23}$$

Let $Q(\varepsilon)=\int_0^{\infty} q(\varepsilon,t)dt$ so that (7A2.0-23) becomes

$$q(\varepsilon,\infty)-q(\varepsilon,0)=2PK_v\frac{\partial}{\partial\varepsilon}\left[g_n(\varepsilon)Q(\varepsilon)\right]+\frac{N_0^{'}}{4}K_v^2\frac{\partial^2}{\partial\varepsilon^2}\left[Q(\varepsilon)\right] \quad \text{(7A2.0-24)}$$

Now it must be true that for the density function to exist that

$$q(\varepsilon,\infty)=0 \qquad q(\varepsilon,0)=\delta(\varepsilon) \quad \text{(7A2.0-25)}$$

Using (7A2.0-25) leads to

$$-\delta(\varepsilon)=\alpha KK_vP\frac{d}{d\varepsilon}\left[q_n(\varepsilon)Q(\varepsilon)\right]+\frac{N_0^{'}}{4}K_v^2\frac{\partial^2}{\partial\varepsilon^2}\left[Q(\varepsilon)\right] \quad \text{(7A2.0-26)}$$

Notice the boundary conditions become

$$Q(\varepsilon_{LL})=Q(-\varepsilon_{LL})=0 \quad \text{(7A2.0-27)}$$

The solution to (7A2.0-26) is called the Green's function [7A2] for the problem. Forming an indefinite integration of ε one obtains

$$-u(\varepsilon)=\alpha KK_vPg_n(\varepsilon)Q(\varepsilon)+\frac{N_0^{'}}{4}K_v^2\frac{d}{d\varepsilon}\left[Q(\varepsilon)\right]-c \quad \text{(7A2.0-28)}$$

where $u(\varepsilon)$ is the unit step function and c is a constant of integration. Rearranging

$$\frac{d}{d\varepsilon}\left[Q(\varepsilon)\right]K_v^2\left(\frac{N_0^{'}}{4}\right)+\alpha KK_vPq_n(\varepsilon)Q(\varepsilon)=c-u(\varepsilon) \quad \text{(7A2.0-29)}$$

again yields

$$\frac{d}{d\varepsilon}\left[Q(\varepsilon)\right]\frac{4\alpha KP}{K_vN_0^{'}}\left[g_n(\varepsilon)Q(\varepsilon)\right]=\frac{c-u(\varepsilon)}{K_v^2\left(\frac{N_0^{'}}{4}\right)} \quad \text{(7A2.0-30)}$$

From [7A3] the solution of (7A2.0-30) is given by

$$Q(\varepsilon)=e^{-\gamma\int^{\varepsilon}g_n(\varepsilon^{'})d\varepsilon^{'}}\int_{-\varepsilon_{LL}}^{\varepsilon}\left[\frac{c-u(\varepsilon")}{\beta}\right]e^{\gamma\int^{\varepsilon}g_n(\varepsilon^{'})d\varepsilon^{'}}d\varepsilon"+De^{-\gamma\int^{\varepsilon}g_n(\varepsilon^{'})d\varepsilon^{'}} \quad \text{(7A2.0-31)}$$

Now the following two conditions apply

$$Q(-\varepsilon_{LL}) = 0 \Rightarrow D = 0$$
$$Q(\varepsilon_{LL}) = 0 \Rightarrow c = 1/2 \tag{7A2.0-32}$$

Using these two conditions leads to

$$Q(\varepsilon) = e^{-\gamma\int^{\varepsilon} g_n(\varepsilon')d\varepsilon'} \int_{-\varepsilon_{LL}}^{\varepsilon} \left[\frac{1/2 - u(\varepsilon")}{\beta}\right] e^{\gamma\int^{\varepsilon} g_n(\varepsilon')d\varepsilon'} d\varepsilon" \tag{7A2.0-33}$$

Denote $G(\varepsilon)$ as the following indefinite integral

$$G(\varepsilon) = \int^{\varepsilon} g_n(\varepsilon')d\varepsilon' \tag{7A2.0-34}$$

such that $G(\varepsilon)$ is an even function of ε. Using (7A2.0-34) in (7A2.0-33) produces

$$Q(\varepsilon) = e^{-\gamma G(\varepsilon)} \int_{-\varepsilon_{LL}}^{\varepsilon} \left[\frac{1/2 - u(\varepsilon")}{\beta}\right] e^{\gamma G(\varepsilon)} d\varepsilon" \tag{7A2.0-35}$$

From (7A2.0-21) and (7A2.0-22) one has

$$\bar{T} = \int_0^{\infty} \int_{-\varepsilon_{LL}}^{\varepsilon_{LL}} q(\varepsilon,t)dtd\varepsilon = \int_{-\varepsilon_{LL}}^{\varepsilon_{LL}} Q(\varepsilon)d\varepsilon \tag{7A2.0-36}$$

Using the definition of $Q(\varepsilon)$ produces

$$\bar{T} = \frac{1}{\beta} \int_{-\varepsilon_{LL}}^{\varepsilon_{LL}} e^{-\gamma G(\varepsilon)} \int_{-\varepsilon_{LL}}^{\varepsilon} \left[1/2 - u(\varepsilon')\right] e^{\gamma G(\varepsilon')} d\varepsilon' d\varepsilon \tag{7A2.0-37}$$

Now using the definitions

$$\gamma = \frac{4\alpha KP}{K_v N_0'} \quad and \quad \beta = K_v^2 \left(\frac{N_0'}{4}\right) \tag{7A2.0-38}$$

and

$$B_L = \frac{A_{eq}}{4} \quad and \quad A_{eq} = \alpha PKK_v \tag{7A2.0-39}$$

Using the value of N_0' from (7A2.0-11) produces for γ

$$\gamma = \frac{2\alpha KP}{K_v\left[4N_0\alpha PR_{PN}^2(\delta T_s)(1-R_{PN}(2\delta T_s)+2N_0^2W\left(1-R_{PN}^2(2\alpha T_s)\right)\right]} \quad (7A2.0\text{-}40)$$

Recognizing that γ is $1/(\sigma^2)$ yields

$$\sigma^2 = \frac{N_0B_L}{K^2(\alpha P)}\left[8N_0R_{PN}^2(\delta T_s)\left[1-R_{PN}(2\delta T_s)\right]+\frac{4N_0^2W}{\alpha P}\left[1-R_{PN}^2(2\delta T_s)\right]\right] \text{ symbols}^2 \quad (7A2.0\text{-}41)$$

which is the tracking error variance result for the first-order bandpass noncoherent code-tracking loop. It is known [7A4] that for an integrate-and-dump filter code-tracking loop that the variance of the code-tracking error is given by

$$\sigma_{I\&D}^2 = \frac{N_0B_L}{K^2P}\left[8R_{PN}^2(\delta T_s)\left(1-R_{PN}(2\delta T_s)\right)+\frac{4N_0\left[1-R_{PN}^2(2\delta T_s)\right]}{PT_b}\right] \quad (7A2.0\text{-}42)$$

again in symbols^2. Equation (7A2.0-42) is used in the slip time estimates with the integrate-and-dump arm filters version of the first-order code-tracking loop. Solving for $1/\beta$ produces

$$\frac{1}{\beta} = \frac{2}{B_L}\frac{\gamma}{8} = \frac{1}{4B_L\sigma^2} \quad (7A2.0\text{-}43)$$

Therefore the final expression for the mean slip time is given by

$$\overline{T} = \frac{1}{4B_L\sigma^2}\int_{-\varepsilon_{LL}}^{\varepsilon_{LL}} e^{-\left[\frac{G(\varepsilon)}{\sigma^2}\right]}\int_{-\varepsilon_{LL}}^{\varepsilon} e^{\left[\frac{G(\varepsilon')}{\sigma^2}\right]}\left[\frac{1}{2}-u(\varepsilon')\right]d\varepsilon' d\varepsilon \text{ (s)} \quad (7A2.0\text{-}44)$$

Thus this double integral determines the mean time to lose lock in seconds for the first-order code-tracking loop and depends on the one-sided closed-loop noise bandwidth B_L, the linearized tracking error variance σ^2, and the integral of the discriminator function curve, $G(\varepsilon)$. This expression holds for any binary valued (± 1) symbol waveform such as NRZ or BOC symbols.

In the theoretical results presented in this chapter, in Section 7.9 for the mean slip time results, the variance of the integrate-and-dump arm filter, tracking error variance (7A2.0-42) was used in (7A2.0-44). However either (7A2.0-41) or (7A2.0-42) can be used to access the mean slip time performance, depending on which loop structure is considered.

Appendix 7A References

[7A1] R. L. Stratonovich, *Topics in the Theory of Random Noise*, Vol. I, Chapter 4, New York: Gordon and Breach, 1966.

[7A2] B. Friedman, *Principles and Techniques of Applied Mathematics*, Chapter 3, New York: John Wiley & Sons. 1956.

[7A3] I. S. Sokolnikoff and R. M. Redheffer, *Mathematics of Physics and Modern Engineering*, New York: McGraw-Hill, 1958.

[7A4] J. K. Holmes and S. Raghavan, "Acquisition and Tracking Performance of NRZ and Squarewave Modulated Symbols for Use in GPS," *The Institute of Navigation*, Denver, CO, June 1–3, 1998.

Appendix 7A Selected Bibliography

L. Biederman and J. K. Holmes, "Delay-Lock Loop Mean Time to Lose Lock," *National Telemetry Conference*, 1977.

W. B. Davenport and W. L. Root, *An Introduction to the Theory of Random Signals and Noise*, New York: McGraw-Hill, 1958.

J. K. Holmes and L. Biederman, "Delay-Lock Loop Mean Time to Lose Lock," *IEEE Trans. on Communications*, November 1978.

A. J. Viterbi, *Principles of Coherent Communication*, New York: McGraw-Hill, 1972.

CHAPTER 8

Tracking of Frequency-Hopped Signals

8.0 INTRODUCTION

In this chapter we will present the tracking performance of a frequency-hopped spread spectrum (SS) signal. We will see that the loop configuration is very similar to the direct sequence code tracking loop, which was considered in Chapter 7.

Frequency-hopped signals are commonly used with differential binary shift keying (DPSK) modulation, *M*-ary frequency shift keying (MFSK), or binary phase shift keying (BPSK). Initially we will limit our analysis to the case of frequency-hopping without data modulation. Later the performance of tracking with BPSK or DPSK data modulation will be presented.

This chapter does not address frequency-hopping acquisition; however, it is worthwhile to note that it is very similar to direct sequence code acquisition. In fact the same procedure that was used in Chapter 6, Section 6.5, for active search can be used to determine the false alarm probability, the effective false alarm probability, and the detection probability. The reason that the acquisition process for frequency-hopping is typically much faster than direct sequence code acquisition is the fact that the frequency hop symbols are usually milliseconds in duration, rather than microseconds as are the direct sequence SS symbols. Thus, when the initial time uncertainty is divided by the search update time (usually 1/2 of one hop time for frequency-hopping or 1/2 a code symbol for direct sequence SS), the number of acquisition hypotheses to be tested is much less for a frequency hop system than for a direct sequence system. And therefore the acquisition time is greatly reduced, since the acquisition time is proportional to the number of hypotheses that have to be tested.

Another possible application of a frequency-hopping signal is its use as an acquisition aid to direct sequence code acquisition. The frequency-hopped signal acquisition aid is a signal that is used in addition to, (and accompanies) the direct sequence signal. This type of signal, for use as an acquisition aid, has been proposed for use in conjunction with the direct sequence signal used in an existing spread spectrum system.

This approach is most useful when the initial timing error is large. The idea is to send a low rate hopping signal, of hop duration on the order of tens of milliseconds, to be acquired initially and then frequency hop tracked. This provides synchronization at the receiver to a millisecond or two of timing error. Then using a short direct sequence SS search, the acquisition of the direct sequence signal system can be completed. The technique of first acquiring the frequency-hopped signal and then the direct sequence signal can greatly decrease the acquisition time when there are large initial time uncertainties. For a small initial time uncertainty, direct sequence acquisition is normally quicker.

8.1 DATALESS FREQUENCY-HOPPED TIME TRACKING LOOP MODEL

Now we consider a model that describes a loop capable of tracking a frequency-hopped signal. Basically is it very similar to the I-Q noncoherent code-tracking loop for direct sequence tracking.

Before we discuss the loop we will first describe the frequency-hopped signal, which is modeled as

$$s(t)=\sqrt{2P}\cos\left[\omega_0 t+\sum_{n=-\infty}^{\infty}\left(\omega_n t+\phi_n\right)p_{T_c}(t-T-nT_h)+\theta_d(t-T)\right] \tag{8.1-1}$$

in which ω_0 is the carrier radian frequency, radians/sec, ω_n is the hop frequency at time n (from a set of many frequencies), ϕ_n is the phase change that occurs at time n, $p_{T_h}(t)$ is a unit amplitude pulse that is nonzero for 0 to T_h seconds, T is the unknown time delay between the transmitter and the receiver in seconds, and $\theta_d(t)$ is the data modulation on the hopped signal, which will be assumed to be either BPSK or DPSK.

The type of modulation on the hopping signal impacts the tracking ability of the frequency-hopping tracking loop (FHTL). For example, if the modulation is BPSK, then in order to demodulate the data, the dehopping signal must be able to coherently dehop the received frequency-hopped signal. In addition, for coherent demodulation the acquisition time of the carrier tracking loop on each hop must be accounted for. It is possible that the first few data bits, while not be demodulated due to the carrier acquisition time, pull in transient. Thus the effective data rate might be reduced due to this phenomenon.

Alternatively if the dehopped signal can provide a stable channel in phase and frequency, then it may be possible to differently encode the data on a low hop rate system by using the first bit or two as the reference bit(s) in the demodulation process. In this case only one or two data bits will not be able to carry useful data.

Figure 8.1-1 illustrates the noncoherent FHTL over one hop time when no data modulation exists on the hopped signal. The four integrators have to be reset at the end of each hop frequency.

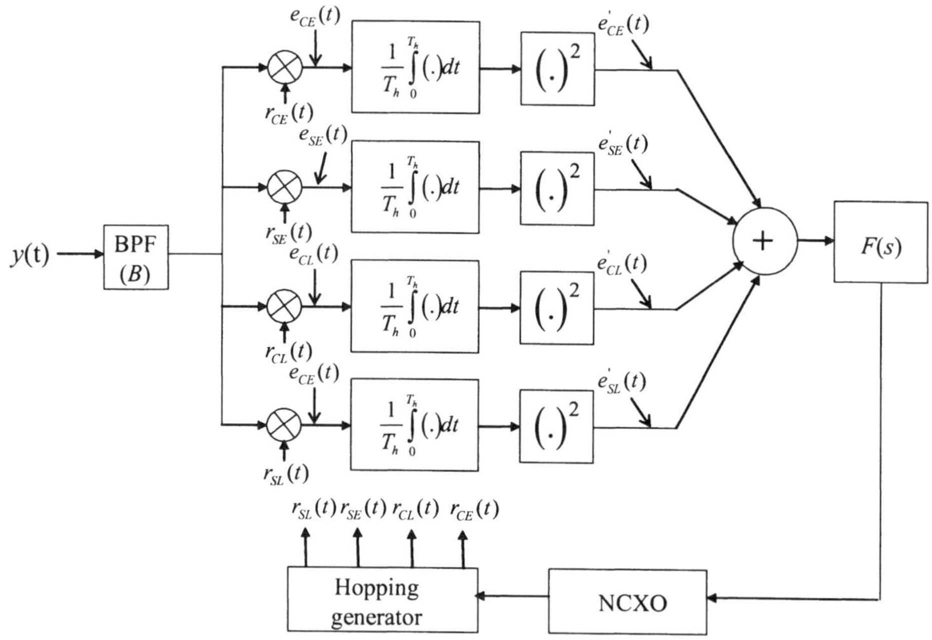

Figure 8.1-1 Noncoherent frequency-hopping tracking loop functional model.

In the figure the four reference signals are given by

$$r_{CE}(t)=\sqrt{2}\cos\left[\omega_0 t+\sum_{n=-\infty}^{\infty}\left(\omega_n t+\phi_n'\right)p_{T_h}(t-\hat{T}-nT_h+T_h/2\right] \tag{8.1-2}$$

$$r_{SE}(t)=\sqrt{2}\sin\left[\omega_0 t+\sum_{n=-\infty}^{\infty}\left(\omega_n t+\phi_n^{'}\right)p_{T_h}(t-\hat{T}-nT_h+T_h/2\right] \tag{8.1-3}$$

$$r_{CL}(t)=\sqrt{2}\cos\left[\omega_0 t+\sum_{n=-\infty}^{\infty}\left(\omega_n t+\phi_n^{'}\right)p_{T_h}(t-\hat{T}-nT_h-T_h/2\right] \tag{8.1-4}$$

$$r_{SL}(t)=\sqrt{2}\sin\left[\omega_0 t+\sum_{n=-\infty}^{\infty}\left(\omega_n t+\phi_n^{'}\right)p_{T_h}(t-\hat{T}-nT_h-T_h/2\right] \tag{8.1-5}$$

where the early (*E*) components have the arguments located early in time, the late (*L*) components are placed late in time, and the (*C*) and (*S*) denote cosine and sine, respectively. In addition $\phi_n^{'}$ is the reference phase, modeled as a random phase, and is unrelated to ϕ_n, by assumption. Finally $\hat{T}$ is the FHTL's estimate of the propagation time delay between the transmitter and the receiver, and is obtained from the initial acquisition process. These reference signals beat the incoming signal down to baseband where it is noncoherently processed.

It is to be noticed that the correlators with the second subscript *E* are both early by one-half of a hop, and the ones with the second subscript *L* are late by one-half of a hop. Thus, the in-phase and quadrature channels (sine and cosine) have a pair of early and a pair of late correlators that generate the error signal. The resultant signals out of the multipliers are given by

$$e_{CE}(t)=\sqrt{P}\cos\left[\sum_{n=-\infty}^{\infty}\left(\overbrace{\left(\omega_n t+\phi_n\right)p_{T_h}(t-T-nT_h)}^{\Phi(t)}-\overbrace{\sum_{n=-\infty}^{\infty}\left(\omega_n t+\phi_n^{'}\right)p_{T_h}(t-\hat{T}-nT_h+T_h/2)}^{\Phi^+(t)}\right)\right]$$
$$+n_c(t)\sin(\Phi^+(t))-n_s(t)\cos(\Phi^+(t))+(2\omega_0) \tag{8.1-6}$$

$$e_{SE}(t)=\sqrt{P}\sin\left[\sum_{n=-\infty}^{\infty}\left(\left(\omega_n t+\phi_n\right)p_{T_h}(t-T-nT_h)-\sum_{n=-\infty}^{\infty}\left(\omega_n t+\phi_n^{'}\right)p_{T_h}(t-\hat{T}-nT_h+T_h/2)\right)\right]$$
$$+n_c(t)\sin(\Phi^+(t))+n_s(t)\cos(\Phi^+(t))+O(2\omega_0) \tag{8.1-7}$$

$$e_{CL}(t)=\sqrt{P}\cos\left[\sum_{n=-\infty}^{\infty}\left(\left(\omega_n t+\phi_n\right)p_{T_h}(t-T-nT_h)-\overbrace{\sum_{n=-\infty}^{\infty}\left(\omega_n t+\phi_n^{'}\right)p_{T_h}(t-\hat{T}-nT_h-T_h/2)}^{\Phi^-(t)}\right)\right]$$
$$+n_c(t)\sin(\Phi^-(t))-n_s(t)\cos(\Phi^-(t))+O(2\omega_0) \tag{8.1-8}$$

$$e_{SE}(t)=\sqrt{P}\sin\left[\sum_{n=-\infty}^{\infty}\left(\left(\omega_n t+\phi_n\right)p_{T_h}(t-T-nT_h)-\sum_{n=-\infty}^{\infty}\left(\omega_n t+\phi_n^{'}\right)p_{T_h}(t-\hat{T}-nT_h-T_h/2)\right)\right]$$
$$+n_c(t)\sin(\Phi^-(t))+n_s(t)\cos(\Phi^-(t))+O(2\omega_0) \tag{8.1-9}$$

Note the three phase terms $\Phi(t),\Phi^{+}(t)$, and $\Phi^{-}(t)$ are defined in (8.1-6) and (8.1-8).

Consider the autocorrelation function of each noise term

$$n_{CE}(t)\triangleq n_c(t)\cos\left(\Phi^{+}(t)\right)-n_s(t)\sin\left(\Phi^{+}(t)\right) \tag{8.1-10}$$

so that

$$R_{n_{CE}}(\tau)=E\begin{bmatrix}\left(n_c(t)\cos\left(\Phi^{+}(t)\right)-n_s(t)\sin\left(\Phi^{+}(t)\right)\right)\\ \times\left(n_c(t)\cos\left(\Phi^{+}(t+\tau)\right)-n_s(t)\sin\left(\Phi^{+}(t+\tau)\right)\right)\end{bmatrix} \tag{8.1-11}$$

Simplifying the above equation leads to

$$R_{n_{CE}}(\tau)=\frac{N_0}{2}\delta(\tau)\cos^2\left(\Phi^{+}(t)\right)+\frac{N_0}{2}\delta(\tau)\sin^2\left(\Phi^{+}(t)\right)=\frac{N_0}{2}\delta(\tau) \tag{8.1-12}$$

In a similar manner it can be shown that

$$R_{n_{CE}}(\tau)=R_{n_{SE}}(\tau)=R_{n_{CL}}(\tau)=R_{n_{SL}}(\tau)=\frac{N_0}{2}\delta(\tau) \tag{8.1-13}$$

Now the early inphase and quadrature phase noise terms are statistically independent, it is also true that the early and late signals are statistically independent since the early and late hops do not overlap when the early late spacing is one hop. Thus all the noise terms are statistically independent of each other. The signals denoted $e_{CE}(t)$ and the corresponding other four can be written as

$$e_{CE}(t)=\sqrt{P}\cos\left[\Phi(t)-\Phi^{+}(t)\right]+n_{CE}(t) \tag{8.1-14}$$

$$e_{SE}(t)=\sqrt{P}\sin\left[\Phi(t)-\Phi^{+}(t)\right]+n_{SE}(t) \tag{8.1-15}$$

$$e_{CL}(t)=\sqrt{P}\cos\left[\Phi(t)-\Phi^{-}(t)\right]+n_{CL}(t) \tag{8.1-16}$$

$$e_{SL}(t)=\sqrt{P}\sin\left[\Phi(t)-\Phi^{-}(t)\right]+n_{SL}(t) \tag{8.1-17}$$

Since the multiplier outputs will produce a signal component that is offset in frequency (usually out of band) when the time difference between the actual delay T and the estimated delay $\hat{T}$ is not zero, we can approximate the correlator outputs by the following approximations with the assumption that signal component is out of band when the timing error is not zero. Thus

$$e_{CE}(t)\cong\sqrt{P}\sum_{n=-\infty}^{\infty}p_{T_h}(t-T_h-nT_h)p_{T_h}(t-\hat{T}-nT_h+T_h/2)\cos\left(\phi_n-\phi_n^{'}\right)+n_{CE}(t) \tag{8.1-18}$$

$$e_{SE}(t)\cong-\sqrt{P}\sum_{n=-\infty}^{\infty}p_{T_h}(t-T_h-nT_h)p_{T_h}(t-\hat{T}-nT_h+T_h/2)\sin\left(\phi_n-\phi_n^{'}\right)+n_{SE}(t) \tag{8.1-19}$$

$$e_{CL}(t) \cong \sqrt{P}\sum_{n=-\infty}^{\infty} p_{T_h}(t-T_h-nT_h)p_{T_h}(t-\hat{T}-nT_h-T_h/2)\cos\left(\phi_n-\phi_n^{'}\right)+n_{CL}(t) \quad (8.1\text{-}20)$$

$$e_{SL}(t) \cong -\sqrt{P}\sum_{n=-\infty}^{\infty} p_{T_h}(t-T_h-nT_h)p_{T_h}(t-\hat{T}-nT_h-T_h/2)\sin\left(\phi_n-\phi_n^{'}\right)+n_{SL}(t) \quad (8.1\text{-}21)$$

Basically these equations are based on the assumption that if the input and reference signals don't overlap in time, they are assumed to be zero. Each reference signal, when in lock, will correlate over about one-half of the hop duration with the same frequency and one-half the time with a previous or subsequent frequency. Thus it is assumed here that the "wrong frequency" doesn't correlate. This is a reasonable assumption when the frequency difference between the two is significant. If, however, the early or late frequency is very close in frequency, then some correlation output will occur. This effect, which should be small, is neglected in what follows. These approximations are actually more justifiable out of the correlator, since then different frequencies would produce a very small output from the correlators, as the correlation would filter out the difference frequencies.

Now consider the output from the correlators with the previous approximations. Over one hop time we have

$$e_{CE}^{'}(T_h)=\sqrt{P}R\left(\varepsilon+T_h/2\right)\cos(\Delta\phi_n)+\frac{1}{T_h}\int_0^{T_h} n_{CE}(t)dt \quad (8.1\text{-}22)$$

$$e_{SE}^{'}(T_h)=-\sqrt{P}R\left(\varepsilon+T_h/2\right)\sin(\Delta\phi_n)+\frac{1}{T_h}\int_0^{T_h} n_{SE}(t)dt \quad (8.1\text{-}23)$$

$$e_{CL}^{'}(T_h)=\sqrt{P}R\left(\varepsilon+T_h/2\right)\cos(\Delta\phi_n)+\frac{1}{T_h}\int_0^{T_h} n_{CL}(t)dt \quad (8.1\text{-}24)$$

$$e_{SL}^{'}(T_h)=-\sqrt{P}R\left(\varepsilon+T_h/2\right)\sin(\Delta\phi_n)+\frac{1}{T_h}\int_0^{T_h} n_{SL}(t)dt \quad (8.1\text{-}25)$$

where $\Delta\phi_n=\phi_n-\phi_n^{'}$ and the tracking error is $\varepsilon=T-\hat{T}$. Equations (8.1-22)–(8.1-25) can be written more succinctly as

$$e_{CE}^{'}(T_h)=\sqrt{P}R\left(\varepsilon+T_h/2\right)\cos(\Delta\phi_n)+N_{CE}(T_h) \quad (8.1\text{-}26)$$

$$e_{SE}^{'}(T_h)=-\sqrt{P}R\left(\varepsilon+T_h/2\right)\sin(\Delta\phi_n)+N_{SE}(T_h) \quad (8.1\text{-}27)$$

$$e_{CL}^{'}(T_h)=\sqrt{P}R\left(\varepsilon+T_h/2\right)\cos(\Delta\phi_n)+N_{CL}(T_h) \quad (8.1\text{-}28)$$

$$e_{SL}^{'}(T_h)=-\sqrt{P}R\left(\varepsilon+T_h/2\right)\sin(\Delta\phi_n)+N_{SL}(T_h) \quad (8.1\text{-}29)$$

where the integrated noise terms are defined over the first hop for convenience, by

$$N_{CE}(T_h) = \frac{1}{T_h}\int_0^{T_h} n_{CE}(t)dt \tag{8.1-30}$$

$$N_{SE}(T_h) = \frac{1}{T_h}\int_0^{T_h} n_{SE}(t)dt \tag{8.1-31}$$

$$N_{CL}(T_h) = \frac{1}{T_h}\int_0^{T_h} n_{CL}(t)dt \tag{8.1-32}$$

$$N_{SL}(T_h) = \frac{1}{T_h}\int_0^{T_h} n_{SL}(t)dt \tag{8.1-33}$$

Now the error correction signal is given by

$$e(t) = \left(e'_{CL}(T_h)\right)^2 + \left(e'_{SL}(T_h)\right)^2 - \left(e'_{CE}(T_h)\right)^2 - \left(e'_{SE}(T_h)\right)^2 \tag{8.1-34}$$

Using (8.1-26)–(8.1-29) in (8.1-34) produces the result

$$\begin{aligned} e(t) &= P\left[R(\varepsilon - T_h/2)^2 - R(\varepsilon + T_h/2)^2\right] - N_{CE}^2(T_h) - N_{SE}^2(T_h) + N_{CL}^2(T_h) + N_{SL}^2(T_h) \\ &+ 2\sqrt{P}\left[N_{CE}(T_h)R(\varepsilon + T_h/2)\cos(\Delta\phi_n) - N_{CL}(T_h)R(\varepsilon - T_h/2)\cos(\Delta\phi_n)\right] \\ &- 2\sqrt{P}\left[N_{SE}(T_h)R(\varepsilon + T_h/2)\sin(\Delta\phi_n) - N_{SL}(T_h)R(\varepsilon - T_h/2)\sin(\Delta\phi_n)\right] \end{aligned} \tag{8.1-35}$$

By assuming that the adjacent hop frequencies are uncorrelated in the hop time correlator, we can describe the normalized autocorrelation function as

$$\begin{aligned} R(\varepsilon) &= 1 - |\varepsilon| \; for \; |\varepsilon| \le 1 \\ R(\varepsilon) &= 0 \; for \; |\varepsilon| > 1 \end{aligned} \tag{8.1-36}$$

Thus the normalized autocorrelation function and the S-curve (tracking function curve) are shown in Figure 8.1-2.

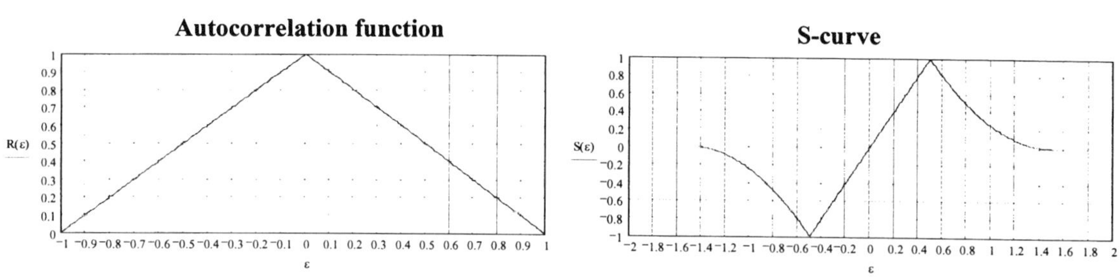

Figure 8.1-2 The autocorrelation function and the S-curve for the frequency-hopping tracking loop.

Within the linear region, (8.1-35) can be written as

$$e(t) = 2P\varepsilon + N_1(T_h) + N_2(T_h) \tag{8.1-37}$$

since the slope of the S-curve is 2, as can be seen from the S-curve in Figure 8.1-2. The two noise terms of (8.1-37) are defined by

$$\begin{aligned} N_1(T_h) &= 2\sqrt{P}\left[N_{CE}(T_h)R(\varepsilon + T_h/2)\cos(\Delta\phi_n) - N_{CL}(T_h)R(\varepsilon - T_h/2)\cos(\Delta\phi_n)\right] \\ &\quad - 2\sqrt{P}\left[N_{SE}(T_h)R(\varepsilon + T_h/2)\sin(\Delta\phi_n) - N_{SL}(T_h)R(\varepsilon - T_h/2)\sin(\Delta\phi_n)\right] \end{aligned} \tag{8.1-38}$$

and

$$N_2(T_h) = N_{CL}^2(T_h) + N_{SL}^2(T_h) - N_{CE}^2(T_h) - N_{SE}^2(T_h) \tag{8.1-39}$$

Consider the variance of the individual noise terms.

$$Var\{N_{CL}(T_h)\} = E\{N_{CL}^2(T_h)\} = E\left\{\frac{1}{T_h^2}\int_0^{T_h}\int_0^{T_h} n_{CL}(t)n_{CL}(u)dtdu\right\} = \frac{N_0}{2T_h} \tag{8.1-40}$$

Furthermore, it can be shown that

$$Var\{N_{CL}(T_h)\} = Var\{N_{SL}(T_h)\} = Var\{N_{CE}(T_h)\} = Var\{N_{SE}(T_h)\} = \frac{N_0}{2T_h} \tag{8.1-41}$$

Recall that we indicated that all four of the noise terms are mutually statistically independent. We can rewrite the error signal as for the time between 0 and T_h seconds as

$$e(t) = 2P\varepsilon + (N_{1i} + N_{2i})p_{T_h}(t) \quad 0 \le t \le T_h \tag{8.1-42}$$

in the linear region ($|\varepsilon| \le T_h/2$). Now the noise samples are defined, as before

$$N_{1i} = 2\sqrt{P}\left[N_{CEi}R(\varepsilon + T_h/2) - N_{CLi}R(\varepsilon - T_h/2)\right] \tag{8.1-43}$$

and

$$N_{2i} = N_{CLi}^2 + N_{SLi}^2 - N_{CEi}^2 - N_{SEi}^2 \tag{8.1-44}$$

at the i-th sample.

8.1.1 Loop Model for the Frequency-Hopping Loop Without Data

Now consider writing the loop equation for the frequency-hopping loop. The delay estimate $\hat{T}$ is formed from the loop filter as

$$\hat{T} = \frac{K_v}{s}F(s)e(t) \tag{8.1-45}$$

Since $\hat{T}$ is related to the error signal,[1] we have, in the linear region for the *i*-th hop,

$$T - \varepsilon_i = \frac{K_v}{s} F(s)\left[2P\varepsilon_i + N_i\right] \tag{8.1-46}$$

where $N_i = N_{1i} + N_{2i}$. Solving for ε_i, we obtain

$$\varepsilon_i = \frac{T_i}{1 + \dfrac{2PK_v}{s}F(s)} - \frac{\dfrac{2PK_v}{s}F(s)N_i}{\left(1 + \dfrac{2PK_v}{s}F(s)\right)2P} \tag{8.1-47}$$

Letting *H*(*s*) be the closed loop transfer function, it is defined as

$$H(s) = \frac{\dfrac{2PK_v}{s}F(s)}{1 + \dfrac{2PK_v}{s}F(s)} \tag{8.1-48}$$

So that the error signal can be written as

$$\varepsilon_i = H(s)T_i - (1 - H(s))\frac{N_i}{2P} \tag{8.1-49}$$

Since we are only concerned with the effect of the thermal noise, we can write ($T_i = 0$)

$$\varepsilon_i = -H(s)\left(\frac{N_i}{2P}\right) \tag{8.1-50}$$

It follows that the tracking error variance is given by

$$\sigma_\varepsilon^2 = \int_{-\infty}^{\infty} |H(f)|^2 \frac{S_N(f)}{4P^2} df \cong \frac{2B_L S_N(0)}{4P^2} \tag{8.1-51}$$

where B_L is the closed loop noise bandwidth of the frequency-hopping tracking loop. Now it is necessary to evaluate the noise spectral density at $f = 0$ in order to obtain the variance of the tracking error.

8.1.2 Evaluation of the Spectral Density of the Noise Terms

The integrate and dump process produces a step-wise process error control process that is known to have a noise process autocorrelation of form

---

[1] In reality $\hat{T}$ should be written as $\hat{T}(t)$, but we don't, for notational simplicity. This is true for other variables also.

$$R_{N_i}(\tau) = E\{N_i^2\}\left[1 - \frac{|\tau|}{T_h}\right]$$
$$R_{N_i}(\tau) = 0 \tag{8.1-52}$$

The corresponding power spectral density of the noise, is given by the Fourier transform of the autocorrelation function, so that

$$S_{N_i}(f) = \left[\text{Var}(N_{1i}) + \text{Var}(N_{2i})\right]T_h \frac{\sin(\pi f T_h)^2}{(\pi f T_h)^2} \tag{8.1-53}$$

Clearly from (8.1-53), since the sinc($\pi f T_h$) function is unity at $f = 0$, we have

$$S_{N_i}(0) = \left[\text{Var}(N_{1i}) + \text{Var}(N_{2i})\right]T_h \tag{8.1-54}$$

Therefore it is necessary to evaluate each variance term to evaluate the noise spectral density at $f = 0$. First consider N_{1i}. We have

$$\text{Var}(N_{1i}) = 4PE\left\{\begin{matrix}(N_{CEi}R(\varepsilon + T_h/2) - N_{CLi}R(\varepsilon - T_h/2)) \\ \times(N_{CEi}R(\varepsilon + T_h/2) - N_{CLi}R(\varepsilon - T_h/2))\end{matrix}\right\} \tag{8.1-55}$$

Since the early and late noise terms are statistically independent, we can simplify the result to

$$\text{Var}(N_{1i}) = 4P\left[\left(\text{Var}(N_{CEi})R^2(\varepsilon + T_h/2) + \text{Var}(N_{CLi})R^2(\varepsilon - T_h/2)\right)\right] \tag{8.1-56}$$

To make the analysis tractable we will approximate the correlation functions for the noise at $\varepsilon = 0$; this leads to the result using (8.1-41)

$$\text{Var}(N_{1i}) = 4P\left[\frac{N_0}{8T_h} + \frac{N_0}{8T_h}\right] = \frac{N_0 P}{T_h} \tag{8.1-57}$$

Now consider the evaluation of the second noise term. Consider the variance of N_{2i}. One has

$$\text{Var}(N_{2i}) = E\left\{\left(N_{CLi}^2 + N_{SLi}^2 - N_{CEi}^2 - N_{CLi}^2\right)\left(N_{CLi}^2 + N_{SLi}^2 - N_{CEi}^2 - N_{CLi}^2\right)\right\} \tag{8.1-58}$$

This can be written as four terms thusly

$$\text{Var}(N_{2i}) = T_1 + T_2 + T_3 + T_4 \tag{8.1-59}$$

where

$$T_1 = E\{N_{SLi}^4\} + E\{N_{SLi}^2 N_{CLi}^2\} - E\{N_{SLi}^2 N_{SEi}^2\} - E\{N_{SLi}^2 N_{CEi}^2\} \tag{8.1-60}$$

$$T_2 = E\{N_{SLi}^2 N_{CLi}^2\} + E\{N_{CLi}^4\} - E\{N_{CLi}^2 N_{SEi}^2\} - E\{N_{CLi}^2 N_{CEi}^2\} \tag{8.1-61}$$

$$T_3 = -E\left\{N_{SEi}^2 N_{SLi}^2\right\} - E\left\{N_{SEi}^2 N_{CLi}^2\right\} + E\left\{N_{SEi}^4\right\} + E\left\{N_{SEi}^2 N_{CEi}^2\right\} \tag{8.1-62}$$

$$T_4 = -E\left\{N_{CEi}^2 N_{SLi}^2\right\} - E\left\{N_{CEi}^2 N_{CLi}^2\right\} + E\left\{N_{CEi}^2 N_{SEi}^2\right\} + E\left\{N_{CEi}^4\right\} \tag{8.1-63}$$

All the T_i yield the same value, so let us consider just T_1. We have

$$T_1 = 3\left(\frac{N_0}{2T_h}\right)^2 + \left(\frac{N_0}{2T_h}\right)^2 - \left(\frac{N_0}{2T_h}\right)^2 - \left(\frac{N_0}{2T_h}\right)^2 = 2\left(\frac{N_0}{2T_h}\right)^2 = \frac{N_0^2}{2T_h^2} \tag{8.1-64}$$

It follows that

$$Var\left(N_{2i}\right) = 4T_1 = \frac{2N_0^2}{T_h^2} \tag{8.1-65}$$

8.1.3 Closed Loop Tracking Loop Performance

In Problem 1 it is shown that the tracking error variance for the frequency-hopping tracking loop, based on the previous calculations, is given by

$$\sigma_\varepsilon^2 = \frac{N_0 B_L}{2P}\left[1 + \frac{2N_0}{PT_h}\right] \text{ (hopping symbols)}^2 \tag{8.1-66}$$

Thus we see that the tracking error variance depends on product of the inverse loop SNR (N_0B_L/P) and a term that is the sum of one plus the inverse hop SNR (N_0/PT_h). Thus, this signal would allow accuracies on the order of a small fraction of a hop duration.

Example 1 Consider a simple example for the tracking standard deviation when $B_L = 2$ Hz and $T_h = 1$ ms, and plot it as a function of the C/N_0 (P/N_0) ratio. Figure 8.1-3 illustrates the tracking standard deviation under these conditions, as a function of the C/N_0 (P/N_0) ratio. Clearly, the greater the C/N_0 ratio (dB-Hz), the smaller the standard deviation of the tracking error.

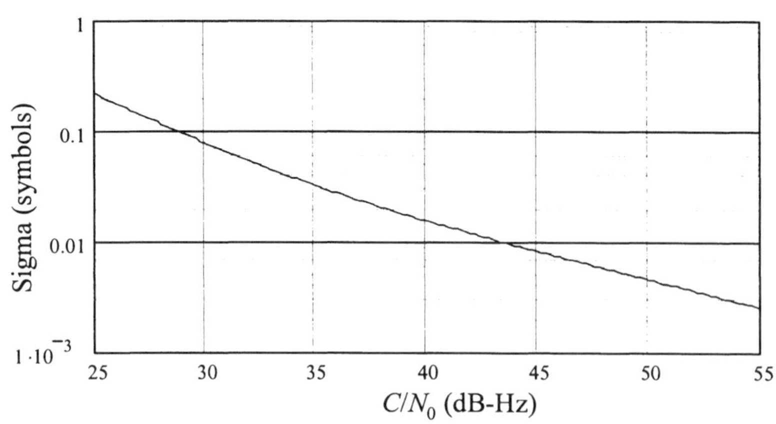

Figure 8.1-3 Frequency-hopping tracking one sigma error as a function of the C/N_0 ratio expressed in dB-Hz.

Based on this model the standard deviation of tracking error is about 0.01 hops at C/N_0 of 43.5 dB-Hz. This corresponds to an rms time tracking error of about 10 microseconds ($0.01 \times 1/1{,}000 = 10^{-5}$).

8.2 FREQUENCY-HOPPING TRACKING WITH BPSK AND DPSK DATA MODULATION

So far we have not discussed data modulation on the frequency-hopping signal. In order to carry data in a useful manner on the frequency-hopped signal, it is desirable to have the hopping frequency synthesizer be frequency synchronous with the data rate. Let the number of data symbols per hop be N. Then the data can be modulated on each frequency hop such that an integer number of N data symbols per hop exist.

This noncoherent hopping tracking loop can be accomplished with the understanding that the correlation time for frequency-hopping tracking must correspond to the data symbol time period, rather than the hop period. This may provide somewhat poorer tracking performance, as measured in symbols^2, than the unmodulated case, since the energy per hop is normally N times better than the energy per data symbol. However, in absolute time (seconds) the data tracker would be expected to be much better (on the order of roughly $1/N$ as large as the frequency-hopping tracking case). Figure 8.2-1 illustrates the frequency-hopping tracking model when BPSK or DPSK data is modulated on the hop frequencies.

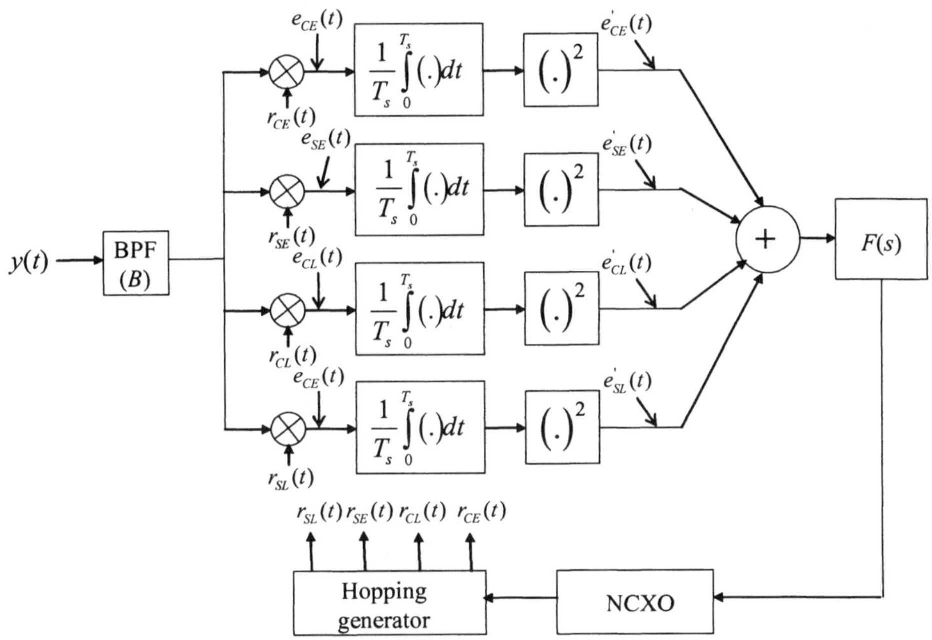

Figure 8.2-1 The tracking loop model for a frequency-hopped signal with BPSK or DPSK data modulation.

Here the reference signals are the same, but the integration is only over the code bit time.

Again it is assumed that the integrators are reset after every coded bit time. If we assume that the time to change any frequency of the frequency-hopper to any other frequency is small compared to a data symbol duration, then the tracking error variance can be obtained for BPSK of DPSK modulation with the understanding that the correlation (integration) is over a coded symbol time, rather than a hop time. The tracking error variance, in the linear region, follows from (8.1-66) and is given by

$$\sigma_\varepsilon^2 = \frac{N_0 B_L}{2P}\left[1 + \frac{2N_0}{PT_s}\right] = \frac{N_0 B_L}{2P}\left[1 + \frac{2NN_0}{PT_h}\right] \quad \text{(data symbols)}^2 \qquad (8.2\text{-}1)$$

where T_s is the symbol duration, in seconds. The loop bandwidth has to be picked so that the bandwidth is large enough to track any expected dynamical disturbances, yet be narrow enough to limit the noise influence on the tracking error. Conversion to time is accomplished by the following equation

$$\sigma_t = \sqrt{\frac{N_0 B_L}{2P}\left[1+\frac{2N_0}{PT_s}\right]}(T_s) \quad \text{(s)} \tag{8.2-2}$$

Further references on the subject of frequency-hopped signal tracking are cited in [1–5].

It should be pointed out that the use of BPSK in frequency hop systems is now feasible in view of the ability of the digital synthesizers to maintain phase coherence from hop to hop.

8.3 SUMMARY

This short chapter has presented an analysis of a frequency-hopped tracking loop performance designed to time track a spread SS frequency-hopping signal that has no data. Frequency hop acquisition was discussed briefly and it was noted that it is quite similar to direct sequence acquisition. A model was presented for tracking the frequency-hopping signal and an analysis of the closed loop noise analysis was accomplished for the variance of the tracking error. The variance of the tracking error was derived and a plot of the tracking performance was presented in an example.

Then the case that the frequency-hopping waveform has either BPSK or DPSK modulation on it was considered. The tracking error performance (i.e., the variance of the tracking error) was obtained, based on the previous analysis of the unmodulated frequency-hopping tracking loop.

References

[1] C. Lo, E. Masry, and L. Milstein, "Performance of Fast Frequency-Hopped DBPSK Spread Spectrum Communication in AWGN," *IEEE Milcom 1990*, September 30–October 3, 1990.

[2] G. Cherubini and L. Milstein, "Performance of Both Hybrid and Frequency-Hopped Phase-Coherent Spread-Spectrum Systems Part II: An FH System," *IEEE Trans. on Communications*, Vol. 37, No. 6, June 1999.

[3] T. Saba, D. Park, and S. Mori, "A Modified Time-Shared Code Tracking Loop for Frequency-Hopping Systems," *1995 IEEE Global Telecommunications conf.* Vol. 3, November 13–17, 1995, pp. 1884–1888.

[4] T. Saba, Y. Sato, and S. Mori, "Performance of the Time-Shared Code Tracking Loop with Extended Tracking Range for Frequency Hopped Signals," *8th Symposium Personal Indoor and Mobile Radio Comm. Conference*, Vol. 2, September 1–4, 1997.

[5] R. E. Ziemer and R. L. Peterson, *Digital Communications and Spread Spectrum Systems*, New York: Macmillan Publishing Company, 1985.

Problem

1. Show that (8.1-66) follows from (8.1-51), (8.1-57), and (8.1-65).

CHAPTER 9

Multiple Access Methods for Digital Wireless Cellular Communications

9.0 INTRODUCTION

In this chapter we will present an introduction to multiple access techniques for digital wireless cellular communications [1–7]. The basic idea of multiple access techniques in general is the simultaneous usage of multiple signals over the same bandwidth. The sharing of the spectrum is necessary to achieve high capacity by simultaneously allocating the available bandwidth to multiple users. The must be done in such a way that each signal does not suffer significant degradation in performance by affecting one another.

9.1 BRIEF HISTORY OF CELLULAR SYSTEMS

The Advanced Mobile Phone Service (AMPS) was an analog cellular phone service that originated in Chicago in 1983. It was an immensely popular 1G (first generation) cellular system and was used in North and South America. Later the Electronic Industry Association Interim Standard IS-54 and later IS-136 provided the opportunity for some single-user analog providers to replace the analog channels (AMPS) with digital channels, which supported three users in the same 30-kHz bandwidth. IS-54, IS-136, IS-95, and Global System for Mobile (GSM) are all 2G (second generation) cellular systems.

A digital cellular system, based on code division multiple access (CDMA), was developed by Qualcomm Inc. and standardized by the Telecommunications Industry Association (TIA) as an Interim Standard (IS-95). It is a second generation (2G) system. This standard supports a variable number of users in a 1.25-MHz bandwidth, using direct sequence CDMA. Digital CDMA systems have been shown to operate at a lower power level, for the same bandwidth, than analog systems. In addition CDMA allows the use of multiple users over the same bandwidth. However IS-95 utilized a variable rate vocoder with voice activity detection, which reduces the required data rate by the cell phone transmitter. IS-95 is 2G. Second generation versions of CDMA include cdmaOne which includes IS-95, IS-95A, and IS-95B. Third generation CDMA systems are called 3G, and include cdma2000, WCDMA, and others. The 3G standards offer increased network capacity for wireless services and high-speed data services.

Another CDMA standard that is not interoperable with these standards is known as wideband CDMA (WCDMA). The world's first WCDMA service, Freedom of Mobile Multimedia Access (FOMA), was launched by NTT DoCoMo in Japan in 2001. It utilizes 5-MHz channels, whereas CDMA standards utilize a 1.25 MHz channel.

The Universal Mobile Telecommunication System (UMTS) is a 3G evolution of the 2G system GSM, which is a time division multiple access (TDMA) system. WCDMA networks are already being operated commercially in Austria, Italy, Sweden, and the U.K. with more to come. There is a 2.5G system based on General Packet Radio Service (GPRS), which is a natural stepping-stone toward UMTS.

Another CDMA-based 3G system is *time division synchronous code division multiple access* (TD-SCDMA). The design was based on the cooperation of Siemens and its Chinese partners. It is to be noted that China is the largest mobile phone market in the world. Via *time division duplex* (TDD), the UMTS transmission mode, traffic is sent and received over the same frequency band, but using different time slots.

The synchronous “S” signifies that TD-SCDMA can master both synchronous circuit-switched services (such as speech or video services) and asynchronous packet-switched services via an Internet access.

9.2 CELLULAR COMMUNICATIONS

A cellular communication system has the ability to connect to the public switched telephone network (PSTN) for any cellular user within the radio range of the mobile base station. A cellular system is capable of accommodating a large number of users with a reasonable bandwidth requirement.

Each mobile station utilizes a separate radio channel to talk to the cell site. The cell site allows communication of many mobile stations at the same time, using one channel per mobile station. Figure 9.2-1 illustrates the mobile links and the connection to the public PSTN.

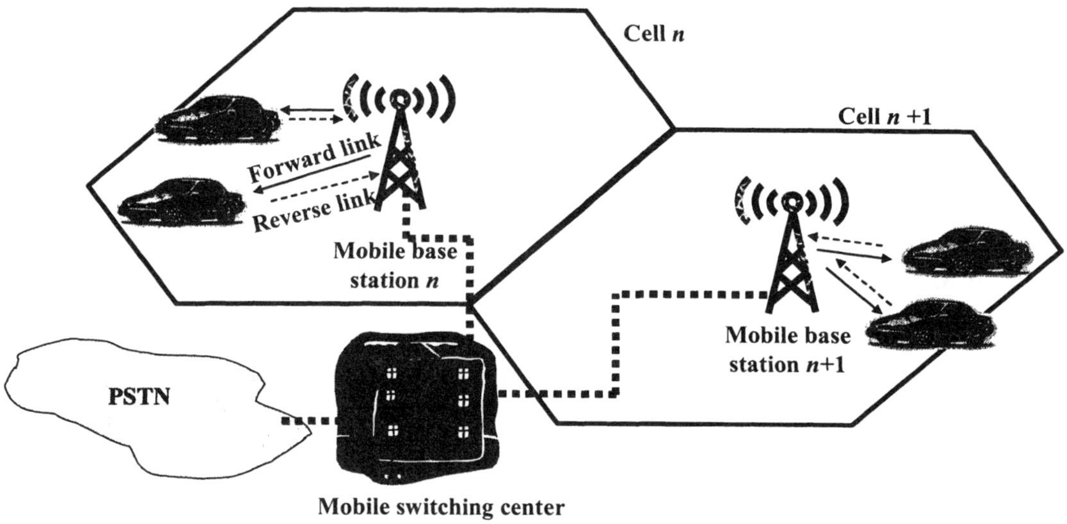

Figure 9.2-1 Basic mobile telephone service network.

The mobile units, or mobile stations (cars in the figure), communicate in both directions to the mobile base station in its cell location. The mobile base station communicates to the mobile switching center (MSC) which communicates to the copper wire telephone system that is the PSTN. The link from base station to the mobile user (mobile) is called the *forward link* or the downlink. The link from the mobile to the base station is called the *reverse link* or the uplink. These terms will be used interchangeably in what follows.

9.2.1 Cellular System Architecture

Increases in demand and poor quality of the then existing services led mobile service providers to obtain ways to improve the capacity of the systems and to increase the quality of service. The communication between the base station and the users is accomplished at two different frequencies, one for transmission and the other for reception.

9.2.2 Mobile Cells

A cell is the basic geographic element of a cellular system. The honey combed shape of the areas into which a coverage region is divided is termed “cellular.” A *cell* is a hexagon shape in which the base station is usually located near its center, and the base station communicates with the mobile stations within the cellular region. Each cell may vary somewhat in size, depending on the landscape. In reality because of the constraints imposed by the natural terrain as well as manmade structures, the actual shape of the cell is not exactly a hexagon. The actual shape of the cell can be determined by field measurements or propagation modeling.

9.2.3 Mobile Clusters

A *cluster* is a group of n cells. Channels are not reused within a cluster. Figure 9.2-2 illustrates a cluster of seven cells.

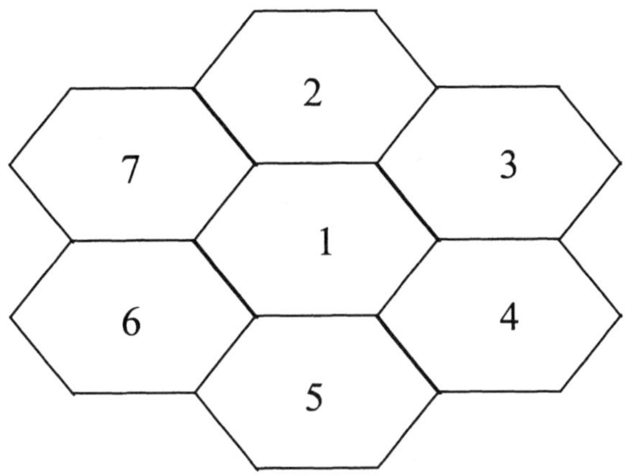

Figure 9.2-2 An example of a cluster of $n = 7$ cells.

9.2.4 Frequency Reuse in a Cellular System

Due to the fact that only a few radio channels were available for mobile systems, the designers had to develop a method to reuse the radio channels in order to increase the capacity of the system. The solution was frequency reuse. *Frequency reuse* was based on the assignment of each cell to a group of radio channels used within a small geographic area. Cells are assigned a group of radio channels that are different from all the adjacent cells. The coverage area of the cells is called the *footprint*. The footprint is roughly limited by the cell boundary, so that the same group of radio channels can be used in different cells that are sufficiently far away from each other, so that radio frequency interference does not occur. Figure 9.2-3 illustrates an example of frequency reuse with three clusters (bold lines).

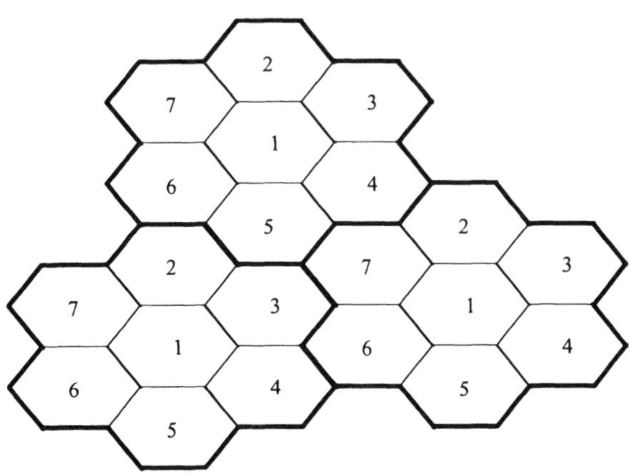

Figure 9.2-3 An example of frequency reuse.

Cells with the same cell number utilize the same set of frequencies. The *frequency reuse factor* is the rate at which the same frequency can be used in the network. It is equal to $1/n$ where n is the number of cells in a cluster. In the previous example the number of cells in a cluster is seven, and the frequency reuse factor is 1/7. In other words each cell is using 1/7 of the available channels.

9.2.5 Cell Splitting

Economic considerations have made the concept of creating systems that work well in heavy traffic areas as well as low traffic areas, but that do not work well because the traffic demand is not homogeneous. To overcome this problem system operators have developed the concept of *cell splitting*. This involves splitting a cell into smaller cells to accommodate the need for high traffic volume. In this manner urban areas can be broken into many small areas as is needed, in order to provide adequate performance in heavy traffic areas. Then low traffic areas (for example, rural areas) can be covered with less expensive larger cells. Figure 9.2-4 illustrates an example of cell splitting in which rural areas have large size cells and urban areas have smaller cells. In the figure the parallel lines denote the ocean and the dark curving line, the shore line. Clusters, small cells, and individual cells are shown in the example.

Another technique that is used to reduce interference is called sectoring. *Sectoring* is a procedure that utilizes, for example, three directive antennas at a cell site. Each antenna then covers about 120°. However sectors are not constrained to 120°. Thus sectoring allows frequency reuse within a cell, or more often, simply narrows the pattern of interference in adjacent cells. Figure 9.2-5 illustrates an example of sectoring one cell. It is seen that there are three sectors in the one cell, numbered 1, 2, and 3. There are actually three cell sizes. (1) *Macrocells* provide the main coverage in a cellular network. These macrocell base stations typically have tens of watts output power. (2) *Microcells* provide radio coverage in areas with a high number of users. Their output power is a few watts. (3) *Picocells* are normally found in buildings where coverage is poor or there are a very large number of users, such as in an airport terminal or shopping center.

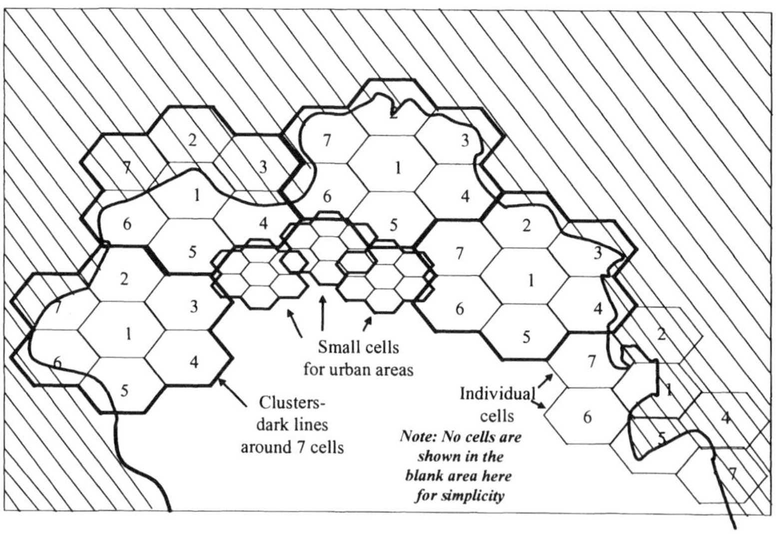

Figure 9.2-4 An example of cell splitting with the small cells used for urban areas.

9.2.6 Handoff

An obstacle in the development of the cellular network involves the problem created when a mobile subscriber (station) travels from one cell to another adjacent cell. Since adjacent cells do not use the same radio channels, a call must either be dropped or transferred from one radio channel to another one when a user crosses the line between adjacent cells. Due to the fact that dropping the call is unacceptable operationally, the process of handoff was developed. A *handoff* occurs when the mobile telephone network automatically transfers a call from one radio channel in the current cell to another radio channel as the mobile station crossed into the adjacent cell. A handoff is normally made when one base station in which the mobile station is entering supplies a stronger signal than that received from the current base station. In second generation systems the decision to make a handoff is decided in the mobile station (cell phone). First generation systems made the decision in the base station.

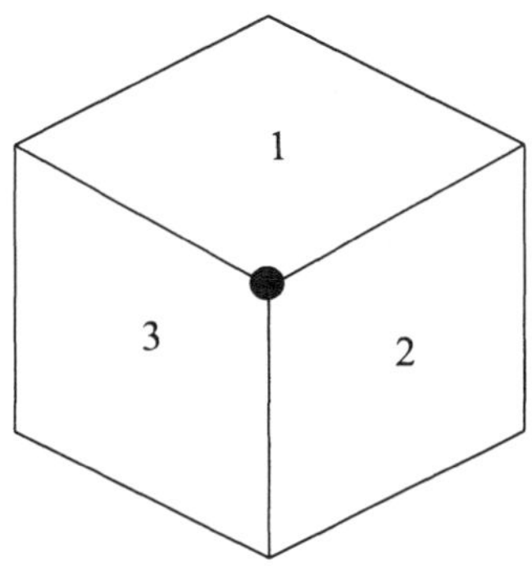

Figure 9.2-5 An example of cell sectoring from one cell to three sectors.

During a call, two parties are on one voice channel. When the mobile unit moves out of the coverage area of a given cell site, the reception becomes progressively weaker. At a preset point, the cell site in use requests a handoff. The system switches the call to a stronger frequency channel in a new cell site without interrupting the call or alerting the users. The call is thus not interrupted when changing cells as the mobile unit travels to its destination, and the users are unaware of the handoff.

In summary a cellular approach offers advantages and disadvantages: Advantages are that: (1) the required transmitter power is reduced, and (2) the capacity through frequency reuse is increased. Disadvantages are that: (1) the network has increased complexity, (2) handoffs must be performed, and (3) if the system is not well designed, cochannel interference can be a problem.

9.2.7 More on Cell Structure

As was noted earlier the cellular concept replaced a large area with a high power transmitter (large cell) with many low power transmitters, using a small area (small cell) with each cell providing coverage to only a small portion of the service area. Each base station is allocated a portion of the total number of channels available to the complete communication system, and the nearby base stations are assigned different groups of channels so that all of the available channels are assigned to a relatively small number of neighboring base stations. By systematically arranging the base stations and their channel groups throughout the market area, the allowed channels are distributed throughout the geographical area and they may be reused as many times as needed, so long as the interference between channels is maintained at permissible levels.

As the demand for more channels increases, the demand can be met by increasing the number of base stations with the decrease in the required transmitter power. This small cell size allows for increased capacity without requiring an increase in the radio spectrum.

Assume that the cellular system of interest has a total of n duplex channels available for use. If each cell is allocated a group of $k \leq n$ channels, and if the n channels are divided among N cells into unique and disjoint channel groups each of which have the same number of channels, the total number of available radio channels is simply

$$n = kN \tag{9.2-1}$$

We have already identified the N cells that collectively use the complete set of frequencies as a cluster. When the cluster is replicated M times within the system, the total number of duplex channels that are available is given by

$$C = MkN = Mn \tag{9.2-2}$$

and is a measure of capacity of the system. As seen from (9.2-2) the capacity of a mobile cellular system is directly proportional to the number of times a cluster is repeated. The factor N is the cluster size and is typically a value of 4, 7, or 12. It can be seen from (9.2-2) that if the cluster size N, is reduced while the cell size is kept constant, more clusters are required to cover the desired area, and therefore more capacity (greater value of C) is achieved. The *frequency reuse factor* is defined as $F_{fr} = 1/N$, because each cell within a given cluster is only assigned $1/N$ of the total available channels in the system.

Due to the hexagonal geometry only certain repeat patterns can *tessellate* (connect without gaps between adjacent cells) [8]. The number N of cells per cluster is given by [9]

$$N = i^2 + ij + j^2 \tag{9.2-3}$$

where i and j are nonnegative integers. The allowable values of N are given by [9]

$$N = 1,\ 3,\ 4, 7, 9, 12, 13, 16, 19, 21, 25... \tag{9.2-4}$$

In order to find the nearest cochannel neighbor one must do the following [1, 10]: (1) move i cells along any chain of hexagons and then (2) turn 60° *counterclockwise* and move j cells. It is to be noted that if the construction is altered slightly (the two lower clusters are rotated counterclockwise to the next position of tessellation) rule (2) must be changed to turn *clockwise* rather than counterclockwise. The former rule is illustrated in Figure 9.2-6 for the case that $N = 7$, based on $i = 2$ and $j = 1$.

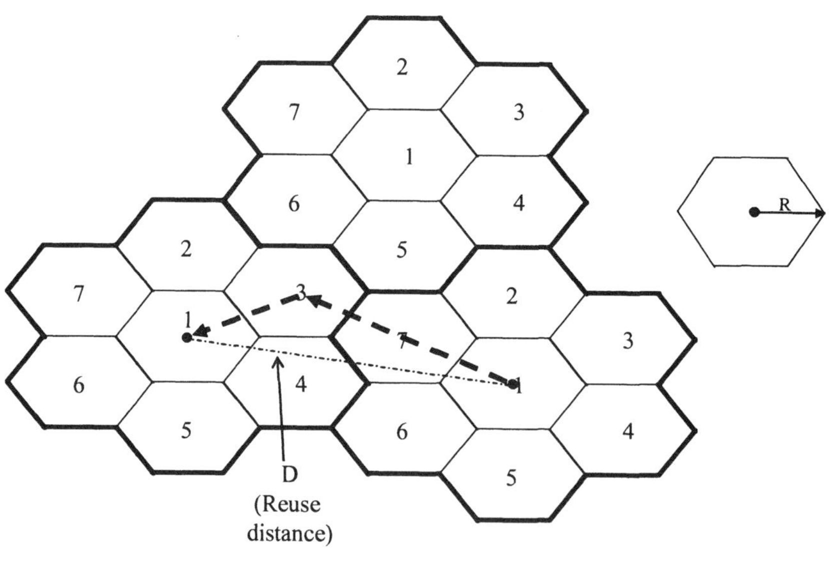

Figure 9.2-6 An example of locating the next cochannel neighbor (1), $N = 7$, $i = 2$, and $j = 1$, in this case.

The *reuse distance* is given by the relationship $D = \sqrt{3N}R$, where R is the *cell radius*.

Example 1 Compute the reuse distance for this example. Using $D = \sqrt{3N}R$ produces the result $D = 4.58 * R$.

Example 2 Assume that a frequency division duplexing (FDD) cellular system has 31 MHz available for telephone usage. Also assume that two 25-kHz channels are used to provide duplex operation. Determine the number of available channels per cell if the system uses (1) 7-cell reuse, (2) 9-cell reuse, and (3) 12-cell reuse. Determine the frequency reuse factor for each case. The FDD channel bandwidth is given by 2(25 kHz) = 50 kHz for each duplex channel. The total number of channels that are available is given by $31(10^6)/50 \cong 620$. For (1) $N = 7$ we have $N_{ch} = 620/7 \cong 88$ channels, and the reuse factor is $F_{fr} = 1/7 = 0.1428$. For (2) we have $N_{ch} = 620/9 \cong 68$ channels, and the reuse factor is $F_{fr} = 1/9 = 0.1111$. Finally for (3) we have $N_{ch} = 620/12 \cong 51$ channels, and the reuse factor is $F_{fr} = 1/12 = 0.0833$.

9.2.8 Assignment Strategies for Channelization

Any channel assignment strategy of the available radio spectrum must consider frequency reuse capacity needs and the minimization of interference when being designed. A number of channel strategies have been devised to achieve these objectives. The channel assignment can be either dynamic or fixed. The choice of the assignment can affect the performance of the system, particularly how the calls are managed in the case of handoff from one cell to another cell [1, 11–13].

In a dynamic channel assignment strategy, voice channels are allocated to difference cells as required; they are not permanent assignments. When a call request is made, the serving base station requests a channel from the MSC. The switch then allocates a channel to the requested cell based on an algorithm that accounts for the possibility of future blocking within a cell, the candidate cell frequency of use, the reuse distance of the cell, and other applicable costs functions.

In a fixed channel assignment strategy, each cell is allocated a predetermined set of voice channels. Any new call attempt with the cell can only be satisfied by the unused channels in that cell. When all the channels are in use, the call is *blocked* and the subscriber will not receive service immediately. There is one variant that allows for borrowing of unused channels in an adjacent cell when all of its channels are occupied. The MSC supervises the borrowing process and the strategy is called a *borrowing strategy*.

Dynamic channel assignment can reduce the chance of blocking, which increases the trunking capacity of the system.

In this section we have attempted to discuss the concepts but have not provided the detail needed to design the system. Numerous books are available that cover considerably more than is possible in this chapter, such as [1, 3, 8, 14].

9.3 MULTIPLE ACCESS TECHNIQUES FOR WIRELESS COMMUNICATIONS

In utilizing wireless communication systems, it is often desirable to allow the mobile units (subscribers) to send information to the base station and concurrently receive information from the base station. In conventional wired telephone systems it is possible to talk and listen at the same time. Communication in both directions simultaneously is known as *duplexing*.

Duplexing can be accomplished with frequency domain techniques or with time domain techniques. FDD provides two distinct frequency bands for every user. The *forward band* provides traffic from the base station to the mobile unit, and the *reverse band* allows traffic from the mobile unit to the base station.

Time division duplexing (TDD) utilizes time to provide a forward and a reverse link. In TDD multiple users each share time slots in the time domain. Each user is assigned one time slot for the forward data flow and one time slot for the reverse data flow.

Figure 9.3-1 illustrates a simple example of FDD and TDD. Both TDD and FDD have advantages and disadvantages. FDD fits well with radio communications systems that allocate individual radio frequencies for each user. Since both transmitter and receiver work simultaneously and there can be a very large difference in the transmitted and received signal power, the forward and reverse channels must be carefully chosen to avoid cross talk from the transmitted signal to the received signal. In addition careful assignment of channels must be employed in order to utilize inexpensive radio frequency components and oscillators.

TDD enables each transceiver (transmitter-receiver) to operate on the same frequency, thereby eliminating the need for separate forward and reverse frequency bands. Because of the very tight timing requirements needed to form precise time slots, TDD is normally limited to cordless phones or for short range portable access. TDD is also effective in a stationary environment in which the propagation delays are fixed.

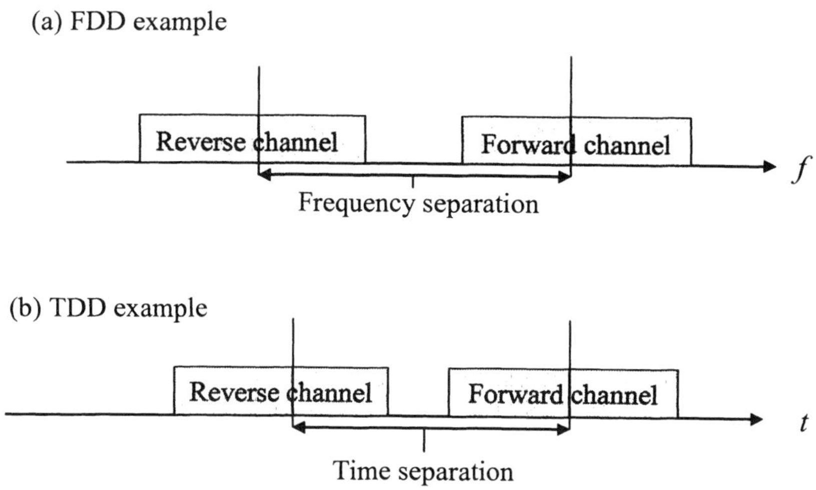

Figure 9.3-1 (a, b) Simplified FDM and TDD duplex channel.

9.3.1 A Brief Introduction to Multiple Access

A number of multiple access techniques have been used for wireless communications: frequency division multiple access (FDMA), time division multiple access (TDMA), and code division multiple access (CDMA). These systems are usually grouped into either narrowband or wideband wireless systems.

Narrowband systems relate to the bandwidth of a single channel to the expected coherence bandwidth of the channel [1]. In a narrowband multiple access system, the available radio frequency spectrum is broken up into a large number of narrowband channels. In this case the channels are usually operated using FDD.

In narrowband FDMA a user is assigned a particular channel, which is not shared by others users, in the near vicinity. If FDD is also used in this case it is known as FDMA/FDD.

In narrowband TDMA usually there are a large number of radio channels allocated with either FDD or TDD, and each channel is shared using TDMA. These systems are known as TDMA/FDD or TDMA/TDD.

Wideband systems have a transmission bandwidth of a single channel that is much larger than the coherence bandwidth of the channel [1]. In wideband multiple access systems, a large number of transmitters are permitted to transmit on the same channel. In contrast to narrowband systems, a wideband system sustaining multipath fading does not cause a large variation in the received power in the channel undergoing multipath fading. TDMA and CDMA systems could be found utilizing either FDD or TDD multiplexing techniques. Table 9.3-1 [1, 2] illustrates the multiple access and duplexing method in various multiple access systems.

Table 9.3-1 Multiple Access and Duplexing Techniques for Different Wireless Systems

Cellular System	Multiple Access and Duplex Technique
Global System for Mobile (GSM)	TDMA/FDD
Pacific Digital Cellular (PDC)	TDMA/FDD
US Digital Cellular (USDC)	TDMA/FDD
Advanced Mobile Phone System (AMPS)	FDMA/FDD
CT2 (Cordless telephone)	FDMA/TDD
Digital European Cordless Telephone (DECT)	FDMA/TDD
IS-95 (U.S. Narrowband Spectrum)	CDMA/FDD
W-CDMA (3G)	CDMA/FDD CDMA/TDD
cdma2000	CDMA/FDD CDMA/TDD

9.3.2 Frequency Division Multiple Access

Frequency division multiple access (FDMA) allocates individual channels to individual users. Each user is assigned a unique frequency band or channel. Users who request the channels are assigned them on a demand basis. The channel is given to that user during that call. When FDD duplexing is employed, the users are allocated as a pair of frequencies, one for the forward channel and one for the reverse channel. Figure 9.3-2 illustrates a conceptual view of FDMA for the case of four FDMA signals. It is seen that each channel is separated in frequency, with a guard band between each channel.

Now consider some salient features of FDMA [1]. (1) There is only one phone circuit on at one time. (2) When a channel is not in use, it cannot be used by other users to share the capacity. (3) When a channel is assigned, the mobile unit and the base station communicate continuously. (4) Normally FDMA channels are narrowband. (5) Since FDMA is on continuously, fewer overhead bits are needed for synchronization, compared to TDMA. (6) Partly due to the need for somewhat costly bandpass filters, the costs of the cell site system is higher than with TDMA. (7) Typically little or no equalization is needed in an FDMA system, since the symbol time duration is large compared to the average delay spread time.

One effect that FDMA users have to deal with is the phenomena of intermodulation frequencies (IM) that occur when many channels are amplified through a common high power amplifier that is operated near saturation, in order to maximize the efficiency of the link. If there are N frequencies in the nonlinear amplifier and the m_i are integers that satisfy the following

$$m_i \in (\ldots -3, -2, -1, 0, 1, 2, 3, \ldots) \tag{9.3-1}$$

then any signal frequency that is produced at the output of the nonlinear amplifier is of the form

$$f_{IM}(m_1, m_2, \ldots m_N) = \sum_{i=1}^{N} (m_i f_i) \tag{9.3-2}$$

for all possible values of the m_i.

Thus many frequencies are generated from the nonlinear amplifier, some of which fall in band, and some fall out of band, and some of which are very weak, while some are strong. Those that fall in band cause interference [3] and appear to be near noise-like when there are a fair number of signals feeding the amplifier.

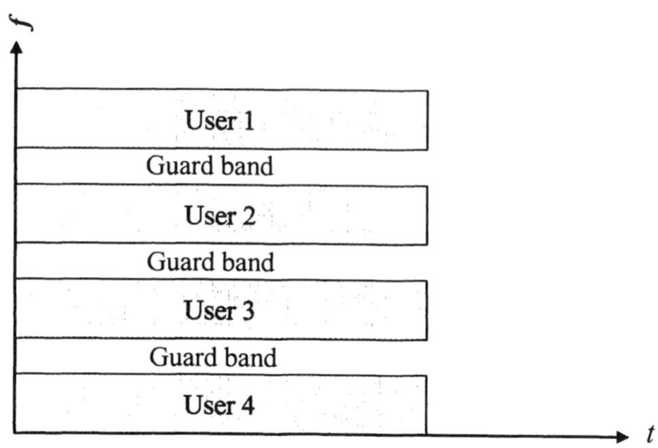

Figure 9.3-2 Conceptual view of a FDMA multiple access communication system.

9.4 TIME DIVISION MULTIPLE ACCESS

Time division multiple access (TDMA) systems partition the radio frequency spectrum into distinct time slots. In each slot, only one user is allowed to transmit or receive at a time. Figure 9.4-1 illustrates four TDMA users.

It is to be noted that each channel has access to the system via a cyclically repeating time slot. The same channel has access to the system via this time slot and has some "protection time" afforded by the guard time slots. TDMA systems transmit data in a *buffer-and-burst* method, and therefore the transmission for any user is noncontinuous. As a consequence of this fact TDMA, unlike FDMA, which accommodates analog modulation, must utilize digital data and digital modulation.

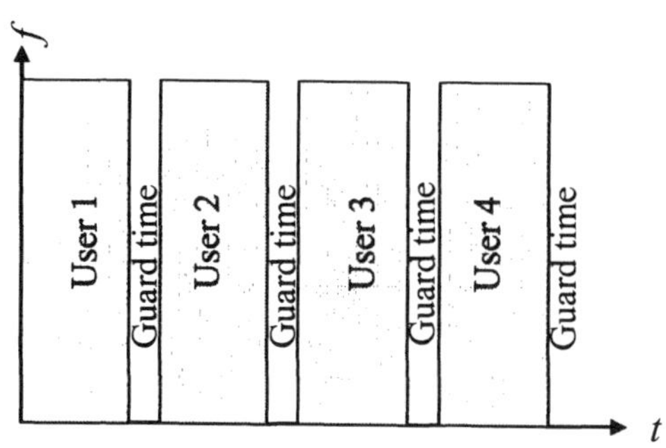

Figure 9.4-1 Conceptual view of a TDMA multiple access communication system.

The transmission from the various users is interlaced into a repeating frame structure as is seen in Figure 9.4-2. Each frame has M slots. The beginning of a frame is formed with a preamble sequence, then an information message, followed by the post amble sequence. The figure indicates the usual functional elements and they may be arranged in a different order. For the case of TDMA/TDD one half of the time slots in the information message would be used for the forward channels and the other half would be used for the reverse channels. However in TDMA/FDD systems, an identical or similar frame structure would be used only for the forward channels or only the reverse channels, but the carrier frequencies would be different for the forward and reverse links. In general TDMA/FDD systems design in several time slots of delay between the forward and reverse time slots for a particular user, in order to not need duplexers in the subscriber unit [1].

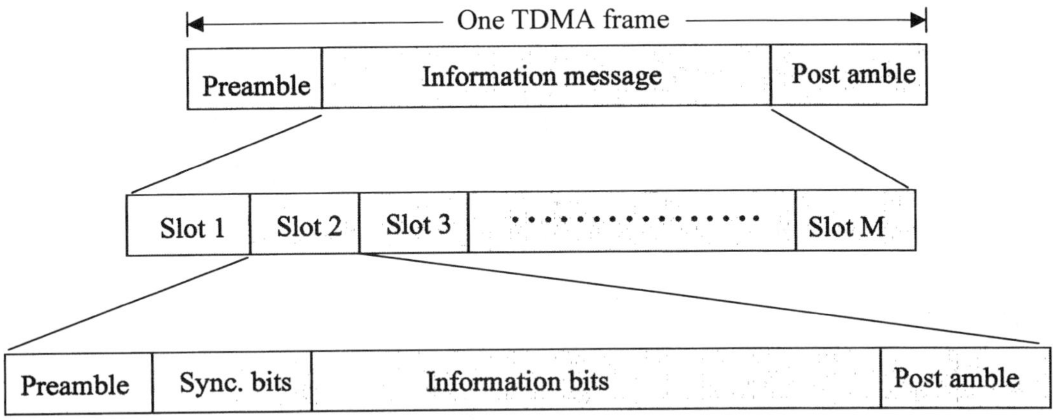

Figure 9.4-2 The TDMA repeating frame structure.

For a TDMA frame structure the preamble contains the synchronization and address information that both the base station and the subscribers use to identify each other. Each TDMA standard may have a somewhat different structure, but the concepts are very similar.

Now consider some salient features of TDMA [1]. (1) It shares several users on one frequency. (2) The data transmission of TDMA is not continuous, but occurs in a burst mode. The result of this is that battery drain of the subscriber unit (user) is smaller since it only needs to be transmitting or receiving during its own time slot. (3) The handoff process is much simpler for TDMA than FDMA since the subscriber unit can listen for other base stations during the slot idle (off) times. (4) The mobile assisted handoff (MAHO) can be carried out by a subscriber listening on an idle slot in the TDMA frame. (5) TDMA uses different time slots for transmission and reception so duplexers are not required. And if FDD is used, a switch rather than a duplexer inside the subscriber unit is sufficient to switch between transmitter and receiver. (6) Unlike FDMA, adaptive equalization is normally necessary in TDMA systems, since the transmission rates are normally very high when compared to FDMA channels rates. (7) A high level of overhead in synchronization bits is needed in TDMA. (8) A major advantage of TDMA is the fact that more than one time slot can be allocated to users, so that bandwidth can be supplied on demand as long as the system has capacity compatible with the need.

9.4.1 The Efficiency of TDMA Systems

The frame efficiency of TDMA is the ratio of the information bits time to the total frame time. If coding is employed the time associated with the data transmission determines the information time. Alternatively one can determine the time in the channel-transmitted bits. Thus if the number of channel bits per frame is given by the product

$$N_{cb} = T_f R_{cb} \tag{9.4-1}$$

where T_f is the frame duration in seconds and R_{cb} is the channel bit rate. The number of overhead channel bits per frame is given by [4]

$$N_{OH} = N_{ref} B_{ref} + N_{tr} B_{preamble} + N_{tr} B_g + N_{ref} B_g \tag{9.4-2}$$

where N_{ref} is the number of reference bursts per frame, N_{tr} is the number of traffic bursts per frame, B_{ref} is the number of overhead channel bits per channel burst, $B_{preamble}$ is the number of channel bits per preamble in each slot, and B_g is the number of equivalent channel bits per guard time slot. Using (9.4-1) and (9.4-2) yields the overall TDMA frame efficiency

$$\eta_{TDMA} = \left(1 - \frac{N_{OH}}{T_f R_{cb}}\right) \times 100\% \tag{9.4-3}$$

9.4.2 The Number of Available Channels in a TDMA System

The number of available TDMA channel slots that can be provided in a TDMA system is found from the equation [1]

$$N_{slots} = \left\lfloor \frac{m\left(BW_{total} - 2BW_{guard}\right)}{BW_{ch}} \right\rfloor \tag{9.4-4}$$

where m is the number of TDMA users supported on each radio channel, BW_{guard} is the guard bandwidth needed at each end, BW_{ch} is the channel bandwidth in Hz, BW_{total} is the total bandwidth in Hz and $\lfloor x \rfloor$ is the greatest integer less than or equal to x.

Example 3 Consider GSM, which is a TDMA/FDD system that employs a 25-MHz forward link. That link is composed of radio channels that have a 200-kHz bandwidth. There are eight speech channels on a single radio channel, and the guard band is 50 kHz. Determine the number of users that can be utilized on this model of GSM. The solution is given by

$$N_{slots} = \left\lfloor \frac{8\left(25\times 10^6 - 2\times 50\times 10^3\right)}{200\times 10^3} \right\rfloor = 996 \tag{9.4-5}$$

Example 4 If a GSM system frame has 8 time slots and contains 6 trail bits, 8.25 guard bits, 26 sync bits, and two groups of 58 data bits each, determine the frame efficiency [1]. The solution is given by first determining the time slot duration of

$$NB_{slot} = 6 + 8.25 + 26 + 2(58) = 156.25 \text{ channel bits} \tag{9.4-6}$$

A frame has therefore

$$NB_{frame} = 8(156.25) = 1{,}250 \text{ bits} \tag{9.4-7}$$

Now the number of overhead bits per frame is given by

$$N_{OH} = 8(6) + 8(8.25) + 8(26) = 322 \text{ channel bits} \tag{9.4-8}$$

It follows from (9.4-3) that the frame efficiency is given by

$$\eta_{frame} = \left(1 - \frac{322}{1250}\right) = 74.24\% \tag{9.4-9}$$

9.5 SPREAD SPECTRUM MULTIPLE ACCESS

Spread spectrum multiple access (SSMA) is a communication method in which the transmission bandwidth is much larger than the minimum bandwidth required to transmit the data rate used in the communication link. A pseudonoise (PN) sequence that runs at a rate much greater than the bit rate modulates the carrier to a wideband signal that appears noise-like to an unsynchronized receiver. SSMA mitigates against multipath effects, but is not bandwidth efficient when only one signal is transmitted, and is almost never operated with only one signal. However, it is quite efficient when multiple signals are transmitted over the same bandwidth, so that they share the bandwidth. This is accomplished by each user having a unique PN code that has low correlation to all other codes in the bandwidth. There are two basic types of SSMA; and the first is frequency-hopped multiple access (FHMA) and the second is code division multiple access (CDMA).

9.5.1 Frequency-Hopped Multiple Access

Frequency-hopped multiple access (FHMA) is a digital multiple access system in which the carrier frequencies of the individual users are hopped in a pseudorandom manner across a wideband channel. Figure 9.5-1 illustrates an example of FHMA. The hopping codes (hopping frequencies) are designed to have zero or minimal occurrences of the same frequencies at the same time (near-zero correlation). The rectangles represent the signal bandwidth at the frequency at the vertical center of each rectangle.

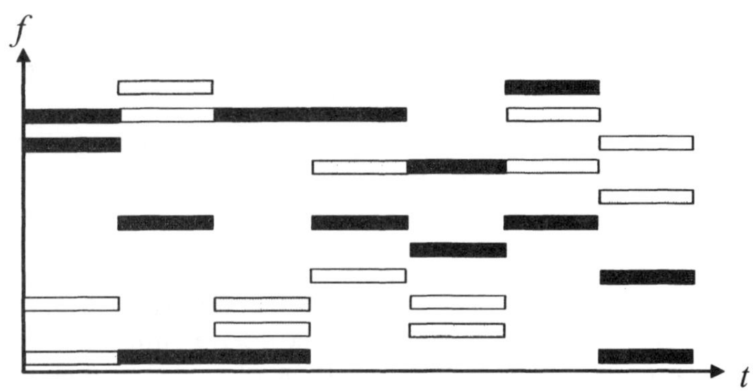

Figure 9.5-1 FHMA example with four hopped signals occupying the same bandwidth.

In the figure the four signals hop to a different frequency at the end of each hop period. For example the black rectangle represents the hop frequency versus time (t) for one FHMA user and the first hop is at the highest frequency, the second at the lowest frequency, and the hop pattern continues in this manner. Each FHMA user has its own unique hopping pattern. They are hopped in such a way that they don't occupy the same frequency very often or at all.

9.5.2 Code Division Multiple Access

Code division multiple access (CDMA) is a modulation scheme that mathematically multiplies the data by a PN code that operates at a much greater rate than the data. This causes the spectrum to be spread much more than the data would actually require by itself. Each user is assigned a particular PN code that is designed to provide low cross-correlation over the code bit time between all the other codes in the band. Thus the receiver, in order to obtain its desired signal, correlates its code against all the received codes, and by design only the selected code associated with that user is demodulated; all the other codes appear as background noise to the desired code, which is sometimes called *CDMA noise*. None of the signals are time synchronized. A code search by the receiver of the desired code provides synchronization to the received signal that has that code. Figure 9.5-2 illustrates conceptually, in the form of a box layers, how the CDMA signals relate to each other.

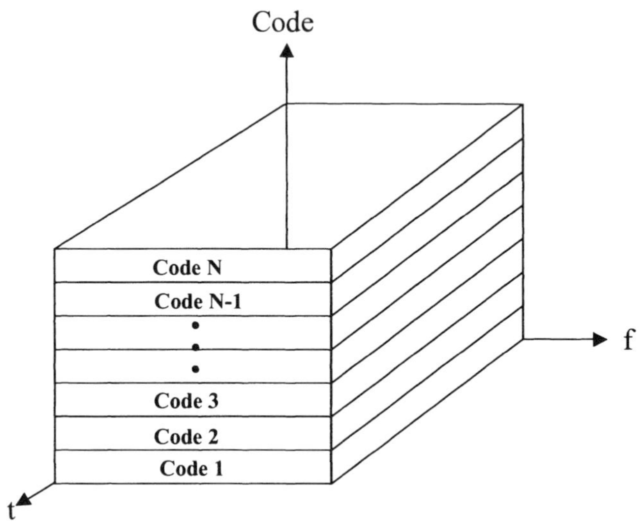

Figure 9.5-2 CDMA SSMA in which each unique code defines a link that coexists with the other codes operating in the same bandwidth and at the same time.

In Figure 9.5-2 it can be seen that each code lies in the same frequency spectrum and operates over the same time, but is identifiable via its own unique CDMA code.

A problem that CDMA users can observe is the *near-far problem*. Basically the near-far problem is the existence of different received powers from the different transmitters. This can be due to (1) the same power level at each transmitter located at various ranges from the receiver, so that the received power is not the same; or (2) from a lack of balance in received power at the receiver, from each transmitter. For a cellular system this can occur when each user within a cell does not control its transmit power well enough so that, at the base station, all the received signal powers are not the same. When the received power is not the same for each signal, the capacity of the channel is diminished because the higher powered signals cause more interference than necessary.

In general power control is used to prevent the near-far problem from occurring and to improve the capacity. When this power control is put into the system, a closer transmitter will not appear to be any stronger than a user farther away from the base station. One particular CDMA cell phone manufacturer maintains the power to a couple of tenths of a dB, to maximize the capacity. Power control is implemented at the base station by rapidly sampling the radio signal strength indicator levels of each mobile unit within a cell. Other users outside the cell (adjacent cells) are not controlled and thus cause some interference to the adjacent cells.

The main features [1] of CDMA are as follows. (1) Either TDD or FDD may be used, and many users share the same frequency band at the same time. (2) CDMA, unlike TDMA or FDMA, has a soft capacity limit (there is no absolute maximum number of users). The CDMA noise rises linearly with the number of power controlled CDMA users. (3) Multipath fading may be reduced due to the fact the signal is spread over a large bandwidth. If the channel is spread greater than the coherence bandwidth[1] the inherent diversity will mitigate the effects of small-scale fading. (4) Commonly the chip rates are very high. Therefore

[1] The coherence bandwidth and channel delay spread will be discussed in the next chapter.

the chip duration is very short and may be less than the channel delay spread (see footnote 1). Since the correlation beyond one chip is normally negligible compared to the on-time correlation, multipath components delayed by more than one chip will appear as some minor additional noise. A RAKE[2] receiver can combine delayed versions of the signal due to multipath to improve receiver performance. (5) Since CDMA uses cochannel cells, it can utilize spatial diversity to provide soft handoff. Soft handoff is realized by the mobile switching center (MSC), which is able to simultaneously monitor a user from one or more base stations. The MSC has the ability to choose the highest quality signal at any time. (6) Since typically most CDMA links are not truly orthogonal, some CDMA noise is present that adds to the background noise and limits performance. (7) As we have noted earlier the near-far problem is an issue in CDMA and the base station received signal power levels must be strictly maintained to yield nearly equal received powers at the receiver.

9.5.3 Hybrid Techniques for Spread Spectrum Signals

In addition to FHMA and CDMA signals, combinations of these can be advantageous in some applications. These hybrid combinations will be discussed in the next section.

9.5.3.1 Hybrid FDMA/CDMA or FCDMA

Hybrid frequency division multiple access/code division multiple access (FCDMA) is useful in obtaining a wider spread spectrum signal than can normally be obtained by CDMA. Basically a different CDMA signal is placed on each FDMA band. There is no hopping in this case. Figure 9.5-3 illustrates both the power spectral density of CDMA and FCDMA with four signals present.

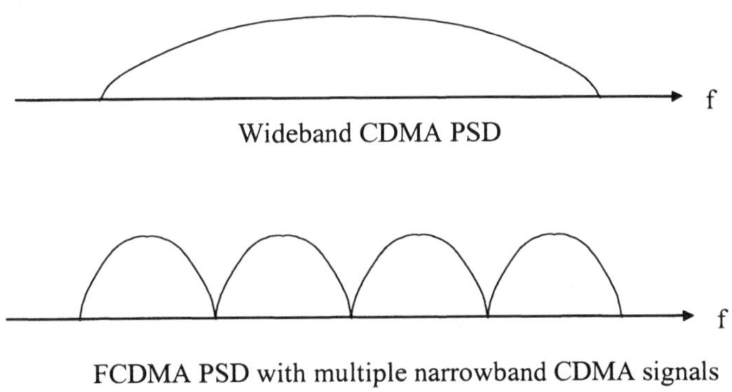

Figure 9.5-3 Power spectral density (PSD) of both WCDMA and FCDMA with four signals.

Each of the smaller FDMA channels (four in the figure) has lower chip rate CDMA signals on them. Since the chipping rate is about 1/4 of the wideband CDMA system its processing gain is about 6 dB less on each one. One advantage of this system is the fact that the bandwidth need not be contiguous and flexibility is offered in that the chip rate and bandwidth can be tailored to the needs of the user. The power spectral density shows only the main spectral response for channelization clarity. In reality these sidebands would be filtered so that they have a negligible impact on the adjacent channels.

[2] The RAKE receiver will be discussed in the next chapter.

9.5.3.2 Hybrid Direct Sequence/Frequency-Hopped Multiple Access

This CDMA approach, direct sequence frequency-hopped multiple access, is abbreviated as DS/FHMA. This hybrid approach is based on the idea that the direct sequence signal can be frequency-hopped in a pseudo-random manner. Figure 9.5-4 illustrates an example of DS/FHMA.

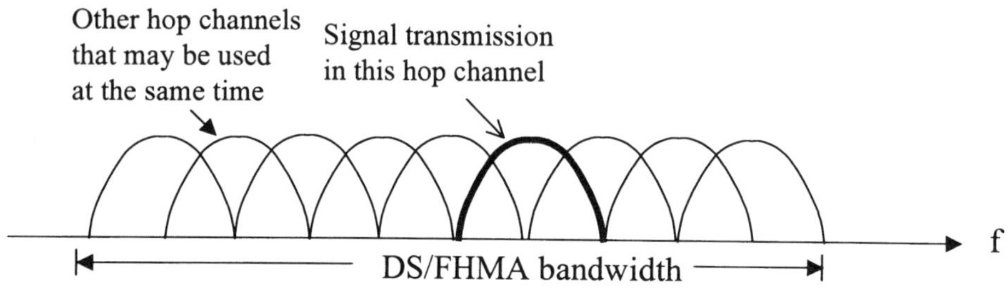

Figure 9.5-4 PSD of DS/FHMA showing one active channel and many inactive channels.

Multiple DS/FHMA signals can operate at the same time over the same bandwidth as long as they are coordinated so that they don't hop to the same frequency, at the same time, very often. Since these hybrid systems operate at different hop frequencies at a given time, the near-far problem is not present, unless multiple CDMA signals are placed on the same hop pattern.

On the other hand the DS/FHMA systems have a disadvantage of not easily being able to process a soft handoff, since it may be challenging to synchronize the frequency-hopped base station receiver to the multiply hopped signals.

9.5.3.3 Time Division Code Division Multiple Access

Time division code division multiple access (TDCDMA) is also sometimes called TDMA/CDMA. In TDCDMA different spreading codes are assigned to different cells. Inside a cell a user is allocated a particular time slot. Hence at any particular time only one CDMA user is transmitting in each cell. In case of a handoff, the CDMA code of the user is moved to that of the adjacent new cell in which the user appears. Like CDMA DS/FHMA, this scheme does not tend to suffer from the near-far problem since only one user transmits at any one given time. A synchronous version of TDCDMA is known as TDSCDMA, which denotes time division synchronous code division multiple access, and is the basis for China's nascent 3G cellular standard [5]. All the uplink signals to the base station are synchronized at the base station.

9.5.3.4 Time Division Frequency-Hopping

Time division frequency-hopping (TDFH), (also known FH/TDMA) is a multiple access scheme in which TDMA and frequency-hopping are combined. The subscriber can hop to a new frequency-hop frequency at the start of a new TDMA frame, thereby avoiding a severe fade or an erasure event on a particular frequency. The hop sets are selected such that hops within the cell are orthogonal and hops in adjacent cells are random [1, 6]. This mode of frequency-hopping is called *mixed mode frequency-hopping*. This scheme (TDFH) is very advantageous to GSM in that it can increase the capacity in fading by a factor of several times [7].

9.6 SPACE DIVISION MULTIPLE ACCESS

Space division multiple access [1] (SDMA) is a multiple access technique that controls the radiated antenna energy for each user in the area of communications. The use of spot beams facilitates the realization of SDMA. This approach utilizes another multiple access method. For example the different areas covered by antenna beam can also be served by the same frequency in CDMA or TDMA or other frequencies when using FDMA. We discussed sectorization earlier. Sectorization may be viewed as a simple version of SDMA. Figure 9.6-1 illustrates a simple example of SDMA.

Usually the reverse link is the most difficult to design for several reasons. The base station controls the power of all the transmitted signals on the forward link. But the path loss from each user to base station link, the powers from each mobile unit, must be dynamically controlled in order to prevent one user from "generating" too much interference when its signal is stronger than necessary. When the base station can utilize spot beams, less mobile power is needed, and the system is more efficient. SDMA is potential future multiple access technique that is just now being explored to make the system more efficient.

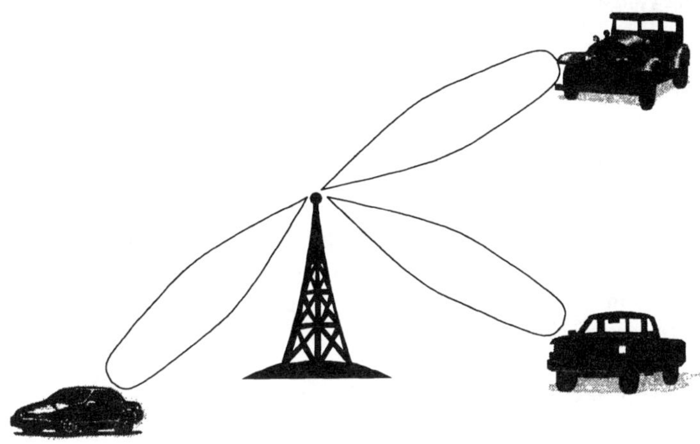

Figure 9.6-1 A base station antenna serving different users by the use of spot beams.

9.7 THE CAPACITY OF CELLULAR CDMA OF A SINGLE CELL

The capacity of a CDMA system [15, 16] is interference limited by the CDMA noise that comes from the collection of other signals occupying the same signal spectrum. One mode of CDMA operation is to operate in the *discontinuous transmission mode* (DTX) in which the transmitter is turned off when there are periods of silence in speech. The observed duty cycle (α) has been estimated at between 35% and 50% [17] for voice. The average capacity of a CDMA system can be increased roughly by $1/\alpha$.

We consider a single cell system for continuous CDMA in a cell using BPSK modulation. The scenario is as follows. The cellular network is composed of a number of mobile stations communicating with the base station. It will be assumed that all the signals on the reverse channel are received at the same power level at the base station. Assume that there are N users in the cell system. Each base station processes a composite received waveform consisting of the desired signal and N–1 interfering signals that also have power P.

Consider the CDMA correlator model illustrated in Figure 9.7-1. Degradation to the C/N_0 ratio of signal 1 (the signal being received) due to the interference by the other CDMA signals is modeled as Gaussian noise, with the same spectral shape as the CDMA signals.

Figure 9.7-1 illustrates the coherent degradation signal correlation model and the interference and thermal noise. The input BPSK signal is modeled as the signal being correlated, which is signal 1, plus thermal noise, plus interference of the other CDMA signals.

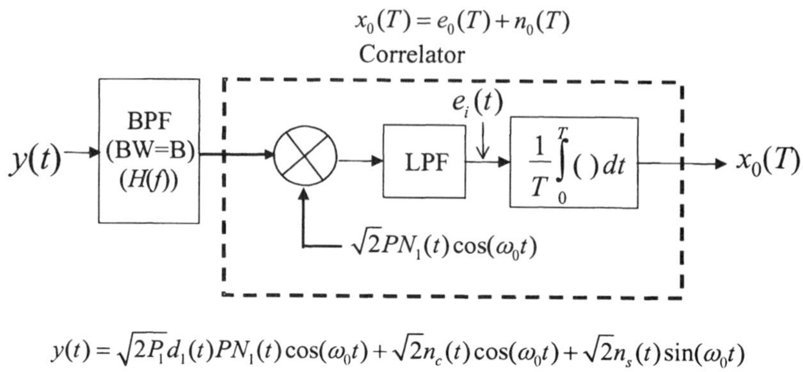

Figure 9.7-1 Model for the evaluation of the capacity of a CDMA link using BPSK.

The received BPSK signal plus interference and thermal noise is modeled as

$$y(t)=\sqrt{2P_1}d_1(t)PN_1(t)\cos\left(\omega_0 t\right)+\sqrt{2}n_c(t)\cos\left(\omega_0 t\right)+\sqrt{2}n_s(t)\sin\left(\omega_0 t\right) \quad (9.7\text{-}1)$$

and P_1 is the power of the received signal being tracked, $d_1(t)$ is the baseband data signal, and $PN_1(t)$ is the spread spectrum code used in signal 1. Also the thermal noise plus the jamming or interference signals are modeled with the last two terms in (9.7-1) as a Gaussian random process that can be decomposed as the sum of the thermal noise and the interfering signals.

After demodulating the signal with the coherent reference signal, the input to the averaging integrator, out of the low-pass filter (removes the double frequency term) is given by

$$e_i(t)=\sqrt{P}d(t)+PN_1(t)n_c(t)\ s \quad (9.7\text{-}2)$$

At the output of the despread and averaged integration (correlator), one obtains the correlated signal plus noise

$$x_o(T)=\sqrt{P}\int_{-\infty}^{\infty}H_{BB}(f)S_{PN_2}(f)df+\frac{1}{T}\int_0^T n_c(t)PN_1(t)dt \quad (9.7\text{-}3)$$

where the correlated signal component [18] depends on the baseband equivalent low-pass filter $H_{BB}(f)$ of the bandpass filter at the input of the correlator system, with RF transfer function $H(f)$ and bandwidth B Hz. The output can be separated into the signal term and the noise term as suggested by (9.7-3)

$$x_0(T)=s_0(T)+n_0(T) \quad (9.7\text{-}4)$$

The variance of the noise term is given by

$$Var(n_o(T))=\frac{1}{T^2}E\left\{\int_0^T\int_0^T n_c(t)n_c(u)PN_1(t)PN_1(u)dtdu\right\} \quad (9.7\text{-}5)$$

Modeling the unit power $PN_1(t)$ code[3] as a stationary random process independent of the signal process and the thermal noise process (9.7-5) yields

$$\text{Var}(n_o(T)) = \frac{1}{T^2}\int_0^T\int_0^T R_{n_c}(t-u)R'_{PN_1}(t-u)dtdu \tag{9.7-6}$$

where $R'_{PN_1}(\tau)$ is the unit power autocorrelation function of the $PN_1(t)$ code process. This can be expressed as a single integration [19] with a change of variables to the form

$$\text{Var}(n_0(T)) = \frac{1}{T^2}\int_{-T}^{T} R_{n_c}(\tau)R'_{PN_1}(\tau)\left[T-|\tau|\right]d\tau \tag{9.7-7}$$

With the assumption that the two processes are wideband relative to $1/T$, (9.7-7) can be well approximated by

$$\text{Var}(n_0(T)) = \frac{1}{T}\int_{-T}^{T} R_{n_c}(\tau)R'_{PN_1}(\tau)d\tau \ S \tag{9.7-8}$$

Approximating the integral with finite T second limits with an integral with unbounded limits, and using Parseval's theorem for real functions [20], leads to the spectral form[4]

$$\text{Var}(n_0(T)) = \frac{1}{T}\int_{-\infty}^{\infty} S_{n_c}(f)S'_{PN_1}(f)df \tag{9.7-9}$$

where $S_{n_c}(f)$ is the sum of white Gaussian (thermal) noise (WGN) and the interfering or jamming of signals, which are modeled as Gaussian random processes, and $S'_{PN_1}(f)$ is the unit psd[5] of the despreading signal. Note that due to the front-end filter, the noise process spectral density $S_{n_c}(f)$ may have filtering on it. Consequently the limits of the integral may be finite for ideal filters.

Now consider the case that the other signals act as interference to signal 1. We can write the noise spectral density as the sum of thermal noise and interfering noise so that

$$S_{n_c}(f) = S_n(f) + S_I(f) \tag{9.7-10}$$

where $S_n(f)$ is the two-sided noise spectral density and $S_I(f)$ is the filtered, two-sided interference spectral density. Accounting for the band-limiting due to the front-end bandpass filter, one has

$$\text{Var}(n_0(T)) = \frac{N_0}{2T}\int_{-\infty}^{\infty}|H_{BB}(f)|^2 S'_{PN_1}(f)df + \frac{1}{T}\int_{-\infty}^{\infty}|H_{BB}(f)|^2 S_I(f)S'_{PN_1}(f)df \tag{9.7-11}$$

where $S_I(f)$ is the interference spectral density, prior to being filtered by the baseband equivalent bandpass filter $H_{BB}(f)$.

Note that the equivalent noise spectral density is obtained from (9.7-11) as

---

[3] We have used the notation $S'_{PN_1}(f)$ as the unit power, power spectral density of $PN_1(t)$ process.

[4] This finite versus infinite time limit is justified if the correlation function becomes essentially zero in a time small compared to T seconds.

[5] All code spectral densities denoted by the form $S_{PN}(f)$ will be assumed to have unit power in this section.

$$N_0^{'} = 2TVar(n_0(T)) \tag{9.7-12}$$

This can be written explicitly in terms of the code and interference spectral densities as

$$N_0^{'} = N_0 \int_{-\infty}^{\infty} |H_{BB}(f)|^2 S_{PN_1}^{'}(f)df + 2\int_{-\infty}^{\infty} |H_{BB}(f)|^2 S_I(f) S_{PN_1}^{'}(f)df \tag{9.7-13}$$

Noting that the baseband equivalent interference process $S_I(f)$ contains $I/2$ watts of interference power, (9.7-13) can be written as

$$N_0^{'} = N_0 \int_{-\infty}^{\infty} |H_{BB}(f)|^2 S_{PN_1}^{'}(f)df + I\int_{-\infty}^{\infty} |H_{BB}(f)|^2 S_I^{'}(f) S_{PN_1}^{'}(f)df \tag{9.7-14}$$

where $S_I^{'}(f)$ denotes the unit power, unfiltered, power spectral density (psd) of the interference, and is related to the nonunit power spectral density by

$$S_I^{'}(f) = \frac{2}{I} S_I(f) \tag{9.7-15}$$

The effective noise spectral density can be written, from (9.7-14), as

$$N_0^{'} = N_0 \int_{-\infty}^{\infty} |H_{BB}(f)|^2 S_{PN_1}^{'}(f)df + (N-1)P\int_{-\infty}^{\infty} |H_{BB}(f)|^2 S_I^{'}(f) S_{PN_1}^{'}(f)df \tag{9.7-16}$$

since there are N–1 interferers and one desired signal, for a total of N signals. This can be simplified to the form

$$N_0^{'} = \nu N_0 + (N-1)PI_0 \tag{9.7-17}$$

where

$$I_0 = \int_{-\infty}^{\infty} |H_{BB}(f)|^2 S_I^{'}(f) S_{PN_1}^{'}(f)df \tag{9.7-18}$$

I_0 is known as the *spectral separation coefficient* (SSC). It is a measure of how much spectral overlap occurs in the received and dispreading spectral densities. The greater the value of I_0, the fewer signals that can exist on the CDMA channel. Now

$$\nu = \int_{-\infty}^{\infty} |H_{BB}(f)|^2 S_{PN_1}^{'}(f)df \le 1 \tag{9.7-19}$$

Often times ν is very close to the value of one; under these conditions (9.7-17) can be well approximated as

$$N_0^{'} = N_0 + (N-1)PI_0 \tag{9.7-20}$$

We have not accounted for the activity factor, α. This factor accounts for the fraction of time that the interference is on. When it is included into the effective noise we have

$$N_0^{'} = \nu N_0 + (N-1)\alpha P I_0 \tag{9.7-21}$$

Now we are in a position to determine the CDMA capacity. Using (9.7-21) to form the effective noise, we have, following Raghavan and Holmes [21, 22]

$$\frac{E_b}{N_0^{'}} = \frac{PT_b}{\nu N_0 + (N-1)\alpha P I_0} \geq \left(\frac{E_b}{N_0^{'}}\right)_{reqd.} \tag{9.7-22}$$

with P the received signal power and the power of the other N–1 CDMA users in the same bandwidth. Dividing the second ratio of (9.7-22) by N_0, produces the result

$$\frac{\left(\dfrac{PT_b}{N_0}\right)_{avail.}}{\nu + (N-1)\dfrac{\alpha P I_0}{N_0}} \geq \left(\frac{E_b}{N_0^{'}}\right)_{reqd.} \tag{9.7-23}$$

Solving for the allowable N produces our main result

$$N \leq 1 + \frac{1}{\alpha}\left[\frac{\left(\dfrac{PT_b}{N_0}\right)_{avail.}}{\left(\dfrac{PT_b}{N_0^{'}}\right)_{reqd.}} - \nu\right]\left(\frac{PI_0}{N_0}\right)^{-1} \tag{9.7-24}$$

It is to be noted that the available SNR must be greater than the required SNR or it is not possible to have more than one user total. In general the greater the available SNR is over the required SNR the more users one can operate simultaneously. From (9.7-24) it is to clear that if ν was approximately one and the available and required SNRs were equal, then N is bounded by one. In fact the system would achieve one, since there would be no interference from the other (zero number) CDMA signals. Another observation is that N decreases as the spectral separation coefficient (or interference (I_0)) increases.

It should also be noted that it is assumed that the signals received at the base station are not time synchronized, so that they are asynchronous in time, in the correlator. If sectorization is employed with three sectors, then the capacity is approximately multiplied by 3. Although we will not address the capacity of multicell CDMA here, the effective noise (accounting for the other signal interference) will increase compared to the single cell case and the capacity will correspondingly reduce.

Example 5 Assume that a CDMA system has the following parameters: NRZ BPSK PN code CDMA signals, a data rate of 500 bps, a required SNR of 9.6 dB, an available SNR of 12.6 dB, a chip rate of 100 Kcps, and a front-end ideal bandpass filter that has a bandwidth of 400 kHz, centered at the carrier frequency. Assume no sectorization is used. Assume that the voice duty factor, α, is 0.4. Determine (1) the spectral separation coefficient, (2) the value of ν, and finally (3) the allowable number N of users. First we compute the value of ν and I_0. Observe that all the spectral densities are the same for the signal and the CDMA interference. We have

$$\nu = \int_{-2\times10^6}^{2\times10^6} T_c \frac{\sin^2(\pi f T_c)}{(\pi f T_c)^2} df = 0.95$$

and

$$I_0 = \int_{-0.2\times10^6}^{0.2\times10^6} \left[T_c \frac{\sin(\pi f T_c)^2}{(\pi f T_c)^2} \right]^2 df = 6.664\times10^{-6}$$

Find P/N_0 from the relationship $P/N_0 = 10^{1.26}(50) = 909.8$. Letting SNRA be the available SNR and SNRR be the required SNR yields SNRA = 18.2 and SNRR = 9.12 absolute numbers (not dB). Using (9.7-24) produces the result $N \le 44.1$, so that $N = 44$ users can operate on this channel with BPSK using CDMA with a voice activity factor of 0.4, in theory. In reality other losses will occur and the actual number will be less.

9.8 PACKET RADIO ACCESS TECHNIQUES

Packet radio access techniques are based upon many subscribers attempting to access a single channel in a partially coordinated or uncoordinated manner. The information is transmitted in bursts. *Collisions* occur when two or more packets overlap in time on the same channel. Either an ACK, acknowledging correct reception at the base station, is sent to all users, or a NACK is sent (from the base station) to the desired user and all other users alerting them of the fact that the previous packet was not received correctly at the base station.

Packet radio is relatively easy to implement but it is not very efficient and there may be significant delays in receiving the desired message. The first computer system to employ radio instead of point-to-point wires for its communication facility was the ALOHA system at the University of Hawaii in 1971 [23].

Packet radio allows numerous users to utilize the channel with basically no overhead. The user transmits his data then the user must wait to listen to see if he received an ACK or a NACK. If the user receives a NACK he must send his data again. If he instead receives an ACK then he can assume that the message was sent. This procedure is an example of *contention*. That is, all users are contending for the use of the channel.

9.8.1 ALOHA Channel

The contention of the usage of the channel can be evaluated by use of the throughput (R) in terms of messages per second and the delay (D) in seconds experienced by a typical user.

In order to analyze the throughput performance it is necessary to determine the *vulnerable period* (VP) for the packets [24]. Figure 9.8-1 illustrates an example of a vulnerable period for packets. If any packet from another source appears within the time denoted by $2T_P$, a collision will occur and the two affected packets will have to be re-sent.

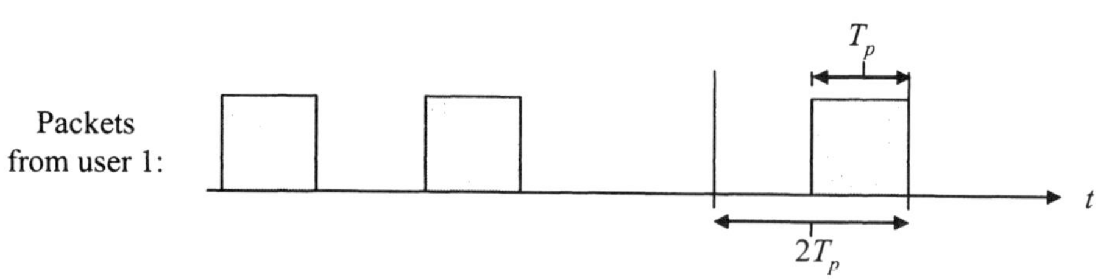

Figure 9.8-1 Vulnerable period for packet interference in an ALOHA channel.

We will assume that all packets that are sent by users have a constant packet length and a fixed channel data rate, and all the other users generate new packets at random time intervals. As is standard in this type of analysis, it will be assumed that the packets occur with a Poisson distribution, having a mean arrival rate of λ packets per second. If τ is the packet duration in seconds, then the *throughput* S of a packet radio network is given by the relationship [1, 24]

$$S = \lambda \tau \ \text{Erlangs} \tag{9.8-1}$$

The normalized throughput is also known as the *traffic intensity*. An Erlang is a measure of traffic volume in one hour in a telecommunication system.

Example 6 Consider an example of how to compute an Erlang. Assume there are 60 calls in one hour, each call lasting 5 minutes. The number of minutes of traffic in one hour is the number of calls × call duration. The minutes of traffic in the hour is 60 × 5 = 300. The corresponding equivalent hours of traffic in one hour is given by 300/60 = 5. Thus the traffic intensity is 5 Erlangs.

The *channel traffic* G is the sum of the new packets and the retransmitted packets (that were involved in collisions in the previous transmission). Thus we have

$$G = \lambda' \tau \tag{9.8-2}$$

where

$$\lambda' = \lambda + \lambda_{retrans.} \tag{9.8-3}$$

is the sum of the new packet rate (λ) and the retransmission rate ($\lambda_{retrans.}$). It follows that

$$\lambda' \geq \lambda \tag{9.8-4}$$

If R is the channel data rate in bits per second, then the *normalized throughput, s*, is given by

$$s = \frac{\lambda \tau}{R} \tag{9.8-5}$$

and the *normalized channel traffic*, g, is defined by

$$g = \frac{\lambda' \tau}{R} \tag{9.8-6}$$

If $s > 1$ the users are generating packets at a rate greater than the channel can handle [24]. Therefore it would be expected that $0 < s < 1$ in normal operations.

For a Poisson arrival process, for statistically independent users, the probability of having exactly K new packets arrive over a time interval of τ seconds is given by

$$P(K) = \frac{(\lambda \tau)^K e^{-\lambda \tau}}{K!} \tag{9.8-7}$$

We noted that the vulnerable period was 2τ seconds. A packet from a user will be successful if new packets are transmitted during the 2τ seconds from the right side of the packet as seen in Figure 9.8-1. Thus we may evaluate (9.8-7) at $K = 0$ for 2τ seconds to obtain the probability

$$P_{succ} = P(K = 0) = \frac{(\lambda' 2\tau)^0 e^{-2\lambda' \tau}}{0!} = e^{-2\lambda' \tau} \tag{9.8-8}$$

Note that λ' was used in the expression and not λ since all the transmissions (that is, the new and the retransmissions) must be included. This result does not fully describe the utilization of the ALOHA channel.

Rosner [24] has shown that this probability is also given by

$$P_{succ} = \frac{channel\ throughput}{channel\ traffic} = \frac{\lambda}{\lambda'} \tag{9.8-9}$$

Now since both measures of probability of success for the same channel must be the same, we can equate them to yield

$$\frac{\lambda}{\lambda'} = e^{-2\lambda'\tau} \tag{9.8-10}$$

or solving for λ

$$\lambda = \lambda' e^{-2\lambda'\tau} \tag{9.8-11}$$

Now multiplying both sides by τ produces the result

$$\lambda\tau = \lambda'\tau e^{-2\lambda'\tau} \tag{9.8-12}$$

Using the definitions of S and G produces

$$S = Ge^{-2G} \tag{9.8-13}$$

This expression relates the useful delivered throughput of the ALOHA channel, S, to the total traffic flowing on the channel, G. Figure 9.8-2 illustrates the relationship.

Example 7 Let us determine what value of G that S is maximized for. Let $f(G) = Ge^{-2G}$. The first derivative yields $f'(G) = e^{-2G} + G(-2)e^{-2G} = 0$. The only finite value of G that satisfies this equation is $G_0 = 1/2$.

It can be seen from the figure that only about 18.4% throughput (maximum value) can be achieved with the ALOHA channel. Another issue that we have not covered here is the ALOHA delay characteristics. When the delay is plotted against the throughput, it can be shown that when the throughput reaches the peak value of 18.4% the delay becomes unbounded [24].

9.8.2 The Slotted ALOHA Channel

Although ALOHA is a marvel in itself, it is limited to about 18% throughput as was seen in the last section. By establishing a slotted channel (slotted ALOHA), which is a channel with discrete time slots in which users may transmit their packets, it is possible to greatly reduce the vulnerable period. This transmission protocol requires that the user wait until the beginning of the time slot to transmit the packet. Hence with this protocol there is either no packet overlap or a complete overlap. Packets in this approach are not destroyed just because a few bits have packet overlap; in other words, only complete overlap occurs and never partial overlap. Figure 9.8-3 illustrates the case in which all slots have no collisions, except for the time between 4τ and 5τ.

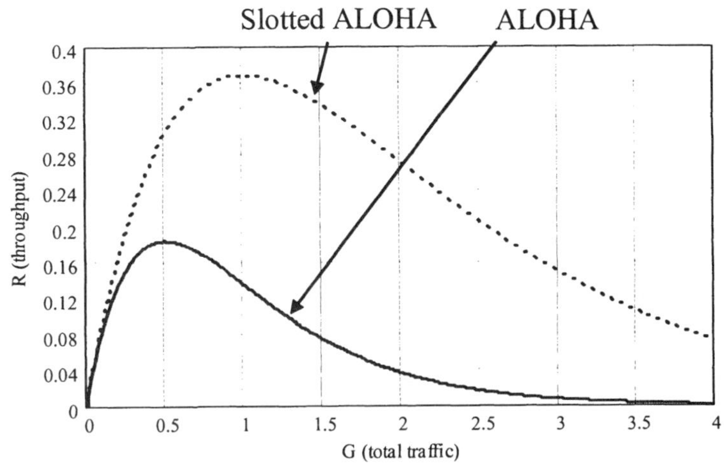

Figure 9.8-2 The relationship of the throughput versus the total traffic for the ALOHA and the slotted ALOHA channels.

The slotted ALOHA channel protocol has two disadvantages. The first is the timing impact on the terminal design, as well as the need to allow for slight time differences from the user to the base station. Each terminal must be able to determine range or distance between the user and the base station and to account for this by adjusting the user time accordingly. The second disadvantage is the fact that the packet duration τ is fixed for all packets no matter if they are of less duration or not. Thus short packets will "consume" one full packet slot but may be much shorter in duration, thereby wasting time that would not be wasted in the original ALOHA channel protocol.

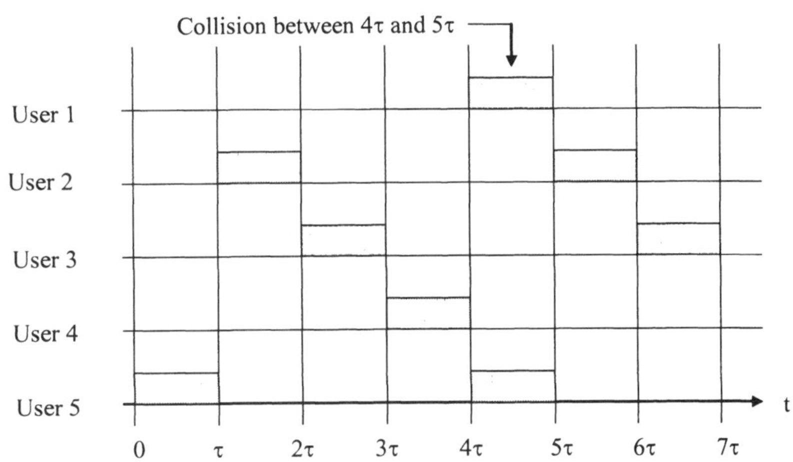

Figure 9.8-3 Slotted ALOHA channel example with five users showing one collision.

It will be assumed that every packet has a fixed duration of τ seconds. This assumption leads to an over bound in the throughput estimate. Assume that

$$\begin{aligned} S_i &= \text{probability that user } i \text{ successfully transmits a packet} \\ G_i &= \text{probability that user } i \text{ transmits any packet} \\ &\quad \text{(that is a successful or unsuccessful packet)} \end{aligned} \tag{9.8-14}$$

In order that user i transmits a successful packet, he must transmit his packet while no other user transmits one. Thus

$$S_i = \frac{G_i}{1-G_i}\prod_{k=1}^{n}(1-G_k) = G_i\prod_{\substack{k=1\\k\neq i}}^{n}(1-G_k) \tag{9.8-15}$$

for the case of n users. It will now be assumed that all of the users of the channel are statistically equivalent in the sense that they have an equally likely probability of transmitting at any instant of time, then

$$\begin{aligned} S_i &= \frac{S}{n} \\ G_i &= \frac{G}{n} \end{aligned} \tag{9.8-16}$$

where S is the channel throughput and G is the channel traffic, defined in the same manner as in ALOHA.

The first expression implies that an individual user's probability of a successful packet, S_i, is $1/n$-th of the total normalized throughput of the channel. In a similar manner his probability of making a transmission is just $1/n$-th of the total traffic transmitted on the channel. From (9.8-15) and using (9.8-16) produces the relationship

$$\frac{S}{n} = \frac{G}{n}\prod_{\substack{k=1\\k\neq i}}^{n}\left(1-\frac{G}{n}\right) = \frac{G}{n}\left(1-\frac{G}{n}\right)^{n-1} \tag{9.8-17}$$

This can be written, after multiplying both sides by n, as

$$S = G\left(1-\frac{G}{n}\right)^{n-1} \tag{9.8-18}$$

If there are a large number of users, then we may write

$$LS = \left(1-\frac{x}{n}\right)^{n-1} \cong e^{-x} = RS = e^{-G} \tag{9.8-19}$$

where LS denotes the left side and RS denotes the right side. Actually this approximation is quite good for $n \geq 10$, up to $x = 10$, which can be seen by plotting the left side and right side versus x for $n = 10$ in Figure 9.8-4. The dashed line is the LS, and the solid line is the RS. It is seen that the approximation is very good for $n = 10$ for x up to 10.

Using the large n approximation leads to the result

$$S \cong Ge^{-G} \tag{9.8-20}$$

as our final result for the slotted ALOHA channel throughput versus the total traffic and is plotted in Figure 9.8-2.

Notice that the slotted ALOHA reaches a peak value of about 37%, which is about twice as large as the ALOHA channel. It can be shown [24] that there is a slight additional delay in the time to deliver the packet. Overall, slotted ALOHA is an improvement over ALOHA.

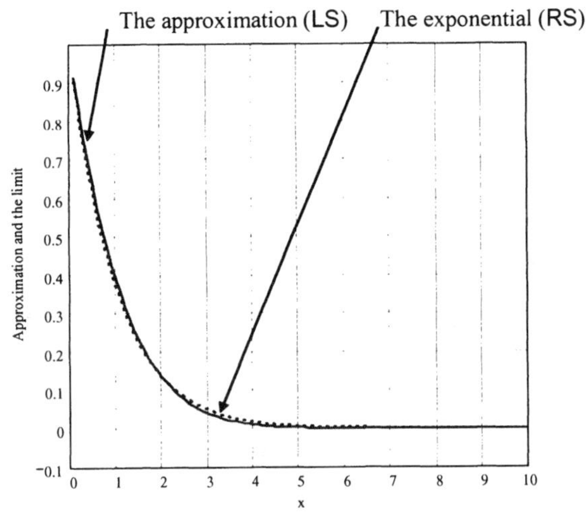

Figure 9.8-4 The approximation (solid line) and the negative exponential limit (dashed line) of (9.8-19) plotted versus x.

9.9 CARRIER SENSE MULTIPLE ACCESS PROTOCOLS

The previously discussed ALOHA protocols did not check the channel before transmitting the packet. By first "sensing" the channel prior to a packet transmission, greater efficiencies can be obtained in packet throughput. Protocols in which the users listen for a carrier (a signal transmission) and act accordingly are *carrier sense protocols* [23]. This section follows Tobagi's work [25] on carrier sense multiple access (CSMA) protocols. It will be assumed that: (1) all packets are the same length; (2) the only errors are due to collisions; (3) the random delay, after a collision, is uniformly distributed and large compared to the packet transmission time; (4) packet generation attempts (new and retransmitted) form a Poisson process with mean G packets per packet time; (5) a user may not transmit and receive simultaneously; (6) each user can sense the transmissions of all other users; (7) the propagation delay is small compared to the packet transmission time and identical for all users; and (8) sensing the state of the channel can be done instantaneously. We will now discuss some of the CSMA protocols.

9.9.1 1-Persistent CSMA

In 1-persistent CSMA when a user has data to send, it first listens to the channel to determine if anyone else is transmitting. If the channel is busy, the station waits until the channel becomes idle (not being used). When the user detects an idle channel, it transmits a packet. If a collision occurs, the user waits a random amount of time and starts all over again. The reason the protocol is called 1-persistent is the fact that the station transmits with a probability of 1 whenever it finds the channel idle.

The effect of propagation delay is important in this protocol. There is a small chance that just after a user begins sending, another user will be ready to send and sense the channel. If the first packet has not reached the second user, the latter will sense that the channel is free and will begin sending its packet, resulting in a collision.

Without propagation delay there will be collisions. If two users wait while a third user sends its packet, they will both begin transmitting at the same time after the third user's packet has finished. Even accounting for this problem we will see that it is considerably better than ALOHA. Finally it should be noted that 1-persistent CDMA should not be used in satellite broadcasting [24].

For zero propagation delay the throughput has been derived by Tobagi [25] and is given by

$$S = \frac{G^{-G}(1+G)}{G+e^{-G}} \tag{9.9-1}$$

in terms of the total traffic. This result is plotted in Figure 9.9-1.

9.9.2 Nonpersistent CSMA

This protocol has a better approach to transmitting a packet than 1-persistent CDMA. Before transmitting a packet the user listens for the idle state. If there is no other packet being transmitted, it then transmits its packet. If, on the other hand, the channel is "busy" the user does not wait until the idle state is found, but rather waits a random period of time and then repeats the algorithm. It would seem that this algorithm will be more efficient than the previous one. In the case of zero propagation delays, both the slotted and the unslotted versions produce the same throughput result as a function of the total traffic. The result [25] is given by

$$R = \frac{G}{1+G} \tag{9.9-2}$$

This result is also plotted in Figure 9.9-1.

9.9.3 p-Persistent CSMA

This protocol applies to slotted channels and operates as follows. When a user is ready to transmit a packet, it senses the channel. If the channel is idle, it transmits with probability p, or waits with probability $q = 1-p$ for the next slot. If that slot is also idle, it either transmits the packet with probability p, or waits for the next slot with probability q. This process is repeated until either the packet has been transmitted or another user has begun transmitting. In the latter case, the user acts as if there has been a collision (that is, it waits a random time and starts again). If the user initially sensed the channel is busy, it waits until the next slot and applies the previous algorithm.

This is a very complicated algorithm but has been analyzed by Tobagi [25] with the result that the throughput is given by

$$R = \frac{Ge^{-G}(1+pGX)}{G+e^{-G}} \tag{9.9-3}$$

and where X is defined as

$$X = \sum_{k=0}^{\infty} \frac{(qG)^k}{\left(1-q^{k+1}\right)k!} \tag{9.9-4}$$

This function is plotted in Figure 9.9-1 where all of the protocols are plotted. It can be seen that 0.01 persistent has the best throughput in the range of $0 \le G \le 9$.

There are other protocols that have been devised but we will not consider them here. One of them is a distributed queuing request update multiple access (DQRUMA). It is a time slotted protocol.

9.9.4 Conceptual Comparison of the Multiple Access Methods

Now consider a tree relating the different types of multiple access schemes. Figure 9.9-2 is based on the work of Mori [26]. Most, but not all, of the schemes were discussed in this chapter.

Basically there are connection-oriented and packet-oriented multiple access methods. Within the connection-oriented methods are TDMA, FDMA, and CDMA. Packet-oriented methods are based on demand base assignment of slotted and nonslotted channels.

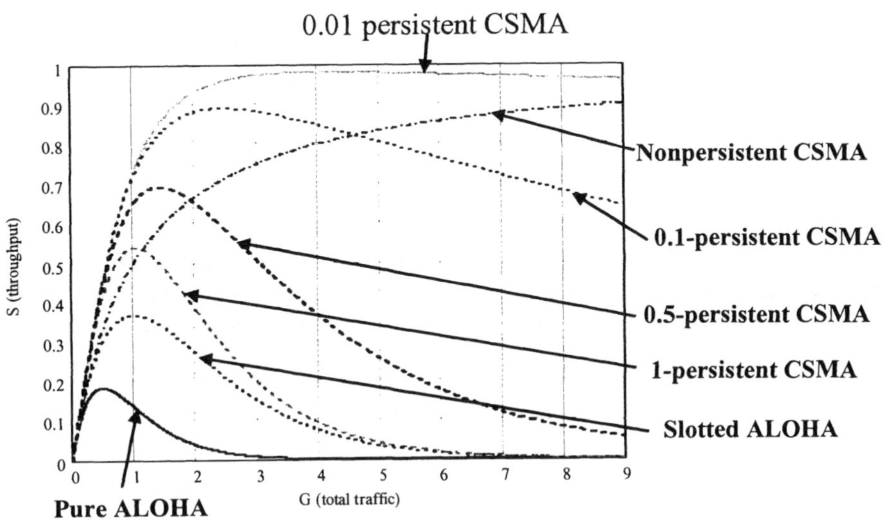

Figure 9.9-1 Comparison of the throughput versus the total traffic for some random access protocols discussed in this chapter.

These can be broken into two types: contention-based and contention-free (conflict free). Among the contention based channel access are (1) random access and (2) scheduled access.

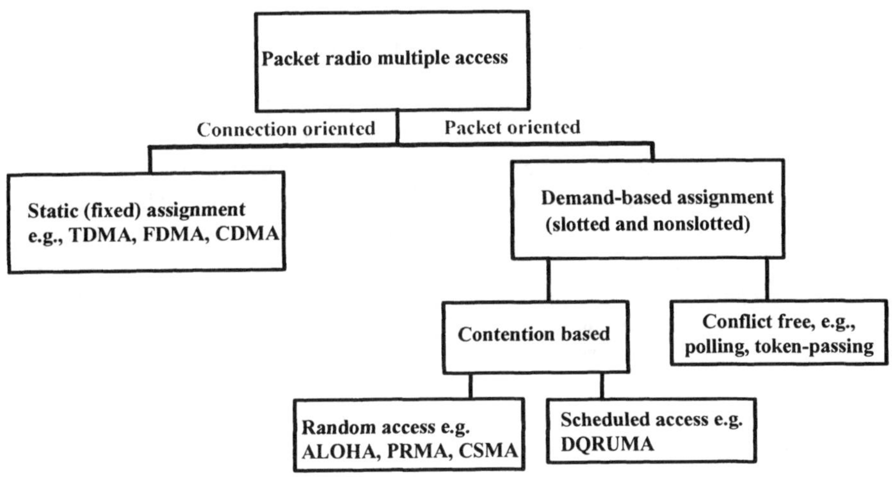

Figure 9.9-2 A comparison of the various multiple access techniques.

9.10 MULTIUSER DETECTION CONCEPTS

Multiuser detection is involved in the demodulation of the mutually interfering digital signals that are found in areas such as wireless communications, satellite communications, digital television, high speed data transmissions, and magnetic recordings. Multiuser detection has been viewed as a very important recent advance in communication technology [27]. We will discuss a few of the detection approaches used in multiuser detection (MD). As mentioned earlier, one example of MD occurs when a wireless communications link is transmitting data on the uplink from the user to the base station. Assume that there are K uplinks. The job of the base station is to demodulate and decode the desired signal when being interfered with by the K–1 interfering signals. There are two possible scenarios, one in which the received signals are *synchronous* and the more general case in which the received signals are *asynchronous.* In the synchronous case there are no differential delays between the received signals. In the asynchronous case differential delays exist. We will concentrate on the synchronous case, since that will illustrate the concepts without being bogged down with the details of the much more complex asynchronous case.

9.10.1 The Matched Filter for CDMA Signals

It will be assumed that the CDMA channel has K simultaneous users. Each user has a *signature waveform.* The signature waveform is a unique waveform that is much wider in bandwidth than the bandwidth needed to convey the data information. For CDMA signals the signature function is a pseudonoise (PN) sequence of L code chips. A CDMA signature waveform that has a duration of T seconds can be expressed as

$$sg_k(t) = \sum_{j=0}^{L-1} PN_k(j)p(t - jT_c)\ , \qquad 0 \le t \le T \tag{9.10.1-1}$$

in which the $PN_k(j)$ denotes the pseudonoise code sequence chips which are binary valued (± 1) for j = 0, 1, 2…L–1, and $p(t)$ is a unit pulse rectangular function of duration T_c seconds. Therefore there are L code chips per signature function symbol, so that LT_c=T. It is assumed that all signature functions, $sg_k(t)$ have unit signal energy, so that

$$\int_0^T (sg_k(t))^2 = 1 \tag{9.10.1-2}$$

Thus for the CDMA case the signature functions are just the unique PN functions, one for each CDMA channel.

There are two types of sequence periods that may be used in CDMA; one has a *short period* sequence that occurs when the code period is equal to, or less than, one symbol (in this case usually an integer multiple of periods will fit in the symbol duration). The other case occurs when the code period is of much larger duration than a symbol and is called the *long period* sequence case. The clear access (C/A) codes in the GPS are an example of the short period sequence in which there are 20 code periods per bit period. An example of the long period sequences is again found in GPS where the P(Y) codes have periods many times the bit period. In the *short period* sequence case the cross correlations are unique; however, in the case of the long period sequences they are not unique (the partial correlations may be small but vary, depending where in the sequence the partial correlations are taken). We will assume in what follows that the codes have short periods for analytical convenience.

As mentioned earlier we will only consider the synchronous case for short period sequences. The asynchronous case can be found in the literature [27, 28]. Define the cross correlations [28] to be given by

$$\rho_{ij}(0) = \int_0^T sg_i(t) sg_j(t)dt \quad i \text{ and } j = 1,\ 2,\ 3,...K \tag{9.10.1-3}$$

It is convenient to assume that binary antipodal signals are utilized to communicate information from each user. Now let the information sequence of the k-th user be denoted by $\{d_k(n)\}$ in which each data bit takes on the value of ± 1. Assume that the data blocks are N bits long (N bit data block). Then the data block from the k-th user is given by the $N \times 1$ column vector

$$\boldsymbol{b}_k = [d_k(1), d_k(2)...d_k(N)]^t \tag{9.10.1-4}$$

where the superscript "t" denotes the transpose. The complex envelope (real in this case) of the transmitted signal is then given by

$$s_k(t) = \sqrt{E_k} \sum_{j=1}^{N} d_k(j) sg_k(t - jT) \tag{9.10.1-5}$$

where E_k is the signal energy per bit of the k-th signal. In the synchronous case the complex envelope of the composite signal for the K users can be written as

$$s(t) = \sum_{k=1}^{K} sg_k(t) = \sum_{k=1}^{K} \sqrt{E_k} \sum_{j=1}^{N} d_k(j) sg_k(t - jT) \tag{9.10.1-6}$$

with the assumption that all the carrier frequencies are the same.

If the asynchronous case were considered there would be various delayed versions of the signals and possible different carrier frequencies for each user. This is the multiuser model for synchronous signals. The asynchronous case will not be considered here, but can be found in many references, such as [27–29], to mention a few.

In the environment of white Gaussian noise we have that the received signal is

$$r(t) = s(t) + n(t) \tag{9.10.1-7}$$

and $n(t)$ represents white Gaussian noise.

9.10.1.1 The Optimum Receiver for Synchronous CDMA Signals

The optimum receiver [28] determines the most probable sequence of bits $\{d_k(n),\ 1 \le n \le N,\ 1 \le k \le K\}$ given the signal over the time interval $0 \le t \le NT$. Assuming the synchronous transmission case, only one data symbol (or bit if uncoded) is affected by the other K–1 symbols. In the case of white Gaussian noise it is sufficient to consider the first complex envelope of the signal symbol interval; that is, consider

$$r(t) = \sum_{k=1}^{K} \sqrt{E_k} d_k(1) sg_k(t) + n(t), \quad 0 \le t \le T \tag{9.10.1-8}$$

where $n(t)$ is white Gaussian noise.

The optimum maximum-likelihood receiver formulates the log-likelihood function

$$\Lambda(\boldsymbol{d}_K) = \int_0^T \left[r(t) - \sum_{k=1}^{K} \sqrt{E_k} d_k(1) sg_k(t - jT) \right]^2 dt \tag{9.10.1-9}$$

and obtains the symbol information sequence $\{d_k(1),\ 1 \le k \le K\}$ that minimizes the likelihood function $\Lambda(\boldsymbol{d}_K)$, where the column vector $\boldsymbol{d}_K$ is defined by

$$\boldsymbol{d}_K = \left(\sqrt{E_1} d_1(1), \sqrt{E_2} d_2(1), ... \sqrt{E_K} d_K(1) \right)^t \tag{9.10.1-10}$$

and again the superscript "t" denotes the vector transpose. Expanding the square in (9.10.1-9) produces

$$\Lambda(\boldsymbol{d}_K) = \int_0^T r^2(t)dt - 2\sum_{k=1}^{K}\sqrt{E_k}d_k(1)\int_0^T r(t)sg_k dt + \sum_{k=1}^{K}\sum_{i=1}^{K}\sqrt{E_k E_i}d_k(1)d_i(1)\int_0^T sg_k(t)sg_i(t)dt \quad (9.10.1\text{-}11)$$

The first term does not have any information about the choice of the likelihood function so that it may be ignored when looking for the optimum selection. Let the integration in the second term be denoted by r_k so that

$$r_k = \int_0^T r(t)sg_k(t)dt, \qquad 1 \le k \le K \quad (9.10.1\text{-}12)$$

Furthermore let the last integral of signature functions be denoted as

$$\rho_{ki}(0) = \int_0^T sg_k(t)sg_i(t)dt \quad (9.10.1\text{-}13)$$

Let the $K \times 1$ column vector $\boldsymbol{r}_K$ be defined by

$$\boldsymbol{r}_K = (r_1,\ r_2,\ r_3, \ldots r_K)^t \quad (9.10.1\text{-}14)$$

With these definitions, (9.10.1-11) can be written in the form of correlation metrics [28] as

$$Cor(\boldsymbol{r}_K, \boldsymbol{d}_K) = 2\sum_{k=1}^{K}\sqrt{E_k}d_k(1)r_k - \sum_{k=1}^{K}\sum_{i=1}^{K}\sqrt{E_k}\sqrt{E_i}d_k(1)d_i(1)\rho_{ki}(0) \quad (9.10.1\text{-}15)$$

The optimum detector chooses the largest value[6] of Cor($\boldsymbol{r}_K$,$\boldsymbol{d}_K$) over $\boldsymbol{d}_K$. An alternate method of describing this correlation is writing it as an inner product of the form [27]

$$Cor(\boldsymbol{r}_K, \boldsymbol{d}_K) = 2(\boldsymbol{d}_K)^t\boldsymbol{r}_K - (\boldsymbol{d}_K)^t\boldsymbol{R}_c\boldsymbol{d}_K \quad (9.10.1\text{-}16)$$

in which $\boldsymbol{R}_c$ is the K by K correlation matrix, having elements $\rho_{ki}(0)$. Figure 9.10-1 illustrates the optimum multiuser (MU) receiver for synchronous transmission.

With the knowledge of all the signal energies, there are 2^K choices of the bits (or symbols) in the sequence of K users. The function of the optimum detector is to compute the correlation metrics of each sequence from (9.10.1-16) and choose the sequence that produces the largest correlation metric from the 2^K possible transmitted information sequences.

We have only discussed the synchronous case here. It is much simpler and applies to very restricted links. The asynchronous case is treated elsewhere [27, 28] and will not be considered here. Complexity is a major issue in obtaining the solution to the correlation metrics and is $O(2^K/K)$ per bit [27]. The time complexity per bit of $f(K)$ is written as $O(g(K))$ if there exists a constant $C > 0$ such that for large enough K, $f(K) \le Cg(K)$. Due to this complexity issue of the optimum detector, less complex and suboptimal schemes are desirable.

[6] Note we desire the minimum of (9.10.1-11) or alternatively the maximum of (9.10.1-15), since the algebraic signs have changed.

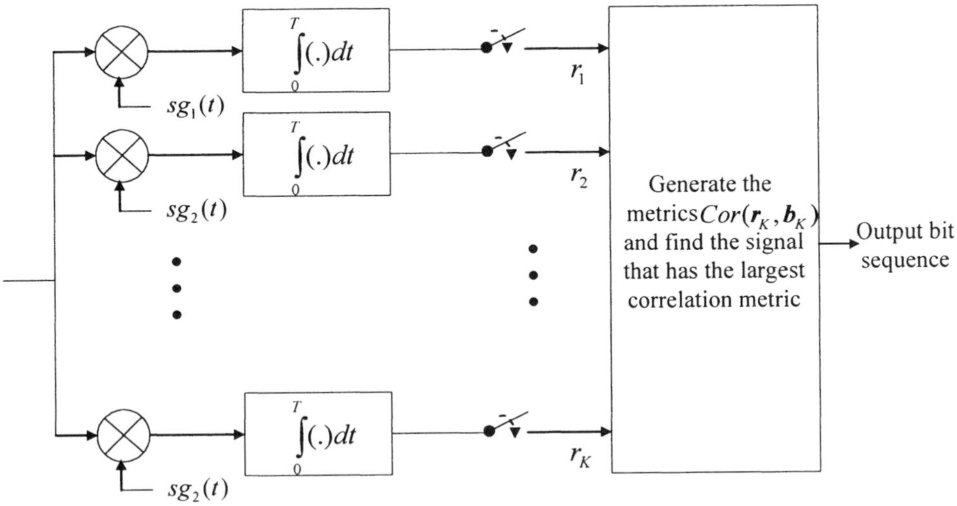

Figure 9.10-1 Optimum multiuser receiver for the case of synchronous transmission.

9.10.2 Conventional Single User Detector in the Synchronous Case

Now that we have presented the optimum detector we shall consider some suboptimum detectors. Conventional receivers have been discussed previously. This is the simplest suboptimum receiver structure that one can consider. The receiver correlates (or uses a matched filter) with the signature waveform against the received signal. The decision is made based on the highest correlation with the desired local signature waveform. Figure 9.10-2 illustrates the receiver model for the *k*-th user. The user simply correlates his signature waveform against the received signals and noise and tracks the carrier that is used to demodulate the data. We are assuming in this analysis that the signals are synchronous.

Consider the first bit in time period $0 \le t \le T$. The output of the correlator shown in Figure 9.10-2 is given by

$$r_k = \int_0^T r(t) sg_k(t) dt \tag{9.10.2-1}$$

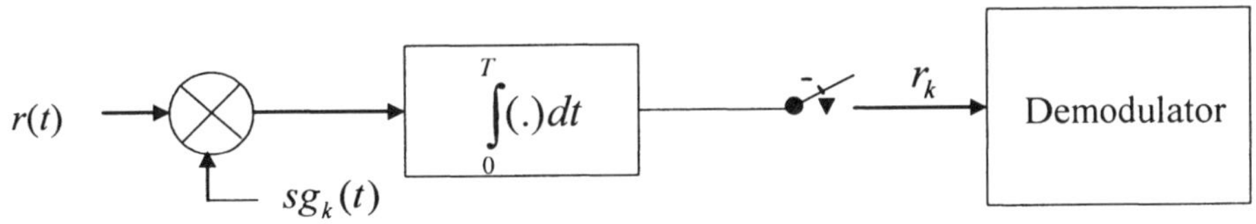

Figure 9.10-2 Conventional single user detector of the signature waveforms.

This can be expanded into a simpler form, using (9.10.1-6) for the first bit duration

$$r_k = \sqrt{E_k} d_k(1) + \sum_{\substack{n=1 \\ n \neq k}}^{K} \sqrt{E_n} d_n(1) \rho_{nk}(0) + n_k(1) \tag{9.10.2-2}$$

in which E_k is the energy of the k-th signal, $d_k(1)$ is the data value of the k-th waveforms first data symbol in the N bit block. In this expression the noise is given by

$$n_k(1) = \int_0^T n(t) sg_k(t) dt \tag{9.10.2-3}$$

Clearly the mean value is zero. The variance is given by

$$E\{n_k^2(1)\} = \int_0^T \int_0^T E\{n(t)n(s)\} sg_k(t) sg_k(s) dt ds \tag{9.10.2-4}$$

This can be evaluated as

$$E\{n_k^2(1)\} = \int_0^T (N_0/2) sg_k^2(t) dt = \frac{N_0}{2} \tag{9.10.2-5}$$

We can write the $K \times 1$ column vector of $n_k(1)$ values as

$$\boldsymbol{n}_K = [n_1(1), n_2(1), ... n_K(1)]' \tag{9.10.2-6}$$

So far we have not addressed the middle term in (9.10.1-2) which contains K–1 terms. If the value of $\rho_{nk}(0)$ were zero then only noise would be affecting the decision, and the receiver would be optimum. However it is not possible to generate orthogonal codes, and thus the middle term is always present and is nonzero in general. In particular if one or more signals that are not the k-th signal are much stronger than the others, its interference will be significant. This has been discussed earlier as the *near-far problem*. We will not address the perhaps more important problem of the *asynchronous case*. However this topic is addressed in Verdu [27] and Proakis [28] and other texts. The goal in this chapter is to provide only an introduction to the techniques of multiuser detection.

9.10.3 Decorrelating Detector

There are numerous suboptimum detectors for the CDMA link we have been studying. The first example was the conventional single-user detector discussed in the last section. The conventional detector has problems in performance when the near-far problem exists. Another type of detector is called the *decorrelating detector*. Again we will consider the simpler case of synchronous symbol transmission. Let the received signal vector $\boldsymbol{r}_K$ represent the K matched filter outputs so that the $K \times 1$ vector $\boldsymbol{r}_K$ is given by

$$\boldsymbol{r}_K = \boldsymbol{R}_c \boldsymbol{d}_K + \boldsymbol{n}_K \tag{9.10.3-1}$$

where $\boldsymbol{d}_K$ is a $K \times 1$ vector given by (9.10.1-10), $\boldsymbol{R}_c$ is given by the $K \times K$ correlation matrix, having elements $\rho_{ki}(0)$, and $\boldsymbol{n}_K$ represents the white Gaussian noise out of the correlators and is given by (9.10.2-6). The noise vector has a $K \times K$ covariance given by

$$E\left(\boldsymbol{n}_K \boldsymbol{n}_K'\right) = \frac{N_0}{2} \boldsymbol{R}_c \tag{9.10.3-2}$$

since the (j,k)-th term is of the form

$$E\left[n_j(1)n_k(1)\right] = E\int_0^T\int_0^T n(t)sg_j(t)n(u)sg_k(u)dtdu = \frac{N_0}{2}\rho_{jk}(0) \tag{9.10.3-3}$$

Thus (9.10.3-3) follows. The thermal noise is modeled a Gaussian random process and $\boldsymbol{r}_K$ is described by a K dimensional Gaussian probability density function of the form [28]

$$p(\boldsymbol{r}_K|\boldsymbol{d}_K) = \frac{1}{\sqrt{(N_0\pi)^K \det(\boldsymbol{R}_c)}}\exp\left[-\frac{1}{N_0}(\boldsymbol{r}_K - \boldsymbol{R}_c\boldsymbol{d}_K)^t\boldsymbol{R}_c^{-1}(\boldsymbol{r}_K - \boldsymbol{R}_c\boldsymbol{d}_K)\right] \tag{9.10.3-4}$$

The optimum linear estimate of $\boldsymbol{d}_K$ is the value of $\boldsymbol{d}_K$ that minimizes the likelihood function [28]

$$\Lambda(\boldsymbol{d}_K) = (\boldsymbol{r} - \boldsymbol{R}_c\boldsymbol{d}_K)^t\boldsymbol{R}_c^{-1}(\boldsymbol{r} - \boldsymbol{R}_c\boldsymbol{d}_K) \tag{9.10.3-5}$$

The value that minimizes the likelihood function is

$$\boldsymbol{d}_K^{opt} = \boldsymbol{R}_c^{-1}\boldsymbol{r}_K \tag{9.10.3-6}$$

where $\boldsymbol{R}^{-1}$ denotes the inverse of the matrix $\boldsymbol{R}$. The detected symbols, by this process, are obtained by making a hard decision on each element of the vector $\boldsymbol{d}_K^{opt}$, so that

$$\hat{\boldsymbol{d}}_K = sgn(\boldsymbol{d}_K^{opt}) \tag{9.10.3-7}$$

Figure 9.10-3 illustrates the receiver of the decorrelating receiver.

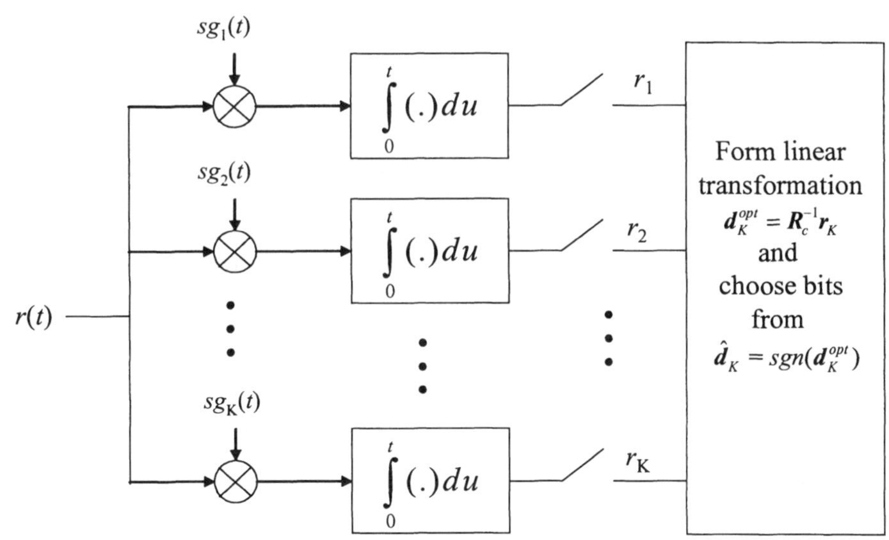

Figure 9.10-3 Decorrelating receiver structure model.

It should be observed that the best (maximum likelihood) linear estimate given by (9.10.3-6) is not the same as the best nonlinear ML sequence detector that finds the best discrete valued (± 1) sequence that maximizes the likelihood function [28].

It is insightful to consider the case that K=2 [28]. First we will form $\boldsymbol{d}_K^{opt}$ (9.10.3-6) and then $\hat{\boldsymbol{d}}_K$ (9.10.3-7). Consider the 2×2 autocorrelation matrix

$$\boldsymbol{R}_c = \begin{bmatrix} 1 & \rho \\ \rho & 1 \end{bmatrix} \quad (9.10.3\text{-}8)$$

in which ρ is defined as

$$\rho = \int_0^T sg_1(t)sg_2(t)dt \quad (9.10.3\text{-}9)$$

The inverse of $\boldsymbol{R}_c$ is obtained when $\rho < 1$, by obtaining the cofactor $R_c(i,j)$ divided by the determinant of the matrix, and the cofactor is signed according to $(-1)^{i+j}$ for the (i,j)-th element. Thus $R_c(1,1)$ is 1 since when the first column and first row are crossed out that is the remaining term in this 2×2 matrix. The determinate is given by $|\boldsymbol{R}_c| = 1-\rho^2$. Thus the inverse is given by

$$\boldsymbol{R}_c^{-1} = \frac{1}{1-\rho^2}\begin{bmatrix} 1 & -\rho \\ -\rho & 1 \end{bmatrix} \quad (9.10.3\text{-}10)$$

The receiver of Figure 9.10-3 needs to form the correlations of the received signal with $sg_1(t)$ and $sg_2(t)$. The received complex envelope of the signal (from (9.10.1-8)) can be written, for K =2, as

$$r(t) = \sqrt{E_1}d_1(1)sg_1(t) + \sqrt{E_2}d_2(1)sg_2(t) + n(t) \quad (9.10.3\text{-}11)$$

so that the projections onto $sg_1(t)$ and $sg_2(t)$ (the correlator outputs) are given by

$$\boldsymbol{r}_2 = \begin{bmatrix} \sqrt{E_1}d_1(1) + \rho\sqrt{E_2}d_2(1) + n_1(1) \\ \rho\sqrt{E_1}d_1(1) + \sqrt{E_2}d_2(1) + n_2(1) \end{bmatrix} \quad (9.10.3\text{-}12)$$

in which $n_1(1)$ and $n_2(1)$ are the noise components out of the two correlators. Hence the best linear estimator is obtained from (9.10.3-6) so that evaluating product of the 2×2 matrix and the 2×1 vector leaves a 2×1 vector. The values are given by

$$\boldsymbol{d}_K^{opt} = \boldsymbol{R}_c^{-1}\boldsymbol{r}_2 = \begin{bmatrix} \sqrt{E_1}d_1(1) + \left(n_1(1) - \rho n_2(1)\right)/(1-\rho^2) \\ \sqrt{E_2}d_2(1) + \left(n_2(1) - \rho n_1(1)\right)/(1-\rho^2) \end{bmatrix} \quad (9.10.3\text{-}13)$$

Thus we see the transformation has uncoupled the interference between the two signals and therefore it is not necessary to be concerned about the near-far problem and it is also not necessary to control the signal level very accurately. Hence the detector is called the *decorrelating detector* since it decorrelates the other signals.

Clearly if the two signature signals are orthogonal there is no cross signal or cross noise (i.e., when ρ = 0). The final decision is made by forming the sign function to the upper and lower components of the vector $\boldsymbol{d}_K^{opt}$, as indicated in (9.10.3-7).

An alternative approach in this problem [28] is to consider the modified signature functions

$$sg_1'(t) = sg_1(t) - \rho sg_2(t)$$
$$sg_2'(t) = sg_2(t) - \rho sg_1(t) \qquad (9.10.3\text{-}14)$$

with the primed signature functions used in the correlation process as shown in Figure 9.10-4. When this receiver structure is used it is not necessary to use the matrix transformation. Then let the vector $\boldsymbol{r}_2'$ be defined as the 2×1 vector

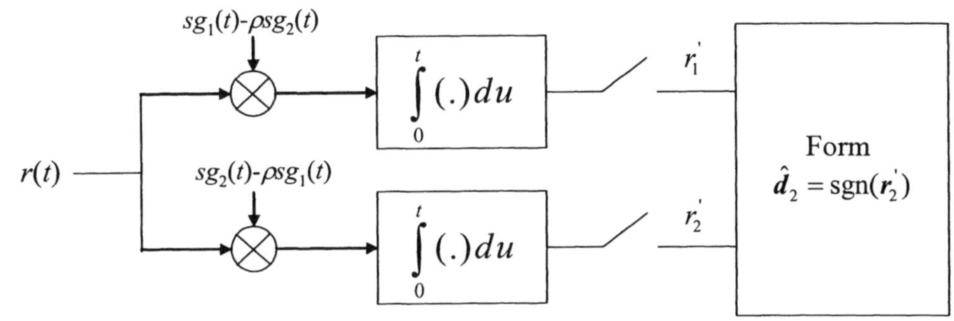

Figure 9.10-4 Alternate model for decorrelation detection when $K = 2$.

$$\boldsymbol{r}_2' = \begin{bmatrix} r_1' \\ r_2' \end{bmatrix} \qquad (9.10.3\text{-}15)$$

Evaluating $\boldsymbol{r}_2'$ after canceling similar terms produces the result

$$\boldsymbol{r}_2' = \begin{bmatrix} \sqrt{E_1} d_1(1)(1-\rho^2) + n_1(1) - \rho n_2(1) \\ \sqrt{E_2} d_2(1)(1-\rho^2) + n_2(1) - \rho n_1(1) \end{bmatrix} \qquad (9.10.3\text{-}16)$$

Thus we see that (9.10.3-16) and (9.10.3-13) are identical except for a scale factor that does not affect the bit decisions.

9.10.4 Minimum Mean Square Error Estimator

In the decorrelating detector we found the best linear maximum-likelihood estimate $\boldsymbol{b}_K$ that minimizes the quadratic likelihood function indicated in (9.10.3-5). The result found the solution based on a linear transformation given by (9.10.3-13) was the solution. Another approach is to search for the linear transformation of the outputs of the correlators so as to minimize the mean squared error (MSE).

The approach is based on seeking the linear transformation $\boldsymbol{d}_K^{opt} = \mathbf{A}\boldsymbol{r}_K$ where $\mathbf{A}$ is a $K \times K$ matrix that is to be determined such that the mean square error (MSE) is minimized, and both $\boldsymbol{d}^{opt}$ and $\boldsymbol{r}_K$ are $K \times 1$ vectors. In a similar manner we can define the noise terms

$$\boldsymbol{n}_K = [n_1(1), n_2(1), \ldots n_K(1)]^t, \quad K \times 1 \text{ matrix} \qquad (9.10.4\text{-}1)$$

Also note that the Gaussian noise vectors are of zero mean and have the autocorrelation function that can be written as

$$E\left[\boldsymbol{n}_K \boldsymbol{n}_K^t\right] = \frac{N_0}{2} \boldsymbol{R}_c \tag{9.10.4-2}$$

and $\boldsymbol{R}_c$ is a $K \times K$ matrix with (j, k)-th elements, given by

$$R_c(j,k) = \int_{-\infty}^{\infty} sg_j(t) sg_k(t) dt \tag{9.10.4-3}$$

Recall that $\boldsymbol{d}_K$ is given by

$$\boldsymbol{d}_K = \left(\sqrt{E_1} d_1(1), \sqrt{E_2} d_2(1), \ldots \sqrt{E_K} d_K(1)\right)^t \tag{9.10.4-4}$$

For the synchronous case the minimum mean squared error is given by

$$MSE(\boldsymbol{d}_K) = E\left[(\boldsymbol{d}_K - \boldsymbol{d}_K^{opt})^t (\boldsymbol{d}_K - \boldsymbol{d}_K^{opt})\right] \tag{9.10.4-5}$$

or using $\boldsymbol{d}^{\text{opt}} = \mathbf{A}\boldsymbol{r}$ in (9.10.4-5) produces

$$MSE(\boldsymbol{d}_K) = E\left[(\boldsymbol{d}_K - \mathbf{A}^{\text{opt}} \boldsymbol{r}_K)^t (\boldsymbol{d}_K - \mathbf{A}^{\text{opt}} \boldsymbol{r}_K)\right] \tag{9.10.4-6}$$

where the expectation is in respect to $\boldsymbol{d}_K$ and $\boldsymbol{n}_K$ from (9.10.2-6). The optimum value of $\mathbf{A}$ can be obtained from requiring the error $\boldsymbol{d}_K - \mathbf{A}\boldsymbol{r}_K$ to be orthogonal to the data vector $\boldsymbol{r}_K$, which is the orthogonality principle [30]

$$E\left[\left(\boldsymbol{d}_K - \mathbf{A}\boldsymbol{r}_K\right)\boldsymbol{r}_K^t\right] = \mathbf{0} \tag{9.10.4-7}$$

or equivalently

$$E\left[\boldsymbol{d}_K \boldsymbol{r}_K^t\right] - \mathbf{A} E\left[\boldsymbol{r}_K \boldsymbol{r}_K^t\right] = \mathbf{0} \tag{9.10.4-8}$$

Now consider the two terms separately. From (9.10.3-1) we have

$$\boldsymbol{r}_K^t = \left(\mathbf{R}_c \boldsymbol{d}_K\right)^t + \boldsymbol{n}_K^t \tag{9.10.4-9}$$

Let matrix $\mathbf{A}$ have dimensions $(m \times p)$ and matrix $\mathbf{B}$ have dimensions $(p \times n)$, then the *reversal law of transposed products* [31] applies; that is,

$$(\mathbf{AB})^t = \mathbf{B}^t \mathbf{A}^t \tag{9.10.4-10}$$

and this matrix is of dimensions $(n \times m)$. First consider the first term of (9.10.4-8). From (9.10.4-9) and (9.10.4-8) we have

$$E\left[\boldsymbol{d}_K \boldsymbol{r}_K^t\right] = E\left[\boldsymbol{d}_K \left(\left(\mathbf{R}_c \boldsymbol{d}_K\right)^t + \boldsymbol{n}_K^t\right)\right] \tag{9.10.4-11}$$

which can be written evaluated as

$$E\left[\boldsymbol{d}_K \boldsymbol{r}_K^t\right] = E\left[\boldsymbol{d}_K \left(\mathbf{R}_c \boldsymbol{d}_K\right)^t\right] = E\left[\boldsymbol{d}_K \boldsymbol{d}_K^t\right] \mathbf{R}_c^t \tag{9.10.4-12}$$

Denote the expectation of the $K \times K$ matrix as $\mathbf{D}$ we have

$$E\left[\boldsymbol{d}_K \boldsymbol{d}_K^t\right] = \mathbf{D} \tag{9.10.4-13}$$

so that (9.10.4-12) can be written as

$$E\left[\boldsymbol{d}_K \boldsymbol{r}_K^t\right] = \mathbf{D}\mathbf{R}_c^t \tag{9.10.4-14}$$

From (9.10.3-1) the second term in (9.10.4-8) can be written as

$$E\left[\boldsymbol{r}_K \boldsymbol{r}_K^t\right] = E\left[\left(\mathbf{R}_c \boldsymbol{d}_K + \boldsymbol{n}_K\right)\left(\mathbf{R}_c \boldsymbol{d}_K + \boldsymbol{n}_K\right)^t\right] \tag{9.10.4-15}$$

Assuming that the data and the noise are statistically independent, we obtain

$$E\left[\boldsymbol{r}_K \boldsymbol{r}_K^t\right] = E\left[\left(\mathbf{R}_c \boldsymbol{d}_K\right)\left(\mathbf{R}_c \boldsymbol{d}_K\right)^t\right] + E\left[\left(\boldsymbol{n}_K\right)\left(\boldsymbol{n}_K\right)^t\right] \tag{9.10.4-16}$$

or using the reversal law of transposed products, we have

$$E\left[\boldsymbol{r}_K \boldsymbol{r}_K^t\right] = E\left[\mathbf{R}_c \boldsymbol{d}_K \boldsymbol{d}_K{}' \mathbf{R}_c^t\right] + E\left[\left(\boldsymbol{n}_K\right)\left(\boldsymbol{n}_K\right)^t\right] = \mathbf{R}_c \mathbf{D} \mathbf{R}_c^t + \frac{N_0}{2}\mathbf{R}_c \tag{9.10.4-17}$$

From (9.10.4-8), (9.10.4-14), and (9.10.4-17), we can write

$$\mathbf{D}\mathbf{R}_c' = \mathbf{A}\left(\mathbf{R}_c \mathbf{D} \mathbf{R}_c^t + \frac{N_0}{2}\mathbf{R}_c\right) \tag{9.10.4-18}$$

To solve for the matrix $\mathbf{A}$ we note that $\rho_{ij}(0) = \rho_{ji}(0)$ so that $\mathbf{R}_c = \mathbf{R}_c^t$. Thus post multiplying (9.10.4-18) by $\mathbf{R}_c^{-1}$ produces

$$\mathbf{D} = \mathbf{A}\left(\mathbf{R}_c \mathbf{D} + \frac{N_0}{2}\boldsymbol{I}\right) \tag{9.10.4-19}$$

where $\boldsymbol{I}$ is the $K \times K$ identity matrix. Post multiply both sides of (9.10.4-19) by $\mathbf{D}^{-1}$, which yields

$$\mathbf{I} = \mathbf{A}\left(\mathbf{R}_c + \frac{N_0}{2}\mathbf{D}^{-1}\right) \tag{9.10.4-20}$$

Solving for the matrix $\mathbf{A}$ produces

$$\mathbf{A} = \left(\mathbf{R}_c + \frac{N_0}{2}\mathbf{D}^{-1}\right)^{-1} \tag{9.10.4-21}$$

is the solution. It follows that the minimum mean square estimator is of the form [28]

$$\boldsymbol{d}_K^{opt} = \left(\mathbf{R}_c + \frac{N_0}{2}\mathbf{D}^{-1}\right)^{-1} \boldsymbol{r}_K \tag{9.10.4-22}$$

and the data bit estimator is given by

$$\hat{\boldsymbol{d}}_K = \text{sgn}(\boldsymbol{d}_K^{opt}) \tag{9.10.4-23}$$

The MMSE estimate is similar to the decorrelating detector, as can be seen by comparing (9.10.3-6) and (9.10.4-22).

9.10.5 Additional Types of Multiuser Detector Systems

So far we have discussed the matched filter for CDMA signals, which is the optimum receiver for synchronous CDMA signals; the conventional single user detector in the synchronous case; the decorrelating detector; and the minimum mean squared error detector in the synchronous case. We have not considered the asynchronous case for any of these schemes since the goal of this section was to introduce the subject but not to provide an exhaustive introduction.

Due to the keen interest in commercial CDMA communication systems, the design of multiuser detection algorithms is an active area of research activity. A number of algorithms that we have not discussed but have been reported in the literature include the situation when the channel is time dispersive and therefore has intersymbol interference (ISI). Another case involves the situation in which the signature waveforms of the other CDMA users are unknown, so that the user receiver has to deal with both ISI and CDMA interference (sometimes called multiple access interference (MAI)). In this environment it is possible to develop adaptive algorithms [28].

Suitable adaptive algorithms that are designed to suppress both ISI and MAI in multiuser CDMA communications are presented in the works of Abdulrahman et al. [32], Honig [33], Miller [34], and Mitra and Poor [35, 36]. Both textbooks [27, 29] are useful in this area also.

In CDMA communication systems the use of transmitting and/or receiving antennas allows each user to employ spatial filtering along with temporal filtering in order to reduce the ISI and MAI. Algorithms that are designed for blind multiuser detection with multiple antenna systems are described by Wang and Poor [36].

In a typical case the CDMA links are encoded, with either a single level of encoding or with concatenated encoding. A more optimum strategy to reduce the ISI and MAI is to use soft-decision metrics from the decoder rather than operating at the demodulator level. Hence in this approach a turbo type iterative demodulating decoding algorithm can be devised for CDMA applications. The papers by Wang and Poor [37], Reed et al. [38], Moher [39], and Alexander et al. [40] address these issues.

9.10.6 Successive Interference Cancellation

Successive interference cancellation (SIC) (also known as stripping, successive decoding, or onion peeling [27]) is another multiuser detection scheme that is based on the idea that if a decision on an interferer's bit has been made, then that interfering signal can be recreated at the receiver and subtracted from the received waveform. If that decision is correct the interference for that signal will be removed from the input; however, if the decision is incorrect the interference will be doubled. Commonly the interfering signal with the highest received power level is chosen to detect first. After the detection of the bit has been made and it is removed from the input, the next interferer is detected and processed in the same way, until all the interference is removed.

In making a decision on the k-th user, it is assumed that the decisions of users k+1, k+2,...K are correct and the remaining interferers (waveforms 1, 2, ...k–1) are disregarded. Consequently the decision on the k-th user information bit, assuming synchronous transmission, is given by [27, 28]

$$\hat{d}_k = \text{sgn}\left[r_k - \sum_{j=k+1}^{K} \sqrt{E_j}\rho_{jk}(0)\hat{d}_j \right] \tag{9.10.6-1}$$

where $\hat{d}_j$ is the decision value or estimate of d_j. Recall that r_k is the output to the correlator corresponding to the k-th users' signature sequence.

Another approach that can be used is to base the demodulation on the order of decreasing received cross-correlation power at the outputs of the cross correlators. The correlation metrics power is given by

$$E\left(\left[\int_0^T g_k(t(r(t)dt\right]^2\right) = E_k + \sum_{j\neq k} E_j \rho_{jk}^2(0) + \frac{N_0}{2} \tag{9.10.6-2}$$

in the time synchronous case. There are some observations that one can make regarding the SIC technique [27, 28]: (1) the delay in demodulating the weakest user increases in a linear manner with the user count, (2) the computational complexity of the demodulation process is linear in the user count, (3) estimation errors result in residual multiuser interference, which causes an increase in the bit error rate (BER), and (4) the interference from users whose signal levels are weaker than the detected users signal level is processed as additive interference.

9.10.7 Multistage Interference Cancellation

Multistage interference cancellation (MIC) is a technique that exploits multiple iterations in the detection process for the user bits and cancellation of the interference. The method is readily described by an example with two users [27, 28].

Example 8 Let us assume that there are two users. This technique utilizes the SIC detector or any of the suboptimum detectors discussed in Section 9.10. Let us use the decorrelating detector in the first stage of the algorithm [28]. The first stage for the decorrelating detector produces

$$\hat{d}_1 = \text{sgn}(r_1 - \rho r_2)$$
$$\hat{d}_2 = \text{sgn}(r_2 - \rho r_1)$$

The corresponding second stage becomes

$$\hat{\hat{d}}_1 = \text{sgn}\left(r_1 - \sqrt{E_2}\hat{d}_2\rho\right)$$
$$\hat{\hat{d}}_2 = \text{sgn}\left(r_2 - \sqrt{E_1}\hat{d}_1\rho\right)$$

In a similar manner the third stage becomes

$$s\hat{\hat{d}}_1 = \text{sgn}\left(r_1 - \sqrt{E_2}\hat{d}_2\rho\right)$$
$$\hat{\hat{d}}_2 = \text{sgn}\left(r_2 - \sqrt{E_1}\hat{d}_1\rho\right)$$

and the process continues until there are no changes in the detected bits. This detector is suboptimum and does not converge to the optimum detector discussed in Section 9.10.1.1.

9.10.8 Bit Error Rate Performance Estimates of the Detectors

Normally the bit error rate (BER) performance is a meaningful metric for channel data quality. As a standard to compare against, we use the single user, with no other users on the channel for BPSK modulation. Thus the k-th user will have a BER given by

$$PE_k(\gamma_k) = Q\left(\sqrt{2\gamma_k}\right) \tag{9.10.8-1}$$

in which γ_k is defined by

$$\gamma_k = \frac{E_k}{N_0} \tag{9.10.8-2}$$

and N_0 is the one-sided noise spectral density of the additive white Gaussian noise (AWGN) channel and E_k is the energy per bit in the k-th signal.

9.10.8.1 Bit Error Rate Performance of the Optimum Detector

The analysis for the BER of the optimum receiver (Section 9.10.1.1) for the synchronous case is very difficult [28] and will not be attempted here. However, (9.10.8-1) serves as a useful lower bound on the BER of the optimum receiver and a suboptimum receiver's performance can be used for an upper bound.

9.10.8.2 Bit Error Rate Performance of the Conventional Synchronous Detector

For this case the output of the correlator for the k-th user is given by (9.10.2-2) which is repeated here:

$$r_k = \sqrt{E_k}d_k(1) + \sum_{\substack{n=1 \\ n\neq k}}^{K} \sqrt{E_n}d_n(1)\rho_{nk}(0) + n_k(1) \tag{9.10.8-3}$$

It follows that the BER of the k-th user [28], conditioned on a sequence $\boldsymbol{d}_k$ (see 9.10.1-12), is given by

$$PE(\boldsymbol{d}_k) = Q\left(\sqrt{2\left[\sqrt{E_k} + \sum_{\substack{j=1 \\ j\neq k}}^{K} \sqrt{E_j}d_j(1)\rho_{jk}(0)\right]^2 \Big/ N_0}\right) \tag{9.10.8-4}$$

where $Q(x)$ is defined by

$$Q(x) = \int_x^{\infty} \frac{1}{\sqrt{2\pi}} e^{-\frac{y^2}{2}} dy \tag{9.10.8-5}$$

Making the assumption that each bit is equally likely, it is clear that the K–1 bits can take on any one of the 2^{K-1} sequences, and the probability of any sequence is 2^{-K+1}. Thus we have that the average BER is given by

$$PE_k = \left(\frac{1}{2}\right)^{K-1} \sum_{j=1}^{2^{K-1}} PE_k(\boldsymbol{d}_j) \tag{9.10.8-6}$$

The minimum signal-to-noise ratio (SNR) occurs when the second term in the argument of (9.10.8-4) is the most negative. This occurs when the algebraic sign is negative for all terms, hence

$$SNR_{\min} = \frac{1}{N_0}\left[\sqrt{E_k} - \sum_{\substack{j=1 \\ j\neq k}}^{K} \sqrt{E_j}\left|\rho_{jk}(0)\right|\right]^2 \tag{9.10.8-7}$$

Since the BER is given by (9.10.8-6) the tightest lower bound occurs by choosing the one largest single term. This occurs when the sequence that yields the minimum SNR occurs, which yields the left-hand side of (9.10.8-8). The tightest upper bound can be obtained by choosing the one term that has the largest sequence BER (i.e., the one with the minimum BER). If all terms are upper-bounded by this term, and the sum indicated in (9.10.8-6) is taken, the right-hand side of (9.10.8-8) is obtained. Thus the average BER is bounded by

$$\left(\frac{1}{2}\right)^{K-1} Q\left(\sqrt{2SNR_{\min}}\right) \leq PE_k \leq Q\left(\sqrt{2SNR_{\min}}\right) \tag{9.10.8-8}$$

9.10.8.3 Bit Error Rate Performance of the Decorrelating Detector

Now consider the decorrelating detector discussed in Section 9.10.3. When a decorrelating detector is utilized all the other user interference terms are completely eliminated. However the noise will increase, as we will see in an extension of an earlier example. The BER can be written as [28]

$$PE_k = Q\left(\sqrt{\frac{E}{\sigma_k^2}}\right) \tag{9.10.8-9}$$

in which σ_k^2 is the variance of the k-th element of the estimate $\boldsymbol{d}_K^{opt}$, where $\boldsymbol{d}_K^{opt}$ is defined in (9.10.3-6).

Example 9 In order to clarify these concepts let us consider an example in the synchronous case for two users indicated in (9.10.3-13). Now consider the BER performance of this detector. From (9.10.3-13) the signal term for the first signal is given by $\sqrt{E_1}$. The noise component of the first signal is given by

$$n_{first} = \frac{n_1 - \rho n_2}{1-\rho^2} \tag{9.10.8-10}$$

in which ρ is the correlation between the $sg_1(t)$ and $sg_2(t)$. The corresponding variance is defined by

$$Var\left(n_{first}\right) = \frac{E\left[\left(n_1 - \rho n_2\right)^2\right]}{\left(1-\rho^2\right)} \tag{9.10.8-11}$$

Each noise term in this equation can be written as

$$n_j(1) = \int_0^T n(t) sg_j(t) dt \quad (9.10.8\text{-}12)$$

And the variance of each noise term can be written as

$$E\left[\left(n_j(1)\right)^2\right] = \frac{N_0}{2}\int_0^T sg_j^2(t)dt = \frac{N_0}{2} \quad (9.10.8\text{-}13)$$

Therefore the numerator of (9.10.8-11) can be evaluated as

$$E\left[\left(n_1 - \rho n_2\right)^2\right] = \frac{\left(\frac{N_0}{2}\right)\left(1-\rho^2\right)}{\left(1-\rho^2\right)^2} = \frac{N_0}{2\left(1-\rho^2\right)} \quad (9.10.8\text{-}14)$$

It follows that the BER from (9.10.8-9) is given by

$$PE_{first} = Q\left(\sqrt{\frac{2E_1}{N_0}\left(1-\rho^2\right)}\right) \quad (9.10.8\text{-}15)$$

Thus The BER has an effective SNR that is $(1-\rho^2)$ smaller than the single user cases. It is easy to show that the second signal has the same variance. In conclusion for the decorrelating detector it can be seen that the cost of removing the inference results in a larger noise term with a resulting degradation in BER when compared to the single user BER performance.

9.10.8.4 Bit Error Rate Performance of the MMSE Detector

The MMSE estimate is similar to the decorrelating detector as can be seen by comparing (9.10.3-6) and (9.10.4-22). In fact when the noise is small compared to the interference, the two detectors are basically the same. Actually for small multiuser interference levels the MMSE detector has smaller noise increase, but also has some residual bias due to the other user's interference.

9.11 AN EXAMPLE OF A CDMA SYSTEM: CDMA2000

Some of the concepts presented in this chapter may well be best understood by reviewing some specific examples that are currently used in practice. To that end we will summarize the architecture and some details of cdma2000 and WCDMA (wideband CDMA, also known as UMTS (universal mobile telecommunication system)).

This section provides some details [41–45] on the CDMA mobile radio standard cell phone system cdma2000 air interface between users in the same area of operations (AOO) and the link layers used in cdma2000. By 2003 there were more than 73 million cdma2000 subscribers in 24 countries.

This section covers the important aspects of the forward link (base station to user) and the reverse links (user to base station). cdma2000 is the third generation (3G) wireless standard developed from the second generation versions of IS-95A and IS-95B and cdmaOne. cdma2000 was designed to be backward compatible with the existing cdmaOne (IS-95) family of standards. References [41, 42] were the primary basis for this section.

cdma2000 uses a CDMA air interface that provides wire-line quality voice and data devices ranging from 144 Kbps for mobile stations to 2 Mbps for stationary ones.

In order to proceed it is useful to review (or perhaps see for the first time) the International Standards Organization (ISO) protocol hierarchy [24]. Figure 9.11-1 illustrates the ISO protocol hierarchy.

The lowest level of the hierarchy, level 1, is the *physical level* where the physical interface is defined. The meaning of physical interface with the network means the pin connections, electrical voltage levels, and signal formatting, such as the data rate.

Level 2 is known as the *data link level.* It controls the data link between the user and the network. This level also defines the data format, error control and recovery procedures, data transparency, and the implementation of certain command sequences. For networks that are not switched or the interface of simple terminals with computers via point-to-point services, generally only levels 1 and 2 are needed.

The *network level* forms level 3. It defines most of the protocol-driven functions of the packet network interface, or the internal network. The flow-control procedures are employed and its switched services are initiated through a data call establishment procedure at this level.

Level 4 is known as the *transport level.* It assures the end-to-end flow of complete messages. If the network requires that messages be broken down into segments or packets at the interface, the transport level assures that the message segmentation takes place and that the message is properly delivered.

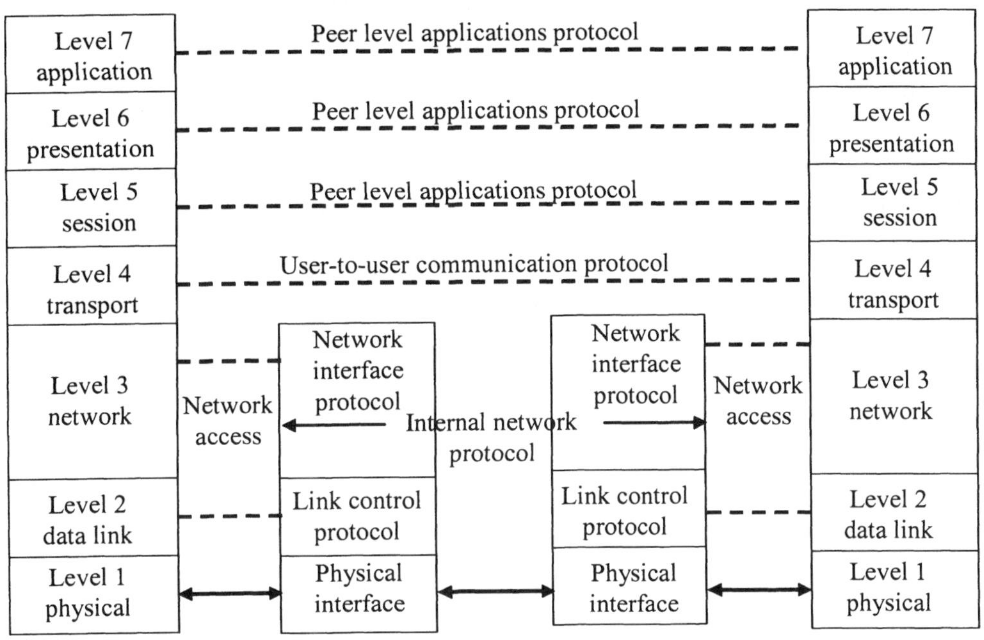

Figure 9.11-1 International Standards Organization (ISO) protocol hierarchy.

Session control is performed in level 5. It controls the interaction of user software, which is exchanging data at each end of the network. Session control includes such activities as log-on, user authentication, and the allocation of ADP resources within the user equipment, and manages and terminates connections (sessions) between cooperating applications.

Level 6 is the *presentation level.* It controls the display formats, data code conversion, and information going to and from peripheral storage devices. It provides independence to the application processes from differences in data presentation (syntax).

Level 7 is the user process or user *application level* and deals directly with the software application programs that interact through the network.

Implementation of software that can operate at the levels of 5, 6, and 7 has been slow. The software at all of these levels (often called peer-level software) tends to be both equipment and application dependent. However, the layered approach to protocol development achieves a degree of isolation and modularity between the various layers. Hence changes in one level can be made without changes in any other level.

With this short outline of the ISO model we can now proceed to the cdma2000 layering structure.

9.11.1 cdma2000 Layering Structure Overview

Figure 9.11-2 illustrates the layering structure for cdma2000. Although the layering structure is not part of the air interface it will be presented briefly to give an overall picture of the system.

The upper layers contain the three basic services: (1) *voice services* including PSTN (public switched telephone network) access mobile-to-mobile voice services, and Internet telephony; (2) *end-user data-bearing services* that deliver any form of data on behalf of mobile end users, including packet data (e.g., IP service), circuit data services (e.g., B-ISDN emulation services), and short message service (SMS); (3) *signaling services* that control all aspects of operation of the mobile unit. "Mobile," "mobile unit," and "mobile station" will be used interchangeably in the next two sections.

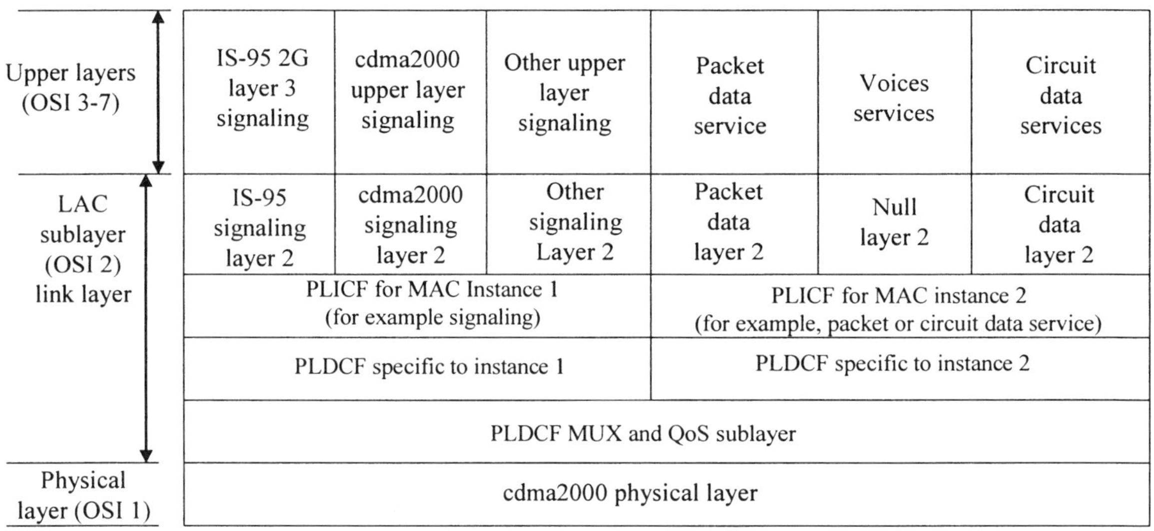

Notes: PLICF denotes "physical layer independent convergence function," PLDCF denotes "physical layer dependent convergence function," and QOS denotes "qualilty of service."

Figure 9.11-2 cdma2000 layering structure.

In Figure 9.11-2 MAC denotes the medium access control, LAC denotes the link access control, PLICF denotes the physical-layer independent convergence function, and PLDCE denotes the physical layer dependent convergence function.

The LAC sublayer manages the point-to-point communications channels between peer upper layer entities and provides framework to support a wide range of different end-to-end reliable link layer protocols. We will not delve into the layering aspects of the link; however, the interested reader can consult [41] for more details.

9.11.2 Forward Link and Reverse Link Channels Overview

The forward physical channels that are used in cdma2000 and IS-95 are shown in Figure 9.11-3.

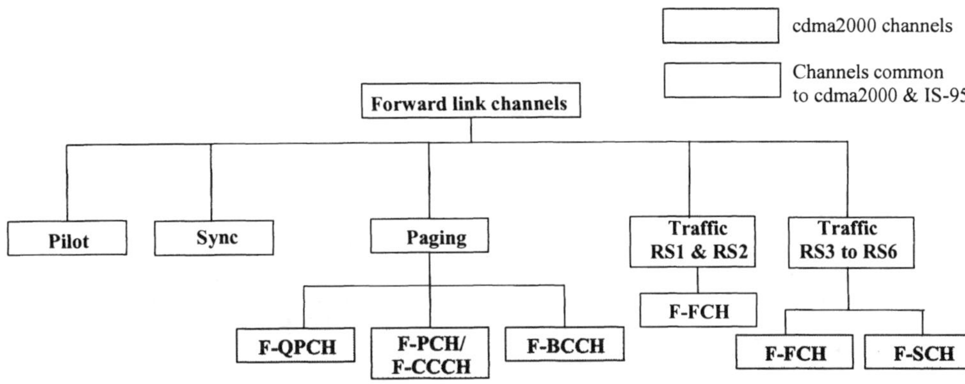

Figure 9.11-3 The forward physical channels for cdma2000 and IS-95.

The following abbreviations are used in Figure 9.11-3: F-FCH is the forward fundamental channel, F-QPCH is the forward quick paging channel, F-PCH is the forward paging channel, F-CCCH is the common control channel, F-BCCH is the forward broadcast common channel, and F-SCH is the forward supplemental channel. RS denotes the rate set. The shaded blocks are the new cdma2000 channels, and the unshaded blocks denote the blocks common to both IS-95 and cdma2000.

The reverse link structure is illustrated in Figure 9.11-4. Both the forward and reverse links will be discussed in more detail in the following.

In Figure 9.11-4 the following definitions have been used. R-ACH is the reverse access channel, R-CCCH is the reverse common control channel, R-FCH is the reverse fundamental channel, and R-SCH is the reverse supplemental channel. Again RS denotes the particular rate set. The shaded blocks are common to both IS-95 and cdma2000.

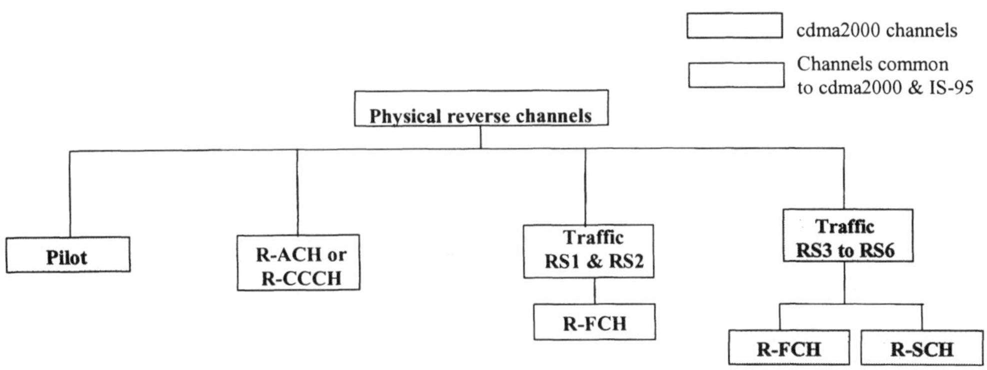

Figure 9.11-4 Reverse physical channels for cdma2000 and IS-95

9.11.3 Physical Layer of cdma2000

The physical layer provides coding and modulation services for a set of logical channels used by physical layer dependent convergence function (PLDCF) MUX and the quality of service (QoS) sublayer. The physical channels are classified as forward/reverse dedicated physical channels (F/R-DPHCH) and forward/reverse common physical channels (F/R-CPHCH). F/R-DPHCH is a collection of all physical channels that carry information in a dedicated, point-to-point manner between the base station and a single mobile unit. F/R-CPHCH is a collection of all physical channels that carry information in a shared access, point-to-multipoint manner between the base station and multiple users. Figure 9.11-5 [42] illustrates the cdma2000 physical channels. The major features of cdma2000 and IS-95 are illustrated in Table 9.11-1. In Table 9.11-1 EVRC denotes enhanced variable rate codec.

Table 9.11-1 Major Features in cdma2000 and IS-95

Characteristics	IS-95	cdma2000
Core network	ANSI-41	ANSI-41
Chip rate	1.2288 Mcps	N*1.2288 Mcps N=1,3,6,9,12
Frame length	20 ms	20 ms
Voice coder	EVRC	EVRC
Modulation	QPSK	QPSK
Data rate	Nominal: 9.6 Kbps	From 9.6 – 921.6 Kbps
Filtered bandwidth	1.23 MHz	N*1.25 MHz N=1,3,6,9,12
Coding	Convolutional with Viterbi decoding	Convolutional and Turbo decoding
Interleaving	20 ms	N*20 ms N=1,2 or 4

The primary differences in cdma2000 from IS-95 are the increased chip rate, channel bandwidth, interleaver depth, Turbo coding, and the data rates.

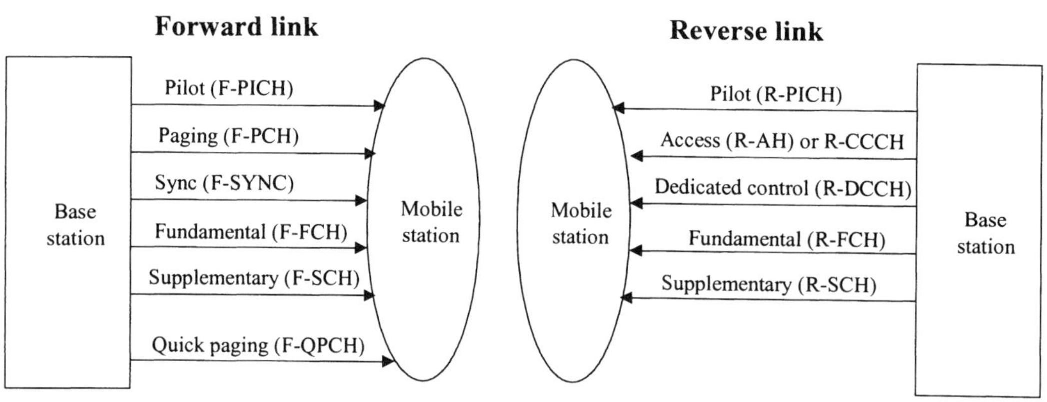

Figure 9.11-5 cdma2000 physical channels.

9.11.4 Forward Link Physical Channels

Forward dedicated channels provide information from the base station to a particular station (mobile). Common channels carry information from the base station to a set of mobile units in a point-to-multipoint manner. Table 9.11-2 indicates these channels. A brief discussion of the forward link channels will now be described.

Table 9.11-2 Forward Link Channels

Forward Channels	Physical Channel	Channel Name
Forward common physical channels (control and overhead hannels)	Forward pilot channel	F-PICH
	Forward paging channel	F-PCH
	Forward sync channel	F-SYNC
	Forward common control channel	F-CCCH
	Forward common auxiliary pilot channel	F-CAPICH
	Forward quick paging	F-QPCH
	Forward broadcast common channel	F-BCCH
Forward dedicated physical channels	Forward dedicated auxiliary pilot channel	F-DAPICH
	Forward dedicated common control channel	F-DCCH
	Forward traffic channel	
	-Fundamental	F-FCH
	-Supplementary	F-SCH

9.11.4.1 The Forward Pilot Channel (F-PICH)

The F-PICH is continuously broadcast throughout the cell to provide phase information and timing. The F-PICH is used by all traffic channels and is used for estimating phase and gain, detecting multipath rays so that the RAKE[7] fingers are efficiently assigned to the strongest multipath component and for cell acquisition and handoff. The F-PICH channel spreads the all-zeros sequence with Walsh code number 0. The pilot is shared between all mobiles in the cell and is used to obtain fast acquisition on new multipath components and obtain, by estimation, the multipath phase and signal strength. This common pilot can be transmitted without a lot of overhead. The common pilot approach is more efficient than providing a pilot for each user. In voice traffic the common pilot can yield better channel estimation with the result of better receiver performance. It can be utilized for improved search and handoff performance.

9.11.4.2 The Forward Paging Channel (F-PCH)

The F-PCH is used to send control information and paging messages from the base station to users and operates at 9.6 or 4.8 Kbps (the same as IS-95). The F-PCH carries overhead messages, pages, and acknowledges channel assignments, status request, and shared secret data (SSD) updates from the base station to the mobile user. There are two types of paging channels: *shared* F-PCH and *wideband* F-PCH. The shared F-PCH provides service to both the IS-95B and cdma2000 when using the F-PCH in an IS-95B underlay channel. The wideband F-PCH is modulated across the entire band. This mode is applicable to both overlay and nonoverlay configurations.

[7] The RAKE receiver will be discussed in Chapter 10.

9.11.4.3 Forward Sync Channel (F-SYNC)

The mobile units, operating within the coverage area of the base station to obtain initial time synchronization, utilize the sync channel. There are two types of F-SYNC channels: (1) *wideband* F-SYNC and *shared* F-SYNC. The wideband F-SYNC is modulated over the entire wideband channel bandwidth. It is modulated as a separate channel within the forward common physical channel (F-CPHCH). This mode is applicable to both the nonoverlay and overlay configurations. The shared F-SYNC channel provides service to both the IS-95B and cdma2000 when using the F-SYNC in an IS-95B underlay configuration.

9.11.4.4 The Forward Common Control Channel (F-CCCH)

The forward common control channel (F-CCCH) is a common channel used for communication of layer 3 and median access control (MAC) messages from the base station to the mobile user. Frame sizes for the F-CCCH are 5, 10, or 20 ms, depending on the current operating environment. The data rates range from 9.6, 19.2, and 38.4 Kbps. It is identical to F-PCH for a 9.6 Kbps rate (20 ms frame). The F-CCCH provides a short burst capability for data communication.

9.11.4.5 The Forward Common Auxiliary Pilot Channel (F-CAPICH)

The F-CAPICH is used with antenna beam-forming application to generate spot beams. As mentioned earlier in this chapter, spot beams can be used to increase coverage in a particular geographical area or to increase capacity in areas that need it. The F-CAPICH can be shared among multiple mobiles in the same spot beam. There is no data on the pilot channels. Auxiliary pilots are code multiplexed with other forward link channels, and they utilize Walsh codes.[8] Auxiliary pilots can be also used for orthogonal diversity transmission in the forward spread spectrum link. If directional or spot beams are supported with an antenna array, it is necessary to provide a separate forward link pilot for channel estimation.

9.11.4.6 The Forward Broadcast Common Channel (F-BCCH)

The F-BCCH is a paging channel dedicated to carrying only overhead messages and possible SMS ((wireless) short message service) broadcast messages from the paging channel to a separate broadcast channel. It removes the overhead messages from the paging channel to a separate broadcast channel. By doing this the mobile initialization time and system access performance is improved. The F-BCCH has a fixed Walsh code that is communicated to the mobile on the F-SYNC.

9.11.4.7 The Forward Quick Paging Channel (F-QPCH)

The F-QPCH is used by the base station and is a new type of paging channel and is used when it needs to contact the mobile user, in the slotted mode. Reduced time needed to be awake is obtained when it is used, which results in decreased battery drain for the mobile user. The F-QPCH contains the *quick page message* contains a single bit message. The F-QPCH page message directs a slotted-mode mobile user to monitor its assigned slot on the paging channel, that will follow immediately. The quick page message is transmitted up

---

[8] Walsh codes are baseband orthogonal codes. An example for the case of four of them is (1,1,1,1), (1,–1,1,–1), (1,1,–1,–1), and (1,–1,–1,1). Each code is orthogonal to each other. For $n = 2^k$ we have $H_{2^k} = \begin{bmatrix} H_{2^{k-1}} & H_{2^{k-1}} \\ H_{2^{k-1}} & \bar{H}_{2^{k-1}} \end{bmatrix}$ where the bar denotes the same matrix with inverted elements and $H_2 = \begin{bmatrix} 1 & 1 \\ 1 & -1 \end{bmatrix}$.

to 80 ms prior to the page message to alert the user to the paging channel. It uses a different modulation, so that will look like a different physical channel.

9.11.4.8 The Forward Fundamental Channel (F-FCH)

The F-FCH is transmitted at a variable rate as in the IS-95B standard and therefore requires data rate detection in the receiver. Each F-FCH is transmitted on a different orthogonal code channel and uses 5 and 20 ms frame sizes. The 20 ms frame supports the data rate corresponding to Rate Set 1 (RS1) and Rate Set 2 (RS2), where the rates are 9.6, 4.8, 2.7, and 1.5 Kbps for RS1 and 14.4, 7.2, 3.6, and 1.8 Kbps for RS2. The N=1 F-FCH for RS1 is shown in Figure 9.11-6. In this case a rate ½ convolutional encoder is used.

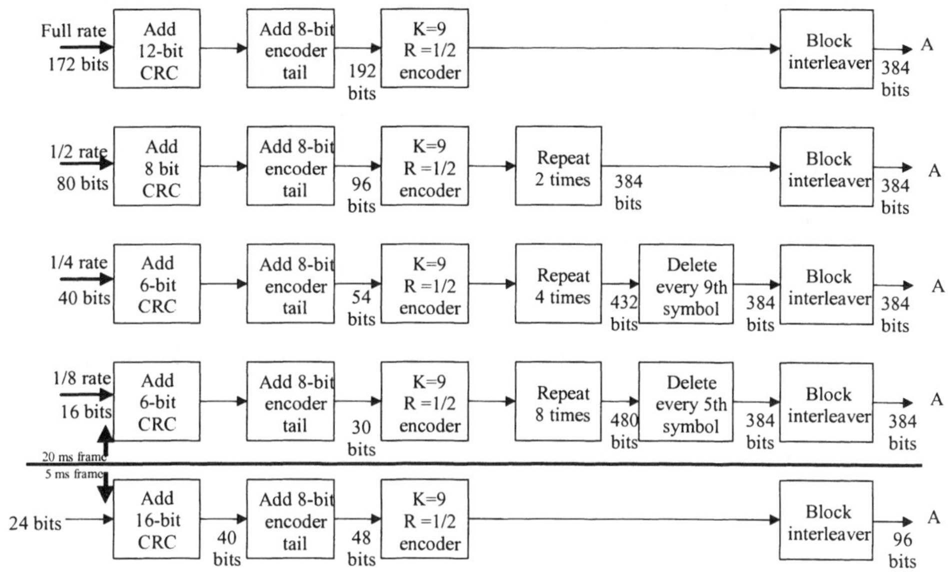

Note: Point A goes to point A on Figure 9.11-13 for the I-Q mapping, Walsh modulation, PN spreading, and frequency modulation, for N=1.

Figure 9.11-6 Forward fundamental channel for N=1 for RS1.

For the case of N= 1 and RS2, a rate 1/3 convolutional code, followed by puncturing every ninth bit effectively provides an effective code rate of 3/8. The case for RS2 and N=1 is illustrated in Figure 9.11-7.

In the figure the asterisk denotes the fact that the effective rate is 3/8. The two cases (RS1 and RS2) are relatively similar, but not identical. The cases for N = 3, 6, 9, and 12 for RS1 and RS2 can be found in Garg [42].

9.11.4.9 Forward Supplemental Channel (F-SCH)

The F-SCH can be operated in two different modes. The first mode limits the data rate to 14.4 Kbps and detects without scheduling or rate information. The variable rates correspond to the IS-95 RS1 and RS2 rates. The variable rate mode structure is identical to the F-FCH for the 20 ms frame case.

In the second mode the rate information is provided to the base station. The high data rates can have either K = 9 convolutional coding or turbo coding with K = 4 component coders. When convolutional coding is used there are 8 tail bits. In the case when turbo coders are used, 6 tail bits and 2 reserve bits are used.

The forward supplementary channel supports 20 ms frame periods. When data rates are derived from RS1, the F-SCH supports data rates starting at 9.6 up to 307.2 Kbps. Figure 9.11-8 illustrates F-SCH for $N = 1$ with a channel bandwidth of 1.25 MHz and Figure 9.11-9 for the case $N = 3$.

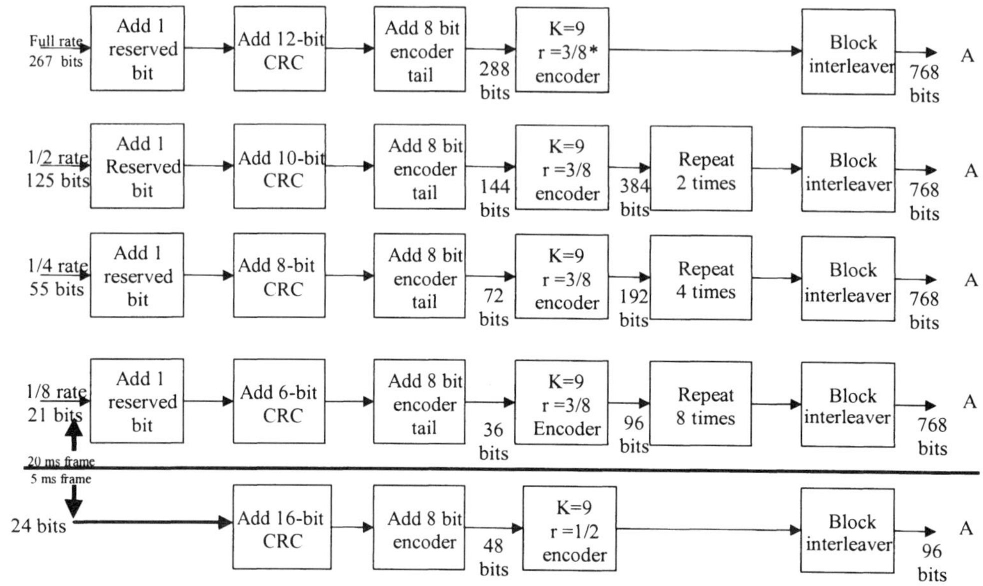

Note: Point A goes to point A on Figure 9.11-13 for the I-Q mapping, Walsh modulation, PN spreading, and frequency modulation for $N = 1$.

Figure 9.11-7 Forward fundamental channel for $N = 1$ for RS2.

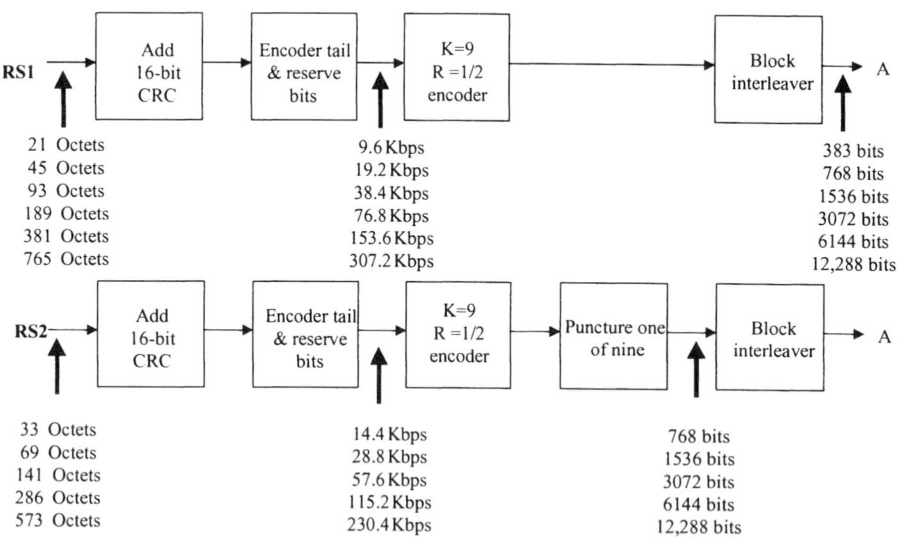

Note: Point A goes to point A on Figure 9.11-13 for the I-Q mapping, Walsh modulation, PN spreading, and frequency modulation for $N = 1$.

Figure 9.11-8 F-SCH for $N = 1$, multicarrier, and a channel bandwidth of 1.25 MHz.

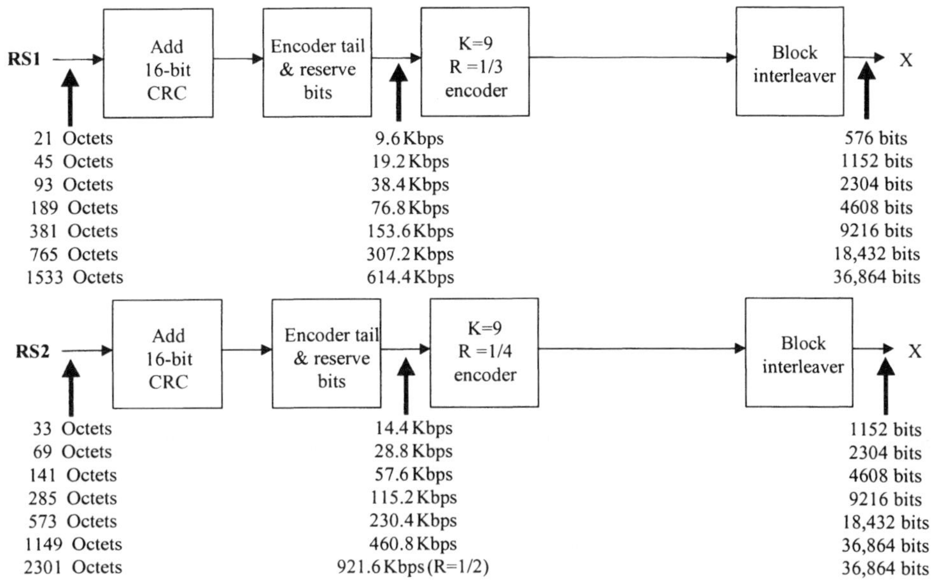

Note: For the case of multicarrier spreading, signal point X feeds into Figure 9.11-14 and for $N>1$ signal point X feeds into Figure 9.11-16.

Figure 9.11-9 F-SCH for $N = 3$ and a channel bandwidth of 3.75 MHz.

9.11.4.10 Forward Dedicated Control Channel (F-DCCH)

The F-DCCH supports both 5 ms and 20 ms frame sizes at the 9.6 Kbps encoder input rate. Sixteen CRC bits are added to the information bits, in the case of 5 ms frames, and 12 CRC bits are added to the 20 ms frames. For both frames the following items follow: 8 tail bits, convolutional coding, interleaving, and scrambling. Figures 9.11-10 and 9.11-11 illustrate the forward dedicated control channel for $N = 1$ and $N \geq 3$.

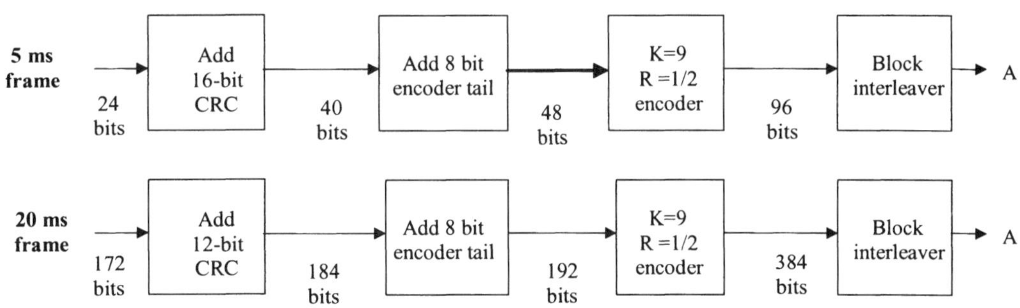

Note: Point A goes to point A on Figure 9.11-13 for the I-Q mapping, Walsh modulation, PN spreading, and frequency modulation for $N=1$.

Figure 9.11-10 Forward dedicated control channel (F-DCCH) for the case $N = 1$.

9.11.4.11 Forward Link Channelization (Physical Layer)

The forward link supports chip rates of 1.228N Mcps for $N = 1, 3, 6, 9$, and 12. For $N = 1$, the spreading is similar to IS-95B; however QPSK modulation and fast closed-loop power control is also used. There are two options for $N > 1$ (1) multicarrier (multiples of 1.25 MHz) and (2) direct spreading as shown in Figure 9.11-12.

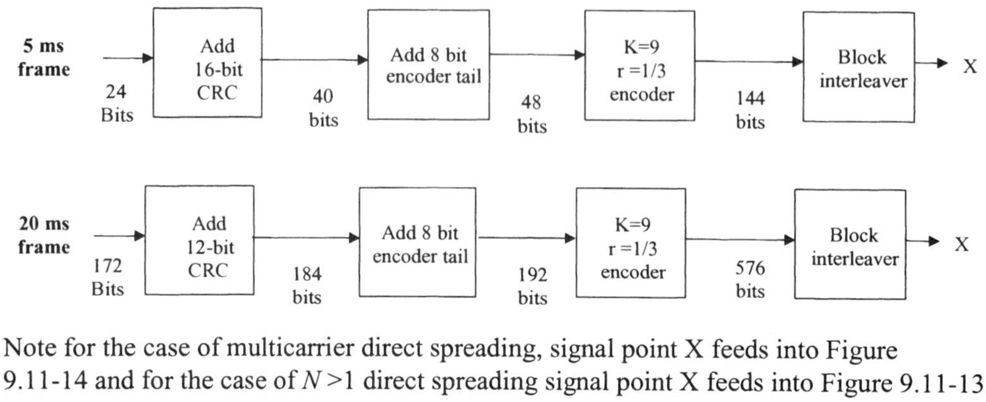

Figure 9.11-11 Forward dedicated control channel (F-DCCH) for $N \geq 3$.

The multicarrier approach demultiplexes modulation symbols onto N separate 1.25-MHz carriers ($N = 3, 6, 9, 12$). Each carrier is spread with a 1.2288-Mcps PN code chip rate. For the case when $N > 1$ the direct spread technique transmits modulation symbols on a single carrier, which is spread with a chip rate of $1.2288 \times N$ Mcps where $N = 3, 6, 9$, and 12. Figure 9.11-12 illustrates the $N = 3$ case.

9.11.4.12 Forward Link Common Pilot (Physical Layer Description)

The cdma2000 system provides a common code multiplexed pilot for all users. The common pilot is an all-zero sequence prior to Walsh code spreading with Walsh code number 0. This pilot is shared by all traffic channels, and thus it provides for an efficient utilization of resources. The pilot channel is used for the following purposes: (1) estimating channel gain and phase, (2) detecting multipath rays such that Rake fingers are efficiently assigned to the strongest multipath components, and (3) cell acquisition and handoff.

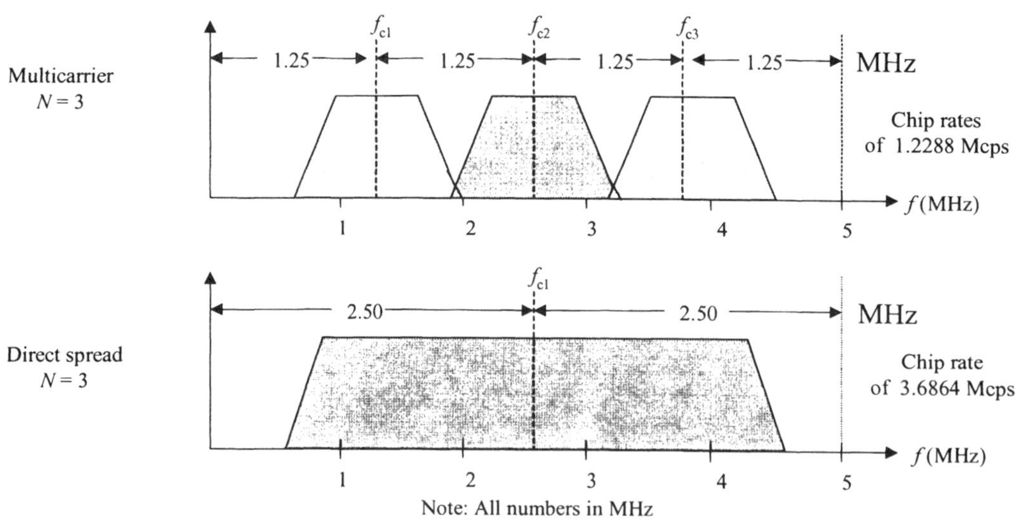

Figure 9.11-12 Multicarrier and direct spread configuration for $N = 3$.

A strong common pilot channel allows for more accurate estimation of the fading channel and faster detection of weak multipath rays than a per-user pilot approach. Furthermore, for a common pilot, it is possible to send the pilot signal without incurring significant overhead for each user. As a result, a system with a common pilot approach can achieve better performance than that in the case of a per-user pilot approach.

9.11.4.13 Forward Link Auxiliary Pilots

Certain applications such as antenna arrays and antenna transmit diversity require a separate pilot for channel estimation and phase tracking. Auxiliary pilots are code multiplexed with other forward link channels and use orthogonal Walsh codes. Since a pilot contains no data (i.e., the all-zeros data case), auxiliary pilots may use a longer Walsh code to lessen the reduction of orthogonal Walsh codes available for traffic. Auxiliary pilots can also be used for orthogonal diversity transmission in the direct spread forward link. Furthermore, if the CDMA system uses a separate antenna array to support directional or spot beams, it is necessary to provide a separate forward link pilot for channel estimation.

9.11.4.14 Forward Link Independent Data Channels

The cdma2000 system provides two types of forward link physical data channels (fundamental and supplemental) that can each be adapted to a particular type of service. The use of fundamental and supplemental channels enables the system to be optimized for multiple simultaneous services. The two separate physical channels are separately coded and interleaved and in general have different transmit power levels and frame error rate set points. Each channel carries a different type of service, depending on the service scenarios.

9.11.4.15 Forward Link Orthogonal Modulation

To reduce or eliminate intracell interference, each forward link physical channel is modulated by a Walsh code. To increase the number of usable Walsh codes, QPSK modulation is employed prior to spreading. Every two information bits are mapped into a QPSK symbol. As a result, the available number of Walsh codes is increased by a factor of two relative to BPSK (prespreading) symbols. Furthermore, the Walsh code length varies to achieve different information bit rates. The forward link may be interference limited or Walsh code limited depending on the specific deployment and operating environment. When a Walsh code limit occurs, additional Walsh-like codes may be created by multiplying Walsh codes by masking functions. The codes created in this way are called *quasi-orthogonal functions*. Other suboptimal codes may also be used.

9.11.4.16 Forward Link Carrier and Multicarrier Transmit Diversity

Transmit diversity can reduce the required transmit power per channel and thus enhance capacity. Transmit diversity can be implemented in two basic different ways: (1) multicarrier transmit diversity and (2) direct spread transmit diversity.

Multicarrier transmit diversity (MCTD) can be implemented with antenna diversity with no impact on the subscriber user (terminal), where a subset of carriers is transmitted on each antenna. The main characteristics of the multicarrier approach are the following: (1) coded information symbols are demultiplexed among multiple 1.25-MHz carriers, (2) frequency diversity is equivalent to spreading the signal over the entire bandwidth, (3) both time and frequency diversity are captured by convolutional coder/symbol repetition and interleaver, (4) the RAKE receiver[9] captures signal energy from all the bands,

[9] The RAKE receiver will be discussed in the following chapter.

(5) an identical Walsh code is used on all the carriers on each forward link channel, and (6) fast power control is used on the forward links.

For direct spread transmit diversity, *orthogonal transmit diversity* (OTD) may be used. A different orthogonal code is used per antenna for spreading. This maintains the orthogonality between the two output streams, and hence self-interference is virtually eliminated in a flat-fading environment. Note that by splitting the coded bits into separate streams, the effective number of spreading codes per user is the same as the case without OTD. An auxiliary pilot signal is introduced with the additional antenna.

9.11.4.17 Forward Link Rate Matching

The cdma2000 system uses several approaches to match the data rates to the Walsh spreader input rates. These approaches include: (1) adjusting the code rate, (2) using symbol repetition with or without symbol puncturing, and (3) sequence repetition. Subrates of speech signals are generated by symbol repetition and by symbol puncturing, when necessary. A supplementary channel rate that is not the same data rate as the main channel data rate is realized by sequence repetition or by symbol repetition with symbol puncturing to match the desired channel data rate. These rate-matching approaches provide flexibility in matching data rates to channel rates.

9.11.4.18 Fast Forward Power Control

The cdma2000 system uses fast closed-loop power control on the forward link dedicated channels with 800 updates per second. The closed-loop power control compensates for medium to fast fading and for inaccuracies in open loop power control. In addition fast forward link power control is effective for adaptation of dynamically changing interference conditions due to the activation and deactivation of high power high data rate users.

9.11.4.19 Frame Length

The cdma2000 system supports both 5 ms and 20 ms frames for control information on the fundamental and dedicated control channels, and utilizes 20 ms frames for other types of data, including voice. Interleaving and sequence repetition are applied over the entire frame interval. This improves the time diversity over systems that use shorter frame sizes.

9.11.4.20 Forward Error Correction

Forward error correction is composed of two basic types: convolutional coding and Turbo coding. The convolutional coding on the forward link uses $K = 9$ convolutional codes for the fundamental channel (F-FCH). The supplemental channel (F-SCH) uses $K = 9$ convolutional codes for rates up to and including 14.4 Kbps. Convolutional codes for higher data rates are optional on the F-SCH and Turbo codes are preferred. The F-SCH uses Turbo codes with $K = 4$, $R = 1/4$, 1/3, and 1/2 for data rates greater than 14.4 Kbps. Turbo codes are preferred for all data rates greater than 14.4 Kbps.

9.11.4.21 Modulation and Spreading for the Forward Link

The case for the $N = 1$ system can be deployed in a new spectrum environment or as a backward-compatible upgrade anywhere an IS-95B forward link is deployed in the same RF channel. The new cdma2000 channels have been designed so that they can coexist in an orthogonal manner with the code channels of the existing IS-95B system. The case for the $N = 1$ spreading, which is filtered quadriphase, is shown in Figure 9.11-13.

The first processing involves the user data scrambling by the user long code generator, followed by the inphase (I) and quadrature (Q) mapping, the setting of the channel gain, the optional power control symbol puncturing, and the Walsh function spreading.

Finally, as shown in Figure 9.11-13, the signal is inphase and quadrature (complex) PN spread, followed by baseband filtering and transmission. "X" denotes the input user data that is coded and has a CRC added (see Figure 9.11-11, for example).

9.11.4.22 Modulation and Spreading for the Multicarrier Forward Link

The multicarrier system can be deployed in a new spectral location or as a backwards-compatible upgrade anywhere a IS-95B can be employed in the same N RF channels.

The overall structure of the multicarrier CDMA channel is shown in Figure 9.11-14. After scrambling with the PN code corresponding to user m, the user data is demultiplexed into N carriers, where N = 3, 6, 9, or 12. On each carrier, the demultiplexed bits are mapped onto I and Q followed by Walsh spreading. When applicable, power control bits (PC bits), for reverse closed-loop power control, may be punctured onto the forward link channel at a rate of 800 Hz. The signal on each carrier is orthogonally spread by the appropriate Walsh code function (W_{nk}) in such a manner as to maintain a fixed chip rate of 1.2288 Mcps per carrier, where the Walsh code may differ on each carrier. The signal on each complex PN spread waveform, as shown in Figure 9.11-14, is followed by baseband filtering and frequency modulation. The box labeled "Complex PN spreading" is shown in Figure 9.11-15, and outputs the I and Q for Figure 9.11-14.

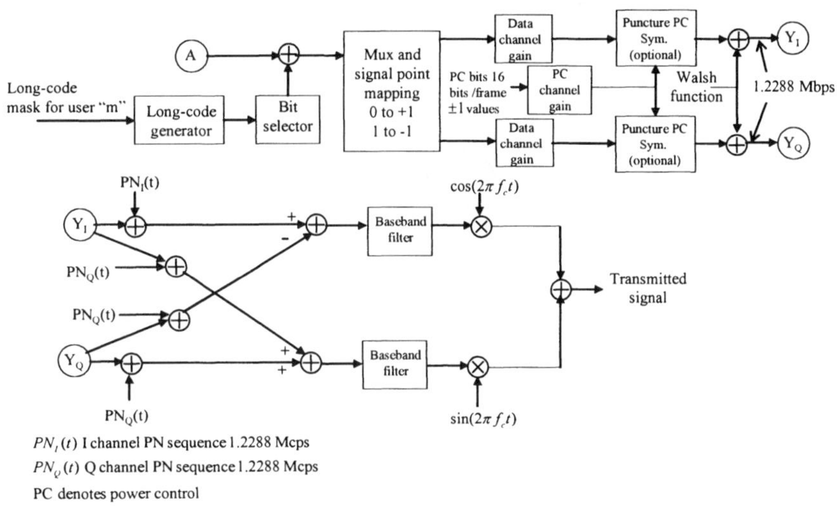

Figure 9.11-13 I and Q mapping, Walsh modulation, PN spreading, and baseband spreading for $N = 1$.

For the cases $N = 1, 3, 6, 9$, and 12 use direct spreading as shown in Figure 9.11-16. The user data is first scrambled by the user long PN code followed by I and Q mapping, channel gain, power control puncturing, and Walsh spreading. The power control bits may not be punctured onto the forward link channel depending on the specific logical-to-physical channel mapping. Next, the signal is complex PN spread, followed by baseband filtering and phase modulation. W_n denotes the n-th Walsh function.

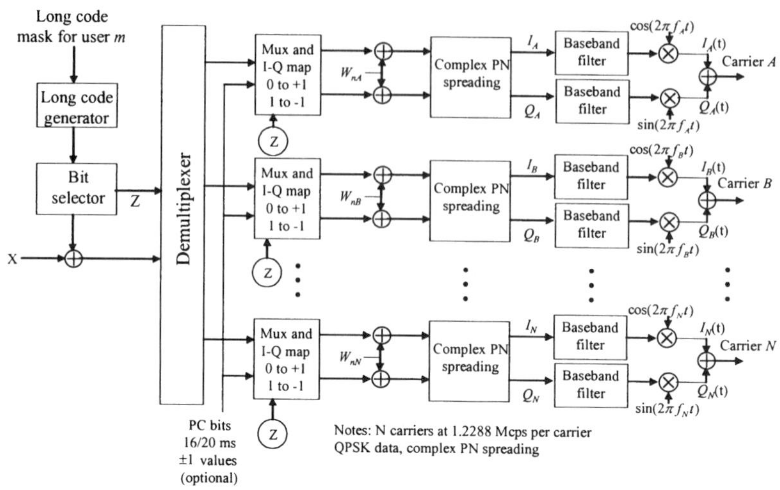

Figure 9.11-14 Multicarrier CDMA forward link structure.

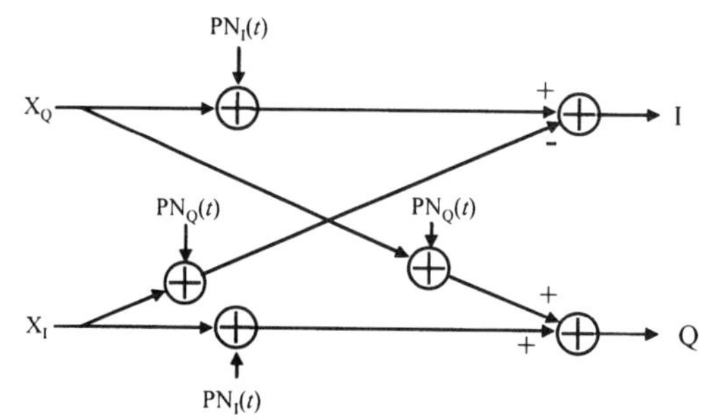

Note: Maps the complex input into a complex output.

Figure 9.11-15 Complex PN spreading for Figure 9.10-14.

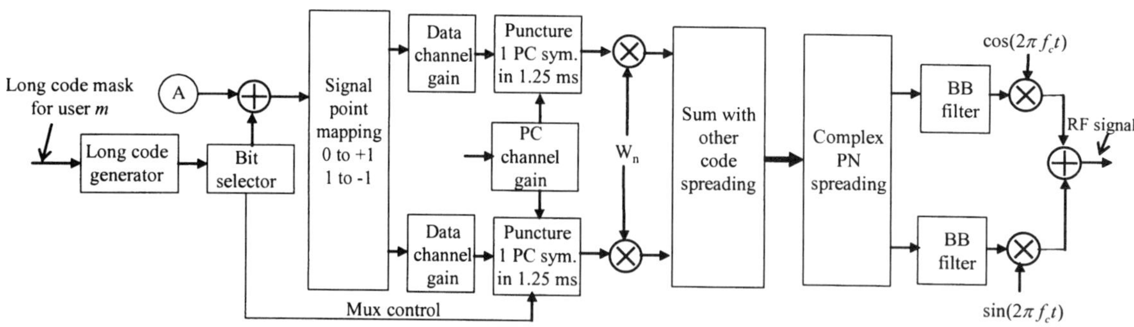

Figure 9.11-16 The case of N = 3, 6, 9, and 12 I-Q mapping and Walsh modulation.

9.11.4.23 Main Characteristics of the cdma2000 Forward Link

The main characteristics of the forward link are summarized in the following [42]:

(1) Channels are orthogonal and use variable length Walsh codes. Different length Walsh codes are used to achieve the same chip rate for different information rates.

(2) QPSK modulation is used before spreading to increase the number of usable Walsh codes.
(3) Forward error correction (FEC) is used.
   - Convolutional codes with $K = 9$ are used for voice and data.
   - Turbo codes with $K = 4$ are used for higher data rates on SCHs.
(4) Supports nonorthogonal forward link channelization.
   - These are used when running out of orthogonal space (not a sufficient number of Walsh codes available).
   - Quasi-orthogonal functions are generated by masking existing Walsh functions.
(5) It has a synchronous forward link.
(6) The forward link provides diversity.
(7) Fast forward power control in a closed loop mode is activated 800 times per second.
(8) The supplemental channel active set is a subset of the fundamental channel active set. The maximum data rate supported for RS3 and RS5 for the supplemental channel is 153.6 Kbps (raw data rate). RS4 and RS6 will be supported only for voice calls with the fundamental channel rates of up to 14.4 Kbps (raw data rate).
(9) Frame lengths are either one of those indicated here:
   - 20 ms frames are used for signaling and user information.
   - 5 ms frames are used for control information.

9.11.5 cdma2000 Reverse Physical Channels

The reverse physical channels (see Figure 9.11-5) provide a continuous waveform for all data rates, including both a continuous data channel and a continuous pilot channel from the mobile user to the base station. The continuous channel provides interleaving to be done over an entire frame. This allows the interleaving to achieve the full benefit of the frame time diversity. Table 9.11-3 lists the reverse channels.

9.11.5.1 cdma2000 Reverse Access Channel (R-ACH) and the Reverse Common Control Channel (R-CCCH)

The R-ACH and R-CCCH are common channels used for communication of layer 3 and the MAC (media access control) messages from the mobile unit to the base station. The R-CCCH is different from the R-ACH in that the R-CCCH provides enhanced capabilities beyond those of the R-ACH. As an example, the R-CCCH supports the lower latency access procedures required for efficient operation of the packet data in a suspended state.

Table 9.11-3 Reverse Physical Channels

Reverse Channel Type	Physical Channels	Channel Abbreviation
Reverse common physical channel	Reverse access channel	R-ACH
	Reverse common control channel (9.6 Kbps only)	R-CCCH
Reverse dedicated physical channel	Reverse pilot channel	R-PICH
	Reverse dedicated control channel	R-DCCH
	Reverse traffic control: -Fundamental -Supplemental	R-FCH R-SCH

Both R-ACH and R-CCCH are multiple access channels as mobile users transmit without explicit authorization by the (ground mobile) base station (GMBS). The R-ACH and R-CCCH use a slotted ALOHA type of operation with higher capture probabilities due to the CDMA properties of the channel (simultaneous transmission of multiple users). It is to be noted that there can be one or more access channels per frequency assignment. The different access channels are distinguished by their different long PN codes. The R-CCCH is identical to the R-ACH for the data rate of 9.6 Kbps, with 20 ms frames. The rates of 19.2 Kbps and 38.4 Kbps and frames of 5 ms and 10 ms will be supported for the R-CCCH. Figure 9.11-17 illustrates the structure of R-ACH and R-CCCH.

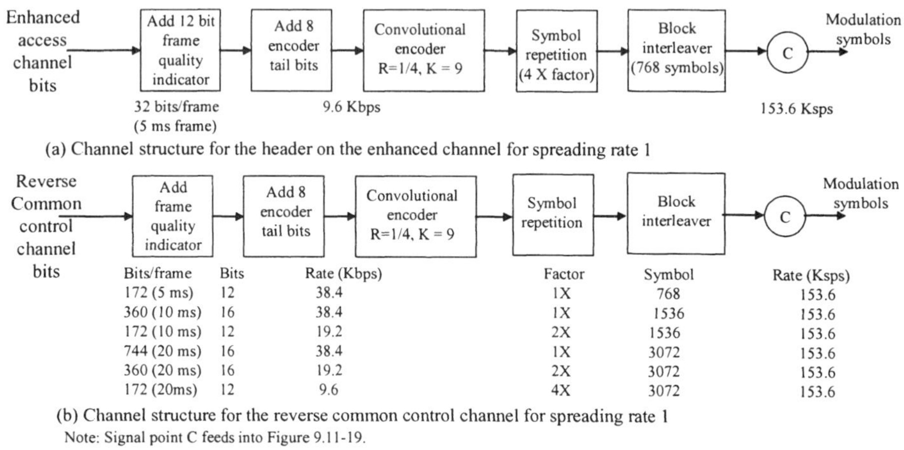

Figure 9.11-17 (a, b) Reverse access and reverse common control channel structure.

9.11.5.2 cdma2000 Reverse Pilot Channel

The reverse dedicated pilot channels (R-PICH) consist of a fixed reference value and multiplexed forward power channel control information, as shown in Figure 9.11-18. This time-multiplexed forward power control information is referred to as the *power control subchannel.* This subchannel provides information on the quality of the forward link at the rate of 1 bit per 1.25 ms power control group (PCG) and the forward link channels adjust their power via this PCG. The power control symbol repetition means that the 1-bit value is constant for that repeated symbol's duration. The power control bit uses the last portion of each PCG. The +1 pilot symbols and multiplexed power-control symbols are all sent at the same power level. The binary power control symbols are represented with ± 1 values in Figure 9.11-18.

The R-PICH signal is used for pilot signal time tracking, initial acquisition, power control measurements, and RAKE-receiver coherent reference recovery, as shown in Figure 9.11-18 for the R-PICH structure.

9.11.5.3 Reverse Dedicated Control Channel

The three channels the reverse dedicated control channel (R-DCCH), the R-FCH, and the R-SCH may or may not be used depending on the service scenario. A Walsh code sequence is used to spread each physical channel and thus provide orthogonal channelization among these physical channels. Both the spread pilot and the R-DCCH are mapped into the inphase (I) data channel.

Both the spread R-FCH and the R-SCH are mapped into the quadrature (Q) data channel. The I and Q data channels are then spread using a complex-type PN spreading approach with filtering, as shown in Figure 9.11-19. See Figure 9.11-20 for reverse the dedicated channel structure and for reverse link I and Q mapping for 1X and 3X.

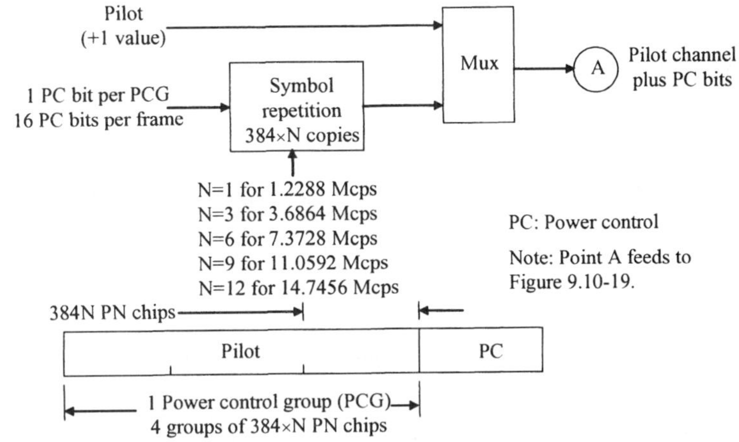

Figure 9.11-18 Reverse pilot channel (R-PICH) structure for reverse dedicated channels

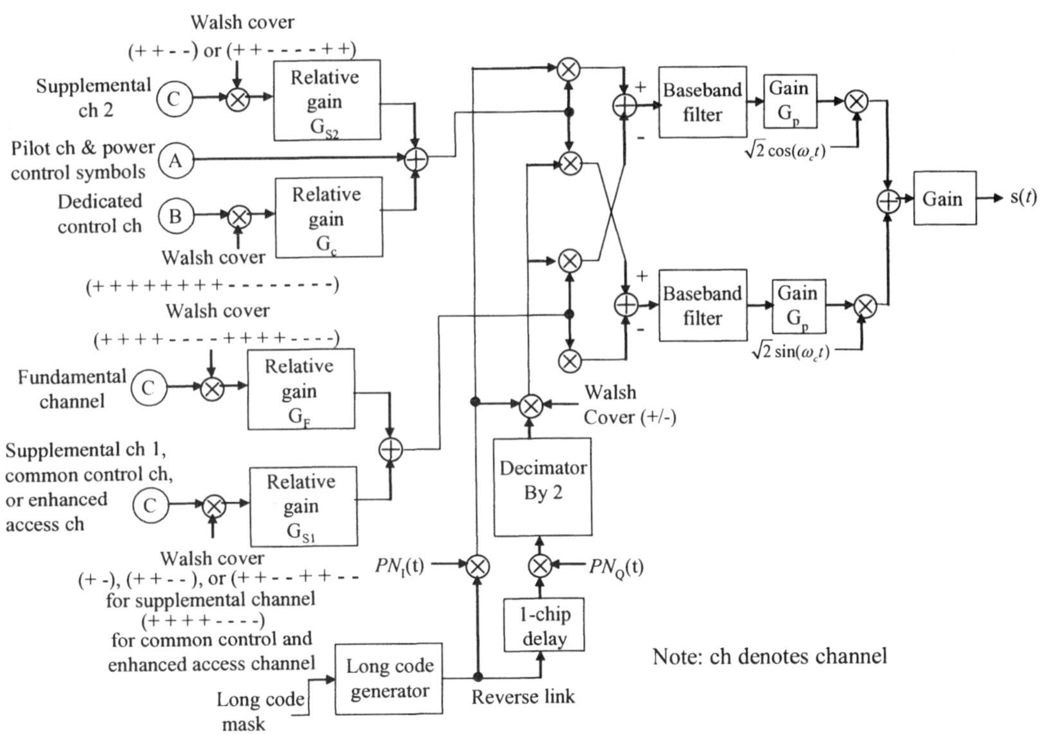

Figure 9.11-19 Reverse link I and Q mapping for 1X and 3X.

9.11.5.4 Reverse Fundamental Channel

The reverse fundamental channel (R-FCH) supports both 5 and 20 ms frames. The 20 ms frame provides rates derived from the IS-95B for RS1 or RS2. Within each 20 ms frame interval, either one 20 ms R-FCH structure, up to four 5 ms R-FCH structure(s), or no information can be transmitted. Also, when a 5 ms R-FCH structure is used, it can be either on or off in each of the four 5 ms segments of a 20 ms frame interval. The R-FCH is transmitted at different rates. The 5 ms frames provide 24 information bits per frame with a 16 bit cyclic redundant code (CRC). The rates supported for the R-FCH are 1.5, 2.7, 4.8, and 9.6 kbps for RS3 and RS5, and 1.8, 3.6, 7.2, and 14.4 Kbps for RS4 and RS6. Figures 9.11-21 and 9.11-22 illustrate R-FCH/R-SCH for 1X, radio configuration (RC) 4 and 3X, RC6.

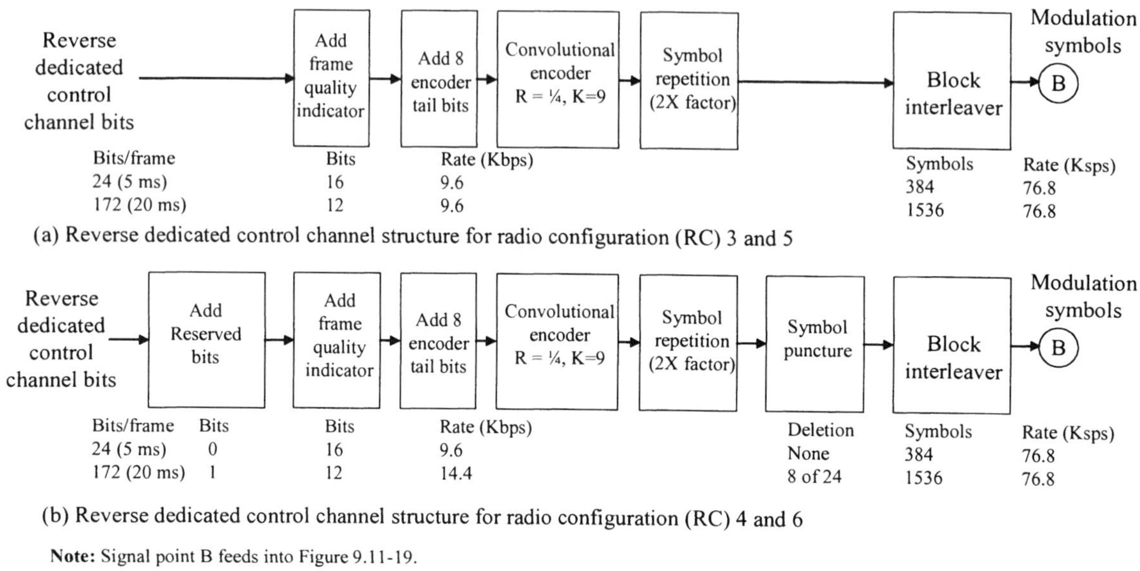

Figure 9.11-20 (a, b) Reverse dedicated channel structure.

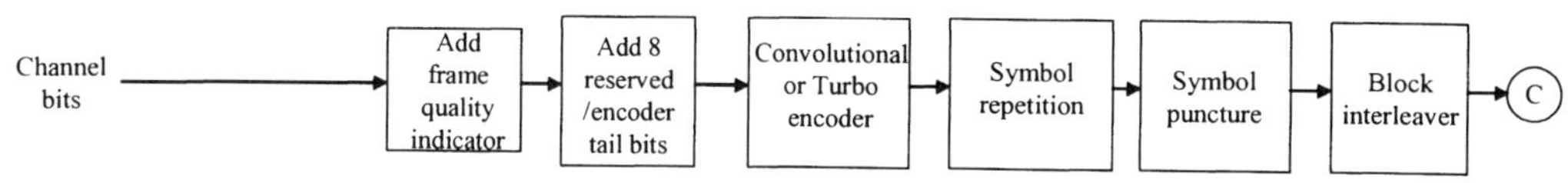

Bits/frame	Bits	Rate (Kbps)	Rate	Factor	Deletion	Symbols	Rate (Ksps)
24 (5 ms)	16	9.6	¼	2X	None	384	76.8
16	6	1.5	¼	16X	1 of 5	1536	76.8
40	6	2.7	¼	8X	1 of 9	1536	76.8
80	8	4.8	¼	4X	None	1536	76.8
172	12	9.6	¼	2X	None	1536N	76.8
360	16	19.2	¼	1X	None	1536N	76.8
744	16	38.4	¼	1X	None	3072N	153.6
1512	16	76.8	¼	1X	None	6144N	614.4
3048	16	153.6	¼	1X	None	12,288N	614.4
6120	16	307.2	¼	1X	None	12,288N	614.4

Notes:

1. The 5 ms frame is used only for the fundamental channels, and only rates of 9.6 Kbps or less are used for the fundamental channels.
2. Turbo coding may be used for the supplemental channels with rates of 19.2 Kbps or more; otherwise the K=9 convolutional coding is used.
3. With convolutional coding, the reserved tail bits provide an encoder tail. With Turbo encoding, the first 2 of these bits are reserved bits that are encoded and the last 6 bits are replaced by an internally generated tail.
4. N is the number of consecutive 20 ms frames which the interleaving is done (that is N =1, 2, or 4).
5. Signal point C feeds into Figure 9.11-19.

Figure 9.11-21 R-FCH/R-SCH for 1X, RC3 and 3X, RC5.

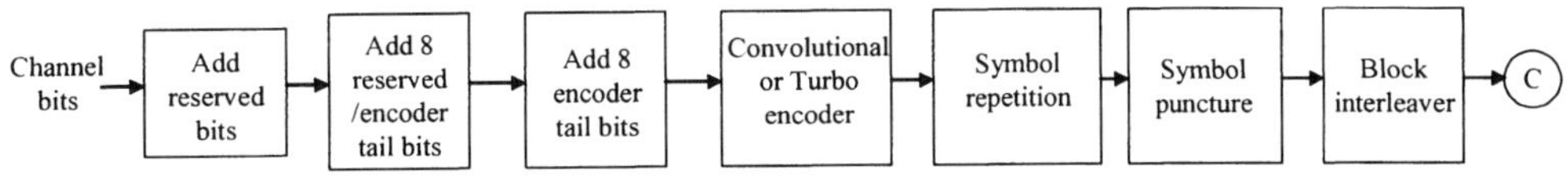

Bits/frame	Bits	Bits	Rate (Kbps)	Rate	Factor	Deletion	Symbols	Rate (Ksps)
24 (5 ms)	0	16	9.6	¼	2X	None	384	76.8
21	1	6	1.8	¼	16X	8 of 24	1536	76.8
55	1	8	3.6	¼	8X	8 of 24	1536	76.8
125	1	10	7.2	¼	4X	8 of 24	1536	76.8
267	1	12	14.4	¼	2X	8 of 24	1536N	76.8
552	0	16	28.8	¼	2X	8 of 24	3072N	153.6
1128	0	16	57.6	¼	2X	8 of 24	5144N	307.2
2280	0	16	115.2	¼	2X	8 of 24	12,288N	614.4
4584	0	16	230.4	¼	1X	8 of 24	12,288N	614.4

Notes:

1. The 5 ms frame is used only for the fundamental channels, and only rates of 14.4 Kbps or less are used for the fundamental channels.
2. Turbo coding may be used for the supplemental channels with rates of 28.8 Kbps or more; otherwise the K=9 convolutional coding is used.
3. With convolutional coding, the reserved tail bits provide an encoder tail. With Turbo encoding, the first 2 of these bits are reserved bits that are encoded and the last 6 bits are replaced by an internally generated tail.
4. N is the number of consecutive 20 ms frames which the interleaving is done (that is N =1, 2, or 4).
5. Signal point C feeds into Figure 9.11-19.

Figure 9.11-22 R-FCH/R-SCH for 1X, RC4 and 3X, RC6.

9.11.5.5 Reverse Supplementary Channel

There are two distinct modes for the reverse supplementary channel (R-SCH). In the first mode the data rates do not exceed 14.4 Kbps, and it uses blind rate detection (i.e., no scheduling or rate information). For the second mode, the base station explicitly knows the rate information. The R-SCH is used for data calls and can operate at different prenegotiated rates. Only RS3 and RS5 are supported for the R-SCH. Since only RS3 is supported on the R-FCH for high-speed packet data calls, the rates supported for the R-SCH are 9.6, 19.2, 38.4, 76.8, and 153.6 Kbps as indicated for R-FCH/R-SCH for 1X, RC4 and 1X, RC6 in Figure 9.11-22. It can be seen that Figures 9.11-21 and 9.11-22 are very similar.

9.11.5.6 Reverse Power Control

The power of the reverse link channels for a specific user is adjusted at a rate of 800 bits per second. The reverse power control bits are punctured onto a dedicated transmitter power. Figure 9.11-4 provides a comparison between the reverse physical channels used in IS-95A/B and cdma2000. This figure is located earlier in the cdma2000 section.

9.11.5.7 Reverse Link Forward Error Correction Coding

For R-FCH the reverse link utilizes a K = 9, R = 1/4 (code rate of 1/4) convolutional code. The distance properties of this code provide performance gains versus higher rate codes in fading and additive white Gaussian noise (AWGN) channel conditions. The constraint length K = 9, R = 1/4 convolutional codes provides a gain of about 0.5 dB over a K = 9, R = 1/2 in AWGN. The R-SCH uses convolutional codes for data rates of up to 14.4 Kbps.

Convolutional codes for higher data rates on the R-SCH are optional, and the use of Turbo codes is preferred, since they can offer even better bit error rate performance. A common constituent code is used for the reverse link. For all R-SCH links Turbo codes of constraint length 4, rate 1/4, 1/3, and 1/2 are used.

9.11.5.8 Main Reverse Link Physical Layer Characteristics for cdma2000

The main reverse link physical layer characteristics are listed next.

(1) **A continuous waveform.** cdma2000 utilizes a continuous pilot and a continuous data-channel waveform, for all data rates. This continuous waveform minimizes interference to external biomedical devices such as pacemakers and hearing aids, and permits a range increase at lower transmission rates. Interleaving can be performed over the entire frame, rather than just the portions that are not gated off, since it is a continuous waveform. The continuous waveform enables the interleaving processing to achieve the benefits of frame time diversity. The pilot signal is used by the base station for multipath searches, tracking, and as a reference for coherent demodulation, as well as to measure the quality of the link for power control purposes. cdma2000 utilizes separate orthogonal channels for the pilot and each of the data channels. Therefore, the relative power levels of the pilot and physical data channels can easily be adjusted, without changing the frame structure or power levels of some symbols of a frame in cdma2000.

(2) **Orthogonal channels using different length Walsh sequences.** cdma2000 uses orthogonal channels for the pilot and the other physical channels, as mentioned earlier. These orthogonal channels are provided with different length Walsh sequences. The higher-rate channels are used with shorter Walsh sequences. The use of shorter Walsh sequence allows high encoder output rates to be accommodated.

(3) **Low spectral side lobes.** The cdma2000 system achieves low spectral side lobes with nonideal mobile (nonlinear) power amplifiers by splitting the physical channels into an I and Q channel and using a complex-multiply-type PN spreading approach, along with filtering.

(4) **Rate matching.** cdma2000 utilizes several approaches to match the data rates to the Walsh function spreader input rates. These include adjusting the code rate using puncturing, symbol repetition, and sequence repetition. The design approach is to first try to use a low rate code, but not to reduce the code rate below $R = 1/4$, since additional coding gains for lower rates would be small, and the decoder implementation complexity would increase very significantly.

(5) **Independent data channels.** cdma2000 provides two types of physical data channels (R-FCH and R-SCH) that are used on the reverse link and that can be adapted to a particular type of service. The system can be optimized for multiple simultaneous services by the use of both R-FCH and R-SCH. These channels are separately coded and interleaved, and may have different transmit power level and frame error rate (FER) set points.

(6) **Reverse signal power control.** cdma2000 utilizes three components for reverse power control from the mobile to the base station: (a) *open loop*, (b) *closed loop*, and (c) *outer loop*. Open-loop power control adjusts the mobile transmitter power, based upon the power that is received at the user (mobile) terminal for the path loss (for a low path loss rate variation) from the mobile to the base station, and handles very slow fading rates. Closed-loop power control consists of an 800 bps feedback loop from the base station to the mobile unit to set the transmit power of the mobile unit. The closed-loop power control compensates for medium to fast fading and inaccuracies in the open-loop power control. The outer loop power control is dependent on the particular implementation, but normally adjusts the closed-loop power control threshold in the ground mobile base station to maintain a desired forward error rate.

(7) **Frame length.** The cdma2000 system uses both 5 and 20 ms frames for control information on fundamental and dedicated control channels. It also uses a 20 ms frame for the other types of data, including voice. Interleaving and sequence repetition are accomplished over the entire frame interval. This provides improved time diversity over systems that utilize shorter frame sizes. The 20 ms frames are used for voice. A shorter frame would reduce the one component of the total voice delay, but it would negatively affect the bit error rate demodulation performance, because of the shorter interleaving span.

(8) **Separate dedicated control channel.** The cdma2000 system reverse link consists of a separate low-rate, continuous, orthogonal, dedicated control channel. This channel provides for a flexible

and dedicated control channel structure, which does not impact the other pilot and physical channel frame structures.

9.11.5.9 Primary cdma2000 Reverse Link Modulation and Coding

Direct sequence spreading is used in the reverse link with an IS-95B chip rate of 1.2288 Mcps (denoted as a 1X chip rate) or chip rates that are 3, 6, 9, or 12 times the IS-95B chip rate. The higher chip rate systems are denoted by 3X, 6X, 9X, and 12X and they are have corresponding chip rates of 3.6863 Mcps, 7.3728 Mcps, 11.0592 Mcps, and 14.7456 Mcps.

The 1X system can be utilized whenever an IS-95B reverse link can be used. An IS-95B reverse link carrier frequency can be also be shared with users transmitting the IS-95B waveform and those transmitting the 1X cdma2000 waveform. In applications where larger bandwidth allocations are available the higher chip rate reverse links can be used.

Users that support a higher chip rate would typically also support the 1X chip rate. This will allow these mobiles to access ground mobile base stations that support only the 1X chip rate and allow operators with larger bandwidths the flexibility of using a mixture of 1X and higher chip rate systems.

The 1X cdma2000 reverse links, within an operator's allocated band, would typically occupy the same bandwidth as the IS-95B reverse link system (i.e., a 1.25-MHz bandwidth), and higher chip rate cdma2000 links would typically occupy a bandwidth that is equal to 1.25 MHz times the chip rate factor. Typically a guard band of 625 kHz (1.25/2 MHz) would be used on both sides of the operator's allocated band, to minimize cross channel interference.

The reverse CDMA channel is composed of reverse common channels and reverse dedicated channels.

The reverse common channel is used by the mobile station to initiate communications with the base station and to respond to forward link paging channel messages. The reverse common channel uses a random access protocol. reverse common channels are uniquely identified by their long code.

The reverse dedicated channel may be used for the transmission of user traffic and signaling information to the base station.

9.11.5.10 Key Characteristics of the Reverse Link

The key characteristics of cdma2000 reverse link are the following:

(1) The channels are primarily code multiplexed.
(2) Different QoS and physical layer characteristics use separate channels.
(3) Signal transmission is provided in a continuous format to avoid electromagnetic interference (EMI).
(4) The employment of Walsh functions renders the channels orthogonal. They are also I/Q split so that performance is equivalent to BPSK (the higher data rate channels imply shorter Walsh sequences).
(5) The links are a hybrid combination of QPSK and BPSK modulation.
(6) There is a coherent reverse link that has a continuous pilot signal.
(7) Forward power control information is time multiplexed with the pilot signal.
(8) Power peaking is reduced by restricting alternate phase changes of the complex scrambling, and also the side lobes are narrowed by this process, yielding a requirement for less channel bandwidth.
(9) Independent fundamental and supplemental channels have different transmit power levels and frame error rate (FER) target levels.
(10) Forward error correction coding schemes: Convolutional codes with $K = 9$ are used for voice and data. Parallel turbo codes with $K = 4$ are used for high data rates on the supplemental channels.
(11) Reverse power control is accomplished with (a) open loop, (b) closed loop, and (c) outer loop approaches. Fast-reverse power control is used at a rate of 800 times per second.
(12) Frame lengths are as follows: (a) 5 ms are used for control information and (b) 20 ms frames are used for signaling and user information.

9.11.6 Data Services in cdma2000

cdma2000 supports simultaneous voice data operations without impacting the voice quality of high-speed data performance. There are two types of data services in cdma2000: (1) packet data services and (2) high-speed circuit data services.

Packet data services support a large number of mobile stations. Dedicated channels for packet services users are allocated based "on demand" and released immediately after the end of the activity period. Short data bursts can be transmitted over a common traffic channel. It uses Internet protocol (IP) to support wireless packet data networking capability.

High speed circuit data services utilize dedicated traffic and control channels that are typically assigned to the mobile unit for extended periods of time during the circuit service sessions. In addition some delay sensitive services, such as video applications, require a dedicated channel for the duration of the call.

Now we turn our attention to another example of a modern mobile wireless system.

9.12 WCDMA

Wideband code division multiple access (WCDMA) is one of the main technologies for implementation for the third generation (3G) cellular radio systems, as is cdma2000. In a report by the GSM Association, it indicated WCDMA is going to be the main standard used for 3G services. It also said Europe and Asia would have 85% of the 3G market using the WCDMA standard. WCDMA is the 3G standard promoted by most European operators and is used by NTT DoCoMo (Japan) in its FOMA services. The other 3G standard is cdma2000, which was discussed earlier; this is being promoted by an American CDMA manufacturer. cdma2000 is beginning to be adopted by telecom operators around the world. Some countries, such as South Korea and the United States, are using both standards, but most prospective 3G operators are opting for the WCDMA standard.

Various aspects of WCDMA will now be presented.

9.12.1 WCDMA Radio Frequency Protocol Architecture

The radio interface protocol stack is shown in Figure 9.12-1. The lowest block is the WCDMA physical layer. The second layer is made up of the medium access layer (MAC), the radio link control (RLC-C for the

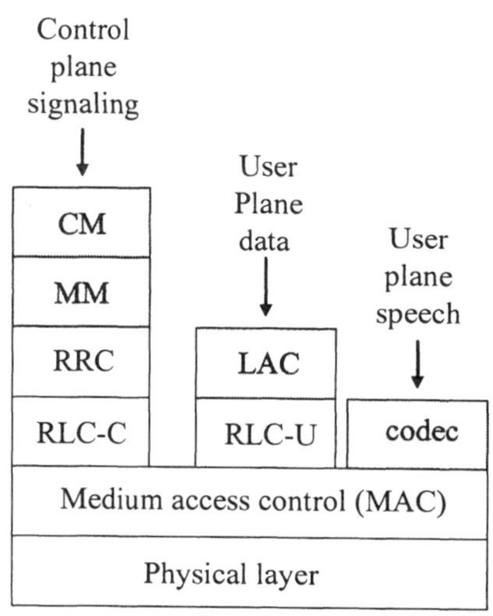

Figure 9.12-1 WCDMA radio interface protocol architecture.

control plane and RLC-U for the user plane) protocols, as well as the link access control (LAC) protocol [46, 47]. MAC and RLC belong to the access stratum. The network layer of the control plane is divided into the radio resource control (RRC) sublayer and the mobility management (MM) and connection management (CM) sublayers. Both CM and MM belong to the access management stratum. The codec layer shown in Figure 9.12-1 can be part of the access or the nonaccess stratums.

9.12.2 WCDMA Channels

We will briefly outline the channels of WCDMA, which include the logical channels, the transport channels, and the physical channels [48]. The logical channels are mapped to the transport channels which are once again mapped to the physical channels. The logical to transport channel conversion occurs in the medium access control (MAC) layer, which is a lower sublayer in the data link layer, which is layer 2. WCDMA has both an uplink (reverse link) and a downlink (forward link).

The logical channels include (1) broadcast channel (BCCH) for downlink broadcast control, (2) paging control channel (PCCH) for downlink paging information, (3) dedicated control channel (DCCH), which is dedicated between the mobile and the network, (4) common control channel (CCCH), which is common between the mobile and the network, (5) dedicated traffic channel (DTCH) for the uplink and the downlink for one mobile, and (6) common traffic channel (CTCH), which is unidirectional one to one and one to many.

The transport channels include: (1) dedicated transport channel (DCH) for the uplink and the downlink, which is mapped to the DCCH and the DTCH, (2) broadcast channel (BCH) for the downlink mapped to the BCCH, (3) forward access channel (FACH) for the downlink, which is mapped to the BCCH, the CCCH, the CTCH, the DCCH, and the DTCH, (4) paging channel (PCH) for the downlink, which is mapped to the PCCH, (5) random access channel (RACH) for the uplink, which is mapped to the CCCH, the DCCH, and the DTCH, (6) uplink common packet channel (CPCH) for the uplink, which is mapped to the DCCH and the DTCH, and (7) downlink shared channel (DSCH) for the downlink, which is mapped to the DCCH and the DTCH.

The physical channels include: (1) primary common physical channel (PCCPCH), which is mapped to the BCH, (2) secondary common control physical channel (SCCPCH), which is mapped to the FACH and the PCH, (3) physical random access channel (PRACH), which is mapped to the RACH, (4) dedicated physical data channel (DPDCH), which is mapped to the DCH, (5) dedicated physical control channel (DPCCH), which is mapped to the DCH, (6) physical downlink shared channel, which is mapped to the DSCH, (7) physical common packet channel (PCPCH), which is mapped to the CPCH, (8) synchronization channel (SCH), (9) common pilot channel (CPICH), (10) acquisition indicator channel (AICH), (11) paging indication channel (PICH), (12) CPCH status indication channel (CSICH), and (13) collision detection/channel assignment indication channel (CD/CA-ICH).

9.12.3 WCDMA Physical Layer

WCDMA has both an uplink (reverse link) that transmits from the mobile to the base station and a downlink (forward link) that transmits from the base station to the mobile. WCDMA is a wideband CDMA direct sequence spread spectrum system that has a basic chip rate of 4.096 Mcps. However, the chip rate can be increased to 8.192 or 16.384 Mcps in order to accommodate user bit rates of 2 Mbps. WCDMA utilizes frequency division multiplexing (FDD) and has a carrier frequency spacing of 4.4–5.0 MHz with a carrier raster of 200 kHz. The reason for the 200 kHz carrier raster is to provide satisfactory coexistence and interoperability with GSM (2G). This section utilizes material from [46].

9.12.3.1 WCDMA Uplink DPDCH and DPCCH

For the uplink (reverse link) the DPDCH signal and the DPCCH signal are code and I Q multiplexed within each radio frame. The uplink DPDCH carries layer 2 data, while the DPCCH carries pilot bits, transmit-power-control (TPC) commands, and an optional transport-format indicator (TFI). The uplink (reverse link) for DPCCH and DPCCH are shown in Figure 9.12-2. Each frame is composed of 10 ms and is divided into 16 slots of length 0.625 ms, each corresponding to one power control period. Therefore the power control frequency is 1,600 Hz. Within each slot, the DPCCH and the DPCCH are transmitted in parallel on the I and Q phases of the QPSK signal, using different codes.

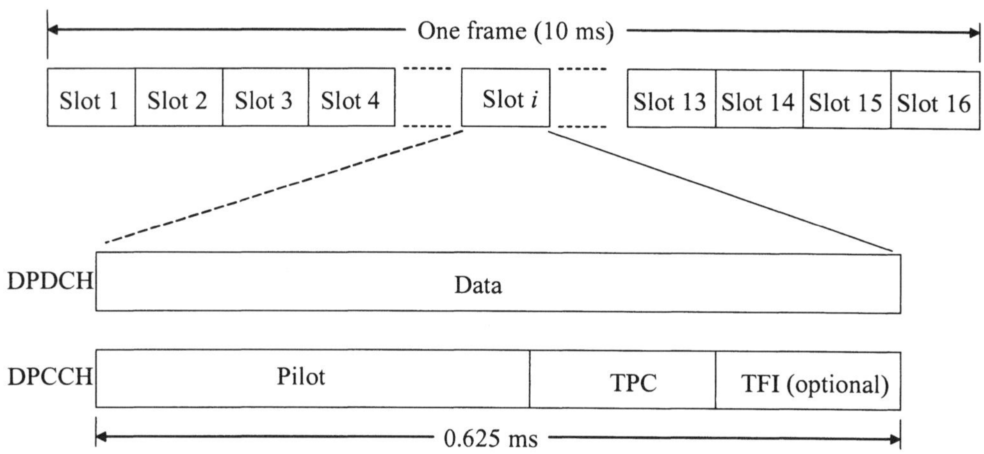

Figure 9.12-2 Uplink frame structure showing DPCCH/DPDCH.

The DPDCH signal and the DPCCH signals are modulated as shown in Figure 9.12-2.

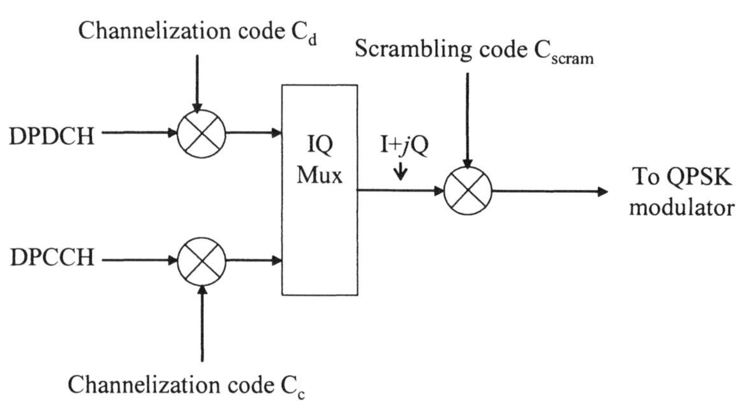

Figure 9.12-3 WCDMA channelization and scrambling for the uplink (reverse link) dedicated physical channels.

The spreading process and the modulation are shown in Figure 9.12-3. The DPDCH and the DPCCH signals are mapped to the I and Q branches, respectively, and spread by two different channelization codes on I and Q. The resulting complex signal is scrambled by the scrambling code and then the result is QPSK modulated with square root cosine pulse shaping with a roll-off factor of 0.22. Square root raised cosine pulse shaping is very commonly used in mobile communications and satisfies the Nyquist criterion to avoid intersymbol interference. The filter is characterized by a flat response out to $f = (1-\alpha)1/(2T)$ where T is the symbol duration, and goes to zero frequency response at a frequency that depends on the roll-off factor (α). When the roll-off factor is zero then the filter has an abrupt drop off at $f = 1/(2T)$. In actuality the square root filter is used on the transmitter side and a similar square root filter is used on the receiver side so that the resulting signal pulses out of the receiver filter appear to have been filtered by a raised cosine filter and don't have intersymbol interference (ideally).

The channelization codes are used to spread the data to the chip rate, while preserving orthogonality between physical channels with different rates and spreading factors. A method called *orthogonal variable spreading factor* (OVSF) codes are used for the channelization. The OVSF codes can be defined in a tree-like manner as shown in Figure 9.12-4.

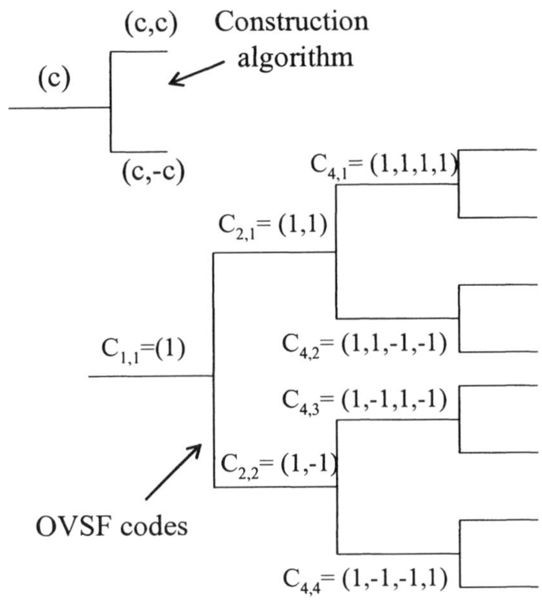

Figure 9.12-4 OVSF code tree showing the code construction.

Each level in the code tree corresponds to a certain spreading factor. A physical channel spread by the code C_a is orthogonal to another physical channel spread by code C_b if and only if code C_b is not on the path to the root of the tree C_a or in the subtree below C_a [46].

Uplink scrambling can be based on a short or long code. The short code is a complex code based on two 256 chips long Gold codes of length 256. The long scrambling code is a 40,960 chip segment of a Gold code that has length $2^{41} - 1$.

The I-Q multiplexing of control and data is used to ensure that electromagnetic compatibility (EMC) problems are diminished in the user equipment. During silent periods in speech transmission no data is transmitted to maximize capacity and minimize interference. However pilot bits and power-control commands are needed to keep the link time synchronized and power controlled.

9.12.3.2 WCDMA Downlink DPDCH and DPCCH

In the downlink (forward link (base station to mobile)) the DPDCH and the DPCCH are time multiplexed within each radio frame. Similar to the uplink, the downlink DPDCH contains layer 2 data, and the DPCCH signal carries pilot bits, TPC commands, and an optional TPI as shown in Figure 9.12-5. In a parallel manner

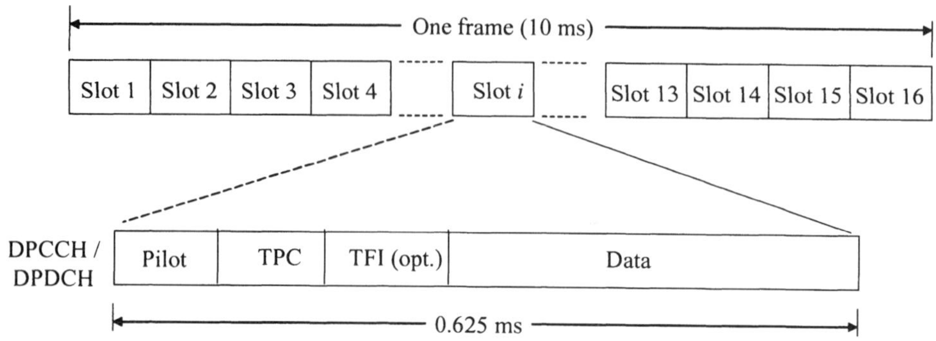

Figure 9.12-5 Downlink frame structure showing DPCCH/DPDCH.

to the uplink, each frame is 10 ms and is divided into 16 slots of length 0.625 ms each. The DPCCH and DPDCH signals are multiplexed and transmitted with the same code on both the I and Q phases. The spreading factor for the DPDCH and the DPCCH signals can vary between 4 and 256. The spreading factor, denoted SF, can be described by $SF = 256 \times 2^{-k}$ where $k = 0, 1, \ldots, 6$. Hence there are a total of 20×2^k bits per slot.

The spreading and modulation of the downlink dedicated physical channels is shown in Figure 9.12-6.

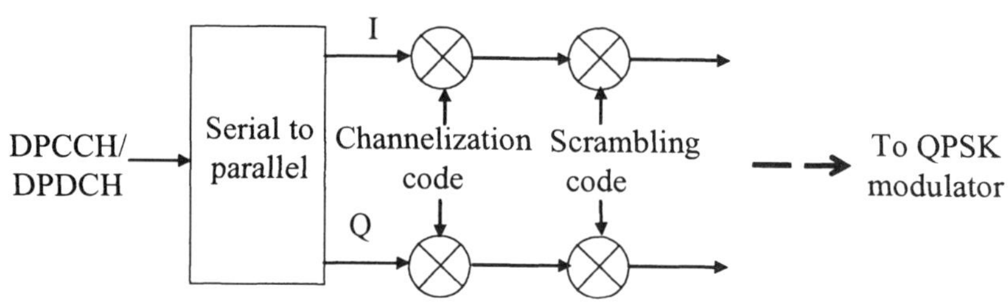

Figure 9.12-6 Channelization and scrambling for the downlink dedicated physical channel.

In reference to Figure 9.12-6 the DPCCH/DPDCH bits are mapped into pairs to the I and Q branches, and spreading to the chip rate is done with the same channelization code on both I and Q branches. The scrambling is done before modulation onto the complex signal onto the carrier. Root raised cosine pulse shaping is also done on the downlink, similar to the uplink.

Channelization is done in the same manner as the uplink, that is, with OVSF codes. The downlink scrambling code is a 40,960 chip segment of a Gold code of length $2^{18}-1$. There are a total of 512 different segments available for downlink scrambling. These codes are divided into 16 groups of 32 codes, each in order to simplify the cell search procedure. Each individual cell is assigned a specific downlink scrambling code at the initial deployment.

Unlike the uplink, time multiplexing of the control and data does not lead to EMC problems on the downlink.

Using pilot bits on the WCDMA dedicated physical channels ensures that adaptive antennas can be introduced on the downlink.

9.12.4 WCDMA Channel Coding

Various coding and interleaving schemes can be utilized for a transport channel (TrCH) depending on the particular needs in terms of the bit error rates, time delay, and so on. The codes include rate 1/2 and 1/3 rate convolutional coding for lower data rate applications, a concatenated rate 1/3 convolutional code concatenated with a Reed-Solomon code, a rate 1/3 Turbo code for use at higher data rates, and a 8-state parallel concatenated convolutional code (PCCC). Channel coding is used in conjunction with transport channel multiplexing as shown in Figure 9.12-7.

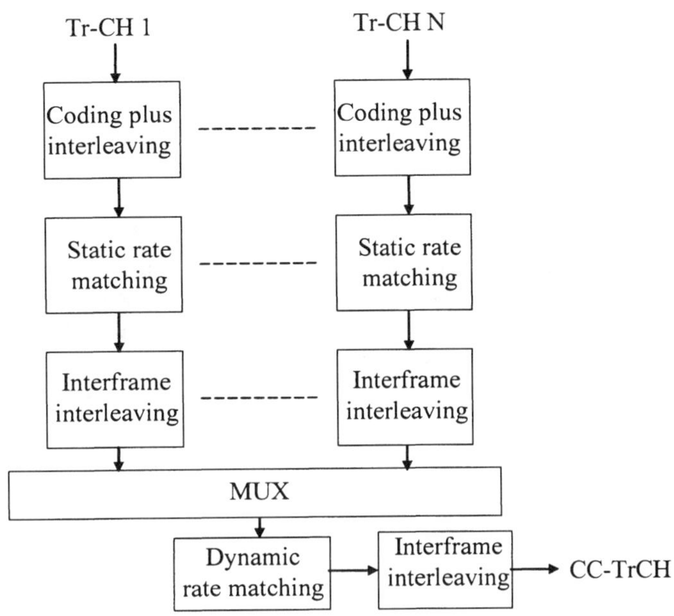

Figure 9.12-7 Block diagram illustrating the channel coding and the transport-channel multiplexing.

Figure 9.12-7 illustrates the method of channel coding and transport-channel multiplexing in WCDMA. N parallel channels are assumed at the input (TrCH 1 to TrCH N). Parallel TrCHs are separately channel coded and interleaved. These coded channels are then time multiplexed into a coded composite TrCH (CC-TrCH) digital bit stream. Final interframe (10 ms duration) interleaving is performed after the channel multiplexing.

Rate matching is used in order to match the bit rate of the of the CC-TrCH data streams to one of the limited sets of bit rates of the uplink or the downlink physical channels. There are two types of rate matching: *static rate matching* and *dynamic rate matching*. In static rate matching puncturing is used to match the data rates and is done on a very slow basis. Dynamic rate matching is carried out every 10 ms, which is on a very fast basis, and uses symbol repetition to obtain exact data rate matching.

9.12.5 WCDMA Power Control

Power control is utilized on both the uplink and the downlink. The function of power control is to remove the near-far effect, mitigate against fading, compensate for changes in the propagation conditions, and at the system level decrease the CDMA noise level and thereby increase the capacity of the system.

WCDMA employs fast closed-loop power control in both the uplink and the downlink signals. The basic control rate is 1600 Hz, and the power control step can be varied according to the user equipment speed and operating environment. Signal-to-interference ratio (SIR) is used to compare the estimate received SIR with a target value. The transmitter is commanded to increase or decrease according to the need to achieve the correct SIR.

The target values of the SIR are controlled by the outer power control loop. The outer loop measures the link quality by estimating the frame and bit error rates and adjusts the SIR targets accordingly. Open-loop power control is used by the random access procedure, in which the uplink path loss is estimated from the downlink path loss. In addition common packet transmissions depend on open-loop power control.

9.12.6 WCDMA Random Access

The random access scheme for WCDMA is based on the slotted ALOHA system. The random access procedures are described in detail in [49] and [50].

9.12.7 WCDMA Initial Cell Search

WCDMA base stations are asynchronous in that there is no universal time reference known to all of the base stations. In order to distinguish different cells, different downlink scrambling codes are used. During the initial cell search, the user equipment (UE) first searches for the strongest base station cell signal. The UE then determines the scrambling code and the frame synchronization of that cell. The cell search consists of three steps: in step (1) the UE acquires the slot synchronization to the strongest base station signal based on searching for the largest output for the system specific code, in step (2) the UE correlates the received slot synchronized signal with the 16 codes used on the secondary synchronization channel (SCH) signal, and in step (3) the UE conducts an exhaustive search of all the scrambling codes in the code group identified in the second step. When the scrambling code can be identified, the cell and system specific broadcast information on the primary CCPCH can be read.

9.12.8 WCDMA Handover

There are three types of handovers in WCDMA for FDD operation. The normal handover in WCDMA is *soft intrafrequency handover*, in which the UE (mobile) is connected to two or more cells simultaneously on the same frequency. The UE continuously searches for new cells, using the cell search technique discussed here; however the search is limited to a list of neighboring cells broadcast from the network. The neighborhood list informs the UE in which order to search for the scrambling codes, and in addition it can also limit the search to a subset of all of the available codes. In a soft handover, the uplink signals are combined in the network, and the downlink combining of the signals is done in the UE's RAKE receiver.

A second handover scheme is the *softer handover* in which the UE is connected to two cells belonging to one base station. In lieu of processing the uplink combining in the network, as is done for soft handover, softer handover combining can be done at the base station. This allows more efficient uplink combining (e.g., maximum ratio combining).

The third handover scheme in WCDMA is *interfrequency handover*. Interfrequency handover in the system is essential in order to support the following scenarios: (1) hot-spot scenarios in which a cell user has more carriers than the surrounding cells, (2) hierarchical cell structures, in which macro, micro, and pico layers are on different frequencies, (3) handovers between different operators, and (4) handovers to other systems, such as GSM.

9.12.9 WCDMA Packet Data Services

This section deals with packet transmission in the uplink. However the downlink is done in a very similar manner. In WCDMA there are three ways that packet data can be transmitted. First, if a layer 3 packet is generated, the UE may elect to transmit it on the RACH. In this case the data is appended to the access burst as shown in Figure 9.12-8. This option is chosen only when the UE has a small amount of data to transmit. In this option no reservation scheme is utilized, so that the overhead necessary to transmit a packet is kept to

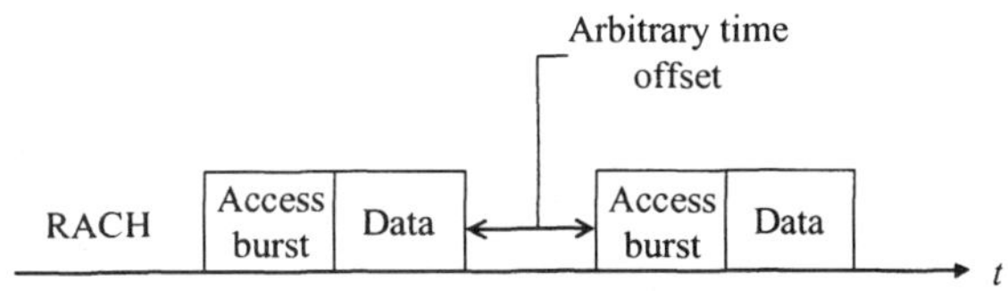

Figure 9.12-8 Uplink packet transmission on the RACH.

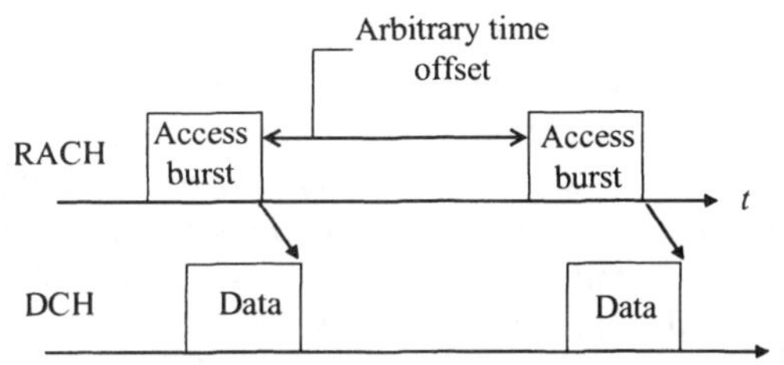

Figure 9.12-9 Uplink packet transmission on a dedicated channel.

a minimum. In this case the UE does not need to get a channel assigned to it, hence the access delay is kept small. The other method for packet transmission is based on a dedicated channel and is shown in Figure 9.12-9. In this option the UE first sends a resource request (Res_Req) message. In this resource request message is an idea of what sort of traffic is to be transmitted. The network checks to see if the UE can be assigned the necessary resources. If the resources are adequate it transmits a resource allocation (Res_All) message on the FACH. A Res_All message is composed of a set of transport formats (TFs). Out of this set the UE will use a TF to transmit its data on the DCH. The particular TF the UE may use and at what time the UE may initiate its transmission is transmitted together with the Res_All message or is indicated in a separate capacity allocation (Cap_All) message at a later time. The first option (on the RACH) would normally be used when the traffic load is low, while the second option would be utilized in cases in which the traffic load is high. A tabular comparison of both cdma2000 and WCDMA is made in Table 9.12-1 [51, 52].

Table 9.12-1 Parameter Comparison of WCDMA and cdma2000

	WCDMA	cdma2000
Channel bandwidth (MHz)	5, 10, and 20	1.25, 5, 10, 15, and 20
Downlink RF channel structure	Direct spread	Direct spread or multicarrier
Chip rate (Mcps)	4.096/8.192/16.384	1.228/3.6864/7.3728/11.0593/ 14.7456 for direct spread
Frame length (ms)	10/20 (optional)	20 ms for data & control/5 ms for Ctrl. Info. on Fund. & Ded. Ctrl. Ch.
Spreading modulation	QPSK (DL) Dual Ch. QPSK (UL)	QPSK (DL) Dual Ch. QPSK (UL)
Data modulation	QPSK (DL) BPSK (UL)	QPSK (DL) BPSK (UL)
Coherent detection	User ded. time Mplx. pilot (DL & UL)	Pilot time Mplx. with PC & EIB (UL) Com Cont. pilot Ch. and aux pilot (DL)
Ch mplx in UL	Control & pilot Ch. time mplx. I&Q mplx. for data and control channel	Control, pilot, fund. & Supl code mplx. I&Q Mplx. for data and control Ch.
Multirate capability	Variable spreading and multicode	Variable spreading and multicode
Spreading factors	4–256 (4.096 Mcps)	4–256 (3.6864 Mcps)
Power control	Open and fast closed loop (1.6 kHz)	Open loop and fast CL (800 Hz)
Spreading downlink	Var. length OVSF for Ch. separation Gold codes for cell & user separation	Var. length Walsh seq. for Ch. separ. M-seq. time shifted for diff. Cells diff. seq. in I & Q Ch.
Spreading uplink	Var. length OVSF for Ch. separ. Gold codes for cell and UE separation	Var. length OVSF for Ch. separate, M sequence UE separation
Handover	Soft handover, interfrequency handover	Soft handover, interfrequency handover

9.13 SUMMARY

In this chapter multiple access methods for digital wireless communications were presented. The chapter started with a brief history of cellular systems. This was followed by a discussion of cellular communications including clusters, frequency reuse, and cell splitting handoff. Next multiple access techniques for wireless communications was discussed. FDMA, TDMA, and various forms of spread spectrum multiple access including FHMA/CDMA, SDMA, and hybrid techniques were also discussed. Then the capacity formulation of cellular CDMA of a single cell was developed, using BPSK modulation. Packet radio access techniques were then discussed. This included the ALOHA channel and the slotted ALOHA channel. Carrier sense multiple access was then presented. This included 1-persistent, nonpersistent, and p-persistent schemes.

The next topic that was presented was multiuser detection concepts. Numerous techniques were discussed: the optimum receiver, the convention single channel receiver, the decorrelating detector, the minimum mean square error detector, successive interference cancellation, multistage interference cancellation, and the BER performance was developed for some of the detectors that were presented.

Finally the chapter concluded with two brief outlines of examples of mobile, spread spectrum, multiple access cellular systems, and cdma2000 and WCDMA—both of which are 3G. The intent of presenting the material on cdma2000 and WCDMA was to provide some familiarity of current systems and provide some of the details of very complex systems. The topic of this chapter has a table that compares some of the salient features of cdma2000 and WCDMA.

References

[1] T. S. Rappaport, *Wireless Communications Principles and Practice*, 2nd ed., Upper Saddle River, NJ: Prentice-Hall, 2002.

[2] R. Peterson, R. Ziemer, and D. Borth, *Introduction to Spread Spectrum Communications,* Upper Saddle River, NJ: Prentice-Hall, 1995.

[3] M. D. Yacoub, *Foundation of Mobile Radio Engineering,* Boca Raton, FL: CRC Press, 1993.

[4] R. E. Ziemer and R. L. Peterson, *Digital Communications*, Englewood Cliffs, NJ: Prentice-Hall, 1992.

[5] TD-SCDMA forum Web site: http://tdscdma-forum.org/EN/resources.

[6] A. Hassan, S. Chennakeshu, and J. Anderson, "Performance of Coded Slow-Frequency-hopped TDMA Cellular Systems," *IEEE 43rd Vehicular Technology Conference*, May 18–20, 1993, pp. 289–292.

[7] B. Gudmundson, J. Skold, and J. K. Ugland, "A Comparison of CDMA and TDMA Systems," *Proc. of 42nd IEEE Vehicular Technology Conference*, Vol. 2, 1992, pp. 732–735.

[8] "Cellular Communications," Web ProForum Tutorials, http://www.iec.org, The International Engineering Consortium.

[9] J. D. Gibson, (ed.), *The Mobile Communications Handbook*, Boca Raton, FL: CRC Press and IEEE Press, 1996.

[10] V. H. MacDonald, "The Cellular Concept," *The Bell Systems Technical Journal*, Vol. 58, No. 1, January 1979, pp. 15–43.

[11] J. Oeting, "Cellular Mobile Radio-An Emerging Technology," *IEEE Communications Magazine*, November 1983, pp. 10–15.

[12] S. Tekinay and B. Jabbari, "Handover and Channel Assignment in Mobile Cellular Networks," *IEEE Communications Magazine*, November 1981, pp. 42–46.

[13] C. W. Sung, and W. S. Wong, "User Speed Estimation and Dynamic Channel Allocation in Hierarchical Cellular System," *Proceedings of IEEE 1994 Vehicular Conference*, Stockholm, Sweden, 1994, pp. 91–95.

[14] S. S. Rapport, "Blocking, Hand-off and Traffic Performance for Cellular Communication Systems with Mixed Platforms," *IEE Proceedings*, Vol. 140, No. 5, October 1993, pp. 389–401.

[15] Raymond Steele, (ed.), *Mobile Radio Communications*, New York: IEEE Press and Pentech Press, 1992.

[16] K. S. Gilhouson et al., "On the Capacity of a Cellular CDMA System," *IEEE Trans. on Vehicular Technology*, Vol. 40, No. 2, 1991.

[17] P. T. Brady, "A Statistical Analysis of On-Off Patterns in 16 Conversations," *Bell Systems Tech. Journal*, Vol. 47, January 1968, pp. 73–91.

[18] J. K. Holmes, *Coherent Spread Spectrum Systems*, Malabar, FL: Kreiger Press, Chapter 8, 1990.

[19] A. Papoulis, *Probability, Random Variables, and Stochastic Processes*, 2nd ed., New York: McGraw-Hill, 1984.

[20] L. Couch II, *Digital and Analog Communication Systems*, Upper Saddle River, NJ: Prentice-Hall, 2001.

[21] S. Raghavan and J. K. Holmes, "Frequency Band Sharing Between NRZ and Split Spectrum Signals–Analysis and Simulation Results," *20th AIAA Inter. Comm. Sat. Systems Conf. and Exhibit*, Montreal, Canada, May 12–15, 2002.

[22] S. Raghavan and J. K. Holmes, "Mixed Modulation Signals For CDMA Capacity Improvement," *IEEE Aerospace Conf.*, 2004, Big Sky, MT, March 6–13, 2004.

[23] A. S. Tanenbaum, *Computer Networks*, Englewood Cliffs, NJ: Prentice-Hall, 1981.

[24] R. Rosner, "Packet Switching Tomorrow's Communication Today," Lifetime Learning Publications, Belmont, CA, 1982.

[25] F. A. Tobagi, "Random Access Techniques for Data Transmission over Packet Radio," Ph.D. Thesis, Computer Science Dept., UCLA, 1974.

[26] Prof. Mori's Web site, University of Tokyo: http://www.mlab.t.u-tokyo.ac.jp/~mori/courses/WirelessMobileComm2000/CDMA.pdf.

[27] S. Verdu, *Multiuser Detection,* Cambridge, U.K.: Cambridge University Press, 1998.

[28] J. G. Proakis, *Digital Communications*, 4th ed., New York: McGraw-Hill, 2001.

[29] P. Castoldi, *Multiuser Detection in CDMA Mobile Terminals*, Norwood, MA: Artech House, 2002.

[30] A. Papoulis, *Probability, Random Variables, and Stochastic Processes*, 2nd ed., New York: McGraw-Hill, 1984.

[31] L. A. Pipes, *Matrix Methods for Engineering*, Englewood Cliffs, NJ: Prentice-Hall, 1963.

[32] A. Abdulrahman, D. D. Falconer, and A. Sheikh, "Decision Feedback Equalization for CDMA in Indoor Wireless Communications," *IEEE J. Select Areas of Comm.*, Vol. 12, May 1994, pp. 678–706.

[33] M. L. Honig, "Adaptive Linear Interference Suppression for Packet DS-CDMA," *European Trans. Telecom. (ETT)*, Vol. 41, July 1998, pp. 944–960.

[34] S. L. Miller, "An Adaptive Direct-Sequence Code-Division Multiple Access Receiver for Multiuser Interference Rejection," *IEEE Trans. on Communications*, Vol. 43, February-March-April 1995, pp. 1746–1755.

[35] U. Mitra and H. V. Poor, "Adaptive Receiver Algorithms for Near-Far Resistant CDMA," *IEEE Trans. on Communications*, Vol. 43, April 1995, pp. 1713–1724.

[36] X. Wang and H. V. Poor, "Blind Equalization and Multiuser Detection for CDMA Communications in Dispersive Channels," *IEEE Trans. on Communications*, Vol. 46, January 1998, pp. 91–103.

[37] X. Wang and H. V. Poor, "Iterative (Turbo) Soft Interference Cancellation and Decoding for Coded CDMA," *IEEE Trans. on Communications,* Vol. 47, July 1999, pp. 1046–1061.

[38] M. C. Reed, C. B. Schlegel, P. D. Alexander, and J. A. Asenstorfer, "Iterative Multiuser Detection for CDMA with FEC: Near Single User Performance," *IEEE Trans. on Communications*, Vol. 46, December 1998, pp. 1693–1699.

[39] M. Moher, "An Iterative Multiuser Decoder for Near-Capacity Communications," *IEEE Trans. on Communications*, Vol. 46, July 1998, pp. 870–880.

[40] P. D. Alexandra, M. C. Reed, J. A. Asenstorfer, and C. B. Schlegel, Iterative Multiuser Interference Reduction: Turbo CDMA," *IEEE Trans. on Communications*, Vol. 47, July 1999, pp. 1008–1014.

[41] The cdma2000 ITU-R RTT Candidate Submission (.18), Steve Dennett Chair, TR45.5.4, 1998.

[42] V. K. Garg, *IS-95 CDMA and cdma2000*, Upper Saddle River, NJ: Prentice-Hall, 2000.

[43] Y. S. Rao and A. Kripalani, "cdma2000 Mobile Radio Access for IMT-2000," *Proc. IEEE International Conference on Personal Wireless Comm.*, 1999, pp. 6–15.

[44] A. Sharma, "WCDMA and cdma2000 Compared," *Proc. IEEE WCNC2002*, Vol. 1, 2002, pp. 23–24.

[45] Mobile Computing and Broadband Networking Laboratory CIS, NCTU, "CDMA2000 Overview," undated, from the Internet.

[46] E. Dahlman, P. Beming, F. Ovesjo, M. Persson, and C. Roobol, "WCDMA—The Radio Interface for Future Mobile Multimedia Communications," *IEEE Trans. on Vehicular Technology*, Vol. 47, November 1998.

[47] Helsinki University of Technology postgraduate course in radio communications, 36 pages, January 26, 2006.

[48] http://umtsworld.com/technology/UMTSChannel.htm.

[49] R. Esmailzadeh and M. Gustafsson, "A New Slotted ALOHA Based Random Access Method for CDMA Systems," *Proc. IEEE Int. Conf. on Universal Personal Comm., ICUPC 97*, San Diego, CA, October 1997, pp. 43–47.

[50] "UMTS Terrestrial Radio Access: Concept Evaluation (UMTS 30.06)," ETSI Tech. report 101 146, version 3.0.0, December 1997.

[51] R. Prasad and T. Ojanpera, "A Survey on CDMA: Evolution Towards Wideband CDMA," *Proc of the IEEE, 5th International Symposium on Spread Spectrum Technology*, 1998.

Problems

1. Determine the next two numbers of cell size clusters after 25. Also determine what values of i and j yield these numbers.

2. Using the same assumptions as Example 5, assume that the available SNR is 15.6 dB. Show that the allowable number of users in this case is 43.

3. Following Example 4, show that for the slotted ALOHA protocol the maximum throughput occurs for $G = 1$.

4. Assume that there are 50 calls in 1 hour, half of them lasting for 5 minutes and the other half for 10 minutes. Determine the traffic intensity in Erlangs for this case.

CHAPTER 10

An Introduction to Fading Channels

10.0 INTRODUCTION

Previous chapters have assumed, for the most part, that channels were disturbed by additive white Gaussian noise (AWGN), which caused the received signal to sometimes be detected in error. In this chapter we will present an introduction to the phenomena of fading channels, which affects spread spectrum signal performance, as well as nonspread spectrum signals. Fading, in general, causes a deleterious effect on the link performance. Understanding its cause and effect will be the goal of this chapter.

Early work in the area of propagation modeling occurred in the period of the 1950s and the 1960s. The methodology developed was applicable to over-the-horizon communications applicable to a large range of frequencies. The high-frequency (HF) band of 3–30 MHz, the very-high-frequency (VHF) band of 30–300 MHz, the ultra-high-frequency (UHF) band of 300 MHz–3 GHz, and the super-high-frequency (SHF) band of 3 GHz–30 GHz are examples of frequency bands in which the channels suffer fading. Even though the early work on ionospheric and tropospheric scatter are somewhat different than what is encountered in mobile radio channels, the early work is useful in predicting the effects of mobile radio channels.

We will discuss various channel models of fading and also discuss some forms of mitigation against fading, such as the RAKE receiver, the equalizer, and diversity methods.

10.1 AN INTRODUCTION TO RADIO PROPAGATION

The physical phenomenon that produces wave propagation is quite diverse and includes reflection, scattering, and diffraction. As an example, most cellular radio systems operate in an urban environment in which direct line-of-sight propagation between the transmitter and the receiver does not normally occur. When line-of-sight is not possible, multiple signals with different amplitudes and phase arrive, at the receiver which results in multipath fading at the location of the receiver. Also, as the distance between the transmitter and receiver increases, the composite signal amplitude decreases.

There are two primary categories of propagation models: *large-scale* models also known as path loss and *small-scale models*, which is also known as *multipath fading* models. Large-scale effects specify a power level averaged over an area of tens or hundreds of meters and is sometimes called the area mean power. Large-scale effects are characterized by the slow variation of the signal level with distance between the transmitter and the receiver. It is the natural expansion of the wave, with distance, from the transmitter.

Small-scale or multipath fading leads to a fast variation of the amplitude and phase of the resulting signal if the receiver moves over a distance of the order of one carrier wavelength or more.

There are three basic propagation mechanisms: reflection, diffraction, and scattering. These mechanisms apply to both large-scale and small-scale propagation.

Refection occurs when a radio wave impacts one medium coming from another medium having different electrical properties. The wave is partially reflected from and partially transmitted through the new medium. If the second medium is a perfect conductor then all the energy is reflected from the conductor. It is required that the medium is very large relative to the radio wave wavelength (λ).

Diffraction occurs when the radio wave is obstructed by a dense body having dimensions that are large compared to λ, causing secondary radio waves to be formed behind the obstruction. This phenomenon also occurs in beyond-the-horizon effects and propagation over the curved surface of the Earth.

Scattering occurs when a radio wave impinges on a rough surface that has dimensions on the order of λ or less. Common sources of diffraction in an urban environment include lamp posts, street signs, foliage, and trees.

10.2 OUTDOOR MODELS FOR LARGE-SCALE EFFECTS

In this section we will present some large-scale effects propagation models. We will start with the free space model and progress from there. Figure 10.2-1 illustrates [1] a sketch of some examples of path loss for large-scale effects.

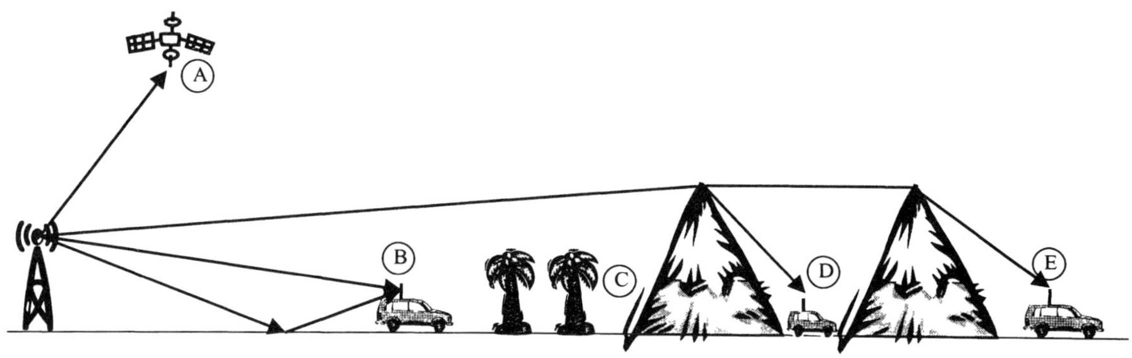

Figure 10.2-1 Figure showing various large-scale effects.

In location A we see an example of free space propagation between the transmitter and the receiver, in which there are no blockages or reflections to combine with the signal. In location B we see that the direct path and a reflected path occur at the receiver. The plane Earth model might be an appropriate model. In location C the path loss will be compounded by the fact that there will probably be significant diffraction losses due to the trees diffracting the signal path. When we consider location D we may accurately use a simple model of diffraction. However in location E the loss to that point would be quite difficult to predict, since multiple diffractions have occurred.

10.2.1 Free Space Path Loss Model

Radio propagation is such a complex subject that only a few cases can be determined by a deterministic analysis. Free space propagation is one such example. Consider Figure 10.2-2, which illustrates the free space model.

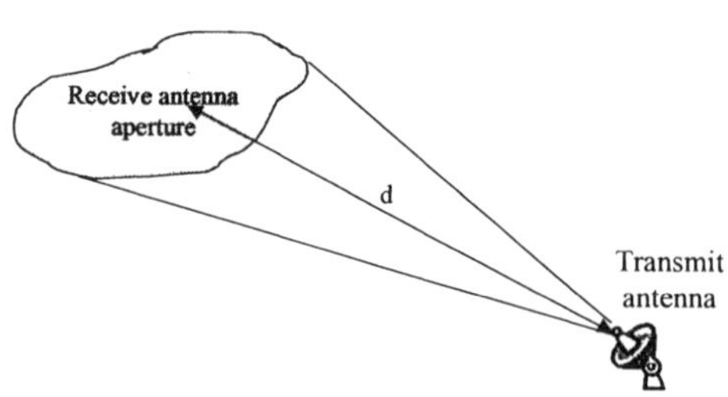

Figure 10.2-2 Free space propagation model.

It is assumed that the antenna is located in free space so that there are no obstructions near the signal path. If the transmitting antenna has gain G_T in the direction of the receiving antenna, the power density, expressed in watts per unit area, at a distance d, is given by

$$W = \frac{P_T G_T}{4\pi d^2} \quad (10.2.1\text{-}1)$$

since the spherical area at distance d is $4\pi d^2$. If the receiving antenna has an effective area A_e, then the received power at the receiver is given by

$$P_R = \frac{P_T G_T A_e}{4\pi d^2} = \frac{P_T G_T}{4\pi d^2}\frac{\lambda^2 G_R}{4\pi} \quad (10.2.1\text{-}2)$$

with G_R denoting the received gain in the direction of the transmitter. Note that the effective area of an antenna is defined as [2] $A_e = \lambda^2 G_R / 4\pi$.

Now (10.2.1-2) can be written as

$$P_R = P_T G_T G_R \left(\frac{\lambda}{4\pi d}\right)^2 \quad (10.2.1\text{-}3)$$

This is known as the *free space* or *Friis equation* [3]. Since $c = f\lambda$, (10.2.1-3) can be written as

$$P_R = P_T G_T G_R \left(\frac{c}{4\pi fd}\right)^2 \quad (10.2.1\text{-}4)$$

Rendering the received power equation in dB and using the base 10 logarithm, yields

$$P_R\text{dB} = 10\log(P_T) + 10\log(G_R) + 10\log(G_T) - 20\log(d) - 20\log(f) + 20\log\left(\frac{3\times 10^8}{4\pi}\right)\ (\text{dB}) \quad (10.2.1\text{-}5)$$

Upon evaluating the last term, one finally obtains

$$P_R\text{dB} = 10\log(P_T) + 10\log(G_R) + 10\log(G_T) - 20\log(d) - 20\log(f) + 147.56\ (\text{dB}) \quad (10.2.1\text{-}6)$$

This expression is often put into the following form

$$P_R\text{dB} = -32.45 + \log(P_T) + 10\log(G_R) + 10\log(G_T) - 20\log(d_{km}) - 20\log(f_{MHz})\ (\text{dB}) \quad (10.2.1\text{-}7)$$

where P_T and P_R are the transmitted power and received power in watts and G_R and G_T are, respectively, the receive antenna gain and the transmit antenna gain in dB. Also the distance is in kilometers and the frequency is expressed in MHz. When statute miles and MHz are used the received power takes the form

$$P_R\text{dB} = -36.58 + \log(P_T) + G_R\text{dB} + G_T\text{dB} - 20\log(d_{sm}) - 20\log(f_{MHz})\ (\text{dB}) \quad (10.2.1\text{-}8)$$

and when expressed in nautical miles and MHz, it becomes

$$P_R\text{dB} = -37.8 + \log(P_T) + G_R\text{dB} + G_T\text{dB} - 20\log(d_{nm}) - 20\log(f_{MHz}) \text{ (dB)} \tag{10.2.1-9}$$

The free space path loss can be obtained from the received signal power and typically is the most common loss parameter cited in the references. From (10.2.1-7) we have

$$L_{path} = 32.45 + 20\log(d_{km}) + 20\log(f_{MHz}) \text{ (dB)} \tag{10.2.1-10}$$

when the separation between the transmitter and the receiver is expressed in *km* and the frequency is expressed in *MHz*. The free space model only applies for values of *d*, which are in the far field of the transmitting antenna. The far field is also called the *Fraunhofer region* at the transmitting antenna. If the antenna source's largest dimension is *D*, then the Fraunhofer region satisfies the three conditions

$$\begin{aligned} &(1)\ d_{fr} = \frac{2D^2}{\lambda} \\ &(2)\ d_{\text{fr}} >> D \\ &(3)\ d_{fr} >> \lambda \end{aligned} \tag{10.2.1-11}$$

Example 1 Assume that the transmitted power is 120 watts, at a frequency of 1.575 GHz, at a range of 10,988 nm and a 12-dB transmitter gain with and an omni receiver antenna gain (0 dB). Since the distance is expressed in nautical miles it is convenient to utilize (10.2.1-9). Thus converting the frequency to MHz and the power transmitted to dBW one has for the path loss

$$L_{path} = 182.56 \text{ dB}$$

and therefore the received power is given by

$$P_R = P_T\text{dBW} + G_R\text{dB} + G_T\text{dB} - L_{path} = -149.77 \text{ dBW}$$

10.2.2 Received Signal Power and the Electric Field Strength

Sometimes it is necessary to relate the received power in watts to the electric field strength in free space [4]. Let P_{den} denote the power flux density, expressed in W/m^2. The received power in watts for the free-space model is given by

$$P_R = P_{den}A_e = \frac{|E|^2}{120\pi}A_e = \frac{|E|^2}{377}A_e = \frac{P_T G_T G_R \lambda^2}{(4\pi)^2 d^2} \tag{10.2.2-1}$$

with the units of the electric field being volt/m.

Now consider the relationship between the received signal power and the voltage applied to receiver antenna terminals. Consider the equivalent circuit shown in Figure 10.2-3 for the voltage seen at the receiver for a given received signal power.

Here the antenna voltage is fed through the equivalent antenna resistance, R_{ant}, and fed to the load R_{out}, and the power delivered to the load (receiver input resistance) is given by

$$P_R = \frac{(V/2)^2}{R_{ant}} = \frac{1}{4}\frac{V^2}{R_{ant}} \tag{10.2.2-2}$$

since half the antenna voltage is input to the receiver input resistance.

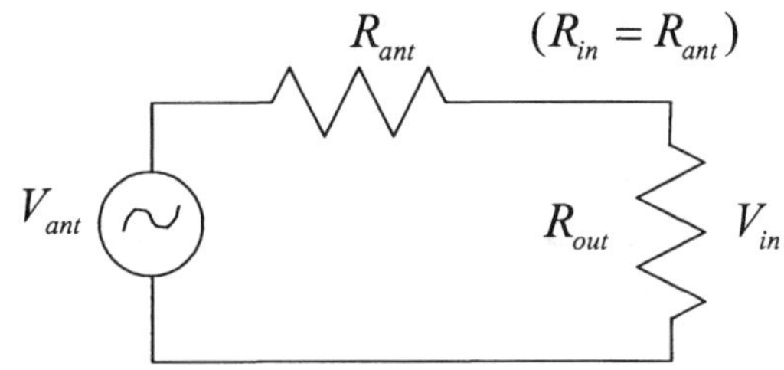

Figure 10.2-3 Equivalent circuit for the voltage applied to the receiver input.

Example 2 Using Example 1, determine (a) the electric field strength and (b) the voltage at the receiver input. Assume that the antenna resistance is 50 ohms. The solution is obtained by utilizing (10.2.2-1) for the electric field and (10.2.2-2) for the voltage. Thus using (10.2.2-1) and the fact that $A_e = G_R(\lambda)^2/(4\pi)$ produces the result

$$A_e = 1(.19)^2/(1.575*10^9) = 2.89*10^{-3}\ \mathrm{m}^2$$

$$|E| = \sqrt{\frac{377P}{A_e}} = 1.173*10^{-5}\ \mathrm{V/m}$$

and from (10.2.2-2) the voltage at the receiver input is given by

$$V = \sqrt{4PR_{ant}} = 4.59*10^{-7}\ \mathrm{V}$$

10.2.3 Plane Earth Propagation Path Loss Model

The plane Earth model [2] is based on a two-ray model of a flat reflecting surface. Figure 10.2-4 illustrates the two-path model geometry.

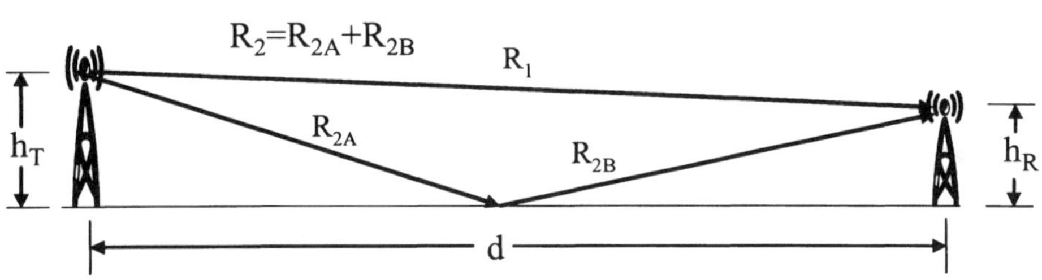

Figure 10.2-4 The two-path model for a flat reflecting surface.

Parsons has shown that the difference in the two path lengths, $\Delta R = R_2 - R_1$ is well approximated by

$$\Delta R = \frac{2h_T h_R}{d} \qquad (10.2.3\text{-}1)$$

He has also shown that the electric field strength at the receiving antenna due to both the direct wave (distance R_1) and the reflected path (R_1+R_2) is given by

$$E = E_d\left[1+\rho\exp(-j\Delta\phi)\right] \tag{10.2.3-2}$$

where ρ is the reflection coefficient, $\Delta\phi$ is the phase difference between the two paths, and E_d is the signal field strength of the direct path at the receiver antenna. For very small angles of incidence ($d >> h_T$ and $d >> h_R$) it is known that $\rho = -1$. Thus (10.2.3-2) can be rewritten as

$$E = E_d\left[1-\exp(-j\Delta\phi)\right] \tag{10.2.3-3}$$

The phase difference for the two signals is given by

$$\Delta\phi = \frac{2\pi}{\lambda}\Delta R = \frac{4\pi h_T h_R}{\lambda d} \tag{10.2.3-4}$$

Taking the magnitude of the electric field strength, one has

$$|E| = |E_d||1-\cos(\Delta\phi)+j\sin(\Delta\phi)| = |E_d|\left[1+\cos^2(\Delta\phi)-2\cos(\Delta\phi)+\sin^2(\Delta\phi)\right]^{1/2} \tag{10.2.3-5}$$

which can be simplified to

$$|E| = 2|E_d|\sin\left(\frac{\Delta\phi}{2}\right) = 2|E_d|\sin\left(\frac{2\pi h_T h_R}{\lambda d}\right) \tag{10.2.3-6}$$

after using (10.2.3-4). The received power is proportional to $|E|^2$ (let K by the proportionality factor) so that

$$P_R = 4K|E_d|^2\sin^2\left(\frac{2\pi h_T h_R}{\lambda d}\right) = 4P_T\left(\frac{\lambda}{4\pi d}\right)^2 G_T G_R \sin^2\left(\frac{2\pi h_T h_R}{\lambda d}\right) \tag{10.2.3-7}$$

When $d >> h_T$ and $d >> h_R$, then (10.2.3-7) can be rewritten as

$$P_R = 4P_T\left(\frac{\lambda}{4\pi d}\right)^2 G_T G_R\left(\frac{h_T h_R}{d^2}\right)^2 \tag{10.2.3-8}$$

Equation (10.2.3-8) is known as the *plane Earth propagation equation*. It is important to note that it is frequency independent. Also it has a fourth power with distance fall with range. Thus, doubling the distance results in a 12-dB loss! This equation only applies when $d >> h_T$ and $d >> h_R$. When this inequality relationship is not met, there are variations with distance that are predicted by the right side of (10.2.3-7). They are oscillatory in nature. Equation (10.2.3-8) can be rewritten in decibel form as

$$P_R = G_T\text{dB} + G_R\text{dB} + 20\log(h_T) + 20\log(h_R) - 40\log(d) \tag{10.2.3-9}$$

10.2.4 Egli's Path Loss Model

Experiments have confirmed that in macrocell links over smooth plane terrain, the received signal power decreases as 40log(d) and the path loss depends on frequency, unlike the plane Earth propagation model developed earlier. Egli proposed the following semi-empirical equation for mean path loss

$$L_{path} = 10\log\left(\left(\frac{40\text{ MHz}}{f}\right)^2 \frac{(h_T h_R)^2}{d^4}\right) \text{ dB} \qquad (10.2.4\text{-}1)$$

From (10.2.4-1) is follows that the received power, with antenna gains of G_T and G_R, respectively, of the transmit and receive antennas, and with a transmitted power of P_T, one has

$$P_R = G_T + G_R + P_T + 10\log\left(\left(\frac{40\text{ MHz}}{f}\right)^2 \frac{(h_T h_R)^2}{d^4}\right) \text{ dB} \qquad (10.2.4\text{-}2)$$

10.2.5 Okumura-Hata Path Loss Model

The path loss model developed by Hata [5] based on empirical curve fitting to data developed by Okumura [6] is quite well known. The *urban Hata model* mean path loss is given by

$$\begin{aligned} L_{urban} &= 69.55 + 26.16\log(f) - 13.82\log(h_T) - a(h_R) \\ &\quad + \left(44.9 - 6.55\log(h_T)\right)\log(d) \text{ dB} \end{aligned} \qquad (10.2.5\text{-}1)$$

in which f is in MHz, h_T and h_R are the elevation in meters, $a(h_R)$ is a correction factor that depends on the city size, and d is the separation of the transmitter and receiver in km. The mobile antenna correction factor for small and medium-sized cities is given by

$$a(h_R) = \left(1.1\log(f) - 0.7\right)h_R - \left(1.56\log(f) - 0.8\right) \qquad (10.2.5\text{-}2)$$

In the case of large cities the parameter takes the form

$$\begin{aligned} a(h_R) &= 8.29\left(\log(1.54h_R)\right)^2 - 1.1 \text{ dB for } f \le 300 \text{ MHz} \\ a(h_R) &= 3.2\left(\log(11.75h_R)\right)^2 - 4.97 \text{ dB for } f > 300 \text{ MHz} \end{aligned} \qquad (10.2.5\text{-}3)$$

The *suburban Hata model* is obtained by applying a correction term to the urban model and is of the form

$$L_{suburban} = L_{urban} - 2\left(\log(f/28)\right)^2 - 5.4 \text{ dB} \qquad (10.2.5\text{-}4)$$

Finally, the rural Hata model modifies the urban model differently and is given by

$$L_{rural} = L_{urban} - 4.78\left(\log(f)\right)^2 + 18.33\log(f) - 40.94 \text{ dB} \qquad (10.2.5\text{-}5)$$

The following ranges are given by Hata: 150 MHz $< f <$ 2000 MHz, 30m $< h_T <$ 200m, 1m $< h_R <$ 10m, and 1 km $< d <$ 20 km.

The Hata models are applicable to large cellular area systems (sometimes called macrocells), but are not very useful for small cell sizes on the order of 2-km diameter (1-km radius).

10.2.6 COST-231 Hata Path Loss Model

A model that is widely used for predicting path loss in mobile wireless systems is known as the COST-231 Hata model [7]. It was developed as an extension of the Hata-Okumura model [5, 6] to extend the upper frequency range to 2000 MHz. It also provides for corrections for urban, suburban, and rural (flat) environments. Its simplicity and the availability of correction factors have seen it widely used for path loss prediction in this frequency band. The path loss equation is given by

$$L = 46.3 + 33.9\log(f) - 13.82\log(h_T) - ah_m + \left(44.9 - 6.55\log(h_T)\right)\log(d) + c_m \qquad (10.2.6\text{-}1)$$

where d is the distance between transmitter and receiver in km, f is the frequency in MHz, and h_T and h_R are the transmit and receiver antenna heights in meters. The parameter c_m is defined to be 0 dB for suburban or open environments and 3 dB for urban environments. The parameter ah_m is defined for urban environments as

$$ah_m = 3.20\left(\log(11.75h_R)\right)^2 - 4.97 \text{ dB}, \quad \text{for } f > 400 \text{ MHz} \qquad (10.2.6\text{-}2)$$

and for suburban or rural (flat) environments it is defined as

$$ah_m = \left(1.1 \times \log(f) - 0.7\right)h_R - \left(1.56\log(f) - 0.8\right) \text{ dB} \qquad (10.2.6\text{-}3)$$

Again the allowable range of parameters is: 1,500 MHz < f < 2,000 MHz, 30m < h_T < 200m, 1m < h_R < 10m, and 1 km < d < 20 km.

10.2.7 ECC-33 Path Loss Model

The original Okumura data was obtained in the suburbs of Tokyo, Japan. The authors divided the urban areas into "large city" and "medium city" categories. In addition they gave corrections for "suburban" and "open areas." Due to the fact that the propagation characteristics of highly built-up areas in Tokyo are rather different from those found in European cities, the "medium city" model is recommended for European cities [2]. Due to the fact that the Hata model is quite accurate at UHF bands, but is known to be less accurate at the higher frequencies, the COST-231 Hata model was developed. The COST model extends the frequency range to 2000 MHz, but was proposed for mobile systems having omnidirectional customer premises equipment (CPE) antennas located less than 3m above ground. Another approach was taken in the ECC-33 path loss model. The results of Okumura's results were modified to more closely represent fixed wireless systems (FWS). The model ECC-33 is based on [1]. The path loss in dB is given by

$$L_{path} = L_{freespace} + L_{bm} - \Delta L_T - \Delta L_R \; dB \qquad (10.2.7\text{-}1)$$

where $L_{freespace}$ is the free space path loss, L_{bm} is the medium path loss, Δ_{LT} is the transmit (base station) gain adjustment in dB, and Δ_{LR} is the gain adjustment in dB for the receiver (CPE). Each term is defined as

$$L_{freespace} = 92.4\text{Log}(f) + 20\text{Log}(d) \; dB \qquad (10.2.7\text{-}2)$$

with f in GHz and d in km. L_{bm} is defined by

$$L_{bm} = 20.41 + 9.83\text{Log}(d) + 7.894\text{Log}(f) + 9.56\left[\text{Log}(f)\right]^2 \qquad (10.2.7\text{-}3)$$

The transmitter correction term is given by

$$\Delta L_T = \text{Log}(h_T / 200)\left[13.958 + 5.8\left(\text{Log}(d)\right)^2\right] \qquad (10.2.7\text{-}4)$$

and for medium city environments by

$$\Delta L_R = [42.57 + 13.7\text{Log}(f)][\text{Log}(h_R) - 0.585] \quad (10.2.7\text{-}5)$$

Here h_R and h_T are the antenna heights in meters. The medium city model is more appropriate for European cities.

10.2.8 Microcell Propagation Models

A microcell [8] is defined to be a relatively small outdoor area in the location of, say, a street, having the transmitter antenna located below the rooftop of the surrounding buildings. The cell's size is smaller than the macrocells mentioned in the previous sections. By installing a microcell structure, an efficient use of the limited frequency spectrum is realized and it produces a lower priced infrastructure. The primary assumption in a microcell is the use of short radio paths on the order of 200m to 1,000m, low height base stations on the order of 3m to 10m, and low transmitter power on the order of 10 mW to 1W. Microcells are commonly used in IS-95, DCS, PCS, DECT, GSM, and other systems.

10.2.8.1 The Kaji and Aleyama Model for Microcells

Many models exist for microcells, and we will review two models here. Kaji and Aleyama [9] have suggested the following empirical model [2] when the base transmitter antenna height is 5–20m, and the range is between 200m and 2 km:

$$S = -20Log\left[d^a(1+d/g)^b\right] + c \quad (10.2.8\text{-}1)$$

where S is the signal level in dBμV (dB relative to a microvolt), d is the distance between transmitter and receiver in meters, a is the basic attenuation exponent for short distances, b is the additional attenuation for distances beyond the *turning point* (the signal drops of faster after the turning point), g is the distance corresponding to the turning point, and c is the offset factor. It is assumed that both transmit and receive antennas are monopoles (nominally omnidirectional)

The model has two limiting cases: (1) for distances less than g the attenuation produces a signal level of

$$S = -20\log(d^a) + c \quad (10.2.8\text{-}2)$$

and (2) for distances greater than g the propagation yields

$$S = -20\log(d^{a+b}) + c + C \quad (10.2.8\text{-}3)$$

where C is a constant. The previously cited authors have fitted (10.2.8-1) to empirical data measured at 900 MHz. Table 10.2-1 illustrates the parameters needed to evaluate the received signal strength equation as a function of the transmitter antenna height.

Example 3 Determine the received signal voltage in dBμV for a 9-m microcell omnidirectional transmitter that has f = 900 MHz, with d being 500m, and assume that the receiver antenna is omnidirectional. Evaluating (10.2.8-1) and using Table 10.2-1, one obtains 36.4 dBμV.

Table 10.2-1 Empirical Curve-Fit Parameters to (10.2.8-1) for *f* = 900 MHz

Antenna height (m)	*a*	*b*	*g*	*c*
5	1.15	– 0.14	148.6	94.5
9	0.74	0.27	151.8	79.8
15	0.20	1.05	143.9	55.5
19	– 0.48	2.36	158.3	37.3

10.2.8.2 The Lee Microcell Model

The Lee model [10] for predicting the electric field strength in microcells is based on the assumption that there is a high correlation between the signal attenuation and the total path length through the buildings in the path of the radio signal. The model presented here follows [8]. This hypothesis is not quite accurate since the signal at the mobile receiver is the sum of the direct path and all multipath components. The model assumes that if more buildings are propagated through then the attenuation is greater. It is possible to utilize an airplane to take aerial photographs so that the total length through buildings can be estimated. The line-of-sight, P_{LOS}, is obtained from the measurement data along the streets in an open line-of-sight environment. It is necessary to add an additional signal attenuation loss, α_B, to account for the building blockage over the direct path. It can be obtained in the following way:

(1) Calculate the total building blockage length by adding the individual building path distances.
(2) Measure the signal strength P_{LOS} for the line-of-sight condition.
(3) Measure the signal strength P_{NLOS} for the nonline-of-sight condition.
(4) If the signal strength at a particular location is P_{NLOS}, the distance from the base station to the mobile user is *d*, and *B* is the blockage length between the transmitter and the receiver, then the value of α_B for a blockage length *B* can be found from

$$\alpha_B(B) = P_{LOS}(d) + P_{NLOS} \tag{10.2.8-4}$$

Table 10.2-2 lists the value of $\alpha_B(B)$ as a function of the respective distance. This table was obtained from the curves in Lee [10], which were obtained experimentally.

Numerous measurements have provided data that allowed him to conclude that the antenna height gain for various antennas is in the range of 20 to 30 dB/decade. Therefore the microcell prediction model has two curves (P_{LOS} and α_B) to predict the signal strength. The received power at the receiver is then given by

$$P_R = P_{LOS} + \alpha_B \quad (dB) \tag{10.2.8-5}$$

It is known that the original Lee Microcell Model has some deficiencies in two cases yielding large errors in path loss estimate [8]: (1) when the prediction point is in the main street, but there is no direct path in the street, and (2) when the prediction point is in a side street close to an intersection, large building blockage exists between the point of prediction and the transmitter, and the side street and the transmitter are on the same side of the street. The accuracy of this model can be considerably improved by some modifications discussed in Neskovic [11] and Neskovic, Neskovic, and Paunovic [12].

Table 10.2-2 Lee's Microcell Prediction Parameters for P_{LOS} and α_B

***d* and *B* (Distance in Feet)**	**α_B(dB)**	**P_{LOS}(dB)**
50	−5	− 42
250	− 14	− 55
500	− 17.5	− 58
750	− 18	− 60
1000	− 18	− 63.5
1500	− 18	− 70
2000	− 18	− 75
2500	− 18	− 77
3000	− 18	− 82
3500	− 18	− 84
4000	− 18	− 87
4500	− 18	− 88.5
5000	− 18	− 91.5

10.3 LARGE-SCALE EFFECTS FOR INDOOR MODELS

Following the introduction of the personal communication systems (PCS) there developed considerable interest in characterizing the propagation characteristics within a building on the same floor and between floors. It will be seen that the typical distances between the transmitter and the receiver are much less, and range of path loss is greater over small distances. In addition the actual path loss depends on the type of building, the layout of the building, and the construction materials used in the building.

Again the same elements of propagation are evident in indoor propagation: reflection, diffraction, and scattering. Moreover, things such as the location of the transmitting antenna and having the doors open or closed affect the signal level received.

It might seem that predicting the signal strength in an indoor environment might be easier than the outdoor environment; however some field measurements [13–15] have shown a variation of over 80 dB. It is known that the indoor model has more fluctuation than the outdoor model. This is presumed to be due to the fact that the indoor model has many more indirect components than in the outdoor environment, and thus it has more fluctuation in signal strength.

10.3.1 Log-Normal Path Loss Model for Indoors

Many researchers have determined that indoor mean path loss obeys the distance power law given by

$$\overline{L}_{path} = L_{path}(d_0) + 10n\log\left(\frac{d}{d_0}\right) \text{ (dB)} \tag{10.3.1-1}$$

where $\overline{L}_{path}$ is the mean path loss in dB at a known distance d (d_0 is usually 1m) and n is the exponent of the distance dependence. Cox [16], Rappaport [17], and Sklar [18] have shown that the path loss is log-normally distributed about the mean denoted in (10.3.1-1). Thus

$$L_{path} = \overline{L}_{path}(d_0) + 10\times\log\left(\frac{d}{d_0}\right) + X_\sigma \text{ (dB)} \tag{10.3.1-2}$$

where X_σ is a zero mean, log-normally distributed random variable in dB, that has a standard deviation of σ dB and represents the uncertainty in the model. Table 10.3-1 illustrates typical values of n and σ in dB for many different types of buildings [4, 15] and the effect of passing through floors.

The averages of all locations are indicated in the first row, and for the same floor on the second row.

Table 10.3-1 Path Loss Exponent and Standard Deviation for Numerous Types of Buildings

Building Type and Number of Floors	n	σ(dB)
All locations	3.14	16.3
Same floor	2.76	12.9
Through one floor	4.19	5.1
Through two floors	5.04	6.5
Through three floors	5.22	6.7
Grocery store	1.81	5.2
Retail floor	2.18	8.7
Office building 1:		
Entire building	3.54	12.8
Same floor	3.27	11.2
West wing 5th floor	2.68	8.1
Central wing 5th floor	4.01	4.3
West wing 4th floor	3.18	4.4
Office building 2:		
Entire building	4.33	13.3
Same floor	3.25	5.2

10.3.2 Floor Attenuation Factor Path Loss Model

We have seen that the path loss in multifloored structures was predicted with a model that gave the mean path loss exponent. Another approach is to add a constant floor attenuation factor (FAF) in dB, which depends on the number of floors and the building type and is added to the mean path loss for the same floor path loss exponent model. Seidel and Rappaport [15] have suggested a model of the form

$$\overline{L}_{path} = L_{path}(d_0) + 10n\log\left(\frac{d}{d_0}\right) + FAF \text{ (dB)} \qquad (10.3.2\text{-}1)$$

where, as before, d_0 is the reference distance (usually 1m), and n is the exponent of the separation from transmitter to receiver distance (d). The additional term FAF is the floor attenuation factor. Hence if the first two terms accurately represent the path loss in one floor, adding the FAF term would make it applicable to different floors. Table 10.3-2 [15] lists some FAFs and σs. Again the term FAF accounts for the difference between the path loss observed at multifloor locations and the mean path loss predicted by the simple path loss model in (10.3.1-1). In Table 10.3-2 the losses were different for the two different buildings. Office building 1 was built within 10 years of the study whereas building 2 was built 20 to 30 years before the study. All the floors in the two office buildings were made of reinforced concrete.

Table 10.3-2 Average Floor Attenuation Factor (*FAF*), in dB, for One to Four Floors in Two Office Buildings [14]

Building	FAF (dB)	σ(dB)
Office building 1:		
Through one floor	12.9	52
Through two floors	18.7	9
Through three floors	24.4	9
Through four floors	27	9
Office building 2:		
Through one floor	16.2	21
Through two floors	27.5	21
Through three floors	31.6	21

Example 4 Using Tables 10.3-1 and 10.3-2, determine the mean path loss through the third floor of office building 2, assuming that the separation between the transmitter and receiver is 30m, and the frequency is 914 MHz. Assume that $L_{path}(1) = 31.7$ dB. From Table 10.3-1 the same floor loss exponent is $n = 3.25$. From Table 10.3-2 the *FAF* value is 31.6 dB. The total mean path loss is estimated from (10.3.2-1), so that $\overline{L}_{path} = 111.3\text{ dB}$. Clearly the mean path loss inside a building is quite significant. Again it is to be noted that this is the mean path loss. Considerable variation is to be expected around the mean value.

10.4 SMALL-SCALE EFFECTS MULTIPATH FADING

At this point in our study of multipath fading it is beneficial to look at the overall picture of path loss and fading. Figure 10.4-1, based on [19], illustrates the large-scale path loss, the variation around the large-scale path loss, and the small-scale fading effects.

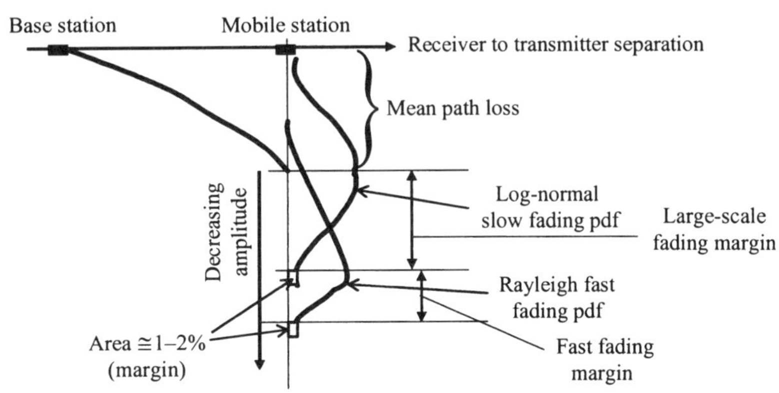

Figure 10.4-1 Link budget components for a fading channel.

We will denote the envelope of the modulation as $R(t)$. We shall use the complex envelope characterization[1] of the received signal. Let $s(t)$ be the received signal, let $v(t)$ be the complex envelope, and let f_c be the carrier frequency expressed in Hz, then

$$s(t) = \mathrm{Re}\left[v(t)e^{j2\pi f_c t} \right] \tag{10.4-1}$$

with Re[.] denoting the real part of operator. The term is the baseband modulation term that contains the information portion of the signal and is complex in general. The complex envelope can be expressed as

$$v(t) = \left| v(t) \right| e^{j\phi(t)} = R(t)e^{j\phi(t)} \tag{10.4-2}$$

and $R(t)$ is the complex envelope magnitude.

In a fading environment $v(t)$ is modified by the multiplicative factor $\alpha(t)e^{-\theta(t)}$ so that $v(t)\alpha(t)e^{-j\theta(t)}$ is the fading induced complex envelope; that is,

$$v(t)\alpha(t)e^{-j\theta(t)} = \alpha(t)\left| v(t) \right| e^{j\phi(t)}e^{-\theta(t)} = R(t)\alpha(t)e^{j[\phi(t)-\theta(t)]} \tag{10.4-3}$$

Thus the magnitude of the signal is $\alpha(t)R(t)$.

We will denote $L(t)$ as the large-scale path loss effects of the signal. Let $r_0(t)$ denote the small scale fading component. Oftentimes $L(t)$ is referred to as the *local mean* or the *log normal fading component.* Furthermore, $r_0(t)$ is sometimes referred to as the multipath or *Rayleigh fading* term. Then [18, 20] the amplitude can be written as

$$\alpha(t)R(t) = L(t)r_0(t)R(t) \tag{10.4-4}$$

Figure 10.3-1 illustrates the mean path loss due to large-scale ($L(t)$) and Rayleigh or small-scale ($r_0(t)$) fading, which are multiplicative components of the amplitude of the received signal. When constructing a link budget in this environment it is necessary to account for the variation around the mean path loss and the variation due to small-scale fading. The shaded areas in Figure 10.4-1 represent an attempt to indicate that the links must provide a high probability that the signals will not drop below some set value, presumably a value that makes the link budget work acceptably well.

Now consider why it is feasible to model the amplitude of the channel effect by (10.4-4). The received signal is modeled as the sum of discrete time delayed and amplitude modified replicas of the transmitted signal, so that

$$r(t) = \sum_n \alpha_n(t)s\left(t - \tau_n(t)\right) \tag{10.4-5}$$

where $\alpha_n(t)$ is the attenuation factor for the n-th multipath component, and $\tau_n(t)$ is the associated delay for the n-th received multipath component. Using (10.4-1), we have

$$r(t) = \sum_n \alpha_n(t)s\left(t - \tau_n(t)\right) = \mathrm{Re}\left[\left\{ \sum_n \alpha_n(t)e^{-j2\pi\,\tau_n(t)}v(t-\tau_n(t)) \right\} e^{j2\pi f_c t} \right] \tag{10.4-6}$$

This can be rewritten as

[1] Chapter 1 discussed the complex envelope characterization of a signal.

$$r(t)=\sum_{n}\alpha_n(t)s\left(t-\tau_n(t)\right)=\mathrm{Re}\left[\left\{\sum_{n}\alpha_n(t)e^{-j\theta_n(t)}v(t-\tau_n(t))\right\}e^{j2\pi f_c t}\right] \tag{10.4-7}$$

From (10.4-1) it is clear that the equivalent low-pass channel is given by

$$v_{eff}(t)=\sum_{n}\alpha_n(t)v\left(t-\tau_n(t)\right)e^{-j\theta_n(t)} \tag{10.4-8}$$

which can be described more succinctly as a multiplicative factor of the envelope in the form

$$v_{eff}(t)=\alpha(t)e^{-j\theta(t)} \tag{10.4-9}$$

in which $\alpha(t)$ is the resultant amplitude and $\theta(t)$ is the resultant phase due to the multipath channel signal combining.

10.4.1 Rayleigh and Rician Fading Models

Now consider *Rayleigh fading* and *Rician fading*[2] models. We will assume that only two components are present at the receiver initially, the direct path and the reflected path. Figure 10.4-2 illustrates the case of two components, the direct path and the reflected path.

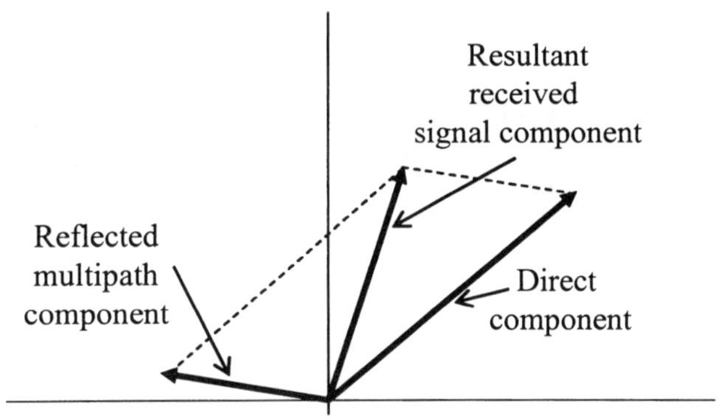

Figure 10.4-2 Resultant signal formed from the direct path and the reflected path.

When the direct path and the reflected path combine, the resultant signal is commonly of a different amplitude and phase, as is shown in Figure 10.4-2. In the general case when many multipath signals are received at the receiver, the resultant signal can be described in rectangular or polar coordinates as

$$x(t)+jy(t)=\alpha(t)e^{-j\theta(t)} \tag{10.4.1-1}$$

Under the assumption that there are many multipath components the two components $x(t)$ and $y(t)$ can be assumed to have statistically independent Gaussian probability density functions (pdf).

The small-scale amplitude is given by

$$r_0(t)=\sqrt{x(t)^2+y(t)^2} \tag{10.4.1-2}$$

[2] Often Rician is spelled as Ricean in the literature.

The resulting pdf of the amplitude $r_0(t)$ is given by

$$p(r_0)=\frac{r_0}{\sigma^2}\exp\left[-\frac{(r_0^2+A^2)}{2\sigma^2}\right]I_0\left(\frac{Ar_0}{\sigma^2}\right)\ ,\ r_0\geq 0, A\geq 0 \tag{10.4.1-3}$$

and is zero elsewhere. $I_0(x)$ denotes the modified Bessel function of the first kind and order zero, at x. The parameter σ^2 is the value of the mean prediction power of the multipath components and $A^2/2$ is the power of the direct component (nonfading). In (10.4.1-3) we have left of the time dependence of $r_0(t)$, as is common in the literature. Also in (10.4.1-3) $I_0(x)$ is the modified Bessel function of the first kind and zero-th order. The Rician pdf is often described by the parameter K, which is the ratio of the power in the direct path to the power in the multipath, so that

$$K=\frac{A^2}{2\sigma^2} \tag{10.4.1-4}$$

When the direct path power goes to zero, the Rician fading model pdf is realized. Hence the Rayleigh pdf for fading channels is given by

$$p(r_0)=\frac{r_0}{\sigma^2}\exp\left[-\frac{r_0^2}{2\sigma^2}\right]\ ,\ r_0\geq 0 \tag{10.4.1-5}$$

The mean and variance of the Rician pdf are $\mu_R=(\sqrt{\pi/2})\sigma$ and $Var_R=\left(2-(\pi/2)\right)\sigma^2$.

The Rayleigh fading model is sometimes called the *random*, *scatter*, or *diffuse* component-fading model. It results when there is no direct path to the receiver. As such, this represents a worst-case model for fading. Figure 10.4-3 illustrates a Rayleigh pdf along with three Rician pdfs for the case that K = 1, 5, and 10, respectively, and σ = 1 in the plots.

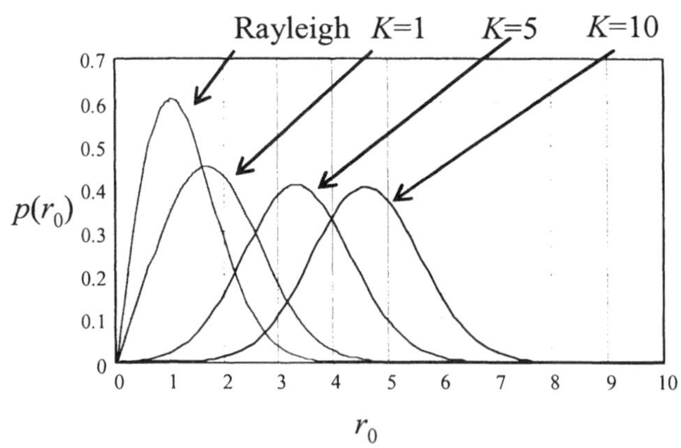

Figure 10.4-3 The Rayleigh and Rician pdfs for three values of K for the Rician pdf.

Figure 10.4-4 illustrates a typical Rayleigh fading (flat fading) of the signal amplitude when the user receiver is moving at a high speed. It is seen that the amplitude can vary significantly (about 45 dB) over time and the amplitude can vary significantly in one half of wavelength.

10.4.2 Small-Scale Fading Types

The effect on the signal due to multipath fading depends on the relative bandwidth of the channel and the signal as well as symbol period of the signal compared to the delay spread of the channel. Multipath delay spread leads to *time dispersion* and *frequency selective fading*. In addition Doppler spread leads to *frequency dispersion* and *time selective fading* [21]. Figure 10.4-5 [21] illustrates the relationship between multipath delay spread and Doppler spread effects. The delay spread and Doppler spread are independent phenomena.

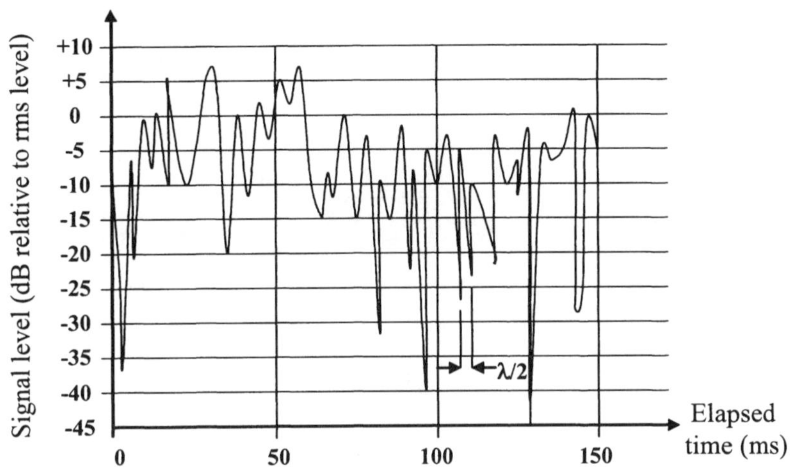

Figure 10.4-4 A representative signal amplitude when the receiver is moving at a high rate of speed.

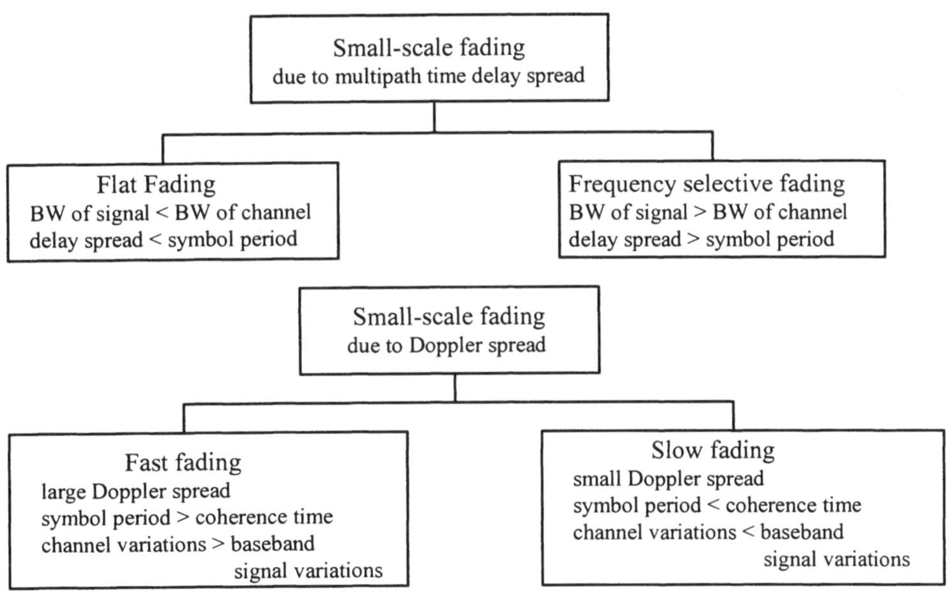

Figure 10.4-5 Multipath delay spread and Doppler spread cases for small-scale fading.

10.4.3 Multipath Time Delay Spread Fading

As is seen from Figure 10.4-5, time delay spread produces either flat fading or frequency selective fading. Delay spread is the phenomena caused by the aggregate of reflected and scattered propagation paths in the radio channel. The *coherence bandwidth*, BW_{coh}, is a parameter obtained from the root mean square (rms) delay spread function (σ_τ). The coherence bandwidth is a statistical measure of the bandwidth in which the channel can be assumed to have equal gain and linear phase (i.e., "flat" fading occurs). Two parameters of delay spread are the mean excess delay (τ_k) and the rms delay spread (σ_k). Let $P(\tau_k)$ denote the received power at delay τ_k; then the mean excess delay is given by

$$\bar{\tau} = \frac{\sum_k \tau_k P(\tau_k)}{\sum_k P(\tau_k)} \tag{10.4.3-1}$$

and the rms delay spread is given by

$$\sigma_\tau = \sqrt{\overline{\tau^2} - (\bar{\tau})^2} \tag{10.4.3-2}$$

where the second moment is given by

$$\overline{\tau^2} = \frac{\sum_k \tau_k^2 P(\tau_k)}{\sum_k P(\tau_k)} \tag{10.4.3-3}$$

It is to be noted that the delay is measured relative to the first detectable signal component, hence the term mean excess delay. In cases where the delay cannot be determined as a discrete process (10.4.3-1)–(10.4.3-3) can be generalized to the continuous case. Rappaport [21] has shown that the coherence bandwidth is related to the rms delay spread by

$$BW_{coh} \approx \frac{1}{5\sigma_\tau} \tag{10.4.3-4}$$

Example 5 Using the power versus delay specified in the plot indicated next determine $\bar{\tau}$, σ_τ, and BW_{coh}. First the power levels are converted to non-dB (arithmetic) values, then (10.4.3-1)–(10.4.3-4) are applied.

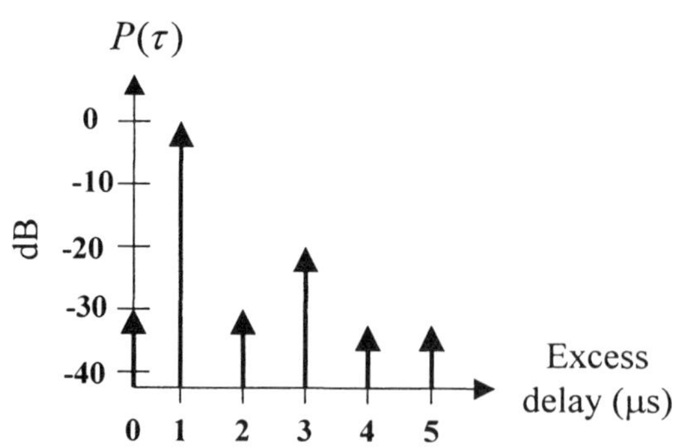

Figure for Example 5 showing the power versus delay relationship.

The results are $\bar{\tau} = 1.027$ μs, $\sigma_\tau = 0.256$ μs, and BW_{coh}=782 MHz.

10.4.3.1 Flat Fading Due to Multipath Time Spread

Flat fading occurs when the channel has approximately constant gain and linear phase over a bandwidth that is larger than the signal bandwidth. Even though the signal is not distorted spectrally in flat fading, the received amplitude may fluctuate considerably at the receiver with time. The characteristics of flat fading are illustrated in Figure 10.4-6, based on [21].

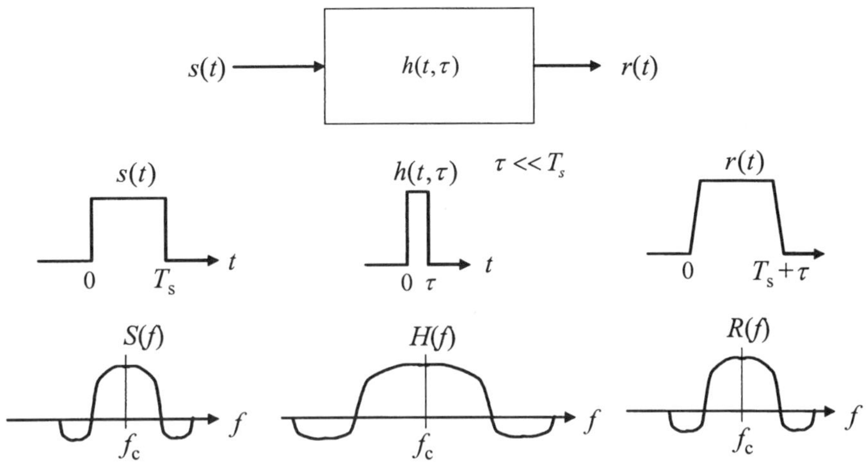

Figure 10.4-6 An example of a flat fading channel characteristic.

In the first row of signals of Figure 10.4-6 the input signal is shown as a baseband pulse modulation and the impulse response of the channel is assumed to be wideband so that $h(t,\tau)$ is approximately an impulse function, having a duration of τ seconds. The resulting output pulse is almost a replica of the input pulse, except for the fact that it has finite rise and fall times. The output pulse has been widened by about τ seconds from its original T_s seconds. In this case $\tau << T_s$ is satisfied, resulting in a "flat" fading channel. The lower set of signals shows the corresponding the signal spectrum, the channel frequency response, and the output signal spectrum. The center frequency is f_c. In summary the following conditions need to apply for flat fading:

$$\begin{aligned} &BW_s << BW_{ch} \\ &\text{and } T_s >> \sigma_\tau \end{aligned} \tag{10.4.3-5}$$

where BW_s is the signal bandwidth, BW_{ch} is the channel bandwidth, T_s is the symbol duration, and σ_τ is the standard deviation of the delay spread of the channel.

10.4.3.2 Frequency Selective Fading Due to Multipath Time Delay Spread

When a channel provides a linear phase and fixed gain over a bandwidth that is smaller than the bandwidth of the signal, the resulting conditions are called a frequency selective fading channel. In this case the impulse response has a delay spread that is greater than the inverse signal bandwidth. In effect some parts of the signal spectrum have different gains than other parts, and distortion occurs. When this type of signal distortion occurs, *intersymbol interference* (ISI) results.

Clearly channels that are frequency selective are more difficult to model than flat fading channels. Figure 10.4-7 illustrates small-scale frequency selective fading channel characteristics. The equivalent linear filter is shown in Figure 10.4-7 [21] for a frequency selective fading channel. In this case the pulse rate is such that the bandwidth of the signal is greater than the bandwidth of the channel. The result is the output pulse is considerably distorted and elongated when compared to the flat fading case.

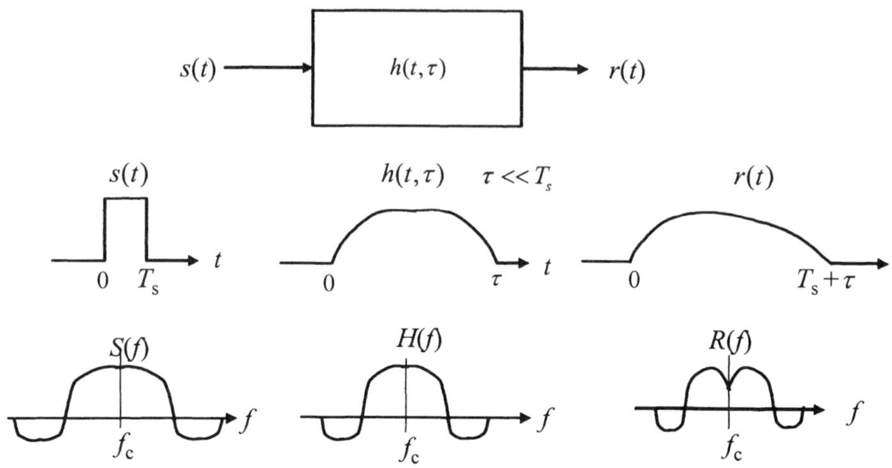

Figure 10.4-7 An example of a frequency selective fading channel characteristic.

Frequency selective channels are also known as *wideband channels*, because the bandwidth of the signal is wider than the bandwidth of the channel. The channel gain and phase vary with time in a wideband channel, producing distortion in the received signal. To summarize [21], frequency selective fading occurs when

$$BW_s > BW_{ch} \quad \text{and} \quad T_s < \sigma_\tau \tag{10.4.3-6}$$

10.4.4 Fading Effects of Multipath Doppler Spread

Now we consider the effects of fading when Doppler spreading occurs on the received signal. There are two classes of Doppler spread: (1) fast fading and (2) slow fading. The measure of the Doppler spread, denoted by BW_{Dop}, is a measure of the spectral broadening of the channel due to the time rate of change of the channel. It denotes the bandwidth over which the Doppler spectrum is not zero. The amount of spectral broadening depends on the velocity of the user and the angle between the user and the arriving scattered waves.

The *coherence time* (T_{coh}) is the time domain dual of Doppler spread and is used to characterize the time varying nature of the frequency spreading of the channel when viewed in the time domain. The coherence time and the Doppler spread are inversely related; that is,

$$T_{coh} = \frac{1}{f_{max}} \tag{10.4.4-1}$$

where f_{max} is the maximum Doppler shift and is given by

$$f_{max} = \frac{v}{\lambda} \tag{10.4.4-2}$$

with v being the velocity in m/sec and λ being the wavelength in meters. A more accurate definition of T_{coh} is given by [21]

$$T_{coh} = \frac{0.423}{f_{max}} \qquad (10.4.4\text{-}3)$$

10.4.4.1 Fast Fading Due to Doppler Spread

In *fast fading* the impulse response changes quickly compared to a symbol time. In other words the channel transmission characteristics changes during the symbol time. The result of this channel effect is *frequency dispersion*, which is also called *time selective fading*; the resulting effect on the signal is signal distortion. Alternatively when viewed in the frequency domain, fast fading causes signal distortion with increasing Doppler spread in relation to the transmitted signal bandwidth. A signal suffers from fast fading [21] when

$$\begin{aligned} T_s &> T_{coh} \\ BW_s &< BW_{Dop} \end{aligned} \qquad (10.4.4\text{-}4)$$

where T_s is the symbol duration in seconds, T_{coh} is the coherent time of the channel in seconds, BW_s is the bandwidth of the signal in Hz, and BW_{Dop} is the Doppler spread bandwidth in Hz.

10.4.4.2 Slow Fading Due to Doppler Spread

In the case of slow fading Doppler spread fading case the fading channel impulse response changes at a rate that is much slower compared to the baseband equivalent of the signal. The fading channel may be considered to be roughly constant over a few inverse signal bandwidths. From the viewpoint of the frequency domain the Doppler spread spans only a small portion of the baseband signal bandwidth. Therefore we may state [21] the conditions for slow fading due to Doppler spread to be

$$\begin{aligned} T_s &<< T_{coh} \\ BW_s &>> BW_{Dop} \end{aligned} \qquad (10.4.4\text{-}5)$$

These conditions are all relative relations. A low baseband data rate might not be considered to be slow fading whereas a high baseband data rate might be considered to have slow fading. Figure 10.4-8 attempts to capture the relationship of the four cases discussed earlier.

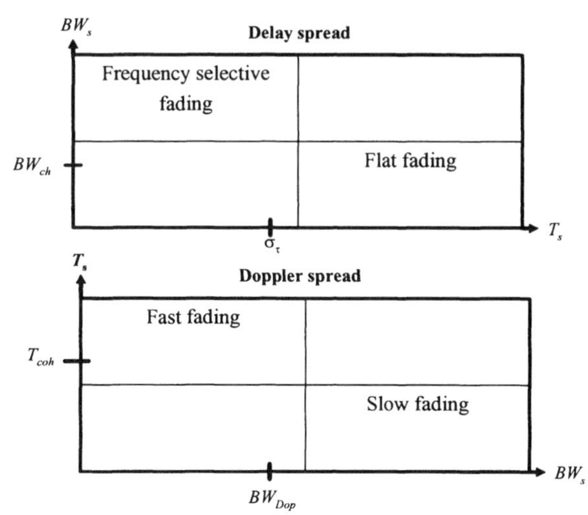

Figure 10.4-8 Summary of conditions for the four fading cases studied so far.

10.5 CHARACTERIZATION OF WIDEBAND CHANNELS

There are two possible characterizations for wideband channels: (1) time varying linear deterministic channels and (2) randomly time-variant linear channels. The theory of time varying linear systems was developed by Zadeh [22] and extended by Bello [23]. This theory applies directly to slowly varying time-variant systems. However, if the channel is rapidly varying, problems occur with these models [24, 25].

10.5.1 Deterministic Models

The deterministic models can be described in both the time domain and the frequency domain. This leads to four possible transmission functions that can be used to describe the channel.

10.5.1.1 The Time Domain Function

In what follows it will be convenient to represent the real bandpass signals in complex form as we have done earlier in this chapter. Thus we write

$$x(t) = \mathrm{Re}\left\{v(t)e^{j2\pi f_c t}\right\} \tag{10.5.1-1}$$

where $\mathrm{Re}\{x\}$ denotes the real part of x, and as before $v(t)$ is the complex envelope of $x(t)$ and f_c is the center frequency of the bandpass signal.

In the time domain characterization of wideband channels is with time varying impulse response functions. Let the complex envelope of the time varying impulse response of the channel equivalent filter be denoted by $h(t,\tau)$, which we have already called the *input delay-spread function* [23], where τ is a delay variable. Then the output complex envelope, $y(t)$, is given by

$$y(t) = \int_{-\infty}^{\infty} v(t-\tau)h(t,\tau)d\tau \tag{10.5.1-2}$$

One interpretation of (10.5.1-2) is a physical representation of the channel as a continuum of fixed (not moving), scintillating scatters, with each elementary scatterer possessing a gain fluctuation of $h(t,\tau)d\tau$ and providing delays in the range $(t,t+\tau)$. Another interpretation of $h(t,\tau)$ is the channel response at time t to an impulse response input τ seconds earlier. Since $h(t,\tau)$ represents a physical channel and cannot have an output before the input, therefore it must be true that $h(t,\tau) =$ for $\tau < 0$.

It is possible to render (10.5.1-2) in discrete time form in order to obtain a tapped delay line model [2]. Writing (10.5.1-2) as a discrete summation produces

$$y(t) \cong \Delta\tau \sum_{n=1}^{N} v(t-n\Delta\tau)h(t,n\Delta\tau) \tag{10.5.1-3}$$

With the assumption that $T = N\Delta\tau$ is the range of time that $h(t,\tau)$ is essentially nonzero. This equation implies a physical representation shown in Figure 10.5-1. The input is the input complex envelope, and the output is complex envelope. The upper row can

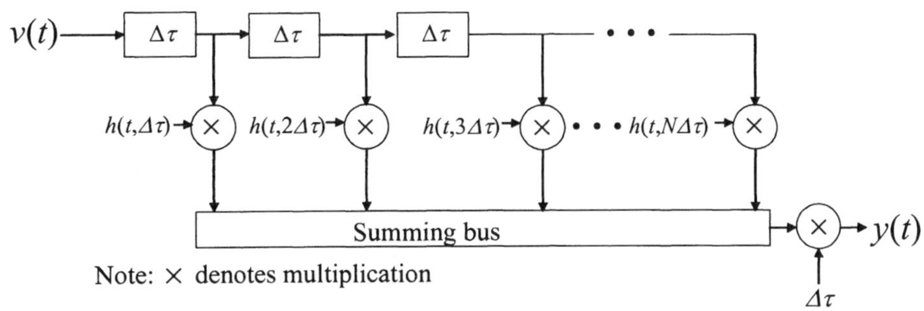

Figure 10.5-1 Time domain representation of a tapped delay line model.

be envisioned as a tapped delay line that is densely spaced. Each boxed $\Delta\tau$ denotes a delay of $\Delta\tau$ seconds; the weighting factors $h(t,n\Delta\tau)$ are the baseband equivalent impulse response for the envelope evaluated at t and $n\Delta\tau$. Finally the output is multiplied by the value $\Delta\tau$ to agree with (10.5.1-3).

10.5.1.2 The Frequency Domain Function

What one can describe in the time domain, one can also describe in the frequency domain through the dual of the time-variant impulse function. This dual channel function $H(f,\upsilon)$ is called the output Doppler spread function, and relates the channel output spectrum to the channel input spectrum in a manner analogous to the way in which $h(t,\tau)$ is related to the input envelope and the output envelope. This characterization relates the output spectrum $Y(f)$ as a superposition of elemental Doppler shifted and filtered replicas of the input spectrum $V(f)$. Bello has referred to the transfer function $H(f,\upsilon)$ as the *output Doppler-spread function.* The output spectrum is then given by [2]

$$Y(f)=\int_{-\infty}^{\infty}V(f-\upsilon)H(f-\upsilon,\upsilon)d\upsilon \tag{10.5.1-4}$$

where $V(f)$ is the Fourier transform of $v(t)$ (it should be noted that v and υ are not the same quantity).

It is possible to write (10.5.1-4) in a summation form [2], as was done to (10.5.1-3), to yield

$$Y(f)=\Delta\upsilon\sum_{m=1}^{M}V(f-m\Delta\upsilon)H(f-m\Delta\upsilon,m\Delta\upsilon) \tag{10.5.1-5}$$

This assumes that the frequency range, in which the functions are essentially nonzero, are over the frequency range of $M\Delta\upsilon$ Hz.

10.5.1.3 The Time-Variant Transfer Function

So far we have discussed in Section 10.5.1 the characterization of linear time-variant channels in terms of the input delay spread function $h(t,\tau)$, which relates output envelope function to the input envelope function. However the output Doppler spread function $H(f,\upsilon)$ relates the output envelope spectrum to the input envelope spectrum. It turns out that another characterization of the channel is possible. This characterization is based on relating the output time envelope function expressed in terms of the input envelope spectrum function to the channel equivalent filter [23]. This function is known as the *time invariant transfer function* $T(f,t)$ and was introduced by Zadeh [22]. The input-output relationship is given by

$$y(t) = \int_{-\infty}^{\infty} V(f)T(f,t)\exp\left(j2\pi ft\right)df \tag{10.5.1-6}$$

in which $V(f)$ is the input envelope spectrum and $T(f,t)$ is the time-variant transfer function. The two variable functions, sometimes called *kernels* [23], are related. For example the time-variant transfer function is the Fourier transform of the input delay-spread function with respect to the delay variable, so that

$$T(f,t) = \int_{-\infty}^{\infty} h(t,\tau)\exp\left(-j2\pi f\tau\right)d\tau \tag{10.5.1-7}$$

In addition, the inverse Fourier transform of the output Doppler spread function with respect to the Doppler variable is the time-variant transfer function, so that

$$T(f,t) = \int_{-\infty}^{\infty} H(f,\upsilon)\exp\left(j2\pi\upsilon t\right)d\upsilon \tag{10.5.1-8}$$

The function $T(f,t)$ can be viewed as the frequency transmission characteristic of the channel. There is one more characterization that will be discussed in the next section.

10.5.1.4 The Delay Doppler Spread Function

A linear time invariant filter (LTV) can be represented by the input delay spread function or the output Doppler spread function as we have already seen. It is advantageous from an engineer's perspective to have a system function that simultaneously provides a description of both the Doppler shift domain and the time delay domain. A characterization that has the time delay operation at the input and the Doppler shift operation at the output can be described as a *delay-Doppler domain* characterization [2].

The approach to representing the delay-Doppler domain function as the input delay spread function $h(t,\tau)$ is by the inverse Fourier transform of its spectrum $S(\tau,\upsilon)$ so that

$$h(t,\tau) = \int_{-\infty}^{\infty} S(\tau,\upsilon)\exp\left(j2\pi\upsilon t\right)d\upsilon \tag{10.5.1-9}$$

Now substitute (10.5.1-9) into (10.5.1-2) produces the relationship

$$y(t) = \int_{-\infty}^{\infty}\int_{-\infty}^{\infty} v(t-\tau)S(\tau,\upsilon)\exp\left(j2\pi\upsilon t\right)d\upsilon d\tau \tag{10.5.1-10}$$

Now (10.5.1-10) indicates that the output envelope function is described by the integral of the delayed and Doppler shifted signals. The signals which have delays in the range of τ to $\tau+d\tau$ and Doppler shifts in the range of υ to $\upsilon+d\upsilon$ have a differential scattering amplitude given by $S(\tau,\upsilon)d\tau d\upsilon$ [2]. Therefore the delay-Doppler spread function describes the dispersive nature of the channel based on both time delays and Doppler shifts. A *nondispersive fading channel* displays no spreading in frequency or time.

10.5.1.5 The Relationship Between the Four System Functions

It is of interest to describe the relationship between the four system functions described in the previous sections. The relationship is depicted in Figure 10.5-2 [2].

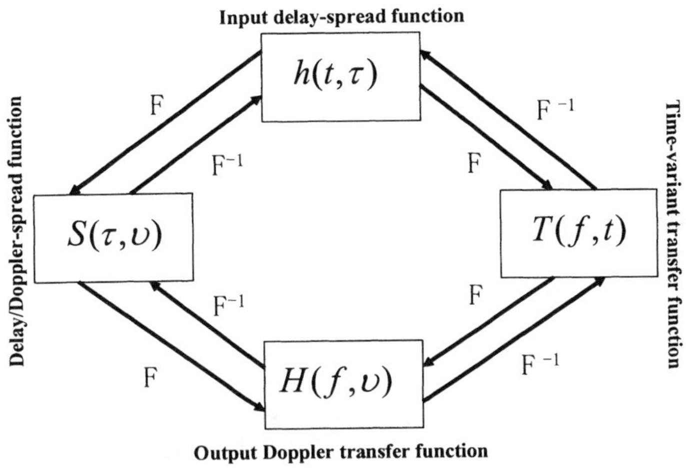

Figure 10.5-2 The relationship between the system functions for fading channels.

The directional arrows indicate that a Fourier transform (F) or an inverse Fourier transform (F^{-1}) is needed to move between the two connected blocks. For example, to describe the channel in terms of $h(t,\tau)$ from $S(\tau,\upsilon)$ one uses the inverse transform as noted in (10.5.1-9). The inverse transform uses the positive exponential, whereas the Fourier transform uses the negative exponential. All system functions have two variables, and any pair of system functions that are adjacent to each other (connected by the arrows) have one common variable. In the Fourier and inverse Fourier transforms the common variable should be considered fixed, when employing the transforms. It is to be noted that a negative exponent (Fourier transform) is used when transforming from a time variable (t, τ) to a frequency variable (f, υ). In addition when transforming from a frequency variable to a time variable, the positive exponent is used (inverse Fourier transform).

The only connection we have not shown is from $H(f,\upsilon)$ to $S(\tau,\upsilon)$. This will be addressed in the following example.

Example 6 Based on Figure 10.5-2, we see that

$$S(\tau,\upsilon) = \int_{-\infty}^{\infty} H(f,\upsilon)\exp\left(j2\pi f\tau\right)df \tag{10.5.1-11}$$

since υ is held constant, f is the variable of integration, and τ is the new time variable.

10.5.2 Stochastic Time-Variant Linear Channels

We have formed a basis of the various channel model described earlier so that we are in a position to extend those models to real radio channels, which are randomly time variant.

In order to have a complete statistical model of the channel a representation of the joint probability function of all the system functions is necessary. However the likelihood of having this information is very low, so a more realistic approach is to obtain a statistical characterization via the correlation functions of the various system functions [2, 23]. If the output of the channel is a Gaussian random process, then the mean and autocorrelation describe the channel completely, in a statistical sense. It will be assumed in what follows that all system functions have zero ensemble average.

10.5.2.1 The Various Channel Correlation Functions

It is convenient to utilize the signal in complex representation as indicated in (10.5.1-1). Thus the signal is represented by

$$x(t) = \mathrm{Re}\left\{v(t)e^{j2\pi f_c t}\right\} \tag{10.5.2-1}$$

Obtaining the autocorrelation function from (10.5.2-1) produces

$$\begin{aligned} E\left[z(t)x(s)\right] &= (1/2)\,\mathrm{Re}\left\{E\left[z(t)z^*(s)\right]\exp\left[j2\pi f_c(s-t)\right]\right\} \\ &\quad + (1/2)\,\mathrm{Re}\left\{E\left[z(t)z(s)\right]\exp\left[j2\pi f_c(s+t)\right]\right\} \end{aligned} \tag{10.5.2-2}$$

in which $E\{z\}$ is the ensemble average of z. We see from (10.5.2-2) that it is necessary to specify two autocorrelation functions when a bandpass process is represented in complex envelope form; that is,

$$\begin{aligned} R_z(t,s) &= E\left[z(t)z^*(s)\right] \\ \tilde{R}_z(t,s) &= E\left[z(t)z(s)\right] \end{aligned} \tag{10.5.2-3}$$

Bello [23] has stated that most bandpass processes possess the property that

$$E\left[z(t)z(s)\right] = 0 \tag{10.5.2-4}$$

If the channel is *wide-sense stationary* (WSS), then this relationship must be true. This follows since the statistics of a WSS process must depend only on the time difference and not on time directly. When this is true it follows that

$$E\left[z(t)z(s)\right] = (1/2)\,\mathrm{Re}\left\{E\left[z(t)z^*(s)\right]\exp(j2\pi f_c(s-t)\right\} \tag{10.5.2-5}$$

The autocorrelation functions that are used for the system functions discussed in Section 10.5.1.2, under the assumption that they are WSS system functions, are given by

$$\begin{aligned} E\left[h(t,\tau)h^*(s,\eta)\right] &= R_h(t,s;\tau,\eta) \\ E\left[H(f,\upsilon)H^*(m,\mu)\right] &= R_H(f,m;\upsilon,\mu) \\ E\left[T(f,t)T^*(m,s)\right] &= R_T(f,m;t,s) \\ E\left[S(\tau,\upsilon)S^*(\eta,\mu)\right] &= R_T(\tau,\mu;\upsilon,\mu) \end{aligned} \tag{10.5.2-6}$$

where τ and η are the time delay variables and υ and μ are the frequency shift variables.

It is possible to obtain the relationships between the autocorrelation function of the output complex envelope and the autocorrelation functions of the system functions through the use of the channel input/output relationships described in (10.5.1-8).

As an example [2], we may use (10.5.1-2), which is reproduced here

$$y(t) = \int_{-\infty}^{\infty} v(t-\tau)h(t,\tau)\,d\tau \tag{10.5.2-7}$$

to obtain the autocorrelation function of $y(t)$. We have, since $y(t)$ is complex,

$$R_y(t,s) = E\left[y(t)y^*(s)\right] \tag{10.5.2-8}$$

Using (10.5.2-7) in (10.5.2-8) produces the equation

$$R_y(t,s) = E\left[\int_{-\infty}^{\infty}\int_{-\infty}^{\infty} v(t-\tau)v^*(s-\eta)h(t,\tau)h^*(s,\eta)d\tau d\eta\right] \tag{10.5.2-9}$$

Since the delay-spread functions are the only stochastic portions of the integral, we may write

$$R_y(t,s) = \int_{-\infty}^{\infty}\int_{-\infty}^{\infty} v(t-\tau)v^*(s-\eta)E\left[h(t,\tau)h^*(s,\eta)\right]d\tau d\eta \tag{10.5.2-10}$$

Recall that the input delay-spread function is defined in (10.5.2-6) and it therefore follows that (10.5.2-10) can be written as

$$R_y(t,s) = \int_{-\infty}^{\infty}\int_{-\infty}^{\infty} v(t-\tau)v^*(s-\eta)R_h(t,s;\upsilon,\eta)d\tau d\eta \tag{10.5.2-11}$$

Thus we see that the autocorrelation of the output envelope can be described in terms of the autocorrelation function of the input delay-spread function. Actual channels can be measured to obtain the approximate value of $R_h(t,s;\upsilon,\eta)$ using impulse response sounding methods [2].

10.5.2.2 The Relationships Between the Four Channel Correlation Functions

Previously we have shown that the four system functions are connected via a Fourier transform or inverse Fourier transforms seen in Figure 10.5-2. It is possible in a similar way to relate the four system functions [2]. From Parsons [2], we have Figure 10.5-3.

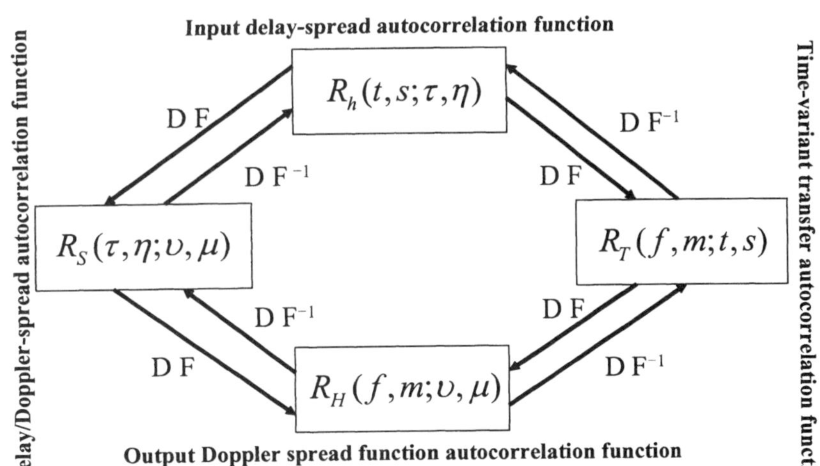

Note: D F double Fourier transform, D F^{-1} double inverse Fourier transform.

Figure 10.5-3 The relationship between the channel autocorrelation functions.

In Figure 10.5-3 it is seen that the notation DF that denotes the double Fourier transform is used on the connections between autocorrelation functions. In addition DF^{-1} denotes the double inverse Fourier transform. In that there are four variables in the functions it is seen that any two functions must be connected by two common variables, which are to be considered constant with regard to the double Fourier transforms. The convention that is used is the following: when transforming from a pair of time variables to a pair of frequency variables, a positive exponential connects the first variable in each pair and a negative connects the second variable in each pair [23]. In addition when transforming from a pair of frequency variables to a pair of time variables, a negative exponential connects the first variable in each pair and a positive exponential connects the second variable in each pair [23]. It is to be noted that f, m, υ, and μ are the frequency variables and s, t, τ, and η are the time variables in Figure 10.5-3. For example, the $R_h(.)$ has all time variables, the $R_T(.)$ function has two frequency variables and two time variables. In addition, $R_H(.)$ has four frequency variables, and RS(.) has two time variables and two frequency variables.

As an example consider the relationship between the autocorrelation functions of $R_H(t,s,\tau,\mu)$ and $R_T(f,m,t,s)$.

Example 7 Determine the relationship between $R_H(t,s,\tau,\mu)$ and $R_T(f,m,t,s)$ in terms of double integrals. From (10.5.1-8) we can write

$$T(f,t)=\int_{-\infty}^{\infty} H(f,\upsilon)\exp\left(j2\pi\upsilon t\right)d\upsilon$$

we can form the product

$$T(m,s)T(f,t)^{*}=\int_{-\infty}^{\infty}\int_{-\infty}^{\infty} H(m,\mu)H^{*}(f,\upsilon)\exp\left(j2\pi\left(s\mu-\upsilon t\right)\right)d\upsilon d\mu$$

Now taking the expected value of both sides produces the result

$$E\left[T(m,s)T(f,t)^{*}\right]=\int_{-\infty}^{\infty}\int_{-\infty}^{\infty} E\left[H(m,\mu)H^{*}(f,\upsilon)\right]\exp\left(j2\pi\left(s\mu-\upsilon t\right)\right)d\upsilon d\mu$$

or

$$R_T(f,m;t,s)=\int_{-\infty}^{\infty}\int_{-\infty}^{\infty} R_H(f,m;\upsilon,\mu)\exp\left(j2\pi\left(s\mu-\upsilon t\right)\right)d\upsilon d\mu$$

Thus we see that $R_H(t,s,\tau,\mu)$ and $R_T(f,m,t,s)$ are connected by a double Fourier transform.

It is to be noted that there are other functions developed by Bello [23], including the output delay-Doppler function, the frequency dependent modulation function, the input Doppler-spread function, and the Doppler-delay spread function. These functions are directly relatable to the functions indicated in Figure 10.5-3 [23]. These functions will not be discussed any further here.

10.5.3 The Wide-Sense Stationary Channels

At this point in the development of fading channels we move toward practical channel models [2].

A set of physical channel fading statistics can be modeled as stationary over short time periods or over small distances. It is therefore convenient to define channels for analytical purposes. A channel that is wide-sense stationary (WSS) has its first two moments independent of time; that is, the second moment only depends on the time difference ($\xi = s - t$) and not absolute time. As an example the input delay-spread autocorrelation function for a WSS channel is of the form

$$R_h(t,s;\tau,\mu)\big|_{WSS} = R_h(t-s;\tau,\mu) \tag{10.5.3-1}$$

The output of this channel may not be WSS, since the output depends on the input process, which may be WSS or not, as can be seen from (10.5.2-11). In similar manner the time-variant transfer function can be written, when it is WSS, as

$$R_T(f,m;t,s)\big|_{WSS} = R_T(f,m;t-s) \tag{10.5.3-2}$$

The Doppler shift domain contains the same basic information as the time domain. Therefore wide-sense stationary in the time variable t must also be realized in the Doppler domain. Consider the correlation function of the delay-Doppler spread function $R_S(t,s;\tau,\mu)$, described in terms of the input delay-spread correlation function. From Figure 10.5-3 we have, using the guidance of the paragraph following the figure,

$$R_S(\tau,\eta;\upsilon,\mu) = \int_{-\infty}^{\infty}\int_{-\infty}^{\infty} R_h(t,s;\tau,\eta)\exp\big(j2\pi(\upsilon t - \mu s)\big)\,dtds \tag{10.5.3-3}$$

Now imposing the fact that for WSS processes, the dependence on t and s is through the difference, say $\xi = s - t$, so that with this assumption (10.5.3-3) can be rewritten as

$$R_S(\tau,\eta;\upsilon,\mu) = \int_{-\infty}^{\infty}\int_{-\infty}^{\infty} R_h(\xi;\tau,\eta)\exp\big(j2\pi(\upsilon t - \mu t - \mu\xi)\big)\,dtd\xi \tag{10.5.3-4}$$

since $ds = d\xi$. Rearranging the t integration and the ξ integration produces

$$R_S(\tau,\eta;\upsilon,\mu) = \int_{-\infty}^{\infty} \exp\big(j2\pi(\upsilon-\mu)t\big)dt \int_{-\infty}^{\infty} R_h(\xi;\tau,\eta)\exp\big(-j2\pi\mu\xi\big)\,d\xi \tag{10.5.3-5}$$

The first integral is just the delta function; that is,

$$\delta(\upsilon-\mu) = \int_{-\infty}^{\infty} \exp\big(j2\pi(\upsilon-\mu)t\big)dt \tag{10.5.3-6}$$

at $\upsilon = \mu$. Further, the second integral of (10.5.3-5) is the *delay-Doppler cross-power spectral density* $P_S(\tau,\eta;\upsilon)$, since it is the Fourier transform of $P_S(\xi;\tau,\eta)$ with respect to the variable ξ. Thus, the second integral can be represented as

$$P_S(\tau,\eta;\upsilon) = \int_{-\infty}^{\infty} R_h(\xi;\tau,\eta)\exp\big(-j2\pi\upsilon\xi\big)\,d\xi \tag{10.5.3-7}$$

It therefore follows that (10.5.3-5) can be rewritten as

$$R_S(\tau,\eta;\upsilon,\mu) = \delta(\upsilon-\mu)P_S\left(\tau,\eta;\upsilon\right) \tag{10.5.3-8}$$

Thus it is seen that for a WSS channel, signals arriving with different Doppler shift values are uncorrelated. Similarly it can be shown, for the output Doppler spread function, that

$$R_H(f,m;\upsilon,\mu) = \delta(\upsilon-\mu)P_H(f,m;\upsilon) \tag{10.5.3-9}$$

where $P_H(f,m;\upsilon)$ is the Fourier transform of $R_T(f,m;\xi)$ with respect to the delay variable ξ. That is to say that

$$P_H(f,m;\upsilon) = \int_{-\infty}^{\infty} R_T(f,m;\xi)\exp(-j2\pi\upsilon\xi)d\xi \tag{10.5.3-10}$$

10.5.4 The Uncorrelated Scattering Channel

An uncorrelated scattering channel (US) is defined as a channel in which the contributions from elemental scatters with different path delays are uncorrelated [2]. The uncorrelated scattering channel is the dual of the WSS channel [23]. From the duality of the channel models one might expect that, analogous to (10.5.3-7), autocorrelation functions of these channels can be expressed as product of delta functions in the time delay variable. In fact we have [2]

$$\begin{aligned} R_h(t,s;\tau,\mu) &= \delta(\eta-\tau)P_h(t,s;\tau) \\ R_S(\tau,\eta;\upsilon,\eta) &= \delta(\eta-\tau)P_S(\tau;\upsilon,\mu) \end{aligned} \tag{10.5.4-1}$$

in which

$$\begin{aligned} P_h(t,s;\tau) &= \int_{-\infty}^{\infty} R_T(\Omega;t,s)\exp(j\pi 2\Omega\tau)d\Omega \\ P_S(\tau;\upsilon,\mu) &= \int_{-\infty}^{\infty} R_H(\Omega;\upsilon,\mu)\exp(j\pi 2\Omega\tau)d\Omega \end{aligned} \tag{10.5.4-2}$$

where $P_h(t,s;\tau)$ is the delay spectral density and $P_S(\tau;\upsilon,\mu)$ is the delay-Doppler cross-power spectral density.

Since the US and the WSS channels are time frequency duals, the US channel can be viewed as having WSS statistics in the frequency variable. Hence the autocorrelation functions depend only on the frequency difference variable Ω $(\Omega = m - f)$. The autocorrelation functions of the output Doppler spread function $H(f,\upsilon)$ and the time-variant transfer function $T(f,t)$ can be written as [2]

$$\begin{aligned} R_H(f,f+\Omega;\upsilon,\mu) &= R_H(\Omega;\upsilon,\mu) \\ R_T(f,f+\Omega;t,s) &= R_T(\Omega;t,s) \end{aligned} \tag{10.5.4-3}$$

Example 8 Let us establish the first of (10.5.4-1). From Figure 10.5-3 we see that we need a double inverse Fourier transform from $R_T(.)$ to $R_h(.)$. We see that the first transformed variable τ has a negative sign and the second variable η has a positive sign in the exponential; thus we can write

$$R_h(t,s;\tau,\eta) = \int_{-\infty}^{\infty}\int_{-\infty}^{\infty} R_T(f,m;t,s)\exp(-j2\pi(f\tau - m\eta))dfdm \tag{10.5.4-4}$$

Recalling for a US channel that the correlation function depends on the frequency variables only through the difference, $\Omega = m - f$. Hence from the second equation in (10.5.4-3) and using the fact that $\Omega = m - f$ allows us to rewrite the previous equation as

$$R_h(t,s;\tau,\eta) = \int_{-\infty}^{\infty}\int_{-\infty}^{\infty} R_T(\Omega;t,s)\exp(-j2\pi(f\tau - f\eta - \Omega\eta))\,df d\Omega \tag{10.5.4-5}$$

Collecting similar terms produces

$$R_h(t,s;\tau,\eta) = \int_{-\infty}^{\infty} \exp(j2\pi f(\eta-\tau))df \int_{-\infty}^{\infty} R_T(\Omega;t,s)\exp(j2\pi\eta\Omega))\,d\Omega \tag{10.5.4-6}$$

The first integral is the delta function $\delta(\eta-\tau)$, and the second integral is delay cross power spectral density $P_h(t,s;\tau)$. Hence

$$\begin{aligned}\delta(\eta-\tau) &= \int_{-\infty}^{\infty} \exp(j2\pi f(\eta-\tau))df \\ P_h(t,s;\tau) &= \int_{-\infty}^{\infty} R_T(\Omega;t,s)\exp(j2\pi\eta\Omega))\,d\Omega\end{aligned} \tag{10.5.4-7}$$

The second equation in (10.5.4-1) can be verified in the same manner.

10.5.5 The Wide-Sense Stationary Uncorrelated Scattering Channel

Perhaps the most useful channel to mobile radio engineering is the *wide sense stationary uncorrelated scattering channel* (WSSUS), which is a hybrid of the US and WSS channels. It has the property that its first and second order statistics are invariant under time or frequency translation. The result of this stationary is the fact that the WSSUS needs to be determined only once, since it applies for all time and frequency.

Under the assumption that the channel is a WSSUS, it has been shown that the autocorrelation functions of the channel system functions is given by [2, 19]

$$\begin{aligned}R_h(t,t+\xi;\tau,\eta) &= \delta(\eta-\tau)P_h(\xi;\tau) \\ R_H(f,f+\Omega;\upsilon,\mu) &= \delta(\upsilon-\mu)P_H(\Omega;\upsilon) \\ R_T(f,f+\Omega;t,t+\xi) &= R_T(\Omega;\xi) \\ R_S(\tau,\eta;\upsilon,\mu) &= \delta(\eta-\tau)\delta(\upsilon-\mu)P_S(\tau;\upsilon)\end{aligned} \tag{10.5.5-1}$$

where the power spectral density terms are defined in (10.5.3-10) and (10.5.4-2).

We can conclude this section by observing the following [2]: The autocorrelation function of the input delay-spread function, $R_h(t,t+\xi;\tau,\eta)$, exhibits uncorrelated scattering in the time delay variable and wide-sense stationarity in the time variable.

The autocorrelation function of the Doppler-spread function, $R_H(f,f+\Omega;\upsilon,\mu)$, exhibits wide-sense stationarity in the frequency variable and uncorrelated scattering in the Doppler shift variable.

The autocorrelation function of the time-variant transfer function, $R_T(f,f+\Omega;t,t+\xi)$, exhibits wide-sense stationarity in both frequency and time variables. This function is related to the correlation between the signal amplitudes between two signal frequencies that are separated by Ω hertz [26–28].

Finally, the autocorrelation function of Doppler-spread function, $R_S(\tau,\eta;\upsilon,\mu)$, possesses uncorrelated scattering in both the Doppler-shift variable and the time-delay variable. In the case of WSSUS channels, the delay-Doppler cross-power spectral density $P_S(\tau;\upsilon)$ is the same as the radar target scattering function $\sigma(\tau;\upsilon)$ [2]. However the radar target scattering function was developed for radar targets modeling [29]. The relationship between the channel correlation functions, or factors of them, is illustrated in Figure 10.5-4.

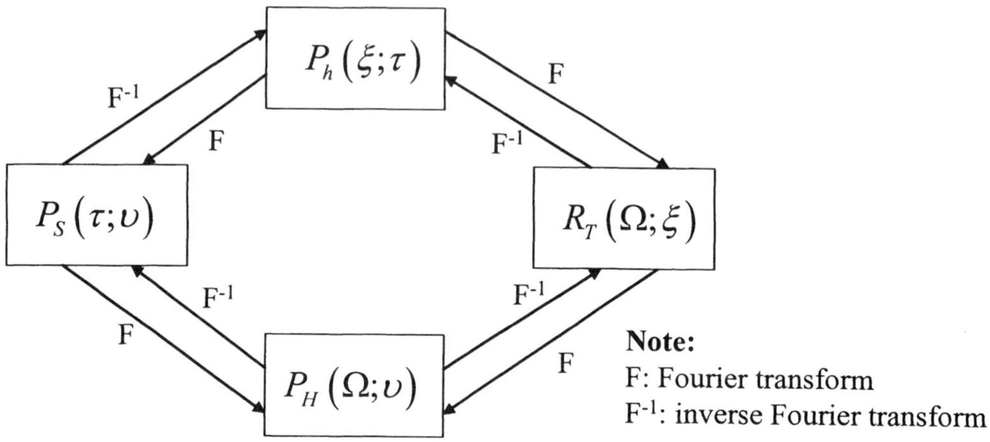

Figure 10.5-4 Fourier transform relationships between correlation functions in WSSUS channels.

Now will briefly summarize the hierarchy of the channel models studied so far. Figure 10.5-5 illustrates the hierarchy of the fading channel models [30] that we have studied.

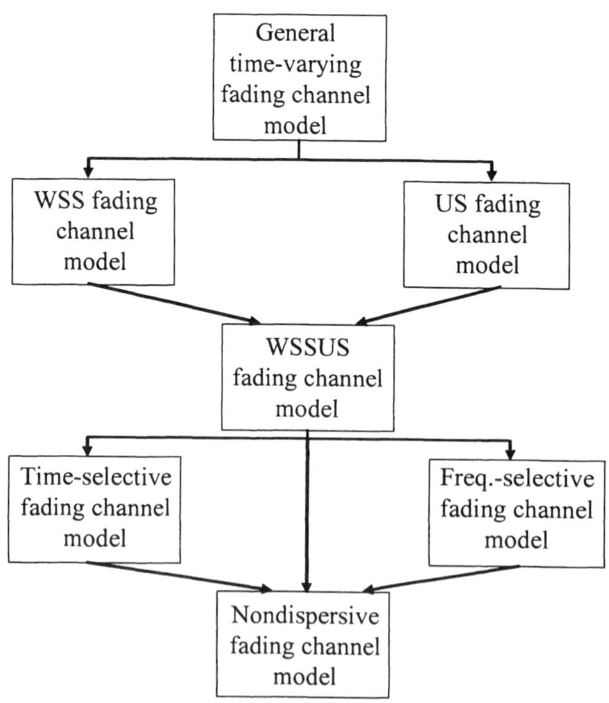

Figure 10.5-5 Hierarchical relationship between the fading random channel models studied so far.

We have started with a randomly time varying channel fading model following the pioneering work of Bello. We found that they could be characterized by correlation functions. Modeled after the deterministic fading channels, we presented the input delay-spread autocorrelation function, the delay/Doppler-spread autocorrelation function, the time-variant transfer autocorrelation function, and output Doppler-spread function autocorrelation function.

Then we moved to a more practical channel model where we introduced the WSS channel, where the input delay-spread autocorrelation function and the time-variant transfer function autocorrelation functions only depend on t and s through their time difference $\xi = s - t$. In addition, for the WSS channels, we showed that delay/Doppler-spread autocorrelation function can be written as a product of $\delta(\upsilon - \mu)$, in the frequency variables υ and μ, and delay-Doppler cross-power spectral density $P_S(\tau, \eta; \upsilon)$. We also showed that the output Doppler-spread function autocorrelation function could be written as the product of $\delta(\upsilon - \mu)$ and $P_H(f, m; \upsilon)$.

Next we considered the uncorrelated scattering (US) model. This channel has the property that the contributions of the elemental scatters with different path delays are uncorrelated. For this model the input delay-spread function autocorrelation function can be written as the product of the delta function in the time variables, $\delta(\eta - \tau)$, and the delay cross power spectral density, $P_h(t, s; \tau)$. In the same manner we found that the Delay/Doppler-spread autocorrelation function can be written as the product of $\delta(\eta - \tau)$ and the delay-Doppler cross-power spectral density, $P_S(\tau, \upsilon; \mu)$. Furthermore, we found that the autocorrelation functions of the output Doppler-spread function and the time-variant transfer function can be written in terms of the frequency difference Ω, $R_H(\Omega; \upsilon, \mu)$ and $R_T(\Omega; t, s)$, respectively.

Finally we reviewed the WSSUS channel. These channels exhibit wide-sense stationarity in the time variable and uncorrelated scattering in the time delay variable. This channel model is nontrivial and displays uncorrelated dispersiveness in both the time-delay and Doppler-shift functions [23]. The singular behavior in both the time-delay and Doppler-shift variables exist as indicated in (10.5.5-1). In Figure 10.5-5 we see that both time selective fading channels and frequency selective fading channels are a simplification of the WSSUS fading channel model. Finally when the fading channel model exhibits no dispersiveness, it results in a *nondispersive fading channel* model. A nondispersive channel model is also known as a flat-flat channel model [31] or a Rayleigh fading channel model, which we have studied earlier in this chapter.

10.6 THE EFFECTS OF A RAYLEIGH FADING CHANNEL ON THE BIT ERROR RATE

In this section we will briefly determine the impact of Rayleigh fading on the bit error rates of some common modulation schemes.

10.6.1 The Effects of a Rayleigh Fading Channel on the BPSK Bit Error Rate

In Section 10.4.1 we presented the Rayleigh channel fading pdf of the received signal amplitude; hence for $\sigma^2 = 1$ we have

$$p(\alpha) = \alpha \exp\left[-\frac{\alpha^2}{2}\right], \quad \alpha \geq 0 \qquad (10.6.1\text{-}1)$$

From Chapter 3 we showed that the bit error rate (BER) for BPSK with multiplicative relative signal amplitude α, is given by

$$PE_b = Q\left(\sqrt{\frac{2\alpha^2 E_b}{N_0}}\right) \tag{10.6.1-2}$$

where E_b is the bit energy, N_0 is the one-sided noise power spectral density, and $Q(x)$ is defined by

$$Q(x) = \int_x^{\infty} \frac{1}{\sqrt{2\pi}} \exp\left(-\frac{y^2}{2}\right) dy \tag{10.6.1-3}$$

In Problem 6 it is shown that the pdf of $\gamma = \alpha^2 E_b / N_0$ is given by

$$p(\gamma) = \frac{1}{\bar{\gamma}} \exp\left(-\frac{\gamma}{\bar{\gamma}}\right) \quad ,\gamma \geq 0 \tag{10.6.1-4}$$

where $\bar{\gamma}$ is the average value of γ. Therefore the average bit error probability is given by

$$\overline{PE_b} = \int_0^{\infty} Q\left(\sqrt{2\gamma}\right) \frac{1}{\bar{\gamma}} \exp\left(-\frac{\gamma}{\bar{\gamma}}\right) d\gamma \tag{10.6.1-5}$$

which can be evaluated to (see Problem 7)

$$\overline{PE_b} = \frac{1}{2}\left[1 - \sqrt{\frac{\bar{\gamma}}{1+\bar{\gamma}}}\right] \tag{10.6.1-6}$$

Figure 10.6-1 plots the probability of bit error for BPSK modulation for the case of a Rayleigh fading channel with additive white Gaussian noise and for the case of the additive white Gaussian noise (AWGN) channel as a function of $\overline{E_b / N_0} = \bar{\gamma}$, for the case of fading and against $\overline{E_b / N_0}$ for the AWGN case.

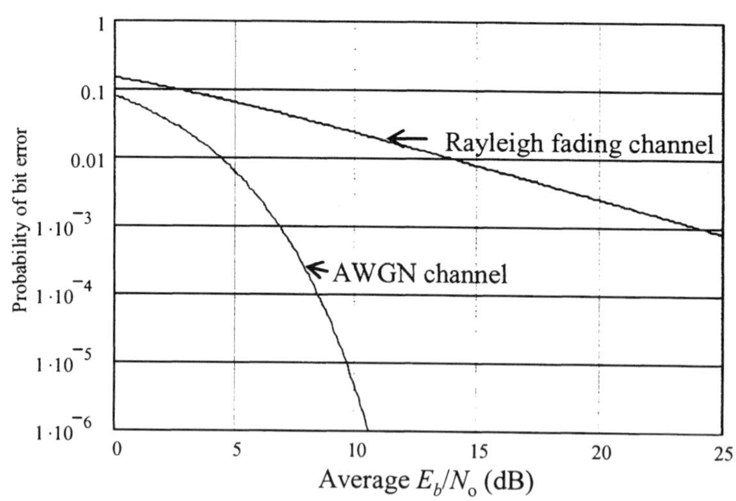

Figure 10.6-1 Bit error probability for BPSK modulation for the AWGN and the Rayleigh fading channel.

As can be seen in Figure 10.6-1 the impact of a Rayleigh fading channel is very considerable, as the power required to achieve, say, a bit error probability of 10^{-3} is about 17.18 dB more for the Rayleigh fading case than compared to the AWGN channel case.

10.6.2 The Effects of a Rayleigh Fading Channel on the DPSK Bit Error Rate

The performance of differential phase shift keying bit error rate is well known. In Chapter 3 we found that the bit error rate for DPSK is given by

$$PE_b = \frac{1}{2}\exp\left(-\frac{E_b}{N_0}\right) = \frac{1}{2}\exp(-\gamma) \quad (10.6.2\text{-}1)$$

To determine the effect of Rayleigh fading on DPSK it is necessary to evaluate the average bit error rate using (10.6.1-4) so that

$$\overline{PE_b} = \int_0^\infty PE_b(\gamma)p(\gamma)d\gamma = \int_0^\infty \frac{1}{2}\exp(-\gamma)\left[\frac{1}{\bar{\gamma}}\exp\left(-\frac{\gamma}{\bar{\gamma}}\right)\right]d\gamma \quad (10.6.2\text{-}2)$$

This can be evaluated as readily with integration tables to arrive at

$$\overline{PE_b} = \frac{1}{2(1+\bar{\gamma})} \quad (10.6.2\text{-}3)$$

Again we see the dependence on γ or $\bar{\gamma}$ is very different for the AWGN case (10.6.2-1) and the Rayleigh fading case (10.6.2-3). The two results are shown in Figure 10.6-2.

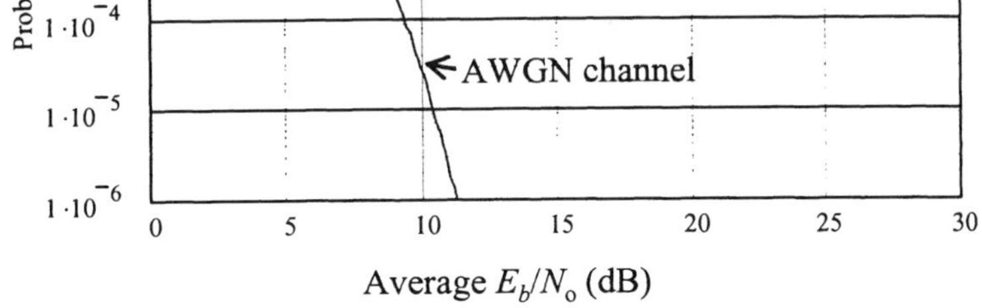

Figure 10.6-2 Bit error probability for DPSK modulation for the AWGN and the Rayleigh fading channel.

The difference in performance for the AWGN channel and the Rayleigh channel, at a bit error rate of 10^{-3}, is 19.05 dB. Again we see that the fading channel is much more deleterious to the bit error rate performance than the white Gaussian noise channel. This is usually the case when Rayleigh fading occurs.

10.6.3 The Effects of a Rayleigh Fading Channel on Noncoherent Orthogonal BFSK Bit Error Rate

The bit error rate of orthogonal noncoherent binary frequency shift keying (BFSK) has been determined [32] to be given by

$$PE_b = \frac{1}{2}\exp(-\gamma/2) \tag{10.6.3-1}$$

Averaging over the Rayleigh channel pdf via the left side of (10.6.2-2) produces

$$\overline{PE_b} = \frac{1}{2+\overline{\gamma}} \tag{10.6.3-2}$$

which is obtained from the results of Problem 7.

We have seen from these three cases that the bit error rate decreases only asymptotically with the average E_b/N_0 value. The amount of power needed to overcome the fading environment is very large and often times cannot be achieved economically. We will see in Section 10.7 that there are ways of overcoming the fading losses to some extent.

10.6.4 Nakagami Fading Channel Model

In (10.6.1-2) we saw that the signal to noise ratio (SNR) E_b/N_0 value is modified by α^2, the fading squared amplitude. So if α described as a Nakagami-*m* pdf, then the random variable $\gamma = \alpha^2 E_b / N_0$ has the pdf

$$p(\gamma) = \frac{m^m}{\Gamma(m)\overline{\gamma}^m}\gamma^{m-1}\exp(-m\gamma/\overline{\gamma}) \tag{10.6.4-1}$$

where $\Gamma(m)$ is the Gamma function evaluated at m and $\overline{\gamma} = E(\alpha^2)E_b/N_0$. Note that when m =1 the pdf degenerates to the Rayleigh fading model for γ, (see (10.6.1-4)). It can be shown [32] that the bit error rate (BER) performance improves with increasing value of m. Also when m is unbound in value (in a positive direction) the BER approaches the AWGN channel with no fading [32]. Now consider the case of binary PSK modulation with AWGN and a Nakagami-*m* fading channel model. The average BER is given by

$$\overline{PE_b} = \int_0^\infty PE_b(\gamma)p(\gamma)d\gamma = \int_0^\infty Q\left(\sqrt{2\gamma}\right)\frac{m^m}{\Gamma(m)\overline{\gamma}^m}\gamma^{m-1}\exp(-m\gamma/\overline{\gamma})d\gamma \tag{10.6.4-2}$$

This result for the average bit error probability with Nakagami fading is shown in Figure 10.6-3, based on numerical integration.

As can be seen from the BER curves the fading effect diminishes as m increases, until at very large m it approaches the unfaded AWGN channel. The case for m = 50 yields a BER = 10^{-5} that occurs at abut 10.2 dB. The AWGN BPSK yields a value of 9.6-dB SNR to achieve a BER = 10^{-5}. Thus, at m =50 the performance is quite close to a channel without fading and with AWGN. Lee [10] and Parsons [2] have obtained the average BER for the Rician channel for binary and other cases.

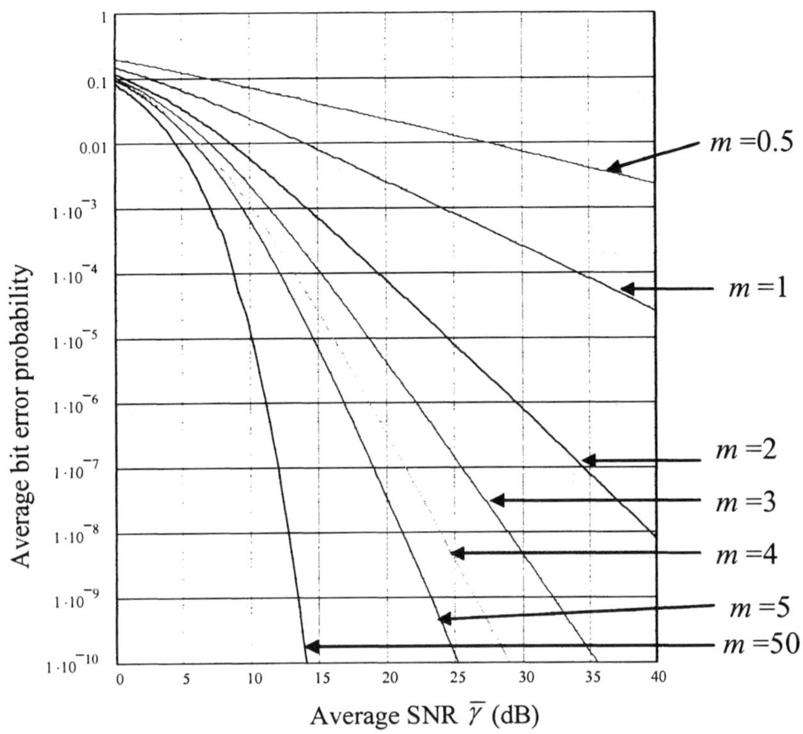

Figure 10.6-3 The average bit error probability of BPSK with a Nakagami-*m* fading channel model.

10.7 MITIGATION METHODS FOR MULTIPATH EFFECTS

There are three basic methods that can be used independently or together to improve signal quality, which can be measured by the bit error rate (BER), for example. They are diversity, equalization, and channel coding.

We have seen in the last section that channel fading caused severe degradation in BER performance if no mitigation was employed. One method to improve BER performance is the use of *redundancy*. For example, when one signal fades it is possible that one or more other independently fading channels might not be in a fade, and therefore could be utilized to improve the bit error probability performance. A means of achieving redundancy is through the use of diversity techniques.

Another approach to mitigate multipath is *equalization*. The role of equalization is to compensate for intersymbol interference (ISI), which is created by multipath within time dispersive channels. When the modulation bandwidth exceeds the coherence bandwidth of the channel, ISI occurs and the modulation symbols are elongated in time. Equalizers compensate for the expected channel delay and amplitude characteristics. Since most channels are unknown to the receiver at the start of operation, the equalizers are adaptive and require some learning time.

The third approach is *channel coding*, which introduces redundancy in the transmitted signal. Commonly interleaving is introduced when channel coding is utilized. The combination provides lower BER for the same transmitted power levels.

10.7.1 Diversity for Multipath Improvement

Redundancy can be achieved in various ways. With *frequency diversity* the same data is carried on the L carriers in which the separation in frequency between successive carriers is equal to or greater than the coherence bandwidth of the channel.

Time diversity is another method of achieving independently fading channels. If the diversity is L then L time slots are to transmit the same signal. In this approach successive time slots must be spaced by the coherence time of the channel.

Another diversity method is *antenna diversity*. This approach utilizes multiple receiver antennas and one transmit antenna. The antennas must be located sufficiently far apart so that the path delays are sufficiently different so as to provide near statistical independence. Commonly a few wavelengths are needed for the antenna separation distance.

Other diversity techniques include *angle of arrival diversity* and *polarization diversity*. Angle of arrival diversity has found applications in troposcatter systems in which large reflector-type of antennas equipped with multiple feeds that produce numerous narrow beamwidth beams pointing in a slightly different direction. It has been observed that the scattered signals related to these directions are uncorrelated [26]. Polarization diversity, for the two orthogonal polarizations, when used for mobile radio communications, exhibit uncorrelated fading statistics. There is an attendant loss of 3 dB on each polarization, however, since the power is split between the two polarizations.

10.7.2 Combining Methods for Fading Mitigation

Four combining (or diversity) approaches have been used to mitigate the effects of fading when separate receiver antennas are used for reception. These methods are known as *space diversity*. Space diversity methods are: (1) selection combining, (2) maximal ratio combining, (3) scanning combining, and (4) equal gain combining.

10.7.2.1 Selection Combining

Selection combining, also called *selection diversity*, is quite simple in principle. Figure 10.7-1 illustrates the selection combining approach. The signal plus noise-to-noise ratio is estimated in the receiver to select the largest ratio, which yields the strongest signal. Assume that each diversity branch in Figure 10.7-1 is uncorrelated and has a Rayleigh distribution function, with mean power P_0.

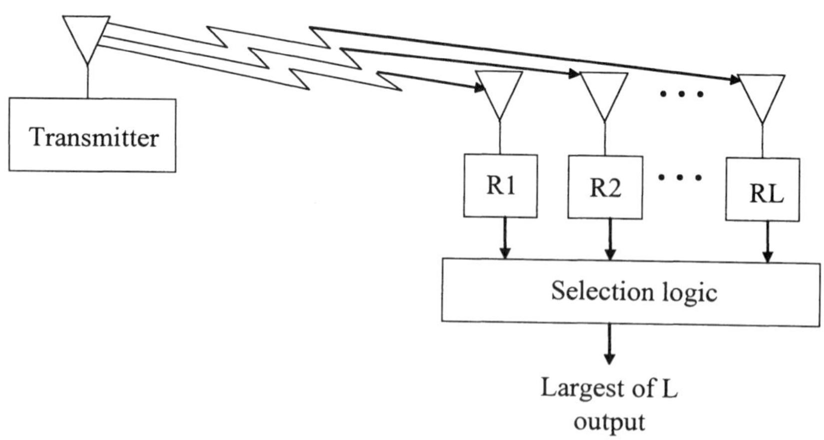

Figure 10.7-1 A block diagram model of selection combining.

The pdf of each diversity branch signal envelope is given by [26]

$$p(r_i) = \frac{r_i}{P_0} \exp\left(-\frac{r_i^2}{2P_0}\right) \quad r_i \geq 0 \tag{10.7.2-1}$$

in which r_i is the signal envelope of the i-th branch and P_0 is the mean power. It is convenient to convert the variable to signal-to-noise ratio (SNR). If r_i is the signal envelope, then the average power is $r_i^2/2$ over one RF cycle. Denote the mean noise power in each branch as $\overline{n_i^2} = N$; then the i-th channel SNR is

$$\gamma_i = \frac{r_i^2}{2N} \tag{10.7.2-2}$$

The average value, over one cycle, is therefore

$$\langle \gamma_i \rangle = \frac{\overline{r_i^2}}{2N} = \frac{P_0}{N} = \Gamma \tag{10.7.2-3}$$

The transformation from r_i to γ_i is obtained from [33]

$$p(\gamma_i) = p(r_i(\gamma_i))\left|\frac{dr_i}{d\gamma_i}\right| \tag{10.7.2-4}$$

In Problem 8 it is shown that

$$p(\gamma_i) = \frac{1}{\Gamma} \exp\left(-\frac{\gamma_i}{\Gamma}\right) \tag{10.7.2-5}$$

From (10.7.2-5) it follows that the probability of the SNR not exceeding some value threshold value γ_{Th} is given by

$$P(\gamma_i \leq \gamma_{Th}) = \int_0^{\gamma_{Th}} p(\gamma_i) d\gamma_i = 1 - \exp\left(-\frac{\gamma_{Th}}{\Gamma}\right) \tag{10.7.2-6}$$

In order to obtain the distribution of the largest, it is convenient to utilize what is called the distribution theory of order statistics [33]. In that theory it is well known that the distribution of the maximum is given by [33]

$$P(all\ of\ the\ observations \leq \gamma_{TH}) = P_L(\gamma_{Th}) = \left[1 - \exp\left(-\frac{\gamma_{Th}}{\Gamma}\right)\right]^L \tag{10.7.2-7}$$

The pdf of the maximum can be obtained by taking the derivative with respect to γ_{Th}. Thus the pdf of the maximum SNR branch satisfies

$$p_L(\gamma_{Th}) = \frac{d}{d\gamma_{Th}}(P_L(\gamma_{Th})) = \frac{L}{\Gamma}\left(1 - \exp\left(-\frac{\gamma_{Th}}{\Gamma}\right)\right)^{L-1} \exp\left(-\frac{\gamma_{Th}}{\Gamma}\right) \tag{10.7.2-8}$$

Now the mean value of the largest is given by

$$\overline{\gamma_{Th}} = \int_0^\infty \gamma_{Th} p_L(\gamma_{Th}) d\gamma_{Th} \tag{10.7.2-9}$$

This can be written as

$$\overline{\gamma_{Th}} = \int_0^\infty \gamma_{Th} \frac{L}{\Gamma}\left(1-\exp\left(-\frac{\gamma_{Th}}{\Gamma}\right)\right)^{L-1} \exp\left(-\frac{\gamma_{Th}}{\Gamma}\right) d\gamma_{Th} \tag{10.7.2-10}$$

Now let $x = \gamma_{th} / \Gamma$ so that

$$\overline{\gamma_{Th}} = L\Gamma \int_0^\infty x\left(1-\exp(-x)\right)^{L-1} \exp(-x)\, dx \tag{10.7.2-11}$$

Expanding the term raised to the L–1 power in a binomial series yields

$$\overline{\gamma_{Th}} = L\Gamma \int_0^\infty x \sum_{r=0}^{L-1} \binom{L-1}{r} (-1)^r \exp(-rx) \exp(-x)\, dx \tag{10.7.2-12}$$

which can be evaluated, after some algebra, as

$$\overline{\gamma_{Th}} = \Gamma \sum_{r=0}^{L-1} \frac{L!}{(L-1-r)!r!} (-1)^r \frac{1}{(r+1)^2} \tag{10.7.2-13}$$

This can be shown to be equal to the following simpler series[3]

$$\overline{\gamma_{Th}} = \Gamma \sum_{i=1}^{L} \frac{1}{i} \tag{10.7.2-14}$$

Therefore it is possible to relate the mean maximum SNR to the individual channel SNR by the relationship

$$\frac{\overline{\gamma_{Th}}}{\Gamma} = \sum_{i=1}^{L} \frac{1}{i} \tag{10.7.2-15}$$

10.7.2.2 Maximal Ratio Combining

This method of combining the received signals, known as *maximal ratio combining*, yields the optimal SNR. Kahn [34] first devised this approach. Each of the L signals are weighted proportionately to their signal voltage to noise power ratios, phase aligned, and then summed.

It is necessary to first phase-align the signals so that they are coherent prior to their being combined. Figure 10.7-2 illustrates the transmitter and receiver block diagram structure.

---

[3] Note that the two series can be shown to be numerically equal for all values of L.

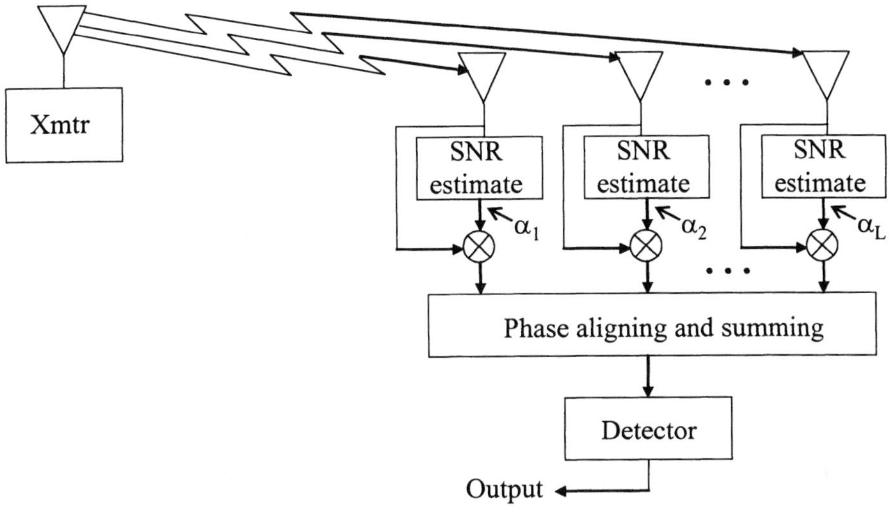

Figure 10.7-2 A block diagram model of maximal ratio combining.

After phase alignment the envelope of the output is given by

$$r = \sum_{k=1}^{L} \alpha_k r_k \tag{10.7.2-16}$$

where α_k are the weights of the k-th channel. Letting the mean noise power of each channel be denoted as N, the weighted sum of the noise is given by

$$N_L = N\sum_{k=1}^{L} \alpha_k^2 \tag{10.7.2-17}$$

The resulting SNR after the combining is then given by

$$\gamma_L = \frac{r^2}{2N_L} = \frac{\sum_{k=1}^{L} \alpha_k r_k}{2N\sum_{k=1}^{L} \alpha_k^2} \tag{10.7.2-18}$$

The resulting SNR can be maximized with the choice $\alpha_k = r_k / N$, based on the Schwarz Inequality [35, 36]. With that choice of α_k, we have

$$\gamma_{Lopt} = \frac{\left(\sum_{k=1}^{L} r_k^2 / N\right)^2}{2N\sum_{k=1}^{L} (r_k / N)^2} = \sum_{k=1}^{L} \frac{r_k^2}{2N} = \sum_{k=1}^{L} \gamma_k \tag{10.7.2-19}$$

Therefore the combined SNR, with the optimum weights, is just the sum of the channel SNRs. The received fading signal envelope for a fading mobile channel can be written as the sum of the squares of two statistically independent Gaussian random variables with zero mean and variance P_0. Thus γ_k^2 can be written as

$$\gamma_k^2 = \frac{r_k^2}{2N} = \frac{1}{2N}\left(x_k^2 + y_k^2\right) \tag{10.7.2-20}$$

It therefore follows that the distribution of γ_{Lopt} is a chi-squared random variable with $2L$ degrees of freedom and has a variance of $P_0 / 2N = \Gamma / 2$. Note that Γ was already defined by (10.7.2-3). The probability density function for the optimized maximal ratio combining of γ_{Lopt} is therefore given by (we have dropped the "opt" for convenience)

$$\begin{aligned} P(\gamma_L) &= \frac{\gamma_L^{L-1}\exp(-\gamma_L/\Gamma)}{\Gamma^L (L-1)!} \quad && for\ \gamma_L \geq 0 \\ P(\gamma_L) &= 0 && for\ \gamma_L < 0 \end{aligned} \tag{10.7.2-21}$$

The probability distribution function of the maximal ratio combiner is given by the integral

$$P_{MR}(\gamma_L) = \frac{1}{\Gamma^L (L-1)!}\int_0^{\gamma_L} u^{L-1}\exp(-u/\Gamma)\,du \tag{10.7.2-22}$$

Evaluating (10.7.2-22) for the maximal ratio (MR) for the SNR produces

$$P_{MR}(\gamma_L) = 1 - \exp(-\gamma_L/\Gamma)\sum_{k=1}^{L}\frac{(\gamma_L/\Gamma)^{k-1}}{(k-1)!} \tag{10.7.2-23}$$

This form of combining is superior to selection combining as will be shortly demonstrated. The mean SNR of the optimally combined signal can be obtained from (10.7.2-19)

$$\overline{\gamma_{MR}}(L) = \sum_{k=1}^{L}\overline{\gamma_k} = \sum_{k=1}^{L}\Gamma = L\Gamma \tag{10.7.2-24}$$

where the explicit dependence on L has been indicated. It is seen that $\overline{\gamma_{MR}}(L)$ varies linearly with the number of combined channels.

It is of interest to mention two approaches to phase alignment for use in maximal ratio combining. Figure 10.7-3 shows two possible methods [26].

The first circuit utilizes a heterodyne process to translate the two signals to baseband. At baseband they are compared in carrier phase and the signal on the upper arm is phase shifted to agree with the phase of the lower signal. The second circuit utilizes a voltage-controlled oscillator (VCO) to adjust the phase of the upper branch signal to match the phase of the lower branch signal. It is assumed that a carrier component is available to track in these implementations.

10.7.2.3 Equal Weight Combining

It is not always easy to obtain the SNR estimates needed for maximal ratio combining. It is much simpler to utilize equal weight combining. Let the gains be set to "1." Figure 10.7-3 still applies, but there no need to estimate the gain, since they are all assumed to be unity. The envelope of the combined (and phase aligned) signal is given by (10.7.2-16) with the $\alpha_k = 1$.

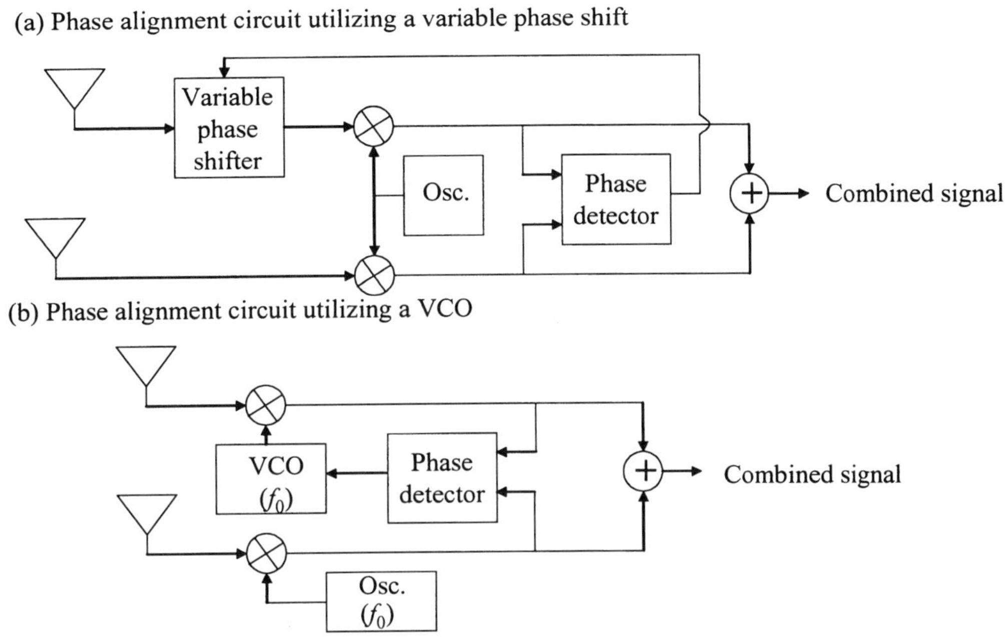

Figure 10.7-3 (a, b) Two possible circuits for aligning the phase of one signal with another.

Thus we have

$$r = \sum_{k=1}^{L} r_k \tag{10.7.2-25}$$

Therefore the SNR of the equal weight combiner is given by

$$\gamma_E = \frac{r^2}{2N\sum_{k=1}^{L} 1^2} = \frac{r^2}{2NL} \tag{10.7.2-26}$$

The distribution function is not available for $L \geq 3$. However, the mean value of γ_E can be obtained. Let the overbar denote the ensemble average; then

$$\overline{\gamma_E} = \frac{\overline{\left(\sum_{k=1}^{L} r_k\right)^2}}{2NL} = \frac{\overline{\sum_{k=1}^{L}\sum_{j=1}^{L} r_k r_j}}{2NL} \tag{10.7.2-27}$$

The double sum can be broken down into the diagonal terms ($k = j$) and the off-diagonal terms ($k \neq j$) so that

$$\overline{\sum_{k=1}^{L}\sum_{j=1}^{L} r_k r_j} = \sum_{k=1}^{L} \overline{r_k^2} + \sum_{j=1}^{L}\sum_{\substack{k=1 \\ j\neq k}}^{L} \overline{r_k}\,\overline{r_j} \tag{10.7.2-28}$$

Now $\overline{r_k^2} = 2P_0$; considering the value of $\overline{r_k} = \sqrt{\pi P_0 / 2}$, we have

$$\overline{r_i} = \int_0^\infty r_i p(r_i) dr_i = \int_0^\infty r_i \left[\frac{r_i}{P_0} \exp\left(-\frac{r_i^2}{2P_0} \right) \right] dr_i = \sqrt{\frac{\pi P_0}{2}} \qquad (10.7.2\text{-}29)$$

from a suitable tables of integration. Using these values in (10.7.2-28) gives

$$\overline{\sum_{k=1}^{L} \sum_{j=1}^{L} r_k r_j} = 2LP_0 + L(L-1)\frac{\pi P_0}{2} \qquad (10.7.2\text{-}30)$$

So

$$\overline{\gamma_E}(L) = \frac{2LP_0 + L(L-1)\frac{\pi P_0}{2}}{2NL} = \Gamma\left(1 + (L-1)\frac{\pi}{4}\right) \qquad (10.7.2\text{-}31)$$

where we have shown the explicit dependence on L. Figure 10.7-4 illustrates the plot of $\overline{\gamma_E}(L)/\Gamma$ in dB. It is easy to show that

$$\left(\frac{\overline{\gamma_E}(L)}{\overline{\gamma_{MR}}(L)} \right)_{Lim\ L \to \infty} = \frac{\pi}{4} \qquad (10.7.2\text{-}32)$$

so that we may conclude that equal weight combining is only 1.05 dB worse than optimal weight combining in the limit as L becomes very large. Figure 10.7-4 compares maximal ratio, equal weight, and selection diversity combining methods. As can be seen maximal ratio offers the best performance.

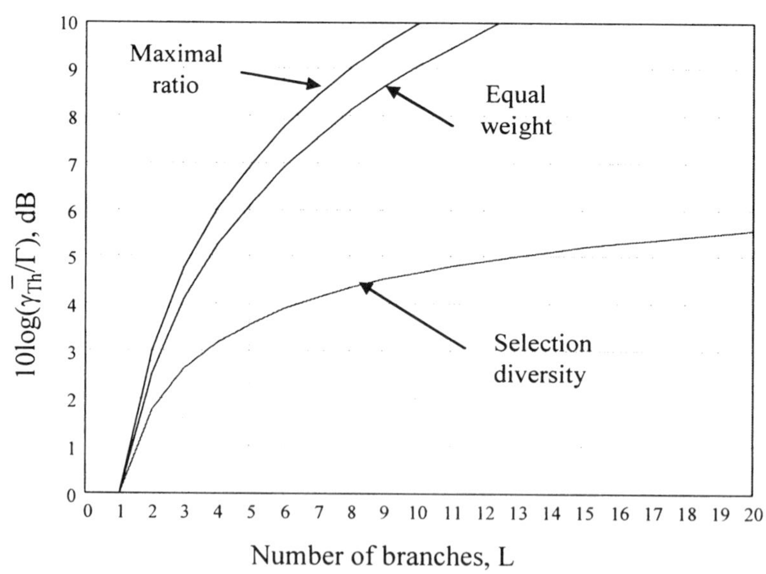

Figure 10.7-4 The improvement of SNR from three diversity combiners: maximal ratio, equal gain combining, and maximum selection.

10.7.2.4 Scanning Diversity

Another very simple type of elementary diversity is known as *scanning diversity*. It is similar to selection diversity; however it differs in the sense that instead of always using the best one of the L signals, it scans the signal sequentially until one is found that is above the threshold. This signal is then used continuously until it falls below the threshold and the search starts anew. The performance is not as good as even selection diversity; however, it is easy to implement and offers reasonable performance. Figure 10.7-5 illustrates the technique.

One advantage of this scheme is the fact that only one receiver is needed to implement it.

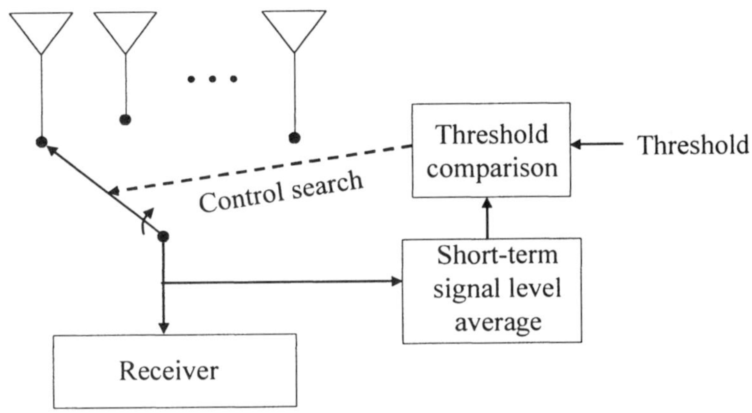

Figure 10.7-5 Model of the scanning diversity approach.

10.8 EQUALIZATION FOR MULTIPATH IMPROVEMENT

As we mentioned earlier in this section there are three basic methods that can be used independently or together to improve the signal quality such as the bit error rate (BER). They are diversity, equalization, and channel coding. We now address equalization.

The function of *equalization* is to compensate for intersymbol interference (ISI) that was caused by multipath within time dispersive channels. Recall when the channel bandwidth is greater than the coherence bandwidth of the channel, ISI occurs, and the baseband equivalent symbols are spread in time. The role of the equalizer is to compensate for the expected range of the channel amplitude and phase characteristics. Furthermore equalizers need to be adaptive in order to compensate for the time varying channel and to determine the unknown channel model parameters.

Conceptually equalization is a terminology that can be used to describe a type of signal processing that minimizes the ISI of the processed received signal. Due to the time varying nature of the fading channel the equalizer must track the changes; consequently, they are called *adaptive equalizers*. Before we discuss adaptive equalizers, let us briefly discuss the transversal symbol rate equalizer model.

10.8.1 Baseband Transversal Symbol Rate Equalizer

Consider a baseband communication system that is represented by samples that are running at the symbol rate of $R_s = 1/T$, which is assumed to be the Nyquist rate. Let the channel be represented by a transversal filter having taps that are the sampled values of the channel filter. This filter represents a distortion to the transmitted signal. Assume that the symbol rate equalizer is also represented by a transversal filter having taps that are the sampled values of the transversal filter.

The combination is depicted in Figure 10.8-1 for a particular channel distortion and equalizer model [37] for the case that the channel has 3 stages ($n = 3$) and the equalizer has 5 stages ($m = 5$).

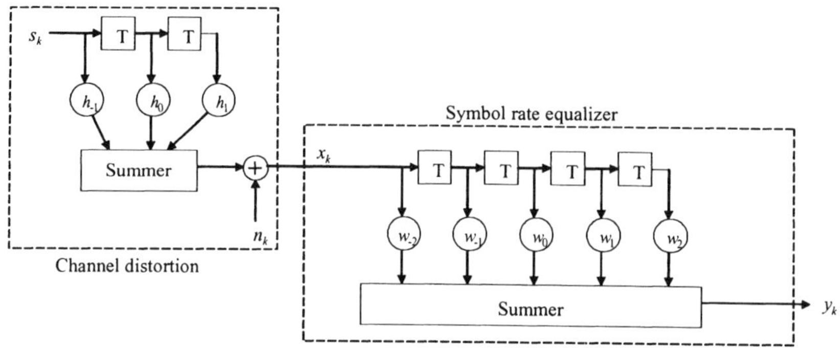

Figure 10.8-1 Channel distortion model followed by a symbol rate equalizer.

Now let $\boldsymbol{h}$ be written in general as

$$\boldsymbol{h} = \left(h_{-\frac{n-1}{2}}, h_{-\frac{n-1}{2}+1}, \ldots h_{-1}, h_0, h_1, \ldots h_{\frac{n-1}{2}-1}, h_{\frac{n-1}{2}} \right) \tag{10.8.1-1}$$

And be the vector of the channel distortion model weights. There are n weights in total. Note that in an undistorted Nyquist channel $\boldsymbol{h}$ reduces to

$$\boldsymbol{h} = (0,0,0,\ldots 1,0,0\ldots 0) \tag{10.8.1-2}$$

with only the center component having unit value, and the rest having the value of zero, whereas a distorted channel, for example, could have the vector (for the case $n = 3$)

$$\boldsymbol{h} = (-0.09,1,0.05) \tag{10.8.1-3}$$

with the center value normalized to unity. The sampled impulse response vector $\boldsymbol{h}$ with n components (n = odd), of the channel, can be represented as a z-transform as

$$h(z) = h_{-\frac{n-1}{2}} z^{\frac{n-1}{2}} + h_{-\frac{n-1}{2}+1} z^{\frac{n-1}{2}+1}, \ldots h_{-1}z + h_0 + h_1 z^{-1}, \ldots h_{\frac{n-1}{2}-1} z^{-\frac{n-1}{2}} + h_{\frac{n-1}{2}} z^{-\frac{n-1}{2}-1} \tag{10.8.1-4}$$

Likewise we can represent the equalizer in the z domain, having m components (m odd) as

$$w(z) = w_{-\frac{m-1}{2}} z^{\frac{m-1}{2}} + w_{-\frac{m-1}{2}+1} z^{\frac{m-1}{2}+1}, \ldots w_{-1}z + w_0 + w_1 z^{-1}, \ldots w_{\frac{m-1}{2}-1} z^{-\frac{m-1}{2}} + w_{\frac{m-1}{2}} z^{-\frac{m-1}{2}-1} \tag{10.8.1-5}$$

Similarly the output $y(z)$ can be written in the z domain also. This output has $n+m-1$ components. The operation of the equalizer on the channel may be expressed in the z domain as

$$h(z)w(z) = y(z) \tag{10.8.1-6}$$

For perfect equalization (no noise) we must have

$$h(z)w(z) = z^{-K} \tag{10.8.1-7}$$

which states that the equalized channel response is a delay of K time units. Perfect equalization (without noise) would suggest that

$$w(z) = z^{-K} / h(z) \tag{10.8.1-8}$$

This result would suggest that an infinite impulse response (IIR) filter would be required as the equalizer, with an infinite number of taps. However, in practice [37], finite impulse response (FIR) filters are normally used. Thus we will consider transversal filter models for equalization.

Let the input to the system of Figure 10.8-1 be a single isolated data symbol. Our filter model causing the distortion has n taps and the equalizer has m taps. When the channel filter has n taps and the equalizer has m taps then the output will have $n+m-1$ symbols. This is true since the first symbol will pass through with no delay. The last symbol will have a delay of $n-1+m-1$ symbols. Hence the range is $[n-1+m-1]-0+1=n+m-1$.

This system can be described by a pair of vectors and a channel matrix. Let the output vector $\boldsymbol{w}$ be described by the $n+m-1$ vector

$$\boldsymbol{y} = \left(y_{-\frac{n+m-1}{2}} \ldots, y_{-2}, y_{-1}, y_0, y_1, y_2, \ldots y_{\frac{n+m-1}{2}} \right) \tag{10.8.1-9}$$

Further, let the equalizer be described by the m component vector

$$w = \left(w_{-\frac{m-1}{2}} \ldots, w_{-2}, w_{-1}, w_0, w_1, w_2, \ldots w_{\frac{m-1}{2}} \right) \tag{10.8.1-10}$$

For notational convenience let the $n = 3$ and $m = 5$ for the H matrix description, so that

$$H = \begin{bmatrix} h_{-1} & h_0 & h_1 & 0 & 0 & 0 & 0 \\ 0 & h_{-1} & h_0 & h_1 & 0 & 0 & 0 \\ 0 & 0 & h_{-1} & h_0 & h_1 & 0 & 0 \\ 0 & 0 & 0 & h_{-1} & h_0 & h_1 & 0 \\ 0 & 0 & 0 & 0 & h_{-1} & h_0 & h_1 \end{bmatrix} \tag{10.8.1-11}$$

In the general case the H matrix is an $m \times (m+n-1)$ convolution matrix of the channel. It then follows that

$$\boldsymbol{y} = \boldsymbol{w}H \tag{10.8.1-12}$$

In order to achieve a useful equalizer it is necessary that the number of taps of the equalizer should exceed the number of taps of the channel. An ideal output would produce

$$\boldsymbol{y}_I = (0001000) \tag{10.8.1-13}$$

for this example of $n = 3$ and $m = 5$. In general the unit value would occur at the $((m+n)/2)$-th position. In the previous example the value is position 4 ((5+3)/2). The difference between the actual output and the ideal distortion-free output, when no noise is present, is given by

$$MSE = \|y - y_I\|^2 = \|wH - y_I\|^2 \tag{10.8.1-14}$$

Thus for this example

$$y = (w_{-2}\ w_{-1}\ w_0\ w_1\ w_2)\begin{bmatrix} h_{-1} & h_0 & h_1 & 0 & 0 & 0 & 0 \\ 0 & h_{-1} & h_0 & h_1 & 0 & 0 & 0 \\ 0 & 0 & h_{-1} & h_0 & h_1 & 0 & 0 \\ 0 & 0 & 0 & h_{-1} & h_0 & h_1 & 0 \\ 0 & 0 & 0 & 0 & h_{-1} & h_0 & h_1 \end{bmatrix} \tag{10.8.1-15}$$

For our example it should be observed that the vector $y\text{–}y_I$ is a 1×7 vector; therefore, there will be 7 squared terms in the expression for the MSE. In Problem 9 it is shown that for this example that the MSE is given by

$$\begin{aligned} MSE = {} & \left(w_{-2}h_{-1}\right)^2 + (w_{-2}h_0 + w_{-1}h_{-1})^2 + (w_{-2}h_1 + w_{-1}h_0 + w_0h_{-1})^2 \\ & + (w_{-1}h_1 + w_0h_0 + w_1h_{-1} - 1)^2 + (w_0h_1 + w_1h_0 + w_2h_{-1})^2 \\ & + (w_1h_1 + w_2h_0)^2 + (w_2h_1)^2 \end{aligned} \tag{10.8.1-16}$$

10.8.2 Baseband Adaptive Equalization

In this section we only consider adaptive equalization at the receiving end of the link; some schemes utilize prechannel equalization, which requires feedback. We again consider the tapped delay line filter, which is *synchronous* in the sense that the symbol duration, T, is the same as the tap spacing. Consider the tapped delay line model shown in Figure 10.8-2.

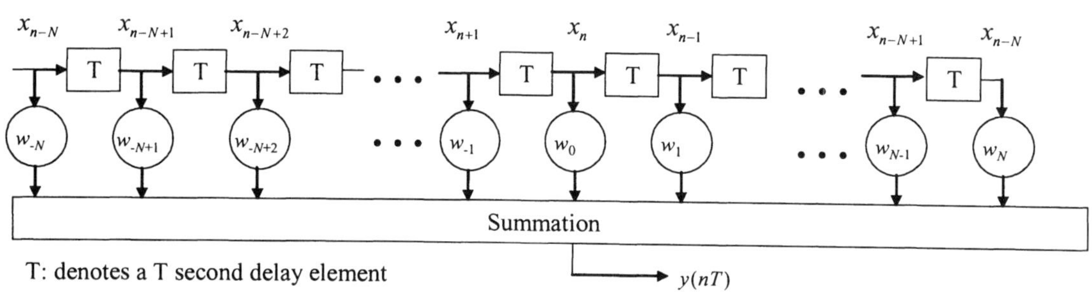

Figure 10.8-2 Baseband adaptive equalizer filter with $2N$+1 components.

10.8.2.1 Least Mean Square Algorithm for the Baseband Adaptive Equalization

Consider the adaptive equalizer shown in Figure 10.8-2 in which the coefficients w_k are adjustable by external control. The input sequence $x(n\text{T})$ is assumed to have passed through an unknown channel that is both noisy and dispersive, and is fed to the equalizer. It is assumed that some form of pulse shaping has been utilized in the design of the transmission system. The adaptive equalizer has the task of correcting for the effects of thermal noise and residual distortion. Let the output of a linear filter, with zero mean additive white Gaussian noise (AWGN), be described by

$$x_n = \sum_{i=1}^{K} h_i a_{n-i} + n_i \tag{10.8.2-1}$$

where the channel filter model is assumed to be represented by a finite impulse response filter (FIR) with values h_i, with the $\{a_n\}$ sequence being the transmitted symbol sequence, and the n_i the zero mean WGN samples.

In addition, let the input of the equalizer and the output of the equalizer be denoted by, respectively,

$$\begin{aligned} x_n &= x(n\mathrm{T}) \\ y_n &= y(n\mathrm{T}) \end{aligned} \tag{10.8.2-2}$$

The output sample sequence of the tapped delay line equalizer, with $2N+1$ taps, to the finite energy input sequence x_n is given by the discrete convolution

$$y_n = \sum_{k=-N}^{N} w_k x_{n-k} \tag{10.8.2-3}$$

The equalizer weight adjustments may be achieved by observing the error between the desired pulse shape and the observed pulse shape at the equalizer output. This error is based on the observing the sampling instants and then processing the error to determine the direction that the tap weights should be changed to obtain the optimum values.

There is more than one criterion to define the "error." For example, the minimization of the peak distortion is one criterion. This approach minimizes the intersymbol interference at the output of the equalizer. This approach is only valid if the intersymbol interference is not too severe.

An alternative approach is to minimize the mean square error [36]. This approach has some advantages and is more general in its application. Thus we will utilize the mean square error as our error measure for the determination of the weights of the equalizer. Let a_n be the desired response and y_n the actual response at time $t = n$. Then the error e_n, at time n, is defined as

$$e_n = a_n - y_n \tag{10.8.2-4}$$

The mean squared error is given by

$$MSE = E\left[e_n^2\right] \tag{10.8.2-5}$$

with $E[\ ]$ denoting the statistical expectation operator. The derivative of the MSE with respect to the weight w_k is given by

$$\frac{\partial MSE}{\partial w_k} = 2E\left[e_n \frac{\partial e_n}{\partial w_k}\right] \tag{10.8.2-6}$$

Using (10.8.2-4) in (10.8.2-6) yields the result

$$\frac{\partial MSE}{\partial w_k} = -2E\left[e_n \frac{\partial y_n}{\partial w_k}\right] \tag{10.8.2-7}$$

From (10.8.2-3) one has

$$\frac{\partial MSE}{\partial w_k} = -2E\left[e_n x_{n-k}\right] \tag{10.8.2-8}$$

The expectation (denoted by $E[.]$) in (10.8.2-8) is the cross correlation between the input signal delayed by n samples and the error signal e_n. That is the partial derivative is given by

$$\frac{\partial MSE}{\partial w_k} = -2ER_{ex}(k) \tag{10.8.2-9}$$

where

$$R_{ex}(k) = E\left[e_n x_{n-k}\right] \tag{10.8.2-10}$$

It therefore follows that the optimality condition for the minimum mean squared error must satisfy the conditions

$$\frac{\partial MSE}{\partial w_k} = 0 \text{ for } k = 0, \pm 1, \pm 2, \ldots \pm N \tag{10.8.2-11}$$

From (10.8.2-11) and (10.8.2-9), it is seen that an equivalent condition is that

$$R_{ex}(k) = 0 \text{ for } k = 0, \pm 1, \pm 2, \ldots \pm N \tag{10.8.2-12}$$

Thus we see that a necessary condition for optimality is that the cross correlation of the input sequence and the error sequence be 0 for all of the $2N+1$ values. This result is known as the *principle of orthogonality.*

It is not hard to show that the MSE is a second order function in the tap weights w_k. Use (10.8.2-3) and (10.8.2-4) in (10.8.2-5) to obtain

$$MSE = E\left[y_{In}^2\right] - 2E\left[y_{In}\sum_{k=-N}^{N} w_k x_{n-k}\right] + E\left[\sum_{k=-N}^{N}\sum_{j=-N}^{N} w_k w_j x_{n-k} x_{n-j}\right] \tag{10.8.2-13}$$

and therefore is a second order function in the variable w_k. This quadratic form generates a bowl-like smooth surface having a unique minimum and no local minima [38, 39]. The goal of the adaptive equalizer algorithm is to seek the minimum value (minimum MSE) of this bowl-like shape by successively adjusting the weights. This algorithm is akin to a steepest descent algorithm for finding the minimum MSE.

It is convenient to formulate the least mean square (LMS) algorithm with a vector and matrix description. Let $\boldsymbol{x}_n$ (vectors in bold) denote the tap inputs of the equalizer so that

$$\boldsymbol{x}_n = \left(x_{n+N}, x_{n+N-1}, \ldots x_{n-1}, x_n, x_{n-1}, \ldots x_{n-N+1}, x_{n-N}\right)^T \tag{10.8.2-14}$$

where the T superscript (in italics and not to be confused with the time delay T in the adaptive equalizer) on the vector $\boldsymbol{x}_n$ denotes the transpose function, so that $\boldsymbol{x}_n$ is actually a vertical vector. Also the tap weights of the equalizer can be written as

$$\boldsymbol{w}_n = \left(w_{-N}(n), w_{-N+1}(n), \ldots w_{-1}(n), w_0(n), w_1(n), \ldots w_{N-1}(n), w_N(n)\right)^T \tag{10.8.2-15}$$

The convolution indicated by (10.8.2-3) can then be written in the compact form

$$y_n = \boldsymbol{x}_n^T \boldsymbol{w}_n \tag{10.8.2-16}$$

as is clear by comparing (10.8.2-16) to (10.8.2-3). The term $\boldsymbol{x}_n^T \boldsymbol{w}_n$ is termed *the inner product* of the vectors $\boldsymbol{x}_n$ and $\boldsymbol{w}_n$. We are now in a position to describe the *least mean square (LMS) algorithm*. It is described as follows

(1) Set the initial value of $\boldsymbol{w}_1$ to be $\boldsymbol{0}$ (that is, all the $2N+1$ tap weights are zeroed out at time $n=1$ or T seconds). The bold $\boldsymbol{0}$ denotes a vector zero of $2N+1$ dimensions.

(2) Compute the first two terms and sequentially determine the third term.

(3)

$$\begin{aligned} y_n &= \boldsymbol{x}_n^T \boldsymbol{w}_n \\ e_n &= a_n - y_n \\ \boldsymbol{w}_{n+1} &= \boldsymbol{w}_n + \alpha e_n \boldsymbol{x}_n \end{aligned} \tag{10.8.2-17}$$

in which α is the step size parameter for the weight vector $\boldsymbol{w}_n$.

(4) Continue the recursion until steady state is obtained.

Figure 10.8-3 illustrates the T-second recursion process that generates the new weights from the previous ones.

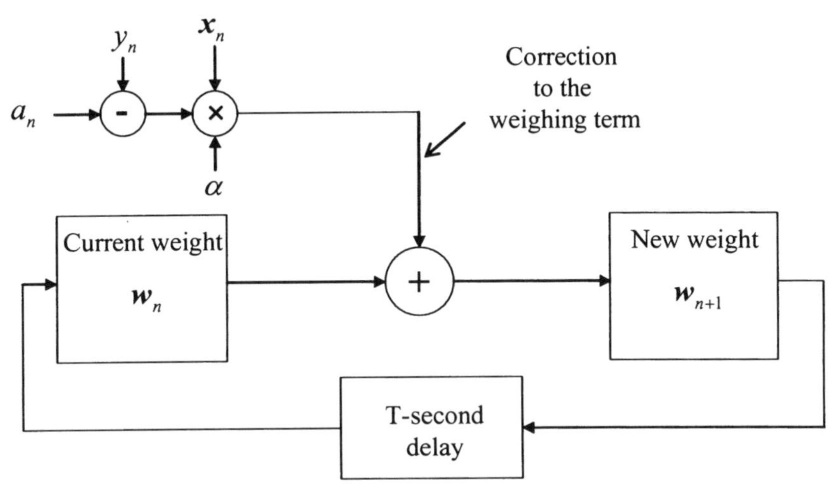

Figure 10.8-3 Vector weight recursion update process of the LMS algorithm.

10.8.2.2 Least Mean Square Algorithm for the Bandpass Adaptive Equalization

When a bandpass function is utilized (the most usual case), it is possible to have one LMS algorithm for each of the I and Q phases of the heterodyned to baseband components. Therefore the above theory applies to each component of the baseband derived signal from the bandpass waveform.

10.8.2.3 Basic Operation of the Adaptive Equalizer

The adaptive equalizer has two modes of operation: (1) initially the *training mode*, followed by (2) the *decision-directed mode* that equalizes the channel. The training mode is based on providing a known sequence that is transmitted and a synchronized replica (with the transmission delay accounted for) of that signal that is generated in the receiver, as the desired response. This training sequence is applied to the receiver (adaptive equalizer) and the tap weights are determined according to the LMS algorithm. The training sequence used in practice is commonly a pseudonoise sequence, which can be viewed has having random symbol polarity characteristics.

At the conclusion of the training process the adaptive equalizer changes over to its second mode of operation, which is the decision-directed mode. In this mode the error is defined by

$$e_n' = \hat{a}_n - y_n \tag{10.8.2-18}$$

where $\hat{a}_n$ is the final (converged, but not necessarily converged to a_n) estimate of y_n and y_n is the equalizer at time nT. In normal operation the decisions made by the adaptive equalizer (receiver) are, with high likelihood, correct, so that the receiver would normally operate properly.

There is a trade-off between a larger value of α and a smaller value of α—the convergence parameter. Too large of a value of α may yield too large a value of the mean square error yielding poor equalizer performance, or the minimum is overshot and convergence never takes place. However, larger values of α allow a faster tracking capability of the equalizer. When α is too small, the equalizer may not be able to track small changes in the channel characteristics. In conclusion there is a balance that must be maintained between fast channel characteristics and reduced mean squared error, obtained by utilizing the proper value of α.

10.8.2.4 Least Mean Square Implementation Options for the Baseband Model

A main advantage of the LMS algorithm is its ease of implementation. There are three general approaches to the implementation: (1) analog, (2) nonprogrammable digital, and (3) programmable digital. A brief description follows.

(1) Analog implementation: This approach is useful in regimes where the symbol rate is too high for digital implementation. It is based on the use of charge coupled device (CCD) technology. The set of adjustable weights are stored in the CCD memory locations, with the multiplications of the analog sample values and the digitized tap weights done in an analog manner.

(2) Hardwired (nonprogrammable) digital implementation: This approach is the most widely used method of constructing adaptive equalizers and is commonly implemented using very large-scale integrated (VLSI) circuits. In this approach the input to the equalizer is sampled and then quantized into a format that is easily stored in shift registers. The set of adjustable weights is stored in shift registers also. The multiply and accumulate functions are accomplished with digital arithmetic. The circuitry that is used for the equalization process is hard-wired.

(3) Programmable digital implementation: This approach is based on the use of a microprocessor that can perform the equalization function. In addition, the microprocessor can time share other processing tasks such as modulation and demodulation when used in a modem (used to transmit digital data over phone lines) and filtering.

Most high-speed phone modems use a form of a LMS algorithm since it is superior to algorithms that minimize the intersymbol interference peak distortion [40].

10.8.3 Baseband Decision Feedback Equalizers

So far we have been considering linear equalizers. Oftentimes a linear equalizer does not perform well when the channel has a spectral null [41]. A decision feedback equalizer (DFE) is a nonlinear equalizer that utilizes previous detector decisions to eliminate unwanted ISI on the symbols that are being detected currently.

To gain a little understanding of the process [36], let the sequence $\{h_i\}$ denote the sampled values of the impulse response with $h_i = h(i\mathrm{T})$. The output of this filter to an input sequence x_n, without noise, is given by the discrete convolution

$$y_n = \sum_i h_i x_{n-i} = \underbrace{h_0 x_n}_{\substack{desired\\data\\symbol}} + \underbrace{\sum_{i>0} h_i x_{n-i}}_{postcursor} + \underbrace{\sum_{i<0} h_i x_{n-i}}_{precursor} \tag{10.8.3-1}$$

where the first term on the right-hand side is proportional to the current symbol, the second term is termed the *postcursors* of the channel response, since all the terms occur after the current symbol x_n, and the third term is the precursor of the channel response, since all of the terms occur before the current symbol x_n.

The concept used in *decision feedback equalization* is to utilize data decisions processed from the precursors of the channel impulse response to accommodate the postcursors. It is necessary that the decisions, based on the precursor, were correct. Under these conditions the DFE can produce an improvement over the performance of the tapped delay line equalizer (linear equalizer). It is to be noted that the decision device that makes a decision on the current symbol is nonlinear. Figure 10.8-4 illustrates a representative model of a DFE.

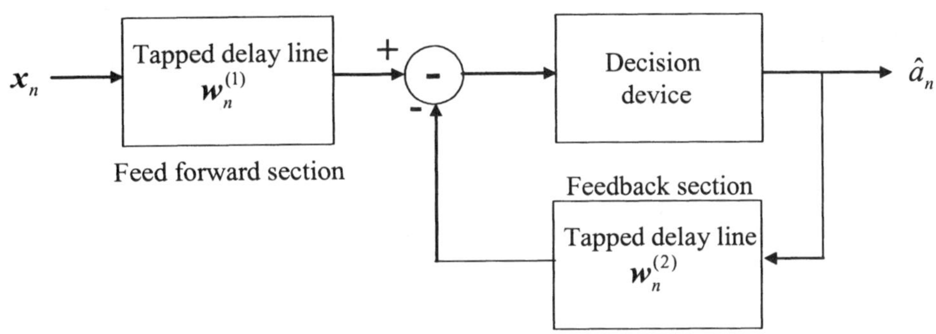

Figure 10.8-4 A decision feedback equalizer block diagram model.

The input is a tapped delay line filter with weights $w_n^{(1)}$. It forms the feed forward section and has its taps separated by the symbol duration. The symbol estimate $\hat{a}_n$ feeds the input of the feedback section, which again is a tapped delay line filter with the taps also spaced apart by the symbol duration. The decision device takes the sum of the feed forward and feedback components and estimates the current symbol, $\hat{a}_n$. The feedback portion attempts to subtract out that portion of the intersymbol interference produced by previously detected symbols, from the estimates of the future samples.

It is possible [36] to utilize a mean square error criterion to obtain the optimum decision feedback equalizer. In fact the LMS algorithm can be used to jointly optimize the feed forward and the feedback tap weights of the equalizers, utilizing a common error signal. Consider the augmented weight vector $\boldsymbol{w}_n$ defined by

$$\boldsymbol{w}_n = \begin{bmatrix} \boldsymbol{w}_n^{(1)} \\ \boldsymbol{w}_n^{(2)} \end{bmatrix} \tag{10.8.3-2}$$

where $\boldsymbol{w}_n^{(1)}$ is the tap weights of the feed forward equalizer and $\boldsymbol{w}_n^{(2)}$ is the tap weights of the feedback equalizer. Let the augmented vector $\boldsymbol{z}_n$ be defined by the vector of the input samples, $\boldsymbol{x}_n$, to the feed forward section and the vector of symbol values, $\hat{\boldsymbol{a}}_n$, input to the feedback section. Thus

$$z_n = \begin{bmatrix} x_n \\ \hat{a}_n \end{bmatrix} \tag{10.8.3-3}$$

The common error signal is defined by

$$e_n = a_n - w_n^T z_n \tag{10.8.3-4}$$

where, again, the T over the vector denotes the transpose operation. The LMS algorithm for the decision feedback equalizer is specified by the update equation pair

$$\begin{aligned} w_{n+1}^{(1)} &= w_n^{(1)} + \alpha_1 e_n x_n \\ w_{n+1}^{(2)} &= w_n^{(2)} + \alpha_2 e_n \hat{a}_n \end{aligned} \tag{10.8.3-5}$$

in which α_1 and α_2 are the step size parameters for the feed forward and feedback segments, respectively.

We have not addressed the rate of convergence in these discussions. Although that topic is not addressed in this book, it has been observed that the number of iterations ranges from about 100 to about 1,000 iterations, depending on a number of parameters. Clearly there is a lot more to this subject than what was presented here. Kurzweil [39] has two chapters devoted to various types of equalizers and can be consulted for more details. In the environment of a fading radio channel in which moderate to severe intersymbol interference occurs, the DFE yields relatively good performance.

10.9 DIVERSITY TECHNIQUES FOR MULTIPATH IMPROVEMENT

Now we will consider the bit error rate performance of a few binary modulation methods when diversity is utilized to improve the fading channel performance. The fading is modeled as frequency nonselective, slowly fading with Rayleigh distributed envelope statistics. It is further assumed that the fading processes on all L channels are mutually statistically independent and the signal is corrupted by additive Gaussian white noise. All the noise processes are assumed to be statistically independent. The means of obtaining diversity are discussed in Section 10.7.1.

10.9.1 Multipath Performance Improvement Via Diversity Techniques for Binary Channels

We will develop a general binary model and then specialize it to a few binary modulation cases. This section follows Proakis [32].

The low-pass equivalent (baseband) signals can be written in the form

$$r_{km}(t) = \alpha_k e^{-j\theta_k} s_{km}(t) + n_k(t) \qquad k = 1,2,3,...L \quad m = 1,2 \tag{10.9.1-1}$$

in which $\left(\alpha_k e^{-j\theta_k}\right)$ includes the attenuation factor $\left(\alpha_k\right)$ and the phase change $\left(e^{-j\theta_k}\right)$ for the k-th channel, $s_{km}(t)$ is the k-th channel signal which is either $m = 1$ or 2, and $n_k(t)$ is the complex baseband equivalent white Gaussian noise process so that the radiofrequency noise is of the form

$$n_k^{rf}(t) = n_{ck}(t)\cos(\omega_0 t + \theta_k) - jn_{sk}(t)\sin(\omega_0 t + \theta_k) = \text{Re}\left[n_k(t)e^{j(2\pi f_0 t + \theta_k)}\right] \tag{10.9.1-2}$$

and $n_{ck}(t)$ and $n_{sk}(t)$ are statistically independent. Furthermore, for white Gaussian noise

$$R_{n_k}(t-u) = \frac{1}{2}E(n_k^*(t)n_k(u)) = N_0\delta(t-u) \tag{10.9.1-3}$$

The optimum demodulator for the received binary signal from the k-th channel has two matched filters with the following impulse responses [42]

$$IR_{k1} = s_{k1}^*(T-t)$$
$$IR_{k2} = s_{k2}^*(T-t) \qquad (10.9.1\text{-}4)$$

where * denotes the complex conjugate. In the case of binary BPSK, only one matched filter is needed since $s_{k1}(t) = -s_{k2}(t)$. The outputs of the matched filters are combined in a combiner and produce the two hypothesis variables. The optimum combiner is one in which each matched filter output is multiplied by the corresponding complex valued channel gain $\alpha_k e^{j\theta_k}$. This multiplication by the complex channel gain compensates for the delay and weights the channels proportionally to the received signal strength. Finally two sums are generated; the first is composed of the real part of weighted outputs from the matched filters with the hypothesis of a "0" being transmitted, and the second is composed of the real part of weighted outputs from the matched filters with the hypothesis of a "1" being transmitted [42]. This combiner has been discussed earlier and is known as a *maximal ratio combiner*. The combiner is based on the assumption that the delays and the signal levels are known exactly. When these parameters are not known exactly, additional degradation will result. Figure 10.9-1 illustrates the digital communication model.

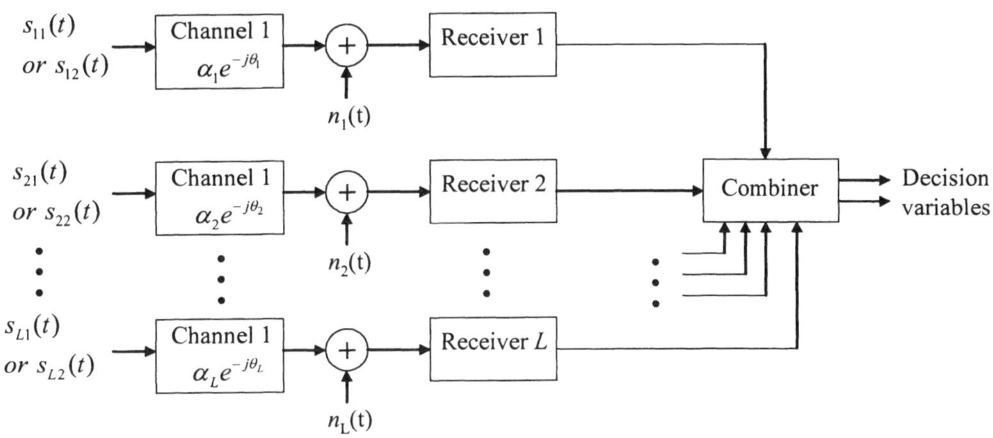

Figure 10.9-1 Binary digital communication system model with diversity.

10.9.1.1 Performance Improvement Via Diversity Techniques for BPSK Channels

Now we are in a position to analyze the performance of binary PSK channels with Rayleigh fading in the case of L-th order diversity. With the maximal ratio combiner model a single statistic for the decision variable is given by

$$U = \mathrm{Re}\left[\sum_{k=1}^{L} \alpha_k e^{j\theta_k} \int_0^T r_{km}(t) s_{k1}^*(t)dt\right] = \mathrm{Re}\left[\sum_{k=1}^{L} \alpha_k e^{j\theta_k} \int_0^T \left(\alpha_k e^{-j\theta_k} s_{k1}(t) + n_k(t)\right) s_{k1}^*(t)dt\right] \qquad (10.9.1\text{-}5)$$

where it has been assumed that $s_{k1}^*(t)$ has been transmitted for convenience. If $s_{k2}^*(t)$ was transmitted, the opposite value of the signal component would occur. Equation (10.9.1-5) can be simplified to

$$U = \mathrm{Re}\left[\sum_{k=1}^{L}\alpha_k^2(2E) + \sum_{k=1}^{L}\alpha_k N_k\right] \tag{10.9.1-6}$$

in which N_k is defined by

$$N_k = e^{j\theta_k}\int_0^T n_k(t)s_{k1}^*(t)dt \text{ and } 2E = \int_0^T s_{k1}(t)s_{k1}^*(t)dt \tag{10.9.1-7}$$

Finally, from (10.9.1-6), our result can be simplified to

$$U = 2E\sum_{k=1}^{L}\alpha_k^2 + \sum_{k=1}^{L}\alpha_k \,\mathrm{Re}\left[N_k\right] \tag{10.9.1-8}$$

The method used to determine the bit error rate (BER) is to compute the BER conditioned on the set of channel attenuation parameters $\{\alpha_k\}$, and then average over the distribution function of the set $\{\alpha_k\}$.

Conditioned on the set $\{\alpha_k\}$, the decision statistic U is a Gaussian random variable with a mean of

$$E(U) = 2E\sum_{k=1}^{L}\alpha_k^2 \tag{10.9.1-9}$$

and a variance of

$$\sigma_U^2 = 2EN_0\sum_{k=1}^{L}\alpha_k^2 \tag{10.9.1-10}$$

which is established in Problem 10. Given that conditional on the set $\{\alpha_k\}$, U is a Gaussian random variable, so that the probability of error is given by

$$PE = Q\left(\sqrt{2\gamma_L}\right) \tag{10.9.1-11}$$

where now γ_L depends on all L channels via

$$\gamma_L = \frac{E}{N_0}\sum_{k=1}^{L}\alpha_k^2 \tag{10.9.1-12}$$

Now E is the energy of one signal and N_0 is the one-sided noise spectral density. Notice that when α_k is unity and $L = 1$, (10.9.1-11) reverts to the well-known result for the bit error rate for BPSK channels in white Gaussian noise. It is convenient to write (10.9.1-12) in the form

$$\gamma_L = \sum_{k=1}^{L}\gamma_k \text{ , and } \gamma_k = \frac{E}{N_0}\alpha_k^2 \tag{10.9.1-13}$$

The parameter γ_k is the instantaneous signal-to-noise ratio (SNR) on the k-th channel. In order to proceed it is necessary to obtain the probability density function of γ_L, which we denote as $p(\gamma_L)$. The easiest way to obtain $p(\gamma_L)$, is to work with the characteristic function (CF). Note that when $L = 1$, $\gamma_L = \gamma_1$. For the Rayleigh fading channel, Problem 9 has shown that the probability density function of the channel SNR is given by

$$p(\gamma_k) = \frac{1}{\Gamma} e^{-(\gamma_k/\Gamma)}, \quad \gamma_k \geq 0 \tag{10.9.1-14}$$

where

$$\Gamma = \frac{E}{N_0} E\left(\alpha_k^2\right) \tag{10.9.1-15}$$

The characteristic function of the channel SNR is given by

$$\Phi(v) = E\left[e^{jv\gamma_k}\right] \tag{10.9.1-16}$$

Using (10.9.1-14) in (10.9.1-16) with the assumption that the average SNR is the same for all channels, so that

$$\Phi(v) = E\left[e^{jv\gamma_k}\right] = \frac{1}{1 - jv\Gamma} \tag{10.9.1-17}$$

for all k. For convenience this assumption of the same channel statistics on all channels will be assumed in the rest of this section. We desire the CF of the sum as indicated in the first equation in (10.9.1-13), so we can write for the sum the L terms

$$\Phi_L(jv) = \frac{1}{\left(1 - jv\Gamma\right)^L} \tag{10.9.1-18}$$

Using the known inverse transform of the CE leads to the probability density of the sum

$$p(\gamma_L) = \frac{1}{(L-1)!\Gamma^L} \gamma_L^{L-1} e^{-\gamma_L/\Gamma} \tag{10.9.1-19}$$

Now it is necessary to average the BER over the density function of γ_L. Thus we need to evaluate

$$PE_{BPSK} = \int_0^\infty Q\left(\sqrt{2\gamma_L}\right) p(\gamma_L) d\gamma_L \tag{10.9.1-20}$$

Proakis [42] has a provided a closed form solution, that can be written as

$$PE = \left(0.5(1-\lambda)\right)^L \sum_{k=0}^{L-1} \binom{L-1+k}{k} \left(0.5(1+\lambda)\right)^k \tag{10.9.1-21}$$

in which λ is defined by

$$\lambda = \sqrt{\frac{\Gamma}{1+\Gamma}} \tag{10.9.1-22}$$

When Γ is, say, larger than 10, $(1-\lambda)/2$ and $(1+\lambda)/2$ can be approximated as

$$0.5(1+\lambda)\simeq 1 \text{ and } 0.5(1-\lambda)\simeq\frac{1}{4\Gamma} \quad (10.9.1\text{-}23)$$

with the latter relationship coming from the approximation

$$\frac{1}{2}\left(1-\sqrt{\frac{\Gamma}{1+\Gamma}}\right)=\frac{1}{2}\left(1-\sqrt{\frac{1}{(1/\Gamma)+1}}\right)\simeq\frac{1}{2}\left(1-\left(1-\frac{1}{2\Gamma}\right)\right)=\frac{1}{4\Gamma} \quad (10.9.1\text{-}24)$$

In addition, the sum

$$\sum_{k=0}^{L-1}\binom{L-1+k}{k}=\binom{2L-1}{L-1} \quad (10.9.1\text{-}25)$$

is established in Problem 11. Using the prior two relationships yields[4]

$$PE\simeq\left(\frac{1}{4\Gamma}\right)^{L}\binom{2L-1}{L-1} \quad (10.9.1\text{-}26)$$

Clearly as the diversity order L increases, the PE decreases proportional to Γ^{-L}.

In some channels in the case that the fade is at a fast rate, it may not be feasible to utilize BPSK modulation and demodulation. In such cases a noncoherent approach such as DPSK or FSK should be utilized.

10.9.1.2 Performance Improvement via Diversity Techniques for DPSK Channels

For differential phase shift keying (DPSK) to be a viable approach for a Rayleigh fading channel, the channel phase variation must be slow enough so that the channel phase does not change significantly over two contiguous channel symbols. The analysis to follow assumes that the channels don't change amplitude or phase over two contiguous symbols.

Assuming that binary DPSK is utilized with 0 degrees and π degrees as the two phases, we can consider the decision statistic [42] for L-th order diversity

$$U=\mathrm{Re}\sum_{k=1}^{L}\left[V_{k2}V_{k1}^{*}\right] \quad (10.9.1\text{-}27)$$

where the complex signal V_{kn} is given by

$$V_{kn}=\left(2E\alpha_k e^{-j\theta_k}+N_{kn}\right) \quad (10.9.1\text{-}28)$$

where the complex noise term is given by the first equation in (10.9.1-7). It follows that U, for the case that the phase does not change between adjacent symbols, is given by

$$U=\mathrm{Re}\left[\sum_{k=1}^{L}\left(2E\alpha_k e^{-j\theta_k}+N_{k2}\right)\left(2E\alpha_k e^{j\theta_k}+N_{k1}^{*}\right)\right] \quad (10.9.1\text{-}29)$$

---

[4] Note that $\binom{2L-1}{L}=\binom{2L-1}{L-1}$, so that this expression may differ with other published results, but they are equivalent.

and the two noise terms are defined in the first equation in (10.9.1-7) and are successive in time. Since (10.9.1-29) is based on the condition that a phase change does not occur, if follows that an error occurs when $U < 0$. Proakis [42] has shown that the BER, conditioned on amplitude parameters $\{\alpha_k\}$, is given by

$$PE(\gamma_L) = \left(\frac{1}{2}\right)^{2L-1} e^{-\gamma_L} \sum_{k=0}^{L-1} c_k (\gamma_L)^k \tag{10.9.1-30}$$

In (10.9.1-30) c_k is defined as

$$c_k = \frac{1}{k!} \sum_{m=0}^{L-1-k} \binom{2L-1}{m} \tag{10.9.1-31}$$

The BER can be obtained by averaging over the Rayleigh fading amplitude parameters $\{\alpha_k\}$ via the parameter γ_L (see (10.9.1-12)) via

$$PE_{DPSK} = \int_0^{\infty} PE(\gamma_L) p(\gamma_L) d\gamma_L \tag{10.9.1-32}$$

Proakis has evaluated this expression with the result

$$PE_{DPSK} = \frac{1}{2^{2L-1}(L-1)!(1+\Gamma)^L} \sum_{k=0}^{L-1} c_k (L-1+k)! \left(\frac{\Gamma}{1+\Gamma}\right)^k \tag{10.9.1-33}$$

with Γ defined in (10.9.1-15). Proakis [24] has shown that (10.9.1-33) can be approximated when $\Gamma >> 1$ as

$$PE_{DPSK} \simeq \left(\frac{1}{2\Gamma}\right)^L \binom{2L-1}{L-1} \tag{10.9.1-34}$$

Again we see that the BER goes as Γ^{-L}.

10.9.1.3 Performance Improvement Via Diversity Techniques for Orthogonal FSK Channels

In this section we consider noncoherent, orthogonal frequency shift keying (OFSK) performance in the maximally combined L channel BER in Rayleigh fading. Again it will be assumed that the channel parameters $\{\alpha_k e^{-j\theta_k}\}$ are constant over the symbol time. For this case we will assume that the combiner of Figure 10.9-1 is square law, since it simplifies some calculations to follow. Under the assumption that signal "1" was transmitted, the two output statistics is given by

$$\begin{aligned} U_1 &= \sum_{k=1}^{L} \left|2E\alpha_k e^{-j\theta_k} + N_{k1}\right|^2 \\ U_2 &= \sum_{k=1}^{L} \left|N_{k2}\right|^2 \end{aligned} \tag{10.9.1-35}$$

It follows that an error occurs when $U_2 > U_1$. Proakis [42] has shown that orthogonal FSK has the same conditional error probability as DPSK, except that γ_L is replaced with $\gamma_L/2$ in (10.9.1-30); thus

$$PE(\gamma_L) = \left(\frac{1}{2}\right)^{2L-1} e^{-\gamma_L} \sum_{k=0}^{L-1} c_k (\gamma_L / 2)^k \tag{10.9.1-36}$$

The average BER for orthogonal, noncoherent FSK is given by the DPSK result with Γ replaced with $\Gamma/2$, so that

$$PE_{NCOFSK} = \frac{1}{2^{2L-1}(L-1)!(1+\Gamma/2)^L} \sum_{k=0}^{L-1} c_k (L/2 - 1 + k)! \left(\frac{\Gamma}{2+\Gamma}\right)^k \tag{10.9.1-37}$$

Finally using the high SNR approximation leads to

$$PE_{NCOFSK} \approx \left(\frac{1}{\Gamma}\right)^L \binom{2L-1}{L-1} \tag{10.9.1-38}$$

The BER performance of BPSK, BDPSK, and noncoherent, orthogonal, BFSK (NCOFSK) is plotted and compared in Figure 10.9-2, using the high SNR approximations, as a function of average combined SNR, $\bar{\gamma}_L = L\Gamma$.

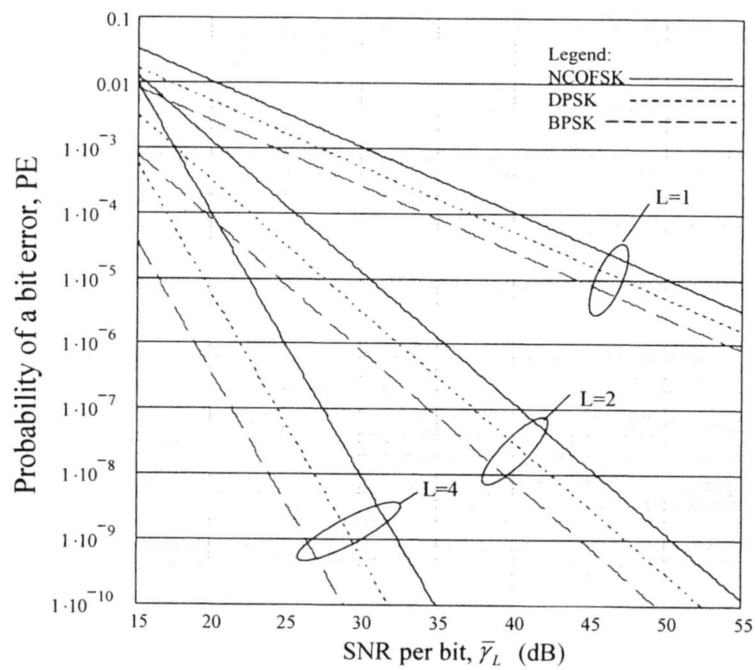

Figure 10.9-2 The bit error rate of three type of modulations with L-th order diversity.

Clearly BPSK is best followed by DPSK with the worst performance obtained by the noncoherent orthogonal FSK modulation. We will not address multiphase signals here, but Proakis [42] has results for M-ary orthogonal signal with square law combining.

10.10 THE RAKE RECEIVER

The original RAKE receiver was first developed by Price and Green [43] around 1958. The RAKE receiver is one method to achieve diversity improvement on a frequency-selective fading channel. The basic idea is that a wideband channel may be represented in terms of a tapped delay line model. To introduce this subject we first consider the tapped delay-line channel model [42, 43].

10.10.1 The Tapped Delay Line Channel Model for a Frequency Selective Slowly Fading Channel

In the case that the channel has a bandwidth (BW) that is greater than the coherent bandwidth (i.e., $BW >> BW_{coh}$), the channel can be subdivided into a number of frequency division multiplexed *subchannels* that have a separation of at least BW_{coh}.

Consider the bandpass signal $s(t)$, which can be written in the form

$$s(t) = \mathrm{Re}\left[v(t)e^{j\omega t}\right] \tag{10.10.1-1}$$

in which $v(t)$ is the low-pass equivalent signal. Assume that the low-pass equivalent signal $v(t)$ is bandlimited to the range $|f| < BW/2$. It is convenient to let $BW = W$ in what follows. It follows that the bandpass version of the signal (the actual signal) has a positive frequency bandwidth of W Hz. Since $v(t)$ is band-limited, it can be represented via the sampling theorem in the time domain

$$v(t) = \sum_{k=-\infty}^{\infty} v\left(\frac{k}{W}\right)\frac{\sin(\pi W(t-k/W))}{\left(\pi W(t-k/W)\right)} \tag{10.10.1-2}$$

The received signal, neglecting noise that has passed through a frequency-selective channel, can be represented in the form (see (10.5.1-6)) as

$$v_0(t) = \int_{-\infty}^{\infty} T(f,t)V(f)e^{j2\pi ft}df \tag{10.10.1-3}$$

where $v_0(t)$ is the output low-pass equivalent function representing the output of the channel and $T(f,t)$ is the time invariant transfer function. The Fourier transform of $v(t)$ is given by

$$V(f) = \frac{1}{W}\sum_{k=-\infty}^{\infty} v(k/W)e^{-j2\pi fk/W} \tag{10.10.1-4}$$

Using (10.10.1-4) in (10.10.1-3) produces

$$v_0(t) = \frac{1}{W}\sum_{k=-\infty}^{\infty} v(k/W)\int_{-\infty}^{\infty} T(f,t)e^{j2\pi f(t-k/W)}df = \frac{1}{W}\sum_{k=-}^{\infty} v(k/W)g(t-kW,t) \tag{10.10.1-5}$$

where

$$g(t-kW,t) = \int_{-\infty}^{\infty} T(f,t)e^{j2\pi f(t-k/W)}df \tag{10.10.1-6}$$

The right-hand side of (10.10.1-5) can be written in the alternative form

$$v_0(t) = \frac{1}{W}\sum_{k=-\infty}^{\infty} v(t-k/W)g(k/W,t) \quad (10.10.1\text{-}7)$$

Denote the time varying channel weights as

$$g_k(t) = \frac{1}{W} g\left(\frac{k}{W}, t\right) \quad (10.10.1\text{-}8)$$

It then follows that (10.10.1-7) can be written as

$$v_0(t) = \sum_{k=-\infty}^{\infty} g_k(t)v(t-k/W) \quad (10.10.1\text{-}9)$$

If the autocorrelation function of the channel is effectively nonzero over the multipath spread of the channel, then this time range is denoted by T_{ms}. ($T_{ms} \cong 3\sigma_\tau$). Then (10.10.1-9) leads to the observation that a discrete representation of the channel applies and can be truncated to $L = \lfloor T_{ms}W \rfloor + 1$ taps, so that we may simplify (10.10.1-9) to

$$v_0(t) = \sum_{k=1}^{L} g_k(t)v(t-k/W) \quad (10.10.1\text{-}10)$$

where $\lfloor x \rfloor$ denotes the *greatest integer function* and is also called the *floor function*. It satisfies $\lfloor 3.2 \rfloor = 3$ and $\lfloor -3.2 \rfloor = -4$, for example. The tapped delay line model of (10.10.1-10) is depicted in Figure 10.10-1. Each element has delay of $1/W$ seconds.

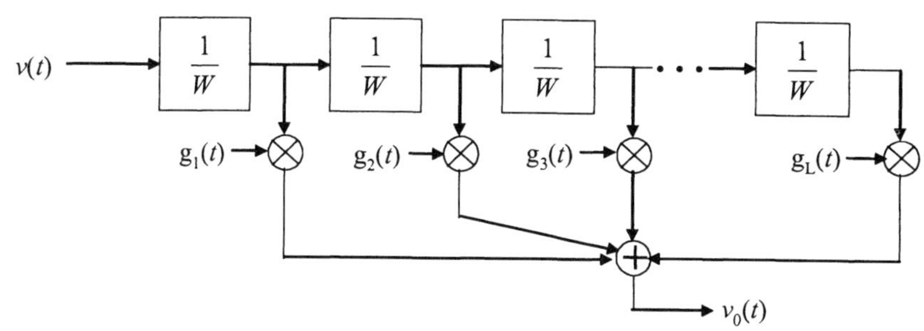

Figure 10.10-1 Tapped delay line model of the complex envelope in a frequency selective channel.

The received signal $v_0(t)$, then has white Gaussian noise added to it to model the received signal plus noise complex envelope, $r(t)$.

10.10.2 The RAKE Receiver

Now that we have established the tapped delay line model of a frequency selective channel, we consider the task of providing digital signals over this channel, assuming that the weighting functions, $g_k(t)$, are statistically independent. The received complex envelope is composed of L weighted replicas of the received (complex envelope of) signal.

Consider binary modulation transmitted over the previously described channel. Let the two low pass equivalent signals be denoted by $v_1(t)$ and $v_2(t)$ and assume that they are either orthogonal or antipodal. It is assumed that the symbol rate (1/*T*) is set so that $T >> T_{ms}$. Therefore it is reasonable to neglect any intersymbol problems due to multipath effects. Based on our channel model, the received signal can be represented as

$$r(t) = \sum_{k=1}^{L} g_k(t) v_n(t - k/W) + n(t) \quad 0 \le t \le T \quad n = 1,2 \tag{10.10.2-1}$$

where $n(t)$ is the complex envelope (baseband equivalent) white Gaussian noise process that has zero mean. It follows that the radio frequency (RF) version of the noise is of the form

$$n_k^{rf}(t) = n_c(t)\cos(\omega_0 t + \theta_k) + jn_s(t)\sin(\omega_0 t + \theta_k) = \mathrm{Re}\left[n(t)e^{j2\pi f_0 t}\right] \tag{10.10.2-2}$$

and $n(t)$ has two-sided noise spectral density N_0, and $n_c(t)$ and $n_s(t)$ are statistically independent Gaussian random processes. Furthermore, for white Gaussian noise

$$E(n(t)n^*(u)) = 2N_0\delta(t-u) \tag{10.10.2-3}$$

The optimum receiver has two filters matched to $v_1(t)$ and $v_2(t)$. The receiver provides samples at the symbol rate from both filters and selects the signal according to the largest output. Thus the two variables are computed

$$U_n = \mathrm{Re}\left[\int_0^T r(t) v_n^*(t) dt\right] = \mathrm{Re}\left[\sum_{k=1}^{L}\int_0^T r(t) g_k^*(t) v_n^*(t - k/W) dt\right] \quad n = 1,2 \tag{10.10.2-4}$$

Figure 10.10-2 illustrates the receiver structure based on the previous equation. The terms (1/*W*) denote a delay of that amount in seconds.

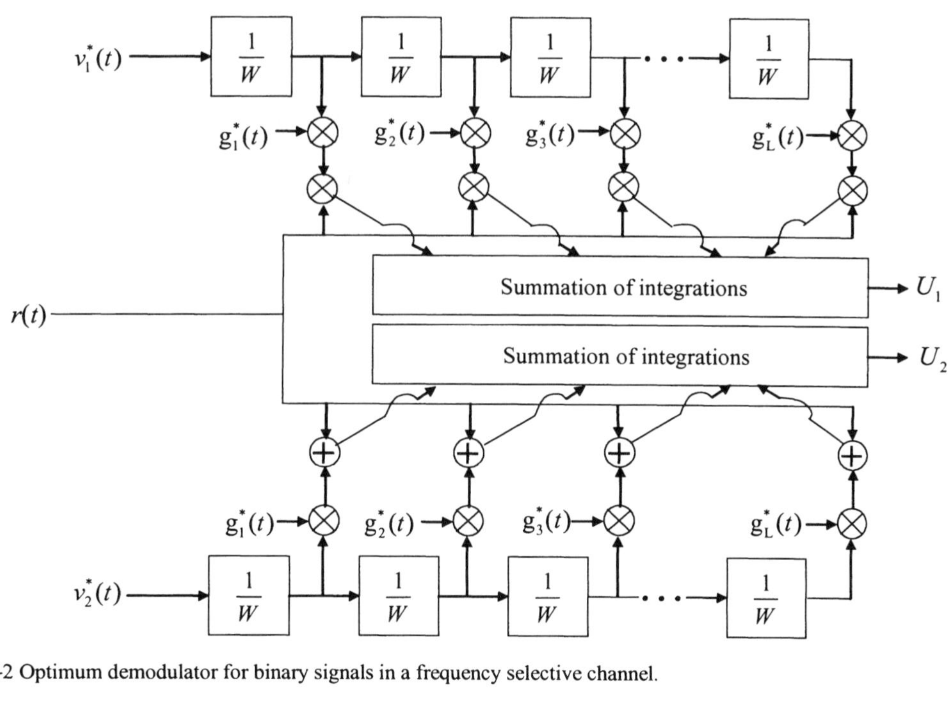

Figure 10.10-2 Optimum demodulator for binary signals in a frequency selective channel.

10.10.3 Performance of the RAKE Receiver

We shall obtain an estimate of the BER of the RAKE receiver under the assumption that the $g_k(t)$ are known exactly. In addition it will be assumed that the set of $g_k(t)$ are constant over one signaling interval (T seconds) and can be represented as g_k. With these assumptions the right-hand side of (10.10.2-4) produces

$$U_n = \mathrm{Re}\left[\sum_{k=1}^{L} g_k^* \int_0^T r(t)v_n^*(t-k/W)dt\right] \quad n=1,2 \tag{10.10.3-1}$$

Let us assume that the transmitted signal is $v_1(t)$ so that the received signal plus noise is given by

$$r_1(t) = \sum_{m=1}^{L} g_m v_1\left(t-m/W\right) + n(t) \tag{10.10.3-2}$$

and $n(t)$ has been described in the last section. Using (10.10.3-2) in (10.10.3-1) produces

$$\begin{aligned} U_n = {} & \mathrm{Re}\left[\sum_{k=1}^{L} g_k^* \sum_{m=1}^{L} g_m \int_0^T v_1(t-m/W)v_n^*(t-k/W)dt\right] \\ & + \mathrm{Re}\left[\sum_{k=1}^{L} g_k^* \int_0^T n(t)v_n^*(t-k/W)dt\right] \quad n=1,2 \end{aligned} \tag{10.10.3-3}$$

When dealing with CDMA wideband signals, $v_1(t)$ and $v_2(t)$ are pseudorandom sequences that are quasi-orthogonal. Under this assumption

$$\int_0^T v_i(t-m/W)v_i(t-k/W)dt \approx 0, \quad k \neq m, \quad i=1,2 \tag{10.10.3-4}$$

Using this assumption for our binary signals, (10.10.3-3) can be simplified to

$$\begin{aligned} U_n = {} & \mathrm{Re}\left[\sum_{k=1}^{L} |g_k|^2 \int_0^T v_1(t-k/W)v_n(t-k/W)dt\right] \\ & + \mathrm{Re}\left[\sum_{k=1}^{L} g_k^* \int_0^T n(t)v_n(t-k/W)dt\right], \quad n=1,2 \end{aligned} \tag{10.10.3-5}$$

In the case that the signals are antipodal a single decision variable is sufficient and (10.10.3-5) then simplifies to

$$U_1 = \mathrm{Re}\left[2E\sum_{k=1}^{L} \alpha_k^2 + \sum_{k=1}^{L} \alpha_k N_k\right] \tag{10.10.3-6}$$

and $\alpha_k = |g_k|$ and in addition

$$N_k = e^{j\theta_k} \int_0^T n(t)v^*(t-k/W)dt \tag{10.10.3-7}$$

which is the same decision variable as given by (10.9.1-6) for the ratio combiner in Rayleigh fading. Therefore the RAKE receiver, having perfect estimates of the channel tap weights, is equivalent to a maximal ratio combiner in a system with L-th-order diversity. Thus the BER performance result indicated in (10.9.1-21) is applicable in the case of the RAKE receiver also.

10.11 BINARY CODED CHERNOFF BER BOUNDS FOR FADING CHANNELS

At this point in this chapter we have presented equalizers and diversity techniques and shown that they improve performance in a multipath fading environment. The other technique, channel coding, will be briefly introduced as a means of improving performance in a fading channel environment. Time and frequency diversity can be viewed as a form of a repetition code. However repetition coding is known not to be as efficient as more efficient coding procedures. We will show that the amount of diversity created by a code is related to the code's minimum distance properties [42]. We shall also see that soft-decision decoding is considerably better than hard-decision decoding.

To extend the concept of diversity providing "independent" copies of the signal to coding, we will require that code symbols (coded bits) fade independently of any other coded symbols that are transmitted. This can lead to inefficient use of the time frequency signaling space, leaving a large unused portion of the time frequency signaling space. To overcome this possible problem a number of code words may be interleaved in time or frequency of both, in such way that coded symbols fade independently. For our work in what follows we assume that the time frequency signal space is separated into nonoverlapping time frequency cells. We will assume that there is negligible intercell interference in what follows.

In addition to the assumption of statistically independent fades we need to add the assumption of the WGN components that are statistically independent from each other and identical for each time-frequency cell.

When a channel suffers from fading, it is difficult to establish a phase reference, unless the channel fades slowly enough to utilize PSK or DPSK modulation. Therefore we restrict our analysis to noncoherent detection of binary orthogonal BFSK modulation with coding.

10.11.1 Chernoff Bound for Binary Linear Block Codes

We will find the use of bounds advantageous for our work to follow. Sometimes it is easier to obtain the bound to PE rather than PE directly. We will discuss the *Chernoff bound* in what follows.

First consider the Chernoff bound [42, 43–45] for orthogonal FSK with L-th order diversity. This will be useful for the BER bound of binary orthogonal FSK signaling, which is noncoherently detected. Recall from (10.9.1-35) that

$$U_1 = \sum_{k=1}^{L} \left|2E\alpha_k e^{-j\theta_k} + N_{k1}\right|^2$$
$$U_2 = \sum_{k=1}^{L} \left|N_{k2}\right|^2 \qquad (10.11.1\text{-}1)$$

U_1 and U_2 are the two decision variables that we used in Section 10.9.1.3. Recall that U_1 represent the combined signal plus noise terms and U_2 represents the square law combined noise terms for L-th order diversity. The probability of error for this binary FSK modulation is denoted by $PE(L)$ and is given by

$$PE(L) = P(U_2 - U_1 > 0) \qquad (10.11.1\text{-}2)$$

If we let Z=U_1–U_2 then we can rewrite the probability of error expression as

$$PE(L) = P(Z > 0) = \int_0^\infty p(z)dz \tag{10.11.1-3}$$

Since the phase (assumed to be constant) does not affect the performance of the square law detector, we will let $\theta_k = 0$ in what follows. With that simplification we have

$$Z = \sum_{k=1}^{L}\left(\left|N_{k2}\right|^2 - \left|2E\alpha_k + N_{k1}\right|^2\right) \tag{10.11.1-4}$$

Now let $U(z)$ denote the unit step function in the variable z, then $PE(L)$ can be written as

$$PE(L) = E\left[U(z)\right] \tag{10.11.1-5}$$

The Chernoff bound is obtained by using an exponential bounding function that upper bounds the unit step function. Thus consider the bound

$$U(Z) \le e^{\rho Z}\ , \quad \rho \ge 0 \tag{10.11.1-6}$$

and ρ is to be optimized to make the bound as tight as possible. Therefore, we have so far

$$PE(L) = E\left[U(Z)\right] \le E\left[e^{\rho Z}\right] \tag{10.11.1-7}$$

Using the definition of Z, from (10.11.1-4), and noting that the random variables in the summation are mutually statistically independent, we have

$$PE(L) \le \prod_{k=1}^{L} E\left[e^{\rho\left|N_{k2}\right|^2}\right] E\left[e^{-\rho\left|2E\alpha_k + N_{k1}\right|^2}\right] \tag{10.11.1-8}$$

The first and second terms in (10.11.1-8) are the characteristic function of Gaussian random variables. From Proakis [42] the characteristic function of the noise only case (first term in the product) is given by

$$E\left[e^{\rho\left|N_{k2}\right|^2}\right] = \frac{1}{1 - 2\rho\left(2EN_0\right)}\ , \qquad \rho < \frac{1}{4EN_0} \tag{10.11.1-9}$$

with the bound on ρ due to the fact that it is only meaningful if the bound is finite. In a similar manner the second characteristic function (second expectation) is given by [42]

$$E\left[e^{-\rho\left|N_{k1} + 2E\alpha_k\right|^2}\right] = \frac{1}{1 + 2\rho\left(2EN_0\left(1 + \bar{\gamma}\right)\right)}\ , \quad \rho > \frac{1}{4EN_0\left(1 + \bar{\gamma}\right)} \tag{10.11.1-10}$$

in which $\bar{\gamma}$ is the average E_s/N_0 ratio per diversity channel. Notice that both expressions do not depend on k. It is assumed that the additive noise terms are statistically independent as well as the fading statistics on the diversity channels. Thus the bit error rate (BER) bound for soft decisions is given by

$$PE_b(L) \le \left[\frac{1}{\left(1 - 2\rho\sigma_N^2\right)\left(1 + 2\rho\sigma_S^2\right)}\right]^L \tag{10.11.1-11}$$

where we have introduced for convenience

$$\begin{aligned}\sigma_N^2 &= 2EN_0 \\ \sigma_S^2 &= 2EN_0(1+\bar{\gamma})\end{aligned} \quad (10.11.1\text{-}12)$$

in which the first term is the variance due to the inphase and quadrature terms due to noise only and the second term is the inphase and quadrature terms due to signal plus noise being present. In order to obtain a tight bound, it is necessary to optimize the value of ρ. This can be done by differentiating the right-hand side of (10.11.1-11), setting to zero and solving for the value of ρ. From the results of Problem 12, the optimum value of ρ, denoted by ρ_0 is given by

$$\rho_0 = \frac{\sigma_S^2 - \sigma_N^2}{4\sigma_S^2\sigma_N^2} \quad (10.11.1\text{-}13)$$

When this value of ρ_0 is inserted into the right hand side of (10.11.1-11), utilizing the results of Problem 12, one has

$$PE_b(L) \le \left[\frac{4(1+\bar{\gamma})}{(2+\bar{\gamma})^2}\right]^L \quad (10.11.1\text{-}14)$$

If we note that from (10.6.3-2) that the probability of error for binary orthogonal FSK signaling (noncoherently detected) is given by

$$p = \frac{1}{2+\bar{\gamma}} \quad (10.11.1\text{-}15)$$

so that (10.11.1-14) can be written as

$$PE(L) \le \left[4p(1-p)\right]^L \quad (10.11.1\text{-}16)$$

10.11.2 Coded Orthogonal FSK Signal Model for Fading Channels

At this point we are ready to address FSK modulated coded waveforms. The model of a coded digital communication system, in which we desire the BER performance, is shown in Figure 10.11-1. We will assume that the encoder is binary. In addition, the encoding type may be block or convolutional coding, or even a combination of block and convolutional such as in a concatenated coding arrangement.

Consider first a linear block code with k information bits and n total bits in the block code. It will be assumed, without loss of generality, that all n code word bits are transmitted simultaneously over the channel on separate frequency cells. Consider the following mapping from the code word to the BFSK modulation. Let the code word C_i have bits $\{c_{ij}\}$, $i \in \{1,\ 2^k\}$ and $j \in \{1,\ n\}$, Now if $c_{ij} = 0$, the tone f_{0j} is transmitted, and if $c_{ij} = 1$, the tone f_{1j} is transmitted. Thus there are $2n$ possible tones that can be sent; however only n tones can be sent in any given signaling interval. Now the bandwidth expansion is given by the ratio ($k/2n$).

The FSK demodulator is based on $2n$ tone detectors, each matched to each a possible tone, and they are detected noncoherently. By assumption the Rayleigh fading and the additive white Gaussian noise in the $2n$ frequency cells are mutually statistically independent and identically distributed processes. The optimum maximum-likelihood detector with soft decisions requires that the filter output be square law detected and properly combined to form the 2^k decision variables. The largest output is selected to be the transmitted code word. When hard decisions are utilized in the decoder, the code word decision is based on choosing the code word with the smallest Hamming distance to the received code word, which is based on the optimum maximum-likelihood decoder.

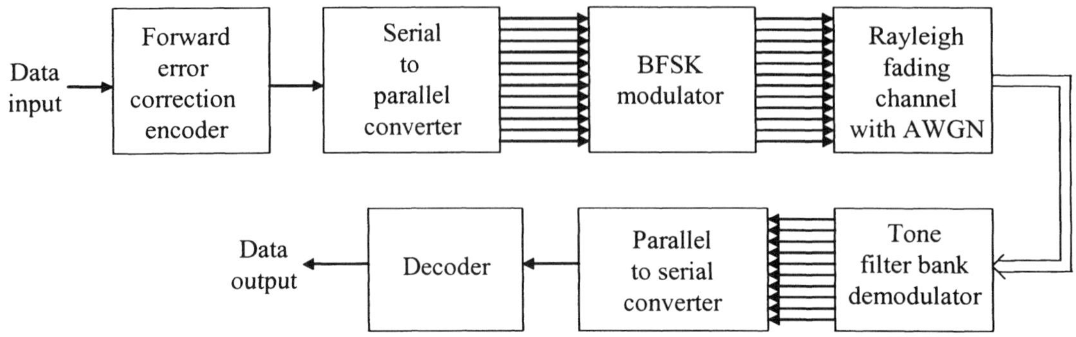

Figure 10.11-1 Communication system model employing BFSK orthogonal modulation and demodulation and forward error correction coding.

10.11.3 BER of Soft-Decision Decoding and FSK Modulation with Linear Binary Block Codes over Rayleigh Fading Channels

Now consider the decoding of a linear (*n*, *k*) block code that is transmitted over a Rayleigh fading channel with additive white Gaussian noise. As described in the previous paragraph the optimum soft-decision decoder forms the 2^k decision variables of the form

$$U_i = \sum_{j=1}^{n}\left[\left(1-c_{ij}\right)\left|v_{0j}\right|^2 + c_{ij}\left|v_{1j}\right|^2\right] = \sum_{j=1}^{n}\left[\left|v_{0j}\right|^2 + c_{ij}\left(\left|v_{1j}\right|^2 - \left|v_{0j}\right|^2\right)\right] \tag{10.11.3-1}$$

in which $\left|v_{ij}\right|^2$ for over $i \in \{1, 2^k\}$ and j =1, 2, 3,…*n*, are the squared received signal envelopes at the outputs of the 2*n* filters, one for each possible frequency. The code word chosen is based on the largest value of U_i over $i \in \{1, 2^k\}$. To obtain the BER of soft-decision linear block codes we will assume, without the loss of generality that the all-zeros code word, C_1 was transmitted, so that $c_{ij} = 0$ for all *j*. The average coded symbol SNR (E_b/N_0) is given by $\overline{\gamma}$ for each tone (cell). The average code word SNR for all *n* symbols is given by

$$\overline{\gamma}_w = n\overline{\gamma} \tag{10.11.3-2}$$

and the corresponding average SNR for a bit is given by

$$\overline{\gamma}_b = \frac{n}{k}\overline{\gamma} \tag{10.11.3-3}$$

since there are *k* bits in *n* coded symbols. The *code rate* is the ratio of bits per word to the coded symbols per word, or

$$R_c = k/n \tag{10.11.3-4}$$

Thus, (10.11.3-3), in combination with (10.11.3-4), can be written as

$$\overline{\gamma}_b = \overline{\gamma}/R_c \tag{10.11.3-5}$$

We have assumed that C_1 was transmitted; therefore, $c_{ij} = 0$ for all j. Consequently the probability of a decision is made for the m-th code word, instead of the correct word C_1, and is given by

$$PE(m) = P(U_m > U_1) = P(U_1 - U_m < 0) \tag{10.11.3-6}$$

Using (10.11.3-1), $PE(m)$ can be written in the form (since the two $|v_{0j}|^2$ cancel for all j, for both terms)

$$PE(m) = P\left(\sum_{j=1}^{n}(c_{1j} - c_{mj})\left(|v_{0j}|^2 - |v_{1j}|^2\right) < 0\right) \tag{10.11.3-7}$$

or, noting that the two codes only differ in w_m places from the all-zeros word, leaves

$$PE(m) = P\left(\sum_{j'=1}^{w_m}\left(|v_{0j}|^2 - |v_{1j}|^2\right) < 0\right) \tag{10.11.3-8}$$

and w_m is the weight between the first and the m-th code word. We may use the Chernoff bound to evaluate (i.e., bound) this expression. From (10.11.1-16) we have

$$PE(m) \le [4p(1-p)]^{w_m} \tag{10.11.3-9}$$

where w_m and L have the same essential meaning of diversity and p is given by

$$p = \frac{1}{2+\bar{\gamma}} = \frac{1}{2+R_c\bar{\gamma}_b} \tag{10.11.3-10}$$

Since the sum of the binary error events over the 2^k–1 nonzero weight code words gives an upper bound on the word error probability, we have for the word error probability bound

$$PE_W \le \sum_{m=2}^{2^k} PE(m) \tag{10.11.3-11}$$

The union bound is then obtained, noting that $d_{\min}$ satisfies

$$d_{\min} = \min_{m, m\neq 1}\{w_m\} \tag{10.11.3-12}$$

so that the word error probability for binary orthogonal FSK noncoherently detected is given by

$$PE_W < (M-1)[4p(1-p)]^{d_{\min}} \tag{10.11.3-13}$$

in which $M = 2^k$ and $p = 1/(2+R_c\bar{\gamma}_b)$, or

$$PE_W < (M-1)\left[\frac{4(1+R_c\bar{\gamma}_b)}{(2+R_c\bar{\gamma}_b)^2}\right]^{d_{\min}} \tag{10.11.3-14}$$

To relate the word error rate to the bit error rate we have the approximate relationship [46–53]

$$PE_b \simeq \frac{d_{\min}}{n} PE_w \tag{10.11.3-15}$$

for group block codes, where $d_{\min}$ is the minimum distance and n is the number of coded bits in the code word to yield the union bound

$$PE_b < \frac{d_{\min}}{n}(M-1)\left[\frac{4(1+R_c\bar{\gamma}_b)}{(2+R_c\bar{\gamma}_b)^2}\right]^{d_{\min}} \tag{10.11.3-16}$$

Example 9 Consider a Rayleigh fading channel in which the Golay (23,12) code is utilized with binary FSK modulation. Plot the word error bound and compare to uncoded binary FSK with $L = 2$ diversity. The weight of this code is 7, based on Table 5.1 of Peterson and Weldon [53]. The extended Golay code has a d_{min} of 8 [53]. The BER for the Golay code is given by (10.11.3-14) and (10.11.3-15).

Consider the bandwidth expansion factors (BEF) for both the coded and uncoded schemes. For the coded scheme $2n = 48$ frequency cells are required, and $k = 12$, so that in the coded case the BEF= 4. For the binary FSK case with $L = 2$, the BEF = 4/1 = 4 also.

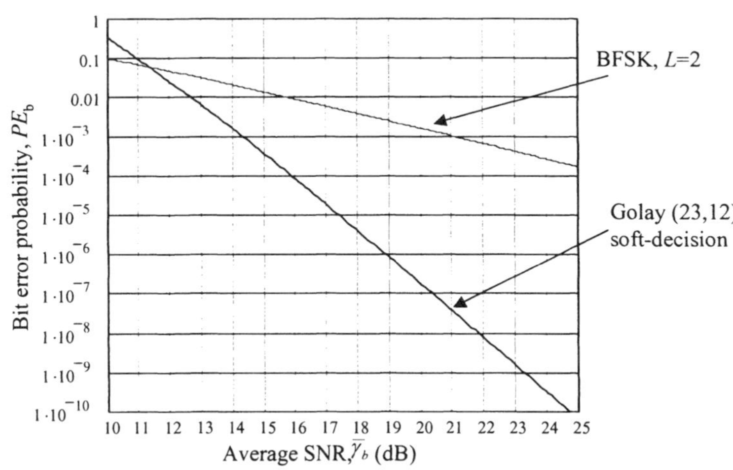

Figure 10.11-2 Union bound BER performance of binary FSK with dual diversity and a Golay (23,12) code using binary FSK modulation, both with noncoherent demodulation and soft decisions.

As can be seen in Figure 10.11-2 the Golay encoded scheme is considerably better than the uncoded scheme.

10.11.4 BER of Hard-Decision Decoding and FSK Modulation with Linear Binary Block Codes over Rayleigh Fading Channels

Proakis [42] has shown that the case of linear binary FSK, noncoherently detected in Rayleigh fading channels with a hard-decision decoding error is of the form

$$PE(m) \le [4p(1-p)]^{w_m/2} \tag{10.11.4-1}$$

The word error probability is given by the union bound

$$PE_W \leq \sum_{m=2}^{M} [4p(1-p)]^{w_m/2} \leq (M-1)[4p(1-p)]^{d_m/2} \tag{10.11.4-2}$$

where $M = 2^k$ and p, for noncoherent detection of binary FSK modulation is given by

$$p = \frac{1}{2 + R_c \bar{\gamma}_b} \tag{10.11.4-3}$$

so that the probability of a word error is given by

$$PE_W < (M-1)\left[\frac{4(1+R_c\bar{\gamma}_b)}{(2+R_c\bar{\gamma}_b)^2}\right]^{d_{\min}/2} \tag{10.11.4-4}$$

Again we use the approximate relationship [44]

$$PE_b \simeq \frac{d_{\min}}{n} PE_w \tag{10.11.4-5}$$

for group block codes, where $d_{\min}$ is the minimum distance, and n is the number of coded bits in the code word. Hence our final union bound for hard decisions on Rayleigh fading channels is given by

$$PE_b < \frac{d_{\min}}{n}(M-1)\left[\frac{4(1+R_c\bar{\gamma}_b)}{(2+R_c\bar{\gamma}_b)}\right]^{d_{\min}/2} \tag{10.11.4-6}$$

Let us now compare the Golay (24,12) code performance with soft and hard decisions. From [45] the minimum distance for the Golay (24,12) linear block code is $d_{\min} = 8$. So plotting (10.11.3-16) and (10.11.4-6), we have Figure 10.11-3.

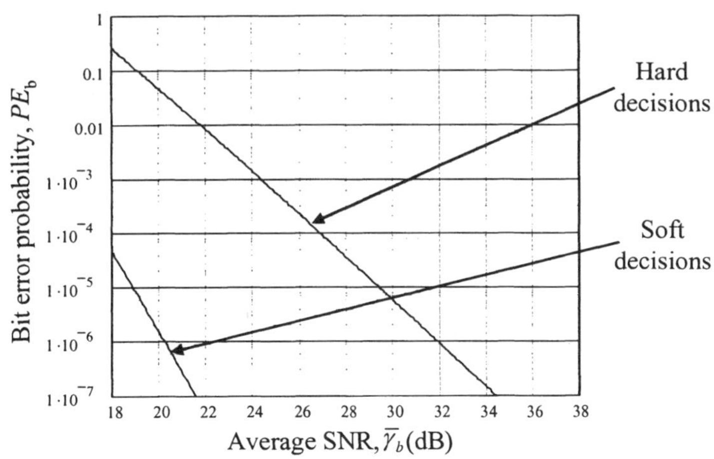

Figure 10.11-3 Union bound BER performance of Golay (24,12) hard and soft decisions in Rayleigh fading.

By comparing the two bounds we see that the performance of soft decisions decoding is greatly superior to hard decision decoding for the example of Golay (24,12) codes on a Rayleigh channel. Typically the difference between a hard-decision and a soft-decision, in white Gaussian noise, is about 2 dB. This is significantly less than the difference seen in Figure 10.11-3 for the Golay (24,12) codes. It should be cautioned that the union bounds used in Figure 10.11-3 are not very tight and a more accurate analysis would yield lower BER values for a given SNR and less difference in the two cases. Nonetheless, these bounds indicate the trend of the differences. The message is, soft decisions are much better than hard decisions, in Rayleigh fading.

10.11.5 Chernoff Bounds for the BER of Convolutional Codes over Rayleigh Fading Channels with Soft and Hard Decisions and Binary FSK Modulation

We will consider both the soft-decision case and the hard-decision case using Chernoff bounds for the BER over Rayleigh fading channels with binary FSK (orthogonal) modulation.

10.11.5.1 Soft Decisions

First we consider the soft-decision case. Consider noncoherent detection of orthogonal FSK modulation with convolutional codes over Rayleigh fading channels with maximum likelihood decoding which is effectively implemented with the Viterbi decoder. Proakis [42] has shown that the bit error probability is bounded by

$$PE_b < \frac{1}{k}\sum_{d=d_{\min}}^{\infty} \beta_d PE(d) \tag{10.11.5-1}$$

where the weighting multipliers (β_d) are determined from the coefficients of the expansion of the first derivative evaluated at $N = 1$, convolutional transfer function $T(D,N)$. That is, the coefficients are obtained from the coefficients of the expansion

$$\left.\frac{dT(D,N)}{dN}\right|_{N=1} \tag{10.11.5-2}$$

The parameter k is the number of bits per symbol and is one in the binary case. The derivative determination of the coefficients will be made clearer via an example. We will use the Chernoff bound (upper bound) for soft decisions with binary (orthogonal) FSK modulation for the evaluation of $PE(d)$. It is given by

$$PE(d) \le \left[\frac{4(1+R_c\overline{\gamma}_b)}{\left(2+R_c\overline{\gamma}_b\right)^2}\right]^d \tag{10.11.5-3}$$

To illustrate the method let us determine an upper bound of the convolutional code described by (10.11.5-1) and obtain the coefficients from the series expansion of (10.11.5-2) for the $K = 3$. The convolutional code illustrated in Figures 4.3-2 or 4.3-3 (they are alternative representations of the same code). In (4.3-7) the transfer function is given by

$$T(D,N) = \frac{ND^5}{1-2ND} \tag{10.11.5-4}$$

Using the series expansion for $(1-x)^{-1}$ produces the series

$$T(D,N) = ND^5\left[1+2ND+4N^2D^2+8N^3D^3+...\right] \tag{10.11.5-5}$$

which can be written as

$$T(D,N) = ND^5+2N^2D^6+4N^3D^7+8N^4D^8+... \tag{10.11.5-6}$$

Computing the derivative at $N=1$ yields

$$\left.\frac{dT(D,N)}{dN}\right|_{N=1} = D^5+4D^6+12D^7+32D^8+... \tag{10.11.5-7}$$

It follows that

$$\beta_5=1,\ \beta_6=4,\ \beta_7=12,\ \beta_8=32,... \tag{10.11.5-8}$$

Thus our bit probability of error is given by

$$PE_b < \left[\frac{4(1+R_c\overline{\gamma}_b)}{(2+R_c\overline{\gamma}_b)^2}\right]^5 + 4\left[\frac{4(1+R_c\overline{\gamma}_b)}{(2+R_c\overline{\gamma}_b)^2}\right]^6 + 12\left[\frac{4(1+R_c\overline{\gamma}_b)}{(2+R_c\overline{\gamma}_b)^2}\right]^7 + 32\left[\frac{4(1+R_c\overline{\gamma}_b)}{(2+R_c\overline{\gamma}_b)^2}\right]^8 + ... \tag{10.11.5-9}$$

which is our bit error rate bound for the convolutional code described in Figure 4.3-2 or 4.3-3, in Rayleigh fading, with binary (orthogonal) FSK modulation and noncoherent detection.

10.11.5.2 Hard Decisions

Now consider hard-decision detection. Proakis [42] has shown that when hard decisions are employed in convolutional codes with Chernoff bounds similar bounds occur with the exception that the effect of the hard decision is to reduce the distances (effective diversity) by half, in comparison to the soft-decision. Therefore the bound of (10.11.5-1) is replaced with

$$PE_b < \frac{1}{k}\sum_{d=d_{\min}}^{\infty} \beta_d PE(d/2) \tag{10.11.5-10}$$

Following the example for the convolutional code of Figures 4.3-2 or 4.3-3, the Chernoff bound for hard decisions is given by (four-term approximation)

$$\begin{aligned} PE_b < &\left[\frac{4(1+R_c\overline{\gamma}_b)}{(2+R_c\overline{\gamma}_b)^2}\right]^{5/2} + 4\left[\frac{4(1+R_c\overline{\gamma}_b)}{(2+R_c\overline{\gamma}_b)^2}\right]^{6/2} \\ &+12\left[\frac{4(1+R_c\overline{\gamma}_b)}{(2+R_c\overline{\gamma}_b)^2}\right]^{7/2} + 32\left[\frac{4(1+R_c\overline{\gamma}_b)}{(2+R_c\overline{\gamma}_b)^2}\right]^{8/2} + ... \end{aligned} \tag{10.11.5-11}$$

Again we see that the soft-decision decoding produces significantly better performance than with hard decisions. Figure 10.11-4 is based on bounds to simply illustrate the difference in hard- and soft-decision decoding for convolutional codes; however, more accurate estimates can be obtained from numerous coding texts and articles available in the literature.

We have only scratched the surface in the area of coding for Rayleigh fading channels, and a number of books and articles delve into much more detail than indicated here. The goal of this section was to present an introduction to coding in Rayleigh fading channels and observe the performance difference between hard and soft decisions.

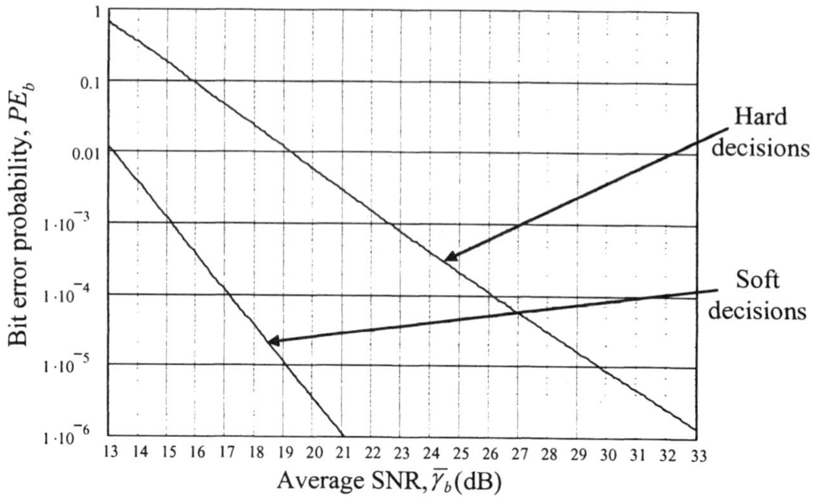

Figure 10.11-4 BER performance of a rate 1/2, $K = 3$, convolutional code with and soft- and hard-decision decoding in Rayleigh fading with noncoherent detection of binary orthogonal FSK modulation.

10.12 SMART ANTENNA SYSTEMS FOR WIRELESS SYSTEMS

A *smart antenna system* is a group of antenna elements with a signal processing capability to optimize its radiation or reception pattern automatically in response to the signal environment [47–54]. Usually the signals received from each antenna element are multiplied by complex weights and then summed up. Smart antenna systems are used in wireless applications. New smart (adaptive) antenna systems can enhance the performance of any CDMA system, including IS-95, IMT-2000, and wideband CDMA.

The utilization of smart antennas will enhance the coverage through range extension, improve the penetration of the buildings, increase system capacity, improve link quality by mitigating multipath, and fill holes in coverage that may occur without smart antennas. Smart antennas can be used in some applications to *spatially separate the signals at the base station.* These enhancements can occur due to the smart antenna providing greater antenna gain in the directions needed. Smart antennas can be installed at the base station or the subscriber receiver; however smart antennas are not normally fitted to mobile subscriber terminals.

10.12.1 Smart Antenna Systems

There are basically two types of smart antennas, adaptive arrays or switched beam. We will consider first the adaptive array and then discuss the switched beam antenna system. Adaptive arrays produce an infinite number of patterns that are adjusted in real time. Switched beam antenna systems produce a finite number of predefined antenna patterns.

10.12.2 Adaptive Array Smart Antennas

Adaptive array smart antennas are based on an "array" of elemental antennas called *elements* that are controlled by a combining network that controls the amplitude and phase of each of the elements. Consider the spherical and the rectangular coordinate systems in Figure 10.12-1 showing some elements.

In this figure ϕ is the azimuth angle, and θ is the elevation angle to the plane wave direction. Some assumptions are used in this analysis [47]. They are: (1) the element array spacing is small enough that the assumption of no amplitude variation between elements is valid, (2) the elements have no mutual coupling, and (3) the signal on the array is narrowband.

Consider a plane wave incident on the array from the direction (θ, ϕ). The signal phase difference between a reference element n and a reference element at the origin is given by the expression

$$\Delta\varphi_m = \Delta k d_m = k\left(x_m \cos\phi \sin\theta + y_m \sin\phi \sin\theta + z_m \cos\theta\right) \qquad (10.12.2\text{-}1)$$

In (10.12.2-1) k is known as the *phase constant* [48] and is given by

$$k = 2\pi / \lambda \qquad (10.12.2\text{-}2)$$

in which λ is the signal wavelength and $\lambda = c / f$, where c is the speed of light (3×10^8 m/s).

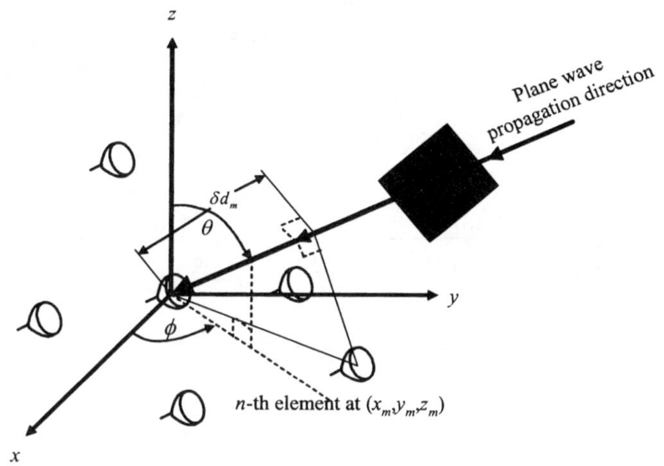

Figure 10.12-1 Coordinate system used in the array model.

Consider now an antenna array model that uses complex weights w_m, as shown in Figure 10.12-2.

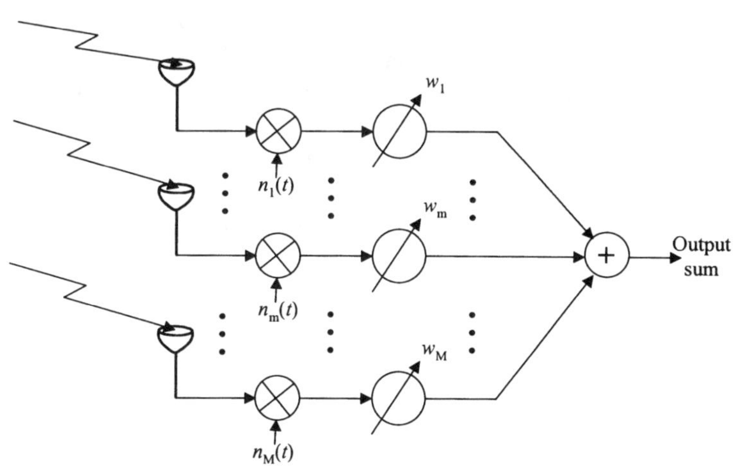

Figure 10.12-2 Array model showing the received noise terms and the weighting factors.

Arrays can be of various shapes [47, 49]. For example a *linear equally spaced* (LES) array is a linear array, equally spaced along a line. Arrays can also be constructed in a circle, which is two-dimensional or they can be arranged in three dimensions. A simple linear array composed of M elements is shown in two dimensions in Figure 10.12-3 with element spacing. The array Δx starts at $x = 0$ and extends to $(M-1)\Delta x$. The upper left corner illustrates the array in three dimensions, showing the M elements and both spherical angles, and the direction of the plane wave phase front.

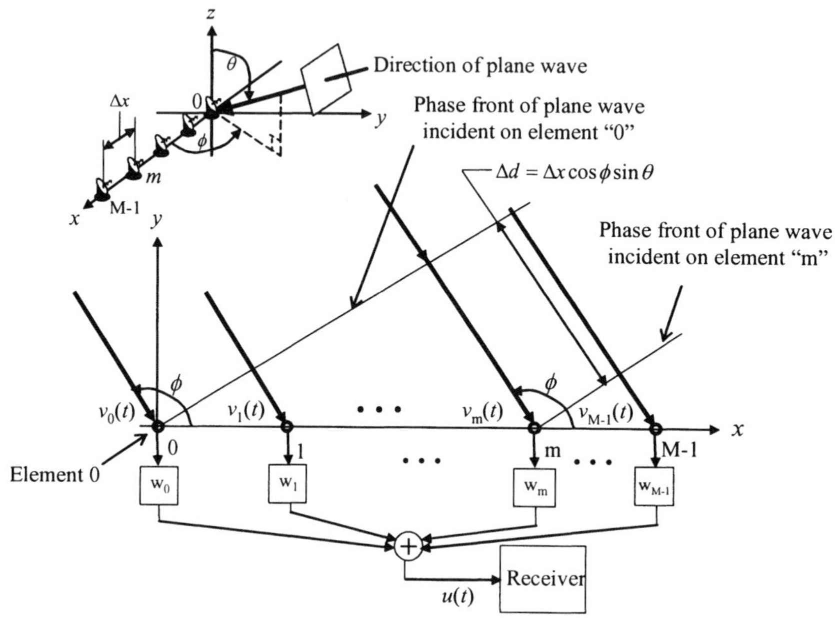

Figure 10.12-3 Antenna array showing the plane wave direction of arrival.

The (baseband) complex envelope of the signals are denoted by $v_m(t)$. Each array weights its respective signal by a complex value, w_m, and sums it with the other weighted elements to form the sum term, $u(t)$.

Let the plane wave phase front impact the array at the spherical angle (θ,ϕ) as seen in the upper left-hand corner of the figure. The plane wave's complex envelope will be denoted by $s(t)$. As noted earlier, the x location of the elements is given by $x_m = m\Delta x$. Therefore the signal received at element m on the array is given by

$$v_m = As(t)e^{-jkm\Delta d} = As(t)e^{-jkm\Delta x\cos\phi\sin\theta} \tag{10.12.2-3}$$

in which A is the received amplitude. The resultant signal out of the array is given by

$$u(t) = \sum_{m=0}^{M-1} w_m v_m(t) = As(t)\sum_{m=0}^{M-1} w_m e^{-jkm\Delta x\cos\phi\sin\theta} = As(t)F_{array}(\theta,\phi) \tag{10.12.2-4}$$

where

$$F_{array}(\theta,\phi) = \sum_{m=0}^{M-1} w_m e^{-jkm\Delta x\cos\phi\sin\theta} \tag{10.12.2-5}$$

$F_{array}(\theta,\phi)$ is called the *array factor*. The array factor determines the ratio of the output complex envelope signal $u(t)$ to the input complex envelope signal $v(t)$, as a function of the plane wave's direction of arrival. The set of complex weights $\{w_m\}$ allow the array to be pointed in any direction (θ_1,ϕ_1). Now consider the array output power. It is given by

$$P_a = \frac{1}{2}|u(t)|^2 = \frac{1}{2}|As(t)|^2 |F_{array}(\theta,\phi)|^2 \tag{10.12.2-6}$$

Consider the following weight assignment to be [47]

$$w_m = e^{jkm\Delta x\cos\theta_1} \tag{10.12.2-7}$$

The array factor becomes

$$F_{array}(\theta,\phi) = \sum_{m=0}^{M-1} e^{-jkm\Delta x(\cos\phi\sin\theta-\cos\theta_1)} \tag{10.12.2-8}$$

Using the summation form of the geometric progression [50] we obtain

$$F_{array}(\theta,\phi) = \frac{1-e^{-jMk\Delta x(\cos\phi\sin\theta-\cos\theta_1)}}{1-e^{-jk\Delta x(\cos\phi\sin\theta-\cos\theta_1)}} \tag{10.12.2-9}$$

Rewriting the array factor as

$$F_{array}(\theta,\phi) = \frac{(e^{jMk\Delta x(\cos\phi\sin\theta-\cos\theta_1)/2} - e^{-jMk\Delta x(\cos\phi\sin\theta-\cos\theta_1)/2})e^{-jMk\Delta x(\cos\phi\sin\theta-\cos\theta_1)/2}}{(e^{jk\Delta x(\cos\phi\sin\theta-\cos\theta_1)/2} - e^{-jk\Delta x(\cos\phi\sin\theta-\cos\theta_1)/2})e^{-jk\Delta x(\cos\phi\sin\theta-\cos\theta_1)/2}} \tag{10.12.2-10}$$

This can be written as

$$F_{array}(\theta,\phi) = \frac{\sin\left(\frac{Mk\Delta x}{2}(\cos\phi\sin\theta-\cos\theta_1)\right)e^{-j\frac{(M-1)k\Delta x}{2}(\cos\phi\sin\theta-\cos\theta_1)}}{\sin\left(\frac{k\Delta x}{2}(\cos\phi\sin\theta-\cos\theta_1)\right)} \tag{10.12.2-11}$$

To obtain a feel for this pattern let $\theta = 90°$, so that the plane wave is incident in the *x-y* plane, shown in Figure 10.12-3 in the upper figure. This is the approximate angle for many cellular antenna applications. Figure 10.12-4 illustrates the magnitude squared amplitude, in dB, of the array factor in the case $\theta = 90°$ as a function of ϕ, when θ_1 = 45° and 75°. As can be seen in the figure the maximum antenna gain is pointed to the value of θ_1 that is input to the array. The dashed line is the 75° case and the solid line corresponds to the 45° case. Notice that the array response is symmetric about the 0° angle point. The linear array was based on seven elements with $\lambda/2$ spacing.

In general if an element has an antenna pattern represented by $F_{element}(\theta,\phi)$ and the linear array consists of identical elements all spaced equally apart on a line, then the resulting pattern is given by

$$F(\theta,\phi) = F_{array}(\theta,\phi)\times F_{element}(\theta,\phi) \tag{10.12.2-12}$$

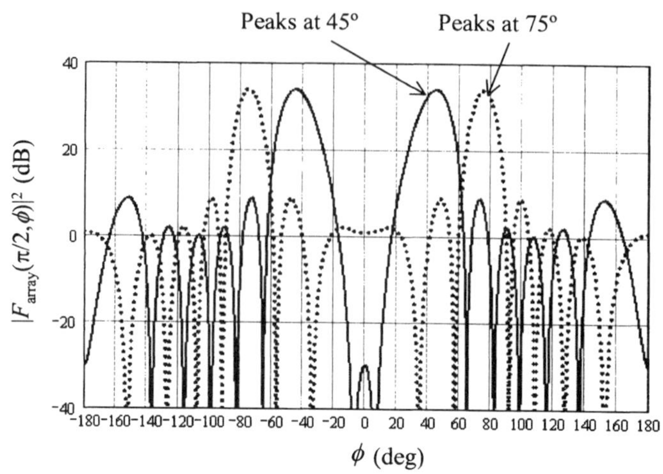

Figure 10.12-4 The magnitude squared of a seven element array factor for the cases in which the steered angles are 45° and 75°.

This pattern factorization is called the principle of pattern multiplication [51]. This result can be extended to rectangular arrays as long as the group of elements, in, say, the *y* direction, can be viewed as super elements when viewed in the *x* direction [51].

Figure 10.12-5 illustrates an example in which *super elements* (*subarrays*) are employed. Here a planar (Figure 10.12-5(a)) has element locations that define the rectangular grid. The rows or columns may be treated as super elements. The number of elements is $N = N_1 \times N_2$. All elements are assumed to be identical. Figure 10.12-5(b) is equivalent [51] to Figure 10.12-5(a).

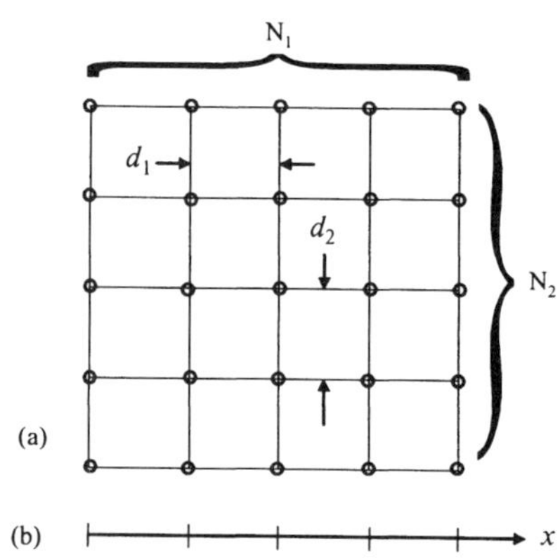

Figure 10.12-5 Equivalent arrays: (a) $N = N_1 \times N_2$ isotropic radiators; and (b) N_1 nonisotropic radiators.

It is convenient when working with arrays to utilize a vector representation of the weights and signals [47]. Let the *weight vector* $\boldsymbol{w}$ be defined by

$$\boldsymbol{w} = \left(w_0, w_1, w_2, w_3 ... w_{M-1}\right)^H \qquad (10.12.2\text{-}13)$$

where the "*H*" denotes the *Hermitian transpose*, which is *a transposition* followed by a *complex conjugation*. The signals from each antenna element defines the *data vector*

$$\boldsymbol{v}(t) = \left(v_0(t), v_1(t), v_2(t), ... v_{M-1}(t)\right)^T \qquad (10.12.2\text{-}14)$$

where the "T" denotes the transpose of the vector. The array output, from the left side of (10.12.2-4), can be written in vector form as

$$u(t) = \boldsymbol{w}^H \boldsymbol{v}(t) \qquad (10.12.2\text{-}15)$$

and it is true that

$$\left(\boldsymbol{w}^H\right)^H = \boldsymbol{w} = \left(w_0, w_1, w_2, w_3 \ldots w_{M-1}\right) \qquad (10.12.2\text{-}16)$$

Consequently (10.12.2-15) indicates a component-by-component multiplication and summation as is defined in the left side of (10.12.2-4). A *steering vector* $\boldsymbol{a}(\theta,\phi)$, has the m-th component (for position x_m), in a direction (θ,ϕ), relative to the element at $x = 0$, is given by

$$a_m(\theta,\phi) = e^{-jk[x_m \cos(\phi)\sin(\theta) + y_m \sin(\phi)\sin(\theta) + z_m \cos(\theta)]} \qquad (10.12.2\text{-}17)$$

The M component steering vector can be written as

$$\boldsymbol{a}(\theta,\phi) = \left(1, a_1(\theta,\phi), a_2(\theta,\phi) \ldots a_{M-1}(\theta,\phi)\right) \qquad (10.12.2\text{-}18)$$

Thus the array factor in direction (θ,ϕ) can be written as

$$F(\theta,\phi) = \boldsymbol{w}^H \boldsymbol{a}(\theta,\phi) \qquad (10.12.2\text{-}19)$$

The angle pair (θ,ϕ) is called the *direction -of-arrival* (DOA) of the received plane wave.

In addition to linear arrays there are rectangular arrays, circular arrays, and elliptical arrays [49]. We have not discussed nonuniformly spaced planar arrays. The main advantage of nonuniformly spaced elements in the array is the fact that the side lobe levels can be reduced [52, 53].

10.12.3 Adaptive Array Spatial Processing

It is possible to implement spatial processing receivers in a number of ways. However, we will present one implementation that gives the general idea of the approach. Figure 10.12-6 illustrates the digital complex baseband weighting and combining processing approach for receivers.

This processing, for a base station, utilizes for each signal path a bandpass filter (BPF), low noise amplifier (LNA), complex mixing (both I and Q components), low-pass filtering (LPF) to $1/M$ of the bandwidth, analog to digital conversion (A/D), along with decimating by the factor M of the original sample rate. Digital down conversion is accomplished by the combination of the complex multiplication, low-pass filtering, and decimating by the factor M. The complex envelops $v_i(t)$ are multiplied by the complex weights w_i and combined in the complex summer to produce a complex (I and Q) output for the receiver to track and demodulate the signal. Adaptive algorithms exist to adjust the weights so as to maximize the signal that is available to the demodulator. This subject will not be discussed here, but is discussed in a number of books [49].

10.12.4 Forming Smart Antennas with Switched Beams

There are two major categories of smart antennas: (1) adaptive arrays and (2) switched beam antennas. Switched beam antennas form multiple fixed beams with increased sensitivity in particular directions. These antenna systems sense signal strength, choose from one or several predetermined, fixed beam antennas, and switch from one beam to another as the mobile user moves about the sector.

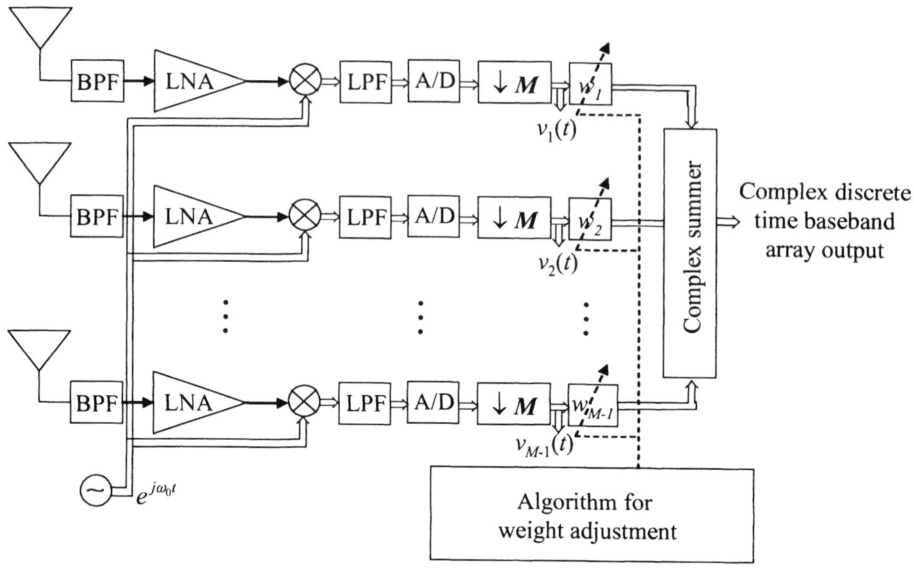

Figure 10.12-6 The digital complex baseband weighting and combining processing approach for receivers.

Figure 10.12-7 illustrates a switched beam network utilizing a beamforming network to form M beams from M array element to allow the processing of K signals. A separate beam selection is made for each of the K receivers.

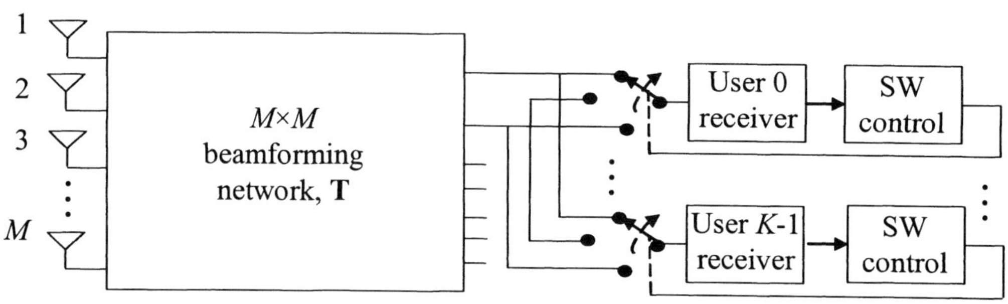

Figure 10.12-7 Structure of switched beam network for forming M beams from M array elements for K receivers.

Depending on what multiple type of access is used, TDMA, FDMA, or CDMA, the mechanism for beam selection is different [47].

10.12.5 MIMO Systems

Up until now we have been discussing only multiple antennas at the base station (normally). However, when multiple antennas are used at both the base station and the user we have a multiple input multiple output (MIMO) link. In applicable channels with the proper signal processing, MIMO can increase effective data rates an order of magnitude or more. MIMO signaling can improve wireless communication in two ways [46, 54]: (1) diversity methods and (2) spatial multiplexing.

Diversity methods improve the robustness of the communication system in terms of BER by exploiting the multiple paths between the transmitter and receiver antennas. Diversity at the receiver can provide performance enhancement similar to the RAKE receiver.

Spatial multiplexing, another MIMO technique, was derived from the fact that in a rich scattering environment it is possible for the receiver to reassemble the transmitted signals simultaneously from the multiple antennas. In effect one is transmitting parallel independent data streams and can achieve dramatic improvement in system capacity. The general MIMO model is shown in Figure 10.12-8.

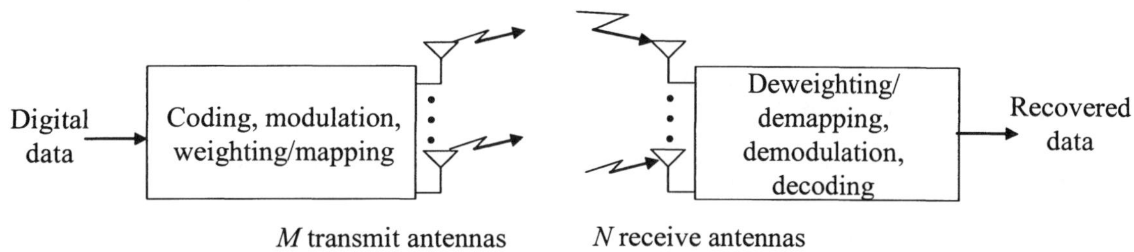

Figure 10.12-8 Model of a MIMO transmission system.

In Figure 10.12-8 baseband digital signals are input into the transmission block. The block is composed of the following functions: error control coding and (possibly combined with) mapping to complex modulation symbols (QPSK, M-QAM and so on). This produces several separate symbol streams that may be independent, partially redundant, or fully redundant. Each of these streams is then mapped onto one of the M transmit antennas. This mapping may be linear spatial weighting of the antenna elements or linear antenna space-time precoding. The signals are transmitted after being filtered, amplified, and frequency translated to the proper frequency.

At the receiver the signals are received by the M multiple antennas, and the inverse process is performed at the receiver: deweighting and demapping, demodulation, and decoding. The appropriate selection of the coding, mapping, weighting, and modulation can offer benefits in performance.

It should be observed that the central idea in MIMO systems is *space-time* signal processing in which the time parameter is augmented with the spatial dimension derived from the use of multiple antennas. As such, MIMO can be viewed as an extension of smart antennas. The use of MIMO [54] has been observed to greatly enhance channel capacity. Recently, MIMO models have been standardized in IEEE 802.16 specifications for fixed broadband wireless access and the third generation partnership project (3GPP) for applications. Liberti and Rappaport [47] discuss smart antennas in considerable depth.

10.13 SUMMARY

In this chapter we presented an introduction to fading channels. To do this we introduced the free space model and other propagation models for various environments as part of the large-scale effects propagation models. Following the large-scale effects we introduced small-scale effects multipath fading. This included the important Rayleigh and Rician channel models. In addition fast and slow fading was introduced. Then wideband channels were explained, which included deterministic models and stochastic time-variant linear channel models. The wide-sense stationary channels and the uncorrelated scattering model followed. Finally, the wide-sense stationary uncorrelated scattering channel model was introduced. The important area of bit error rate degradation due to Rayleigh fading was presented for a few modulation cases. Then the mitigation of multipath effects was introduced through signal diversity and signal combining methods. Next the important area of equalization for multipath improvement was presented. This led to the RAKE receiver. This was followed by the advantage of soft-decision coding over hard-decision coding in a Rayleigh channel fading environment shown for both block and convolutional codes. Finally smart antenna systems and MIMO systems were briefly introduced.

References

[1] Data obtained from the Harvard Web site of Professor Victor Jones, a Robert Wallace research professor of applied physics in the Division of Engineering and Applied Sciences Harvard University. The Web site is the following: http://people.deas.harvard.edu/~jones/es151/prop_models/propagation.html.

[2] J. D. Parsons, *The Mobile Radio Propagation Channel*, 2nd ed., Chichester, England: John Wiley & Sons, 2000.

[3] E. C. Jordan and K. G. Balmain, *Electromagnetic Waves and Radiating Systems*, Englewood Cliffs, NJ: Prentice-Hall, 1968.

[4] T. S. Rappaport, *Wireless Communications Principles and Practice*, New York: IEEE Press & Prentice-Hall, 1996.

[5] M. Hata, "Empirical Formula for Propagation Loss in Land Mobile Radio Services," *IEEE Trans. on Vehicular Technology*, Vol. VT-29, No. 3, August 1980, pp. 317–325.

[6] Y. Okumura, "Field Strength and Its Variability in VHF and UHF Land-Mobile Radio Services," *Review of the Electrical Communications Laboratory,* Vol. 16, September–October 1968.

[7] COST Action 231, "Digital Mobile Radio Towards Future Generation Systems, Final Report," Tech. Rep. European Communities, EUR 18957, 1999.

[8] A. Neskovic, N. Neskovic, and G. Paunovic, "Modern Approaches in Modeling of Mobile Radio Systems Propagation Environment," an IEEE tutorial, 2000, found at http://www.comsoc.org/livepubs/surveys/public/3q00issue/neskovic.html.

[9] M. Kaji and A. Akeyama, "UHF-Band Propagation Characteristics for Land Mobile Radio," *AP-S International Symposium Digest*, Vol. 2, 1985, pp. 835–838.

[10] W. Y. C. Lee, *Mobile Communications Design Fundamentals*, 2nd ed., New York: Wiley-Interscience, 1993.

[11] N. Neskovic, "The Field Strength Prediction for Urban Microcells (Band 900 MHz)," Master's Thesis, Faculty of Electrical Engineering, Belgrade, 1997.

[12] N. Neskovic, A. Neskovic, and D. Paunovic, "A New Microcell Prediction Model Based on the Arrangement of Their Streets and Types," *Proc. 10th IEEE Med. Electrotech. Conf. MELECON*, Limassol, Cyprus, May 2000.

[13] A. Neskovic, "Artificial Neural Network Indoor Radio Propagation Model (Band 900 MHz)," Master's Thesis, Facility of Engineering, Belgrade, 1997.

[14] J. B. Andersen, T. S. Rappaport, and S. Yoshida, "Propagation Measurements and Models for Wireless Communication Channels," *IEEE Comm. Mag.*, January 1995, pp. 42–49.

[15] S. Y. Seidel and T. S. Rapport, "914 MHz Path Loss Prediction for Wireless In-Building Personal Communication System Design," *IEEE Trans. on Antennas and Propagation*, Vol. 40, No. 2, February 1992, pp. 207–217.

[16] D. C. Cox, R. R. Murray, and A. W. Norris, "800 MHz Attenuation Measured in and Around Suburban Houses," *AT&T Bell Labs. Tech. Journal*, Vol. 63, July/August 1984, pp. 921–954.

[17] T. S. Rappaport, "Characteristics of UHF Multipath Radio Channels in Factory Buildings," *IEEE Trans. on Antennas and Propagation*, Vol. 37, August 1989, pp. 1058–1069.

[18] B. Sklar, *Digital Communications Fundamentals and Application*, 2nd ed., Chapter 15, Englewood Cliffs, NJ: Prentice-Hall, 2001.

[19] R. Steele (ed.), *Mobile Radio Communications*, Chapter 2, New York: IEEE Press and Pentech Press, 1994.

[20] W. Y. C. Lee, "Elements of Cellular Mobile Radio Systems," *IEEE Trans. on Vechicular Technology*, Vol. V-35, No. 2, May 1986, pp. 48–56.

[21] T. S. Rappaport, *Wireless Communications Principles and Practice*, Upper Saddle River, NJ: IEEE Press and Prentice Hall, 1996.

[22] L. A. Zadeh, "Frequency Analysis of Variable Networks," *Proc. IRE*, Vol. 38, March 1995, pp. 291–299.

[23] P. Bello, "Characterization of Randomly Time-Variant Linear Channels," *IEEE Trans. on Communication Systems*, Vol. CS-11, December 1963, pp. 360–393.

[24] W. Kozek, "On the Transfer Function Calculus for Underspread LTV Channels," *IEEE Trans. on Signal Processing*, Vol. 45, 1997, pp. 215–223.

[25] W. Kozek and A. F. Molisch, "On the Eigenstructure of Underspread WSSUS Channels," *Proc. First Processing Workshop Advances in Wireless Communications (SPAWC'97)*, 1997, pp. 325–328.

[26] W. C. Jakes, (ed.), *Microwave Mobile Communications*, New York: John Wiley, 1974.

[27] D. C. Cox and R. P. Leck, "Correlation Bandwidth and Delay Spread Multipath Propagation Statistics for 910 MHz Urban Mobile Radio Channels," *IEEE Trans. on Communications*, Vol. 23, No. 11, 1975, pp. 1271–1280.

[28] M. J. Gans, "A Power-Spectral Theory in the Mobile-Radio Environment," *IEEE Trans. VT*, Vol. 21, No. 1, 1972, pp. 27–38.

[29] R. Price and P. E. Green, ""Signal Processing in Radar Astronomy," Report 234, MIT Lincoln Lab., Lexington, MA, 1960.

[30] R. Peterson, R. Ziemer, and D. Borth, *Introduction to Spread Spectrum Communications*, Upper Saddle River, NJ: Prentice-Hall, 1995.

[31] P. A. Bello and B. D. Nelin, "The Effect of Frequency Selective Fading on the Binary Error Probabilities of Incoherent and Differentially Coherent Matched Filter Receivers," *IEEE Trans. on Communications Systems*, Vol. CS-10, June 1962, pp. 160–168.

[32] J. G. Proakis, *Digital Communications*, 4th ed., New York: McGraw-Hill, 2001.

[33] B. W. Lindgren, *Statistical Theory*, Chapter 3, New York: Macmillan Co., 1962.

[34] L. R. Kahn, "Ratio Squarer," *Proc. IRE (Correspondence)*, No. 42, November 1954, p. 1704.

[35] L. Franks, *Signal Theory*, Englewood Cliffs, NJ: Prentice-Hall, 1969.

[36] S. Haykin, *Communication Systems*, New York: John Wiley & Sons, 1994.

[37] J. Kurzweil, *An Introduction to Digital Communications*, Chapter 10, New York: John Wiley & Sons, 2000.

[38] M. Salhav, A. Wiesel, and Y. C. Eldar, "Robust Peak Distortion Equalization," *ICASSP*, 2004.

[39] J. Kurzweil, *An Introduction to Digital Communications*, Chapter 11, New York: John Wiley & Sons, 2000.

[40] S. U. H. Qureshi, "Adaptive Equalization," *Proc. IEEE*, Vol. 73, No. 9, September 1985, pp. 1340–1387.

[41] A. Feuer and E. Weinstein, "Convergence Analysis of LMS Filters with Uncorrelated Gaussian Data," *IEEE Trans. on Acoustics, Speech, and Signal Processing*, Vol. V-33, 1985, pp. 220–230.

[42] J. G. Proakis, *Digital Communications*, 4th ed., New York: McGraw-Hill, 2001.

[43] R. Price and P. E. Green, "A Communication Technique for Multipath Channels," *Proc. of the IRE*, Vol. 46, 1958, pp. 555–570.

[44] G. C. Clark Jr. and J. B. Cain, *Error Correction Coding for Digital Communications*, New York: Plenum Press, 1981.

[45] W. W. Peterson and E. J. Weldon, *Error-Correcting Codes*, 2nd ed., Cambridge, MA: MIT Press, 1972.

[46] S. Sanayei and A. Nosratina, "Antenna Selection in MIMO Systems," *IEEE Comm. Magazine*, October 2004.

[47] J. C. Liberti, Jr. and T. S. Rappaport, *Smart Antennas for Wireless Communications: IS-95 and Third Generation CDMA Applications*, Upper Saddle River, NJ: Prentice-Hall, 1999.

[48] S. Ramo and J. R. Whinnery, *Fields and Waves in Modern Radio*, 2nd ed., New York: John Wiley & Sons, 1962.

[49] R. A. Monzingo and T. W. Miller, *Introduction to Adaptive Arrays*, Raleigh, NC: SciTech Publishing Company, 2004.

[50] L. B. Jolley, *Summation of Series*, 2nd ed. (rev.), New York: Dover Publications, 1961.

[51] B. D. Steinberg, *Principles of Aperture and Array System Design*, New York: Wiley-Interscience Publications, 1976.

[52] R. F. Harrington, "Sidelobe Reduction by Nonuniformly Spaced Element Spacing," *IRE Trans. on Antennas and Propagation*, March 1961, pp. 187–192.

[53] F. Hodjat and S. A. Hovanessian, "Nonuniformly Spaced Linear and Planar Array Antennas for Sidelobe Reduction," *IEEE Trans. on Antennas and Propagation*, Vol. AP, No. 2, March 1978, pp. 198–204.

[54] D. Gesbert, M. Shafi, Da-shan Shiu, P. Smith, and A. Naguib, "From Theory to Practice: An Overview of MIMO Space-Time Coded Wireless Systems," *IEEE Journal on Selected Areas in Communication*, Vol. 21, No. 3, April 2003.

Problems

1. Use the parameters of Example 1, but assume that the receive gain is 3 dB, the transmitted power is 200W, and everything else is the same. Determine the path loss and received power in dBm (this denotes dB value relative to an mW).

2. Using Hata's model determine the path loss for a medium-sized suburban city with h_T = 30m, h_R = 10m, f = 1,500 MHz, and d = 10 km. Also determine the path loss assuming that the propagation is in free space.

3. Recompute Example 4 when $P(0) = 0$ dB, $P(1) = -10$ dB, $P(2) = 0$ dB, $P(3) = -20$ dB, $P(4) = 0$ dB, and $P(5) = -20$ dB.

4. Extrapolate the results of (10.4.3-1)–(10.4.3-3) to the case that the delay spread can be interpreted to be continuous so that the three equations can be written as

$$\bar{\tau} = \frac{\int \tau P(\tau)d\tau}{\int P(\tau)d\tau}$$
$$\overline{\tau^2} = \frac{\int \tau^2 P(\tau)d\tau}{\int P(\tau)d\tau} \quad \text{(P10-1)}$$

where $P(\tau)$ is the power density at excess delay τ.

5. Show that the Fourier transform relationship going from $S(\tau,\upsilon)$ to $H(f,\upsilon)$ is given by

$$H(f,\upsilon) = \int_{-\infty}^{\infty} S(\tau,\upsilon)\exp(-j2\pi f\tau)d\tau \quad \text{(P10-2)}$$

since υ is held constant and τ is the variable of integration and f is the new "frequency variable."

6. Show that using (10.6.1-1) with the substitution $\gamma = \alpha^2 E_b / N_0$ yields the pdf of γ, which is given by $p(\gamma) = \frac{1}{\bar{\gamma}}\exp\left(-\frac{\gamma}{\bar{\gamma}}\right)$ where $\bar{\gamma}$ is the average value of γ. Next, using the fact that $\int_0^{\infty} \Phi(\sqrt{t})\exp(-pt)\,dt = \frac{1}{p}\frac{1}{\sqrt{1+p}}$ and the fact that $Q(x) = \frac{1}{2}\left[1-\Phi\left(\frac{x}{\sqrt{2}}\right)\right]$ show that the average probability of error is given by $\overline{PE_b} = \frac{1}{2}\left(1-\sqrt{\frac{\bar{\gamma}}{1+\bar{\gamma}}}\right)$.

7. Show that for noncoherent, orthogonal BFSK, the average BER is given by

$$\overline{PE_b} = \frac{1}{2+\overline{\gamma}} \tag{P10-3}$$

8. Starting with the Rayleigh probability density function of (10.7.2-1) and the definition of γ_i, (10.7.2-2) and using (10.7.2-3), show that

$$p(\gamma_i) = \frac{1}{\Gamma}\exp\left(-\frac{\gamma_i}{\Gamma}\right) \tag{P10-4}$$

9. Establish the expression for the MSE given in (10.8.1-16) using (10.8.1-14) in (10.8.1-16). Observe that $\boldsymbol{y}_l$ has a one in the fourth position and zeros elsewhere.

10. Establish the expression for the variance using the following: (a) (10.9.1-3), (10.9.1-7), and the fact that Re[z]=$\left(\frac{z+z^*}{2}\right)^2$.

11. Establish the summation

$$\sum_{k=0}^{L-1}\binom{L-1+k}{k} = \binom{2L-1}{L-1} = \binom{2L-1}{L} \tag{P10-5}$$

by noting that

$$\binom{n-m}{0}+\binom{n-m+1}{1}+\binom{n-m+2}{2}+\ldots\binom{n}{m} = \binom{n+1}{m} \tag{P10-6}$$

from a result in M. Abramowitz and I. Stegun, *Handbook of Mathematical Functions*, Delaware, NJ: Dover Publishers Inc., 1968, p. 822.

12. By differentiating the right-hand side of (10.11.1-11) show that the optimum value of ρ is given by

$$\rho_0 = \frac{\sigma_S^2 - \sigma_N^2}{4\sigma_S^2\sigma_N^2} \tag{P10-7}$$

In addition show that when ρ_0 is used in the right-hand side of (10.11.1-11) that the Chernoff bound is of the form

$$PE_b(L) \le \left[\frac{4(1+\overline{\gamma})}{(2+\overline{\gamma})^2}\right]^L \tag{P10-8}$$

13. Establish that the summation in (10.12.2-8) can be simplified to

$$F_{array}(\theta,\phi) = \frac{\sin\left(\frac{Mk\Delta x}{2}(\cos\phi\sin\theta - \cos\theta_1)\right)e^{-j\frac{(M-1)k\Delta x}{2}(\cos\phi\sin\theta-\cos\theta_1)}}{\sin\left(\frac{k\Delta x}{2}(\cos\phi\sin\theta - \cos\theta_1)\right)} \tag{P10-9}$$

CHAPTER 11

Low Probability of Detection Systems

11.0 INTRODUCTION

This chapter will provide an introduction to the area of the techniques used by "adversaries" to detect covert communication systems. These include the radiometer and its variations: the rate line detectors including the chip rate and the carrier line rate detectors, and transform techniques. The three types of LPI/classification techniques are shown in Table 11.0-1.

Table 11.0-1 The Three Main Types of LPI Detection Devices

Transform Techniques	Energy Detectors	Rate Line Detectors
Spectrum analyzer	Radiometers	Chip rate detectors
Compressive receiver	Channelized receivers	Carrier frequency detectors
Bragg cell		
FFT		

The first class of detectors is based on transform techniques. They are based on computing the power spectral density of the received signal plus noise. Common implementations of this class of detector are the spectrum analyzer, the compressive receiver (such as chirp transformers using surface acoustic devices), the fast Fourier transform (FFT), and the Bragg cell, which is an acoustic/optic device used to obtain the power spectral density of an incoming signal.

The second class of detector is the energy detector. The radiometer and the channelized radiometer are examples of this class of detector. They can be used to detect either direct sequence or frequency-hopped signals. Since they only detect energy, they are not capable of signal classification and signal extraction.

The third class of detector is the rate line detectors, which include the delay and multiply detector, and the square law detector followed by narrowband filters to extract the line spectra. Other types of rate line detectors include carrier frequency extraction from squaring and quadrupling the input signal plus noise. These devices extract harmonics of either the carrier frequency or the chip rate, from the signal plus noise.

From the user's point of view, the signal must be low power; if possible, a very short burst transmission; not repeated often, if at all; have a design free of any appreciable spurs (which can be detected by a spectrum analyzer); and hopefully operate far enough away from the interceptor as to not be detectable. Also, the transmitted signal must not allow line spectra, such as a multiple of the carrier frequency or the chip or hop rate, to be generated easily in the interceptors' receiver.

Emission covertness is almost never absolute in that radiated signals that are above the noise level in the unintended receiver can easily be detected. Careful design of the user's transmitter can minimize the probability of the unintended receiver detecting the transmission. The primary measure of transmitter covertness is normally taken to be the maximum range at which the interceptors' receiver can successfully accomplish detection. To be meaningful at all, this range depends on the detection and false alarm probabilities, as well as the transmission path that the signal must undergo to be detected. Generally speaking the air-to-air, or air-to-ground, detection range is much greater than is the ground-to-ground detection range.

If we let r denote the range between the transmitter and the receiver, then the ground to ground path loss commonly drops as r^{-4}, and as an air to air path the loss drops as r^{-2}.

This chapter will discuss the various types of detectors and present their performance.

11.1 LOW PROBABILITY OF INTERCEPT (LPI)

In this section we consider some terminology used in covert communications, followed by a discussion of the LPI scenario, and then we'll discuss a few propagation models that are useful in LPI link calculations.

11.1.1 Covert Communications

In this chapter the following concepts will be used [1]. *Covert transmission* is a radio signaling technique that is designed so that an agent can communicate with other friendly units with a low likelihood of disclosing his presence or his mission.

Two terms are used in the area of covert communication. The first is *COMSEC*, or communication security, which refers to the security of the message. The most common way to protect the message is to encrypt it. If one does not have the proper key to decrypt the message, then it cannot normally be read. Ideally, the use of COMSEC will not degrade the data bit error performance (in practice it sometimes does), but will prohibit the unauthorized listener from deciphering the data.

The other term that is used in covert communications is *TRANSEC*, or transmission security, which refers to use of the transmitted signal for security, including modulation and the use of spread spectrum techniques. Ideally the use of spread spectrum techniques for transmission security should not diminish the performance of the communication link, but would greatly diminish the ability of the unauthorized interceptor to detect the signal transmission and would diminish the effect of any jamming signal to interfere with the communication link.

In the military it is desirable to be able to communicate without the enemy being aware of the transmission, especially if one is behind enemy lines. Designing signals that have a low probability of being detected is a continual challenge. The more noise-like the signal, the wider the bandwidth of the transmitted signal, the lower the transmitted power of the signal, and the shorter[1] the time it takes to send the communication message, the less likely it is to be detected by the undesired listener. Other techniques include employing very narrow beam antenna beamwidths with low sidelobes (high antenna gain) and communicating in the same band as well-established signals, so that the covert signal is "hidden" among the well-established signals (which are hopefully stronger).

A *low probability of detection* (LPD) system is one that transmits a signal that is difficult to detect (be aware that there is any transmission), without additional information that is known only by the desired user (receiver). Detection means that a determination that the receiver decides that something other than noise is present. Once detected, it may be intercepted or exploited.

Low probability of interception (LPI) means that the signal will be difficult to recognize as a signal of interest (SOI). An LPI signal, although not escaping detection, should be difficult to identify and recognize.

A signal that has a *low probability of exploitation* (LPE) means that the signal, if detected, will be difficult to exploit. Exploitation includes reading of plaintext, and identifying the transmitter unit and its mission. LPE is generally provided to an overt signal by encrypting the message content (COMSEC), and by providing TRANSEC cover of the signal structure.

11.1.2 The LPI Scenario

Now consider the relationship between the transmitter, the intended receiver, and the intercept receiver. The basic model for LPI transmission is shown in Figure 11.1-1.

[1] However, there are numerous factors that affect detectability. See Problem 2 for an idealized viewpoint.

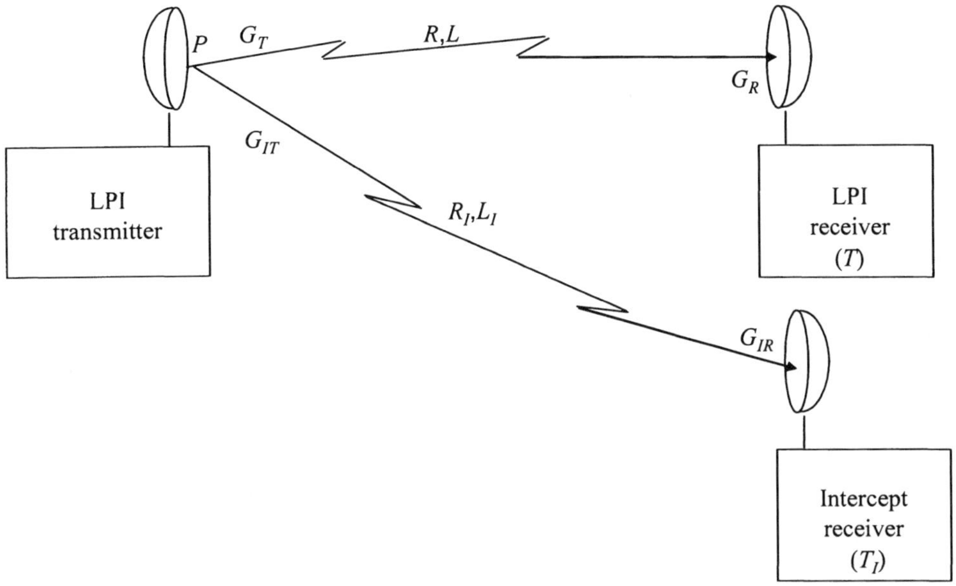

Figure 11.1-1 Communications and the intercept links.

The transmitter sends the signal to the receiver with transmitted power P, and with transmit gain G_T, along a path of range R, which corresponds to a path loss L. The friendly receiver, with antenna gain G_R, having a noise figure of T, receives the signal with path loss R. In addition to the friendly receiver, there is an intercept receiver used by an unauthorized user, whose job it is to detect enemy transmissions. Its noise temperature is assumed to be T_I , it suffers a path loss of R_I, and the transmit gain in the intercept direction is G_{IT} along a path loss of L_I. The intercept receiver has antenna gain G_{IR} in the direction of the transmitter.

Sometimes the figure of merit or *quality factor* (*QF*), defined as the ratio of the maximum range to the communication receiver to the maximum range to the unauthorized user, is used as a measure of LPI effectiveness. The factors that affect that range ratio include the type of terrain, the relative antenna gain from the transmitter to the receiver, the gain from the transmitter to the interceptor, the relative noise figures of the communication receiver and the interceptor receiver, relative atmospheric losses, communication and interceptor antenna gains, the communication and spread spectrum modulation type, and the interceptor type.

The friendly receiver receives a signal power of P_R watts given by

$$P_R = \frac{G_T P G_R}{L} \tag{11.1-1}$$

where the antenna gains are for the path between the transmitter and the desired receiver, and the range loss to the friendly receiver is L. The received power at the intercept receiver is similarly given by

$$P_I = \frac{G_{IT} P G_{IR}}{L_I} \tag{11.1-2}$$

It is to be noted that G_{IT} and G_T may not be the same, as the transmitter may have higher gain in the direction of the intended receiver than to the intercept (unintended) receiver. Furthermore the intercept receiver antenna gain may be much higher if it can be pointed, so that G_R and G_{IR} may not be the same either. If we assume that the path loss is from free space propagation we can write the received power at the intended receiver as

$$P_R = \frac{G_T P G_R}{KR^2} \tag{11.1-3}$$

and at the intercept receiver, as

$$P_I = \frac{G_{IT} P G_{IR}}{KR_I^2} \tag{11.1-4}$$

and K is a proportionality constant. The one-sided noise spectral density is related to the noise temperature T (kT) at the LPI receiver, and T_I at the intercept receiver. The ratio of received power to one-sided noise power spectral density is, for the friendly receiver, given by $P/(kT)$. In a similar manner the equivalent ratio for the intercept receiver is given by $P_I/(kT_I)$ The ratio of received power to one-sided noise spectral densities is given by

$$\frac{P/(kT)}{P_I/(kT_I)} = \left(\frac{G_T P G_R}{KR^2}\right)\left(\frac{KR_I^2}{G_{IT} P G_{IR}}\right)\left(\frac{T_I}{T}\right) = \frac{G_T G_R R_I^2 T_I}{G_{IT} G_{IR} R^2 T} \tag{11.1-5}$$

Solving for the ratio of the ranges from the left- and right-hand sides of (11.1-5) yields the quality factor for free space

$$QF = \frac{R}{R_I} = \sqrt{\frac{(P_I/N_0)\, T_I G_T G_R}{(P/N_0)\, T G_{IT} G_{IR}}} \tag{11.1-6}$$

where P/N_0 is proportional to P/T for both the desired signal and the intercept signal. Thus we see the quality factor improves when the transmit and receive gains increase for the friendly receiver and decrease when the transmit gain in the direction of the intercept receiver increases and the intercept gain increases. Also, if the desired receiver has a lower P/N_0 threshold than the intercept receiver (P_I/N_0) threshold, the quality factor increases also.

11.1.3 Brief Signal Propagation Summary

In order to determine the range that an interceptor can detect the communication signal, it is useful to have a model of the path loss for air-to-ground and ground-to-ground propagation. A few will be summarized here; additional results are presented in Chapter 10. The path loss for free space propagation is given by

$$L_p = 32.4 + 20\log(f_{MHz}) + 20\log(d_{km}) \tag{11.1-7}$$

where L_p is the path loss in dB, the frequency, f_{MHz}, is in megahertz, and the distance, d_{km}, is in kilometers.

For ground-to-ground propagation, when the antenna heights are 10 or more meters, and for frequencies above 40 MHz, one has the median path loss equation in dB from Palmer and Egli [2, 3]:

$$L_{50} = 85.9 + 20\log(f_{MHz}) + 40\log(d_{km}) - 20\log(h_t) - 20\log(h_r) \tag{11.1-8}$$

where h_t and h_r are, respectively, the transmit height in meters and the receive height in meters.

For ground-to-ground propagation [3], when the antenna heights are less than 10 meters, and above 40 MHz, one has the median path loss equation in dB

$$L_{50} = 76.3 + 20\log(f_{MHz}) + 40\log(d_{km}) - 20\log(h_t) - 10\log(h_r) \quad (11.1\text{-}9)$$

The Hata-Okumura model [4–6] an empirical model for losses between base station and mobile user, is based on the work of the two aforementioned authors. For urban areas the mean path loss, denoted by L_{50}, is given in dB by

$$L_{50} = 69.55 + 26.16\times\log(f_{MHz}) - 13.82\log(h_t) - a(h_r) + (44.9 - 6.55\log h_t)\times\log(d_{km}) \quad (11.1\text{-}10)$$

where the parameters $a(h_r)$ will be defined shortly, and the other parameters must satisfy $150 \le f_{MHz} \le 1{,}500$ MHz, $30 \le h_t \le 200$m, $1 \le h_r \le 10$m, $1 \le d_{km} \le 20$ km. The parameter $a(h_r)$ is prescribed for the following cases:

(1) For a small or medium-sized city:

$$a(h_r) = (1.1\times\log(f_{MHz}) - 0.7)h_r - (1.56\times\log(f_{MHz}) - 0.8)\ \text{dB} \quad (11.1\text{-}11)$$

(2) For a large city:

$$a(h_r) = 8.29(\log(1.54\times h_r))^2 - 1.1\ \ dB,\quad f_{MHz} < 200\ \text{MHz} \quad (11.1\text{-}12)$$

or

$$a(h_r) = 3.2\times(\log(11.75\times h_r))^2 - 4.97\ \ dB,\quad f_{MHz} > 400\ \text{MHz} \quad (11.1\text{-}13)$$

(3) For suburban areas:

$$L_{50} = L_{50}(urban) - 2\left[(\log\left(\frac{f_{MHz}}{28}\right)^2 - 5.4\right]\ \ \text{dB} \quad (11.1\text{-}14)$$

(4) For open areas:

$$L_{50} = L_{50}(urban) - 4.78\times(\log(f_{MHz}))^2 + 18.33\times\log(f_{MHz}) - 40.94\ \ \text{dB} \quad (11.1\text{-}15)$$

Clearly in all the ground-to-ground path loss equations, the assumption is made that line-of-sight transmission is physically possible, that is the path distance is not so great that line of sight is obstructed by the Earth's curvature and is therefore not geometrically possible. It is also clear that transmitting the LPI signal close to the ground will minimize the chance of being detected, but it will also increase the path loss of the signal received by the intended receiver as well.

It should be noted that there are various other models for ground-to-ground propagation, most of which have close to fourth power dependence on range.

Example 1 Consider an example that illustrates the difference between free space path loss and path loss in ground-to-ground propagation. Let the distance between the transmitter and the receiver be 20 km in free space and the frequency of transmission be 250 MHz. Let the transmitter height be 20m, and the receiver height be 10m. (1) Assume that the propagation is free space. Evaluation produces the result that the free space loss is 106.4 dB. (2) Assume ground-to-ground propagation. Using (11.1-2), it is determined that the path loss is 139.8, about 33 dB larger than the free space case! Other propagation conditions could lead to even greater path loss.

11.2 AN INTRODUCTION TO RADIOMETRIC DETECTORS

In this section some different types of radiometric detectors will be analyzed to determine their performance. This is the second class of LPI devices indicated in Table 11.0-1. A significant portion of the material to follow in this chapter is based on the unclassified sections of a report by Betz and Rifkin [7].

11.2.1 The Radiometer

In this section an idealized radiometer is treated. Consider Figure 11.2-1 as the model of the radiometer that is considered here. The signal plus noise is passed through a bandpass filter (BPF), which is square law detected and integrated for T seconds and compared to a threshold (derived from the noise process) that is designed to operate on background noise primarily.

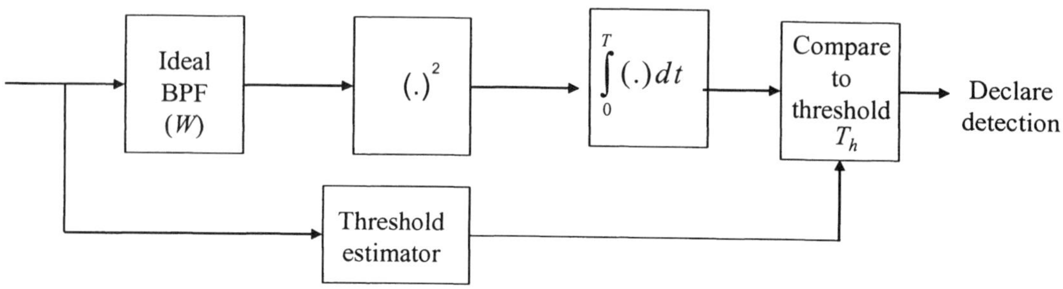

Figure 11.2-1 The classic radiometer.

The radiometer can be shown to be the optimum structure for detecting a signal that is a Gaussian random process and is immersed in a white Gaussian noise background, in which the power spectral density of the signal is small, compared to the thermal noise [8, 9]. This optimum radiometer has a bandpass filter that is matched to the signal bandwidth [8]. The radiometer considered here does not have a matched filter at the input, since it is assumed that the signal to be detected is unknown in spectral shape. The effect of the threshold estimator is not included in this analysis, but is covered in Section 11.2.2.

The input filter is assumed to be an ideal bandpass filter having positive frequency bandwidth of W Hz, which is, by intent, equal to the signal bandwidth. In reality a multipole filter is often used, so the ideal filter assumption provides a simplification in the analysis. The output of the filter feeds the square law detector that feeds the T second integrator. This is the classic integrate-and-dump radiometer. In the analysis it will be assumed that the signal exists for T seconds and that the radiometer is synchronized in time with the signal at the receiver. This assumption leads to the best possible performance one can hope for, since in reality the timing will not be known in advance (even the bandwidth may not be known in advance). The threshold detector compares the integrator output with a threshold and makes a decision on whether the signal is present. The received signal plus noise is modeled by

$$y(t) = \sqrt{2}s_c(t)\cos\omega_0 t - \sqrt{2}s_s(t)\sin\omega_0 t + \sqrt{2}n_c(t)\cos\omega_0 t + \sqrt{2}n_s(t)\sin\omega_0 t \tag{11.2-1}$$

where the first two terms comprise the signal, which is expressed in its quadrature components and is assumed to have signal energy E. The third and fourth terms compose the band-limited white Gaussian receiver noise process, expressed in its in-phase and quadrature components, and each baseband noise term has two-sided spectral density of $N_0/2$. It is assumed that essentially all the power of the signal is passed through the ideal bandpass filter. Since this problem has been treated in Section 6.5 in a slightly less general form, we will utilize the results for the detection and false alarm probabilities derived there for the specific case of a BPSK signal. It can be shown [10, 11] that the result holds for all well-behaved signals that can be expressed in the form of the first two terms of (11.2-1) and have finite energy. From Section 6.5.4 the false alarm and detection probabilities are given by

$$P_{FA} = \int_{\delta}^{\infty} \frac{1}{2^M \Gamma(M)} z^{M-1} e^{\frac{-z}{2}} dz \tag{11.2-2}$$

and

$$P_D = \frac{1}{2}\int_{\delta}^{\infty}\left(\frac{z}{2R}\right)^{\frac{M-1}{2}} e^{\frac{-z}{2}} e^{-R} I_{M-1}(\sqrt{2Rz})dz \tag{11.2-3}$$

where M is equal to $\underline{WT}$, which denotes the closest integer to the product WT, $R = E/N_0$, and δ is the normalized threshold, $\delta = (T_h/\sigma^2)$, with σ^2 the variance of a noise sample of the output of the BPF. The integrand of (11.2-2) is called the (central) chi-squared probability density function with $2M$ degrees of freedom. The integrand of the detection probability in (11.2-3) is called the noncentral, chi-squared probability density function with $2M$ degrees of freedom [11].

In Problem 1 it is shown that P_D can be written in terms of the generalized Marcum Q function of the form

$$P_D = Q_M(\sqrt{2R}, \sqrt{\delta}) \tag{11.2-4}$$

where the generalized Marcum Q function is defined by

$$Q_M(a,b) = \int_b^{\infty} x\left(\frac{x}{a}\right)^{M-1} e^{-\frac{x^2}{2}} e^{-\frac{a^2}{2}} I_{M-1}(ax)dx \tag{11.2-5}$$

It should be pointed out that the false alarm rate, which is often specified along with the probability of detection, is just the false alarm probability divided by the time period (T) between decisions, so that

$$FAR = P_{FA}/T \tag{11.2-6}$$

Note that both equations can be evaluated with the use of a commercial software package since the equations are written in a normalized form.[2] Only the parameters M, R, δ need be specified in order to evaluate the two probabilities. Before we discuss the limitations of these performance equations for the radiometer, let us consider a Gaussian approximation for these results. When the parameter M becomes large, the output density functions tend to Gaussian random variables. Thus, following Section 6.5.5, one has the result

$$P_D = Q\left[\frac{\beta - \sqrt{WT}SNR}{\sqrt{1+2SNR}}\right] \tag{11.2-7}$$

and

$$P_{FA} = Q(\beta) \tag{11.2-8}$$

where

$$\beta = \left(\frac{T_h - N_0 WT}{N_0\sqrt{WT}}\right) \tag{11.2-9}$$

and where the $Q(x)$ function is defined by

$$Q(x) = \int_x^{\infty} \frac{1}{\sqrt{2\pi}} \exp\left(-\frac{t^2}{2}\right)dt \tag{11.2-10}$$

[2] Actually evaluating P_D is best done with a series expansion; see Chapter 6, Section 6.5.11, for an example.

and the SNR is the signal power (P) divided by the noise power (N_0W); that is,

$$\text{SNR} = \frac{P}{N_0 W} \tag{11.2-11}$$

This result applies as a reasonable approximation to all well-behaved signals having power P, including QPSK and MSK. This Gaussian probability calculation, in normalized form, is very simple to use, since choosing a false alarm probability implies β, which then allows the evaluation of the detection probability, when the SNR is known. Alternatively, if one knows the desired value of the detection probability then the argument of the Q function of (11.2-7) is known and the SNR can be solved to achieve the desired detection probability. It should be noted that if the signal to be detected and the radiometer have a nonnegligible filtering loss due to the bandpass filter having bandwidth W, then the SNR should be modified to

$$\text{SNR} = \frac{\alpha P}{N_0 W} \tag{11.2-12}$$

where α is the filtering loss, due to the effect of the ideal bandpass filter on the signal. If $S(f)$ is the one sided signal power spectral density centered at f_0, then

$$\alpha P = \int_{f_0 - B/2}^{f_0 + B/2} S(f)df \tag{11.2-13}$$

Problem 2 shows that under the constraint the pulse energy and the probability of false alarm are fixed, the pulse type that minimizes the probability of detection, from the transmitter's point of view, is to send a very low-power, long duration pulse, rather than a high-power, short pulse. This conclusion is based on the assumption that the radiometer has a large WT product and that the radiometer is synchronized with the pulsed signal and the pulse energy is held constant.

Example 2 Determine the minimum (synchronous) integration time needed to detect a signal that has power level of –160 dBW through an omni antenna (0-dB gain), assuming that the Gaussian approximation is adequate. Assume that the noise level is –200 dBW/Hz and the bandwidth of the signal is 10 MHz. Set the integration time to achieve a $P_D = 0.95$ and $P_{FA} = 10^{-2}$. By evaluating (11.2-8) and (11.2-7), it can be shown that 1.575 seconds are needed to achieve the desired P_D and P_{FA}. The corresponding value of β is 2.32. The actual threshold (T_h) can be found from (11.2-9) and knowing W, $T\,N_0$, and β.

11.2.2 Limitations of the Radiometer Performance Results

There are a number of limitations in this analysis that need to be pointed out. First, for sensitive radiometers the background temperature variation and interfering signals limit the performance of the radiometer. The reason for this is the fact that the threshold of the radiometer must be derived from the receiver noise process without the signal present. It follows that the threshold is a statistical quantity, and therefore has some variation to it, and depends upon whether the input is noise only or has other additional interference signals in addition to the thermal noise.

If the threshold estimator is heavily filtered then it cannot respond to noise temperature and interference signal variations that are faster than the threshold estimator filter response. If on the other hand, the threshold estimator filter is not very narrowband (to give it a faster response) it will have a larger output variation on it, which will also change the detection and false alarm probabilities quite dramatically. Antenna background temperature in the tactical environment depends on a number of factors, including elevation angle, azimuth angle, temporal variation, and location of the detector. All these factors can influence the threshold setting. It is to be noted that background interference comes from all types of signals including television, radio, radio relay links, microwave links, multipath, and so on.

Now let us quantify the sensitivity to the radiometer threshold variation. This formulation follows Sonnenschein and Fishman [12].

From (11.2-7) and (11.2-8) one can write the detection probability as

$$P_D = Q\left[\frac{Q^{-1}(P_{FA}) - \sqrt{WT}\text{SNR}}{\sqrt{1+2\text{SNR}}}\right] \quad , \text{SNR}=\frac{P}{N_0 W} \qquad (11.2\text{-}14)$$

where P is the signal power and N_0 is the one-sided noise spectral density, with B the positive frequency noise bandwidth. Solving for the SNR from (11.2-14) produces

$$\text{SNR} = \frac{B}{\sqrt{WT}} + \frac{A}{WT}\left[A - \sqrt{A^2 + WT + 2\sqrt{WT}B}\right] \qquad (11.2\text{-}15)$$

where the parameters A and B are given by

$$A = Q^{-1}(P_D) \quad \text{and} \quad B = Q^{-1}(P_{FA}) \qquad (11.2\text{-}16)$$

When WT is large, (11.2-15) can be approximated by

$$\text{SNR} \cong \frac{Q^{-1}(P_{FA}) - Q^{-1}(P_D)}{\sqrt{WT}} \qquad (11.2\text{-}17)$$

Thus one can compute the required SNR needed to obtain the required P_D and P_{FA} from either (11.2-15) or (11.2-17), depending on whether or not WT is large.

From (11.2-8) and (11.2-9) the ideal interceptor threshold value for P_{FA0} is

$$T_{h0} = N_0\left[TW + \sqrt{TW}Q^{-1}(P_{FA0})\right] \qquad (11.2\text{-}18)$$

Hence the interceptor's actual threshold would be given by

$$\hat{T}_{h0} = \hat{N}_0\left[TW + \sqrt{TW}Q^{-1}(P_{FA0})\right] \qquad (11.2\text{-}19)$$

where $\hat{N}_0$ is the estimate of the one-sided noise spectral density.

Now, since the threshold must be determined from the input noise process, it will be proportional to N_0, at the output of the detector. Since this estimate will not be perfect, it will be modeled by $\hat{N}_0$, and will be assumed to satisfy the following constraints

$$(1 - \varepsilon_1)N_0 \le \hat{N}_0 \le (1 + \varepsilon_2)N_0 \qquad (11.2\text{-}20)$$

where N_0 is the actual value and the parameters ε_1 and ε_2 satisfy

$$0 \le \varepsilon_1 < 1 \quad \text{and} \quad \varepsilon_2 \ge 0 \qquad (11.2\text{-}21)$$

Since the threshold is proportional to N_0 it follows that

$$(1-\varepsilon_1) \le \frac{\hat{T}_h}{T_{h0}} \le (1+\varepsilon_2) \tag{11.2-22}$$

where $\hat{T}_h$ is the estimate of the threshold (T_{h0}) for the ideal case (which occurs when $\hat{N}_0 = N_0$). If the threshold estimate satisfies the left-hand side of (11.2-22) with equality, then the observed P_{FA} will be greater than the desired value, in the ideal case. Thus, the threshold estimate will have to be increased by the factor $1/(1-\varepsilon_1)$ to obtain the correct P_{FA}. If, on the other hand, the noise level satisfies the right-hand side of (11.2-20) with equality then the threshold must be increased by the factor $(1+\varepsilon_2)$ to ensure the achievement of the desired P_{FA}. It is concluded that the worst-case threshold must satisfy the relationship

$$(T_h)_{wc} = U\,T_{h0} \tag{11.2-23}$$

where U is the peak-to-peak uncertainty defined as

$$U = \frac{1+\ \varepsilon_2}{1-\ \varepsilon_1} \tag{11.2-24}$$

If we define the SNR needed to achieve a $P_D > P_{D0}$ and $P_{FA} < P_{FA0}$, over the entire range uncertainty of U, using (11.2-23) and (11.2-18), and

$$P_D = Q\left[\frac{UT_h - N_0 WT(1+\text{SNR})}{N_0\sqrt{WT}\sqrt{(1+2\text{SNR})}}\right] \tag{11.2-25}$$

leads to the solution for SNR as

$$\text{SNR} = (U-1) + U\left(\frac{B}{\sqrt{WT}}\right) + \frac{A}{WT}\left[A - \sqrt{A^2 + (2U-1)WT + 2\sqrt{WT}B}\right] \tag{11.2-26}$$

If the estimate is perfect ($U = 1$) then (11.2-26) degenerates to (11.2-15). When $WT >> 1$ (11.2-26) becomes

$$\text{SNR} \cong (U-1) + \frac{UB - A\sqrt{2U-1}}{\sqrt{WT}} \tag{11.2-27}$$

Note that for large WT the minimum value of the SNR satisfies

$$(\text{SNR})_{\min} \cong U - 1 \tag{11.2-28}$$

The parameter $(\text{SNR})_{\min}$ is the minimum value of SNR that is needed to overcome the threshold level uncertainty irrespective of the detector parameters, P_D and P_{FA}, and the integration time T. This situation is in direct contrast to the ideal case where a longer integration time yields a lower required SNR! Equation (11.2-26) is plotted in Figure 11.2-2 with $P_{FA}=1-P_D = 0.01$ and in Figure 11.2-3 with $P_{FA}= 1-P_D =10^{-6}$.

The plot is for the SNR required versus the WT product parameterized by the parameter U, expressed in dB. At the larger values of WT the two figures do not differ much. However at smaller values of WT the respective curves differ by about 5 dB. Notice that as the peak-to-peak uncertainty increases, the minimum SNR increases. From (11.2-28) it is clear that when U is 3 dB, U–1 is equal to 0 dB as the value of the minimum value of the SNR, and is shown in both figures.

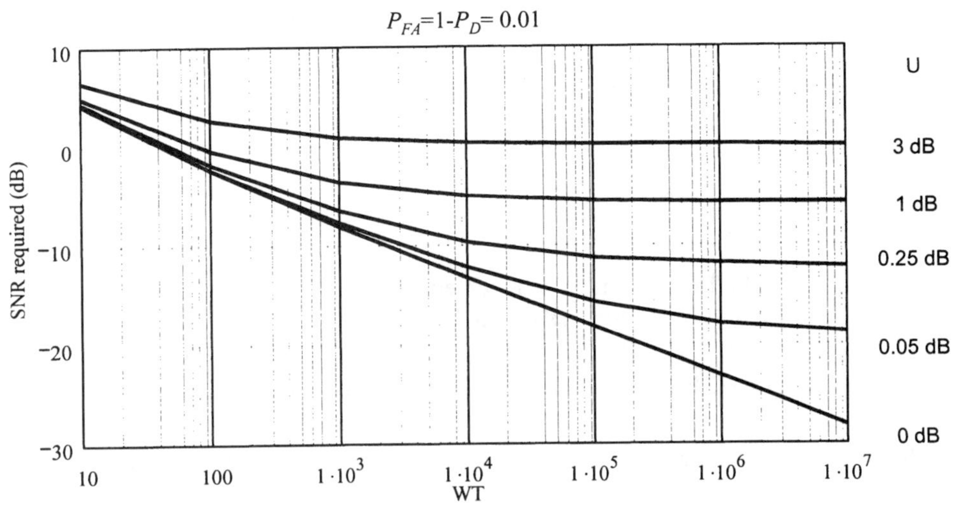

Figure 11.2-2 SNR required for a radiometer for various values of U (dB) versus WT.

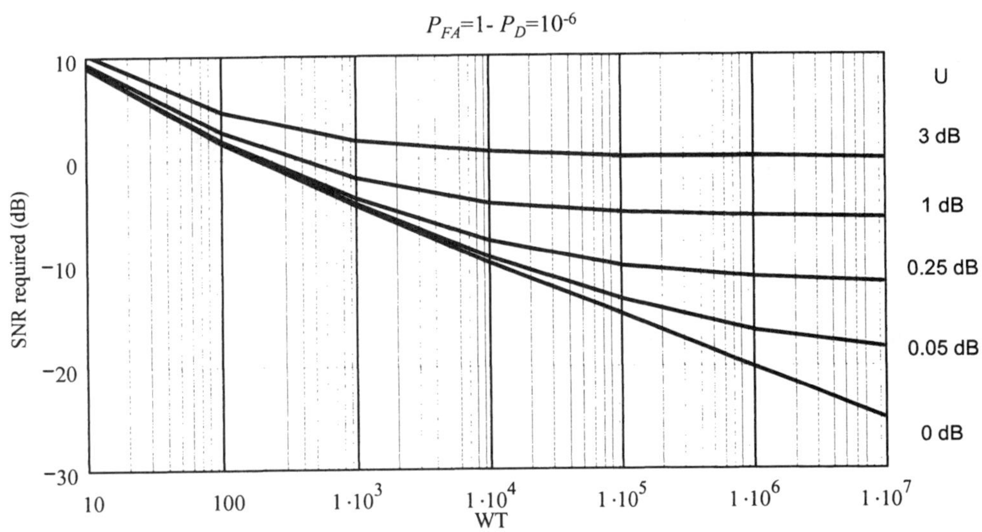

Figure 11.2-3 SNR required for a radiometer for various values of U (dB) versus WT.

11.2.3 Low-Pass Filter Radiometer

Another type of radiometer [13], which does not require an integrate-and-dump filter, is the low-pass filter radiometer (LPFR). Timing of the integrator is not required in this radiometer. This radiometer is similar to the standard radiometer, except that the integrator is replaced by a one-pole RC low-pass filter, with time constant $\tau = RC$. Figure 11.2-4 illustrates the LPFR.

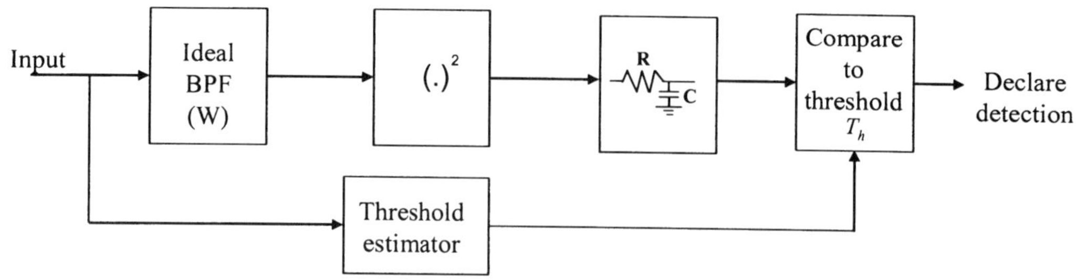

Figure 11.2-4 The low-pass filter radiometer.

The signal out of the bandpass filter is modeled as a tone of duration T seconds, plus band-limited white Gaussian noise to yield the input process $y(t)$ to the radiometer, where $y(t)$ is given by

$$y(t) = \sqrt{2P}\sin(\omega_0 t + \theta) + \sqrt{2}n_c(t)\cos(\omega_0 t) + \sqrt{2}n_s(t)\sin(\omega_0 t) \tag{11.2-29}$$

where P is the power of the signal, ω_0 is the angular frequency, and θ is the random phase, assumed to be random over 0 to 2π. The two-sided noise power spectral density of the input white noise process is assumed to be $N_0/2$. It can be easily be shown that the two-sided noise spectral density of the two quadrature noise terms ($n_c(t)$ and $n_s(t)$) is $N_0/2$ also. In Problem 3 it is shown that the probability of detection can be approximated by

$$P_D = Q\left(\frac{\beta - \sqrt{2B\tau}\ \text{SNR}\left[1 - e^{-T/\tau}\right]}{\sqrt{1 + 2\text{SNR}}}\right) \tag{11.2-30}$$

when T/τ is large and where β is the effective threshold, as indicated next. The parameter SNR is the input signal-to-noise ratio and is given by

$$\text{SNR} = \frac{P}{N_0 W} \tag{11.2-31}$$

and the false alarm probability is given by

$$P_{FA} = Q(\beta) \tag{11.2-32}$$

The parameter β is given by

$$\beta = \frac{\delta - N_0 W}{\sqrt{N_0^2 W\left(\dfrac{1}{2\tau}\right)}} \tag{11.2-33}$$

In order to compare the detection performance with the ordinary (integrate-and-dump) radiometer at low SNR, it can be seen that the effective ratio of input SNRs (at low input SNR) satisfies

$$L = \frac{\text{SNR}_{RC}}{\text{SNR}_{\text{int}}} = \sqrt{2}\sqrt{\frac{\tau}{T}}\left[1 - e^{-T/\tau}\right] \tag{11.2-34}$$

It can be shown by direct evaluation that L is maximum when $\tau/T = 0.795$. The value of L at this optimum ratio is 0.9. Thus, L is reduced by $10\times\log(0.9) = 0.5$ dB. It is concluded that the LPFR is about 0.5 dB (input SNR) worse in performance than the integrate-and-dump detector, but does not require timing information! The results presented here also apply to constant envelope modulated signals as long as the spectra are constrained by the bandpass filter of bandwidth W. For the optimized case, the probability of detection can be written

$$P_D = Q\left[\frac{\beta - 0.9\sqrt{WT}\text{SNR}}{\sqrt{1+2\text{SNR}}}\right] \tag{11.2-35}$$

When modulation is present, SNR should be replaced by αSNR as discussed in the text adjacent to (11.2-13).

11.2.4 The Correlation Radiometer

Another type of radiometer device is the correlation radiometer [14] and is shown in Figure 11.2-5. The analysis approach in this section differs from that in [14]. The basic idea of the device is based on the fact that the signal is highly correlated and the noise of each channel, when due to receiver noise, is uncorrelated. Therefore the device should perform better than the radiometer. However common noise such as a "hot spot" seen by the antennas would be correlated and appear to be a signal. Another concern with this type of device is the fact that if the signal is received from off-bore sight the phase difference of the received signal could cause a loss in detection capability, since there would be a phase shift between the signals. If the phase difference is 90° and the signal is BPSK, the signal would be totally suppressed.

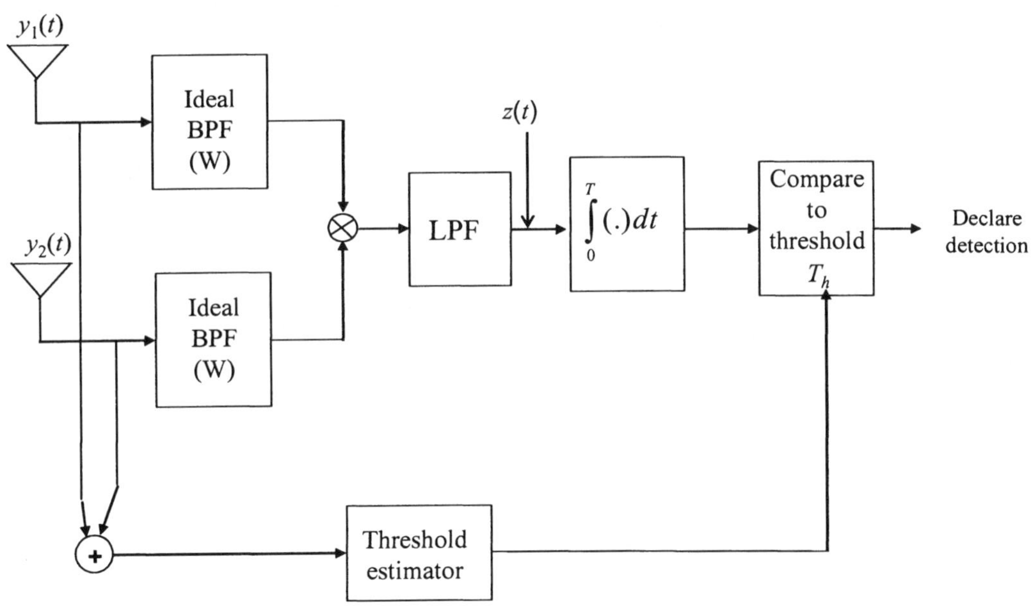

Figure 11.2-5 The correlation radiometer.

One solution is to place the two antennas as close together as possible and to use relatively high gains so that signals detected in the sidelobes would be very weak. Then, only the main beam signals would tend to be detected.

Let the two inputs be represented as

$$y_1(t) = \sqrt{2P}\sin(\omega_0 t) + n_1(t)$$
$$y_2(t) = \sqrt{2P}\sin(\omega_0 t) + n_2(t) \tag{11.2-36}$$

where P is the signal power, T is the integration time, and $n_1(t)$ and $n_2(t)$ are the two statistically independent noise processes, originating from the two receiver front ends. It will be assumed that WT is large so that the Gaussian assumption for the output of the integrate-and-dump filter can be assumed. The noise terms can be written in their quadrature band-limited form

$$n_1(t) = \sqrt{2}n_{c1}(t)\cos(\omega_0 t) + \sqrt{2}n_{s1}(t)\sin(\omega_0 t) \tag{11.2-37}$$

$$n_2(t) = \sqrt{2}n_{c2}(t)\cos(\omega_0 t) + \sqrt{2}n_{s2}(t)\sin(\omega_0 t) \tag{11.2-38}$$

where it is assumed that the two noise processes are statistically independent of each other. The product of $y_1(t)$ and $y_2(t)$, after filtering the $2\omega_0$ terms, is given by (call it $z(t)$)

$$z(t) = y_1(t)y_2(t)\big|_{LPF} = P + \sqrt{P}n_{s1}(t) + \sqrt{P}n_{s2}(t) + n_{c1}(t)n_{c2}(t) + n_{s1}(t)n_{s2}(t) \tag{11.2-39}$$

The mean output is given by

$$\mathrm{E}\left\{y_1(t)y_2(t)\big|_{LPF}\right\} = P \tag{11.2-40}$$

since the two noise terms are uncorrelated. To obtain the variance of the output we determine the autocorrelation function of the output process of the filtered product

$$R_z(\tau) = PR_{n_{s1}}(\tau) + PR_{n_{s2}}(\tau) + R_{n_{c1}}(\tau)R_{n_{c2}}(\tau) + R_{n_{s1}}(\tau)R_{n_{s2}}(\tau) \tag{11.2-41}$$

Since all the noise terms are statistically independent and of equal power, (11.2-41) simplifies to

$$R_z(\tau) = 2PR_{n_{s1}}(\tau) + 2R_{n_{c1}}^2(\tau) \tag{11.2-42}$$

It is assumed that the input bandwidth signal duration time (WT) is large, and the spectral density is nearly constant over the integrate-and-dump bandwidth, it is sufficient to determine the spectral density at $f = 0$, thus

$$S(0) = 2P\frac{N_0}{2}\int_{-\infty}^{\infty}\delta(\tau)d\tau + 2\int_{-\infty}^{\infty}\left(\frac{N_0}{2}\right)^2 |H_{LPF}(f)|^4\, df \tag{11.2-43}$$

where $H_{LPF}(f)$ is the low-pass filter equivalent of the ideal bandpass filter (i.e., $H_{LPF}(f) = 1$, $|f| \le \frac{W}{2}$) and zero otherwise). Evaluating the power spectral density at $f = 0$ produces

$$S(0) = PN_0 + \frac{N_0^2 W}{2} \tag{11.2-44}$$

Using Parseval's theorem, it follows that the variance of the output variable is given by integral of the spectral density and the transfer function of an integrate-and-dump filter; thus

$$\sigma_z^2 = \int_{-\infty}^{\infty} S(0)T^2 \left[\frac{\sin(\pi f T)}{\pi f T} \right]^2 df \qquad (11.2\text{-}45)$$

Evaluating this, one obtains

$$\sigma_z^2 = \frac{N_0^2 WT}{2} + PN_0 T, \quad \sigma_N^2 = \frac{N_0^2 WT}{2} \qquad (11.2\text{-}46)$$

Hence, when the signal is not present, $P = 0$ in (11.2-40), so that the probability of false alarm and detection are given by

$$P_{FA} = \int_{\delta}^{\infty} \frac{1}{\sqrt{2\pi}\sigma_N} e^{-\frac{u^2}{2\sigma_N^2}} du$$

$$P_D = \int_{\delta}^{\infty} \frac{1}{\sqrt{2\pi}\sigma_Z} e^{-\frac{(u-\mu_Z)^2}{2\sigma_Z^2}} du \qquad (11.2\text{-}47)$$

Evaluating one obtains

$$P_{FA} = Q(\beta) \quad \text{and} \quad \beta = \frac{\delta}{\sqrt{\frac{N_0^2 WT}{2}}} \qquad (11.2\text{-}48)$$

where β is a normalized threshold, δ is the unnormalized threshold, and the denominator is the standard deviation of the noise when the signal is absent (σ_N). The detection probability is given by

$$P_D = Q\left[\frac{\beta - \sqrt{2}\mathrm{SNR}\sqrt{WT}}{\sqrt{1+2\mathrm{SNR}}} \right] \qquad (11.2\text{-}49)$$

with SNR (SNR = $P/(N_0 W)$ the input signal-to-noise ratio, and β a constant that determines the false alarm probability. Notice that at low input SNRs (where these devices are normally used) the effective input SNR is 1.5 dB (10 log($\sqrt{2}$) = 1.5 dB) higher when compared to the classic radiometer (11.2-7).

11.2.5 Relationship of the Output SNR and the Deflection

The output SNR, denoted by SNR_0, can be defined as

$$\mathrm{SNR}_0 \cong \frac{(\mu_{S+N} - \mu_N)^2}{\sigma_N^2} = d^2 \qquad (11.2\text{-}50)$$

where μ_{S+N} is the mean value of the output, under the hypothesis that the signal plus noise is present, and μ_N is the mean output value under the assumption that only the noise is present. The parameter d is also known as the *deflection* [15], and the square of d is the output SNR at low input SNR values. Consider the two cases: (1) when the signal plus noise are present and (2) when only the noise is present. Let the mean signal and variance for both cases be denoted by

$$\begin{aligned}&\text{Case 1:}\\&\text{mean} = \mu_N \quad \text{variance} = \sigma_N^2\\&\text{Case 2:}\\&\text{mean} = \mu_{S+N} \quad \text{variance} = \sigma_{S+N}^2\end{aligned} \qquad (11.2\text{-}51)$$

If the threshold is δ, then the false alarm can be written as (δ is the unnormalized threshold)

$$P_{FA} = Q\left(\frac{\delta - \mu_N}{\sigma_N}\right) = Q(\beta) \qquad (11.2\text{-}52)$$

where

$$Q(x) = \int_x^{\infty} \frac{1}{\sqrt{2\pi}} e^{-\frac{y^2}{2}} dy \qquad (11.2\text{-}53)$$

Now it is easy to show that the probability of detection is given by

$$P_D = Q\left(\frac{\delta - \mu_{S+N}}{\sigma_{S+N}}\right) \qquad (11.2\text{-}54)$$

Now from (11.2-52) it follows that

$$\delta = \beta\sigma_N + \mu_N \qquad (11.2\text{-}55)$$

Using (11.2-55) in (11.2-54) produces the result

$$P_D = Q\left(\frac{\beta\sigma_N + \mu_N - \mu_{S+N}}{\sigma_{S+N}}\right) \qquad (11.2\text{-}56)$$

Under the assumption that $\sigma_{S+N} \cong \sigma_N$, it is also true that $\mu_N < \mu_{S+N}$ at low input SNR yields the result

$$P_D = Q[\beta - d] \qquad (11.2\text{-}57)$$

From (11.2-57) it follows that for a fixed β (fixed P_{FA}) the larger the output SNR, the greater the value of P_D. It also follows that the negative of the second term in the expression for P_D is the square root of the output SNR at low input SNRs. This result applies to all radiometric type calculations in which the Gaussian assumption applies at low input SNRs. Note that for the correlation radiometer the output SNR is 3 dB better in output SNR than the classic (integrate-and-dump) radiometer, as is seen by comparing (11.2-49) and (11.2-7). It follows that it requires 1.5 dB less input SNR than the classic radiometer. This assumes that the phases are aligned perfectly.

11.2.6 The Optimum Detector for Frequency-Hopped Waveforms

The maximum-likelihood detection derivation of the optimum detector for frequency-hopped waveforms has been determined in [16, 17]. Figure 11.2-6 illustrates the optimum detector for the case that the hopping frequency separation is equal to the modulation bandwidth. The parameter M is the number of hops that are observed.

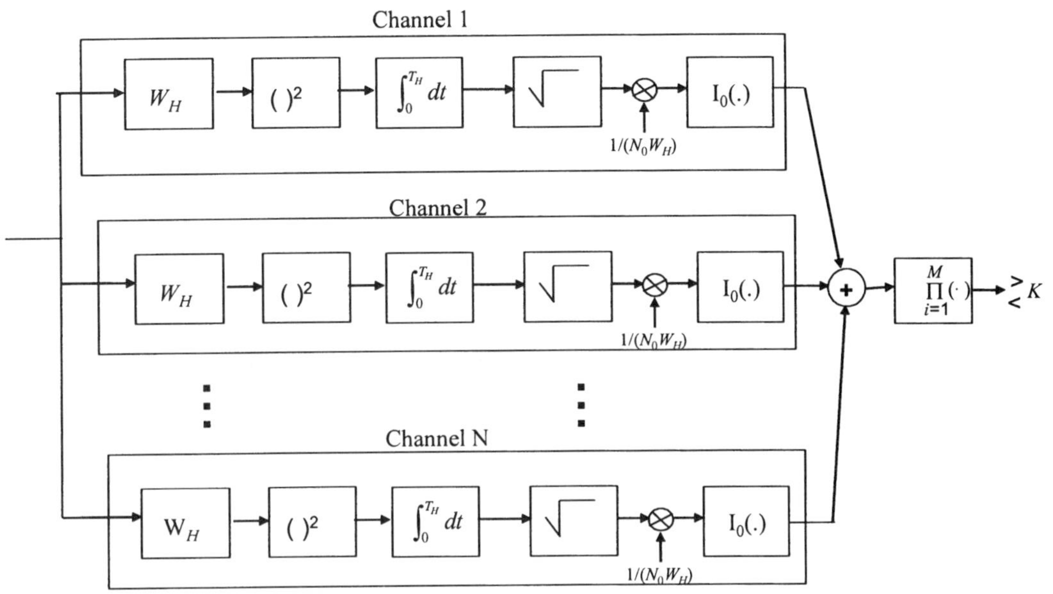

Figure 11.2-6 Optimum FH radiometer detector.

The optimum detector has a hop radiometer for each of the N hops in the hopping pattern. Each channel has a front-end radio frequency (RF) bandwidth W_H Hz. The bandwidth and time are matched in the radiometer. The square root of each radiometer output is taken, weighted and scaled by the modified Bessel function of the first kind, and then added to the other N–1 channels for each hop. Then, on a hop-by-hop basis, the product is formed over each hop. At the end of the hops, for that detection period the result is compared to a threshold, K. The analysis is very difficult to accomplish except in the case when N is a large number (at least 100) when Gaussian statistics can be assumed [17].

Woodring [17] has shown that the required P/N_0 ratio needed, based on a Gaussian assumption, is given by

$$\left(\frac{P}{N_0}\right)_{reqd} \cong \frac{W_H}{2} I_0^{-1}\left[1+N\left(e^{(\rho^2/M)}-1\right)\right] \tag{11.2-58}$$

where $I_0^{-1}(x)$ is the inverse modified Bessel function of the first kind evaluated at x, and ρ^2 is given by

$$\rho^2=\left[Q^{-1}(P_{FA})-Q^{-1}(P_D)\right] \tag{11.2-59}$$

and depends on the false alarm and detection probabilities desired. The Q function is defined by (11.2-10). A plot of $I_0(x)$ allows one to obtain the inverse function readily, and one is shown in Figure 11.2-7.

11.2.7 The Filter Bank Combiner

The filter bank combiner (FBC) [14, 16–18] is a radiometer that lends itself to detecting frequency-hopped signals and is shown in Figure 11.2-8. Again M is the number of hop intervals processed.

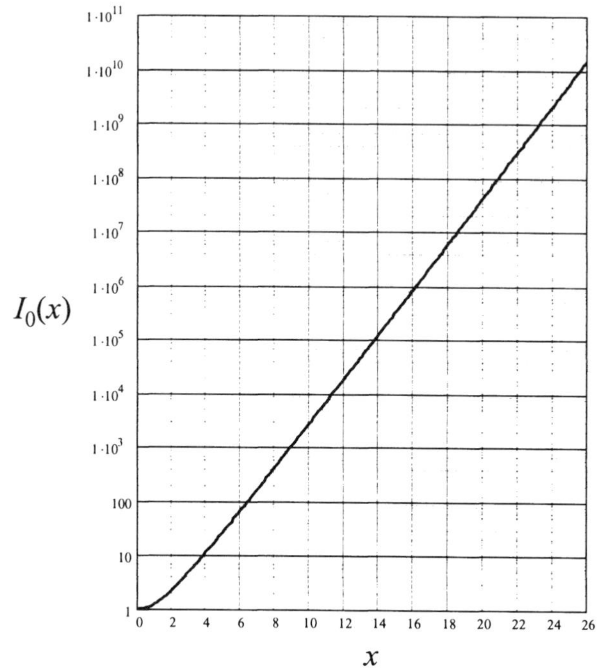

Figure 11.2-7 Plot of $I_0(x)$ versus x.

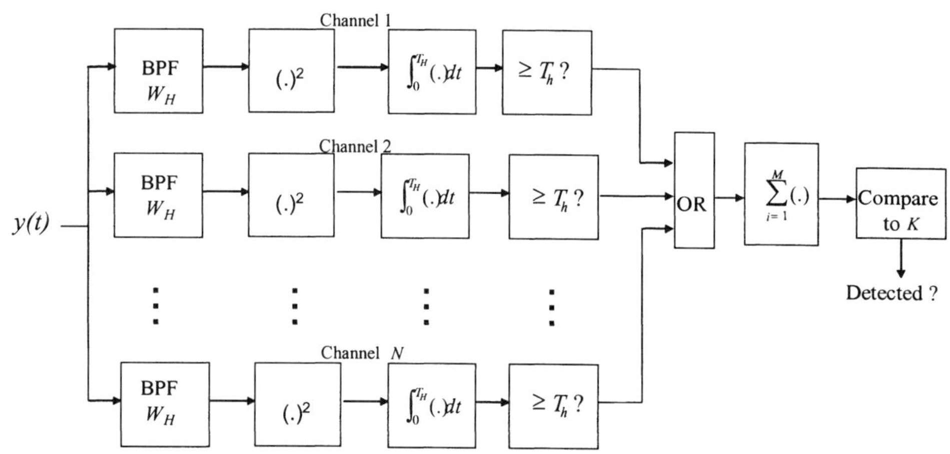

Figure 11.2-8 The filter bank combiner radiometer (FBC).

Dillard [19] calls a very similar detection system the binary moving window detector (BMWD). The main difference is the fact that the comparison is made after every hop and not at the end of the test period as in the FBC. Assume that there are N possible hop frequencies and M hops with the integration time set equal to the hop time, which is T_H seconds. Thus the total message time is MT_H. It is assumed that the bandpass filters are centered at each hop frequency and the filters do not overlap in frequency. Each filter is W_H Hz wide. The total spread bandwidth is W Hz. Hence $NW_H = W$. After the integration of T_H seconds, the outputs of the integrators are compared to the thresholds, which are set at T_h.

At each hop time a decision on each radiometer is made on whether the signal is present and those decisions are logically OR'd for each hop; and then the summation of the M hops are made and compared to an integer valued threshold which we denote as "K." The threshold on successful hops, K, is less than M in general. Note that there are two thresholds, the individual radiometer threshold and the threshold on the output of the summer of the "OR" function. It is to be noted that if several radiometer subsystem outputs cross their individual thresholds simultaneously, only a single "1" would be transferred to the accumulator for the final threshold crossing.

The filter bank combiner radiometer is potentially effective against frequency-hopped signals with DPSK modulation or MFSK modulation in which the hops are spread apart by W_H Hz and the modulation is about W_H Hz wide. It would be necessary, in order to use this scheme, to know a priori the hop frequencies and the hopped bandwidth. It would also be necessary to know the hop times, which is very unlikely. In a more practical version either a low pass filter would have to replace the integrator (with some loss of performance) or a dual set of integrators offset in time by $T_H/2$ seconds would have to be employed. The analysis to follow provides the best performance that this approach can hope to obtain. This FBC may also be effective against a narrowband direct sequence signal known to be about W_H Hertz wide.

To do the analysis we note that for adjacent ideal bandpass filters that do not overlap, the outputs are statistically independent. Now let P_{DI} and P_{FAI} denote the individual radiometer detection probabilities for one particular radiometer and for one hop. Then the probability of a false alarm during any one hop time is given by

$$P_{FAH} = 1-(1-P_{FAI})^N \tag{11.2-60}$$

where P_{FAH} is the probability of one or more false alarms, in one hop time, from the N different radiometers. This is the same probability as one minus the probability of no false alarms from the N radiometers. Similarly the probability of a correct detection is equal to 1 minus the probability of N–1 correct rejections (noise only present) times the probability of a missed detection. Thus

$$P_{DH} = 1-(1-P_{FAI})^{N-1}(1-P_{DI}) \tag{11.2-61}$$

where P_{DI} is the probability of a detection in the channel that the hopped signal appears.

Assume that the decisions from hop to hop are statistically independent. The overall probability of false alarm for the message, after the M hops, is given by [16]

$$P_{FA} = \sum_{i=K}^{M}\binom{M}{i}(P_{FAH})^i(1-P_{FAH})^{N-i} \tag{11.2-62}$$

and the overall detection probability is given by [14]

$$P_D = \sum_{i=K}^{M}\binom{M}{i}(P_{DH})^i(1-P_{DH})^{M-i} \tag{11.2-63}$$

In order to set the threshold on the individual radiometer threshold, note from (11.2-60), one has

$$P_{FAH} = 1-\left[1-P_{FAI}\right]^N \cong NP_{FAI} \tag{11.2-64}$$

for small values of P_{FAI}. Then knowing the radiometer threshold one can evaluate the final false alarm probability from (11.2-62), when K is picked. Knowing the radiometer threshold allows the computation of P_{DI} and the overall P_D from (11.2-63). Reference [17] has compared the radiometer and the FBC and found that optimum detector and the filter bank combiner requires less C/N_0 than the radiometer to achieve the same P_D and P_{FA}.

The individual values of P_{DI} and P_{FAI} can be determined from Section 11.2.4.

$$P_{DI} = Q\left[\frac{\beta - \sqrt{WT}\text{SNR}}{\sqrt{1+2\text{SNR}}}\right] \tag{11.2-65}$$

and

$$P_{FAI} = Q(\beta) \tag{11.2-66}$$

where

$$\beta = \left(\frac{T_h - N_0 WT}{N_0\sqrt{WT}}\right) \tag{11.2-67}$$

and the SNR is the signal power (P) divided by the noise power (N_0W); that is,

$$\text{SNR} = \frac{P}{N_0 W} \tag{11.2-68}$$

In conclusion the greater complexity of the FBC or the optimum detector offers better performance when the timing is known, however when timing is not known (as it normally is not) it is not clear there is an advantage when compared to the radiometer. Clearly the radiometer is not sensitive to the hop time; and is only sensitive to the burst transmission time, which is the advantage of the radiometer.

11.3 SPECTRUM ANALYZERS

A spectrum analyzer is probably the simplest (and most readily available or constructed) device that an adversary can use to look for unexpected signals. Typically a portion of the received signal is processed by a discrete Fourier transform (DFT). Figure 11.3-1 illustrates the model for the spectrum analyzer considered here.

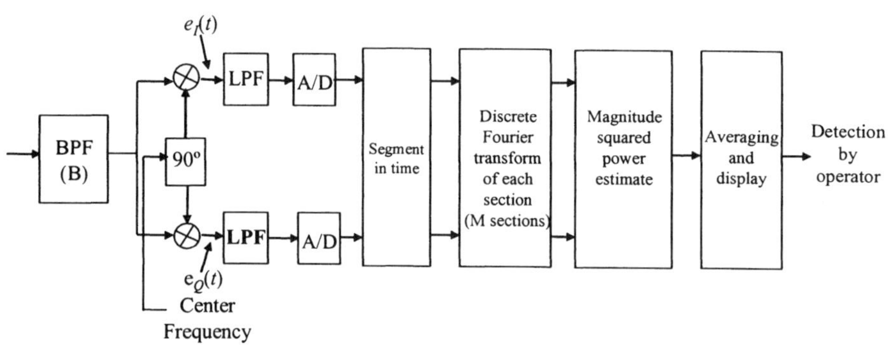

Figure 11.3-1 The spectrum analyzer model.

It is assumed that the bandpass filter is an ideal bandpass filter of bandwidth B Hz. The I and Q baseband components are formed by low-pass filters, which eliminate the $2\omega_0$ terms. The A/D converters sample the signal streams, which are then segmented in time before the DFTs are performed. From the DFTs, the magnitudes are formed for the N frequency bins, which are then averaged and displayed.

This section follows the approach of the work of Betz and Rifkin [7]. A total of NM samples of the filtered signal segment $\{x(k)\}_{k=0}^{MN-1}$ is segmented into M sequences of N samples each. The samples are spaced apart by Δt seconds, where $\Delta t = 1/B$ so that the noise samples are statistically independent (see the appendix in this chapter for more details on the statistical independence). Thus the total time span is $NM\Delta t$ seconds. The sequences are considered to consist of contiguous samples, without overlap. The j-th subsequence is denoted by

$$\{x_j(m)\}_{m=0}^{N-1} = \{x(k)\}_{k=jN}^{(j+1)N-1} \tag{11.3-1}$$

Taking the DFT of the j-th subsequence one obtains

$$X_j(n) = \sum_{k=0}^{N-1} x_j(k) e^{-i2\pi nk/N} \tag{11.3-2}$$

It will be assumed that the spectrum analyzer displays averaged power estimates of the form

$$\frac{1}{M}\sum_{j=0}^{M-1}\left|X_j(n)\right|^2 \tag{11.3-3}$$

for each frequency bin "n." The frequency resolution of the spectrum analyzer is approximately $1/(N\Delta t) = B/N$, since the bins are separated by B/N Hz. Thus, if the averaging is done for each frequency bin "n," then the test statistic is given by the set of N frequencies

$$\frac{1}{M}\sum_{j=0}^{M-1}\left|X_j(n)\right|^2 \tag{11.3-4}$$

It will be assumed initially that the input signal is an unmodulated tone and additive white Gaussian noise, so that

$$y(t) = \sqrt{2}A\cos[(\omega_0 + \Delta\omega)t] + \sqrt{2}n_1(t)\cos(\omega_0 t + \theta) + \sqrt{2}n_2(t)\sin(\omega_0 t + \theta) \tag{11.3-5}$$

The low-passed versions of the I and Q signals, along with the noise components for the tone, are given by

$$e_I(t)\big|_{LPF} = A\cos(\Delta\omega t) + n_1(t)\cos\theta + n_2(t)\sin\theta \tag{11.3-6}$$

and

$$e_Q(t)\big|_{LPF} = -A\sin(\Delta\omega t) - n_1(t)\sin\theta + n_2(t)\cos\theta \tag{11.3-7}$$

Now the sum of the two noise terms in (11.3-6) and (11.3-7) can be shown equal to two uncorrelated noise terms with essentially the same statistics as either $n_1(t)$ or $n_2(t)$. Thus, the low-passed versions can be rewritten as

$$e_I(t)\big|_{LPF} = A\cos(\Delta\omega t) + n_I(t) \tag{11.3-8}$$

and

$$e_Q(t)\big|_{LPF} = -A\sin(\Delta\omega t) + n_Q(t) \tag{11.3-9}$$

where $n_Q(t)$ and $n_I(t)$ are statistically independent white Gaussian noise processes, each of which have zero means and power spectral densities of $N_0/2$, and are independent from sample to sample (the autocorrelation function for each noise term in (11.3-5) is given by $2R_{n_c}(\tau)\cos(\omega_0\tau)$. Since the spectrum analyzer is modeled with a human operator making detection decisions, there are no thresholds used in the determination of detection.

11.3.1 Narrowband Signal Spectrum Analyzer Performance

For the case of a unmodulated tone, the I-Q samples written as a complex number are given by

$$x_j(k) = A\cos(\Delta\omega t_k) + n_I(t_k) + i\left(-A\sin(\Delta\omega t_k) + n_Q(t_k)\right) \tag{11.3-10}$$

From (11.3-2) and (11.3-10) the $X_j(n)$ are seen to be complex Gaussian random variables with mean given by

$$E\left[X_j(n)|H_1\right] = \sum_{k=0}^{N-1} Ae^{-i2\pi\Delta f t_k}e^{-i2\pi nk/N} \tag{11.3-11}$$

when the signal is present and by

$$E\left[X_j(n)|H_0\right] = 0 \tag{11.3-12}$$

when the signal is not present. The condition H_0 denotes the fact that the signal is absent and the condition H_1 denotes the fact that the signal is present. The variance of the complex random variable, under hypothesis H_0, is given by

$$E\left[\left|X_j(n)\right|^2|H_0\right] = E\left[\sum_{k=0}^{N-1}\sum_{l=0}^{N-1}\left(n_I(t_k) - in_Q(t_k)\right)\left(n_I(t_l) + in_Q(t_l)\right)e^{-i2\pi n(k-l)/N}\right] \tag{11.3-13}$$

Since the $n_I(t)$ and the $n_Q(t)$ are statistically independent, and since each sample of each process is statistically independent, one has under H_0

$$E\left[\left|X_j(n)\right|^2|H_0\right] = NN_0B \tag{11.3-14}$$

where B is the bandwidth of the ideal bandpass filter preceding the spectrum analyzer. Thus $X_j(n)$ is a Gaussian complex random variable with mean given by either (11.3-11) with signal present or is zero with the signal absent; the variance of the signal with the signal absent is given by NN_0B. Two cases will be considered: (1) the narrowband case and (2) the wideband case.

For signals that are narrowband, there have been limited experimental detection experiments for binary spectrograms. Webster [20] has determined that the minimum detectable narrowband signal for visual detection of a spectrogram produces an output SNR (the deflection) of approximately 0 dB per DFT when $M = 100$. The output SNR at the n-th frequency bin is defined by the deflection[2]

$$\text{SNR}_0(n) = \frac{\left\{ E\left(\frac{1}{M}\sum_{j=0}^{M-1} |X_j(n)|^2 \,|H_1 \right) - E\left(\frac{1}{M}\sum_{j=0}^{M-1} |X_j(n)|^2 \,|H_0 \right) \right\}^2}{Var\left(\frac{1}{M}\sum_{j=0}^{M-1} |X_j(n)|^2 \,|H_0 \right)} \tag{11.3-15}$$

Based on [20], the minimal detectable condition is given by the case when M = 100 and $\text{SNR}_0(n)$ = 0 dB. Now the variable

$$y = |X_j(n)|^2 \tag{11.3-16}$$

has mean NN_0B from (11.3-14), when the signal is not present, and the variance, when the signal is not present, is given by [13]

$$\sigma_y^2 = 4\left(\frac{N_0NB}{2} \right)^2 = (NN_0B)^2 \tag{11.3-17}$$

In the case that the signal is present, and is in one of the DFT bins, the mean output is given by

$$E\left[y|H_1 \right] \cong NN_0B + N^2P \tag{11.3-18}$$

where P is the signal power. If the narrowband signal is between the DFT bins the effective input SNR could be reduced as much as about 3.9 dB. Hence the output SNR, or deflection, is given from (11.3-14), (11.3-17), and (11.3-18) to be

$$\text{SNR}_0 = \frac{\left(N^2P\right)^2}{\frac{1}{M}N_0^2N^2B^2} = MN^2\left(\text{SNR}_i\right)^2 \tag{11.3-19}$$

where SNR_i is the input SNR ($\text{SNR}_i = P/N_0B$) in the bandwidth B. Solving for the input SNR needed to obtain detection using a spectrum analyzer (with $\text{SNR}_0 = 1$) produces (M =100)

$$SNR_i = \frac{1}{10N} \tag{11.3-20}$$

11.3.2 Wideband Signal Spectrum Analyzer Performance

In the wideband spectral analyzer case, it will be assumed that the signal power spectral density is spread over a bandwidth that is large compared to the resolution bandwidth and is essentially flat. If the signal occupies Q frequency bins in spectral extent then the output power will be distributed over Q frequency bins so that each bin will be $1/Q$ as large as before. Therefore, from the deflection, the output SNR in each bin will be approximately given by

$$\text{SNR}_0 = \frac{MN^2\left(\text{SNR}_i\right)^2}{Q^2} \tag{11.3-21}$$

Thus the input SNR needed to achieve an output SNR of 1 in the wideband case is given by (M =100)

$$(SNR_i) = \frac{Q}{10N} \tag{11.3-22}$$

11.4 SECOND-ORDER CYCLOSTATIONARY FEATURE DETECTION

In this section nonlinear transformations applied to the received signal plus noise are considered. These transformations will generate spectral lines, one or more of which can be detected. Typically these spectral lines are multiples of the chip rate or the carrier frequency in the case of direct sequence signals, or the hop rate or modulation rate in the case of frequency-hopped signals. These spectral lines can be used to detect the LPI signal just as a radiometer can be used for this purpose. Depending on how much is known a priori about the signal, the performance can rival or exceed the performance of a radiometer.

11.4.1 Cyclostationary Processes

We will be dealing with wide-sense cyclostationary processes. Therefore, a brief introduction to cyclostationary processes will be presented now. A *wide-sense cyclostationary process* [21] has the property that the autocorrelation function and the mean are periodic; that is,

$$\begin{aligned} R(t_1+T, t_2+T) &= R(t_1, t_2) \\ E[X(t+T)] &= E[X(t)] \end{aligned} \tag{11.4-1}$$

where T (= $2\pi/\omega_0$) is the period of the periodicity. These processes are cyclic in the period T, and are not wide-sense stationary. The fact that they are periodic implies that they possess a Fourier series expansion. The tones (harmonics) in the expansion are what typically are detected in cyclostationary detectors. The frequencies of these tones are called the *cycle frequencies* and the set of frequencies is called the *cycle spectrum* [7, 21].

11.4.2 The Baseband and Carrier Cyclostationarity

The narrowband signal will be modeled as

$$s(t) = \text{Re}\left\{v(t)e^{i\theta}e^{i\omega_0 t}\right\} \tag{11.4-2}$$

where ω_0 is the angular center frequency and θ is the phase angle of the signal. It is assumed that both parameters are constants. The baseband signal $v(t)$ is the complex envelope of the spreading code and is complex in general for any signal but BPSK. It will also be assumed that the spreading code random variables in $v(t)$ will take on the values $\pm$ 1 with equal likelihood and are statistically independent from one chip to any other chip. Let us consider the autocorrelation function for the signal $s(t)$. We have, taking the expectation

$$R_{ss}(t, t-\tau) = E\left\{\text{Re}\left[v(t)e^{i(\omega_0 t+\theta)}\right]\text{Re}\left[v(t-\tau)e^{i(\omega_0(t-\tau)+\theta)}\right]\right\} \tag{11.4-3}$$

which can be written as

$$E\left\{\left[\frac{1}{2}v(t)e^{i(\omega_0 t+\theta)}+\frac{1}{2}v^*(t)e^{-i(\omega_0 t+\theta)}\right]\left[\frac{1}{2}v(t-\tau)e^{i(\omega_0(t-\tau)+\theta)}+\frac{1}{2}v^*(t-\tau)e^{-i(\omega_0(t-\tau)+\theta)}\right]\right\} \quad (11.4\text{-}4)$$

Equation (11.4-4) can be written as

$$R_{ss}(t,t-\tau)=\frac{1}{2}\text{Re}\left[R_{vv}(t,t-\tau)e^{+i(2\omega_0 t-\omega_0\tau+2\theta)}\right]+\frac{1}{2}\text{Re}\left[R_{vv^*}(t,t-\tau)e^{+i\omega_0\tau}\right] \quad (11.4\text{-}5)$$

in which the two code spreading autocorrelation functions are given by

$$R_{vv}(t,t-\tau)=E\{v(t)v(t-\tau)\} \quad (11.4\text{-}6)$$

$$R_{vv^*}(t,t-\tau)=E\{v(t)v^*(t-\tau)\} \quad (11.4\text{-}7)$$

The covariance of (11.4-6) in this book will be called the *carrier covariance function* and the covariance of (11.4-7) will be called the *baseband covariance function*. The baseband covariance function deals with tones that arise at or near baseband, while the carrier covariance function deals with the tones that are near twice the carrier frequency. The carrier covariance function has also been called the conjugate covariance, the pseudo covariance, the zero-covariance, and the carrier cyclostationary. The baseband covariance has also been called the *unconjugate covariance*, the *covariance*, and the *one-conjugate covariance* [7].

Evaluating the complex envelope ($v(t)$) in (11.4-5) determines whether the output process has periodic spectral components in the output or not.

11.4.3 BPSK Through a Filter and Squarer Circuit

Consider now a model for tone detection based on a filter and squarer circuit for BPSK that produces the line spectra and followed by a phase locked loop (PLL) with a lock detector, as shown in Figure 11.4-1. If the PLL can lock up to a line component from the squaring operation, the lock detector will produce an output declaring lock, indicating that a BPSK signal has been found. The low-pass filters (LPF) after both multipliers are there to remove the sum frequency terms. It will be assumed that the PLL has sufficient tone signal level to remain in lock. This means that $P_{tone}/N_0' B_L > 10$ dB, where P_{tone} is the power of the generated tone, and N_0' is the one-sided (positive frequency) effective noise spectral density, and B_L is the one-sided loop noise bandwidth expressed in Hz.

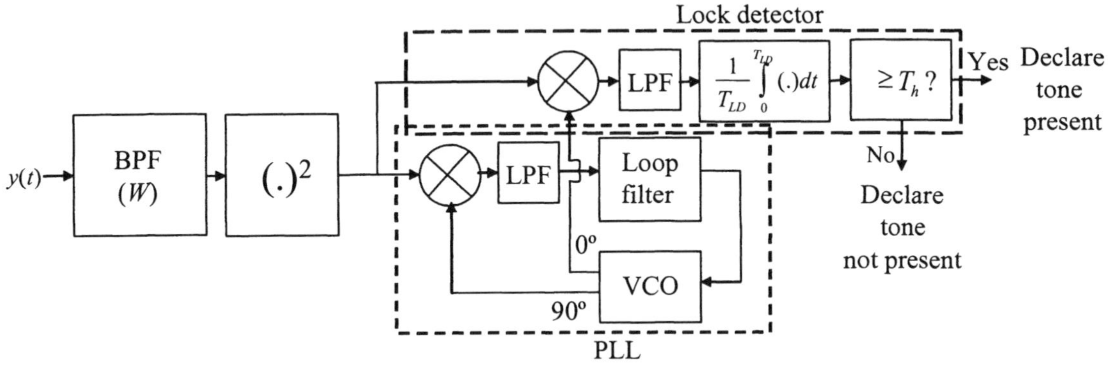

Figure 11.4-1 The filter and square tone detector with PLL and lock detector.

The complex envelope for the code process is given by

$$v(t) = \sqrt{2}A\sum_k a_k \tilde{p}_T(t-kT) \tag{11.4-8}$$

where the tilde on the pulse sequence denotes that it is filtered by the input filter of bandwidth W. Chapter 1 discusses the relationship between the unfiltered pulse and the filtered pulse ($\tilde{p}_T(t)$) to the input pulse ($p_T(t)$) convolved with the low-pass equivalent filter impulse response $h_L(t)$. Considering only the signal as the input, the output will be just $s^2(t)$. The signal portion is defined to be the expected value of the output, so that the signal output is given by

$$E\left[s^2(t)\right] = R_{ss}(t,t) \tag{11.4-9}$$

From (11.4-5) it follows that the signal output is given by

$$R_{ss}(t,t) = \frac{1}{2}\mathrm{Re}\left[R_{vv}(t,t)e^{+i(2\omega_0 t+2\theta)}\right] + \frac{1}{2}\mathrm{Re}\left[R_{vv^*}(t,t)\right] \tag{11.4-10}$$

which can be written as two terms

$$R_{ss}(t,t) = R_{car}(t,t) + R_{BB}(t,t) \tag{11.4-11}$$

with the first term being the one that produces a double carrier frequency component, and the second term, the one that produces the baseband terms. That is,

$$R_{car}(t,t) = \frac{1}{2}\mathrm{Re}\left[R_{vv}(t,t)e^{+i(2\omega_0 t+2\theta)}\right] \tag{11.4-12}$$

and

$$R_{BB}(t,t) = \frac{1}{2}\mathrm{Re}\left[R_{vv^*}(t,t)\right] \tag{11.4-13}$$

Evaluating $R_{vv}(t,t)$ and $R_{vv^*}(t,t)$ produces

$$R_{vv}(t,t) = 2A^2\sum_{k=-\infty}^{\infty}[\tilde{p}_T(t-kT)]^2 \tag{11.4-14}$$

$$R_{vv^*}(t,t) = 2A^2\sum_{k=-\infty}^{\infty}[\tilde{p}_T(t-kT)][\tilde{p}_T^*(t-kT)] \tag{11.4-15}$$

Both of these equations can be shown to be nonzero. In fact since the first one has a dc value, from (11.4-10) there will be a term at twice the carrier frequency. In addition since (11.4-15) is periodic there will be spectral lines at baseband. To evaluate the "twice the carrier" term, expand $(1/2)(R_{vv}(t,t))$ in a complex exponential series to produce

$$A^2\sum_{k=-\infty}^{\infty}\left[\tilde{p}_T(t-kT)\right]^2 = \sum_{n=-\infty}^{\infty}c_n e^{i2\pi nt/T} \tag{11.4-16}$$

whereupon letting $P = A^2$, one has

$$c_n = P\left[\frac{1}{T}\int_{-T/2}^{T/2}\sum_{k=-\infty}^{\infty}[\tilde{p}_T(t-kT)]^2 e^{-i2\pi nt/T}dt\right] \tag{11.4-17}$$

and the term in the brackets is the Fourier coefficient, for the unit power expansion. Equation (11.4-17) can be evaluated as

$$c_n = \frac{P}{T}\int_{-\infty}^{\infty}[\tilde{p}_T(t)]^2 e^{-i2\pi nt/T}dt \tag{11.4-18}$$

Note that $\tilde{p}_T(t)$ is defined by the inverse Fourier transform

$$\tilde{p}_T(t) = \int_{-\infty}^{\infty} H_L(f)P_T(f)\exp(i2\pi ft)df \tag{11.4-19}$$

and furthermore $P_T(f)H_L(f)$ is the Fourier transform of $\tilde{p}_T(t)$ so that

$$P_T(f)H_L(f) = \int_{-\infty}^{\infty}\tilde{p}_T(t)\exp(-i2\pi ft)dt \tag{11.4-20}$$

In addition, $p_T(t)$ is the inverse Fourier transform of $P_T(f)$.

In Problem 5, using (11.4-19) in (11.4-18), it is shown that c_n can be written in the frequency domain, as

$$c_n = \frac{P}{T}\int_{-\infty}^{\infty}H_L(f)P_T(f)H_L(n/T-f)P_T(n/T-f)df \tag{11.4-21}$$

where $H_L(f)$ is the low-pass equivalent of the bandpass filter. The spectral line component at $2\omega_0$, for example, is obtained when $n = 0$ in (11.4-21) and its value is given by

$$c_0 = \frac{P}{T}\int_{-\infty}^{\infty}H_L(f)P_T(f)H_L(-f)P_T(-f)df \tag{11.4-22}$$

If the low-pass equivalent filter is *complex conjugate symmetric* (that is, $H_L(-f) = H_L^*(f)$), which implies that $h_L(t)$ is real, then

$$c_0 = \frac{P}{T}\int_{-\infty}^{\infty}\left|H_L(f)\right|^2\left|P_T(f)\right|^2 df \tag{11.4-23}$$

The spectral line terms near twice the carrier frequency are given by (11.4-12) as

$$e_{2f_0}(n,t) = \mathrm{Re}\left[c_n e^{+i(2\omega_0 t+2\pi n/T+2\theta)}\right] \tag{11.4-24}$$

Writing c_n has a magnitude and a phase, in the form

$$c_n = \left|c_n\right|e^{i\theta_{c_n}} \tag{11.4-25}$$

one can express (11.4-24) as

$$e_{2f_0}(n,t) = |c_n|\cos(2\pi(2f_0 + n/T)t + 2\theta + \theta_{c_n}) \tag{11.4-26}$$

where θ_{c_n} is the phase associated with c_n. The power of this tone near $2f_0$ is given by

$$P_{2f_0+\frac{n}{T}} = \frac{|c_n|^2}{2} \tag{11.4-27}$$

Clearly there is always a tone present at exactly twice the carrier frequency for any reasonable filter (if $c_0 \neq 0$). The double carrier frequency can always be detected with a squarer circuit. In general, for BPSK out of the squarer, there are spectral lines generated at the following frequencies near $2f_0$:

$$\text{terms at } 2f_0 = 2f_0 \pm n/T \tag{11.4-28}$$

where n = 0, 1, 2, 3, 4,... Normally, the larger the value of n, the weaker the strength of the n-th spectra line.

In addition to the cyclostationary production of spectral lines, in any real system the ones and zeros will not be perfectly balanced, and the mixers will have leakage, so that both ω_0 and $2\omega_0$ will be present to some extent in "real" systems. In a well-designed system, leakages and imbalances would normally be minimized to enhance LPI transmission.

Now consider the baseband terms, which are derived from the second term in (11.4-10). To evaluate the baseband terms, expand $(1/2)(R_{vv^*}(t,t))$ in a complex exponential series to produce

$$A^2 \sum_{k=-\infty}^{\infty} [\tilde{p}_T(t-kT)][\tilde{p}_T^*(t-kT)] = \sum_{k=-\infty}^{\infty} d_n e^{i2\pi nt/T} \tag{11.4-29}$$

where ($P = A^2$) the Fourier coefficients of the baseband spectral lines are given by

$$d_n = \frac{P}{T}\int_{-T/2}^{T/2} \sum_{k=-\infty}^{\infty} [\tilde{p}_T(t-kT)]\left[\tilde{p}_T^*(t-kT)\right] e^{-i2\pi nt/T} dt \tag{11.4-30}$$

In the same manner that c_n was obtained, it can be shown that d_n is given by

$$d_n = \frac{P}{T}\int_{-\infty}^{\infty} H_L(f)P_T(f)H_L^*(n/T - f)P_T^*(n/T - f)df \tag{11.4-31}$$

Writing d_n as

$$d_n = |d_n| e^{i\theta_{d_n}} \tag{11.4-32}$$

from (11.4-29) it follows that the tone at n/T Hz is given by

$$e_{n/T}(t) = |d_n|\cos(2\pi nt/T + \theta_{d_n}) \tag{11.4-33}$$

which has power

$$P_{n/T} = \frac{|d_n|^2}{2} \tag{11.4-34}$$

It is clear from (11.4-31) that there are spectral lines at all multiples of $1/T$; that is,

$$\text{baseband terms} = \pm n/T \quad \text{where } n = 1,2,3,4,... \tag{11.4-35}$$

which does not include the dc term, since it contains no information about the chip rate.

One of the first things to recognize from (11.4-31) is that if the received pulse spectrum ($H_L(f)P_T(f)$) is spectrally limited to $\pm 1/(2T)$ then there will be no line component at $f = 1/T$ or at multiples of $1/T$, since from (11.4-31) the product of disjoint spectra produces a zero value of the integral. Thus a signal designer could spectrally limit the transmitted signal by filtering to a total baseband (positive frequency) bandwidth of $1/2T$; then an adversary could not detect a rate line at $f = 1/T$, or multiples, since none would exist.[3]

Equation (11.4-31) applies to the product of the receiver filtering $H_L(f)$ and the transmitted pulse, which may also be filtered. Thus d_n may be interpreted as an integration dependent on the unfiltered pulse $P(f)$, and the composite filter due to both the transmitter and receiver filters via $H_L(f)$, as long as the channel is linear (a TWTA would have to be backed off to its linear region, for example). With heavy filtering on the transmitted signal, the receiver would be forced to detect the second line at $f = 2/T$, which is normally much weaker than a mildly filtered tone at $f = 1/T$, or use a fourth power detector which has lower output SNR and is therefore harder to detect.

Another thing to notice from (11.4-18), (11.4-26), (11.4-30), and (11.4-34) is that the power output of these spectral lines are proportional to the square of the input power, P. A fourth power detector output power would be proportional to the fourth power of the input power.

Example 3 As an example of the evaluation of (11.4-31) consider unit pulses of duration T seconds and a low-pass equivalent filter that is an ideal unit response filter of bandwidth $2\alpha/T$. Note that if $\alpha = 1$ then the filter would cover the null-to-null bandwidth and if $\alpha = 1/2$ the bandwidth would be 1/2 the null-to-null bandwidth. It is shown in Problem 6 that the value of d_1 is given by

$$d_1(\alpha)\big|_{\alpha \geq 0.5} = \frac{P}{\pi}\int_{(1-\alpha)\pi}^{\alpha\pi} \frac{\sin(x)}{x}\frac{\sin(x-\pi)}{(x-\pi)}dx, \text{ and } d_1(\alpha)\big|_{\alpha \leq 0.5} = 0 \tag{11.4-36}$$

Figure 11.4-2 illustrates the signal amplitude for the first spectral line component (d_1) at $f = 1/T$ versus the parameter α for $P = 1$.

Note that when the ideal low-pass filter equivalent is $1/T$ wide ($\alpha = 0.5$), the amplitude goes to zero. For values of α less than 0.5 the amplitude is zero. The maximum occurs at about $\alpha = 1$. However, the wider the filter bandwidth, the more noise or interference is accepted by the rate line detector. This affects the detection and false alarm performance. In general for most reasonable filters and signals, increasing n decreases the output signal strength. Hence, normally the first spectral line is the one chosen for detection, since it is the strongest.

11.4.3.1 BPSK Through a Filter and Squarer Circuit Detection and False Alarm Probabilities

Now let us consider the noise performance of the double carrier frequency term at low input signal-to-noise ratios, based on the PLL lock detector. Figure 11.4-1 illustrates the detector for the lines generated by the squaring process. It is composed of a phase lock loop (PLL), which tracks the desired line component, and the lock detector, which signifies when the PLL is in lock. Thus the PLL lock detector is the device that

---

[3] In practice one cannot build a real filter to respond exactly like an ideal filter; however, it is possible to approximate it by a high order filter and cause the output level of the "filter and square" circuit to be greatly reduced.

indicates that the BPSK signal has been detected. It is to be noted that typically the PLL will require that the VCO be frequency swept over the expected tone frequency range in order to achieve PLL lock. Thus in the search mode the frequency sweeping will occur until the PLL acquires the signal and the lock detector indicates a lock that would then turn off the sweep function. If the lock detector indicates lack of lock, the sweep process would be initiated again.

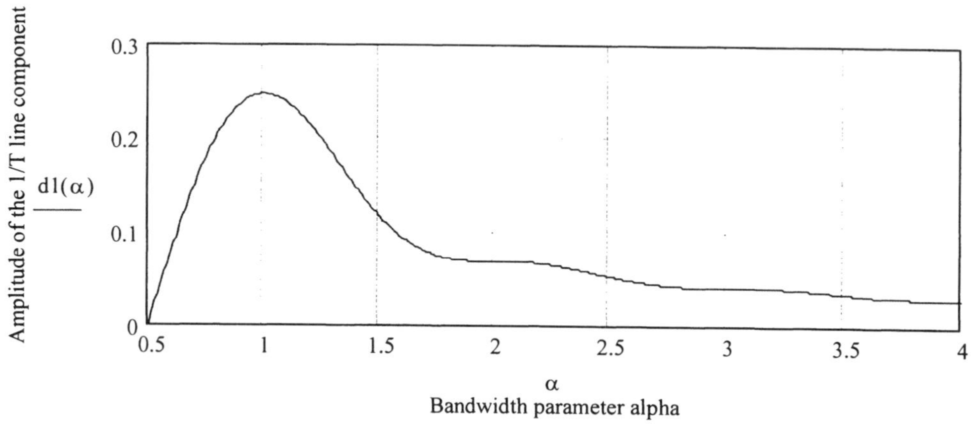

Figure 11.4-2 Signal amplitude of the first spectral line at $f=1/T$.

We can model the received signal plus white Gaussian thermal noise with two-sided noise power spectral density of $N_0/2$ as

$$y(t)=S(t)+\sqrt{2}n_c(t)\cos(\omega_0 t)+\sqrt{2}n_s(t)\sin(\omega_0 t) \tag{11.4-37}$$

Out of the squaring function we have, generically speaking, the three terms

$$S\times S+n\times n+S\times n \tag{11.4-38}$$

The first term (the signal component) out of the squarer has been evaluated in (11.4-22) and has power given by

$$P_{c_0}=\frac{|c_0|^2}{2} \tag{11.4-39}$$

To simplify the analysis we will assume that the input noise power is much greater than the input signal power. This is a reasonable assumption in this type of unauthorized detection work, where the signal term is very weak typically and allows us to neglect the $S\times n$ term in what follows. The $n\times n$ term can be easily evaluated by using a result for the $n\times n$ spectral density out of the squarer, given in Davenport and Root [22]. The $n\times n$ spectral density out of the squarer in the region of $2\omega_0$ is given by [22]

$$N_0^{2\omega_0}=N_0^2 W \tag{11.4-40}$$

Under the assumption that $WT_{LD} >> 1$ and $W/B_L >> 1$, where B_L is the one-sided loop noise bandwidth of the PLL in Hz, we may assume that the two-sided noise spectral density out of the squarer is approximately flat and has value

$$\frac{N_0^{2\omega_0}}{2} = \frac{N_0^2 W}{2} \tag{11.4-41}$$

We will assume that the PLL has sufficient input tone power to noise spectral density to maintain PLL lock.[4] We may model the input line component plus squared thermal noise into the coherent amplitude detector (the input to the lock detector multiplier) as

$$e_i(t) = \sqrt{2}\left|c_0\right|\sin(2\omega_0 t+\varphi)+\sqrt{2}n_s(t)\sin(2\omega_0 t+\varphi)+\sqrt{2}n_c(t)\cos(2\omega_0 t+\varphi) \tag{11.4-42}$$

Modeling the inphase reference signal (at 0°) as

$$\sqrt{2}\sin(2\omega_0 t+\varphi) \tag{11.4-43}$$

produces the input to the lock detector integrator as

$$e_{inLD}(t) = \left|c_0\right| + n_s(t) \tag{11.4-44}$$

The detected line component out of the integrator is given by

$$e_{outLD} = \left|c_0\right| \tag{11.4-45}$$

Under the assumption that $WT_{LD} >> 1$ we will assume that the baseband noise term $n_s(t)$ is approximately white, so that

$$\sigma^2 = \frac{1}{T_{LD}^2} E\left\{\int_0^{T_{LD}}\int_0^{T_{LD}} n_s(u)n_s(t)dtdu\right\} = \frac{1}{T_{LD}^2}\int_0^{T_{LD}}\int_0^{T_{LD}} E\left\{n_s(u)n_s(t)\right\}dtdu \tag{11.4-46}$$

Now this can be evaluated by noting that

$$\sigma^2 = \frac{1}{T_{LD}^2}\int_0^{T_{LD}}\int_0^{T_{LD}} \frac{N_0^2}{2}\delta(u-t)dtdu = \frac{N_0^2}{2T_{LD}^2}\int_0^{T_{LD}} dt = \frac{N_0^2 W}{2T_{LD}} \tag{11.4-47}$$

Since it is Gaussian, with the mean value given in (11.4-45) and with the variance given by (11.4-47), it can be shown in Problem 7 that the detection and false alarm probabilities are given by

$$P_D \cong Q\left(\frac{T_h - \left|c_0\right|}{N_o\sqrt{\dfrac{W}{2T_{LD}}}}\right) \qquad P_{FA} \cong Q\left(\frac{T_h}{N_o\sqrt{\dfrac{W}{2T_{LD}}}}\right) \tag{11.4-48}$$

and the units of $\left|c_0\right|$ and T_h must be in input power units, as can be seen from (11.4-18). The value of σ (the denominator in the probability expressions) is also in units of input noise power. The value of c_0 was given in (11.4-22). It was not necessary to pick the second harmonic as our choice in the line component to detect; lines near the baseband could as well have been detected. In general, it is only necessary to analyze a few components to determine which are most detectable in noise, in the sense of having the best SNR.

---

[4] Clearly both the PLLs must maintain lock in order for the lock detector to function properly.

11.4.4 Balanced QPSK Through a Filter and Squarer Circuit

Let us now consider balanced (equal power in the inphase and quadrature components) QPSK chip modulation through a filter and squarer to investigate the spectral lines at the output. Recall the QPSK chip modulated signal has a complex envelope of the form

$$v(t) = A\sum_{k}\left[a_k \tilde{p}_T(t-kT) + ib_k \tilde{p}_T(t-kT)\right] \tag{11.4-49}$$

where the tilde on the pulse sequences again denotes the fact that the pulse process is filtered by the input bandpass filter. Considering only the signal as the input, by neglecting the noise, the output will again be just $s^2(t)$. The signal portion of the output is defined to be the expected value of the output, so that the signal output is given by

$$E\left\{s^2(t)\right\} = R_{ss}(t,t) \tag{11.4-50}$$

From (11.4-5) it follows that, as before, the signal output is given by

$$R_{ss}(t,t) = \frac{1}{2}\mathrm{Re}\left[R_{vv}(t,t)e^{+i(2\omega_0 t+2\theta)}\right] + \frac{1}{2}\mathrm{Re}\left[R_{vv^*}(t,t)\right] \tag{11.4-51}$$

with the first term producing a double carrier frequency component and the second term producing the baseband terms. Evaluating the baseband and carrier covariance functions produces

$$R_{vv}(t,t) = A^2\sum_{k=-\infty}^{\infty}[\tilde{p}_T(t-kT)]^2 - A^2\sum_{k=-\infty}^{\infty}[\tilde{p}_T(t-kT)]^2 = 0 \tag{11.4-52}$$

$$R_{vv^*}(t,t) = 2A^2\sum_{k=-\infty}^{\infty}[\tilde{p}_T(t-kT)][\tilde{p}_T^*(t-kT)] \tag{11.4-53}$$

From the first term of (11.4-51) and (11.4-52) it is seen that there will be no double carrier frequency term, or any frequencies near $2f_0$ (i.e., $R_{vv}(t,t)$ is zero). In addition, since (11.4-53) is periodic and nonzero, there will be spectral lines at baseband. Since (11.4-53) is the same expression as for BPSK, the same spectral line results apply to the QPSK case. Thus, using $P = A^2$, one obtains

$$d_n = \frac{P}{T}\int_{-\infty}^{\infty} H_L(f)P_T(f)H_L^*(f-n/T)P_T^*(f-n/T)df \tag{11.4-54}$$

Again writing d_n as in (11.4-32) the spectral lines can be written in the form

$$e_{n/T}(t) = |d_n|\cos(2\pi nt/T + \varphi_n) \tag{11.4-55}$$

which has power

$$P_{n/T} = \frac{|d_n|^2}{2} \tag{11.4-56}$$

Therefore, just as in the BPSK case, spectral lines at baseband for QPSK are generated at the following frequencies

$$\text{baseband terms} = \pm\, n/T \tag{11.4-57}$$

where n = 1, 2, 3, 4, ... Again the dc term was not included since it does not provide any information about the chip rate. It should be noticed that if QPSK were not power balanced, then lines would also appear near $2f_0$.

Now consider the detectability of a line spectral component near baseband. Consider one of the line components (d_n) near zero frequency, but still excluding zero frequency. We again assume the model of Figure 11.4-1, which is assumed to be centered at $\pm\, n/T$ Hz. Again the $n \times n$ spectral density out of the squarer in the region of n/T Hz is given by [22]

$$N_0^{'} = N_0^2 W \tag{11.4-58}$$

if $WT >> 1$. The corresponding noise power out of the lock detector averaging-filter is given by

$$\sigma^2 = \frac{N_0^2 W}{2T_{LD}} \tag{11.4-59}$$

The mean and variance of the power out of the averaging-filter is given by

$$\begin{aligned} \mu &= \frac{|d_n|^2}{2} \\ \sigma^2 &= \frac{N_0^2 W}{2T_{LD}} \end{aligned} \tag{11.4-60}$$

When the signal is not present, μ becomes zero. The value d_n is given by (11.4-54) for all nonzero n. Following the approach in Section 11.4.3.1 we find that the approximate (Gaussian approximation) detection and false alarm probabilities are given by

$$P_D \cong Q\left(\frac{T_h - |d_n|}{N_0\sqrt{\dfrac{W}{2T_{LD}}}}\right) \qquad P_{FA} \cong Q\left(\frac{T_h}{N_0\sqrt{\dfrac{W}{2T_{LD}}}}\right) \tag{11.4-61}$$

where d_n is given by (11.4-54), and $n = \pm 1, \pm 2, \pm 3, \ldots$

11.4.5 Balanced OQPSK Through a Filter and Squarer Circuit

Let us now consider balanced offset quadriphase shift keying (OQPSK) chip modulation (also known as quadrature phase shift keying) through a filter and squarer, for spectral lines at the output. Recall the OQPSK chip modulated signal has a complex envelope of the form

$$v(t) = A\sum_k \left[a_k \tilde{p}_T(t-kT) + ib_k \tilde{p}_T(t-kT-T/2)\right] \tag{11.4-62}$$

In Problem 8 it is shown that the spectral lines near $2f_0$ are of the form

$$e_{2f_0}(n,t) = |c_n|\cos(2\pi(2f_0 - n/T)t + 2\theta + \theta_{c_n}) \tag{11.4-63}$$

where $P=A^2$ and

$$c_n = \frac{P}{2T}\left[1-e^{i\pi n}\right]\int_{-\infty}^{\infty} H_L(f)P_T(f)H_L(n/T-f)P_T(n/T-f)df \qquad (11.4\text{-}64)$$

Notice that when n is even, $c_n = 0$ and therefore the double carrier frequency spectral line is not present. However frequency terms near $2f_0$ are

$$\text{near by } 2f_0 \text{ terms} = 2f_0 \pm (2n-1)/T \qquad (11.4\text{-}65)$$

when $n = 1, 2, 3, 4,\ldots$ That is to say, the spectral lines for OQPSK occur at odd multiples of twice the chip rate, near twice the carrier frequency. The results of Problem 8 show that the spectral lines near baseband are of the form

$$e_{n/T}(t) = |d_n|\cos(2\pi nt/T + \varphi_n) \qquad (11.4\text{-}66)$$

and their Fourier coefficients are given by

$$d_n = \left(1+e^{-i\pi n}\right)\frac{P}{2T}\int_{-\infty}^{\infty} H_L(f)P_T(f)H_L^*(f-n/T)P^*(f-n/T)df \qquad (11.4\text{-}67)$$

From these Fourier coefficients it is seen that only even values of n produce nonzero values of d_n. Thus the baseband frequencies that exist for OQPSK are

$$\text{baseband terms} = \pm 2n/T \quad \text{with} \quad n = 1,2,3,4\ldots \qquad (11.4\text{-}68)$$

which is due to the fact that the RF phase of an OQPSK signal can change every one-half chip time, rather than every chip time as in QPSK. Thus it is seen that OQPSK and QPSK are very different in their spectral line characteristics, as far as detectability.

It is not difficult, assuming that baseband spectral lines are to be detected, to have with a Gaussian approximation the following approximate detection and false alarm probabilities

$$P_D \cong Q\left(\frac{T_h - |d_n|}{N_0\sqrt{\dfrac{W}{2T_{LD}}}}\right) \qquad P_{FA} \cong Q\left(\frac{T_h}{N_0\sqrt{\dfrac{W}{2T_{LD}}}}\right) \qquad (11.4\text{-}69)$$

and n takes on the values $\pm 2, 4, 6,\ldots$ Also d_n is given by (11.4-67).

11.4.6 MSK Through a Filter and Squarer Circuit

Let us now consider MSK chip modulation through a filter and squarer for spectral lines, at the output. Recall from Chapter 1 that the MSK chip modulated signal has a complex envelope of the form

$$v(t) = \sqrt{2}A\sum_k\left[a_k\tilde{p}_T(t-kT) + ib_k\tilde{p}_T(t-kT-T/2)\right] \qquad (11.4\text{-}70)$$

In a manner similar to the solution of Problem 8 it is shown that the spectral lines near $2f_0$ are of the form

$$e_{2f_0}(n,t) = |c_n|\cos(2\pi(2f_0 - n/T)t + 2\theta + \theta_{c_n}) \quad (11.4\text{-}71)$$

where

$$c_n = \frac{P}{T}\left[1 - e^{i\pi n}\right]\int_{-\infty}^{\infty} H_L(f)P_T(f)H_L(n/T - f)P_T(n/T - f)df \quad (11.4\text{-}72)$$

and $P = A^2$. Notice that when n is even, $c_n = 0$ and therefore the double carrier frequency term is not present. However frequency terms at

$$2f_0 \ terms = 2f_0 \pm (2n-1)/T \quad (11.4\text{-}73)$$

do occur, that is to say the spectral lines for MSK occur at odd multiples of twice the chip rate near twice the carrier frequency, just as in OQPSK. In a manner analogous to the solution of Problem 8 it may be shown that the spectral lines near baseband are of the form

$$e_{n/T}(t) = |d_n|\cos(2\pi nt/T + \varphi_n) \quad (11.4\text{-}74)$$

and their Fourier coefficients are given by

$$d_n = \left(1 + e^{-i\pi n}\right)\frac{P}{T}\int_{-\infty}^{\infty} H_L(f)P_T(f)H_L^*(f - n/T)P^*(f - n/T)df \quad (11.4\text{-}75)$$

From these Fourier coefficients it is seen that only even values of n produce nonzero values of d_n. Thus the baseband frequencies that exist for MSK are

$$\text{baseband terms} = 2n/T \quad \text{with} \quad n = 1, 2, 3, 4... \quad (11.4\text{-}76)$$

Again it is not difficult to show that the approximate detection and false alarm probabilities are given by

$$P_D \cong Q\left(\frac{T_h - |d_n|}{N_0\sqrt{\dfrac{W}{2T_{LD}}}}\right) \qquad P_{FA} \cong Q\left(\frac{T_h}{N_0\sqrt{\dfrac{W}{2T_{LD}}}}\right) \quad (11.4\text{-}77)$$

and d_n is given by (11.4-75) and n takes on the values $\pm 2, 4, 6, \ldots$

In Table 11.4-1 the cycle spectrum are summarized for these four spreading modulation types.

Table 11.4-1 Direct Sequence Cycle Spectrum for the Filter and Square Circuit

Spreading Modulation	Baseband Cycle Spectrum	Carrier Cycle Spectrum
BPSK	$\pm n/T$ $(n = 1, 2, 3,...)$	$2f_0 \pm n/T$, $(n = 0,1, 2, 3,...)$
Balanced QPSK	$\pm n/T$ $(n = 1, 2, 3,...)$	None exists
Balanced OQPSK	$\pm 2n/T$ $(n = 1, 2, 3,...)$	$2f_0 \pm (2n-1)/T$, $(n = 1, 2, 3,...)$
MSK	$\pm 2n/T$ $(n = 1, 2, 3,...)$	$2f_0 \pm (2n-1)/T$, $(n = 1, 2, 3,...)$

11.4.7 Frequency-Hopped Signals with MFSK Through a Filter and Squarer Circuit

Typically frequency-hopped signals use either a form of MFSK or DPSK data modulation. First we consider a frequency-hopped signal with MFSK data modulation. Initially we consider slow hopping in which there are one or more data code bits per hop.

11.4.7.1 Slow Frequency Hopped Signals with MFSK Data Modulation Through a Filter and Squarer Circuit

The complex envelope of a slow frequency-hopped signal with MFSK data modulation is given by

$$v(t) = \sqrt{2}A\sum_k e^{i(2\pi f_k t+\theta_k)} p_{T_h}(t-kT_h)\sum_j e^{i(2\pi F_j t+\vartheta_j)} p_{T_d}(t-jT_d) \tag{11.4-78}$$

where T_d is the duration of the data frequency modulation (MFSK) with frequencies F_j, assumed to be M_d equally spaced frequencies. Also $NT_d = T_h$ is the duration of the hop frequencies (the duration of N data modulation frequencies equals the duration of one hop). It is assumed that the occurrence of the M_d frequencies, F_j, is randomly chosen. It is also assumed that the M_h equally spaced hopping frequencies f_k are also randomly chosen, and the two processes are independent. The random variables θ_k and φ_j are statistically independent, and are uniformly distributed over 2π radians as well as from hop to hop and from modulation frequency to modulation frequency. Again, the signal component of the output is defined as

$$E\left\{s^2(t)\right\} = R_{ss}(t,t) \tag{11.4-79}$$

Evaluating $R_{vv}(t,t)$ produces

$$R_{vv}(t,t) = E\left[v(t)v(t)\right] = 0 \tag{11.4-80}$$

since the random variables θ_k and φ_j are assumed to be statistically independent, uniformly distributed, and each one has zero mean value; that is,

$$E\left[e^{i\theta_k}\right] = E\left[e^{i\vartheta_j}\right] = 0 \tag{11.4-81}$$

It is therefore concluded that there are no frequencies present near the carrier double frequency. Consider $R_{vv^*}(t,t)$ now, in order to investigate the baseband cycle frequencies. Evaluating $R_{vv^*}(t,t)$ produces

$$R_{vv^*}(t,t) = 2P\sum_k\sum_j p_{T_d}(t-kT_d)p_{T_h}(t-jT_h)p^*_{T_d}(t-kT_d)p^*_{T_h}(t-jT_h) \tag{11.4-82}$$

In this model it is assumed that the pulse functions are not filtered since the analysis would be difficult because the filtering would depend on where the hop was located. Expanding in a Fourier series produces

$$\frac{1}{2}R_{vv^*}(t,t) = \sum_{n=-\infty}^{\infty} d_n e^{i2\pi nt/T} \tag{11.4-83}$$

Before evaluating $1/2R_{vv}*(t,t)$ note that $R_{vv}*(t,t)$ is dominated by the short pulse duration of the modulation so that

$$\frac{1}{2}R_{vv*}(t,t) \cong P\sum_k p_{T_d}(t-kT_d)p^*_{T_d}(t-kT_d) \quad (11.4\text{-}84)$$

and therefore the d_n are given by

$$d_n \cong \frac{P}{T_d}\int_{-\infty}^{\infty} P_{T_d}(f)P^*_{T_d}(f-n/T_d)df \quad (11.4\text{-}85)$$

where $P_{T_d}(f)$ is the Fourier transform of the time pulses $p_{T_d}(t)$. For rectangular pulse shapes (which is the case in frequency hopping), $P_{T_d}(f)$ would have a sin(*x*)/*x* shape, which means that d_n would decrease monitonically as *n* increases, as seen via (11.4-85). The tones at $f = n/T_d$ (at multiples of the FSK data rate) are given by

$$e_{n/T}(t) = |d_n|\cos(2\pi nt/T_d + \theta_{d_n}) \quad (11.4\text{-}86)$$

with the associated average power

$$P_{n/T} = \frac{|d_n|^2}{2} \quad (11.4\text{-}87)$$

where as noted earlier T_d is the MFSK data modulation time duration. Thus tones can be detected at

$$\text{baseband tones} = \pm n/T_d \quad \text{where } n = 1, 2, 3... \quad (11.4\text{-}88)$$

Using the model of Figure 11.4-1 it is not difficult, assuming that baseband spectral lines are to be detected, to show that, using a Gaussian approximation, the following detection and false alarm probabilities result

$$P_D \cong Q\left(\frac{T_h - |d_n|}{N_0\sqrt{\dfrac{W}{2T_{LD}}}}\right) \qquad P_{FA} \cong Q\left(\frac{T_h}{N_0\sqrt{\dfrac{W}{2T_{LD}}}}\right) \quad (11.4\text{-}89)$$

For the values of $n = \pm 1, \pm 2, \pm 3,\ldots$ and d_n is given by (11.4-85).

11.4.7.2 Fast Frequency-Hopped Signals with MFSK Data Modulation Through a Filter and Squarer Circuit

The complex envelope of a fast frequency-hopped signal (the hop rate is greater than the data rate) with MFSK is given by

$$v(t) = \sqrt{2}A\sum_k e^{i(2\pi f_k t+\theta_k)}p_{T_h}(t-kT_h)\sum_j e^{i(2\pi F_j t+\vartheta_j)}p_{Td}(t-jT_d) \quad (11.4\text{-}90)$$

where T_d is the duration of the data frequency modulation (MFSK) with frequencies F_j, (assumed to be M_d equally spaced frequencies) and T_h is the duration of the frequency hops (N frequency hops time periods equals one data modulation frequency period). It is assumed that the M_d frequencies, F_j, are randomly chosen. The random variables θ_k and φ_j are statistically independent random variables, uniformly distributed over 0 to 2π, which are independent from each other and from hop to hop and from modulation frequency to modulation frequency. Again the signal component of the output is defined as

$$E\{s^2(t)\} = R_{ss}(t,t) \tag{11.4-91}$$

Evaluating $R_{vv}(t,t)$ produces the same result as in the slow frequency hopping case

$$R_{vv}(t,t) = E[v(t)v(t)] = 0 \tag{11.4-92}$$

since the expected value of each phase term is zero, that is

$$E\left[e^{i\theta_k}\right] = E\left[e^{i\vartheta_j}\right] = 0 \tag{11.4-93}$$

It is thus concluded that there are no spectral lines at twice the carrier (hop) frequency.

Note that $R_{vv}*(t,t)$ is dominated by the short pulse duration of the hopping waveform so that in a manner to the slow hopping case one obtains for the Fourier coefficients the result

$$d_n \cong \frac{P}{T_h}\int_{-\infty}^{\infty} P_{T_h}(f)P_{T_h}^*(f - n/T_h)df \tag{11.4-94}$$

for the frequencies at $f = n/T_h$, where T_h is the frequency hop time duration and $1/T_h$ is the hop rate. Thus the hop rate lines can be detected at

$$\text{baseband tones} = \pm n/T_h \quad \text{for } n = 1,2,3... \tag{11.4-95}$$

In a similar manner to the slow frequency-hopping case the baseband tones have a lock detector probability of false alarm and detection given by

$$P_D \cong Q\left(\frac{T_h - |d_n|}{N_0\sqrt{\dfrac{W}{2T_{LD}}}}\right) \qquad P_{FA} \cong Q\left(\frac{T_h}{N_0\sqrt{\dfrac{W}{2T_{LD}}}}\right) \tag{11.4-96}$$

For the values of $n = \pm 1, \pm 2, \pm 3, \ldots$ and d_n is given by (11.4-94).

11.4.8 Slow Frequency-Hopped Signals with DPSK Data Modulation Through a Filter and Squarer Circuit

The other common data modulation used with frequency-hopped signals is differential phase shift keying (DPSK). The complex envelope of the frequency hopped signal with DPSK data modulation is given by

$$v(t) = \sqrt{2}A\sum_k e^{i(2\pi f_k t + \theta_k)} p_{T_h}(t - kT_h)\sum_j d_j p_{T_d}(t - jT_d) \qquad (11.4\text{-}97)$$

where f_k are the hop frequencies, T_d is the DPSK data symbol duration, and T_h is the frequency hop duration. The θ_k are assumed to be uniform random variables, independent from sample to sample. In addition the d_j are assumed to equally likely random variables taking on the value of +1 or –1 with equal likelihood. It is also assumed that the data, phase, and frequencies are statistically independent. It will also be assumed that the data rate is much larger than the hop rate. Again the signal component of the output is defined as

$$E\{s^2(t)\} = R_{ss}(t,t) \qquad (11.4\text{-}98)$$

Since the frequency-hopping phase is independent and uniformly distributed over 0 to 2π, we conclude that

$$R_{vv}(t,t) = 0 \qquad (11.4\text{-}99)$$

since the phases are random and equally likely. Again we see that there is no double the carrier hop frequencies generated in this system. Consider the baseband covariance function

$$R_{vv^*}(t,t) = 2A^2 E\left\{\sum_k d_k p_{T_d}(t - kT_d)\sum_l p_{T_h}(t - lT_h)\sum_j d_j p^*_{T_d}(t - jT_d)\sum_m p^*_{T_h}(t - mT_h)\right\} \times E\left\{e^{i2\pi(f_l - f_m)}\right\} E\left\{e^{i(\theta_l - \theta_m)}\right\} \qquad (11.4\text{-}100)$$

where $E\{.\}$ is the mathematical expectation operator. Now because of the statistical independence of the data sequence and the phase sequence, the summation over the four indices reduces to a summation over two indices yielding

$$R_{vv^*}(t,t) = 2A^2 \sum_k p_{T_d}(t - kT_d) p^*_{T_d}(t - kT_d)\sum_m p_{T_h}(t - mT_h) p^*_{T_h}(t - mT_h) \qquad (11.4\text{-}101)$$

If now one neglects the transitions at every hop time (assuming that the data rate is much larger than the hop rate), then (11.4-101) is well approximated by

$$R_{vv^*}(t,t) \cong 2A^2 \sum_k p_{T_d}(t - kT_d) p^*_{T_d}(t - kT_d) \qquad (11.4\text{-}102)$$

Expanding (½)$R_{vv}*(t,t)$ in a complex Fourier series and solving for the Fourier coefficients yields (using $A^2 = P$)

$$d_n = \frac{P}{T}\int_{-T/2}^{T/2} \sum_{k=-\infty}^{\infty} p_T(t - kT) p^*_T(t - kT) e^{-i2\pi kt/T} \qquad (11.4\text{-}103)$$

This result can again be simplified, as before, to

$$d_n = \frac{P}{T_d}\int_{-\infty}^{\infty} P_{T_d}(f) P^*_{T_d}(f - n/T_d)\,df \qquad (11.4\text{-}104)$$

Thus, spectral lines occur at multiples of $1/T_d$, with the case $n = 1$ being the strongest, under reasonable conditions. Again recall that T_d is the data time duration. It follows that the lines appear at the following frequencies out of a filter and squarer

$$\text{baseband frequencies} = \pm n/T_d \quad n = 1, 2, 3... \qquad (11.4\text{-}105)$$

In a similar manner to the previous frequency-hopping cases, the baseband tones have a lock detector probability of false alarm and detection given by

$$P_D \cong Q\left(\frac{T_h - |d_n|}{N_0\sqrt{\dfrac{W}{2T_{LD}}}}\right) \qquad P_{FA} \cong Q\left(\frac{T_h}{N_0\sqrt{\dfrac{W}{2T_{LD}}}}\right) \qquad (11.4\text{-}106)$$

where the line spectra at baseband occur as shown in (11.4-105) and d_n is given by (11.4-104).

Table 11.4-2 summarizes the spectral lines that are present in frequency-hopped waveforms.

Table 11.4-2 The Cycle Spectrum in a Frequency-Hopping System Through a Filter and Squarer Circuit

Frequency-Hop Type	Baseband Cycle Spectrum	Carrier Cycle Spectrum
Slow freq. hop/FSK	n/T_d (n = 1, 2, 3, 4,...)	None exist
Fast freq. hop/FSK	n/T_h(n = 1, 2, 3, 4,...)	None exist
Freq. hop/DPSK	n/T_d (n = 1, 2, 3, 4,...)	None exist

Table note: T_d is the FSK and DPSK symbol time and T_h is the frequency-hop time.

11.4.9 Delay and Multiply Chip Rate Detectors with Balanced QPSK

As another technique to generate line spectra from a QPSK signal, consider the delay and multiply chip rate detector indicated in Figure 11.4-3 for QPSK chipping (direct sequence) signals.

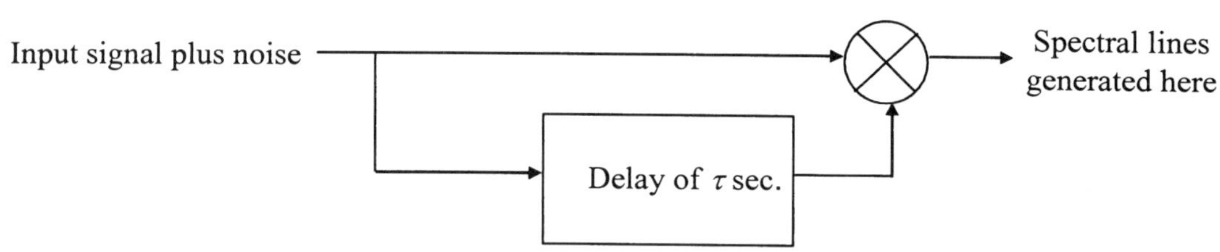

Figure 11.4-3 The delay and multiply chip rate detector.

It will be seen in the development to follow that there is a problem with the delay and multiply circuit in terms of its sensitivity to frequency errors. It will be assumed that the signal is not filtered significantly in this section. From Figure 11.4-3 it is seen that the output due to the signal times the signal term is given by

$$e_{ss}(t) = s(t)s(t - \tau) \qquad (11.4\text{-}107)$$

Define the output signal component as the following mean value

$$E\{s(t)s(t-\tau)\} = R_{ss}(t,t-\tau) \tag{11.4-108}$$

From (11.4-5) it follows that

$$R_{ss}(t,t-\tau) = \frac{1}{2}\mathrm{Re}\left[R_{vv}(t,t-\tau)e^{+i(2\omega_0 t-\omega_0\tau+2\theta)}\right] + \frac{1}{2}\mathrm{Re}\left[R_{vv^*}(t,t-\tau)e^{+i\omega_0\tau}\right] \tag{11.4-109}$$

From Chapter 1 the QPSK complex envelope signal is given by

$$v(t) = A\sum_k\left[(a_k + ib_k)p_T(t-kT)\right] \tag{11.4-110}$$

Based on (11.4-109) it is necessary to evaluate the carrier and the baseband autocovariance functions for this modulation. Just as in the filter and square detector it is easy to show that

$$R_{vv}(t,t-\tau) = 0 \tag{11.4-111}$$

So that the carrier cycle spectrum does not exist. Now consider one half the baseband autocorrelation function

$$(1/2)R_{vv^*}(t,t-\tau) = P\left[\sum_k p_T(t-kT)p_T^*(t-\tau-kT)\right] \tag{11.4-112}$$

since $A^2 = P$. It follows, using (11.4-5), that the signal autocorrelation function is given by

$$R_{ss}(t,t-\tau) = P\sum_k p_T(t-kT)p_T^*(t-\tau-kT)\cos(\omega_0\tau) \tag{11.4-113}$$

Note that $p_T(t)$ is, in fact, real in this case, when the pulse is unfiltered. Again, since this autocorrelation function is cyclostationary, it can be expanded in a complex Fourier series of the form

$$P\sum_k p_T(t-kT)p_T^*(t-\tau-kT)\cos(\omega_0\tau) = \sum_{n=-\infty}^{\infty} d_n e^{i2\pi nt/T} \tag{11.4-114}$$

In Problem 9 it is shown that the coefficients of the Fourier series are given by

$$d_n = \frac{P}{T}\cos(\omega_0\tau)e^{-i2\pi n\tau/T}\int_{-\infty}^{\infty} P(f)P^*(f-n/T)e^{i2\pi f\tau}df \tag{11.4-115}$$

Thus it is seen that the same spectral lines (frequencies) appear at the frequencies $\pm n/T$ for $n = 1, 2, 3\ldots$ This is the same set of spectral lines that occurred for the filter and squarer circuit. However the intensity of the spectral lines is not the same. The intensities are the same as the filter and squarer circuit, only in the case when τ is zero.

There is one important point that should be emphasized: the angular carrier frequency time delay product must be equal to $2\pi n$, with n an integer, or the amplitude of the spectral line could be considerably reduced in intensity. Since the frequency is not normally known in advance exactly, and the product is very sensitive to variations in frequency or delay, with the delay and multiply circuit it is not very feasible for line rate detection, because it is impossible to guarantee that the spectral line will not be suppressed. When this circuit is used, τ is usually set to be $0.5T$, since it maximizes the $1/T$ rate line spectral intensity component [23].

Again it is possible to obtain Gaussian approximations to the false alarm probability and the detection probability for the lines appearing at baseband. They are given by

$$P_D \cong Q\left(\frac{T_h - |d_n|}{N_0\sqrt{\dfrac{W}{2T_{LD}}}}\right) \qquad P_{FA} \cong Q\left(\frac{T_h}{N_0\sqrt{\dfrac{W}{2T_{LD}}}}\right) \tag{11.4-116}$$

where the parameters have been already defined, d_n is given by (11.4-115), and n can take on values of $n = \pm 1, \pm 2, \pm 3, \ldots$

11.5 PERFORMANCE OF A CHIP RATE DETECTOR FOR BPSK

In this section the approximate detection performance of a BPSK signal, in additive noise, will be considered. Figure 11.5-1 illustrates the model that will be considered here for the performance analysis.

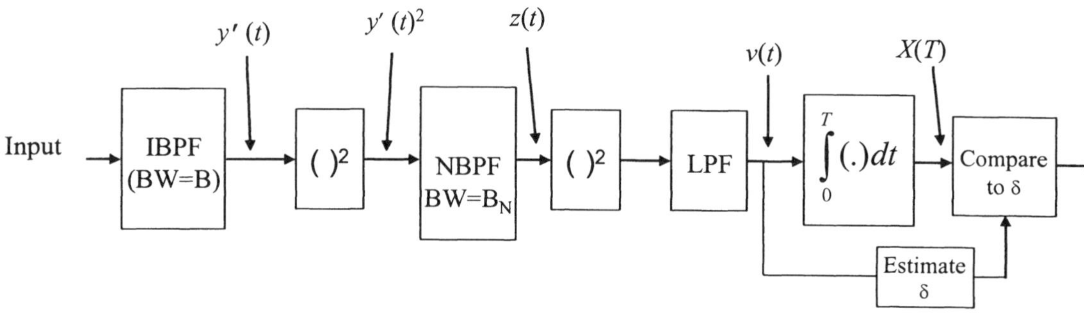

Figure 11.5-1 Chip rate detector for BPSK modulation.

First, the signal plus noise is filtered by the ideal bandpass filter (IBPF) of bandwidth B, and then the result is detected in a square law detector and narrowband filtered (NBF) to a bandwidth of B_N centered at $f = 1/T$. The tone that is generated at this point is still immersed in noise, and it is detected in another square law detector. This result is low pass filtered to pass the dc terms proportional to signal power at $f = 1/T$ and the low-pass filtered noise.

These filtered terms are then integrated in an integrator and the output is compared to a threshold, which was based on the input to the integrator when no signal was present.

The input is composed of a BPSK signal and band-limited noise with a two-sided spectral density of $N_0/2$. Thus the input can be modeled as

$$y(t) = s(t) + n(t) \tag{11.5-1}$$

After filtering by the ideal bandpass filter the output is given by

$$\tilde{y}(t) = \tilde{s}(t) + \tilde{n}(t) \tag{11.5-2}$$

which is the filtered version of (11.5-1). The signal is modeled as

$$\tilde{s}(t) = \mathrm{Re}\left[v(t)e^{i(\omega_0 t+\theta)}\right] \tag{11.5-3}$$

and the complex envelope is given by

$$v(t) = \sqrt{2}A\sum_k a_k \tilde{p}_T(t-kT) \tag{11.5-4}$$

where $\tilde{p}_T(t)$ is the filtered version of $p_T(t)$, with the filter being the low pass equivalent of the ideal bandpass filter. The noise can be expressed in the quadrature form, as

$$\tilde{n}(t) = \sqrt{2}\tilde{n}_c(t)\cos(\omega_0 t+\theta) + \sqrt{2}\tilde{n}_s(t)\sin(\omega_0 t+\theta) \tag{11.5-5}$$

The signal component is defined to be the mean value of the signal squared; thus

$$E\left[\tilde{s}^2(t)\right] = R_{ss}(t,t) \tag{11.5-6}$$

Based on (11.5-3) and (11.5-4) along with (11.4-5) out of the ideal bandpass filter, one obtains the signal component

$$E\left\{(\tilde{y}(t))^2\right\} = \frac{1}{2}\mathrm{Re}\left[R_{vv^*}(t,t)\right] + \frac{1}{2}\mathrm{Re}\left[R_{vv}(t,t)e^{-i(2\omega_0 t+2\theta)}\right] \tag{11.5-7}$$

It follows that the signal plus noise out of the squarer is given by

$$\begin{aligned}(\tilde{y}(t))^2 &= \frac{1}{2}\mathrm{Re}\left[R_{vv^*}(t,t)\right] + \frac{1}{2}\mathrm{Re}\left[R_{vv}(t,t)e^{-i(2\omega_0 t+2\theta)}\right] \\ &+ \text{n.p.s.t.} + \tilde{n}(t)\mathrm{Re}\left[v(t)e^{i(\omega_0 t+\theta)}\right] + (\tilde{n}(t))^2\end{aligned} \tag{11.5-8}$$

where n.p.s.t. denotes nonperiodic signal terms. These terms will not be considered since they are only relevant at high SNRs, and the analysis will be based on the assumption of low SNRs. The narrowband bandpass filter (NBPF) will filter out the entire signal components, except the line spectra generated by the squaring operation. Thus only the signal component located at $f = 1/T$, where f is the chip rate, will be considered, plus the noise that passes through the NBPF. Now consider the baseband terms, which are derived from the first term in (11.5-8). Expand $(1/2)R_{vv^*}(t,t)$ in a complex exponential series

$$A^2\sum_{k=-\infty}^{\infty}[\tilde{p}_T(t-kT)][\tilde{p}_T^*(t-kT)] = \sum_{k=-\infty}^{\infty} d_n e^{i2\pi nt/T} \tag{11.5-9}$$

The Fourier coefficients are given by

$$d_n = \frac{P}{T}\int_{-T/2}^{T/2}\sum_{k=-\infty}^{\infty}[\tilde{p}_T(t-kT)]\left[\tilde{p}_T^*(t-kT)\right]e^{-i2\pi nt/T}dt \tag{11.5-10}$$

which again can be written as

$$d_n = \frac{P}{T}\int_{-\infty}^{\infty} H_L(f)P_T(f)H_L^*(f-n/T)P_T^*(f-n/T)df \tag{11.5-11}$$

Writing d_n as

$$d_n = |d_n| e^{i\theta_{d_n}} \tag{11.5-12}$$

The steady state value of the signal component at $f = 1/T$ out of the NBPF is given

$$e_{1/T}(t) = 2|d_1|\cos(2\pi t/T + \theta_{d_1}) \tag{11.5-13}$$

which has already been derived in (11.4-32). With the assumption that α = BT/2, and α = 1 (the maximum value of d_1 occurs at α =1) or B = 2/T the expression for d_1 ((11.4-36) is given by

$$d_1 = \frac{P}{\pi}\int_0^{\pi} \frac{\sin(x)}{x}\frac{\sin(x-\pi)}{(x-\pi)}dx = 0.247P \tag{11.5-14}$$

where P is the power of the input direct sequence signal. The tone power is given by

$$P_1 = (2d_1)^2/2 = 0.122P^2 \tag{11.5-15}$$

where P in the input signal power. The signal times noise term and the noise times noise terms of (11.5-8) have been analyzed by Davenport and Root [22] and with a minor correction to account for modulation, the results at $f = 1/T$ are given by

$$\begin{aligned} S_n(1/T) &= 2\gamma P N_0 \\ S_{nn}(1/T) &= N_0^2(B - 1/T) \end{aligned} \tag{11.5-16}$$

where N_0 is the input noise spectral density and γ is the filter loss. The parameter γ is given by

$$\gamma = \frac{\int_{-B/2}^{B/2} S_s(f)df}{\int_{-\infty}^{\infty} S_s(f)df} \tag{11.5-17}$$

Thus the two-sided noise spectral density at $f = 1/T$ is given by

$$\frac{N_0^{'}}{2} = N_0^2(B - 1/T) + 2\gamma P N_0 \tag{11.5-18}$$

The spectral density, at $f = 1/T$, is not flat over the bandwidth and the process is not Gaussian (when $B_N \ll B$ it is approximately Gaussian). However in order to proceed with the analysis it will be assumed that the process is band-limited white Gaussian noise, with one-sided equivalent spectral density $N_0^{'}$. Therefore, the signal plus noise at the point in the Figure 11.5-1, labeled $z(t)$, is given by

$$z(t) = \sqrt{2}a\cos(2\pi t/T + \theta_{d_1}) + \sqrt{2}\tilde{n}_c(t)\cos(2\pi t/T) + \sqrt{2}\tilde{n}_s(t)\sin(2\pi t/T) \tag{11.5-19}$$

where $a = \sqrt{2}\,d_1$. The noise terms have density $N_0^{'}$ and each have two-sided bandwidth B_N. The output of the squarer and the low-pass filter (which removes the $2/T$ frequency terms) is given by, after some simplifications,

$$z^2(t)\big|_{LPF} = v(t) = a^2 + 2a\tilde{n}_c(t) + (\tilde{n}_c(t))^2 + (\tilde{n}_s(t))^2 \tag{11.5-20}$$

At this point (11.5-20) can be written in the form

$$v(t) = (a + \tilde{n}_c(t))^2 + (\tilde{n}_s(t))^2 \tag{11.5-21}$$

Thus it is seen that $v(t)$ is the composed of two independent noise processes; one has zero mean and one has mean a. The signal $v(t)$ is integrated and compared with the threshold, which was estimated from the input when only noise was present. The integral can be approximated by the sum of discrete steps of the form

$$X(T) = \int_0^{T_I} v(t)dt \cong \sum_{i=1}^{M} v(t_i)\Delta t \tag{11.5-22}$$

where M is the least integer greater than $B_N T_I$ and $\Delta t = 1/B_N$, with T_I the integration time. It is shown in the appendix that samples of white Gaussian noise are independent every B_N seconds, where B_N is the bandpass (positive frequency) bandwidth of the ideal bandpass filter. Following Proakis [11], the random variable $X(T)$ can be described by the following probability density function in the case of noise only as

$$p_X\left(x|H_0\right) = \frac{1}{\sigma_N^n 2^{n/2}\Gamma(n/2)} x^{\frac{n}{2}-1} e^{-\frac{x}{2\sigma_N^2}} \tag{11.5-23}$$

and when the signal is present by

$$p_X(x|H_1) = \frac{1}{2\sigma_{S+N}^2}\left(\frac{x}{S^2}\right)^{\frac{n-2}{4}} e^{-\left(\frac{(S^2+x^2)}{2\sigma_{S+N}^2}\right)} I_{\frac{n}{2}-1}\left(\sqrt{x}\frac{S}{\sigma_{S+N}^2}\right) \tag{11.5-24}$$

where

$$n = 2\lfloor B_N T_I \rfloor \tag{11.5-25}$$

and $\lfloor x \rfloor$ denotes the least integer greater than or equal to x. The S parameter is given by

$$S^2 = \frac{n}{2}a^2 \tag{11.5-26}$$

and the variance is given the product of the two-sided noise spectral density times the two-sided noise bandwidth; thus,

$$\sigma_{S+N}^2 = N_0^{'} B_N = \left[2N_0^2(B - 1/T) + 4\gamma P N_0\right] B_N \tag{11.5-27}$$

and

$$\sigma_N^2 = N_0^{'} B_N = \left[2N_0^2(B - 1/T)\right] B_N \tag{11.5-28}$$

The false alarm and the detection probabilities are given by

$$P_{FA} = \int_{\delta}^{\infty} p_X(x|H_0)dx$$
$$P_D = \int_{\delta}^{\infty} p_X(x|H_1)dx \qquad (11.5\text{-}29)$$

Letting $u = y/S^2$ the false alarm and detection probabilities can be put in a more convenient form to yield

$$P_{FA} = \frac{1}{2^{n/2}\Gamma(n/2)}\int_{\frac{\delta}{\sigma_N^2}}^{\infty} u^{\frac{n}{2}-1} e^{-u/2} du \qquad (11.5\text{-}30)$$

and[5]

$$P_D = \frac{1}{2}\int_{\frac{\delta}{\sigma_{S+N}^2}}^{\infty} \left(\frac{\sigma_{S+N}^2 u}{S^2}\right)^{\frac{n-2}{4}} e^{-\left(\frac{S^2}{2\sigma_{S+N}^2}\right)} e^{-u} I_{\frac{n}{2}-1}\left(\sqrt{u}\frac{S}{\sigma_{S+N}}\right) du \qquad (11.5\text{-}31)$$

Consider the quantity $\dfrac{S^2}{\sigma_{S+N}^2}$. Denoting P_i as the input power it can be shown that

$$\frac{S^2}{\sigma_{S+N}^2} = \frac{\lfloor 2B_N T_I \rfloor \left(2d_1(\alpha)^2 P_i^2\right)}{\left[2N_0^2(B-1/T)+4\gamma P_i N_0\right]2B_N} \qquad (11.5\text{-}32)$$

where γ is the filtering loss due to the first bandpass filter of bandwidth B and is given by (11.5-17) and $d_1(\alpha)$ is the integral

$$d_1(\alpha)\big|_{\alpha\ge 0.5} = \frac{1}{\pi}\int_{(1-\alpha)\pi}^{\alpha\pi} \frac{\sin(x)}{x}\frac{\sin(x-\pi)}{(x-\pi)}dx, \quad d_1(\alpha)\big|_{0\le\alpha\le 0.5} = 0 \qquad (11.5\text{-}33)$$

From (11.5-17) and (11.5-27) to (11.5-33) it is possible to compute the detection and false alarm probabilities for this rate line detector. It should be pointed out that the narrower the narrowband filter and the longer the integrator integration time, the better the performance, at least in theory. However, phase noise and the lack of exact rate line (frequency) knowledge due to Doppler effects and oscillator drift limit the practical bandwidth B_N and therefore the performance of this rate line detector.

11.6 FREQUENCY ESTIMATION OF AN UNMODULATED TONE WITH A LIMITER DISCRIMINATOR

In the field of covert communications any parameter that can be estimated to provide some information about the signal under investigation is a worthwhile piece of information. Instantaneous frequency measurement (IFM) systems provide carrier frequency information and possibly the spectral shape information of the signal under investigation. Consider Figure 11.6-1, which illustrates the discriminator frequency detector.

[5] Note that P_D is best evaluated by the series of Schnidman discussed in Section 6.5.4.

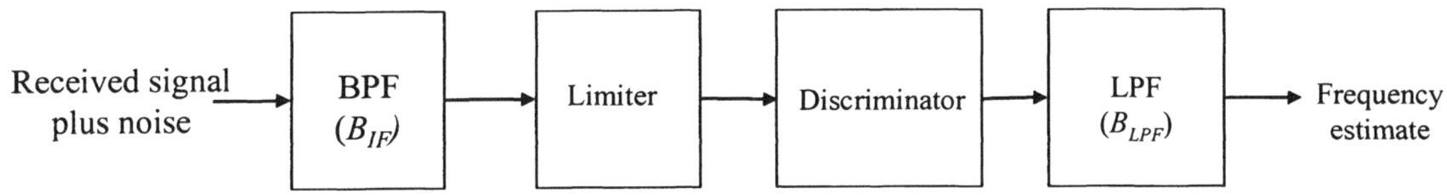

Figure 11.6-1 Limiter discriminator frequency estimator.

The first element of the system is the bandpass filter having bandwidth B_{IF}. It will be assumed that this filter is an ideal bandpass filter to simplify the approximate analysis to follow. The second element is a limiter, followed by a discriminator, and in turn followed by an ideal low-pass filter having bandwidth B_{LPF}. The unmodulated tone plus band-limited white Gaussian noise out of the BPF can be written as

$$y(t) = \sqrt{2}A\cos(\omega_c t + \theta) + \sqrt{2}n_c(t)\cos(\omega_c t + \theta) + \sqrt{2}n_s(t)\sin(\omega_c t + \theta) \tag{11.6-1}$$

where it has been assumed, without a loss in generality, that the phase of the signal and the noise terms are the same. The noise power out of the bandpass filter is given by $N_0 B_{IF}$. Each low-pass noise term has a spectral density of $N_0/2$, and a two-sided bandwidth of B_{IF}. In what follows, a simplified analysis will be presented. Writing (11.6-1) in the envelope and phase form produces

$$y(t) = \sqrt{2}\sqrt{(A + n_c(t))^2 + (n_s(t))^2}\cos\left(\omega_c t + \theta + \varphi(t)\right) \tag{11.6-2}$$

where the phase function is given by

$$\varphi(t) = \tan^{-1}\left[\frac{n_s(t)}{A + n_c(t)}\right] \tag{11.6-3}$$

After the limiter, the approximate fundamental component of the output is given by

$$y(t) \cong A_L \cos\left(\omega_c t + \theta + \varphi(t)\right) \tag{11.6-4}$$

where A_L is the amplitude of the limiter output level. Assuming that the signal-to-noise ratio is reasonably high, the total phase of (11.6-4) can be approximated by

$$\phi(t) \cong 2\pi f_c t + \theta + \left[\frac{n_s(t)}{A}\right] \tag{11.6-5}$$

The instantaneous frequency, in Hz, is given by

$$f(t) = \frac{\dot{\phi}(t)}{2\pi} = f_c + \frac{\dot{n}_s(t)}{2\pi A} \tag{11.6-6}$$

The noise-induced frequency error is seen to be the quantity

$$\Delta f = \frac{\dot{n}_s(t)}{2\pi A} \tag{11.6-7}$$

It is well known that the power spectral density of the derivative of a noise process is given by

$$S_{\Delta f}(f)=\frac{(2\pi f)^2 S_{n_s}(f)}{(2\pi A)^2}=\frac{f^2}{A^2}\left(\frac{N_0}{2}\right) \qquad |f|\le\frac{B_{IF}}{2} \tag{11.6-8}$$

It therefore follows that the variance of the frequency error, for this high SNR case, is given by

$$\sigma_{\Delta f}^2=\frac{N_0}{A^2}\int_0^{B_{LPF}} f^2 df \tag{11.6-9}$$

where it is assumed that the low-pass filter bandwidth is less than one half of the bandpass filter bandwidth. Evaluating (11.6-9) produces the result

$$\sigma_{\Delta f}^2=\frac{(B_{LPF})^3}{3(A^2/N_0)}=\frac{(B_{LPF})^3}{3B_{IF}\rho} \quad \text{Hz}^2 \tag{11.6-10}$$

where $\rho = A^2/(N_0B_{IF})$ and is the signal-to-noise ratio out of the bandpass filter. Thus the center frequency of a tone has been estimated with a one sigma error of

$$\sigma_{\Delta f}=\sqrt{\frac{(B_{LPF})^3}{3B_{IF}\rho}} \quad \text{Hz} \tag{11.6-11}$$

Hence the IFM could be used to detect the line components from a filter and squarer or delay and multiply detector.

A major problem with IFM receivers is the fact that they do not work well with two or more spectral lines present at one time. If one tone is considerably stronger than the other tones, the others will degrade its estimate. If the powers of the stronger signals are comparable then the estimate will be normally useless.

All the rate line detectors have assumed to have only one signal present at the input. However this may not be the case in the "real world," where multiple signals may be present, and thus multiple lines might appear. If a high gain antenna can be utilized then it may be possible to "separate" the different signals.

11.7 SUMMARY

This chapter introduced some of the types of low probability of intercept receivers that are used to surreptitiously detect signals. Covert communications were discussed along with some propagation models that are applicable to LPI detection. The radiometer and its performance were introduced along with its variants. This was followed by a discussion of spectrum analyzers.

Narrowband signals were introduced and applied to various chip modulations. With this background, second order cyclostationary feature detection was presented via the filter and square circuit. The intensity and the frequency of the lines that generated from this process are developed for BPSK, QPSK, OQPSK, MSK, and frequency hopping, with both frequency shift keying modulation and DPSK modulation.

Based on a Gaussian approximation, an estimate of both the detection probability and the false alarm probability were obtained for a PLL-based tone detection scheme for the filter and squarer detector, followed by a PLL and its lock detector. BPSK, balanced QPSK, balanced OQPSK, and MSK were assumed as the direct sequence modulation type. This same scheme was then applied to slow and fast frequency hopping with MFSK modulation and slow frequency hopping with DPSK data modulation.

In addition, the delay and multiply detector was briefly looked at for balanced QPSK data modulation.

Finally an instantaneous frequency estimator was analyzed, assuming high signal-to-noise ratio for the standard deviation of the frequency error of the carrier frequency. Obviously this brief introduction doesn't do justice to the many techniques used in covert communications work, but at least it has provided a brief introduction to some of the techniques used.

References

[1] D. L. Nicholson, *Spread Spectrum Signal Design*, New York: Computer Science Press, 1988.

[2] F. H. Palmer, "Measurements of VHF/UHF Propagation Characteristics over Arctic Paths," *IEEE Trans. on Antennas and Propagations,* Vol. AP-28, No. 6, November 1980.

[3] J. J. Egli, "Radio Propagation Above 40 MHz over Irregular Terrain," *Proc. IRE*, 1957, pp. 1383–1391.

[4] M. Hata, "Empirical Formula for Propagation Loss in Mobile Radio Services," *IEEE Trans. on Vehicular Technology*, VT-29, 1980, pp. 317–325.

[5] Y. Okumura, E. Ohmori, T. Kawana, and K. Fukuda, "Field Strength and Its Variability in VHF and UHF Land-Mobile Radio Service," *Rev. Elec. Communication Lab.*, 1968, pp. 825–873.

[6] V. K. Garg, K. Smolik, and J. E. Wilkes, *Applications of CDMA in Wireless/Personal Communications*, Englewood Cliffs, NJ: Prentice-Hall, 1997.

[7] J. Betz and R. Rifkin, "Intercept Detection Performance," Mitre Report MTR 94B0000092, September 1994, Secret (only unclassified portions used in this book).

[8] H. Van Trees, *Detection, Estimation, and Modulation Theory*, Vol. III, Section 12.2, New York: John Wiley & Sons, 1971.

[9] R. Price and P. E. Green, "Signal Processing in Radar Astronomy-Communication Via Fluctuating Multipath Media," MIT, Lincoln Lab., TR 234, October 1960.

[10] H. Urkowitz, "Energy Detection of Unknown Deterministic Signals," *Proceedings of the IEEE*, Vol. 55, No. 4, April 1967.

[11] J. G. Proakis, *Digital Communications*, New York: McGraw-Hill, 1983.

[12] A. Sonnenschein and P. Fishman, "Limitations on the Detectability of Spread Spectrum Signals," *Milcom 89*, Boston, MA, October 15–18, 1989.

[13] J. K. Holmes, *Coherent Spread Spectrum Systems*, Malabar, FL: Krieger Publishing, 1990.

[14] Probe Systems Inc. "Principles of Spread Spectrum Intercept," *Teal Wing Program*, PSI-ER-5419-05, February 1978, Secret (unclassified sections only).

[15] W. A. Gardner and C. M. Spooner, "Signal Interception: Performance Advantages of Cyclic-Feature Detectors," *IEEE Trans. on Communications*, Vol. 40, No. 1, January 1992.

[16] J. D. Edell, "Wideband, Noncoherent, Frequency-Hopped Waveforms and Their Hybrids in Low-Probability-of-Intercept Communications," Naval Research Lab. Report NRL 8025, Washington, D.C., November 8, 1976.

[17] D. G. Woodring, "Performance of Optimum and Suboptimum Detectors for Spread Spectrum waveforms," Naval Research Lab. Report NRL 8432, Washington, D.C., December 30, 1980.

[18] D. Woodring and J. Edell, "Detectability Calculation Technique," U.S. Naval Research Laboratory, Code 5480, September 1, 1977.

[19] R. Dillard, "Detectability of Spread-Spectrum Signals," *IEEE Trans. on Aerospace and Electronic Systems*, Vol. AES-15, No. 4, July 1979.

[20] R. J. Webster, "Minimum Detectable Signal of Spectrogram Displays," *IEEE Trans. on Signal Processing*, Vol. 41, No. 4, April 1993, pp. 1724–1726.

[21] W. A. Gardner, *Introduction to Random Process*, 2nd ed., New York: McGraw-Hill, 1989.

[22] W. B. Davenport and W. L. Root, *An Introduction to the Theory of Random Signals and Noise*, New York: McGraw-Hill, 1958.

[23] R. E. Ziemer and R. L. Peterson, *Digital Communications and Spread Spectrum Systems*, New York: Macmillan Publishing Company, 1985.

Selected Bibliography

R. A. Dillard and George M. Dillard, *Detectability of Spread-Spectrum Signals*, Norwood, MA: Artech House, 1989.

R. A. Scholtz, "The Spread Spectrum Concept," *IEEE Trans. on Communications*, Vol. COM-25, No. 8, August 1977.

D. J. Torrieri, *Principles of Secure Communication Systems*, Dedham, MA: Artech House, 1985.

Problems

1. Show that (11.2-3) can be written in the form of (11.2-4), where $Q_M(a,b)$ is defined by (11.2-5).

2. Investigate the optimum burst time of a signal to minimize the probability of detection for a fixed false alarm probability under the assumption of fixed signal burst energy, at low input SNR. Using (11.2-7) to (11.2-9) show that for a fixed false alarm probability, P_{FA}, and fixed energy ($E = PT$), with the assumption that the radiometer has fixed integration time T and the signal burst is T seconds long, and the radiometer is synchronized in time with the pulse transmission, that the optimum burst time (minimum detection probability) is as long as possible (and therefore the signal power is as low as possible). To establish this result it is necessary to adjust the threshold in such a way that as T varies E and β are held constant and P will vary to satisfy the energy constraint.

3. Consider the one-pole LPF radiometer that was discussed in Section 11.2.3 and assume that the time constant is τ.
   (a) Show that the mean output at the end of the signal pulse of T seconds is given by

$$\mu = N_0 B\left[1 + SNR\left(1 - e^{-T/\tau}\right)\right] \tag{P11-1}$$

   where SNR is the input signal-to-noise ratio (SNR = $P/(N_0B)$).
   (b) Show that the steady state variance of the output noise is given by

$$\sigma^2 = \left(N_0^2 B + 2PN_0\right)\left(\frac{1}{2\tau}\right) \tag{P11-2}$$

   where the two-sided noise bandwidth of the low-pass filter is $1/(2\tau)$.
   (c) Show that the false alarm probability, with an assumption that the output variable is approximately Gaussian, that

$$P_{FA} = Q(\beta) \quad \text{where } \beta = \frac{\delta - N_0 B}{\sqrt{N_0^2 B\left(\dfrac{1}{2\tau}\right)}} \tag{P11-3}$$

   and where

$$Q(x) = \int_x^{\infty} \frac{1}{\sqrt{2\pi}} e^{-\frac{t^2}{2}} dt \tag{P11-4}$$

   (d) Show that the under the same approximation that the probability of detection is given by

$$P_D = Q\left[\frac{\beta - \sqrt{2B\tau}\ SNR\left(1 - e^{-T/\tau}\right)}{\sqrt{1 + 2SNR}}\right] \quad \text{(P11-5)}$$

(e) Show that the value of τ that makes the detection probability greatest occurs when $\tau = 0.795T$.

(f) Show that at low SNR the loss relative to the integrate-and-dump radiometer is about 0.5-dB on the input SNR.

4. Show that the complex envelopes satisfy the same relationships as the actual signals themselves; that is, show that

$$\begin{aligned} &(1)\quad v_o(t) = v_i(t) * h_L(t) \quad \text{and} \\ &(2)\quad V_o(f) = V_i(f)H_L(f) \end{aligned} \quad \text{(P11-6)}$$

where $V_o(f)$ and $V_i(f)$ are defined in Chapter 1.

5. Starting with (11.4-18), that is, if

$$c_n = \frac{P}{T}\int_{-\infty}^{\infty} [\tilde{p}_T(t)]^2 e^{-i2\pi nt/T} dt \quad \text{(P11-7)}$$

show that c_n can be written, in the frequency domain, in the form

$$c_n = \frac{P}{T}\int_{-\infty}^{\infty} H_L(f)P(f)H_L(n/T - f)P_T(n/T - f)df \quad \text{(P11-8)}$$

where $H_L(f)$ is the low-pass equivalent filter, of the bandpass filter preceding the squarer.

6. Establish that the value of d_1 can be expressed as (11.4-36) by doing the following:

(a) Evaluate the region that the product of the filter at baseband and the filter at $1/T$ Hz is nonzero over the frequency $(1-\alpha)/T$ to α/T so that from (11.4-31) d_1 becomes

$$d_1(\alpha)\big|_{\alpha \ge 0.5} = PT\int_{(1-\alpha)/T}^{\alpha/T} \frac{\sin(\pi fT)}{\pi fT}\frac{\sin(\pi fT - \pi)}{(\pi fT - \pi)}df, \text{ and } d_1(\alpha)\big|_{\alpha \le 0.5} = 0 \quad \text{(P11-9)}$$

(b) Make a change of variables and achieve the result indicated in (11.4-36).

7. (a) Evaluate the carrier covariance Fourier series expansion coefficients for OQPSK to obtain

$$c_n = \frac{P}{2T}\left[1 - e^{i\pi n}\right]\int_{-\infty}^{\infty} H_L(f)P_T(f)H_L(n/T - f)P_T(n/T - f)df \quad \text{(P11-10)}$$

(b) Evaluate the baseband covariance to obtain

$$d_n = \left(1 + e^{-i\pi n}\right)\frac{P}{2T}\int_{-\infty}^{\infty} H_L(f)P_T(f)H_L^*(n/T - f)P^*(n/T - f)df \quad \text{(P11-11)}$$

Conclude that the carrier spectral lines occur only for n even and the baseband spectral lines occur only for n odd.

8. Show that the coefficients of the Fourier series for (11.4-115) is given by

$$d_n = \frac{P}{T}\cos(\omega_0 t)\int_{-\infty}^{\infty} p(t)p^*(t-\tau)e^{-i2\pi nt/T_c}dt \tag{P11-12}$$

$$d_n = \frac{P}{T}\cos(\omega_0 \tau)e^{-i2\pi n\tau/T}\int_{-\infty}^{\infty} P(f)P^*(f-n/T)e^{i2\pi f\tau}df \tag{P11-13}$$

Hint: Assume that the second pulse function in (11.4-113) is complex.

9. Show that (11.2-15) follows from (11.2-14).

APPENDIX 11A
SAMPLES FROM A BANDPASS FILTERED GAUSSIAN RANDOM PROCESS

In this appendix ideal bandpass filtered Gaussian noise will be sampled and shown that all samples taken at the rate of every $1/B$ seconds will be statistically independent.

Let the noise be represented in narrowband bandpass form

$$n(t) = \sqrt{2}n_c(t)\cos(\omega_0 t + \theta) + \sqrt{2}n_s(t)\sin(\omega_0 t + \theta) \tag{11A-1}$$

where θ is a random variable uniform from 0 to 2π. The autocorrelation function is given by

$$R_n(\tau) = E\left[n(t)n(t+\tau)\right] \tag{11A-2}$$

Evaluating one obtains

$$R_n(\tau) = 2R_{n_c}(\tau)\cos(\omega_0 t)\cos(\omega_0(t+\tau)) + 2R_{n_s}(\tau)\sin(\omega_0 t)\sin(\omega_0(t+\tau)) \tag{11A-3}$$

Since both quadrature autocorrelation functions are the same, this simplifies to

$$R_n(\tau) = 2R_{n_c}(\tau)\cos(\omega_0 \tau) \tag{11A-4}$$

To evaluate the quadrature autocorrelation function consider

$$R_{n_c}(\tau) = \int_{-\infty}^{\infty} S(f)e^{i2\pi f\tau}df \tag{11A-5}$$

$$R_{n_c}(\tau) = \frac{N_0 B}{2}\frac{\sin(\pi B\tau)}{\pi B\tau} \tag{11A-6}$$

since the low-pass noise terms have bandwidth B Hz extending from $-B/2$ to $B/2$ Hz. The autocorrelation of the noise process satisfies

$$R_n(\tau) = N_0 B\frac{\sin(\pi B\tau)}{\pi B\tau}\cos(\omega_0 \tau) \tag{11A-7}$$

It is clear that samples of $n(t)$ are uncorrelated, and therefore statistically independent when τ satisfies

$$\tau = k/B, \quad k=\pm 1, \pm 2, \pm 3, ... \tag{11A-8}$$

Thus samples are independent every $1/B$ seconds.

CHAPTER 12

Lock Detector Theory and Absorbing Markov Chains

12.0 INTRODUCTION

This chapter deals with the performance of lock detectors, which are normally found in conjunction with almost all synchronization devices. The analytical approach taken in this chapter is to treat the lock detectors as absorbing Markov chains. The results of which, with the use of simple matrix operations, allow one to obtain the mean and variance of the time to indicate the out-of-lock state. The same theory can be used to determine performance for the case of acquisition lock detection and for out-of-lock detection as well. Lock detectors are a very important part of any system design. They are used with PN code acquisition and PN code-tracking loops, frequency-hopping acquisition and tracking, carrier acquisition and tracking loops, frequency-hopping acquisition and tracking loops, as well as bit synchronizers in acquisition and tracking.

Lock detectors are used for both acquisition indication and tracking lock status of code acquisition circuits, code-tracking circuits, and bit synchronization circuits. When the system is in lock, it is desirable that the lock detector indicates the in-lock condition as soon as possible, after the system locks. Conversely, if the system suddenly drops out of lock, it is highly desirable that the lock detector indicate that fact, as soon as possible.

Acquisition time is normally determined by the time the system takes to acquire the signal (be it PN code, hop pattern, carrier, and so on) and to indicate that it is in lock. Thus the lock detector's response time is a factor in the overall acquisition time. Also the time to indicate the out-of-lock state is an important consideration when designing systems, since reacquisition can't start until the out-of-lock indication occurs.

The method developed here follows the work of Holmes [1], whose matrix approach was based on Kemeny and Snell's [2] textbook and was first applied to a lock detector problem by Hopkins [3]. First, some background material will be presented, and then the theory will be applied to some lock detectors configurations.

First we will present the mathematical background for lock detectors and then analyze different algorithms for lock detector performance via the derived theory and then follow that with some different lock detector models for various modulation types.

12.1 ABSORBING MARKOV CHAINS

In this section a few definitions will be introduced to set the stage for the forthcoming analysis. The states of a Markov chain can be divided into transient states and persistent states (note: the terms "recurrent" and "ergodic" are sometimes used in place of persistent). *Transient states*, once left, can never be entered again, while *persistent states*, once entered, can never be left. A set of persistent states with only one element is called an *absorbing state*, which is characterized by a transition probability of $p_{ii} = 1$. In this case all the other entries in the i-th row of the transition matrix must be zero. In general p_{ij} denotes the probability of going from state i to state j in one step.

A Markov chain, consisting of a single persistent set, is called a *persistent Markov chain*. A Markov chain, all of whose nontransient states are absorbing, is called an *absorbing Markov chain*.

Every Markov chain must have a persistent set of states, but it does not have to have a set of transient states. If a Markov chain contains only one state, then it must be an absorbing chain (this is not a very interesting chain, however!). As an example to clarify these definitions, consider a Markov chain characterized by the state transition matrix of the form

$$P_1 = \begin{array}{c} \\ 1 \\ 2 \\ 3 \\ 4 \\ 5 \end{array} \begin{array}{c} \begin{array}{ccccc} 1 & 2 & 3 & 4 & 5 \end{array} \\ \begin{bmatrix} 1 & 0 & 0 & 0 & 0 \\ p & 0 & q & 0 & 0 \\ 0 & p & 0 & q & 0 \\ 0 & 0 & p & 0 & q \\ 0 & 0 & 0 & 0 & 1 \end{bmatrix} \end{array} \tag{12.1-1}$$

where q is the probability of the process going to one higher state and p the probability that of going to one lower state. Notice once the process reaches state 1 or 5, it remains there forever; hence states 1 and 5 are absorbing states. States 2, 3, and 4, are transient states, since eventually the process leaves them and remains in the persistent states (1 and 5). Note that this matrix describes an absorbing Markov chain. Note also that the sum of the row entries equals unity. This means that the sum the probabilities going from state i to any of the states j is one.

If we modify our example to have "reflecting boundaries" at states 1 and 5, the transition matrix becomes

$$P_2 = \begin{array}{c} \\ 1 \\ 2 \\ 3 \\ 4 \\ 5 \end{array} \begin{array}{c} \begin{array}{ccccc} 1 & 2 & 3 & 4 & 5 \end{array} \\ \begin{bmatrix} 0 & 1 & 0 & 0 & 0 \\ p & 0 & q & 0 & 0 \\ 0 & p & 0 & q & 0 \\ 0 & 0 & p & 0 & q \\ 0 & 0 & 0 & 1 & 0 \end{bmatrix} \end{array} \tag{12.1-2}$$

Now there are no transient states; they are all persistent. Since they all *communicate* (i.e., any state can be reached by any other state) there is one persistent set. Hence we have a persistent Markov chain. The rest of this section we will be concerned with absorbing Markov chains.

Theorem 1 In any Markov chain the probability that the process is in a persistent state tends to 1 as the number of steps, n, tends to infinity, no matter where the starting point is.

Proof Once a persistent set is reached it can never be left (by definition of a persistent set of states). Suppose that the process starts in a transient state. Suppose that from any transient state it is possible to reach an ergodic (persistent) state in not more than n steps. There is positive probability p of entering a persistent state from any transient state in n steps. Therefore the probability of not reaching a persistent state, in n steps, is $1-p < 1$. Consequently, the probability of not reaching a persistent state in mn steps in less than or equal to $(1-p)^m$, which approaches zero, as $m \to \infty$. Hence, it must reach a persistent state.

From this theorem we see that if the chain is absorbing, the process will eventually end up in an absorbing state with probability of 1. Most of the lock detector problems that we will be concerned with will be modeled as absorbing chains, with a single absorbing state.

In order to utilize our matrix results it is convenient to utilize a *canonical form* of the transition matrix P, which we will call P_C. First, group all the persistent states together, and then all the transient states. Assume there are n total states and t transient states, and therefore $n-t$ persistent states. Rearranging the matrix according to the two types of states yields

$$P_c = \begin{array}{c} \\ n-t\} \\ t\} \end{array} \overset{\begin{array}{cc} n-t & t \end{array}}{\begin{bmatrix} S & \boldsymbol{O} \\ R & Q \end{bmatrix}} \tag{12.1-3}$$

where the submatrix S is concerned with the process after it has reached the persistent states, $\boldsymbol{O}$ is an $(n-t) \times (t)$ all-zero submatrix, the t by $(n-t)$ submatrix R is concerned with the transition from transient states to persistent states, and the submatrix Q deals with the process as long as it stays in the transient states.

Theorem 1 allows us to conclude that the powers of Q tend to a t by t zero matrix, and consequently all the elements of the last t columns of $P_c^n \rightarrow 0$ as $n \rightarrow 0$.

Now consider an absorbing chain. For an absorbing Markov chain it is obvious that the *canonical form* is given by

$$P_c = \begin{array}{c} \\ n-t\} \\ t\} \end{array} \overset{\begin{array}{cc} n-t & t \end{array}}{\begin{bmatrix} I & \boldsymbol{O} \\ R & Q \end{bmatrix}} \tag{12.1-4}$$

since, by definition, if we have $n - t$ absorbing states, then I represents the absorbing states as an $(n-t) \times (n-t)$ unit submatrix. With probability one, the chain will enter an absorbing state and stay there (i.e., it will be "absorbed"). In what follows, the unit matrix I, when it is used in this chapter for addition, subtraction, and multiplication, will be assumed to be of the same dimensions as the matrix it is being added to or subtracted from, or multiplied by.

Example 1 Let us consider our first example, which has the transition matrix given by

$$P_1 = \begin{array}{c} \\ 1 \\ 2 \\ 3 \\ 4 \\ 5 \end{array} \overset{\begin{array}{ccccc} 1 & 2 & 3 & 4 & 5 \end{array}}{\begin{bmatrix} 1 & 0 & 0 & 0 & 0 \\ p & 0 & q & 0 & 0 \\ 0 & p & 0 & q & 0 \\ 0 & 0 & p & 0 & q \\ 0 & 0 & 0 & 0 & 1 \end{bmatrix}} \tag{12.1-5}$$

Since states 1 and 5 are absorbing, the canonical form is given by

$$P_{c1} = \begin{array}{c} \\ 1 \\ 5 \\ 2 \\ 3 \\ 4 \end{array} \overset{\begin{array}{ccccc} 1 & 5 & 2 & 3 & 4 \end{array}}{\begin{bmatrix} 1 & 0 & 0 & 0 & 0 \\ 0 & 1 & 0 & 0 & 0 \\ p & 0 & 0 & q & 0 \\ 0 & 0 & p & 0 & q \\ 0 & q & 0 & p & 0 \end{bmatrix}} \tag{12.1-6}$$

where the states 1 and 5 are arranged to form a two-dimensional I matrix, and all states agree with the original matrix in the sense that the same elements occur at the same row and column number. Hence it is clear from (12.1-6) that

$$
I=\begin{bmatrix}1 & 0\\ 0 & 1\end{bmatrix} \quad \boldsymbol{O}=\begin{bmatrix}0 & 0 & 0\\ 0 & 0 & 0\end{bmatrix}
$$
$$
R=\begin{bmatrix}p & 0\\ 0 & 0\\ 0 & q\end{bmatrix} \quad Q=\begin{bmatrix}0 & q & 0\\ p & 0 & q\\ 0 & p & 0\end{bmatrix} \tag{12.1-7}
$$

If we do not care at which state the absorbing set of states is entered, we may combine the two absorbing states together to obtain

$$
P_{c1}^{'}=\begin{array}{c} \\ 1\text{ or }5\\ 2\\ 3\\ 4\end{array}\begin{array}{c}\begin{array}{cccc}1\text{ or }5 & 2 & 3 & 4\end{array}\\ \begin{bmatrix}1 & 0 & 0 & 0\\ p & 0 & q & 0\\ 0 & p & 0 & q\\ q & 0 & p & 0\end{bmatrix}\end{array} \tag{12.1-8}
$$

When we are interested in what happens as we go from state j to state k we may make state k an absorbing state. We will find that absorbing chains answer the type of questions we have regarding the time to drop out of lock, as well as the time to enter lock, for lock detectors.

12.2 THE FUNDAMENTAL MATRIX

A very important matrix will be used in the following development. First we prove a theorem required to prove that the matrix exists.

Theorem 2 If Q^n tends to **0** (the zero matrix) as n tends to infinity, then $[I-Q]$ has an inverse, and further

$$
[I-Q]^{-1}=\sum_{k=0}^{\infty}Q^k \qquad Q^0=I \tag{12.2-1}
$$

Proof Consider the following equality:

$$
(I-Q)(I+Q+Q^2+Q^3+...+Q^{n-1})=I-Q^n \tag{12.2-2}
$$

We shall show that $I-Q$ has a nonzero determinant since this is a sufficient condition for the matrix to have an inverse. By hypothesis, the right side of this equation converges to the unit matrix I. Hence the determinant of the right side is nonzero for sufficiently large n. Now the determinant of the product of two matrices is the product of the two determinants. Therefore $I-Q$ cannot have a zero determinant, which is what we set out to show; that is, $I-Q$ has an inverse. Now premultiply both sides of the equation by $[I-Q]^{-1}$ to yield

$$
\left[I+Q+Q^2+Q^3+...Q^{n-1}\right]=[I-Q]^{-1}\left[I-Q^n\right] \tag{12.2-3}
$$

Therefore, as n becomes unbounded, one obtains from (12.2-3)

$$
[I-Q]^{-1}=\sum_{k=0}^{\infty}Q^k \tag{12.2-4}
$$

This leads to the following result.

Corollary 1

$$[I-Q]^{-1}=\sum_{k=0}^{\infty}Q^k \qquad Q^0=I \tag{12.2-5}$$

Definition 1 We define the *fundamental matrix* for an absorbing Markov chain to be the matrix, where:

$$N=[I-Q]^{-1}, \qquad t\times t \tag{12.2-6}$$

and I and Q are of the same dimensions, and Q is as defined in (12.1-4).

12.3 MEAN AND VARIANCE OF THE NUMBER OF TIMES A PROCESS IS IN A TRANSIENT STATE

In this section we develop the mean and variance of the number of times a Markov process, starting in a transient state, remains in a transient state.

Definition 2 We define n_j to be the total number of times that the process is in state j (denoted by S_j), where S_j is a transient state. Also, define u_j^k as a function that is equal to 1 if the process is in state S_j after k steps, and is zero otherwise.

In this chapter we will denote $\{a_{ij}\}$ as a matrix whose ij component is a_{ij}. Similarly we write $\{a_j\}$ for a row vector and $\{a_i\}$ for a column vector.

Theorem 3 Letting T denote the set of transient states, we have

$$\{E_i[n_j]\}=N=[I-Q]^{-1} \qquad t\times t \qquad S_i,S_j\in T \tag{12.3-1}$$

where $\{E_i[n_j]\}$ denotes the $t\times t$ dimensioned matrix, composed of elements formed by the mean number of total times the process is in state j, starting in state i. Hence this result shows that the average number of times that the process is in a particular transient state j, starting in state i, is the ij-th element of the matrix N. Furthermore it establishes the fact that the mean number of times the process is in a particular state is finite.

Proof First we note that by definition

$$n_j=\sum_{k=0}^{\infty}u_j^k \tag{12.3-2}$$

Therefore

$$\{E_i[n_j]\}=\left\{E_i\left[\sum_{k=0}^{\infty}u_j^k\right]\right\}=\left\{\sum_{k=0}^{\infty}E_i\left[u_j^k\right]\right\} \tag{12.3-3}$$

Now since $p_{ij}^{(k)}$ is defined as the probability of going from state i to state j in k steps, we have, since u_j^k is either one or zero, the result

$$\left\{E_i\left[u_j^k\right]\right\} = \left\{\sum_{k=0}^{\infty}(1 - p_{ij}^{(k)})\times 0 + p_{ij}^{(k)}\times 1\right\} = \left\{\sum_{k=0}^{\infty} p_{ij}^{(k)}\right\} \tag{12.3-4}$$

But $\left\{p_{ij}^{(k)}\right\} = \left[P_c\right]^k$, for all states $i, j \in T$. Recall that

$$\left[P_c\right] = \begin{bmatrix} S & \boldsymbol{O} \\ R & Q \end{bmatrix} \tag{12.3-5}$$

It is clear by matrix multiplication that

$$\left[P_c\right]^n = \begin{bmatrix} S^n & \boldsymbol{O} \\ R' & Q^n \end{bmatrix} \tag{12.3-6}$$

where R' is in general different from R, but is of no concern here, and S^n and Q^n are just the n-th powers of their respective matrices. But since the states i and j were restricted to transient states, we have

$$\left\{\sum_{k=0}^{\infty} p_{ij}^{(k)}\right\} = \sum_{k=0}^{\infty} Q^k \qquad S_i, S_j \text{ transient states} \tag{12.3-7}$$

By our corollary 1, we have

$$\left\{\sum_{k=0}^{\infty} p_{ij}^{(k)}\right\} = \left[I - Q\right]^{-1} = N \qquad t \times t \text{ matrix} \tag{12.3-8}$$

This completes the proof.

Example 2 To illustrate the application of the matrix $[N]$ consider the canonical form of the transition matrix

$$P_c = \begin{array}{c} \\ 1 \\ 2 \\ 3 \end{array}\begin{array}{c} \begin{array}{ccc} 1 & 2 & 3 \end{array} \\ \begin{bmatrix} 1 & 0 & 0 \\ 1/2 & 0 & 1/2 \\ 1/4 & 1/4 & 1/2 \end{bmatrix} \end{array} \tag{12.3-9}$$

Therefore it follows that

$$Q = \begin{array}{c} \\ 2 \\ 3 \end{array}\begin{array}{c} \begin{array}{cc} 2 & 3 \end{array} \\ \begin{bmatrix} 0 & 1/2 \\ 1/4 & 1/2 \end{bmatrix} \end{array} \tag{12.3-10}$$

Hence the fundamental matrix is given by

$$N = [I - Q]^{-1} = \begin{matrix} & 2 & 3 \\ 2 & 4/3 & 4/3 \\ 3 & 2/3 & 8/3 \end{matrix} \qquad (12.3\text{-}11)$$

It follows that the mean number of times the process is in state 2 starting in state 3 before entering state 1 is 2/3, whereas the mean number of times the process is in state 3, starting at state 3, before entering state 1, is 8/3. Note that the mean number of times the process is in a nonabsorbing state, starting in state 3, is 10/3 (2/3+8/3), and the mean number of times the process is in a nonabsorbing state, starting in state 2, is 8/3 (4/3+4/3).

We are now at a point in our development that we can prove a slightly less general form of our main results. The results are found in Kemeny and Snell [2].

Definition 3 Let n be the function giving the time, including the original position, in which the process is in a transient state. Note that n is an integer-valued function.

If a process starts in an ergodic state, then $n = 0$. If the process starts in a transient state, then n gives the total number of steps needed to reach an ergodic set. It is to be noted that it is assumed that the time (or dwell in each state) is unity. In an absorbing chain this is the time to absorption.

Theorem 4 The mean value of n, given that the process starts in state i, is given by

$$\{E_i[n]\} = [N]\mathbf{1} \qquad S_i \in T \text{ and } \mathbf{1} \text{ is a } t \times 1 \text{ column vector} \qquad (12.3\text{-}12)$$

where $[N]$ is the matrix $[I\text{-}Q]^{-1}$ and $\mathbf{1}$ is a t component unit column vector (recall that Q is a $t \times t$ matrix), each component of which represents the time in that state. Note that both the left side and the right side of (12.3-12) are column vectors with t components, as is required.

Proof From our definition of n_j we have

$$n = \sum_{S_j \in T} n_j \qquad (12.3\text{-}13)$$

with T denoting the set of transient states, so that

$$\{E_i[n]\} = \left\{ \sum_{S_j \in T} E_i[n_j] \right\} \qquad (12.3\text{-}14)$$

But in Theorem 3 we have showed that

$$\{E_i[n_j]\} = N \qquad (12.3\text{-}15)$$

Hence it follows that

$$\{E_i[n]\} = N\mathbf{1} \qquad (12.3\text{-}16)$$

Since this expression gives the row sums of the matrix N.

Theorem 5 The variance of the time that the process is in a transient state, starting in state i, is given by

$$\text{var}_i[n] = (2N - I)N\mathbf{1} - (N\mathbf{1})_{\text{sq}} \qquad S_i \in T \tag{12.3-17}$$

where again $(N\mathbf{1})_{\text{sq}}$ is a t dimensional column vector having components that are the square of each component of $N\mathbf{1}$. Thus this equation gives the variance of the number of times (n) that the process is in transient states before being absorbed, given that the process started in state i. This result holds for unit occupation times (alternatively dwell times).

Proof First we note that by definition

$$\{\text{var}_i[n]\} = \{E_i[n^2]\} - \{E_i[n]\}^2 \tag{12.3-18}$$

Hence it remains to evaluate $E_i[n^2]$, since the second term has been evaluated in (12.3-16). Denote $\tilde{T}$ as the set of nontransient states (persistent states). Then

$$\{E_i[n^2]\} = \left\{\sum_{S_k \in \tilde{T}} p_{ik} \cdot 1 + \sum_{S_k \in T} p_{ik} E_k[(n+1)^2]\right\} \tag{12.3-19}$$

so that

$$\{E_i[n^2]\} = \left\{\sum_{S_k \in T \cup \tilde{T}} p_{ik} \cdot 1 + \sum_{S_k \in T} p_{ik} E_k[n^2] + 2\sum_{S_k \in T} p_{ik} E_k[n]\right\} \tag{12.3-20}$$

where $S_k \in T \cup \tilde{T}$ denotes the states that are in T or $\tilde{T}$ (i.e., the whole space). Now since the probability of going from state i to any of the possible states is 1, we have

$$\{E_i[n^2]\} = \left\{\mathbf{1} + \sum_{S_k \in T} p_{ik} \, \text{E}_i[n^2] + 2Q(N\mathbf{1})\right\} \tag{12.3-21}$$

where the last term on the right was obtained by noting $\sum_{S_k \in T} p_{ik} E_k[n] = Q(N\mathbf{1})$. Finally, we have

$$\{E_i[n^2]\} = \mathbf{1} + 2QN\mathbf{1} + QE_i[n^2] \tag{12.3-22}$$

or

$$\{E_i[n^2]\} = (I - Q)^{-1}(2QN\mathbf{1} + \mathbf{1}) \tag{12.3-23}$$

or since $N = (I-Q)^{-1}$ it follows that

$$\{E_i[n^2]\} = 2NQN\mathbf{1} + N\mathbf{1} \tag{12.3-24}$$

From the definition of N it is not difficult to show that

$$QN = NQ = N - I \tag{12.3-25}$$

so that

$$\{E_i[n^2]\} = 2[N - I]N\mathbf{1} + N\mathbf{1} = (2N - I)N\mathbf{1} \tag{12.3-26}$$

Finally, we have the relationship

$$\{Var_i[n]\} = (2N - I)N\mathbf{1} - \{E_i[n]\} = (2N - I)N\mathbf{1} - (N\mathbf{1})^2_{sq} \tag{12.3-27}$$

This is the result that we wanted to prove. Note that $(N\mathbf{1})_{sq}$ is the column vector in which the elements are the square of the elements in $(N\mathbf{1})$.

12.4 MEAN AND VARIANCE OF THE NUMBER OF TIMES A PROCESS IS IN A TRANSIENT STATE—GENERAL CASE

In this section the concept of n, the number of times that the system is in a transient state, is generalized to the case that the dwell times are not equal. A more general result for the mean and the variance of the "time" that a system is in a transient state is presented with the proofs.

Definition 4 Let t_n be the function giving the time, including the time in the original position, that the process is in a transient state. Note that t_n generalizes the variable n, of Definition 3, to noninteger and unequal values for different states. The function t_n may be written as

$$t_n = \sum_{s_j \in T} n_j T_j \tag{12.4-1}$$

where n_j is the number of times the process is in state j and T_j is the duration or "dwell" of state j. Now the mean time to reach an absorbing state, starting in state i, is given by

$$\{M_i(t_n)\} = \left\{ \sum_{s_j \in T} M_i(n_j T_j) \right\} \tag{12.4-2}$$

Theorem 6 The mean time the process is in a transient state including the original position, starting in state i, is given by

$$\{E_i[t_n]\} = N\boldsymbol{\tau} \qquad \boldsymbol{\tau} = \begin{Bmatrix} T_1 \\ T_2 \\ \vdots \\ T_t \end{Bmatrix} \tag{12.4-3}$$

where $\boldsymbol{\tau}$ is a t-component column vector, whose components are the respective "dwell" times and t_n denotes time. Note that $\boldsymbol{\tau}$ generalizes the $\mathbf{1}$ vector to unequal (and not necessarily unity) components. This result generalizes (12.3-16) to the case that the dwell times are not unity and not necessarily equal.

Proof Starting in state "i", one has for the mean time to reach an absorbing state, the expression

$$\{M_i(t_n)\} = \left\{ \sum_{s_j \in T} n_j T_j \right\} \tag{12.4-4}$$

which is a $t \times 1$ column vector. Again use the definition of n_j given in Definition 2, and placing the expectation inside the summation produces

$$\{M_i(n_j T_j)\} = \left\{ \sum_{k=0}^{\infty} M_i\{u_j^k T_j\} \right\} \tag{12.4-5}$$

Evaluating the mean value produces

$$\{M_i(n_j T_j)\} = \left\{ \sum_{k=0}^{\infty} \left[(1 - p_{ij}^{(k)})0 + p_{ij}^{(k)} T_j \right] \right\} \tag{12.4-6}$$

where $p_{ij}^{(k)}$ is the transition probability of going from state i to state j in k steps. Hence we have

$$\{M_i(n_j T_j)\} = \left\{ \sum_{k=0}^{\infty} \left[p_{ij}^{(k)} T_j \right] \right\} = \sum_{k=0}^{\infty} Q^k \boldsymbol{\tau} \tag{12.4-7}$$

From (12.2-4) and (12.2-6) it follows that

$$\{M_i(n_j T_j)\} = \sum_{k=0}^{\infty} Q^k \boldsymbol{\tau} = N \boldsymbol{\tau} \tag{12.4-8}$$

Therefore Theorem 6 follows. Hence the mean time to go from state i, to an absorbing state, is given by the result of Theorem 6.

Theorem 7 The variance of the time that the process is in a transient state, including the initial state, given that it starts in state i (a transient state), is given by

$$\{\mathrm{var}_i(t_n)\} = 2[N][T]Q[N]\boldsymbol{\tau} + [N](\boldsymbol{\tau}_{sq}) - ([N]\boldsymbol{\tau})_{sq} \tag{12.4-9}$$

with $(a)_{sq}$ defined for the column vector a as the column vector with each component squared. In addition the matrix T is defined by

$$T = \begin{bmatrix} T_1 & 0 & 0 & 0 & \cdots & 0 \\ & T_2 & & & \cdots & 0 \\ & & T_3 & & \cdots & 0 \\ & & & T_4 & \cdots & 0 \\ \vdots & \vdots & \vdots & \vdots & \ddots & \vdots \\ 0 & 0 & 0 & 0 & 0 & T_t \end{bmatrix} \tag{12.4-10}$$

Note that the T_i in this matrix is the same parameters as defined in the definition of $\boldsymbol{\tau}$.

Proof Now consider a recursive expression of the variance of the time to reach an absorbing state. Consider

$$\left\{M_i(t_n^2)\right\} = \left\{\sum_{s_k \in \tilde{T}} p_{ik}T_i^2 + \sum_{s_k \in T} p_{ik}M_k\left\{((t_n + T_i)^2)\right\}\right\} \tag{12.4-11}$$

where T denotes the transient states and $\tilde{T}$ denotes the absorbing states. Expanding the squared term on the right-hand side of (12.4-11) and combining the T_i^2 terms produces

$$\left\{M_i(t_n^2)\right\} = \left\{\sum_{s_k \in \tilde{T} \cup T} p_{ik}T_i^2 + \sum_{s_k \in T} p_{ik}M_k\left\{(t_n)^2\right\} + 2\sum_{s_k \in T} p_{ik}T_iM_k\left\{(t_n)\right\}\right\} \tag{12.4-12}$$

Recognizing the definition of the Q matrix in the second term produces

$$\left\{M_i(t_n^2)\right\} = \{\sum_{s_k \in \tilde{T} \cup T} p_{ik}T_i^2\} + Q\left\{M_k(t_n)^2\right\} + 2\left\{\sum_{s_k \in T} p_{ik}T_iM_k(t_n)\right\} \tag{12.4-13}$$

Solving for $\left\{M_i(t_n^2)\right\}$ yields

$$[I - Q]\left\{M_i(t_n^2)\right\} = \{\sum_{s_k \in \tilde{T} \cup T} p_{ik}T_i^2\} + 2\left\{\sum_{s_k \in T} p_{ik}T_iM_k(t_n)\right\} \tag{12.4-14}$$

Consider the first term on the right side of (12.4-14) and denote it as T_I

$$T_I = \left\{\sum_{s_k \in \tilde{T} \cup T} p_{ik}T_i^2\right\} = row\ sum\ of \begin{bmatrix} p_{11}T_1^2, p_{12}T_1^2, \ldots p_{1s}T_1^2 \\ p_{21}T_1^2, p_{22}T_1^2, \ldots p_{2s}T_1^2 \\ \vdots \\ p_{s1}T_1^2, p_{s2}T_1^2, \ldots p_{ss}T_1^2 \end{bmatrix} \tag{12.4-15}$$

Since the probability of going from state i to one of the s states is unity, it follows that

$$T_I = \begin{Bmatrix} T_1^2 \\ T_2^2 \\ \vdots \\ T_s^2 \end{Bmatrix} = \boldsymbol{\tau}_{sq} \tag{12.4-16}$$

Now consider the second term, call it T_{II}.

$$T_{II} = 2\left\{\sum_{s_k \in T} p_{ik}T_iM_k(t_n)\right\} \tag{12.4-17}$$

Reordering the transient states as 1 through s–1, assuming only one absorbing state, and denoting $M_k(t_n) = M_k$ for notational convenience yields

$$T_{II} = 2\times \text{ row sum of} \begin{bmatrix} p_{11}T_1M_1, p_{12}T_1M_2, \dots p_{1(s-1)}T_1M_{s-1} \\ p_{21}T_2M_1, p_{22}T_2M_2, \dots p_{2(s-1)}T_2M_{s-1} \\ \vdots \\ p_{(s-1)1}T_{s-1}M_1, p_{(s-1)2}T_{s-1}M_2, \dots p_{(s-1)(s-1)}T_{s-1}M_{s-1} \end{bmatrix} \tag{12.4-18}$$

This can be rewritten as

$$T_{II} = 2\begin{bmatrix} p_{11}T_1, p_{12}T_1, \dots p_{1(s-1)}T_1 \\ p_{21}T_2, p_{22}T_2, \dots p_{2(s-1)}T_2 \\ \vdots \\ p_{(s-1)1}T_{s-1}, p_{(s-1)2}T_{s-1}, \dots p_{(s-1)(s-1)}T_{s-1} \end{bmatrix} \begin{Bmatrix} M_1 \\ M_2 \\ \vdots \\ M_{s-1} \end{Bmatrix} \tag{12.4-19}$$

Now consider the matrix A defined by

$$A = \begin{bmatrix} p_{11}T_1, \dots & p_{1(s-1)}T_1 \\ p_{21}T_1, \dots & p_{2(s-1)}T_2 \\ \vdots & \\ p_{(s-1)1}T_1, \dots p_{(s-1)(s-1)}T_{s-1} & \end{bmatrix} \tag{12.4-20}$$

Also define the matrix T by

$$T = \begin{bmatrix} T_1, 0, 0, \dots 0 \\ 0, T_2, 0, \dots 0 \\ 0, 0, T_3, \dots, 0 \\ \vdots \\ 0, 0, 0, \dots T_{s-1} \end{bmatrix} \tag{12.4-21}$$

Then it is clear that

$$\begin{bmatrix} T_1, 0, 0, \dots 0 \\ 0, T_2, 0, \dots 0 \\ 0, 0, T_3, \dots, 0 \\ \vdots \\ 0, 0, 0, \dots T_{s-1} \end{bmatrix} \begin{bmatrix} p_{11}, p_{12}, \dots p_{1(s-1)} \\ p_{21}, p_{22}, \dots p_{2(s-1)} \\ \vdots \\ p_{(s-1)1}, p_{(s-1)2}, \dots p_{(s-1)(s-1)} \end{bmatrix} = \begin{bmatrix} T_1, 0, 0, \dots 0 \\ 0, T_2, 0, \dots 0 \\ 0, 0, T_3, \dots, 0 \\ \vdots \\ 0, 0, 0, \dots T_{s-1} \end{bmatrix} [Q] = [A] \tag{12.4-22}$$

It follows from (12.4-14), (12.4-16), (12.4-19), (12.4-20), and (12.4-22) that

$$[I - Q] = \{M_i(t_n^2)\} = \boldsymbol{\tau}_{sq} + 2[T][Q]\{M\} \tag{12.4-23}$$

where $\{M\}$ is the column vector having components $\{M_i(t_n)\}$. Since $\{M\}$ has been evaluated as

$$\{M\} = [N]\boldsymbol{\tau} \tag{12.4-24}$$

we can evaluate the second moments by

$$\{M_i(t_n^2)\} = 2[I-Q]^{-1}[T][Q][I-Q]^{-1}\boldsymbol{\tau} + [I-Q]^{-1}\boldsymbol{\tau}_{sq} \tag{12.4-25}$$

Now the variance is given by

$$\sigma_i^2 = M_i(t_n^2) - M_i(t_n)^2 \tag{12.4-26}$$

or

$$\{\sigma_i^2\} = 2[N][T][Q][N]\boldsymbol{\tau} + [N]\boldsymbol{\tau}_{sq} - \{[N]\boldsymbol{\tau}\}_{sq} \tag{12.4-27}$$

As an application of these theorems, let us consider an example. We will consider the time at each state in the rest of the chapter as the "dwell time."

Therefore the variance of the time it takes to reach the absorbing state from state i is given by the i-th component of $\{\sigma_i^2\}$.

Example 3 Obtain the mean and variance of the time in a transient state using the transition matrix of Example 2. Recall that the transition matrix is in canonical form and is represented by

$$P_c = \begin{matrix} & 1 & 2 & 3 \\ 1 & 1 & 0 & 0 \\ 2 & 1/2 & 0 & 1/2 \\ 3 & 1/4 & 1/4 & 1/2 \end{matrix} \tag{12.4-28}$$

From this transition matrix, Q and N were obtained as

$$Q = \begin{matrix} & 2 & 3 \\ 2 & 0 & 1/2 \\ 3 & 1/4 & 1/2 \end{matrix} \tag{12.4-29}$$

$$N = [I - Q]^{-1} = \begin{matrix} & 2 & 3 \\ 2 & 4/3 & 4/3 \\ 3 & 2/3 & 8/3 \end{matrix} \tag{12.4-30}$$

It will be assumed that the dwell time in each state is one second. Using Theorem 4 or 6 with $T_i = 1$, one obtains

$$\{E_i[t_n]\} = N\mathbf{1} = \begin{matrix} & 2 & 3 \\ 2 & 4/3 & 4/3 \\ 3 & 2/3 & 8/3 \end{matrix} \begin{Bmatrix} 1 \\ 1 \end{Bmatrix} = \begin{matrix} 2 \\ 3 \end{matrix} \begin{Bmatrix} 8/3 \\ 10/3 \end{Bmatrix} \tag{12.4-31}$$

Thus we see that the mean time in a transient state before being absorbed, starting in state 2, is 8/3 seconds and, starting in state 3, is 10/3 seconds. To compute the variance, one must use (12.3-27) or (12.4-27). Hence, one has, using (12.3-27),

$$\{\text{var}_i[t_n]\} = (2N - I)N\mathbf{1} - (N\mathbf{1})_{sq} \tag{12.4-32}$$

Therefore

$$\{\text{var}_i[t_n]\} = \frac{2}{3}\left\{\begin{bmatrix} 8/3 & 8/3 \\ 4/3 & 16/3 \end{bmatrix} - \begin{bmatrix} 1 & 0 \\ 0 & 1 \end{bmatrix}\right\}\begin{Bmatrix} 8/3 \\ 10/3 \end{Bmatrix} - \begin{Bmatrix} 64/9 \\ 100/9 \end{Bmatrix} \tag{12.4-33}$$

Combining leads to the final result

$$\{\text{var}_i[t_n]\} = \frac{2}{3}\begin{Bmatrix} 56/9 \\ 62/9 \end{Bmatrix} \tag{12.4-34}$$

Consequently the variance of the number of times (or time) the process is in transient states is slightly larger starting in state 3 than starting in state 2.

This theory allows us to readily compute the mean and variance of the number of steps (including the original position) and the time it takes to be absorbed in an absorbing state.

In general to study the mean time (or variance of the time) it takes to reach a particular state, it is convenient to modify the transition matrix or state transition diagram so that the state under study is an absorbing one. The resulting process is a Markov process with one absorbing state. The behavior of the process, before absorption, is exactly the same as the original process.

12.5 THE PROBABILITY OF STARTING IN A TRANSIENT STATE AND ENDING IN A PERSISTENT STATE

Another interesting result due to Kemeny and Snell [2] concerns the probability of starting in a transient state and ending up in a persistent state. This result yields no time information. Furthermore, with only one absorbing state the probabilities are all 1! Therefore the result is primarily useful when there are two or more absorbing states.

Theorem 8 The probability that the Markov process, starting in a transient state i, ends up in an absorbing state j, is b_{ij}, where

$$\{b_{ij}\} = B = NR \qquad (t \times (n-t) \text{ matrix}) \tag{12.5-1}$$

where R is the submatrix defined in (12.1-3) and deals with the transition from transient to absorbing states.

Proof Starting in state i, the process may be absorbed in state j in one or more steps. For a single step the probability of absorption is p_{ij}. If the process is not absorbed, the process can move to either another absorbing state (and cannot reach state j) or to a transient state k. In this latter case the probability of being absorbed in state j, from state k, is b_{kj} for any transient state k. Therefore we can write

$$b_{ij} = p_{ij} + \sum_{S_k \in T} p_{ik} b_{kj} \tag{12.5-2}$$

which, in matrix form is

$$B = R + QB \tag{12.5-3}$$

Since the submatrix R contains the transition probabilities of the transient states (the set T) to the absorbing states (the set $\widetilde{T}$), and the submatrix Q contains the transition probabilities of transient states to transient states. Solving (12.5-3) yields

$$B = [I - Q]^{-1} R = NR \tag{12.5-4}$$

where $N = [I - Q]^{-1}$ from (12.2-6).

As can be seen from this problem, this theory can be applied to an acquisition process, as well as to a lock detection process, besides a myriad of other problems in other fields.

Example 4 Consider the following transition matrix in canonical form

$$P = \begin{bmatrix} 1, & 0, & 0, & 0, & 0 \\ 0, & 1, & 0, & 0, & 0 \\ 1/3, & 0, & 0, & 2/3, & 0 \\ 0, & 0, & 1/3, & 0, & 2/3 \\ 0, & 2/3, & 0, & 1/3, & 0 \end{bmatrix} \tag{12.5-5}$$

If follows from the definition of R that

$$R = \begin{bmatrix} 1/3, 0 \\ 0, \quad 0 \\ 0, 2/3 \end{bmatrix} \tag{12.5-6}$$

And N can be shown to be given by

$$N = [I - Q]^{-1} = \begin{bmatrix} 7/5, 6/5, 4/5 \\ 3/5, 9/5, 6/5 \\ 1/5, 3/5, 7/5 \end{bmatrix} \tag{12.5-7}$$

It therefore follows that

$$B = NR = \begin{matrix} \\ S_2 \\ S_3 \\ S_4 \end{matrix} \overbrace{\begin{bmatrix} 7/15, 8/15 \\ 1/5, \quad 4/5 \\ 1/15, 14/15 \end{bmatrix}}^{S_1 \qquad S_5} \tag{12.5-8}$$

Thus the probability of going from state 2 to state 1 is 7/15, and the probability of going from state 4 to state 5 is 14/15.

12.6 LOCK DETECTOR PERFORMANCE

Now that we have built up sufficient theory to address lock detector problems, let us consider some lock detector algorithms.

Lock detectors are devices that indicate when a synchronizing device has succeeded in obtaining synchronization or lock. The synchronizing device could be a code tracking loop, a code acquisition system, a carrier-tracking loop, a frequency-tracking loop, or a bit synchronization loop. Virtually any synchronization function that requires an acquisition process or a tracking process also requires a lock detector to indicate when lock occurs. The results that are developed here are applicable to lock detectors that can be characterized as Markov chains, with the transition probabilities being independent of time. In the theory to follow it is assumed that the transition probabilities are known. These probabilities have to be determined from the lock detection devices themselves. This section will utilize the results of Section 12.4.

The basic function of a lock detector is to indicate the in-lock condition when the signal is "locked" and to indicate the out-of-lock condition when the system is out of lock. Unfortunately, no lock detector is perfect and thus there is a probability that when the system is in lock the lock detector will declare the out-of-lock state and vice versa. Normally this probability is very small, by design. The mean time to declare the out-of-lock condition, when the system is in lock, is a useful measure of the capabilities of a lock detector. Clearly a very long time is desired.

Commonly, lock detectors utilize a long integration time, or a series of integrations based on a counter that declares the system to be in lock until some number of below-threshold indications occur, at which time the system is declared out of lock. One possible problem with simply extending the integration time to ascertain the lock state is due to the fact that if a signal drop-out occurs due, for example, to antenna switching or deep fades, a single integration might become unreliable. Thus it, may indicate that the system is in lock, when it is in fact out of lock. Counter-type lock detectors will be the subject of this section, since a single drop out does not seriously affect their performance. The actual implementation can be done in hardware or software, or a combination of both.

We shall utilize the results that were developed in this chapter to solve a very simple type of lock detector based on an integrator with an integration time of T seconds. We shall assume that when the signal is present, the probability of lock detector detection is p, and the probability of a missed detection is q. It follows that $p + q = 1$. One missed detection causes the system to declare the out of lock condition, even if the system is in lock. Also a single detection puts the system in the locked state, directly from the unlocked state. The state transition diagram of this single count lock detector is shown in Figure 12.6-1.

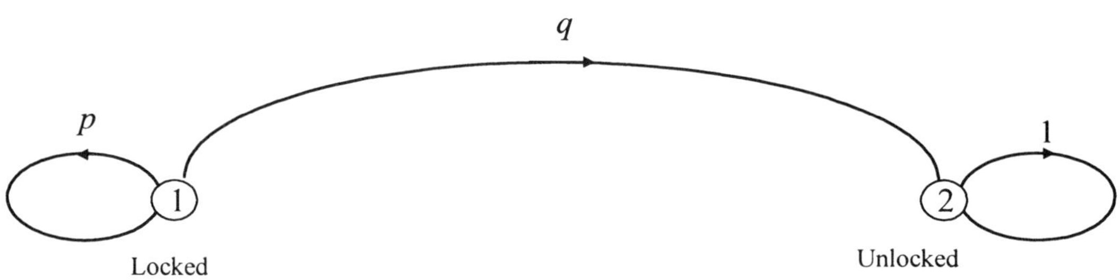

Figure 12.6-1 A simple single-count lock detector state transition diagram.

The process starts in state 1 (the locked state) for T seconds, then the system is tested again and either remains in state 1 for T more seconds or is "absorbed" in state 2. This is lock state monitoring.

The state transition matrix for the detection of the unlocked condition, given that the system is in the locked condition, is given by

$$P = \begin{matrix} & \begin{matrix} 1 & 2 \end{matrix} \\ \begin{matrix} 1 \\ 2 \end{matrix} & \begin{bmatrix} p & q \\ 0 & 1 \end{bmatrix} \end{matrix} \tag{12.6-1}$$

In canonical form the matrix becomes

$$P_c = \begin{matrix} & \begin{matrix} 2 & 1 \end{matrix} \\ \begin{matrix} 2 \\ 1 \end{matrix} & \begin{bmatrix} 1 & 0 \\ q & p \end{bmatrix} \end{matrix} \tag{12.6-2}$$

The mean time to absorption is given by (using Theorem 6 with $\tau = T$)

$$E_1[t_n] = N\tau = (1-p)^{-1}T = \frac{T}{q} \tag{12.6-3}$$

since Q is a one-dimensional matrix. Thus, the mean time to declare that the system is out of lock, when it is in lock, is given by (12.6-3). As an example, if $p = 0.99$, then the mean time to falsely indicate that the system is out of lock, when it is in lock, is $100T$. Note that the results could be applied to the case when the signal is absent (for example, when the signal drops out). In this case p would normally be near zero, and q near one, so that the mean time to correctly indicate the out of lock state, when the signal is out of lock, is about T seconds, since $(1-p)^{-1}T \cong T$ seconds. It is to be noted that (12.6-3) is the mean time to declare that the system is out of lock when it is in fact in lock. Thus p must be designed to be very large when the system is in lock, and p should be very small when the system is out of lock.

Thus there are two conflicting requirements for lock detectors: (1) the mean time to declare out of lock, when the system is in lock, should be very large, and (2) the mean time to detect the out of lock condition, when the system is out of lock, should be as small as possible!

Thus, from this example it is seen that the basic modeling method is to consider the state or states that one is operating in to be transient states and then characterize the state that is to be investigated as an absorbing state. In the previous example the unlocked state was modeled as an absorbing state and the in-lock state (state 1) as the transient state. This simple example for the mean time to incorrectly indicate that the system is out of lock can be solved in a straightforward manner without resorting to the matrix theory presented here. Let P_k be the probability of being absorbed on the k-th trial. Also $p = 1-q$. The mean time to be absorbed is given by the following expression

$$\bar{T} = \sum_{k=1}^{\infty} kTP_k = \sum_{k=1}^{\infty} k(1-q)^{k-1}qT \tag{12.6-4}$$

Note that in $k-1$ trials, it is not absorbed (with probability $(1-q)^{k-1}$), but in the k-th trial it is absorbed (with probability q), so that (12.6-4) follows. Note that

$$\sum_{k=1}^{\infty} k(1-q)^{k-1} = \frac{1}{q^2} \tag{12.6-5}$$

Inserting (12.6-5) into (12.6-4) results in the expression

$$\bar{T} = \frac{T}{q} = \frac{T}{1-p} \tag{12.6-6}$$

as was obtained form the matrix theory approach indicated earlier. However, as the number of states increases, this direct enumeration of the k-th absorption probability becomes much more difficult to formulate, and therein lies the advantage of the matrix method. Although not computed, the variance can be easily obtained in the same manner.

As a more complex example of a lock detector consider Example 5.

Example 5 Consider a lock detector that requires that it must have three consecutive missed detections before the system is declared to be out of lock. It is clear that the state transition diagram associated with this detector is given in Figure 12.6-2.

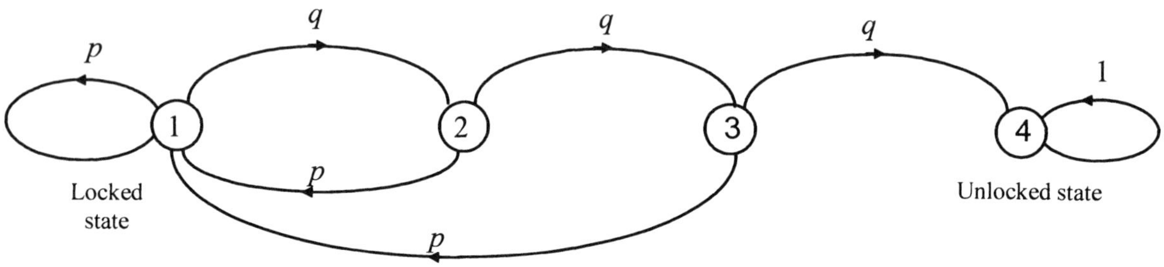

Figure 12.6-2 A "three consecutive counts" lock detector state transition diagram.

The state transition matrix is given by

$$P = \begin{array}{c} \\ 1 \\ 2 \\ 3 \\ 4 \end{array}\begin{array}{c} \begin{array}{cccc} 1 & 2 & 3 & 4 \end{array} \\ \begin{bmatrix} p & q & 0 & 0 \\ p & 0 & q & 0 \\ p & 0 & 0 & q \\ 0 & 0 & 0 & 1 \end{bmatrix} \end{array} \tag{12.6-7}$$

In canonical form the matrix is of the form

$$P_c = \begin{array}{c} \\ 4 \\ 1 \\ 2 \\ 3 \end{array}\begin{array}{c} \begin{array}{cccc} 4 & 1 & 2 & 3 \end{array} \\ \begin{bmatrix} 1 & 0 & 0 & 0 \\ 0 & p & q & 0 \\ 0 & p & 0 & q \\ q & p & 0 & 0 \end{bmatrix} \end{array} \tag{12.6-8}$$

since the fourth state is the absorbing one. The mean value vector is given by

$$\{E_i[t_n]\} = \left\{ \begin{bmatrix} 1 & 0 & 0 \\ 0 & 1 & 0 \\ 0 & 0 & 1 \end{bmatrix} - \begin{bmatrix} p & q & 0 \\ p & 0 & q \\ p & 0 & 0 \end{bmatrix} \right\}^{-1} \begin{Bmatrix} 1 \\ 1 \\ 1 \end{Bmatrix} T \tag{12.6-9}$$

where T is the duration of each dwell time. Evaluating this expression yields

$$\{E_1[t_n]\} = \frac{1+q+q^2}{q^3} \tag{12.6-10}$$

It has been proven by Holmes [4] that the n consecutive count lock detector has a mean false dismissal time of

$$\{E_1[t_n]\} = \frac{1+q+q^2+\cdots q^{n-1}}{q^n} \tag{12.6-11}$$

where there are $n+1$ nodes on the state transition diagram. Kordosz [5] has also obtained the variance for this lock detector and some bounds on the probability distribution function for the time to lose lock.

Example 6 Another type of detector is the up-down counter that starts in state 1 and remains there unless a miss detection occurs, at which time the counter drops to state 2. A hit returns the counter to state 1, whereas a miss detection sends the counter to state 3 from state 2. Finally when the fourth state is reached, the lock detector declares the system is out of lock.

Consider the state transition diagram for a three-state, up-down counter, lock detector shown in Figure 12.6-3.

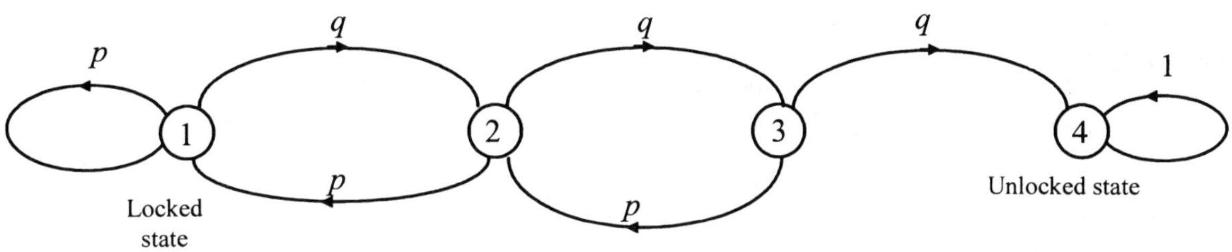

Figure 12.6-3 "Three state up-down counter" lock detector state transition diagram.

The canonical form of the matrix is given by

$$P_c = \begin{array}{c} \\ 4 \\ 1 \\ 2 \\ 3 \end{array}\!\!\begin{array}{c} \begin{array}{cccc} 4 & 1 & 2 & 3 \end{array} \\ \begin{bmatrix} 1 & 0 & 0 & 0 \\ 0 & p & q & 0 \\ 0 & p & 0 & q \\ q & 0 & p & 0 \end{bmatrix} \end{array} \tag{12.6-12}$$

It can be shown (Problem 5) that the mean time to declare a false lock, starting in state 1, is given by

$$\{E_1[t_n]\} = \frac{1+2q^2}{q^3}T \tag{12.6-13}$$

which does not differ significantly from the lock detector shown in Figure 12.6-2. Note that as a check, if $q = 1$ (and therefore $p = 0$), that

$$\{E_1[t_n]\} = 3T \tag{12.6-14}$$

as is obvious from Figure 12.6-3. Note also that (12.6-13) holds when the signal is not present also. In general there is not much to choose between the two types of three-count lock detectors.

The design of lock detectors must balance the desire not to declare the out-of-lock state, when the signal is present, against the desire to quickly declare the out-of-lock state when the signal is not present.

12.7 LOCK DETECTOR SYSTEM MODELS

In this section we will indicate some typical lock detector system models, which can be used for lock detection and monitoring. We will consider residual carrier, suppressed carrier, and a PN code lock detector, in the following. When the lock detector statistics are known, the probabilities p and q can be determined and applied to the lock detector algorithms as discussed in this chapter. This enables one to determine the lock detector performance.

In order to utilize the theory presented here, it is necessary to: (1) determine the values of the probabilities p and q, (2) choose a lock detector algorithm, and (3) then apply these probabilities to the algorithm to evaluate the mean time to an event, to acquire lock, for example.

12.7.1 Residual Carrier Loop Lock Detector Block Diagram Model

This lock detector model for coherent detection of an unmodulated tone plus noise is shown in Figure 12.7-1. It is assumed that the modulated sidebands do not possess any significant spectral composition at the carrier frequency. It is assumed that the tone plus white Gaussian noise are fed into the phase lock loop. The multiplier, VCO, along with the quadrature (input) multiplier, forms the phase locked loop. The lock detector is formed from inphase multiplier, the low pass filter (LPF), and the integrator. The LPF removes the double frequency term, and the integrator does the filtering of the signal plus noise. The integrated output is then compared to a threshold, which determines the lock status.

In this scheme it is necessary to provide for an AGC circuit to prevent the noise level from changing with the absence or presence of the signal. Typically both a noncoherent and a coherent AGC are used in coherent systems.

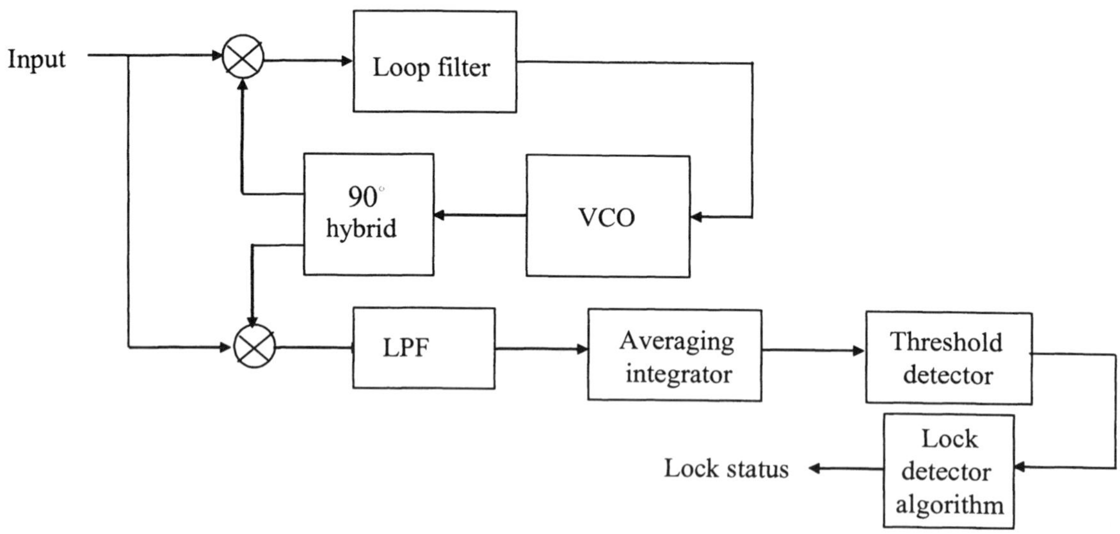

Figure 12.7-1 Carrier loop lock detector block diagram model.

12.7.1.1 Residual Carrier Loop Lock Detector Performance

Let us consider, as an example, how to compute the values of p and q for the simple lock detector of Figure 12.7-1. Let us model received signal plus thermal noise as

$$y(t)=\sqrt{2P}\cos(\omega_0 t+\theta)+\sqrt{2}n_c(t)\cos(\omega_0 t+\theta)+\sqrt{2}n_s(t)\sin(\omega_0 t+\theta) \quad (12.7\text{-}1)$$

where P is the received signal power in watts, ω_0 is the center frequency in radians/sec, and the white Gaussian noise is represented by the in-phase and quadrature baseband components as seen in the last two terms in (12.7-1). The input noise is assumed to have a one-sided noise spectral density of $N_0/2$. The reference signal $r(t)$ is given by

$$r(t)=\sqrt{2}\cos\left(\omega t+\hat{\theta}\right) \quad (12.7\text{-}2)$$

It will be assumed that the carrier phase lock loop is tracking perfectly, so that $\hat{\theta}=\theta$, for simplicity in the analysis to follow. Out of the lower mixer, just before the low pass filter, in Figure 12.7-1, the following signal and noise components exist

$$u(t)=\sqrt{P}+n_c(t)+O(2\omega_0) \quad (12.7\text{-}3)$$

where $O(2\omega_0)$ is the sum term which is located at twice the carrier frequency. After removing $O(2\omega_0)$ and averaging over T seconds, the output is given by

$$v(T)=\frac{1}{T}\int_0^T\left[\sqrt{P}+n_c(t)\right]dt \quad (12.7\text{-}4)$$

The mean and variance can be computed as

$$\mu=\frac{1}{T}\int_0^T\sqrt{P}dt=\sqrt{P} \quad (12.7\text{-}5)$$

and

$$\sigma^2=\frac{1}{T^3}\int_0^T\int_0^T \mathrm{E}\left[n_c(t)n_c(u)\right]dtdu=\frac{N_0}{2T} \quad (12.7\text{-}6)$$

where we have used the fact that the white noise component has expectation given by $(N_0/2)\delta(t-u)$. Let the threshold be δ; then the probability p is given by

$$p=\int_\delta^\infty\frac{1}{\sqrt{2\pi\sigma^2}}\exp\left(-\frac{(y-\mu)^2}{2\sigma^2}\right)dy \quad (12.7\text{-}7)$$

Now if the threshold is chosen halfway between zero and $\sqrt{P}$, then p can be expressed as

$$p = Q\left(-\sqrt{\frac{PT}{N_0}}\right) = Q\left(-\sqrt{R}\right) \text{ and } q = 1 - p \tag{12.7-8}$$

where $Q(x)$ is defined by

$$Q(x) = \frac{1}{\sqrt{2\pi}} \int_x^{\infty} \exp\left(-\frac{t^2}{2}\right) dt \tag{12.7-9}$$

and $R = \sqrt{PT/N_0}$. The results are shown in Figure 12.7-2. The upper curve (solid curve) is the value of p plotted against the signal-to-noise ratio parameter, R in dB. The lower curve, which is dashed, represents the value of q versus R in dB.

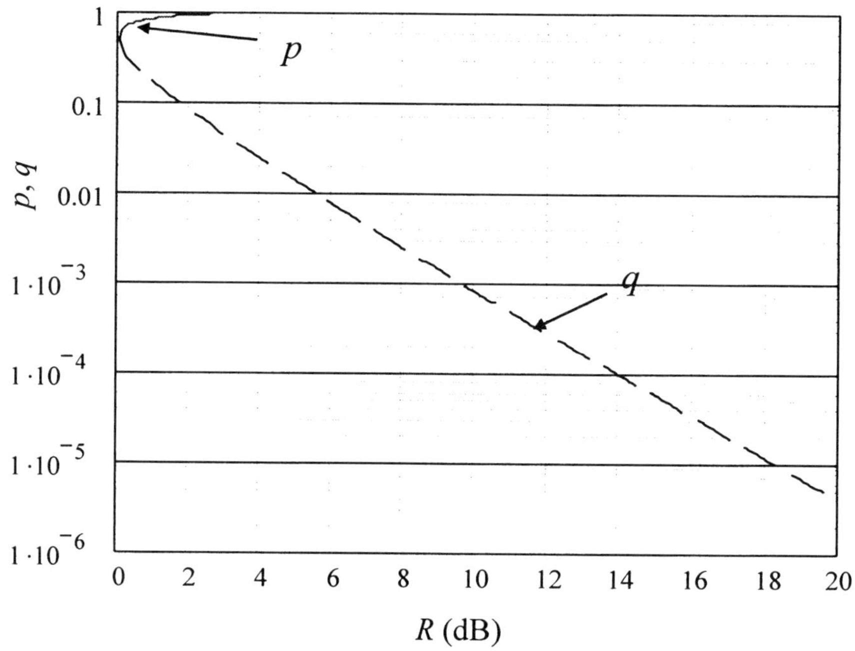

Figure 12.7-2 The values of p and q for a carrier loop lock detector versus the SNR, R in dB.

12.7.2 Suppressed Carrier Lock Detector

A suppressed carrier signal would typically be demodulated by either a Costas loop or a squaring loop [1]. Figure 12.7-3(a) and Figure 12.7-3(b) contain representative models for the lock detector's block diagram models for a Costas loop and a squaring loop, respectively.

Figure 12.7-3(a) illustrates the lock detector model and a Costas loop carrier tracking loop. The left side of the system is the Costas loop up to the multiplier feeding the loop filter. The right side of the diagram is the lock detector for this loop. It forms $I^2 - Q^2$, which is filtered by the integrator and compared to a threshold. The lock detector decisions are made after every integration time.

Figure 12.7-3(b) is the model for the squaring loop lock detector. Basically it is the same as the carrier lock detector with the addition of a squaring circuit after a bandpass filter.

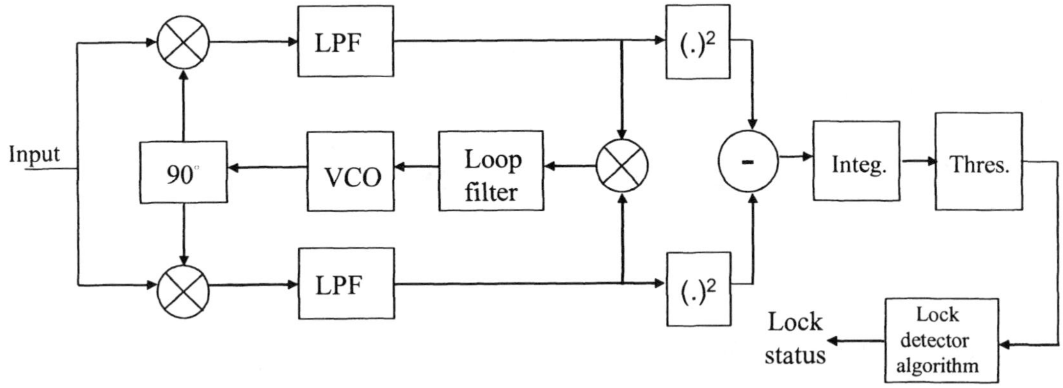

Figure 12.7-3 (a) Costas loop lock detector block diagram model.

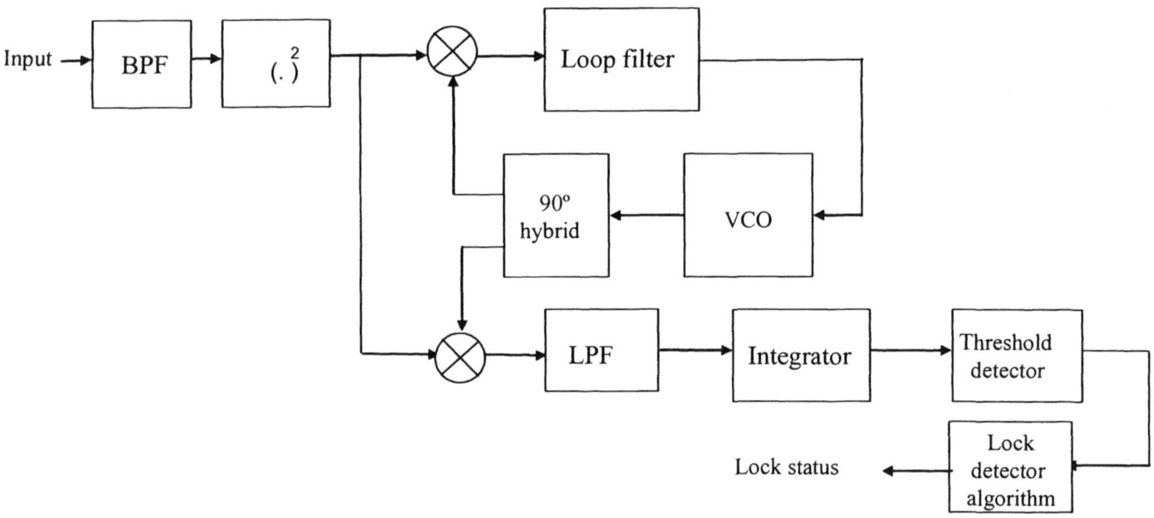

Figure 12.7-3 (b) Squaring loop lock detector block diagram model.

12.7.3 PN Code Acquisition Lock Detector

A code lock detector is shown in Figure 12.7-4. It may be used for code acquisition and code track monitoring. Consider Figure 12.7-4(a).

The on-time code generator despreads the incoming code, which is then filtered in the bandpass filter and detected and integrated. The integrated process is then compared to the threshold and either the signal plus noise is above the threshold, or it is not. These decisions are then fed into the lock detector algorithm that determines if the system is still in synchronization. The baseband version is shown in Figure 12.7-4(b). It basically does the same operation as Figure 12.7-4(a), except at baseband.

12.7.4 A Frequency-Hopping Lock Detector for SFH/DPSK

In Figure 12.7-5 a block diagram model of a lock detector for a frequency hopping tracking system is illustrated. The tracking system utilizes an early and a late hopping pattern to generate an error signal to feed back to the voltage controlled crystal oscillator (VCXO) that drives the hopping code generator. This closed loop process maintains the tracking function to keep the on-time (OT) hopping pattern in time with the received hopping pattern.

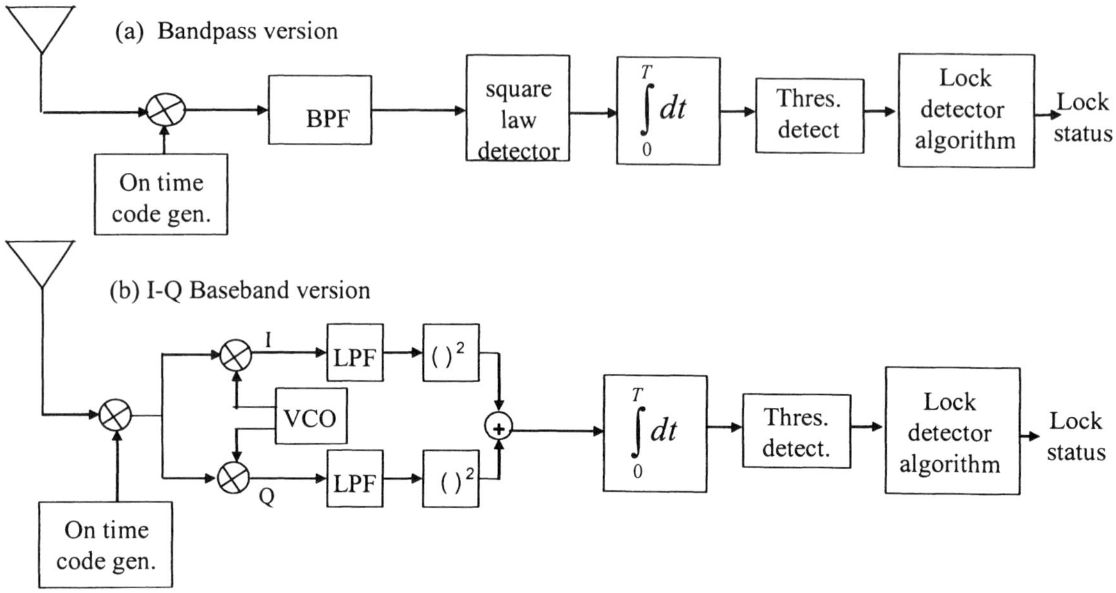

Figure 12.7-4 (a, b) A direct sequence code lock detector for PN code acquisition for PN code acquisition and tracking.

The lock detector is composed of an on-time dehopped signal, a bandpass filter, a square law detector, a threshold detector, followed by a lock detector algorithm. Basically, the DPSK signal is filtered to limit the noise level, and then is detected and integrated. The integrated signal is then compared to a threshold and the binary decision is fed to the lock detector algorithm, which indicates lock status.

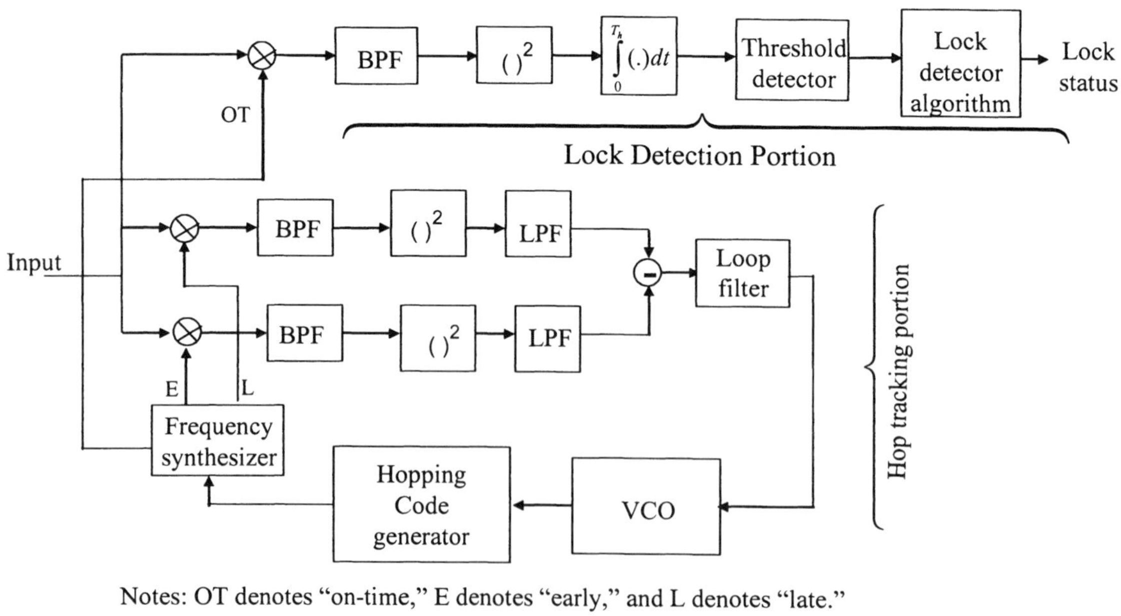

Figure 12.7-5 A slow hop FH/DPSK lock detector.

12.8 SUMMARY

In this chapter absorbing Markov chains were introduced from a matrix point of view. Some theorems that were pertinent to lock detector theory were presented and proven. The main result of this chapter was the mean and variance of the time it takes to start in a transient state and end in a persistent state (Theorems 6 and 7), since this relates to lock detectors directly. In particular it models both the case when the signal is present and computes the mean and the variance of the time it takes to falsely declare out of lock; it also allows modeling the case in which the signal has dropped out and allows the determination of the mean and variance of the time it takes to declare that the signal is not present.

Then lock detector performance was developed for a few lock detectors, in terms of the probabilities of being above and below a threshold, p and q. It was assumed that the relevant probabilities of being above and below the threshold were known. A methodology was developed so that virtually any type of counting lock detector could be analyzed by the methods presented in this chapter. One analysis was developed that determined the values of p and q for the case of a carrier loop lock detector. The values of p and q were obtained in terms of the threshold level and the signal to noise ratio.

Finally some block diagram models of lock detectors for both suppressed and residual carrier tracking, PN code tracking, and frequency hopping were presented and discussed.

As another note, we have used the convention that q is the probability of moving toward the persistent state, and p is the probability of moving away from it.

References

[1] J. K. Holmes, *Coherent Spread Spectrum Systems*, Malabar, FL: Krieger Publishing Co., 1990.

[2] J. Kemeny, and J. Snell, *Finite Markov Chains*, New York: Van Nostrand, 1960.

[3] P. Hopkins, "A Spread Spectrum Synchronization Scheme for the Space Shuttle Orbiter," Johnson Space Center, Tracking and Communications Development Division Internal Note (Rev A), September 1975.

[4] J. K. Holmes, "Code Tracking Lock Detector Mean Time to Declare Out-of-Lock," TRW IOC IUS-110, May 1979.

[5] R. A. Kordosz, "Useful New Results on the Time to Lose Lock for a Simple Markov Tracking Scheme," *Milcom 96*, McLean, VA, October 21–24, 1996.

Selected Bibliography

R. Howard, *Dynamic Programming and Markov Processes*, New York: Technology Press and Wiley, 1960.

W. H. Huggins, "Signal Flow Graphs and Ransom Signal," *Proc. IRE*, January 1957.

S. J. Mason, "Feedback Theory-Some Properties of Signal Flow Graphs," *Proc. IRE*, Vol. 41, No. 9, September 1953.

S. J. Mason, "Feedback Theory-Further Properties of Signal Flow Graphs," *Proc. IRE*, Vol. 44, No. 7, July 1956.

R. W. Sittler, "Systems Analysis of Discrete Markov Processes," *IRE Trans. Circuit Theory*, Vol. CT-3, December 1956.

Problems

1. Show that N is as indicated in (12.3-11) of Example 1. That is to say, evaluate N from Q by performing the appropriate matrix inverse.

2. Show that (12.4-8) collapses to (12.3-12) when τ equals **1**.

3. Consider the following state transition diagram shown in the figure. Here q_d is the probability of going to a higher state and p_d is the probability of going to a lower state.

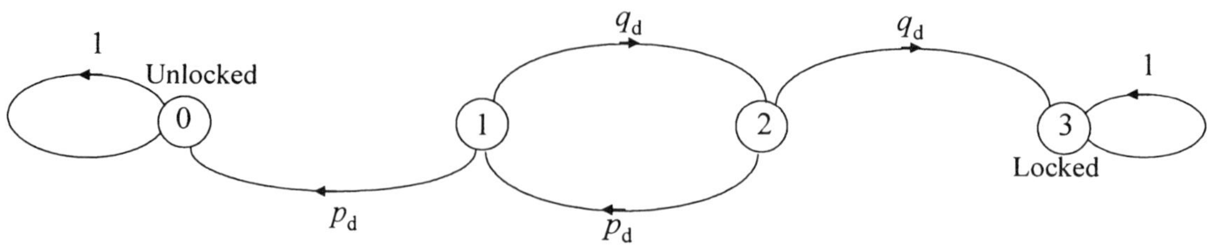

Figure for Problem 3 The state transition diagram.

The diagram represents a simplified model to determine the probability of lock, where $q_d = P(\text{hit}|\text{search mode})$ and $p_d = P(\text{no hit}|\text{search mode})$.

(a) Show that the probability of reaching state 3, starting from state 1, is given by

$$P_L = \frac{q_d^2}{1 - p_d q_d} \tag{P12-1}$$

(b) Show that the probability of reaching state 3 or state 0 from state 1 is unity!

4. Show that the variance of the previously described lock detector for unit dwell times is given by

$$\sigma^2 = \frac{q_d}{p_d^2} T^2 \tag{P12-2}$$

where T is the dwell time in seconds. Use both the matrix method and the direct evaluation method to obtain the variance.

5. Referring to Figure 12.6-3, which illustrates the three-state up-down lock detector counter state transition diagram, show that the mean time to indicate the out-of-lock state, when the system is in lock, is given by

$$\{E_1[t_n]\} = \frac{1+2q^2}{q^3} T \tag{P12-3}$$

In this equation q is the probability of the T-second integrator being above the threshold and p is the probability of it being below the threshold. Note that this applies in the case when the signal is present and to the case when the signal is absent.

6. Consider the count-to-five lock detector shown in the figure following. This lock detector requires that 5 consecutive above-threshold values occur before acquisition is declared. Let q_d denote the probability of a correlation indicating an above threshold value and p_d the probability of a correlation being below the threshold. Show that the mean time to indicate acquisition (locked state) is given by

$$\overline{T}_{Declare\ lock} = \frac{1 + q_d + q_d^2 + q_d^3 + q_d^4}{q_d^5} T \tag{P12-4}$$

Therefore if q_d is very near unity, then $\overline{T}_{Declare\ lock} \cong 5T$.

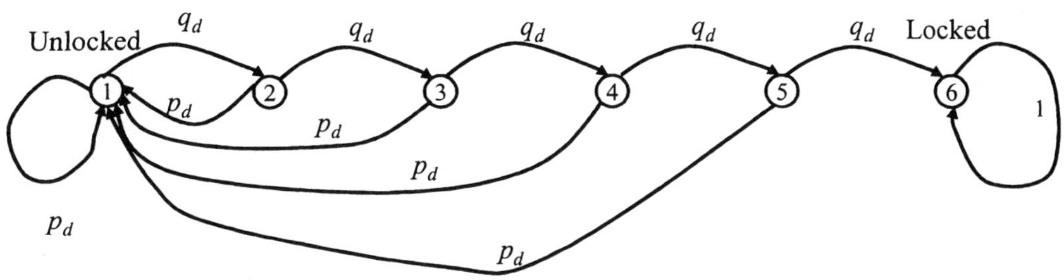

Figure for Problem 6 An illustration of the "five-counts-in-a-row" lock detector.

About the Author

Jack K. Holmes has a B.S., an M.S., and a Ph.D. from UCLA in electrical engineering. He is the author of the book *Coherent Spread Spectrum Systems* (Wiley Interscience, 1982; Krieger Press, 1990). Dr. Holmes has published over 50 papers in journals and conferences in the area of spread spectrum systems, communications, navigation, and synchronization. He has been employed by a number of aerospace companies in the Los Angeles area as well as operating as a consultant. He has taught numerous short courses and tutorials, as well as graduate courses, at both UCLA Extension School and California State College, Northridge.

Index

The GNSS Technology and Applications Series

Elliott Kaplan and Christopher Hegarty, Series Editors

Applied Satellite Navigation Using GPS, GALILEO, and Augmentation Systems, Ramjee Prasad and Marina Ruggieri

Digital Terrain Modeling: Acquisition, Manipulation, and Applications, Naser El-Sheimy, Caterina Valeo, and Ayman Habib

Geographical Information Systems Demystified, Stephen R. Galati

GNSS Markets and Applications, Len Jacobson

GNSS Receivers for Weak Signals, Nesreen I. Ziedan

Introduction to GPS: The Global Positioning System, Second Edition, Ahmed El-Rabbany

Spread Spectrum Systems for GNSS and Wireless Communications, Jack K. Holmes

Understanding GPS: Principles and Applications, Second Edition, Elliott Kaplan and Christopher Hegarty, editors

For further information on these and other Artech House titles, including previously considered out-of-print books now available through our In-Print-Forever® (IPF®) program, contact:

Artech House Publishers
685 Canton Street
Norwood, MA 02062
Phone: 781-769-9750
Fax: 781-769-6334
e-mail: artech@artechhouse.com

Artech House Books
46 Gillingham Street
London SW1V 1AH UK
Phone: +44 (0)20 7596 8750
Fax: +44 (0)20 7630 0166
e-mail: artech-uk@artechhouse.com

Find us on the World Wide Web at: www.artechhouse.com